M_v
42	2.01 00	2.45 00	1.42 01	1.80 01	1.94 01	1.83 01	1.63 01	1.48 01	1.29 01	1.15 01
45	1.32 00	3.27 00	3.16 00	1.87 01	2.49 01	2.46 01	2.23 01	1.96 01	1.71 01	1.53 01
48	1.16 01	3.57 00	4.30 00	4.35 00	2.93 01	2.65 01	2.69 01	2.56 01	2.24 01	2.01 01
51	2.18 01	4.12 00	5.39 00	5.69 00	5.27 00	3.42 01	2.74 01	2.57 01	2.64 01	2.49 01
54	3.85 01	2.90 00	6.38 00	7.03 00	6.96 00	6.48 00	4.95 01	3.12 01	2.69 01	2.91 01
57	9.53 00	2.35 00	6.69 00	8.23 00	8.18 00	7.78 00	7.19 00	6.67 00	3.24 01	2.89 01
60	4.92 00	6.86 00	8.28 00	1.08 01	1.03 01	9.71 00	8.90 00	8.20 00	7.49 00	3.05 01
63	6.27 00	1.67 01	1.24 01	1.36 01	1.27 01	1.18 01	1.08 01	9.94 00	9.06 00	8.37 00
66	1.26 01	1.99 01	1.95 01	1.79 01	1.63 01	1.50 01	1.35 01	1.24 01	1.12 01	1.03 01
69	1.16 01	1.81 01	1.89 01	1.98 01	2.03 01	1.88 01	1.69 01	1.54 01	1.39 01	1.28 01
72	8.93 00	1.81 01	2.13 01	2.15 01	2.28 01	2.25 01	2.03 01	1.85 01	1.68 01	1.54 01
75	7.16 00	1.66 01	2.33 01	2.45 01	2.57 01	2.56 01	2.41 01	2.21 01	2.01 01	1.85 01
78	8.93 00	1.49 01	2.22 01	2.78 01	2.77 01	2.93 01	2.75 01	2.68 01	2.44 01	2.25 01
81	1.12 01	9.40 00	2.21 01	2.87 01	3.09 01	3.03 01	3.19 01	3.02 01	2.90 01	2.72 01
84	2.48 01	6.11 00	1.81 01	2.78 01	3.53 01	3.44 01	3.26 01	3.42 01	3.17 01	3.07 01
87	3.48 01	9.10 00	9.00 00	2.39 01	3.20 01	3.54 01	3.77 01	3.56 01	3.55 01	3.46 01
90	7.19 01	6.82 00	1.17 01	1.93 01	2.98 01	3.38 01	3.88 01	3.80 01	3.59 01	3.37 01

HANDBOOK ON SYNCHROTRON RADIATION

Series editors

ERNST-ECKHARD KOCH
Fritz-Haber Institut der Max-Planck-Gesellschaft, Faradayweg 4–6, 1000 Berlin 33/Dahlem

TAIZO SASAKI
Department of Material Physics, Faculty of Engineering Science, Osaka University, Toyonaka-shi, Osaka 560, Japan

HERMAN WINICK
Stanford Synchrotron Radiation Laboratory, SLAC, Bin 69, P.O. Box 4349, Stanford, CA 94305, USA

HANDBOOK ON SYNCHROTRON RADIATION

VOLUME 2

Edited by

GEOFFREY V. MARR

Department of Physics, Natural Philosophy Building,
Aberdeen University, Aberdeen AB9 2UE, Scotland

NORTH-HOLLAND
1987   AMSTERDAM · OXFORD · NEW YORK · TOKYO

© Elsevier Science Publishers B.V., 1987

All rights reserved. No part of this publication may be reproduced, stored in a retrieval system, or transmitted, in any form or by any means, electronic, mechanical, photocopying, recording or otherwise, without the prior permission of the publisher, Elsevier Science Publishers B.V. (North-Holland Physics Publishing Division), P.O. Box 103, 1000 AC Amsterdam, The Netherlands. Special regulations for readers in the USA: This publication has been registered with the Copyright Clearance Center Inc. (CCC), Salem, Massachusetts. Information can be obtained from the CCC about conditions under which photocopies of parts of this publication may be made in the USA. All other copyright questions, including photocopying outside of the USA, should be referred to the publisher.
USA, should be referred to the publisher.

ISBN: 0 444 87046 6

Published by:

North-Holland Physics Publishing
a division of
Elsevier Science Publishers B.V.
P.O. Box 103
1000 AC Amsterdam
The Netherlands

Sole distributors for the USA and Canada:
Elsevier Science Publishing Company, Inc.
52 Vanderbilt Avenue
New York, NY 10017
USA

Library of Congress Cataloging in Publication Data

(Revised for Vol. 2)

Handbook on synchrotron radiation

    Vol. 2 edited by Geoffrey V. Marr.
    Includes bibliographies and indexes.
    1. Synchrotron radiation—Handbooks, manuals, etc.
I. Koch, Ernst-Eckhard, 1943–   . II. Marr, Geoffrey V.
QC793.5.E627H37   1983   539.7'2112   83-12116
ISBN 0-444-86425-3 (Set)
ISBN 0-444-86709-0 (U.S. : V. 1a)
ISBN 0-444-86710-4 (U.S. : V. 1b)

Printed in The Netherlands

PREFACE

Volume 1 of this series contained an extensive list of topics which attempted to review the characteristics of Synchrotron Radiation, the instrumentation and principles of various research applications for a branch of scientific activity which, following two decades of vigorous research, shows no sign of reducing its rate of development in the foreseeable future. Volume 2 of this series concentrates on the use of synchrotron radiation which covers that region of the electromagnetic spectrum which extends from about 10 eV to 3 keV in photon energy, and is essentially the region where the radiation is strongly absorbed by atmospheric gases, and so has to make extensive use of a high vacuum to transport the radiation to the workstation, and where the presence of hard X-rays can cause extensive damage to both the optics and the targets used in the experimental rigs. In order to keep the volume to a reasonable size, a selection of topics has been chosen and attention has been limited essentially to the disciplines of physics and chemistry.

The first three chapters are concerned with the synchrotron radiation source and its beam lines, with attendant facilities of data collection needed to exploit the VUV to soft X-ray radiation. The following three chapters concentrate on the problems of some free atom and molecule studies, while the final five chapters cover surface and solid state phenomena. It is hoped that this material will be of interest to the research student or newcomer to the techniques of exploiting the radiation from a large multi-user facility, as well as providing material of use to the researcher who wishes to consider new developments or the use of techniques, as yet unexplored, where synchrotron radiation may be significant.

Each of the 21 authors has made a substantial contribution to the volume and I would like to take this opportunity of expressing my appreciation to each and every one of them. I hope that they will feel that the final result represents a suitable vehicle for their reviews and for their comments on the future use of Synchrotron Radiation.

As volume editor I am indebted to the general editors to the series for their help in getting this volume under way. In particular, it is a pleasure to express my appreciation to Ernst Koch, who as editor of Volume 1 and subsequently series editor with Dr. Sasaki and Dr. Winick, did much to support early efforts in the assembly of topics for Volume 2.

It is a fact that this volume would not have been possible without the very capable help received from my secretary Carole Faulkner, who in the preparation of manuscript and through the efficient handling of the secretarial work involved, ensured that material reached the publishers on time. North-Holland, as ever, have continued to handle this contribution to the Handbook Series in their usual competent way. Their patience and understanding in dealing with this volume editor is much appreciated.

Geoffrey V. Marr
Aberdeen 1987

CONTENTS

Preface v
Contents vii
Contributors to Volume 2 xi

1. Synchrotron radiation sources. 1
      I.H. Munro and G.V. Marr

    1. Introduction 3
    2. Synchrotron radiation sources in the soft X-ray (SXR) and vacuum ultraviolet (VUV) regions 6
    3. The prospects for increasing the brilliance of SR sources in the soft X-ray region 14
    4. Synchrotron radiation source installations 19
    References 20

2. Optical engineering 21
      J.B. West and H.A. Padmore

    1. Introduction 23
    2. The synchrotron radiation source 24
    3. Mirrors 35
    4. Diffraction grating optics 56
    5. Beam line layout 84
    6. Optical components 99
    7. Beam line auxiliary equipment 109
    8. Conclusion 116
    References 117

3. Data aquisition and analysis systems 121
      P.A. Ridley

    1. Introduction 123
    2. Data aquisition electronics 124
    3. Local computer systems 148
    4. Data processing facilities 157
    References 170

4. High resolution spectroscopy of atoms and molecules including Faraday rotation effects 175
      J.P. Connerade and M.A. Baig

    1. Introduction 177
    2. High resolution spectra of atoms 177

3. High resolution spectra of molecules 212
4. SR experiments with polarised radiation 230
5. Conclusion 236
References 237

5. Resonances in molecular photoionization 241
   J.L. Dehmer, A.C. Parr and S.H. Southworth

   1. Introduction 243
   2. Shape resonances 245
   3. Autoionization 263
   4. Triply differential photoelectron measurements – experimental aspects 275
   5. Case studies 280
   6. Survey of related work 330
   7. Prospects for future progress 334
   Appendix. A bibliography on shape resonances in molecular photoionization through early 1985 336
   References 342

6. Molecular photodissociation and photoionization 355
   I. Nenner and J.A. Beswick

   1. General introduction 357
   2. Experimental methods 361
   3. Selected examples 382
   4. Miscellaneous and future trends 455
   References 458

7. Surface science with synchrotron radiation 467
   I.T. McGovern, D. Norman and R.H. Williams

   1. Introduction 469
   2. Techniques 469
   3. Clean surfaces 476
   4. Surfaces with adsorbates 500
   5. Interfaces 521
   6. Conclusions 533
   References 533

8. Metal–semiconductor interface studies by synchrotron radiation techniques 541
   L.J. Brillson

   1. Introduction 543
   2. Metal–semiconductor interactions and electronic properties 544
   3. Soft X-ray photoemission spectroscopy for microscopic interface characterization 547
   4. Chemical bonding at metal–semiconductor interfaces 549
   5. Metal–semiconductor interdiffusion 576
   6. Fermi level pinning and semiconductor band bending 582

7. Other synchrotron radiation techniques for interface analysis 594
8. Conclusions and future directions 600
References 603

9. **Inner shell photoelectron process in solids** 611
   A. Kotani

   1. Introduction 613
   2. Fundamental theory of many-body effects in inner shell photoelectron process 614
   3. Simple metals 623
   4. Rare earth metals and compounds 627
   5. Transition metals and compounds 639
   6. Related topics 650
   7. Concluding remarks 656
   Appendix. Derivation of basic photoemission formulae 657
   References 660

10. **Surface core level shift** 663
    Y. Jugnet, G. Grenet and Tran Minh Duc

    1. Introduction 665
    2. Surface–bulk core level shift for clean surfaces 666
    3. Applications 693
    4. Conclusion 718
    References 719

11. **Optical constants** 723
    D.W. Lynch

    1. Introduction 725
    2. Definitions 726
    3. Sample characteristics 738
    4. Measurement methods 742
    5. Summary 762
    6. Some examples 764
    7. Comments on data collections and recent literature 770
    Appendix. Recent literature reference to optical data for $E \geqslant 6 \, \mathrm{eV}$ 772
    References 775

Author index 785
Subject index 829

CONTRIBUTORS TO VOLUME 2

M.A. Baig, *Physikalisches Institut, Universität Bonn, Nussallee 12, Bonn, Fed. Rep. Germany*.

J.A. Beswick, *LURE, Centre National de la Recherche Scientifique et Université de Paris Sud, 91405 Orsay, France.*

L.J. Brillson, *Xerox Webster Research Center, 800 Phillips Road W114, Webster, NY 14580, USA.*

J.P. Connerade, *The Blackett Laboratory, Imperial College, London SW7 2AZ, UK.*

J.L. Dehmer, *Argonne National Laboratory, Argonne, IL 60439, USA.*

G. Grenet, *Institut des Sciences de la Matière and Institut de Physique Nucléaire de Lyon (and I N2 P3), Université Claude Bernard Lyon-1, 43 Bd. du 11 Novembre 1918, F69622 Villeurbanne Cédex, France.*

Y. Jugnet, *Institut des Sciences de la Matière and Institut de Physique Nucléaire de Lyon (and I N2 P3), Université Claude Bernard Lyon-1, 43 Bd. du 11 Novembre 1918, F69622 Villeurbanne Cédex, France,*

A. Kotani, *Department of Physics, Faculty of Science, Tohoku University, Sendai 980, Japan*

D.W. Lynch, *Department of Physics and Ames Laboratory-USDOE, Iowa State University, Ames, IA 50011, USA*

G.V. Marr, *Department of Physics, Natural Philosophy Building, Aberdeen University, Aberdeen AB9 2UE, Scotland.*

I.T. McGovern, *Department of Pure and Applied Physics, Trinity College, Dublin, Republic of Ireland.*

I.H. Munro, *Science and Engineering Research Council, Daresbury Laboratory, Daresbury, Warrington WA4 4AD, UK.*

I. Nenner, *Commisariat à l'Energie Atomique, IRDI, Department de Physico-Chimie, Centre d'Etudes Nucléaires de Saclay, 91191 Gif sur Yvette, France and LURE, Centre National de la Recherche Scientifique et Université de Paris Sud, 91405 Orsay, France.*

D. Norman, *Science and Engineering Research Council, Daresbury Laboratory, Warrington, WA4 4AD, UK.*

H.A. Padmore, *Science and Engineering Research Council, Daresbury Laboratory, Daresbury, Warrington WA4 4AD, UK.*

A.C. Parr, *Synchrotron Ultraviolet Radiation Facility, National Bureau of Standards, Gaithersburg, MD 20899, USA.*

---

* Present address: Physics Department, Quaid-i-Azam University, Islamabad, Pakistan.

P.A. Ridley, *Science and Engineering Research Council, Daresbury Laboratory, Daresbury, Warrington WA4 4AD, UK.*

S.H. Southworth, *Los Alamos National Laboratory, Los Alamos, NM 87545, USA.*

Tran Minh Duc, *Institut des Sciences de la Matière and Institut de Physique Nucléaire de Lyon (and I N2 P3), Université Claude Bernard Lyon-1, 43 Bd. du 11 Novembre 1918, F69622 Villeurbanne Cédex, France.*

J.B. West, *Science and Engineering Research Council, Daresbury Laboratory, Daresbury, Warrington WA4 4AD, UK.*

R.H. Williams, *Physics Department, University College, Cardiff, UK.*

CHAPTER 1

SYNCHROTRON RADIATION SOURCES

I.H. MUNRO
Science and Engineering Research Council, Daresbury Laboratory, Daresbury, Warrington WA4 4AD, England

G.V. MARR
Department of Physics, Natural Philosophy Building, Aberdeen University, Aberdeen AB9 2UE, Scotland

Contents

1. Introduction	3
2. Synchrotron radiation sources in the soft X-ray (SXR) and vacuum ultraviolet (VUV) regions	6
2.1. The properties of storage ring sources	6
2.2. Collimation of synchrotron radiation and its polarisation	8
2.3. The radiated spectrum	9
2.4. The photon source in practice	10
2.5. The time modulation of the source	14
3. The prospects for increasing the brilliance of SR sources in the soft X-ray region	14
3.1. Insertion devices	15
3.2. Undulator radiation	17
3.3. Coherence	18
4. Synchrotron radiation source installations	19
References	20

Handbook on Synchrotron Radiation, Vol. 2, edited by G.V. Marr
© *Elsevier Science Publishers B.V., 1987*

1. Introduction

The synchrotron is a machine which is used by physicists to produce high energy (GeV) charged particles. This type of accelerator confines the particles (usually electrons) by magnetic fields so as to cause them to move over approximately circular orbits. It then accelerates them to speeds approaching that of light before they are extracted for collision experiments. The centripetal force acting on the relativistic electrons causes them to radiate electromagnetic radiation predominantly in the vacuum ultraviolet and soft X-ray regions. The radiation is emitted in a narrow cone in the direction of the fast moving electrons so that for a closed circular orbit the radiation appears as a narrow fan emitted tangential to the electron orbit and centred about the orbit plane.

Many individuals have been involved in the development of the classical electrodynamical theory of synchrotron radiation. It effectively began with the early work of Lienard (1898), was developed by Schott (1912) in an attempt to solve the radiation problem of the hydrogen atom and is now to be found in standard texts (Jackson 1975) as well as being the subject of detailed reviews (see Handbook on Synchrotron Radiation, Vol. 1: Koch et al. 1983, Krinskey et al. 1983). It is not the purpose of the present chapter to repeat these reviews, but to attempt a brief summary of the features of modern synchrotron radiation sources so that intended users of these facilities may be aware of their properties and so that it may serve as an introduction to the following chapters in the present volume where the exploitation and use of the vacuum ultraviolet to soft X-ray radiation is discussed.

The past 30 years have seen synchrotron radiation sources and their user communities "come of age". During the 1960's and 1970's, each novel experimental result had a good chance of leading to the generation of a whole new area of experimental science. The literature contains several important examples including the development of mirror and grating systems for the vacuum region, cross section related studies of atoms and molecules, topography, bulk and surface EXAFS, small angle diffraction, soft X-ray imaging and many others. The significance of these advances was often enhanced at the time because of their interdisciplinary nature, drawing together the rather different skills of the physicist, chemist and bioscientist in the solution of problems of mutual interest.

A somewhat different but rather more obvious transition has occurred in the properties of the sources themselves. Almost all synchrotron radiation research until well into the 1970's was conducted using electron accelerators (synchrotrons) whose primary objective was to carry out research into the properties of elementary particles. This "parasitic" exploitation of the unwanted electromagnetic radiation produced by the unavoidable energy loss from circular accelerators, gradually devolved into a relationship between the exploiters of synchrotron radiation and the high-energy physicists, which is now called "symbiotic". As a consequence, the past decade has seen the appearance of a community of accelerator physicists who have been actively encouraged to design machines to maximise the output of synchrotron radiation from accelerators whose emittance is designed to be as small as is physically compatible with a source which is expected to run for more than 5000 hours per year. The costs of such a machine, together with the very high level of demand for access from a large number of scientific groups, decrees that predictable and reliable operation of the source is the paramount requirement of the designers. In addition, many of the user scientists originally had little or no understanding of the source, nor of working at a large institutionalized facility. With 24 hour usage required, shift working of technical staff simply to maintain a functioning source is necessary and to maintain the experimental workstations in good working order a scientific support staff must be always on call. Support staff are also needed to develop new workstations and to provide the software essential for equipment control, data collection and analysis. The synchrotron radiation facility is a large, expensive, and complex organisation, devoted to the provision of electromagnetic radiation to a wide range of experimental rigs, and service a community with diverse scientific backgrounds. Differing methods for tackling these problems have been established at the various facilities around the world, and discussions of cost effective exploitation is increasingly prominent in the planning and modus operandi at each synchrotron radiation facility.

The first generation of purpose built sources of synchrotron radiation were designed through the 1970's as electron storage rings with the synchrotron radiation coming from the use of dipole magnets to bend the stored electron beam into a closed loop. There has been a steady increase in the stable circulating beam current which can be sustained by a storage ring for long stored beam periods. Present day accelerators are normally expected to support circulating currents of at least 100 mA (often <500 mA is achieved) for periods of several hours. This represents probably a tenfold increase in "ampere hours" in as many years. Less dramatic has been the improvement in control of the electron beam position and direction at the photon beam source points. The essential requirement for all photon beam lines from such storage rings is for direction sensors to interactively define the operating parameters of the storage ring. Source control simultaneously at many positions around each storage ring is difficult due to the interdependence of electron beam dynamics on conditions prevailing at different magnets round the ring. Nevertheless, beam stability for the majority of present day users is an essential requirement.

At a very early stage in the development of synchrotron radiation sources it was realised that the ability to harness radiation from a linear array of dipole magnets could provide a means of maximising the photon flux at a workstation and would be more feasible than attempting to store extremely high circulating ring currents (see Kincaid 1977). The transformation of the design criteria associated with the European Synchrotron Radiation Facility (the ESRF) illustrates rather dramatically this evolution in source design which has taken place between approximately 1977 and 1987. The original ESRF design was for a storage ring of fairly high current and small electron beam cross section with synchrotron radiation being derived principally from dipole magnets and wigglers (wavelength shifters). The present (Buras and Tazzari 1985) design however is based exclusively on "insertion devices" where dipole magnets are used essentially only to transfer the circulating beam from one undulator or wiggler device to the next.

The specificity of undulator radiation (in terms of spectral output, beam emittance and polarisation) has led to the present day philosophy in which the source parameters are tailored as fully as possible to match the specific needs of a particular experiment. There is the immediate example of synchrotron radiation used in lithography where the storage ring is exclusively designed, constructed and operated in a manner dedicated to a single set of technical requirements.

The evolution of synchrotron radiation sources as sources of hard and soft X-rays is summarised in fig. 1. It incorporates conventional X-ray sources which can operate only with very poor efficiency. The total radiated (photon) power is typically only 0.1% of the input power in such devices which, because of Joule heating, are unable to exploit input power levels of greater than approximately 100 kW. All sources of this kind radiate approximately isotropically and the unpolarised radiated spectrum contains spectral lines superimposed on a back-

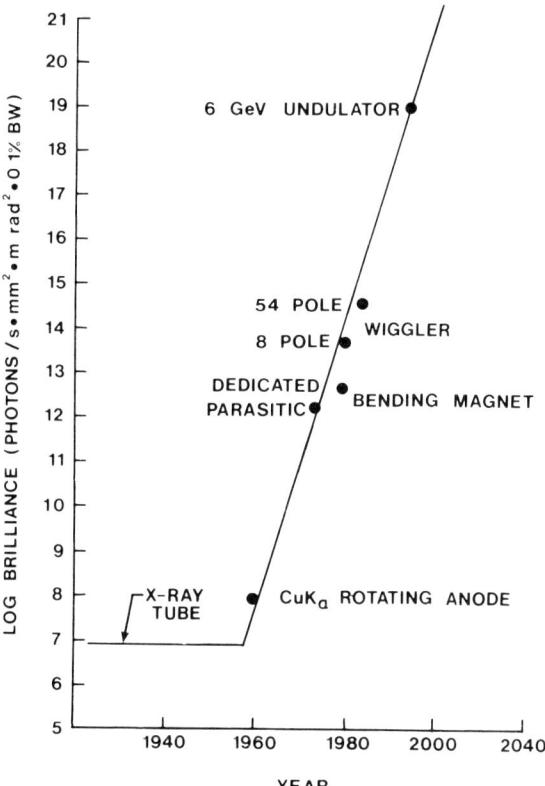

Fig. 1. History of X-ray sources.

ground continuum. The source size also is limited by thermal load considerations to a value approximately a few 10^{-8} to 10^{-5} m^2. The source brilliance is therefore fundamentally limited to less than 10^9.

The first generation of synchrotron radiation sources (that is storage ring designs approximately up to the present time) have a high efficiency for conversion of input power into radiated power in a broadband (white) spectrum. The maximum input power is restricted mainly by the cost of producing and coupling a high level RF source (the klystron) to a storage ring. The total input power to present day rings is usually greater than 100 kW and up to 1 MW and beyond will be perfectly feasible in high energy storage rings. The source size (the electron beam cross section) remains comparable with the best conventional X-ray microfocus source in the region from approximately 10^{-8} m^2 to 10^{-6} m^2. However, the beam divergence and degree of polarisation (usually approximately 95%) which are dictated by the relativistic properties of the source, push up the source brilliance by at least 10^4 times above that of non-synchrotron-radiation sources.

The evolutionary leap *in brilliance* associated with undulator sources is not based on greater source input power (which remains approximately <2 MW) nor on increased circulating current, but on the reduction in source emittance and the increase in undulator source coherence. By the mid-1990's, linear or circularly polarised radiation should be produced from undulators containing from 100 to 1000 periods. The source size of 10^{-8} m^2, source divergence of $\sim 10^{-5}$ radians and intrinsic spectral purity from 1% to 0.1% will yield an unprecedentedly high brilliance approaching 10^{19} photons per s mm^2 mrad^2 0.1% bandwidth. This projection must surely tremendously expand the boundaries in science and for example could lead on to the study of exceedingly small samples measured in short time periods. Very high momentum resolution in scattering and high energy resolution in spectroscopic experiments, together with the potential for high spatial resolution overall for imaging studies are other features of future undulator radiation which will be exploited in future.

2. Synchrotron radiation sources in the soft X-ray (SXR) and vacuum ultraviolet (VUV) region

2.1. The properties of storage ring sources

The optical cross section and effective length of a Synchrotron Radiation Source (SRS), together with its spectral characteristics, angular divergence and time structure are each defined by the parameters of the ring, which will have been selected in each case to optimise the match between user experimental needs, minimum source construction and operation costs. Although a full discussion of the physics of storage ring sources and the technological problems in their construction is reviewed elsewhere (see for example Krinsky et al. 1983 and Sands 1970) it is important to summarise the relationships between the storage ring parameters and the properties of the emitted radiation.

Radiation will be emitted when any charged particle is accelerated. The case of synchrotron radiation is restricted (normally) to electrons or positrons confined in a circular accelerator where they go around the ring at a fixed frequency (ν) given by

$$\nu = \frac{v}{2\pi R}, \tag{1}$$

where R is the radius of the electron orbit and v is the particle speed.

In practice, v will be very close to the speed of light, c. Looking back along a soft X-ray beam line into any synchrotron radiation source, a laboratory-based observer will "see" the electron or positron emitting radiation from a frame of reference moving towards him with velocity v relative to his own (hopefully stationary) reference frame. Transposing from the reference frame of the source to that of the observer results in dramatic changes in the form of the emitted radiation pattern and predicts many unusual results all of which have been shown to be in agreement with observation.

The properties of a charged particle in the ring will be governed by its instantaneous momentum, which is equal to the product of the magnetic field strength B (in tesla) the orbit radius R (in metres) and the particle charge.

In fact the radius of curvature of the particle orbit is given by

$$R = 3.3\ E/B, \tag{2}$$

where E is the electron energy in Gev. Using iron electromagnets, where B_{max} is typically 1.2 T, the bending radius will be 5.6 m for example for a 2 GeV storage ring. For higher energy rings (producing "harder" X-rays) this radius will be increased. Of course the average radius of a storage ring is always greater than the particle bending radius because of the extra lengths of orbit (straight sections) used to connect dipole magnets.

A typical mean radius for a 2 GeV machine (the Daresbury SRS) is 15 m. When a specially constructed high field bending magnet is used (such as the 5 T superconducting wiggler at the Daresbury Laboratory SRS) then the electron orbit radius will be correspondingly reduced – in this case to ~1.3 m in the vicinity of the wiggler.

The energy lost by each electron in a storage ring per single revolution is given by

$$\Delta E = \frac{88.5\ E^4}{R}\ \text{keV}, \tag{3}$$

which corresponds to about 250 keV for the SRS, or several photons per electron per revolution, where the energy of each photon ($h\nu$) is given by $h\nu$ (eV) = $12.4/\lambda$ (Å). If a corresponding amount of energy were not to be fed into the storage ring via the RF system then the electrons would rapidly lose energy and spiral inwards to collide with the vacuum vessel or a target "scraper".

The total power, P, radiated by a storage ring is given by the total number of electrons contained in the ring (measured by the mean circulating current I in amperes) and is given by

$$P = 88.5\, E^4 I/R \text{ (kW)}. \tag{4}$$

The total power radiated by each relativistic electron is in fact

$$P = \frac{2}{3} \frac{e^2 c}{R^2} \left(\frac{E}{m_0 c^2}\right)^4 \left(\frac{1}{4\pi\varepsilon_0}\right) \text{ (W)}, \tag{5}$$

where m_0 is the rest mass of the charged particle. This expression reveals why it is that protons radiate far less power than electrons when accelerated to the same energy E. While electrons and positrons are equally effective as synchrotron radiation sources, a proton at the same energy will radiate only 10^{-13} as much power. Alternatively, to extract as much synchrotron radiation from a proton as from an electron storage ring would require 1836 times the energy!

For the Daresbury SRS, operating at above 100 mA circulating current (that is with approximately 10^{12} electrons in the ring), at least 50 kW of radiated power will be produced, necessitating the use of a 500 MHz klystron with an output power of at least 100 kW.

The instantaneous synchrotron radiation beam divergence is set by the electron energy (i.e., the source velocity) and is given roughly by the angle $1/\gamma$ where $\gamma = (1 - v^2/c^2)^{-1/2}$. For a medium (2 GeV) energy storage ring such as the SRS this will correspond to a cone angle of somewhat less than one milliradian (where 1 mrad $\sim$ 200 seconds of arc).

It is important to establish the radiative power loading on a target which results from the highly directional nature of synchrotron radiation. The power emitted in the horizontal plane may be shown to be

$$P_{\text{Hor}} = 4 \times 10^{-3}\, E^3 B I \text{ (W/rad)}. \tag{6}$$

For the SRS operating at 2 GeV and approximately 250 mA this yields approximately 10 W/mrad from a dipole magnet. For the X-ray region, the vertical collimation of the beam is also $\sim$1 mrad and the power will fall on an area of $\sim$10 mm^2 at 10 m from the source. In many experiments, the bulk of this power will fall on a mirror, crystal or grating which obviously must be radiation resistant. For multipole wiggler devices and undulators, the power loading on the first element can be enormous (up to approximately 100 W/mm^2) and radiation and thermal damage or distortion effects on the instrumentation or on the sample may ultimately limit the benefits from sources planned to have exceptionally high brilliance.

2.2. Collimation of synchrotron radiation and its polarisation

The detailed calculation of radiated power from a storage ring allows the intensity emitted to be expressed in terms of the radiation polarised with either the E

vector perpendicular or parallel to the electron orbit plane. Viewed in the plane of the orbit, the radiation is 100% plane polarised with the E vector in the plane of the orbit. Out of this plane both components of polarisation are seen and the degree of polarisation is wavelength-dependent. Nevertheless, the degree of polarisation is at least 75% for radiation integrated over all wavelengths and all vertical angles. Radiation from an electron storage ring is therefore either plane polarised or elliptically polarised, depending on the beam line conditions and the wavelength region of the spectrum.

The collimation of emitted radiation can be related to λ_c, the so-called "critical wavelength" which divides into two equal parts the total radiated power (i.e., half at wavelengths greater than λ_c and half at wavelengths less than λ_c). For a given λ_c, the radiation divergence angle decreases as wavelength is reduced. However, for a particular wavelength, the divergence angle reduces as λ_c is increased and, since λ_c varies inversely as the electron energy, lower energy accelerators produce less divergent radiation than high energy accelerators.

The intensity distribution in the vertical direction can be satisfactorily described by a Gaussian over a wide wavelength range. The actual vertical opening angle is less than $1/\gamma$ below λ_c, roughly equal to $1/\gamma$ at λ_c and increases to several times $1/\gamma$ for $\lambda \sim 100\,\lambda_c$.

2.3. The radiated spectrum

The unique "white" continuum associated with all synchrotron radiation sources results from the extremely high electron velocity with respect to the laboratory experimental station. The observer will gather radiation only from a small opening angle ($\sim 1/\gamma$) which corresponds to a length of orbit about R/γ. The radiation pulse produced for a time Δt will appear to the laboratory-based user to be relativistically reduced, corresponding to $R/c\gamma^3$. A frequency analysis of such a short duration pulse reveals harmonics of the orbit frequency (usually ~ 1 MHz) up to $1/\Delta t$, i.e., the pulse analysis reveals angular frequencies up to γ^3 times the orbit frequency. In practice, the individual harmonics cannot be resolved at such high frequencies and so the radiation appears to have a continuous spectrum extending from the microwave region to very short wavelengths about $2\pi R/\gamma^3$. In practice, the overall spectral profile rather resembles that of a black-body radiator at an extremely high temperature: 10^7 K or so.

The detailed spectral profile calculations incorporate the critical wavelength, λ_c (Å), which is defined to be the wavelength which divides the power spectrum into two equal parts:

$$\lambda_c = 19(BE^2)^{-1}. \tag{7}$$

The radiated power as a function of wavelength in a 0.1% bandwidth (where $\Delta\lambda = 10^{-3}\,\lambda$) per milliradian of horizontal angle is given by

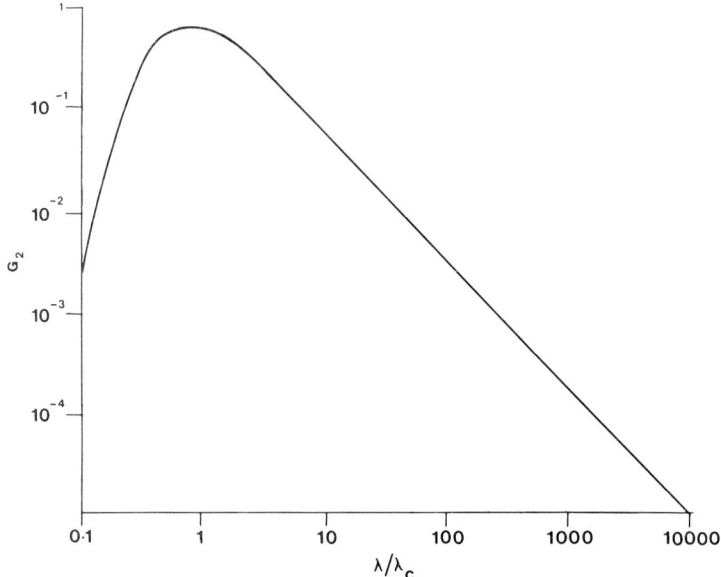

Fig. 2. Radiated power as a function of wavelength in a 0.1% bandwidth per milliradian of horizontal angle.

$$P(\lambda) = 8.7 \times 10^{-3} \frac{E^4}{R} G_2 \frac{\lambda}{\lambda_c} I, \qquad (8)$$

where $G_2(\lambda/\lambda_c)$ is a Universal function (Krinsky et al. 1983) and is shown in fig. 2. The peak power is radiated at $\lambda = 0.7 \lambda_c$. Almost all the radiated power is contained within the range $0.2 \lambda_c < \lambda < 10 \lambda_c$.

For users of synchrotron radiation, the radiated spectrum is almost always presented as a photon flux, $N(\lambda)$, rather than power. Figure 3 shows the Universal distribution of photon flux in synchrotron radiation given by

$$N(\lambda) = 2.5 \times 10^{14} EG_1 \frac{\lambda}{\lambda_c} I. \qquad (9)$$

Useful flux is normally assumed to extend from $\lambda > 0.1 \lambda_c$ with the maximum occurring at $\lambda = 4 \lambda_c$. The flux then falls off extremely slowly to longer wavelengths. For the Daresbury SRS, with $\lambda_c = 3.9$ Å, the maximum photon flux at 2 GeV and 300 mA is $\sim 3 \times 10^{13}$ photons s^{-1} mrad^{-2} in 0.1% bandpass and is typical for a dipole magnet source at this current.

2.4. The photon source in practice

The properties of an actual radiation source, viewed from the laboratory reference frame, are determined partly by the synchrotron radiation characteristics

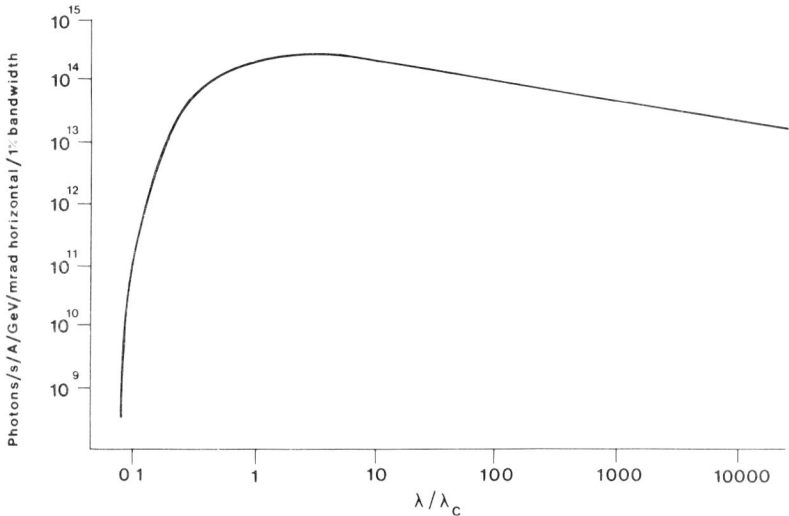

Fig. 3. Universal distribution of photon flux in synchrotron radiation.

associated with a single electron at the energy of the accelerator and partly by the statistical spread in position, direction and time of the large number of electrons ($\sim 10^{12}$ or so for $I > 100$ mA) which constitute the circulating beam.

A fundamental constant for any storage ring is its emittance which is conserved at positions around the orbit and is equal to the product of the overall electron beam dimension and divergence in the horizontal (radial) and vertical directions. In principle, the vertical and horizontal emittances are independent although in practice the source properties can be modified by altering the degree of coupling between the two components. The emittances are determined by complex functions which describe the damping and focussing processes in the electron beam around the ring. A good indication of the range of storage ring emittance values is given in table 1. For example, the emittance at the Daresbury SRS results in a full width at half maximum intensity source size of approximately 1 mm^2. Obviously, this value will define the size of the image at a sample or will limit the minimum slit size to be used in a monochromator. Similarly, the electron beam divergence will contribute to the intrinsic divergence of the synchrotron radiation from a source. In present day storage rings, however, the small values for electron beam emittance yield electron beam divergence values which are <1 mrad and are therefore significant only at the very shortest wavelengths.

For the future, it will be important for the vertical and horizontal source sizes to be as small as possible (<100 μm) in order to allow good source size limited resolution to be achieved with entrance slitless monochromators and to permit monochromators to be developed to exploit both polarisation components of radiation to be used from the storage ring. Obviously, the exploitation of such very small source sizes also imposes the requirement that the beam position can be reproducibly controlled to within the dimensions of the beam and that no oscillations or long term drifts in position should be permitted.

Table 1
Summary of synchrotron radiation sources (worldwide). (D – dipole magnet, W – wiggler, U – undulator, and FEL – free electron laser.)

	λ_c (Å)	Approx. no. expt. stations	Emittance ($10^{-8}\pi$ mrad)	Insertion devices	Operating energy (GeV)	Operating current (mA)	Annual hours operation (h)
Brazil							
Campinas – LNRS					2	100	–
China							
Beijing – BEPC	2.6	4 (X-ray only)	66	–	2.8	150	Part dedicated, under construction: 1st beam end 1987
Hefei – HESYRL	20	–	9	Undulator planned	0.8	–	Dedicated
European SR Facility							
Grenoble – ESRF	1 (D), 0.86 (U), 0.1 (W)	Approx. 30 by year 2000	0.7	28–30 W or U, as required	6	100–200	Design phase
France							
LURE, Orsay – ACO	39	13	15	1 FEL	0.54	150	1700
– DCI	4 (D), 1.1 (W)	14–20	150	1 W	1.75	200	1300
– SUPERACO	18.5 (D)	19	3	1 W, 1 FEL, 4U	0.8	500 (goal)	Operation commenced 1987
India							
Indore – Ring 1 (approved, under construction)		Design studies in progress...			0.45	100	–
– Ring 2 (planned)		Design studies in progress...			0.8	–	
Italy							
Frascati – ADONE (part dedicated)	8.3 (D), 4.4 (W)	7	22.5	2 W	1.5	100	1500
Trieste – CARS/AFRODITE (approved)	6.6 (D), 600–6 (U)	10–20	~1	10 U (max.)	1.5	100	Design phase
Japan[*]							
Tokyo – INSOR	112	6	30	–	0.38	250	1820
Tsukuba (KEK) – Photon Factory	3.1 (D), 0.6 (W)	33	50	1 W, 1 U	2.5	150	2000
– Accumulator Ring	~1 (D)	4	~50	3 (planned)	6	~20	–
– 1.5 GeV ring	250–40	10–20	small	10–13	1.5	150	Planned
Tsukuba (ETL) – TERAS	~10	~10	–	–	0.6	~100	–
Okazaki – UVSOR	29 (D), 8.3 (W)	12	12	1 W, 1 U	0.75	100	–
Konsai – 6–8 GeV ring	~1 (D)	many	small	–	6–8	~100	Proposed
– 2.5 GeV ring		Low emittance ring for soft X-ray studies					

Synchrotron radiation sources

Facility							
Sweden							
Lund – MAX	~40		3	–	0.55	–	Operation commenced 1987
Taiwan							
Hsinchu – TLS	Design studies in progress...				1.5	200	Partially approved, dedicated
United Kingdom							
Daresbury – SRS	3.9 (D), 0.93 (W), 10–100 (U)	28	11	1 W, 1 U	2.0	300	6000
USA							
Brookhaven – NSLS I	25 (D)	25–30	~14	–	0.75	600	–
– NSLS II	2.5 (D)	30–40	8	1 W, 1 U	2.5	~100	–
Stanford – SPEAR	2.6 (D), 1.2 (8-PW) 1.7 (54-pole)	20–25	45	5	3	80–100	2500
– PEP	–	10–20	15	1 U	14.5	–	–
– SXRL (FEL ring)	–	4	1	1 FEL, 4 for SR	1.0	–	–
Wisconsin – TANTALUS	256	11	23	–	0.24	100	2000
– ALADDIN	12	5–17	11	–	1	~100	2000 (max.)
Ithaca – CESR (CHESS facility)	2.1 to 0.35	6	20	1 W	5.5	40	–
Gaithersberg (NBS) – SURF II	>150	11	27	–	0.28	50 (130 max.)	–
Berkeley – ALS	6.6 (D), ~6–1000 (U)	30–50	1	10 U	>1.5	>100	Design phase
Argonne – 6 GeV ANLS	1 (D), >0.62 (U), >0.1 (W)	30–50	0.8	30 (max.)	6	200	Design phase
USSR							
Novosibirsk – VEPP-2M	–	8	–	2	0.7	–	–
– VEPP-3	–	9	–	2	2.2	–	–
Fed. Rep. Germany							
Hamburg – HASYLAB – DORIS	1.34 (D), 2.3 (W)	32	27	2 W	3.7	150	1950
Berlin – BESSY I	19 (D), 28 (U)	31	4	1 U	0.8	600	2700
– BESSY II	5.5 (D), 600–6 (U)	8–16	2	8 U max.	1.5	100	Under construction
– COSY	~15	10–20	250	–	0.56	–	Operational
Bonn – Two synchrotrons	–	–	–	–	0.5, 2.5	–	–

*Other synchrotron radiation source projects are under consideration (or in some cases under construction) at Sendai, Yamagata, Tsukuba (SORTEC), Atsugi (at least 2 rings for N.T.T.) and Hiroshima.

Finally, a storage ring source will appear to an observer to have a very significant depth, L, from which photons are collected and which is simply related to the horizontal aperture, θ, defined to be $L \sim \theta R$. The length of the source apparently will be further extended by virtue of the finite area (horizontal size) of the electron beam with a further contribution arising from the opening angle of the radiation. For a typical source such as the Daresbury SRS, the source length for approximately 5 mrad aperture at the experiment can be as much as 50 mm, necessitating careful representation of the source volume in the ray tracing package used to define the radiation volume in the image plane.

2.5. The time modulation of the source

The electron beam in a storage ring is intrinsically modulated in a longitudinal direction because of the alternating RF field which is used to accelerate the beam. A wide range of accelerating frequencies are used extending from about 1 MHz to 500 MHz. Electrons will be accelerated only if they are in phase with the accelerating field applied when they traverse the RF cavity in the ring. Electrons which are out of phase with the field will receive either too much or too little energy, resulting in an electron orbit which cannot be contained within the vacuum chamber – they collide with the chamber walls and are lost. In phase (or synchronous) electrons are typically confined within a time window which is roughly 10% of the period of the RF and orbit the ring in a series of well defined and regularly spaced bunches.

For accelerators with 500 MHz accelerating fields the bunches are therefore spaced in time by the period of the field (=2 ns) and are roughly of 200 ps duration (corresponding to a bunch length of ~6 cm f.w.h.m.). In practice, the length of the bunch may be determined by the complex impedance of the orbit chamber in which the electron beam, modulated at high frequency, is contained. The bunch length also tends to increase as the circulating current is increased and short bunch operation is usually limited therefore to currents of about 10 mA or so. The number of electron bunches in a storage ring will have a maximum value given by the ratio of the ring orbit period divided by the RF period. For the SRS this value is 160 bunches. It is possible also to operate many storage rings with electrons effectively confined solely to a single bunch which will then provide all the characteristic properties of synchrotron radiation in terms of its spectral profile, polarisation, etc., but, in the case of the SRS, in ~200 ps wide pulses and with a duty cycle of 1/1600!

3. The prospects for increasing the brilliance of SR sources in the soft X-ray region

The properties of synchrotron radiation sources are usually discussed in connection with radiation produced at magnet dipoles and determined at different points around the ring.

The spectral brilliance (defined to be the number of photons per second, per 100 mA circulating current, per mm² electron beam source cross section, per mrad², per 0.1% $\Delta\lambda/\lambda$) is expected to have a maximum value of $\sim 10^{13}$ for realistic values of present circulating current and storage ring emittance. The next major step in the quest for high source brilliance will be to increase the number of source points (i.e., magnets), to further reduce the electron beam emittance and the photon beam divergence. Note that the spectral brilliance is normally considered in terms of a cw source. However, when a storage ring is operating in the single bunch mode, the peak (instantaneous) brilliance may be already up to 10^3 times higher than for cw operation.

3.1. Insertion devices

There are three basic types of insertion devices, each of which has some advantage over a normal dipole bending magnet in terms of spectral range, greater flux or brilliance. Insertion devices are appealing to accelerator physicists since they can be incorporated and operated in a storage ring in a largely independent manner. This should offer much more flexibility in terms of the number and types of insertion devices which can be used without serious interaction with the basic beam properties, such as the emittance of the electron beam.

The first device (attempted as early as about 1970 in Wisconsin) was called a wavelength shifter. This is normally a three-pole arrangement based on a central dipole usually operating at a very high field strength. The effect is to reduce the value of λ_c for a particular machine energy and is used at the SRS at Daresbury to provide high flux at ~ 1.5 Å for X-ray diffraction and scattering experiments.

The second type of insertion device, called a multipole wiggler, yields an increase in output flux simply by using a magnetic field which changes periodically in polarity along the length of the device. The multipole wiggler will then act in effect as a series of N independent dipole sources. Unlike a dipole magnet source, however, both the wiggler and multipole wiggler produce a modification of the electron source for a given storage ring current and energy in which the divergence of the synchrotron radiation in the orbit plane is restricted to some extent, according to the deflection of the electron beam.

The distinction between the radiation pattern produced from a dipole and that from a multipole wiggler is shown clearly in fig. 4. Assuming the magnetic field distribution to be sinusoidal along the direction of electron motion with a peak field amplitude B_0(T) and a spatial field period (of λ_0 cm) it can be shown that the angular deflection of the electron beam, θ_w (mrads), is given by

$$\theta_w \sim 0.05\, B_0 \lambda_0 / E , \tag{10}$$

and the lateral displacement of the beam (mm),

$$a_w \sim 10^{-3}\, \theta_w \lambda_0 / \pi . \tag{11}$$

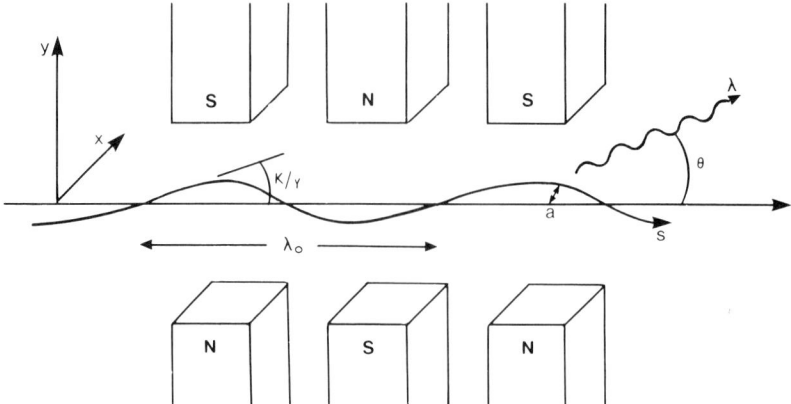

Fig. 4. Radiation pattern from a dipole and a multipole wiggler.

The important parameter, K, is called the deflection parameter, and

$$K \sim 93 \, B_0 \lambda_0 \tag{12}$$

giving a horizontal angular deflection of $\sim K/\gamma$.

The value of K for wigglers is normally rather large – for example, $K \sim 230$ for the three pole 5 T wiggler on the SRS at Daresbury. The maximum deflection of the electron beam is $\theta_w \sim 60$ mrad, although the radiation divergence in the vertical direction remains $\sim 1/\gamma$, as for conventional synchrotron radiation. The deflection of the electron beam through the SRS wiggler ($a_w = 9$ mm) is also relatively large. Wiggler magnets are generally used as wavelength shifters, that is, to reduce the effective λ_c of the storage ring at the wiggler magnet without recourse to the much more expensive solution which would be to increase the energy $E(\text{GeV})$ of the electron beam. The multipole wigglers therefore simply produce from N poles, N times the intensity from a dipole magnet radiating into about the same solid angle.

In practice, the opening angle for the emitted radiation θ_{ph} will contain a contribution both from the electron beam divergence and from the synchrotron radiation divergence:

$$\theta_{ph}^2 = \theta_w^2 + \left(\frac{1}{\gamma}\right)^2, \tag{13}$$

which can be rearranged to give

$$\theta_w \sim \frac{1}{\gamma}(1 + \tfrac{1}{2}K^2). \tag{14}$$

Clearly, when K is large ($\gg 1$) the photon emission angle is dominated by the electron trajectory. However, as K is reduced, for example by reducing the field strength B_0 or the period (λ_0) of the magnet array, a regime will be reached where all radiation is confined within an angular range $\sim 1/\gamma$ in both vertical and horizontal planes. Devices which operate in this mode are called undulators.

3.2. Undulator radiation

When the insertion device causes the electron beam to perform a series of wiggles by using a large number ($N \sim 100$) equally spaced alternately polarised dipole magnets, it is possible to increase the photon flux seen at the detector by N, the number of single period wiggles. If, however, the spacing λ_0 between the single period wiggles is reduced there comes a time when the radiation is best described by considering the relativistic oscillation of the electron transverse to the motion of the electron beam. In this situation we would expect the electrons to radiate at the oscillation wavelength and the radiation to be more nearly monochromatic. The radiation received at the detector would of course be shifted due to the Doppler effect, so the radiation is shifted from λ_0 to

$$\lambda \sim \frac{\lambda_0}{2\gamma^2}. \tag{15}$$

Taking into account the angular spread of the radiation over an angle θ with respect to the electron beam axis we get for $K < 1$:

$$\lambda = \frac{\lambda_0}{2\gamma^2}(1 + \tfrac{1}{2}K^2 + \gamma^2\theta^2), \tag{16}$$

so that for $\theta = 0$ a 2 GeV electron traversing a magnetic field configuration with $\lambda_0 = 10$ cm would emit radiation at ~ 3.3 nm, provided K is sufficiently low to be neglected. The power radiated is peaked in the forward direction as is the radiation for the conventional synchrotron source but because of the oscillating motion transverse to the radiation, the radiation pattern is more complex than conventional synchrotron radiation (Hoffman 1980).

As the value of the magnetic field B_0 is increased, the transverse force on the electron increases so that it moves faster and it is then possible to have relativistic transverse motion of the electron superimposed on its relativistic forward motion. Under these conditions any increase in the transverse momentum must be accompanied by a decrease in the longitudinal momentum and vice versa. In the moving frame of reference of the average relativistic drift velocity along the longitudinal axis, the electron executes a figure of eight type of motion instead of the simple harmonic oscillation previously assumed. This motion now contains harmonics of the oscillation frequency, and transformation into the laboratory frame gives a wavelength spectrum,

$$\lambda = \frac{\lambda_0}{2k\gamma^2}(1 + \tfrac{1}{2}K^2 + \gamma^2\theta^2), \tag{17}$$

where k is the odd harmonic number (1, 3, 5, etc.) which occurs in the forward direction. Increasing K has the effect of smearing out the harmonics and ultimately returning to the continuum of the conventional synchrotron radiation source.

Whether the insertion device is called a multipole wiggler or an undulator depends on the particular configuration of parameters. Wigglers have large K values and essentially are used to increase the power emitted in a particular wavelength region. Undulators have low K values and provide a more peaked selection of quasi-monochromatic radiation. Both devices have their places in the synchrotron radiation source and since, in principle, they can be switched in or out as the experimenters desire, without altering the running of the whole facility, they are economically desirable as well as providing new versatile high flux sources. The total power radiated by an undulator is relatively small (~20–30 W), although for the wavelength of the fundamental the central brightness is extremely high (~10^4 that of a bending magnet). Restriction of the central cone of radiation about the electron drift axis by a pinhole could be used to produce essentially monochromatic radiation. In practice, the radiation pattern is also governed by the angular spread of the electron beam so that for example at the SRS, prior to the introduction of the high brightness lattice, there was considerable smoothing out of the spectrum and the undulator served to increase the photon flux over a wide energy band rather than to provide quasi-monochromatic radiation.

3.3. Coherence

Coherence is usually regarded as an exclusive property of the laser. It is of interest in that it refers to the ability to form interference patterns when wave fronts are recombined from a given source. In fact, sources which exhibit partial coherence can be used to produce clear interference fringes provided the displacement of the wave front is kept within limits set by the coherence length,

$$l_c = \frac{\lambda^2}{\Delta\lambda}, \qquad (18)$$

where λ is the wavelength and $\Delta\lambda$ the spectral line width of the radiation.

For radiation at 10 nm a value of l_c which is 1–10 μm would be quite useable for many purposes so that the use of undulators as partially coherent sources is now feasible with modern low emittance storage rings. For wave front separations of less than l_c it is possible to observe clear interference patterns and with some improvement in current conditions, soft X-ray interferometry and holography should be feasible.

The radiation from an undulator discussed in section 3.2 would be expected to appear as a group of harmonic lines each of spectral purity: $\lambda/\Delta\lambda \sim kN$. Consequently, the longitudinal coherence length is given by

$$l_c = \frac{\lambda^2}{\Delta\lambda} \sim kN\lambda,\qquad(19)$$

which is essentially the relativistically contracted length of the undulator. For $\lambda = 3.3$ nm, $N = 100$ and $k = 3$, $l_c = 1$ µm. Additionally, by using a monochromator following the undulator, $\lambda/\Delta\lambda$ could be increased by another order of magnitude, with the penalty of course that the power available would be reduced. With coherence lengths of tens of micrometres holographic microscopy with a relatively deep depth of field should be possible.

4. Synchrotron radiation source installations

Around the world, the number of dedicated storage ring sources for synchrotron radiation research is continuing to increase at a rapid rate. This increase is notable in view of the major capital costs associated with the design, construction and operation of any new accelerator (~£100 M for high energy 5 GeV storage rings). Although these source costs can present considerable problems to the science community in any country, they are followed in due course by the equally considerable costs associated with the construction and operation of ports, beam lines and experimental stations. Recent years have seen many international collaborative activities in the field of synchrotron radiation research of which the largest and best example is that of the European Synchrotron Radiation Facility (the ESRF) which is approaching an advanced stage in its design at Grenoble in France.

To introduce economies into capital funding, the resources required to develop new beam lines and experimental stations are sometimes provided wholly by the future user of the beam line. This approach has led to the creation of participating research teams or of industrial consortia who are exclusively responsible for an experimental station and actually pay the science community for use of source beam time. The most recent development has been the evolution of specific sources which will be exclusively dedicated to the large scale replication of large area integrated circuit devices with submicron features using soft X-ray radiation. Projects of this kind are at a very advanced stage of development already in the Federal Republic of Germany, the UK, USA and Japan.

Table 1 presents a catalogue of source properties (derived in part from Winick and Watson 1986). The most striking feature of the table is that approximately 700 experimental beam lines and stations will be operating world wide by the beginning of the next decade (1990). This represents a capital investment in experimental equipment alone which will greatly exceed a £1000 M. Collectively, the sources and their associated stations will incorporate several thousands optical elements (mirrors, gratings and crystals) of the very highest optical quality obtainable and this has already led to the creation of a small industry of specialist ultra-high vacuum optical and instrument making companies.

References

Buras, B., and S. Tazzari, 1985, European Synchrotron Radiation Facility, Rep. ESRP c/o CERN (CERN, Switzerland).
Hoffman, A., 1980, Phys. Rep. **64**(5), 253.
Jackson, J.D., 1975, Classical Electrodynamics, 2nd Ed. (Wiley–Interscience, New York) p. 672ff.
Kincaid, B.M., 1977, J. Appl. Phys. **48**, 2648.
Koch, E.-E., D.E. Eastman and Y. Farge, 1983, in: Handbook on Synchrotron Radiation, Vol. 1, ed. E.-E. Koch (North-Holland, Amsterdam) ch. 1.
Krinsky, S., M.L. Perlman and R.E. Watson, 1983, in: Handbook on Synchrotron Radiation, Vol. 1, ed. E.-E. Koch (North-Holland, Amsterdam) ch. 2.
Lienard, A., 1898, L'Eclairage Electron. **16**, 5.
Sands, M., 1970, Stanford Linear Accelerator Rep. SLAC (SLA, Stanford, CA) p. 121.
Schott, G.A., 1912, Electromagnetic Radiation (Cambridge).
Walker, R.P., 1986, Daresbury Laboratory SCI (Daresbury Laboratory, Daresbury, Warrington) p. 513A.
Winick, H., and R.E. Watson, 1986, Nucl. Instrum. Methods **222**, 373.

CHAPTER 2

OPTICAL ENGINEERING

J.B. WEST and H.A. PADMORE

Science and Engineering Research Council, Daresbury Laboratory, Daresbury, Warrington WA4 4AD, UK

Contents

1. Introduction	23
2. The synchrotron radiation source	24
2.1. Source emittance	24
2.2. The calculation of synchrotron radiation intensities	28
3. Mirrors	35
3.1. Focusing properties and surface equations of common optical elements	36
3.1.1. The parabola	36
3.1.2. The ellipse	37
3.1.3. Toroids	39
3.2. Reflectivity of mirrors	41
3.3. Separated function mirror systems	52
4. Diffraction grating optics	56
4.1. Application of the grating equation	56
4.1.1. Diffraction	56
4.1.1.1. Spectrograph geometry	57
4.1.1.2. Fixed grating-deviation-angle geometry	57
4.1.1.3. On-blaze geometry	59
4.1.1.4. Plane grating constant-image-distance geometry	61
4.1.1.5. In-focus Miyake geometry	61
4.1.2. Dispersion	62
4.1.3. Entrance and exit slit-width-limited resolution	62
4.2. The geometric aberration theory of diffraction gratings	63
4.2.1. The optical path function	64
4.3. The toroidal grating monochromator	68
4.4. Design of TGM pre-optics	77
4.5. The astigmatically corrected spherical grating in a TGM mounting; the SGM	78
4.6. Diffraction efficiency	82

Contents continued overleaf

Handbook on Synchrotron Radiation, Vol. 2, edited by G.V. Marr
© Elsevier Science Publishers B.V., 1987

Contents continued

5. Beam line layout	84
5.1. General design principles for a multiple instrument beam line	85
5.2. Details of a four instrument beam line at the Daresbury SRS	88
5.2.1. The normal incidence monochromator (Seya)	88
5.2.2. The high resolution normal incidence monochromator	90
5.2.3. The toroidal grating monochromator	92
5.2.4. The soft X-ray monochromator	93
5.3. Post-focusing optics	97
5.3.1. Remote exit slit	98
5.2.2. Capillary light guide	98
5.3.3. Post-focusing mirrors	98
6. Optical components	99
6.1. The mounting and adjustment of optical components	100
6.2. Mirror materials	105
6.3. Optical coatings	108
7. Beam line auxiliary equipment	109
7.1. Laser alignment system	109
7.2. Vacuum equipment	111
7.3. Pressure measuring and monitoring devices	113
7.4. Beam line controls and safety systems	114
7.4.1. Safety aspects	114
7.4.2. Vacuum protection	115
7.4.3. Radiation damage protection	115
8. Conclusion	116
References	117

1. Introduction

Of the many aspects involved in constructing a synchrotron radiation facility, one which is not widely met in other research areas is the close link between large scale mechanical engineering and precision optics. The only other example which comes readily to mind is the construction of large astronomical telescopes, and these have the advantage in general of being single purpose instruments. In contrast, a synchrotron radiation source will be used for a wide variety of experiments covering research areas in physics, chemistry, biology, and applied areas such as material science and technology. The means of bringing the radiation from the orbiting electrons onto the experimental sample with maximum efficiency is therefore highly important, and the experimental requirements vary considerably in this respect. The "beam line" is the means of bringing the "white" radiation to the monochromator and/or experiment, and in general is far from being just a straight piece of pipe between the experiment and the synchrotron radiation source. It may have to feed three or more experiments covering a wide spectral range from the soft X-ray to the near ultraviolet, in addition to being the interface between the experimental vacuum environment and the ultrahigh vacuum of the source. It must also provide users with protection against damaging radiation, particularly in those cases where the electron beam energy in the source exceeds a few hundred MeV. Thus a beam line is often a very complex assembly; due allowance should always be made for the fact that the capital cost in providing a complete set of beam lines on an electron storage ring designed for synchrotron radiation work will far exceed the capital cost of the accelerator itself.

It is the purpose of this chapter to identify the main problems in constructing beam lines for synchrotron radiation and to show how precision optics can be integrated into the general beam line design; hence the term "optical engineering". It is not our intention to review all the beam lines and monochromators on facilities throughout the world, since the literature already contains a wealth of such information. We set out to identify fundamental concepts and good design practice, and hope that these, together with examples of beam line designs presently used, will be helpful to those in the process of designing beam lines for future synchrotron radiation sources.

In many respects, the monochromatising element is an integral part of beam line optical design and the two should be considered as one system. As can be seen from the chapter on VUV (vacuum ultraviolet) monochromators in Volume 1 of this Handbook (R.L. Johnson 1983) there are many designs presently available, and, contrary to the situation a few years ago, any experimental requirement can probably be met by one of these designs. Some of these instruments have been well characterised (see a summary by Ederer 1982, also Saile and West 1983) and a few are available commercially as complete working monochromators. The view taken here is that, as a starting point the monochromators required for the experimental programme in mind should be designed first and the beam line, with all the preoptics involved, designed to incorporate them. As will be seen later, when fitting three or four monochromators into a beam line compromises are inevitable and restraints on the optical design can arise when considering the construction of the whole assembly. There are also applications, (e.g. extremely high aperture, or very large resolving power) where one instrument will dominate beam line design. We shall be concerned primarily with beam lines on which more than one experimental facility is located, although design details for single experiment lines will be considered.

Another factor which enters into beam line design is the wavelength range to be covered, in particular whether experiments using wavelengths shorter than 4 Å, the longest practical wavelength at which a beryllium window can be used, are to be mixed with longer wavelength experiments on the same beam line. For low-energy synchrotron radiation sources this does not apply, but on higher energy or "compromise" sources there are good practical reasons to separate beam lines which have beryllium windows and whose experiments run in air or helium atmospheres from those which run in vacuum. One primary reason is the need for shielded hutches around experiments where there are X-ray paths in air; it is a considerable simplification not to mix shielded experiments with unshielded ones. In addition, so-called X-ray beam lines are simpler in construction and less demanding from the vacuum point of view. This chapter will concentrate on design principles and practical considerations for VUV/soft X-ray beam lines.

Thus we begin with a description of basic source requirements, followed by methods of calculating the light flux from the source. Focusing properties of optical elements and aberration theory will then be considered, since a knowledge of these is essential before a final beam line design, in which the instruments are integrated with the source, can be reached. For general information purposes, a section on the behaviour and choice of optical coatings is included, since, although this has no direct effect on beam line layout, it has a marked effect on beam line performance.

2. The synchrotron radiation source

2.1. Source emittance

There can be little doubt that a prime requirement of a VUV source is small emittance, where emittance is defined by considering the behaviour of the electron beam in phase space. If we define the direction along which the electrons move as the s direction, x and y perpendicular to this direction in and perpendicular to the orbit plane, respectively, then phase space coordinates are x and x', y and y', where $x' = dx/ds$ and $y' = dy/ds$.

Thus x' is the angle the electrons make with the straight ahead direction in the orbit plane, and y', correspondingly, perpendicular to the orbit plane. In general

Optical engineering

these angles are small enough that the approximation $\tan^{-1} dx/ds = dx/ds$, $\tan^{-1} dy/ds = dy/ds$ can be made. Figure 2.1 shows examples of electron beam cross sections in configuration and phase space, and fig. 2.2 shows the phase space ellipses at $s = 0$ chosen to be at a waist of the electron beam cross section, and at a point distance s along the orbit from it, referring to the y direction. The ellipses in figure 2.2 are "one σ" contours, and thus the "one σ" emittance is $\sigma_y \sigma_{y'}$, the equation of the ellipse being

$$\frac{\sigma_{y'}}{\sigma_y} y^2 + \frac{\sigma_y}{\sigma_{y'}} y'^2 = \sigma_y \sigma_{y'}.$$

Similar relations hold for the x direction.

Thus the emittances in the x and y directions are a combination of the electron beam dimension and the angular divergence of the electrons, both of which vary with position along the direction of travel, s. In reality one observes a finite length of arc of the electron trajectory rather than a single point on the trajectory. It is

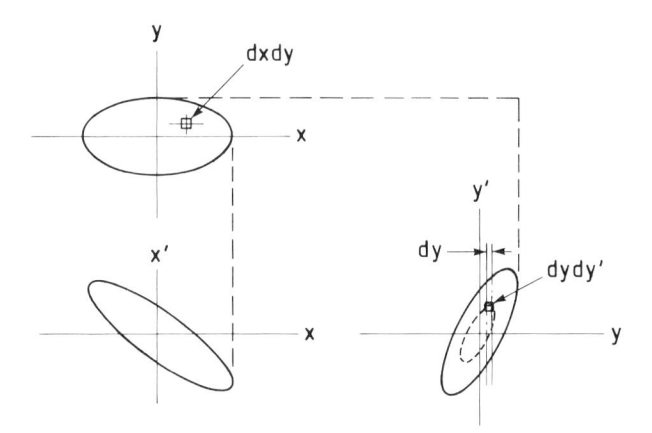

Fig. 2.1. Electon beam cross sections in phase space.

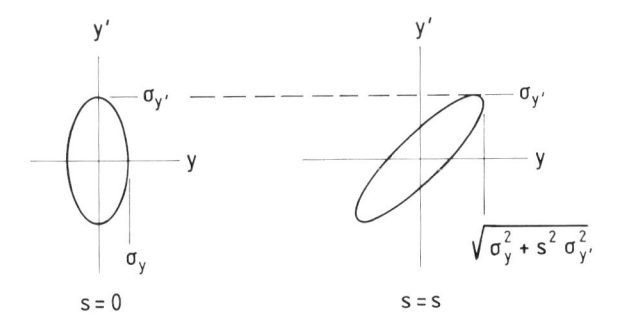

Fig. 2.2. Phase space ellipse at $s = 0$ (waist).

then reasonable to select a midpoint on the arc and project the radiation from points either side of this onto a plane perpendicular to the electron trajectory at this midpoint, the midpoint lying in this plane. This is shown diagrammatically in fig. 2.3 for the yy' phase space. The midpoint ($s = 0$) is chosen to be at an electron beam waist. Hence the phase ellipse (1) is vertical. Moving to s, and assuming no magnetic focusing, the ellipse transforms to a slanted ellipse (2) where $\sigma_{y'}$ remains constant but σ_y increases. The divergence angle of the radiation emitted, an angle which varies with optical wavelength, then increases σ_y, to give the real source at $s = s$ (3). This is then transformed back onto $s = 0$ (4), where the angles are preserved, but the source has increased in apparent size. $\sigma_{y'}$ has become $(\sigma_{y'}^2 + \sigma^2)^{1/2}$, where σ refers to the *photon* angular distribution.

Full mathematical details of all these transformations, and calculations of photon flux from a real source, are to be found in two compreshensive documents by Green (1976, 1977). Clearly, the equations have to be integrated over the full source length (i.e. from $s = -s$ to $s = +s$) in order to map all points of the orbit seen by the beam line optical system onto the plane at $s = 0$. The reader is therefore referred to Green's papers for full mathematical details; his notation has been used here. It is sufficient for present purposes, having presented the basic ideas of the beam optics, to consider which are the dominant source parameters. Good source design should ensure minimum emittance, and provide beam line access to those points in the storage ring lattice where this is a minimum. A factor which influences the vertical beam size is the coupling between horizontal and vertical oscillations of the electron beam; when kept to a minimum this gives the characteristic letter box shape of the electron beam, and thus matches vertically dispersing instruments well. The coupling between the horizontal and vertical oscillations can be kept down to the 1% level with proper design and precise installation of the magnetic components; fig. 2.4 shows the standard deviations for

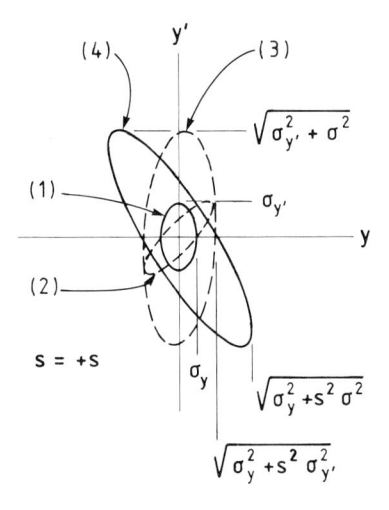

Fig. 2.3. Projection onto a plane at the extended source midpoint.

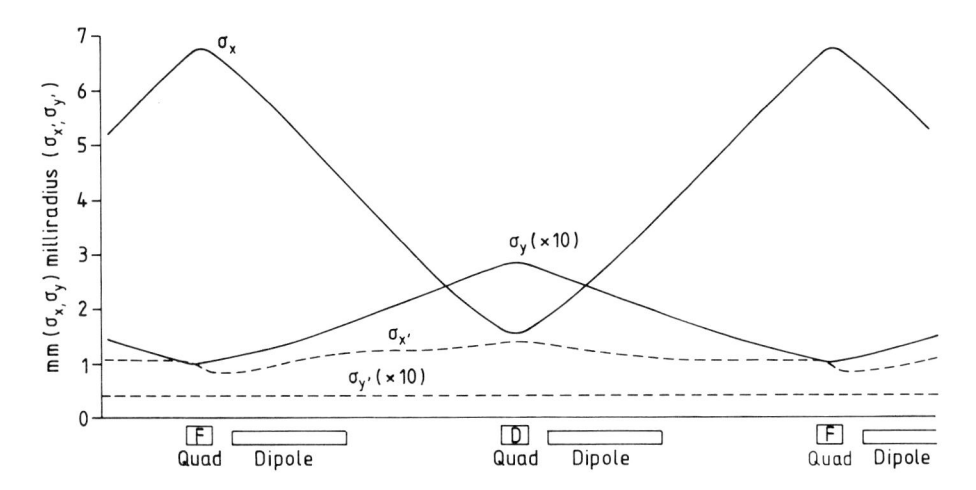

Fig. 2.4. Source size and emittance in the Daresbury SRS.

the source size and emittance for the present magnet lattice in the Daresbury synchrotron radiation source. In fact this is not a low emittance source compared to the VUV rings at Brookhaven and Berlin, where the emittances are at least a factor of 10 smaller, but the point to realise here is that the major contribution to the emittance figure comes from the physical source size; the contribution from the different angular trajectories is substantially less than the divergence of the emitted radiation as far as the VUV and soft X-ray region is concerned. Figure 2.5 shows the improved emittance expected from the modified SRS (Walker

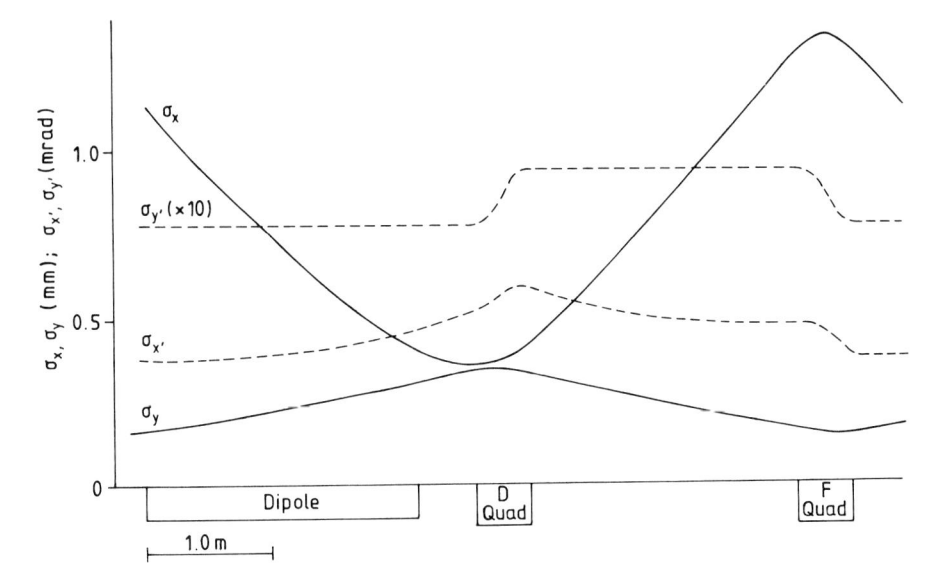

Fig. 2.5. As fig. 2.4, but after the high-brightness lattice modification.

1985). Thus in calculating the apertures (and thus sizes) of collimating or focusing mirrors to collect radiation from the source and feed it to monochromators, the main parameters to consider are source size and the vertical angular divergence resulting from the properties of the radiation itself. The former is provided by the source designer; the latter can be found in published calculations, though not always in a useful form. Thus the next section is devoted to deriving the most useful equations for calculation of the synchrotron radiation flux.

2.2. The calculation of synchrotron radiation intensities

The basic equations describing synchrotron radiation have been given by Schwinger (1949), Sokolov and Ternov (1968), and Tomboulian and Hartman (1956), the latter giving a quantitative comparison between measured synchrotron radiation spectra and the basic theory. Several review papers have been published detailing the properties of synchrotron radiation and their numerical evaluation, of which those by Mack (1966) and by Green (1976, 1977) are the most useful. Several other more general papers have been given by Winick (1980), Codling (1973) and Rowe (1973). The most useful forms of the equations given in the above references are considered here.

The number of photons per second radiated from a tangent point can be expressed in several forms, the most useful being the following.

$$N_k(\lambda) = 1.256 \times 10^{10} k \gamma y G_0(y) \text{ photons}/k\lambda, \text{ s, mA, mrad } \theta . \tag{2.1}$$

Equation (2.1) gives the number of photons per unit bandwidth k.

$$N_{\Delta\lambda}(\lambda) = 0.2998 \frac{\gamma^4}{\rho} y^2 G_0(y) \text{ photons}/\text{Å, s, mA, mrad } \theta . \tag{2.2}$$

Equation (2.2) gives the number of photons per angstrom.

$$N_{\Delta E}(\lambda) = 1.013 \times 10^6 \gamma \lambda_c G_0(y) \text{ photons}/\text{eV, s, mA, mrad } \theta . \tag{2.3}$$

Equation (2.3) gives the number of photons per electron volt photon energy.

$$P(\lambda) = 1.421 \times 10^{-26} \frac{\gamma^7}{\rho^2} y^3 G_0(y) \text{ W}/\text{Å, mA, mrad } \theta . \tag{2.4}$$

Equation (2.4) gives the power radiated per angstrom. These four equations apply to the case where the radiation spectrum is integrated over all vertical angles φ. The terms used in eqs. (2.1)–(2.5) are as follows:

ρ: bending magnet radius (m),
θ: horizontal opening angle (mrad),
V: machine energy (GeV),

φ: vertical angle from the storage ring plane (mrad),
λ: wavelength (Å),
γ: 1957 V (GeV),
λ_c: critical wavelength; equal power is radiated above and below the critical wavelength. $\lambda_c = 5.59\rho/V^3$ (Å),
$y = \lambda_c/\lambda = E_c/E$.

Function G_0 is defined as

$$G_0(y) = \int_y^\infty K_{5/3}(x)\, dx, \tag{2.5}$$

where $K_{5/3}(x)$ is a modified Bessel function of the second kind. Bessel functions of this type can be related to normal modified Bessel functions of non-integral order by the following relation:

$$K_\nu(x) = \frac{\pi}{2} \frac{[I_{-\nu}(x) - I_\nu(x)]}{\sin \nu\pi}. \tag{2.6}$$

For small arguments (e.g. $\nu = 5/3$ and $1/3 < x < 3.0$) $I_\nu(x)$ can be evaluated using a series approximation,

$$I_\nu(x) = \sum_{s=0}^\infty \frac{1}{s!(s+\nu)!} \left(\frac{x}{2}\right)^{2s+\nu},$$

$$I_{-\nu}(x) = \sum_{s=0}^\infty \frac{1}{s!(s-\nu)!} \left(\frac{x}{2}\right)^{2s-\nu}. \tag{2.7}$$

The non-integer factorials can be evaluated for large arguments using Stirling's approximation and extended to lower values using the factorial difference relation

$$s! = \sqrt{2\pi}\, s^{s+1/2}\, e^{-s}(1 + 1/12s) \quad \text{Stirling's approximation}, \tag{2.8}$$

$$(s-1)! = \frac{s!}{s} \quad \text{difference relation}. \tag{2.9}$$

An alternative method is to calculate $s!$ in the range $0 \leq s \leq 1$ from a fitted polynomial and extend to higher values using the difference relation

$$s! = 1 + \sum_{n=1}^8 b_n s^n. \tag{2.10}$$

A polynomial fit of this type has been made by Hastings (1955) with an error of less than 3×10^{-7}. Negative factorials can be found using the relation

$$-s! = \frac{1}{s!}\left(\frac{\pi s}{\sin \pi s}\right). \tag{2.11}$$

For arguments of $K_\nu > 3$, an asymptotic approximation can be used:

$$K_\nu(x) = \left(\frac{\pi}{2x}\right)^{1/2} e^{-x}\left[1 + \frac{(4\nu^2 - 1)}{8x} + \frac{(4\nu^2 - 1)(4\nu^2 - 9)}{2!(8x)^2} + \cdots\right]. \quad (2.12)$$

The series form for $x \leq 3$ and the asymptotic form for $x \geq 3$ has been used by Poole (1975, 1976a, b) for the calculation of synchrotron radiation spectra. A more convenient form of $K_\nu(x)$ is given by Arfken (1970) and by Abramowitz and Stegun (1972):

$$K_\nu(x) = \int_0^\infty e^{-x \cosh t} \cdot \cosh \nu t \, dt. \quad (2.13)$$

The form has been directly evaluated using a numerical integration technique by Williams and Weisenbloom (1979). A rapidly converging series form of $K_\nu(x)$ which can be computed faster than the previously given forms has been given by Kostroun (1980):

$$K_\nu(x) = h\left[\tfrac{1}{2} e^{-x} + \sum_{r=1}^{\infty} e^{-x \cosh(rh)} \cdot \cosh(\nu r h)\right], \quad (2.14)$$

where h is a small interval. For the computation of G_0 (eq. 5) required in the calculation of synchrotron radiation spectra we require the integral of $K_\nu(x)$ from λ_c/λ to infinity. This can be obtained by numerical integration of $K_\nu(x)$, or using an expression related to eq. (2.14) given by Kostroun (1980):

$$\int_y^\infty K_\nu(x) = h\left[\tfrac{1}{2} e^{-x} + \sum_{r=1}^{\infty} e^{-x \cosh(rh)} \frac{\cosh(\nu r h)}{\cosh(rh)}\right]. \quad (2.15)$$

For $\nu = 1/3$ or $2/3$ the series can be truncated at $r = 20$. The expressions given in eqs. (2.1)–(2.4) give the vertically integrated photon flux and radiated power. As many beam line optical systems cannot vertically integrate and in some cases have a variable aperture (e.g. a grating of fixed length scanning in wavelength), it is important to know the variation of flux with vertical angle φ. For a given wavelength, the variation of parallel and perpendicular components of the electric vector (with respect to the storage ring plane) can be given by

$$F_\parallel(\varphi) = [1 + (\gamma\varphi)^2]^2 K_{2/3}^2\left\{\frac{\lambda_c}{2\lambda}[1 + (\gamma\varphi)^2]^{3/2}\right\}, \quad (2.16)$$

$$F_\perp(\varphi) = (\gamma\varphi)^2[1 + (\gamma\varphi)^2]K_{1/3}^2\left\{\frac{\lambda_c}{2\lambda}[1 + (\gamma\varphi)^2]^{3/2}\right\}. \quad (2.17)$$

The degree of linear polarisation can then be found directly from

$$P_{\text{lin}} = \frac{F_{\parallel} - F_{\perp}}{F_{\parallel} + F_{\perp}}, \qquad (2.18)$$

and the degree of circular polarisation can then be found from

$$P_{\text{circ}}^2 + P_{\text{lin}}^2 = 1. \qquad (2.19)$$

In general, a beamline will accept a vertical aperture (which might be wavelength dependent) which is symmetric about the orbit plane. To evaluate the number of photons collected at a particular wavelength, the vertical distribution functions have to be integrated up to the desired vertical aperture,

$$N_{k,\gamma}(\lambda) = 1.256 \times 10^{10} k \gamma y G_0 \frac{\int_0^w F_{\parallel}(\varphi) \, d\varphi}{\int_0^\infty [F_{\parallel}(\varphi) + F_{\perp}(\varphi)] \, d\varphi}, \qquad (2.20)$$

where w is the desired vertical angle from the orbit plane. In this case the parallel component has been calculated, but by replacing $F_{\parallel}(\varphi)$ in the numerator by $F_{\perp}(\varphi)$, the perpendicular component can be calculated. In the example given the number of photons per unit bandwidth emitted into a full angle of $2w$ symmetric about the orbit plane is given, but any of the functions given in eqs. (2.1)–(2.4) could be used if units of (unit bandwidth)$^{-1}$ were not desired. An alternative way to calculate the vertical distribution of power or photon flux is to calculate the function in the orbit plane and then use $F_{\parallel}(\varphi)$ and $F_{\perp}(\varphi)$ to scale the function out of plane. Note that $F_{\parallel}$ has a maximum at $\varphi = 0$ and $F_{\perp}(\varphi)$ has a minimum at $\varphi = 0$. The in-plane functions analogous to eqs. (2.1)–(2.4) have been given by Green (1976, 1977) as

$$N_k(\lambda, 0) = 3.461 \times 10^6 k \gamma^2 y^2 K_{2/3}^2(y/2) \text{ photons}/k\lambda, \text{ s, mA, mrad } \theta, \qquad (2.21)$$

$$N_{\Delta\lambda}(\lambda, 0) = 8.263 \times 10^{-5} \frac{\gamma^5}{\rho} y^3 K_{2/3}^2(y/2) \text{ photons}/\text{Å, s, mA, mrad } \theta, \qquad (2.22)$$

$$N_{\Delta E}(E, 0) = 3.951 \times 10^{19} \frac{E\rho^2}{\gamma^4} K_{2/3}^2(y/2) \text{ photons}/\text{eV, s, mA, mrad } \theta, \qquad (2.23)$$

$$P(\lambda, 0) = 3.918 \times 10^{-30} \frac{\gamma^8}{\rho^2} y^4 K_{2/3}^2(y/2) \text{ W}/\text{Å, mA, mrad } \theta. \qquad (2.24)$$

The out-of-plane functions can now be found by scaling, as $F_{\perp}(0)$ and therefore $N_{\perp}(\lambda, 0)$ is zero,

$$N_\|(\lambda, \varphi) = N_\|(0) F_\|(\varphi)/F_\|(0) , \qquad (2.25)$$

$$N_\perp(\lambda, \varphi) = N_\|(0) F_\perp(\varphi)/F_\|(0) . \qquad (2.26)$$

$N_\|(0)$ is representative of any of the functions given by eqs. (2.21)–(2.24). For a particular beam line aperture $2w$ we therefore have to integrate up to the limit. In the case of photon flux we would have

$$N_{\|,w}(\lambda) = \frac{N_\|(0)}{F_\|(0)} \int_0^w F_\|(\varphi) \, d\varphi , \qquad (2.27)$$

and

$$N_{\perp,w}(\lambda) = \frac{N_\|(0)}{F_\|(0)} \int_0^w F_\perp(\varphi) \, d\varphi . \qquad (2.28)$$

In designing beam line components such as masks and collimators which absorb all the synchrotron radiation power over their aperture, it is desirable to know both the total power to be dissipated and the vertical power distribution. From the latter, the power density can be found as a function of φ so that the thermally induced stress in the device can then be calculated. The total power can be given by

$$P = 14.08 V^4 \frac{I}{R} \text{ W/mrad} , \qquad (2.29)$$

where V is the machine energy (GeV), I is the machine current (A), R is the radius of bending magnet (m), and the vertical power distribution is given by

$$P = 1.440 \times 10^{-18} \frac{\gamma^5}{\rho} \left\{ (1 + \gamma^2\varphi^2)^{-5/2} \left[\frac{7}{16} + \frac{5}{16} \frac{\gamma^2\varphi^2}{(1+\gamma^2\varphi^2)} \right] \right\}$$

W/mrad θ, mrad φ, mA .

To illustrate the form of the synchrotron radiation spectrum, fig. 2.6 shows $N_k(\lambda)$ ($k = 0.1\%$) plotted for two machines of different λ_c. The upper curve is for the SRS with $\lambda_c = 3.88$ Å and the lower curve is for BESSY with $\lambda_c = 19.4$ Å. It can be seen that in the low-energy limit the flux is the same, but that BESSY peaks at 200 eV compared to the SRS peak of 900 eV. The extension of the spectrum to high photon energies causes heating problems on XUV beam lines which do not occur on lower energy (higher λ_c) machines such as BESSY.

Figures 2.7 and 2.8 show the vertical functions $F_\|$, $F_\perp$ and their sum for 3 eV and 300 eV for the SRS ($V = 2$ GeV, $\rho = 5.56$ m). It can be seen that the vertical function is much more localised near the orbit plane at 300 eV than at 3 eV. In both spectra it is useful to note the large perpendicular component localised out of plane.

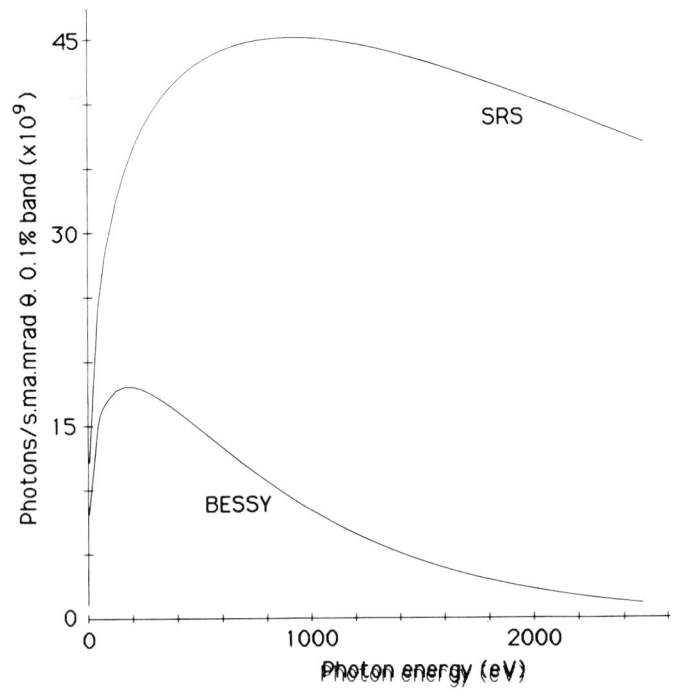

Fig. 2.6. Vertically integrated photon fluxes for the Daresbury SRS and for BESSY as a function of photon energy (0.1% bandpass).

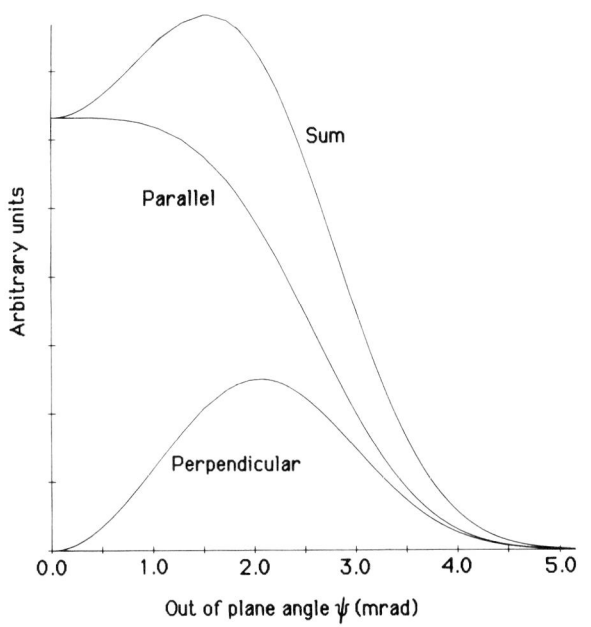

Fig. 2.7. The parallel and perpendicular components to the electric vector and their sum for 3 eV photon energy from the SRS (2 GeV, $\rho = 5.56$ m).

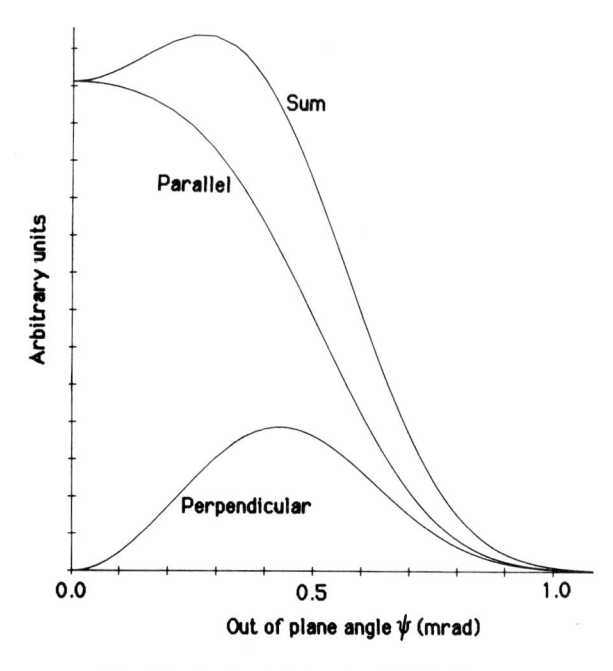

Fig. 2.8. As fig. 2.7, but for 300 eV photons.

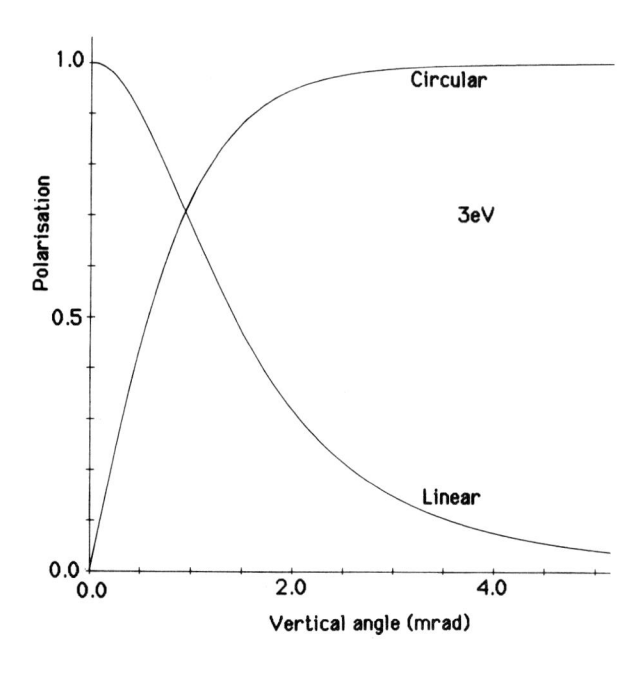

Fig. 2.9. Linear and circular polarisation as a function of the vertical emission angle φ for 3 eV photons from the SRS.

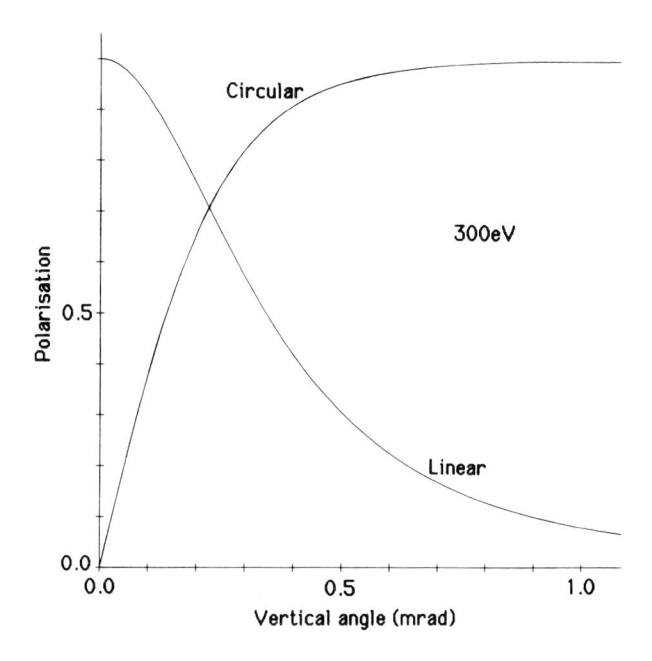

Fig. 2.10. As fig. 2.9, but for 300 eV photons.

For experiments requiring a high degree of linear polarization at low (<100 eV) photon energies, the perpendicular component can be removed by utilising the differential reflectivity of parallel and perpendicular polarized light from grazing incidence mirrors (see later) or by operating at smaller apertures. In the soft X-ray region, however, the reflectivity is very similar for the two components and so highly linearly polarised light can only be obtained by operating at small vertical apertures. Figures 2.9 and 2.10 show the linear and circular polarization, with the use of eqs. (2.18) and (2.19), as a function of the vertical angle φ for 3 eV and 300 eV, respectively. It can be seen that in the limit of high φ, the radiation becomes circularly polarized. This property of synchrotron radiation has been utilised in some very elegant experiments by Eyers et al. (1983) and Heckenkamp et al. (1984) in which off-plane radiation is selected by a vertically moveable baffle. By selecting radiation from above or below the storage ring plane the helicity of the light can be changed (σ^+, σ^-).

3. Mirrors

In order to use the radiation from the source it first has to be monochromatized and focused. The collection and focusing of VUV-SXR light is accomplished using glancing incidence mirrors. The purpose of this section is to examine the general focusing characteristics of single and multiple element systems and to describe the fundamental relations which affect X-ray reflectivity. In general there are many possible combinations of element type and coating that could satisfy a particular application and it is the designer's task to find the simplest and most cost effective solution. The requirement for the simplest solution is twofold: firstly, the solution should have a minimum number of components so that the overall transmission is maximised; secondly, complex mirror shapes are expensive and often require complex mounting and alignment systems, e.g. paraboloids, ellipsoids, and so should only be used where really necessary. A further consideration is that aspheric elements are difficult to manufacture and involve complex metrology. The result is that a designer will have less confidence in these elements where used in a critical application than if cylindrical or spherical elements had been used. The penalty for using the easily fabricated cylinder or sphere is that more elements would have to be used to give the same properties as the aspheric mirror, e.g. double focusing or coma-corrected imaging in one plane. The design procedure should therefore consider many possible solutions so that detailed comparison of the cost of the optical and associated mechanical systems can be made together with a comparison of their theoretical optical performance, sensitivity to source movement or misalignment and sensitivity to possible errors in manufacture.

3.1. Focusing properties and surface equations of common optical elements

3.1.1. The parabola

The geometry of the parabola is given in fig. 3.1. The parabola has the property that rays travelling parallel to the symmetry axis OX are all focused to a point at A. Conversely, the parabola will collimate rays emanating from a point situated at the focus of the parabola A. If focusing in one plane only is required the mirror surface would be plane in the direction perpendicular to the plane of the diagram. If the angle of incidence on the off-axis segment is θ, f is the distance from the pole of the segment to the focus and a is the distance from the focus to the pole of the parabola, then the line equation for the parabola is

$$Y^2 = 4aX, \tag{3.1}$$

where

$$a = f \cos^2 \theta. \tag{3.2}$$

The position of the pole can be given by

$$X_0 = a \tan^2 \theta, \tag{3.3}$$

$$Y_0 = 2a \tan \theta. \tag{3.4}$$

If double focusing is required a paraboloidal shape could be used and can be obtained by rotating a parabola around its axis of symmetry OX. The equation of

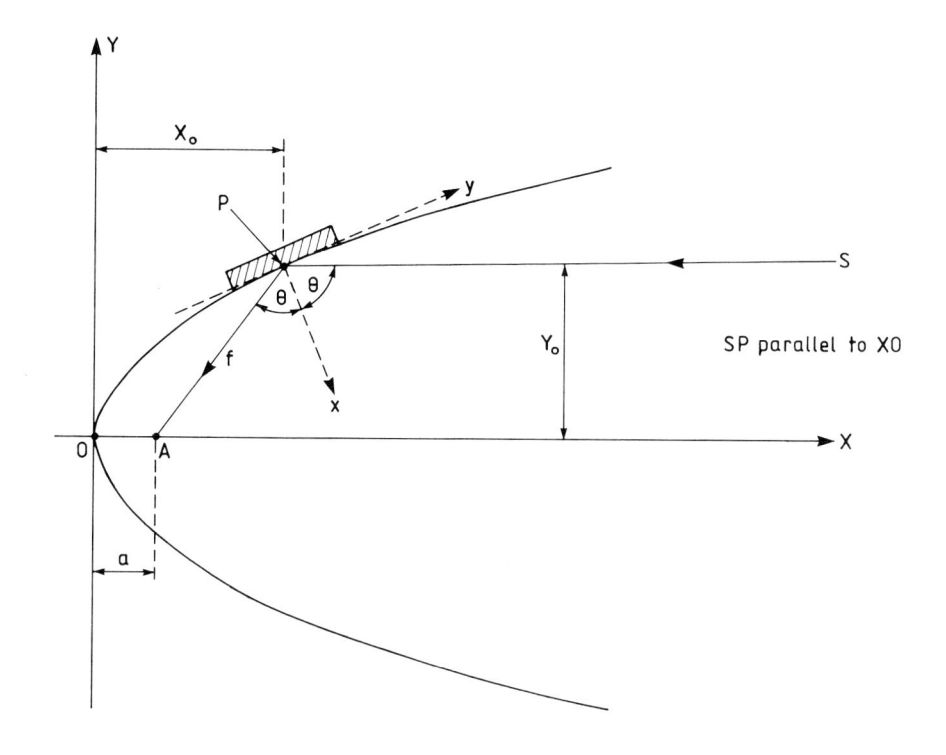

Fig. 3.1. Geometry of the parabola.

the surface of the paraboloid would then be

$$Y^2 + Z^2 = 4aX. \tag{3.5}$$

These equations use a Cartesian coordinate system centred on the pole of the parabola (X, Y, Z). It is often necessary, however, to use a coordinate system centred on the centre of the optical segment under consideration. In fig. 3.1, the off-axis coordinate system has the x axis normal to the centre of the element with the y axis being the tangent plane and the z axis passing through the centre of the element and perpendicular to the plane of the paper. In this coordinate system, the equation of the paraboloid becomes

$$x^2 \sin^2 \theta + y^2 \cos^2 \theta + z^2 - 2yx \sin \theta \cos \theta - 4ax \sec \theta = 0 \tag{3.6}$$

3.1.2. The ellipse

The geometry of the ellipse is given in fig. 3.2. The ellipse has the property that rays from one point focus F_1 will always be perfectly focused to the second point focus F_2. In a Cartesian system (XYZ) centred on the centre of the ellipse O, the equation of the ellipse is given by

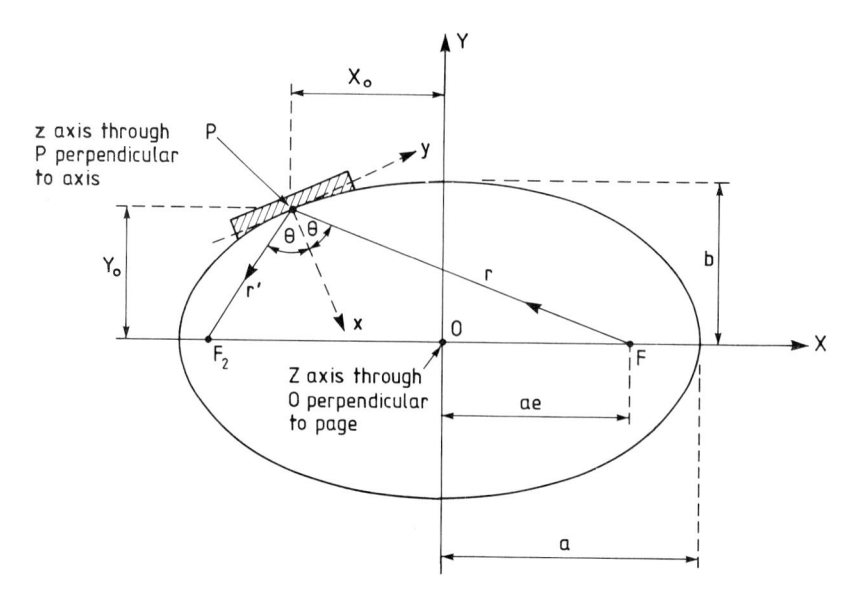

Fig. 3.2. Geometry of the ellipse.

$$\frac{X^2}{a^2} + \frac{Y^2}{b^2} = 1. \tag{3.7}$$

The parameter of the ellipse a and b can be defined in terms of the object and image distances r and r' and the angle of incidence θ by

$$a = (r + r')/2, \tag{3.8}$$

$$b = [a^2(1 - e^2)]^{1/2}, \tag{3.9}$$

where e is the eccentricity and is given by

$$e = \frac{1}{2a} [r^2 + r'^2 - 2rr' \cos 2\theta]^{1/2}. \tag{3.10}$$

The position of the pole of the element is given by

$$Y_0 = \frac{rr' \sin 2\theta}{2ae}, \tag{3.11}$$

$$X_0 = a \left\{ 1 - \frac{Y_0^2}{b^2} \right\}^{1/2}. \tag{3.12}$$

If double focusing is required an ellipsoidal shape could be used. This would be obtained by rotating the ellipse around the major axis OX. The equation of the surface formed would be given by

$$\frac{X^2}{a^2} + \frac{Y^2}{b^2} + \frac{Z^2}{b^2} = 1 . \tag{3.13}$$

In a coordinate system x, y, z with the origin at the centre of the optical element, as given in fig. 3.2, the surface equation of the ellipsoid can be given by

$$x^2 \left(\frac{\sin^2 \theta}{b^2} + \frac{1}{a^2} \right) + y^2 \left(\frac{\cos^2 \theta}{b^2} \right) + \frac{z^2}{b^2} - x \left(\frac{4f \cos \theta}{b^2} \right)$$
$$- xy \left[\frac{(e^2 - \sin^2 \theta)^{1/2} 2 \sin \theta}{b^2} \right] = 0 , \tag{3.14}$$

where

$$f = \left(\frac{1}{r} + \frac{1}{r'} \right)^{-1} .$$

3.1.3. Toroids

Toroidal elements can be classified into two types, the "bicycle tyre" toroid and the "apple core" toroid. These are shown in fig. 3.3. The geometry for reflection is given in fig. 3.4. The bicycle tyre toroid is generated by rotating a circle of radius ρ in an arc of radius R. The apple core toroid is generated by rotating a circle of radius R around a chord which lies a maximum distance ρ from the circle along a radial line. Referring to fig. 3.4, the conditions for a meridian focus are

$$\left(\frac{1}{r} + \frac{1}{r'} \right) \frac{\cos \theta}{2} = \frac{1}{R} , \tag{3.15}$$

and for a sagittal focus are

$$\left(\frac{1}{r} + \frac{1}{r'} \right) \frac{1}{2 \cos \theta} = \frac{1}{\rho} . \tag{3.16}$$

To obtain a stigmatic image of a point source we therefore have the relationship between R and ρ,

$$\frac{\rho}{R} = \cos^2 \theta . \tag{3.17}$$

In the case of a sphere therefore, a stigmatic image can only be obtained at normal incidence. In the coordinate system x, y, z with the origin at the centre of the optical element, the surface equations can be given as

bicycle type toroid:

$$x^2 + y^2 + z^2 = 2Rx - 2R(R - \rho) + 2(R - \rho)[(R - x)^2 + y^2]^{1/2} , \tag{3.18}$$

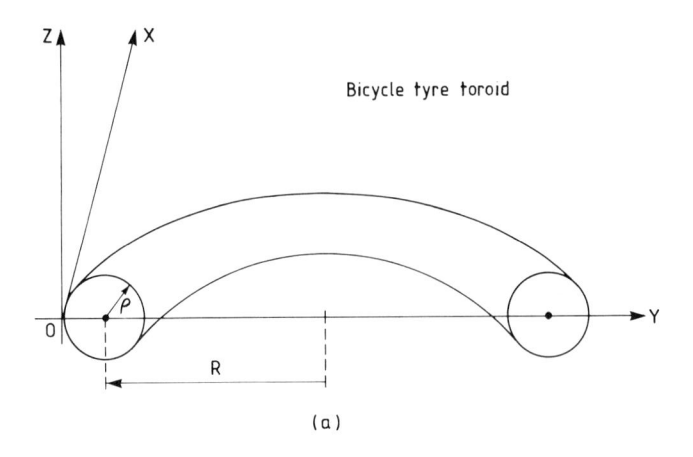

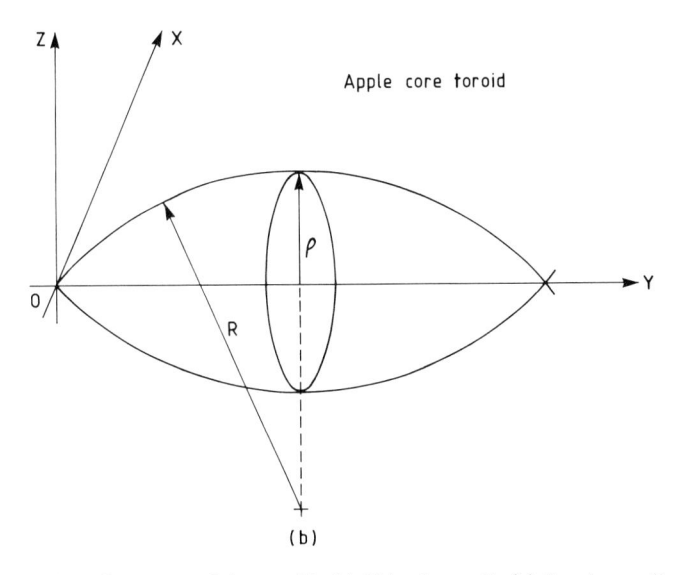

Fig. 3.3. Geometry of the toroid: (a) "bicycle tyre", (b) "apple core".

apple core toroid:

$$x^2 + y^2 + z^2 = 2\rho x + 2R(R - \rho) - 2(R - \rho)[R - y^2]^{1/2}. \tag{3.19}$$

The magnification is given simply by the ratio of the image to object distances and is the same in both the sagittal and meridian directions. Often in beam line optical systems it is required to have different sagittal and meridian magnifications and therefore separate elements are used for focusing in each direction. Different combinations of systems are examined in a later section.

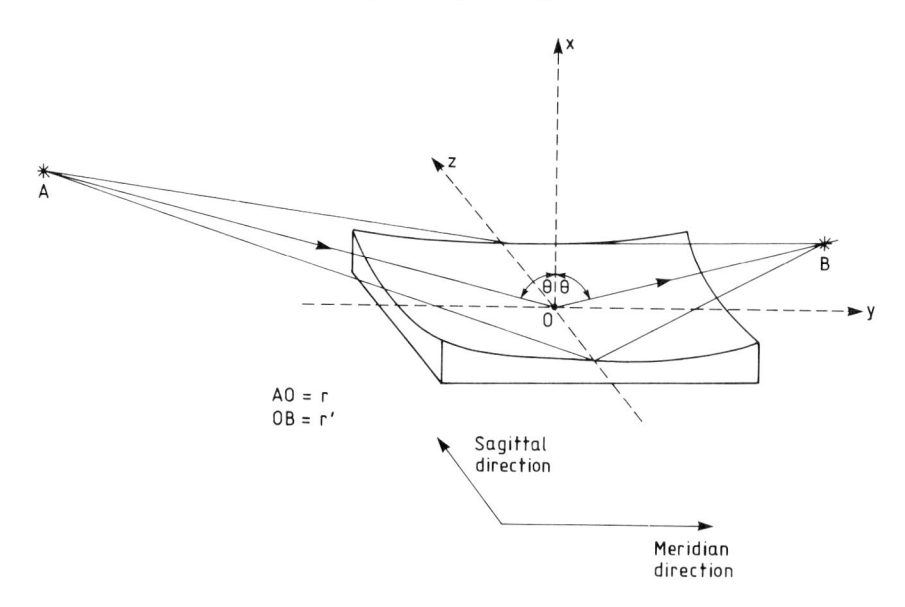

Fig. 3.4. Reflection geometry.

3.2. Reflectivity of mirrors

The interaction of electromagnetic radiation with mirror surfaces can be expressed in terms of a wavelength dependent complex constant, the real part representing velocity and the imaginary part representing absorption. The complex constant can either be expressed as the complex refractive index $\tilde{N}$,

$$\tilde{N} = n - \mathrm{i}k, \tag{3.20}$$

or as the complex dielectric constant $\tilde{K}$,

$$\tilde{K} = 1 - \alpha - \mathrm{i}\gamma. \tag{3.21}$$

In the soft X-ray and X-ray region the real part of the complex refractive index can become close to unity and so is expressed in the form

$$\tilde{N} = 1 - \delta - \mathrm{i}\beta. \tag{3.22}$$

The complex refractive index is related to the complex dielectric constant by

$$\tilde{N}^2 = \tilde{K}, \tag{3.23}$$

and so we can relate the individual components of the complex numbers

$$\alpha = 2\delta - \delta^2 + \beta^2, \tag{3.24}$$

$$\gamma = 2\beta(1 - \delta).\tag{3.25}$$

As both δ and β are small in the soft X-ray and X-ray regions expressions (3.24) and (3.25) are often approximated to

$$\alpha \simeq 2\delta, \quad \gamma \simeq 2\beta.\tag{3.26}$$

The complex dielectric constant can be calculated from the complex atomic scattering factors as given by Henke (1982) and by Henke et al. (1981, 1982):

$$\tilde{F} = f_1 + if_2,\tag{3.27}$$

where

$$\alpha = Df_1, \quad \gamma = Df_2,$$

and

$$D = \frac{r_0 \lambda^2 N_A \rho}{\pi A}.\tag{3.28}$$

r_0 is the classical radius of the electron ($=2.8179 \times 10^{-15}$ m), N_A is Avogadro's number, A is the atomic weight, ρ is the density. These expressions for α and γ are for a single-element reflector but similar simple expressions can be found for compound reflectors (Henke et al. 1981, 1982). The atomic scattering factors are related to measurable parameters by the Kramers–Kronig relation (James 1965),

$$f_1 = Z + C \int_0^\infty \frac{\varepsilon^2 \mu_a(\varepsilon) \, d\varepsilon}{E^2 - \varepsilon^2},\tag{3.29}$$

$$f_2 = (\pi/2) C E \mu_a(E),\tag{3.30}$$

where E is the incident photon energy, $C = 1/(\pi r_0 hc)$, with r_0 the classical radius of the electron, μ_a is the atomic photoabsorption cross section, and Z represents the number of electrons per atom. The integral in eq. (3.29) represents anomalous dispersion and can vary rapidly close to ionization thresholds. The atomic scattering factors can be uniquely determined from atomic photoabsorption data using relations given in eqs. (3.29) and (3.30). A compilation of atomic scattering factors from 100 eV to 2000 eV photon energy for the elements $z = 1$ to $z = 94$ has been given by Henke et al. (1982) together with a review of the methods of calculation. This work also includes calculated reflectivities at several selected angles of incidence.

The generalised Fresnel equations (Mahan 1956) can now be used to calculate the reflectivity of a mirror surface for s and p polarized light. Figure 3.5 gives the geometry of polarized light reflection. The notation normally adopted is s, σ or

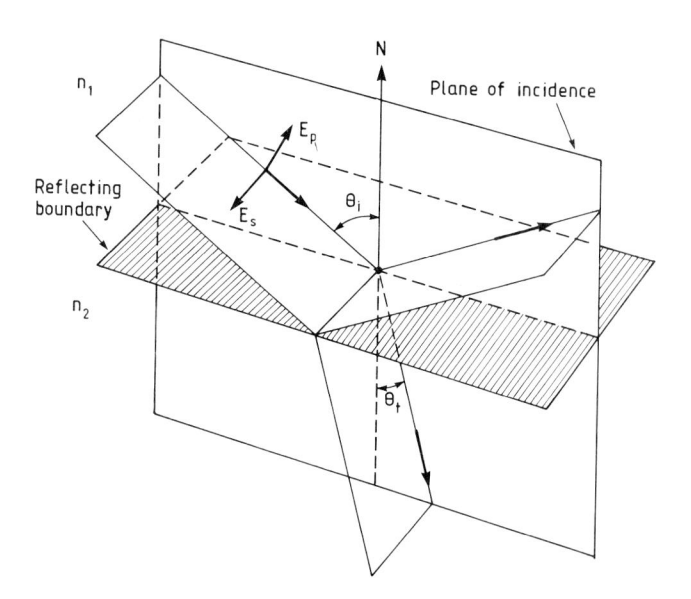

Fig. 3.5. Geometry for polarized light reflection.

TE polarization: electric vector perpendicular to the plane of incidence; p, π or TM polarization: electric vector parallel to the plane of incidence. The generalized Fresnel equations are usually given in the form

$$R_s = \frac{a^2 + b^2 - 2a \cos \theta + \cos^2 \theta}{a^2 + b^2 + 2a \cos \theta + \cos^2 \theta}, \tag{3.31}$$

$$R_p = R_s \left(\frac{a^2 + b^2 - 2a \sin \theta \tan \theta + \sin^2 \theta \tan^2 \theta}{a^2 + b^2 + 2a \sin \theta \tan \theta + \sin^2 \theta \tan^2 \theta} \right), \tag{3.32}$$

where

$$2a^2 = [(n^2 - k^2 - \sin^2 \theta)^2 + 4n^2 k^2]^{1/2} + (n^2 - k^2 - \sin^2 \theta), \tag{3.33}$$

$$2b^2 = [(n^2 - k^2 - \sin^2 \theta)^2 + 4n^2 k^2]^{1/2} - (n^2 - k^2 - \sin^2 \theta). \tag{3.34}$$

By equating the complex refractive index and the complex dielectric constant using eqs. (3.23) and (3.33), eq. (3.34) can be stated in terms of α and γ and hence the atomic scattering factors,

$$1 - \alpha = n^2 - k^2, \tag{3.35}$$

$$\gamma^2 = 4n^2 k^2. \tag{3.36}$$

The degree of polarization P can be defined as

$$P = \frac{R_s - R_p}{R_s + R_p}, \qquad (3.37)$$

where R_s and R_p are the reflectivities for incident light polarized with the electric vector perpendicular and parallel to the plane of incidence, respectively. Using the optical constant data of Hagemann et al. (1974) ($n = 1.07$, $k = 0.84$, 21.2 eV), the reflectivity and degree of polarization for a gold surface at 21.2 eV photon energy (He I) is given in fig. 3.6 as a function of the angle of incidence. For a non-absorbing material (e.g. quartz at 21.2 eV) the p reflectivity reaches zero at the well-known Brewster angle given by

$$\theta_B = \tan^{-1}\left(\frac{n_2}{n_1}\right). \qquad (3.38)$$

For an absorbing material, the p reflectivity now goes through a minimum as shown in fig. 3.6 for gold at 21.2 eV. By using two or three reflections in series a high degree of polarization can be achieved in the low-energy VUV region. Multi-element reflection polarizers are widely used with unpolarized discharge sources and have been extensively reviewed by Samson (1967) and Vodar and Romand (1974). Humphreys-Owen (1961) has given an expression for the angle of incidence, as a function of n and k, at which R_p is a minimum and Damany (1965) has given an expression for the angle of incidence at which the polarization is a maximum. The polarization of VUV radiation can be found using the method

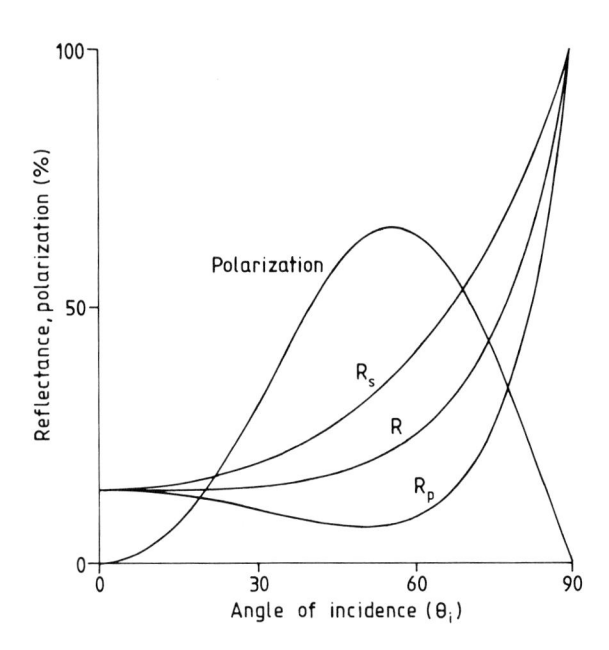

Fig. 3.6. Reflection and polarization characteristics of gold at 21.2 eV.

of Abelès (1950). This uses the relation

$$R_s = R_p^2 \quad \text{for} \quad \theta_i = 45°. \tag{3.39}$$

If R_1 represents the absolute reflectivity with the plane of incidence perpendicular to the plane in which the electric vector is maximum and R_2 represents the absolute reflectivity with the plane of incidence parallel, then the polarization is given by Samson (1976, 1978) as

$$P = \frac{R_2 - R_1}{1 + R_1 + R_2 - [1 + 4(R_1 + R_2)]^{1/2}}. \tag{3.40}$$

This method requires no knowledge of the optical constants and the validity of this expression has been verified by Rabinovitch et al. (1965). A similar method but using multiple mirror reflections has been given by Hamm et al. (1965). The design criteria for reflection polarizers and analysers together with an analysis of the errors involved in measuring polarization have been given by Hunter (1978) and by Hass and Hunter (1978).

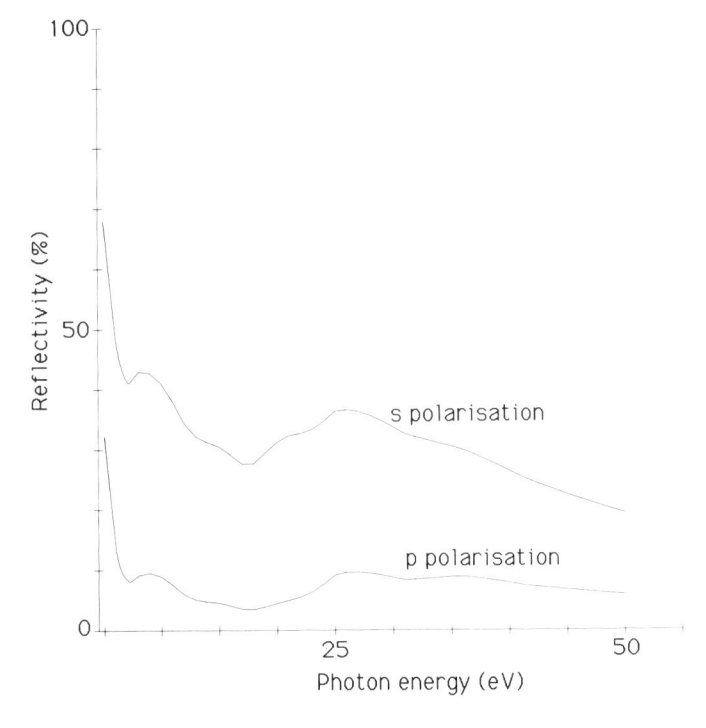

Figure 3.7 shows the reflectivity of gold for s and p polarization at an angle of incidence of 55° calculated from the optical constant data of Hagemann et al. (1974). Clearly, the reflectivity for s polarization is far higher than for p

Fig. 3.7. Reflectivity of gold for s and p polarization at an angle of incidence of 55°.

polarization in the VUV region. Consequently, as bending magnet sources of synchrotron radiation are polarized in the orbit plane with the electric vector parallel to the plane (horizontal), vertical reflection or diffraction geometry should be used, where possible, in the VUV energy range. The alternative to this arrangement is to use a glancing p reflection (horizontal deflection) as the s and p reflectivities converge at small glancing angles. If a high degree of linear polarization is required by a VUV experiment, s-polarization geometry is to be preferred. Most VUV monochromators collect all the available light in the vertical direction and from basic synchrotron radiation theory it can be seen that vertically integrated VUV light is elliptically polarized with a significant component of the electric vector being perpendicular to the orbit plane. To obtain a high degree of polarization therefore the beam-line optical system should be designed to be all s reflecting to act as a polarizer.

Figure 3.8 shows the reflectivity for s polarization of gold as a function of the angle of incidence calculated using the optical constant data of Hagemann et al. (1974) over the photon energy range of 5–100 eV. This demonstrates the strong dependence of the reflectivity on the angle of incidence and the complex structure due to core excitations.

The reflectivities also do not show a sharp cut-off at higher energies even for large angles of incidence, but rather a gradual fall in efficiency. As synchrotron

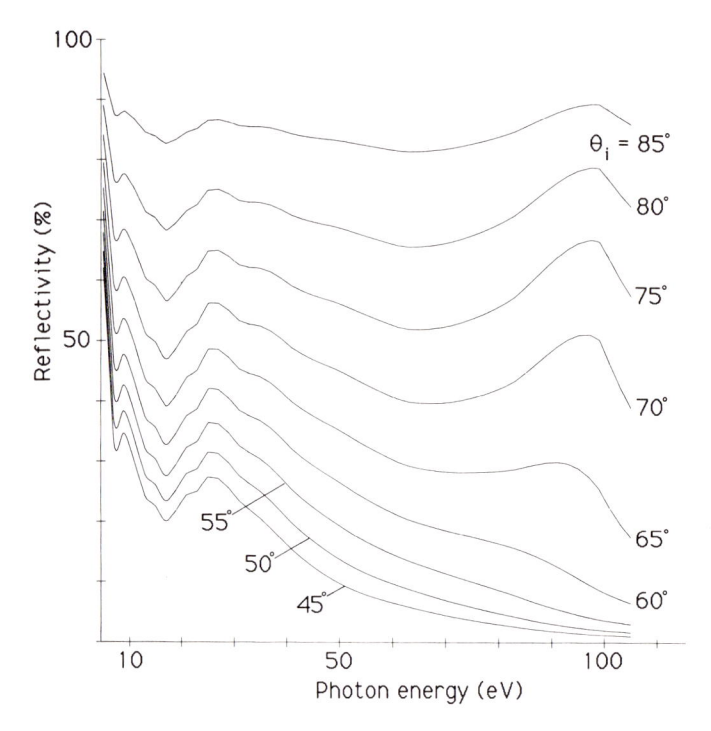

Fig. 3.8. s polarized reflectivity of gold as a function of the angle of incidence θ_i in the photon energy range 5 to 100 eV.

light is a continuum it is often desirable to use mirrors as low-pass filters. From the form of the reflectivities given in fig. 3.8 gold is not a suitable reflector for a low-pass filter in this energy region.

Figures 3.9 and 3.10 show reflectivities for gold, calculated from the atomic scattering factors of Henke et al. (1982, 1981). The convergence of the s and p reflectivities in the soft X-ray region can be seen in fig. 3.10 ($\theta_i = 86°$). Beam lines operating over this energy range can therefore be divided by horizontal (p) reflections with only very small loss of efficiency compared to an s reflection and only a very small loss of linear polarization. Figure 3.9 shows the calculated reflectivity for s reflection in the soft X-ray region over a range of incidence angles from 80° to 89°. Above 300 eV the curves for glancing angles less than 4° are characterized by gradually decreasing reflectivity up to a photon energy higher than which the reflectivity falls sharply. This "cut-off" becomes sharper towards higher energy. In the energy range 800 eV to 4 keV (2 to 4 keV not known) gold reflectors can be used as low-pass filters. This is particularly useful on a high-energy storage ring where delicate soft X-ray optical elements have to be shielded from excessive power loading by a glancing incidence premirror. For a monochromator covering up to 1600 eV, a 2° glancing angle gold-coated premirror could be used with high efficiency, but provide almost zero reflectivity above 2500 eV. The phenomenon leading to a high-reflectivity plateau followed by a rapid drop in

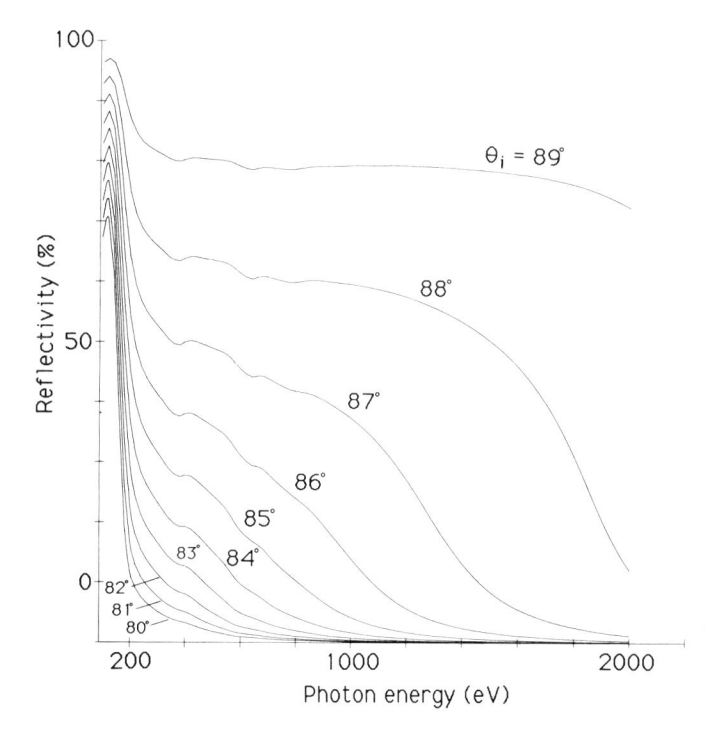

Fig. 3.9. s polarized reflectivity of gold as a function of the angle of incidence θ_i in the photon energy range 100 to 2000 eV.

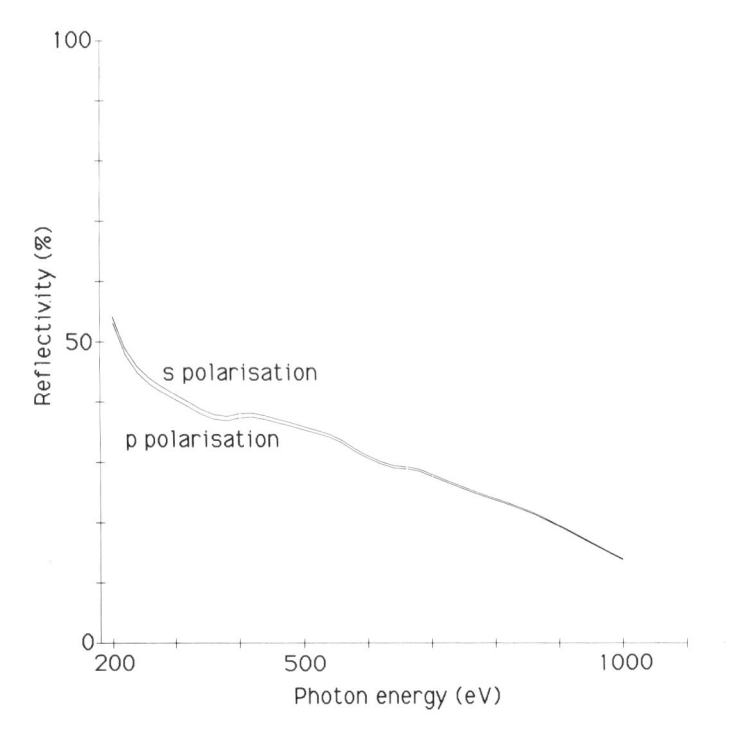

Fig. 3.10. s and p polarized reflectivity of gold as a function of photon energy in the range 200 to 1000 eV at an angle of incidence of 86°.

reflectivity towards higher energy is known as total external reflection. Although very high reflectance can be achieved over a small energy region total reflection can never occur. In the limit of the absorption being zero ($K \rightarrow 0$) from eqs. (3.31), (3.33) and (3.34) we have

$$b = 0, \quad a = (n^2 - \cos^2 \phi)^{1/2}, \tag{3.41}$$

where ϕ is the glancing angle. Substituting eq. (3.41) in the expression for R_s reflectivity in the limit of $R_s = 1$ leads to

$$n = 1 - \delta = \cos \phi, \tag{3.42}$$

and expanding $\cos \phi$ for small angles gives

$$\phi_c \simeq (2\delta)^{1/2}, \tag{3.43}$$

where ϕ_c is the critical glancing angle.

The regions in which total reflection occurs for various materials throughout the VUV region has been examined by Rehn (1981) (from 10 eV to several keV) and by Bilderback (1981) and Bilderback and Hubbard (1982a,b) for hard X-rays

(above 3.8 keV). Rehn (1981) gives ranges of total reflection for gold as
- photon energy: 87 to 197 eV and 800 to 3000 eV;
- critical angle: 30° to 11° and 4.1° to 2.6°.

These ranges can be seen in fig. 3.9. The loss of total reflectance behaviour between 200 and 800 eV is due to contributions to the oscillator strength from the $N_{4,5}$ shells in gold. The conclusion of Rehn (1981) was that for effective total reflection behaviour, light materials with widely spaced core levels should be used so that core level excitations would not contribute strongly in the region of interest. The much better total reflection behaviour is of course achieved at the expense of much lower critical angles (via the density term in eq. 3.28) leading to longer, more expensive mirrors for the same aperture. Although gold has a good performance as a low-pass mirror filter between 800 and 3000 eV, in the region down to 200 eV (covering the carbon, nitrogen and oxygen K edges) other materials have to be used. As synchrotron radiation is a continuum, harmonic contamination from monochromators can be a severe problem. In some measurements such as total yield SEXAFS or fluorescent SEXAFS, harmonic content less than 1% is necessary. For energy dispersive techniques such as photoemission, although harmonic contamination is less of a problem, it can make experiments much more difficult to perform. Bartlett et al. (1984) and Rehn (1981) have examined many mirror filter systems for the range 20 to 1200 eV. In the work of Bartlett et al. (1984) materials and glancing angles were selected on the basis of reflectivity calculations using the atomic scattering factors of Henke et al. (1981, 1982) to fulfill the requirements of 50% transmission 50:1 harmonic rejection, and 25% transmission 100:1 harmonic rejection. It was found that Al, Al_2O_3, Be and C coatings were necessary to cover the range from 80 to 1300 eV. Rehn (1981) has also considered one- and three-mirror filters using total reflectance behaviour, and also filters using the strong absorption around the K and L shell excitations in several light elements. A simple mechanical rotation mechanism which minimizes the vertical deviation of the output beam for a two-mirror filter has been described by Gluskin et al. (1978). A mechanism which would give constant deviation for a two-mirror system has been described by Golovchenko et al. (1981).

The penetration of the photon field into a reflecting surface has been calculated by Henke et al. (1981, 1982) to be given by

$$E = E_0 \exp(-2\pi\gamma z/\lambda\rho), \qquad (3.44)$$

and therefore we can define the depth d at which the energy flow has fallen to $1/e$ of its initial value as

$$d = \lambda\rho/2\pi\gamma, \qquad (3.45)$$

where d is measured perpendicular to the surface.

Figure 3.11 demonstrates that in regions where near total reflectance behaviour is found, the penetration is typically only 10–30 Å. In the region near the cut-off

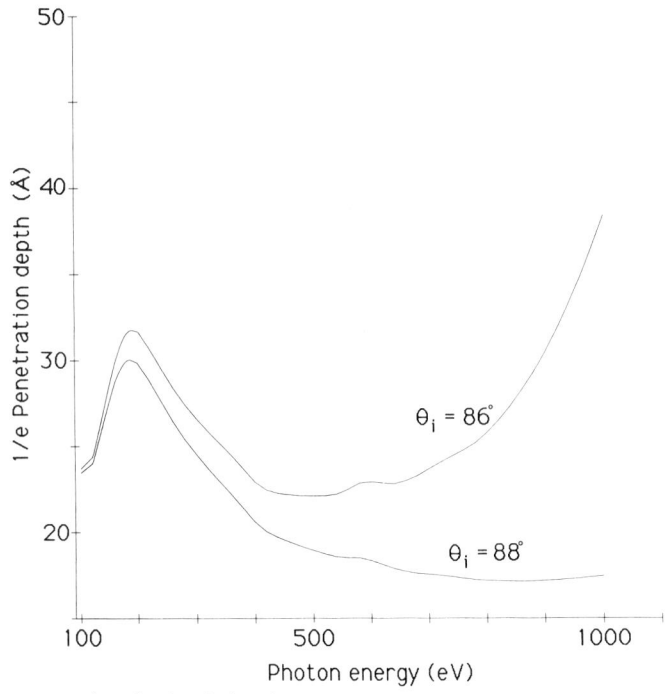

Fig. 3.11. 1/e penetration depth of the photon field into a gold surface at 86° and 88° angles of incidence in the soft X-ray region.

energy for $\theta_i = 86°$, the penetration increases sharply as it also does for $\theta_i = 86°$ and 88° around 200 eV, due to core-level excitation. This behaviour is also seen at low photon energies where large penetration depths readily occur at photon energies far less than the cut-off energy. An example of this behaviour is shown in fig. 3.12 for $\theta_i = 55°$ in the photon energy range 5–55 eV. As the interaction is only with relatively few layers of atoms the cleanliness and order of the surface is important on an atomic scale.

The variation of interaction depth with incidence angle and wavelength has been used by Parratt (1954) in the investigation of the oxidation and crystalline state of evaporated layers of copper, gold, silver, and aluminium. Parratt extended the Fresnel equations for the general case of N layers of stratified reflecting material for the study of mirror surfaces consisting of a bulk density substrate with a surface layer of oxide. Using the same method, the reflectivity of overcoated or contaminated mirror surfaces can be calculated if the density and chemical composition of the contaminant (e.g. graphitic carbon) are known. Several models have been proposed for the variation of reflectivity with surface roughness the simplest being given by Beckman and Spizzichino (1963), Bennett and Porteus (1961) and Bennett (1978). This is based on a scalar scattering theory applicable at optical wavelengths but has been successfully used at XUV wavelengths by Jark and Kunz (1985) and by Haelbich et al. (1979) at X-ray wavelengths (multi-layer mirror at 1.54 Å). The model represents the roughness as a Gaussian function of depth about the mean of the roughness profile. A

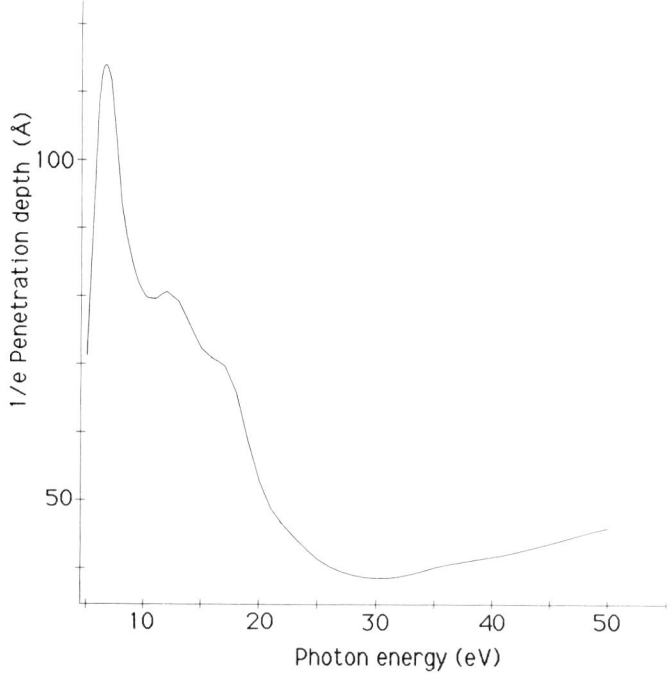

Fig. 3.12. 1/e penetration depth of the photon field into a gold surface at 55° angle of incidence in the photon energy range 5 to 50 eV.

Debye–Waller form is used for the attenuation of the reflectivity,

$$R = R_0 \exp[-(4\pi\sigma \sin \phi/\lambda^2)] , \tag{3.46}$$

where R_0 is the smooth-surface reflectivity, R is the attenuated reflectivity, σ is the rms roughness and ϕ is the glancing angle. This expression gives reasonable agreement with experiment but in general overestimates the attenuation in the region of the critical angle (Bilderback 1981).

A more realistic model has been used by Névot and Croce (1975, 1980) in which the surface roughness introduces a local variation of the refractive index near the surface (fig. 3.13). The surface roughness gives rise only to a specular reflectance and not to any scattering as the variation in refractive index is only normal to the surface.

The index of refraction can be expressed as

$$n(z) = 1 + (n_1 - 1) F(z) , \tag{3.47}$$

where

$$F(z) = \frac{1}{\sigma\sqrt{2\pi}} \int_{-\infty}^{z} \exp(-z^2/2\sigma^2) \, dz ,$$

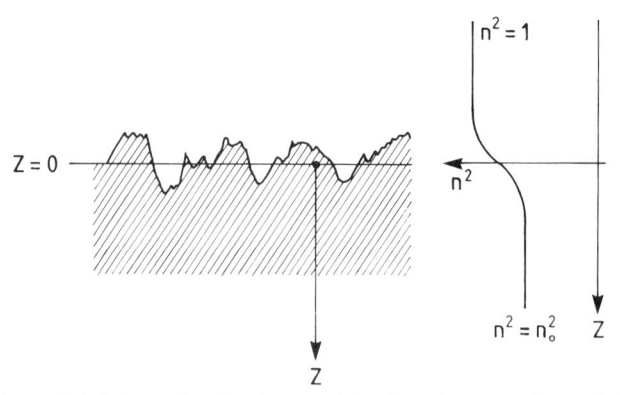

Fig. 3.13. Depth graded index of refraction model of surface roughness (Névot and Croce 1976, 1980).

where z is the distance perpendicular to the mean roughness plane and σ represents the rms variation in surface height. This formalism leads to a model of reflectivity which is fairly accurately reproduced by experiment. The model has been applied by Névot and Croce to the analysis of compression layers near the surface of polished glass and by Bilderback (1981) to the analysis of platinum-coated float glass and etched float glass X-ray mirrors in the photon energy range 3.8 to 50 keV. Above the critical angle the photon field can penetrate deeply into the mirror surface and if the mirror consists of a thin coating on top of a substrate, some of the radiation can be reflected from the substrate–coating interface. The wave reflected from the substrate and the wave reflected from the coating can now interfere leading to strong modulations in the reflectivity versus photon energy characteristics. To match the reflectivity between 3.8 and 50 keV for a 250 Å platinum layer on glass, Bilderback found it necessary to have the air–platinum interface much rougher (29 Å) than the glass–platinum interface (7 Å rms) although the bulk density was used for platinum. The fit with the experimental data including modulations was very good.

In the model of X-ray reflectivity due to Smirnov (Smirnov 1977, Smirnov et al. 1979) the reflecting surface is assumed to consist of a transition layer on the surface of a solid reflector of bulk density. The density in the transition layer changes as a function of depth with a smooth matching at the bulk–transition layer interface. The density profile was assumed to be Gaussian although any arbitrary profile could be used. The model was tested against experimental measurements of reflectivity as a function of glancing angle for copper on glass at 1.54 Å and 2.28 Å; reasonable agreement was found. The experiments were conducted on very thick evaporated layers and it was therefore found necessary to make the transition layer thick (360 Å) to account for the large surface roughness. No actual measurement of surface roughness (e.g. Talystep) was conducted in order to check this model.

3.3. Separated function mirror systems
The principle use of separated function mirror systems is for the pre-optical

Optical engineering

systems of monochromators where the source has to be imaged onto the entrance slit of the monochromators. In view of the fact that single-element double focusing devices are available (section 3.1) and from the previous section it is obviously necessary to keep the number of elements to a minimum, the question might be asked, why use a separated function mirror system. The principle reason is that it is possible to independently adjust the horizontal and vertical magnifications, or have an astigmatic arrangement where for example the vertical focus would be on the entrance slit and the horizontal focus would be at the sample (e.g. the SGM; see section 4.5). It is often necessary, for example in the case of toroidal grating monochromators, to strongly demagnify the source onto the entrance slits in the dispersive direction, but it is highly undesirable to demagnify in the horizontal direction as the horizontal aperture determines one of the principle resolution determining aberrations of the monochromator.

Figure 3.14 shows the two basic types of separated function systems. The arrangement in fig. 3.14a is deflecting in only one plane, the first mirror being

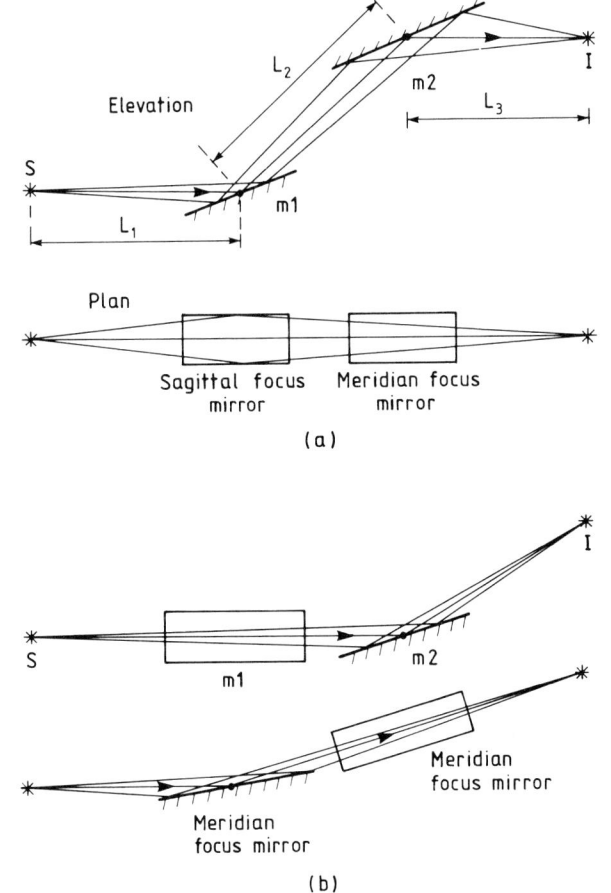

Fig. 3.14. (a) Separated function mirror system employing coplanar sagittal and meridian focusing mirrors; (b) mirror system employing orthogonal meridian focusing mirrors.

sagittally focusing and the second meridionally focusing. This arrangement is particularly convenient where the monochromator has to be situated on a platform above other beam lines and is therefore often used for low-energy instruments such as Seya or low-energy toroidal grating monochromators. For those instruments that collect a large vertical aperture, this arrangement also has the desirable feature that it can substantially enhance the polarization. Mirror m1 would be a sagittal cylinder (radius given in eq. 3.16) and mirror m2 could be either a meridian cylinder (radius given in eq. 3.15) or a plane elliptic cylinder. For large vertical apertures, the plane elliptic cylinder can produce a better focus but at a much higher cost. It appears possible for the off-axis elliptic shape to be fabricated to sufficient accuracy for VUV use and such a scheme has been used in the design of a low-energy TGM for beam line 1 of the SRS.

The arrangment shown in fig. 3.14b consists of two mirrors deflecting in different planes. Mirror m1 would normally be a meridian cylinder and mirror m2 could be either a meridian cylinder or a plane elliptic cylinder if large apertures were required. An attractive feature of this arrangement is that if m1 is a horizontally deflecting mirror the focused light is deflected away from the main fan of radiation allowing more room for both the monochromator and experiment. This scheme is often used for soft X-ray beam lines where the depolarization caused by m1 is small. If this scheme is used for VUV beam lines the deflection angle must be made small so that the depolarization is not too severe. For soft X-ray beam lines where the mirrors are used at extreme glancing incidence, if the required demagnification is not high, the two cylinder mirrors may be replaced by spheres. This is because the sagittal focusing is very small at glancing incidence for the sphere. The advantage of this arrangement is that the sphere is cheap to manufacture and can be figured to high accuracy. The degradation in image quality due to the finite sagittal curvature is usually very small but any specific application should be ray-traced in detail.

For soft X-ray applications it is difficult to obtain elliptic cylinders that are figured accurately enough and at a reasonable price. An alternative is to use a double spherical mirror coma-corrected system as described by Namioka et al. (1983) and extensively used at the Photon Factory (e.g. Maezawa et al. 1986b).

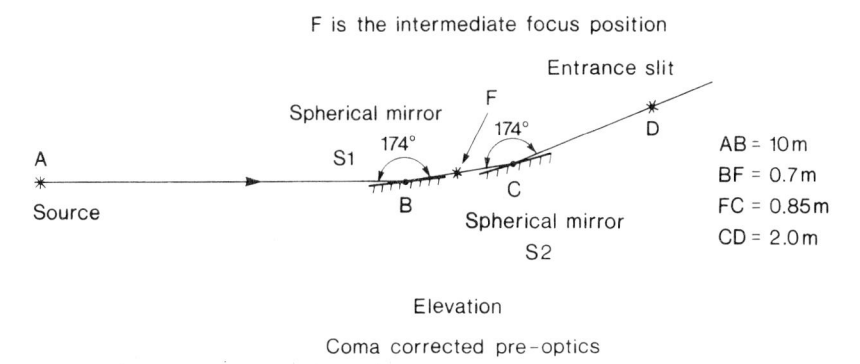

Fig. 3.15. Coma-corrected meridian focus mirror system.

The layout of an example system is shown in fig. 3.15. The first mirror forms an intermediate focus between the mirrors at F which is then used as the object for the second mirror. For a pre-determined first mirror object and image distance there is only one combination of object and image distance for the second mirror that gives coma correction. The conditions for coma correction for mirrors can be derived from the general condition for two gratings given by Namioka et al. (1983),

$$(r_1' \sec \theta_1)^3 f_1 + (r_2 \sec \theta_2)^3 f_2 = 0 ,\qquad (3.48)$$

where

r_i = object distance of mirror i,

r_i' = image distance of mirror i,

$$f_i = \frac{T_i}{r_i} \sin \theta_i - \frac{T_i^*}{r_i'} \sin \theta_i ,$$

$$T_i = \frac{\cos \theta_i}{R_i} - \frac{\cos^2 \theta_i}{r_i} ,$$

$$T_i^* = \frac{\cos \theta_i}{R_i} - \frac{\cos^2 \theta_i}{r_i'} ,$$

R_i = radius of mirror i.

Having chosen the object distance, the intermediate image distance and the two incidence angles the overall magnification is chosen thus defining the magnification of the second mirror. The actual values of r_2 and r_2' are changed but keeping the same magnification (r_2/r_2' = const.) and adjusting R_2 for a focus condition (eq. 3.15). At some value of r_2 and r_2' the second term will cancel the first term in eq. (3.48). The sign of θ_2 in the example shown in fig. 3.15 is positive, but would be negative if the deflection direction had been opposite to the first mirror. Figure 3.16a shows a ray trace from a 3×3 matrix of points representing a source 0.4 mm high in the focusing direction. The light diverges in the non-focusing direction but is almost perfectly focused in the other direction. The conditions used are fairly extreme ($\theta = 87°$, demagnification = 6.07, vertical aperture 1 mrad) in order to demonstrate the coma correction. A single mirror of the same demagnification gives the ray trace shown in fig. 3.16b demonstrating the large coma tail.

This final example aptly demonstrates the decisions which must be made in the design progress. For a double focusing system we would end up with three mirrors. The designer would need to calculate the reflectivity of the system and weigh that against the near perfect focusing of the arrangement. In this example it may well be better for a pre-optical system to accept imperfect focusing from a one- or two-mirror system but with the benefit of higher overall reflectivity. Apart from the purely optical performance the tolerance of the system to manufacturing

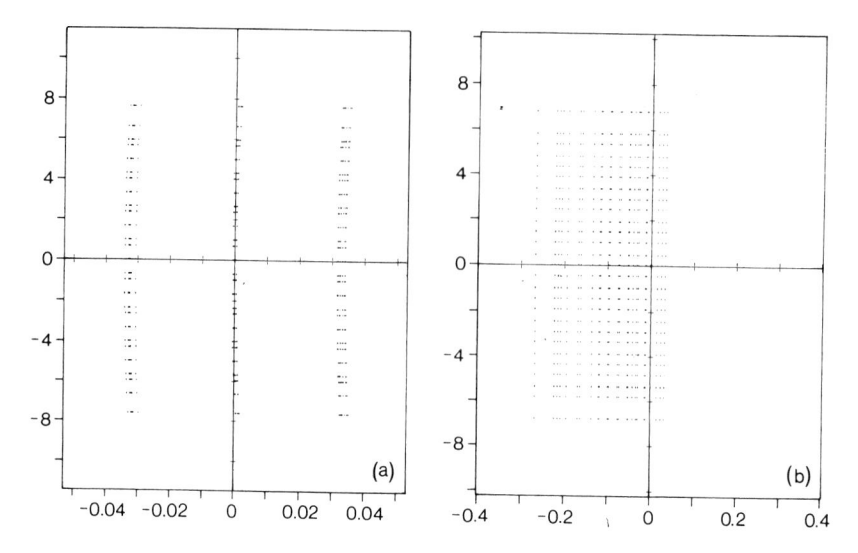

Fig. 3.16. (a) Raytrace of a double spherical mirror coma-corrected system: $AB = 10$ m, $BF = 0.7$ m, $FC = 0.85$ m, $CD = 2$ m; $\theta = 87°$; demagnification = 6.07; 1 mrad vertical aperture; (b) Raytrace of a single spherical mirror: $\theta = 87°$; demagnification = 6.07; 1 mrad vertical aperture.

error and misalignment should also be carefully considered. Finally, the cost of the associated mechanical systems should be considered. In the example given the advantage of having cheap optics may be outweighed by the additional vacuum chambers and adjusting mechanisms necessary in comparison to the more expensive single aspheric mirror alternative.

4. Diffraction grating optics

In this section the basic theory necessary for the design of grating instruments is presented. The theory is then applied to the practical case of the toroidal grating monochromator although the same method can be simply applied to mirrors, plane grating monochromators, Rowland circle monochromators and normal incidence monochromators. It is not the intention of this section to provide a review of monochromator designs, but to state the basic design methods which should be used for all monochromators. The design of separated function grating monochromators employing separate dispersive and non-dispersive focusing elements is also examined.

4.1. Application of the grating equation

4.1.1. Diffraction

Figure 4.1 shows the standard notation used. Radiation of wavelength λ incident on the grating at an angle α to the surface normal is either reflected (zero order)

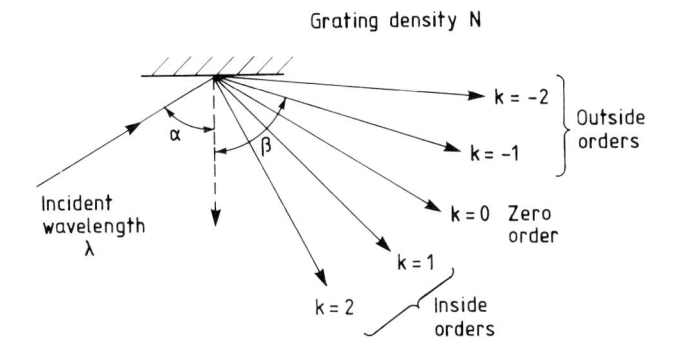

Fig. 4.1. Standard notation for a diffraction grating. Note: α and β have opposite signs when on opposite sides of the surface normal. β is shown for the first negative order.

or diffracted at an angle β. The relative amounts of light of wavelength λ reflected or diffracted into a particular order k is determined by the groove shape and dielectric constant. The angle of diffraction β is related to the angle of incidence α, the groove density N and the wavelength λ by the grating equation,

$$Nk\lambda = \sin \alpha + \sin \beta . \tag{4.1}$$

α and β are of opposite sign if on opposite sides of the surface normal. In fig. 4.1, α is positive and β negative. From eq. (4.1) it can be seen that α and β are free parameters in that diffraction of a ray of wavelength λ is possible over a range of incident and diffracted angles. The various types of common optical configurations, however, constrain the relationship between α and β.

4.1.1.1. Spectrograph geometry. In a spectrograph, α is fixed and different wavelengths are selected by varying the viewed diffraction angle β. This would be done by scanning a slit along the focal line in a particular order, or by having an imaging detector (e.g. photodiode array) placed on part of the focal line.

4.1.1.2. Fixed grating-deviation-angle geometry. Two types of fixed grating-deviation-angle geometry are shown in fig. 4.2. The most common and simplest geometry is the TGM as shown in fig. 4.2a. The entrance and exit slits fix the deviation angle 2θ giving the relation

$$\alpha - \beta = 2\theta . \tag{4.2}$$

The negative β is due to the angle sign convention. A similar arrangement using a plane grating and a focusing mirror is shown in fig. 4.2b. As the focusing mirror is fixed eq. (4.2) is still valid. By combination of eqs. (4.2) and (4.1) we have the definitions of α and β,

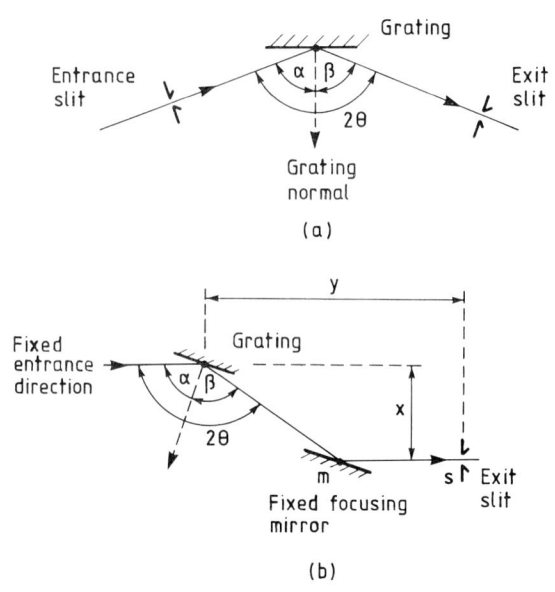

Fig. 4.2. Fixed grating-deviation-angle geometry: (a) single element, e.g. TGM; (b) two elements, e.g. plane grating and fixed paraboloid.

$$\alpha = \sin^{-1}\left(\frac{Nk\lambda}{2\cos\theta}\right) + \theta, \qquad (4.3)$$

$$\beta = \sin^{-1}\left(\frac{Nk\lambda}{2\cos\theta}\right) - \theta, \qquad (4.4)$$

and by rearrangement of eq. (4.4) we have a simple definition of the wavelength scale,

$$\lambda = \frac{2}{Nk}\cos\theta\sin(\theta + \beta). \qquad (4.5)$$

As $(\theta + \beta)$ is simply the angle turned from zero order ϕ, the wavelength is simply proportional to $\sin\phi$. By using a sine bar drive as shown in fig. 4.3 the wavelength is therefore proportional to the perpendicular separation L, which can be measured easily using a commercial linear encoder. A consequence of a fixed deviation angle geometry is that the long wavelength end of the spectrum is

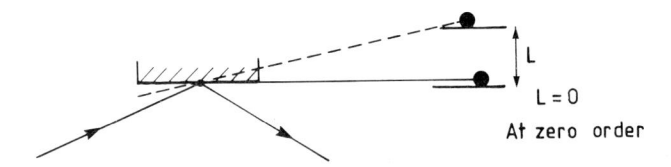

Fig. 4.3. Simple sine bar drive. For a fixed grating-deviation-angle geometry, $\lambda \propto L$.

limited in the limit of $\alpha \to 90°$ in positive order, or $\beta \to 90°$ in negative order. The limiting or horizon wavelength is given by

$$\lambda_H = \frac{2}{Nk} \cos^2 \theta . \qquad (4.6)$$

4.1.1.3. On-blaze geometry. Figure 4.4 shows the different forms of groove profile, lamellar, sinusoidal and blazed. The geometry for the blazed grating is shown in fig. 4.5. The grating grooves are in the form of facets with the facet inclined at the blaze angle γ to the macroscopic surface. The grating is on-blaze when the incident ray and the diffracted ray make equal angles with the surface normal of the groove facet. Under these conditions the efficiency of the grating is maximum and is higher than that obtainable from either the lamellar or sinusoidal

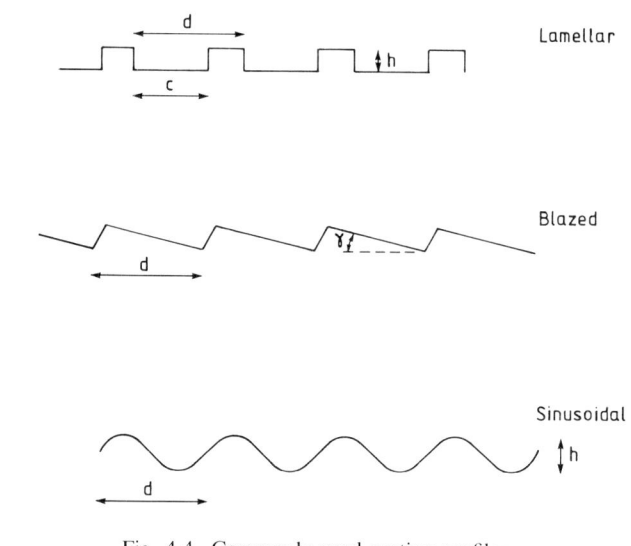

Fig. 4.4. Commonly used grating profiles.

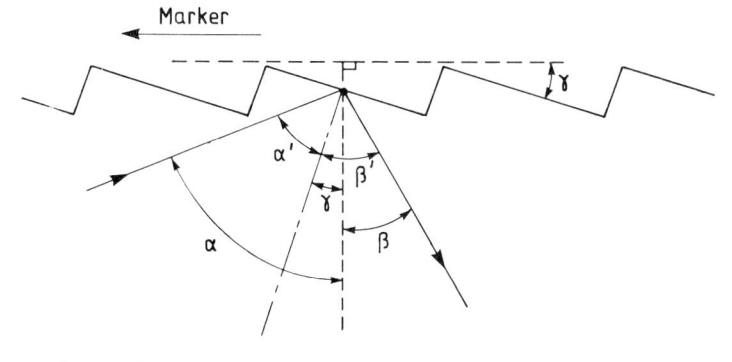

Fig. 4.5. Standard notation for a blazed grating; γ is the blaze angle.

gratings. For the on-blaze condition the following relation must be satisfied:

$$\alpha + \beta = 2\gamma. \tag{4.7}$$

The grating equation is constrained by this condition and therefore becomes

$$\lambda = \frac{2}{Nk} \sin\gamma \cos(\gamma + \beta). \tag{4.8}$$

As the groove profile is asymmetric, blazed gratings are usually marked with an arrow pointing towards the blaze angle as shown in fig. 4.5; conventionally, this arrow points towards the entrance slit with the grating used in positive order.

In order to satisfy the condition given by eq. (4.7) a complex mechanical linkage is necessary. Figure 4.6a shows the arrangement used by Dietrich and Kunz (1972) and an approximate arrangement has been given by Petersen (1982) and was used in the BESSY SX700. This arrangement differs from the simple Miyake (1969) arrangement shown in fig. 4.2b in that a plane premirror is free to translate and rotate along the incident light axis allowing a free choice of α and β. In this way eq. (4.7) can be satisfied. A similar arrangement has been used by Hunter et al. (1982) except that the light is first collimated by a paraboloidal mirror (fig. 4.6b).

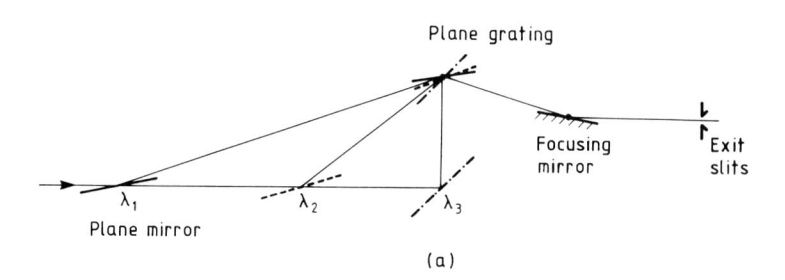

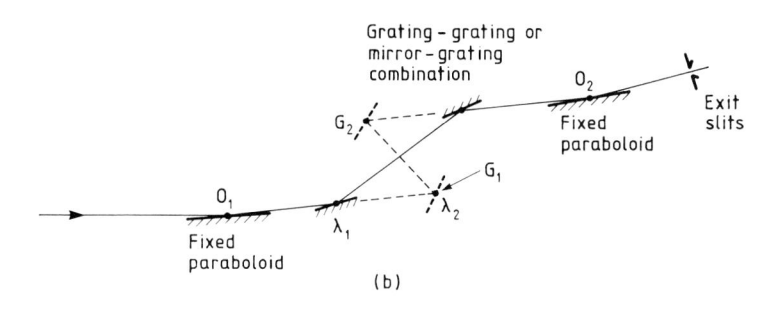

Fig. 4.6. Mechanisms allowing selection of the deviation angles: (a) Dietrich and Kunz (1972), Petersen (1982); (b) Hunter et al. (1982).

4.1.1.4. Plane grating constant-image-distance geometry.
For a source of light at a finite distance r from a plane grating, an image is formed at a distance r' given by Murty (1962) as

$$r' = -r \frac{\cos^2 \beta}{\cos^2 \alpha} = -rK . \tag{4.9}$$

For a fixed exit slit and focusing mirror arrangement as shown in fig. 4.2b a correct focus will only be obtained at the exit slit if the virtual image distance r' and hence $\cos^2 \beta / \cos^2 \alpha$ is a constant K. The grating equation is therefore constrained by this condition and becomes

$$(Nk\lambda - \sin \beta)^2 = 1 - (1 - \sin^2 \beta)/K . \tag{4.10}$$

The arrangement given in fig. 4.2b can therefore not satisfy simultaneously the on-blaze and in-focus conditions. The arrangement of Hunter et al. (1982) uses a paraboloidal collimating premirror and therefore produces parallel light for the grating. This therefore allows a free choice of α and β and so the on-blaze and in-focus conditions can be satisfied together.

4.1.1.5. In-focus Miyake geometry.
The simple Miyake et al. (1969) geometry is shown in fig. 4.2b. In most monochromators of this type, the spherical focusing mirror is free to translate along MS parallel to the entrance direction between the two focal positions possible for a single mirror (West et al. 1974, Howells et al. 1978). For a given deviation angle 2θ the mirror–slit distance MS is defined by the geometry

$$\text{MS} = y + x \cot 2\theta = l_b \tag{4.11}$$

The required object position is therefore defined from the paraxial focus equation

$$l_a = \frac{R' l_b}{2 l_b - R'} , \tag{4.12}$$

where $R' = R \cos \theta$, R is the radius of the mirror, l_b is the image distance (eq. 4.11), and l_a is the required object distance.

The object distance is the sum of the grating mirror distance and the virtual object distance given by eq. (4.9),

$$l_a = \frac{x}{\sin 2\theta} + \frac{r \cos^2 \beta}{\cos^2 \alpha} . \tag{4.13}$$

It is clear therefore that at a fixed deviation angle only one wavelength gives a focus at the exit slits. This wavelength can be found by equating eqs. (4.13) and (4.12) (Padmore 1986a):

$$\lambda = \frac{2}{Nk} \cos\theta \, \frac{H}{(H^2 + J^2)^{1/2}}, \tag{4.14}$$

where $H = \cos\theta(1 - G)$, $J = \sin\theta(1 + G)$, and

$$G = \left[\frac{1}{r} \left(\frac{R' l_b}{2 l_b - R'} - \frac{x}{\sin 2\theta} \right) \right]^{1/2}.$$

By altering the deviation angle, the in-focus wavelength can be moved to the required value. For each wavelength and deviation angle the mirror–slit distance has to be changed in accordance with eq. (4.11).

4.1.2. Dispersion

The angular dispersion can be obtained directly from the grating equation

$$\left(\frac{d\lambda}{d\beta} \right)_{\alpha = \text{const.}} = \frac{1}{Nk} \cos\beta. \tag{4.15}$$

The dispersion at the exit slit perpendicular to the principal ray for a single element monochromator can be found from eq. (4.15) by substituting ds/F for $d\beta$ where F is the monochromatic image distance,

$$\left(\frac{d\lambda}{ds} \right)_{\alpha = \text{const.}} = \frac{1}{NkF} \cos\beta. \tag{4.16}$$

For a particular slit dimension, this relation also defines the exit slit limited resolution. For a two-element system such as a plane grating and focusing mirror separated by a distance P the situation is more complex. If the virtual source–grating distance is r' (eq. 4.9) and the monochromatic image distance is F it is simple to prove that for a spherical focusing mirror,

$$\left(\frac{d\lambda}{ds} \right)_{\alpha = \text{const.}} = \frac{1}{Nk} \left(\frac{1}{F} + \frac{P}{Fr'} \right) \cos\beta. \tag{4.17}$$

In the limit of the grating and spherical mirror merging ($P \to 0$) or parallel incident light ($r' = \infty$) the extra term vanishes. The monochromatic magnification is also similarly modified (Padmore 1986a).

4.1.3. Entrance and exit slit-width-limited resolution

The finite size of the entrance and exit slits in the dispersive direction sets a limit to the achievable resolution. The slit-width-limited resolution can be obtained directly from the differentiated grating equation with $d\alpha = S_1/r$ and from the dispersion (eq. 4.17) with $dS = S_2/F$, where S_1 and S_2 are the entrance and exit slit widths, respectively.

Entrance slit:

$$\Delta\lambda_1 = \frac{1}{Nk} \frac{S_1}{r} \cos\alpha, \quad (4.18)$$

where r is the source or entrance slit–grating distance.

Exit slit:

$$\Delta\lambda_2 = \frac{1}{Nk} \frac{S_2}{F} \cos\beta, \quad (4.19)$$

where F is the monochromatic image distance.

Equation (4.19) is for the single-element case; for a separate focusing mirror the form of eq. (4.17) should be used. The magnification can be obtained by substituting the entrance slit or source limited resolution (4.18) into the expression for the dispersion (4.17). For a single-element case this simplifies to

$$M(\lambda) = \frac{\cos\alpha}{\cos\beta} \frac{F}{r}. \quad (4.20)$$

4.2. The geometric aberration theory of diffraction gratings

Figure 4.7 shows the grating geometry under consideration. The x axis is normal to the grating surface and the y–z plane is the tangent plane at O. The z axis is

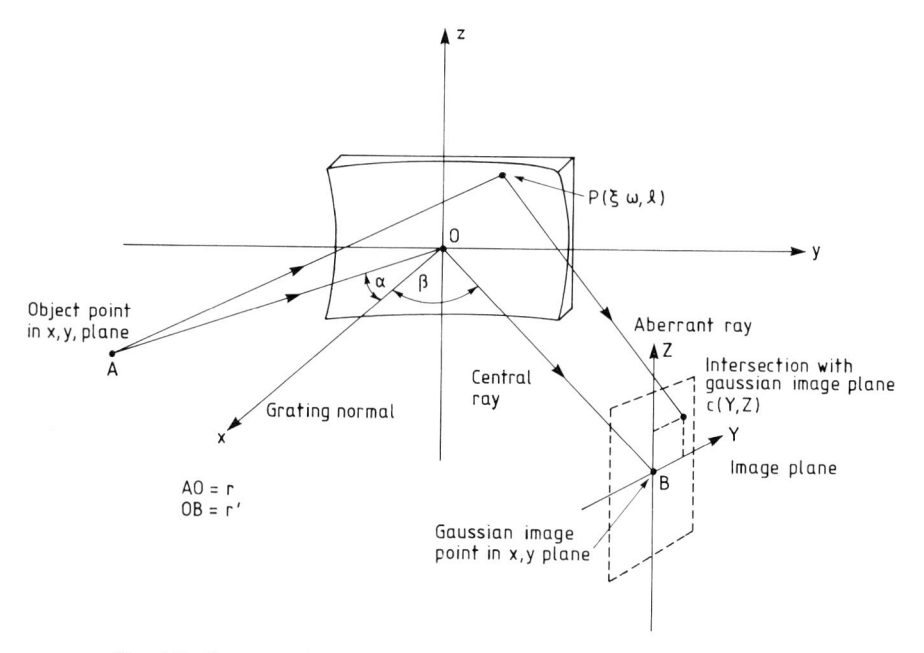

Fig. 4.7. Geometry for calculation of the optical path of a diffracted ray.

parallel to the grating rulings and the source point is constrained to lie in the x–y plane. This constraint is imposed for simplicity and is not a general restriction on the theory. The object of an aberration theory for a diffraction grating is to produce a simple expression for the intersection points in the image plane produced by rays diffracted from different points on the grating surface. This expression can be reduced to a form consisting of a summation in powers of the aperture coordinates of the grating (y, z), each term representing a particular geometrical aberration. Although they are not separate, the influence of particular types of aberrations can be seen in diffracted images and by iterating adjustable parameters, such as substrate shape, object distance, image distance and groove density, the sum of the aberrations can be reduced. Aberrations can be reduced further by using grating grooves that are curved and have a variable spacing as can be produced using holographic techniques. Geometric aberration theory has been applied to the general case of holographic gratings by Noda et al. (1974) and Velzel (1976, 1977). The following discussion of the geometrical aberration theory is restricted to straight ruled symmetrical gratings for simplicity as given by Beutler (1945), Haber (1950) and Namioka (1959a,b). Geometrical aberration theory is of prime importance in the design of monochromators and therefore the equations describing the principal aberrations are described in detail.

4.2.1. The optical path function

The surface under consideration given in fig. 4.7 is symmetric about the x–y plane. The equation of the surface can then be expressed as a power series,

$$x = \sum_{i=0}^{\infty} \sum_{j=0}^{\infty} a_{ij} y^i z^j , \qquad (4.21)$$

where $a_{00} = a_{10} = 0$ and the summation is only over even powers of j. These conditions arise from a consideration of the position of the pole and the plane of symmetry. A particular shape such as a toroid will therefore have a particular array of a_{ij} coefficients describing its surface, although the optical path function will be derived without reference to particular a_{ij}'s.

A ray from an object point A in the x–y plane is incident on the grating at point $P(\xi, w, l)$ and diffracted to point $C(Y, Z)$ in the image plane. The optical path function can then be written (Noda et al. 1974) as

$$F = \langle AP \rangle + \langle PC \rangle + \frac{m\lambda}{d} , \qquad (4.22)$$

where m is the order of diffraction, λ is the wavelength and d is the grating spacing. The pathlengths $\langle AP \rangle$ and $\langle PC \rangle$ can be expressed as

$$\langle AP \rangle = [(x - \xi)^2 + (y - w)^2 + l^2]^{1/2} ,$$
$$\langle PC \rangle = [(x' - \xi)^2 + (y' - w)^2 + (z' - l)^2]^{1/2} , \qquad (4.23)$$

where x', y', z' are the Cartesian coordinates of the intersection in the image plane. By substituting eqs. (4.21) and (4.23) into eq. (4.22) and by power series expansion in the aperture coordinates we obtain

$$F = F_{00} + wF_{10} + \tfrac{1}{2}w^2F_{20} + \tfrac{1}{2}l^2F_{02} + \tfrac{1}{2}w^3F_{30} + \tfrac{1}{2}wl^2F_{12}$$
$$+ \tfrac{1}{8}w^4F_{40} + \tfrac{1}{4}w^2l^2F_{22} + \tfrac{1}{8}l^4F_{04} + \cdots. \quad (4.24)$$

The F_{ij} terms are grouped in terms of powers in w and l. Each F_{ij} term represents a particular form of aberration and can be grouped as follows ($AO = r$; $OB = r'$):

$$F_{00} = r + r', \quad (4.25)$$

$$F_{10} = \frac{m\lambda}{d} - (\sin\alpha + \sin\beta), \quad (4.26)$$

$$F_{20} = (\cos^2\alpha/r) + (\cos^2\beta/r') - 2a_{20}(\cos\alpha + \cos\beta), \quad (4.27)$$

$$F_{02} = (1/r) + (1/r') - 2a_{02}(\cos\alpha + \cos\beta), \quad (4.28)$$

$$F_{30} = [T(r,\alpha)/r]\sin\alpha + [T(r',\beta)/r']\sin\beta - 2a_{30}(\cos\alpha + \cos\beta), \quad (4.29)$$

$$F_{12} = [S(r,\alpha)/r]\sin\alpha + [S(r',\beta)/r']\sin\beta - 2a_{12}(\cos\alpha + \cos\beta), \quad (4.30)$$

$$F_{40} = [4T(r,\alpha)/r^2]\sin^2\alpha - [T^2(r,\alpha)/r]$$
$$+ [4T(r',\beta)/r'^2]\sin^2\beta - [T^2(r',\beta)/r']$$
$$- 8a_{30}[(\sin\alpha\cos\alpha/r) + (\sin\beta\cos\beta/r')]$$
$$- 8a_{40}(\cos\alpha + \cos\beta), \quad (4.31)$$

$$F_{22} = [2S(r,\alpha)/r^2]\sin^2\alpha + [2S(r',\beta)/r'^2]\sin^2\beta$$
$$- [T(r,\alpha)S(r,\alpha)/r] - [T(r',\beta)S(r',\beta)/r']$$
$$+ 4a_{20}a_{02}(1/r + 1/r') - 4a_{22}(\cos\alpha + \cos\beta)$$
$$- 4a_{12}[(\sin\alpha\cos\alpha/r) + \sin\beta\cos\beta/r')], \quad (4.32)$$

$$F_{04} = 4a_{02}^2(1/r + 1/r') - 8a_{04}(\cos\alpha + \cos\beta) - [S^2(r,\alpha)/r] - [S^2(r',\beta)/r'], \quad (4.33)$$

where

$$T(r,\alpha) = (\cos^2\alpha/r) - 2a_{20}\cos\alpha,$$
$$S(r,\alpha) = (1/r) - 2a_{02}\cos\alpha,$$

and α and β are respectively the angles of incidence and diffraction (in the x–y plane). The signs of α and β are opposite if the incident and diffracted rays are on opposite sides of the surface normal. In fig. 4.7 α is positive and β negative.

According to Fermat's principle the path function F should have a stationary value such that $\partial F/\partial w = 0$ and $\partial F/\partial l = 0$. Applications to wF_{10} yields

$$m\lambda/d = \sin\alpha + \sin\beta , \qquad (4.34)$$

which is the grating equation. The path function can now be seen to consist of two parts, one representing the Gaussian path function and the other representing aberrations:

$$F = F_0 + F_1 , \qquad (4.35)$$

where

$$F_0 = r + r' + \frac{m\lambda}{d} - (\sin\alpha + \sin\beta) .$$

Each F_{ij} term in the F_1 group can be associaited with a particular type of imaging error:

F_{20} ... defocus

F_{02} ... astigmatism

F_{30} ... coma

F_{12} ... astigmatic coma

F_{40} ...

F_{22} ... } generalised spherical aberration

F_{04} ...

By iterating the variable parameters in a design, the sum of the aberrations may be reduced to a minimum. In some designs individual terms in the path function (such as defocus) may be minimised without affecting the other aberrations but in general a global optimisation is necessary. The aberration terms in the path function can be related directly to the deviation of the aberrated ray from the Gaussian image point (Madden and Ederer 1972, Namioka 1959a, Howells 1980a). The deviation in the dispersive direction (ΔY) and non-dispersive direction (ΔZ) can be expressed as

$$\Delta Y = \frac{r'}{\cos\beta}\frac{\partial F_1}{\partial w} , \qquad (4.36)$$

$$\Delta Z = r'\frac{\partial F_1}{\partial l} . \qquad (4.37)$$

The deviation in terms of wavelength error can be found by substituting the expression for reciprocal linear dispersion in eq. (4.36):

$$\Delta\lambda = \frac{1}{Nk}\frac{\partial F_1}{\partial w}. \tag{4.38}$$

From the expression of the path function given in eq. (4.24), the aberration-limited wavelength resolution can now be found using eq. (4.38),

$$\Delta\lambda = \frac{1}{Nk}(wF_{20} + \tfrac{3}{2}w^2 F_{30} + \tfrac{1}{2}l^2 F_{12} + \tfrac{1}{2}w^3 F_{40} + \cdots). \tag{4.39}$$

The way in which each aberration affects the aberration limited resolution as a function of aperture can readily be seen. The defocus term is linear in the ruled length of the grating ($\pm w$) and gives an error which is symmetric about the Gaussian image point corresponding to rays from the top and bottom of the grating. If a monochromator is only limited by defocus a decrease in dispersive aperture will cause a corresponding linear decrease in the dispersive error. In contrast, the coma term is proportional to w^2 giving a dispersive error which only occurs on one side of the Gaussian image point for rays from both the top and bottom of the grating ($\pm w$). If the monochromator is only limited by coma a reduction in dispersive aperture by a factor of 2 will cause the error from the most aberrated rays to reduce by a factor of 4. The astigmatic coma term is proportional to l^2 giving a wavelength error which again only occurs on one side of the Gaussian image point. The form of these errors will be considered in detail in the next section.

These conclusions have been reached without the need to consider the form of the substrate. The shape of the substrate will alter the magnitude of the F_{ij} terms and their wavelength dependence but not their aperture dependence. To obtain a value for the aberration limited resolution, the a_{ij} terms of the power series expressing the form of the substrate (given in eq. 4.21) must be substituted into the expressions for F_{ij} (eqs. 4.27 to 4.33). These have been given by Howells (1980b) for paraboloidal, ellipsoidal and toroidal substrates and are listed below.

Substrate surface power series coefficients:

paraboloid

$$a_{02} = \frac{1}{4f\cos\theta}, \qquad a_{20} = \frac{\cos\theta}{4f},$$

$$a_{12} = -\frac{\tan\theta}{8f^2}, \qquad a_{30} = -\frac{\sin\theta\cos\theta}{8f^2},$$

$$a_{40} = \frac{5\cos\theta\sin^2\theta}{64f^3}, \qquad a_{22} = \frac{3\sin^2\theta}{32\cos\theta\,f^3},$$

$$a_{04} = \frac{\sin^2\theta}{64f^3\cos^3\theta}. \tag{4.40}$$

ellipsoid

$$a_{02} = \frac{1}{4f \cos \theta}, \qquad a_{20} = \frac{\cos \theta}{4f},$$

$$a_{12} = \frac{\tan \theta}{8f^2} \frac{(e^2 - \sin^2 \theta)^{1/2}}{\cos \theta}, \qquad a_{30} = \frac{\sin \theta}{8f^2} (e^2 - \sin^2 \theta)^{1/2},$$

$$a_{40} = \frac{b^2}{64 f^3 \cos \theta} \left(\frac{5 \sin^2 \theta \cos^2 \theta}{b^2} - \frac{5 \sin^2 \theta}{a^2} + \frac{1}{a^2} \right),$$

$$a_{22} = \frac{\sin^2 \theta}{16 f^3 \cos^3 \theta} \left[\frac{3}{2} \cos^2 \theta - \frac{b^2}{a^2} \left(1 - \frac{\cot^2 \theta}{2} \right) \right],$$

$$a_{04} = \frac{b^2}{64 f^3 \cos^3 \theta} \left(\frac{\sin^2 \theta}{b^2} + \frac{1}{a^2} \right). \tag{4.41}$$

bicycle tyre toroid

$$a_{02} = \frac{1}{2\rho}, \qquad a_{20} = \frac{1}{2R},$$

$$a_{40} = \frac{1}{8R^3}, \qquad a_{22} = \frac{1}{4R\rho^2},$$

$$a_{04} = \frac{1}{8\rho^3}. \tag{4.42}$$

apple core toroid

$$a_{02} = \frac{1}{2\rho}, \qquad a_{20} = \frac{1}{2R},$$

$$a_{40} = \frac{-3}{8R^3}, \qquad a_{22} = \frac{-1}{4R\rho^2},$$

$$a_{04} = \frac{-3}{8\rho^3}. \tag{4.43}$$

4.3. The toroidal grating monochromator

The optical path function of the toroidal grating has been extensively studied by Haber (1950) and the optimisation of toroidal grating monochromators using straight ruled gratings has been given in very useful papers by Howells (1980a) and McKinney and Howells (1980). The optimisation of holographic toroidal gratings has been given by Lepère (1975). The principle features of the optimisation process described by McKinney and Howells (1980) will be used to illustrate the use of the optical path function in monochromator design.

The normal toroidal substrate is the bicycle tyre type shown in fig. 3.3a. The power series coefficients of this shape given in eq. (4.42) [which were derived from an expansion (Maclaurins series) of the surface equation transformed to a coordinate system centred on the optic] can now be substituted into the aberration coefficients F_{ij} given by eqs. (4.27)–(4.33). As an example this is done for the defocus term F_{20}:

$$F_{20} = (\cos^2 \alpha / r) + (\cos^2 \beta / r') - 2a_{20}(\cos \alpha + \cos \beta).$$

From eq. (4.42) we have $a_{20} = 1/(2R)$, which gives

$$F_{20} = \left(\frac{\cos^2 \alpha}{r} - \frac{\cos \alpha}{R} \right) + \left(\frac{\cos^2 \beta}{r'} - \frac{\cos \beta}{R} \right). \tag{4.44}$$

Using the notation of Howells (1980a) we have

$$\frac{R}{2} F_{20} = C_{20} = \frac{1}{2} \left(\frac{\cos^2 \alpha}{r^*} - \cos \alpha \right) + \left(\frac{\cos^2 \beta}{r'^*} - \cos \beta \right), \tag{4.45}$$

where the asterisk denotes division by R, i.e.,

$$r^* = r/R, \qquad r'^* = r'/R. \tag{4.46}$$

In terms of the aperture coordinates x, y the aberrant part of the path function becomes

$$C = C_{20} \frac{y^2}{R} + C_{30} \frac{y^3}{R^2} + C_{02} \frac{z^2}{R} + C_{12} \frac{yz^2}{R^2} + C_{40} \frac{y^4}{R^3} + C_{04} \frac{z^4}{R^3} + C_{22} \frac{y^2 z^2}{R^3}$$
$$+ \cdots . \tag{4.47}$$

Howells (1980a) has shown that the first few terms in the aberrant part of the path function dominate and it is unnecessary to consider the generalised spherical aberration (C_{40}, C_{04}, C_{32}) or higher terms. Substituting the appropriate power series coefficients into the F_{ij} terms yields

defocus coefficient:

$$C_{20} = \frac{1}{2} \left[\left(\frac{\cos^2 \alpha}{r^*} - \cos \alpha \right) + \left(\frac{\cos^2 \beta}{r'^*} - \cos \beta \right) \right], \tag{4.48}$$

coma coefficient:

$$C_{30} = \frac{1}{2} \left[\frac{\sin \alpha}{r^*} \left(\frac{\cos^2 \alpha}{r^*} - \cos \alpha \right) + \frac{\sin \beta}{r'^*} \left(\frac{\cos^2 \beta}{r'^*} - \cos \beta \right) \right], \tag{4.49}$$

astigmatic coma coefficient:

$$C_{12} = \frac{1}{2} \left[\frac{\sin \alpha}{r^*} \left(\frac{1}{r^*} - \frac{\cos \alpha}{\rho^*} \right) + \frac{\sin \beta}{r'^*} \left(\frac{1}{r'^*} - \frac{\cos \beta}{\rho^*} \right) \right], \tag{4.50}$$

astigmatism coefficient:

$$C_{02} = \frac{1}{2}\left[\left(\frac{1}{r^*} - \frac{\cos\alpha}{\rho^*}\right) + \left(\frac{1}{r'^*} - \frac{\cos\beta}{\rho^*}\right)\right]. \tag{4.51}$$

The principle advantage of a toroidal grating over a spherical grating is that by a correct choice of the sagittal radius ρ, the monochromator can be made stigmatic at one wavelength. From eq. (4.37) the deviation in the non-dispersive direction is set to zero giving

$$\Delta z = r'\frac{\partial}{\partial z}\left(C_{02}\frac{Z^2}{R}\right) \equiv C_{02} = 0.$$

Setting $C_{02} = 0$ in eq. (4.51) gives

$$\rho^* = \left(\frac{\cos\alpha + \cos\beta}{1/r^* + 1/r'^*}\right). \tag{4.52}$$

Astigmatism can therefore be corrected at one wavelength. As originally pointed out by Madden and Ederer (1972), however, the astigmatism is approximately corrected over a large range of wavelengths as $\cos\alpha + \cos\beta$ is nearly constant for deviation angles greater than 140° and for a reasonable choice of groove density.

The deviations in the dispersive direction in the image plane can similarly be found by applying eq. (4.36). Before numerically evaluating the aberration-limited resolution it is instructive to examine the form of the defocus term. This is shown schematically in fig. 4.8 for two cases, one in which the entrance and exit arms are of equal length and one in which the exit arm is longer than the entrance arm. In the former case the design is symmetrical and may be used in positive or negative order. For the asymmetrical case with $r < r'$ the design can only be used in positive order. For the reverse case with $r > r'$ the design will only focus in negative order. By moving from a symmetrical to an asymmetrical configuration

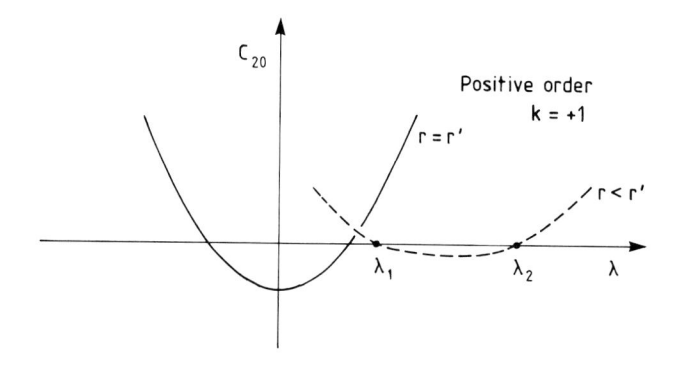

Fig. 4.8. Defocus coefficient C_{20} for a symmetric and asymmetric TGM.

the two zeros of C_{20} are moved into the same order with the result that there is approximate focusing over a reasonably wide wavelength range. The values of entrance and exit arm length to give $C_{20} = 0$ at the chosen wavelengths λ_1, λ_2 can be found by setting $C_{20} = 0$ in eq. (4.48) and simultaneously solving at λ_1 and λ_2,

$$\frac{\cos^2 \alpha_1}{r^*} - \cos \alpha_1 + \frac{\cos^2 \beta_1}{r'^*} - \cos \beta_1 = 0, \qquad (4.53)$$

$$\frac{\cos^2 \alpha_2}{r^*} - \cos \alpha_2 + \frac{\cos^2 \beta_2}{r'^*} - \cos \beta_2 = 0. \qquad (4.54)$$

The entrance and exit arm lengths can then be given by

$$r = R\left(\frac{GA - CE}{BG + DG - FC - HC}\right), \qquad (4.55)$$

$$r' = R\left(\frac{GA - CE}{FA + HA - BE - DE}\right), \qquad (4.56)$$

where $A = \cos^2 \alpha_1$, $B = \cos \alpha_1$, $C = \cos^2 \beta_1$, $D = \cos \beta_1$, $E = \cos^2 \alpha_2$, $F = \cos \alpha_2$, $G = \cos^2 \beta_2$ and $H = \cos \beta_2$.

The selection of the two in-focus positions λ_1, λ_2 and hence r and r' is the first step in defining the configuration of a TGM. The selected entrance and exit arm lengths can now be substituted in the expression for astigmatism correction given in eq. (4.52). A design in which the defocusing term is minimised, however, may not lead to an optimum design as coma and astigmatic coma may significantly contribute to the aberration-limited resolution. Although selection of the C_{20} zeros defines the starting point for r^* and r'^* the overall resolution must then be minimised by iterating r^* and r'^*.

By differentiating the aberrant path function given by eq. (4.47) with respect to y, the dispersive length of the grating, the overall resolution can be calculated:

$$\Delta \lambda = \frac{1}{Nk}\left(\frac{2y}{R} C_{20} + \frac{3y^2}{R^2} C_{30} + \frac{z^2}{R^2} C_{12} + \cdots\right). \qquad (4.57)$$

To assess $\Delta \lambda$ for a real situation, the angular properties of the light source need to be evaluated. For an isotropic source, $\Delta \lambda$ would be assessed for a large number of points equally distributed over the grating surface. $\Delta \lambda$ would then be the FWHM of the $\Delta \lambda_n$ distribution. For synchrotron radiation, the vertical distribution is highly anisotropic and a function of wavelength and so an accurate calculation would reflect this in a wavelength-dependent distribution of points on the grating surface. It is usually sufficient, however, for the purposes of optimising the resolution to take a matrix of 3×3 points covering the grating surface. Each of the points will contribute a particular $\Delta \lambda$ which may be positive, zero or negative. The overall $\Delta \lambda$ for that wavelength and configuration is the difference between the most aberrant positive and negative rays, i.e.,

$$\Delta\lambda_{\text{FWZH}} = \Delta\lambda_+ - \Delta\lambda_- .$$

In eq. (4.52) it can be seen that the coma and astigmatic coma will always be single-signed for both positive and negative y and z, respectively due to the squared term. The defocus term will contribute equal positive and negative values corresponding to the top and bottom of the grating.

If the half-length and half-width of the grating is represented by a and b, respectively then the absolute contributions from the various aberrations are

$$\text{defocus:} \quad D = \frac{1}{Nk} \cdot \frac{2a}{R} C_{20}, \tag{4.58}$$

$$\text{coma:} \quad C = \frac{1}{Nk} \cdot \frac{3a^2}{R^2} C_{30}, \tag{4.59}$$

$$\text{astigmatic coma:} \quad A = \frac{1}{Nk} \cdot \frac{b^2}{R^2} C_{12}, \tag{4.60}$$

and so

$$\Delta\lambda = [(D + C); (-D + C); 0.0; (A); (D + C + A); (-D + C + A)]_{\min}^{\max}, \tag{4.61}$$

where max/min indicates that $\Delta\lambda$ is the difference between the maximum term and the minimum term. This is shown schematically in fig. 4.8. $\Delta\lambda$ can then be evaluated at intervals over the desired wavelength range. The figure of merit of the design is then the inverse of the integral of $\Delta\lambda$ from the lower wavelength limit λ_L to the upper wavelength limit λ_H.

$$\frac{1}{M_\lambda} = \int_{\lambda_L}^{\lambda_H} \Delta\lambda(\lambda, r^*, r'^*). \tag{4.62}$$

As α and β for a particular wavelength are fixed (N, the groove density, is fixed by a correct choice of the horizon wavelength and 2θ, the deviation angle, by the reflectivity), the only remaining variables are r^* and r'^*. Starting from the zero defocus values given by eqs. (4.55) and (4.56), r^* and r'^* are iterated in a grid search until the figure of merit is maximised. At each new r^*, r'^* position a new value for the sagittal curvature is found from the expression for the astigmatism correction given by eq. (4.52), using a wavelength at the middle of the wavelength range. The figure of merit given by eq. (4.62) would give a design that was optimised for wavelength resolution. If the design was to be optimised for energy resolution or resolving power, the following figures of merit would be used respectively.

$$\Delta E, \qquad \frac{1}{M_E} = \int_{\lambda_L}^{\lambda_H} \frac{\Delta\lambda(\lambda, r^*, r'^*)}{\lambda^2} ; \qquad (4.63)$$

$$\text{Resolution}, \quad \frac{1}{M_R} = \int_{\lambda_L}^{\lambda_H} \frac{\Delta\lambda(\lambda, r^*, r'^*)}{\lambda} . \qquad (4.64)$$

An example of the results of an energy optimization for a 150°, 710 lines/mm TGM is given in fig. 4.9. Figures 4.9a–4.9c show the individual aberration components given by D, C and A in eqs. (4.58)–(4.60) which represent the contributions from defocus, coma and astigmatic coma, respectively. These are then combined in the manner given by eq. (4.61) to give the energy difference between the most aberrant rays. This is plotted in fig. 4.9d. The optimization range was from 10 to 40 eV and yielded $r^* = 0.1931$, $r'^* = 0.3273$ and $\rho^* = 0.0628$ for a dispersive aperture of 12.0 mrad and a non-dispersive aperture of 16.0 mrad.

It can be seen from figs. 4.9a–4.9c that the dominant aberration at low energy is defocus. If provision were made in the design for a translating exit slit a better overall resolution could be achieved by adjusting r' so that a better balance between the aberrations could be found. The optimum r' would not give zero defocus for the wavelength required but would be reached by a minimisation procedure similar to the one previously described; however, only r'^* would be changed. The starting position for the optimization would be the in-focus r'^* given by

$$r'^* = \frac{\cos^2 \beta}{\cos \alpha + \cos \beta - (1/r^*) \cos^2 \alpha} . \qquad (4.65)$$

Any analytic optimisation procedure should be checked in detail by ray tracing. In order to demonstrate the nature of coma and astigmatic coma, the previous example of a 150° TGM is used with the photon energy set to 35 eV so that the monochromator is in focus (fig. 4.9a). From figs. 4.9b and 4.9c it can be seen that coma and astigmatic coma at this energy are of the same sign and will therefore give rise to aberrations on the same side of the dispersive axis. In order to demonstrate coma, the monochromator was ray-traced for 20 mrad dispersive aperture by 1 mrad non-dispersive aperture. The result is shown in fig. 4.10 using the ray-tracing program of Hubbard and Pantos (1983). In order to demonstrate astigmatic coma the apertures were reversed with the result given in fig. 4.11. As expected, the aberrations occur on only one side of the axis and have the same sign in each case. In order to demonstrate defocus, the monochromator was ray-traced at 24.7 eV, where from figures 4.9a–4.9c it can be seen that $C_{30} = 0$ and C_{20} is the dominant aberration. The result is shown in fig. 4.12. As expected, the

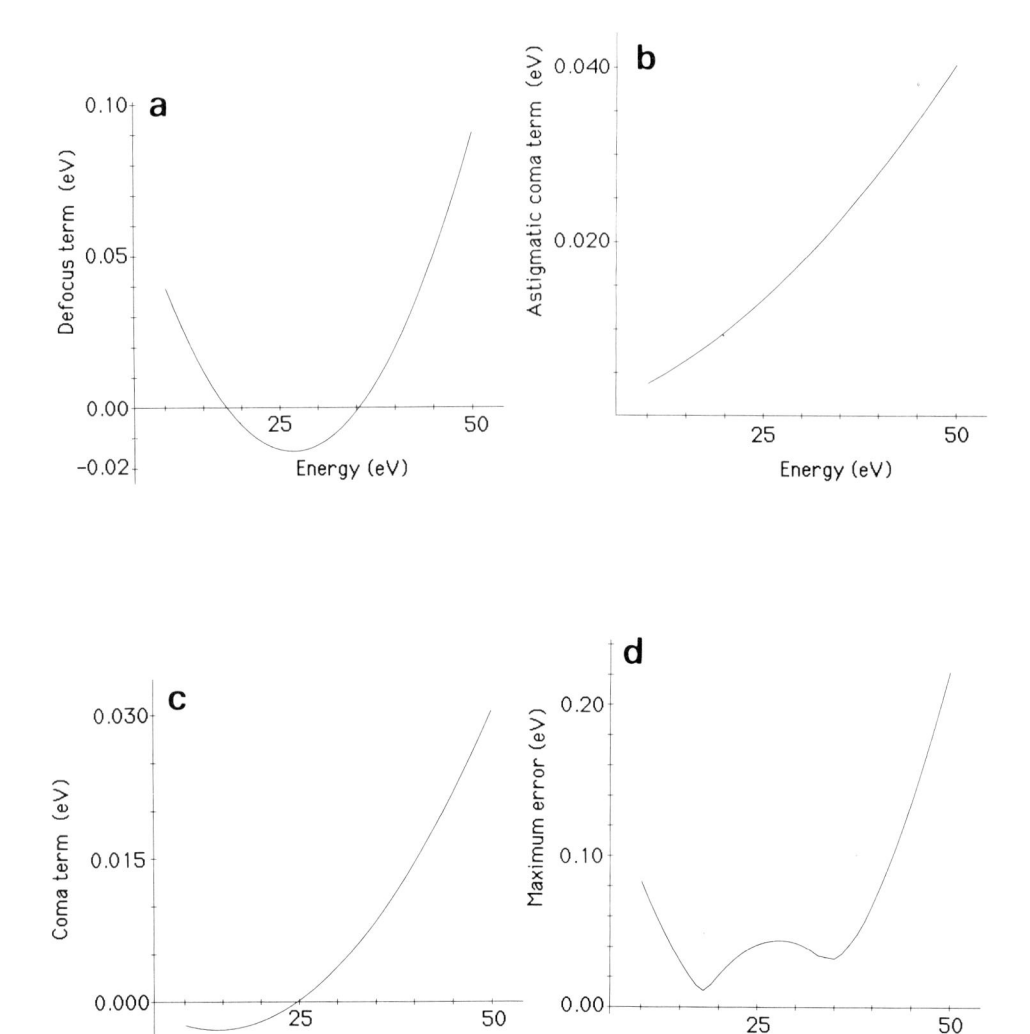

Fig. 4.9. Defocus (a), astigmatic coma (b) and coma (c) contributions given by eqs. (4.58)–(4.60). The resultant energy separation of the most aberrated rays is shown in (d) following eqs. (4.63) and (4.61).

dispersive aberration is approximately symmetrically distributed about the dispersive axis. In order to remove defocus at this energy, the exit arm length was adjusted to $r'^* = 0.3233$ in accordance with eq. (4.65). The result of this is shown in fig. 4.13. The defocus has clearly been removed leading to a large reduction in the dispersive aberrations although coma has now appeared in the image. In order to reduce the aberrations further r' was iterated until the image size was minimised with the results shown in fig. 4.14 for $r'^* = -0.3238$.

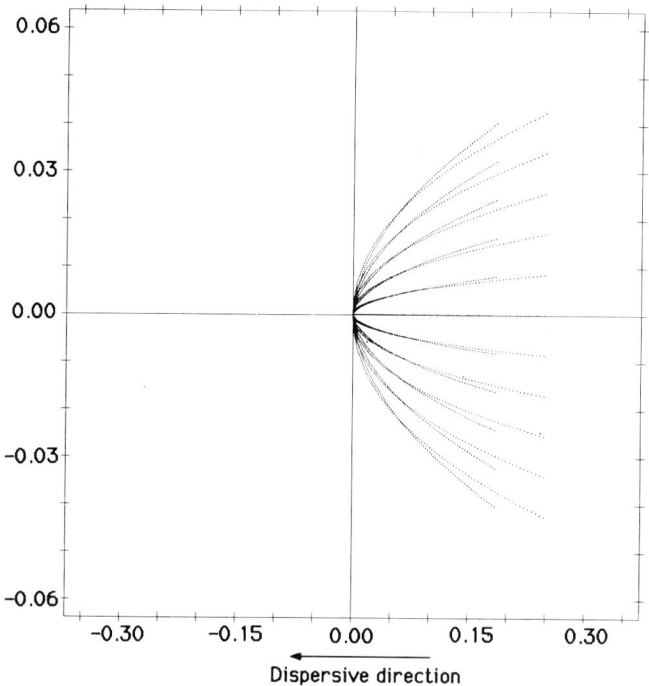

Fig. 4.10. Raytrace of 150° TGM with groove density 710 lines/mm, first positive order, $r^* = 0.1931$, $r'^* = 0.3273$ and $\rho^* = 0.0628$. Aperture; dispersive 20 mrads, non-dispersive 1 mrad. Dispersive aberration principally coma. Photon energy 35 eV, $C_{20} = 0$.

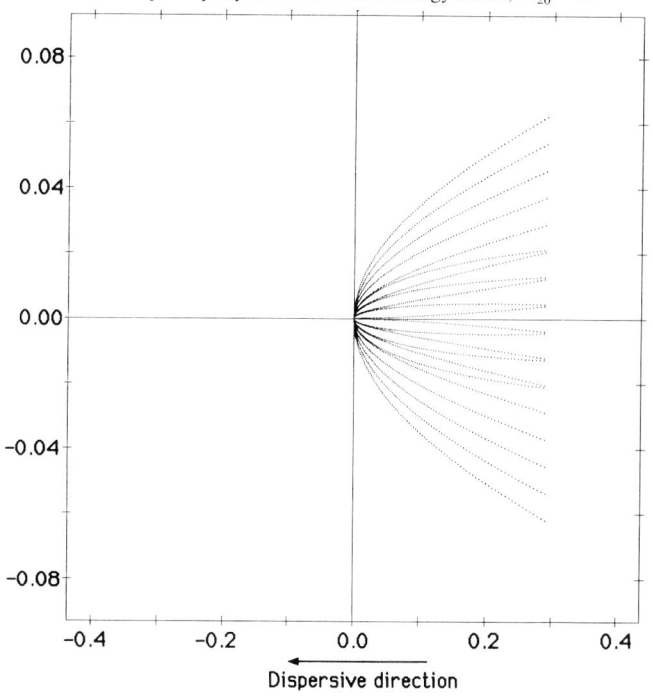

Fig. 4.11. Parameters as for fig. 4.10 except aperture; dispersive 1 mrad, non-dispersive 20 mrad. Dispersive aberration principally astigmatic coma.

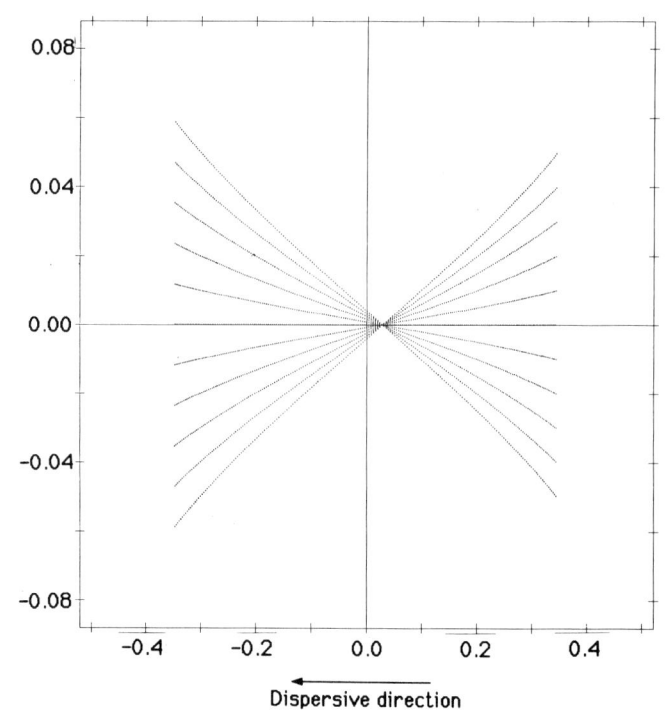

Fig. 4.12. Parameters as for fig. 4.10 except photon energy 24.7 eV, $C_{30} = 0$. Aperture; dispersive 20 mrads, non-dispersive 1 mrad. Dispersive aberration principally defocus.

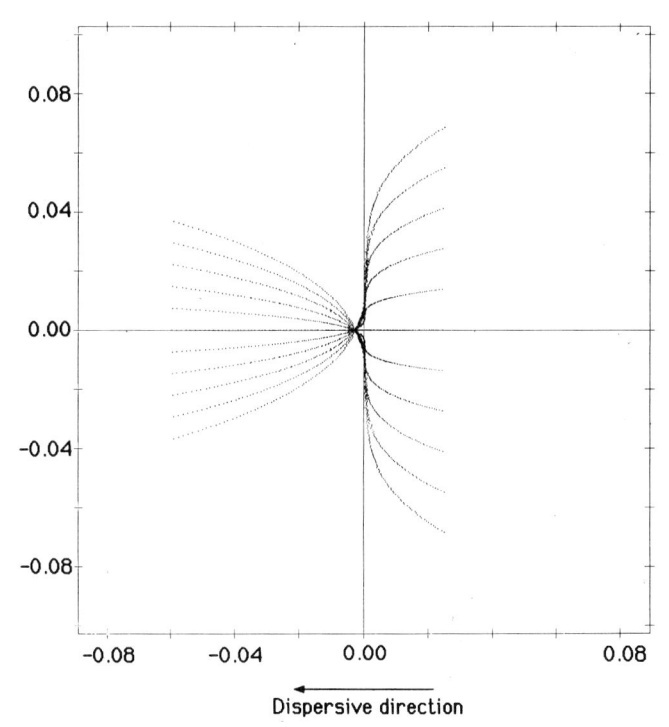

Fig. 4.13. Parameters as for fig. 4.12 except $C_{30} \neq 0$; $r'^* = 0.3233$, $C_{20} = 0$. Photon energy 24.7 eV. Dispersive aberrations coma (lhs) and astigmatic coma (rhs).

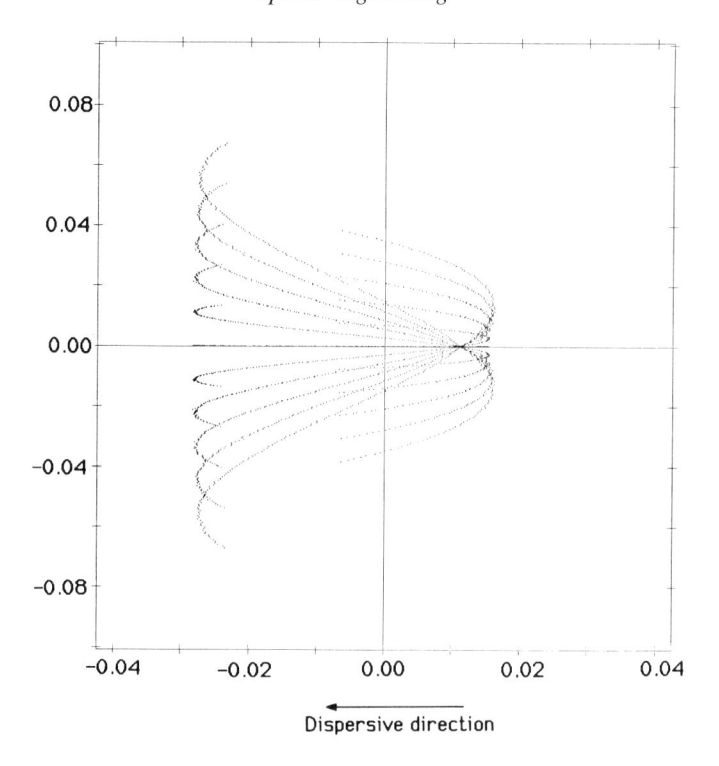

Fig. 4.14. Parameters as for fig. 4.13 except $r'^* = 0.3238$. Fully optimised r'^* at 24.7 eV. Dispersive aberration is a combination of defocus, coma and astigmatic coma.

4.4. Design of TGM pre-optics

A selection of the main types of pre-optical systems has been described in sections 3.1 and 3.3. It is important to note that the design of the pre-optical system can severely affect both the throughput and resolution of the monochromator. As the aberration limited resolution of a TGM is independent of the overall size, the parameter which determines the size of the monochromator is the source size limited resolution. From eq. (4.18) it can be seen that the resolution is proportional to s_1/r, where s_1 is the entrance slit size in the dispersive direction and r is the entrance arm length. For maximum efficiency, it would be desirable that the size of the demagnified source on the entrance slits was small enough to give the maximum required resolution without substantial loss at the entrance slit. In general therefore the monochromator is built as large as possible for high-resolution applications. For many applications, the maximum size of monochromator allowed by the physical constraints of the experimental area does not give sufficient resolution without demagnification of the source. Assuming the grating can be made as large as necessary the question arises as to what limits the desirable demagnification. The image quality of the pre-optical system will degrade as a function of increasing demagnification, but usually a more severe restriction is the aperture dependence of the various aberrations described in the previous section. For excessive demagnification, the defocus and coma aberration will dominate the demagnified source size limited resolution. Clearly therefore a correct balance must be made between source-limited and aberration-limited resolution.

One of the advantages of using a two-mirror pre-optical system is that the magnifications in each direction may be adjusted independently. As described in the previous section, one of the principal aberrations in the TGM is astigmatic coma which depends on the non-dispersive aperture. The pre-mirror focusing in the non-dispersive direction should therefore be designed to operate at near-unity magnification. For the range of horizontal source sizes common in storage rings, the system should in general not magnify as higher order aberrations become more important as the non-dispersive object for the TGM becomes excessively large.

4.5. The astigmatically corrected spherical grating in a TGM mounting: the SGM

The combined pre-optics described in the previous section and the toroidal grating monochromator are designed to produce stigmatic foci, i.e., the pre-optics form a stigmatic image at the TGM entrance slits and the grating forms a monochromatic stigmatic image at the exit slits. The question therefore arises as to whether the focus of the pre-optics in the non-dispersive direction needs to be at the entrance slits or whether by making the focus non-stigmatic and therefore using a different sagittal radius for the grating, the overall performance can be improved. Such a non-stigmatic TGM has been described by Chen et al. (1984) in which the non-dispersive focus was placed close to the grating and the overall design was found to have an improved aberration limited resolution. An extension of this idea is to completely remove the sagittal radius of the grating so that all the non-dispersive focusing is done by the pre-optics. Figure 4.15 shows the layout of such a system. In practice the cylindrical grating is approximated by a sphere, as the very small sagittal curvature at glancing incidence produces little focusing. For VUV use, where the included angle of the TGM could be as low as 140°, the effective sagittal curvature has to be compensated by focusing after the exit slits with the non-dispersive direction pre-optics arranged so that the sagittal focusing of the sphere in the middle of the wavelength range brings the overall focus to the exit slits. In the soft X-ray region, where the deviation angle is much larger, this effect is negligible.

An alternative to the entrance slitted SGM is shown in fig. 4.16. This monochromator uses the electron beam at the tangent point as the entrance slit and is broadly based on the FHI high-energy TGM at BESSY (Dietz et al. 1985). Two millirad of the horizontal fan of radiation is collected by a 2° glancing angle CVD SiC cylinder premirror and focused onto the exit slits of the monochromator. The grating is spherical and is mounted in a simple sine bar scanning mechanism. The light from the exit slits is refocused by a fixed ellipsoidal mirror onto the sample. Due to the large amount of shielding necessary on a high-energy storage ring, the optical components can only be placed a large distance from the

Optical engineering

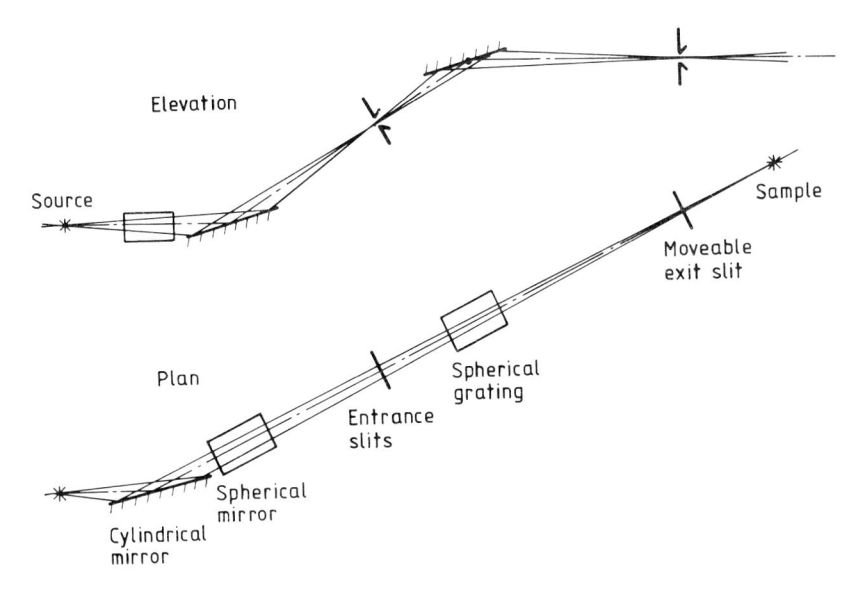

Fig. 4.15. Layout of a positive order entrance slitted SGM.

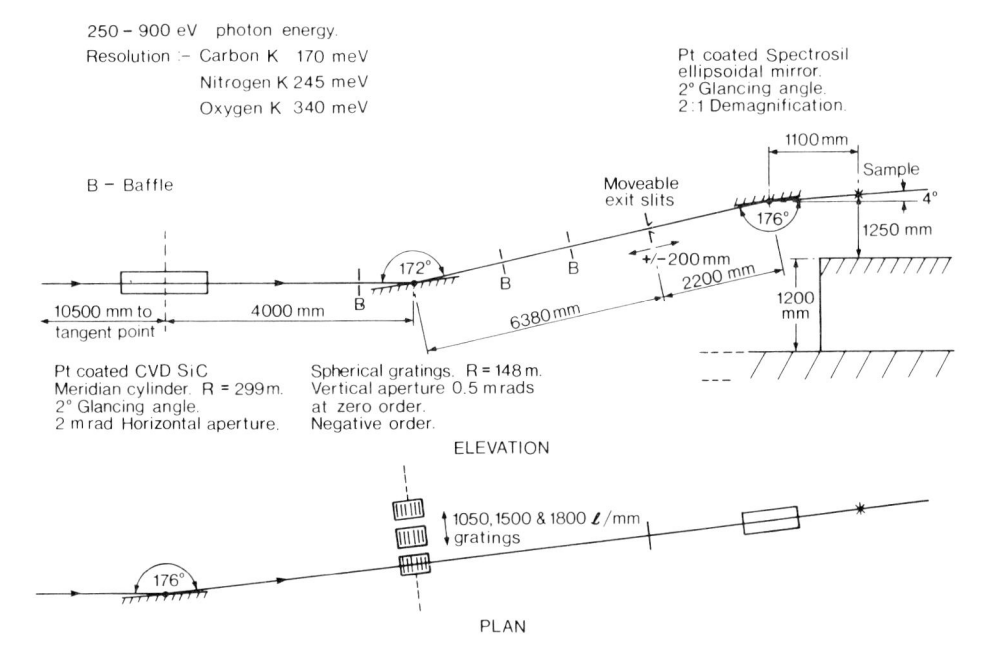

250 – 900 eV photon energy.
Resolution :– Carbon K 170 meV
                Nitrogen K 245 meV
                Oxygen K 340 meV

Pt coated Spectrosil ellipsoidal mirror.
2° Glancing angle.
2:1 Demagnification.

B – Baffle

Pt coated CVD SiC
Meridian cylinder. R = 299 m.
2° Glancing angle.
2 mrad Horizontal aperture.

Spherical gratings. R = 148 m.
Vertical aperture 0.5 mrads at zero order.
Negative order.

ELEVATION

1050, 1500 & 1800 ℓ/mm gratings

PLAN

Fig. 4.16. Layout of SRS BL1 HESGM. Negative order entrance slitless soft X-ray monochromator.

tangent point and so a negative order design ($r > r'$) was adopted. The aperture collected is 2 mrad (horizontal) by 0.5 mrad (vertical) at zero order rising to typically 0.7 mrad in the spectrum. The wavelength-dependent image distance of the gratings (eq. 4.65) can be tracked in this design by computer controlled translation of the exit slits. The exit arm length as a function of photon energy is given in fig. 4.17. Three interchangeable gratings (1050, 1500 and 1800 lines/mm) are used to give the same image distances at the carbon, nitrogen and oxygen K-edges, to optimise the diffracted efficiency in each region of interest. The focal curve is arranged so that for near-edge spectroscopy the defocus broadening is minimised. For a vertical source size of 0.4 mm FWHM resolutions of 170, 245 and 340 meV should be achieved at the carbon, nitrogen and oxygen K-edges, respectively. The result of removing the sagittal curvature and using a separated function focusing system is demonstrated in fig. 4.18. The entrance slitless design shown in fig. 4.16 has been ray-traced using a 3×3 matrix representation of the source (0.4×2 mm) for a 1050 lines/mm grating at 280, 340, 450 and 600 eV photon energy. The defocus has been eliminated in each case by moving the slits to the correct focal position (fig. 4.17). It can be seen that the only dispersive aberration appears to be a small coma tail and that the normally dominant aberration of astigmatic coma is missing. It can be seen also that at 340 eV where the closest approach to the Rowland circle is made that the coma from the central part of the source has vanished. The increase and decrease in the horizontal image quality is simply caused by the movement of the image plane about the fixed horizontal focus of the premirror. The optical design as with all types of SGM is essentially aberration-free and the resolution will only be limited by the source (or slit) size and the slope errors in the optical elements. An attractive feature of the design shown in fig. 4.16 is that the slope errors in the premirror will, essentially, only contribute in the non-dispersive direction; the quality of the spherical grating is the only parameter that will affect the resolution. The sphere

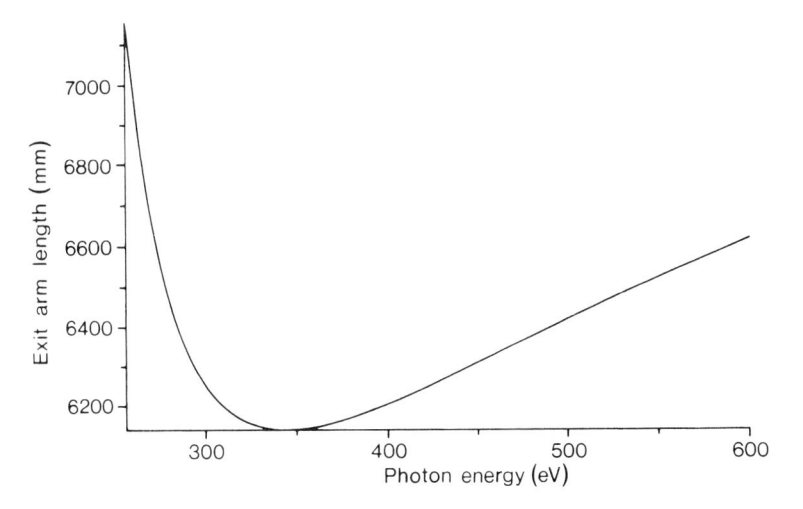

Fig. 4.17. Exit arm length as a function of photon energy HESGM, 1050 lines/mm.

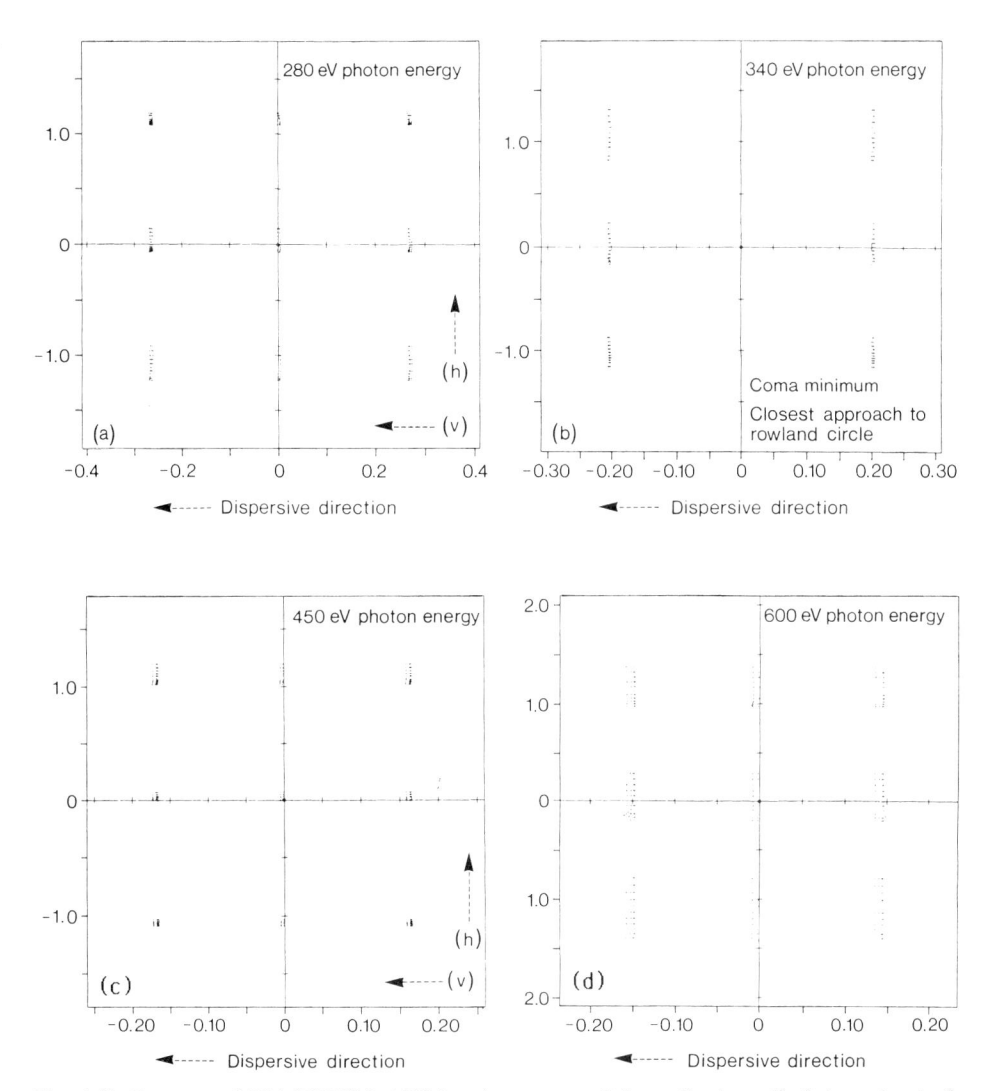

Fig. 4.18. Raytrace of BL1 HESGM; 1050 lines/mm, source 2.0 mm (horizontal), 0.4 mm (vertical); (a) 280 eV, (b) 340 eV, (c) 450 eV and (d) 600 eV photon energy. Exit slits at focus position for each energy.

is the simplest focusing optic to produce and it is currently believed that slope errors down to 0.2 arc seconds can be achieved. This implies that for entrance slitted SGM's, resolutions down to 20 meV at the carbon K-edge should be achievable. To conduct useful experiments at this resolution will require a high brightness source such as storage ring undulators. A further benefit of removing the sagittal focusing from the normal TGM is that the SGM resolution is highly insensitive to the non-dispersive aperture or to source movements in the non-dispersive direction. As the sagittal curvature is very small, the gratings can be translated across the beam line without affecting the resolution. This affords the opportunity of making the gratings several times wider than necessary so that they may be translated to expose fresh coating areas when the used grating area becomes contaminated. An alternative would be to ion-etch the different ruling densities into adjacent stripes on the one optical element. The entrance slitless monochromator in fig. 4.16 is currently (July 1986) being installed in BL1 of the SRS (Padmore 1986b) and several entrance slitted SGM's are in the planning or construction stage at other SR facilities [e.g. Brookhaven: one monochromator on a VUV ring bending magnet, one monochromator on the X-ray ring undulator (C.T. Chen); Stanford: one monochromator on an undulator (Hogrefe et al. 1986)].

4.6. Diffraction efficiency

The form of the various groove profiles is shown in fig. 4.4. In order to design an optimised monochromator, the grating efficiency must be calculated as a function of incident and diffracted angles, groove spacing and groove depth. It may also be important to calculate higher order diffracted efficiencies. A rigorous calculation would involve solution of Maxwell's electromagnetic field equations using the grating profile as a periodic boundary. Such calculations have been made for VUV and soft X-ray gratings (e.g. Petit 1980, Neviere 1980, Neviere et al. 1982, Neviere and Flamand 1980, Flamand et al. 1986, Jark and Neviere 1986). The main features of the efficiency behaviour can be predicted by simpler scalar theories although the predictions of absolute diffraction efficiency and higher order content are not accurate. For lamellar gratings Hellwege (1937) has shown that the diffracted efficiency in odd orders m can be given by

$$E_m = \left(\frac{2}{m\pi}\right)^2 \sin^2\left(\frac{\delta}{2}\right), \tag{4.66}$$

where $\delta = 2\pi h(\cos \alpha + \cos \beta)/\lambda$.

This leads to an oscillatory form for the diffraction efficiency due to constructive and destructive interference of waves diffracted at the top and bottom of the grooves. This is shown in fig. 4.19. It is clear that to obtain a high diffraction efficiency over a large wavelength range the maximum marked A should be used. The position of the odd-order maximum A is clearly a function of groove depth and the correct depth for the maximum to occur at a chosen "blaze" wavelength λ is given by

$$\left(\frac{1}{h} - \frac{1}{h_0}\right)^2 = \left(\frac{8 \sin \alpha}{\lambda d}\right)m + \frac{1}{h_0^2}, \tag{4.67}$$

where $h_0 = \lambda/2 \cos \alpha$.

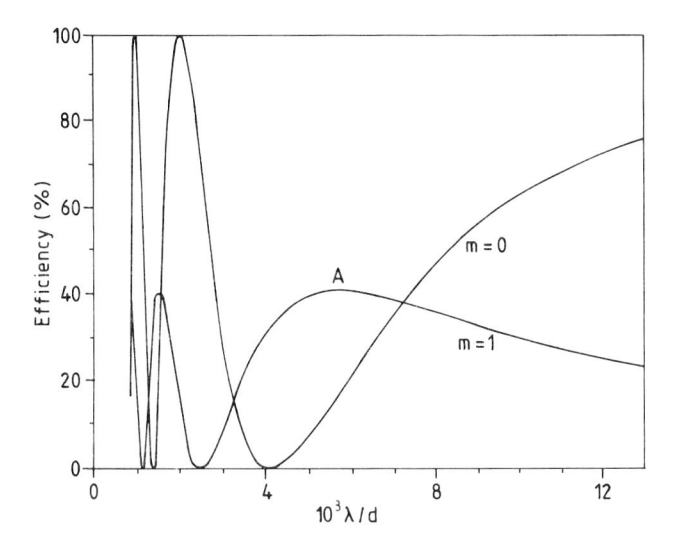

Fig. 4.19. Zero- and first-order diffraction efficiencies using the model of Hellwege (1937) for a groove depth to grating pitch ratio (h/d) of 0.01 (Franks et al. 1975).

From eq. (4.66), the position of the primary maximum is also a function of the deviation angle and so a wider range of wavelengths can be used if the deviation angle is changed for different regions. The basic theory has been extended by Bennett (1971), Johnson (1975) and Franks et al. (1975) to include the effects of shadowing and penetration of the X-rays through the grating lands. Neviere et al. (1982) have investigated the applicability of the simple scalar theory and the optimisation of the efficiency by altering the groove width (c) to groove spacing (d) ratio. By reducing the land width the shadowing is reduced and so more of the bottom of the groove can contribute to the diffraction. In the soft X-ray region, the grazing incidence geometry can lead to completely or heavily shadowed grooves for $c/d = 0.5$ and therefore to low diffraction efficiency. The simple geometrical argument for the optimum c/d ratio that the unshadowed groove length should be equal to the land length has been investigated by Neviere et al. (1982), who have shown that for soft X-ray gratings with groove densities above 600 lines/mm this simple view breaks down. The diffraction efficiencies of blazed, lamellar and sinusoidal gratings optimised at 109 Å are shown in fig. 4.20.

The blazed grating has the highest efficiency with the other two being comparable. The advantage of the lamellar grating is that the even-order contribution is strongly suppressed. By optimising the c/d ratio to give the highest efficiency at a given wavelength the suppression of even order radiation is reduced. The minimum even order content will always occur at $c/d = 0.5$, but the highest ratio between first order and even order light will occur at a value between 0.5 and the optimised efficiency c/d value. For many experiments, the minimization of higher order content is critical and this aspect of grating design should not be neglected.

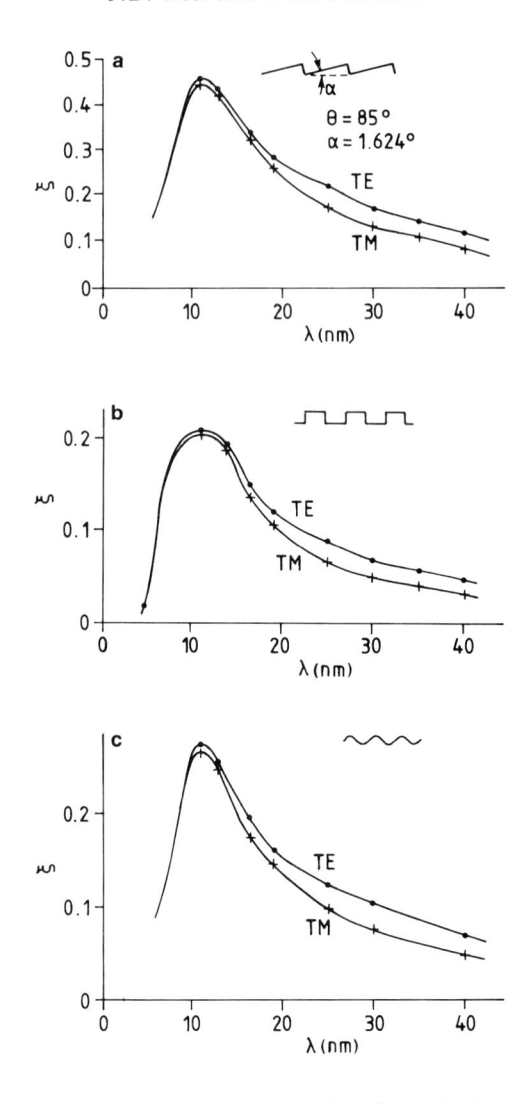

Fig. 4.20. Diffraction efficiency for blazed (a), lamellar (b) and sinusoidal (c) gratings. N = 600 lines/mm, $\theta = 85°$. Blazed grating blaze angle 1.624°: lamellar, $h = 198$ Å: sinusoidal, $h = 302$ Å. (Courtesy M. Neviere.)

5. Beam line layout

Using the information in the previous sections, it should now be possible to design the instruments required and calculate the light flux and resolution expected from them. We describe in this section the major considerations relevant to the layout, and give an example of a four instrument beam line at the Daresbury SRS.

5.1. General design principles for a multiple instrument beam line

Ideally, one would assume that the first optical component feeding a monochromator should be as close as possible to the source, in order to gather the maximum amount of radiation. Other factors dominate this situation, however; generally the beam line must clear the storage ring components as it emerges from the bending magnet, and move sufficiently far from the tangent point to make space for mirror boxes to contain the collecting optics. This depends on the bending radius of the electrons and thus goes back to the original machine design and the chosen electron energy. For VUV spectroscopy, a low-energy, small radius ring is feasible and the distance to the first optical component can be kept to two metres or less (see e.g. Ederer et al. 1980). There is an added advantage that the shielding required against high-energy radiation is minimal during stored beam conditions. Quite large apertures (50 mrad or more) can be taken from such sources; however, it should be borne in mind that the source is an extended one *along* the optic axis, and as described earlier there will be an apparent increase in source size which in practice will become serious for high emittance sources for apertures above ~ 30 mrad. The final decision will depend on the wavelength resolution required, whether or not a prefocusing mirror system is needed and the demagnification which can be used bearing in mind the aberration performance of the system, discussed in the earlier section.

In the more general case, where the storage ring energy is higher to give useful flux in the soft X-ray region at a few angstrom, the closest component is rarely less than 5 metres away, to avoid collisions with storage ring components. With careful shielding design it is possible to provide access for optical adjustments to this component while synchrotron radiation is falling on it, and in general this is a desirable feature. The design approach from this point onwards depends upon whether the beam line is to serve a single experimental station or many such stations. A single experiment beam line is obviously a simplification, and can easily be incorporated on a machine where the bending radius is small and thus it is physically possible to arrange two or more separate beam lines from the same bending magnet. This allows single, large aperture instruments and gathers maximum flux, but in general these will be medium/low wavelength resolution, for the reasons described above. The optical design principles for such beam lines are the same as for multiple experiment beam lines, but the latter have additional constraints. What follows summarises the major decisions to be made concerning beam line layout:

(i) Common versus individual mirror boxes.
  The choice amounts to the following:
  (a) (individual mirror line) separate the full aperture of the beam at its front end into separate beam lines, thereby losing the radiation in the "crotch" areas between beam lines, and then lead each individual branch to its optics;
  (b) (common mirror box line) use mirror boxes the initial one of which takes the complete fan of radiation, and contains a mirror to divert part of it to an experiment passing the rest on to the next mirror box.

The advantage of the first technique is that breakdown/mechanical work on one branch does not affect the other branches. However, up to 30% of the available aperture can be lost. The second technique uses all the aperture available by ensuring that mirror mounting plates and mechanisms are always in the "dark" region at the edges of the beam, and thus reflect the radiation across the beam passing straight through. Any problem with the first mirror box, however, renders the whole line unuseable if the vacuum must be broken to rectify it.

Considering the pressure on synchrotron radiation sources nowadays, the common mirror box approach should be considered seriously. There is, however, a major consideration which favours the individual mirror box design. Any branch leading to an instrument which works in the 280–600 eV region, where carbon contamination has its most serious effects, may need frequent access. Particularly on high-energy storage rings, it is almost impossible to prevent gradual carbon build-up on the first optical component, thus it would be wise to make this branch independent of the rest so that its mirror can be cleaned or changed without disturbing the rest of the beam line.

(ii) Decide on the aperture available.

This aperture is generally dominated by the requirement that no surface which stray electrons from the beam can hit is visible directly by an observer looking down the beam line. By placing the centre of the arc emitting radiation into the beam line near the centre of the bending magnet, it has been possible to build a "reverse-tangent" recess into the "tangent vessels" at the Daresbury SRS and thereby ensure the above requirement is met and this should hold true for other machines. The reason is to reduce the high-energy bremsstrahlung intensity down the beam line, which would be hazardous to experimenters. Careful analysis shows that bremsstrahlung from residual gas molecules is a more serious problem, but this can be solved by careful collimator design in the beam line, and by the use of beam stops so that no one can "look" down the beam line aperture without "looking" through a substantial thickness (~ 20 cm in the case of the SRS) of lead. The aim is for the maximum possible aperture without needing heavy beam line shielding; the maximum value will also be limited by space considerations.

(iii) Differential pumping at the beam line front end.

The aim of this is to cope with gas loads produced from irradiation of new mirrors in the beam line. Despite bakeout, UV light on a new mirror will cause substantial outgassing. In the early stages of monochromator installation it can also be used for testing purposes since the monochromator can be opened up to the beam at relatively high pressure ($\sim 10^{-7}$ mbar). However, this practice should be kept to an absolute minimum because of the danger of contaminating the optical components; this, in its turn, can be minimised by being scrupulous with cleanliness of all components in the vacuum system, pre-baking them where possible. In this way the hydrocarbon content of the residual vacuum can be kept to a minimum, and it is the partial pressure of these hydrocarbons, rather than the ultimate total pressure, which should be minimised.

Once the beam line is in routine use, the differential pumping system is less important, and there may be no case for fitting it in single instrument beam lines where the aperture is large. It should *not* be used to cope with gas loads from experiments; this should be dealt with by differential pumping at the experiment itself.

(iv) All optical components accessible for manual adjustment with the beam on.

This decision has important implications for the general beam line design, since the distance of the first optical component from the tangent point must allow for radiation shielding between this component and the storage ring. In addition, the differential pumping section must be included. On the Daresbury SRS the distance to the first optical component is ~6 m, resulting in a mirror of ~500 mm in length to accept 10 mrad and deflect it through 10°. This is practical, and has advantages for mirror design and heat loading. The requirement for access to this mirror is paramount during the early stages of setting up; the vacuum chamber which contains it should be liberally fitted with windows so that the way the mirror is being illuminated and the position of the reflected beam can be seen clearly. As the beam line settles down these requirements are far less important, but are so desirable in the early stages that the design of the beam line should be constrained to accommodate them. On low-energy rings where radiation shielding is minimal the constraint does not, of course, arise.

(v) A laser alignment system to aid the setting up of monochromators and experiments.

A facility of this kind is an invaluable aid during the early stages of setting up a beam line. It should be accessible all the time, i.e., placed outside the shielding area. It should be adjustable so that it can illuminate any instrument on the beam line and simulate the direction of the synchrotron radiation beam, i.e., it should appear to originate at the theoretical tangent point. An angular accuracy of ~0.1 mrad is required, and thus the assembly must be very stably anchored to overcome vibration and differential vacuum forces. Such a system is useful for locating position and direction for components in the beam line, but *not* for focus adjustment. A more sophisticated system which has an auto-collimation facility built in has been described by Williams and Howells (1983) for the Brookhaven VUV ring. Details of the Daresbury system will be given in a later section. They are intended for initial alignment, and to act as a reference datum for beam *direction*. Final adjustments, particularly focus, must be carried out using the synchrotron radiation beam.

Thus, with the total beam line aperture known, the instruments (and their pre-optics if required) chosen, it should now be possible to design the beam line as a whole. At this stage space constraints will affect the final design, but will not, hopefully, seriously alter the optical layout of any instrument. In the next section, we describe for the sake of example the layout of a four instrument beam line now in use at the Daresbury SRS.

88   *J.B. West and H.A. Padmore*

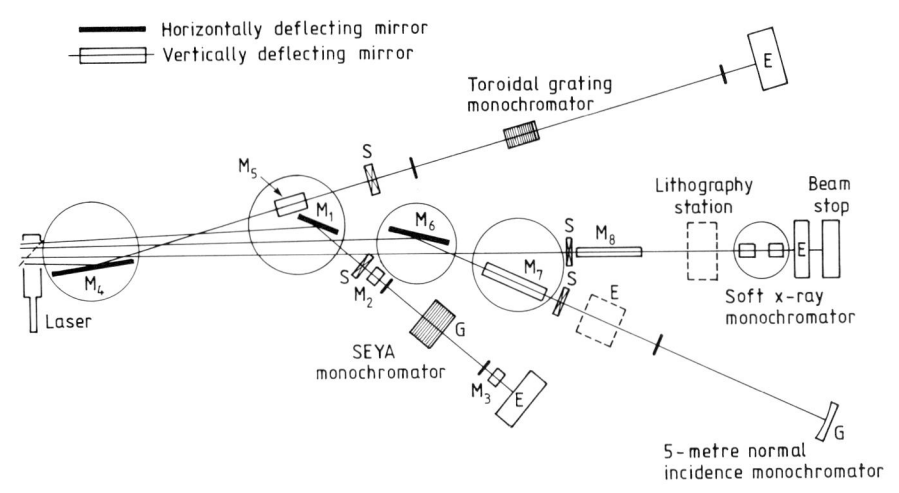

Fig. 5.1. Plan layout of 4-instrument beam line; not to scale. Mirror details: M_1: cylindrical, SiC, 4:1 demagnification, grazing angle 10°, aperture 7.5 mrad horizontal, 2.5 mrad vertical. M_4: cylindrical, SiC, 2:1 demagnification, grazing angle 5°, aperture 10 mrad horizontal, 2.5 mrad vertical. M_6: plane, SiC, grazing angle 5°, 5 mrad horizontal and vertical aperture. M_2, M_3: see fig. 5.2. M_5, M_7: see fig. 5.3. M_8: see fig. 5.4. E: experimental chamber; G: diffraction grating; TP: tangent point.

5.2. Details of a four instrument beam line at the Daresbury SRS

This beam line layout is shown, in plan view, in fig. 5.1, and its instruments cover the photon energy ranges 5–120 eV and 800 eV–6.5 keV. Since there is no instrument on it covering the energy range near the carbon K-edge a common mirror box philosophy has worked quite well, and has enabled four instruments to be fitted with reasonable apertures. Even so, the efficiency of the primary optical components has deteriorated by ~25% over two years, presumably due to slow build-up of carbon contamination, despite care with regard to cleanliness during construction. This would have had a far more dramatic effect near the carbon K-edge. Thus other lines planned for use in the 250 eV–1 keV region should incorporate facilities for easy removal of mirrors for recoating, since it seems impossible to prevent slow contamination where there is a substantial flux of X-rays absorbed by the mirrors. On multiple instrument beam lines, this would mean a separated mirror box design, as discussed earlier.

This beam line has a total aperture of 28 mrad and is fitted with a laser system to facilitate initial alignment; the source size is 12 mm (horizontal) by 0.4 mm (vertical). The primary mirrors are fitted to reflect *across* the beam line centre, as shown, except the last one. This allows the unused radiation to pass on to the next mirror with the minimum obstruction by the mirror mount or mirror edge. Outline details of the four instruments are as follows.

5.2.1. The normal incidence monochromator (Seya)
This was already available as a 1 metre spectrometer, with cylindrical mirrors at entrance and exit slits to reduce astigmatism and is a variation of the principle

described by Rehfeld et al. (1973). It contains two gratings interchangeable under vacuum, whose ruled areas are 40 × 80 mm, the 80 mm dimension being perpendicular to the rulings and thus parallel to the dispersion plane. The gratings are blazed at 500 and 1500 Å. As a first approximation, one should attempt to fill the grating. With a vertical acceptance of 2 mrad from the source, the first cylindrical mirror demagnifies the source 40 times and thus fills the 80 mm dimension (fig. 5.2), with the instrument at 8 metres from the ring. This loses some light by not accepting the full angular divergence from the source, but in order to improve this the instrument must either be moved closer to the source or the cylindrical mirror geometry changed. The position of the small cylindrical mirror (M_2 in fig. 5.2) should not be varied if the astigmatism correction is to be maintained, and to move the spectrometer closer to the source by more than 1 metre was impractical and thus little could be gained in this respect.

With a source size of ~0.4 mm in the vertical direction at the SRS, a demagnification of 40 would produce an image 10 μm wide at the entrance slit. In practice, the performance of the cylindrical mirror M_2 is limited by coma and this image is more like 100 μm wide. This provides a source size limited resolution of 0.8 Å. This can be reduced to 0.25 Å with reduction in flux by about an order of magnitude.

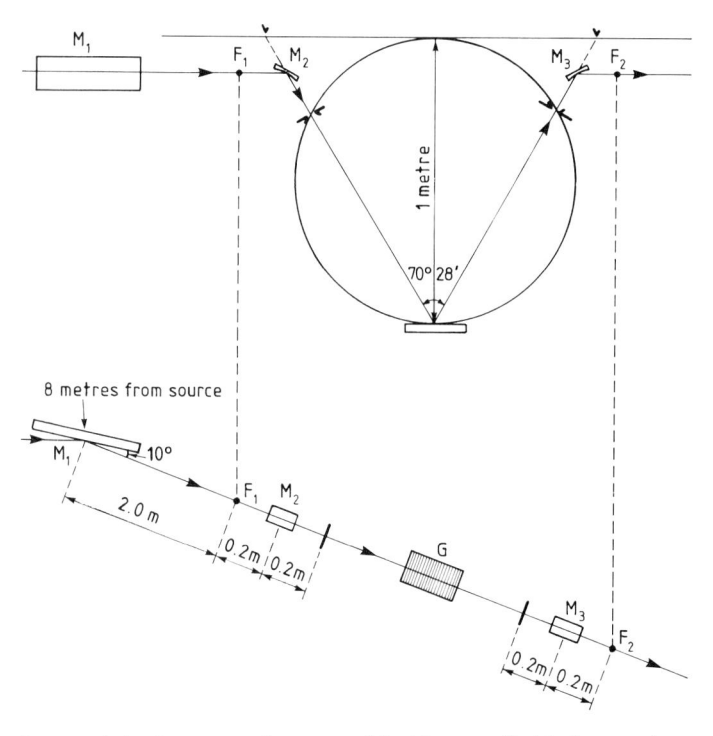

Fig. 5.2. The layout of the Seya monochromator. M_2, M_3 are cylindrical pre-end post-mirrors; see section 5.2.1 for further details.

In the horizontal plane, the 40 mm dimension of the grating should be filled, and this is done with the cylindrical premirror M_1 (fig. 5.2), which also serves to deflect the beam sideways away from the main beam line. It must also bring the horizontal dimension of the source ($\sim$12 mm in this case) to a focus at F_1 (fig. 5.2) to satisfy the condition for astigmatism correction (for details see Rehfeld et al. 1973). Thus, with a demagnification of 4:1 and an acceptance from the source of 10 mrad, the 40 mrad aperture perpendicular to the dispersion plane of the instrument is also filled.

Finally, a deflection angle of 20° (i.e. 10° grazing angle onto the mirror) is chosen to allow sufficient clearance for the spectrometer and its experiment from the main line, a compromise between good reflectivity and accessibility.

At the point F_2 an image 3 mm wide and equal to the exit slit width in height should appear. The horizontal dimension is 4–5 mm in practice, and equal to the image size of F_1 as expected. Again, coma is the primary aberration, this time in mirror M_1. For an instrument of this type, not operating at high resolution, the coma losses are not severe. The primary loss is the reflectivity of the two cylindrical mirrors M_2 and M_3, which are at angles of incidence of 62.5° where the reflection efficiency is not high. With the grating, there are thus three reflections with low efficiency which gives a low overall efficiency figure. The advantage of this instrument is its simplicity of construction, high aperture and the fact that its output is better than 98% plane polarised. This simplifies the analysis of the data in many experiments, for example, angle resolved photo-electron spectroscopy and fluorescence spectroscopy.

5.2.2. The high resolution normal incidence monochromator

For high resolution molecular spectroscopy a scanning monochromator with a resolving power of around 20 000 was considered adequate and was provided by the McPherson Instrument Company in the form of a 5-metre normal incidence spectrometer with two 1200 lines/mm diffraction gratings, ruled areas 150 × 75 mm, the ruling length being along the 75 mm dimension. These are blazed at 600 Å and 1500°, as with the Seya above. For stability this was mounted on concrete blocks and arranged to have a horizontal exit beam. It is in general useful to arrange that exit beams from monochromators are horizontal; it certainly aids interchange of experiments from one monochromator to another, and simplifies the design of experimental equipment.

Because of the support structure for this monochromator, it proved necessary to deflect the entrance beam sideways, using a plane mirror, thus making room for the straight-through beam to be used for the soft X-ray spectrometer. It then remained to focus the source onto the entrance slit of the monochromator. The horizontal aperture available is 5 mrad, equal to the vertical aperture. Thus a demagnification of 3:1 filled the grating horizontally and this was provided by an ellipsoidal mirror deflecting the beam vertically. The layout is shown in fig. 5.3a. The vertical aperture was thus underfilled by a factor of 2, leading to a loss of $\sim$40% in the optical throughput at the peak wavelength in the output of this spectrometer, a consequence of the differential pumping section early on in this

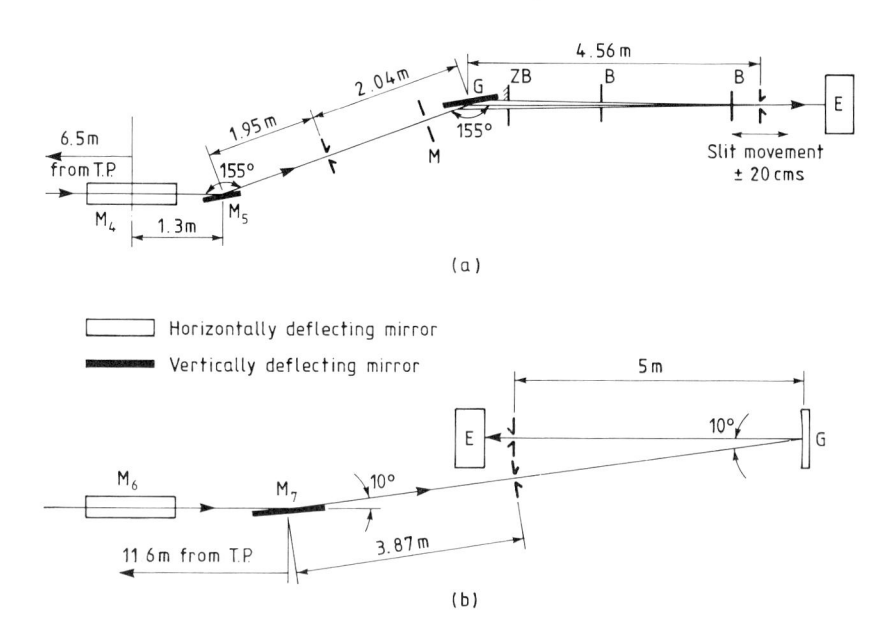

Fig. 5.3. (a) The layout of the TGM, (b) the layout of the 5-metre normal incidence. M_5: cylindrical mirror, fused silica; 4:1 demagnification. M_7: ellipsoid, fused silica; 3:1 demagnification. B: exit arm baffles; ZB: zero order trap; TP: tangent point; E: experimental chamber; G: diffraction grating; M: adjustable grating mask.

beam line. The source size limited resolution of the monochromator was 0.2 Å for this demagnification, and could have been improved by greater demagnification, offset however by the loss due to overfilling the horizontal aperture. In addition to this, the off-axis aberrations of the ellipsoidal mirror become worse as the demagnification increases resulting in the gain expected not being achieved in practice.

To achieve the high resolution of which this instrument is capable, it is essential to orient the grating rulings correctly with respect to the slits. The instrument was levelled with the centre of the exit slit vertically above the centre of the entrance slit, and the beam line laser used to arrange for the direction of the instrument to be correct, i.e., the whole spectrometer was rotated about the vertical axis through the slit centres, and moved horizontally, until the laser beam went through the centre of the entrance slit and hit the centre of the grating. At zero order, the grating was then rotated about a vertical axis until the reflected beam came through the centre of the exit slit. A window is provided to allow the diffracted beam (for a He–Ne laser) to be viewed. The grating rulings were then rotated until this beam lay in the vertical plane which also contained the slit centres and grating centre. A theodolite was used to do this. The lower slit jaws were then set horizontal, and a mercury vapour lamp placed in front of them. The upper slit orientation was adjusted parallel to the lower slit by scanning the zero order image to achieve the minimum width. Final adjustment to the grating ruling orientation was made by scanning through the 2537 Å line from a mercury lamp. This instrument also contains two focus adjustmens, to accommodate diffraction gratings of slightly differing radii; these were adjusted for minimum width at the zero order position and at 2537 Å.

This is the kind of procedure required for such instruments, and although most of the adjustments can be done with the instrument off-line they may have to be checked with the spectrometer in position. It is as well to provide for this possibility when the detailed beam line plans are made.

5.2.3. The toroidal grating monochromator

This instrument was intended to cover the wavelength range 100–1200 Å (120–10 eV), using two, interchangeable gratings with ruling densities of 710 lines/mm and 2400 lines/mm to cover the ranges 300–1200 Å and 100–300 Å, respectively. The main radius was 15 m, and included an angle of 155°. Since the theory and design procedure of these instruments has been covered in section 4.2, with particular reference to a 150° instrument, only a brief description is given here. The image properties are very similar to the 150° instrument described earlier, and a resolution of 20 meV was achieved at energies below 60 eV without the need to mask the projected grating aperture, which was 20 mrad horizontal, 8 mrad vertical at zero order. The grating ruled area was 90×50 mm. As with the Seya, orthogonal cylindrical mirrors were used as the prefocusing system, giving different demagnifications for the two directions to fill the grating aperture in both directions, as shown in fig. 5.3. Mechanically, the instrument is very simple, being scanned in positive order by simply rotating the grating. A precision and reproducibility in this drive of ±10 seconds of arc is required and easily achieved using a lever arm and precision lead screw driven by a stepper motor. The gratings are interchanged by mounting them on parallel bars attached to the rotation mechanism, and moving them sideways, i.e. perpendicular to the incoming beam and parallel to the rotation axis. Thus either grating can be chosen to intercept the incoming light. For better than 50 meV resolution at energies above 60 eV, the grating aperture must be masked, primarily its 90 mm dimension, to reduce the defocus aberration. The vertical masking jaws are shown in fig. 5.3; corresponding horizontal jaws are also fitted. As described in section 4.2, the defocus term can also be reduced, at a particular wavelength, by moving the exit slit, and provision is made to do this over a range of ±10 cm. It is important to reduce stray light using a zero order trap baffle and further baffles down the exit arm, as shown.

Grating alignment in these instruments is not difficult provided the gratings have been manufactured with their rulings perpendicular to the main toroid axis within 1 minute of arc. It is then sufficient to rotate the toroid about an axis perpendicular to its front surface, and through its pole, to achieve a symmetrical image in the exit beam. (See the section on grating mounting later.) This is best done at zero order, using synchrotron radiation, at a point about three quarters of the way down the exit arm where the image is out of focus. The effect of this "yaw" adjustment being wrong is then easily seen in the shape of a skewed image.

Normally there is no need to provide this adjustment external to the vacuum, unless it is suspected that the gratings have not been manufactured within the above tolerance. It is assumed here, of course, that window valves have been provided which allow the use of synchrotron radiation to make this adjustment with the instrument at atmospheric pressure. The above procedure becomes insufficiently accurate as the angle of incidence decreases; in high resolution normal incidence instruments precise ruling orientation adjustment is required as described in the previous section.

5.2.4. The soft X-ray spectrometer

The remaining 5 mrad on this beam line have been used for a soft X-ray crystal spectrometer working over the wavelength range 3–15 Å. The instrument is preceded by a long (~750 mm) premirror which deflects the beam vertically, and whose grazing angle is variable over the range 0–5° to vary the high energy cutoff of the whole optical system. The soft X-ray spectrometer is placed ~6 m from this mirror, which is itself 14.5 m from the tangent point. Thus the instrument is mounted on a platform which has to move vertically to accept an entrance beam moving over the range 0–10° from the horizontal. This has been done by mounting the whole of the beam line from the mirror to the spectrometer on a long platform which pivots about a horizontal axis through the mirror centre. This is shown as the pivot P in fig. 5.4. The platform angle is varied using the hoist H, and the mirror is separately adjusted. Since the mirror box itself is stationary, a large bellows is required connecting it to the beam line leading up to the soft X-ray spectrometer, to accommodate the 0–10° range of movement. Two options exist for the mirror:

(i) It can be bent along its axis to focus the beam vertically.
(ii) A flat or cylindrical mirror is available, *not* interchangeable under vacuum.

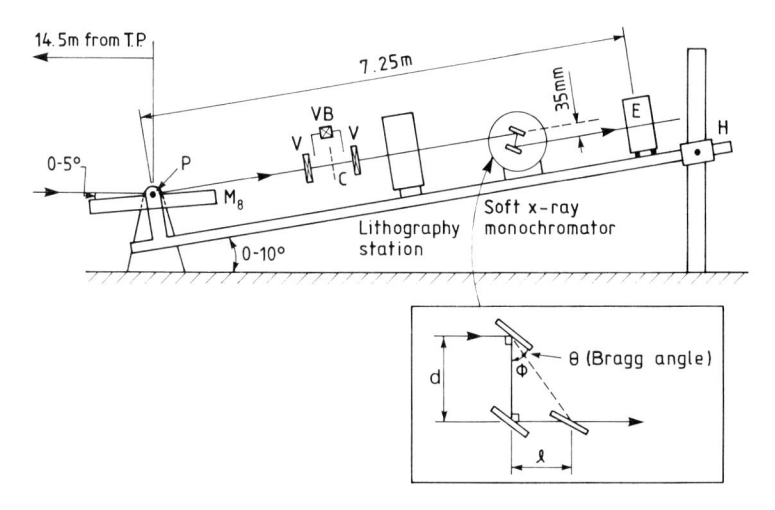

Fig. 5.4. The layout of the soft X-ray monochromator: side view. C: carbon film; V: isolating vacuum valves; VB: bypass valve; H: hoist, P: pivot.

If cylindrical, the curvature is along the direction perpendicular to the beam direction and thus provides horizontal focusing, but this varies with mirror angle, of course. Thus, combined with option (i) a toroid can be generated; at a grazing angle of 15 mrad, the horizontal focus is in the experimental chamber.

For the Daresbury SRS, in the soft X-ray region, a vertical aperture of less than a milliradian is sufficient to accept most of the radiation. Thus with a cylindrical mirror the horizontal aperture is dominated by the cylindrical radius, and since this is calculated to provide focusing at very grazing angles, it can soon be shown that the horizontal aperture will be 5 mrad or less.

This beam line is constructed to include a section where either carbon or beryllium windows can be used to isolate the UHV of the main beam line from the high vacuum of the spectrometer. The pumping arrangements are such that a large differential cannot be placed across these windows whose main function is to prevent contamination reaching the main mirror and also to act as a low-energy cutoff. Thus the window thickness is in the range 2000–4000 Å; and its aperture is 70×10 mm. In fact two windows are fitted; the first absorbs the heat from the beam, and the second provides the differential. They are held in an isolated section, and to prevent a large pressure difference building up across them a bypass valve is fitted as shown. Control of this valve is left to the vacuum control system, which monitors the pressure difference across the differential window. Care must still be taken during initial pump down, however, to prevent their destruction. The beam line can also serve as a wide bandpass monochromator, since the mirror, with its variable angle of incidence, can be used as a high-energy cutoff. In this mode the beam line is used for lithography experiments.

The crystal spectrometer itself is of standard layout, similar to the "JUMBO" spectrometer described by Cerino et al. (1980) in principle, but is rather different in mechanical detail. It is an example of precision engineering for soft X-ray optics.

The aim is to provide a mechanical linkage which keeps two crystals parallel, within 5 seconds of arc, over the range of Bragg angles 10–80°. In addition, the exit beam is to remain stationary during scanning within a similar angular tolerance. A linear drive system is used to scan the spectrometer, and is specified such that the crystal angle is within 10 seconds of arc of that calculated from the known position of the linear drive. The linear drive position is encoded using a Heidenhain linear encoder, and a rotary encoder fitted to the rotation axis of the lower crystal to check this. The top crystal is water-cooled; without this, crystal heating would cause the exit beam to deviate as the instrument is scanned.

The linkage between the two crystals, designed by D.K. Tole of Bird and Tole Ltd., is shown in fig. 5.5. In this diagram, $F_1M_1 = F_2M_2$, $F_1F_2 = F_2C_1$, $M_1M_2 = M_2C_2$ and $F_1C_1C_2M_1$ and $F_2C_1C_2M_2$ are parallelograms. The bearings at C_1 and C_2 are constrained to remain in contact with the surfaces on which they roll by springs, and these surfaces are ground accurately flat. F_1 and F_2 are fixed pivots, M_1 and M_2 are moving pivots. Suppose the upper crystal is rotated through an angle θ; the upper plate is confined by the geometry to rotate through 2θ about

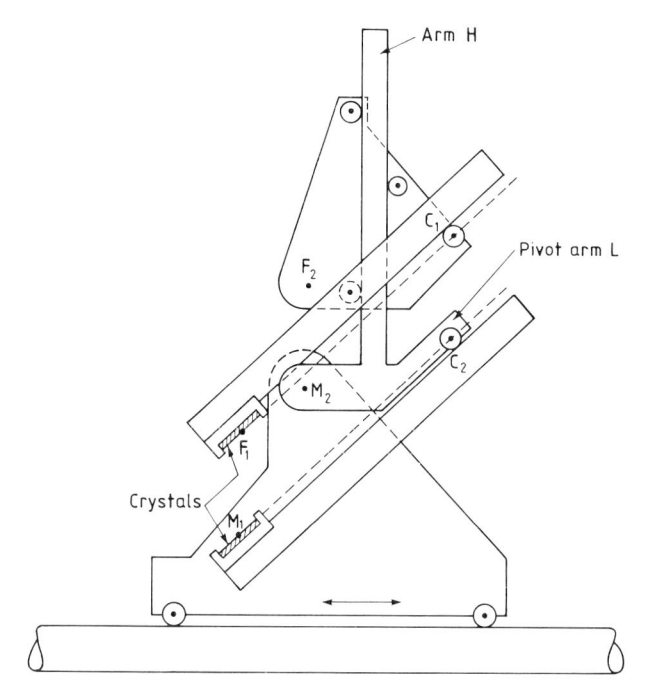

Fig. 5.5. The soft X-ray spectrometer mechanism.

F_2. The arm H, free to move along its length and rigidly connected to pivot arm L, will also rotate through 2θ, and will therefore drive the lower carriage along to follow a beam which has rotated through an angle 2θ about F_1, i.e., this angle-doubling mechanism provides the required lateral translation of the lower crystal. The lower crystal itself, however, using a similar angle doubling mechanism via the pivot arm L, which is rigidly attached to arm H, is constrained by the geometry to rotate through θ. The mechanism is driven by a linear translation of the lower crystal, and this movement is encoded.

This design places high demands on accurate positioning of the components, and in particular in arranging for the geometry of the triangles $F_1F_2C_1$ and $M_1M_2C_2$ to be identical. Jig boring can achieve this, but the instrument contains preset adjustments so that the position of bearing C_1 and the parallelism of the shafts which hold the crystals can be trimmed when the instrument is finally aligned. When it comes to calibrating the drive system, it is important to know the linear encoder reading when the crystals are at the 45° Bragg-angle position. The relationship between Bragg angle and linear encoder reading can then be calculated, using the simple relationship, $\tan \phi = l/d$ (inset fig. 5.4); l is the distance moved from the 45° position, $\frac{1}{2}\phi$ is the angle through which the crystal rotates. Thus the new Bragg angle is $45° - \frac{1}{2}\phi$.

Of course, this relies on knowing the encoder reading for the 45° position accurately, and also the value of d. In general, if the encoder reading at the 45°

position is L, and the lower crystal is moved such that the new encoder reading is L', then the new Bragg angle is given by

$$\theta = 45° - \tfrac{1}{2}\phi, \quad \text{where} \quad \phi = \tan^{-1}\frac{(L-L')}{d}.$$

If there is an error δL in L and δd in d, this becomes

$$\phi = \tan^{-1}\frac{(L \pm \delta L - L')}{d \pm \delta d}. \tag{5.1}$$

The procedure adopted for this instrument was to program eq. (5.1) into the computer used to drive the spectrometer, and then check the correspondence between the angle through which the lower crystal moved as determined by the angular shaft encoder, and the angle through which it was supposed to move according to eq. (5.1), initially with $\delta L = \delta d = 0$. By trial and error with the values of δL and δd, the difference between the calculated value and actual value was reduced to ± 30 arc seconds over the whole range. It is helpful to realise that errors in d are most important at small values of ϕ and errors in L at large values of ϕ; this can be deduced by differentiating eq. (5.1). Thus the trial and error procedures can be shortened.

Preset adjustments are of course required in the mechanism, to ensure the crystals remain parallel as they are scanned. Referring to fig. 5.5, the directions F_1C_1 and M_1C_2 must be made parallel. If the crystals are located at the front surface, they can be replaced by mirrors and autocollimation techniques used to check their parallelism as the instrument is scanned.

Fine tuning of the upper crystal is provided by a solenoid mechanism, which provides a range of movement of ± 200 arc seconds for the top or first crystal. This is used to tune onto the Bragg peak, or to tune off it for higher order rejection, and can be controlled by peak sensing hardware. A coarse adjustment is provided with it to account for the fact that the Bragg planes may not be parallel to the front surface of the crystal; for the same reason, a rotation about an axis perpendicular to the main rotation axis is also provided. All the three adjustments above work with the instrument under vacuum.

The energy calibration of the spectrometer is done using absorption edges in the soft X-ray region. Equation (5.1) above assumes that the Bragg planes lie along the F_1C_1 and M_1C_2 direction in fig. 5.5. This will rarely be the case, and can be accommodated by including a $\delta\phi$ error term in eq. (5.1) once the best values of δL and δd have been established. The energy calibration then determines the value of $\delta\phi$.

The outline specification described above included cooling of the first crystal, and this is provided using a gallium interface between the crystal and a water-cooled copper block. It works well with crystals such as silicon, germanium and indium antimonide which have good conductivity. The original angular tolerance specified (±5 arc seconds) has been met over the range of Bragg angles 10–60°. Since the angular resolution requirement of such an instrument is less important for large Bragg angles, through the dispersion relationship obtained by differentiating the Bragg equation, this is not a serious limitation.

Of necessity, the description given here is somewhat abbreviated and is included to bring out the main features of the design. It has demonstrated the practicality, and limitations of an instrument of this kind. It would be beyond the scope of this chapter to give a detailed description of soft X-ray spectrometers and their performance; this particular instrument is described in greater detail by MacDowell et al. (1987).

All the instruments described in this section can take beam simultaneously, and there is a further feature which is usefully incorporated into the general design. Where possible to do so, the exit beams from the monochromators should be horizontal, and at the same height above floor level. The exit beams of the TGM and 5-metre normal incidence monochromators are well above floor level, as can be seen from fig. 5.3, and mezzanine floors have been fitted, with an area large enough to accommodate all the experimental equipment and ancillaries, maintaining the same floor-to-slit distance. This feature simplifies the transfer of experiments between monochromators, though as in the case of the soft X-ray branch where the whole beam line is mounted on a moving base-plate, it cannot always be accommodated.

Except for the soft X-ray mirror all the other primary premirrors are made of silicon carbide, of the REFEL variety. Because these mirrors deflect the beam horizontally, the heat loading amounts to ~ 10 W/cm^2 in the worst case, and this material has been able to withstand this without damage or deterioration in use. The soft X-ray mirror deflects vertically and thus the main heat load is spread over a greater area. It amounts to $\sim 1/8$ W/cm^2 and thus fused silica is satisfactory. None of these mirrors is cooled and the temperature of the SiC rises to ~ 240°C with the SRS running at 2 GeV, 300 mA. Some thermal effects are evident; alignment drift has been observed as the mirrors warm up, over a period of a few minutes. Thus they are probably on the limit of satisfactory performance without cooling.

Finally, we address a topic sometimes neglected by instrument designers: how to arrange for the photons which leave the exit slit of a monochromator to reach the sample. In addition, a later section will deal with details of mirror and grating mounting mechanisms and the way the adjustments are provided.

5.3. Post-focusing optics

There are occasions when it is possible to place an experimental chamber close to the exit slit of a monochromator, but this is not usually the case. The following three methods have been used to reduce flux losses between monochromator exit slit and experimental sample.

5.3.1. Remote exit slit

This method has been used and described elsewhere by Howells (1979). It has a scissor-like structure, the "handle" of the scissors being driven by a precision micrometer, the "points" carrying the slit jaws. The pivot is of the flexural type. It is thus simple to construct, but rather limited in its application. It cannot support a differential pressure across it, and if the slit is in an experimental chamber, where for example electrons and/or fluorescent photons are being detected, there may be problems with scattering from the slit jaws. Nevertheless, it is an efficient way of bringing light to a sample if the above problems are unimportant.

5.3.2. Capillary light guide

A capillary light guide is a very efficient way of transporting light from the exit slit of a monochromator, or indeed from a laboratory light source as has been shown by Dehmer and Berkowitz (1974). Since the angular divergence of the radiation emerging from monochromators fitted to synchrotron radiation sources is ~ 20 mrad or less in most cases, lengths up to a metre do not involve many reflections. A 2-mm bore in ordinary pyrex glass is adequate and easily obtainable, and work at the NBS storage ring and the Daresbury SRS has shown that light guides up to 750 mm in length are very efficient (i.e. $\sim 75\%$) for photon energies up to 40 eV and do not affect the polarisation properties of the radiation unless they become contaminated. They are also ideal where a pressure differential is required. Optically, they transfer, in effect, the exit slit and beam divergence from the entrance end of the guide to the exit end, and thus appear to put the exit slit at the sample. The main disadvantage is that they must become less efficient at high photon energies, although they have been used at the Daresbury SRS up to 100 eV. In addition, clearly there cannot be a vacuum valve between the exit slit of the monochromator and the sample chamber without interrupting the light pipe and thereby severely reducing its efficiency. This feature is a disadvantage where, for example in a UHV surface-science experiment, the experimental chamber must be baked and yet has to be connected to a monochromator which in general is not baked. They also suffer from the same problems of scatter, which can be reduced using an earthed cap assembly, but never eliminated. Thus it is their differential pumping ability which is their asset, and they find most use in gas phase experiments.

5.3.3. Post-focusing mirrors

The method permitting greater flexibility in matching the experiment to the monochromator is to use a post-focusing mirror or mirrors. Where a small spot size is required on a sample, as in most surface science experiments, demagnification of the image at the exit slit is usually required. This can be achieved with a toroidal or ellipsoidal mirror, operating at a grazing angle of $\sim 10°$ or more if possible. As usual, there has to be a compromise between reflection efficiency and minimising optical aberration. Very often, the demagnification required for the slit width dimension is less than that required for the slit length dimension and a

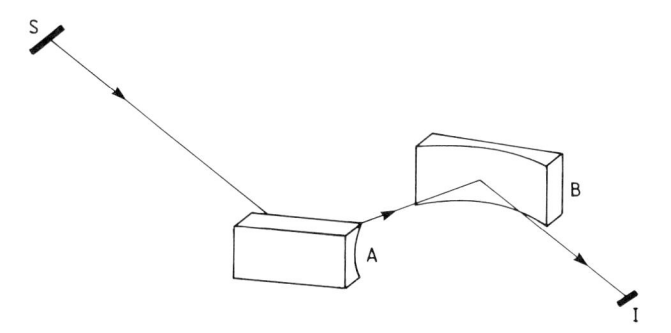

Fig. 5.6. "Crossed" cylindrical mirror assembly for post-focusing.

crossed cylinder arrangement can be used. These have been discussed in section 3.2 and also in detail by Hubbard (1982), who found that a cylindrical mirror with sagittal focusing, followed by a plane cylinder, can be arranged to provide an exit beam parallel to the input beam, and has good focusing properties. A sketch of such an optical layout is shown in fig. 5.6; it is relatively cheap to build and alignment of the unit can be preset. The actual geometry will depend on the demagnifications required and the exact experimental layout. In fig. 5.6, S represents the horizontal exit slit of a monochromator, whose vertical (narrow) dimension is brought to a focus at I with unity magnification by the sagittal focusing mirror A. Mirror B brings the horizontal dimension of S to a focus I with a demagnification of 3:1.

There is some disadvantage in using the above system because of the reflectivity loss of the two mirrors, but if these are used at grazing angles of 10° or less this need not be severe. In cases where the divergence of the beam from the exit slit of a monochromator is low, there may not be any need for post-focusing optics, since the collecting mirror before the monochromator can be used to provide focusing perpendicular to the dispersion plane. Provided the sample is fairly close to the exit slit, the divergence in the dispersion plane as the light leaves the exit slit can be tolerated. For crystal monochromators, as described for example in section 5.2.4, where there is no exit slit, the premirror can provide focusing in both directions, at the expense of energy resolution.

The above methods should cover most applications, and end this section on beam line layout. The aim has been to highlight problems and techniques in designing a multiple instrument beam line, and some of the compromises that must be made.

6. *Optical components*

This section examines practical methods of mounting and adjusting beam line components; aspects of mirror materials and coatings are included.

6.1. The mounting and adjustment of optical components

As a general rule, optical components should be mounted in contact with a reference face, so that they can be removed and repositioned accurately without elaborate realignment. In practice, for reflection optics this means front surface location on three points, positioned in such a way that the assembly is stable when the optical component is pushed from the back. In fig. 6.1 we show, for the sake of example, a mount which incorporates the desired features. The spindle on the axis A defines the axis of rotation, and would normally be located in bearings fixed to the main structure of the monochromator. These spindles would also be attached to an accurate rotation system such as a sine bar driven by a precision lead screw. The three screws B are used to set the front surface of the optical component coincident with this axis. This is usually done on a surface plate using

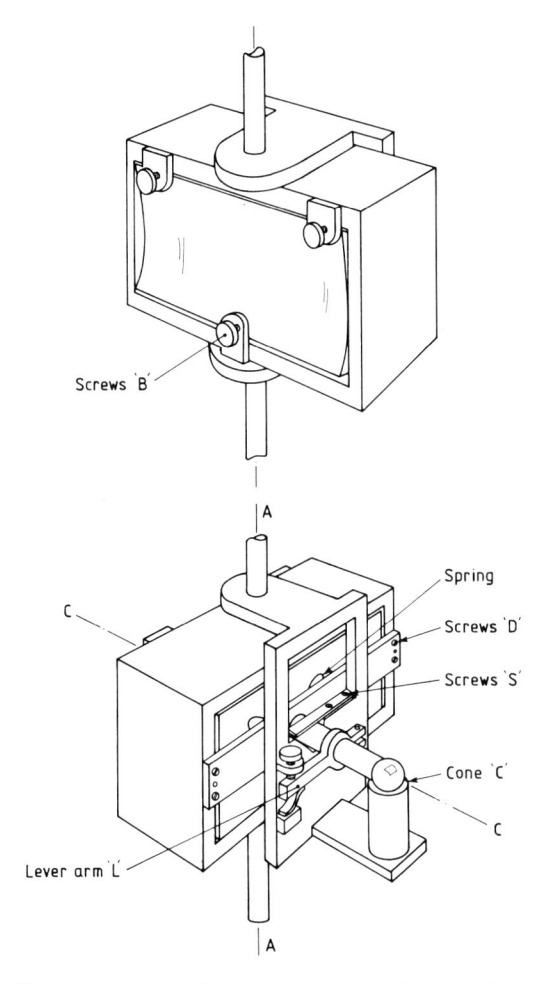

Fig. 6.1. An example of a mirror or grating mount.

a dial gauge indicator, first setting the axis A parallel to the surface plate, using the spindles on this axis as supports, and a known height above it. The optical component should be replaced by a lapped flat with the same physical dimensions, the setting up procedure accounting for the difference between a flat surface and any curvature the optical element may have (sagittal correction). One further adjustment is useful on such a mount, and is provided by the lever arm L and a set-screw. This rotates the component about the axis C, which with accurate machining will be sufficiently perpendicular to the axis A. This adjustment, for which $\pm 2°$ is adequate, will serve to orient grating rulings or the main axis of aspheric components, e.g. toroids, paraboloids or ellipsoids. It is worth noting, at this point, that ruling alignment in monochromators is greatly facilitated by incorporating a small area of coarser ruling density ($\sim$200 lines/mm) at the edge of the blank accurately parallel to the main ruling, where it is not normally illuminated. This can then be used for laser alignment. Normally there is insufficient rotation in the system to bring first order diffracted light from a helium–neon laser onto an exit slit, particularly where the ruling density is high. This technique would overcome this, but care should be taken to ensure that this area is normally masked off. Bausch and Lomb have included such a facility in their high resolution gratings for some time; now that gratings are interferometrically ruled, it should not be difficult to incorporate $\sim$3 mm depth of ruling (i.e. dimension of ruled area perpendicular to the grooves) at the edge of the blank. The alternative is to provide sufficient rotation in the grating assembly to reach the visible, but this is not always consistent with a high precision rotation for the VUV whose range is limited, and there may be other mechanical constraints. Alternatively, of course, it may be possible to arrange for the diffracted light from the helium–neon laser to emerge through a window built into the vacuum chamber specifically for this purpose. Adjustment can then proceed as described in section 5.2.2.

Finally, to facilitate replacement of the optics without the need for realignment, the grating mount itself should be located kinematically in the main mount. In fig. 6.1 the mount is released by undoing screws S and removing the spring clamp which holds the shaft on the C axis in the vee-groove and cone C. The grating holder can then be withdrawn complete with its lever arm L. The vee-groove and cone provide kinematic relocation; the grating itself is removed by undoing the screws D and taking off the back bar on which the spring is attached, this spring pushing the grating forward onto the screws B. Normally, the grating (or mirror) would be restrained from moving around in the mounting by pips in contact with its edges, springs pushing the optical component onto these pips. In this way the orientation of the grating rulings, or of the axis of a toroid, ellipse or parabola, is maintained relative to the mount when it is replaced. Of course, if the blank is cylindrical rather than rectangular, this will not work and realignment will be necessary after it has been removed.

It should be pointed out that the mount shown in fig. 6.1 has been included to demonstrate the desirable features in such a mount, and will not necessarily be appropriate for all applications.

A mount of the above type is useful in spectrometers where generally the requirement is just to rotate a component accurately about one axis after it has been set in position. For mirrors in mirror boxes requirements are slightly different. Front surface location is again desirable, particularly since beam line mirrors may need to be removed for recoating at hopefully not too frequent intervals and it should be possible to replace them without the need for extensive realignment. This may be complicated by the need to cool such mirrors. Generally water/vacuum joints should be avoided, which means that cooling pipes run continuously through the vacuum wall onto the component to be cooled and out again. Thus these pipes must be made sufficiently flexible not to overcome the mirror kinematic location. It is clearly much simpler just to attach the mirror to a cooled block mounted rigidly to the adjustment system, but this temptation should be avoided. A little careful design effort taking into consideration the differential forces acting between the kinematic mechanism and the cooling arrangement will be repaid in the long term. The most difficult case to accommodate is that in which metal mirrors are used, and thus the cooling pipes are attached to the back of the mirror itself. It is clearly only worthwhile cooling mirrors which are themselves good thermal conductors, which means they are made of metal or certain ceramics. Where the cooling pipes are not integral with the mirrors good thermal contact has to be established through an interface. Liquid gallium metal has been used for this purpose, since it is liquid at 30°C, has a very low vapour pressure ($\sim 10^{-16}$ mbar) at this temperature and "wets" metals quite well. Tests on the use of this metal to conduct heat away from silicon carbide mirrors onto water-cooled copper blocks have now been carried out at the Daresbury SRS. MacDowell, West and Koide (1986) have shown that heat load up to a total of 300 W can be tolerated without damage to the optics.

The decision which has to be made on mirror adjustments for VUV lines is: which adjustments are needed outside the vacuum and which inside? One way of achieving the movements required is to adjust the whole vacuum chamber in which the mirror is situated, in those cases where an individual mirror box design is adopted. These adjustments are always opposed by large vacuum forces, which may make precision adjustment difficult. The approach described here assumes that, for normal running adjustments, the vacuum chamber is stationary, having first been positioned correctly while at atmospheric pressure.

For mirrors which are an integral part of the optical system of a spectrometer, the adjustments would normally be inside the vacuum and not accessible while the instrument is in use; it is assumed they have been accurately pre-set. For beam line mirrors, however, external adjustments are essential to ensure the image of the source is correctly placed, both for position and direction, in line with the requirement of the monochromator concerned. Six different adjustments are possible: translation along the x, y, z orthogonal axes and rotation about these axes. Figure 6.2 shows these schematically, for two cylindrical mirrors H and V whose reflection planes are orthogonal. The z, z' axes are parallel and vertical, but the x', y' axes are rotated with respect to the x, y axes by an angle ϕ, the angle through which the incoming beam is deflected by the first mirror. Thus the

Optical engineering

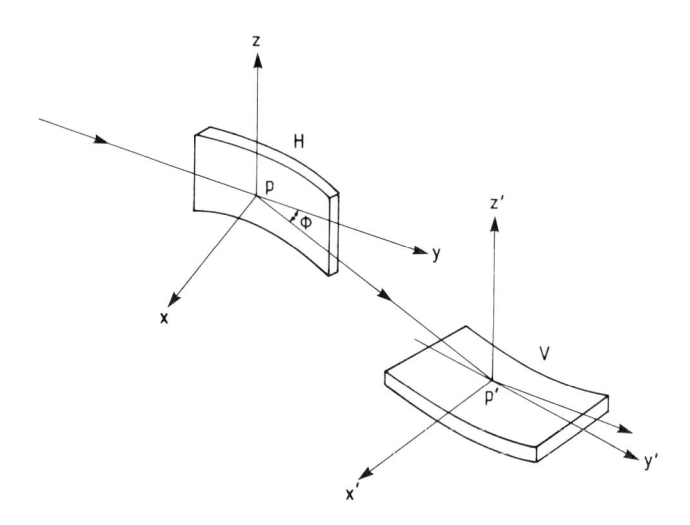

Fig. 6.2. Cross cylindrical mirrors for the purposes of adjustment procedure (see text).

beam incident on the second mirror lies in the $y'-z'$ plane. The y axis is coincident with the incoming beam direction, and the x, y, z origin is at the pole P of mirror H. The y' axis is horizontal and is tangent to mirror V at its pole P'.

To demonstrate how these movements are used in practice, consider the orthogonal mirror system in fig. 6.2, used to focus the beam from the storage ring. Initially, the mirrors are positioned correctly using the x, y, z linear movements. Thus mirror H will be set so that

(a) it takes the correct slice of available beam (x coordinate);

(b) its centre is on the median plane of the storage ring (z coordinate);

(c) its lateral position is correct to maintain the angular geometry of the system, i.e. ensure that angle ϕ and the distance PP' are correct (y coordinate);

(d) the mirror top surface is levelled. For the cylindrical mirror shown here, the cylindrical axis must be ground accurately perpendicular to the long top edge. Thus this adjustment, rotation about the x axis, aligns the cylinder correctly.

For the second mirror V, (a) to (c) apply in just the same way. Adjustment (d) becomes rotation about the y' axis, and can be achieved by placing parallel bars along the mirror y' dimension. Using the x' rotation an accurate (± 5 arc seconds) spirit level set along the y' direction is used to set the mirror horizontal in this direction. The spirit level is then rotated through 90°, still supported by the parallel bars, and the mirror is levelled in the x' direction using the rotation about y'. It is obvious that great care must be taken not to damage the mirror surface, and all objects in contact with it must be scrupulously cleaned. If mirror V were flat, this technique cannot be used; instead the laser alignment system can be used if a correctly pre-aligned target in the $y'-z'$ plane is available, on which the reflected beam can fall.

The above adjustments are those made with the mirrors at atmospheric pressure. For remote adjustments made through the vacuum interface, in general the following will suffice:

(1) for a horizontally deflecting mirror, rotations of ±2° about the y and z axes;

(2) for a vertically deflecting mirror, rotations of ±2° about the x' axis. For non-symmetrical optical components, such as the cylinders depicted in fig. 6.1, on an ellipsoidal mirror, a rotation about z' is desirable and would be set by examining the symmetry of the focused image.

All the other adjustments should only be accessible when the mirrors are at atmospheric pressure. It is a great mistake to provide too many external adjustments, since they interact and utter confusion could result. In addition, it encourages one to use such adjustments in an attempt to overcome misalignment errors, thereby introducing further errors. Since opening up a UHV system is expensive in time, it is well worthwhile ensuring that the x, y, z coordinates are correct initially, and that the mirrors are properly set as described above.

Assuming, then, that the mirrors have been correctly set up at atmospheric pressure, and are now under vacuum, the final adjustments would proceed as follows, using the laser alignment system, set along the incoming beam direction:

Mirror H is rotated about its z axis until the laser beam hits mirror V at its centre with *respect to the x' direction*. The beam may not be central along the y' dimension of mirror V. To achieve this rotate mirror H about its y axis.

This will alter the first adjustment, and these two should be iterated until correct. It is possible to do this using the synchrotron radiation beam, but this is much broader so it could be difficult to define its position on the second mirror.

Mirror V should now be rotated about its x' axis until the image falls on the entrance slit of the monochromator concerned. Ideally, if this mirror has been properly levelled, adjustment about its y' axis should not be needed. If it were provided it would rotate the image of the source (i.e. the electron beam at the tangent point) and thus provide the option of making it parallel to a monochromator entrance slit, for example. Since the mirror can be accurately levelled, however, it would probably be wise to assume the image is horizontal and set the monochromator accordingly.

Thus the system is now set. If a laser was used, there will inevitably be differences between the laser beam direction and that of the synchrotron radiation. Provided these are not so large that the mirrors are not being properly illuminated, they should be accommodated as follows:

(1) An error in horizontal position of the source can be corrected by rotating mirror H about its z axis.

(2) A vertical error, if very small, i.e. ≤0.1 mrad, can be corrected by rotating mirror V about its x' axis. Any more than this should be corrected by steering the electron beam to bring it back onto the median plane, using beam height monitors to assist in this process.

As stated earlier, where non-symmetrical optical components are used, a third rotation (sometimes known as "yaw") is desirable, particularly for vertically deflecting mirrors. For horizontally deflecting mirrors, it is not so important, because the main axis of symmetry is usually specified to be parallel, within a minute of arc or so, to the edge of the blank and this can be set accurately horizontal with a spirit level. For vertically deflecting mirrors, this edge must be set parallel to the incoming beam direction, and is difficult to do accurately. It can be done by optimizing the image from the mirror for maximum symmetry, and it may be helpful to do this at a point where the image is actually out of focus.

The above general principles can be applied to single mirrors or more complex systems containing many mirrors, the latter requiring step by step adjustment along the mirror chain. Finally, there are one or two guidelines to be borne in mind when providing adjustments through the vacuum interface. Of course, the rotational accuracy required will need to be calculated, but this will generally be in the range 1–10 seconds of arc. This can be achieved using a precision linear drive through the vacuum interface with a lever system inside the vacuum. Where space is restricted, a double lever system can be used, though this will reduce the range of movement available. The use of rotary vacuum feedthroughs where high precision is required is not recommended. Their mechanical advantage is in the wrong direction, and this results in wind-up effects as friction in the lever system inside the vacuum is overcome. This friction can be minimised, but never completely eliminated. The use of flexural pivots rather than bearings helps in this respect; also, where parts of a lever system unavoidably rub together, dissimilar metals such as stainless steel and phosphor bronze, should be used.

6.2. Mirror materials

In the past, fused silica has been widely used as the substrate for optical surfaces. It can be obtained in large sizes and is sufficiently hard to enable it to be polished to a very high degree of surface finish, sufficient in fact for X-ray reflection. There is a wealth of literature on this subject, and those interested in obtaining more detail should refer to the work of the National Physical Laboratory, UK (e.g. Franks et al. 1975) or of the Naval Weapons Centre, China Lake, USA (e.g. Rehn et al. 1980).

The major problem arising with synchrotron radiation sources is that of radiation damage, caused by the absorption of high-energy X-rays in the peak of the synchrotron radiation spectrum. Of course, this does not apply to those facilities where a lower energy ring has been built specifically for experiments in the VUV, but most rings are intended to serve the whole spectrum. In the past it has been solved by using metal mirrors, water-cooled or thermoelectrically cooled. It is difficult to achieve good surface figure combined with a low surface roughness on such mirrors, stable over a long period of time, although coating them using the electroless nickel process and polishing overcomes this somewhat. Nevertheless, their use has been restricted to the normal incidence region of the VUV at the Daresbury SRS. For the soft X-ray region, where grazing angles of around 1° or less are used, in *vertical* deflection, fused silica can still be used since the power absorbed/unit area is small (≤ 1 W/cm^2), but power densities larger than this will distort the optical surface, since the thermal conductivity of fused silica is poor. Distortion also occurs with copper mirrors because of their lower tensile strength.

Ideally, a material with high conductivity and tensile strength is needed, and such a material is silicon carbide; Rehn et al. (1977) have found that its chemically vapour deposited (CVD) form has particularly good optical properties. Unfortunately, this cannot be made in pieces larger than about 10 cm^2, and collection/collimating mirrors for synchrotron radiation are often larger than this. Other forms, however, can be used and measurements on their suitability as VUV reflectors, have been made (see, e.g. Kelly et al. 1981; also, for measurements on the CVD form, Rehn and Choyke 1980). In fact, for the region between 600 and 1000 Å, the reflectivity of silicon carbide, at normal incidence, rises to ~40% and is superior to most other surfaces.

The form of silicon carbide which can be produced in larger sizes is the reaction bonded type known as REFEL. Details of the REFEL process are described by Popper and Davies (1961). Lindsey and Morrell (1981) and Franks et al. (1983) have shown that the REFEL material can be polished to give an excellent surface finish, suitable for use as an X-ray reflector. It does, however, contain about 10% free silicon, which shows up as "islands" of silicon a few angstrom high. While this does not detract from its performance as a mirror material in the VUV, it does preclude its use as a diffraction grating substrate if the ion-etching process is used, where the etching rates would be different for the silicon "islands" and the silicon carbide itself. Thus the CVD material is the only one really suitable for diffraction gratings, and for use at very short wavelengths. Worldwide, there is considerable effort being invested in making large area CVD coatings on substrates such as graphite or REFEL.

As described in section 5, several REFEL mirrors are in use at the Daresbury SRS and are coping, uncooled, with heat loads of a few watts/cm^2. There is also one mirror, close in, cooled using liquid gallium as an interface, which has so far survived without deteriorating. The parameters for these mirrors are summarised in table 1, for a storage ring current of 300 mA at 2 GeV; all these mirrors are horizontally deflecting. Other materials have been tried using the SRS test port; table 2, taken from MacDowell, West and Kiode (1986) summarises the results.

Table 1
REFEL SiC mirrors at the Daresbury SRS.

Size L × H × T (mm)	Heat load (SRS at 100 mA)		Grazing angle (deg)	Type
	Total (W)	Per unit area (W/cm^2)		
700 × 50 × 10 (Pt-coated)	120	3	5	2 cylinders, 1 flat
336 × 50 × 10	100	5.5	10	1 cylinder
190 × 50 × 10	120	9	20	1 cylinder
156 × 50 × 10	140	100	45	flat; water-cooled with gallium interface

Table 2
Various mirrors at the Daresbury SRS (MacDowell et al. 1986).

Material	Average current (mA)	Energy SRS (GeV)	Maximum thermocouple temperature (°C)	Exposure time (h)	Observation
$Si_3N_4(6)$[a]	150	2	247	12	No apparent physical damage except for slight brown mark (carbon?) on X-ray axis
SiC (CVD) on graphite[a]	150	2	313	8	No visible damage
WC[a]	170	2	536	10	No visible damage
Fused silica	228	1.8	90	2	Brown mark on X-ray axis at 1.8 GeV (carbon)
		2	150	18	At 2 GeV, no brown deposits; (carbon desorbed)-surface
SiC, hot pressed (5)[b]	100	1.8	175	3	No visible damage
Zerodur	90	1.8	122	9	Surface destroyed and colour centres local to X-ray axis
SiC, hot pressed[b]	100	2	213	2	No visible damage
SiC (REFEL)[b]	100	2	235	4	No visible damage
SiC, on REFEL[b]	100	2	214	4	No visible damage; green fluorescence on X-ray axis
Aluminium (machined surface)	100	2	213	2	No visible damage
Molybdenum	80	2	400	16.5	No visible damage
Copper OFHC	100	2	470	11	Ripples present over entire surface
Invar	110	2	300	11	No visible damage
Borosilicate glass	120	2	–	1 min	Broke up into several pieces; surface melting; colour centres throughout sample
SiC, CVD on REFEL (8)[c]	200	1.8	371	48	No visible damage; did not fluoresce on X-ray axis (unlike CVD sample from BNF)
Float glass	119	1.8	–	1 min	Broke up into several pieces; surface damage/melting on X-ray axis
Sapphire	150	1.8	–	3	Broke into several pieces; no surface melting
Pyrex	200	1.8	–	27	Broke into several pieces; surface melting on X-ray axis

[a] Toshiba Ceramics Co., Japan.   [b] British Nuclear Fuels.   [c] Astron Ltd., UK.

6.3. Optical coatings

Section 3.3 gives a detailed analysis of the behavior of optical coatings; thus only a few practical points will be given here. A detailed description of the preparation of optical coatings, and measurements made on the variation of performance with thickness is beyond the scope of this chapter. The reader is referred to the references given below for such information.

The material used for the coating, and the thickness required, will depend on the wavelength range to be covered. Generally gold or platinum are used for the energy region above 50 eV, on grazing incidence optics. For details of the optical behaviour and preparation of gold films, see Canfield et al. (1964); for platinum, see Jacobus et al. (1963). Knowing the optical constants, the equations in section 3.3 allow the expected performance to be calculated. In the normal incidence region (10–50 eV) data for the optical constants are more abundant, and other materials and coatings are used here with a better performance than gold or platinum. Those with the highest reflectance in the 10–20 eV region are osmium (see Cox et al. 1973) and iridium (see Hass et al. 1967). Cox et al. (1971) have also investigated rhodium films in this region, which have a performance similar to that of gold and platinum. At energies around 10 eV, Hunter et al. (1971) have measured the reflectivity of aluminium films overcoated with MgF_2; a very high efficiency of 80% at normal incidence can be obtained. At lower energies still, aluminium alone is effective.

The robustness of coatings, and their deterioration on exposure to air, are important considerations and one reason why gold has been chosen in the past. Aluminium oxidises rapidly in air, and an overcoat of MgF_2 not only improves the optical performance but also serves to protect the aluminium. But osmium and platinum deteriorate slowly in air and thus care should be taken to minimise such exposure before installation. During the coating process itself a high degree of cleanliness is important; coating under UHV conditions is now being adopted, though it has not yet been proved essential. With the carbon contamination problem apparently unavoidable on high-energy rings, despite extensive efforts to investigate its origins and thus remove it (see, for example, Boller et al. 1983), attention is now being focused on in situ cleaning of optical components by glow discharge. A system for doing this has been described recently by Koide et al. (1986). In certain cases, it may also be possible to use in situ evaporation if the coating required is aluminium or gold, depending on the geometry of the system concerned. For other materials, requiring high-temperature evaporation or RF sputtering, this will probably not be practicable.

The maintenance of clean optical coatings is vital for the performance of any optical system, and it is clearly inconvenient and time-consuming to have to remove components for recoating from a UHV system. It seems likely that in situ cleaning will have to be adopted on synchrotron radiation beam lines in future; in the meantime, care should be taken to expose them for the minimum time required to actually run the experiment concerned. Thus the vacuum control system should always include masks or shutters to protect the primary mirrors when they are not in use.

7. Beam line auxiliary equipment

In principle, one might expect the connection between the first mirror box on a beam line and the light source itself to be a simple piece of pipe with perhaps a valve and a pump. In practice, particularly on larger machines, this is not usually the case. For radiation and vacuum protection, vacuum and beam direction diagnostics, and for general alignment requirements many other components are needed. These are considered in this section.

7.1. Laser alignment system

During the initial assembly of a beam line and its various components, normal surveying techniques will be used to locate the equipment accurately. Ideally, when mounting the optical components in their final positions, synchrotron radiation should be used, but since these components will be at atmospheric pressure a problem arises. It is under these circumstances that a laser alignment system is useful. Figure 7.1 shows a plan view layout for such a system at the Daresbury SRS. The table with the three leveling screws, L, L' supports a prism with two reflecting faces, accurately ground and polished flat, such that angle θ is $135° \pm 5''$ and both faces are perpendicular to the prism base within 2 seconds of arc. The prism table is driven along the accurately ground guide bars G, perpendicular to the beam centre line, and all this part of the mechanism is in

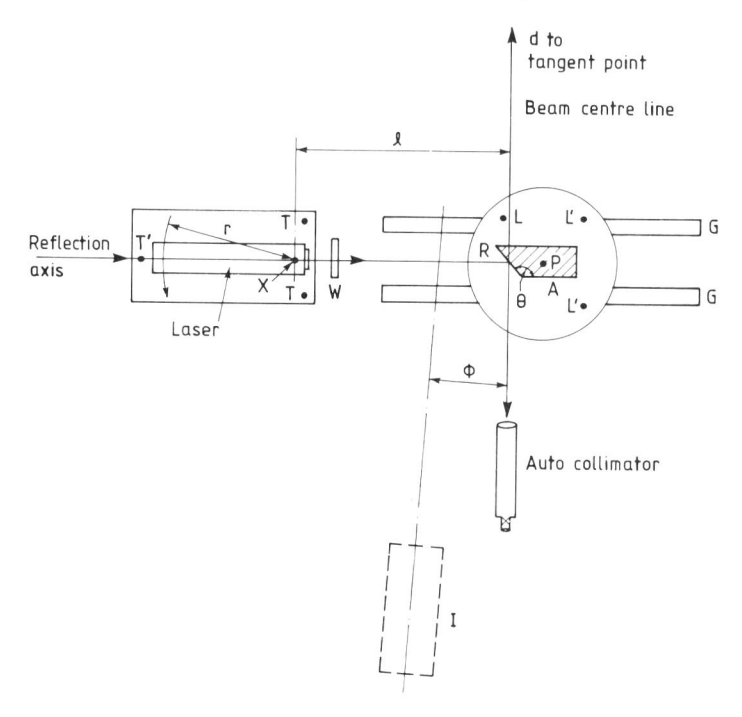

Fig. 7.1. Plan layout for a laser alignment system at the Daresbury SRS (see text).

vacuum. A laser is positioned as shown on a table with three levelling screws T, and can be rotated in a horizontal plane about a vertical axis through X. The laser light enters the vacuum through the optical flat W.

The setting-up procedure is as follows. An autocollimator must be set up along the beam centre line and the prism moved so that the autocollimator's target is seen by reflection in face A. Using the levelling screws L' and a precision rotation about the vertical axis through P, face A is set accurately vertical, and the middle of this face is also set, approximately, on beam height. The prism is then moved such that a target, previously set on beam height, can be viewed by the autocollimating telescope by reflection in face R; the prism should have been moved so that the beam centre lines run approximately through the centre of face R. Screw L is then used to set this reflected image of this target on beam height. The target should then be moved horizontally so that, again using the autocollimating telescope, it can be positioned so that it lies on the "reflection axis" (see fig. 7.1). It remains to set the laser beam axis; the laser should be of the type contained in a cylindrical case with its beam running accurately along the cylindrical axis. It should be mounted on the table shown in fig. 7.1 in Vee-blocks. Using the target just positioned on the "reflection axis", and one on the beam centre line in place of the autocollimator, the laser is adjusted using the levelling screws T, T' and its X-axis rotation until it illuminates one of the targets centrally. It is then reversed in its mount and adjustment is repeated for the other target, this procedure being repeated until both targets are centrally illuminated. A laser provided with a reverse beam would simplify this procedure, and would not need the Vee-block type of mounting. Finally, screws T are used to ensure the rotation axis X is vertical, using them differentially so as not to disturb the previous adjustment.

Two movements are needed to use the laser for illuminating any instrument on the beam line:

(1) Laser rotation about the vertical axis X, provided by a micrometer adjusting the position of the rear end of the laser.

(2) Prism translation provided by a linear drive system attached to the prism table through the vacuum.

Having set the prism and laser as above, the readings on these two drives must be noted, since they represent beam centre line. The distances l and d must also be known. Then, to illuminate an instrument I whose centre line is at an angle ϕ radians to the left of the beam centre line in fig. 7.1, the prism must be moved by an amount $d\phi$ to the left, and the laser rotated anticlockwise through ϕ i.e. its rotation micrometer moved by an amount $r\phi$ (ϕ is assumed sufficiently small for the small angle approximation to be used). Because the laser has not been rotated about the reflection point in the face R, the beam will "walk" along this face and a correction $(l - d\phi) \tan 45° = l - d\phi$ must be made to the prism translation setting. (This ignores the "walk" of the beam towards the tangent point, since this is small compared with d.) Thus, in the example given here, the prism would have to move by an amount

$$d\phi - (l - d\phi)\phi .$$

It will be clear that such a system must be rigidly mounted to the floor if it is to be used as a reference standard. Its main application is to locate optical components accurately, and for this it is better than the synchrotron radiation source because it has a smaller divergence. Thus the locating of component centres in their correct positions is much better carried out using the laser. On higher energy machines there is generally a heavy shield wall between the experiments and the source. Thus the laser system would have to be located outside this shield wall for ease of access. On smaller machines it may be possible to fit a reverse tangent window in the machine vacuum vessel. In such cases a reflecting prism may not be needed and the system can be simplified. It is worth stating at this point that a reverse tangent window is a great benefit for alignment purposes and should be provided wherever possible.

Reliability of and confidence in the laser system depends critically on the way it is mounted and the precision of its linear drive. This drive must be accurate enough not to tilt the prism by more than 2 seconds of arc over the whole range of movement. It would be wise, therefore, to install a surveyed target in the beam line to check the laser settings, or alternatively arrange that the laser beam can emerge from the end of the beam line through a window or window valve to illuminate a survey target or mark set up in the experimental area. All such windows must, of course, be parallel flats of optical quality.

The system described above is in use on the VUV lines at the Daresbury SRS and on one of the X-ray lines. It can be seen that it is a very useful alignment device and can be fitted at no great cost, despite the precision required. A more sophisticated, though no more expensive, system has been designed by Williams and Howells (1983) for the VUV ring at the National Synchrotron Light Source, Brookhaven Laboratory. In this VUV ring the synchrotron radiation source and laser are made to coincide using an autocollimating mirror, and the laser is then used for optical adjustment of the monochromator attached to that beam line. Such a system is ideal for single instrument beam lines, but is not so easily adaptable for multiple instrument beam lines, and would be difficult to use as a general surveying tool.

On X-ray lines where beryllium windows are the rule, a laser system is of little use, unless the windows are fitted in gate valves. Even so, the laser would probably be useful only in the initial construction stage.

7.2. Vacuum equipment

The predominant expense in any VUV beam line is the vacuum equipment and the control system which monitors it. It follows that it will also be the most troublesome, particularly if it is complex. Some complexity is inevitable if it is to protect the electron accelerator at the one end and fragile experiments at the other, and is also to be used safely by many different operators. Nevertheless, it should be kept as simple as possible. Because of the adverse effects of hydrocarbon contamination on reflecting optics, and to a lesser extent on the components of the storage ring, the whole system must operate under UHV conditions. This places high demands on the components and seals to be used, and also means that the whole system must be bakeable.

On beam lines where there is not a heavy gas load, ion pumps can be used extensively with considerable simplification in the vacuum system. Where gas loads are present, however, cryopumps will have to be used, or alternatively, turbo-molecular pumps. Both of these will need to have isolation valves above them, which become large and expensive for the larger pumping speeds. Modern cryopumps are certainly preferable to turbo-molecular pumps; they are cheaper per unit pumping speed, oil-free, and mechanically simpler. They need a valve above them to isolate them when they are warmed up to release the adsorbed gas. They do not, however, pump the rare gases neon and helium very well, unless they are the liquid helium-cooled type, as distinct from those which use a closed-cycle helium compressor and whose operating temperature is around 12 K. For pumping helium and neon, gases often used in atomic and molecular experiments as calibrants, the only alternative is a turbo-molecular pump, unless a liquid helium supply system is readily available.

Most turbo-molecular pumps contain oil-lubricated bearings, although backstreaming is at a very low level, and oil contamination probably originates from the rotary backing pump. Despite the assurances of some manufacturers, it is essential to instal backing line zeolite traps where a rotary pump is used, and these traps *must* be regularly maintained. The exception used to be air-bearing pumps made by Alcatel, which, once below ~0.1 mbar, can exhaust to atmosphere. Thus such a system is entirely oil-free; these pumps, however, are no longer made. Magnetic bearings pumps, in a 500 l/s size, are still made by Leybold–Heraeus, but these require a backing pump.

Turbo pumps require a well thought out control system and regular maintenance. In particular, their bearings are highly stressed and prone to failure, so proper lubrication is essential and should be monitored. Their excellent "lack of contamination" performance depends critically upon their functioning correctly. To be more specific, they should never be opened to a UHV system unless running at full speed. Thus the control system should monitor pump speed and if this falls below normal close the valve at the pump inlet and after allowing it to slow down to about half speed, vent it through a dry nitrogen supply to prevent oil creeping back to the UHV side of the pump. There should be a magnetic valve to the rotary pump, isolating this pump when the power goes off. The control system should also monitor oil flow and temperature, again shutting off the pump if faults or errors occur in this system.

Because ion pumps, cryopumps, and turbo-molecular pumps can only be opened to a vacuum $\sim 10^{-4}$ mbar, to avoid contamination, a rough pump system is required. These are conveniently constructed on a mobile trolley and consist of two or three sorption pumps, cooled by liquid nitrogen, and backed by a rubber diaphragm pump, or preferably, because of its greater speed, a carbon vane rotary pump. The sorption pumps, if carefully used, will easily pump down large chambers onto the 10^{-3} mbar range. The purpose of having more than one pump is to use one for high pressure, and the other one (or two) for low pressures. Thus the high pressure pump is cooled down first, and can be used to remove gas in the other two while they are still warm. They are then cooled down, ready for use in the final stages of pumpdown. The carbon vane pumps have an ultimate vacuum around 100 mbar, depending on the make, somewhat inferior to rubber diaphragm types. Nevertheless, their vastly superior pumping speed more than offsets this disadvantage in practice, despite placing a greater load on the "high pressure" sorption pump. Such a system can be wheeled and bolted on to any chamber where a small metal valve (38 mm bore is sufficient) has been fitted for the purpose, and thus ensures that contamination is minimised.

After a decision has been made on which pumps and where to install them, it remains to decide on the types of pressure gauges and other diagnostic devices.

7.3. Pressure measuring and monitoring devices

The purpose of the vacuum gauges is to provide the user with pressure readings and also to provide a system of interlocks and protection against vacuum accidents or failures. For the latter purpose, Pirani/Penning combinations are suitable since such gauges will be on all the time and are not subject to filament failure. For reliable pressure measurements ionisation gauges should be fitted, and every beam line should be equipped with residual gas analysers. It is more important to ascertain the components present in the residual gas, rather than to achieve the best possible ultimate vacuum. This arises because of the sensitivity of the beam line optics to hydrocarbons. These manifest themselves as peaks in the 54–58 mass number region, and are fragments of heavier hydrocarbons such as rotary pump oil and organic matter. As a rule of thumb, these should only be measurable on the 10^{-13} mbar scale. Thus a background vacuum of 10^{-10} mbar is not at all satisfactory if it is composed primarily of residual hydrocarbons. Summarising, the Pirani/Penning combination is, primarily because it is robust, ideal for control and protection purposes; any changes or uncertainty in the Penning calibration will not affect this function. Ionisation gauges should be used for quantitative pressure measurements and residual gas analysers primarily to identify contaminants and also for leak detection.

Most beam lines will require protection against catastrophic vacuum failure, the most typical being the fracture of a ceramic feedthrough on ion pumps or other electrical feedthroughs on UHV systems. This protection usually takes the form of a fast closing valve at the front end of the beam line, and monitors to detect sudden pressure rises in those areas where an accident is most likely to occur. Fast closing valves, generally of the flap-type, are now available with ~5 ms closing time; the major issue is providing a monitor with a sufficiently fast rise-time, preferably less than a millisecond, to operate the valve. At the Daresbury SRS, small, so called appendage, ion pumps have been used successfully for this purpose. One other form of protection often incorporated is an acoustic delay line to disperse the pressure wave which occurs after a major vacuum failure. A well quantified design for these has been published by Sato et al. (1983) who also include performance figures. In many cases, however, the experimental station, which is where an accident is most likely to occur since most work is in progress there, is many metres from the tangent point. Thus this distance, plus the fact that the gas must pass through the slits of a preceding monochromator in most cases, makes an acoustic delay line unnecessary. However, it may have an application where beryllium windows are in use, as on some X-ray lines, with no preceding monochromator or slit assemblies.

7.4. Beam line controls and safety systems

This item needs careful thought since it interacts closely with the user, and yet should be designed so that it is completely transparent. For reliability, the control system should be kept simple; however, there are important features which it must incorporate:

(1) Protection of users against irradiation.
(2) Protection of the storage ring from vacuum failure in the beam line, and vice-versa.
(3) Protection of beam line components from contamination and radiation damage.

7.4.1. Safety aspects

The control system must ensure that the line is safe, from a radiation hazard point of view, to operate, and on VUV lines will interact with the vacuum system. There are two major hazards: bremsstrahlung radiation from residual gas molecules and possibly metal surfaces in the line of sight in the storage ring itself, and soft X-rays scattered from beam line components which intercept the synchrotron radiation beam. On VUV lines the soft X-rays are contained within the stainless steel vacuum envelope; extra local shielding, e.g. lead glass over windows and thin lead cladding over bellows may be required. Thus the system is safe from this point of view, as long as the vacuum is present; vacuum switches interlocked to the main beam shutter protect the user from irradiation when working on equipment in the beam line at atmospheric pressure. At the Daresbury SRS a second safeguard is included in that the beam shutter is backed up by a vacuum valve constructed to absorb soft X-rays when closed. Referring back to fig. 5.1 such valves are marked S, so that if work is being done at atmospheric pressure on the monochromators, these valves will be closed. For work on mirrors inside the main beam line aperture, the main shutter must be closed to protect against high-energy bremsstrahlung. On branch lines, however, it is sufficient just to close the safety valve, to protect against soft X-ray scatter. It is valuable to provide such valves, on branch lines, with lead glass windows to allow optical alignment using synchrotron radiation. A beam stop, situated at the end of the beam line, serves the purpose of absorbing high-energy bremsstrahlung, and is interlocked to a radiation monitor at the end of the beam line outside the vacuum system, but in the line of sight from the tangent point. Sufficient lead is built into the beam line to absorb such radiation before it can emerge into regions where people may be working, but the monitor acts as an extra safeguard.

7.4.2. Vacuum protection

Protection against catastrophic vacuum failure was covered earlier; however, there is still a need to protect against slow rises of pressure or to interlock valves and prevent them opening "good" vacuum to "poor" vacuum. This is best done with purpose built "valve modules" which accept signals from pressure gauges either side of the valve, compares them and according to criteria set by the user, permits the valve to be opened or not. The unit should not only examine the pressure differential, but also the absolute pressure, with preset adjustments which can be set to the desired conditions. There are few commercial valve operating modules which do this; the Daresbury SRS uses a system where valve modules are a standard feature of the vacuum control system. Penning gauges are used for pressure measurement, backed up by Pirani gauges. The penning gauge controllers switch on automatically when they receive a signal from the Pirani gauge that the pressure is below 5×10^{-3} Torr.

Where a turbo pump is incorporated into the vacuum system, further protection is needed to prevent contamination in the event of pump failure. Signals from the pump-set which indicate low turbine speed or overheating need to be fed to the module which controls the valve presumably fitted above the pump. This valve would then automatically close and shut down the pump-set. After the turbo pump has been allowed to slow down for a few minutes, its control system should automatically vent the turbo-pump to atmospheric pressure using dry nitrogen. This prevents contamination from rotary pump oil which would occur if the rotary were left connected to the turbo-pump with the latter stationary. Before the turbo-pump can be used to pump the main system it must be repaired and brought back to full speed.

Cryo-pumps also need protection in the event of the cooling circuit failing, although they should not be vented! Thus, a pressure differential measurement would normally be sufficient protection, a temperature sensor connected to the valve control module is useful added protection, particularly if noxious gases have been pumped which would damage the vacuum system if released in large quantities. Ion pumps, inherently the safest of all in use but unfortunately totally inadequate for continuous pumping of rare gases and unsatisfactory for coping with heavy gas loads, need no protection in the form of a valve over them. Their controllers generally incorporate sufficient protection against overload.

7.4.3. Radiation damage protection

On sources where high X-ray fluxes would be incident on valves it is wise to protect the valves with a preceding cooled absorber. The valve control module would "look" to see whether this was in position before closing the associated valve, and move the absorber into position if necessary. Depending on the control system philosophy, the absorber, if separately controlled, could not be moved unless the valve which it protects were open. Alternatively, opening the valve could automatically be followed by moving the absorber. One advantage of having the absorber separately controlled is that it can be used as a beam shutter, rather than a vacuum valve, which is a more complicated device and prone to failure if used repeatedly. Other components in the beam line may also require protection from the direct beam, and in general it is good policy to ensure that radiation falls *only on those areas which require it*. This can be achieved by using water-cooled masks or apertures surveyed into position in the beam line and which allow only the front surface of the mirrors to be illuminated. From experience, it is preferable to overdo this than risk allowing radiation onto other components in the vacuum chamber. This avoids small distortions and other long-term heating effects which make precision alignment of the optics pointless. Thus great care in the initial beam line layout is necessary and will be repaid later in terms of optical stability.

Lastly, a word on the control system itself. A modular design is desirable for valve and pump controls, and a standard system of pressure measurement has obvious advantages. It is essential for this to be transparent to the user; in general, all he requires to do is press a button to open, or close, a valve or associated absorber. A panel showing the beam line layout, possibly just the user's branch of it, with lamps showing the status of the components, is a valuable facility; the required control buttons should be mounted on this panel. In the event of a fault, a minimum set of fault indications should appear on this panel, beyond which specialist help is needed. The user should also be able to examine pressure at those points which concern him "at a glance", on analogue meters provided for this purpose. The system may, or may not be computer controlled, depending upon the number of facilities needed and whether vacuum data logging is required. To keep the system as simple as possible will clearly add to its reliability; it should be designed to be consistent with the protection required, and ensure that the user does not require detailed knowledge of the system to operate it. Finally, the performance of the beam line optics depends crucially on the vacuum control system protecting those optics from contamination, radiation damage and overheating. Thus careful thought must go into this aspect of beam line design if the optical components are to perform satisfactorily.

8. Conclusion

We have now covered many of the aspects which affect beam line and optical instrument design. We have concentrated on multiple-instrument beam lines; however, the way ahead for the future rests with the use of undulators which enhance the photon flux output in specific wavelength regions. Brown et al. (1983) have described the properties of undulators, and many modern storage rings have fitted them. By passing the electrons through a periodic magnetic field of 10 periods or more, it is possible to enhance the synchrotron radiation output by constructive interference in a region of the spectrum chosen by the magnetic field strength, its periodicity and electron energy. Synchrotron radiation sources are now being designed, such as the European Synchrotron Radiation Facility and the Advanced Light Source at Berkeley which will use undulators to provide nearly all the radiation for experiments; fluxes three orders of magnitude above that presently available are possible. The divergence of the beam from the undulator is ~1 mrad or even considerably less and can have a bandpass of a few percent. Monochromators will still be needed, therefore, for most experiments, and the information provided in this chapter is relevant to undulator beam line construction. This will be much simplified since single-instrument beam lines will generally be the case, but the optics will have to cope with thermal loads of at least 100 W, and more than this, on high-energy machines. However, the bulk of the radiation will at least be in the region of interest, and with efficient optics should end up on the sample, thereby reducing the absorbed energy which is a problem on present sources. In addition, the optical apertures will be much smaller and thus aberration performance far superior.

Undulators have been installed on storage rings in the USSR, Japan and USA and have been shown to work effectively (see, for example, Maezawa et al. 1986a). At the Daresbury SRS a plane grating grazing incidence monochromator has been installed on an undulator designed to work over the wavelength range 20–120 Å, with an optimum resolution around 0.1 Å. The flux output is in the region of 10^{12} photons/s within this bandpass, and the premirror still has to cope with a heat load of typically 100 W. The mirror, however, is at grazing incidence (~2°) and silicon carbide is used. Apart from the undulator, the monochromator and beam line optics are conventional; the undulator provides the opportunity of making a good match to the optics, and, as discussed above, reduces the problems associated with radiation damage.

It has been our intention throughout this chapter to include details of instrumentation and techniques, which have been collected over years of working on synchrotron radiation sources. This has been done by way of example, but does not mean that the systems described are perfect or even the best available; the aim was to highlight the problems and suggest ways of solving them so that those involved in building beam lines on future machines will be in a good position to build the best design for their own source and experimental requirements. There are a few straightforward rules and procedures to follow in matching an experiment and monochromator to the source, but in general a compromise has to be made among conflicting requirements. Reference has been made to detailed information on monochromator designs and storage ring optics, but this stage follows after the initial purpose of the instrumentation has been decided. We hope this chapter will help the reader to define the objectives clearly, which will then lead to the decisions on how those objectives can best be met.

References

Abelès, F., 1950, Compt. Rend. **230**, 1942.
Abramowitz, M., and I. Stegun, 1972, Handbook of Mathematical Functions (Dover, New York).
Arfken, G., 1970, Mathematical Methods for Physicists, 2nd Ed. (Academic Press, New York).
Bartlett, R.J., D.R. Kania, R.H. Day and E. Kallne, 1984, Nucl. Instrum. Methods **222**, 95.

Beckmann, P., and A. Spizzichino, 1963, The Scattering of Electromagnetic Waves from Rough Surfaces (Macmillan, New York).
Bennett, H.E., 1978, Opt. Eng. **17**, 480.
Bennett, H.E., and J.O. Porteus, 1961, J. Opt. Soc. Am. **51**, 123.
Bennett, J.M., 1971, Ph.D. Thesis (Univ. of London).
Beutler, H.G., 1945, J. Opt. Soc. Am. **35**, 311.
Bilderback, D.H., 1981, in: Reflecting Optics for Synchrotron Radiation, Proc. SPIE **315**, 90.
Bilderback, D.H., and S. Hubbard, 1982a, Nucl. Instrum. Methods **195**, 85.
Bilderback, D.H., and S. Hubbard, 1982b, Nucl. Instrum. Methods **195**, 91.
Boller, K., R-P. Haelbich, H. Hogrefe, W. Jark and C. Kunz, 1983, Nucl. Instrum. Methods **208**, 273.
Brown, G., K. Halbach, J. Harris and H. Winwick, 1983, Nucl. Instrum. Methods **208**, 65.
Canfield, R.L., G. Hass and W.R. Hunter, 1964, J. Phys. (France) **25**, 124.
Cerino, J., J. Stohr, N. Hower and R.Z. Bachrach, 1980, Nucl. Instrum. Methods **172**, 227.
Chen, C.T., E.W. Plummer and M.R. Howells, 1984, Nucl. Instrum. Methods **222**, 103.
Codling, K., 1973, Rep. Prog. Phys. **36**, 541.
Cox, J.T., G. Hass and W.R. Hunter, 1971, J. Opt. Soc. Am. **61**, 360.
Cox, J.T., G. Hass, J.B. Ramsey and W.R. Hunter, 1973, J. Opt. Soc. Am. **63**, 435.
Damany, H., 1965, J. Opt. Soc. Am. **55**, 1558.
Dehmer, J.L., and J. Berkowitz, 1974, Phys. Rev. A **10**, 484.
Dietrich, H., and C. Kunz, 1972, Rev. Sci. Instrum. **43**, 434.
Dietz, E., W. Braun, A.M. Bradshaw and R.L. Johnson, 1985, Nucl. Instrum. Methods **239**, 359.
Ederer, D.L., 1982, Nucl. Instrum. Methods **195**, 191.
Ederer, D.L., B.E. Cole and J.B. West, 1980, Nucl. Instrum. Methods **172**, 185.
Eyers, A., Ch. Heckenkamp, F. Schäfers, G. Schönhense and U. Heinzmann, 1983, Nucl. Instrum. Methods **208**, 303.
Flamand, J., A. Thevenon, B. Touzet and M. Neviere, 1986, in: Abstracts Int. Conf. on Vacuum Ultraviolet Radiation Physics (Lund, Sweden) p. 285.
Franks, A., K. Lindsey, J.M. Bennett, R.J. Speer, D. Turner and D.J. Hunt, 1975, Philos. Trans. R. Soc. **277**, 503.
Franks, A., B. Gale, K. Lindsey, M. Stedman and W.P. Bailey, 1983, Nucl. Instrum. Methods **208**, 223.
Gluskin, E.S., E.M. Trakhtenberg and A.S. Vinogradov, 1978, Nucl. Instrum. Methods **152**, 133.
Golovchenko, J.A., R.A. Levesque and P.L. Cowan, 1981, Rev. Sci. Instrum. **52**(4), 509.
Green, G.K., 1976, Brookhaven National Laboratory Rep. 50522 (BNL, Long Island, NY).
Green, G.K., 1977, Brookhaven National Laboratory Rep. 50595 (BNL, Long Island, NY).
Haber, H., 1950, J. Opt. Soc. Am. **40**, 153.
Haelbich, R.P., A. Segmuller and E. Spiller, 1979, Appl. Phys. Lett. **34**(3), 184.
Hagemann, H.J., W. Gudat and C. Kunz, 1974, Deutsches Elektronen Synchrotron Int. Rep. DESY SR74-7.
Hamm, R.N., R.A. MacRea and E.T. Arakawa, 1965, J. Opt. Soc. Am. **55**, 1460.
Hass, G., and W.R. Hunter, 1978, Appl. Opt. **17**, 76.
Hass, G., G.F. Jacobus and W.R. Hunter, 1967, J. Opt. Soc. Am. **57**, 758.
Hastings, C., 1955, Approximations for Digital Computers (Princeton Univ. Press, Princeton, NJ).
Heckenkamp, Ch., F. Schäfers, G. Schönhense and U. Heinzmann, 1984, Phys. Rev. Lett. **52**, 421.
Hellwege, K.H., 1937, Z. Phys. **106**, 588.
Henke, B.L., 1982, Phys. Rev. A **6**, 94.
Henke, B.L., P. Lee, T.J. Tanaka, R.L. Shimabukuro and B.K. Fujikawa, 1981, in: Proc. Conf. on Low Energy X-Ray Diagnostics, Monterey, Vol. 75, eds D.T. Attwood and B.L. Henke (AIP, New York) p. 340.
Henke, B.L., P. Lee, T.J. Tanaka, R.L. Shimabukuro and B.K. Fujikawa, 1982, Atomic Data and Nuclear Data Tables **27**, 1.
Hogrefe, H., M.R. Howells and E. Hoyer, 1986, Application of Spherical Gratings in Synchrotron Radiation Spectroscopy, in: Proc. SPIE Meeting on Soft X-Ray Optics and Technology, eds E.-E. Koch and G. Schmahl, in press.
Howells, M.R., 1979, Brookhaven National Laboratory Rep. 26027 (BNL, Long Island, NY).

Howells, M.R., 1980a, Nucl. Instrum. Methods **172**, 123.
Howells, M.R., 1980b, Some Geometrical Considerations Concerning Grazing Incidence Reflectors, unpublished.
Howells, M.R., D. Norman, G.P. Williams and J.B. West, 1978, J. Phys. E **11**, 199.
Hubbard, D.J., 1982, Ph.D. Thesis (University of Reading, UK).
Hubbard, D.J., and E. Pantos, 1983, Nucl. Instrum. Methods **208**, 319.
Humphreys-Owen, S.P.F., 1961, Proc. Phys. Soc. London **77**, 949.
Hunter, W.R., 1978, Appl. Opt. **17**, 1259.
Hunter, W.R., J.F. Osantowski and G. Hass, 1971, J. Opt. Soc. Am. **10**, 540.
Hunter, W.R., R.T. Williams, J.C. Rife, J.P. Kirkland and M.N. Kabler, 1982, Nucl. Instrum. Methods **195**, 141.
Jacobus, G.F., R.P. Madden and L.R. Canfield, 1963, J. Opt. Soc. Am. **53**, 1084.
James, R.W., 1965, The Optical Principles of Diffraction of X-Rays (Cornell Univ. Press, Ithaca, NY).
Jark, W., and C. Kunz, 1985, Deutsches Elektronen Synchrotron Int. Rep. DESY SR85-09.
Jark, W., and M. Neviere, 1986, in: Abstracts Int. Conf. on Vacuum Ultraviolet Radiation Physics (Lund, Sweden) p. 307.
Johnson, R.L., 1975, Ph.D. Thesis (Univ. of London).
Johnson, R.L., 1983, Grating Monochromators and Optics for the VUV and Soft X-Ray Region, in: Handbook on Synchrotron Radiation, Vol. 1, ed. E.-E. Koch (North-Holland, Amsterdam) ch. 3.
Kelly, M.M., J.B. West and D.E. Lloyd, 1981, J. Phys. D **14**, 401.
Koide, T., S. Sato, T. Shidara, M. Niwano, M. Yanajihara, A. Yamada, A. Fujimori, A. Mikuni, H. Kjato and T. Miyahara, 1986, Nucl. Instrum. Methods **246**, 215.
Kostroun, V.O., 1980, Nucl. Instrum. Methods **172**, 371.
Lepère, D., 1975, Nouv. Rev. Opt. **6**, 3, 173.
Lindsey, K., and R. Morrell, 1981, in: Reflecting Optics for Synchrotron Radiation, Proc. SPIE **315**, 140.
MacDowell, A.A., J.B. West and T. Koide, 1986, Nucl. Instrum. Methods **246**, 219.
MacDowell, A.A., G. Van der Laan and J.B. West, 1987, to be published.
Mack, R.A., 1966, Spectral and Angular Distributions of Synchrotron Radiation, Int. Rep. CEAL-1027 (Cambridge Electron Accelerator).
Madden, R.P., and D.L. Ederer, 1972, J. Opt. Soc. Am. **62**, 722A.
Maezawa, H., Y. Suzuki, H. Kitamura and T. Sasaki, 1986a, Nucl. Instrum. Methods **246**, 82.
Maezawa, H., S. Nakai, S. Mitami, H. Noda, T. Naikoka and T. Sasaki, 1986b, KEK Preprint 85-74 (Photon Factory, Tsukuba, Japan).
Mahan, A.I., 1956, J. Opt. Soc. Am. **46**, 913.
McKinney, W.R., and M.R. Howells, 1980, Nucl. Instrum. Methods **172**, 149.
Miyake, K.P., R. Kato and H. Yamashita, 1969, Sci. Light **18**, 39.
Murty, M.V.R.K., 1962, J. Opt. Soc. Am. **52**, 768.
Namioka, T., 1959a, J. Opt. Soc. Am. **49**, 440.
Namioka, T., 1959b, J. Opt. Soc. Am. **49**, 951.
Namioka, T., H. Noda, K. Goto and T. Katayama, 1983, Nucl. Instrum. Methods **208**, 215.
Neviere, M., 1980, in: Periodic Structures, Gratings, Moiré Patterns and Diffraction Phenomena, Proc. SPIE **240**.
Neviere, M., and J. Flamand, 1980, Nucl. Instrum. Methods **172**, 273.
Neviere, M., J. Flamand and J.M. Lerner, 1982, Nucl. Instrum. Methods **195**, 183.
Névot, L., and P. Croce, 1975, J. Appl. Crystallogr. **8**, 304.
Névot, L., and P. Croce, 1980, Rev. Phys. Appl. **15**, 761.
Noda, H., T. Namioka and M. Seya, 1974, J. Opt. Soc. Am. **64**, 8, 1031.
Padmore, H.A., 1986a, Daresbury Technical Memorandum DL/SCI/TM45E, (Daresbury, Warrington, UK).
Padmore, H.A., 1986b, Application of a Simple Rotational Grating Mounting to High Resolution Soft X-Ray Spectropscopy, in: Proc. SPIE Meeting on Soft X-Ray Optics and Technology, eds E.-E. Koch and G. Schmahl, in press.
Parratt, L.G., 1954, Phys. Rev. **95**(2), 359.

Petersen, H., 1982, Opt. Commun. **40**, 402.
Petit, R., ed., 1980, Electromagnetic Theory of Gratings, Topics in Current Physics, Vol. 22 (Springer, Berlin).
Poole, J.H., 1975, Daresbury Laboratory Internal Rep. DL/SRF/TM1, (Daresbury, Warrington, UK).
Poole, J.H., 1976a, Daresbury Laboratory Internal Rep. DL/SRF/TM4, (Daresbury, Warrington, UK).
Poole, J.H., 1976b, Daresbury Laboratory Internal Rep. DL/SRF/TM6, (Daresbury, Warrington, UK).
Popper, P., and D.G.T. Davies, 1961, Powder Metallurgy **8**, 113.
Rabinovitch, K., L.R. Canfield and R.P. Madden, 1965, Appl. Opt. **4**, 1005.
Rehfeld, N., U. Gerhardt and E. Dietz, 1973, Appl. Phys. **1**, 229.
Rehn, V., 1981, in: Proc. Conf. on Low Energy X-Ray Diagnostics, Monterey, Vol. 75, eds D.T. Attwood and B.L. Henke (AIP, New York) p. 162.
Rehn, V., and W.J. Choyke, 1980, Nucl. Instrum. Methods **177**, 173.
Rehn, V., J.L. Stanford, A.D. Baer, V.O. Jones and W.J. Choyke, 1977, Appl. Opt. **16**, 1111.
Rehn, V., V.O. Jones, J.M. Elson and J.M. Bennett, 1980, Nucl. Instrum. Methods **172**, 307.
Rowe, E.M., 1973, IEEE Trans. Nucl. Sci. **NS-20**, 973.
Saile, V., and J.B. West, 1983, Nucl. Instrum. Methods **208**, 199.
Samson, J.A.R., 1967, Techniques of Ultraviolet Spectroscopy (Wiley, New York) p. 296.
Samson, J.A.R., 1976, Rev. Sci. Instrum. **47**, 859.
Samson, J.A.R., 1978, Nucl. Instrum. Methods **152**, 225.
Sato, S., T. Koide, Y. Morioka, T. Ishii, H. Sugawara and I. Nagakura, 1983, Nucl. Instrum. Methods **208**, 31.
Schwinger, J., 1949, Phys. Rev. **75**, 1912.
Smirnov, L.A., 1977, Opt. Spectrosc. **43**, 3.
Smirnov, L.A., T.D. Sotnikova, B.S. Anokhin and B.Z. Taibin, 1979, Opt. Spectrosc. **46**, 3.
Sokolov, A.A., and I.M. Ternov, 1968, Synchrotron Radiation (Pergamon Press, New York).
Tomboulian, D.H., and P.L. Hartman, 1956, Phys. Rev. **102**, 1423.
Velzel, C.H.F., 1976, J. Opt. Soc. Am. **66**, 4, 346.
Velzel, C.H.F., 1977, J. Opt. Soc. Am. **77**, 8, 1021.
Vodar, B., and J. Romand, 1974, in: Some Aspects of Vacuum Ultraviolet Radiation Physics, ed. N. Damany (Pergamon Press, New York) p. 31.
Walker, R.P., 1985, Daresbury Laboratory, private communication.
West, J.B., K. Codling and G.V. Marr, 1974, J. Phys. E **1**, 137.
Williams, G.P., and M.R. Howells, 1983, Nucl. Instrum. Methods **208**, 37.
Williams, G.P., and J.F. Weisenbloom, 1979, Brookhaven National Laboratory Rep. 26974 (BNL, Long Island, NY).
Winick, H., 1980, in: Synchrotron Radiation Research, eds H. Winick and S. Doniach (Plenum Press, New York) p. 11.

CHAPTER 3

DATA ACQUISITION AND ANALYSIS SYSTEMS

P.A. RIDLEY

Science and Engineering Research Council, Daresbury Laboratory,
Keckwick Lane, Daresbury, Warrington WA4 4AD, UK

Contents

1. Introduction	123
2. Data acquisition electronics	124
2.1. Introduction	124
2.2. Detector signal processing	125
2.2.1. The NIM system	125
2.2.2. Single channel counting	126
2.2.3. Multiscaling	129
2.2.4. Multichannel counting and coincidence techniques	130
2.2.5. Other methods	136
2.3. Motor driving	137
2.3.1. Stepping motors	137
2.3.1.1. Advantages	137
2.3.1.2. Disadvantages	137
2.3.1.3. Performance parameters	138
2.3.1.4. The resonance problem	138
2.3.1.5. Multi- and micro-stepping	138
2.3.1.6. Stepping motor system components	138
2.3.1.7. Backlash	140
2.3.1.8. Stepping motor variants	140
2.3.1.9. Do stepping motor systems lose steps?	140
2.3.2. DC servo motors	140
2.3.2.1. Advantages	140
2.3.2.2. Disadvantages	141
2.3.2.3. Performance parameters	141
2.4. Digital interfacing systems	141

Contents continued overleaf

Handbook on Synchrotron Radiation, Vol. 2, edited by G.V. Marr
© *Elsevier Science Publishers B.V., 1987*

Contents continued

    2.4.1. Requirements 141
    2.4.2. CAMAC. 143
      2.4.2.1. CAMAC basics. 143
      2.4.2.2. CAMAC strengths and weaknesses 145
    2.4.3. Other interfacing standards 146
      2.4.3.1. ANSI/IEEE-488 146
      2.4.3.2. Small system buses 146
      2.4.3.3. High performance buses. 147
3. Local computer systems 148
 3.1. Introduction. 148
 3.2. Dedicated or shared? 148
 3.3. Computer system requirements 149
    3.3.1. Introduction. 149
    3.3.2. Basic hardware considerations. 150
      3.3.2.1. A typical system 150
      3.3.2.2. Peripherals 150
      3.3.2.3. Word sizes and addressability 151
      3.3.2.4. Programmed I/O and DMA 152
      3.3.2.5. Interrupts 153
    3.3.3. Basic software considerations 153
      3.3.3.1. Program preparation 153
      3.3.3.2. Run-time support and real-time considerations . . . 155
4. Data processing facilities 157
 4.1. Introduction 157
 4.2. Required resources 159
    4.2.1. Data storage 159
    4.2.2. Graphics terminals 160
    4.2.3. Plotters and printers 160
    4.2.4. Computation 161
    4.2.5. Access to a wide area network 161
 4.3. Distributed computer systems 162
    4.3.1. Introduction 162
    4.3.2. Technical issues 163
    4.3.3. Organisational issues 164
    4.3.4. Summary 165
 4.4. Local area networks 165
    4.4.1. Introduction 165
    4.4.2. Ethernet 166
    4.4.3. Other LANs 168
References 170

1. Introduction

This chapter is intended to provide a basic introduction to data acquisition and analysis systems, particularly as they are applied to vacuum ultraviolet (VUV) and soft X-ray experiments using synchrotron radiation (SR). It does not attempt great depth or detail on any topic. References are provided where possible to enable the reader to explore interesting topics further. The presentation is aimed towards the needs of two main categories of reader: (a) young research workers who, while they may not need to implement a data acquisition system themselves, would be aided in planning their experiments by the examples and descriptions of systems, and (b) those beginning a career in a data acquisition and computing support role who would benefit from an overview of the subject.

The computing requirements of VUV and soft X-ray experiments are not very different from those of other SR experiments (or any other scientific data acquisition and analysis requirements). It is hoped, therefore, that those concerned with other sorts of experiments will also find some useful material here. However, it is worth noting that there are significant differences of emphasis between *scientific* data acquisition and *industrial* data acquisition. The latter is generally concerned with the measurement of a large number of slowly varying analogue signals for process monitoring and control. The data acquisition systems considered in this chapter are more concerned with handling pulses from photon and electron detectors. Those readers requiring an introduction to analogue signal measurement are referred to VanDoren (1982).

In practice, the role of a data acquisition system is usually much broader than that of merely acquiring the data. For the purposes of this chapter the term is considered to cover a system that performs the following functions.

(a) Control of instruments (monochromators and spectrometers).
(b) Digitising the output of detectors and measuring instruments.
(c) Recording the digitised data.
(d) Providing feedback on experiment progress and data quality to the experimenter.
(e) The coordination of the above activities into an appropriate sequence of measurements.

It is assumed in the above that data are brought to a digital form. There are rare cases where this is unnecessary and, for example, direct recording on an X–Y chart recorder may be adequate. Such cases are not considered in this chapter.

Figure 1 illustrates a data acquisition system and its connections to experimental apparatus. This chapter contains sections on detector signal processing, motor driving, computer interfacing, and computer systems. Detectors, spectrometers and monochromators are treated extensively elsewhere in Volumes 1 and 2 of this Handbook.

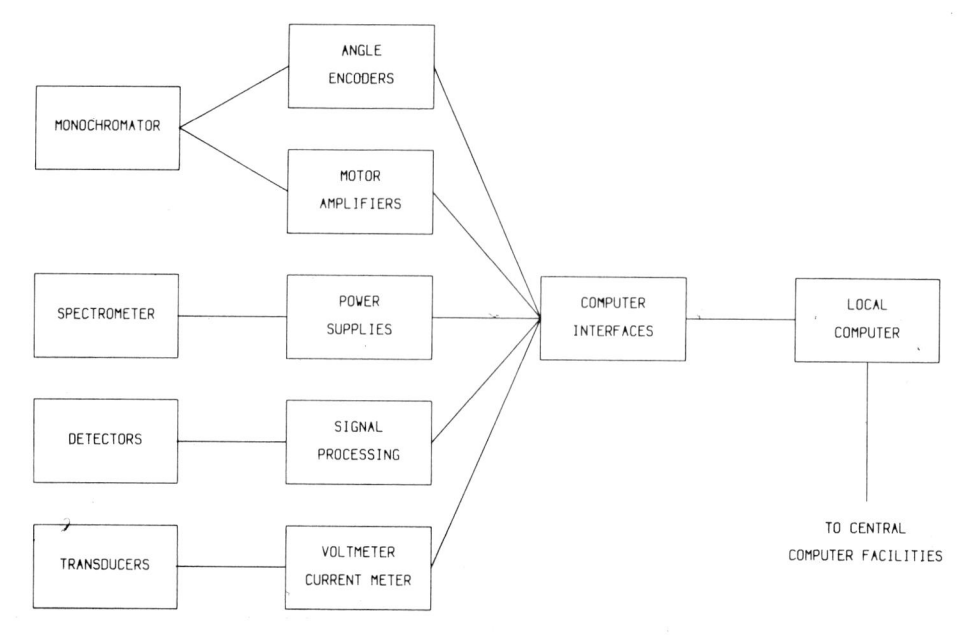

Fig. 1. Schematic of data acquisition system.

2. Data acquisition electronics

2.1. Introduction

This section deals with electronics from a system viewpoint rather than at the level of circuits. It aims to provide the reader with an introduction to the basic electronic units commonly used in SR experiments, mainly by means of simple examples.

In principle, most of the electronics for an experiment, one might imagine, could be neatly provided within the computer interfacing system, with attractive potential savings in cost (eliminating individual power supplies, for example), cabling, and space. However, the practice in SR laboratory systems is very different. The major reasons are the following.

(a) *Modularity*. The demands of an experimental facility are rarely static for long. With a modular system, changes can be made relatively easily. For example, one amplifier can be readily substituted by another with a different specification, or a new signal channel added by using "off-the-shelf" modules. Modularity helps in rapid fault location and elimination. The use of standard modules for several different experiments allows common spares to be held. When the apparatus is no longer required the standard modules can be re-used for other experiments.

(b) *Sensitivity*. It may be necessary to place some electronics very close to the experimental apparatus in order to minimise cable lengths to the apparatus. This is the case where small analogue signals are to be protected against noise interference, in which case the signals may be amplified or converted to digital form close to their source before transmission over a longer path to the computer interfacing subsystem (see, for example, Murphy 1976). Futhermore the computer interfacing system itself is sometimes too noisy an environment for handling precision analogue signals and prior conditioning may be required. Problems of noise pickup are treated by Brookshier (1969) and by Morrison (1967).

(c) *Availability*. It is rarely sensible to develop a specialised interface if there is a commercial instrument available for which a standard interface exists. There are obvious penalties in terms of development costs and timescales, and long term maintenance provision. Exceptions may occur, however, if a large number of copies of a unit are required and sufficient economies and technical improvements can be achieved. The decision to do this, therefore, would normally be taken at the laboratory level rather than at the level of an individual experiment.

(d) *Format*. Some items are very inconvenient to build in to the computer interfacing sub-system because of their size or power dissipation. Space in the computer interfacing sub-system is often limited and/or expensive. The use of external packaging for the unit may overcome this problem and provide more front panel space for visual indicators and manual controls. While the computer expert may regard the latter as an unnecessary and mildly insulting deviation from the spirit of a computer-controlled system, the user frequently finds them invaluable aids to setting up the experiment and locating problems.

Electronics for detector signal processing and stepping motor driving are to be found in nearly all SR experiments and are therefore usually organised into modular sub-systems and standardised throughout the laboratory.

2.2. Detector signal processing

Some useful review articles on the topics of this section are by Timothy and Madden (1983), Granneman and van der Wiel (1983) (both from Volume 1 of this Handbook), Munro and Schwentner (1983) and by Smith and Kevan (1982).

2.2.1. The NIM system

In order to reduce the problem of noise pickup in the detector output signal, it is usual to provide a preamplifier in the form of a "free-standing" unit as close as possible to the detector. The output of the preamplifier may be in the form of voltage pulses, where individual particles have been detected, or as a DC voltage proportional to the rate of particles detected. The subsequent processing of such signals is usually performed by units in the NIM format.

The NIM (Nuclear Instrument Modules) standard (see ref. NIM) is based on the use of a "bin" which provides slots and regulated power supplies on specified connector pins for plug-in modules. A normal bin fits in a 19-inch wide rack and houses up to 12 single-width units. Half-width bins with 6 slots designed for portability and small systems are also available. The version in wide-spread use provides for modules of 8.75 inches height (the NIM specification also defines a 5.25 inch standard).

The NIM standard defines "preferred practices" with regard to "linear signals" (analogue signals, with information in their amplitude) and "logic signals". Linear signals should be in the range 0 to 10 V. There are two types of logic signal defined, positive and negative. Positive logic signals are for slow- or medium-rate applications (to 20 MHz repetition rate): logic 1 levels are in the range +4 to +12 V and logic 0 from +1 to −2 V. When fast-rise time signals are required (e.g. for timing applications or high rates), negative logic signals would be used. These signals are defined to drive a 50 ohm impedance with −14 to −18 mA for a logic 1 and −1 to +1 mA for a logic 0. The rise time for these signals is typically 2 ns (though this is not defined in the NIM standard). In handling fast rise time signals it is necessary to beware of reflections, so 50 ohm cables with correct termination must be used.

The use of these preferred signal levels enables the various modules from different sources in a system to be interconnected easily. A common fault of the inexperienced or unwary user is to confuse positive and negative logic levels so that, for example, a positive logic level is connected to an input designed for negative logic signals. NIM positive logic levels correspond to "TTL" logic levels of digital electronics. However, NIM negative logic levels do *not* correspond to TTL "negative logic" (where the low-voltage level represents a logic 1 and the high level to logic 0, that is, the *inverse* of "positive logic"). Standard NIM units are available to convert between TTL and NIM (fast logic) signals and to produce their logical inverses.

Several companies manufacture NIM modules and a very large selection of units is commercially available. Units commonly used in SR experiments include pulse amplifiers (some specialised for pulse-height analysis and others for fast timing), discriminators, ratemeters, time-to-amplitude converters, single-channel analysers, voltage and current-to-frequency converters, and detector high-voltage power supplies. It is often convenient, too, to use the NIM format for "homemade" electronics. Examples of the use of NIM units are given in the following sections.

In spite of the age (it was first issued in 1964) and title of this standard it remains very relevant to SR research and there is no sign of any successor. New NIM units are continually emerging, and recently a supplementary standard has been established to define the way NIM units should be interfaced to the IEEE-488 General Purpose Interface Bus (GPIB) (see ref. ANSI/IEEE 488).

Some general information on the design of detector signal processing electronics may be found in work by Kowalksi (1970) and Herbst (1970).

2.2.2. Single channel counting

The simplest form of detector signal processing is a single channel counting system, as illustrated in fig. 2.

The experiment might be one in which the signal is a train of pulses corresponding to detected photoelectron from a channeltron. A spectrum of electron yield versus incident photon energy would be obtained by counting these pulses for a fixed time interval at each wavelength selected by a monochromator; the monochromator being driven to a new setting by a motor system (not shown) at the end

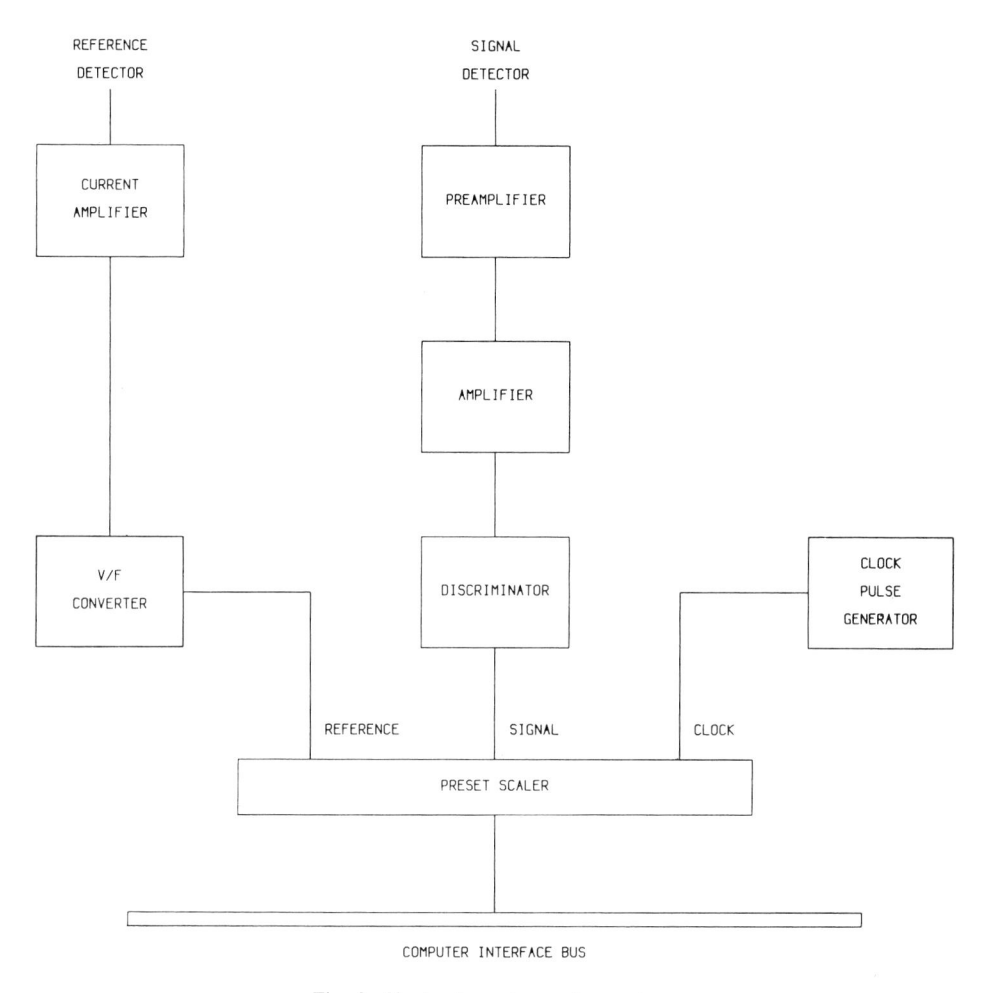

Fig. 2. Single-channel counting system.

of each interval. In the figure, the time interval is governed by a clock pulse generator, producing pulses at perhaps 1 kHz. It is convenient to count both clock and signal pulses in a preset scaler, that is, one that will produce an "overflow" signal when the programmed count of clock pulses is reached. The overflow would be used to inhibit further counting and simultaneously signal to the controlling computer by an interrupt (see section 3.3.2.5) to read the counts from the scaler, reset the monochromator and enable counting for the next channel of the spectrum. Using the overflow signal directly to inhibit counting helps ensure that the actual counting interval is not subject to variation caused by varying delays in the response of the computer to the interrupt. Similarly precautions may be necessary to ensure that all the channels of the scaler are simultaneously enabled at the start of a counting interval.

An improvement on counting for a fixed period is to count for a fixed number of pulses generated by a "reference" detector. The reference detector would be monitoring the beam flux that is incident on the sample. Thus normalising the signal count at each point in the spectrum to the incident beam compensates for variations in SR intensity (which decays, of course, with time) and monochromator transmission efficiency. It is still useful to record the clock pulses, however, so that fluctuations in beam intensity that might effect the validity of the measurement can be detected. In case the beam should vanish or be drastically reduced, the scalar can be preset to produce the overflow signal on either the reference count limit or after an abnormally long time, so that the computer system need not be waiting forever for a measurement that will never complete, but can warn the experimenter of the state of affairs.

In the figure, the signal channel is shown as containing a preamplifier, an amplifier and a discriminator. The preamplifier would normally be "charge-sensitive", converting the charge pulse released by the detector into a voltage pulse. In being charge sensitive rather than voltage sensitive, such a preamplifier avoids the output variations that arise from the variation of detector capacitance with bias voltage (see Kowalski 1970). The output amplitude, normally a fraction of a volt, is proportional to the charge input, and the rise time, typically of the order of 20 ns, is approximately proportional to the width of the input current pulse: thus a charge-sensitive preamplifier may be used for both particle energy analysis and timing applications. The preamplifier is placed close to the detector to minimise noise-pickup on the input.

The type of main amplifier used depends on whether optimum energy resolution or timing resolution and rates are required. Amplifiers suitable for high energy resolution produce shaped output pulses of a few microseconds in width and a few volts in amplitude, suitable for pulse-height analysis. Thus they are limited to handling counting rates of substantially less than 1 MHz. In single-channel counting experiments, count rates are more significant than energy resolution, since the pulse height is not analysed. A timing amplifier has selectable time constants down to the order of 10 ns, permitting count rates of several megahertz.

The basic function of the discriminator is to produce a logic pulse whenever the input voltage pulse rises above a selected level. In a simple counting arrangement, this level would be set high enough to exclude noise and unwanted low-energy particle pulses.

For accurate comparisons of counting rates it is necessary to correct for "dead-time". Each unit in the system requires a finite time to process a pulse, within which period it will be insensitive to the arrival of another pulse at the input. In a simple counting arrangement such as shown in fig. 2, the dead-time of the system would be governed by the dead-time of the slowest element. This might be the recovery time of the discriminator and be of the order of 50 ns. Calling this dead-time per processed pulse, τ, the fraction of the time the system is not sensitive to pulses is $n\tau$, where n is the observed count rate. The corrected count rate is then simply given by $n/(1 - n\tau)$.

The reference channel in fig. 2 illustrates the case when the rate is too high to count pulses from individual particles. From the above discussion of time constants in amplifiers and discrimination it can be seen that this arises for pulse rates of a few tens of megahertz upwards. For higher rates a current amplifier may be used. This instrument (also called an electrometer) is placed close to the detector and converts currents ranging from picoamperes to a milliampere to a voltage of up to 10 V. A convenient way of interfacing this into a counting system is to use a voltage-to-frequency converter to produce a pulse train that may be delivered to a scalar and therefore treated in the same way as pulses delivered by a discriminator. The number of pulses from the voltage-to-frequency converter during a counting interval is proportional to the number of particles detected.

2.2.3. Multiscaling

In a simple scanning system a spectrum is acquired by starting at one end of the range of the scanned parameter (photon wavelength or electron energy, for example), and counting for perhaps several seconds at each position in the scan, proceeding systematically to the other end of the range. A system which makes only a single scan through the spectrum, dwelling a long time at each position, may be vulnerable to long-term drifts in counting efficiency arising from drifts in such parameters as gas jet pressure and channeltron gain. Conditions may be significantly different for data taken at the two ends of the spectrum.

A solution to this problem is to accumulate the spectrum by a series of repeated scans, dwelling for a short time at each position in each scan (see e.g. Parr et al. 1980), thus exposing each channel to approximately the same variations of conditions. The overhead associated with resetting the scan parameter between counting intervals must now be sustained many times, thereby increasing the time taken to accumulate the spectrum. To some extent the overhead is unavoidable, containing the time necessary to drive a stepping motor, for instance, and to allow voltages to settle. The overhead may also contain the response time of the computer to the "end-of-interval" interrupt unless the computer is dedicated to a continuous loop, checking to see if it is time to move to the next scan point (see section 3.3.3.2).

A system designed at Daresbury Laboratory avoids the computer response problem by employing a fast "bit-slice" microprocessor controller in the CAMAC computer interface system (CAMAC is described in section 2.4.2). The controller is used to transfer data between CAMAC modules. The controller responds to each "end-of-interval" pulse from the scalar by reading the scalar contents and updating the count for that spectrum channel held in a memory module in the CAMAC crate. The previously accumulated value for the next channel is then read from the memory module and loaded into the preset scalar ready to accumulate further during the next counting interval. The memory module also holds the reference count limit used to preset the scalar for the next interval and the list of values to be written to a DAC (digital-to-analogue converter) to select the electron energy for the next channel. The processing dead-time is thus reduced to a few microseconds per channel (see fig. 3).

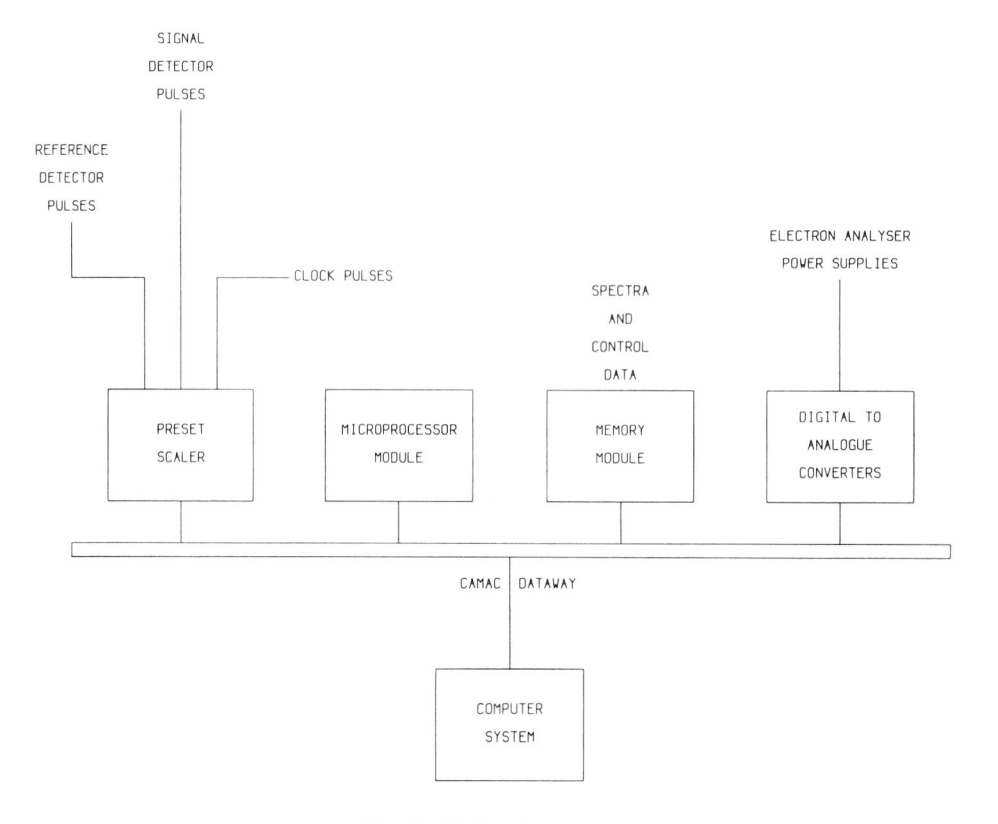

Fig. 3. Multiscaling system.

2.2.4. Multichannel counting and coincidence techniques

Experiments with several parameters to vary may be very time consuming if only one combination of the parameters can be processed at a time. Drift, contamination or sample deterioration effects may become important. Thus multichannel detection systems, in which data for all the desired values of a parameter can be taken in parallel, may be required.

Figure 4 shows a basic "delayed coincidence" counting arrangement. This arrangement is used in time-resolved spectroscopy experiments for fluorescence lifetimes studies (Munro and Schwentner 1983). In these experiments, the synchrotron is typically operated in "single-bunch" mode, providing excitation photons at intervals of the order of a microsecond. The time-to-amplitude converter (TAC) converts the time interval between the excitation pulse and the detected fluorescent photon into a voltage pulse which is analysed and the spectrum accumulated in a pulse-height analyser (also commonly known as a multichannel analyser (MCA)).

The timing resolution achieved depends on the detectors and electronics used. Timing amplifiers have been mentioned in section 2.2.2. There are several types of discriminator with differing performances with respect to timing applications

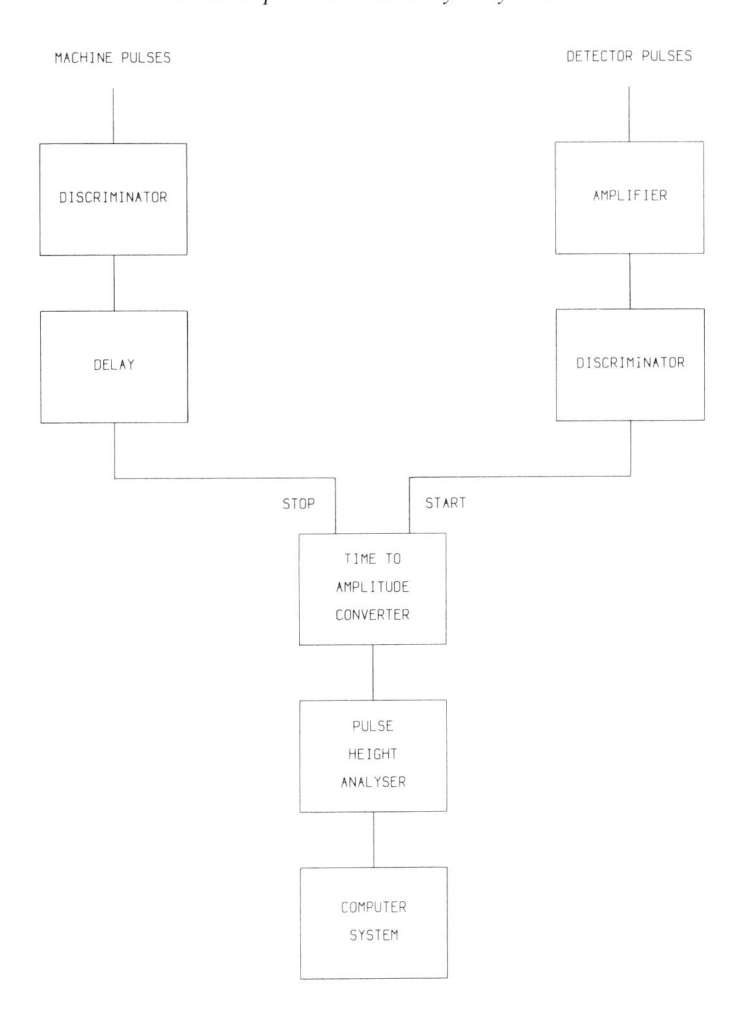

Fig. 4. Delayed coincidence system.

(see ref. EG&G Ortec AN-42). The "constant-fraction discriminator" (CFD) has time resolution down to about 200 ps (Bedwell and Paulus 1979). Munro and Schwentner (1983) reported that using standard commercial detectors and electronics, resolutions of better than 1 ns are usually obtainable, while 50 ps is possible with special photomultiplier tubes and channel plate detectors.

The same technique is applied to "time-of-flight" (TOF) spectroscopy. Hastings and Kostroun (1983) describe a system used to record the charge state distribution of ions. A high voltage is used to accelerate the ions, giving them, since they are produced with very small kinetic energy, essentially the same kinetic energy as they enter a drift tube.

Their velocities differ for different charge-to-mass ratios, $v = k(q/m)^{1/2}$, thus they are detected at the end of the drift tube at different times relative to the signal from the synchrotron bunch responsible for the ionisation. If the rate of detected ions is lower than the synchrotron bunch repetition rate, as is the case for the system Hastings and Kostroun describe, it is advantageous to invert the spectrum by using the ion detector signal to start the TAC conversion, since this will result in fewer unnecessary conversion cycles and thus less dead-time.

TOF spectrometry is also used for the multichannel energy analysis of photo-electrons (see e.g. White et al. 1979: timing resolution of 70 ps contributed to an energy resolution of better than 5%).

Electron–ion (or electron–electron) coincidence experiments can be performed with essentially the same arrangement of electronics as shown in fig. 4. In this case the start signal for the TAC is derived from an electron detector and the stop signal from the ion detector. If complete TOF spectra are stored for each of the settings of electron angle, for example, or other experimental parameter, very large quantities of data may be accumulated. This may be unavoidable if there are several peaks in the spectrum or complex background removal and peak shape analysis is required. For simple situations where so much detail is not required,

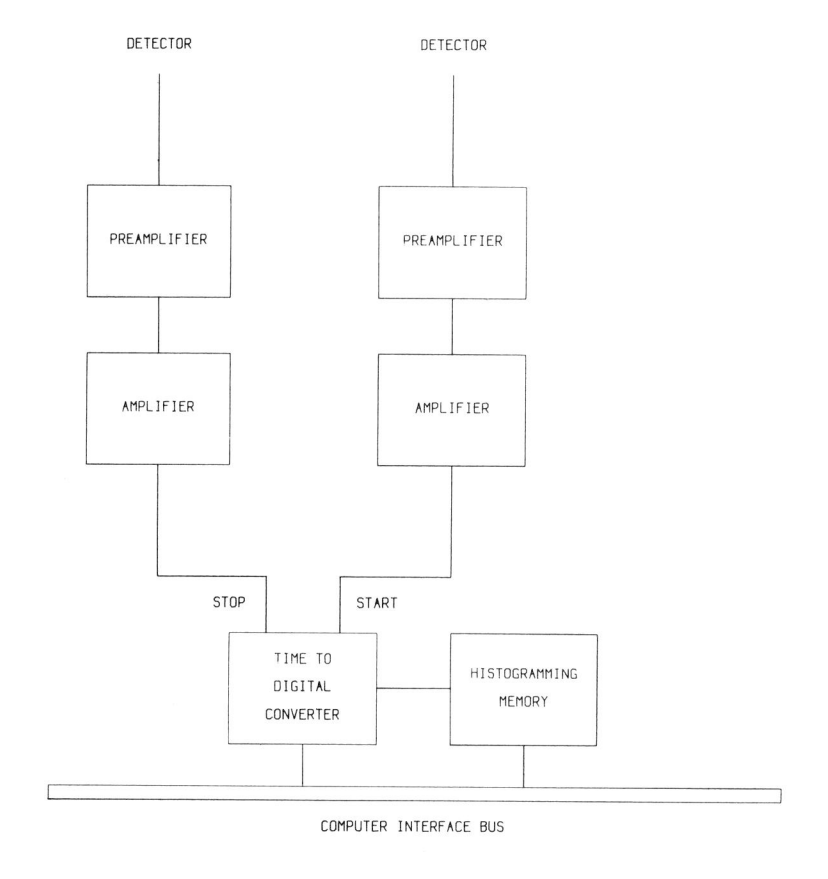

Fig. 5. Coincidence counting system with TDC and histogramming memory.

Granneman and van der Wiel (1983) show how timing single-channel analysers and coincidence (AND) gates may be used to reduce the data at source.

Given that the data will be read out into, and the experiment controlled from, a local computer system, it is sometimes convenient to substitute the combination of a time-to-digital converter (TDC), and local histogramming memory for the TAC and pulse height analyser, as shown in fig. 5.

Commercial MCAs have the advantage that they are (usually) easier to control via their front panel switches than computer systems. TDCs with computer systems, however, are useful where flexibility and expandability are required. For multi-parameter experiments (several detectors to provide angular resolution, for example) such a system can be readily extended to have several TDC channels,

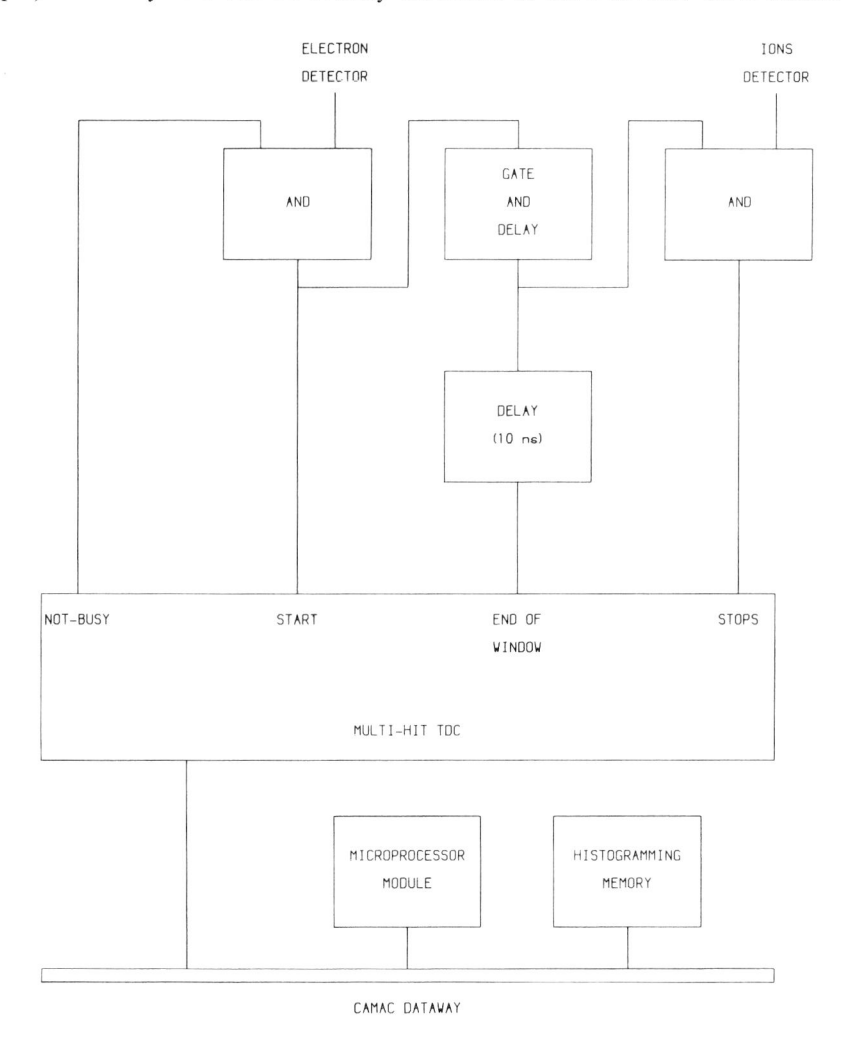

Fig. 6. Multi-hit TDC used for electron–ion coincidence.

accumulating data in a large histogramming memory, supported by appropriate grey-levels or pseudocolour display system. Commercial CAMAC TDCs exist with timing resolution down to 160 ps and conversion times of less than 1 microsecond. Turko (1978) describes a TDC with 9.76 ps resolution.

TDCs exist with a multi-hit capability, able to digitise up to 8 separate STOP pulses following in series through a single input to the TDC. Such an arrangement has been used at Daresbury Laboratory for electron–ion coincidence experiments and is illustrated in fig. 6 (Enderby and Holland 1985). An equivalent arrangement employing several TACs and ADCs with routing circuitry would be considerably more clumsy to achieve.

A more complex example of this application of coincidence circuitry and TDC is described by Butler et al. (1985), referring to an electron–ion coincidence spectrometer constructed at the NBS. They used a well stabilised pulse generator to draw out ions for detection only when threshold electrons have been detected. An elaborate system of AND and OR gates reduces the sensitivity of the system to noise pulses and enables convenient measurement of random coincidence rates.

One- and two-dimensional detector systems have been developed, based on microchannel plate (MCP) detectors. Read-out systems for these have been described by Timothy and Madden (1983) and by Grannemann and van der Wiel (1983). As mentioned in these articles, a position-sensitive detector can be read-out by digitising the time difference between pulses seen at the two ends of a resistive anode (Wiza 1979).

Such a detector can form part of a spectrometer for the energy analysis of soft X-ray fluorescence spectra (MacDowell, Hillier and West 1983). The X-rays are converted to photoelectrons whose energy information is translated to positional information on the detector by an electrostatic field. The read-out system (fig. 7a) has much in common with the basic delayed coincidence system of fig. 4. In this case a prompt pulse is derived from the detector and used to enable the TAC only when the pulse is within limits established by a single channel analyser (SCA), thus discriminating against noise. A refinement employed by MacDowell et al. is shown in fig. 7b. This is to compensate for non-linearities in the translation of electron energies to positions on the detector. The spectrum is scanned across the detector so that response variations are averaged out. To achieve this, a ramp voltage is used to drive the scanning electronics of the spectrometer and also to generate a proportional pedestal voltage to be summed with the output of the TAC. This enables the spectrum to be accumulated in fixed channel addresses in the MCA, while the TAC output varies in response to the movement of the spectrum across the detector. Multiscaling (see section 2.2.3) to reduce drift effects is achieved by repeating the voltage ramp.

A successor read-out system has been developed at Daresbury Laboratory for multiscaling with resistive anode encoding (fig. 7c). The TAC and MCA are replaced by a TDC and companion histogramming memory. At the end of each counting interval a fast microprocessor CAMAC controller adjusts analyser voltages obtained from lists held in another CAMAC memory module through digital-to-analogue converters and adjusts an offset word in the TDC to keep the

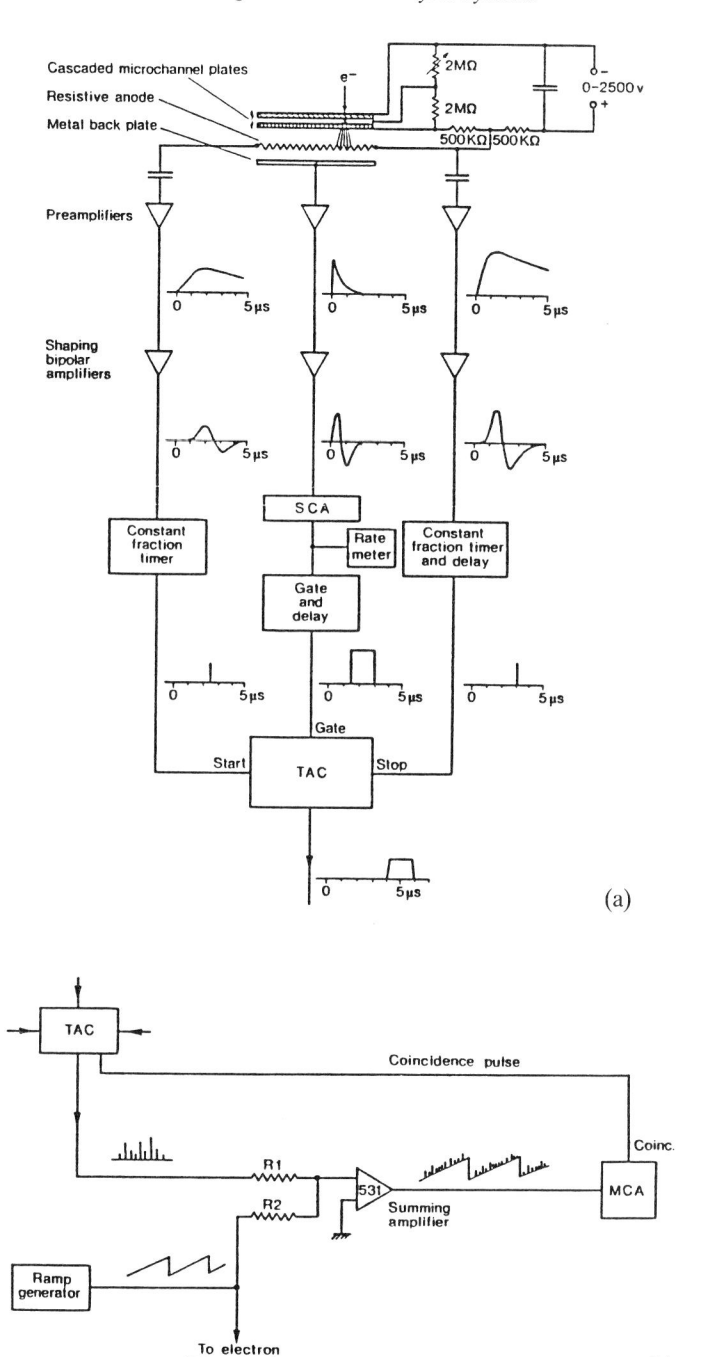

Fig. 7. Resistive anode encoding (TAC-based): (a) read-out system, (b) refinement of (a) employed by MacDowell et al. (1983), (c, overleaf) successor developed at Daresbury Laboratory.

Fig. 7c.

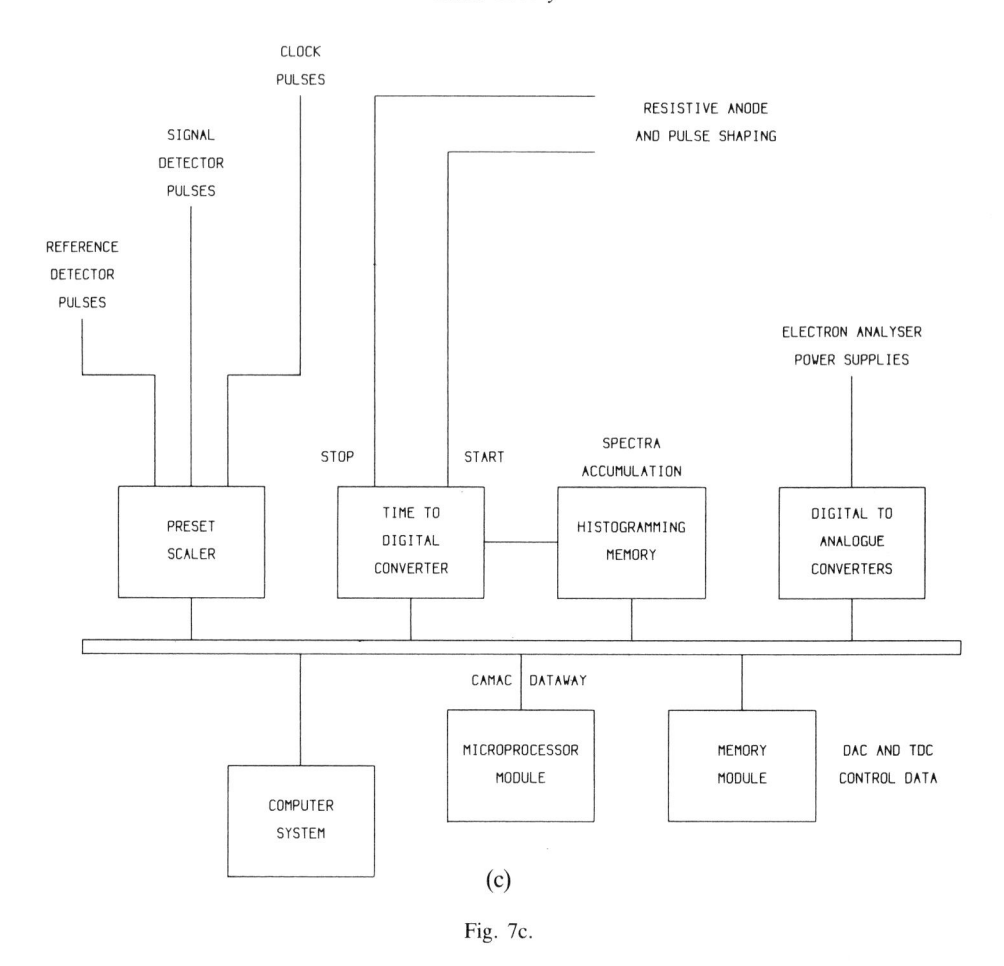

spectrum fixed in memory. This is the multichannel equivalent of the multiscaling system described in section 2.2.3 and fig. 3.

2.2.5. Other methods

The preceeding sections deal almost exclusively with examples of systems handling pulses from detectors. This section makes very brief mention of some other instruments used for detector signal processing.

Streak cameras are used to translate time-varying intensities of SR into positional dispersion, which can then be recorded by a position-sensitive detector. The basic principle is to focus the SR onto a slit in front of a photocathode, and subject the emitted photoelectrons to a high frequency deflecting voltage while accelerating them onto a luminescent screen. The output is recorded on photographic film or by photodiodes. Munro and Schwentner (1983) describe the application of streak cameras to time-resolved spectroscopy. Resolutions of 50–100 ps can be achieved.

Lock-in amplifiers enable low-level AC signals to be measured. An example of their use is described by Hormes et al. (1983), for magnetic circular dichroism spectroscopy. Phase-sensitive lock-in amplifiers may also be used for making phase comparisons between high frequency signals. This method can be applied to the measurement of fluorescent lifetimes, as described by Munro and Schwentner (1983).

Video cameras followed by video digitising systems have been used as read-out sytems for 2D microchannel plate detectors (Weeks et al. 1979). The system has been applied to angle-resolved photoelectron spectroscopy. Photoelectrons are detected and amplified by the microchannel plate and the signal converted to photons by a fluorescent screen. The photons are imaged by the video camera, whose output is then digitised. The digitising may be performed by a commercial image processing system (Rieger et al. 1983, Schnell et al. 1984) so that, for example, the signal-to-noise ratio can be improved by averaging successive frames.

2.3. Motor driving

Almost all synchrotron radiation experiments employ several motors for setting and adjusting monochromator gratings, mirrors and collimators (see, for example, Puester and Thimm 1978, Cerino et al. 1980), and in some cases also for detector positioning. Generally, stepping motors are used on grounds of economy and simplicity, though DC servo motors may be used with advantage in some applications. A brief summary of the relative merits is given below.

2.3.1. Stepping motors

2.3.1.1. Advantages
(a) Readily compatible with digital control systems. A pulse from the control system causes the motor to rotate by one step.
(b) "Open loop" control, no position or speed feedback required.
(c) Inexpensive.

2.3.1.2. Disadvantages
(a) Mechanical resonances occur at particular speeds (but see below on microstepping).
(b) Resolution is determined by the step angle (and gearing). A standard motor has 200 full steps per revolution.
(c) Inefficient overall power consumption. Power is dissipated even when the motor is at a standstill. A reduced "standby" current can be used, but this reduces the ability of the motor to hold the load in position.
(d) Limited maximum speed and acceleration (likely to cause loss of steps if these are exceeded).
(e) No response to changes in load—the torque is roughly constant over the motor speed range.

2.3.1.3. Performance parameters. Stepping motors are normally selected by consulting the speed, torque and inertia data supplied in manufacturers' catalogues. Typical torques range from 0.2 to 15 Nm and are relatively constant with changing speed up to about 1000 steps per second, falling off at higher speeds. The maximum speed and acceleration achievable depend on the type of drive used (see section 2.3.1.6). With appropriate drive, maximum speeds range up to 10 000 steps per second (though at those speeds there is very little torque). The acceleration depends on the motor and load inertia, and friction. The larger these are, the lower the acceleration must be to avoid losing steps. A typical acceleration rate is about 1000 steps per second.

2.3.1.4. The resonance problem. The principle of operation of the stepping motor is that the moving element (the rotor) is driven between sharply defined angular positions by the switching of current from one set of motor coils to another. The inertia of the rotor causes it to overshoot and, in normal operation, produce a damped oscillation. At a step rate close to the natural resonant frequency, the oscillation can become severe and cause the loss or gain of steps. Also, this may generate unacceptable vibration in the experimental equipment (e.g. sensitive monochromator).

2.3.1.5. Multi- and micro-stepping. A way of minimising the resonance problem and at the same time increasing the angular resolution is to employ the technique of multi-stepping or micro-stepping. Whereas in a simple driving system the current in a given coil takes only two values, on or off, multi-stepping and micro-stepping drives provide a finer variation in current to generate intermediate motor positions. Thus, positional change is achieved in less violent jumps. In some examples (called here "multi-stepping") a full step is subdivided into 5 smaller steps. In a micro-stepping drive, each step may be divided into perhaps 50 micro-steps. The current values for each micro-step (sine and cosine values) are held in read-only memory (ROM) and delivered to the coils by digital-to-analogue converters (Galwey 1977, Clout and Johnson 1978). The motor may now require 10 000 pulses per revolution. However, non-linearities inherent in the mechanical construction are such that if micro-stepping is required to yield high angular resolution accuracy, the motor used must be calibrated. A factor to consider with a micro-stepping system is the micro-stepping pulse rate. A typical application might require a travel of 400 revolutions, thus of 80 000 full steps, or 4 M micro-steps with a step division of 50. A pulse generator matched to a standard system of up to 10 000 steps per second may then prove to be intolerably slow, and 100 000 steps per second may be more appropriate.

2.3.1.6. Stepping motor system components. The organisation of the components of a stepping motor driving system is illustrated in fig. 8.

The motor controller is the digital interface between the controlling computer and the motor driver. For example, it might contain registers to be set by the computer for the number of steps to drive (clockwise or counterclockwise) from

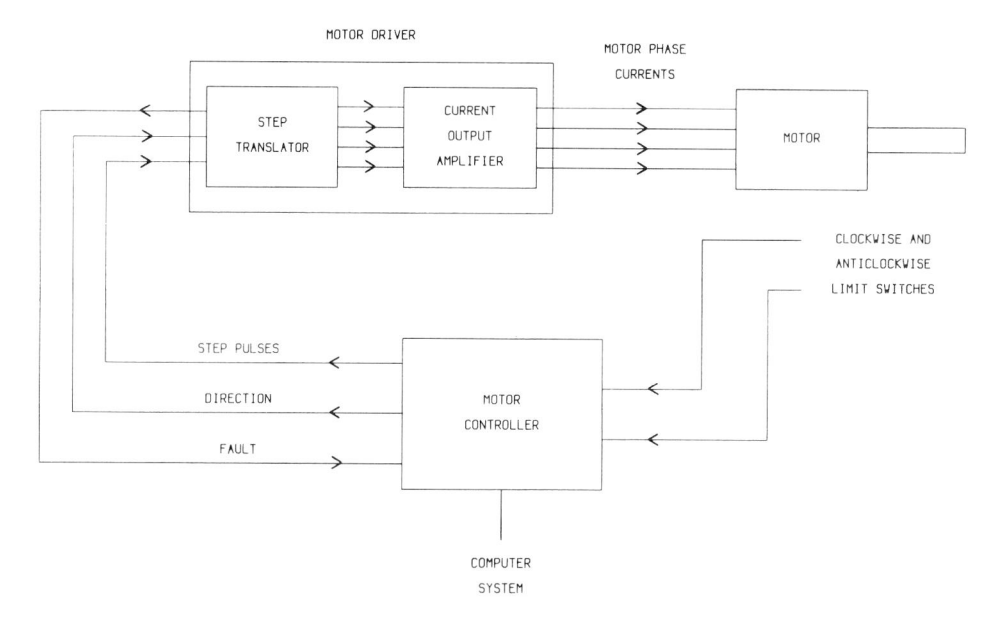

Fig. 8. Stepping motor driving components.

the current motor position, the maximum stepping rate required, the acceleration and deceleration rate, and the starting and stopping speed. Microprocessor-based controllers are now available that can store parameters for a complex series of motions and drive several axes. Such a controller typically communicates with a computer through an RS-232 interface (i.e. through a standard terminal port). Microswitches may be fitted to the driven apparatus to indicate by an open circuit that the limit of desirable (or tolerable) clockwise or counterclockwise travel has been reached. Other switches may be used (placed before the limit switches) to indicate that deceleration must start so that the speed is not too high and steps are not lost if a limit switch is encountered. These signals, together with a fault signal that would indicate a failed or unpowered motor driver, are routed back to the controller. Limit and fault conditions cause the controller to stop sending stepping pulses and interrupt the controlling computer system to demand attention.

The motor driver contains a "stepping motor translator" and a current output stage. The translator has the task of converting the step and direction signals received from the controller into the appropriate phase signals for the current output stage.

Several types of driver output stage are commercially available, including those that contain a micro-stepping facility (Clout and Johnson 1978). The simplest form is the unipolar, resistance-limited drive (a ballast resistor is placed in series with the coil winding). Bipolar drives allow current reversal in the motor winding and improve torque over unipolar drives by about 30%. Chopper drives pulse the current on and off and dissipate much less power.

2.3.1.7. Backlash. Mechanical gears are often placed between the motor shaft and the driven apparatus to improve the angular resolution or translate the rotational into linear motion. This mechanical linkage will in general have a certain slack, so that, for instance, the first few steps of motor rotation may be required to take up the slack before the final element starts to move. The discrepancy between the actual final position and the target position depends in detail on the relative contributions of friction and inertia in the system (see Kuo and Tal 1978, ch. 2). This effect, "backlash", causes the final position of the apparatus to be slightly different for clockwise and counterclockwise motion for the same target position. It is often necessary to remove this effect by writing motor driver software so that the final position is always approached from the same direction. This is simply accomplished by deliberately "overshooting" when approaching from the "wrong" direction so that the final approach is always from the "right" direction at a slow speed.

2.3.1.8. Stepping motor variants. There are several variants in stepping motors and control systems, for instance 5-phase motors and closed-loop drive systems (Clout and Johnson 1978, Kuo 1979) that offer varying advantages in resonance control, efficiency, and so on. They are not in widespread use and are not considered further here.

2.3.1.9. Do stepping motor systems lose steps? In practical systems, stepping motors do lose steps. In principle, a properly configured system would not do so, but in experimental systems a variety of reasons may conspire to spoil the ideal. The inertial load the system is being required to drive may be larger than originally planned, there may be more friction in the system than anticipated, and the motor may be asked to accelerate too fast or sustain too high a speed for the load. It is in the nature of experimental apparatus that requirements change frequently. If lost steps must be detected and corrected then it is advisable to use a shaft encoder. Software can then implement a form of closed-loop correction by comparing the actual with the target position when the movement has finished and make successive iterations until the system has arrived sufficiently close to target. Introductory material on sensors and encoders used with motor systems may be found in the article by Kuo (1978).

2.3.2. DC servo motors
DC servo motors are widely used in industry, though not frequently in SR experiments and so are only briefly treated here. More information may be found in the work by Kuo and Tal (1978).

2.3.2.1. Advantages
(a) No resonance.
(b) Resolution is not dependent on the motor.
(c) Efficient in power consumption. The linear speed/torque characteristic of the

DC motor allows it to "adapt" to the applied load and only draw the current required.
(d) Very high maximum speed and acceleration possible with no loss of position.

2.3.2.2. Disadvantages
(a) Requires digital-to-analogue conversion for control by a digital system. Appropriate electronics can hide this, however, and present a stepping motor-like interface to the control system.
(b) Closed-loop control is required, involving speed (tachometer) and position (encoder) feedback transducers and associated electronics, hence complexity.
(c) Relatively expensive, because of the extra hardware noted under (b).

In situations where an encoder would anyway be required for a stepping motor system (see section 2.3.1.9) the overhead of closed-loop control in a DC servo motor system may prove to be of little disadvantage. Since the DC servo system continuously corrects itself through hardware, it is more efficient in terms of time and software at doing so than the equivalent stepping system.

2.3.2.3. Performance parameters. Typical DC servo motors have rated speeds of up to 100 revolutions per second and torques of up to 3 Nm.

2.4. Digital interfacing systems

The digital interfacing system typically consists of one or more *buses*. A bus is a set of data, address and control signal paths shared by the interface modules and controlling processors. The use of the word "digital" rather than "computer" in the title of this section is intended to emphasise that more than one controller, not all of which need be fully-fledged computers, may be present.

This section reviews the factors to consider in choosing a digital interfacing system. Since CAMAC is the most widely, indeed almost universally, used system in SR and similar laboratories, CAMAC is briefly described. New "standard" systems are emerging, and there are signs of a slow move away from CAMAC towards them. Some of these systems are briefly described.

2.4.1. Requirements
The first requirement of an interfacing system is *modularity*. The advantages of modularity are outlined in section 2.1, and will not be repeated here.

Closely coupled with this is the requirement for *standardisation*. All the experiments should use the same interfacing system, enabling them to use the same module for the same purpose. The benefits of this are that modules may be interchanged between experiments, and so can elements (e.g. standard subroutines) of data acquisition software. A pool of spare modules can be held to support a number of experiments in case of failures and to quickly meet a new or temporary requirement. The ability to share software, in particular, leads to a considerable saving in time and effort.

Having concluded that a standard must be adopted within the laboratory, the choice of standard will depend on a number of interrelated factors. The standard should be *international, specified in detail, computer-independent* and *manufacturer-independent*. These factors combine to ensure that a large number of suppliers can exist, making available a wide range of interfaces, and that the standard will be relatively stable (not subject to change at the whim of the manufacturer), and sufficiently well-specified that modules from different manufacturers can be mixed freely in a system. The choice of standard will depend heavily on the current and anticipated *availability* of a wide range of modules suitable for the laboratory applications envisaged.

Ideally, the hardware standard should be developed in parallel with a corresponding software standard for the (software) interface between an application program and the basic system software. Developing these two together will result in a better, more useful standard. The existence of a standard software interface will promote the portability of software between systems, indeed between laboratories.

The system must have sufficient *speed* and *flexibility* to meet the needs of the range of the applications envisaged. Among the factors to consider are the following.

(a) The data transfer rates—what are the widths of the data paths (i.e. how many bits are transmitted in parallel) and what frequency of transfers is possible?

(b) The addressing structure—how wide are the address paths, and is it possible to address individual data elements in a module (e.g. a single word in a mass memory) directly?

(c) Support for multiple processors—what mechanisms exist for processors to gain control of the bus ("arbitration" mechanisms)? What mechanisms are provided for processors to communicate with each other? Can any processor access any interface module in the system?

(d) Maximum size—what is the maximum number of modules the system can have? How far apart can the modules in the system be?

The criteria against which the answers to these questions are to be judged naturally depend on the field of application. Most, but by no means all, VUV and soft X-ray experiments are relatively undemanding in data rates. Some exceptions have already been mentioned in sections 2.2.4 and 2.2.5. Examples of the use of an auxiliary CAMAC crate controller (thus simple multiple processor systems) have been given in sections 2.2.3 and 2.2.4.

If the benefits of standardisation are to be realised, it must be the exceptional case that an alternative interfacing system is required to satisfy an application. Thus it must be accepted from the outset that the power of the system will be more than is required in many applications, in order that the standard may be maintained while the more demanding applications are satisfied.

The *cost* of units in the standard is clearly an important factor, but a realistic appraisal must take into account not only the *direct* costs of the units (purchase prices) but also all the *indirect* costs associated with in-house design, documentation, engineering, maintenance, and software effort [Clout (1976) explores some of these points in a quantitative manner].

Some compromise must be reached between these various related, and to some extent competing, factors. A single standard does not exist that will meet *all* the requirements within a reasonable budget and without significant in-house design effort. Different standards have strengths in different areas, therefore, in practice, laboratories use a range of standards, overlapping as little in functionality as possible. There are usually "bridging" units between systems (see e.g. CAMAC-IEEE 488, VME-CAMAC, CAMAC-FASTBUS) in order to allow sub-systems to communicate and co-exist to a limited degree in the same system.

2.4.2. CAMAC

CAMAC is worthy of special mention because of its continued very widespread use in medium and large research laboratories throughout the world, including SR laboratories. The reader may find, therefore, an introduction to CAMAC jargon and properties useful.

2.4.2.1. CAMAC basics. CAMAC is defined in a series of documents describing different aspects of the system. One series of documents is published in Europe, under the aegis of the ESONE (European Standards on Nuclear Electronics) Committee, and a parallel series with identical content in the USA, published by the US ERDA (Energy Research and Development Association) NIM Committee. The major documents have also been published by the IEC (International Electrotechnical Commission) in Europe and IEEE (Institute of Electrical and Electronic Engineers) in the USA. These documents are listed in the reference section at the end of the chapter. A collection of the major documents has been published in book form and there are several excellent tutorial articles published.

CAMAC modules are housed in a "crate" with up to 25 "stations", each station having connection to the "dataway" (backplane bus). Station 25 (or station 24 in a crate with 24 stations) is specially wired, so that it and the adjacent normal station can be occupied by the "crate controller" (see ref. CAMAC-3). The special wiring of the end station gives the crate controller access to signals that are specific to each station rather than bus lines, shared by all. At each normal station there is an 86-contact connector that supplies the module with power and gives it access to the signal lines. Two of these lines are for private communication with the module (i.e. inform the module that it must perform the operation defined by the bus signals) and the other (L) for the module to signal an interrupt (see section 3.3.2.5 for an explanation of interrupts) to the controller. Interrupts in a CAMAC system are known as LAMs ("Look At Me"). Most of the other lines are shared by all modules and include 24 write data lines, 24 read data lines, 5 function (F) lines (providing for 32 different read, write and control operations) and 4 sub-address (A) lines used to address features (such as specific registers) within a module.

Dataway transfers are synchronised by two strobe signals S1 and S2, whose timing is such that cycles (i.e. successive operations) can be executed at about 1 microsecond intervals (non-standard short cycles of about 400 nanoseconds are also possible for specially designed modules). Thus CAMAC data throughput can be as high as 3 Mbytes/s (24 bits = 3 bytes).

The crate may contain more than the one source of control. If so, the controllers ("auxiliary crate controllers") are connected to the unit occupying the control station (the "end-station controller") by a supplementary bus, the ACB (Auxiliary Controller Bus, see ref. CAMAC-9). Examples of the use of an auxiliary controller are given in sections 2.2.3 and 2.2.4.

A large system may contain many CAMAC crates. Two standard highways for linking crates via their controllers are defined. The parallel highway (see ref. CAMAC-5) is suitable for short distances (less than 100 metres) and high data rate applications (the basic CAMAC cycle including highway overheads becomes about 1.5 microseconds). Up to 7 crates may be interconnected on this highway, and a given computer may drive more than one such highway if more crates are required. Each highway would be driven by a computer-specific "branch highway" driver module. The crate controllers on the highway, "type A", are defined by the standard.

The CAMAC serial highway system (see ref. CAMAC-7) is able to drive up to 62 crates, and is suitable for systems with long inter-crate separation. This is not frequently a requirement for experimental data acquisition systems, however, but serial highway systems are commonly used in the accelerator control systems of large laboratories. As usual with serial transmission systems, there is a compromise between data rates and distance. For the serial highway, distance may be more or less unlimited since data communication modems and telephone lines may form part of the route. Serial highways may be "byte-serial" (8 bits in parallel per clock period) or "bit-serial" (one bit per clock period), and the clock rate may be up to 5 MHz. As for the parallel highway, the specification defines a standard crate controller, "type L" (see ref. CAMAC-8), the highway being driven by a computer-specific branch driver.

For small systems (up to two or three crates, say) it is common to avoid the complexity and costs associated with the combination of branch driver and computer-independent crate controllers by coupling the controllers directly to the computer I/O bus. These controllers are known as "type U" (for "undefined").

For systems with auxiliary crate controllers, ref. CAMAC-9 defines their organisation within the crate but does not define a method for an auxiliary controller to access modules in another crate. Daresbury Laboratory has developed a multicrate link system to overcome this limitation (Alexander and Howson 1983).

Standards for CAMAC-driving software have also been defined. IML (see ref. CAMAC-10) defines the primitive functions for performing CAMAC operations in a computer- and language-independent way. Real-time Basic for CAMAC (see ref. CAMAC-11) is an implementation of ANSI standard real-time Basic, intended for process control applications, where the process control hardware is in CAMAC. Fortran is the dominant language in the scientific community, however, and for data acquisition programs written in Fortran it is common to use the Fortran implementation defined in "Subroutines for CAMAC" (ref. CAMAC-12).

2.4.2.2. CAMAC strengths and weaknesses. The sucess and longevity of CAMAC (the basic standard was first published in 1969) is a most striking example of successful international collaboration in the scientific community, and greatly to the credit of its pioneers. It was the first truly international computer- and manufacturer-independent standard for computer interfacing. The standards were developed for users by users and many of the very wide range of modules available from the many manufacturers were specified wholly or in part by users. Many CAMAC modules, then, are available to meet the specific needs of scientific research. To give a quantitative context to this, and a measure of what would be demanded of an alternative, in mid-1985 Daresbury Laboratory was using about 700 CAMAC modules of which there were 60 functionally distinct types on SR experiments alone. Approximately one quarter were designed in-house.

CAMAC has evolved complementary, subsidiary standards to meet changing and broader needs (e.g. serial highway, auxiliary controller bus). These "add-on" standards may lack the elegance and generality that would be expected of facilities designed-in from the start, but they work and enable laboratories to build on their past work and investment.

With the benefit of hindsight, it can be seen that certain aspects were not defined early enough or in sufficient detail, leaving too much freedom of choice for the module designer. Many of the details of interrupt handling, for example, identifying the source of interrupt within a module, must be left to the application where it would be preferable to do more at the system software or hardware level. There are many different computer interfaces (type-U controllers and branch drivers), even for the same computer, with differing features and philosophies. The requirements of software, and eventually the software standards, were not fully considered until long after the basic hardware standards were developed. These factors have inhibited the portability of software.

For very demanding applications, CAMAC shows its age in being unable to match the capabilities of modern high performance microprocessor systems. The data paths are 24 bits wide, rather than the 32 bits now becoming common. The addressing structure, in particular, is very restricted (up to 23 modules in a crate, with 16 sub-addresses per module). A processor needing to read from a specific location in a histogramming memory, for example, must first perform a write operation to set up an address pointer within the module before executing the read cycle. An attempt to develop a subsidiary standard (COMPEX) to overcome this restriction by operating the dataway in another mode (principally using the 24 write lines as address lines) has not met with widespread support. The CAMAC data transfer rate of 3 Mbytes/s (24 bits at 1 MHz) is also inferior to that of modern high performance bus systems. Though the auxiliary controller bus provides for multiple controllers in a CAMAC crate, the facilities for interprocessor communication etc. are rudimentary by comparison with those designed into VME and Multibus-II.

In practice, the shortcomings of CAMAC are often by-passed by building more capability into the interfaces used. For example, a TDC may communicate with a histogramming memory by a "private" front panel bus, rather than over the dataway. The emergence of alternative high-performance bus systems (see next section) now makes this less attractive than the use of a bridging module (e.g. CAMAC to FASTBUS), if the system requires CAMAC anyway, to a system better able to meet the requirement directly. It seems unlikely that new CAMAC-related standards will emerge to extend its realm of application.

For undemanding applications requiring simple cards common in process control applications (e.g. 12-bit ADCs, DACs, low-speed digital I/O), CAMAC modules are relatively expensive. This is particularly striking in small microprocessor-based systems where the cost of the basic CAMAC crate plus controller may dominate the cost of the system.

ESONE recognised this problem and began the development of a new bus standard, but abandoned the effort as it became clear that competing buses were emerging from industry that were more likely to attract the cost and availability benefits of larger scale production.

Scientific experiments, however, often require rather specialised modules, such as the multichannel, multihit TDC mentioned in section 2.2.4, for example. The cost of such modules will be relatively high in whatever interfacing system is chosen, since the market will be small and development costs correspondingly spread over few units. It will be several years before the many modules available in CAMAC for scientific data acquisition are readily available in other bus standards.

2.4.3. Other interfacing standards

2.4.3.1. ANSI/IEEE-488. This bus, pioneered by Hewlett-Packard under the more familiar name of the General Purpose Interface Bus (GP–IB or HP–IB), has been compared to CAMAC on a number of occasions (Horelick 1975, Merritt 1976, Clout 1976, and Sachs 1979). The bus is a cable interconnecting up to 15 devices that can communicate with each other in a character by character manner. Typical units are bench-top programmable measuring instruments, such as digital voltmeters. The bus lacks the data rate and expansion capability of CAMAC, and since it carries no power, units generally have the overhead of their own power supply. Of course, a unit on the bus can be a crate with plug-in modules. One such unit is a CAMAC crate: thus at least one CAMAC crate controller exists that may be used by any computer with a GP–IB interface to control CAMAC. Similarly CAMAC interfaces exist to drive the GP–IB and thus to take advantage of digital voltmeters etc. available on the GP–IB.

2.4.3.2. Small system buses. There is a large number of bus systems suitable for small scale 8- or 16-bit microprocessor control with inexpensive cards oriented towards process control and monitoring applications; over 60 in Europe alone. Most of these buses are manufacturer dependent. Of those that are not, S100, STD and G64 stand out as the most widely used. S100 started as a simple system but has changed radically in becoming an IEEE standard (IEEE 696), and is now a high-performance bus, with corresponding expense. Many non-standard boards exist. STD is very popular in the USA and exists in two mechanical forms, one of which is in Eurocard (STE bus). (Eurocard is a popular mechanical packaging system for digital electronics and many suppliers exist.) Standards IEEE P961 and P1000 are proposed for STD and STE.

In Europe, G64 has attracted particular interest in laboratories, especially since CERN and the UK SERC (Science and Engineering Research Council) have adopted it as a standard (ref. G64-Bus).

2.4.3.3. High performance buses. Several bus systems are emerging for high performance applications. This section will refer to FASTBUS, VME-Bus and MULTIBUS-II, as these are currently the best established. A far more comprehensive review is given by Muller (1985). FASTBUS has arisen in the high-energy physics community, and VME-Bus and MULTIBUS-II are gathering momentum as buses developed and promoted by groups of manufacturers. In addition to these, a manufacturer and processor independent bus called FUTUREBUS is being proposed for standardisation as IEEE-P896, but at present it is unclear what commercial support this will receive.

High-energy physics experiments have frequently demanded performance beyond that available from conventional industrial systems. As far back as 1977 it was recognised that multiprocessor, high-speed data transfer systems would be needed and that CAMAC would not meet that need (e.g. Sendall 1985, von Ruden 1985). The FASTBUS system, to be published as IEEE STD 960, has two standard buses: the backplane segment (defined within a crate for housing modules) and a cable segment (used to interconnect crates). The interconnections can be complex, and multiple processors are supported. The bus carries 32 bits of address and 32 bits of data. For single transfers, where address and data must be specified, the data rate can be up to 40 Mbytes/s and for block transfers (data only), up to 80 Mbytes. Introductions to FASTBUS may be found in the work by Rimmer (1985) and by Wadsworth (1980).

VME and MULTIBUS-II each consist of a number of related buses, each bus within the group having a distinctive function. Thus VME has 8, 16, 24, and 32 bit transfers, up to 57 Mbytes/s for general purpose use. VMX bus is for fast (up to 80 Mbytes/s), transfers intended for communication between a subset of modules, for instance a processor addressing memory, without competing for the VME bus. VMS bus provides a serial highway (3.2 M bits/s), for interprocessor communication (see ref. VME-Bus).

MULTIBUS-II, similarly consists of the iPSB (8, 16, and 32 bits at 10 MHz) for general purpose operation, iLBX (48 Mbytes/s for high speed access (e.g. to memory) for groups of modules, iSSB (2 Mbits/s) for interprocessor communication, and the iSBX and Multichannel DMA I/O buses carried forward from its predecessor, MULTIBUS-I (IEEE 796).

At the time of writing (1985), VME bus, being available earlier, has more products available. However, for both system, the emphasis seems to be on *system* devices (processor boards, disc interfaces, communications boards), plus a few analogue I/O boards suitable for industrial process control. This helps to fuel the enormous growth in the availability of "super-micro" computer systems, but as yet offers little of direct interest to scientific data acquisition. Both systems, too, in their present form are restricted to single crate sytems and define no way of linking crates, unlike CAMAC and FASTBUS.

3. Local computer systems

3.1. Introduction

In a large laboratory the experimenter is likely to encounter a variety of computer systems, ranging from small microcomputers to large mainframes, used for a variety of purposes and often organised in a hierarchical network of systems (see, for example, Zacharov 1976a). This section is concerned with computer systems that are local to the experiment and are responsible for the experiment control and data acquisition functions.

Computer systems have been used since the earliest days of SR experiments. The main reasons for their use are

(a) They enable data to be collected much faster than by manual means. Without this speed, some experiments may not be feasible.

(b) They can be programmed to provide complex control functions and measuring sequences.

(c) They are readily adaptable to changes in procedures and apparatus.

(d) They can reduce and display the data in meaningful terms as they are required. Appropriate feedback to experimenters can eliminate bad data and guide them in making efficient use of beam-time.

3.2. Dedicated or shared?

The local computer system is normally dedicated to a single experimental station rather than shared between stations. There are several advantages to this arrangement. The users of a station may work (at least as far as the data acquisition system is concerned) completely independently from the users of other stations. Thus, they do not disturb each other by turning the power off the system to install or replace interfaces, or by inadvertently disconnecting the wrong cable, or by "crashing" the system with faulty software. One experiment does not compete with another for the available computer power. With a dedicated and reasonably mobile computer system, major new experimental apparatus or a new experiment can be developed and tested off-line (sometimes off-site) before being brought to the beam line. It can be removed for further development or for use with a laboratory source without disruption to other stations.

3.3. Computer system requirements

3.3.1. Introduction

This section will attempt to identify the features of local computer systems that are significant to VUV and soft X-ray data acquisition systems. It does not pretend to provide a condensed course in computer science, though some basic concepts and terminology are introduced, nor would it be feasible to present a thorough discussion of the merits of different types of system. Computing technology changes rapidly, and so correspondingly do the facilities that may be afforded for data acquisition systems. Not many years ago, systems in use were first generation minicomputers with 8 K words of memory, paper tape for data output and program input, with programs, including a home-built operating system executive, laboriously written in assembler language (for example, Clout 1975, Wall and Stevenson 1978). Today, professional microcomputers offer several hundred kilobytes of memory, hard and exchangeable disc systems, high-level languages and manufacturer-supplied operating systems at less cost than those earlier systems. It is reasonable to assume that a similar dramatic increase in facilities will take place over the next decade. Also, though the demands of VUV and soft X-ray experiments are not normally very great, this may very well change. The use of dedicated SR sources and devices, such as undulators, will no doubt place increasing emphasis on higher data rates, area detectors in place of single-channel counting and so on, with a corresponding increase in demands on computer systems.

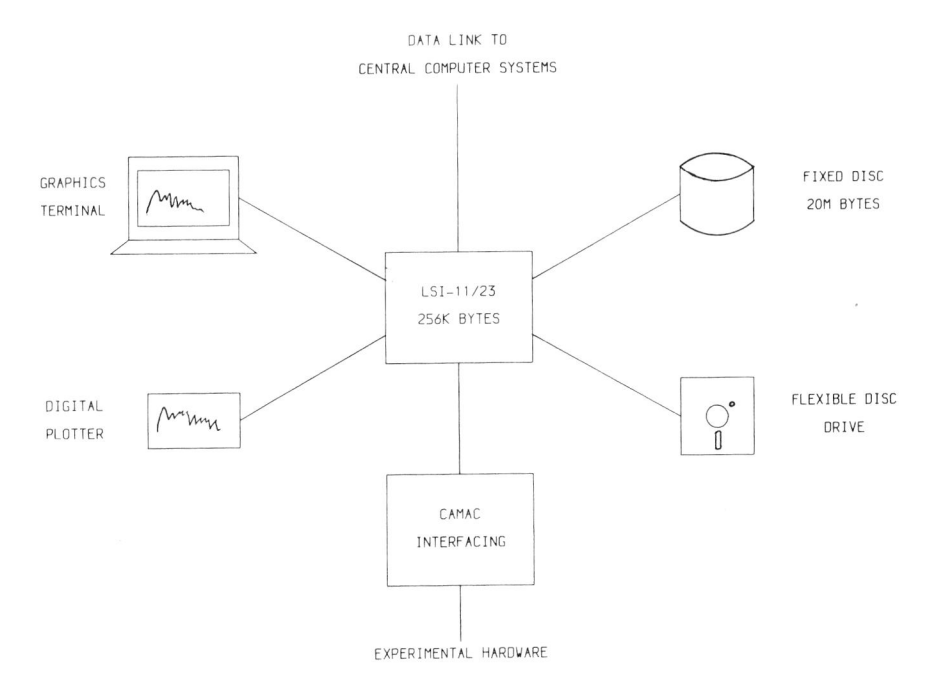

Fig. 9. Local computer system for data acquisition.

3.3.2. Basic hardware considerations

3.3.2.1. A typical system. Figure 9 shows a typical local computer system. It will be used to illustrate a number of aspects of local computer systems discussed in this section.

The system shown is a "single-user" system with a single terminal which is used to control the system and display the acquired data. If there is more than one experimenter present at the station (and most experiments are manned by one or two people at a time), then it is desirable for facilities to be available for at least a second experimenter to be examining and analysing recently acquired data. "16-bit" minicomputers (the term is explained in section 3.3.2.3), such as shown in fig. 9, do not have much capacity for multi-user access and the demands of data analysis at the same time as providing efficient data acquisition. For the system illustrated, the need for members of the team to be analysing data would be met by the provision of a graphics terminal located by the station, accessing a more powerful, remote multi-user computer system to which the data have been transferred over a mini/micro-computer systems with greater direct memory addressing capabilities, and the ability to share expensive peripherals over a high-speed local area network (see section 4.4) increasingly make it attractive to distribute this function to computers located at the experimental station.

3.3.2.2. Peripherals. Most VUV and soft X-ray experiments are spectroscopic in nature, collecting one-dimensional spectra of from 200 to, say, 1000 channels. For these applications, a graphics terminal with a resolution of 1024×768 points is adequate, interfaced by a serial line (RS-232) to the computer. A simple digital plotter is useful for obtaining hard-copy plots of raw data. Some experiments are more demanding, however. For instance, Schell et al. (1984) (also Rieger et al. 1983) describe a data acquisition system used in photo-electron spectroscopy where the data is two-dimensional, electron intensity and angle for a fixed kinetic energy. Here an on-line image-processing system signal-averages the incoming data to increase the signal-to-noise ratio and displays processed images of 256×256 points on a video monitor.

A typical disc configuration would consist of a 20 Mbyte "Winchester" fixed disc drive and a flexible (better known as a "floppy") disc drive. A Winchester disc drive is a sealed unit, protected against the entry of dust into the area around the disc surface and read/write heads, thus considerably reducing the incidence of catastrophic "head-crash" damage to the unit. This is an important consideration in the dusty environment often found at an experimental station. The disc would be used to store the operating system software, user programs, and experimental data. In VUV and soft X-ray experiments, data files are not usually very large, in the region of 2 to 20 Kbytes. However, at SR laboratories files may be acquired very quickly and it is convenient to be able to hold a large number locally. The second main reason for using a hard disc drive is that the speed of access and data transfer rate from such a disc is very much greater, of the order of a factor of 10, than from a low-performance device such as a floppy disc drive. This speed of access is needed if program overlays are required (see section 3.3.2.3). The process of program preparation also requires many disc transfers and is far more efficient on a hard disc system. A further benefit of hard disc capacity is the availability of sufficient space to hold useful text information. This would include details of operating system commands supplied by the manufacturer (usually called HELP text) and experimental station software and general documentation. The role of the floppy disc is to keep "backup" copies of programs in case of their accidental loss from the fixed disc, and to provide a transfer medium between systems that is not completely dependent on network availability. It is not normally necessary to provide a "backup" medium of a size comparable to the fixed disc, since most data files are temporary and soon transferred to another system over the local network. However, it is necessary for the system to have an exchangeable medium sufficiently large to accomodate enough of the operating system to load itself in the event of its loss from the fixed disc, and on large systems this may be several megabytes.

3.3.2.3. Word sizes and addressability. Computer systems are often described by a data length, for example 8-bit (small microcomputers), 16-bit (small minicomputers), 24-bit and 32-bit ("supermicros", "super-minicomputers", mainframes). This characterisation is oversimple and ambiguous, but some appreciation of its meanings and related topics is helpful in considering the type of machine needed for a given application. The topics relate to the internal "architecture" of small computer systems. Some references are Dobinson (1985) and Lippiatt (1978).

The basic unit of computer information storage is the *byte*, a group of 8 binary digits (bits). Most modern computer systems are able to address bytes easily, transferring them as an entity to and from memory, CPU (central processing unit) and peripherals. Computers differ in the widths of the transfer paths, or *buses*, they have, that is to say, in the number of bytes they are able to transfer in parallel in a single operation. A machine instruction, including the operands it requires such as memory addresses, in general consists of several bytes, and is thus slow to fetch and execute on an "8-bit" system, where bytes are obtained in series. A "16-bit" system would have data paths of at least 16-bits wide. In general, a computer system has several internal data paths, of various widths. Thus a system advertised as being "32-bits" may have certain paths within the processor of 32 bits width, but perhaps a path to memory of only 24 bits or 16 bits, with a consequent smaller memory addressing capability.

The number of bits in an instruction available to specify the address of an operand in memory may be greater than, equal to, or less than the width of the pathway between CPU and memory. A program running in the system shown in fig. 9, for instance, can generate only 16-bit addresses and, since each address corresponds to a byte of memory, can only directly access up to 64 Kbytes (1 K = 1024 in computer parlance). Programs that have larger memory requirements must be overlaid (see below), with consequent complications in program preparation and penalties in run-time performance. In this system the number of address bits on the bus is 18 or 22 (depending on model), so the system as a whole can have more than 64 Kbytes. The extra memory might be used for data arrays, as a program overlay medium, or for other programs. The hardware responsible for converting the addresses specified in the program to addresses on the memory bus is called the memory management unit.

On a system such as that in fig. 9, if the program is larger than the memory it can address directly, it must be *overlaid*. An overlaid program is divided into portions, some of which will be on disc and not in memory at any instant during execution. When a portion that is not currently in memory is required, it must be read from disc and the portion now temporarily not required removed (copied to disc if modified from the previous copy on disc). The overlay structure of the program must be planned and implemented explicitly by the programmer.

Larger computer systems (24-bit and 32-bit) may have have *virtual storage* capability (see Lister 1984). Such systems are designed to enable programs to be written as though a very large memory is available, even though the quantity of main memory present is limited, without the need for the programmer to explicitly partition the program into overlay sections. The system views the program as consisting of a number of pages, each page being of the order of 1 Kbytes, and the operating system and hardware together ensure that pages are brought in from disc when required by the program in place of pages not currently in use (*demand paging*). This may have undesirable performance consequences for a real-time system (see section 3.3.3.2) unless the operating system provides appropriate facilities.

Since in practice data acquisition programs typically exceed 64 Kbytes in size, computers with direct memory addressing ranges of greater than 16 bits, now inexpensive, would be a more natural choice for data acquisition computer systems.

A *word* is, in modern computers, a unit of storage consisting of a number of bytes, usually from 2 to 8, and some computers will support a number of word lengths. The internal data paths should be wide enough to handle the entire word in one transfer, thus speeding up arithmetic operations, for example. Counting rates in SR experiments are frequently such that more than 16 bits are needed to represent the contents of a spectrum channel, and a system with only 16-bit words tends to be clumsy in handling such spectra. Floating-point arithmetic is normally performed on 32- or 64-bit words and its performance will again depend partly on the widths of internal data paths.

3.3.2.4. Programmed I/O and DMA. The simplest form of I/O (input and output operations) between a computer and a peripheral is via "programmed I/O". The program initiates the transfer of each word and may require to execute several instructions for each word transferred. A DMA (direct memory access) device, however, has sufficient logic to generate successive memory addresses and perform the transfer of a block of data, software being required only to initiate the transfer of the block. This is in general a much faster method of transferring blocks of data, leaving the CPU free to execute other software. Disc drive controllers, for example, are normally DMA devices. DMA is useful between experimental interfaces and the computer if large blocks of experimental data are to be transferred, for instance, a two-dimensional spectrum from a dedicated histogramming memory. This is only exceptionally the case in present day VUV and soft X-ray experiments. Sachs (1979) gives an expanded discussion of the use of DMA in laboratory computer systems.

A related facility is DMI (direct memory increment), in which the experimental hardware (analogue to digital converter, for example) may present a number (e.g. representing pulse height) that the DMI controller will use to specify an address in computer memory to which 1 must be added. Thus a pulse height spectrum may be directly accumulated in computer memory.

3.3.2.5. Interrupts. Interrupts are the primary method whereby a data acquisition program synchronises itself to events in the experiment. An interrupt is a signal on the computer I/O bus that has the effect of causing the CPU to suspend execution of the current program and begin executing a different piece of software. This new program may do all that is necessary for responding to the interrupt, or may send a message to the data acquisition program to do so as soon as it resumes execution. Interrupts are used extensively to signal between the standard computer peripherals (discs, terminals and so on) and the CPU. In data acquisition systems they are typically used to signal the end of a counting interval, for example, or that a motor movement, which may take a long time, has completed, or that an alarm condition has arisen (a motor limit switch encountered, for example). Further discussion of interrupts and their handling may be found in the article by Baumann (1972).

3.3.3. Basic software considerations
This section discusses aspects of the local computer operating system, that is, the collection of basic system software, utilities, and run-time support that is needed before an application such as a data acquisition program can be prepared and run.

3.3.3.1. Program preparation
Basic elements. The major software elements needed for program development are the following.

(a) An editor. An editor enables the programmer to type in a program, store it in a file, and subsequently modify the text. Simple text editors work in a line-oriented mode, whereby the editor handles one line at a time and must step on, or be guided, to the next line for modification. Increasingly, full-screen editors are available, enabling the user to move the terminal input cursor freely around the text displayed over the whole screen, with considerable gains in user efficiency.

(b) Compilers. A compiler translates a program written in a high-level language, for example FORTRAN, into "object code" suitable for input to the linker (see below) (or exceptionally into assembler input; see below). An application program such as a data acquisition or analysis program would almost invariably be written in a high-level language.

(c) An assembler. An assembler translates a program written in a low-level language (assembler language), an alphanumeric and mnemonic form which has a mainly one-to-one correspondence with the basic machine instruction set, into a binary code representation of the instructions to be executed. The output is then in a form suitable for input to the linker (see below). Normally, only system programs would be written in assembler language, and even then there is an increasing tendency towards the use of a suitable high- (or medium-) level language (e.g. the C programming language).

(d) A linker. A linker accepts an object program, combines it with the object code of any previously compiled or assembled routines required by the program, and outputs a file that is in a form ready for program execution (an "executable image").

Interpreters. Some high-level languages are "interpreted" rather than compiled, the most common example being BASIC. In these cases, the interpreter reads the sources program statements and, usually after some small degree of transformation for the sake of efficiency, translates and executes each statement as it is encountered in the program flow. Thus a statement may be translated many times in the course of program execution, with a consequent penalty (of the order of a factor of ten) in execution time by comparison with a compiled program. Though the speed of an interpreted program may be unsatisfactory for production use, some interpreters are supported by a compatible compiler. Effective use may then be made of the interpreter's main strength—the development cycle of the new program is short, consisting only of editing, interpreting and testing phases (cf. editing, compiling, linking, running and testing). If a compatible compiler exists, then this would be used when the program is well tested and ready for production use.

"*Cross-*"*software*. There is a choice to be made between preparing programs on the local computer on which they will run, and using a separate system. For instance, a multi-user "host" computer system may be used to edit, cross-compile (i.e., use a compiler that produces code suitable for the target computer rather than the host), cross-link (combine it with other object code for the target computer), test the program using software that simulates the target machine and interface hardware, and transmit the resulting program to the local computer for continued testing and execution (see Pyle 1976).

Some years ago, the use of a powerful host system with cross-software was particularly attractive because of the lack of adequate program preparation software on minicomputer systems. This is no longer the case on systems of the capabilities relevant to SR data acquisition systems. Rather, it is preferable to be able to develop programs on the local system to shorten the development cycle of editing, compiling, linking and testing, and to retain the flexibility to develop software when the system starts at, or is moved to, some location (see the preceding section) where communication with the host computer is difficult. However, the local system is frequently unavailable for development purposes because it is in use taking data. A large laboratory is likely to have similar systems reserved for program development, but if these are single-user systems then there may well be competition for their use.

A facility that has proved useful at Daresbury Laboratory takes advantage of the availability of a multi-user central computer and the fact that data acquisition programs are written in a high-level language (FORTRAN). A library of data acquisition simulation or "dummy" routines exists on the central machine so that programs may be developed and partially tested there. The FORTRAN source files are then transferred to the local system for the later stages of testing and refinement.

3.3.3.2. Run-time support and real-time considerations. Data acquisition systems are examples of "real-time" computer systems. A "real-time" system is normally defined as one for which the response to external events must take place rapidly if the system requirements are to be satisfactorily met, though how rapidly this must be, and the consequences of failing to meet the time constraints (lost data, for example), naturally vary from system to system. In a data acquisition system, the external events are signals from the experimental hardware, or a user typing commands at a keyboard. From the viewpoint of the computer, these signals are interrupts (described in section 3.3.2.5). VUV and soft X-ray data acquisition systems are not usually very demanding of response to interrupts. From the system examples given in section 2.2, it can be seen that most of the requirements for rapid response to incoming data can be met by the hardware interfacing system.

Real-time system requirements demand real-time features in the operating system (Clout 1982). Operating systems that do not claim to be "real-time" may nevertheless possess the necessary features required by an undemanding experiment. However, systems designed for business and terminal-intensive applications or for "personal" use do not necessarily have the required facilities and need to be carefully examined. The variants of the wide-spread UNIX operating system illustrate this point. Some have been designed to incorporate real-time features while retaining a UNIX appearance to the terminal user, but most implementations lack the concepts of high priority real-time processes and "locking" of a process in memory (see below). (For further comments on the use of UNIX in a real-time environment, see Lee and Wiegandt 1983).

Interrupts. For the purposes of a data acquisition program, it is necessary that the operating system should allow a user program to "connect" to an interrupt. This means that the program must be able to instruct the system to execute a particular user-supplied routine in response to an interrupt from a specific interface in the data acquisition hardware. In a single-user system it is feasible to dispense with interrupts from experimental hardware if the experiment is sufficiently simple. The data acquisition program may "poll" the hardware, meaning that it may, when waiting for work, continuously examine the status registers of the devices from which signals are expected, each in turn. This may be advantageous in terms of speed, since the time the system would take to activate a waiting program after an interrupt may be several hundred microseconds. In a multi-user or multi-tasking system (see below) however, this would be unsatisfactory. A high-priority data acquisition process looping round testing status registers would be unnecessarily occupying the CPU when other users or processes could be doing useful work, and a low-priority program may be slow to detect an event since other interrupt-driven tasks may be executing.

Multi-user systems. In a multi-user operating system, it is usually necessary to access the data acquisition hardware through a piece of system software called a device driver. The device driver would probably not be provided by the operating system supplier, though he may provide a useful "skeleton" driver as a basis for development. The driver need be written only once for a given combination of computer system, data acquisition hardware and operating system.

Data acquisition programs send messages to the driver in order to have I/O operations performed. This prevents the chaos that might otherwise ensue if several user programs all "simultaneously" attempt to perform operations on the hardware. In such systems, the device driver may implement a "booking" facility whereby programs may request exclusive access to particular interfaces. Since all operations are performed by the device driver, it can ensure that all such reservations are respected. However, the sending of messages between a data acquisition program and a device driver can dramatically increase the time taken to perform I/O, in some cases to some milliseconds (Davies 1976) where direct access from the application might take only a few microseconds. As a consequence, to minimise the effect of this overhead the device driver should have the capability to accept and execute a list of I/O operations on the instruction of one message (e.g. de Laat and Kromme 1985).

Multi-tasking systems. Many operating systems offer "multi-tasking" facilities. The operating system shares the hardware resources (CPU, memory and so on) amongst a number of "tasks" (also known as processes). Some will be system processes and some user processes. A given user may organise his data acquisition software into a number of tasks, which are able to share data easily, and which can run more or less independently, in parallel with each other. On a conventional machine with only one CPU for application processes, only one process can be executing at any instant, so the impression of parallel activity is somewhat of an illusion. The operating system suspends a process that is waiting for some event (an interrupt, for instance, or the completion of a read operation from disc) or has exceeded its "time-slice". *Time-slicing* ensures that a CPU-bound process (one that performs relatively little I/O and so rarely needs to wait for an event) is nevertheless suspended after a fixed interval so that other processes may obtain a share of CPU time.

Some simple systems select the next process to execute in a "democratic" way, choosing the next process in the list in a cyclic fashion. In a real-time system, however, a "priority" system must be implemented, so that a time-critical process will be run as soon as possible when it is required to respond to an event. Thus it would be selected ahead of lower priority processes and any lower priority process that is executing would be suspended immediately to make way for the real-time process. The part of the operating system responsible for process selection and the implementation of the priority scheme is usually called the *scheduler*.

The advantage of a multi-tasking system to a complex data acquisition system is that the software may be written in a modular fashion, each task having a well-defined and relatively simple role within the complex whole. The price to be paid for this simplification is the increased complexity arising from the need to communicate between and synchronise the co-operating tasks. The principal criterion for determining whether the benefits of multi-tasking are likely to outway the disadvantages is the degree of concurrent activity required (Pyle 1976), for instance the number of different sources of interrupt that must be handled. VUV and soft X-ray experiments are not normally so demanding as to require this division, and in the author's experience it is usually satisfactory to have data acquisition and user feedback accomplished within a single task.

Swapping and paging overheads. Multi-user and multi-tasking systems are usually designed to run programs whose combined memory requirements exceed the quantity of main memory available. In such cases, processes are either "swapped" or "paged" onto disc. "Paging", a property of "virtual storage" systems, has been explained in section 3.3.2.3. In a "swapping" system, an entire process is copied onto disc in order to make way for another. A process (or process page) that happens to be absent from memory ("swapped out" or "paged out") when an interrupt occurs, may take a long time to make the necessary response since it must first be read from disc. To be useful in a real-time environment, the operating system must enable time-critical programs to be "locked in" to memory, and thus not be subject to swapping or paging.

Summary. In summary, some important questions to ask about an operating system to be used for real-time, data acquisition applications are the following.

(a) Can a user program connect to an interrupt?

(b) What is the interrupt response time of the systems? In other words, how long will it be after an interrupt occurs until the user interrupt handling software starts to execute?

(c) Does the scheduler implement a priority scheme such that time-critical processes may be given priority over other processes?

(d) Can a process be locked into memory so that it is immediately available for execution after an interrupt?

Further details about real-time aspects of operating systems may be found in work by Davies (1976), Pyle (1972), and Clout (1982). A useful textbook on operating systems is by Lister (1984) .

4. Data processing facilities

4.1. Introduction

An SR laboratory must provide facilities for at least the "on-line" data analysis needs of the experiments. The terms "on-line" and "off-line" are sometimes understood to mean the distinction between interactive (immediate, terminal controlled) and non-interactive (batch) processing. In the present context of data acquisition, however, "on-line" processing is the analysis that must be performed while the experiment is in progress, and "off-line" analysis at some later time.

The first aim of on-line analysis is to enable the experimenter to establish if valid and useful data are being taken, so that he or she can intervene if not. The second aim is to reveal interesting features in the data, so that the experimenter can establish experimental parameters in such a way as to collect the most useful data. The essential point, of course, is to enable the experimenter to make efficient use of what is often limited beam time. In many cases, a simple (usually graphical) display of the raw data is sufficient to achieve these aims. In others, considerable computation may be required, involving back-ground removal, Fourier transformation, or fitting functions to the data (e.g. exponential decay functions in fluorescence life-times measurements).

In general, on-line analysis is a subset of the full analysis procedure, most of which would be performed some time after the experiment (off-line). The SR laboratory may or may not provide sufficient facilities for off-line data analysis. The principal benefits of such facilities are

(a) The visiting experimenter's home institute may not provide the necessary facilities, especially where specialised graphics systems or high performance processors are required.

(b) In-house scientists need computing facilities in order to collaborate effectively with visitors and pursue their own research.

(c) Programming is, and is likely to continue to be for the foreseeable future, a labour intensive activity. Sufficient differences exist between systems (programming language implementations, graphics software, peripherals) to demand programming effort even to move analysis programs from one computer to another. Thus there is considerable value in pooling and focusing programming resources through a centralised library of programs and subroutines. An example is the SRS Program Library described by Pantos (1983), which contains over 100 programs covering general spectroscopic data reduction, specialised application packages, file manipulation and programming aids.

The local computer system used for data aquisition is usually not equipped to provide full data analysis facilities. Firstly, since the process of off-line analysis typically takes place over a period of several months, the local computer would need to become a substantial multi-user system with large quantities of disc space and so on. For the reasons already discussed in section 3.2, this is undesirable on operational grounds, conflicting with requirements of data acquisition in an experimental environment. Secondly, on grounds of expense, it is rarely justifiable to provide the full range of peripherals required on every experimental station computer. Some degree of sharing of peripherals will be needed for economic reasons. Thirdly, the environment of an experimental station is often too hostile for the reliable operation of computer peripherals such as magnetic discs and tape. The number of such units housed in experimental areas is usually kept to a minimum.

As a result of these considerations, while it is advantageous to perform as much of the on-line analysis on the data acquisition computer as possible without impeding data acquisition, non-trivial on-line and off-line analysis facilities are normally provided on computer systems remote from the experimental station, communicating with the local computer system via data links.

For on-line analysis purposes, data links need to be direct and delay-free, fast enough to transfer the data in a time short compared with data acquisition so that effective feedback of results can be made in time to influence data taking. This avoids the delays associated with copying data onto a demountable medium at the local system and manually transporting it to the analysis systems. The links from connecting station computer systems to analysis computers constitute a "local area network". These are discussed in section 4.4.

The remote data processing facilities may be provided by a single, large, central computer or, increasingly, by a "distributed system" of smaller systems. Distributed computer systems are discussed in section 4.3.

4.2. Required resources

4.2.1. Data storage

The data storage capacity at the experimental station will usually be the minimum necessary for efficient use of the station for the reasons already indicated, that is, the unsuitability of the environment of the experimental area for large disc and tape drives, and costs. The cost of a large disc is significantly less than that of a number of medium-sized discs of the same total capacity. Thus somewhere remote from the station, in a computer room environment, there should be one or more "file servers". A file server is a computer system with a large data storage capacity to which the local computer systems send their data files. Apart from raw data files, there are many other files for the system to hold, including processed data, programs in source and executable forms, backup for programs on local computer systems and text files for documentation.

A file server would also have magnetic tape drives. These would be used for archival of experimental data (releasing the relatively expensive on-line disc space for new data) and the copying of data for external users to take home. The 1/2 inch, 9-track tape written at 1600 or 6250 bits per inch remains the most reliable medium for data exchange between dissimilar computers. A 2400-feet, 6250 bpi tape can hold 100–150 Mbytes of data, the exact amount depending on such factors as the size of data blocks used in the writing process.

Demountable optical disc cartridges with capacities in the region of 1 Gbyte are (in 1985) starting to emerge in commercially available systems and may take over from magnetic tape as a data archiving medium, though as yet there is no sign that they will be standardised sufficiently for data interchange purposes.

Mass storage systems that provide for automatic (i.e. without operator intervention) transfer of data between disc and special tape cartridges exist, but for the present are too expensive and large scale for the needs of a typical SR laboratory alone.

4.2.2. Graphics terminals

Most of the graphics requirements of VUV and soft X-ray experiments, typically the display of one-dimensional spectra, can be met with relatively simple terminals, offering a resolution of the order of 1000×1000 points and connected to the analysis computer by a standard serial interface as used for simple alphanumeric terminals. Since the line speed requirement for simple graphics terminals is not high (say 4800 bits/s) they can be a considerable distance from the computer (several hundred metres, or further with the use of modems). It is feasible and highly desirable to locate such a graphics terminal at the experimental station for use in any on-line analysis executing on a remote computer system.

Two-dimensional data, however, are very much more demanding of the display system. A typical display terminal would offer 512×512 points spatial resolution, representing intensities by variations in "grey-level" (brightness of picture elements, "pixels", in a monochrome display) with values in the range 0–255 (i.e. by an 8-bit value) or colour (this colour coding of intensities is often called pseudo-colour or false colour). An image on such a display would contain 256 Kbytes of information, which takes a very long time to transmit down a typical serial line. Such displays would instead be connected to computers by parallel interface, typically with a DMA (see section 3.2.4) interface. The distance such an interface can drive is usually very short, of the order of a few metres. If such a display is needed for on-line analysis, and therefore at the experimental station, it must either be connected directly to the experimental data acquisition computer or to an adjacent graphics workstation computer. There is a rapid growth in the availability of "single-user systems" (see Willers 1985) typically containing a 1–2 Mip (1 Mip = 1 million instructions per second) 32-bit microprocessor, tens of megabytes of local disc storage, a high-speed graphics display and communications via a high-speed local area network (LAN, see section 4.4).

A significant recent development is the establishment of an international standard for graphics software, GKS (graphics kernel system). The standard defines subroutines for 2D drawings (not to be confused, however, with the images of 2D data discussed above) and handling of interactive devices. Related standards are in development for 3D drawings and standard output files ("graphics metafiles"). For further information on GKS, see Hopgood et al. (1983). GKS, however, is a relatively low-level package, so an application program would typically access GKS via an unspecified subroutine library. Also, GKS provides little assistance to image handling on grey-level and pseudo-colour displays. Nevertheless, GKS and its extensions will make an important contribution to the portability and hardware-independence of application programs. It is becoming commonplace for graphics terminals to be designed to implement GKS functions within them. Osland (1984) reviews the current state of graphics software standardisation.

4.2.3. Plotters and printers

It is convenient to locate a cheap and simple digital plotter or "dot addressable" serial printer at the experimental station to provide an immediate hard-copy of graphs plotted on either a terminal on the local computer system or one connected to the on-line analysis computer. Good quality (and more expensive) plotters suitable for publication-quality graphs would be a shared resource, sited in a computer room. It is one of the functions of the graphics metafile of a graphics subroutine package to enable graphs first displayed on a terminal to be then copied to another device such as a plotter, possibly on a different computer by file transfer over the local area network.

Generally, there is little need for text hard-copy at the experimental station (except during intensive program development phases), so printers are normally also shared resources in the computer room. The bulk of text output would be handled by line-printers, printing several hundred lines per minute without particular concern for visual quality.

Printers are also required to produce high-quality text on A4 sheets for documentation and papers. Inexpensive serial printers are available for low-volume requirements of this kind. Most modern multi-user computer systems provide "text-formatting" software, which will process a file produced by the system text editor containing embedded text formatting commands. Increasingly, however, desk-top microcomputer systems with specialised word-processing software will be used for this application.

4.2.4. Computation
Conventional processors can be supplemented by processors for heavy computational work, e.g. for 2D Fourier transforms. Attached processors (connected to the host computer like a peripheral) are available in a wide range of architectures, performance and cost, and in general it is necessary to match the architecture to the problem to be solved. These processors are often called array processors. This term usually refers to their capability for processing arrays of numbers, but in some instances also to their internal design as arrays of processors operating in parallel. It is feasible for a low-cost attached processor to be used on the experimental station computer system to boost on-line data analysis, able to perform of the order of 10 Mflops (1 Mflop = 1 million floating-point instructions per second). Programs for such systems are not easy to write if a good proportion of their speed is to be realised, though a well-supported system will have a FORTRAN compiler and a subroutine library for common operations.

The dominant language for data analysis programming is FORTRAN and, especially in its current standard form FORTRAN 77, it is relatively portable between differing computer systems. Space does not permit a satisfactory review here of the merits of FORTRAN and its alternatives, and the reader is referred instead to Metcalf (1984). It is reasonable to predict, however, that FORTRAN will remain the obvious choice for data analysis programming, though its form will evolve to improve on its weaknesses through new standards such as FORTRAN 8X (in development).

4.2.5. Access to a wide area network
In several countries, computer networks exist that link major laboratories,

universities and other academic institutions. For example, there is ARPANET in the USA and JANET (Joint Academic Network) in the UK. Networks such as these, spanning many kilometres, are called wide area networks (WANs). They offer the visiting experimenter the possibility of performing interactive data analysis on computer systems at the SR laboratory via the network from a terminal at his or her own institute. This enables the experimenter to reap the advantages of a central data analysis program library, as outlined in section 4.1. Alternatively, the network can be used to transfer files to a computer system at his or her institute. The network can also be used to send "electronic mail" messages, a valuable facility for enabling members of a collaboration to communicate, and to access national computer resources such as very powerful computer systems ("supercomputers") used in modelling calculations (see, for example, Rosner 1985).

The conventions established for message formats to enable computer systems to communicate are called protocols. International standardisation of protocols is proceeding in terms of the "OSI Reference Model" (see Day and Zimmermann 1983), which defines a seven-layer model for the structure of network software. For the present, networks with different protocols (e.g. JANET and ARPANET) can nevertheless communicate to some degree through "gateway" computers that perform protocol conversions.

4.3. Distributed computer systems

4.3.1. Introduction

The computing facilities in a large laboratory almost invariably constitute a distributed computer system in some sense; however, there are many different senses of the term and it is necessary to make some distinctions for this discussion (see, for example, Zacharov 1976b, Lorin 1980, Chambers et al. 1984, Stankovic 1984).

The basic concept of any distributed computer system is that more than one processor is employed to fulfil a given task, where the word task is used here in a very general sense. Considering the overall task of acquiring the data from an experiment, the preceding sections have already outlined a distributed system in which there are local computer systems dedicated to data acquisition, supported by computers specialising in subsequent data processing.

The processors in a distributed system are linked in some fashion. The link may be by computer backplane bus, local area network or wide area network. Several processors may be present in a single computer, co-operating in the execution of a single program. This form of distributed system, though an active and interesting research and development area and relevant to the design of powerful and highly reliable computers, is not directly the concern of this section. Instead, we are concerned only with the organisation of the data processing facilities within a laboratory, in which the computers are linked by a local area network. In particular, we are concerned with the comparison between systems with a single powerful central computer shared by all users of the laboratory and those employing a number of smaller machines, divided in some way between the users. Even for the "single central computer" model, a degree of distribution arises, since large computer systems typically employ satellite processors to improve their efficiency by off-loading certain system functions. Examples are processors dedicated to driving peripherals and processors for interfacing the central computer to a computer network. The distinction to be drawn here is that *users* see only one processor for the execution of their programs in that case.

Most modern computer systems, then, are distributed, but they differ in the degree and nature of the distribution, and the extent to which *application* programs are distributed. The reasons for and against distributed processing are partly technical and economic, and partly organisational.

4.3.2. Technical issues
The major technical issues, expressed in terms of advantages and disadvantages of distributed systems, are

(a) Advantages:

(*i*) *Separation of function.* In a distributed system, each component computer of the system can be selected and tailored for its function in an optimum fashion. Thus, for example, high performance graphics facilities may be provided by single-user microcomputer workstations specialised for the task (Willers 1985). Operating systems on single-user systems are able to avoid the overheads of multi-user operating systems that are counter to providing rapid interactive response.

(*ii*) *Increased power.* Recent advances in computer technology have greatly reduced the cost of processors and memories. A distributed computer system is a means by which these may be made available to smaller groups of users while the relatively more expensive peripherals remain shared. Thus, as in (i), the aim is to improve response times, in this case through a reduction in the degree of time-sharing on a given processor.

(*iii*) *Incremental, flexible growth.* Individual components of the system are not expensive, relative to the cost of a large "mainframe" computer. The system can grow according to need in a smooth fashion, and can take advantages of advances in technology and falling prices as it does so. In contrast, the upgrade of a large computer system is usually a disruptive and expensive event.

(*iv*) *Reliability.* The failure of an individual element of the system should not, given a sensible design, lead to a total loss of facilities as is typically the case with single central computer systems. Most activities should be able to proceed, or, if the failure takes place in a component of which there is more than one copy, work may be able to continue with only a degradation in response times or throughput.

(b) Disadvantages:

(*i*) *Inefficiency and cost.* A distributed computer system is relatively inefficient in its use of computer resources. Except in the case described below, the system must be "over-provided" with computer power at each processor in order to meet fluctuating demand. A single large processor sees proportionally less fluctuation since it has a larger user population and need make less "over-provision".

The exceptional case mentioned above is one in which the distributed system is under the control of a single operating system, which can attempt to automatically share the overall load between a number of available processors (see Wang and Morris 1985), thus avoiding the situation where one processor is congested while another is lightly loaded. Such a system, it should be noted, is overwhelmingly likely to be homogeneous with regard to machine type and so to exclude some of the flexibility advantages listed earlier.

A distributed system is also inefficient in the sense that there is duplication of hardware and software. Each computer in the system will have its own operating system, consuming a certain proportion of memory, disc space and CPU power, and funds for licenses and maintenance. At the present time, few suppliers have licensing policies that are appropriate to distributed systems.

(*ii*) *Chaos*. Unless there is strict central control in the choice of elements, especially software, a chaos of incompatible systems may arise. Following Sendall (1985), the way to minimise this problem is to place emphasis on modular system design, focussing on the good definition of system interfaces (especially software interfaces), to avoid unnecessary variety in the system (choice of computer and operating system), and to use standards (e.g. FORTRAN 77, CAMAC) wherever possible.

4.3.3. Organisational issues

Organisational issues are at least as important as technical issues in evaluating distributed systems and in determining their design and operation (see King 1983, Franz et al. 1984, and Ahituv and Sadan 1985). A detailed evaluation of organisational issues can only be made in terms of the nature and objectives of the specific organisation concerned. Some of the general aspects are

(*i*) *Control*. A frequently voiced complaint against central computer systems and their management is that they are not responsive to the specific needs of individual user groups sharing the facility. In a shared system there are inevitable conflicts of interest among users. For instance, there may be conflicts concerning the scheduling of withdrawal of service for maintenance periods, the sharing out of disc space, the relative priorities of batch and interactive working and the merits and timing of major changes to the system. In a distributed system, user groups may be separated onto different computer systems and they will want to exercise some degree (possibly total) of managerial control over their systems.

This may extend to budgetary control, with the user group determining the growth of their facilities according to the need that they perceive. This has the benefit in a scientific laboratory that cases for funding can be expressed more clearly in terms of the scientific benefits anticipated for that user group. The case for funding upgrades to a large multi-user system, in combining the needs of many users, becomes generalised and unspecific, and is harder to weigh against other scientific priorities.

(*ii*) *Accessibility*. Elements of the system may be located where they are most convenient for the users. For instance, a graphics workstation may be placed at an experimental station for on-line analysis purposes. Users may wish to operate certain facilities themselves, for instance to mount their own magnetic tapes and collect line-printer output from nearby peripherals, rather than wait for operator attention, as in a traditional computer centre.

(*iii*) *Development and maintenance*. There is a natural tendency for user groups to wish to become completely independent of other groups in order to better control their own destiny by establishing their own operations and development staff. It is here that the risks of duplication of effort and the development of incompatible and unmaintainable systems are greatest. Central coordination is essential if chaos is to be avoided.

4.3.4. Summary

The debate of centralised versus decentralised computing can be seen as one concerning efficiency versus effectiveness (King 1983). Centralised systems are more efficient in their use of computer resources. However, it may be argued that the greater effectiveness of a distributed sytem leads to an improved efficiency for the overall operation of the organisation. The arguments for distributed systems generally start from the assumption that the objective is not to minimise the cost of the computer resources, but to provide them with a distribution that improves their effectiveness for the organisation. Thus the trend away from centralised computing to decentralised computing is fueled in part by the reduction in the cost of computer systems (hardware at least), which has reduced the emphasis on minimising costs through centralisation. A distributed sytem is *not* likely to be less expensive to buy, or to operate if all the effort, including user effort, is correctly costed. It should, however, be capable of providing a better service, given proper management.

4.4. Local area networks

4.4.1. Introduction

Local area networks (LANs) are distinguishable from wide area networks (WANs, see section 4.2.5), obviously enough, by their short range—a few kilometres or less—and their speed—1 Mbits/s or more. Because they are short range and within the control of the laboratory or department, the error rate on their links is much lower than for WANs. LANs take advantage of this by employing less elaborate error recovery mechanisms, and thus simpler protocols, leading to further performance advantages.

The primary role of a LAN in an SR laboratory is to connect data acquisition systems to data processing systems. If the data processing system is a distributed system, as discussed in section 4.3, then a LAN would also be used to interconnect the computers of the distributed system. If there is a single LAN covering both these requirements, then data acquisition systems are able to access individual elements of the distributed system.

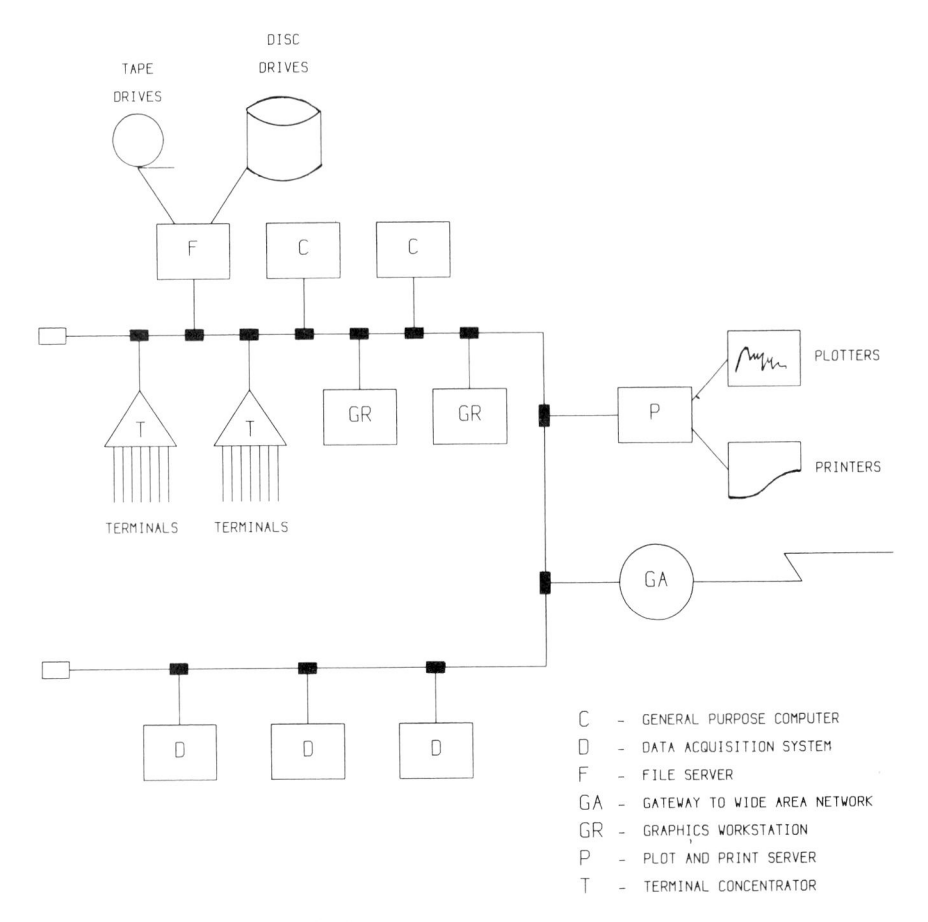

Fig. 10. Distributed computer system on a LAN.

An example of such a distributed computer system based on an Ethernet LAN is shown in fig. 10.

Ethernet is currently the most widespread LAN system and is outlined in section 4.4.2. Further information on LANs can be found in work by Stallings (1984), Tanenbaum (1981), Hutchison (1983) and Rosner (1985).

4.4.2. Ethernet

The first commercial Ethernet systems were based on a specification jointly issued by Digital, Intel and Xerox in the USA. Recently, standard ISO DIS 8802/3 (and equivalent IEEE 802.3) has been approved, which differs from the commercial Ethernet in minor respects. Since the changes are minor, 8802/3 systems are informally known as Ethernets. At the time of writing (1985), some care needs to be exercised in selecting equipment that is compatible with the standard, but it is expected that non-standard implementations will soon disappear.

A basic Ethernet consists of a coaxial cable of up to 500 metres in length, to which systems connect by a device called a *transceiver*, and a cable to the transceiver that may be up to 50 meters long. Power may be applied to, and be removed from, the transceiver without interrupting traffic on the Ethernet, and suitable transceivers may be attached to the cable while the network is running. Five segments of coaxial cable may be connected together by *repeaters*, leading to an overall maximum distance between systems of 2.5 kilometers. There may be up to 1024 systems connected to the Ethernet. Figure 11 illustrates a possible Ethernet structure.

This figure also illustrates the use of a *bridge* between Ethernet segments, and the use of a fibre-optic cable. A bridge is used to filter network traffic so that all segments need not be flooded with all messages. A bridge has tables of addresses of systems on either side, so that it can determine which messages should cross the bridge on the basis of address. In a large system, it may be necessary to electrically isolate Ethernet segments. This may be done by linking repeaters by fibre-optic cable.

Data are transmitted serially along the cable at 10 Mbits/s. In a real application, however, the data rate for a file transfer, for instance, from a disc on one

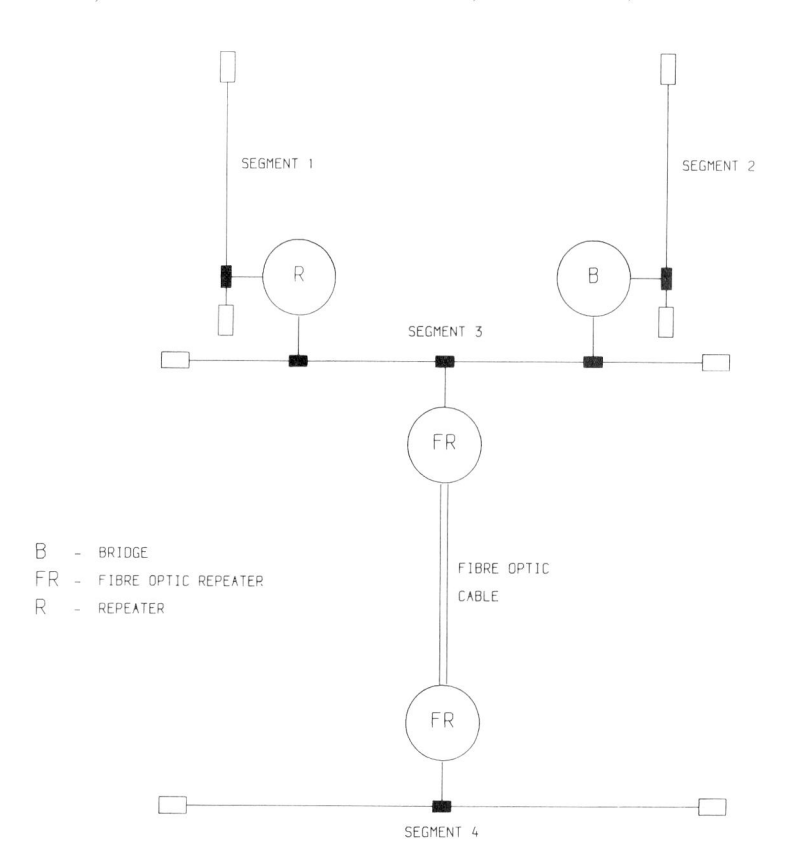

Fig. 11. Ethernet structure.

system to a disc on another will be determined by the complexity of the protocols used and the speed of the filing systems on the two computers. Tasker et al. (1985) observed 60 Kbytes (approximately 0.5 Mbits/s) between LSI-11/23 and VAX-11/750 computers for a system implemented at Daresbury Laboratory for the SRS data acquisition system.

Ethernet is a member of the class of LANs known as CSMA/CD, standing for carrier-sense multiple-access with collision detection. In such systems, a station wishing to transmit must first sense the communication medium and wait until it finds that there are no messages already on the network. When it does transmit, there is a finite chance that another system has "simultaneously" started to transmit. In this case the systems discover that there are "colliding" messages on the network by comparing the data on the cable with the data they transmitted. If they are different, they must transmit a "jamming" signal to ensure that all other transmitting stations see a collision, and then "back-off" for a random interval before trying again. This procedure for competing for the use of the network has the merit of simplicity, but can lead to performance problems on a very heavily loaded network. However, the measurements by Shoch and Hupp (1980) show that the data throughput matches the demand almost exactly up to about 90% of the transmission bandwith for large packet sizes, levelling out at about 96% as demand increases (83% for small packets).

An Ethernet packet contains a fixed header of 22 bytes followed by data, and finally a 4-byte "frame check sequence". The maximum allowable packet size is 1518 bytes and the minimum is 64 bytes. The Ethernet packet (or "frame") is thus very inefficient for sending very short messages (e.g. for character-at-a-time handling for terminals), but becomes quite efficient for long messages, such as for file transfers.

4.4.3. Other LANs

The other major LAN standards likely to compete with Ethernet for popularity and commercial support are the Token Ring (ISO DIS 8802/5, IEEE 802.5) and the Token Bus (ISO DIS 8802/4, IEEE 802.4) (see Stallings 1984 and Hutchison 1983). In the Token Ring, illustrated in fig. 12, a node is allowed to transmit when it sees a special bit-pattern known as the token. It must remove the token (by modifying it) and then send its message before creating a new token and passing it on to its neighbour. Special precautions are taken to deal with lost tokens. Two different transmission rates are defined, 1 Mbit/s and 4 Mbits/s.

The major advantages of this system over Ethernet, achieved at the cost of some additional complexity, are that the problem of collisions does not arise (only one node can have the token), so the system should perform better under high loads, and that nodes are guaranteed the right to transmit within a certain time since at the end of message (of defined maximum length) the token is passed to the next node. Thus such a system is more appropriate for implementing distributed real-time control systems where unpredictable and unlimited delays cannot be tolerated. The Token Ring is favoured by IBM, and thereby ensured commercial support.

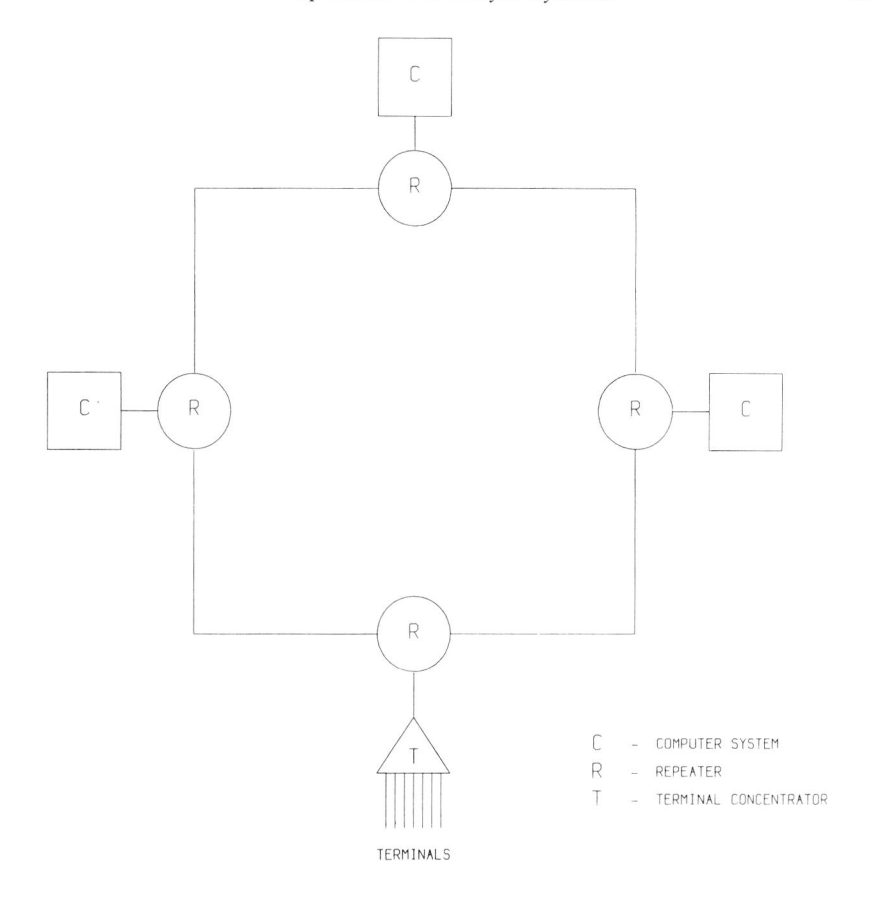

Fig. 12. Ring structure.

The Token Bus has the same approach to network access as the Token Ring, but with a topology similar to Ethernet. The collaboration of a number of large users and suppliers in the USA is leading towards the creation of a de facto standard for protocols over the Token Bus: MAP (manufacturing automation protocols).

There are many other LANs in use, mainly proprietary, some for lower speed, lower cost communication among microcomputers, and others for higher speed (e.g. Hyperchannel, 50 Mbits/s) communication amongst large computer systems. Readers requiring further information on Hyperchannel, Cambridge Ring or broadband systems are referred to Stallings (1984).

Acknowledgements

I gratefully thank my colleagues David Holland, John Howson, Ted Owen, Manolis Pantos, and Frances Rake for reading and suggesting improvements to this chapter; John Alexander, John Bateson and Mark Enderby for supplying information I have used; John West for permission to use diagrams 7a and 7b; and Geoff Mant for the use of his computer program for drawing diagrams.

References

NIM

Standard Nuclear Instrument Modules (NIM), TID-20893 (Rev. 4), July 1974, U.S. Department of Energy, Physical and Technological Research Division, Office of Health and Environmental Research, Washington, DC 20545, USA.

ANSI/IEEE 488

IEEE Standard Digital Interface for Programmable Instrumentation, ANSI/IEEE Std. 488-1978, The Institute of Electrical and Electronic Engineers, 345 East 47 Street, New York, NY 10017, USA.

CAMAC

In this list, equivalent documents are grouped: thus a, b, c and d of CAMAC-3 are equivalent.

CAMAC-1. CAMAC Instrumentation and Interface Standards, The Institute of Electrical and Electronic Engineers, New York, 1982. (Contains references 3, 4, 5, 7, 9, 11, 12 and 13 below.)

CAMAC-2. ERDA Report TID-26618, CAMAC Tutorial Articles, October 1976.

CAMAC-3. CAMAC – A Modular Instrumentation System for Data Handling, Description and Specification.
  (a) EURATOM Report EUR 4100, 1972.
  (b) ERDA Report TID-25875, July 1972, *plus* TID-25877, December 1972.
  (c) ANSI/IEEE Std. 583-1975.
  (d) IEC Publication 516, 1975.

CAMAC-4. Block Transfers in CAMAC Systems.
  (a) EURATOM Report EUR 4100 Supplement.
  (b) ERDA Report TID-26616, February 1976.
  (c) ANSI/IEEE Std. 683-1976.
  (d) IEC Publication 677, 1980.

CAMAC-5. CAMAC – Organisation of Multicrate Systems (Parallel Highway).
  (a) EURATOM Report EUR 4600, 1972.
  (b) ERDA Report TID-25876, March 1972, *plus* TID-25877.
  (c) ANSI/IEEE Std. 596-1982.
  (d) IEC Publication 552, 1976.

CAMAC-6. CAMAC – Specification of Amplitude Analogue Signals Within a 50 Ohm System.
  (a) EURATOM Report EUR 5100, 1974.
  (b) ERDA Report TID-26614, October 1974.

CAMAC-7. CAMAC Serial Highway Interface System.
  (a) EURATOM Report EUR 6100, 1976.
  (b) ERDA Report TID-26488.
  (c) ANSI/IEEE Std. 595-1982.
  (d) IEC Publication 640.

CAMAC-8. Recommendations for Serial Highway Drivers and LAM Graders for the SCC-L2.
  (a) ESONE/SD/02.
  (b) ERDA/DOE Report DOE/EV-0006.

CAMAC-9. Multiple Controllers in a CAMAC Crate.
  (a) EURATOM Report EUR 6500, 1978.
  (b) ERDA/DOE Report DOE/EV-0007, 1978.
  (c) ANSI/IEEE Std. 675-1982.
  (d) IEC Publication 729, 1978.

CAMAC-10. The Definition of IML, A Language for Use in CAMAC Systems.
  (a) ESONE/IML/01, October 1974.
  (b) ERDA Report TID-26615, January 1975.
CAMAC-11. Real-Time Basic for CAMAC.
  (a) ESONE/RTB/03.
  (b) ERDA Report TID-26619.
  (c) ANSI/IEEE Std. 726-1982.
CAMAC-12. Subroutines for CAMAC.
  (a) ESONE/SR/01.
  (b) ERDA/DOE Report DOE/EV-0016.
  (c) ANSI/IEEE Std. 758-1979 (Reaff. 1981).
CAMAC-13. Definition of CAMAC Terms.
  (a) ESONE/GEN/01.
  (b) ERDA/DOE Report DOE/ER-0104.
  (c) IEC Publication 678.

The documents above are available from the following sources:
EURATOM – Office of Official Publications of the European Communities, P.O. Box 1003, Luxembourg.
ESONE – ESONE Secretariat, Commission of the European Communities, CGR-BCMN, B-2440 Geel, Belgium.
ERDA and DOE Reports – L. Costrell, National Bureau of Standards, Washington, DC 20234, USA.
ANSI – Sales Department, American National Standards Institute, 1430 Broadway, New York, NY 10018, USA.
IEEE – IEEE Service Center, 445 Hoes Lane, Piscataway, NJ 08854, USA.
IEC – International Electrotechnical Commission, 1 rue de Varembe, CH-1211 Geneve 20, Switzerland.

FASTBUS
FASTBUS Modular High Speed Data Acquisition and Control System, DOE/ER-0189, U.S. Dept. of Energy, 1983. (Also to be published as IEEE Std. 960.)

G-64 Bus
G-64 Bus Specification, Gespac, s.a., 3 ch. des Auix, CH-1228, Geneve/Plan-les-Ouates, Switzerland.

VME Bus
VME Bus Specification Manual, VME Bus Int. Trade Assoc., Scotsdale, AZ, USA (in process of standardisation as IEEE P1014 and IEC 821).

General references

Ahituv, N., and B. Sadan, 1985, Datamation, September 15th, p. 139.
Alexander, J.R., and J.M. Howson, 1983, Interfaces Computing **1**, 171.
ANSI/IEEE 488, see special reference section above.
Baumann, R., 1972, Interrupt Handling in Real-Time Control Systems, in: Computing with Real-Time Systems, Vol. 2, eds I.C. Pyle and P. Elzer (Transcripta Books, London).
Bedwell, M.O., and T.J. Paulus, 1979, IEEE Trans. Nucl. Sci., **NS-26**(1), 422.
Brookshier, W.K., 1969, Nucl. Instrum. Methods **70**, 1.
Butler, J.J., D.M.P. Holland, A.C. Parr, R. Stockbauer and R. Buff, 1985, J. Phys. E **18**, 286.
CAMAC, see special reference section above.
Cerino, J., J. Stohr, N. Hower and R.Z. Bachrach, 1980, Nucl. Instrum. Methods **172**, 227.
Chambers, F.B., D.A. Duce and G.P. Jones, 1984, Distributed Computing (Academic Press, New York).
Clout, P.N., 1975, Data Acquisition Systems at the SRF, Daresbury Laboratory Technical Memorandum DL/TM 144 (Daresbury Laboratory, Warrington, UK).

Clout, P.N., 1976, Interfacing, in: On-Line Computing in the Laboratory, eds R.A. Rosner, B.K. Penney and P.N. Clout (Advance, London).
Clout, P.N., 1982, Interfaces Comput. **1**, 3.
Clout, P.N., and P.N. Johnson, 1978, Nucl. Instrum. Methods **152**, 151.
Davies, H., 1976, Operating Systems for Experimental Physics, in: Proc. of the 1976 CERN Summer School of Computing, CERN 76-24 (CERN, Geneva, Switzerland).
Day, J.D., and H. Zimmermann, 1983, The OSI Reference Model, in: Special Issue on Open Systems Interconnection – Standard Architecture and Protocols, Proc. IEEE **71**(12), 1334.
de Laat, C.T.A.M., and J.G. Kromme, 1985, Nucl. Instrum. Methods A **239**, 556.
Dobinson, R.W., 1985, Microprocessors: From Basic Chips to Complete Systems, in: Proc. of the 1984 CERN Summer School of Computing, CERN 85-09, ed. C. Verkerk (CERN, Geneva, Switzerland).
EG&G Ortec, AN-42, Principles and Applications of Timing Spectroscopy.
Enderby, M.J., and D.M.P. Holland, 1985, private communication.
FASTBUS, see special reference section above.
Franz, L.S., A. Sen and T.R. Rakes, 1984, Information and Management **7**, 263.
G-64 Bus, see special reference section above.
Galwey, R.K., 1977, Vibrationless High Resolution Stepping Motor Drive, IBM Research Laboratory Report RJ1987(27937) 4/15/77 (IBM, San José, CA).
Granneman, E.H.A., and M.J. van der Wiel, 1983, Transport, Dispersion and Detection of Electrons, Ions and Neutrals, in: Handbook on Synchrotron Radiation, Vol. 1, ed. E.-E. Koch (North-Holland, Amsterdam) ch. 6.
Hastings, J.B., and V.O. Kostroun, 1983, Nucl. Instrum. Methods **208**, 815.
Herbst, L.J., ed., 1970, Electronics for Nuclear Particle Analysis (Oxford Univ. Press, Oxford).
Hopgood, F.R.A., D.A. Duce, J.R. Gallop and D.C. Suttcliffe, 1983, Introduction to the Graphics Kernel System (Academic Press, London).
Horelick, D., 1975, IEEE Trans. Nucl. Sci. **NS-22**, 488.
Hormes, J., A. Klein, W. Krebs, W. Laaser and J. Schiller, 1983, Nucl. Instrum. Methods **208**, 849.
Hutchison, D., 1983, Software & Microsyst. **2**(4), 87.
King, J.L., 1983, Computing Surveys **15**(4), 319.
Kowalski, E., 1970, Nuclear Electronics (Springer, Berlin).
Kuo, B.C., 1978, Sensors and Encoders, in: Incremental Motion Control, Vol. 1: DC Motors and Control Systems, eds B.C. Kuo and J. Tal (SRL Publishing Company, Champaign, IL).
Kuo, B.C., and J. Tal, eds, 1978, Incremental Motion Control, Vol. 1: DC Motors and Control Systems (SRL Publishing Company, Champaign, IL).
Kuo, B.C., ed, 1979, Incremental Motion Control, Vol. 2: Step Motors and Controls (SRL Publishing Company, Champaign, IL).
Lee, G., and D. Wiegandt, 1983, Interfaces Comput. **1**, 329.
Lippiatt, A.G., 1978, The Architecture of Small Computer Systems (Prentice-Hall, London).
Lister, A.M., 1984, Fundamentals of Operating Systems, 3rd. Ed. (Macmillan, London).
Lorin, H., 1980, Aspects of Distributed Computer Systems (Wiley, New York).
MacDowell, A.A., I.H. Hillier and J.B. West, 1983, J. Phys. E **16**, 487.
Merritt, R., 1976, Instrum. Technol. **August**, 29.
Metcalf, M., 1984, Has Fortran a Future?, CERN Report DD/84/7 (CERN, Geneva, Switzerland).
Morrison, R., 1967, Grounding and Shielding Techniques in Instrumentation (Wiley, New York).
Muller, K.D., 1985, IEEE Trans. Nucl. Sci., **NS-32**(1), 262.
Munro, I.H., and N. Schwentner, 1983, Time-Resolved Spectroscopy using Synchrotron Radiation, Nucl. Instrum. Methods **208**, 819.
Murphy, E.L., 1976, Instrum. Control Syst. **June**, 35.
NIM, see special reference section above.
Osland, C.D., 1984, Interfaces Comput. **2**, 1.
Pantos, E., 1983, Nucl. Instrum. Methods **203**, 449.
Parr, A.C., R. Stockbauer, B.E. Cole, D.L. Ederer, J.L. Dehmer and J.B. West, 1980, Nucl. Instrum. Methods **172**, 357.

Puester, G., and K. Thimm, 1978, Nucl. Instrum. Methods **152**, 95.
Pyle, I.C., 1972, Basic Supervisor Facilities for Real-Time, in: Computing with Real-Time Systems, Vol. 2, eds I.C. Pyle and P. Elzer (Transcripta Books, London).
Pyle, I.C., 1976, Software, in: On-Line Computing in the Laboratory, eds R.A. Rosner, B.K. Penney and P.N. Clout (Advance, London).
Rieger, D., R.D. Schnell, W. Steinmann and V. Saile, 1983, Nucl. Instrum. Methods **208**, 777.
Rimmer, E.M., 1985, Interfaces Comput. **3**, 1.
Rosner, R.A., 1985, Data Networks and Open Systems, in: Proc. of the 1984 CERN Summer School of Computing, CERN 85-09, ed. C. Verkerk (CERN, Geneva, Switzerland).
Sachs, M.W., 1979, Nucl. Instrum. Methods **162**, 719.
Schnell, R.D., D. Rieger and W. Steinmann, 1984, J. Phys. E **17**, 221.
Sendall, D.M., 1985, CERN Report DD/85/17 (CERN, Geneva, Switzerland).
Shoch, J.F., and J.A. Hupp, 1980, Communications **23**(12), 711.
Smith, N.V., and S.D. Kevan, 1982, General Instrumentation Considerations in Electron and Ion Spectroscopies using Synchrotron Radiation, Nucl. Instrum. Methods **195**, 309.
Stallings, W., 1984, Local Networks – An Introduction (Macmillan, New York).
Stankovic, J.A., 1984, IEEE Trans. Comput. **C-33**(12), 1102.
Tanenbaum, A.S., 1981, Computer Networks (Prentice-Hall, Englewood Cliffs, NY).
Tasker, R., F.M. Rake, P.S. Kummer and D.P. Hines, 1985, Interfaces Comput. **3**, 153.
Timothy, J.G., and R.P. Madden, 1983, Photon Detectors for the Ultraviolet and X-Ray Region, in: Handbook on Synchrotron Radiation, Vol. 1, ed. E.-E. Koch (North-Holland, Amsterdam) ch. 5.
Turko, B., 1978, IEEE Trans. Nucl. Sci. **NS-25**(1) 75.
VanDoren, A.H., 1982, Data Acquisition Systems (Reston Publ. Company Inc., Reston, VI).
VME Bus, see special reference section above.
von Ruden, W., 1985, in: Proc. of the 1984 CERN Summer School of Computing, CERN 85-09, ed. C. Verkerk (CERN, Geneva, Switzerland).
Wadsworth, B.F., 1980, IEEE Trans. Nucl. Sci. **NS-27**(1), 612.
Wall, W.E., and J.R. Stevenson, 1978, Nucl. Instrum. Methods **152**, 141.
Wang, Y.-T., and R.J.T. Morris, 1985, IEEE Trans. Comput. **C-34**(3), 204.
Weeks, S.P., J.E. Rowe, S.B. Christman and E.E. Chaban, 1979, Rev. Sci. Instrum. **50**(10), 1249.
White, M.G., R.A. Rosenberg, G. Gabor, E.D. Poliakoff, G. Thornton, S.H. Southworth and D.A. Shirley, 1979, Rev. Sci. Instrum. **50**(10), 1268.
Willers, I., 1985, Single User Systems, in: Proc. of the 1984 CERN Summer School of Computing, CERN 85-09, ed. C. Verkerk (CERN, Geneva, Switzerland).
Wiza, J., 1979, Microchannel Plate Detectors, Nucl. Instrum. Methods **162**, 587.
Zacharov, B., 1976a, The Role of the Small Computer in the Laboratory, in: On-Line Computing in the Laboratory, eds R.A. Rosner, B.K. Penney and P.N. Clout (Advance, London).
Zacharov, B., 1976b, Distributed Processor Systems, in: Proc. of the 1976 CERN Summer School of Computing, CERN 76-24 (CERN, Geneva, Switzerland).

CHAPTER 4

HIGH RESOLUTION SPECTROSCOPY OF ATOMS AND MOLECULES, INCLUDING FARADAY ROTATION EFFECTS

J.P. CONNERADE

The Blackett Laboratory, Imperial College, London SW7 2AZ, UK

M.A. BAIG

Physikalisches Institut, Universität Bonn, Nussallee 12, Bonn, Fed. Rep. Germany[*]

Contents

1. Introduction 177
2. High resolution spectra of atoms 177
    2.1. High Rydberg states 177
    2.2. Ionisation potential determinations 181
    2.3. Autoionisation effects 185
    2.4. Inner shell spectroscopy of atoms 194
    2.5. Configuration mixing in inner shell spectra 200
    2.6. "Giant resonances" in atomic spectra 210
3. High resolution spectra of molecules 212
    3.1. Application of synchrotron spectroscopy to the study of rovibronic structure . 212
    3.2. The asymmetric top problems 219
    3.3. High Rydberg states of molecules 226
4. SR experiments with polarised radiation 230
    4.1. External magnetic fields: Faraday rotation spectroscopy . 230
    4.2. Spin polarisation spectroscopy 235
5. Conclusion 236
References 237

[*] Present address: Physics Department, Quaid-i-Azam University, Islamabad, Pakistan.

Handbook on Synchrotron Radiation, Vol. 2, edited by G.V. Marr
© *Elsevier Science Publishers B.V., 1987*

1. Introduction

Synchrotron radiation has a special part to play in the pursuit of high resolution atomic and molecular spectroscopy because of (i) its extremely wide wavelength coverage, extending over a domain not previously accessible for studies at high dispersion, (ii) its high intensity without which high dispersion would not even be feasible over much of the range, and (iii) its unique polarisation properties which have so far been exploited mainly in photoelectron and magneto-optical spectroscopy.

High resolution must always be a relative term and, as such, soon out of date. In the region of interest for the present chapter, namely the vacuum ultraviolet (VUV), tunable coherent sources are becoming available (Wallenstein 1983) down to at least 900 Å in wavelength, and these are capable of much higher resolution than is available with conventional spectrographs equipped with gratings. However, the tunability is restricted to narrow ranges, and the combination of a synchrotron source with a spectrograph of high dispersion is likely to remain the dominant tool over much of the VUV for the forseeable future, because of its wide coverage, its speed and its flexibility.

A practical definition of high resolving power was given by Ginter (1980) as anything above 100 000 for the VUV down to 1200 Å or so. Recent progress with holographic gratings suggests that this number could be revised upwards to 200 000. Below 1200 Å, a convenient rule of thumb is that resolving power falls off as λ. If we take high resolving power as anything above $100 \times \lambda$ (λ in angstroms) below 1000 Å, then we have a definition which joins on to that of Ginter (1980) around 1000 Å and which is also realistic towards 100 Å for grazing incidence spectrographs.

Having given this definition we do not intend to survey any of the well-known techniques of high resolution spectroscopy. Rather, we concentrate on a variety of physical problems for which high resolution is indispensable because of the insight one gains into processes of excitation or decay. Such problems will be illustrated by examples drawn from recent synchrotron radiation experiments at high resolving power. Also, we avoid duplicating recent reviews on soft X-ray spectroscopy of atoms and molecules by restricting ourselves to wavelengths longer than about 100 Å.

2. High resolution spectra of atoms

2.1. High Rydberg states

The present section illustrates the simplest application of high resolution spectroscopy to atoms, namely the observation of long Rydberg series to high principal quantum numbers, by some recent examples obtained with synchrotron radiation.

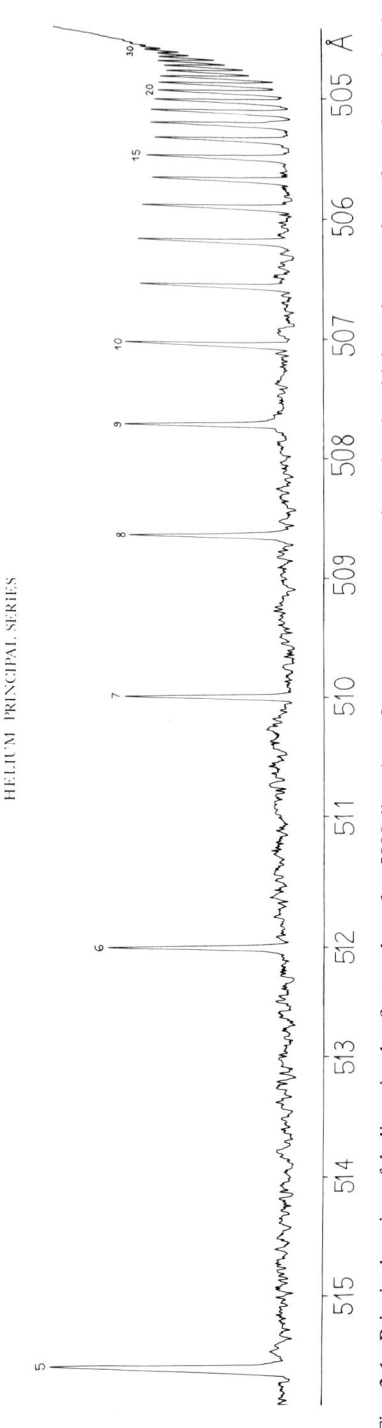

Fig. 2.1. Principal series of helium in the first order of a 5000 lines/mm 3-metre concave grating, showing high series members first observed using synchrotron radiation spectroscopy.

High resolution spectroscopy of atoms and molecules 179

HIGH RESOLUTION ABSORPTION SPECTRUM OF MGI.

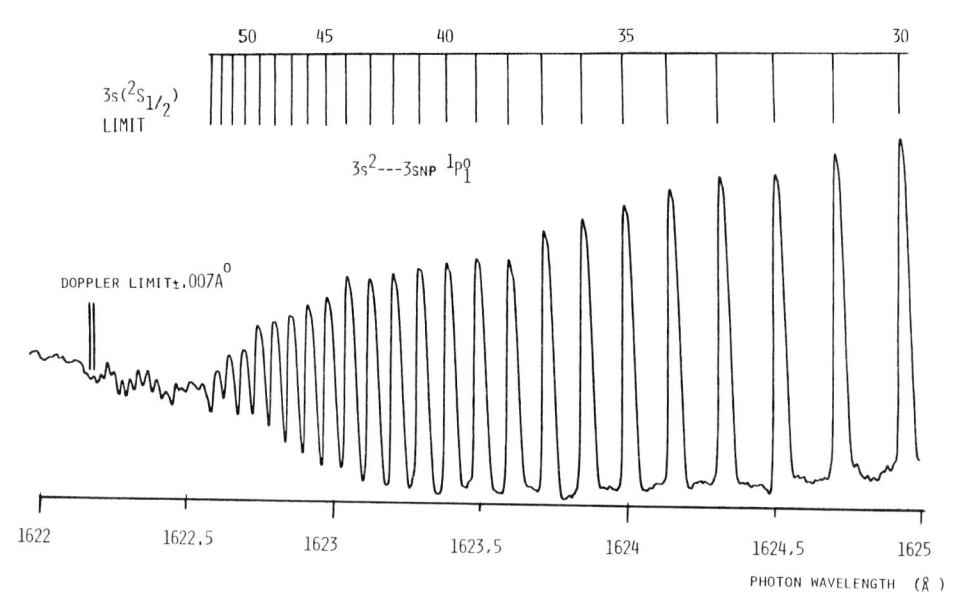

Fig. 2.2. High members of the principal series of magnesium, showing how close the experimental resolution is to the theoretical Doppler width, as calculated from the temperature of the absorption cell.

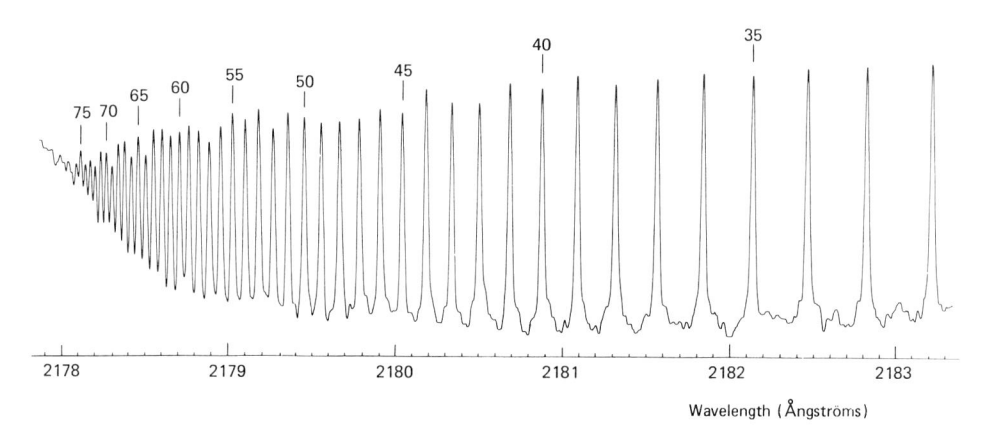

Fig. 2.3. High members of the principal series of strontium, a series for which Faraday rotation studies have recently been carried out (see section 4.1).

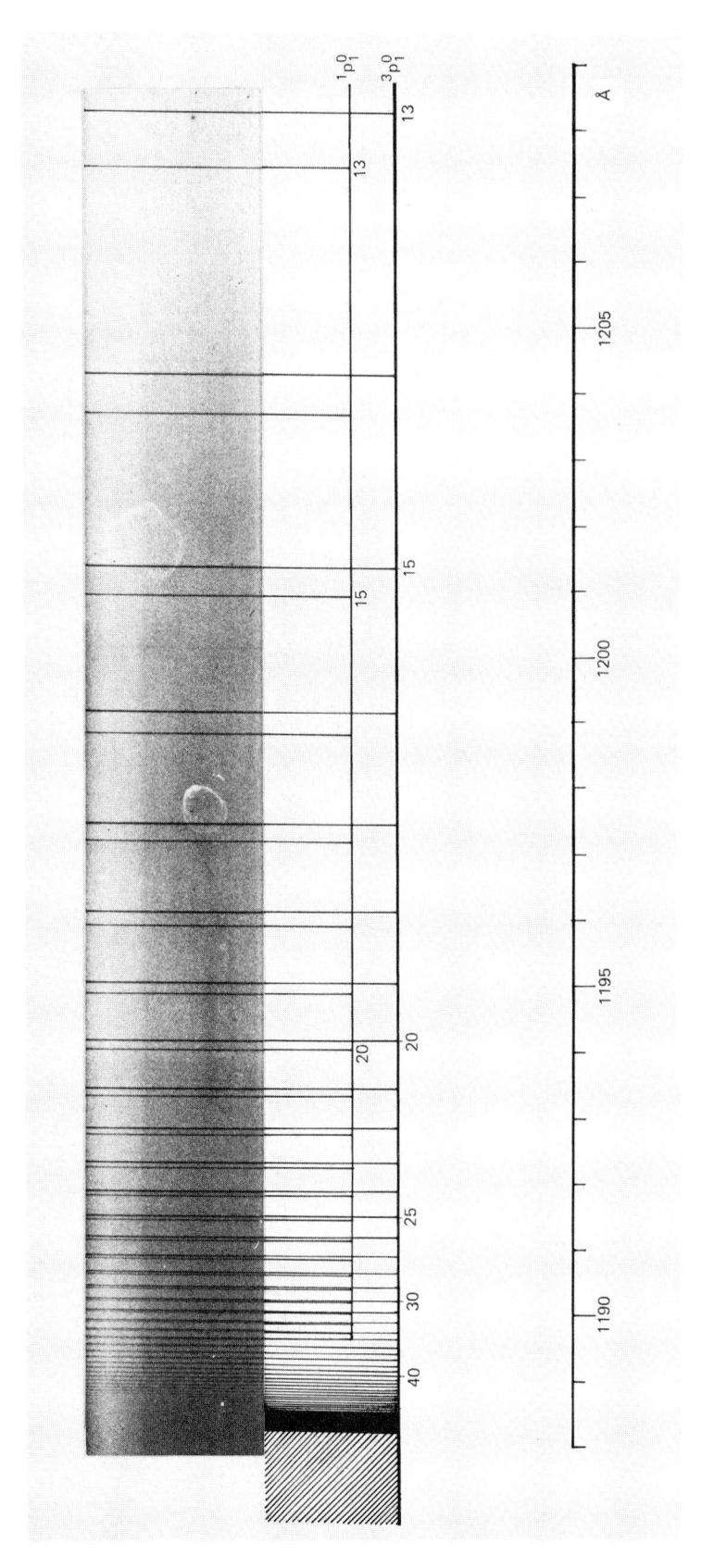

Fig. 2.4. High members of the principal series of mercury, showing also the intercombination series, which gains intensity towards the series limit as a result of the gradual change from LS to jj coupling with increasing n.

The most obvious application of high resolution synchrotron spectroscopy to atoms is the study of high Rydberg states in the vacuum ultraviolet, at energies higher than are readily available using tunable lasers. The highest ionisation potential of a neutral atom is of course that of helium. In fig. 2.1, we show the principal series of He to high quantum as recently determined in a synchrotron radiation experiment involving a 3-metre spectrograph equipped with a holographic grating of 5000 lines/mm (Baig et al. 1984a). Surprisingly, although there has been much work on the spectrum of helium using a variety of sophisticated techniques, the spectrum of fig. 2.1 appears to be the first direct observation of the principal series above $n = 14$. When one consults the best available tabulation of the energy levels of ^{4}He (Martin 1960, 1973), it emerges on close inspection that, although some early work was reported up to $n = 14$ (Suga 1937), the experimentally based energies above $n = 10$ are regarded as so unreliable that a series formula is used in place of observations. The work of Suga (1937) was performed in emission at high pressures. Some recent work (Katayama et al. 1979) on transitions from the metastable excited state $2s\ ^1S_0$ to the same upper levels has also yielded quite high Rydberg states (up to $n = 36$) but involved detection by optogalvanic techniques, requiring pressures of 0.4–1 Torr and the operation of an electrical discharge in the sample. Not surprisingly, the lines observed by Katayama et al. (1979) are broadened at high n. The data of fig. 2.1 were obtained using a path length of roughly 6 metres and pressures of a few microns. Although the resolving power achieved at 500 Å in the synchrotron experiment is not as high as that of the dye laser used by Katayama et al. (1979) at longer wavelengths, the data obtained with synchrotron radiation extend to $n = 35$ and are more reliable because of the more suitable conditions in the sample chamber.

Further examples of high Rydberg states in atoms with closed outer shells are shown in figs. 2.2 (Mg I), 2.3 (Sr I) and 2.4 (Hg I). In the case of Mg I (Connerade and Baig, 1981), the atom is light and the temperature in the cell high enough that the interval between successive series members at high n becomes close to the Doppler limit, so that little further structure would emerge by further improvements in instrumental resolution. To observe the series to yet higher members would require some form of Doppler-free spectroscopy, a technique which is better suited to lasers than to synchrotron radiation. In Sr I, Doppler broadening is small, and the series can be followed to $n = 85$. In Hg I, as for most heavy atoms, departures from LS coupling becomes important, and the intercombination lines are observed to high n, gaining intensity at the expense of the singlet lines as n increases (Baig 1983). In practice, perturbations of the course of intensity often reduce the number of detectable series members near the series limit.

2.2. Ionisation potential determination

The observation of high Rydberg states allows ionisation potentials to be determined accurately.

Consider first the case of an unperturbed series. We then apply the following Rydberg relation:

$$E_n = I_\infty - \frac{R}{(n - \mu_n)^2}, \tag{1}$$

where I_∞ is the ionisation potential, E_n is the transition energy, R is the mass-corrected Rydberg constant, n is the principal quantum number and μ_n is the quantum defect. Using a trial value of I_∞, one calculates quantum defects and seeks a trend towards constant quantum defects at high n-values. One then applies the Ritz formula:

$$E_n = I_\infty - \frac{R}{\nu^2}, \quad \nu = n - \mu + \beta\bar{\Delta}, \tag{2}$$

where the term value,

$$\bar{\Delta} = I_\infty - E_n.$$

Plotting term values against quantum defects yields a straight line of slope β and intercept μ. Unperturbed series, such as that in fig. 2.2 are readily handled in this way.

However, Seaton (1966) remarked that adjusting only the ionisation potential to achieve constant quantum defect at high n is not a good procedure, because the error in quantum defects $\delta\mu_n$ due to an error δE_n in the experimental energy is

$$\delta\mu_n = -\frac{1}{2}\left(\frac{R}{I_\infty - E_n}\right)^{3/2} \frac{\delta E_n}{R}, \tag{3}$$

which increases rapidly as n increases. He therefore proposed that one should simultaneously adjust the ionisation potential and a parameter a_i, where $i = 0, 1, \ldots, N$ so as to obtain a least squares fit of observed and calcuated term values:

$$E_n(\text{calc}) = I_\infty - \frac{R}{\{n - \mu(E_n)\}^2}, \tag{4}$$

where

$$\mu(E_n) = \sum_{i=0}^{N} a_i \left(\frac{E}{R}\right)^i.$$

This procedure was applied very successfully in accurate determinations of the ionisation potential of helium (see e.g. Baig et al. 1984a), of interest in the determination of the Lamb shift for two-electron systems (see Herzberg 1958).

Often, the simple approach outlined above is inapplicable when the Rydberg series is perturbed, and the assumption of constant quantum defects at high n breaks down. One then resorts to Multichannel Quantum Defect Theory (MQDT) (Seaton 1966). It is not our purpose here to describe this method of analysis in detail, since excellent reviews exist (e.g. Seaton 1983), but some brief remarks may be useful.

The theory introduces the concept of a *channel*, meaning an entire Rydberg series with its adjoining continuum, and the bound and low energy scattering spectrum are treated in a unified way. It is perhaps most meaningful, in the context of the reduction of data, to regard MQDT as a method of reducing a vast amount of experimental information to just a few relevant parameters (a minimum number) which can then be compared with truly ab initio theoretical predictions.

In the MQDT approach, the Hamiltonian of the atom is divided into two parts: (i) the Coulomb interaction between the outermost electron and the core at $r > r_0$, which is termed the "collision or loose-coupling" channel i, and (ii) the short range interactions at $r < r_0$ which define the "close-coupling" channel α.

An excited electron at large distance from the core experiences only the Coulomb attraction, and its wavefunction can be treated analytically. The regularity of the Rydberg series is determined by the behaviour of the wavefunctions at $r > r_0$. Perturbation effects, intensity variations or irregularities of level spacings result from the short range interactions which affect the wavefunction inside the core region. If short-range interactions are small, the energy eigenvalues are given by the Rydberg formula, and the quantum defect is a measure of non-Coulombic effects.

In terms of the regular and irregular Coulomb functions $f(r)$ and $g(r)$, the wavefunction is written as

$$\Psi(r) = f(r) \cos \pi\mu - g(r) \sin \pi\mu . \tag{5}$$

At energies E below the ionisation threshold, the condition for a bound state leads to the simultaneous equations

$$\sin \pi(\nu_i + \mu) = 0 \tag{6}$$

with

$$\nu_i = n - \mu , \qquad E = I_\infty - \frac{R}{\nu_i^2}$$

The roots E of these equations recover the Rydberg formula:

$$E_n = I_\infty - \frac{R}{(n - \nu_i)^2}$$

In the *two-channel* case, two ionisation thresholds I_1 and I_2 must be considered.

The bound states which lie below both thresholds are conveniently handled by the graphical approach to MQDT of Lu and Fano (1970). In this method, each discrete level is referred to *both* ionisation potentials, because interchannel mixing can occur as a result of perturbations, and all levels acquire some character from each channel. The fractional parts of the two effective quantum numbers are plotted, modulo one, in a unit square. Were there no interchannel interaction, the resulting graph for constant quantum defect in each channel would exhibit intersecting horizontal and vertical lines. In the presence of interactions, the resulting curve describes the periodic nature of the coupling between the series. Any horizontal line will intersect this curve as many times as there are series converging on the limit of the channel plotted vertically, and vice versa. The periodicity imposes that, for every branch of the curve which exits the Lu–Fano graph at a given point, with a given slope on one side, another branch enters the square with the same slope, from the corresponding point on the opposite side of the square. The curvature at the avoided crossings is determined by the magnitude of the interchannel interaction.

Thus, one can determine from the Lu–Fano graphs how many series converge to a given limit, the purity of channel labels, and the degree of interchannel interaction. One can also readily calculate the positions of any missing levels from the intersection between the curve described above and the functional relation between the two effective quantum numbers:

$$\nu_1 = \left[\frac{1}{(\nu_2)^2} - \frac{(I_2 - I_1)}{R} \right]^{1/2}. \tag{8}$$

The close- and loose-coupled channels are coupled by a matrix $U_{i\alpha}$. One can show that non-trivial solutions occur only if the secular condition:

$$\det | U_{i\alpha} \sin(\pi \nu_i + \mu_\alpha) | = 0 \tag{9}$$

is satisfied. Between them, these two equations determine the energies of the bound states. The analysis consists of adjusting $U_{i\alpha}$ and μ_α until the term values agree with experimental data. Notice that the first equation involves the ionisation potential, while the second describes interchannel interactions.

As a very simple example of this procedure, we give the $5sns\ ^1S_0$ series of Sr I, for which experimental data are due to Esherick (1977). The series is perturbed by the $5p^2$ interloper between $n = 6$ and $n = 8$. Using $I_s = 45932.10\ \text{cm}^{-1}$ and $I_p = 70048.11\ \text{cm}^{-1}$, the Lu–Fano plot of fig. 2.5 is obtained, where the solid line is a hand-drawn curve through the data points and the dashed line is the functional relation between the two effective quantum numbers above, with $\Delta = 0.21976$. The eigenquantum defects μ_1 and μ_2 of the close-coupled eigenstates are determined from the intersections of the full curve with the diagonal, as shown in the figure. The values here are $\mu_1 = 0.325$ and $\mu_2 = 0.145$: One can write $U_{i\alpha}$ as a two-dimensional rotation through an angle θ:

$$U_{i\alpha} = \begin{pmatrix} \cos\theta & \sin\theta \\ -\sin\theta & \cos\theta \end{pmatrix}, \tag{10}$$

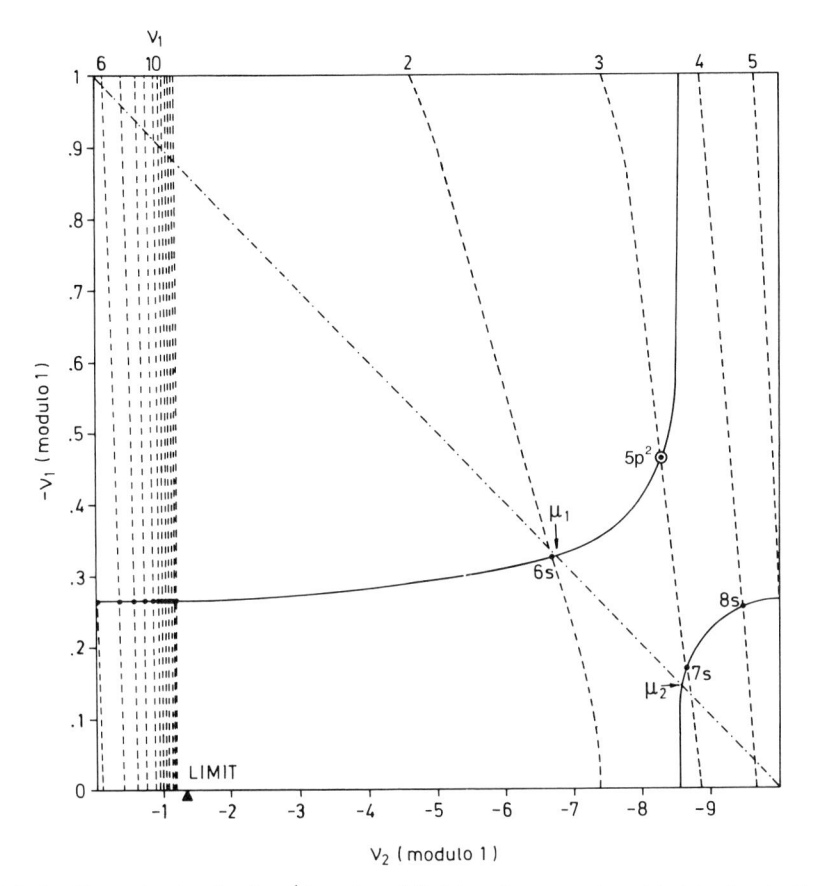

Fig. 2.5. Lu–Fano plot for the $5sns\ ^1S_0$ series of Sr I based on experimental data due to Esherick (1977), illustrating the technique of analysis described in the text.

then θ is determined by the slope of the full curve at the avoided crossing. In our example, it is 0.36.

Once parameters are known for the bound states, the connection between quantum defects and phase shifts in the continuum can be exploited to compute autoionisation profiles above the first threshold.

Of course, only simple couplings can be plotted in two dimensions. Often, many channels are involved, and computer fitting is required. Thus, Robaux and Aymar (1982) have developed a computer code to determine semi-empirical MQDT parameters for problems involving up to 12 channels, with an iterative optimisation.

2.3. Autoionisation effects

The study of line profiles in the autoionising range is also a well-established application of vacuum ultraviolet spectroscopy and often requires high resolution, particularly if the profiles are to be studied up to high principal quantum numbers. Again, the combination of the synchrotron with instrumentation of high dispersion is currently the most versatile, if not the most powerful technique over the full range of the vacuum ultraviolet.

In what follows, we take as read the celebrated paper by Fano (1961) on autoionisation, and we assume the reader is familiar with the Fano parametrisation:

$$\sigma(\varepsilon) = \frac{(q+\varepsilon)^2}{1+\varepsilon^2}, \quad \varepsilon = \frac{2(E-E_0)}{\Gamma}. \tag{11}$$

A good example of the diversity of autoionisation effects to be encountered in the vacuum ultraviolet is presented by the spectra of the rare gases above the first ionisation limits. In the case of Ne (Baig and Connerade 1984a), which lies before the transition sequences begin in the Periodic Table, the series converging on the upper p^5 limit remain fairly sharp even in the autoionising range (see fig. 2.6). Further down in the Periodic Table, the d subshells build up, as a result of the effective potential becoming more binding for d electrons, and the autoionisation rate for the nd series, which depends on the spatial overlap between the d electrons and the core, increases rapidly. Thus, by the time Xe is reached, the broadening of the nd series to the upper p^5 limit has been enhanced dramatically (see fig. 2.7).

The increasing penetration of d-wavefunctions into the core is illustrated from Hartree–Fock single configuration calculations (Baig and Connerade 1984a) in fig. 2.8. There is some further discussion of the centrifugal barrier effects responsible for the changes in section 2.5 below.

The broad nd series in Xe is a particularly good example of autoionisation because it can be approximated with only two interacting channels. This situation is simple enough to be parametrised for all the Rydberg members in the autoionising range using an MQDT formula proposed by Dubau and Seaton (Seaton 1983, Dubau and Seaton 1984) which involves three shape parameters. A straightforward technique for extracting the parameters from experimental data is described by Connerade (1983) and leads to the theoretical curve in fig. 2.7. The formula can be written as

$$\sigma(\nu) = A^2 \frac{\tan^2 \pi\nu + 2B \tan \pi\nu + B^2}{\tan^2 \pi\nu + 2C \tan \pi\nu + D^2} \tag{12}$$

in the notation used by Connerade (1983b), where B, C and D are adjustable shape parameters and A is an amplitude. This formula can be regarded as a generalisation of the classic expression given by Fano (1961) for an isolated autoionising line, in that it represents a full Rydberg series above the ionisation potential. Figure 2.9 shows a comparison between the profile obtained for the lowest autoionising member of the Xe series using both the Dubau–Seaton (curve a) and the Fano (curve b) formulae. The difference between the two, as one might expect, is largest in the region between the resonances.

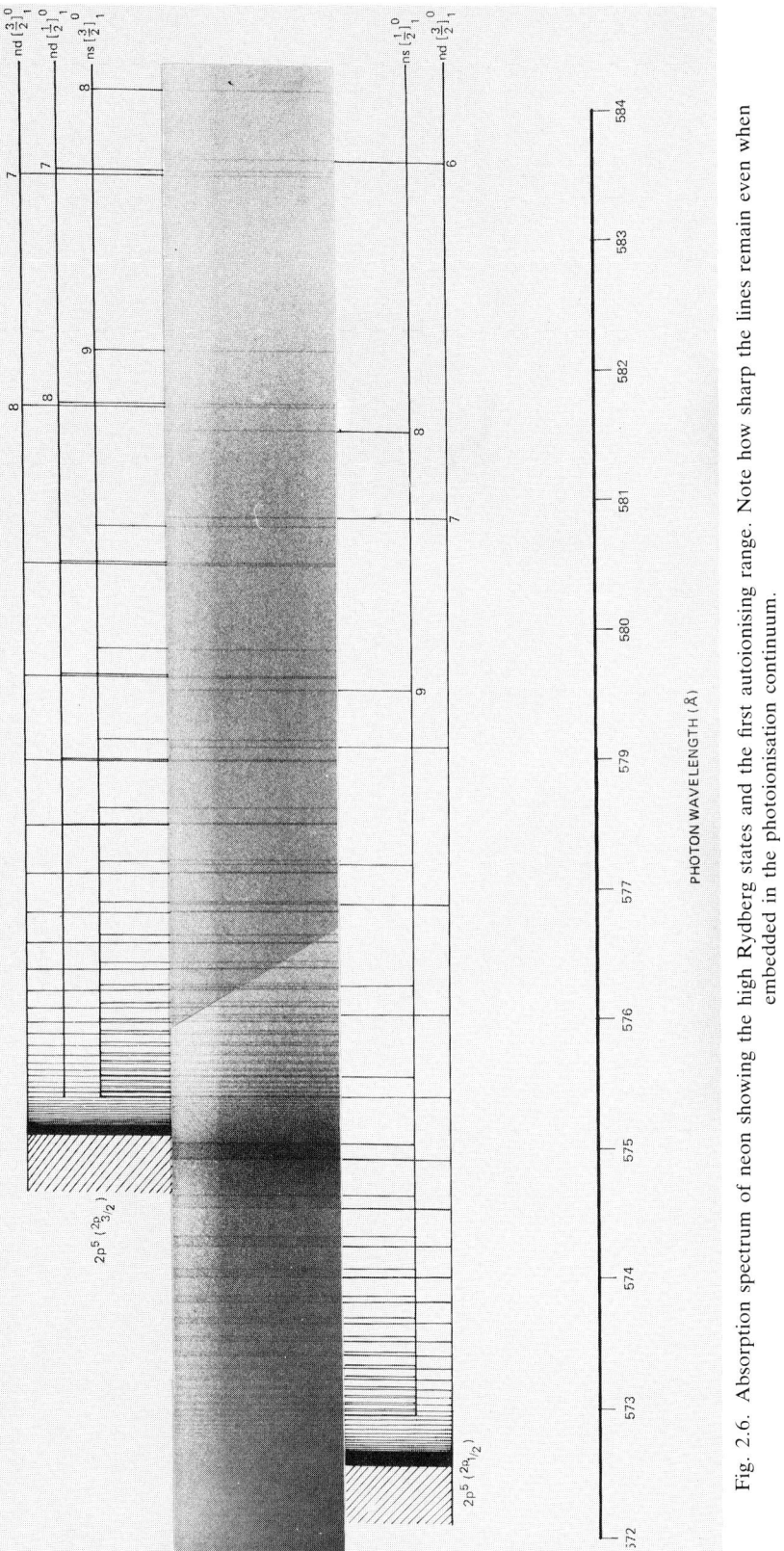

Fig. 2.6. Absorption spectrum of neon showing the high Rydberg states and the first autoionising range. Note how sharp the lines remain even when embedded in the photoionisation continuum.

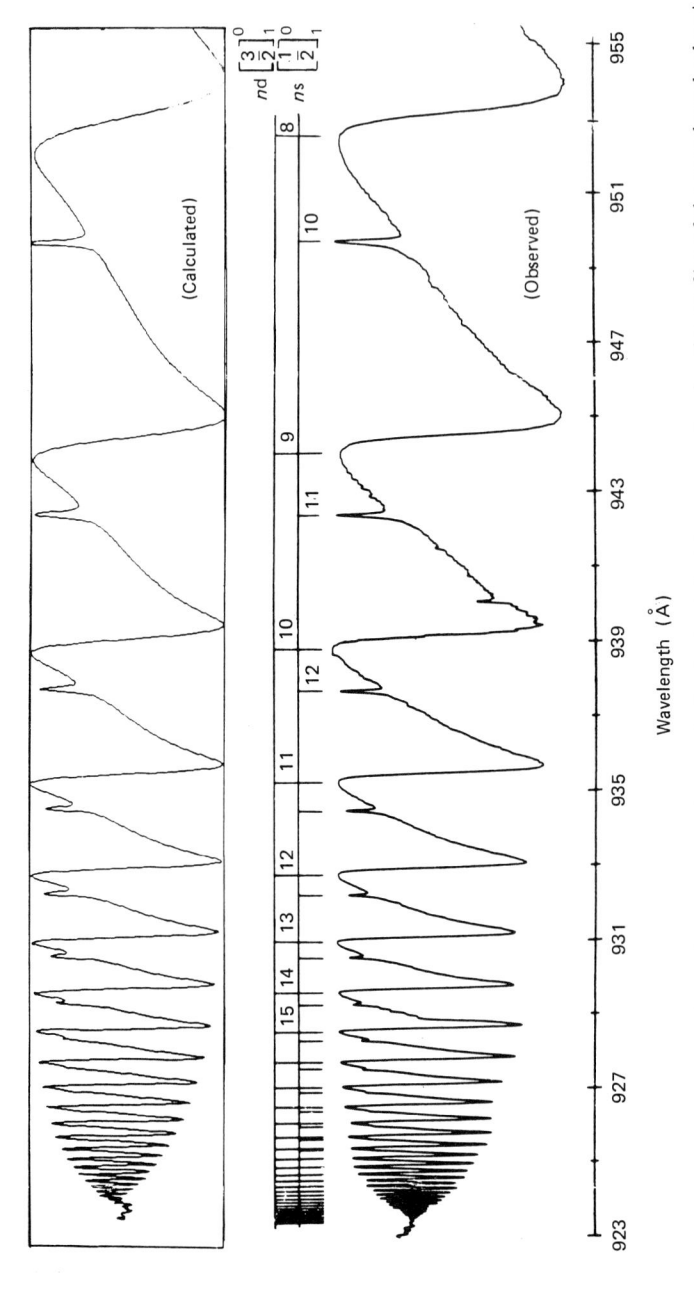

Fig. 2.7. Densitometer trace of the absorption spectrum of xenon in the first autoionising range, showing the profiles of the very broad nd series above the ionisation potential. A theoretical curve obtained from the Seaton–Dubau formula by adjusting four parameters reproduces the data well and is shown alongside the experimental spectrum.

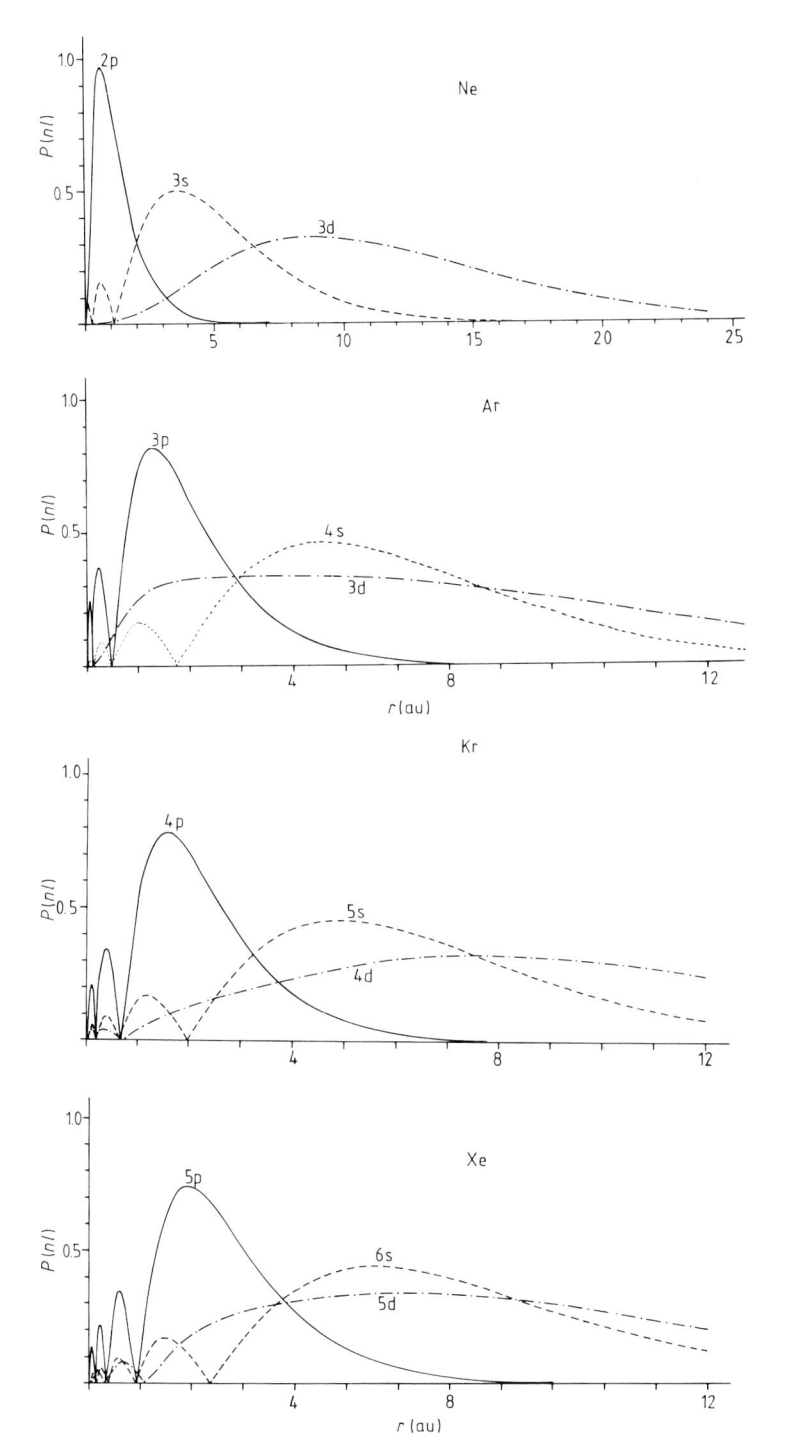

Fig. 2.8. Hartree–Fock single configuration calculations of the outermost s and d wavefunctions in the full sequence of the rare gases (Baig and Connerade 1984a). Note the sudden increase in spatial overlap between the d wavefunction and the p core between Ne and Ar.

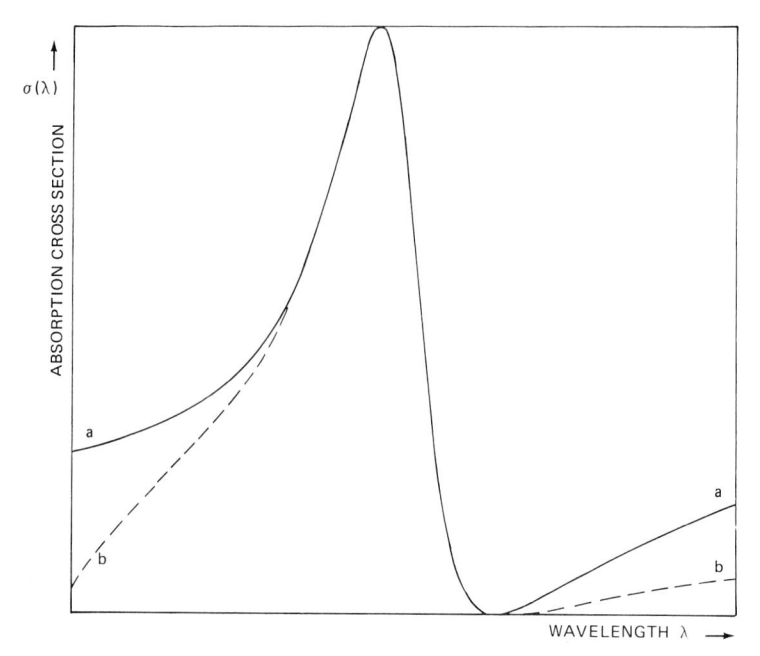

Fig. 2.9. Comparison between (a) the Seaton–Dubau profile, and (b) the Fano profile for parameters adjusted to fit the lowest energy autoionising resonance in the nd series of Xe I.

The connection between eq. (12) and the Fano (1961) formula is readily exhibited by rearranging it thus:

$$\sigma(\nu) = A^2 \frac{(\tan \pi\nu + B)^2}{(\tan \pi\nu + C)^2 + (D^2 - C^2)}$$

$$= A^2 \frac{\left(\dfrac{B - C}{\sqrt{D^2 - C^2}} + \dfrac{\tan \pi\nu + C}{\sqrt{D^2 - C^2}}\right)^2}{1 + \left(\dfrac{\tan \pi\nu + C}{\sqrt{D^2 - C^2}}\right)^2}$$

$$\approx A^2 \frac{(q_n + \varepsilon_n)^2}{1 + \varepsilon_n^2}, \quad q_n = \frac{B - C}{\sqrt{D^2 - C^2}}. \tag{13}$$

For isolated resonances, we thus extract a Fano q-parameter which remains constant for a whole Rydberg series and a given choice of B, C, D. However, while the shape is generally preserved in the full series, the identification of Fano parameters is only valid for isolated resonances. In general, as Connerade (1985) has shown, Dubau–Seaton profiles depart strongly from the Fano shape (see fig. 2.10) and there exist two types of autoionisation profile which are readily distinguished by a geometrical construction (see fig. 2.11). Experimental examples of both types have been found.

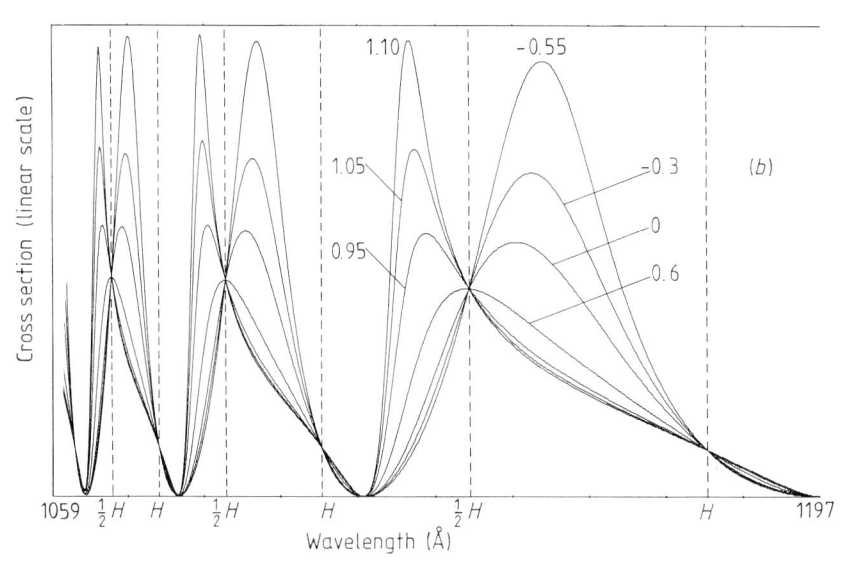

Fig. 2.10. Families of Dubau–Seaton profiles (Connerade 1985) showing the evolution away from the Fano (1961) shape as the parameter C is varied, while B and D are held constant. Hydrogenic points (labelled H) are those with $\tan \pi \nu = 0$, while those with $\tan \pi \nu \to \pm\infty$ are labelled $\frac{1}{2}H$.

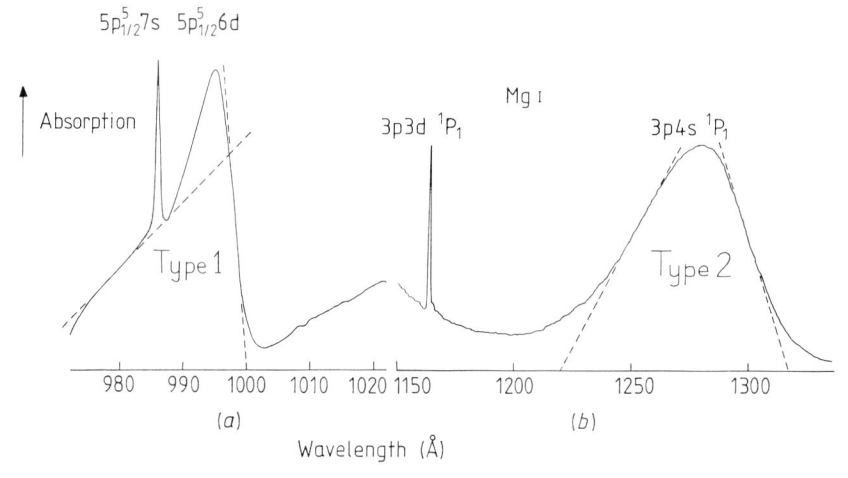

Fig. 2.11. Distinction between autoionising profiles of type 1 and type 2 (Connerade 1985), illustrated from experimental data.

It is of course rare that Rydberg series of autoionising resonances can be approximated by the two channel Dubau–Seaton formula. Often, changes in coupling scheme (Baig and Connerade 1978) or interlopers from other configurations complicate the picture. An especially interesting example of a Rydberg series of fairly narrow autoionising resonances traversing a very broad line and interacting with it occurs at the crossover between the 6s and 5d excitation spectra

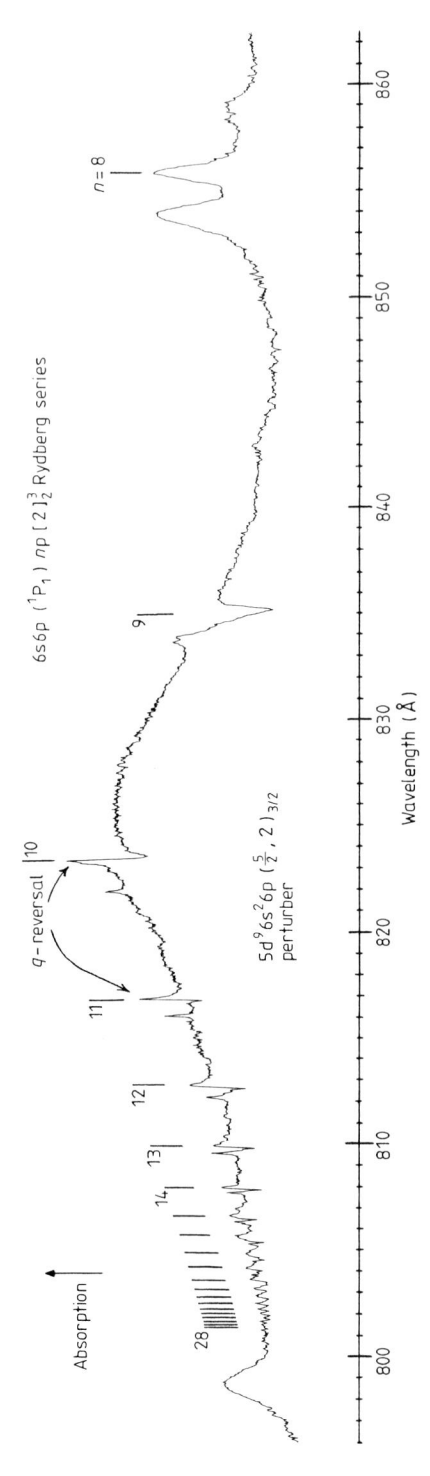

Fig. 2.12. A rare occurrence in the autoionisation range is the case displayed here of a series of autoionising resonances traversing another very broad resonance with which a strong interaction occurs: the profiles of the series members change shape and the profiles "reverse" (the q index changes sign) as the series passes through the energy of maximum interchannel coupling.

of Tl I. The profiles of the series members are modified and interesting changes of quantum defect occur as the point of maximum interaction is traversed. The resulting spectrum, displayed in fig. 2.12, was originally interpreted within the theoretical framework due to Cohen-Tannoudji and Avan (1977). It exhibits one of the finest examples in atomic physics of the 'q-reversal' effect (Connerade 1978a), namely a reversal in the asymmetry of an autoionising profile about the point of maximum coupling with a continuum of finite bandwidth. Very recently, the K-matrix formalism of Lane (1984, 1985) has been extended (Connerade et al. 1985) yielding comparatively simple analytical expressions for the line profile of a full Rydberg manifold interacting with a broad perturber. These formulae include such effects as the 'skewed q-reversal', shown in the calculated curve of fig. 2.13. A simple way of summarising the algebra involved is to write:

$$\sigma(\varepsilon) = A^2 \frac{\left\{\varepsilon + \frac{q_B \Gamma_B}{2} + N(\varepsilon) \sum_n \frac{H_n^2}{E_n - E}\right\}^2}{\left\{\varepsilon + \sum_n \frac{H_n^2}{E_n - E}\right\}^2 + \left\{D(\varepsilon) \sum_n \frac{H_n^2}{E_n - E} - \frac{\Gamma_B}{2}\right\}^2}, \quad (14)$$

where ε is now the detuning from E_B the broad level, q_B, Γ_B are Fano parameters for the broad level, H_n is an interaction strength coupling the Rydberg series of fine levels E_n to the broad level, and $N(\varepsilon)$, $D(\varepsilon)$ are numerator and denominator forms which are linear in ε (for more exact details, see Connerade et al. 1985). In addition to the q-reversal effect, these formulas predict 'vanishing widths' (Connerade et al. 1985) at certain energies, an effect well-known in nuclear physics as part of the subject of Robson asymmetries (Connerade and Lane 1985), but new to atomic physics. Recent, very high resolution laser experiments confirm this conclusion (Rinneberg et al. (1985). It is rather interesting and remarkable that the interaction with a perturber can actually "cancel" the broadening due to autoionisation, and stabilise levels high in the continuum.

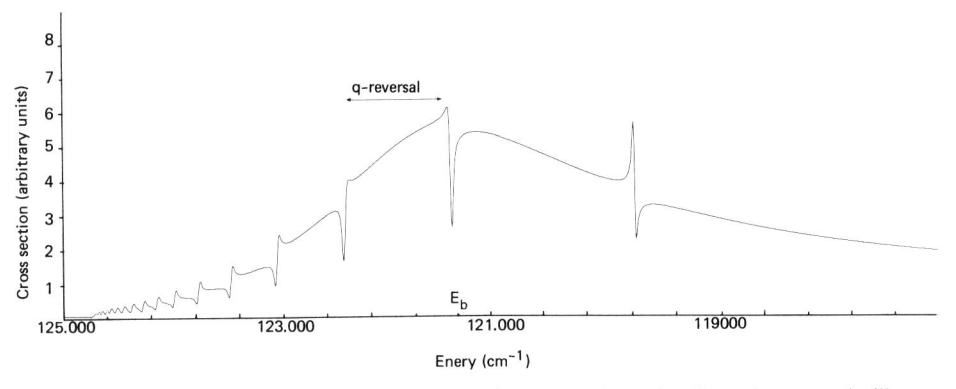

Fig. 2.13. Theoretical curve (Connerade et al. 1985) obtained from the K-matrix approach, illustrating the "skewed q-reversal" effect exhibited by the data in fig. 2.11.

Even this last example is a comparatively simple one in that few interacting channels are involved. As one proceeds upwards in energy through the excitation of inner shells (see below), the description of the broadening mechanisms becomes rapidly more complicated.

The study of autoionising profiles is an important theme of high resolution spectroscopy, because it yields very useful information on interchannel interactions. Further examples are discussed below in the context of inner shell excitation and molecular Rydberg series.

2.4. Inner shell spectroscopy of atoms

Proceeding to energies still higher than considered in sections 2.1 and 2.3, it eventually becomes possible to break into the inner shells of the atom. The outermost of the inner shells are the *inner valence* or *subvalence* shells of the atom; as the energy increases yet further, one breaks into the *deep* inner shells. More and more excitation channels become available with increasing energy, and we therefore begin by considering the outermost inner shells.

The simplest long series in inner shell excitation occur through excitation of an s subvalence electron in an atom with a closed outer shell. A good example is the series

$$3s^2 3p^6\ ^1S_0 \rightarrow 3s 3p^6 np\ ^1P_1$$

in argon, first reported up to $n = 20$ by Madden et al. (1969) and recently extended to $n = 28$ (Connerade et al. 1984, see fig. 2.14). This series, a good example of the "window resonances" discussed by Fano (1961), illustrates the fact that excitation of the inner shells can give rise to high Rydberg states, even though energies well above the first ionisation potential are involved.

Sharp inner shell transitions also occur when high l series are excited, provided the excited states are held out from the core by the repulsive centrifugal potential. An example occurs for 4f-subshell excitation in Yb I (see fig. 2.15), where transitions to nd and ng states are involved. The latter are the highest l values achievable for excitation from the ground state. When the inner shell series are as sharp as in this example, a Lu–Fano analysis as described in section 2.2 is feasible to extract interchannel interaction parameters and has in fact been performed (Baig and Connerade 1984b).

As remarked above, with higher energies one breaks into new excitation channels, and atomic spectra increase rapidly in complexity. It turns out that inner shell excitation energies can lie either (i) below or (ii) above the threshold for simultaneous removal of *two* electrons by absorption of a single photon (photo-double ionisation), and that, as pointed out by Connerade (1977), the appearance of the resulting spectrum depends markedly on which of (i) or (ii) occurs. Figure 2.16 shows some plots of excitation energies for d-shell spectra for elements in a given row of the Periodic Table, together with double ionisation thresholds, to illustrate the crossings. Similar plots can be drawn for each of the subvalence shells.

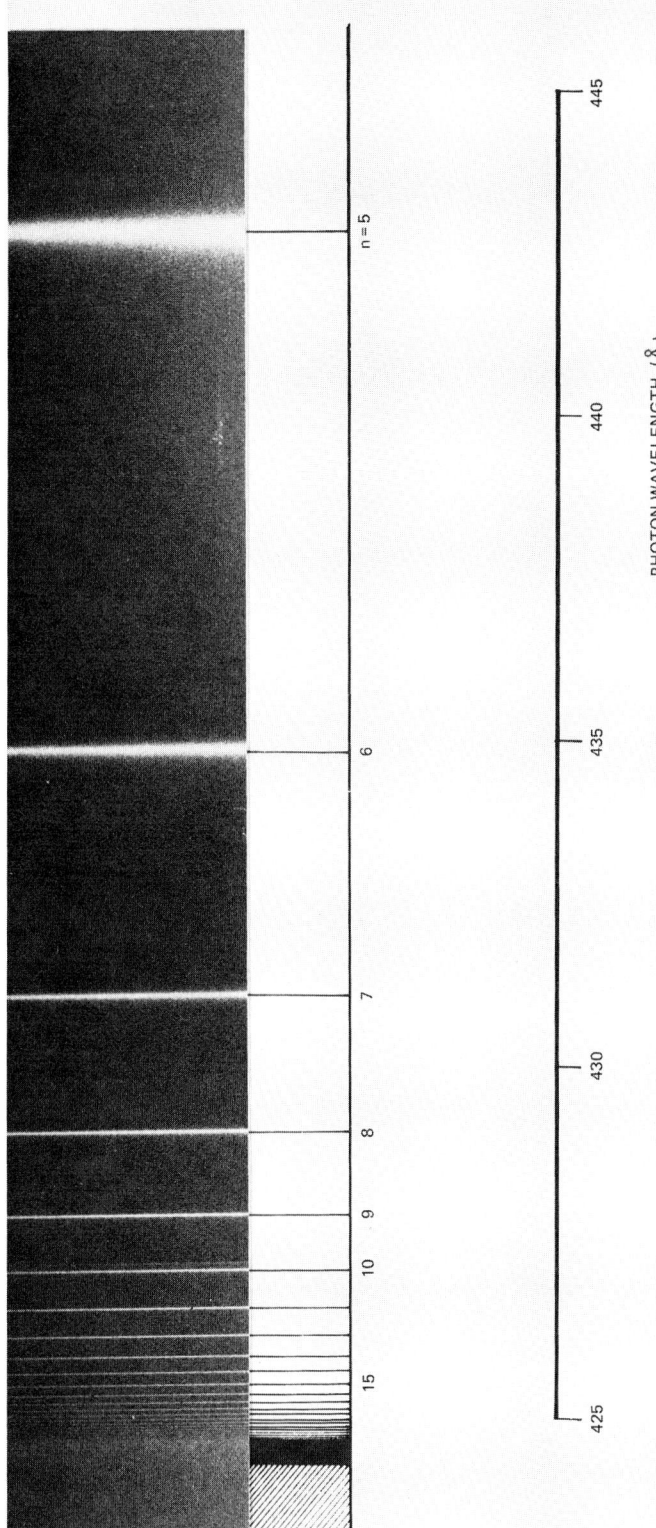

Fig. 2.14. Inner shell excitation series from the 3s subshell of argon, showing that long Rydberg series are by no means confined to excitation of optical electrons.

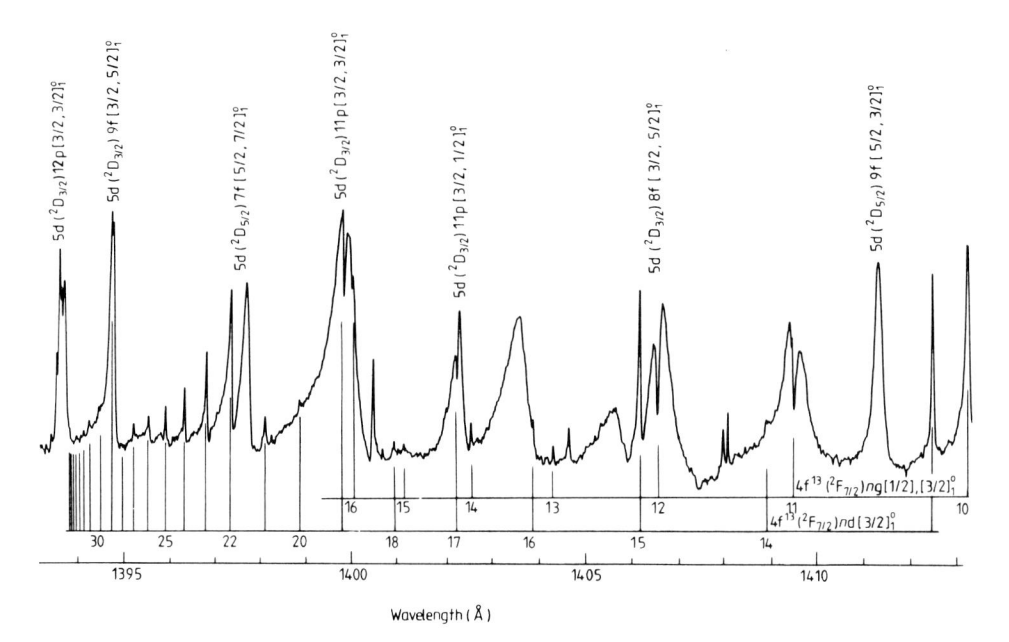

Fig. 2.15. Inner shell spectrum of Yb I, showing series due to the excitation of a single 4f electron to ng final states. These are the highest angular momenta attainable in photoabsorption from ground states.

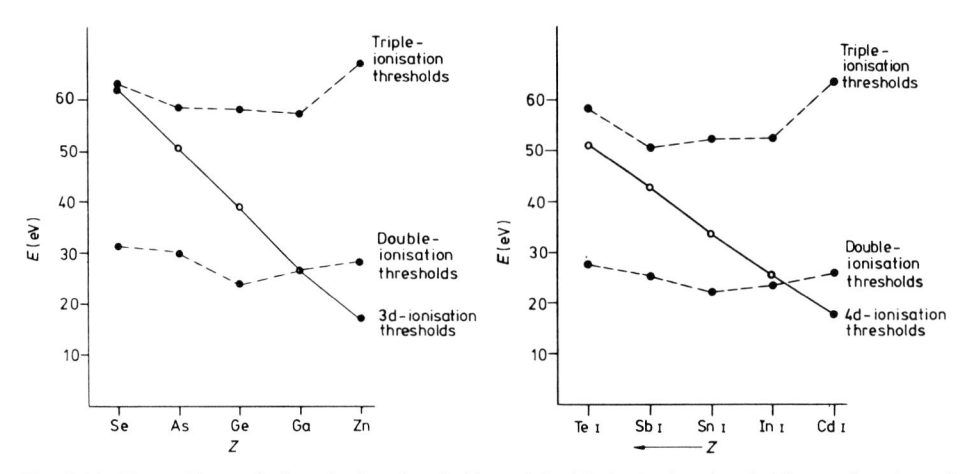

Fig. 2.16. Plots of inner shell excitation thresholds and double ionisation thresholds as a function of atomic species, which illustrate the existence of "crossings" between the curves. Near these "crossings", the interaction between single and double excitation/ionisation channels rises to a maximum.

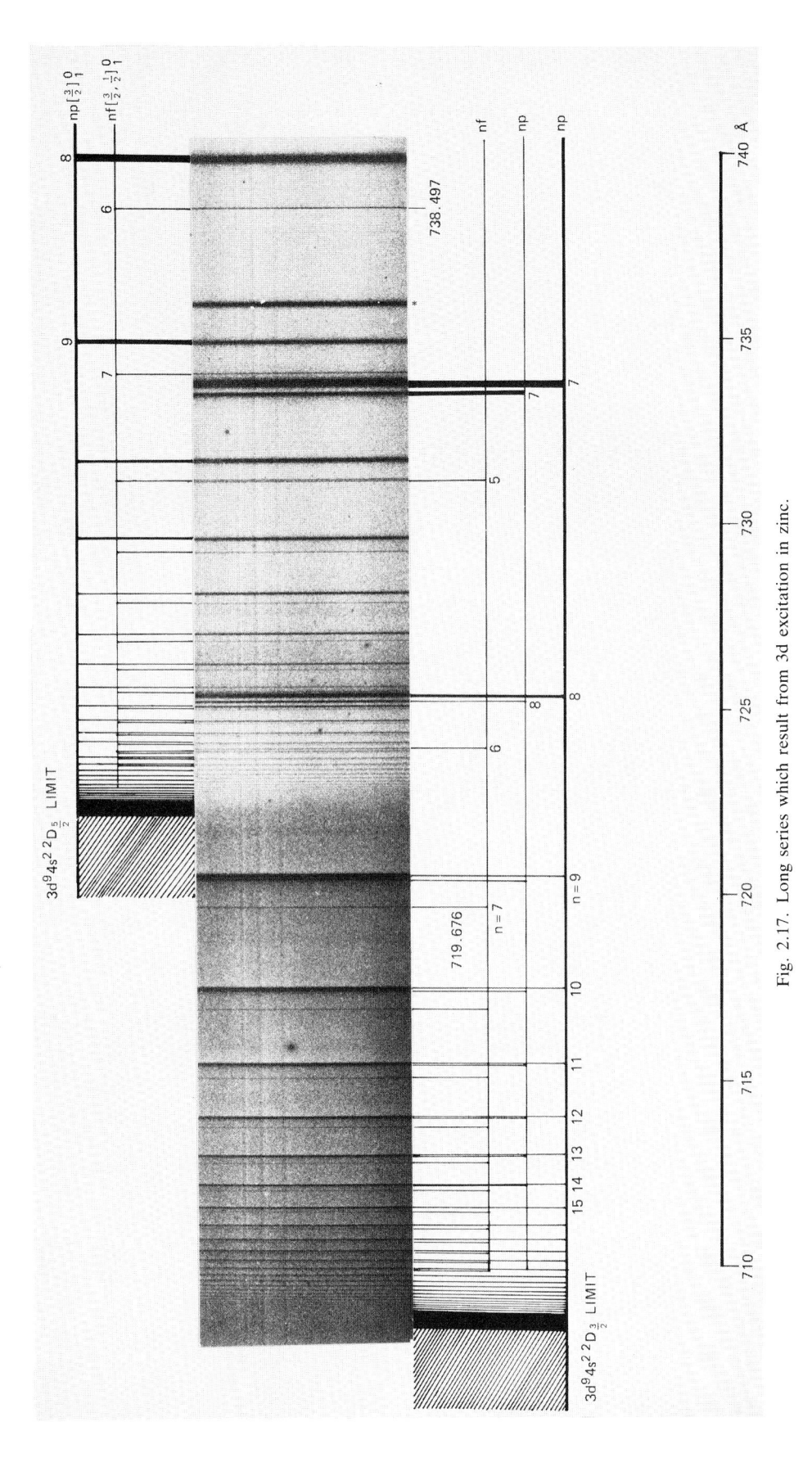

Fig. 2.17. Long series which result from 3d excitation in zinc.

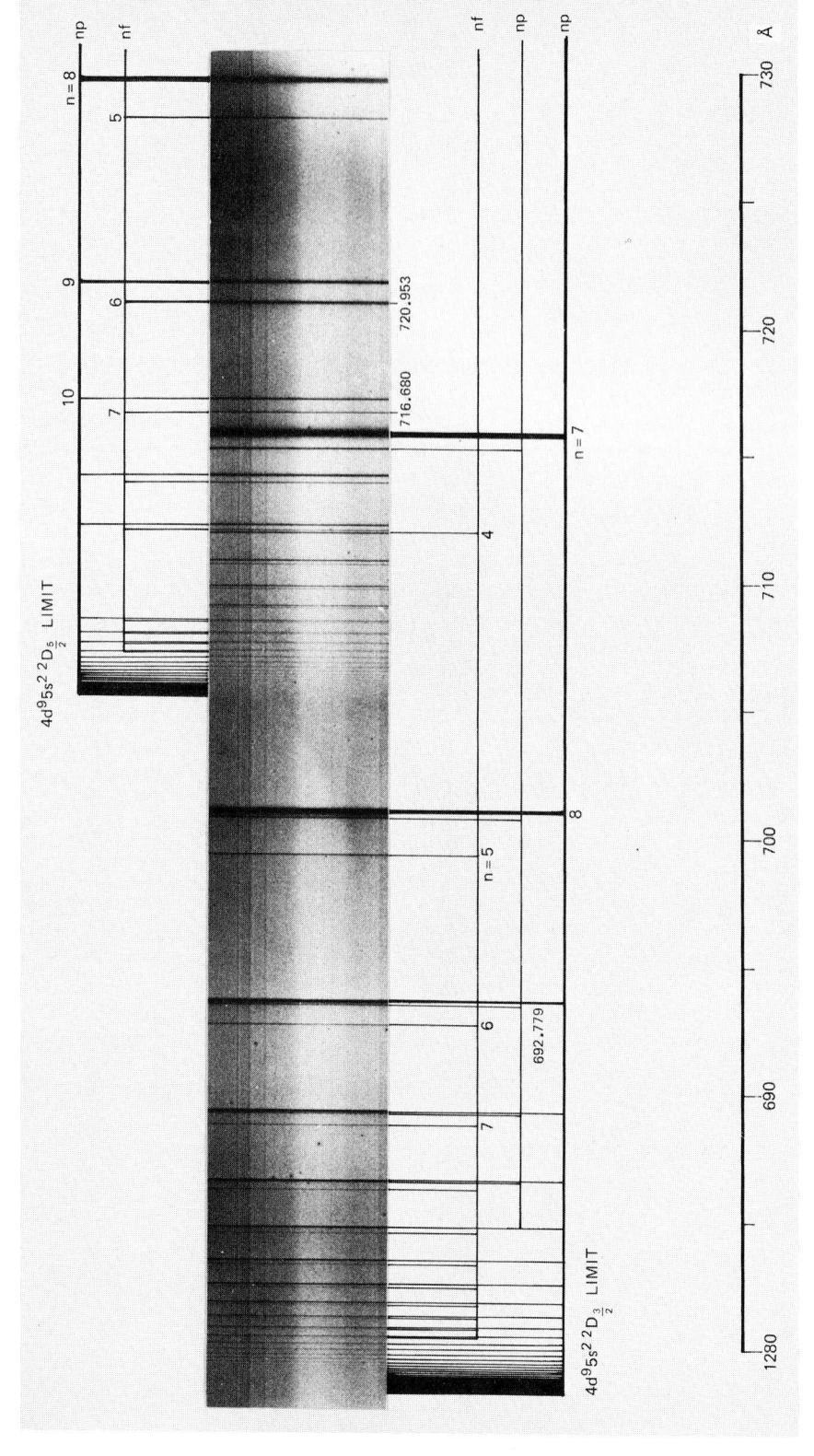

Fig. 2.18. Long series which result from 4d excitation in cadmium.

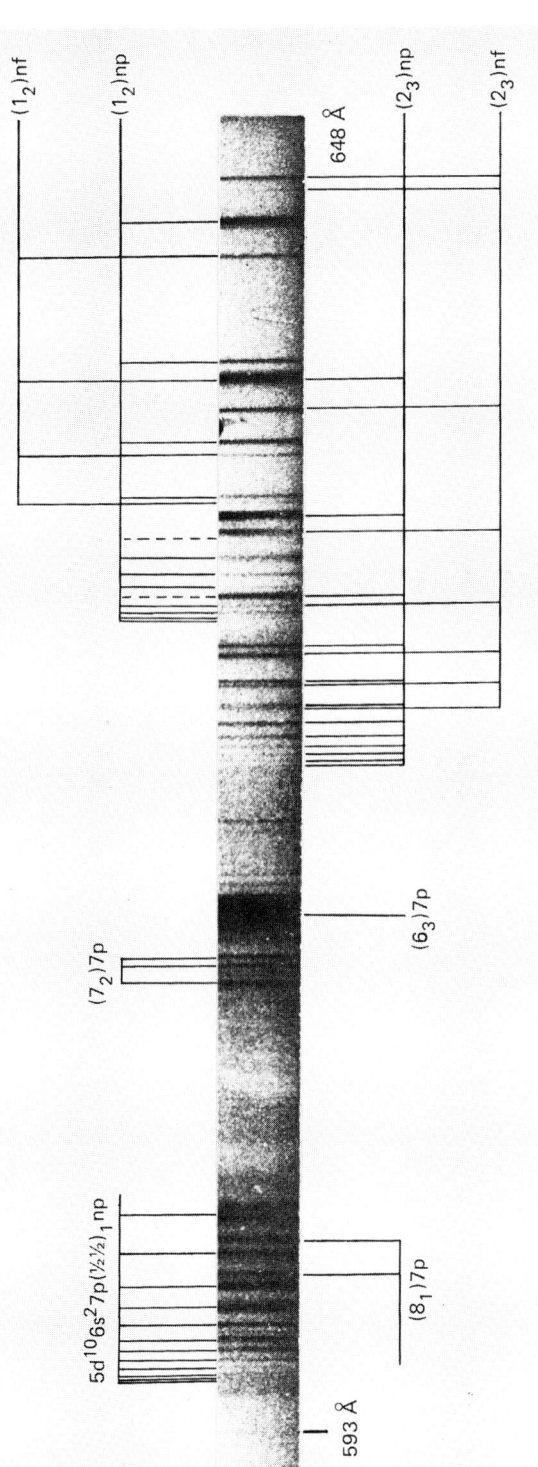

Fig. 2.19. Section of the 5d absorption spectrum of thallium in which fairly simple structures are observed. As the double ionisation threshold is approached (not shown) the spectrum becomes much more complex than shown here.

From these plots, we can see that, when the inner shell spectrum lies below the double ionisation limit, as do the outermost d-shell spectra of Zn and Cd, then the series limits correspond to bound states of the ion, and the only broadening mechanism which occurs is autoionisation, with a $1/\nu^3$ dependence which does not reduce the number of observable series members. Thus, although the lines are broadened at low n and the spectrum lies above the first ionisation threshold, long series are observed, as illustrated for both Zn and Cd in figs. 2.17 and 2.18 (Sommer et al. 1984).

As the inner shell spectrum sweeps up in energy towards the double ionisation threshold, degeneracies between single and double excitation energies become more frequent, and the latter gain oscillator strength at the expense of the former, so that the spectra gain in complexity. Even in such cases, the portion of the spectrum which lies lowest in energy (furthest from the double ionisation limit) may remain comparatively simple, as illustrated in fig. 2.19 for a small section of the Tl I 5d excitation spectrum. High resolution spectroscopy is of value in the analysis of such spectra, for which it becomes important to extract precise values of the ionisation limits in order to sort out which series are attributable to single and which to double excitations.

2.5. Configuration mixing in inner shell spectra

The subject touched upon in the last section, namely the interaction between single and double excitation channels in inner shell spectra, is a vast one, and an adequate review would require more than a single chapter. Rather than attempt to cover it in any detail, we confine our remarks to those aspects which relate directly to high resolution spectroscopy and give a few examples of relevant data obtained with synchrotron radiation.

Double excitation series occur throughout the vacuum ultraviolet spectra of atoms. One must distinguish between various types. In some cases, such as the classic series in He studied originally by Madden and Codling (1963), the doubly excited series stands clear of the singly excited spectrum. This spectrum has recently been observed at normal incidence with a high dispersion grating and is shown in fig. 2.20.

In other cases, double excitations fall near or within a singly excited spectrum. They are then usually enhanced by the resulting interchannel interaction. Another possibility, originally diagnosed in the inner shell spectra of the alkalis (Connerade 1970a,b) is that double or even multiple excitations should actually attach themselves to an inner shell single excitation spectrum as a result of breakdown in the characterisation of the core vacancy.

Perhaps the most spectacular example of the latter is provided by the 5p excitation spectrum of Ba I (discussed below), in which the excitation of a 5p vacancy simultaneously shatters the $6s^2$ outer subshell. Before considering this complicated case, it is useful to give examples drawn from the alkali spectra, which demonstrate the underlying principles.

When an electron is excited from the outermost p subshell of an alkali, there is

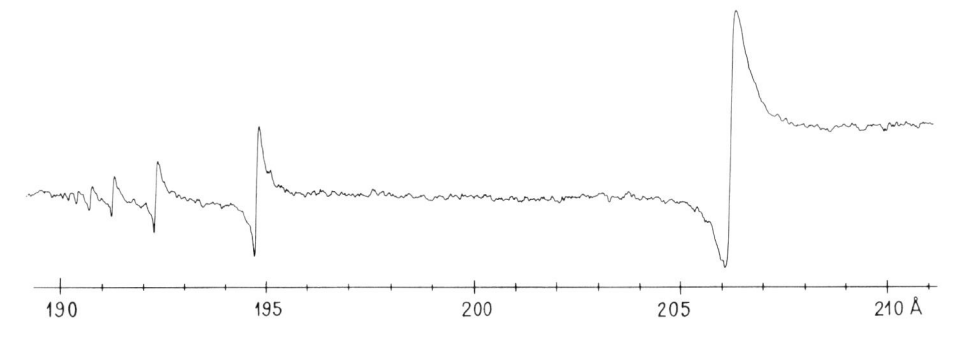

Fig. 2.20. Double excitation series of helium first observed by Madden and Codling (1963) using synchrotron radiation. The observations shown here were made using a 5000 lines/mm grating at normal incidence (Baig et al. 1984).

a fundamental distinction between those alkalis (Li, Na) which do not preceed a transition sequence in the Periodic Table and those which do (K, Rb, Cs). This distinction is apparent in many ways from the observed spectra, and is closely similar to the one made in section 2.3 between Ne and heavier rare gases. For example, the excitation scheme of Na gives rise to series built on the parent ion configuration $2p^53s$ which is comparatively "pure", whereas the corresponding ion state for Rb is $3p^54s$ which lies close in energy to $3p^53d$ and contains a good deal of admixture from the latter configuration. Consequently, strong configuration mixing effects set in between Na and Rb, giving rise to supernumerary Rydberg series in the inner shell spectrum.

Formally, the additional lines appear as double excitations. However, the character of the alleged single excitations is not clear either, and when mixing in the parent ion is severe, the distinction between single and double excitations becomes blurred. In practice, the supernumerary series have all the appearance of the spectrum into which they intrude. In a sense, they appear as the result of a back reaction of the Rydberg excitation on the core, which is "shattered" by correlation effects. This kind of double excitation is in complete contrast with the double excitations described as double Rydberg states, which are Rydberg states involving two running electrons (see for example the theoretical discussion by Rau 1983).

Another consideration in discussing p-subshell excitation in the alkali elements is the importance of centrifugal effects. These have been the subject of several reviews (e.g., Connerade 1978, 1984), and only brief comments will be made here.

Potential barrier effects for d electrons have been an explicit subject of study in recent years, following the detailed ab initio calculations by Griffin et al. (1969), but the importance of the centrifugal term for d electrons in the radial Schrödinger equation was implicitly recognised long before 1969 in a number of empirical statements such as the qualitative treatment of alkali atoms by penetrating and non-penetrating orbits (cf. White 1934). The latter model leads to the statement that μ_l, the quantum defect, depends strongly on l but is almost independent of the principal quantum number. To explain the detailed dependence on l is of course beyond this simple model and requires ab initio calculations.

To make the connection between alkali spectra and centrifugal barrier effects transparently clear, consider the fractional parts of the experimental quantum defects for ns and nd electrons from Li and Cs as deduced, for example, from Moore (1958). (See table 1.) The largest change in each sequence occurs between elements with empty d subshells (Li and Na) and those with filled d subshells (K, Rb and Cs). For the d electrons, the wavefunctions are nearly hydrogenic for Li and Na, where the centrifugal barrier excludes them from the core, but become non-hydrogenic for K, Rb and Cs as a result of "collapses" into the attractive inner region of the potential.

For the s electrons, the change is in some sense an opposite one: as a consequence of the filling of the d subshell, the outer reaches of the potential become more fully screened, that is, more nearly Coulombic, and the s wavefunctions therefore become more closely hydrogenic for K, Rb and Cs.

Additional information on the extent of spatial overlap between the core and ns or nd wavefunctions can be extracted by inner shell spectroscopy. The properties of the autoionising resonances which result from excitation of the p subshells can be related to the optical spectra of the alkalis via the potential barrier model. Although the behaviour described here is quite general down any homologous sequence of the Periodic Table (see Baig and Connerade (1984a) for the rare gases or Ruščić et al. (1984) for the halogens), the alkalis provide the best family of elements for this intercomparison between ns and nd states as: (i) bound states below the first ionisation potential, and (ii) autoionising states built on an excited core.

In fig. 2.21, we reproduce the spectra of Na, K and Cs, showing the autoionising resonances due to excitation from the outermost p-subshells. Both ns and nd series are excited according to the scheme:

$$m\mathrm{p}^6(m+1)\mathrm{s}\,{}^2\mathrm{S}_{1/2} \to m\mathrm{p}^5(m+1)\mathrm{s}({}^1\mathrm{P}_1)n\mathrm{s}, nd,$$

where $m = 2, 3, 4, 5$ respectively for Na, K, Rb and Cs.

The nd resonances, which start out comparatively sharp in Na where the d subshell is empty, rapidly become broad in going to K, Rb and Cs, where the d subshell is filled. This is due to the large increase in spatial overlap between the nd wavefunction and the core once "collapse" has occurred, as explained in the

Table 1
Fractional parts of the experimental quantum defects for ns and nd electrons from Li to Cs.

Quantum defects	Li	Na	K	Rb	Cs
μ_s	0.400	0.350	0.182	0.140	0.059
μ_d	0.003	0.014	0.259	0.330	0.475

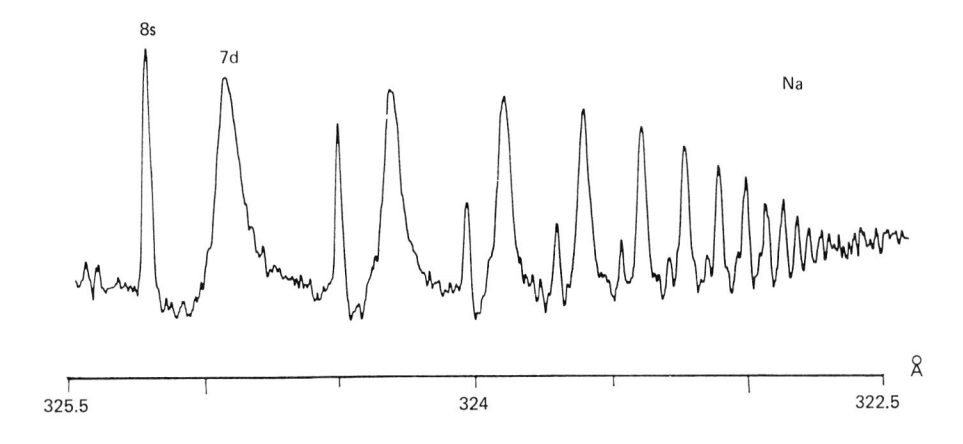

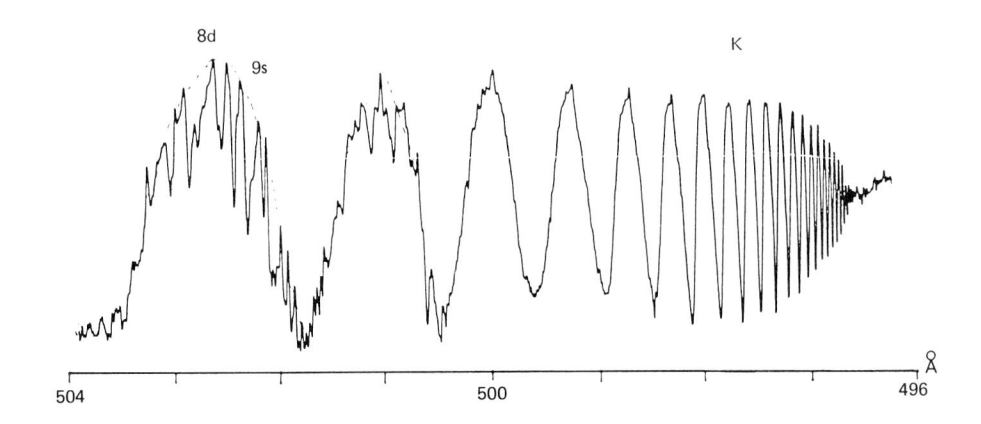

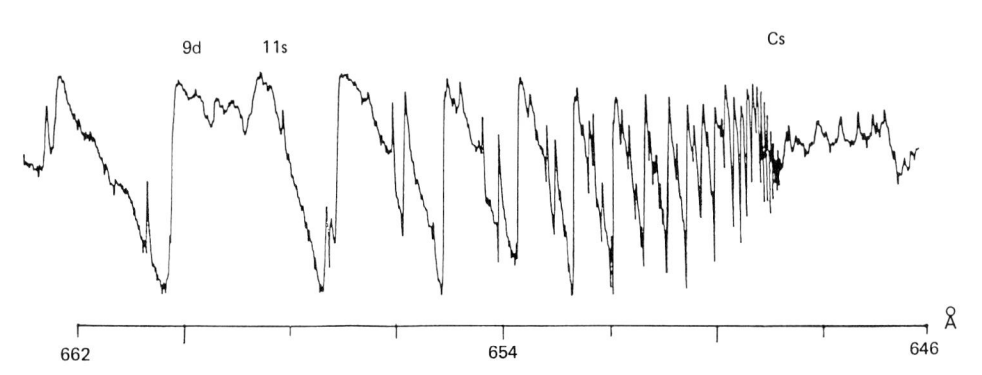

Fig. 2.21. Outermost inner shell absorption spectra of sodium, potassium and caesium: although superficially one might expect a detailed correspondence between these spectra, they are in fact quite different owing to configuration mixing effects in the parent ion.

case of Ne and the other rare gases by Baig and Connerade (1984a). Following the same sequence, ns and nd states which form the nearest pair in sodium cross over in energy in going towards Cs, becoming energy degenerate around K, again in precise analogy with the rare gases. There is, however, a slight distinction between the situation in the alkalis and in the rare gases: for the rare gases, the ns states can be considered as energy "markers", allowing the contraction of the d orbitals to be followed. In the alkalis, the s states also become more hydrogenic towards Cs and thus move oppositely in energy to the d states.

The reason for the selective contraction of the d orbitals with respect to the s orbitals as the atomic number increases is a quantum mechanical effect, which can be traced to the gradual deepening of an inner well in the effective radial potential (including the centrifugal term). The effect, somewhat loosely described as "wavefunction collapse", is present for values of $l > 1$, but is less dramatic for d electrons, in which there is only a shallow barrier between the inner and outer reaches of the potential, than for f electrons, which possess a well characterised double valley potential. For more details on the behaviour of the d wavefunctions in relation to centrifugal barrier phenomena and to the wavefunction collapse problem, see e.g. Connerade (1978b, 1982, 1984).

In table 2, we collect typical quantum defects for the ns and nd states of core-excited Na to Cs for comparison with table 1. We have also performed (Baig and Connerade, unpublished) single configuration Hartree–Fock calculations using the Froese-Fischer (1972) computer code, which show that the spatial overlap between the core and the excited nd wavefunction increases drastically between Na and K. Wavefunctions obtained in these calculations are plotted in fig. 2.22. The situation closely parallels that in the rare gases, for which a similar example of progressively increasing overlap of d-states with the core, as computed from the Hartree–Fock theory, is given in section 2.3 above.

A complication in the detailed interpretation of the spectra of K, Rb and Cs actually results from the "collapse" of the d wavefunctions: in the presence of a core vacancy, the $(m + 1)$s and md orbitals become degenerate in energy, leading to strong mixing between different configurations of the parent ion. As a result, the limiting structure is vastly enriched, and the superposition of many more Rydberg series than would be expected from the one-electron excitation scheme is observed.

The anomalously rich experimental structure of these alkalis was first observed by Beutler and Guggenheimer (1934), who were unable to account for the large number of transitions. An explanation for the enhanced structure in terms of

Table 2
Typical quantum defects for the ns and nd states of core-excited Na to Cs.

Quantum defects	Na	K	Rb	Cs
μ_s	0.72	0.55	0.40	0.32
μ_d	0.35	0.55	0.68	0.75

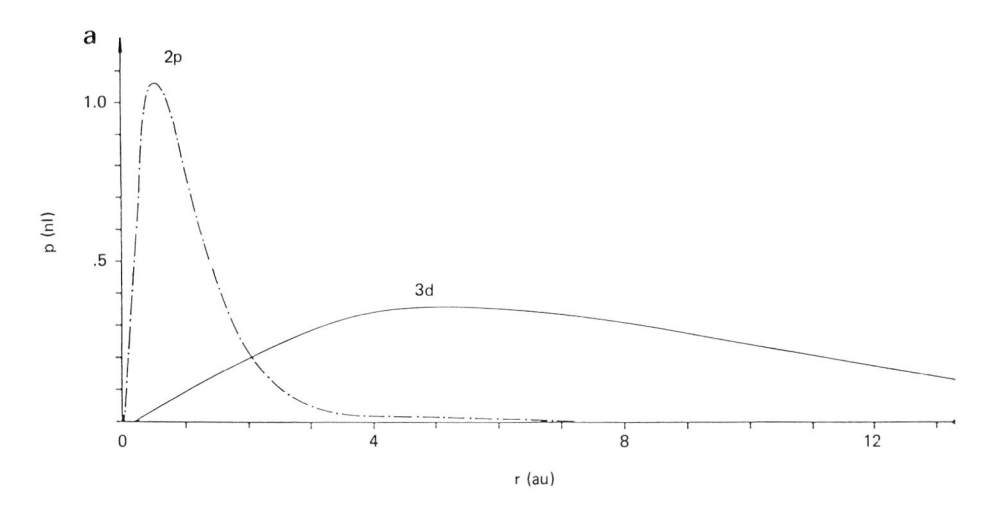

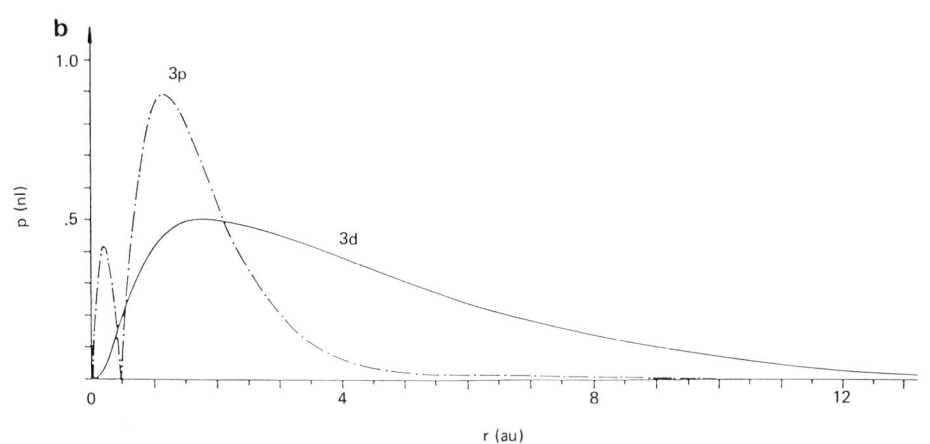

Fig. 2.22. Wavefunctions of the outermost d-electrons in (a) Na and (b) K obtained from single configuration Hartree–Fock calculations. Notice how the spatial overlap between the d-electron and the core increases dramatically between Na and K.

parent ion configuration mixing ("s × d mixing") was first proposed by Connerade (1970a,b) and this interpretation has since been confirmed by a number of more recent experiments (e.g. Süzer et al. 1976) and calculations (e.g. Rose et al. 1980) for not only the alkalis but also the rare earth elements.

The situation in the alkaline earth sequence is very similar and, if anything, even more complex. Here, in absence of mixing, the spectrum would be very simple indeed, with just the two $p^5 s^2 (^2P_{1/2})$ and $(^2P_{3/2})$ limits, as in the rare gases. However, the mixing involves the full $p^5 [s^2 \times sd \times d^2]$ manifold of the ion, leading to structures of great complexity. Figure 2.23 shows the spectra of Ca I and Sr I (Baig and Connerade 1987), and fig. 2.24, a small portion of the corresponding spectrum of Ba I (Baig et al. 1984b). In the latter spectrum, more than fourteen

Ca

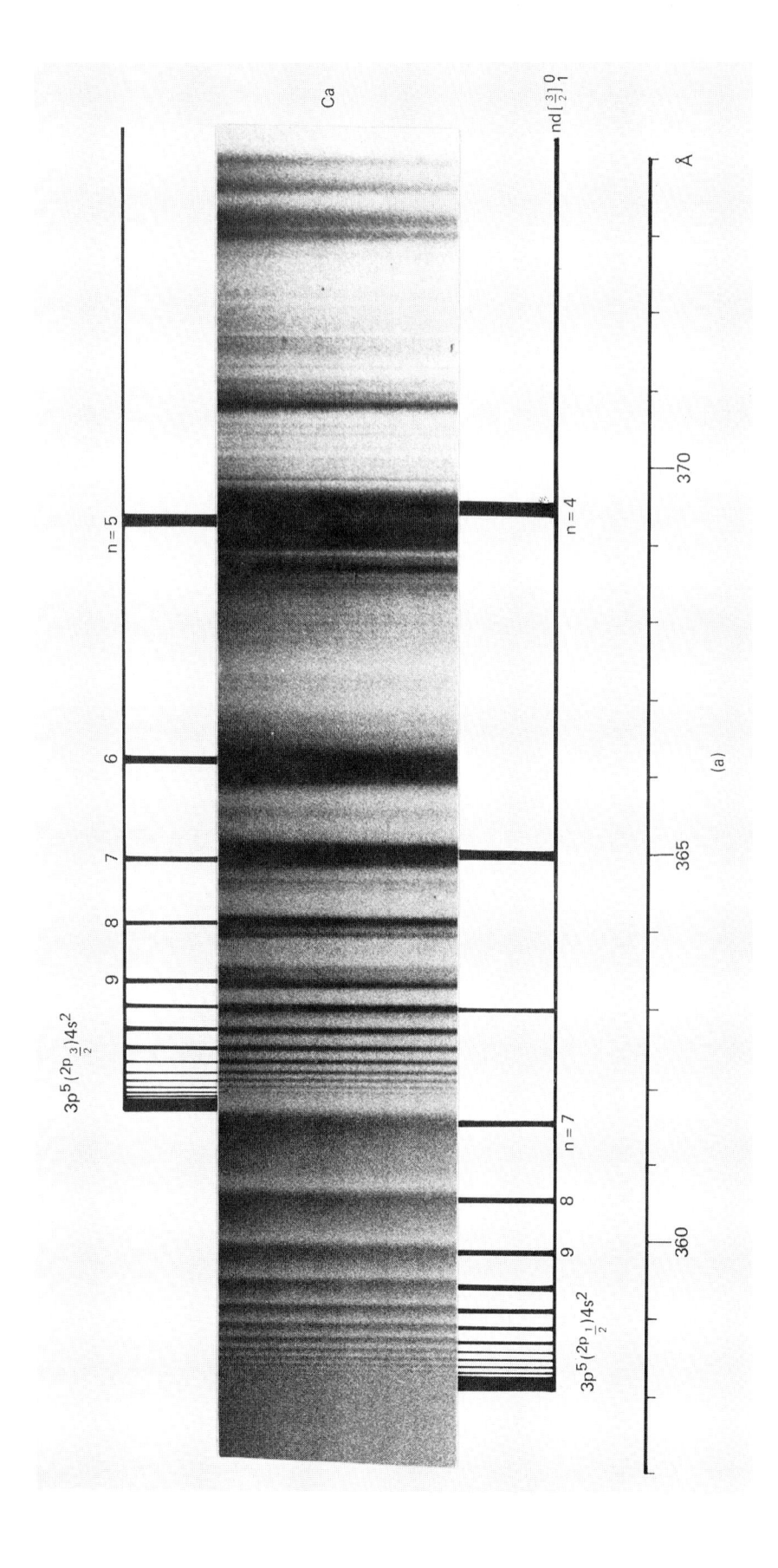

$3p^5(2p_{3/2})4s^2$ n=5 6 7 8 9

$nd[3/2]^0_1$ n=4

n=7 8 9 $3p^5(2p_{1/2})4s^2$

360 365 370 Å

(a)

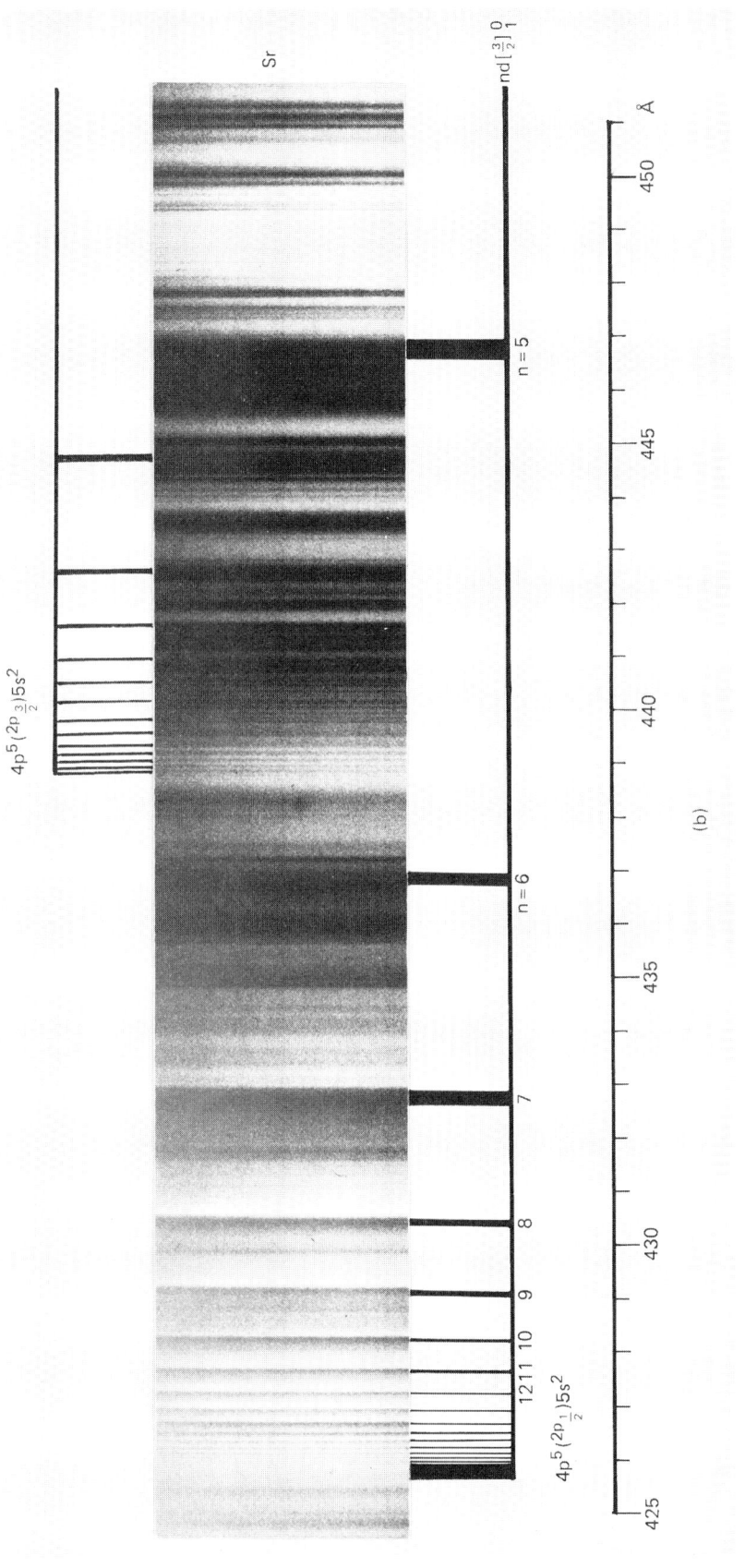

Fig. 2.23. Outermost inner shell spectra of: (a) calcium and (b) strontium, both of which exhibit complexities due to configuration mixing in the parent ion.

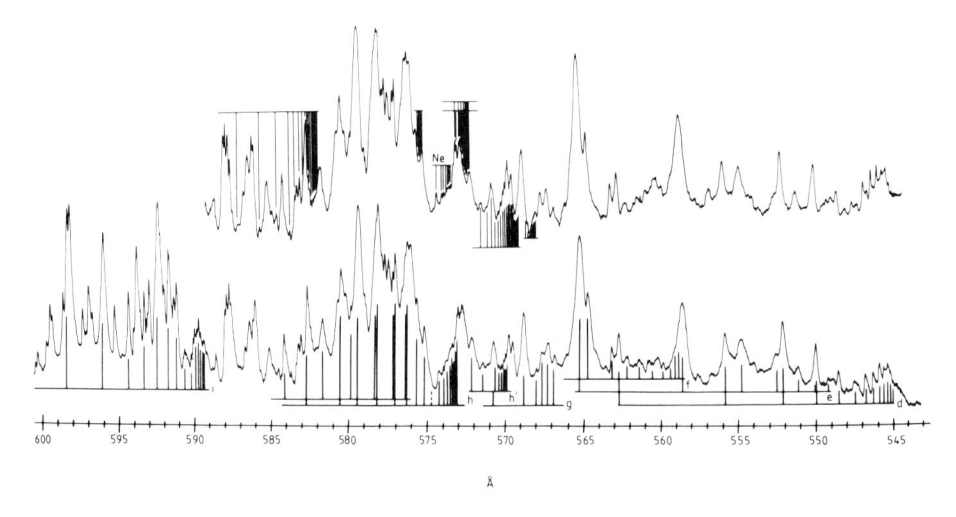

Fig. 2.24. Portion of the 5p absorption spectrum of barium, perhaps the most extreme case of parent ion mixing ever encountered. It has been described as "shattering" of the $6s^2$ subshell by excitation of the 5p electron.

series converging on at least 12 limits have been found, and the highest resolution is necessary to unravel the details of the classification.

For 5p excitation in Ba, it is fair to say that the $6s^2$ subshell is completely shattered and that all possible fragments are found. The resulting spectrum is a spectacular display of correlation effects, and a very elaborate theoretical approach with proper inclusion of relativistic effects has been required to make sense of it (Rose et al. 1980).

The 5p spectrum of Barium also provides a beautiful example of the value of intercomparisons between different high resolution spectroscopies. It has been known for a long time (see Hotop and Mahr 1975 and refs. therein) that there is an anomalously high level of photo-double ionisation from 5p-excited Ba. Figure 2.25 shows the basic process currently believed to be responsible, derived from an original suggestion by Hansen (1975). The process is known as two-step autoionisation and gives a detailed insight into the Auger effect at higher energies. Figure 2.25b shows how the ejected electrons from each of the two steps have been detected in photoelectron spectroscopy under excitation at a single wavelength (Hotop and Mahr 1975). A proper analysis of the absorption spectrum proved critical in establishing that two-step autoionisation is indeed energetically viable (Connerade et al. 1979, Connerade and Martin 1983). Two-step autoionisation was also confirmed in photoelectron spectroscopy by Rosenberg et al. (1979, 1980).

High resolution photo-double ion spectroscopy (Lewandowski et al. 1981) has been successfully applied to determine the spectral structure of Ba^{2+} photoproduction. Figure 2.26 shows a comparison between the photo-double ion

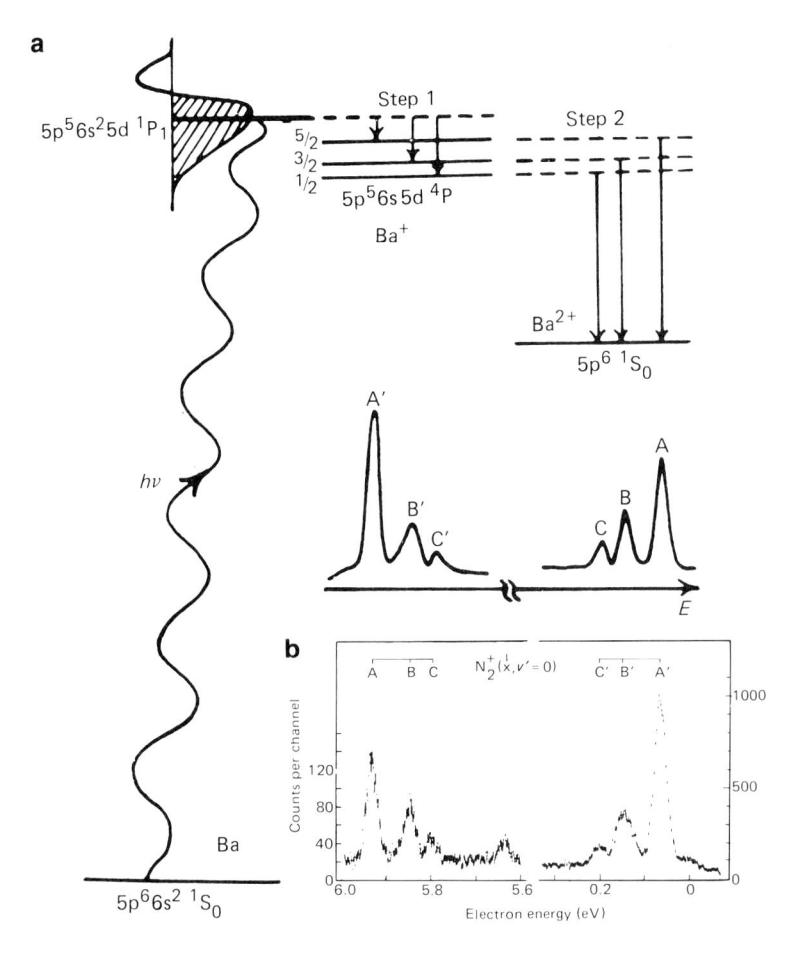

Fig. 2.25. (a) Schematic representation of two-step autoionisation, (b) the spectrum of electrons ejected in the two-step autoionisation process.

spectrum and high resolution photoabsorption data for neutral Ba. This portion of the Ba photoabsorption spectrum has now been observed in the second order of a 3-metre 6000 lines/mm grating (Connerade et al. 1984). The comparison has enabled a detailed analysis of the two-step autoionisation mechanism to be achieved. It is generally true that a complete interpretation of photoabsorption spectra due to inner shell excitation requires more than photoabsorption data, and that information on the decay channels after the original photoexcitation process is crucial to a proper understanding. Thus, the most fruitful approach to high resolution synchrotron spectroscopy of atoms is a careful intercomparison between data obtained from different kinds of experiment, e.g. photoabsorption, photoion, photoelectron, etc.

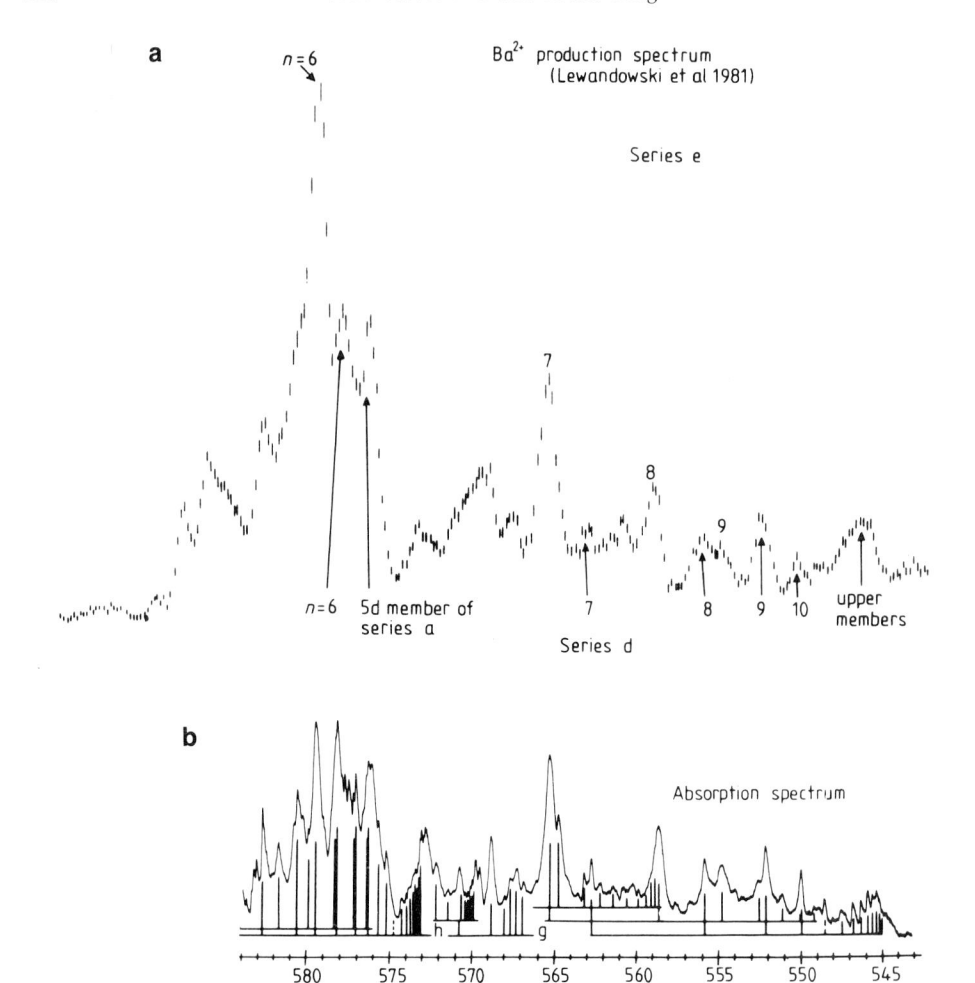

Fig. 2.26. Comparison between (a) the Ba^{2+} photoproduction spectrum, and (b) the photoabsorption spectrum in the range of two-step autoionisation.

2.6. "Giant resonances" in atomic spectra

The subject of "giant resonances" in atomic spectra is one which deserves mention in this account of photoabsorption spectroscopy, because of the important role synchrotron radiation has played in their study. A complete discussion, including for example the ramifications in solid state physics, resonant photoemission spectroscopy and photo-stimulated desorption from surfaces, far exceeds the scope of the present chapter. For the atomic effects, we refer to some reviews (Connerade 1978b, 1982, 1983c), and confine our comments here to some more recent work.

The name "giant resonance" in atomic physics is given to very broad and intense features of the photoabsorption spectrum, in which all the available oscillator strength of a given photoabsorption channel is concentrated, to the exclusion of the usual Rydberg series. These "giant resonances" are normally features of the continuum, and possess the distinctive property that they survive relatively unchanged from the atom to the molecule to the solid. Indeed, the latter property can be used as an experimental test to characterise "giant resonances".

In terms of Hartree–Fock calculations, "giant resonances" are explained by including the centrifugal term in the effective radial potential, which is then found to possess a double-well for the relevant elements. The inner, short-range well is capable of trapping the innermost loop of a continuum state of appropriate energy. This effect produces a resonance, the properties of which (intensity, breadth, energy) are somewhat insensitive to environment, because the inner well lies deep inside the atom, shielded from the outer reaches of the potential.

Although "giant resonances" may be rather insensitive to environment, they are a sensitive measure of how binding is the inner well, and thus to the particular model used in ab initio calculations. They have been found to depend strongly on many-body terms and on the inclusion of relativistic effects in theoretical models (cf. Band 1981), and have therefore served as a testing ground for atomic theory.

High resolution spectroscopy contributes mainly through the study of their profiles, and this is therefore a good place to discuss a two-parameter profile formula specific to "giant resonances". This is obtained from quantum scattering theory by invoking a result due to Schwinger, namely that the low energy scattering spectrum of a short-range well is independent of the precise form of the well. We therefore choose a square well with angular momentum, for which analytic solutions in spherical Bessel functions are readily available. We thus obtain the phase shift in terms of the wavevectors k' and k inside and outside the well and of the radius a of the well. This has the form:

$$\chi = \tan \delta = \frac{zj_l(z') \, j_{l-1}(z) - z'j_l(z) \, j_{l-1}(z')}{zj_l(z') \, j_{-l}(z) + z'j_{-l-1}(z) \, j_{l-1}(z')}, \tag{15}$$

where the angular momentum is l and j_l are spherical Bessel functions of the first kind, of order l, while $z = ka$, $z' = k'a$

Partial wave theory then yields a cross section:

$$\sigma(k) = A^2 \frac{\chi^2}{1 + \chi^2}. \tag{16}$$

The resulting profiles depend, for each l, on two shape parameters, namely the depth and width of the well. The limitation of this approach is implicit in Schwinger's theorem: since the profile depends only on how binding is the well and not on its shape, the parameters have no separate physical interpretation. Nevertheless, the formula gives a very good representation of a pure "giant resonance" as shown in fig. 2.27 for the 4d→f excitation of gadolinium.

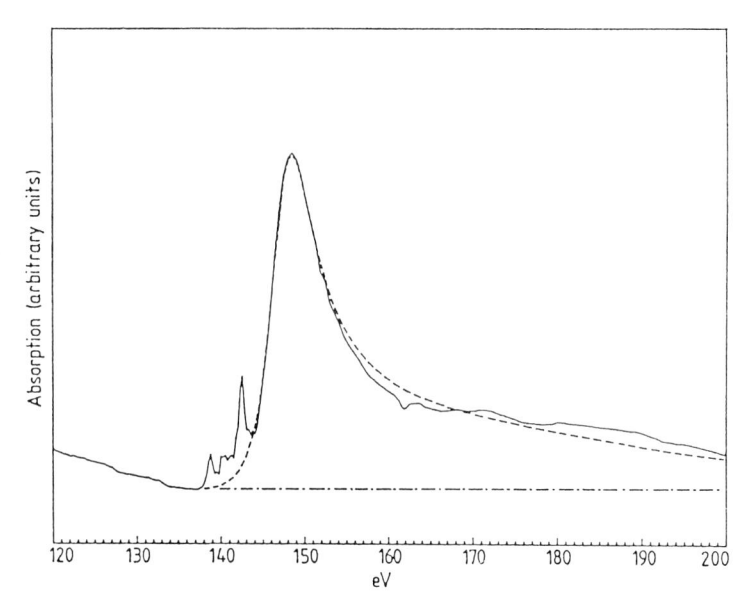

Fig. 2.27. Illustration of the fit obtained from the spherical square well formula for the 4d→f "giant resonance" in gadolinium (Connerade and Pantelouris 1984).

The formula also brings out some general features: the profile is not the same as the Fano profile. In particular, as long as only one channel is involved, the zero is tied not to the resonance energy but to the threshold. This implies that, as the resonance sweeps up through the continuum with decreasing binding strength of the inner well, it becomes progressively broader. In terms of the uncertainty principle, this can be understood if we consider the "giant resonance" as a virtual state embedded in the continuum.

The approximately linear relationship between width and energy above threshold of "giant resonances" has been verified experimentally for free atoms (Connerade 1984).

3. High resolution spectra of molecules

3.1. Application of synchrotron spectroscopy to the study of rovibronic structure

The prime application of high resolution spectroscopy to molecules lies in the determination of rotational constants (see e.g. Herzberg 1950) and vibrational data from which the potential curves can be calculated and, ultimately, the symmetries of the excited states can be determined.

Until recently, the resolution available at synchrotron radiation laboratories has not been high enough to achieve this aim, and rotationally resolved spectra of molecules were mainly studied using conventional sources at longer wavelengths. With the advent of holographic gratings of high ruling density, the situation has changed: rotationally resolved spectra of molecules can now be obtained with synchrotron radiation at energies which were previously inaccessible, and which approach the ionisation potentials of many interesting systems. It is no exaggeration to claim that a new domain of molecular spectroscopy has been opened up.

The simplest rotational structure occurs in diatomic molecules. We begin with the case of H_2, which of course has attracted most interest. In spite of the great wealth of previous work, synchrotron radiation has enabled some extensions to the high resolution spectrum in the VUV.

There have been many investigations, both theoretical and experimental (Huber and Herzberg 1979, and references therein) of the Lyman band system of H_2, which is built on the lowest energy electronic excitation of the H_2 molecule, namely $B\,^1\Sigma - X\,^1\Sigma$. The two most recent high resolution absorption studies (Wilkinson 1968 and Dabrowsky and Herzberg 1974) have relied on conventional sources for background continuum radiation, so that several rotational lines were obscured by extraneous emission features.

Wilkinson (1968) used the 5th order of a 6.6-metre 1200 lines/mm grating, achieving 0.4 Å/mm dispersion, in conjunction with an Ar discharge continuum source. He observed and analysed all the bands, from $(0, 0)$ to $(4, 0)$, but the $(3, 0)$ and $(4, 0)$ bands were partly obscured. Similarly, Dabrowski and Herzberg (1974) used the 3rd order of a 10 metre grating, achieving a resolution slightly lower than that of Wilkinson (1968), in conjunction with a hydrogen tube as background source. They reported and analysed the bands from $(5, 0)$ to $(17, 0)$, but again a number of lines were obscured due to emission lines from the continuum source. Thus, the absorption data were incomplete. Of course, it is possible to obtain more complete data by observing emission spectra, as done by Herzberg and Howe (1959), but photoabsorption data are more reliable for rotational analysis, provided they are recorded at low pressure, so that there is no extraneous broadening mechanism. In Dabrowsky's (1984) flash absorption experiments, a discharge at 12 kV was operated in the absorption cell to populate high vibrational and rotational levels of the $X\,^1\Sigma_g$ state. Pressures of 1 Torr were used in the cell. Dabrowsky (1984) comments that many of the lines are broad, not as a result of predissociation, but as a result of conditions in the absorption cell.

In fig. 3.1, we show a portion of the H_2 photoabsorption spectrum recorded at very low pressure with path lengths of around 10 metres using synchrotron radiation. The experiment covered the bands from $(0, 0)$ to $(9, 0)$. Figure 3.1 shows the $(7, 0)$ band. Within the same wavelength range, the $(0, 0)$ band of the $C\,^1\Pi - X\,^1\Sigma$ system can be seen quite clearly, and there are intensity perturbations as a result of the energy overlap of the two systems.

Most of the bands studied were partially obscured in earlier measurements, and a distinct advantage of synchrotron radiation as compared to conventional sources is the featureless continuum which provides a clean background for high resolution spectroscopy.

Rotational analyses of the Lyman absorption bands have been performed on the basis of the new data. We do not find any drastic disagreements with earlier

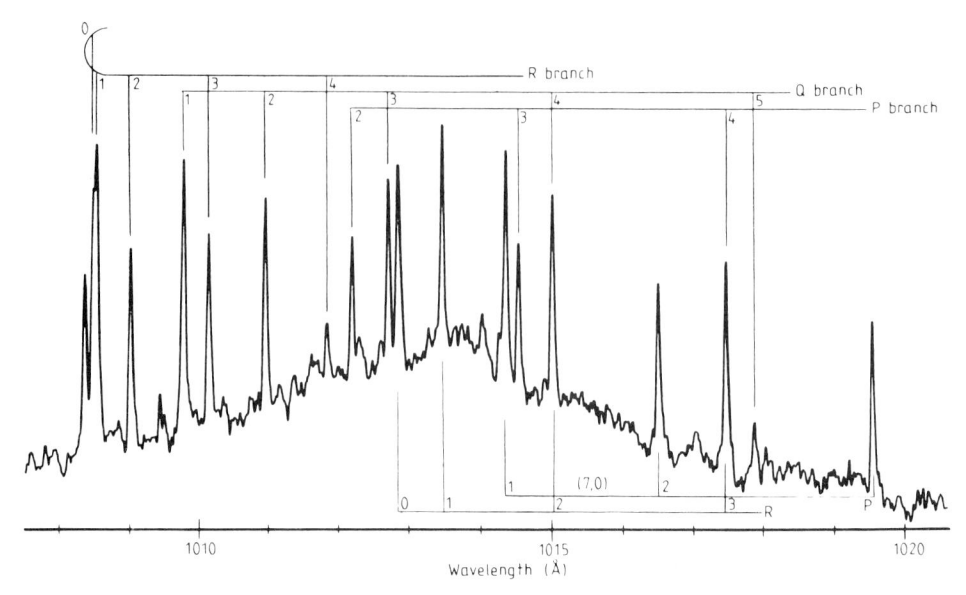

Fig. 3.1. Portion of the photoabsorption spectrum of H_2 recorded at low pressure and long path length, showing a perturbation described in the text.

determinations, but the values we obtain for the rotational constants are more consistent and more reliable, since they are based on data recorded under more favorable conditions.

We turn now to the hydrogen halides which have also received much attention in the vacuum ultraviolet using conventional sources (see e.g., Ginter and Tilford 1970, 1971, Ginter et al. 1980). In fig. 3.2, we show an overall view of the HBr absorption spectrum in the range 1100–1200 Å recorded with synchrotron radiation (Baig et al. 1981). The rotational structure in this spectrum is rendered more complicated by the energy overlap between different electronic states. To see how this arises, consider a p complex, i.e., the full manifold of molecular excited states which arises from the excitation of a p orbital: we expect to find both Σ and Π electronic states. For a d complex, we expect to find Σ, Π and Δ states, and the number of overlapping systems will, in general, increase rapidly with increasing angular momentum. (For a discussion of complexes in molecular spectra, see Kovacs 1969.) Thus, simple spectra only occur for molecules with a filled shell ground configuration (e.g. CO, SiO, GeO, etc.) yielding a $^1\Sigma$ state and for excitation to an s molecular orbital which yields only a Σ state, as observed for example in the F $^1\Sigma$ systems of SiO and GeO (Lagerqvist and Renhorn 1974, Baig and Connerade 1979, 1980).

Such a $\Sigma - \Sigma$ transition occurs in HBr and DBr for the band system shown in fig. 3.3. The comparison between isotopes is useful because (i) the observed differences in rotational constants confirm the rotational analysis, (ii) the isotopic shift is smallest for a (0, 0) transition which yields the vibrational assignment within an

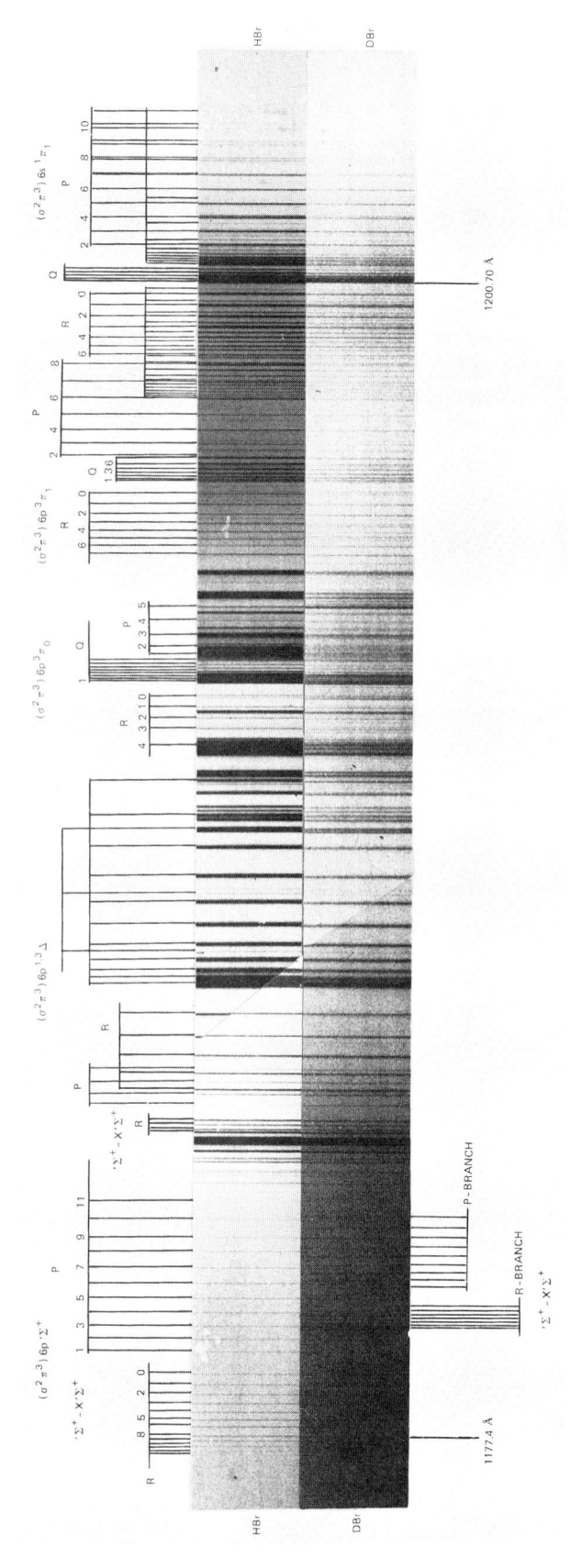

Fig. 3.2. Absorption spectra of HBr and DBr showing complex structure in the 1100 to 1200 Å range.

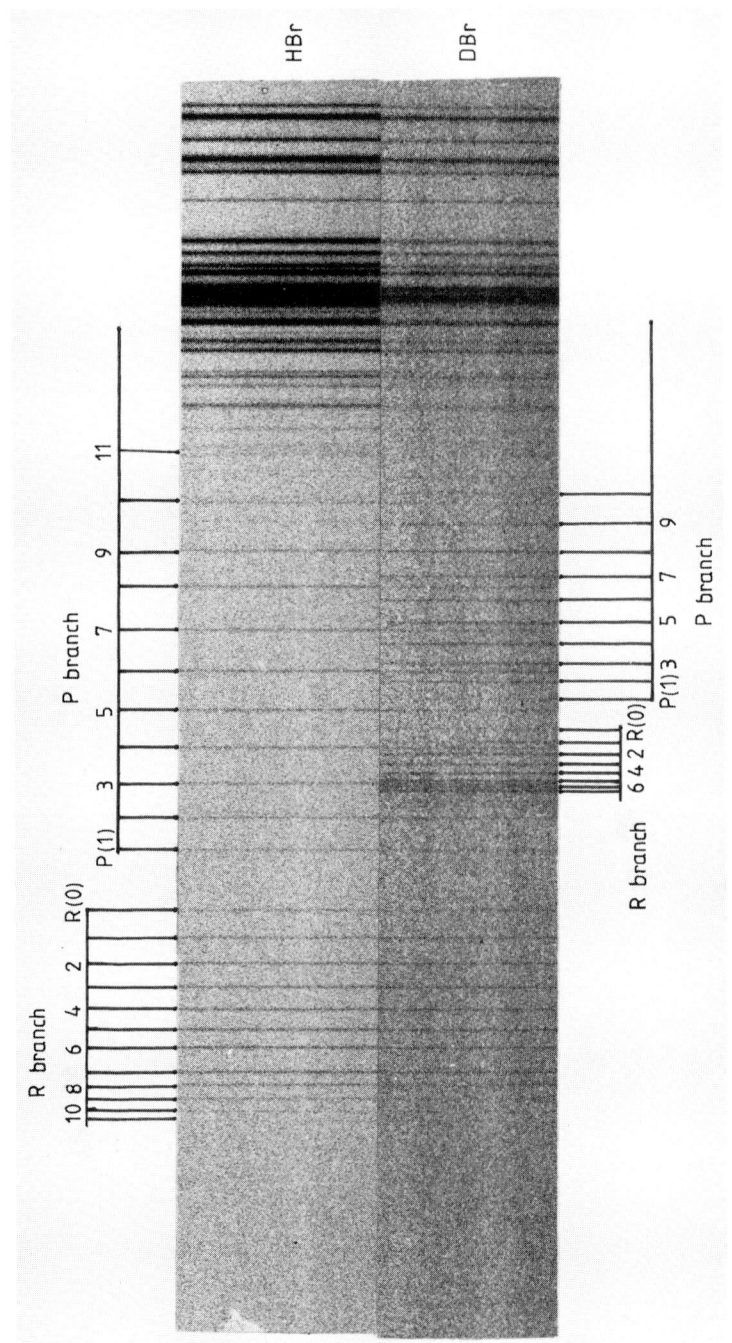

Fig. 3.3. Change in rotational structure between corresponding bands of HBr and DBr due to the isotope effect.

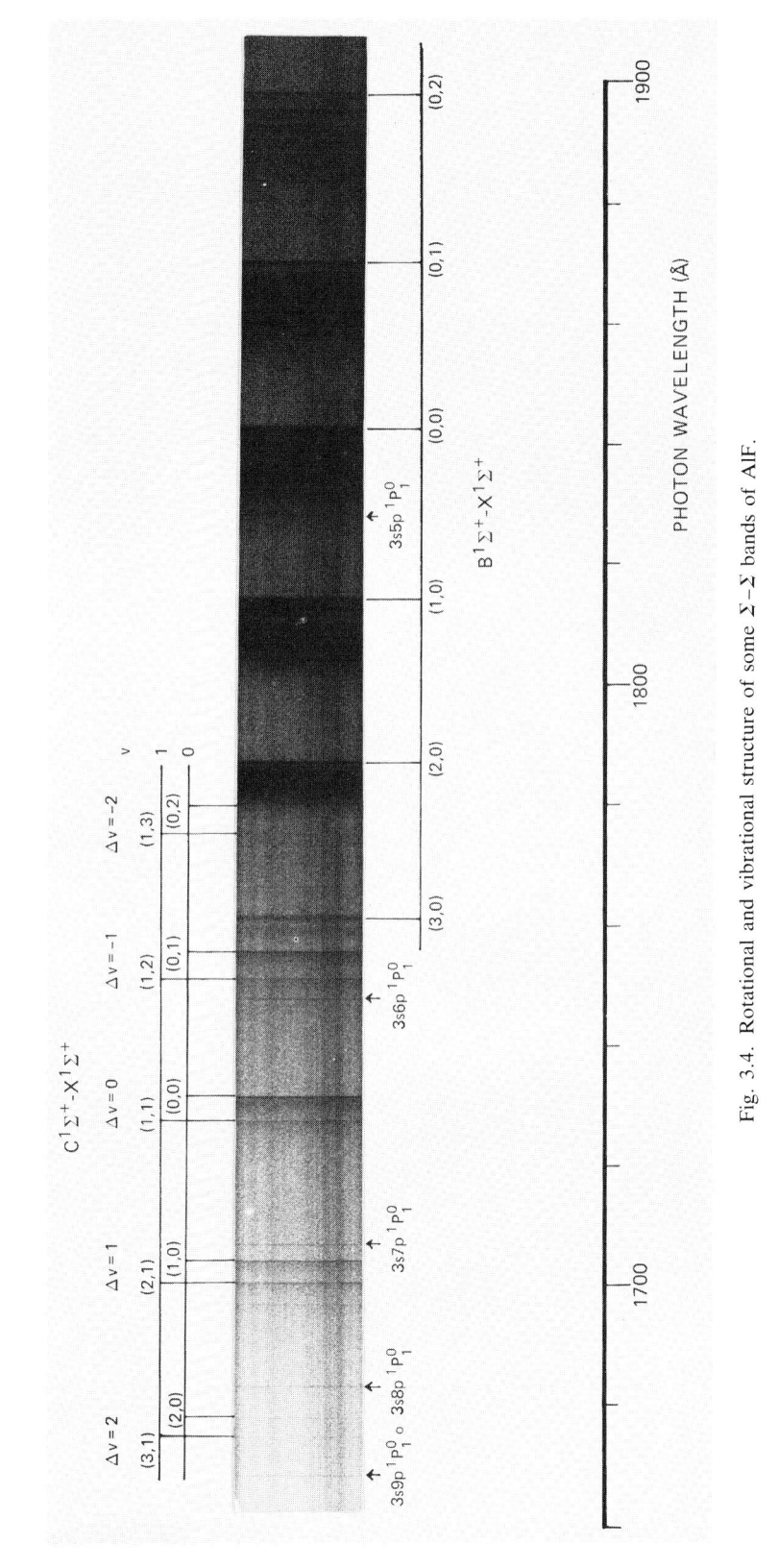

Fig. 3.4. Rotational and vibrational structure of some $\Sigma-\Sigma$ bands of AlF.

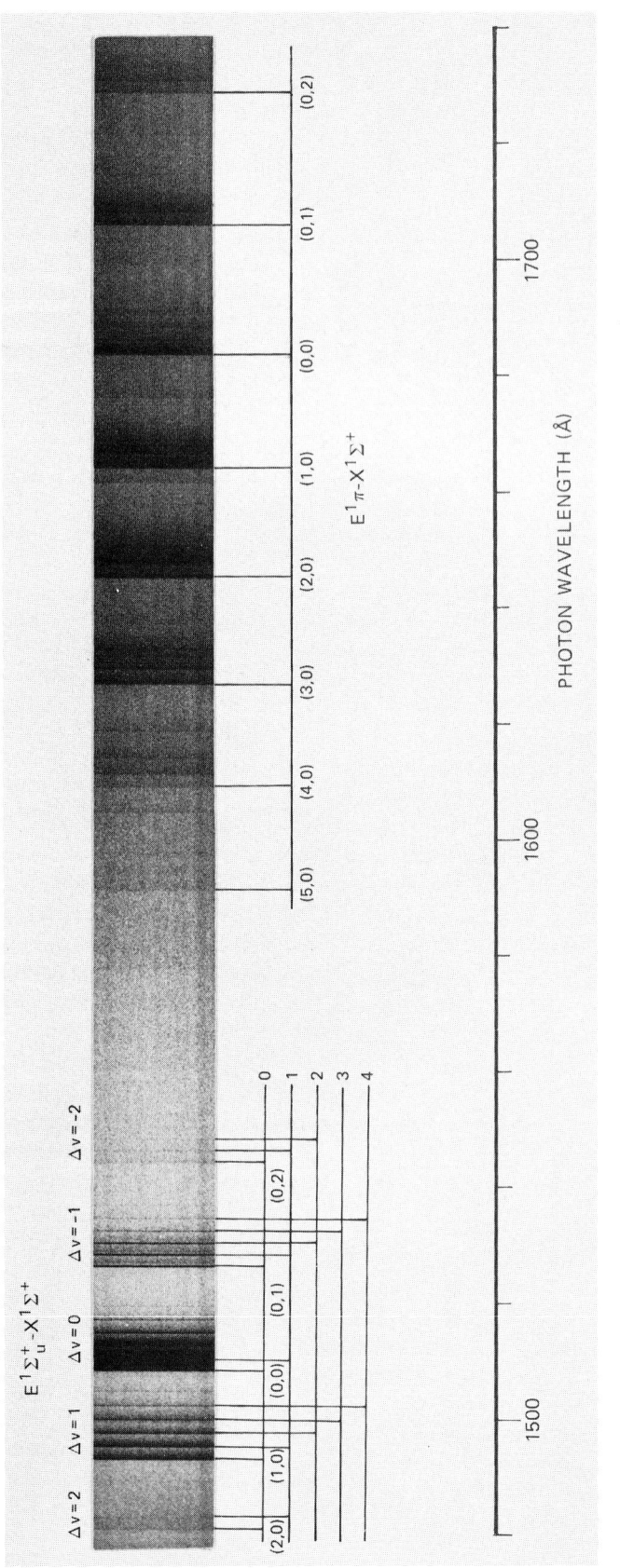

Fig. 3.5. Rotational and vibrational structure of some $\Sigma-\Sigma$ and $\Sigma-\Pi$ bands of P_2.

electronic transition directly, and (iii) the pure electronic transition energy can be extracted from the rovibronic analysis and then correlated with united atom estimates to determine electronic configuration assignments.

Yet another example of $\Sigma-\Sigma$ transitions is shown in fig. 3.4, which displays the B and C systems of the AlF molecule (Baig and Connerade 1979b). The AlF molecule was generated using the chemical reaction:

$$MgF_2 + 2Al \rightarrow 2AlF + Mg$$

and the presence of Mg atoms was recognised by the observation of the principal series, which lies in the same wavelength range. Both the B and C systems are well developed using this method of generation and exhibit interesting vibrational and rotational detail. Here, each of the two electronic states possesses many vibrational levels. For each vibrational level, there is a complete rotational structure (unresolved in fig. 3.4). The intensities of the transitions from the (0, 0) ground state to upper vibrational levels then fall off with increasing vibrational quantum number in the excited state – a clear manifestation of the Franck–Condon Principle.

Another example involving $\Sigma-\Sigma$ transitions is shown in fig. 3.5. A column of the P_2 molecule was generated by the thermal decomposition:

$$P_2N_5 \rightarrow P_2 + 2N_2 + PN$$

and, in addition to the absorption spectrum of P_2 shown in the figure (Baig and Connerade 1979b) the spectrum of N_2 and a Rydberg series of the PN molecule were also observed. The $E\ ^1\Pi$ and $G\ ^1\Sigma$ band systems of P_2 are displayed. Again, the (0, 0) transitions are the most intense, and all possible combinations of the v', v'' progressions were observed in the $G\ ^1\Sigma$ system. The bands are degraded to the red, which shows that the rotational constants are smaller in the excited state than in the ground state.

3.2. The asymmetric top problem

In the previous section on rovibronic structure, the examples chosen were for diatomic molecules. These are, of course, by far the simplest systems, because there is only one value of the moment of inertia to be determined from the rotational constant $B = h/4\pi^2 I$. For most molecules the situation is much more complex, but polyatomic molecules often possess an axis of symmetry with reference to which principal moments of inertia are conveniently defined. One defines three mutually perpendicular axes about which the moments of inertia are I_A, I_B, I_C. One further assumes $I_C > I_B > I_A$. In order of increasing complexity, we then have the following cases:

(1) $I_A = I_B = I_C$. This is known as the spherical top, but no example of electronic spectra has yet been analysed.

(2) $I_A = I_B \neq I_C$ or $I_A \neq I_B = I_C$. This is known as the symmetric top, and subdivides as:
  (a) $I_A = I_B$ symmetric oblate, and
  (b) $I_B = I_C$ symmetric prolate.
(3) $I_A \neq I_B \neq I_C$. This is known as the asymmetric top, which not surprisingly yields the most complex rovibronic structure.

The case of the linear molecule, in this notation, corresponds to $I = I_A = I_B$ if one allows $I_C = 0$ and is therefore a limiting case of (2), with a rotational structure very similar to that of a diatomic molecule.

An important example of the asymmetric top is provided by H_2O, which possesses a rich and complex spectrum in the vacuum ultraviolet, leading up to the ionisation potential at about 900 Å. Each one of the electronic states is split into a multitude of sharp lines with no apparent regularity. The 1240 Å band was analysed by Johns (1963), and more recent data for both H_2O and D_2O obtained with synchrotron radiation are shown in fig. 3.6. Until 1980, there existed little or no data towards higher energies. There are two experimental techniques which have been used successfully to break into the higher energy range, namely synchrotron spectroscopy (Connerade et al. 1980) up to the first ionisation potential, and three-photon laser spectroscopy for the 1130 Å band (Ashfold et al. 1983), but the latter is more restricted in wavelength coverage.

In fig. 3.7, we show some examples of higher energy band systems in H_2O the structure of which has not yet been analysed. A useful approach to this complicated problem is to study corresponding structures along a homologous molecular sequence (i.e. H_2O, H_2S, H_2Se and H_2Te where the substituted atoms all belong to a column of the Periodic Table). The reason for which this leads to a simplification of the structure can be understood as follows. The atomic orbitals responsible for bonding the hydrogen atoms to the substituted atoms X are the p wavefunctions of X which would lead to a 90° bond angle were there no electrostatic repulsion between the two hydrogen atoms or hybridization.

Clearly, electrostatic forces tend to increase the bond angle (see fig. 3.8a). Similarly, it is easy to see how hybridization allows the angle to increase. Consider H_2O, and construct the hybrid orbitals

$$h_1 = N_1(s + \lambda_1 p_1), \qquad h_2 = N_2(s + \lambda_2 p_2),$$

where p_1 is along the x axis and

$$p_2 = \cos\theta_{12} p_x + \sin\theta_{12} p_y.$$

Orthogonality implies that

$$\int h_1 h_2 \, d\tau = N_1 N_2 \left\{ 1 + \lambda_1 \int p_1 s \, d\tau + \lambda_2 \int s p_2 \, d\tau + \lambda_1 \lambda_2 \int p_1 p_2 \, d\tau \right\} = 0$$

that is

$$1 + \lambda_1 \lambda_2 \cos\theta_{1,2} = 0.$$

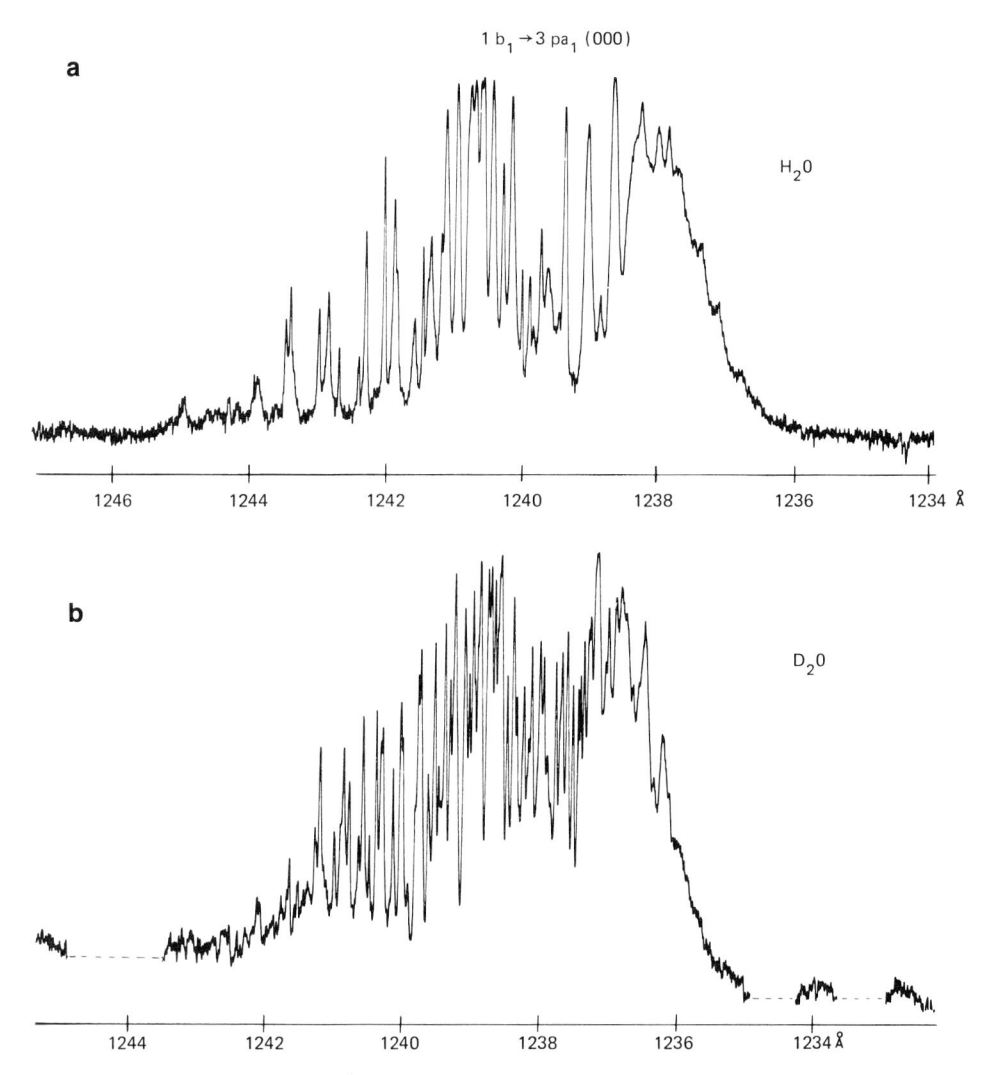

Fig. 3.6. Complexity of the 1240 Å band of (a) H_2O and (b) D_2O as recorded in the first order of a 6000 lines/mm holographic grating using synchrotron radiation. These bands, which were first observed and analysed by Johns (1963) are characteristic of an asymmetric top molecule.

By symmetry $\theta_1 = \theta_2$, so $\cos\theta_{1,2} = -1/\lambda^2$. Since $0 < \lambda_1, \lambda_2 < 1$, this implies $\theta_{1,2} > 90°$, which increases the overlap and decreases the repulsion between the bonds. The third hybrid of the trio must be orthogonal to the other two, so $\cos\theta_{1,3} = -1/\lambda\lambda_3$, and $\cos\theta_{2,3} = -1/\lambda\lambda_3$, i.e. $\cos\theta_{1,3} = \cos\theta_{2,3}$, which means that the lone pair sticks out to the rear of the oxygen atom. The unpaired electrons lead to a measured intrinsic dipole moment. Also, the dissociation energy of H_2O differs by only 10% from twice the dissociation energy of HO, which confirms the picture that the bonds are independent.

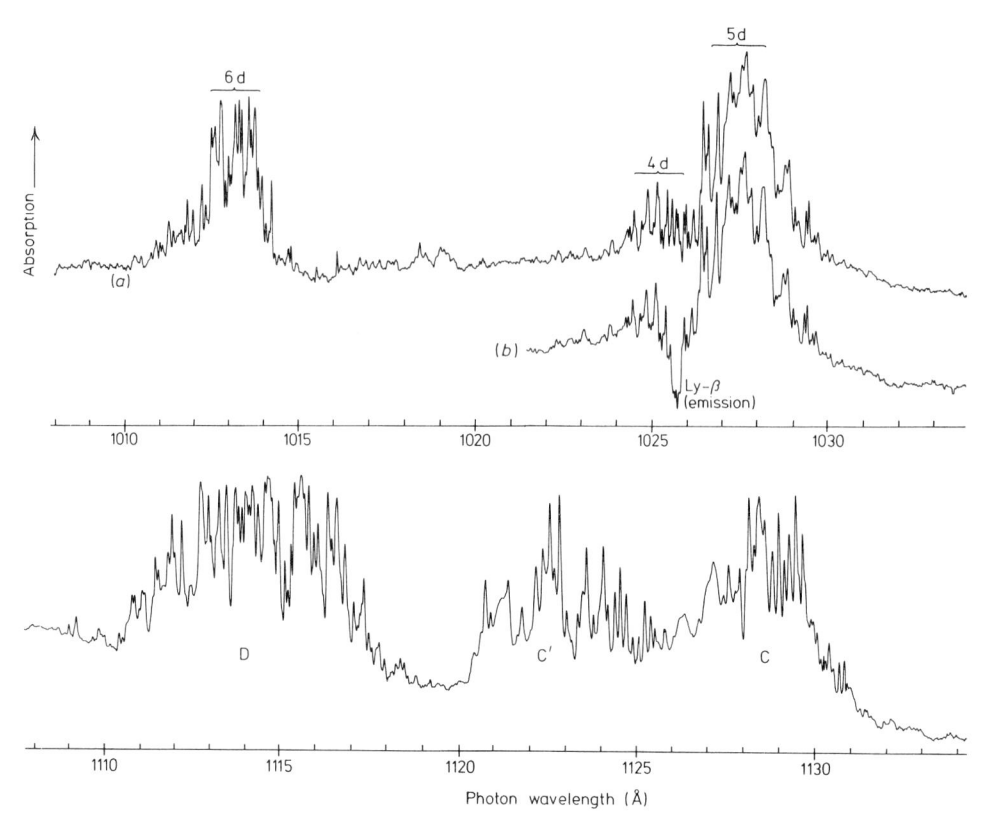

Fig. 3.7. Further band systems of H_2O at shorter wavelengths, showing the persistent complexity of the spectrum towards higher energies. Many such systems have been observed converging on the $^1B_1$ limit of H_2O.

More quantitative considerations lead to the conclusion that about 5° of the increase above 90° in H_2O is attributable to Coulombic repulsion between the two hydrogen atoms, while the rest may be due to hybridization.

In an intuitive way, one can also "guess" that the bond angle will approach 90° as the X atom of H_2X becomes very large. However, as Coulson (1982) remarks: "There is as yet no satisfactory explanation of why, except for H_2O, all bond angles are close to 90°, implying p-bonds with little or no hybridization".

This dramatic decrease of hybridization is a crucial factor in the simplification of the structure referred to above.

It turns out that, the more massive atom X and the closer the bond angle to 90°, the simpler the resulting rovibronic structure will become. To see this, consider fig. 3.8b, which shows what angular momenta are expected for an infinitely massive atom X and a bond angle exactly equal to 90°. From the figure, one readily deduces that $I_A = I_B = I_{c/2}$, the condition for a symmetric oblate top. It turns out, as expected from this model, that H_2S and H_2Se are closer to this

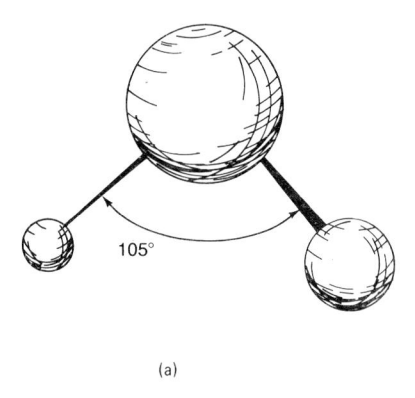

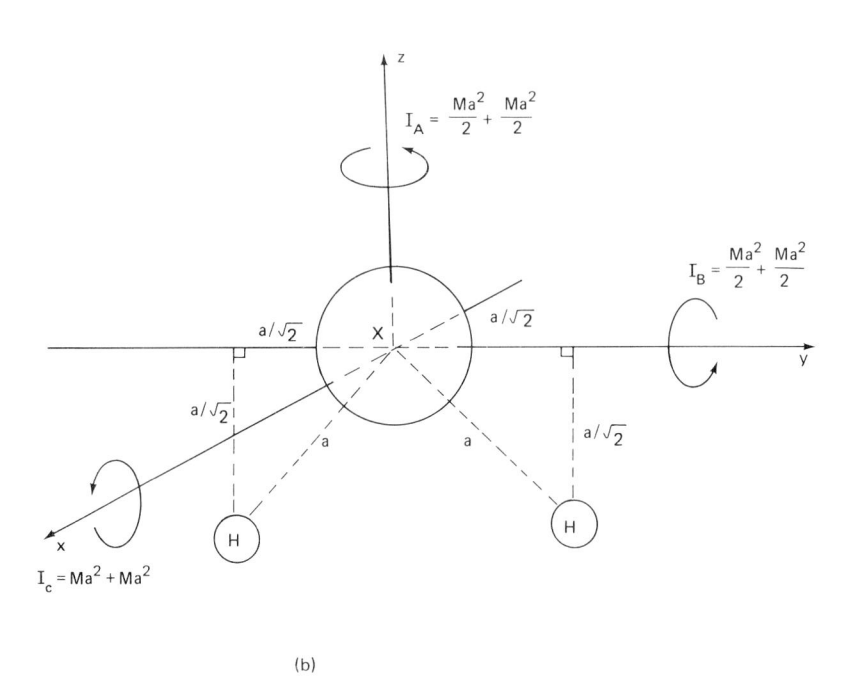

Fig. 3.8. (a) Structure of the H$_2$O molecule; (b) rotations about different principal axes for the special case of an H$_2$X molecule in the limit where X is infinitely massive and the bond angle becomes 90°.

condition than H$_2$O, and indeed that H$_2$O is the only *strongly* asymmetric top in the group.

As a result of the departure from symmetry, the K selection rule is broken (cf. Herzberg 1950) and the bands are generally much more complex in H$_2$O than the corresponding bands in other members of the sequence. Examples of the simpler band systems which occur in H$_2$X molecules heavier than H$_2$O are illustrated in fig. 3.9. All the data for this figure were obtained using synchrotron radiation

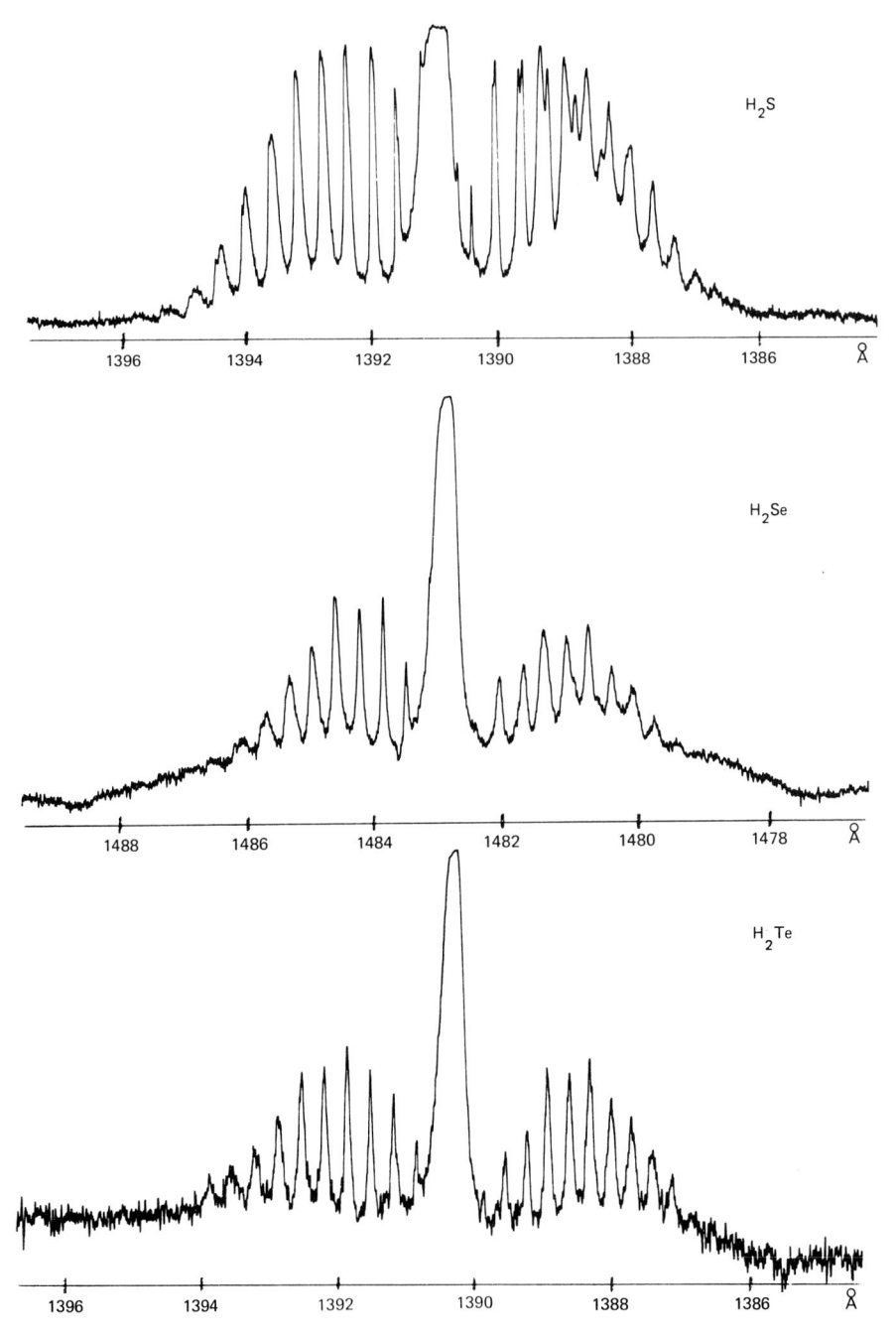

Fig. 3.9. Comparison between observed rotational structures for corresponding type C transitions in H_2S, H_2Se and H_2Te. The data for H_2S shows evidence of saturated absorption in the central peak. The band in H_2Te is assigned by comparison with the corresponding bands in the other molecules, for which band contour analyses are available. A detailed analysis of H_2Te is in progress.

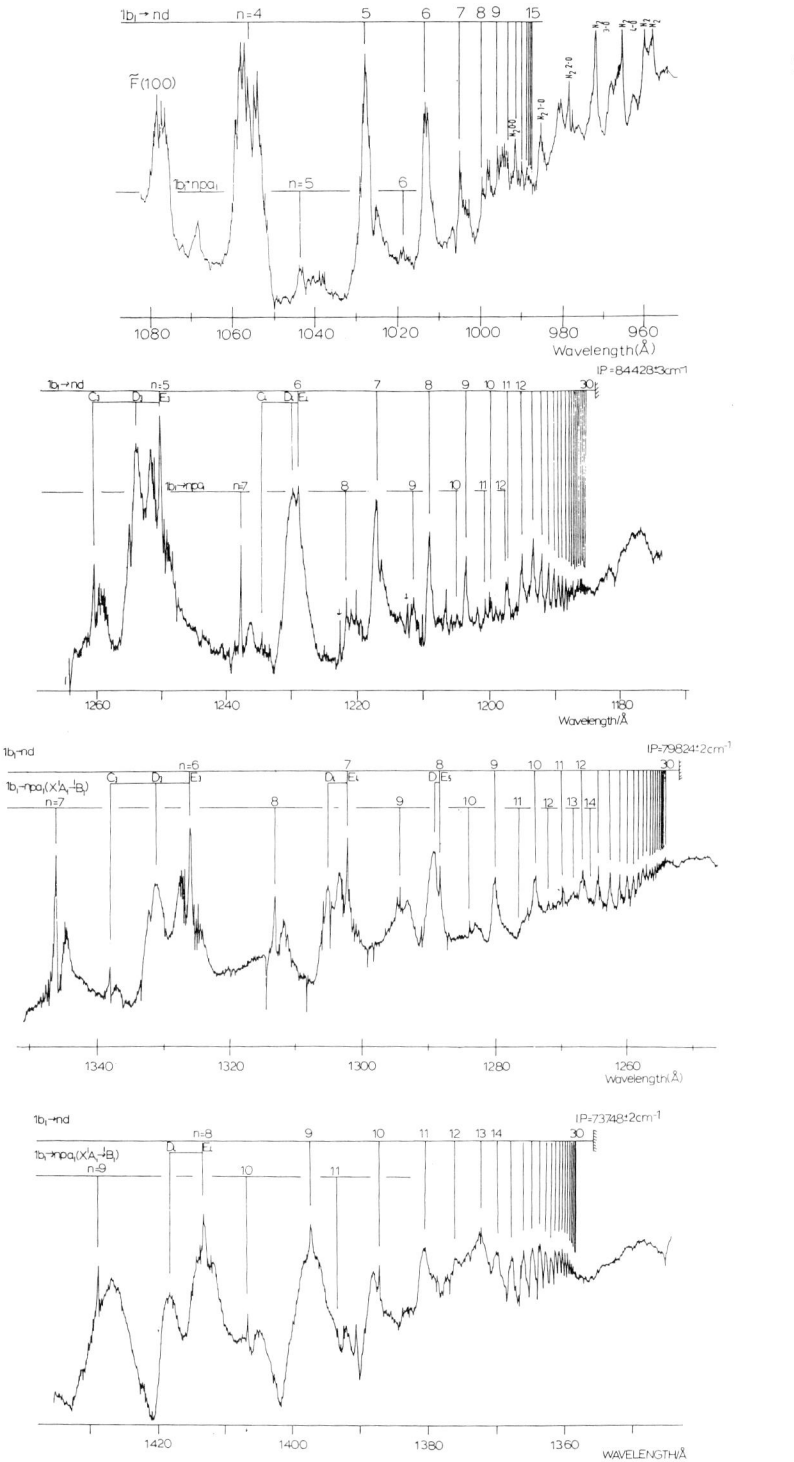

Fig. 3.10. Intercomparison between the Rydberg series to the $^1B_1$ limit in the homologous sequence of H_2X molecules, showing effects due to increasing atomicity as the molecules become heavier (see text).

(Mayhew and Connerade 1986a,b, Mayhew et al. 1986) and the analysis is still in progress at time of writing.

A further consequence of the increase in mass down a homologous sequence is the greater atomicity of the electronic excitations built on the substituted atom. As a result, quasi-atomic Rydberg states of the molecule emerge, and the observation of their spectra provides further data for detailed intercomparisons. Figure 3.10 shows the Rydberg series to the $^1B_1$, limits of H_2O, H_2S and H_2Se due to excitation of an electron from a non-bonding orbital. The upper members of the series are more conspicuous in H_2S (Baig et al. 1981) and H_2Se (Mayhew et al. 1983) than in H_2O (Connerade et al. 1980) for several different reasons: (i) because the change in the bond angle between the ground and the excited states is smaller for H_2S and H_2Se than for H_2O. The Franck–Condon principle thus favours the $(0,0)$ transitions in the heavier molecules of the sequence, leading to less pronounced vibrational structure and clearer series; (ii) because of greater atomicity, as noted above; and (iii) because the heavier molecules have a more compact rotational structure, leading to a better definition in energy of the upper members.

As a general rule, all of (i), (ii) and (iii) must be satisfied for the observation of clear Rydberg series in molecules, and it is also necessary that the residual vacancy on which the series is built should be localised on one site, i.e. that the excited electron should originate from a non-bonding orbital.

3.3. High Rydberg states of molecules

Often, in an introductory discussion on molecular states, the problem of an atom in an electric field is used as an illustrative example: the energy degeneracy of the atomic orbital is lifted by an electric field in a similar way to the formation of different Λ states in molecules (see e.g. Herzberg 1950). Taking the comparison further, the distortions of atomic wave functions by very intense external fields are an important theme of current research (cf. Connerade et al. 1983) in which the high Rydberg states become more and more distorted with increasing n, whereas the low n wavefunctions are nearly those of the central field. There are thus two aspects of perturbations by external fields: (i) the lifting of degeneracy, and (ii) the distortion of radial wavefunctions. The latter can, in extreme cases, lead to breakdown of the Rydberg formula, an effect which grows rapidly with increasing n.

If we consider this last case, Rydberg series of molecules can, in some sense, be thought of as the opposite situation: at high n, the wavefunctions tend more and more closely to a quasi-atomic limit as the excited electron spends more and more time in the Coulombic outer reaches of the potential. In the opposite extreme, the excited states become more and more perturbed by the molecular field with decreasing n.

Rydberg series in molecules, apart from the obvious application of measuring ionisation potentials accurately, provide us, when they occur, with a unique opportunity to study a progressive transition from quasi-atomic to more fully molecular behaviour. When Rydberg series are absent, or when they fade out without reaching high members, this can also provide indirect information on the nature of the vacancy, i.e., whether it is localised (atomic) or distributed (molecular), or on the energy span of rovibronic structure.

A starting point in this area was made many years ago in the pioneering investigations of Price (1936) who used conventional sources and did not have the benefit of the high resolution presently available for the study of Rydberg states, but worked in the vacuum ultraviolet and used the quasi-atomic nature of the excitations to establish electronic assignments by working downwards from the limit and assuming the constancy of quantum defects in the Rydberg formula for a given channel.

At present, with high resolution and synchrotron radiation sources available in the vacuum ultraviolet, it is obviously desirable to renew this approach. In the meantime, atomic theory has of course evolved well beyond the simple approximation quoted above, with the advent of multichannel quantum defect theory (MQDT) (Seaton 1966). There are several ways in which MQDT is applicable to molecular systems (Giusti-Suzor and Lefebvre-Brion 1984). For example, it can be used for intensity calculations in molecular band systems (cf. Jungen 1985). Another possibility is to apply it to the analysis of quasi-atomic Rydberg states in molecules (Dagata et al. 1981) an approach which can be regarded as a refinement of the original attempts by Price and by Mulliken.

The standard tool of atomic MQDT in recent years has become the Lu–Fano graph (Lu and Fano 1970; see also section 2.2 above for a discussion of this method), which plots the interaction between two channels by treating their quantum defects on an equal footing and exploiting the cyclic recurrence with increasing n contained in the Rydberg formula. The example of fig. 3.11 is an important one in the development of MQDT for atoms (the 5p spectrum of XeI).

The width of the avoided crossings in an MQDT diagram and the size of their departure from the diagonal measures the strengths of interchannel interactions. This information can be extrapolated into the continuum by using the relation between quantum defects and phase shifts, and the appearance of the Lu–Fano graph therefore implies the form of the autoionizing spectrum. A simple qualitative rule is that narrow avoided crossings (weak interactions) imply sharp resonances above the threshold. The Lu–Fano graph of Ba^{2+} (isoelectronic with Xe) has been studied by Hill et al. (1982) and this connection exemplified.

Turning back to molecules, a spectrum analogous to 5p excitations in Xe should occur in each molecule of the sequence HI, CH_3I, C_2H_5I, etc. In HI (a light molecule) the rovibronic structure is very open, and the resulting spectrum is too complex for a Rydberg series to be extracted from the data. However, CH_3I (fig. 3.12) provides a fine example of quasi-atomic molecular excitations, and the correspondence with Xe is illustrated by the molecular quantum defect plot in fig. 3.13. It will be noticed that both the avoided crossings and the autoionisation widths are narrower for CH_3I than for Xe, but broader than for Ba^{2+}.

By the same argument, one can relate the spectrum of Kr to that of CH_3Br

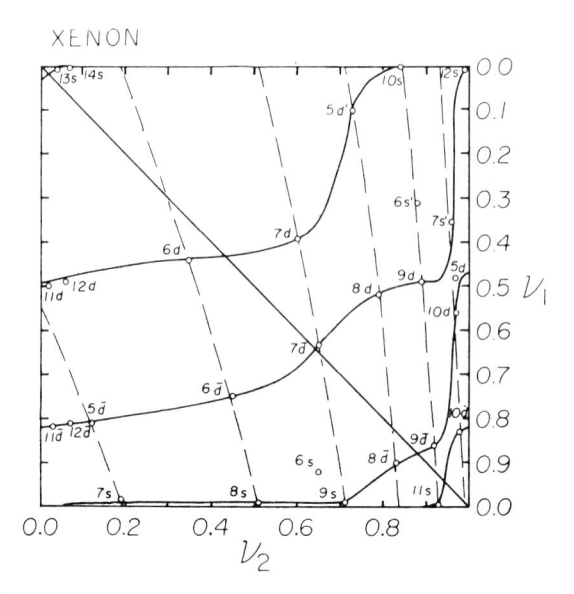

Fig. 3.11. Quantum defect plot for the 5p series of xenon. Xenon is the united atom corresponding to methyl iodide, and this plot should be compared with fig. 3.13.

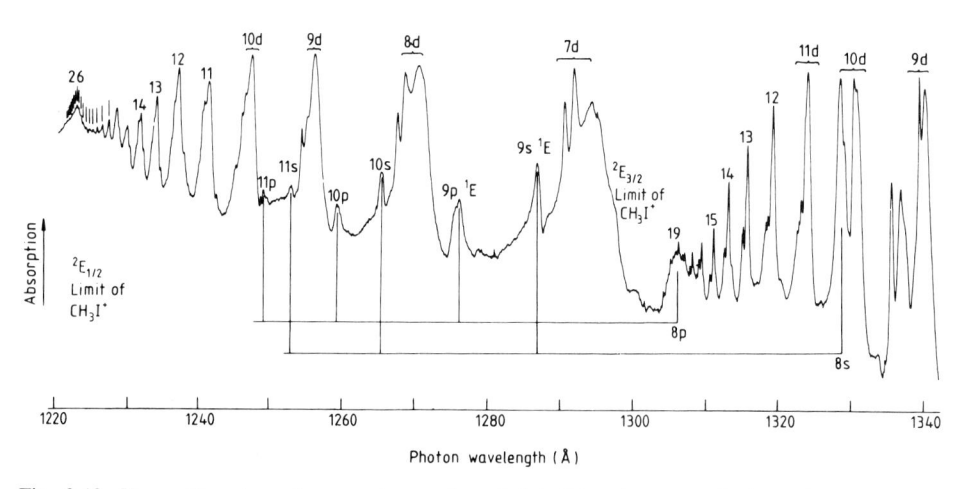

Fig. 3.12. Xenon-like absorption spectrum of methyl iodide, due to excitation of the lone pair electron.

(Baig et al. 1982), and the high Rydberg states of the latter are closely similar to those of CH_3I (Baig et al. 1981).

There are, of course, differences between MQDT for atoms and molecules which become apparent when the Lu–Fano graphs are studied in detail. In fig. 3.13, the emphasis is on quasi-atomic states, but a more complete approach (Dagata 1982) includes extensions to more specifically molecular transitions.

METHYL IODIDE

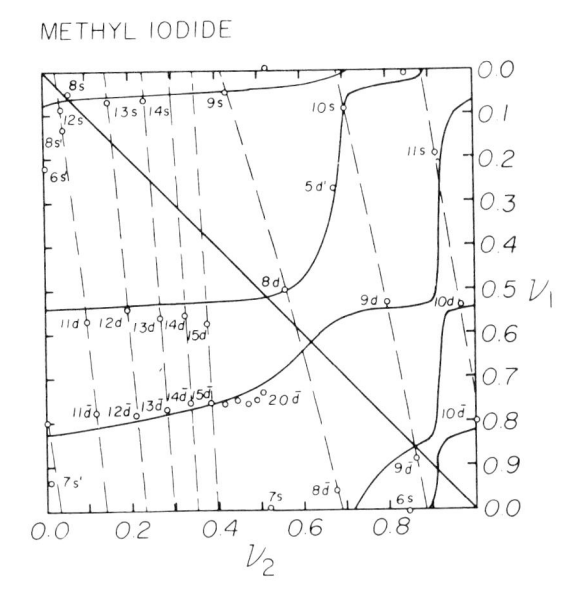

Fig. 3.13. Quantum defect plot for the xenon-like series of methyl iodide, illustrating the correspondence between the atomic and molecular spectra (cf. fig. 3.11).

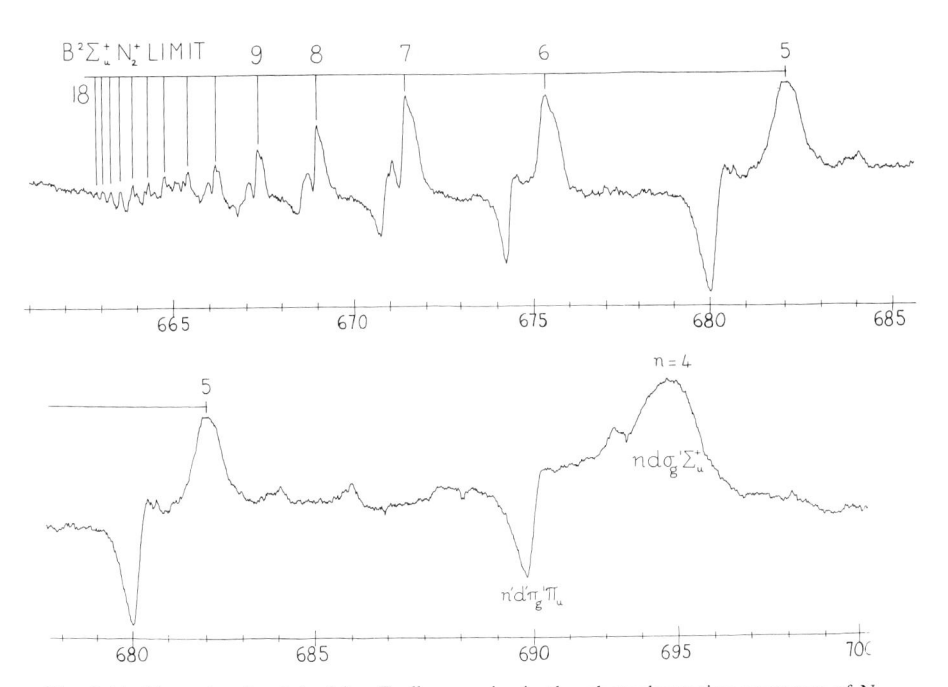

Fig. 3.14. Example of autoionising Rydberg series in the photoabsorption spectrum of N_2.

The approach has been extended as far as C_2H_5I (Baig et al. 1986), and we have also employed other MQDT tools such as the Dubau–Seaton formula (see section 2.3 above) to interpret the autoionisation range (fig. 3.12) of these rather heavy molecular systems.

By applying MQDT, one is thus able to refine and to update the method of electronic assignment pioneered by Price. Again, high resolution techniques are crucial in this approach, since the analysis of perturbations by MQDT depends sensitively on the behaviour of Rydberg series close to the ionisation limit.

Finally, we show in fig. 3.14 an example of autoionising resonances (both transmission windows and asymmetric absorption lines) in the Rydberg series converging to the $B\,^2\Sigma_u^+$ state of N^{2+}. This series is of particular interest because the profiles have recently been calculated ab initio by Raoult et al. (1983).

4. SR experiments with polarised radiation

Synchrotron spectroscopy is ideally suited to the study of external magnetic field effects, because of its unique polarisation properties.

It was first pointed out by Schott (1912), whose excellent treatise on the electromagnetic radiation emitted by charged particles on circular orbits has been unjustly neglected, that what was later to be known as synchrotron light is polarised linearly in the plane of the orbit, circularly along the axis, and elliptically at a general angle to the plane.

Although its polarisation properties constitute one of the most important advantages of synchrotron radiation, progress in exploiting them has been comparatively slow. One should mention here studies of the MCD spectra of benzene and ethylene (Snyder et al. 1981), investigations of the Fano effect in electron spin polarisation spectroscopy (Heinzmann 1980) and MCD spectroscopy of matrix isolated species (Hormes et al. 1983).

4.1. External magnetic fields: Faraday rotation spectroscopy

In the present section, we give a brief description of an experiment involving the high resolution study of Faraday rotation in atoms. This measurement, carried out on the 500 MeV electron accelerator in Bonn, involves plane polarised radiation and is therefore ideally suited to a synchrotron source.

The experimental configuration required (Garton et al. 1983) is essentially the classical setup for Faraday rotation studies in the visible, with crossed polariser and analyser, a continuum source and dispersive optics. A sketch of the apparatus is given in fig. 4.1. The experimental innovations which were introduced are as follows:

(i) The synchrotron was used as the polariser and the grating of the spectrograph as the analyser, thus eliminating the need for any transmitting optics

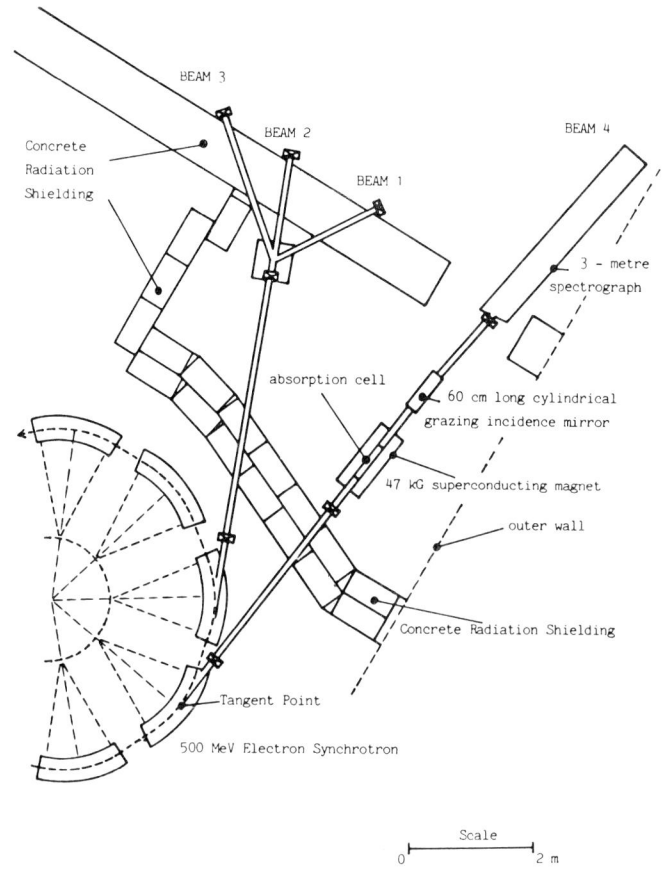

Fig. 4.1. Experimental setup for the observation of magneto-optical spectra of atomic or molecular gaseous samples using synchrotron radiation.

(which limit wavelength coverage) or three-reflection polarisers (which lose much intensity).

(ii) The experiment was performed in a very high magnetic field (4.7 Tesla) with a path length of some 50 cm, allowing rotation angles of many π close to absorption lines. This provides an absolute calibration of the angles measured by the simple expedient of observing intensity maxima and minima in the oscillations of the magneto-optical patterns (see fig. 4.2). It also could open the way to studies of high field effects on atomic f-values, a novel area of research.

(iii) The experiments were performed at high spectrographic resolution, thereby allowing intensity oscillations in the magneto-optical patterns to be followed close in to the lines.

A typical magneto-optical spectrum is shown in fig. 4.2.

The interpretation of the data rests on a theoretical technique (Connerade 1983a) described as the magneto-optical Vernier (MOV) method. The algebra

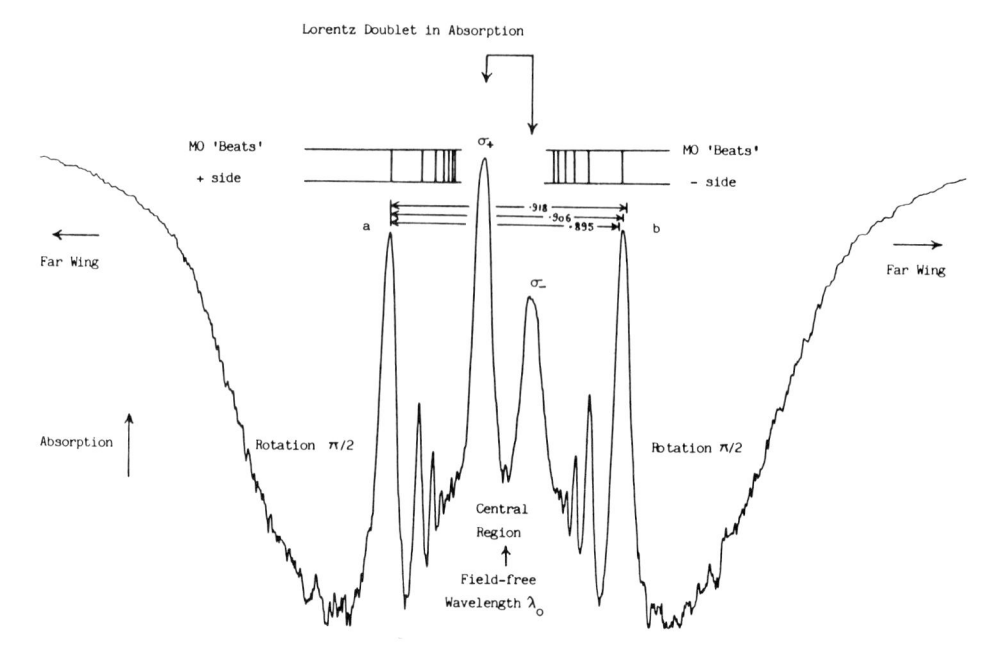

Fig. 4.2. Typical magneto-optical rotation pattern, recorded for strontium, showing well-resolved intensity oscillations on either side of the Lorentz doublet. The magnetic field was 4.7 Tesla.

involved is given fully in the reference just cited, and since it is rather lengthy we content ourselves with a brief description of the principles.

The fundamental equation for the transmitted intensity is

$$I = \frac{I_0}{4}\frac{P}{100}\left\{\left[\exp\left(-\frac{a_+ z}{2}\right) - \exp\left(-\frac{a_- z}{2}\right)\right] + 4\exp\left[-(a_+ + a_-)\frac{z}{2}\right]\sin^2\varphi\right\}$$
$$+ \frac{100-P}{100}I_0\{\exp[-(a_+ + a_-)z]\},$$

where I_0 is the intensity of the incident plane wave, P is the polarisation efficiency of the grating, a_+ and a_- are the absorption coefficients for right- and left-hand circularly polarised radiation, z is the length of the absorption cell and φ is the Faraday angle.

The quantities a_+, a_- and φ are functions of the detuning $\nu - \nu_0$ and of the magnetic field strength B. The simplest relations to use are those of classical dispersion theory, for example in the notation of Mitchell and Zemanski (1971):

$$a_\pm = \frac{e^2 Nf}{mc}\frac{(\Gamma/4\pi)}{(\nu_0-\nu \pm \alpha)^2 + (\Gamma/4\pi)^2}$$

for singlet terms, where

$$\alpha = \frac{eB}{4\pi mc},$$

while φ is related to the refractive indices n_+ and n_- as follows:

$$n_\pm - 1 = \frac{e^2 Nf}{4\pi m} \frac{1}{(\nu_0 \pm \alpha)} \frac{\nu_0 - \nu \pm \alpha}{(\nu_0 - \nu \pm \alpha)^2 + (\Gamma/4\pi)^2}$$

$$\varphi = \frac{1}{2} \frac{\omega z}{c} (n_+ - n_-).$$

The object of the analysis is to extract from the data values for the relative oscillator strengths of the absorption lines. The rotation angle, on which the transmitted intensity depends, is directly related to the f-value through dispersion theory. It turns out that the patterns contain two regimes, namely: (a) a regime of large detuning, where a small change in f-value would result in a slow change in the wavelength of the intensity maximum, and (b) a regime of small detuning, where small changes in the f-value precipitate large oscillations of transmitted intensity.

The theoretical technique consists in computing the full pattern from the complete equations including both birefringence and dichroism (Connerade 1983c), for an estimated f-value based on regime (a). The computed patterns are then "tuned" by varying the f-value in steps until the computed pattern matches experiment in both (a) and (b) which can be thought of as the coarse and fine adjustments of a Vernier measurement.

The method combines high sensitivity with high accuracy and is capable of giving relative f-values with accuracies better than the percent level. It also seems ideally suited to investigate the influence of very high fields on f-values, although the latter capability has not yet been explored experimentally.

Recently, theoretical and computational basis of the method have been considerably improved (Stavrakas 1983) and it has been demonstrated that experimental magneto-optical patterns which overlap in energy can be unravelled by the MOV technique. In fig. 4.3, we illustrate by an example the analysis of overlapping patterns. The material described in the present section is clearly of a preliminary nature: Faraday rotation studies with synchrotron radiation are in their infancy and the area is likely to develop rapidly as new and better synchrotron sources become available.

One should also mention the relevance of high resolution synchrotron radiation studies to the area of atomic and molecular physics close to ionisation thresholds in high external fields (Connerade et al. 1982). To date, this area has been dominated by spectroscopy with conventional continuum sources or with lasers, both in the near ultraviolet or in the visible, but as interest extends to the vacuum ultraviolet, synchrotron radiation will also have an important role to play.

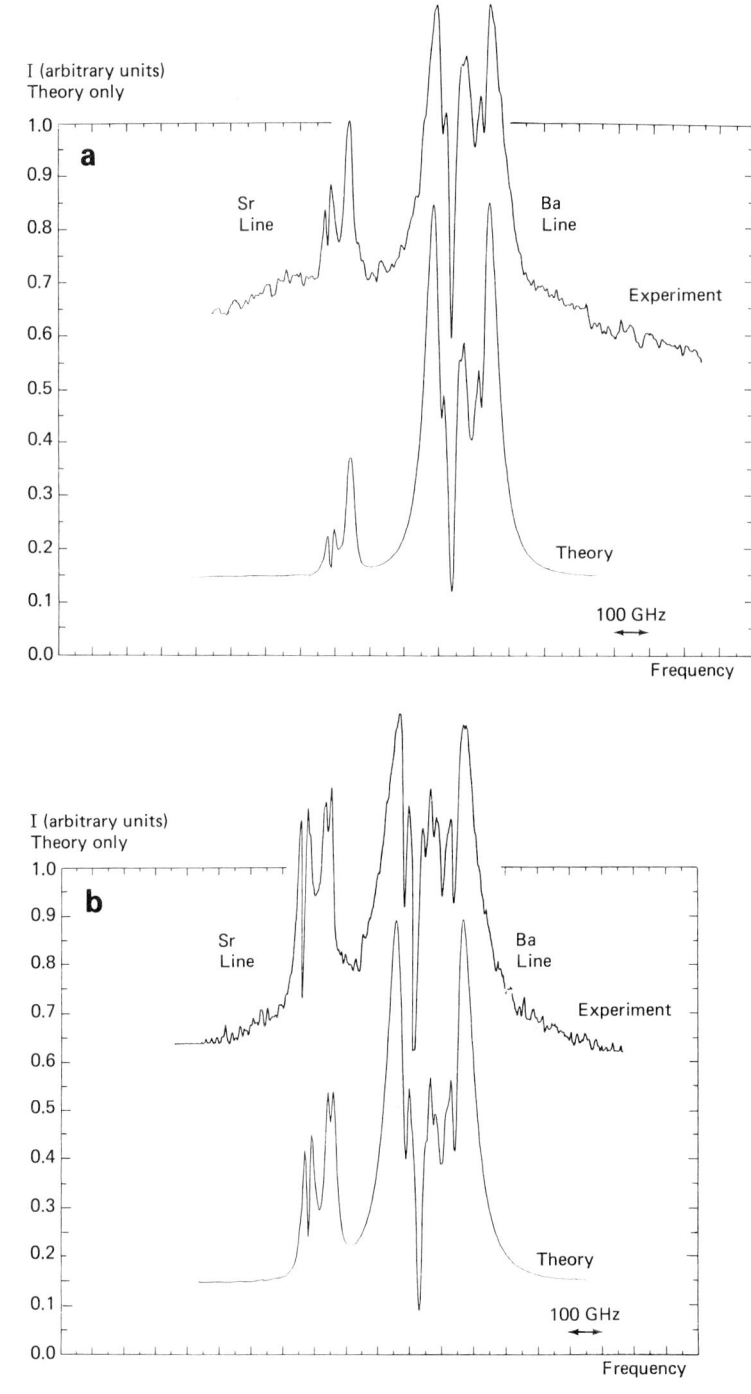

Fig. 4.3. Example of overlapping magneto-optical patterns analysed by the magneto-optical Vernier technique. Theoretical profiles are shown alongside the data for a range of experimental conditions. The transitions involved are two lines, one in barium and the other in strontium, which happen to lie close together, and conditions were chosen such that both patterns are well developed. The field strengths are: (a) 2.0 Tesla, (b) 2.5 Tesla and (c, facing page) 3.0 Tesla.

Fig. 4.3c.

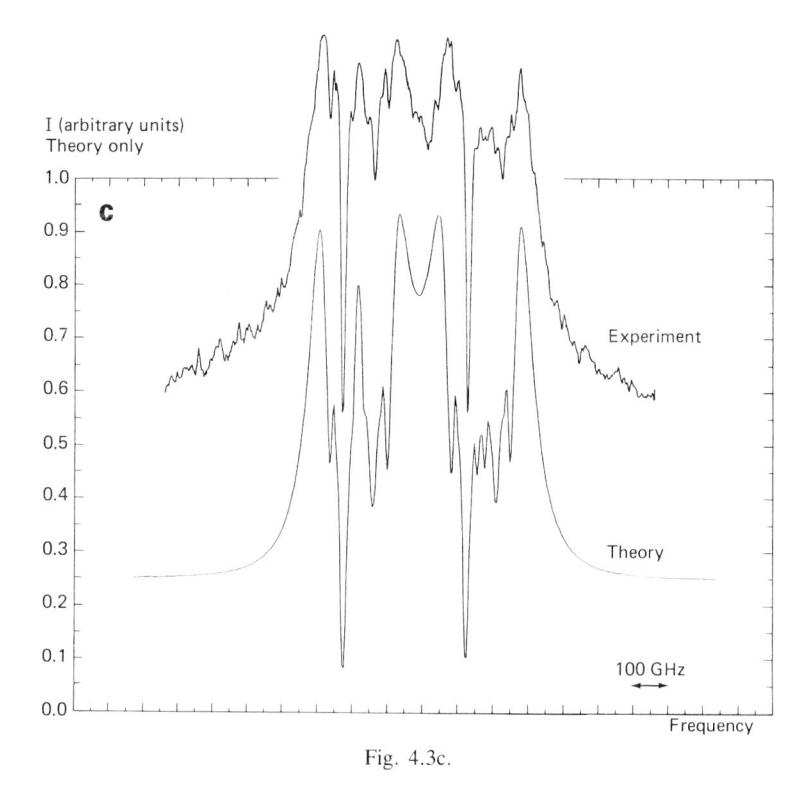

Finally, it is worth noting that Faraday rotation spectroscopy has been applied by Kitagawa et al. (1978) to trace determinations of atomic cadmium, where a detection limit of 5×10^{-13} g was demonstrated at a wavelength of 2288 Å. This shows that the technique may possess practical applications.

4.2. Spin polarisation spectroscopy

In the previous section, the application of synchrotron radiation to spin polarised photoelectron spectroscopy was mentioned. This is not the place to attempt a description of what is now an extensive area of activity in its own right. For a comprehensive discussion, the reader is referred to the review by Heinzmann (1980). The present section is intended merely to emphasise that the techniques of spin polarisation spectroscopy and high resolution photoabsorption spectroscopy complement each other in a fruitful way.

To illustrate their complementary nature, we shall quote two examples, one atomic and one molecular.

The 4d absorption spectrum of Ag I was studied by photoabsorption spectroscopy (Connerade et al. 1978, Connerade and Baig 1979) yielding a fairly complex structure, the details of which were compared with calculations based on ab initio theory. Such calculations are rather complex: first, a Hartree–Fock code is used to solve the radial equation and to obtain all the Slater–Condon parameters.

Next, one sets up the energy matrix in intermediate coupling, which then has to be diagonalised to determine eigenvalues and eigenvectors. Usually, the results must be adjusted somewhat in energy to agree with experiment. In the present case, the configuration average approximation was found from the theory itself to be inadequate, and strong mixing between $4d^9 5s5p$ and the double-hole configuration $4d^8 5s5p$ was also suspected. Thus, an independent method of checking the assignments is highly desirable.

The spin polarisation spectrum of the 4d subshell of Ag I was investigated independently by Heinzmann et al. (1980), who were able to show that their data is consistent with the J values given by the analysis of the photoabsorption spectrum.

Another example is afforded by the spectrum of methyl bromide in the first autoionisation range, which was mentioned in section 3.3 as an instance of quasi-atomic excited states in molecules.

The photoabsorption spectrum of CH_3Br has been obtained using synchrotron spectroscopy at a resolution two orders of magnitude higher than achieved by spin polarisation spectroscopy (Schäfers et al. 1983). On the other hand, the photoabsorption measurement does not yield absolute values, which are readily obtained from the spin polarised photoelectron spectrum. Schäfers et al. (1983) therefore combined both approaches, convolving the high resolution spectrum with the apparatus function of the spin polarisation experiment, to determine both absolute and partial photoionisation cross sections for excitation into three energy degenerate continua. Further experimental work is required to check out the reliability of this approach, but it clearly posesses great potential for the determination of partial cross sections.

5. Conclusion

In the present chapter, we have attempted to present an overview of the wide variety of problems to which high resolution synchrotron spectroscopy has been applied. It is clear that high resolution work depends both on instrumental developments (e.g. progress in the manufacture of holographic gratings) and the development of sources (e.g. the spectral brilliance of available synchrotrons and storage rings). There are also more subtle points to be considered. Photoabsorption spectroscopy at high resolution often necessitates long paths through samples at comparatively high pressures. Attempts to carry out such experiments with storage rings lead to elaborate differential pumping arrangements or windows, with attendant loss of light and/or wavelength coverage. In practice, electron synchrotrons, which operate well with less stringent vacuum requirements, may give as good a photon throughput and enable a simpler arrangement to be used. From this point of view, it is clear that the synchrotron source should ideally be chosen to match the experiment.

Future developments should therefore aim not only at providing high flux, but also at a greater flexibility in operating conditions. Among developments one would like to see on the source side is the provision of helical undulators for the very high magnetic field work mentioned in section 4.1: this would provide a high flux of circularly polarised radiation in the VUV which is essential for future progress.

On the instrumental side, there are physical limits to the useful size of spectrometers and to the ruling density of gratings, but we are still far from being Doppler-limited in most cases. This suggests the value of a radically new approach, namely the development of interferometry into the vacuum ultraviolet. An interferometer can be a comparatively small instrument, much more compact than a spectrograph of comparable resolution, and is also ideally matched to the radiation emitted by a storage ring. An instrument of this kind is already under development at Imperial College (Thorne, unpublished) and it seems therefore quite reasonable to expect significant advances in high resolution synchrotron spectroscopy over the coming years.

Acknowledgements

We thank our colleagues and collaborators, both past and present, who have generously contributed material for this review. In particular, the d-shell spectra of Zn and Cd and the p-shell spectra of Na, K, Ca and Sr form part of the Doktorarbeit of Klaus Sommer, of the University of Bonn. Likewise, the spectra of the H_2X and D_2X molecules form part of the Ph.D. thesis of Chris Mayhew, of Imperial College. We thank both of them for their valuable contributions.

References

Ashfold, M.N.R., J.M. Bailey and R.N. Dixon, 1983, J. Chem. Phys. **79**, 4080.
Baig, M.A., 1983, J. Phys. B **16**, 1511.
Baig, M.A., and J.P. Connerade, 1978, Proc. R. Soc. London Ser. A **364**, 33.
Baig, M.A., and J.P. Connerade, 1979a, J. Phys. B **12**, 2309 (SiO).
Baig, M.A., and J.P. Connerade, 1979b, unpublished results [quoted in: M.A. Baig, 1979, Ph.D. Thesis (Univ. of London)].
Baig, M.A., and J.P. Connerade, 1980, J. Mol. Spectrosc. **83**, 31 (GeO).
Baig, M.A., and J.P. Connerade, 1984a, J. Phys. B **17**, 1785 (Ne).
Baig, M.A., and J.P. Connerade, 1984b, J. Phys. B **17**, L469 (Yb).
Baig, M.A., and J.P. Connerade, 1987, to be published.
Baig, M.A., J.P. Connerade, J. Dagata and S.P. McGlynn, 1981a, J. Phys. B **14**, L25 (MQDT).
Baig, M.A., J. Hormes, J.P. Connerade and S.P. McGlynn, 1981b, J. Phys. B **14**, L725 (H_2F).
Baig, M.A., J. Hormes, J.P. Connerade and W.R.S. Garton, 1981c, J. Phys. B **14**, L147 (HBr).
Baig, M.A., J.P. Connerade and J. Hormes, 1982, J. Phys. B **15**, L5 (CH_3Br).
Baig, M.A., J.P. Connerade, C. Mayhew, G. Noeldeke and M.J. Seaton, 1984a, J. Phys. B **17**, L383 (He).
Baig, M.A., J.P. Connerade, C. Mayhew and K. Sommer, 1984b, J. Phys. B **17**, 371 (Ba).
Baig, M.A., J.P. Connerade and J. Hormes, 1986, J. Phys. B **19**, L343.
Band, I.M., 1981, J. Phys. B **14**, 1649.
Band, I.M., V.I. Fomichev and M.B. Trhavskovskaia, 1981, J. Phys. B **14**, 1103.
Bauer, S.H., G. Herzberg and J.W.C. Johns, 1964, J. Mol. Spectrosc. **13**, 256.

Beutler, H., 1934, Z. Phys. **91**, 132.
Beutler, H., and K. Guggenheimer, 1934, Z. Phys. **88**, 25.
Cohen-Tannoudji, C., and P. Avan, 1977, Etats Atomiques et Moleculaires Couplés a un Continuum, in: Atomes et Molecules Hautement Excités, CNRS Int. Colloq. No. 273, eds S. Feneuille and J.C. Lehmann (Editions CNRS, Paris) p. 93.
Connerade, J.P., 1970a, Astrophys. J. **59**, 685 (Cs).
Connerade, J.P., 1970b, Astrophys. J. **59**, 695 (Rb).
Connerade, J.P., 1977, J. Phys. B **10**, L239 (Double Ionisation).
Connerade, J.P., 1978a, Proc. R. Soc. London Ser. A **362**, 361 (q-reversals).
Connerade, J.P., 1978b, Contemp. Phys. **19**, 415 (Centrifugal).
Connerade, J.P., 1982, The Physics of Non-Rydberg States, in: Les Houches 1982 Session XXXVIII, New Trends in Atomic Physics, eds G. Grynberg and R. Stora (North-Holland, Amsterdam) p. 643 (Centrifugal).
Connerade, J.P., 1983a, J. Phys. B **16**, 399 (Magneto-Optical Vernier Method).
Connerade, J.P., 1983b, J. Phys. B **16**, L329 (Dubau–Seaton Formula).
Connerade, J.P., 1983c, J. Less Common Met. **93**, 171 (Centrifugal).
Connerade, J.P., 1984, J. Phys. B **17**, L165 (Giant Res Profiles).
Connerade, J.P., 1985, J. Phys. B **18**, L367 (Dubau–Seaton Formula).
Connerade, J.P., and M.A. Baig, 1979, Proc. R. Soc. London Ser. A **365**, 253 (Ag).
Connerade, J.P., and M.A. Baig, 1981, upublished results.
Connerade, J.P., and A.M. Lane, 1985, J. Phys. B **18**, L605.
Connerade, J.P., and M.A.P. Martin, 1983, J. Phys. B **16**, L577.
Connerade, J.P., and M. Pantelouris, 1984, J. Phys. B **17**, L173.
Connerade, J.P., M.A. Baig, M.W.D. Mansfield and E. Radtke, 1978, Proc. R. Soc. London Ser. A **361**, 379 (Ag).
Connerade, J.P., S.J. Rose and I.P. Grant, 1979, J. Phys. B **12**, L53.
Connerade, J.P., M.A. Baig, S.P. McGlynn and W.R.S. Garton, 1980, J. Phys. B **13**, L705 (H_2O).
Connerade, J.P., J.C. Gay and S. Liberman, eds, 1982, Physique Atomique et Moléculaire près des Seuils d'Ionisation en Champs Intenses, CNRS International Colloquium Aussois 7–11 June 1982, J. Phys. (France) **43**, C-2.
Connerade, J.P., J.C. Gay and S. Liberman, 1983, Comments At. & Mol. Physics **XIII**, 189.
Connerade, J.P., A.M. Baig and A.M. Lane, 1984, J. Phys. B **18**, 3507.
Connerade, J.P., A.M. Lane and M.A. Baig, 1985, J. Phys. B **18**, 3507.
Coulson, C.A., 1982, The Shape and Structure of Molecules (Clarendon Press, Oxford).
Dabrowsky, I., 1984, Can J. Phys. **62**, 1639.
Dabrowsky, I., and G. Herzberg, 1974, Can J. Phys. **52**, 1110.
Dagata, J.A., 1982, Ph.D. Thesis (Louisiana State University, Baton Rouge, LA).
Dagata, J.A., G.L. Findley, S.P. McGlynn, J.P. Connerade and M.A. Baig, 1981, Phys. Rev. A **24**, 2485.
Dubau, J., and M.J. Seaton, 1984, J. Phys. B **17**, 381.
Esherick, P., 1977, Phys. Rev. A **15**, 1920.
Fano, U., 1961, Phys. Rev. **124**, 2485.
Froese-Fischer, C., 1972, Comput. Phys. Commun. **4**, 107.
Garton, W.R.S., J.P. Connerade, M.A. Baig, J. Hormes and B. Alexa, 1983, J. Phys. B **16**, 389.
Ginter, M.L., and S.G. Tilford, 1970, J. Mol. Spectrosc. **34**, 206.
Ginter, M.L., and S.G. Tilford, 1971, J. Mol. Spectrosc. **37**, 159.
Ginter, M.L., D.S. Ginter and C.M. Brown, 1980, Appl. Opt. **19**, 4015.
Giusti-Suzor, A., and H. Lefebvre-Brion, 1984, Phys. Rev. B **30**, 3057.
Griffin, D.C., R.D. Cowan and K.L. Andrew, 1969, Phys. Rev. A **3**, 1233.
Hansen, J.E., 1975, J. Phys. B **8**, L403.
Heinzmann, U., 1980, J. Phys. B **13**, 4353.
Heinzmann, U., A. Wolcke and J. Kessler, 1980, J. Phys. B **13**, 3149.
Heinzmann, U., B. Osterheld and F. Schafers, 1984, Nucl. Instrum. Methods **195**, 395.
Herzberg, G., 1950, Molecular Spectra and Molecular Structure, in: The Spectra of Diatomic Molecules, Vol. 2 (Van Nostrand, New York).

Herzberg, G., 1958, Proc. R. Soc. London Ser. A **248**, 328.
Herzberg, G., and L.L. Howe, 1959, Can. J. Phys. **37**, 636.
Hill, W.T., K.T. Cheng, W.R. Johnson, T.B. Lucatorto, T.J. McIlrath and J. Sugar, 1982, Phys. Rev. Lett. **49**, 1631.
Hormes, J., A. Klein, W. Krebs, W. Laaser and J. Schiller, 1983, Nucl. Instrum. Methods **208**, 849.
Hotop, H., and D. Mahr, 1975, J. Phys. B **8**, L301.
Huber, K.P., and G. Herzberg, 1979, Molecular Spectra and Molecular Structure, in: The Spectra of Diatomic Molecules, Vol. 4 (Van Nostrand, New York).
Johns, J.W.C., 1963, Can. J. Phys. **41**, 209.
Jungen, C., 1985, 3rd European Workshop on Molecular Spectroscopy and Photon-Induced Dynamics (May 20–25, Seillac, France) unpublished.
Katayama, D.H., J.M. Cook, V.E. Bondybey and T.A Miller, 1979, Chem. Phys. Lett. **62**, 542.
Kitagawa, K., T. Shigeyasu and T. Takeuchi, 1978, Analyst. **103**, 1021.
Kovacs, J., 1969, Rotational Structure in the Spectra of Diatomic Molecules (Hilger, London).
Lagerqvist, A., and I. Renhorn, 1974, J. Mol. Spectrosc. **49**, 157.
Lane, A.M., 1984, J. Phys. B **17**, 2213.
Lane, A.M., 1985, J. Phys. B **18**, 2339.
Lewandowski, B., J. Ganz, H. Hotop and M.W. Ruf, 1981, J. Phys. B **14**, L803.
Lu, K.T., and U. Fano, 1970, Phys. Rev. A **2**, 81.
Madden, R.P., and K. Codling, 1963, Phys. Rev. Lett. **10**, 516.
Madden, R.P., D.L. Ederer and K. Codling, 1969, Phys. Rev. **177**, 136.
Martin, W.C., 1960, J. Res. Natl. Bur. Stand. **64**, 19.
Martin, W.C., 1973, J. Phys. & Chem. Ref. Data **2**, 257.
Mayhew, C., M.A. Baig and J.P. Connerade, 1983, J. Phys. B **16**, L757 (H_2Se).
Mayhew, C.A., and J.P. Connerade, 1986a, J. Phys. B **19**, 3493.
Mayhew, C.A., and J.P. Connerade, 1986b, J. Phys. B **19**, 3502.
Mayhew, C.A., J.P. Connerade and M.A. Baig, 1986, J. Phys. B **19**, 4149.
Mitchell, A.C.G., and M.W. Zemanski, 1971, Resonance Radiation and Excited Atoms (Cambridge University Press, Cambridge).
Moore, C.E., 1958, in: Atomic Energy Levels, Circular of the National Bureau of Standards (U.S. Government Printing Office, Washington, DC).
Price, W.C., 1936, J. Chem. Phys. **4**, 147.
Raoult, M., H. Le Rouzo, G. Rafeev and H. Lefebvre-Brion, 1983, J. Phys. B **16**, 4601.
Rau, R., 1983, J. Phys. (France) **43**, C2-211.
Rinneberg, H., G. Jonsson, J. Neukammer, K. Vietzke, H. Hieronymus, G. Konig and W.E. Cooke, 1985, in: Proc. 2nd European Conf. on Atomic and Molecular Physics, eds A.E. de Vries and M.J. van der Wiel (Vrije Universiteit, Amsterdam).
Robaux, O., and M. Aymar, 1982, Comput. Phys. Commun. **25**, 223.
Rose, S.J., I.P. Grant and J.P. Connerade, 1980, Philos Trans. R. Soc. London A **296**, 527.
Rosenberg, R.A., M.G. White, G. Thornton and D.A. Shirley, 1979, Phys. Rev. Lett. A **21**, 132.
Rosenberg, R.A., S.T. Lee and D.A. Shirley, 1980, Phys. Rev. A **21**, 132.
Ruščić, B., J.P. Greene and J. Berkowitz, 1984, J. Phys. B **17**, 1503.
Schafers, F., M.A. Baig and U. Heinzmann, 1983, J. Phys. B **16**, L1.
Schott, G.A., 1912, Electromagnetic Radiation (Cambridge Univ. Press, Cambridge).
Seaton, M.J., 1966, Proc. Phys. Soc. **88**, 801.
Seaton, M.J., 1983, Rep. Prog. Phys. **46**, 97.
Snyder, P.A., P.A. Lund, P.N. Schatz and E.M. Rowe, 1981, Chem. Phys. Lett. **82**, 546.
Sommer, K., M.A. Baig and J. Hormes, 1984, 16th EGAS Conference Paper A 1-3.
Stavrakas, T.A., 1983, Ph.D. Thesis (Imperial College, London).
Suga, T., 1937, Sci. Pap. Inst. Phys. & Chem. Res. (Japan) **34**, 7.
Suzer, S., S.T. Lee and D.A. Shirley, 1976, Phys. Rev. A **31**, 1842.
Wallenstein, R., 1983, in: Proc. of the VIIth Int. Conf. on VUV Radiation Physics, eds A. Weinreb and A. Ron (Hilger, Bristol).
White, H.W., 1934, Introduction to Atomic Spectra (McGraw-Hill, New York).
Wilkinson, P.G., 1968, Can. J. Phys. **46**, 1225.

CHAPTER 5

RESONANCES IN MOLECULAR PHOTOIONIZATION*

J.L. DEHMER
Argonne National Laboratory, Argonne, IL 60439, USA

A.C. PARR
Synchrotron Ultraviolet Radiation Facility, National Bureau of Standards, Gaithersburg, MD 20899, USA

S.H. SOUTHWORTH
Los Alamos National Laboratory, Los Alamos, NM 87545, USA

Contents

1. Introduction 243
2. Shape resonances 245
   2.1. Overview 245
   2.2. Basic properties 248
   2.3. Eigenchannel plots 255
   2.4. Connections between shape resonances in electron–molecule scattering and in molecular photoionization 260
3. Autoionization 263
   3.1. Overview 263
   3.2. MQDT treatment of H_2 photoionization 268
4. Triply differential photoelectron measurements – experimental aspects 275
5. Case studies 280
   5.1. Shape-resonance-induced non-Franck–Condon effects in N_2 $3\sigma_g$ photoionization 280
   5.2. Autoionization via the Hopfield series in N_2 287
   5.3. Continuum–continuum coupling effects in N_2 $2\sigma_u$ photoionization . . 294
   5.4. Resonance effects in photoionization of the $1\pi_u$ level of C_2H_2 297

Contents continued overleaf

---

* Work supported in part by the U.S. Department of Energy and the Office of Naval Research.

Handbook on Synchrotron Radiation, Vol. 2, edited by G.V. Marr
© *Elsevier Science Publishers B.V., 1987*

Contents continued

5.5. Valence-shell photoionization of SF_6 305
5.6. Valence-shell photoionization of BF_3. 318
6. Survey of related work 330
7. Prospects for future progress 334
Appendix. A bibliography on shape resonances in molecular photoionization through early 1985 336
References 342

1. Introduction

Molecular photoionization is a rich source of information on fundamental intramolecular interactions. This is apparent when photoionization is viewed as a half-collision in which a collision complex, prepared by dipole excitation, decays by ejection of an electron from the field of the target. In the molecular case, the escaping electron must traverse the anisotropic molecular field and can undergo interactions with its nuclear modes. Hence, the photoelectron carries to the detector dynamical information on the two central aspects of molecular behavior – motion of an electron in a multicenter field and interplay among rovibronic modes.

Attention is invariably drawn to resonant photoionization mechanisms, such as shape resonances and autoionization. These resonant processes are important probes of photoionization for various reasons, the most obvious one being that they are usually displayed prominently against nonresonant behavior in such observables as the total photoionization cross section, photoionization branching ratios, and photoelectron angular distributions. More importantly, resonances temporarily trap the excited complex in a quasibound state, causing the excited electron to traverse the molecular core many times before its escape by tunneling or by exchange of energy with the core. In this way, resonances amplify the subtle dynamics of the electron–core interactions for more insightful analysis.

The last decade has witnessed remarkable progress in characterizing dynamical aspects of molecular photoionization. From among the great variety of successful streams of work, one can identify four broad classes which together have propelled the recent activities in this field. *First*, the extensive measurements of total photoabsorption/photoionization cross sections from the VUV to the X-ray range by a variety of means (see, e.g., Koch and Sonntag 1979 and a bibliography of inner-shell spectra by Hitchcock 1982, and original literature cited below) have continually provided fresh impetus to account for novel features displayed in molecular oscillator strength distributions. *Second*, shape resonances have emerged as a major focal point in the study of molecular photoionization dynamics. Initially stimulated by observations of intense, broad peaks in inner-shell spectra, beginning in the late sixties, the study of shape resonances in molecular photoionization has grown into a vigorous subfield. (A bibliography of papers discussing shape resonances in molecular photoionization is presented in the Appendix, along with an indication of the molecule(s) treated in each.) Benefitting greatly from the timely development of realistic, independent-electron models (Dehmer and Dill 1979b, Langhoff 1979, Raseev et al. 1980, Lucchese et al. 1982, Levine and Soven 1983, Collins and Schneider 1984, Levine and Soven 1984, Lynch et al. 1984b, Schneider and Collins 1984, Dill and Dehmer 1974, Lucchese et al. 1980, Lucchese and McKoy 1981c, Richards and Larkins 1984) for treating molecular photoionization, studies in this area have not only accounted for the features in the total photoionization spectra of both inner and outer shells, but also have predicted and confirmed several manifestations in other physical observables as discussed below. *Third*, multichannel quantum defect theory (MQDT) was adapted (Fano 1970, Dill 1972, Herzberg and Jungen 1972, Atabek et al. 1974, Fano 1975, Jungen and Atabek 1977, Dill and Jungen 1980, Guisti-Suzor and Lefebvre-Brion 1980, Jungen 1980, Jungen and Dill 1980, Raoult et al. 1980, Jungen and Raoult 1981, Raoult and Jungen 1981, Giusti-Suzor 1982, Lefebvre-Brion and Giusti-Suzor 1983, Raoult et al. 1983, Raseev and Le Rouzo 1983, Giusti-Suzor and Fano 1984a,b, Giusti-Suzor and Jungen 1984, Giusti-Suzor and Lefebvre-Brion 1984, Jungen 1984a,b, Mies 1984, Mies and Julienne 1984, Lefebvre-Brion et al. 1985) to molecular photoionization, providing a framework for the quantitative and microscopic analysis of autoionization phenomena. This powerful theoretical framework has been successfully applied to a number of prototype diatomic molecules, yielding both insight into the detailed dynamics of resonant photoionization and some specific predictions for experimental testing by means discussed in the next item. *Fourth*, technical advances, especially the development of intense synchrotron radiation sources (Kunz 1979, Winick and Doniach 1980, Koch 1983), have made it feasible to perform triply differential photoelectron measurements (see, e.g., Marr et al. 1979, White et al. 1979, Parr et al. 1980, Krause et al. 1981, Derenbach et al. 1983, Morin et al. 1983, Parr et al. 1983, 1984) on gas phase atoms and molecules. By this we mean that photoelectron measurements are made as a function of three independent variables – incident photon wavelength, photoelectron energy, and photoelectron ejection angle. Variable wavelength permits the study of photoionization at and within spectral features of interest. Photoelectron energy analysis permits separation and selection of individual (ro)vibronic ionization channels. Measurement of photoelectron angular distributions accesses dynamical information, i.e., relative phases of alternative degenerate ionization channels, that is not present in integrated cross sections. This level of experimental detail approaches that at which theoretical calculations are done and, hence, permits us to isolate and study dynamical details which are otherwise swamped in integrated or averaged quantities. We emphasize that, although the current trend is toward use of synchrotron radiation for variable wavelength studies, a variety of light sources have been successfully used to study photoionization dynamics. For example, in the shape resonance literature cited in the Appendix, many of the pioneering measurements were carried out with laboratory sources. Likewise, although most current measurements of vibrational branching ratios and angular distributions within autoionizing resonances employ synchrotron radiation (see, e.g., Morin et al. 1982a,b, Carlson et al. 1983b, Marr and Woodruff 1976, Woodruff and Marr 1977, Baer et al. 1979, Codling et al. 1981, Ederer et al. 1981, Parr et al. 1981, West et al. 1981, Parr et al. 1982b, Truesdale et al. 1983b, Hubin-Franskin et al. 1984), many early and ongoing studies with traditional light sources have made significant observations of the effects of autoionization on vibrational branching ratios (Doolittle and Schoen 1965, Price 1968, Berkowitz and Chupka 1969, Collin and Natalis 1969, Blake et al. 1970, Bahr et al. 1971a,b,

Carlson 1971, Collin et al. 1972, Kleimenov et al. 1972, Gardner and Samson 1973, Tanaka and Tanaka 1973, Gardner and Samson 1974a,b, Caprace et al. 1976, Natalis et al. 1977, Gardner and Samson 1978, Eland 1980, Kumar and Krishnakumar 1981, 1983) and angular distributions (Carlson 1971, Carlson and Jonas 1971, Morgenstern et al. 1971, Carlson and McGuire 1972, Carlson et al. 1972, Niehaus and Ruf 1972, Hancock and Samson 1976, Mintz and Kuppermann 1978, Katsumata et al. 1979, Kibel et al. 1979, Sell et al. 1979, Kreile and Schweig 1980).

Here we review recent progress in this field with emphasis on resonant mechanisms and on the interplay between experiment and theory. Sections 2 and 3 discuss elementary aspects of shape resonances and autoionization, respectively. Section 4 describes experimental aspects of triply differential photoelectron measurements which are currently the major source of new data in this field. Section 5 describes particular case studies of molecular photoionization, chosen to focus on a variety of basic resonant mechanisms that are both under active study currently and likely to form main themes in the future. Whereas these case studies draw heavily upon results of the authors' program at the National Bureau of Standards' SURF-II Facility, section 6 surveys related work with a much broader perspective, stressing aspects which are often unique to other groups. Finally, section 7 offers some thoughts about future directions of research within and beyond the present limitations of this field.

2. Shape resonances

2.1. Overview

Shape resonances are quasibound states in which a particle is temporarily trapped by a potential barrier, through which it may eventually tunnel and escape. In molecular fields, such states can result from so-called "centrifugal barriers", which block the motion of otherwise free electrons in certain directions, trapping them in a region of space with molecular dimensions. Over the past few years, this basic resonance mechanism has been found to play a prominent role in a variety of processes in molecular physics, most notably in photoionization and electron scattering. As discussed more fully in later sections, the expanding interest in shape resonant phenomena arises from a few key factors:

First, shape resonance effects are being identified in the spectra of a growing and diverse collection of molecules and now appear to be active somewhere in the observable properties of most small (nonhydride) molecules. Examples of the processes which can exhibit shape resonant effects are X-ray and VUV absorption spectra, photoelectron branching ratios and photoelectron angular distributions (including vibrationally resolved), Auger electron angular distributions (Dill et al. 1980), elastic electron scattering (Bardsley and Mandl 1968, Schulz 1973, 1976, Lane 1980, Shimamura and Takayanagi 1984), vibrational excitation by electron impact (Dehmer and Dill 1980, Bardsley and Mandl 1968, Schulz 1973, 1976,

Lane 1980, Shimamura and Takayanagi 1984), and so on. Thus concepts and techniques developed in any of these contexts can be used extensively in molecular physics.

Second, being quasibound inside a potential barrier on the perimeter of the molecule, such resonances are localized, have enhanced electron density in the molecular core, and are uncoupled from the external environment of the molecule. This localization often produces intense, easily studied spectral features, while suppressing the nearby continuum and/or Rydberg structure and, as discussed more fully below, has a marked influence on vibrational behavior. In addition, localization causes much of the conceptual framework developed for shape resonances in free molecules to apply equally well (Dehmer and Dill 1979a) to photoionization and electron scattering and to other states of matter such as adsorbed molecules (Davenport 1976a,b, Dill et al. 1976, Davenport et al. 1978, Gustafsson et al. 1978b, Gustafsson 1980b, Stöhr and Jaeger 1982, Gustafsson 1983, Stöhr et al. 1983, 1984, Koestner et al. 1984, Carr et al. 1985), molecular solids (Blechschmidt et al. 1972, Dehmer 1972, Lau et al. 1982, Fock 1983, Fock et al. 1984, Fock and Koch 1984, 1985), and ionic crystals (Åberg and Dehmer 1973, Pulm et al. 1985).

Third, resonant trapping by a centrifugal barrier often imparts a well-defined orbital momentum character to the escaping electron. This can be directly observed, e.g., by angular distributions of scattered electrons (Bardsley and Mandl 1968, Schulz 1973, 1976, Lane 1980, Shimamura and Takayanagi 1984) or photoelectron angular distributions from oriented molecules (Davenport 1976a,b, Dill et al. 1976, Gustafsson et al. 1978b, Gustafsson 1980b, 1983), and shows that the centrifugal trapping mechanism has physical meaning and is not merely a theoretical construct. Recent case studies have revealed trapping of $l = 1$ to $l = 5$ components of continuum molecular wavefunctions. The purely molecular origin of the great majority of these cases is illustrated by the prototype system N_2 discussed in section 2.2.

Fourth, the predominantly one-electron nature of the phenomena lends itself to theoretical treatment by realistic, independent-electron methods (Dehmer and Dill 1979b, Langhoff 1979, Raseev et al. 1980, Lucchese et al. 1982, Levine and Soven 1983, 1984, Collins and Schneider 1984, Lynch et al. 1984b, Schneider and Collins 1984, Dill and Dehmer 1974, Lucchese et al. 1980, Lucchese and McKoy 1981c, Richards and Larkins 1984), with the concomitant flexibility in terms of complexity of molecular systems, energy ranges, and alternative physical processes. This has been a major factor in the rapid exploration in this area. Continuing development of computational schemes also holds the promise of elevating the level of theoretical work on molecular ionization and scattering and, in doing so, to test and quantify many of the independent-electron results and to proceed to other circumstances, such as weak channels, coupled channels, multiply-excited states, etc., where the simpler schemes become invalid.

The earliest and still possibly the most dramatic examples of shape resonance effects in molecules are the photoabsorption spectra of the sulfur K- (LaVilla and Deslattes 1966, LaVilla 1972) and L-shells (Zimkina and Fomichev 1966, Zimkina and Vinogradov 1971, Blechschmidt et al. 1972, LaVilla 1972) in SF_6. The sulfur L-shell absorption spectra of SF_6 and H_2S are shown in fig. 1 to illustrate the type of phenomena that originally drew attention to this area. In fig. 1 both spectra are plotted on a photon energy scale referenced to the sulfur L-shell ionization potential (IP) which is chemically shifted by a few eV in the two molecular environments, but lies near $h\nu \sim 175$ eV. The ordinate represents relative photo-absorption cross section and the two curves have been adjusted so that the integrated oscillator strength for the two systems is roughly equal in this spectral range, since absolute calibrations are not known. The H_2S spectrum is used here as a "normal" reference spectrum since hydrogen atoms normally do not contribute appreciably to shape resonance effects and, in this particular context, can be regarded as weak perturbations on the inner-shell spectra of the heavy atom. Indeed, the H_2S photoabsorption spectrum exhibits a valence transition, followed by partially resolved Rydberg structure, which converges to a smooth continuum. The gradual rise at threshold is attributable to the delayed onset of the "$2p \rightarrow \varepsilon d$" continuum which, for second row atoms, will exhibit a delayed onset prior to the occupation of the 3d subshell. This is the qualitative behavior one might well expect for the absorption spectrum of a core level.

In sharp contrast to this, the photoabsorption spectrum of the same sulfur 2p subshell in SF_6 shows no vestige of the "normal" behavior just described. Instead, three intense, broad peaks appear, one below the ionization threshold and two above, and the continuum absorption cross section is greatly reduced elsewhere.

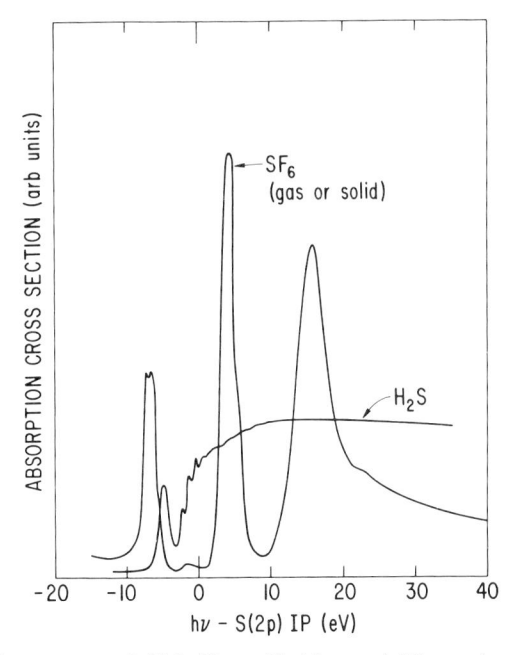

Fig. 1. Photoabsorption spectra of H_2S (from Zimkina and Vinogradov 1971) and SF_6 (from Blechschmidt et al. 1972) near the sulfur $L_{2,3}$ edge.

Moreover, no Rydberg structure is apparent, although an infinite number of Rydberg states must necessarily be associated with any positively charged molecular ion. Actually, Rydberg states superimposed on the weak bump below the IP were detected (Nakamura et al. 1971) using photographic detection, but obviously these states are extremely weak in this spectrum. This radical reorganization of the oscillator strength distribution in SF_6 was interpreted (Nefedov 1970, Dehmer 1972) in terms of potential barrier effects, resulting in three shape-resonantly enhanced final-state features of a_{1g}, t_{2g}, and e_g symmetry, in order of increasing energy. Another shape resonant feature of t_{1u} symmetry is prominent in the sulfur K-shell spectrum (LaVilla and Deslattes 1966) and, in fact, is believed to be responsible for the weak feature just below the IP in fig. 1 owing to weak channel interaction. Hence, four prominent features occur in the photoexcitation spectrum of SF_6 as a consequence of potential barriers caused by the molecular environment of the sulfur atom. Another significant observation (Blechschmidt et al. 1972) is that the SF_6 curve in fig. 1 represents both gaseous and solid SF_6, within experimental error bars. This is definitive evidence that the resonances are eigenfunctions of the potential well inside the barrier, and are effectively uncoupled from the molecule's external environment.

2.2. Basic properties

The central concept in shape resonance phenomena is the single-channel, barrier–penetration model familiar from introductory quantum mechanics. In fact, the name "shape resonance" means simply that the resonance behavior arises from the "shape", i.e., the barrier and associated inner and outer wells, of a local potential. The basic shape resonance mechanism is illustrated schematically (Child 1974) in fig. 2. In the figure an effective potential for an excited and/or unbound electron is shown to have an inner well at small distances, a potential barrier at intermediate distances, and an outer well (asymptotic form not shown) at large separations. In the context of molecular photoionization, this would be a one-dimensional abstraction of the effective potential for the photoelectron in the field of a molecular ion. Accordingly, the inner well would be formed by the partially screened nuclei in the molecular core and would therefore be highly anisotropic and would overlap much of the molecular charge distribution, i.e., the initial states of the photoionization process. The barrier, in all well-documented cases, is a so-called centrifugal barrier. (Other forces such as repulsive exchange forces, high concentrations of negative charge, etc., may also contribute, but have not yet been documented to be pivotal in the molecular systems studied to date.) This centrifugal barrier derives from a competition between repulsive centrifugal forces and attractive electrostatic forces and usually resides on the perimeter of the molecular charge distribution where the centrifugal forces can compete effectively with electrostatic forces. Similar barriers are known for d- and f-waves in atomic fields (Fano and Cooper 1968), however, the l (orbital angular momentum) character of resonances in molecular fields can be higher than those of constituent atoms owing to the larger spatial extent of the molecular charge

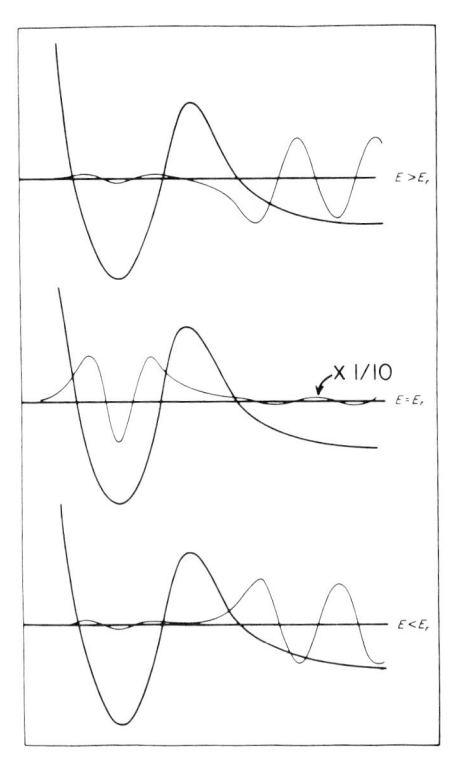

Fig. 2. Schematic of the effect of a potential barrier on an unbound wavefunction in the vicinity of a quasibound state at $E = E_r$ (adapted from Child 1974). In the present context, the horizontal axis represents the distance of the excited electron from the center of the molecule.

distribution, e.g., see the discussion in connection with N_2 photoionization below. The outer well lies outside the molecule where the Coulomb potential ($\sim -r^{-1}$) of the molecular ion again dominates the centrifugal terms ($\sim r^{-2}$) in the potential. We stress that this description has been radically simplified to convey the essential aspects of the underlying physics. In reality effective barriers to electron motion in molecular fields occur for particular l components of particular ionization channels and restrict motion only in certain directions. Specific examples which illustrate alternative types of centrifugal barriers in molecular fields are discussed below in connection with N_2, BF_3, and SF_6.

Focusing now on the wavefunctions in fig. 2, we see the effect of the potential barrier on the wave mechanics of the photoelectron. For energies below the resonance energy, $E < E_r$ (lower part of fig. 2), the inner well does not support a quasibound state, i.e., the wavefunction is not exponentially decaying as it enters the classically forbidden region of the barrier. Thus the wavefunction begins to diverge in the barrier region and emerges in the outer well with a much larger amplitude than that in the inner well. When properly normalized at large r, the amplitude in the molecular core is very small, so we say this wavefunction is essentially an eigenfunction of the outer well although small precursor loops extend inside the barrier into the molecular core.

At $E = E_r$ the inner well supports a quasibound state. The wavefunction exhibits exponential decay in the barrier region so that if the barrier extended to $r \to \infty$, a true bound state would lie very near this total energy. Therefore the antinode that was not supported in the inner well at $E < E_r$ has traversed the barrier to become part of a quasibound waveform which decays monotonically until it re-emerges in the outer well region, much diminished in amplitude. This "barrier penetration" by an antinode produces a rapid increase in the asymptotic phase shift by $\sim \pi$ radians and greatly enhances the amplitude in the inner well over a narrow band of energy near E_r. Therefore at $E = E_r$ the wavefunction is essentially an eigenfunction of the inner well although it decays through the barrier and re-emerges in the outer well. The energy halfwidth of the resonance is related to the lifetime of the quasibound state and to the energy derivative of the rise in the phase shift in well-known ways. Finally, for $E > E_r$ the wavefunction reverts to being an eigenfunction of the outer well as the behavior of the wavefunction at the outer edge of the inner well is no longer characteristic of a bound state.

Obviously this resonant behavior will cause significant physical effects: the enhancement of the inner-well amplitude at $E \sim E_r$ results in good overlap with the initial states which reside mainly in the inner well. Conversely, for energies below the top of the barrier but not within the resonance halfwidth of E_r, the inner amplitude is diminished relative to a more typical barrier-free case. This accounts for the strong modulation of the oscillator strength distribution in fig. 1. Also, the rapid rise in the phase shift induces shape resonance effects in the photoelectron angular distribution. Another important aspect is that eigenfunctions of the inner well are localized inside the barrier and are substantially uncoupled from the external environment of the molecule. As mentioned above, this means that shape resonant phenomena often persist in going from the gas phase to the condensed phase (e.g., fig. 1), and, with suitable modification, shape resonances in molecular photoionization can be mapped (Dehmer and Dill 1979a) onto electron-scattering processes and vice versa. Finally, note that this discussion has focussed on total energies from the bottom of the outer well to the top of the barrier, and that no explicit mention was made of the asymptotic potential, which determines the threshold for ionization. Thus valence or Rydberg states in this energy range can also exhibit shape resonant enhancement, even though they have bound state behavior at large r, beyond the outer well.

We will now turn, for the remainder of this section, to the specific example of the well-known σ_u shape resonance in N_2 photoionization, which was the first case for which shape resonant behavior was demonstrated (Dehmer and Dill 1975) in a diatomic molecule and has since been used as a prototype in studies of various shape resonance effects as discussed below. To identify the major final-state features in N_2 photoionization at the independent-electron level, we show the original calculation (Dehmer and Dill 1975, 1976a) of the K-shell photoionization spectrum performed with the multiple-scattering model. This calculation agrees qualitatively with all major features in the experimental spectrum (Wight et al. 1972/73, 1976, Kay et al. 1977, Hitchcock and Brion 1980a), except a narrow band of double excitation features, and with subsequent calculations, using more accurate techniques (Langhoff 1984, Lynch et al. 1984b, Schneider and Collins 1984). The four partial cross sections in fig. 3 represent the four dipole-allowed channels for K-shell (IP = 409.9 eV) photoionization. Here we have neglected the localization (Bagus and Schaefer 1972, Lozes et al. 1979) of the K-shell hole since it does not greatly affect the integrated cross section, and the separation into u and g symmetries both helps the present discussion and is rigorously applicable to the subsequent discussion of valence-shell excitation. (Note that the identification of shape resonant behavior is generally easier in inner-shell spectra, since the problems of overlapping spectra, channel interaction, and zeros in the dipole matrix element are reduced relative to valence-shell spectra.)

The most striking spectral feature in fig. 3 is the first member of the π_g sequence, which dominates every other feature in the theoretical spectrum by a factor of ~30. (Note that the first π_g peak has been reduced by a factor of 10 to fit in the frame.) The concentration of oscillator strength in this peak is a centrifugal

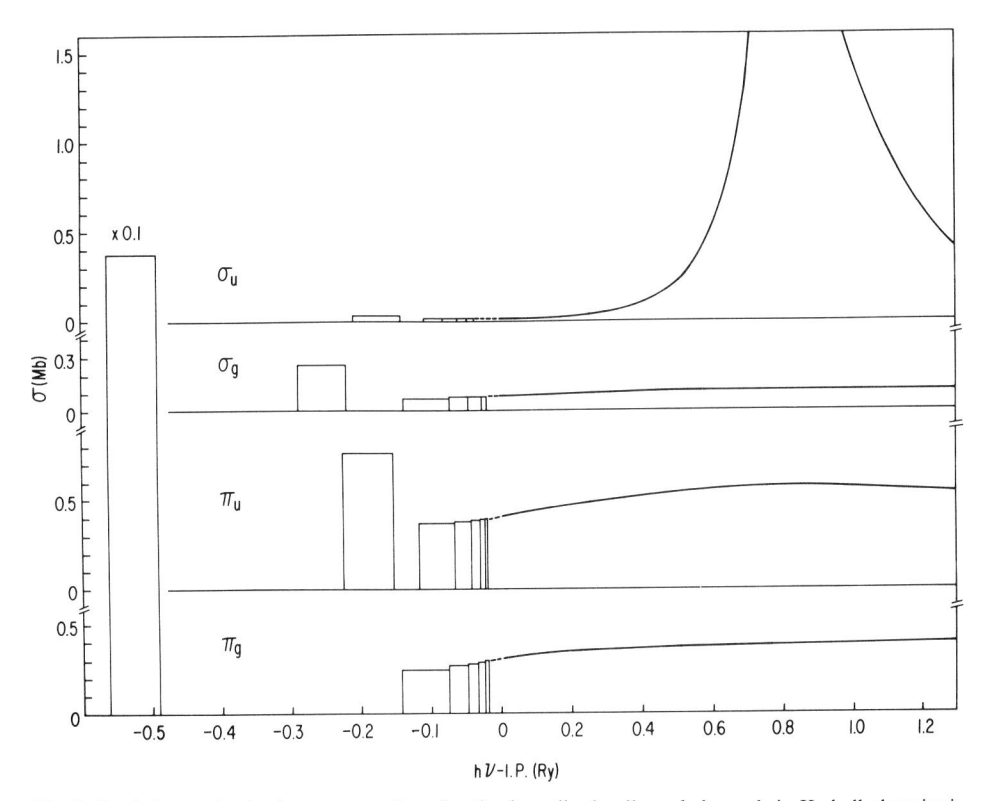

Fig. 3. Partial photoionization cross sections for the four dipole-allowed channels in K-shell photoionization of N_2. Note that the energy scale is referenced to the K-shell IP (409.9 eV) and is expanded twofold in the discrete part of the spectrum.

barrier effect in the d-wave component of the π_g wavefunction. The final state in this transition is a highly localized state, about the size of the molecular core, and is the counterpart of the well-known (Bardsley and Mandl 1968, Schulz 1973, 1976, Lane 1980, Shimamura and Takayanagi 1984) π_g shape resonance in e-N_2 scattering at 2.4 eV. For the latter case, Krauss and Mies (1970) demonstrated that the effective potential for the π_g elastic channel in e-N_2 scattering exhibits a potential barrier due to the centrifugal repulsion acting on the dominant $l = 2$ lead term in the partial-wave expansion of the π_g wavefunction. In the case of N_2 photoionization, there is one less electron in the molecular field to screen the nuclear charge so that this resonance feature is shifted (Dehmer and Dill 1979a) to lower energy and appears in the discrete. It is in this sense that we refer to such features as "discrete" shape resonances. The remainder of the π_g partial cross section consists of a Rydberg series and a flat continuum. The π_u and σ_g channels both exhibit Rydberg series, the initial members of which correlate well with partially resolved transitions in the experimental spectrum below the K-shell IP.

The σ_u partial cross section, on the other hand, was found to exhibit behavior rather unexpected for the K shell of a first-row diatomic. Its Rydberg series was extremely weak, and an intense, broad peak appeared at ~ 1 Ry above the IP in the low-energy continuum. This effect is caused by a centrifugal barrier acting on the $l = 3$ component of the σ_u wavefunction. The essence of the phenomena can be described in mechanistic terms as follows: the electric dipole interaction, localized within the atomic K shell, produces a photoelectron with angular momentum $l = 1$. As this p-wave electron escapes to infinity, the anisotropic molecular field can scatter it into the entire range of angular momentum states contributing to the allowed σ and π ionization channels ($\Delta \lambda = 0, \pm 1$). In addition, the spatial extent of the molecular field, consisting of two atoms separated by 1.1 Å, enables the $l = 3$ component of the σ_u continuum wavefunction to overcome its centrifugal barrier and penetrate into the molecular core at a kinetic energy of ~ 1 Ry. This penetration is rapid, a phase shift of $\sim \pi$ occurring over a range of ~ 0.3 Ry. These two circumstances combine to produce a dramatic enhancement of photoelectron current at ~ 1 Ry kinetic energy, with predominantly f-wave character.

The specifically molecular character of this phenomenon is emphasized by comparison with K-shell photoionization in atomic nitrogen and the united-atom case, silicon. In contrast to N_2, there is no mechanism for the essential p–f coupling, and neither atomic field is strong enough to support resonant penetration of high-l partial waves through their centrifugal barriers. (With subsitution of "d" for "f", this argument applies equally well to the d-type resonance in the discrete part of the spectrum.) Note that the π_u channel also has an $l = 3$ component but does not resonate. This underscores the directionality and symmetry dependence of the trapping mechanism.

To place the σ_u resonance in a broader perspective and show its connection with high energy behavior, we show, in fig. 4, an extension of the calculation in fig. 3 to much higher energy. Again, the four dipole-allowed channels in $D_{\infty h}$ symmetry are shown. The dashed line is two times the atomic nitrogen K-shell

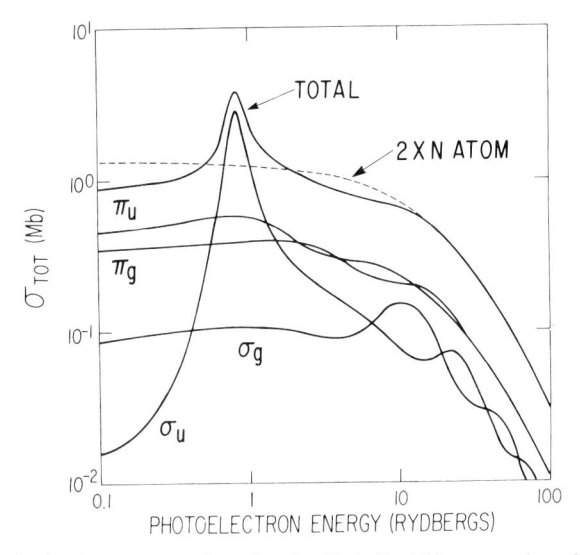

Fig. 4. Partial photoionization cross sections for the K-shell of N_2 over a broad energy range. The dashed line represents twice the K-shell photoionization cross section for atomic nitrogen, as represented by a Hartree–Slater potential.

cross section. Note that the modulation about the atomic cross section, caused by the potential barrier, extends to ~100 eV above threshold before the molecular and atomic curves seem to coalesce.

At higher energies, a weaker modulation appears in each partial cross section. This weak modulation is a diffraction pattern, resulting from scattering of the photoelectron by the neighboring atom in the molecule, or, more precisely, by the molecular field. Structure of this type was first studied over 50 years ago by Kronig (Kronig 1931, 1932, Azaroff 1963) in the context of metal lattices. It currently goes by the acronym EXAFS (extended X-ray absorption fine structure) and is used extensively (Kunz 1979, Winick and Doniach 1980, Koch 1983, Teo and Joy 1981, Lee et al. 1981) for local structure determination in molecules, solids, and surfaces. The net oscillation is very weak in N_2, since the light atom is a weak scatterer. More pronounced effects are seen, e.g., in K-shell spectra (Kincaid and Eisenberger 1975) of Br_2 and $GeCl_4$. Our reason for showing the weak EXAFS structure in N_2 is to show that the low-energy, resonant modulation (called "near-edge" structure in the context of EXAFS) and high-energy EXAFS evolve continuously into one another and emerge naturally from a single molecular framework, although the latter is usually treated from an atomic point-of-view.

Figure 5 shows a hypothetical experiment which clearly demonstrates the l character of the σ_u resonance. In this experiment, we first fix the nitrogen molecule in space and orient the polarization direction of a photon beam, tuned near the nitrogen K-edge, along the molecular axis. This orientation will cause photoexcitation into σ final states, including the resonant σ_u ionization channel. The figure shows the angular distribution of photocurrent as a function of both

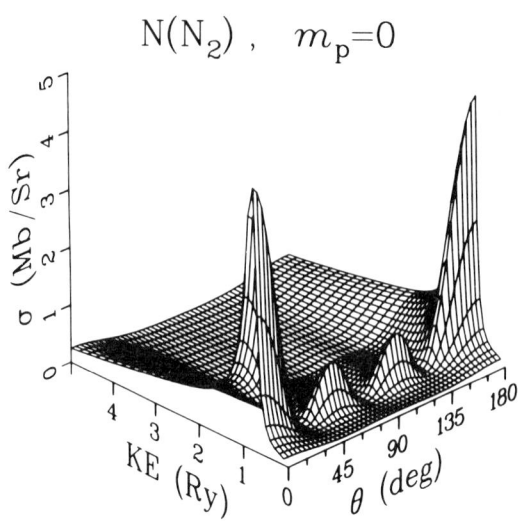

Fig. 5. Fixed-molecule photoelectron angular distribution for kinetic energies 0–5 Ry above the K-shell IP of N_2. The polarization of the ionizing radiation is oriented along the molecular axis in order to excite the σ continua and the photoelectron ejection angle, θ, is measured relative to the molecular axis.

excess energy above the K-shell IP and angle of ejection, θ, relative to the molecular axis. Most apparent in fig. 5 is the enhanced photocurrent at the resonance position, KE ~ 1 Ry. Moreover the angular distribution exhibits three nodes, with most of the photocurrent exiting the molecule along the molecular axis and none at right angles to it. This is an f-wave ($l = 3$) pattern and indicates clearly that the resonant enhancement is caused by an $l = 3$ centrifugal barrier in the σ_u continuum of N_2. Thus the centrifugal barrier has observable physical meaning and is not merely a theoretical construct. Note that the correspondence between the dominant asymptotic partial wave and the trapping mechanism is not always valid, especially when the trapping is on an internal or off-center atomic site where the trapped partial wave can be scattered by the anisotropic molecular field into alternative asymptotic partial waves, e.g., BF_3 (Swanson et al. 1981a) and SF_6. Finally, note that the hypothetical experiment discussed above has been approximately realized by photoionizing molecules adsorbed on surfaces. The shape resonant features tend to survive adsorption and, owing to their observable l-character, can even provide evidence (Gustafsson et al. 1978b, Gustafsson 1980b, 1983) as to the orientation of the molecule on the surface. A related family of measurements on adsorbed molecules record total absorption/ionization measurements as a function of the polarization direction of the light (see, e.g., Stöhr and Jaeger 1982, Stöhr et al. 1983, 1984, Koestner et al. 1984, Carr et al. 1985). This utilizes the symmetry properties of shape-resonance-enhanced absorption features to indicate the relative orientation of the molecular axis and the polarization direction. In the example cited above, the large enhancement above the K-edge of N_2 will occur only when the polarization is parallel to the molecular axis.

In this section, we have utilized the σ_u ($l = 3$) shape resonance in N_2 to illustrate the basic concepts underlying shape resonance phenomena. This resonance is supported by a barrier on the perimeter of the molecular charge distribution, which acts on the $l = 3$ component of the σ_u continuum wavefunction. It is important to realize that potential barriers in molecular fields can also take different forms. For instance, the t_{2g} and e_g shape resonances in SF_6 (sections 2.1 and 5.5) result from the trapping of $l = 2$ waves on the central sulfur atom. Although the trapping is associated with an atomic site, the molecular field plays a crucial role in modifying the potential in the vicinity of the barrier, relative to the free atom. This is manifested in two ways. First, in an isolated sulfur atom, the $l = 2$ wave will penetrate its potential barrier over a much broader energy range centered at higher kinetic energy, thus greatly diminishing the resonance effect. Second, the symmetry of the molecular field splits the $l = 2$ resonance into the crystal-field pair of t_{2g} and e_g quasibound states. In such a case, the d-wave trapping may not be clearly manifested in the asymptotic wave function, i.e., at the detector, since the departing d-wave may be rescattered into other partial waves by the anisotropic molecular field containing six fluorine atoms. Another type of potential barrier in molecules is illustrated by the e' shape resonance in BF_3 (section 5.6). In this case, the essential trapping mechanism was found (Swanson et al. 1981a) to involve the $l = 1$ component on the fluorine site. This off-center trapping site also causes rescattering of the trapped wave before it reaches the detector. In addition, the off-center trapping mechanism permits the trapping of a p wave in photoionization, for which the Coulomb potential would dominate the $l = 1$ centrifugal potential, were they centered on the same origin. These are only three examples, intended to create a broader perspective with which to approach new cases, which are likely to produce yet other types of barriers to photoelectron motion in molecular fields.

Finally, we would like to emphasize an intimate connection which exists between shape resonances and unoccupied valence states in quantum chemistry language (Langhoff 1984). This was dramatically demonstrated over ten years ago, when Gianturco et al. (1972) interpreted the shape resonances in SF_6 photoionization using unoccupied virtual orbitals in an LCAO–MO calculation. This connection is a natural one since shape resonances are localized within the molecular charge distribution and therefore can be realistically described by a limited basis set suitable for describing the valence MOs. However, the scattering approach used in the shape resonance picture is necessary for analysis of various dynamical aspects of the phenomena discussed above.

2.3. Eigenchannel plots

The next topic in the discussion on basic properties of shape resonances involves eigenchannel contour maps (Loomba et al. 1981), or "pictures" of unbound electrons. This is the continuum counterpart of contour maps of bound-state electronic wavefunctions which have proven so valuable as tools of quantum chemical visualization and analysis. Indeed, the present example helps achieve a physical picture of the σ_u shape resonance, and the general technique promises to be a useful tool for analyzing resonant trapping mechanisms and other observable properties in the future (see also Hermann and Langhoff 1981). The key to this visualization, given in eq. (5) below is the construction of those particular combinations of continuum orbital momenta that diagonalize the interaction of the unbound electron with the anisotropic molecular field. These combinations, known as eigenchannels, are the continuum analogues of the eigenstates in the discrete spectrum, i.e., the bound states.

The electronic eigenvalue problem in the molecular continuum is inhomogeneous, i.e., there is a solution at every energy. Moreover, there are, in general, alternative solutions possible at a given energy, depending on how the inhomogeneity is chosen. Typically, calculations are done in terms of real, oscillatory radial functions for the alternative possible orbital momenta. The result is the K-matrix-normalized partial-wave expansion of the continuum electronic wavefunction, which takes the asymptotic form (Dehmer and Dill 1979b, Dill and Dehmer 1974, Newton 1966)

$$\Psi_L \sim (\pi k)^{-1/2} r^{-1} \sum_{L'} (\sin \theta_l \delta_{LL'} + K_{LL'} \cos \theta_{l'}) Y_{L'}(\hat{r}), \qquad (1)$$

where $L = (l, m)$ is the photoelectron orbital momentum and its projection along the molecular z axis, and k^2 is the electron kinetic energy in rydberg. The electron distance r from the molecular center is given in bohr; $\theta_l = kr - l\pi/2 + \omega$, where the Coulomb phase, $\omega = -(Z/k) \ln(2kr) + \arg \Gamma[l + 1 - i(Z/k)]$ and the molecular ion charge, $Z = 1$. The coefficients $K_{LL'}$ of the cosine terms form a real, symmetric matrix known as the K matrix. The K matrix reflects the coupling between different angular momenta due to the nonspherical molecular potential, and, as such, summarizes in a compact way the electron–molecule interaction. That single orbital momentum L (we refer here to l and m collectively as orbital momentum) for which there occurs a sine term specifies the inhomogeneity, and each choice of L gives a row of the K matrix. By determining all rows in this way we obtain the full K matrix and thereby a complete set of functions at the given energy.

The wavefunction (1) specifies what might be called a calculational boundary condition. Its form allows us to obtain the K matrix while working in terms of real radial functions. There are two other types of boundary conditions (Dehmer and Dill 1979b, Dill and Dehmer 1974, Newton 1966) however: physical boundary conditions, appropriate for representing physical observables, and eigenchannel boundary conditions, appropriate for analysis of the continuum wavefunction itself. The sets of wavefunctions for these alternative three types of boundary conditions are interrelated by unitary transformations.

Physical boundary conditions are introduced to obtain physical observables. One transforms to complex radial functions so that the directional character of the continuum electron at large distances, where it is detected, can be represented. This so-called S matrix boundary condition and its use in representing electron–molecule scattering and molecular photoionization is discussed by, e.g., Dill and

Dehmer (1974) and by Dehmer and Dill (1976b). Physical boundary conditions are not well suited, however, for analysis of the continuum wavefunction itself. What is needed, rather, are the "normal modes" of the interaction of the continuum electron with the molecule, the eigenchannels of the electron–molecule complex. These are obtained by diagonalizing the K matrix, viz.

$$\tan(\pi\mu_\alpha)\delta_{\alpha\alpha'} = \sum_{LL'} U_{\alpha L} K_{LL'} U_{L'\alpha'}, \tag{2}$$

to give the eigenvectors Ψ_α and eigenphases μ_α. The coefficients $U_{L\alpha}$ give the composition of the alternative eigenvectors Ψ_α in terms of the K matrix normalized functions Ψ_L, viz.

$$\Psi_\alpha = \sum_L \Psi_L U_{L\alpha}. \tag{3}$$

Using eq. (2) and the fact that the matrix U is unitary, i.e.,

$$\delta_{LL'} = \sum_\alpha U_{L\alpha} U_{\alpha L'}, \tag{4}$$

we can rewrite the eigenvectors as (Newton 1966)

$$\Psi_\alpha \sim (\pi k)^{-1/2} r^{-1} \sum_L [\sin\theta_l + \tan(\pi\mu_\alpha)\cos\theta_l] Y_L(\hat{r}) U_{L\alpha}. \tag{5}$$

This equation is the key result of this discussion. Because molecules are not spherical, an electron of a particular angular momentum is in general "rescattered" into a range of angular momenta. This rescattering is indicated in eq. (1) by the sum over L', and the coefficients $K_{LL'}$ give the relative amplitudes of the various rescatterings. The eigenchannel functions (5), on the other hand, correspond to those special combinations of incident angular momenta which are unchanged by the anisotropic potential of the molecule, i.e., the normal (eigen)-modes of the electron–molecule interaction.

Comparison of eq. (5) with eq. (1), then, shows why the eigenchannel representation (5) is more suitable for analysis of the continuum molecular electronic wavefunction. *First*, as we have seen, the mixing of different orbital momenta is greatly simplified. *Second*, the radial wavefunctions for different angular momenta all have the same mixing coefficient $\tan(\pi\mu_\alpha)$. *Third*, if an eigenchannel is dominated by a particular orbital momentum, then the eigenchannel wavefunction (5) has the characteristic angular pattern of the corresponding spherical harmonic. *Last*, and perhaps most important, because quasibound shape-resonant states generally resonate in a single eigenchannel α, the eigenchannel representation gives us the most direct image of these resonant states.

As discussed earlier, the σ_u resonance is accompanied by a corresponding rise by about π radians in one component of the eigenphase sum,

$$\mu_{sum} = \sum_\alpha \mu_\alpha. \tag{6}$$

This resonant component is in turn composed almost entirely of the single partial wave $l = 3$,

$$U_{L\alpha_{\text{res}}} \simeq \delta_{l3} \,. \tag{7}$$

At these kinetic energies only one other orbital momentum, $l = 1$, contributes appreciably to the photoelectron wavefunction. (Orbital momenta $l = 0, 2, 4$, etc. do not contribute because they are of even parity, and $l = 5$ and higher odd orbital momenta are kept away from the molecule by centrifugal repulsion.) This means that there is only one other appreciable eigenchannel. Its eigenphase component, primarily $l = 1$, is nearly constant throughout the resonant region. Thus, within about 20 eV above the ionization threshold, the N_2 odd-parity continuum can be analyzed in terms of just two eigenchannels, a nonresonant p-like channel and a resonant f-like channel.

In fig. 6 we have plotted the p-like eigenchannel wavefunction for two kinetic energies, one below (top) and one at (bottom) the resonance, which in this calculation falls at ~1.2 Ry. The molecule is in the y–z plane, along the z axis. The surface contours have been projected onto the plane to show more clearly the angular variation of the wavefunction. The single nodal plane characteristic of p waves is clearly seen in this projection. It is remarkable that, despite the complex l-mixing induced by the anisotropic molecular potential, this eigenchannel has such a well-defined ($l = 1$) orbital-momentum character. The cusps in the surface mark the positions of the nuclei. The only apparent change from one surface to the other is the slight shortening of wavelength as the kinetic energy increases. This smooth contraction of nodes continues monotonically through the resonance energy to higher energies. We conclude from fig. 6 that the p eigenchannel is indeed nonresonant.

The f-like eigenchannel, plotted in fig. 7, shows a strikingly different behavior. Now the surfaces whose contours have at large distance the three nodal planes characteristic of f orbitals, show clearly the resonant nature of the f eigenchannel. Again note the clear emergence of a single ($l = 3$) orbital-momentum character over the whole wavefunction. Below and above (not shown) the resonance energy, the probability amplitude is roughly similar to that of the p-wave-dominated eigenchannel. But at the resonance energy there is an enormous enhancement of the wavefunction in the molecular interior; the wavefunction now resembles a molecular bound-state probability amplitude distribution. It is this enhancement, in the region occupied by the bound states, that leads to the very large increase in oscillator strength indicative of the resonance, and to the other manifestations discussed earlier and in subsequent sections.

These eigenchannel plots are discussed more fully elsewhere (Loomba et al. 1981); however, before leaving the subject, several points should be noted. *First,* the N_2 example that we have chosen is somewhat special in that there is a near one-to-one correspondence between the eigenchannels and single values of orbital angular momentum. Orbital angular momentum is, however, not a "good" quantum number in molecules and more generally we should not always expect

N_2 σ_u, 0.9Ry, $l=1$

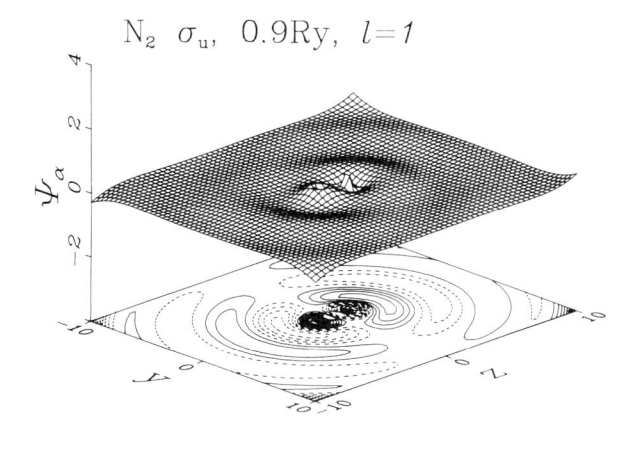

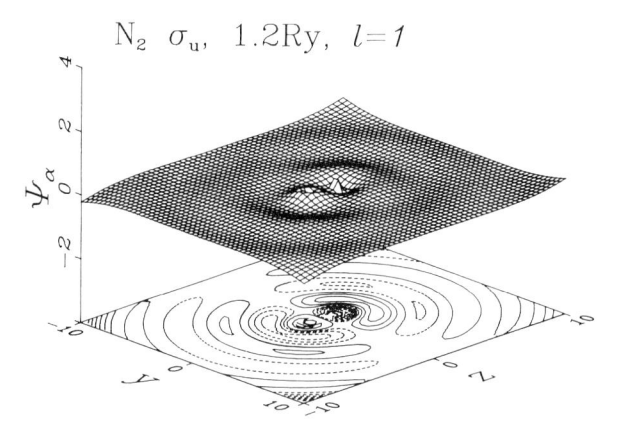

N_2 σ_u, 1.2Ry, $l=1$

Fig. 6. The p-wave-dominated eigenchannel wavefunctions for two electron kinetic energies in the σ_u continuum of N_2. The molecule is in the y–z plane, along the z axis, centered at $y = z = 0$. Contours mark steps of 0.03 from 0.02 to 0.29; solid: positive, dashed: negative.

such clear nodal patterns. Frequently, several angular momenta contribute to the continuum eigenchannels (although a barrier in only one l component will be primarily responsible for the temporary trapping that causes the enhancement in that and coupled components) and this means that the resulting eigenchannel plots will be correspondingly richer. *Second*, eqs. (1) and (5) are asymptotic expressions. The orbital momentum composition of these wavefunctions is more complicated in the molecular interior, as seen, e.g., in figs. 6 and 7. Nonetheless, continuity and a dominant l may, as in the case of N_2, cause the emergence of a distinct l pattern, even into the core region. *Third*, while these ideas were developed (Loomba et al. 1981) in the context of molecular photoionization, the continuum eigenchannel concept carries over without any fundamental change to electron–molecule scattering. This is a further example of the close connection (Dehmer and Dill 1979a) between shape resonances in molecular photoionization

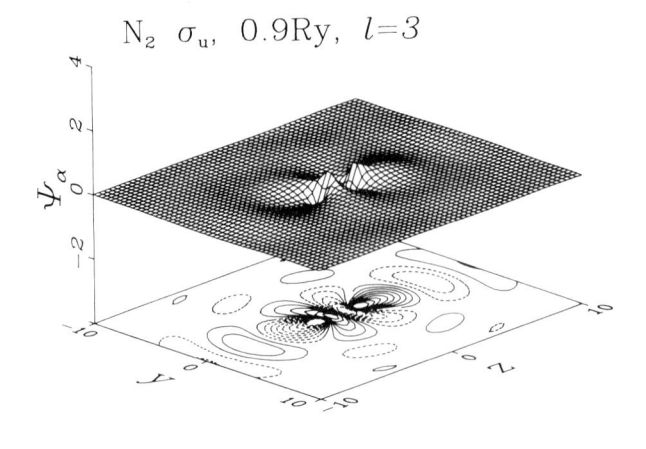

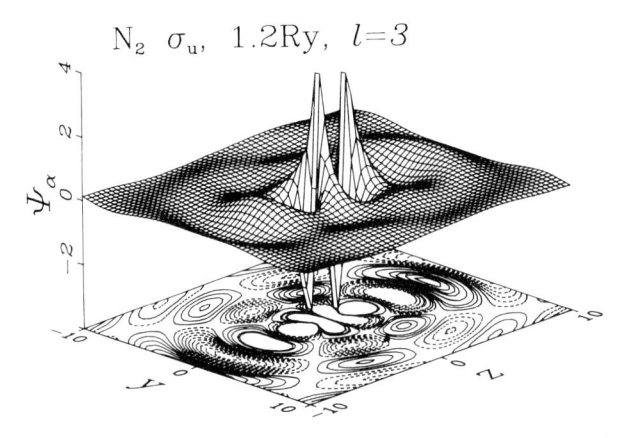

Fig. 7. The f-wave-dominated eigenchannel wavefunctions for nonresonant (top) and resonant (bottom) electron kinetic energies in the σ_u continuum of N_2. The molecule is in the y–z plane, along the z axis, centered at $y = z = 0$. Contours mark steps of 0.03 from 0.02 to 0.29; solid: positive, dashed: negative. The lack of contour lines for 1.2 Ry near the nuclei is because of the 0.29 cutoff.

and electron–molecule scattering. *Finally*, while we have used one-electron wavefunctions here, obtained with the multiple-scattering model, the eigenchannel concept is a general one and we may look forward to its use in the analysis of more sophisticated, many-electron molecular continuum wavefunctions.

2.4. *Connections between shape resonances in electron–molecule scattering and in molecular photoionization*

At first glance, there is little connection between shape resonances in electron–molecule scattering (e + M) and those in molecular photoionization ($h\nu$ + M). The two phenomena involve different numbers of electrons and the collision velocities are such that all electrons are incorporated into the collision complex.

Hence, we are comparing a neutral molecule and a molecular negative-ion system. However, although the long-range part of the scattering potential is drastically different in the two cases, the strong short-range potential is not drastically different since it is dominated by the interactions among the nuclei and those electrons common to both systems. Thus, shape resonances which are localized in the molecular core substantially maintain their identity from one system to another, but are shifted in energy owing to the difference caused by the addition of an electron to the molecular system. This unifying property of shape resonances thus links together the two largest bodies of data on the molecular electronic continuum: $h\nu + M$ and $e + M$, and although these resonances shift in energy in going from one class to another and manifest themselves in somewhat different ways, this link permits us to transfer information between the two. This can serve to help interpret new data and even to make predictions of new features to look for experimentally. Actually, this picture (Dehmer and Dill 1979a) was surmised empirically from evidence contained in survey calculations on $e + M$ and $h\nu + M$ systems and, in retrospect, from data. These observations can be summarized as follows: By and large, the systems $h\nu + M$ and $e + M$ display the same manifold of shape resonances, only those in the $e + M$ system are shifted ~10 eV to higher electron energy. Usually, there is one shape resonance per symmetry for a subset of the symmetries available. The shift depends on the symmetry of the state, indicating, as one would expect, that the additional electron is not uniformly distributed. Finally, there is substantial proof that the l-character is preserved in this process, although interaction among alternative components in a scattering eigenchannel can vary and thus alter the l mixing present.

There are several good examples available to illustrate this point – N_2, CO, CO_2, BF_3, SF_6, etc. In general, one can start from either the neutral or the negative ion system, but, in either case, there is a preferred way to do so: In the $h\nu + M$ case, it is better to examine the inner-shell photoabsorption and photoionization spectra. Shape resonances almost invariably emerge most clearly in this context. Additional effects, discussed briefly at the end of this section, frequently make the role of shape resonances in valence-shell spectra more complicated to interpret. In the $e + M$ case, a very sensitive indicator of shape resonance behavior is the vibrational excitation channel. Vibrational excitation is enhanced by shape resonances (Bardsley and Mandl 1968, Schulz 1973, 1976, Lane 1980, Shimamura and Takayanagi 1984) and is typically very weak for nonresonant scattering. Hence, a shape resonance, particularly at intermediate energy (10–40 eV) (Dehmer and Dill 1980, Dill et al. 1979b) may be barely visible in the vibrationally and electronically elastic scattering cross section, and yet be displayed prominently in the vibrationally inelastic electronically elastic cross section.

Two examples will help illustrate these points. In $e-SF_6$ scattering, the vibrationally elastic scattering cross section has been calculated theoretically (Dehmer et al. 1978) and shown to have four shape resonances of a_{1g}, t_{1u}, t_{2g}, and e_g symmetry at approximately 2, 7, 13, and 27 eV, respectively. The absolute total cross section measured by Kennerly et al. (1979) shows qualitative agreement, although the evidence for the e_g resonance is marginal. (This resonance might be more evident in a vibrational excitation spectrum, which is not available.) Hence, using the guidelines given above, one would expect shape resonance features in the $h\nu + M$ case at -8, -3, 3, and 17 eV (on the kinetic energy scale) to a very crude, first approximation. Indeed, the K- and L-shell photoabsorption spectra of SF_6 show such intense features, as discussed in an earlier section. This correlation is indicated (Fock and Koch 1985) in fig. 8, along with approximate resonance positions in the valence-shell spectra, for which the evidence is more fragmentary (see section 5.5).

Using N_2, we reverse the direction of the mapping, and start with $h\nu + N_2$, which was discussed extensively in earlier sections. Here a "discrete" shape resonance of π_g symmetry and a shape resonance of σ_u symmetry are apparent in the K-shell spectrum (Wight et al. 1972/73, 1976, Kay et al. 1977, Hitchcock and Brion 1980a) (see fig. 3). These occur at ~ -9 and 10 eV on the kinetic energy scale (relative to the ionization threshold). Hence, one would look for the same set of resonances in e–N_2 scattering at ~ 1 and ~ 20 eV incident electron energies. The well-known π_g shape resonance (Bardsley and Mandl 1968, Schulz 1973, 1976, Lane 1980, Shimamura and Takayanagi 1984) is very apparent in the vibrationally elastic cross section; however, there is only a very broad bump at ~ 20 eV (Kennerly 1980). As noted above, the vibrationally inelastic cross section is much more sensitive to shape resonances, and, indeed, the σ_u shape resonance in e–N_2 scattering has been established theoretically and experimentally by looking in this channel (Dehmer and Dill 1980, Pavlovic et al. 1972, Truhlar et al.

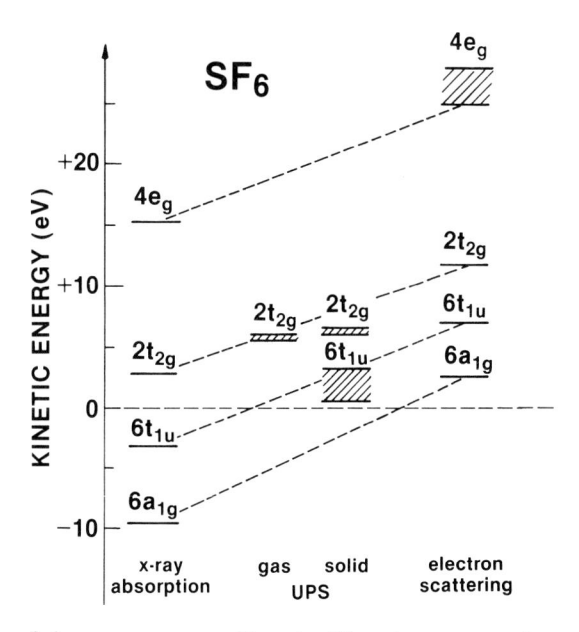

Fig. 8. Systematics of shape resonance positions in different measurements on SF_6. (Adapted from Fock and Koch 1984, 1985.)

1972, Dehmer et al. 1980, Rumble et al. 1981). Several other excellent examples exist, but we will conclude by pointing out that the connections between e–CO_2 and $h\nu + CO_2$ resonances have been recently discussed (Dittman et al. 1983) in detail, including a study of the eigenphase sums in the vicinity of the σ_u shape resonance in the two systems.

Finally, we note similar connections and additional complications upon mapping from inner-shell to valence-shell $h\nu + M$ spectra. On going from deep inner-shell spectra to valence-shell spectra, shape resonances in $h\nu + M$ also shift approximately 1–4 eV toward higher kinetic energy, due to differences in screening between localized and delocalized holes as well as other factors. As mentioned above, several complications arise in valence-shell spectra which can tend to obscure the presence of a shape resonance compared to their more straightforward role in inner-shell spectra. These include greater energy dependence of the dipole matrix element, interactions with autoionizing levels (Morin et al. 1982a, Collins and Schneider 1984), strong continuum–continuum coupling (Dehmer et al. 1982, Stephens and Dill 1985) between more nearly degenerate ionization channels, strong particle–hole interactions (Krummacher et al. 1980, Langhoff et al. 1981a, Krummacher et al. 1983, Bagus and Viinikka 1977, Cederbaum and Domcke 1977, Cederbaum et al. 1977, 1978, 1980, Schirmer et al. 1977, Wendin 1981, Schirmer and Walter 1983) etc. So, for the most transparent view of the manifold of shape resonance features in $h\nu + M$, one should always begin with inner-shell data.

3. Autoionization

3.1. Overview

Autoionization is an intrinsically multichannel process in which a resonantly excited discrete state from one channel couples to the underlying electronic continua of one or more other channels to effect ionization. It has been known since Fano's original work (Fano and Cooper 1968, Fano 1935, 1961) almost fifty years ago, that this process produces characteristic asymmetric Fano–Beutler profiles in the photoionization cross section. Since then, there have been extensive studies of autoionization structure in the total photoionization cross sections of atoms (Fano and Cooper 1968) and molecules (see, e.g., Koch and Sonntag 1979, Hayaishi et al. 1982, Dibeler and Walker 1967, Dibeler and Liston 1968, Chupka and Berkowitz 1969, Dibeler and Walker 1973, McCulloh 1973, Dehmer and Chupka 1975, 1976, Berkowitz and Eland 1977, Gurtler et al. 1977, Berkowitz 1979, Ono et al. 1982, Wu and Ng 1982, P.M. Dehmer et al. 1984). In addition, the manifestations of autoionization in such dynamical parameters as photoionization branching ratios and photoelectron angular distributions have been recognized and have recently developed into a major focal point for current studies of molecular photoionization dynamics (see, e.g., Morin et al. 1980,

Tabché-Fouhailé et al. 1981, Unwin et al. 1981, Hayaishi et al. 1982, Keller et al. 1982, Morin et al. 1982a,b, Parr et al. 1982a, Carlson et al. 1983b, Levine and Soven 1983, Morin 1983, Collins and Schneider 1984, Levine and Soven 1984, Fano 1970, Dill 1972, Herzberg and Jungen 1972, Atabek et al. 1974, Fano 1975, Jungen and Atabek 1977, Dill and Jungen 1980, Giusti-Suzor and Lefebvre-Brion 1980, Jungen 1980, Jungen and Dill 1980, Raoult et al. 1980, Jungen and Raoult 1981, Raoult and Jungen 1981, Giusti-Suzor 1982, Lefebvre-Brion and Giusti-Suzor 1983, Raoult et al. 1983, Raseev and Le Rouzo 1983, Giusti-Suzor and Fano 1984a,b, Giusti-Suzor and Jungen 1984, Giusti-Suzor and Lefebvre-Brion 1984, Jungen 1984a,b, Mies 1984, Mies and Julienne 1984, Lefebvre-Brion et al. 1985, Marr and Woodruff 1976, Woodruff and Marr 1977, Baer et al. 1979, Codling et al. 1981, Ederer et al. 1981, Parr et al. 1981, West et al. 1981, Parr et al. 1982b, Truesdale et al. 1983b, Hubin-Franskin et al. 1984, Doolittle and Schoen 1965, Price 1968, Berkowitz and Chupka 1969, Collin and Natalis 1969, Blake et al. 1970, Bahr et al. 1971a,b, Carlson 1971, Collin et al. 1972, Kleimenov et al. 1972, Gardner and Samson 1973, Tanaka and Tanaka 1973, Gardner and Samson 1974a,b, Caprace et al. 1976, Natalis et al. 1977, Gardner and Samson 1978, Eland 1980, Kumar and Krishnakumar 1981, 1983, Carlson and Jonas 1971, Morgenstern et al. 1971, Carlson and McGuire 1972, Carlson et al. 1972, Niehaus and Ruf 1972, Hancock and Samson 1976, Mintz and Kuppermann 1978, Katsumata et al. 1979, Kibel et al. 1979, Sell et al. 1979, Kreile and Schweig 1980, Berry and Nielsen 1970a,b, Duzy and Berry 1976).

A more physical description of the autoionization process is helpful in discussing the alternative decay mechanisms possible in molecules: In most cases, autoionizing states consist of an excited Rydberg electron bound to an excited ion (also called core) primarily by Coulomb attraction. [The case of two highly correlated electrons bound to an ion is another important case which requires special treatment (Fano 1983) and will not be discussed here.] A necessary condition for decay of this state by ionization is that the excitation energy of the ion must be greater than the binding energy of the Rydberg electron. Then, barring alternative decay paths, autoionization will take place by means of a close collision, between the Rydberg electron and the ion, in which excitation energy of the ion is transferred to the excited electron to overcome its binding energy and permit its escape from the ionic field. Notice that, although a Rydberg electron spends only a very small fraction of time within the molecular ion, such close encounters are essential for autoionization since, only when the Rydberg electron is nearby can it participate fully in the dynamics of the core and exchange energy efficiently with it.

A molecular ion core can store the energy needed to ionize a Rydberg electron in any of its three modes – electronic, vibrational, or rotational. The most direct means of storing electronic energy is to produce a hole in a molecular orbital (MO) other than the outermost occupied MO, e.g., by promoting one of the inner electrons into a Rydberg orbital. In addition, various degrees of vibrational and rotational excitation can accompany photoexcitation of Rydberg states converging to any state of the ion. It is the existence and interplay among the alternative energy modes which lead to the unique properties of molecular autoionization.

As a concrete example of rotational and vibrational autoionization we will discuss photoionization of H_2. The alternative rovibrational ionization channels for para-H_2 ($J = 0$) are shown schematically in fig. 9. The ground ionic state, $H_2^+ X\,^2\Sigma_g^+$, is the only bound electronic state in this spectral range so that the possibility of electronic autoionization is eliminated. In fig. 9, the vertical, shaded bars represent various vibrational channels of $H_2^+\,^2\Sigma_g^+$, labeled by $v^+ = 0$–5. Pairs of continua are associated with each v^+ reflecting the two rotational continua $N^+ = 0, 2$ produced by photoionization of para-H_2. Converging to each of these (and higher) rovibrational thresholds are Rydberg series, supported by the Coulomb field of the H_2^+ ion. A small subset of these Rydberg states is indicated, for later reference, by horizontal lines at their observed spectral location and placed directly under the threshold to which they converge. Any of these optically

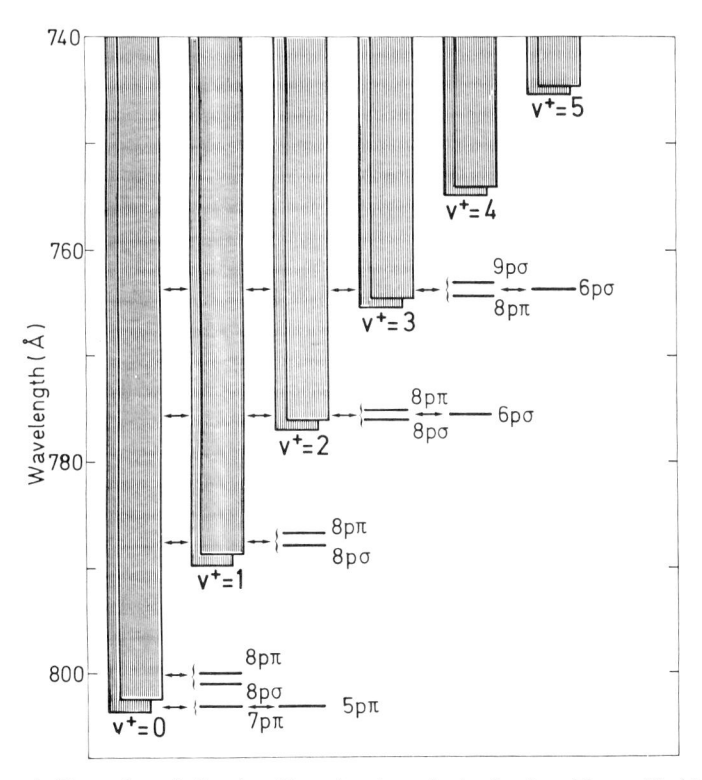

Fig. 9. Schematic illustration of vibrational/rotational autoionization in cold para-H_2 ($J = 1$, negative parity final states). Continua are indicated by vertical hatching. For each given v^+ of the ion H_2^+ there are two continua corresponding to rotational quantum number $N^+ = 0$ and 2 of the ion ($J = 1$). Selected discrete Rydberg levels are indicated below the vibrational ionization limit with which they are associated. (From Raoult and Jungen 1981.)

allowed Rydberg states can autoionize by coupling with accessible open channels. In this case, autoionization would proceed by transferring energy stored in rotation or vibration of the ion core to the photoelectron. If more than one continua is available the decay will proceed into each with a branching ratio determined by the detailed dynamics of the decay process. Moreover, the angular distribution of the photoelectrons escaping in each channel will reflect further

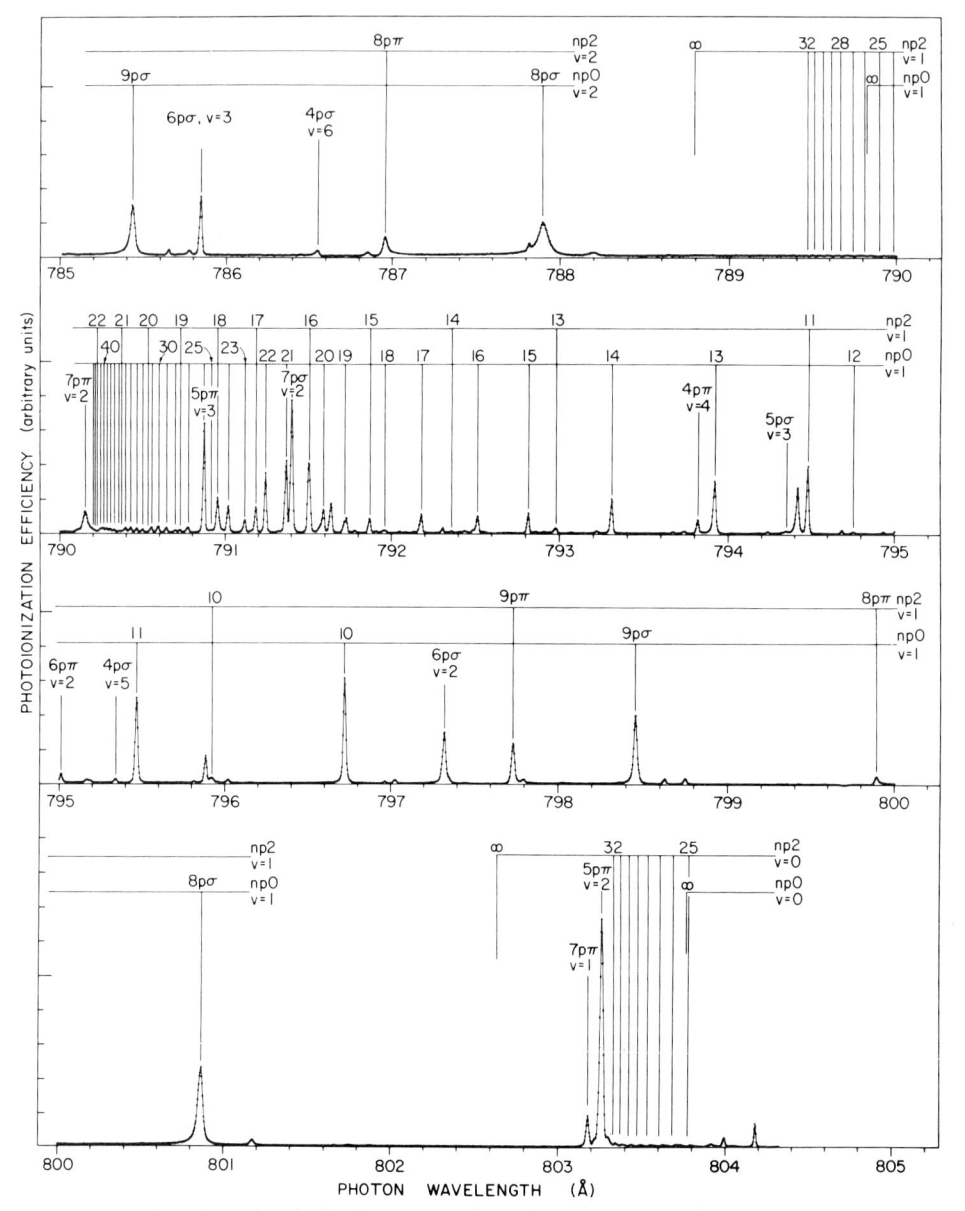

Fig. 10. A portion of the photoionization cross section of para-H_2 at 78 K. (From Dehmer and Chupka 1976.)

details of the dynamics including relative phases of degenerate photoelectron wavefunctions.

Based on the picture, so far, of a set of rovibrational thresholds and Rydberg series converging to each, one might expect a dense pattern of autoionizing levels, but one which would straightforwardly yield to spectroscopic analysis in terms of characteristic rovibrational spacings and known behavior of Rydberg series. However, this simple picture of a rich but fundamentally uncomplicated spectrum ignores all interaction between the Rydberg electron and molecular core, and is

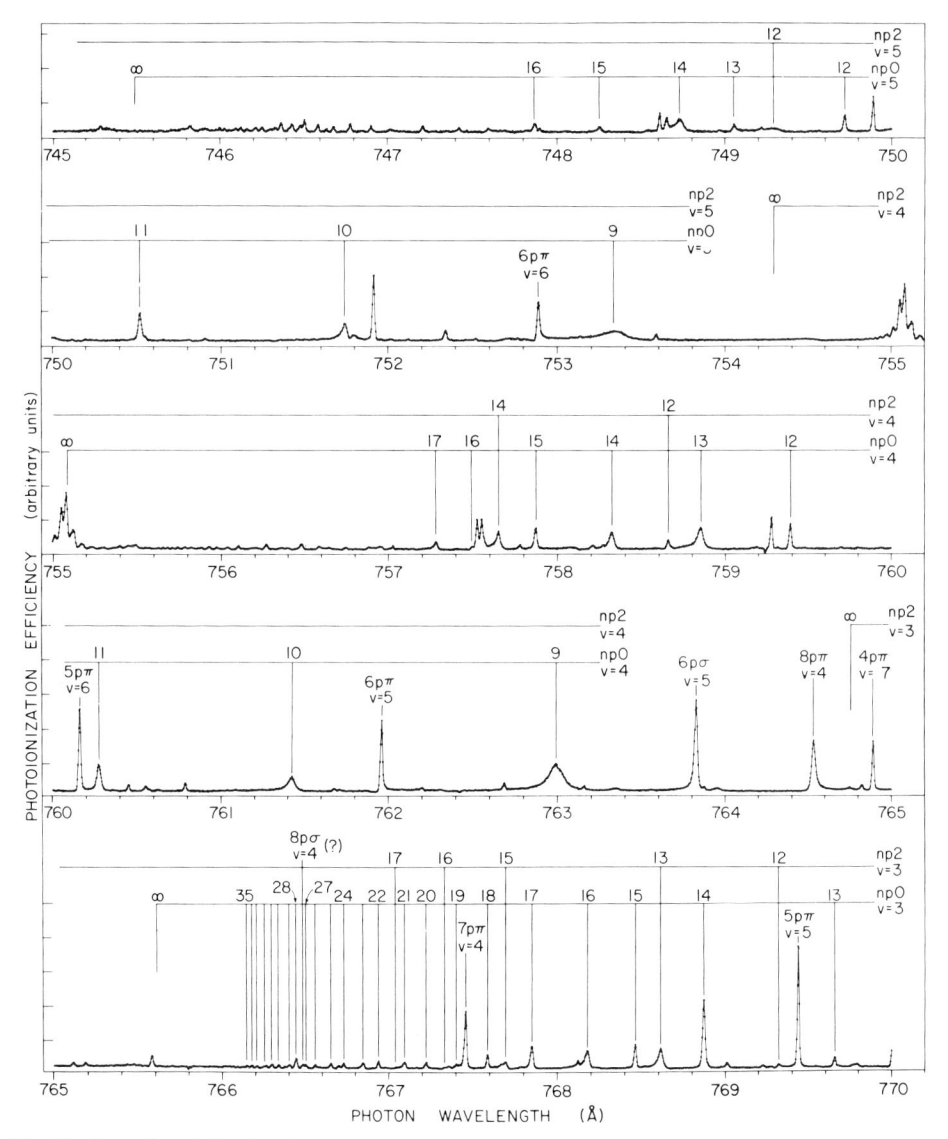

Fig. 11. A portion of the photoionization cross section of para-H_2 at 78 K. (From Dehmer and Chupka 1976.)

wrong. This is shown dramatically in figs. 10 and 11 which show the total photoionization spectrum (Dehmer and Chupka 1976) in two spectral regions (chosen for later discussion) covered in fig. 9, one from the first IP to 785 Å and the other covering 770–745 Å. Careful inspection will reveal that, for any Rydberg series, the spacings, intensities, and profiles will deviate strongly from a simple Rydberg pattern. This is especially true near "interlopers", i.e., Rydberg states falling in the midst of a Rydberg series, but converging to a higher limit. For such cases level shifts and intensity redistribution frequently modify the entire host Rydberg series. These modifications arise from mutual interactions mediated by short range forces and have been accounted for in detail in this prototype system.

In the following two sections we use two examples which represent the state-of-the-art in theoretical and experimental studies of molecular autoionization dynamics. The most accurate and penetrating theoretical analysis (Fano 1970, Dill 1972, Herzberg and Jungen 1972, Dill and Jungen 1980, Jungen 1980, Jungen and Dill 1980, Raoult et al. 1980, Jungen and Raoult 1981, Raoult and Jungen 1981, Giusti-Suzor 1982, Raseev and Le Rouzo 1983, Jungen 1984a,b) has been carried out on parts of the H_2 spectrum using MQDT. Two representative cases will be discussed including rotational autoionization and a prediction of vibrational branching ratios and photoelectron angular distributions resulting from vibrational autoionization above the $v^+ = 3$ limit at ~ 764.8 Å. These predictions have not yet been tested, although equivalent experiments on electronic autoionization in N_2 have recently been performed (Parr et al. 1981, West et al. 1981). The latter case will be discussed in section 5.2.

3.2. MQDT treatment of H_2 photoionization

Multichannel quantum defect theory and its application to molecular photoionization have been described in detail elsewhere (Fano 1970, Dill 1972, Herzberg and Jungen 1972, Atabek et al. 1974, Fano 1975, Jungen and Atabek 1977, Dill and Jungen 1980, Giusti-Suzor and Lefebvre-Brion 1980, Jungen 1980, Jungen and Dill 1980, Raoult et al. 1980, Jungen and Raoult 1981, Raoult and Jungen 1981, Giusti-Suzor 1982, Lefebvre-Brion and Giusti-Suzor 1983, Raoult et al. 1983, Raseev and Le Rouzo 1983, Giusti-Suzor and Fano 1984a,b, Giusti-Suzor and Jungen 1984, Giusti-Suzor and Lefebvre-Brion 1984, Jungen 1984a,b, Mies 1984, Mies and Julienne 1984, Lefebvre-Brion et al. 1985). Hence we will only briefly summarize the important attributes of MQDT and will then turn to two examples of its application to photoionization of H_2.

MQDT is a theoretical framework which *simultaneously* treats the interactions between and within *whole* excitation channels. The input to an MQDT calculation consists chiefly of a small set of physically meaningful parameters (quantum defects and dipole amplitudes) which characterize the short range interactions between the excited electron and the core, are slowly varying functions of energy relative to rovibronic structure in the spectrum, and can generally be obtained from the positions and intensities of low-lying states in the spectrum. Also used are known transformation properties of molecular wavefunctions, e.g., between Hund's coupling cases, and the asymptotic boundary conditions pertinent to a particular spectral range. Given these, straightforward matrix mechanics yields, at each excitation energy, the spectral composition of the total final state wavefunction in terms of the short-range, body-frame basis set, known dipole strengths, and the asymptotic eigenphase shifts of the observable ionization channels. These quantities are then related to such observables as the total photoionization cross section, vibrational branching ratios, and photoelectron angular distributions by now standard formulas.

We wish to emphasize that this theoretical framework is not only elegant, but also reflects very accurately the internal mechanics of the excited complex. Hence, the quality of the computed observable depends solely upon the quality of the input. When accurate empirical quantum defects are used, adiabatic and non-adiabatic corrections to the Born–Oppenheimer approximation are included automatically, to all orders. Hence, given physical input from low-lying excited states, one can use MQDT to generate accurate predictions for experiments throughout the extremely complex high-excitation regions. This is to be contrasted to the normal perturbation approach which would require treating each state separately, with explicit adiabatic and non-adiabatic corrections, and with little hope of treating higher order interactions within and between whole excitation channels.

A striking example of the power and accuracy of MQDT in a complex situation is provided by the rich rotational/vibrational autoionization structure in the 174.3 cm^{-1} spectral range between the $N^+ = 0$ and 2 rotational thresholds associated with the lowest $v^+ = 0$ ionization potential of H_2 (Dehmer and Chupka 1976). The results of a calculation (Jungen and Dill 1980) of this structure is given in fig. 12. Across the top of the figure are indicated the level positions one obtains with discrete boundary conditions, i.e., by eliminating open channels from the linear system. This level of analysis determines the initial assignments of spectral features. Alternatively, photoabsorption directly into the continuum is shown across the bottom of fig. 12. This is the single-channel level of approximation used throughout the discussion of shape resonances in section 2. If rotational autoionization only is introduced, then the levels $np2$ of the Rydberg series converging to the upper ($N^+ = 2$) threshold of $H_2^+ X\,^2\Sigma_g^+$ ($v^+ = 0$) autoionize and distort the continuum into a Rydberg series of Fano–Beutler profiles. This is shown in the middle frame in fig. 12. Finally, if vibrational autoionization channels are also introduced, then the $5p\pi$, $v = 2$ and $7p\pi$, $v = 1$ levels autoionize and strongly distort the rotationally autoionizing levels as well, shifting intensity from above to below the vibrational interlopers.

It is seen that the effect of vibrational autoionization on the ionization cross section is profound and that it affects the whole range shown, corresponding to about 100 cm^{-1}. Indeed, if the fine variations of the cross section are neglected, the whole spectrum can be viewed as one "giant" resonance of about 50 cm^{-1} width which causes a global transfer of intensity from the high-energy to the low-energy side of the vibrational peaks. This transfer leads to further modifica-

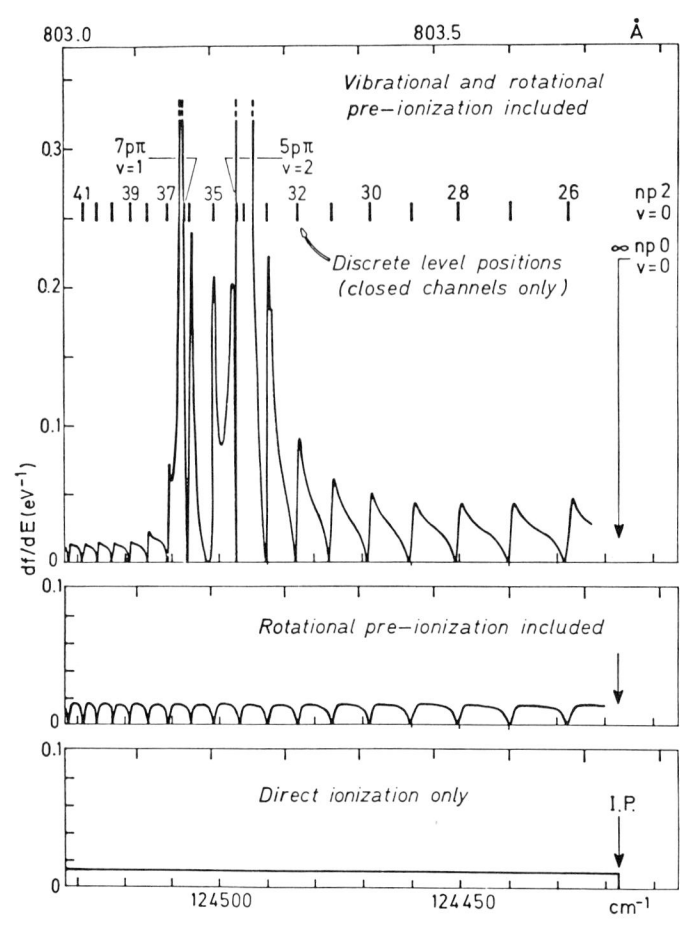

Fig. 12. MQDT calculation of photoionization of $H_2 X\,^1\Sigma_g^+$ ($J'' = 0$, $v'' = 0$) near the ionization threshold. (From Jungen and Dill 1980.)

tions of the fine structure. For example, for $n = 26$ and 27 the intensity minima still correspond nearly to the discrete np2 levels, but for higher n the profiles become progressively distorted until near $n = 32$–35, it is the intensity maxima which coincide with the discrete level positions. In other words, there exists no longer a simple relationship between the extrema in the ionization curve and the positions of the autoionizing levels. In view of these complexities it is clear that vibrational and rotational autoionization cannot be meaningfully treated as separate processes in this spectral region. A key feature of the MQDT is that it is based on no such assumed separability, i.e., it is applicable independently of coupling strengths between alternative decay mechanisms.

In fig. 13 the calculated spectrum (Jungen and Dill 1980) is compared with the high-resolution photoionization spectrum (Dehmer and Chupka 1976). The calculated spectrum from fig. 12 was convoluted with a triangular apparatus function of

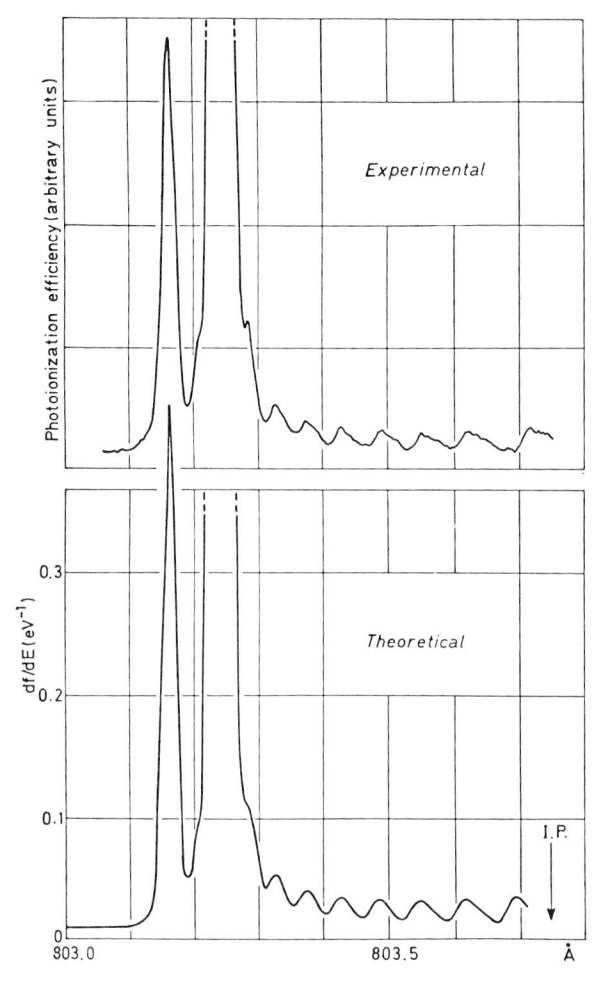

Fig. 13. MQDT results from fig. 12 broadened to a resolution of 0.022 Å and compared with data from Dehmer and Chupka (1976). (Figure from Jungen and Dill 1980.)

halfwidth 0.022 Å to mimic the finite experimental resolution. The comparison in fig. 13 shows essentially exact agreement and reflects more clearly than words the state-of-the-art in computational simulation of detailed photoionization dynamics.

Triply differential cross sections have also been computed (Raoult and Jungen 1981) near each of the levels explicitly shown in fig. 9. We will skip to the highest set of levels, above the $v^+ = 3$ limit, to discuss MQDT predictions of vibrational branching ratios and β's. This spectral range, between 762.5 and 765 Å in fig. 11, is most attractive for future experimental examination since it produces four photoelectron peaks, i.e., corresponding to $v^+ = 0-3$, with sufficiently large photoelectron energies to be measured with existing electron anergy analyzers.

Figure 14 shows the calculated total and vibrational partial cross sections (the

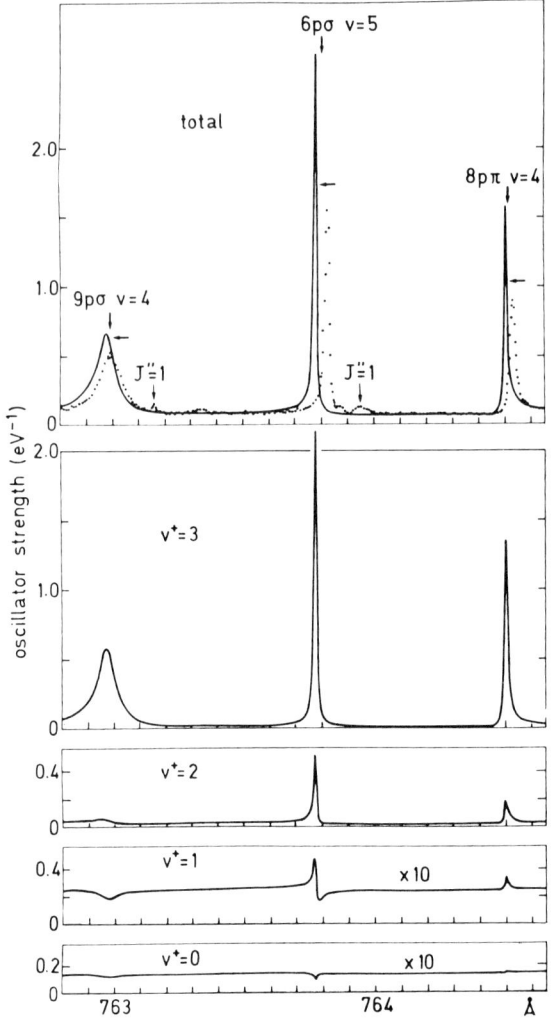

Fig. 14. Total and partial oscillator strengths for photoionization of H_2 near the $v^+ = 3$, $N^+ = 2$ ionization threshold (764.755 Å). (Experimental points from Dehmer and Chupka 1976; figure adapted from Raoult and Jungen 1981.)

rotational sublevels have been summed over) in this spectral range, together with total photoionization data (Dehmer and Chupka 1976). In this case the calculations are not folded with the instrument function and are, accordingly, sharper and higher. The horizontal arrow indicates the height that would be obtained from such a convolution. Also the vertical arrows show more precise peak positions from the high-resolution absorption spectrum (Herzberg and Jungen 1972). Given these qualifications, the agreement in the total cross section is, again, quite satisfactory. The vibrational partial cross sections, for which no experimental data is available is shown in the lower frames of fig. 14. There, we see that the widths of the vibrational autoionization peaks are the same as in the total cross section, as expected, but that the profiles vary drastically. Other details of the behavior of these partial cross sections are displayed more clearly in the vibrational branching ratios, discussed below.

Figure 15 again shows the total photoionization cross section, together with the total β (summed over v^+) and the vibrational branching ratio and β_{v^+} for each vibrational ionization channel v^+. The total β curve is observed to dip strongly

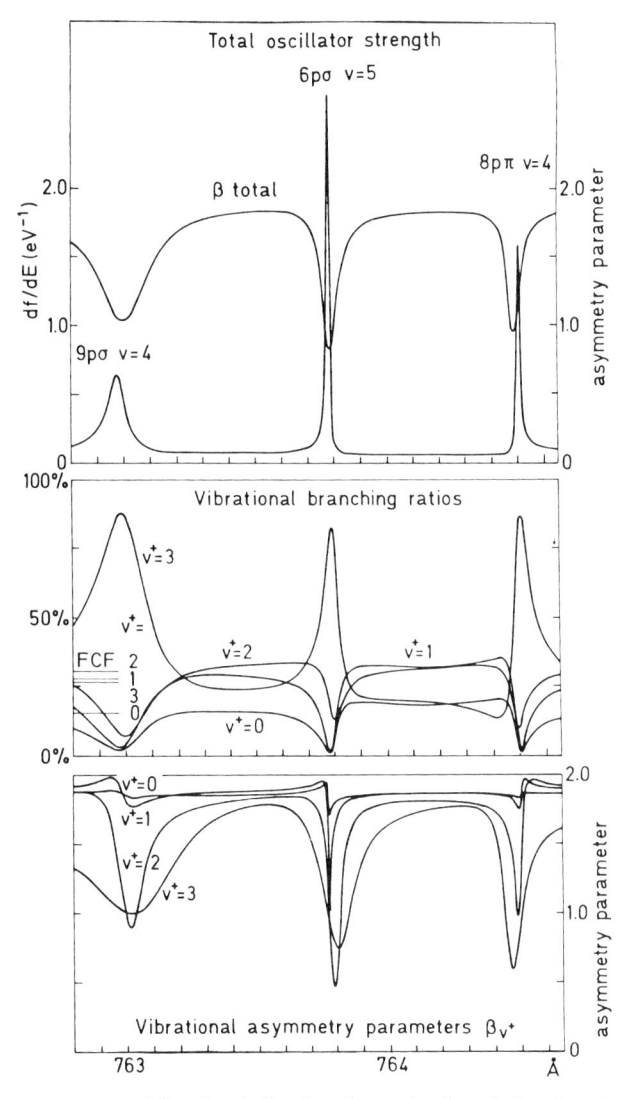

Fig. 15. Asymmetry parameter β (total and vibrationally resolved) and vibrational branching ratios for photoionization of H_2 near the $v^+ = 3$, $N^+ = 2$ ionization threshold (764.755 Å). (Figure adapted from Raoult and Jungen 1981.)

within the resonances. This is a consequence of the discrete wavefunction component mixing strongly into the ionization continuum. Classically speaking, one would say that the "quasibound" photoelectron spends more time near the core so that angular momentum exchange is enhanced. In this case the $N^+ = 2$ ionization channel becomes dominant near the center of the resonance and the value of β is depressed correspondingly, showing that directional information carried by the incoming photon is largely transferred to molecular rotation in the subsequent electron–core collisions. This general behavior is also reflected in the partial β_{v^+} curves although, in addition, a strong v^+ dependence in the magnitudes and shapes of the β_{v^+} is also observed. Note that the spectral extent of the variations induced in the β_{v^+} by vibrational autoionization is considerably larger than the halfwidths of the resonances themselves. This is a significant advantage in triply differential experimental studies which are difficult to perform with narrow photon bandwidth, for intensity reasons, and are only now being attempted with bandwidths of 0.1–0.2 Å.

The middle frame in fig. 15 shows the vibrational branching ratios in this region, along with the FC factors for direct ionization. As in the case of shape resonant photoionization (see, e.g., section 5.1), striking non-FC behavior is observed in the vicinity of the vibrationally autoionizing states. As in the case of β_{v^+}, the spectral extent of the autoionization effect is greater than the autoionizing resonance halfwidth when displayed as vibrational branching ratios, a significant consideration in the context of experimental tests of these predictions. These calculations also predict that the branching ratio in the open channel with the highest v^+, corresponding to autoionization with the lowest possible $|\Delta v|$, is strongly enhanced at the expense of all other channels which are strongly depressed from their FC factors. Thus, the well-known propensity rule (Berry and Nielsen 1970a,b) stating that a vibrationally autoionized level decays preferentially with the smallest possible change of vibrational quantum number, is globally confirmed in these calculations, although for certain wavelengths (e.g., 764.4 Å in fig. 15) the exact opposite may be true.

The results just presented tend to lull one into the feeling that we understand photoionization of H_2 completely, needing only to extend the range of the above MQDT treatment to any region of interest. Nevertheless, it is imperative to perform experimental tests at the triply differential level, as it is in the more detailed quantities such as vibrational branching ratios and angular distributions that we are most likely to observe shortcomings in our detailed understanding of this most important prototype system. That such measurements have not been performed is often surprising to some. To emphasize the dearth of detailed data on this point, we show in fig. 16 a recent summary (Southworth et al. 1982b) of β measurements on H_2 (and D_2) together with other theoretical treatments. The gap between figs. 15 and 16 is enormous. Indeed, the measurements are difficult; however, optimization of current technology should make this goal attainable. This is one of the main motivations for the new generation instrument described in section 4. There is also another dimension to the problem, reflected in fig. 16: autoionization aside, there exists a glaring disagreement between theory and

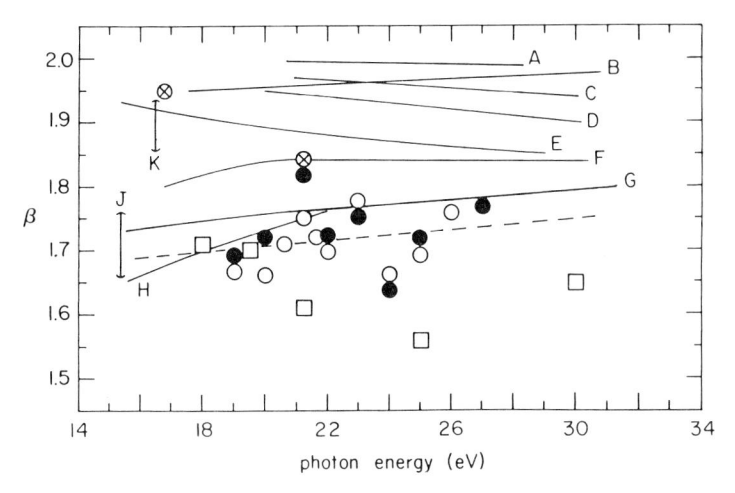

Fig. 16. Various experimental and theoretical results for the asymmetry parameter, β, for photoionization of H_2 and D_2. (Complete citations for the various data are given in the article by Southworth et al. 1982b, from which this figure was taken.)

experiment for the β values in the open continuum of H_2. The theoretical results tend to lie significantly (~ 0.2 β units) higher than measured values. Subsequent theoretical and experimental work shows that this difference is very persistent (see e.g., Itikawa et al. 1983, Hara and Ogata 1985, Raseev 1985, Hara 1985, Richards and Larkins 1986, and references therein), in spite of improvements on both sides. Thus, although great strides have been made during the last decade, the goal of understanding the photoionization dynamics of this most fundamental molecule presents several very contemporary challenges.

4. Triply differential photoelectron measurements – experimental aspects

In order to fully examine dynamical aspects of the resonant photoionization processes discussed above, it is essential to perform measurements of photoelectron intensity as a function of three independent variables – wavelength of the incident light, to select the spectral features; photoelectron kinetic energy, to select the ionization channel of interest; and ejection angle, to measure angular distributions. For convenience, we refer to this level of experiment as "triply differential" photoelectron measurements. Over the last few years, angle-resolved electron spectroscopy has been combined with synchrotron radiation sources to achieve successful triply differential measurements in molecules, including vibrational state resolution. Presently several groups (e.g., Marr et al. 1979, White et al. 1979, Parr et al. 1980, Krause et al. 1981, Derenbach et al. 1983, Morin et al. 1983, Parr et al. 1983, 1984) are involved in this type of experiment, each with their own specific experimental configuration and special emphasis, but each fulfilling the requirements for full triply differential studies. Here we will review, as an example, the experimental aspects of a new instrument (Parr et al. 1983, 1984) presently at the Synchrotron Ultraviolet Radiation Facility (SURF-II) at the National Bureau of Standards, in order to focus on some of the experimental considerations in triply differential photoionization studies.

The new triply differential electron spectrometer system at NBS consists of a high-throughput normal-incidence monochromator (Ederer et al. 1980) (fig. 17) and a pair of 10 cm mean-radius hemispherical electron spectrometers in an experimental chamber (fig. 18). To avoid later confusion we note that an earlier configuration using the same monochromator with a single, rotatable 5 cm mean-radius spectrometer (Parr et al. 1980) has been used for the past several years in several triply differential photoionization studies and is, in fact, the instrument used to obtain the data presented later in sections 5.1, 5.2, and 5.4–5.6. The data in section 5.3 was taken with the new instrument (Parr et al. 1983). The special emphasis with both generations of instruments has been the same, namely, to optimize the photon and electron resolution in order to probe detailed dynamics within shape resonance and autoionization structure in the near-normal-incidence range ($h \leq 35$ eV). The new instrument further optimizes several aspects of the electron spectrometer system to greatly extend the sensitivity and/or resolution compared to earlier measurements, for reasons discussed in section 7.

The high-flux, 2 meter monochromator shown in fig. 17 has been specifically

HIGH THROUGHPUT NORMAL INCIDENCE MONOCHROMATOR AT SURF-II

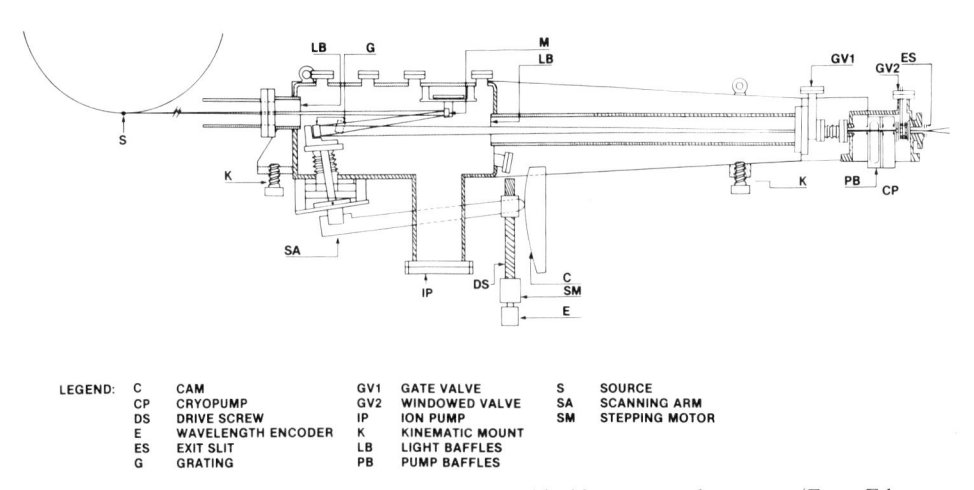

LEGEND:	C	CAM	GV1	GATE VALVE	S	SOURCE
	CP	CRYOPUMP	GV2	WINDOWED VALVE	SA	SCANNING ARM
	DS	DRIVE SCREW	IP	ION PUMP	SM	STEPPING MOTOR
	E	WAVELENGTH ENCODER	K	KINEMATIC MOUNT		
	ES	EXIT SLIT	LB	LIGHT BAFFLES		
	G	GRATING	PB	PUMP BAFFLES		

Fig. 17. Schematic diagram of high-throughput, normal-incidence monochromator. (From Ederer et al. 1980.)

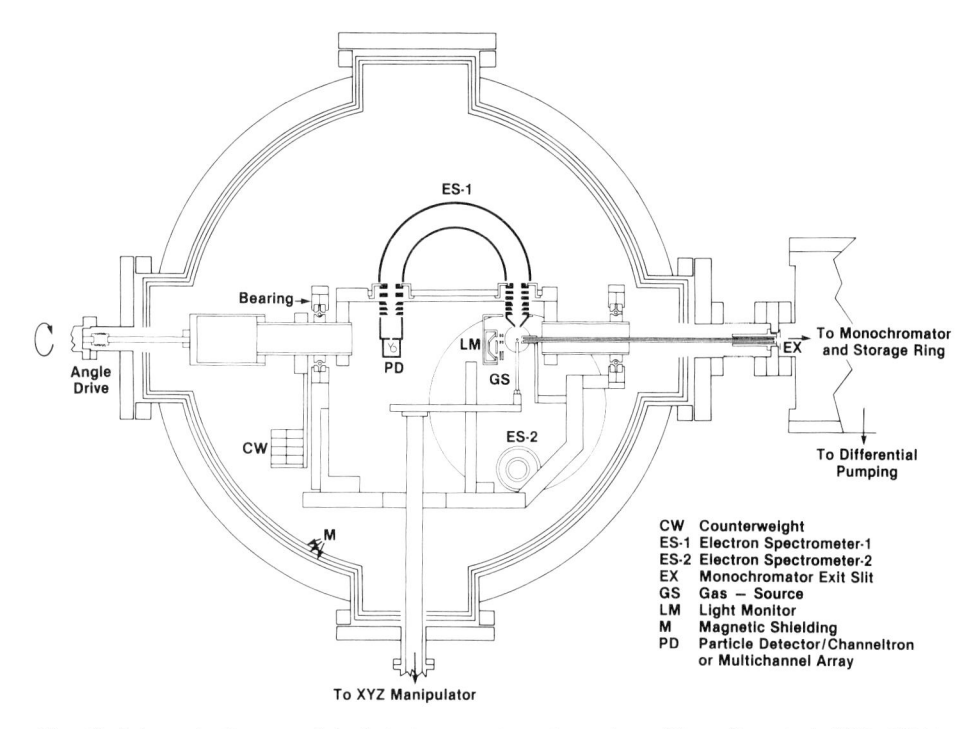

Fig. 18. Schematic diagram of dual electron spectrometer system. (From Parr et al. 1983, 1984.)

Labels in figure:
CW Counterweight
ES-1 Electron Spectrometer-1
ES-2 Electron Spectrometer-2
EX Monochromator Exit Slit
GS Gas – Source
LM Light Monitor
M Magnetic Shielding
PD Particle Detector/Channeltron or Multichannel Array

matched to the characteristics of the SURF-II storage ring: First and foremost, it uses the small vertical dimension of the stored electron beam (~100 μm) as the entrance aperture of the monochromator, an important feature which eliminates loss of incident flux on the entrance slit. Second, it is attached directly to the exit port of the storage ring, resulting in the very large capture angle of 65 mrad in the horizontal plane. This arrangement produces approximately 10^{10} photons s^{-1} $Å^{-1}$ per mA of circulating current (typical initial current presently ~50 mA) at 1000 Å and has been used for triply differential measurements out to ~375 Å. Together with a 1200 line/mm grating and a 200 μm (100 μm) exit slit, this configuration yields a photon resolution of ~0.8 Å (0.4 Å). Plans are made to improve the resolution with a higher dispersion grating. The dispersed light is channeled by a 2-mm-i.d. capillary tube for a distance of ~40 cm into the interaction region of the experimental chamber. The low pumping conductance of this capillary tube is very effective in reducing the gas load on the monochromator and storage ring during experiments in which the gas pressure in the experimental chamber can be as high as 10^{-4} Torr.

The new electron spectrometer system is shown schematically in fig. 18. The chamber is a 76 cm diameter, 92 cm long stainless steel vacuum chamber. It is pumped by a 500 liter/s turbomolecular pump and an 8000 liter/s closed-cycle helium cryopump to provide maximum flexibility in studying the whole range of gaseous targets. Low magnetic fields of <500 μG are maintained throughout the chamber by three layers of high-permeability magnetic shielding.

The system is designed to operate with either one rotating (ES-1) or two stationary (ES-1 and ES-2) electron spectrometers. In either configuration the electron spectra can be recorded as a function of ejection angle relative to the principle axis of polarization. This leads to the determination of the photoelectron branching ratios and angular distributions according to the following expression which applies to dipole excitation of free molecules with elliptically polarized light (Parr et al. 1973, Samson and Starace 1975):

$$\frac{d\sigma}{d\Omega} = \frac{\sigma}{4\pi}\left[1 + \frac{\beta}{4}(3P\cos 2\theta + 1)\right], \qquad (8)$$

where β is the asymmetry parameter, σ is the integrated cross section, P is the polarization of the light with the horizontal component being the major axis, and θ is the angle of ejection of the photoelectron with respect to the horizontal direction. In general, each resolved ionization channel (each photoelectron peak) will have a characteristic set of dynamical parameters, β and σ.

The number of electrons ejected per unit light flux per unit solid angle, $dn/d\Omega$, is proportional to the differential cross section; hence we can write

$$\frac{dn}{d\Omega} = N\left[1 + \frac{\beta}{4}(3P\cos 2\theta + 1)\right]. \qquad (9)$$

Measurement of P, θ and the number of electrons as a function of θ enables a determination of β and N. The relative quantity N, when normalized over a relevant set of possible alternative ionization channels, gives the branching ratio for the particular transition. The measurement of P is accomplished with a triple reflection polarizer based upon the considerations of Horton et al. (1969). The incoming light flux is monitored by a 90% transparent tungsten photocathode on the input aperture of the polarization analyzer. After three reflections the light is intercepted by a second tungsten photodiode. The ratios of these two photodiode signals at 0° and 90° with respect to the major polarization axis determine the polarization.

Each electron analyzer is a 10 cm mean-radius version of our previous instrument (Parr et al. 1980) and utilizes the same electron lens system – a three-aperture "zoom" lens (Harting and Read 1976) to focus the electrons into the hemispherical dispersive element, and a similar one to refocus the energy-analyzed electrons on the exit slit. There are no entrance or exit apertures in the equatorial plane of the hemispheres and therefore the aperture in the entrance cone determines the basic resolution. The resolution obtainable while yet maintaining good signal is expected to be on the order of 20 meV. Measurement of sub-10 meV resolution has been demonstrated. Thus, the resolving power of this instrument is a significant improvement over that typically used now with synchrotron radiation, and will allow for the extension of studies of non-Franck–Condon effects to small polyatomic molecules.

The electrical aspects of the electron spectrometers are generally similar to those of the previously described apparatus (Parr et al. 1980). Briefly, the fixed voltages are controlled by highly regulated conventional power supplies and the variable voltages are under computer control. A 16-bit digital to analog converter (DAC) with a basic increment of 0.0005 V controls the ramping of offset voltages. The variable focus voltages are controlled by isolated power sources run by the computer. The computer (LSI-11/23) is interfaced to a CAMAC crate through which it controls the grating drive, angular position, light detection system, electron counting system, and other experimental chores. Both analyzers are ramped off the 16-bit DAC but have their own separately controlled power supplies for lens voltages. The two identical analyzers allow a determination of the branching ratios and asymmetry parameters without rotation, i.e., the electron intensity at two angles can be measured simultaneously. The instrument is calibrated by reference to gases with known cross sections (Marr and West 1976) and asymmetry parameters (Kreile and Schweig 1980, Dehmer et al. 1975, Holland et al. 1982). The calibration features for the two analyzers are incorporated into a computer program that corrects and analyzes the data. Area detectors have been purchased and will soon be integrated into the instrument, thus significantly increasing its sensitivity.

The gas jet is mounted on an XYZ manipulator in order to optimize signal intensity and resolution by external adjustment. Gas nozzles for the system are interchangeable and provide both effusive beams and supersonic jets by use of pinhole apertures of diameter 7–50 μm. The positioning of the supersonic source is of particular importance and necessitates the positioning capability of this inlet system. With the larger hemispherical dispersive element, the use of two analyzers, better gas source technology, incorporation of area detectors, and enhanced pumping, we expect a significant improvement in the basic sensitivity of the instrument (a very conservative estimate would be $>100\times$) as compared to our previous 5 cm radius single analyzer system (Parr et al. 1980).

The same LSI-11 computer that is used for automation is also used for the data reduction. The basic data consists of electron counts as a function of wavelength, ramp voltage, and angle. The ramp voltage is converted, using known quantities, to electron kinetic energy. The electron counts are then normalized for correction factors that depend upon kinetic energy, such as the transmission functions of the instruments and a small angular correction. Upon obtaining a suitably normalized set of data, the photoelectron spectra are typically fitted to a Gaussian basis set using spectroscopic values for vibrational energy spacings, while treating peak height, peak width, and overall position as free parameters. The calculated curve and normalized data are plotted to aid in the evaluation of the quality of fit. In addition, the fitting program outputs statistical parameters which can be used to estimate the accuracy of the fit. Finally, the areas of the respective peaks are used to infer the values of the branching ratios and asymmetry parameters which contain the dynamical information for the process.

5. Case studies

5.1. Shape-resonance-induced non-Franck–Condon effects in N_2 $3\sigma_g$ photoionization

Molecular photoionization at wavelengths unaffected by autoionization, predissociation, or ionic thresholds has been generally believed to produce Franck–Condon (FC) vibrational intensity distributions within the final ionic state and v-independent photoelectron angular distributions. We now discuss the prediction (Dehmer et al. 1979, Dehmer and Dill 1980) and confirmation (Carlson et al. 1980, Raseev et al. 1980, West et al. 1980, Lucchese and McKoy 1981b, Leal et al. 1984) that shape resonances represent an important class of exceptions to this picture. These ideas are illustrated with a calculation of the $3\sigma_g \to \varepsilon\sigma_u$, $\varepsilon\pi_u$ photoionization channel of N_2, which accesses the same σ_u shape resonance discussed above at approximately $h\nu \sim 30$ eV, or ~ 14 eV above the $3\sigma_g$ IP. The potential energy curves for N_2 are shown in fig. 19 in order to orient this discussion and for later reference. The process we are considering involves

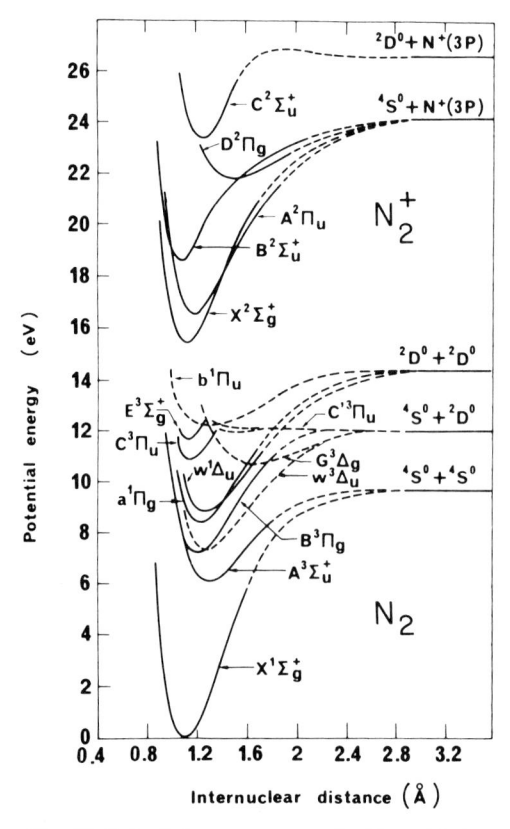

Fig. 19. Potential energy curves for N_2 and N_2^+.

photoexcitation of N_2 $X\,^1\Sigma_g^+$ in its vibrational ground state with photon energies from the first IP to beyond the region of the shape resonance at $h\nu \sim 30$ eV. This process ejects photoelectrons leaving behind N_2^+ ions in energetically accessible states. Figure 20 shows a typical photoelectron spectrum taken at $h\nu = 21.2$ eV which exhibits photoelectron peaks corresponding to production of N_2^+ in its X, A, and B states, i.e., the three lowest ionic states in fig. 19. As we are interested in the ionization of the $3\sigma_g$ electron, which produces the $X\,^2\Sigma_g^+$ ground state of N_2^+, we are concerned with the photoelectron band in the range $15.5\,\text{eV} \leq \text{IP} \leq 16.5\,\text{eV}$ in fig. 20. The physical effects we seek involve the relative intensities and angular distributions of the $v = 0$–2 vibrational peaks in the $X\,^2\Sigma_g^+$ electronic band, and, more specifically, the departures of these observables from behavior predicted by the FC separation.

The breakdown of the FC principle arises from the quasibound nature of the shape resonance, which, as we discussed in section 2, is localized in a spatial region of molecular dimensions by a centrifugal barrier. This barrier and, hence, the energy and lifetime (width) of the resonance are sensitive functions of internuclear separation R and vary significantly over a range of R corresponding to the ground-state vibrational motion. This is illustrated in the upper portion of figs. 21 and 22 where the dashed curves represent separate, fixed-R calculations of the partial cross section and asymmetry parameter for N_2 $3\sigma_g$ photoionization over the range $1.824\,a_0 \leq R \leq 2.324\,a_0$, which spans the N_2 ground-state vibrational wavefunction.

Of central importance in fig. 21 is the clear demonstration that resonance positions, strengths, and widths are sensitive functions of R. In particular, for larger separations, the inner well of the effective potential acting on the $l = 3$ component of the σ_u wavefunction is more attractive and the shape resonance shifts to lower kinetic energy, becoming narrower and higher. Conversely, for lower values of R, the resonance is pushed to higher kinetic energy and is

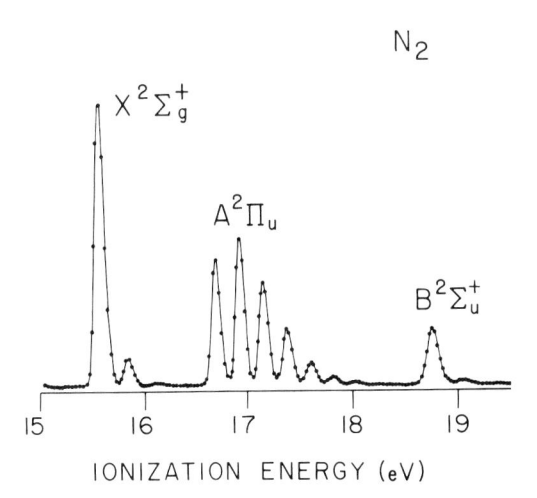

Fig. 20. Photoelectron spectrum of N_2 at $h\nu = 21.3$ eV.

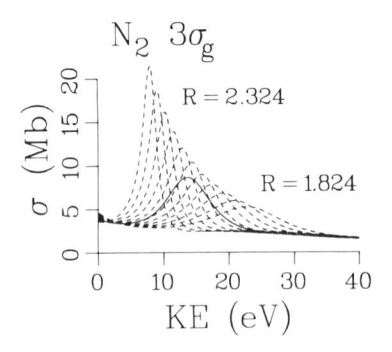

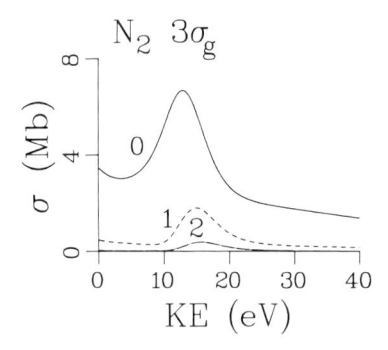

Fig. 21. Cross sections σ for photoionization of the $3\sigma_g$ ($v_i = 0$) level of N_2. Top: fixed-R (dashed curves) and R-averaged, vibrationally unresolved (solid curve) results. Bottom: results for resolved final-state vibrational levels, $v_f = 0$–2.

weakened. This indicates that nuclear motion exercises great leverage on the spectral behavior of shape resonances, since small variations in R can significantly shift the delicate balance between attractive (mainly Coulomb) and repulsive (mainly centrifugal) forces which combine to form the barrier. In the present case, variations in R, corresponding to the ground-state vibration in N_2, produce significant shifts of the resonant behavior over a spectral range several times the fullwidth at half maximum of the resonance calculated at $R = R_e$. By contrast, nonresonant channels are relatively insensitive to such variation in R, as was shown by results (Wallace 1980) on the $1\pi_u$ and $2\sigma_u$ photoionization channels in N_2.

Thus, in the vicinity of a shape resonance, the electronic transition moment varies rapidly with R. This produces non-FC (parametric) coupling that was estimated (Dehmer et al. 1979, Chase 1956) by computing the net transition moment for a particular vibrational channel as an average of the R-dependent dipole amplitude, weighted by the product of the initial- and final-state vibrational wavefunctions at each R,

$$D^-_{v_f v_i} = \int dR \, \chi^\dagger_{v_f}(R) \, D^-(R) \, \chi_{v_i}(R). \tag{10}$$

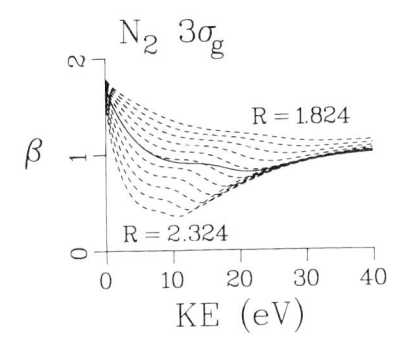

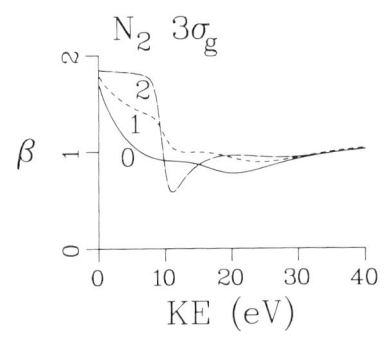

Fig. 22. Asymmetry parameters β for photoionization of the $3\sigma_g$ ($v_i = 0$) level of N_2. Top: fixed-R (dashed curves) and R-averaged, vibrationally unresolved (solid curve) results. Bottom: results for resolved final-state vibrational levels, $v_f = 0$–2.

The vibrational wavefunctions were approximated by harmonic-oscillator functions and the superscript minus denotes that incoming-wave boundary conditions have been applied and that the transition moment is complex. Note that even when the final vibrational levels v_f of the ion are unresolved (summed over), vibrational motion within the initial state $v_i = 0$ can cause the above equation to yield results significantly different from the $R = R_e$ result, because the R-dependence of the shape resonance is highly asymmetric. This gross effect of R averaging can be seen in the upper half of fig. 21 by comparing the solid line (R-averaged result, summed over v_f) and the middle dashed line ($R = R_e$). Hence, even for the calculation of gross properties of the whole, unresolved electron band, it is necessary to take into account vibrational effects in channels exhibiting shape resonances. As we stated earlier, this is generally not a critical issue in nonresonant channels.

The resulting behavior of individual vibrational levels is shown in the bottom half of figs. 21 and 22. Looking first at the partial cross sections in fig. 21, we see that the resonance position varies over a few volt depending on the final vibrational state, and that higher levels are relatively more enhanced at their resonance position than is $v_f = 0$. This sensitivity to v_f arises because transitions to alternative final vibrational states preferentially sample different regions of R. In particular, $v_f = 1, 2$ sample successively smaller R, governed by the maximum overlap with the ground vibrational state, causing the resonance in those vibrational channels to peak at higher energy than that for $v_f = 0$. The impact of these effects on branching ratios is clearly seen in fig. 23, where the ratio of the higher v_f intensities to that of $v_f = 0$ is plotted in the resonance region. There we see that the ratios are slightly above the FC factors (9.3%, $v_f = 1$; 0.6%, $v_f = 2$) at zero kinetic energy, go through a minimum just below the resonance energy in $v_f = 0$, then increase to a maximum as individual $v_f > 0$ vibrational intensities peak, and finally approach the FC factors again at high kinetic energy. Note that the maximum enhancement over the FC factors is progressively more pronounced for higher v_f, i.e., 340% and 1300% for $v_f = 1, 2$, respectively, in this calculation.

Equally dramatic are the effects on $\beta(v_f)$ shown in the lower portion of fig. 22. Especially at and below the resonance position, β varies greatly for different final vibrational levels. The $v_f = 0$ curve agrees well with the solid curve in the upper half, since the gross behavior of the vibrationally unresolved electronic band will be governed by the β of the most intense component. The $R = R_e$ curve has been found to agree well with wavelength-dependent measurements (Marr et al. 1979, Carlson et al. 1980), and the agreement is improved by the slight damping caused by R averaging. More significant for the present purposes is the v_f-dependence of β. Carlson first observed (Carlson 1971, Carlson and Jonas 1971) that, at 584 Å, the $v_f = 1$ level in the $3\sigma_g$ channel of N_2 had a much larger β than the $v_f = 0$ level even though there was no apparent autoionizing state at that wavelength. This is in semiquantitative agreement with the results in fig. 22 which give $\beta(v_f = 0) \sim 1.0$ and $\beta(v_f = 1) \sim 1.5$. Although the agreement is not exact, we feel this demonstrates that the "anomalous" v_f-dependence of β in N_2 stems mainly from the σ_u shape resonance which acts over a range of the spectrum many times its own ~5 eV width. The underlying cause of this effect is the shape-resonance-enhanced

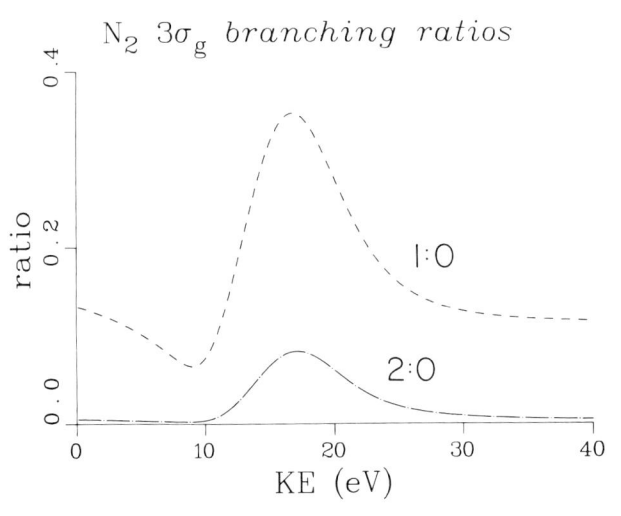

Fig. 23. Vibrational branching ratios $\sigma(v_f)/\sigma(v_f = 0)$ for photoionization of the $3\sigma_g$ level of N_2.

R-dependence of the dipole amplitude, just as for the vibrational partial cross sections. In the case of $\beta(v_f)$, however, both the R-dependence of the phase and of the magnitude of the complex dipole amplitude play a crucial role, whereas the partial cross sections depend only on the magnitude.

This prototype study demonstrated several new aspects of photoionization channels exhibiting shape resonances: First, the localization and delay in photoelectron escape associated with a shape resonance enhances the sensitivity of photoelectron dynamics to nuclear separation, invalidating the FC factorization of the two modes. The resulting asymmetric and non-monotonic dependence of the transition amplitude on internuclear separation requires the folding of the transition amplitude with the vibrational motion of the molecule, at least at the level of eq. (10). Under certain circumstances, interference effects, analogous to those exhibited by the 2.4 eV π_g resonance in e–N_2 scattering (Bardsley and Mandl 1968, Schulz 1973, 1976, Lane 1980, Shimamura and Takayanagi 1984, Herzenberg and Mandl 1962, Birtwistle and Herzenberg 1971, Chandra and Temkin 1976, Schneider et al. 1979) may be important as well. Second, the effects are large in both vibrational intensities and angular distributions, but were largely overlooked in earlier work because shape resonance effects tend to lie in an inconvenient wavelength range for laboratory light sources. Synchrotron radiation has since solved this problem. Third, it is significant to note that the effects of the shape resonance described above act over tens of volts of the spectrum, several times the halfwidth of the resonance, and that σ and β probe the effects differently, i.e., have maximal effects in different energy regions. This underscores the well-known difference in dynamical information contained in the two physical observables. Fourth, a long-standing "anomalous" v_f-dependence in the photoelectron angular distributions of the $3\sigma_g$ channel of N_2 has been resolved. Finally, the phenomena described here for one channel of N_2 should be very widespread, as shape resonances now appear to affect one or more of the inner- and outer-shell channels in most small (nonhydride) molecules.

These theoretical predictions were soon tested in two separate experiments as indicated in figs. 24 and 25. In fig. 24, the branching ratio for production of the $v = 0$ and 1 vibrational levels of N_2^+ X $^2\Sigma_g^+$ is shown. The dash–dot curve is the original prediction (Dehmer et al. 1979) from fig. 23. The solid dots are the measurements (West et al. 1980) in the vicinity of the shape resonance at $h\nu \sim 30$ eV. The conclusion drawn from this comparison is that the observed variation of the vibrational branching ratio relative to the FC factor over a broad spectral range qualitatively confirms the prediction; however, subsequent calculations (Raseev et al. 1980, Lucchese and McKoy 1981b, Leal et al. 1984) with fewer approximations have achieved better agreement based on the same mechanism for breakdown of the FC separation. The dashed and solid curves are results based on a Schwinger variational treatment (Lucchese and McKoy 1981b) of the photoelectron wavefunction. The two curves represent the length and velocity form of the transition matrix element, both of which are in excellent agreement with the data. This is an outstanding example of interaction between experiment and theory, proceeding as it did from a novel prediction, through

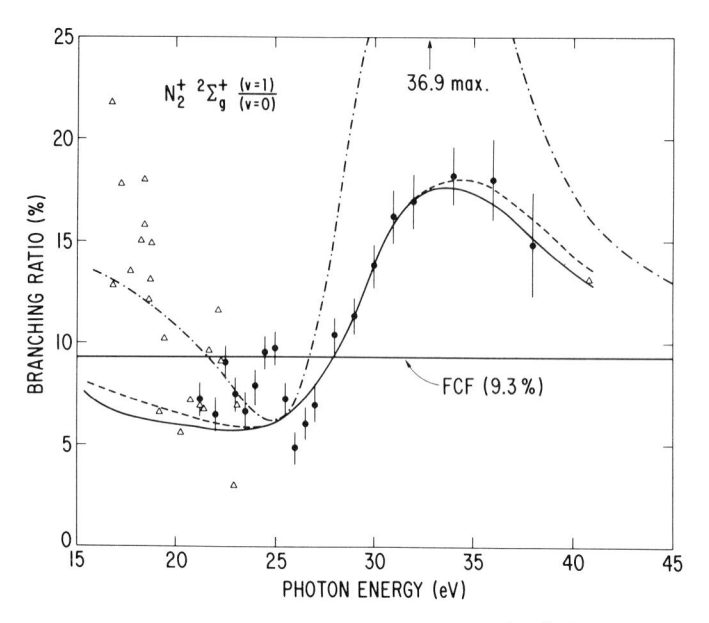

Fig. 24. Branching ratios for production of the $v = 0, 1$ levels of N_2^+ X $^2\Sigma_g^+$ by photoionization of N_2: ●, from West et al. (1980); △, from Gardner and Samson (1978); — · — · —, multiple scattering model prediction from Dehmer et al. (1979); ———, frozen-core Hartree–Fock dipole length approximation from Lucchese and McKoy (1981b); – – –, frozen-core Hartree–Fock dipole velocity approximation from Lucchese and McKoy (1981b).

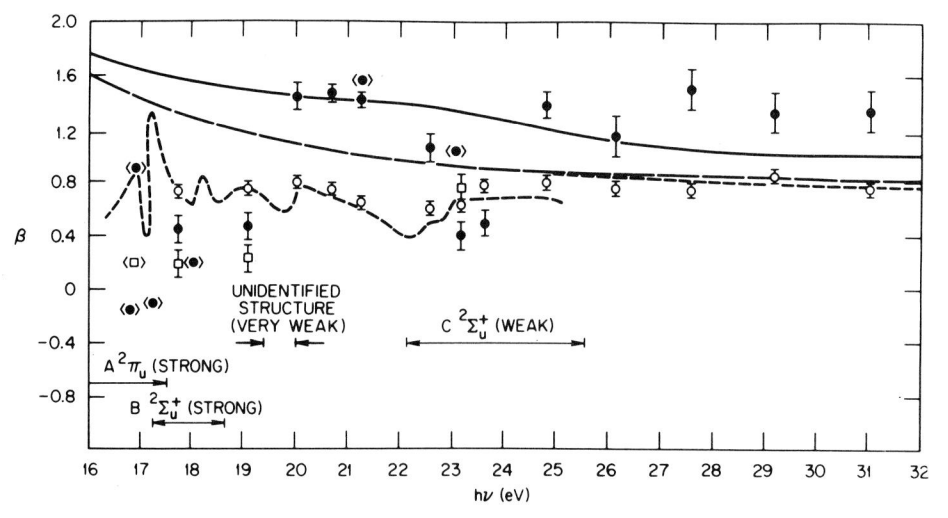

Fig. 25. Photoelectron asymmetry parameters for the $v = 0, 1$ levels of N_2^+ X $^2\Sigma_g^+$: ○, $v = 0$ data from Carlson et al. (1980); ●, $v = 1$ data from Carlson et al. (1980); long-dashed curve, $v = 0$ prediction from Dehmer et al. (1979); solid curve, $v = 1$ prediction from Dehmer et al. (1977); other data described by Carlson et al. (1980), from which this figure was taken.

experimental testing, and final quantitative theoretical agreement in a short time. Also shown in fig. 24 are data in the 15.5 eV ≤ $h\nu$ ≤ 22 eV region which are earlier data (Gardner and Samson 1978) obtained using laboratory line sources. The apparently chaotic behavior arises from unresolved autoionization structure, whose detailed study is discussed in the next section.

Figure 25 shows angular distribution asymmetry parameters (β) for the $v = 0, 1$ levels of N_2^+ X $^2\Sigma_g^+$ over roughly the same energy region. These data were taken at the Synchrotron Radiation Center at the University of Wisconsin by Carlson and coworkers (Carlson et al. 1980). In the region above $h\nu \sim 25$ eV, this data also shows qualitative agreement with the predicted (Dehmer et al. 1979) v-dependence of β caused by the σ_u shape resonance. In this case the agreement is somewhat improved in later calculations (Lucchese and McKoy 1981b), mainly for $v = 1$; however, the change is less dramatic than for the branching ratios.

5.2. Autoionization via the Hopfield series in N_2

Recall the scattered data in the 15.5 eV ≤ $h\nu$ ≤ 20 eV region of the vibrational branching ratio data for N_2 shown in fig. 24. As indicated in the discussion of that data, the scatter was produced by unresolved autoionization structure leading to the A and B states of N_2^+ (see fig. 19). The high-resolution total photoionization spectrum (P.M. Dehmer et al. 1984) of N_2 in that spectral range is given in figs. 26 and 27, showing the rich autoionization structure in the N_2 spectrum. This serves to emphasize the difference in precision needed in studying the two types of resonant photoionization processes, i.e., shape and autoionizing resonances. The relatively broad Hopfield absorption and emission Rydberg series (Hopfield 1930a,b) converging to the $v = 0$ level of N_2^+ B $^2\Sigma_u^+$ at 661.2 Å was chosen for the initial triply differential study (Parr et al. 1981) of molecular autoionization structure. The choice was guided by two considerations: First, the series represented a relatively isolated series whose lower members were sufficiently broad to permit systematic investigation within autoionization profiles with available instrumentation. Second, this was considered a good prototype system wherein the full rotational–vibrational–electronic autoionization process is in play. In this case, the autoionizing states consist of a Rydberg electron bound to an electronically-excited N_2^+ B $^2\Sigma_u^+$, $v = 0$ core. Ionization occurs when the Rydberg electron collides with the core, enabling the exchange of the large electronic excitation energy from the core together with smaller amounts of energy to or from the nuclear modes.

Although several members of the Hopfield absorption and emission series have been studied (West et al. 1981) by triply differential photoelectron spectroscopy, we will focus here on the lowest (and broadest) $m = 3$ members of these series at $\lambda = 723.3$ and 715.5 Å, respectively. The spectroscopic assignments for these series are summarized elsewhere (Parr et al. 1981). Here we note that the window and main absorption series, together with a weaker absorption series between them seem to be most consistent with the designation $nd\pi_g$, $nd\sigma_g$, and $ns\sigma_g$.

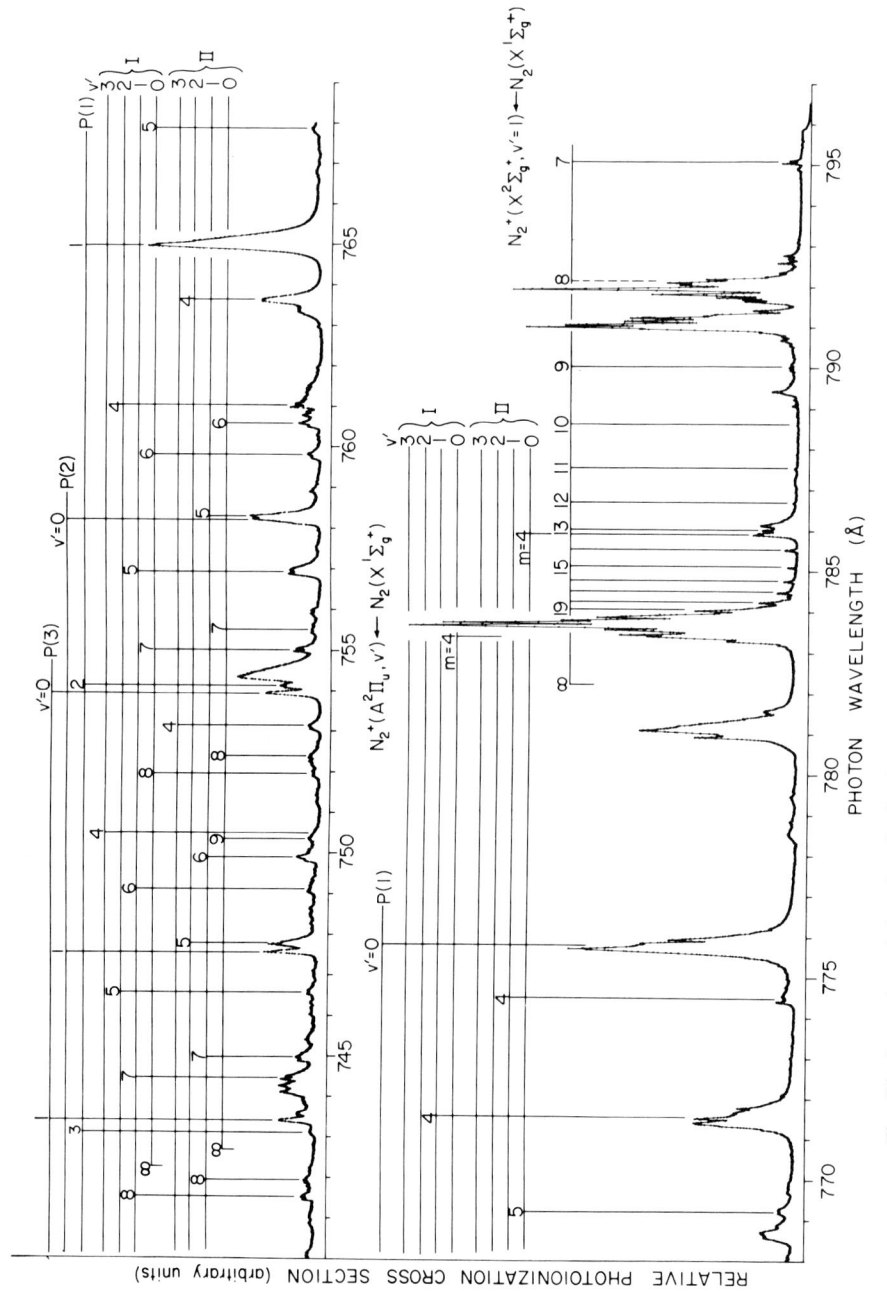

Fig. 26. A portion of the photoionization spectrum of N_2 at 78 K. (Taken from P.M. Dehmer et al. 1984.)

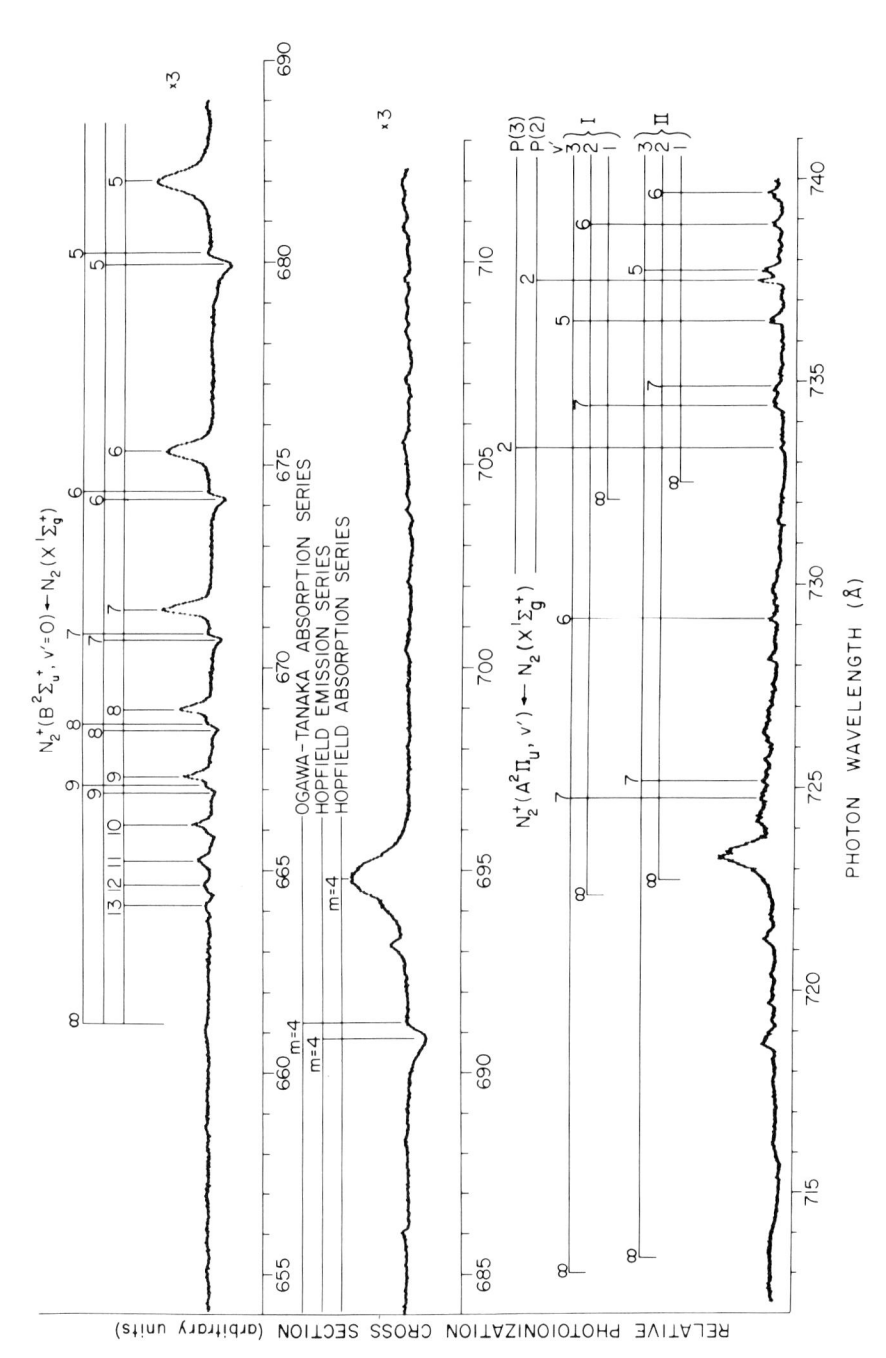

Fig. 27. A portion of the photoionization spectrum of N_2 at 78 K. (From P.M. Dehmer et al. 1980.)

These assignments have been confirmed in recent theoretical work (Raoult et al. 1983).

In fig. 28, we present the vibrational branching ratios for formation of the ground-state ion N_2^+ $X\,^2\Sigma_g^+$ by photoionization in the range $710\,\text{Å} \leq \lambda \leq 730\,\text{Å}$. Here we define the vibrational branching ratio as the ratio of the intensity of a particular vibrational level to the sum over the whole vibrational band. In fig. 29 the asymmetry parameter β is given for the same processes. In both figures the positions of the Hopfield emission and absorption features at 715.5 and 723.3 Å, respectively, are indicated by solid lines joining the upper and lower frames. In the vicinity of these features, a hand-drawn dashed curve is constructed only to guide the eye, and should not be taken too seriously. In both figures, typical error bars for the data in each frame are shown on the last point. Duplicate branching ratio measurements (the open circles were taken at the magic angle and the solid dots were deduced from the angular distribution measurements) show the reproducibility of the data. Note that an early branching-ratio study of this region of the N_2 photoionization spectrum was reported by Woodruff and Marr (Marr and Woodruff 1976, Woodruff and Marr 1977) but without angle dependence and with insufficient wavelength resolution to characterize the profiles of the Hopfield resonances.

Focusing first on the vibrational branching ratios in fig. 28, we see three major qualitative features:

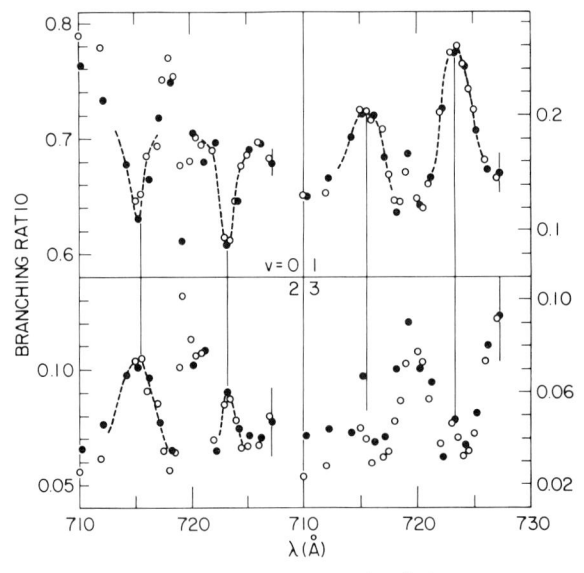

Fig. 28. Vibrational branching ratios for production of N_2^+ $X\,^2\Sigma_g^+$ ($v = 0$–3 in the range $710\,\text{Å} \leq \lambda \leq 730\,\text{Å}$. Vertical lines at 715.5 and 723.3 Å denote the positions of the first members of the Hopfield "emission" and absorption series approaching the N_2^+ $B\,^2\Sigma_u^+$ ($v = 0$) limit. Typical error bars are indicated on the last point in each frame. The dashed line is hand drawn to guide the reader's eye. Open and closed circles represent two independent runs. (Figure taken from Parr et al. 1981.)

(1) The $v = 0$ branching ratio exhibits pronounced dips at the locations of the two major autoionization features, whereas the higher vibrational channels, most notably $v = 1$, show enhancements. Hence, the quasibound autoionizing states mediate a transfer of dipole amplitude from the $v = 0$ channel to the much weaker $v = 1$, 2, and 3 channels, a transfer which involves simultaneous electronic de-excitation and vibrational excitation of the ion core. This transfer is primarily directed to the $v = 1$ channel and is much diminished by $v = 3$. This enhancement of vibrational channels with small Franck–Condon factors relative to the most intense channel is reminiscent of the effects of shape resonances in those few cases studied so far. Comparison with vibrational autoionization in H_2, where ionization channels with the minimum (negative) Δv usually dominate (Raoult and Jungen 1981, Berry and Nielsen 1970a,b) is not straightforward since $\Delta v = 0$ is permitted in this case; and, anyhow, transitions in which vibrational excitation occurs are favored. Establishing the systematics of this diverse set of observations is obviously a most timely problem.

(2) Despite the great contrast between the window and absorption profiles in the photoabsorption and photoionization spectra, the profiles in fig. 28 are of similar shape and both exhibit either an enhancement or depletion, depending upon the channel.

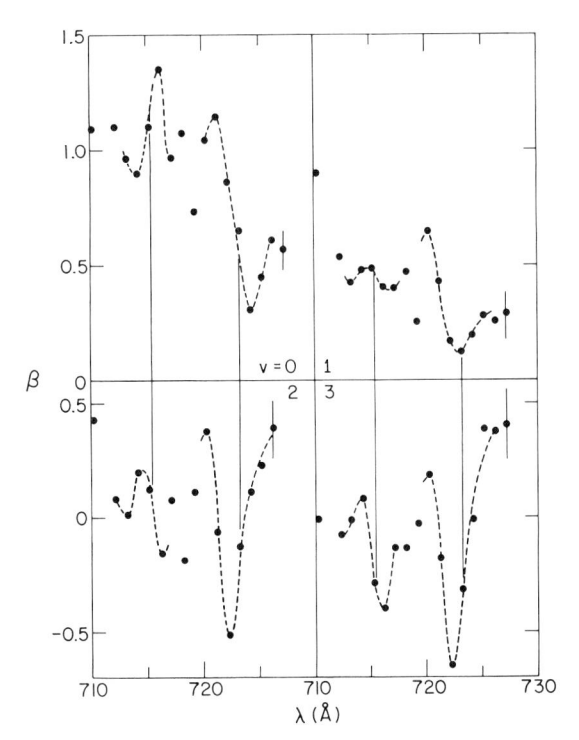

Fig. 29. Photoelectron asymmetry parameters corresponding to the production of N_2^+ $X\,^2\Sigma_g^+$ ($v = 0$–3) in the range $710\,\text{Å} \leqslant \lambda \leqslant 730\,\text{Å}$. Other conventions as in fig. 28. (Figure taken from Parr et al. 1981.)

(3) Definite "interloper" structure occurs between the two major resonances, with variable shape and strength. Both the weak absorption peak near the window resonance and other weak structures (one peak in the photoionization spectrum at 718.8 Å correlates well with the main interloper structures in fig. 28) may play a role here.

The angular distribution results in fig. 29 also exhibit structure at the positions of the two major resonances and in between. Implicit in the spectral variations in β is information on both the vibrational branching ratios and the relative phases of the alternative vibrational ionization channels. Specifically, the competition between asymptotic phases produces large asymmetric variations in β at the resonance positions, which vary from one final vibrational level to another. For instance, the β curve near the Hopfield "emission" line exhibits a peak for $v = 0$ that evolves into a dip for $v = 3$. Near the Hopfield absorption profile, the position of the minimum in β, although not extremely well defined by these data, shifts from the long-wavelength side of the resonance position to the short-wavelength side.

Following these measurements and the related partial cross section measurements by Morin et al. (Morin 1983, Morin et al. 1987, see also Nenner and Beswick 1987), Raoult et al. (1983) applied a two-step formulation of MQDT to the $m = 3$ member of the Hopfield series in N_2. This represented the first *ab initio* study of electronic autoionization profiles in molecules and was very successful in obtaining reasonable agreement with experiment, especially in light of the complexity of the calculation, which involved forty-seven electronic quantities corresponding to five electronic channels for Σ symmetry and six for Π symmetry.

Several approximations were utilized in order to treat such a complicated excited complex (e.g., fixed internuclear separation, neglect of correlation in the final continuum state). Nevertheless, this pioneering calculation succeeded in three important areas: It clearly established the assignments of the Hopfield absorption and emission series. It achieved reasonable agreement with the partial cross sections between 700 and 730 Å, measured by Morin et al. (Morin 1983, Morin et al. 1987, see also Nenner and Beswick 1987) and rationalized the cause for remaining differences. Finally, it achieved semiquantitative agreement for the asymmetry parameters measured by West et al. (1981) for the $X\,^2\Sigma_g^+$ and $A\,^2\Pi_u$ states in this same wavelength region.

Here we will highlight the calculation of the β results as they were found to be consistent with the partial cross section results and, furthermore, proved to be useful in revealing details which are hardly seen in the partial cross sections. This stresses the complementarity of the two types of dynamical parameters.

Figure 30 displays the variation of the asymmetry parameter with photon energy. The calculated β value is purely electronic and has been obtained for $R = 2.068$ a.u. Raoult et al. (1983) compared their calculation to the vibrationally resolved experimental β values for the $v = 0$ component of both the $X\,^2\Sigma_g^+$ and $A\,^2\Pi_u$ N_2^+ states. Indeed, the effect of electronic autoionization is expected to be the largest in the $X\,^2\Sigma_g^+$, $v = 0$ and $A\,^2\Pi_u$, $v = 0$ continua, because the corresponding vibrational wavefunctions have the largest overlap with the $B\,^2\Sigma_u^+$, $v = 0$

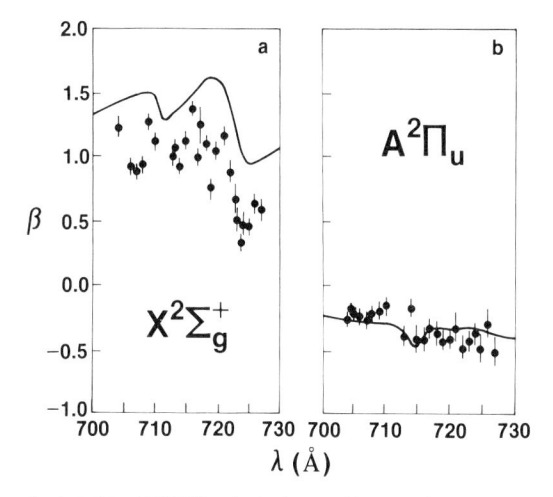

Fig. 30. Comparison of *ab initio* MQDT calculations with experimental results for the asymmetry parameter resulting from photoionization of N_2 in the 700 Å $\leq \lambda \leq$ 730 Å region. Data points are from West et al. (1981) and the theoretical curve (solid line) and figure are taken from Raoult et al. (1983).

vibrational wavefunction. (One assumes that the Rydberg states and the corresponding ionic core have the same vibrational functions.)

The asymmetry parameter β calculated for the $X\ ^2\Sigma_g^+$ core state is globally in agreement with the experimental results (see fig. 30a). The calculated off-resonance value results from an interference between outgoing p and f waves. If only the p wave (f wave) is considered, with the transition moments given by Raoult et al. (1983), the β value should be 1.7 (0.73). At 700 Å, the calculated β value is 1.33, intermediate between these two single-wave values. The two dips around 712 and 725 Å reproduce well the observed features. The β energy variation may be understood as follows. If a resonant state decays mainly in one specific channel $l\lambda$ at the energy corresponding to the maximum of the resonance in the cross section, only one term dominates the β expression, and β takes a geometrical value, $\beta_{l\lambda}$. These geometrical values are listed in table 1 in the article by Thiel (1982). Thus, the dip at 725 Å may be understood when one looks at the different $l\lambda$ contributions to the $X\ ^2\Sigma_g^+$ partial cross section. The $(B\ ^2\Sigma_u^+)3'd'\sigma_g$ Rydberg state autoionizes preferentially in the $(X\ ^2\Sigma_g^+)\varepsilon f\sigma_u$ continuum and at 724 Å, the maximum in the cross section, the β value must be $\beta_{f\sigma} = 0.53$. Indeed, this value is attained experimentally (see fig. 30a). The calculated value is larger because the resonance in the $f\sigma$ cross section was evaluated to be too small (Raoult et al. 1983). Nevertheless, this proves that the calculations are qualitatively consistent with experiment, since the continuum–continuum interaction $(B\ ^2\Sigma_u^+)\varepsilon d\sigma_g/(X\ ^2\Sigma_g^+)\varepsilon f\sigma_u$ is computed to be greater than the $(B\ ^2\Sigma_u^+)\varepsilon d\sigma_g/(X\ ^2\Sigma_g^+)\varepsilon p\sigma_u$ one (Raoult et al. 1983). The dip at 712 Å is more difficult to interpret because the β value results from the combined effects of the $3d\pi_g$ and $4s\sigma_g$ resonances. Indeed, no single $l\lambda$ contribution is dominant in this region. Two experimental features are not reproduced. The feature at 719 Å corresponds to the peak which is seen in the total cross section of fig. 27. This state was previously assigned to the $(A\,^2\Pi_u,\,v^+=3)10s\sigma_g$ state by Ogawa and Tanaka (1962). It is difficult to be confident in this assignment, because the vibrational overlap between $A\,^2\Pi_u$, $v=3$ and $X\,^2\Sigma_g^+$, $v^+=0$ is very weak. However, the appearance of this feature in fig. 30a confirms that this unknown state autoionizes electronically in the $(X\,^2\Sigma_g^+)\varepsilon l\lambda$ continua as argued in the partial cross section discussion in the article by Raoult et al. (1983). Similarly, the dip around 705 Å is not assigned to a $(A\,^2\Pi_u,\,v^+=4)nl\lambda$ Rydberg state but rather to the $(B\,^2\Sigma_u^+,\,v^+=1)4's'\sigma_g$ and $(B\,^2\Sigma_u^+,\,v^+=1)3d\pi_g$ states which are expected in this spectral range. Comparison (Raoult et al. 1983) of the partial cross sections and asymmetry parameters shows that, for equivalent energy resolution, features which appear very weakly in the partial cross sections can be seen in the β measurements.

In contrast to the β for the $X\,^2\Sigma_g^+$ channel, the energy variation of β for the $A\,^2\Pi_u$, $v=0$ channel is flat (fig. 30b). The MQDT calculation reproduces this behavior very well. The detailed dynamics in this channel are discussed further by Raoult et al. (1983).

In this section we have focused our attention on the Hopfield series. In principle, with *ab initio* electronic quantities, it should be straightforward to calculate, by an MQDT treatment, the photoionization spectrum throughout the spectral range between the $X\,^2\Sigma_g^+$ and $A\,^2\Pi_u$ ionization limits. But, in this region, in addition to the vibrational and dissociation effects, it would be important to include the perturbation by the valence states and the intensity borrowing of the $(A\,^2\Pi_u)\,^3\Pi_u$ states from the $^1\Pi_u$ states. Simultaneously with these improvements in the calculations, new measurements of the partial cross sections and of the photoelectron angular distributions would be useful to clarify the assignments of numerous unidentified bands in this region. In particular, the $(A\,^2\Pi_u)nd\delta_g$ series has not yet been identified in spite of its expected strong transition moment. Clearly, the results of this case study have marked the beginning of a significant advance in understanding the detailed dynamics of electronic autoionization in small molecules while indicating the need for further advances in both *ab initio* calculations and high-resolution triply differential photoelectron measurements.

5.3. Continuum–continuum coupling effects in N_2 $2\sigma_u$ photoionization

Our third case study concerns photoionization of the $2\sigma_u$ subshell in N_2, i.e., the third band in fig. 20 with an IP of ~ 18.8 eV. This channel is presently the clearest example in molecular photoionization of continuum–continuum coupling and has been studied very recently by both theoretical (Stephens and Dill 1985) and experimental means (Southworth et al. 1986).

Coupling between molecular photoelectrons and residual electrons is usually weak, because continuum electrons are so diffuse that they have negligible amplitude in the molecular interior (see, e.g., section 2.3) where strong interaction can take place. By the same token, prominent structure in photoionization spectra can signal special dynamical circumstances which enhance electronic amplitude in the core region, and which thereby may amplify many-electron effects. It was proposed (Stephens and Dill 1983) that the marked deviations from one-electron predictions seen in the 25–35 eV photon energy range in the photoionization spectra of the $2\sigma_u$ level of N_2 reflect just such amplification. The deviations coincide with the shape-resonant enhancement of $3\sigma_g^{-1}\varepsilon\sigma_u$ f-wave ($l = 3$) photoelectrons in the molecular core, (see, e.g., section 2.2) and they are particularly striking in the photoelectron asymmetry parameter β (Marr et al. 1979, Wallace et al. 1979, Lucchese et al. 1982, Adam et al. 1983). This correlation in energy with the σ_u shape resonance in the $3\sigma_g$ channel, together with strong angular distortion characteristic of high orbital momenta, is the basis of the surmise that the departure of N_2 $2\sigma_u$ photoionization from independent-electron model predictions is due to electronic continuum configuration mixing with the $3\sigma_g^{-1}\varepsilon\sigma_u$ f-wave shape resonance. The results of a recent *ab initio* K-matrix study (Stephens and Dill 1985) showed directly that this mechanism is an important part of N_2 $2\sigma_u$ photoionization dynamics.

Shape-resonance-enhanced interchannel coupling has been known for many years in atomic photoionization (Starace 1979). For example, one-electron (Herman–Skillman and Hartree–Fock) rare-gas, s-subshell photoionization cross sections and angular distributions may be substantially modified by interchannel coupling with strong, shape-resonant amplitude in other photoionization channels (Starace 1979, Johnson and Cheng 1979). Shape-resonant enhancement, mediated by continuum–continuum coupling, has also been proposed recently to account for resonant activity seen in SF_6 photoionization channels that would not support shape resonances in the independent-electron approximation (see, e.g., Dehmer et al. 1982, and section 5.5).

As inferred from atomic interchannel photoionization studies, what Stephens and Dill included in their description is the probability amplitude of forming a quasibound excited complex composed predominantly of an electron excited out of the $3\sigma_g$ subshell into the $\varepsilon\sigma_u$ continuum. Within this complex, the $\varepsilon\sigma_u$ electron may collide with an electron in the $2\sigma_u$ subshell. Thereby a $2\sigma_u$ electron may be ejected by simultaneous de-excitation of the $\varepsilon\sigma_u$ electron back into the $3\sigma_g$ subshell. The $2\sigma_u$ orbital is more compact than the outer valence levels, and for an electron ejected from the outermost $3\sigma_g$ orbital to experience close-collision with this inner-shell electron with significant probability, and with observable consequences, an amplification mechanism must exist. The mechanism here is the molecular shape resonance, in which electronic amplitude in the $3\sigma_g^{-1}\varepsilon\sigma_u$ channel is quasibound by a potential barrier in the molecular field over a narrow range of electron kinetic energy. In this way, the $\varepsilon\sigma_u$ electron maintains a large amplitude in the core region, and hence provides favorable conditions for electronic collision.

Various results for the photoelectron asymmetry parameter of the $2\sigma_u$ channel of N_2 are shown in figs. 31 and 32 to illustrate the effect of continuum–continuum coupling. The dashed curve is the single-channel multiple-scattering model (MSM) result (Wallace et al. 1979) indicating the predicted behavior in the independent-electron, fixed-R, local exchange approximation. The solid line

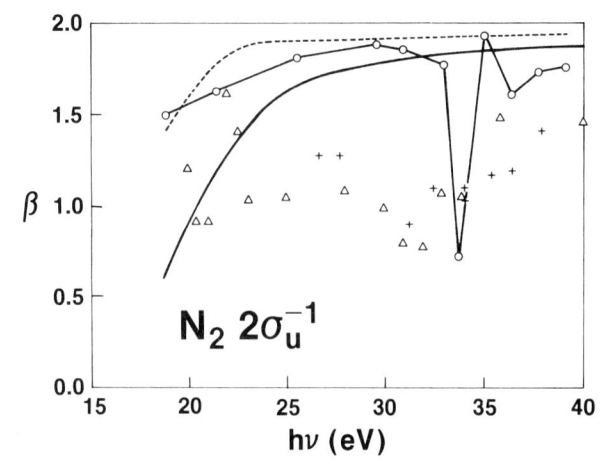

Fig. 31. Photoelectron asymmetry parameter for the $2\sigma_u$ level of N_2: ○, K-matrix results of Stephens and Dill (1985); ———, Hartree–Fock results of Lucchese et al. (1982); ---, multiple-scattering model results of Wallace et al. (1979); △, experimental results of Marr et al. (1979); +, experimental results of Adam et al. (1983).

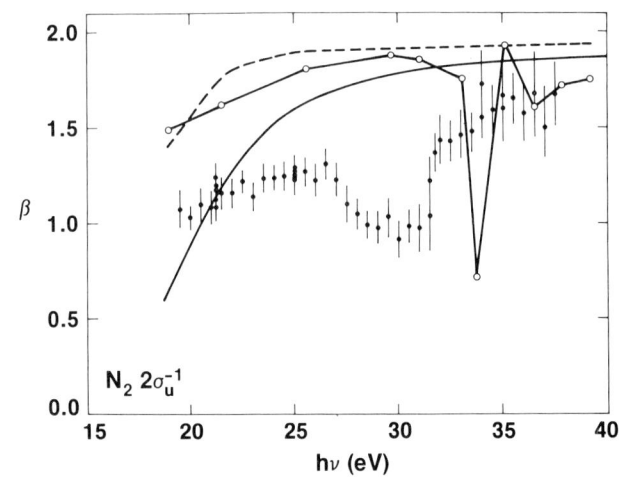

Fig. 32. Photoelectron asymmetry parameter for the $2\sigma_u$ level of N_2: same conventions as fig. 31 plus data of Southworth et al. (1986), given by solid dots.

represents the Hartree–Fock result of Lucchese et al. (1982) which includes initial state correlation in addition to intrachannel coupling in the final state. This calculation improves both the initial state and the final state but still neglects coupling between alternative ionization channels. Early data on this channel taken independently by Marr et al. (1979) and by Adam et al. (1983) are shown in fig. 31. Although sparse, the data clearly shows a large systematic deviation from the one-electron calculations.

This deviation stimulated the K-matrix calculation of Stephens and Dill (1985). By adding electron–electron correlation in an MSM basis, they examined the effects of coupling between the σ_u shape resonance in the $3\sigma_g$ ionization channel and the nominally nonresonant $2\sigma_u$ channel. Their results are indicated by the open circles, connected by a solid line in figs. 31 and 32. These results show that coupling the two channels induces a dramatic dip in the β parameter at about 34 eV. This dip occurs at the photon energy of the σ_u shape resonance in the $3\sigma_g$ channel *in the MSM model calculation* used to generate the one-electron basis for the K-matrix calculation. The experimental energy of the shape resonance in the $3\sigma_g$ channel is, of course, near $h\nu = 30$ eV, as discussed above. (The MSM calculation could have been altered to reproduce this energy exactly, but this is usually not done as no new information is obtained and there is no point in exaggerating the accuracy of the method.)

At this point, it became apparent that better quality data was needed to interpret the significance of the pronounced dip in the K-matrix/MSM calculation. Hence, using second generation instrumentation, described in section 4, Southworth et al. (1986) reexamined the β for the $2\sigma_u$ channel of N_2. The new data is shown as solid dots in fig. 32. This data shows the shape of the β curve much more clearly and allows us to draw several conclusions. First, the β curve exhibits a clear minimum at the photon energy of the σ_u shape resonance in the $3\sigma_g$ channel. Also, the width is close to the σ_u shape resonance width. These observations qualitatively confirm the continuum–continuum coupling mechanism evoked by Stephens and Dill. Second, the minimum is at lower energy and is broader than the calculation. The shift traces mainly to the location of the shape resonance in the independent-electron model calculation. The differences in shape, and to some extent the location, also arise from other approximations inherent in the prototype calculation, for example, neglect of vibrational motion. Third, the data converge to the HF results at energies above and below the shape resonance, indicating the return to essentially single-channel behavior in the absence of shape-resonance-enhanced interchannel coupling.

As in the other case studies emphasized here, this prototype study is but the tip of the iceberg in that this basic mechanism is expected to have widespread effects in molecular photoionization dynamics which can be explored in other circumstances now that the basic mechanism is understood.

5.4. Resonance effects in photoionization of the $1\pi_u$ level of C_2H_2

The fourth case study concerns the photoionization of the outermost $1\pi_u$ orbital of C_2H_2 in the region from the IP up to $h\nu \sim 25$ eV. This process has been under intense study, both experimentally (Kreile et al. 1981, Langhoff et al. 1981b, Unwin et al. 1981, Hayaishi et al. 1982, Keller et al. 1982, Parr et al. 1982a) and theoretically (Kreile et al. 1981, Langhoff et al. 1981b, Hayaishi et al. 1982, Keller et al. 1982, Machado et al. 1982, Levine and Soven 1983, Levine and Soven 1984, Lynch et al. 1984a), over the last few years. This keen interest results from three main interconnected questions which are posed by this spectrum: First,

C_2H_2 is isoelectronic with N_2, which is rather well understood, so that the study of C_2H_2 represents a logical extension of previous work to a polyatomic molecule. In particular, how will the well-established π_g and σ_u shape resonance features influence the photoionization dynamics of this closely related molecule? Second, the total photoionization spectrum (Hayaishi et al. 1982, Berkowitz 1979, Botter et al. 1966, Collin and Delwiche 1967) of C_2H_2 displays a very prominent double hump structure with a deep minimum at $h\nu \sim 14\,\text{eV}$. Strong non-Franck–Condon effects are also observed (Kreile et al. 1981, Langhoff et al. 1981b, Unwin et al. 1981, Keller et al. 1982, Parr et al. 1982a) in the vibrational branching ratios and v-dependent βs. The interpretation (Kreile et al. 1981, Langhoff et al. 1981b, Unwin et al. 1981, Hayaishi et al. 1982, Keller et al. 1982, Machado et al. 1982, Parr et al. 1982a, Levine and Soven 1983, 1984, Lynch et al. 1984a) of these structures has been a central topic of the recent work on C_2H_2. Third, a key point in the photoionization dynamics of polyatomics is the existence of alternative vibrational modes. We discussed the shape-resonance-enhanced non-Franck–Condon effects in section 5.1 and the autoionization-induced non-Franck–Condon effects in section 5.2 for the case of N_2. Having understood that case rather well, it is necessary to look for and understand the ramifications of resonances in other modes which arise in polyatomics. These issues represent increasing complexity relative to the previous case studies and are only partially answered here; however, this case was chosen to underscore the importance of extending our prototypical ideas to more complex molecules.

As mentioned above, the total photoionization spectrum (Hayaishi et al. 1982, Berkowitz 1979, Botter et al. 1966, Collin and Delwiche 1967) of C_2H_2 displays two prominent peaks in the $13\,\text{eV} \leqslant h\nu \leqslant 25\,\text{eV}$ region, one at $\sim 13.3\,\text{eV}$ (930 Å) and another at $15.3\,\text{eV}$ (810 Å), with a dip centered at $\sim 14\,\text{eV}$. In order to examine the dynamics of this process in more detail, triply differential photoelectron measurements have been made on C_2H_2 in this region using synchrotron radiation (Keller et al. 1982, Parr et al. 1982a). We begin by introducing these data and emphasizing the effects of the two resonant features on vibration, including previously unobserved enhancement of weak bending modes below $h\nu \sim 16\,\text{eV}$ (Parr et al. 1982a).

The crux of the analysis of the data is an appreciation that the importance of alternative vibrational modes of the ion can vary drastically in different parts of the spectrum. In particular, predication of the analysis on the fact that the previously published HeI photoelectron spectrum (Baker and Turner 1968) of the $X\,^2\Pi_u$ band of $C_2H_2^+$ exhibits clearly only excitation of the C–C stretching mode would lead to an erroneous analysis of the spectra for $h\nu < 16\,\text{eV}$, where resonant excitation has been found (Parr et al. 1982a) to lead to substantial excitation of other modes, i.e., bending modes. In fact, this observation, documented below for $C_2H_2^+$, is probably the rule rather than the exception for variable-wavelength studies of polyatomics, for which allowance must always be made for enhanced excitation of vibrational modes which have very small Franck–Condon factors and hence are often difficult to detect in photoelectron spectra at nonresonant wavelengths.

To illustrate this, we show angle-resolved photoelectron spectra of $C_2H_2^+$ $X\,^2\Pi_u$ at $\lambda = 563.6$ Å (22.0 eV) and at $\lambda = 885.6$ Å (14.0 eV) in figs. 33 and 34, respectively. At $h\nu = 22$ eV (fig. 33) a single 0.222 eV vibrational spacing, corresponding to the C–C stretch mode, adequately describes the observed spectrum. Use of this single mode, together with a constant (instrumental) peak width of ~120 meV and a background level determined at the extremities of the data in fig. 33, results in an excellent least-squares fit and is in agreement with the He I photoelectron spectrum (Baker and Turner 1968). (In fig. 33 the solid line is the spectrum generated by the least-squares fit and the vertical solid bars indicate the position and relative strengths of the members of the C–C stretch progression.) Below $h\nu = 16$ eV, however, this analysis procedure was found (Parr et al. 1982a) to be inadequate: The peak positions of the higher members of the progression appeared shifted, the valleys between the peaks appeared to fill in more than the resolution or background level warranted, and the quality-of-fit parameter degraded significantly. The key to this puzzle was supplied by a recent high-resolution, high-sensitivity He I photoelectron spectrum (Dehmer and Dehmer 1982) which indicated the location of previously unobserved bending modes of the ground state of $C_2H_2^+$ with intensities at the ~1% level. The observed vibrational spacings were 0.036 eV, 0.086 eV, and 0.172 eV, and are believed to correspond to a trans-bending mode and a cis-bending mode and its harmonic. The first of these is too close to the main C–C stretch progression to be separated with the

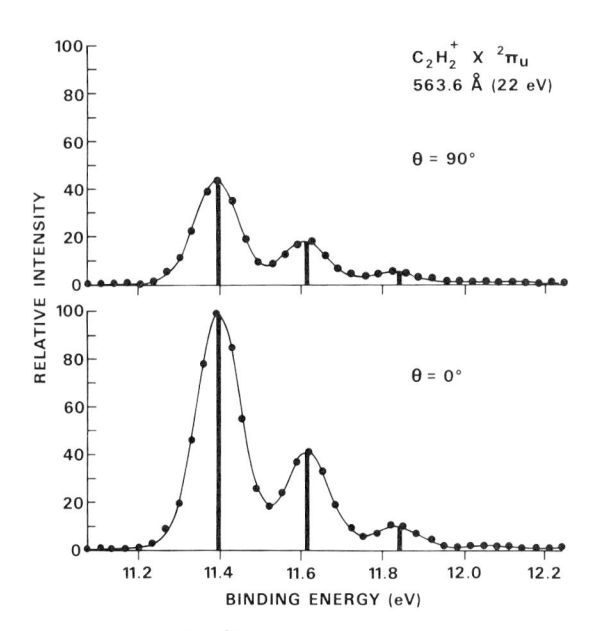

Fig. 33. Photoelectron spectra of $C_2H_2^+$ $X\,^2\Pi_u$ at $h\nu = 22.0$ eV and at $\theta = 0°$ and $90°$. Both spectra are normalized so that the maximum counts in the $\theta = 0°$ spectrum equals 100. The data points (●) and nonlinear least-squares-fit curve (———) are indicated. The amplitudes and positions of the C–C stretch vibrational components yielded by the fit are represented by the vertical solid bars.

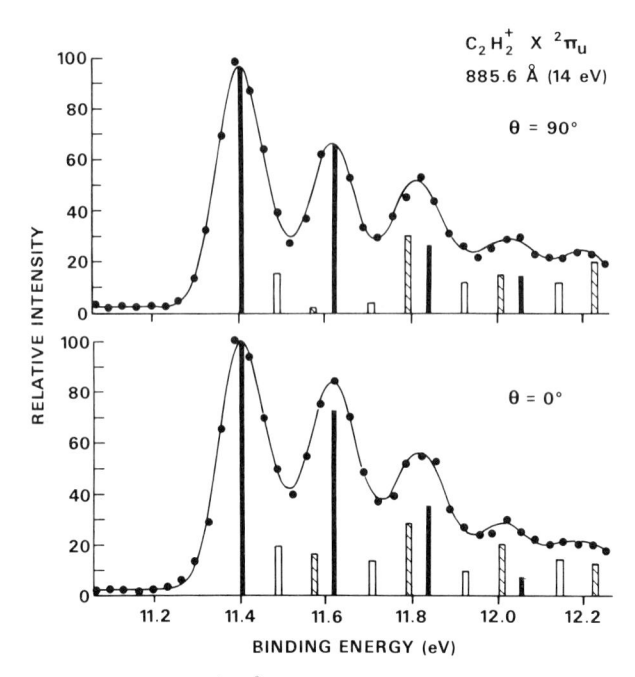

Fig. 34. Photoelectron spectra of $C_2H_2^+$ $X\,^2\Pi_u$ at $h\nu = 14$ eV and at $\theta = 0°$ and $90°$. Both spectra are normalized so that the maximum counts in the $\theta = 0°$ spectrum equals 100. The data points (●) and nonlinear least-squares-fit curve (———) are indicated. The amplitudes and positions of the vibrational components yielded by the fit are as follows: solid bars: C–C stretch; clear bars: 0.086 eV bending mode; cross-hatched bar: 0.172 eV harmonic of bending mode.

resolution of the present experiment; however, when the 0.086 eV and 0.172 eV modes were added to the fitting procedure, the fit quickly converged with a quality-of-fit parameter equal to the high-energy ($h\nu > 16$ eV) fits. The result is illustrated in fig. 34 for $h\nu = 14.0$ eV, where the 0.222 eV, 0.086 eV, and 0.172 eV vibrational progressions are indicated by solid, clear, and hatched bars, respectively. Again, the vertical bars indicate the spectral position and relative intensities of the various vibrational components, and the solid line is the spectral shape of the band generated by the least-squares fit. In comparing figs. 33 and 34, note particularly the reduced peak-to-valley ratio and the shift of the experimental peaks away from the C–C stretch components in fig. 34. The cause for the changes in the spectral shape and peak positions is clearly attributable to the enhanced excitation of the bending modes in the low-energy portion of the excitation spectrum. In fact, transitions involving resonant excitation of bending vibrations become comparable to or even dominate higher members of the main progression at certain wavelengths. Another qualitative observation made clear by figs. 33 and 34 is that the lower-energy spectrum is more isotropic than the higher-energy spectrum, i.e., has a lower β value.

In figs. 35 and 36 we present the spectral variation of the vibrational branching ratios and v-dependent β's for the dominant C–C stretch mode of $C_2H_2^+$ $X\,^2\Pi_u$ in

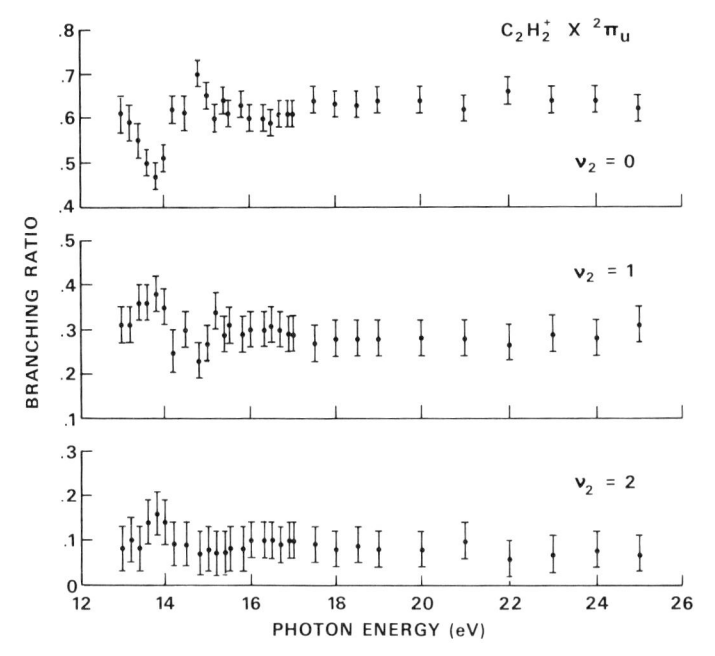

Fig. 35. Vibrational branching ratios for the $v_2 = 0$, 1, and 2 symmetric stretch components of $C_2H_2^+$ $X\,^2\Pi_u$.

the range $13\,\text{eV} \leqslant h\nu \leqslant 25\,\text{eV}$. Although fig. 34 demonstrates the importance of the bending modes for $h\nu < 16\,\text{eV}$, the branching ratios and β's deduced from the least-squares fit were not of sufficient quality to clearly establish the spectral variation of these parameters in the $h\nu < 16\,\text{eV}$ region where they were excited with appreciable intensity. That is, they were usually of less statistical quality than the $v_2 = 2$ C–C stretch data in figs. 35 and 36. Also, they are excited almost exclusively by resonant excitation and tend to vary more sharply than the FC-allowed C–C stretch mode. Hence, they should be mapped on a finer energy mesh. Therefore, we confine further discussion of resonance effects to the stronger C–C stretch mode, whose analysis has nevertheless been improved by including the bending frequencies in the fit for $h\nu < 16\,\text{eV}$. We stress, therefore, that future higher resolution work remains to be done for $h\nu < 16\,\text{eV}$ to completely characterize the photoionization dynamics of this channel in C_2H_2.

Focussing on the spectral behavior of the C–C stretch mode (referred to simply as $v_2 = 0, 1, 2$ from now on) in figs. 35 and 36, we can make the following general observations. First, in fig. 35, the vibrational branching ratios exhibit different profiles in the $h\nu < 16\,\text{eV}$ region and converge to constant values for $h\nu > 16\,\text{eV}$. Thus, they exhibit non-Franck–Condon behavior below $h\nu = 16\,\text{eV}$, and FC behavior above. For example, the $v = 0$ curve exhibits a local dip at $h\nu \sim 13.8\,\text{eV}$, whereas the $v_2 = 1$ and 2 curves show an enhancement; and the $v_2 = 0$ is enhanced near $h\nu = 15\,\text{eV}$, at which energy the $v_2 = 1$ curve dips and the $v_2 = 2$ curve stays flat. Second, the spectral variation of the non-FC branching ratios in fig. 35 are

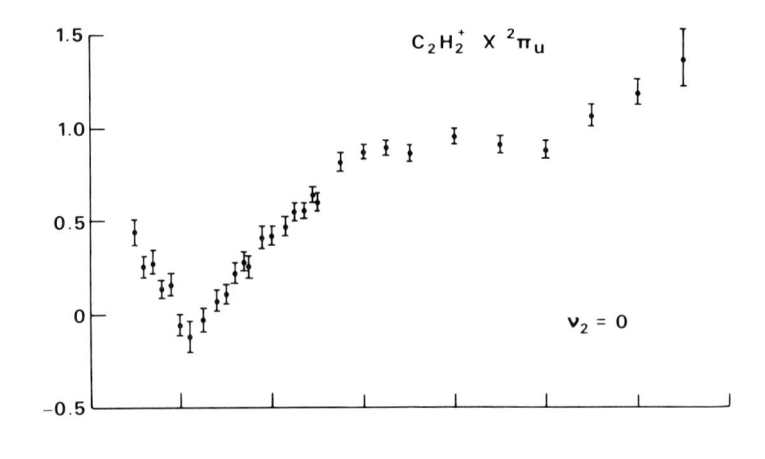

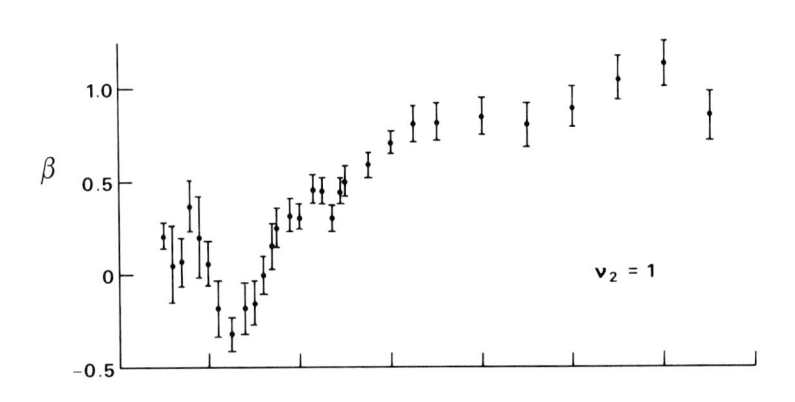

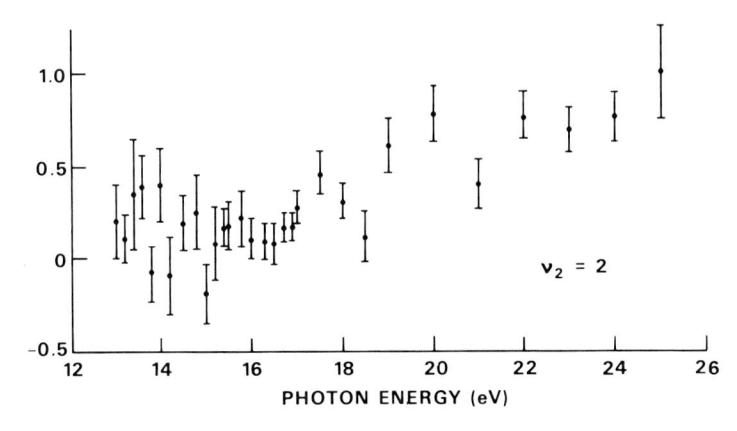

Fig. 36. Photoelectron asymmetry parameters for the $v_2 = 0$, 1, and 2 symmetric stretch components of $C_2H_2^+$ $X\,^2\Pi_u$.

generally similar to the spectral variation of the "peak" intensities reported by Unwin et al. (1981); however, the present branching ratios are believed to be representative of a simple C–C stretching mode (with some admixture of the weak 0.036 eV bending mode) since we have separated it from the other vibrational modes which we have determined to be non-negligible in this spectral range. The $v_2 = 1$ and 2 data of Unwin et al. (1981) could be expected to be affected by this consideration. Above $h\nu \sim 16$ eV, where the excitation of bending modes and other manifestations of resonant excitation are no longer apparent, the branching ratios in fig. 35 agree well with those measured by Kreile et al. (1981) at the Ne I (16.67 eV and 16.85 eV), He I (21.22 eV), and Ne II (26.81 eV and 26.91 eV) resonance lines, and with branching ratios derived from FC factors (Dibeler and Walker 1973, McCulloh 1973) ($v_2 = 0$, 0.69; $v_2 = 1$, 0.21; $v_2 = 2$, 0.09). Third, in fig. 36, the β curves for $v_2 = 0$ and 1 show a distinct broad dip to below $\beta = 0$, centered at $h\nu \sim 14.25$ eV and $h\nu \sim 14.5$ eV, respectively, while the $v_2 = 2$ curve fluctuates near $\beta = 0$, although the statistical uncertainty prohibits a clear picture of the variation of $v_2 = 2$ for $h\nu < 16$ eV. Keep in mind that peaks with small FC factors like $v_2 \geq 2$ and the bending modes will tend to be populated mainly by resonant processes and, hence, may show much sharper variation within Rydberg series than do the FC-allowed channels. Fourth, all three β curves in fig. 36 rise at higher energy with grossly similar spectral shape, within error limits, to a value near $\beta \sim 1$. Note that photoionization of the $1\pi_u$ level of N_2 and the 1π level of CO, both isoelectronic with C_2H_2, also result in a vibrationally-summed β which exhibits a general rise from $\beta = 0$ to $\beta \sim 1$ over the same kinetic energy range (Wallace et al. 1979, Holmes and Marr 1980). This rough correlation among similar orbitals in the three isoelectronic molecules for the nonresonant part of the spectrum seems reasonable. Fifth, comparison of the β values in fig. 36 with the resonance line work of Kreile et al. (1981) confirms the overall accuracy of the independent measurements: The agreement is excellent at all overlapping wavelengths, even at Ar II (13.30 eV and 13.48 eV), for which comparison is hazardous owing to differences in photon bandpass and the failure to take bending vibrations into account in the work by Kreile et al. Furthermore, measurements at the Ne II (26.8 eV and 26.9 eV) and He II (40.8 eV) resonance lines give (roughly v_2-independent) β values of ~ 1.4 and ~ 1.6, respectively, indicating that the trend observed above is continued to higher energy.

The main issue to be resolved en route to full elucidation of the dynamics reflected in this data is the nature of the resonant mechanism(s) responsible for the double hump structure below $h\nu \sim 16$ eV. This has been discussed extensively in the literature (Kreile et al. 1981, Langhoff et al. 1981b, Unwin et al. 1981, Hayaishi et al. 1982, Keller et al. 1982, Machado et al. 1982, Parr et al. 1982a, Levine and Soven 1983, 1984, Lynch et al. 1984a); and, at last, it is now possible to summarize a fairly clear picture of the mechanisms at play. First, there is virtual unanimity regarding the nature of the intense peak at 15.31 eV. Several different calculations place an intense $2\sigma_u \rightarrow 1\pi_g$ transition, converging to the $2\sigma_u^{-1}$ IP, very near this energy. Hence, the resonantly enhanced $1\pi_g$ appears well into the $1\pi_u$ continuum in C_2H_2, partly because the $1\pi_u$ IP is lowered relative to that in N_2. Second, the calculation of Levine and Soven (1983, 1984) has incorporated intrachannel and interchannel interaction via a time-dependent local-density approximation (TDLDA) showing that the cross section in the vicinity of the $2\sigma_u \to 1\pi_g$ transition is accurately reproduced, thus confirming this interpretation. Third, that same calculation indicates that the lower energy peak owes half of its intensity to the $2\sigma_u \to 1\pi_g$ autoionization profile and the other half to a local maximum in the $1\pi_u \to \varepsilon\pi_g$ continuum. This implies that, relative to the independent-particle approximation (IPA), inclusion of intrachannel interactions, within the RPA-type theory used, redistributes the nominally discrete $1\pi_u \to 1\pi_g$ oscillator strength to higher frequencies, some of which appears above the IP as part of the 13.3 eV peak. This is the first clear case in a molecular context of a mechanism known in atomic physics for many years (Starace 1979, Dehmer et al. 1971). Fourth, in a somewhat different framework, Lynch et al. (1984a) also argued that $1\pi_u \to 1\pi_g$ strength contributes to the low-energy feature. Some evidence also exists for a minor contribution from a $3\sigma_g \to 3\sigma_u$ autoionizing transition in the vicinity of the 13.3 eV peak.

The main contributions to the double hump structure would therefore seem to arise from both direct and indirect transitions involving the $1\pi_g$ state. This state is a resonantly enhanced bound state, in the independent-particle approximation, and is analogous to the $1\pi_g$ so prominent in the N_2 K-shell spectrum (see, e.g., section 2.2). To demonstrate that the interpretation summarized above is essentially correct, we show in fig. 37 the vibrationally unresolved β value computed using the TDLDA (Levine and Soven 1983) which is the only calculation on C_2H_2 incorporating the interactions due to electron correlation within and between IPA channels. The TDLDA results (solid line) agree semiquantitatively with the

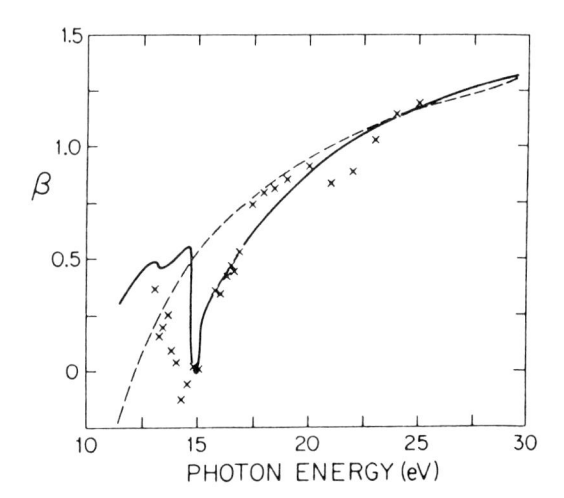

Fig. 37. Photoelectron asymmetry parameter for the $1\pi_u$ level of acetylene. Data from Parr et al. (1982a) have been averaged over vibrationally resolved levels using experimental branching ratios. Experimental uncertainty is typically ± 0.07 β units. Dashed curve, IPA calculation (Levine and Soven 1983); solid curve, TLDA calculation (Levine and Soven 1983).

vibrationally-averaged data (Parr et al. 1982a), clearly reflecting the prominent dip centered at $h\nu \sim 14$–15 eV, totally absent from the IPA calculation. The partial cross section is not shown here, but likewise shows semiquantitative agreement with the measured profile (Levine and Soven 1984). Other measurements are also consistent with these conclusions (Keller et al. 1982).

Hence, to summarize the results of this case study: (i) The prominent double peak has been successfully interpreted; (ii) the $1\pi_g$ state has a major impact, but is reflected differently than in N_2, partly due to its spectral location relative to its own and other IPs (examination of the σ_u resonance is best carried out by examining the photoionization of the σ_g subshells); and (iii) bending vibrations are greatly enhanced in the resonance region, must be included in the analysis, but have not yet been characterized well. This is a notable challenge for future work.

5.5. Valence-shell photoionization of SF_6

The photoionization of SF_6 has been avidly studied over the past 10–15 years. One inducement has been its octahedral symmetry, which should render the study of its spectral properties more tractable than for other large polyatomics. A second, less trivial, motivation has been the central role played by SF_6 in the elucidation of shape-resonance effects in molecules. The four shape-resonant features (a_{1g}, t_{1u}, t_{2g}, and e_g) in the sulfur K-shell (LaVilla and Deslattes 1966, LaVilla 1972) and L-shell (Zimkina and Fomichev 1966, Zimkina and Vinogradov 1971, Blechschmidt et al. 1972, LaVilla 1972) spectra remain the most striking examples of potential barrier effects in molecular spectra, as noted in section 2.1. Third, SF_6 has important practical uses, most notably in gaseous electronics.

For these and other reasons, a large amount of information has been generated on SF_6 photoionization and related excitation processes: On the theoretical side, several groups have calculated the electronic structure (Gianturco et al. 1971, Connolly and Johnson 1971, von Niessen et al. 1975, Hay 1977, von Niessen et al. 1979) of SF_6, and others have calculated partial photoionization cross sections (Gianturco et al. 1972, Sachenko et al. 1974, Levinson et al. 1979, Wallace 1980) and photoelectron angular distributions (Wallace 1980) for all the subshells of SF_6. In addition, the elastic e–SF_6 scattering cross section (Dehmer et al. 1978, Benedict and Gyemant 1978) has been calculated indicating the role of the above-mentioned shape resonances and the close connection (Dehmer and Dill 1979a) between shape resonances in electron scattering and photoionization contexts. An even larger collection of experimental work includes: (i) X-ray absorption and emission cross sections from core levels (LaVilla and Deslattes 1966, Zimkina and Fomichev 1966, Zimkina and Vinogradov 1971, Blechschmidt et al. 1972, LaVilla 1972, Ågren et al. 1978), (ii) VUV absorption by valence levels (Nakamura et al. 1971, Blechschmidt et al. 1972, Sasanuma et al. 1978, Codling 1966, Lee et al. 1977), (iii) photoelectron spectra using X-rays (Gelius 1974) and VUV resonance lines (Gustafsson 1978, Gelius 1974, Potts et al. 1970, Sell and Kuppermann 1978), (iv) photoelectron angular distributions with He I radiation (Sell and Kuppermann 1978), (v) partial photoionization cross sections (Gustafsson 1978, Dehmer et al. 1982, Ferrett et al. 1986) and photoelectron angular distributions (Dehmer et al. 1982, Ferrett et al. 1986) using synchrotron radiation, (vi) photoionization mass spectrometry and ionization yield measurements (Dibeler and Walker 1966, Sasanuma et al. 1979), and (vii) electron scattering measurements of total scattering (Kennerly et al. 1979) differential elastic scattering (Rohr 1979) and inelastic scattering in the low-energy (Simpson et al. 1966, Trajmar and Chutjian 1977), pseudo-optical limit (Hitchcock and Brion 1978a, Hitchcock et al. 1978, Hitchcock and van der Wiel 1979) and (e,2e) (Giardini-Guidoni et al. 1979) configurations.

Despite this great body of information, however, there remained major questions concerning the spectroscopy and dynamics of SF_6 photoionization. The two issues of concern here are the ordering of the valence levels of SF_6 and the role of the t_{2g} shape resonance in valence-shell spectra. Concerning the ordering of valence levels, several ground-state configurations have been proposed on the basis of different types of evidence. This issue is complicated by the near degeneracy of two valence levels (which two is one of the central questions), resulting in the occurrence of six photoelectron peaks in the ionization potential (IP) range $16\,eV \leqslant IP \leqslant 30\,eV$, where seven valence levels are known to lie. The study of shape resonance effects in valence-shell spectra depends very much on establishing the ground-state configurations as, *in the independent-electron approximation*, dipole selection rules govern which orbitals will make transitions to particular shape resonances. The significance of the qualification in italics will be discussed later.

In this case study, we discuss recent evidence (Dehmer et al. 1982) and review the previous literature in an attempt to resolve these problems. In particular, we discuss measurements (Dehmer et al. 1982) of partial photoionization cross sections, branching ratios, and photoelectron angular distributions for the valence levels of SF_6 in the photon range $16\,eV \leqslant hv \leqslant 30\,eV$. The partial cross sections and branching ratios agree well with those measured earlier by Gustafsson (1978), where the two sets of data overlap. In addition, we compare this data with multiple-scattering calculations (Wallace 1980) of the same quantities. They are used here in a form chosen specifically for extracting the needed information from the data. Namely, they have been convoluted with the experimentally observed peak shapes and are plotted in alternative ways to illustrate the consequences of adopting various valence-level orderings. The discussion of these results suggests that the most plausible valence configuration is

$$5a_{1g}^2 4t_{1u}^6 1t_{2g}^6 3e_g^4 (1t_{2u}^6 + 5t_{1u}^6) 1t_{1g}^6 \,{}^1A_{1g},$$

although some small uncertainty still exists regarding the location of the $1t_{2u}$ level. We conclude by proposing further experimental and theoretical work to test these conclusions and to study the strong channel interaction effects implied by them.

There are seven occupied valence states of SF_6 with IP's less than 30 eV (Gelius 1974, Potts et al. 1970, Sell and Kuppermann 1978), all derived from the fluorine

2p and sulfur 3s, 3p, and 3d atomic orbitals. Their approximate ordering, starting with the least tightly bound, is $1t_{1g}^6$, $5t_{1u}^6$, $1t_{2u}^6$, $3e_g^4$, $1t_{2g}^6$, $4t_{1u}^6$, and $5a_{1g}^2$ combining to give a closed-shell ground state with $^1A_{1g}$ symmetry. The photoelectron spectrum (Gelius 1974, Potts et al. 1970, Sell and Kuppermann 1978) covering this range of IP's exhibits six peaks with vertical IPs of 15.7, 17.0, 18.6, 19.7, 22.5, and 26.8 eV. In the following, we will refer to these as peak 1 through peak 6, respectively, in order of increasing IP. Clearly, one of the peaks encompasses two IP's and the probable candidates are peak 2 and peak 3. This will be the main focus of the following discussion. Peak 6 is easily and unanimously assigned to the $5a_{1g}$ molecular orbital (MO) and is not discussed further here since its branching ratio is always <2% for $h\nu < 30$ eV (Gustafsson 1978), and it was not well characterized in the measurements discussed here.

The experimental results are presented in figs. 38–42 for peaks 1–5, respectively, along with corresponding theoretical results which will be described below. In each figure, the top frame contains the photoelectron asymmetry parameter β from the IP up to $h\nu = 29.2$ eV. The β values tend to gravitate around $\beta = 0$, and the resulting nearly isotropic distribution was easily measured with good precision. The average uncertainty was ±0.03 with the largest being ±0.1. Including the uncertainty in the calibration procedure, we assign an overall accuracy of approximately ±0.05 to the β values. Differences of up to 0.15 were noted relative to earlier measurements (Sell and Kuppermann 1978) of β at 584 Å; however, for all but peak 1, the β's in figs. 38–42 lay in the range of β's measured across the bands in the article by Sell and Kupperman (1978). In view of the likelihood of autoionization near 21.2 eV (Codling 1966), one need not be greatly concerned with the differences observed using the medium-resolution synchrotron radiation light source and the narrow-band resonance line.

The middle frames in figs. 38–42 give the branching ratios for peaks 1–5, relative to the sum of their intensities. Uncertainties in these quantities are typically ±0.01. Agreement with earlier measurements by Gustafsson (1978) is generally good, although local differences of 0.05 are observed. Differences in energy mesh, transmission function calibrations for low kinetic energies, and the β-dependence of the earlier measurements (Gustafsson 1978) probably contribute to this, although the differences do not significantly affect the following discussion.

The bottom frames in figs. 38–42 contain the partial cross sections for peaks 1–5, obtained by multiplying the measured branching ratios times the total ionization cross section (total photoabsorption cross section multiplied by the ionization efficiency) reported by Hitchcock and van der Wiel (1979). Again semiquantitative agreement was observed with the analogous analysis by Gustafsson (1978), who used total absorption data by Lee et al. (1977). The total absorption cross sections by Hitchcock and van der Wiel and Lee et al. are in good agreement throughout the range discussed here. The only significant issue is the assumption by Gustafsson (1978) that the ionization efficiency is unity throughout this range. The ionization efficiency measured by Hitchcock and van der Wiel was greater than 90% for $h\nu > 20$ eV, but fell off toward the ionization threshold to a value of 25% at $h\nu = 16$ eV. This accounts for some, but by no means all, of the quantitative differences between the data sets. The observed semiquantitative agreement is considered satisfactory for this type of measurement at this time, and, although it would be desirable to remove the remaining minor discrepancies, they pose no significant problem vis-à-vis the issues discussed below. There, the occurrence of peaks in the partial cross sections represents the most significant aspects of the data; and, on this point, there is no qualitative disagreement.

The continuous curves in figs. 38–42 are theoretical results for each measured quantity. They have been presented (Dehmer et al. 1982) in the following manner to try to aid in resolving assignments in valence-shell photoionization of SF_6 by use of dynamical evidence. First, the partial cross section and photoelectron asymmetry parameter for each valence state of SF_6 was calculated (Wallace 1980) using the multiple-scattering model by now standard procedures (Dehmer and Dill 1979b). Second, three sets of theoretical curves were derived as described in table 1, each set corresponding to one of the possible valence configurations discussed later, each differing in the assignment of the $5t_{1u}$, $1t_{2u}$, and $3e_g$ initial states to peaks 2 and 3. Third, each set of dipole matrix elements and asymmetry parameters was combined with the corresponding experimental IP's, consistent with the assignments in that set, and was then folded with Gaussian line shapes with the halfwidths in the experimental spectrum (0.4, 0.6, 0.7, 0.7, and 0.4 eV for peaks 1–5, respectively). This avoided sudden jumps in the branching ratios at higher IP's and ensured that the comparison between experiment and theory was not confused by the intrinsic and instrumental widths of the photoelectron peaks.

To simplify the discussion of the complex body of data bearing on this subject, we will proceed by presenting a recommended valence-level structure, followed first by evidence supporting this conclusion and second by a discussion of various aspects of the assignment, including a tentative rationalization of seemingly contradictory evidence. Accordingly, we will consider the assignment of photoelectron peaks 1–5 to ionization from $1t_{1g}$, $5t_{1u} + 1t_{2u}$, $3e_g$, $1t_{2g}$, and $4t_{1u}$ MO's, respectively. This is also the ordering arrived at in earlier theoretical work by Connolly and Johnson (1971), Hay (1977), and von Niessen et al. (1975, 1979), and in experimental work by Gelius (1974).

General support for the above ordering is provided by the results of the two

Table 1
Trial assignments used to construct theoretical curves in figs. 38–42.

Peak number	Solid curve	Dashed curve	Dashed–dotted curve
1	$1t_{1g}$	$1t_{1g}$	$1t_{1g}$
2	$1t_{2u} + 5t_{1u}$	$5t_{1u}$	$1t_{2u}$
3	$3e_g$	$1t_{2u} + 3e_g$	$5t_{1u} + 3e_g$
4	$1t_{2g}$	$1t_{2g}$	$1t_{2g}$
5	$4t_{1u}$	$4t_{1u}$	$4t_{1u}$

most sophisticated calculations of SF_6 electronic structure, namely, the many-body Green's-function calculations of von Niessen et al. (1975, 1979) and the generalized valence-bond calculation of Hay (1977). Both types of calculation yield the above sequence of MO's, and the ordering predicted by the calculations by von Niessen et al. was found (von Niessen et al. 1979) to be extremely stable with respect to large changes of basis functions and other conditions. In fact, in all calculations, only the relative ordering of the $5t_{1u}$ and the $1t_{2u}$, and the question of their quasidegeneracy varies.

Important experimental evidence is provided by X-ray emission, resulting from filling holes in the sulfur 1s and 2p subshells by dipole transitions from the valence shells. Sulfur K X-ray emission spectra reported by LaVilla (1972) locate the $3t_{1u}$, $4t_{1u}$, and $5t_{1u}$ valence levels and provide concrete confirmation of the $4t_{1u}$ and $5t_{1u}$ assignments given above. More recent, high-resolution X-ray emission spectra (Ågren et al. 1978) for the sulfur L-shell likewise locate the $5a_{1g}$, $1t_{2g}$, and $3e_g$ levels and provide concrete confirmation of the $3e_g$ and $1t_{2g}$ assignments given above. To summarize the X-ray emission evidence, the $5t_{1u}$, $3e_g$, $1t_{2g}$, and $4t_{1u}$ levels are associated with peaks 2, 3, 4, and 5, respectively.

This leaves the placement of the $1t_{1g}$ and $1t_{2u}$ orbitals. We conform to the assignment of the $1t_{1g}$ orbital to peak 1. There is little qualitative evidence for this (the t_{1g} does not have a dipole allowed component on the sulfur atom which would be active in the sulfur L emission spectrum); however, there is nearly universal agreement on this assignment. The placement of the $1t_{2u}$ orbital, on the other hand, is the most controversial assignment. We assign it to peak 2 mainly on the evidence of the X-ray photoelectron spectrum, as interpreted by Gelius (1974). At $h\nu = 1.25$ keV, the second peak in the photoelectron spectrum is approximately twice as large as the first, third, or fourth. As all of these are derived mainly from fluorine p orbitals, this suggests that the second peak consists of two overlapping bands. This is a simplified version of Gelius' more detailed analysis in terms of net atomic populations. This concludes the main arguments supporting the recommended assignment. Note that it relies heavily on X-ray data, which as we shall suggest later, is valuable in that it should be free from gross channel interaction effects that are believed to significantly modify valence-shell dynamics. Note also that this conclusion means that the slight splitting in peak 2 probably arises from the superposition of two peaks (although other causes could also distort the photoelectron peak) and that peak 3 is split at certain wavelengths and angles by the Jahn–Teller effect. See the work by Gustafsson (1978), Dehmer et al. (1982), Gelius (1974), Potts et al. (1970), and Sell and Kuppermann (1978) for discussions of the doublet structure in peaks 2 and 3.

The major difficulty with the above picture arises from work by Gustafsson (1978), who used partial cross section measurements, and earlier evidence (Dehmer 1972) that a shape resonance occurs at ~ 5 eV kinetic energy in the t_{2g} continuum, to conclude that peaks 2 and 3 corresponded to ionization from the $5t_{1u} + 3e_g$ and $1t_{2u}$ MO's, respectively, with the possible interchange of the two odd-parity MO's. This was later discussed in connection with multiple-scattering calculations with the same general conclusions (Levinson et al. 1979), although the $5t_{1u}$ and $1t_{2u}$ were switched in that work. The reasoning was the following. Since a final-state shape resonance of t_{2g} symmetry is known to lie at $\sim 5\,\mathrm{eV}$ kinetic energy, photoelectron peaks which are significantly enhanced approximately 5 eV above their respective IP's will be odd levels which couple to the t_{2g} resonance in a dipole transition. Peaks 2, 3, and 5 were observed to resonate between 5 and 6 eV above threshold, and therefore they would be assigned to the odd levels $5t_{1u}$, $1t_{2u}$, and $4t_{1u}$ with some ambiguity concerning the first two. Gustafsson used the X-ray photoelectron intensity arguments employed above to conclude that peak 2 contained two peaks and therefore the $3e_g$. It is now fairly clear from X-ray emission data that the $5t_{1u}$ and $3e_g$ levels are associated with peaks 2 and 3, respectively. However, the argument that peak 3 resonates at 5 eV kinetic energy, and therefore contains the $1t_{2u}$ peak is a serious contradiction to the assignment proposed earlier, particularly since the data in figs. 38–42 confirm the resonant behavior, and the existence of the t_{2g} shape resonance is well established.

Dehmer et al. (1982) tentatively resolve this dilemma by attributing the resonant activity of peak 3 to some form of continuum–continuum coupling whereby peak 3 shares in the huge resonant enhancement of peak 2 at $h\nu \sim 23$–$24\,\mathrm{eV}$. The coupling could be direct Coulomb coupling between the nearly degenerate channels, since both have the same excited complex (ion plus photoelectron) symmetry, or possibly vibronic coupling. The latter may be enhanced (relative to typical direct molecular photoionization) since the electron is resonantly trapped, and hence delayed in its escape, and the SF_6^+ ion is known to be unstable relative to fragments of lower symmetry. The conjecture of continuum–continuum coupling is nebulous and would require more theoretical study to demonstrate its validity. However, it is supported by the following observations. First, strong channel interaction resulting in intensity borrowing in the vicinity of the strong resonant enhancement at $h\nu \sim 23\,\mathrm{eV}$ in the total cross section would tend to occur near this photon energy. Indeed, peaks 2 and 3 reach a maximum at $h\nu \sim 23\,\mathrm{eV}$ and are better aligned than on a kinetic energy scale. Second, peak 1, almost surely involving an even initial state, also peaks at $h\nu \sim 23\,\mathrm{eV}$ when the sloping background is taken into account (see fig. 38). In fact, the local enhancement at $h\nu \sim 23\,\mathrm{eV}$ in peak 1 is of the same magnitude (~ 15 Mb) as that in peak 3. The enhancement is more clearly displayed in fig. 38 than in Gustafsson's (1978) data, but both exhibit a clear rise at $h\nu \sim 22$–$23\,\mathrm{eV}$, which is totally absent from the one-electron calculations. Third, the appearance of symmetry-forbidden transitions to shape resonant features has already been noted (Dehmer 1972) in inner-shell absorption spectra for SF_6, e.g., the t_{1u} shape resonance aligns with the weak bump in the sulfur 2p absorption spectra, and the a_{1g}, t_{2g}, and e_g resonant features align with weak features in the sulfur 1s spectra, when the spectra's IP's are aligned. The coupling in the X-ray spectra is weak, only a few percent, whereas one would have to postulate coupling on the order of 20% in the valence shell; but the qualitative trend is reasonable owing to the quasidegeneracy in the valence spectra. Moreover, later work has documented the continuum–continuum coupling mechanism for the $2\sigma_u$ channel of N_2, as described in the article by Stephens and Dill (1985) and section 5.3.

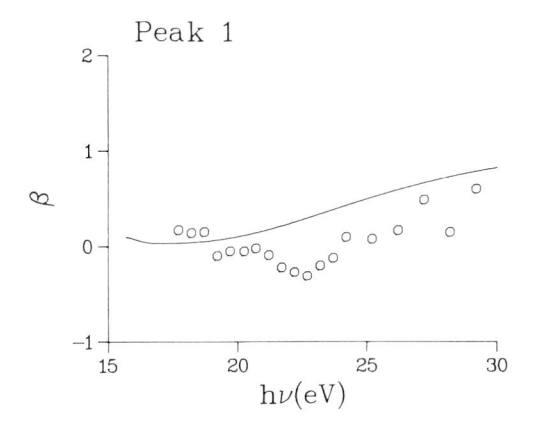

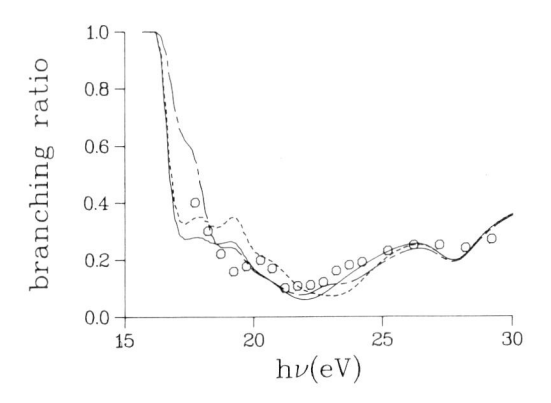

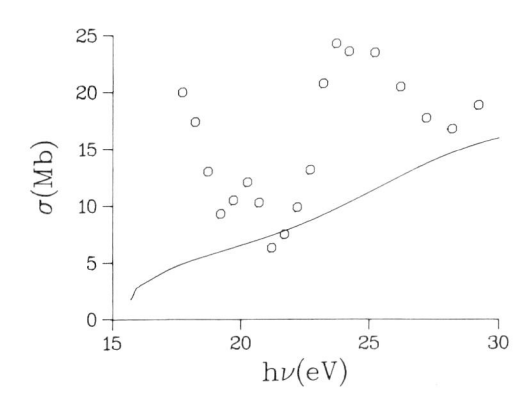

Fig. 38. Photoelectron asymmetry parameter, branching ratio, and partial cross section for peak 1 (IP = 15.7 eV) of the photoelectron spectrum of SF_6. Open circles: data from Dehmer et al. (1982). Curves are theoretical calculations, as described in the text.

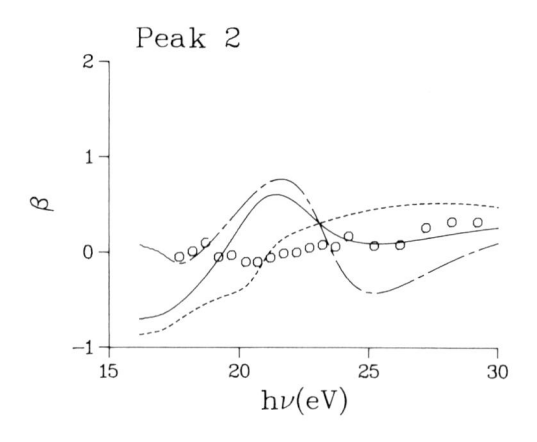

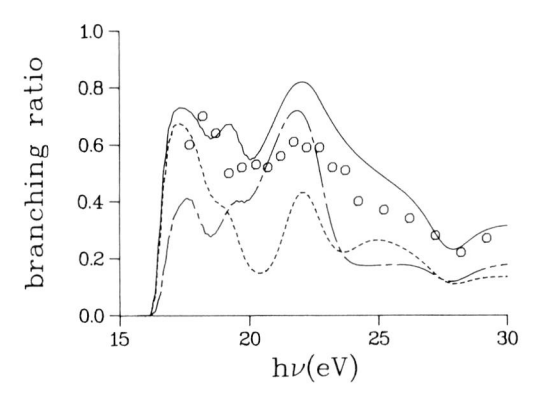

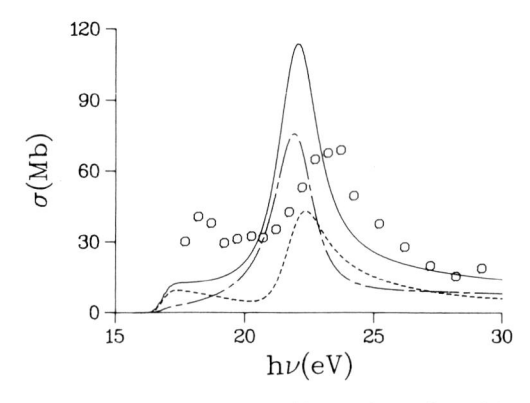

Fig. 39. Photoelectron asymmetry parameter, branching ratio, and partial cross section for peak 2 (IP = 17.0 eV) of the photoelectron spectrum of SF_6. Open circles: data from Dehmer et al. (1982). Curves are theoretical calculations, as described in the text.

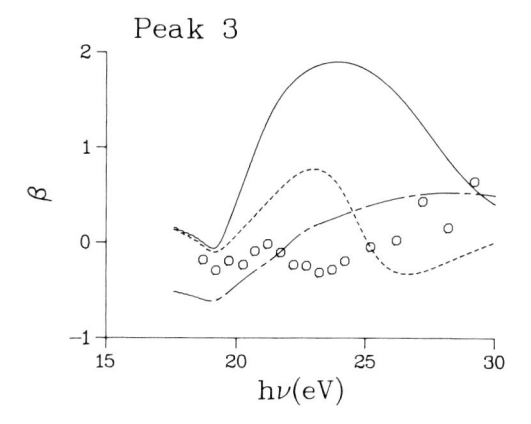

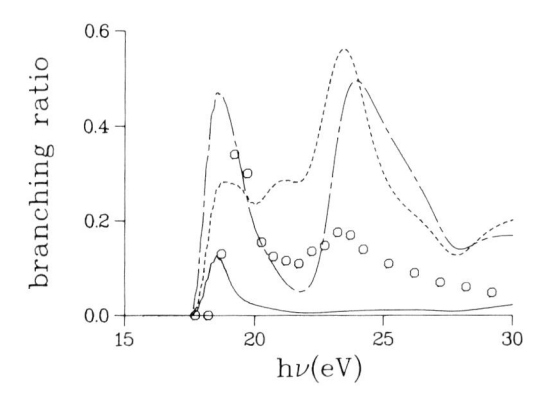

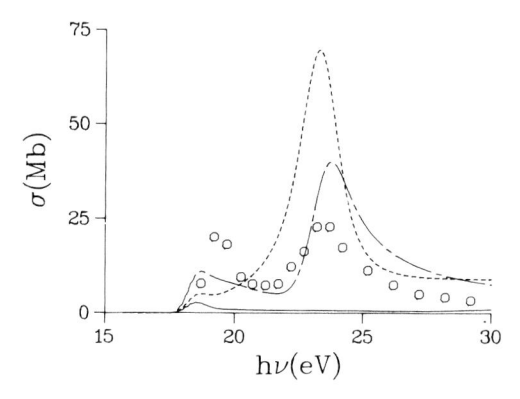

Fig. 40. Photoelectron asymmetry parameter, branching ratio, and partial cross section for peak 3 (IP = 18.6 eV) of the photoelectron spectrum of SF_6. Open circles: data from Dehmer et al. (1982). Curves are theoretical calculations, as described in the text.

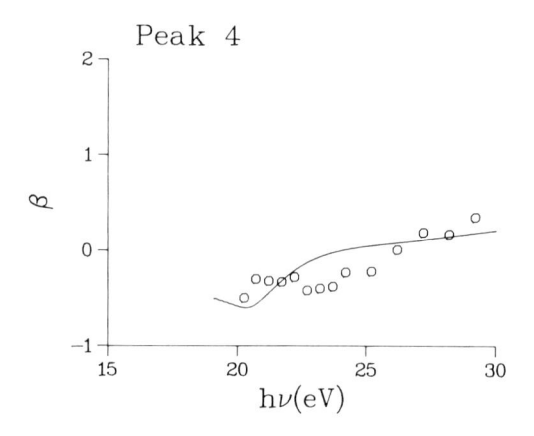

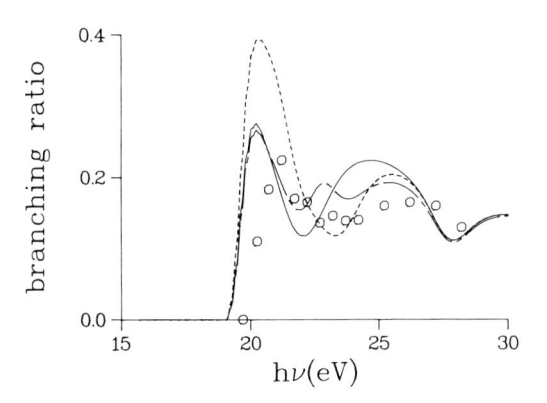

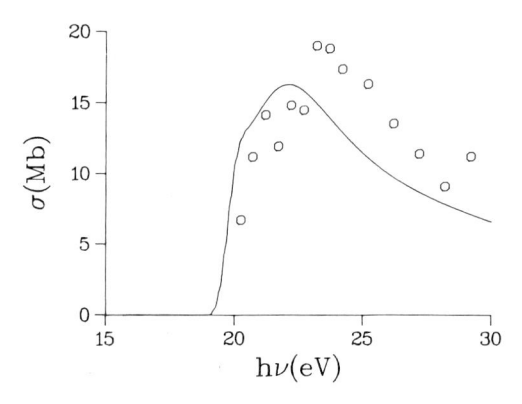

Fig. 41. Photoelectron asymmetry parameter, branching ratio, and partial cross section for peak 4 (IP = 19.7 eV) of the photoelectron spectrum of SF_6. Open circles: data from Dehmer et al. (1982). Curves are theoretical calculations, as described in the text.

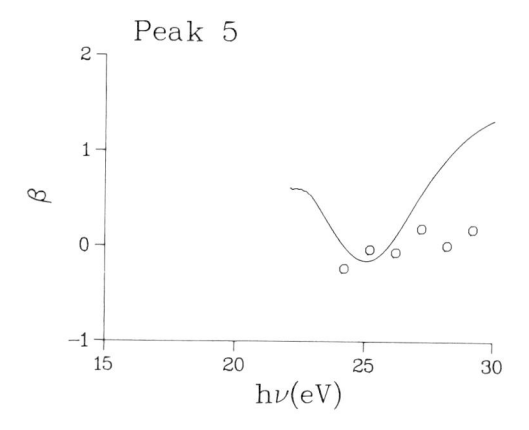

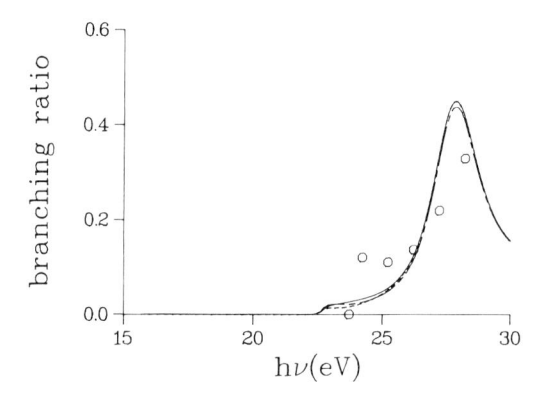

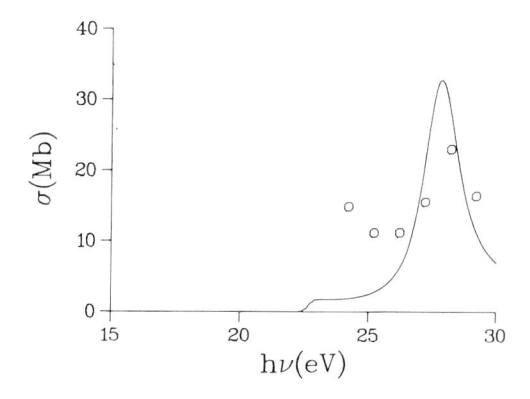

Fig. 42. Photoelectron asymmetry parameter, branching ratio, and partial cross section for peak 5 (IP = 22.5 eV) of the photoelectron spectrum of SF_6. Open circles: data from Dehmer et al. (1982). Curves are theoretical calculations, as described in the text.

Another possible source for deviations from the independent-electron reasoning regarding the appearance of resonant enhancements in the partial cross sections, is autoionization structure, particularly that involving the shape-resonantly-enhanced antibonding $6a_{1g}$ and $6t_{1u}$ MO's, known to cause strong features below inner-shell thresholds. If one of these states occurred near $h\nu \sim 23$ eV, this could perturb the simplified shape-resonance picture. The two most likely candidates are the $5a_{1g} \rightarrow 6t_{1u}$ and the $4t_{1u} \rightarrow 6a_{1g}$ transitions. Taking the kinetic energy of the t_{2g} shape resonance as ~ 5.7 eV and the $6a_{1g}$–t_{2g} and $6t_{1u}$–t_{2g} spacings from X-ray absorption data, we arrive at transition energies of $h\nu \sim 17.2$ and 26.7 eV for the $5a_{1g} \rightarrow 6t_{1u}$ and the $4t_{1u} \rightarrow 6a_{1g}$ transitions, respectively. (Note that shape-resonant features shift by ~ 1–4 eV toward higher kinetic energy in going from inner-shell to valence-shell spectra due to different screening and other differences in relaxation effects. Therefore, we approximate relative energies from X-ray spectra, but normalize to the t_{2g} in the valence-shell spectra.) Neither matches the position of the main resonance peak at $h\nu \sim 23$ eV; however, we note in passing that the total ionization cross section has an unidentified peak at $h\nu \sim 17$ eV. This may be caused by the $5a_{1g} \rightarrow 6t_{1u}$, which in turn could account for the rises in the partial cross sections in figs. 38–40. A similar observation was made by Fock and Koch (1985). Hence, the $6a_{1g}$ and $6t_{1u}$ excited states do not appear to bear on the present discussion of the $h\nu \sim 23$ eV feature. Other weaker, nonresonantly enhanced autoionization states are known to lie in this region (Codling 1966) and may cause departures from a one-electron framework of interpretation. However, the observed structures are weak, relative to the magnitude of the resonant enhancements in peaks 1 and 3; and, therefore, the importance of autoionizing Rydberg states in this connection is tentatively discounted.

Against this background, we now examine the experimental and theoretical results presented in figs. 38–42. In assessing the agreement between experiment and theory, recall that, in most diatomic and triatomic cases studied, the independent-electron multiple-scattering model achieves qualitative to semiquantitative agreement with shape-resonant and nonresonant photoionization (Wallace 1980). The β's are usually within 0.25 of a β unit and have the same general shape as the data. The partial cross sections exhibit most known shape resonances, although the theoretical resonance line shape tends to be too intense and narrow relative to the data and may be shifted by a few eV. Nuclear motion and electron correlation tend to smear out these sharp features. We might expect good agreement for SF_6 owing to the favorable close-packed geometry, which should minimize the impact of assumptions inherent in the multiple-scattering potential. However, anticipating our results, we find qualitative departures in the vicinity of the major resonance at $h\nu \sim 23$ eV and better agreement away from this main resonant peak, which tends to support the idea that the one-electron channels are exhibiting strong channel interaction enhanced near the t_{2g} shape resonance.

In fig. 38, the β computed for the $1t_{2g}$ channel agrees satisfactorily with the data. The measured branching ratio also agrees well with the calculations, regardless of how the assignments for peaks 2 and 3 are chosen. The base level of the partial cross section also agrees well with the calculated curve, although significant enhancements exist at threshold and at $h\nu \sim 23\text{--}24$ eV, i.e., where large peaks occur in the total cross section, as noted above. We therefore conclude, in the context of the above discussion, that the dynamical information is consistent with the assignment of peak 1 to ionization of the $1t_{1g}$ valence orbital with significant coupling near the strong t_{2g} resonance in the $1t_{2u} + 5t_{1u}$ channel. Note that the branching ratio and partial cross section give rather different overall impressions about the agreement between experiment and theory. This arises since a small difference in the branching ratio can be amplified in the partial cross section by a large peak in the total ionization cross section. Moreover, differences in wavelength scale and bandwidth between the total-ionization and photoelectron measurements can produce artificial structure, although this is not believed to be a problem with the broad structures involved here.

In fig. 39, the comparisons with different assignments do not immediately suggest that the recommended solid curve $(5t_{1u} + 1t_{2u})$ agrees better with the data. However, the following points offer some support. First, beyond the resonance peak, $h\nu > 26$ eV, the data fall closest to the solid line. Second, this is also true at higher energies, e.g., $h\nu \sim 50$ eV, where peaks 2 and 3 have cross sections of ~ 16 Mb and ~ 6 Mb, respectively (Gustafsson 1978) which is reasonably in agreement with the solid line which goes to 13 and 5.5 Mb for peaks 2 and 3, respectively (Levinson et al. 1979, Wallace 1980). Third, only the solid curve exceeds the experimental peak which, as stated above, is most often found in such comparisons. It should be mentioned that the $1t_{2u}$ is responsible for $\sim 2/3$ of the cross section in the peak, as indicated by the dash–dot curve so that its presence in the most intense channel (peak 2) is strongly suggested. Note also that the excess of theoretical cross section over experimental cross section roughly equals the magnitude of the resonant enhancement in peaks 1 and 3 at $h\nu \sim 23$ eV.

In fig. 40, the calculations all badly fail to account for major aspects of the data. The β, branching ratio, and partial cross section data depart qualitatively from the solid curves, particularly near $h\nu \sim 23$ eV. At the highest energy, however, they begin to converge with the solid curves, and at $h\nu \sim 50$ eV the $3e_g$ cross section is ~ 6 Mb (Gustafsson 1978), in good agreement with the calculated value of 5.5 Mb (Levinson et al. 1979, Wallace 1980). We therefore ascribe the enhanced cross section of peak 3 at $h\nu \sim 23$ eV to intensity borrowing from the intense, nearly degenerate channel represented by peak 2. This assignment is made difficult by the fairly good agreement between the dash–dot curve and the cross section data and the β data in fig. 40; however, adoption of the dash–dot convention is in direct opposition to LaVilla's decisive argument based on X-ray emission data (LaVilla 1972).

In fig. 41, the overall agreement with the solid curve is very good, lending dynamical support to the assignment of peak 4 to ionization of the $1t_{2g}$ MO. In fig. 42, the appearance of the resonant enhancement at ~ 5.5 eV kinetic energy indicates the action of the t_{2g} shape resonance in this channel and supports its assignment to $4t_{1u}$ ionization as suggested by Gustafsson (1978) and others.

To summarize, having examined diverse evidence concerning valence-shell photoionization of SF_6, one is led to conclude that the valence configuration

$$5a_{1g}^2 4t_{1u}^6 1t_{2g}^6 3e_g^4 (1t_{2u}^6 + 5t_{1u}^6) 1t_{1g}^6 \, ^1A_{1g}$$

is most consistent with the most definitive evidence. We note that apparent contradictions, such as the comparison of the data and the solid curve in fig. 40, challenge these conclusions. However, these contradictions are based on an independent-electron picture of valence-shell photoionization in SF_6. For reasons stated above, the evidence leads one to postulate (Dehmer et al. 1982) strong channel interaction in the vicinity of the very intense t_{2g} shape resonance in the $5t_{1u} + 1t_{2u}$ channel (peak 2) at $h\nu \sim 23$ eV. If this interpretation is correct, it reconfirms that shape-resonant features can be most easily identified in inner-shell spectra, whereas their role in valence-shell spectra can be significantly affected not only by the increased energy dependence of the dipole matrix element, but also by the possibility of strong channel interaction between the more closely spaced optical channels. This discussion should not be taken as conclusive on these issues as stressed by Dehmer et al. (1982). Clearly, more work tailored to this problem area needs to be carried out: Experimentally, it would be beneficial to extend triply differential measurements such as those reported here into the soft X-ray range, say up to $h\nu \sim 150$ eV, in order to avoid the strong channel interactions at lower energy. Gustafsson (1978) reported partial cross sections up to $h\nu \sim 50$ eV, which do, in fact, tend to support most of these conclusions, although they also raise additional interesting questions concerning the failure to clearly observe the strong e_g shape resonance at ~ 15 eV kinetic energy. Similar arguments to those discussed above may apply to this problem as well. In addition, high-energy, narrow shape resonances have been found (Swanson et al. 1980, 1981b, Lucchese et al. 1982) to be significantly smeared out by nuclear motion, which would be especially important for this resonance in SF_6. All these interesting aspects notwithstanding, it would be very useful to move into a region where such effects were absent in order to confirm important underlying assignments. On the theory side, it is imperative to begin examining channel interaction and vibrational effects in this and similar systems. Owing to the complexity of SF_6, this is probably only feasible at this time in connection with extensions (Stephens and Dill 1985) of the multiple scattering model, used here for independent-electron, fixed nuclei results. In any case, this case study should help stimulate some of this much needed advancement of present capabilities, since issues such as those raised here will surely be frequently encountered in the growing body of work in valence-shell photoionization of polyatomics using synchrotron radiation.

5.6. Valence-shell photoionization of BF_3

In this section, we present angle-resolved photoelectron data (J.L. Dehmer et al. 1984) for the valence shells of BF_3 to investigate valence-shell photoionization dynamics in this highly-symmetric polyatomic molecule. This case was chosen because an e' shape resonance is firmly established (Fomichev 1967, Fomichev and Barinskii 1970, Hayes and Brown 1971, Dehmer 1972, Mazalov et al. 1974, Swanson et al. 1981a, Ishiguro et al. 1982) to occur at ~2.2 eV kinetic energy in the boron K-shell spectrum and because comprehensive independent-electron calculations (Swanson et al. 1981a) have been carried out for all the subshells of BF_3 using the multiple-scattering model. Furthermore, as the previous case study on SF_6 clarified the role of the t_{2g} resonance in valence-shell photoionization in SF_6, the comparative study (J.L. Dehmer et al. 1984) of these two highly symmetric fluorides seemed promising. In fact, agreement between experiment and theory is very reasonable in many of the comparisons discussed below, indicating a realistic, first-order theoretical description (Swanson et al. 1981a). However, a predicted shape resonance feature in the branching ratio for the 4e' channel is absent, possibly due to some of the reasons touched upon in the last few sections. These results are discussed in the context of the analogous study on SF_6 and future measurements are suggested to clarify the role of the e' resonance in valence-shell spectra of BF_3.

Figure 43 shows a typical set of data taken at a photon energy of $h\nu = 23$ eV. All three spectra are normalized so that the largest peak (third peak in the $\theta = 0°$ spectrum) has a value of 100. In the top frame, the six peaks are labeled by the symmetry of the orbital being ionized, based on the well-established valence

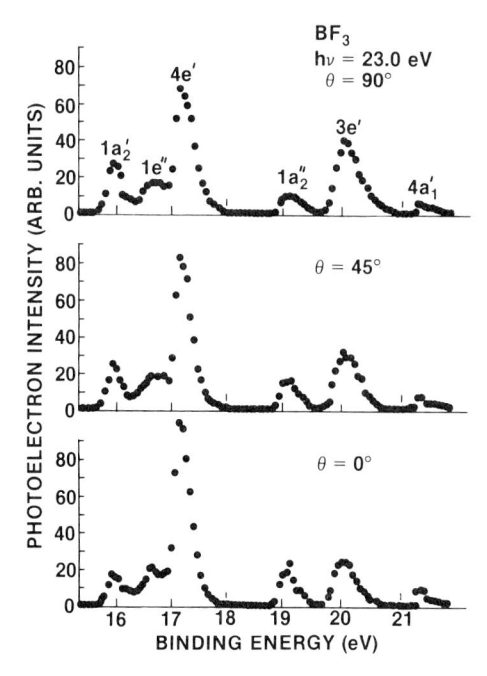

Fig. 43. Photoelectron spectra of BF_3 at $h\nu = 23$ eV and $\theta = 0°$, 45°, and 90°. The normalization of the three spectra is internally consistent and set so that the maximum count rate (third peak of the $\theta = 0°$ spectrum) is equal to 100.

configuration (Potts et al. 1970, Batten et al. 1978, Kimura et al. 1981, Åsbrink et al. 1981, Haller et al. 1983). By careful inspection of fig. 43, one can see that, at this wavelength, the β's for peaks 1 and 5 are negative, the β for peak 2 is nearly isotropic, and the β's for peaks 3 and 4 are positive.

At each angle of observation, the net counts in each photoelectron peak were summed, and the integrated counts were corrected for the transmission function of the electron spectrometer and a small <4% angular correction factor based on the aforementioned electron spectrometer angular calibration. The asymmetry parameter β was then determined for each peak, followed by the determination of the photoionization branching ratios from the measured intensities and β values. Note that in fig. 43, the 1e″ and 4e′ photoelectron bands (peaks 2 and 3) are not clearly resolved. In fact, recent theoretical work by Haller et al. (1983) indicates that a tail from the 1e″ band runs under the 4e′ band. As the actual shapes of the two bands are not known, the data is presented in two ways: In one approach, the partially resolved band has been deconvoluted simply by separating the two peaks at the point of minimum intensity. In the second approach, the sum of the two peaks is reported as a composite photoelectron band. This will be discussed below.

There are six occupied valence orbitals in BF_3 with ionization potentials in the energy range discussed in this case study. Starting with the outermost orbital, from the left in fig. 43, the symmetries and vertical IP's are as follows: $1a_2'$ (15.96 eV), 1e″ (16.70 eV), 4e′ (17.12 eV), $1a_2''$ (19.14 eV), 3e′ (20.12 eV), and $4a_1'$ (21.4 eV). Here the symmetry assignments are taken from Haller et al. (1983) and agree with most previous assignments (Potts et al. 1970, Batten et al. 1978, Kimura et al. 1981, Åsbrink et al. 1981). The vertical IP's are an average of several independent measurements (Potts et al. 1970, Batten et al. 1978, Kimura et al. 1981, Åsbrink et al. 1981) all of which are in close agreement.

The results are presented in figs. 44–50 for each photoelectron peak and for a combination of the partially resolved 1e″ and 4e′ peaks (fig. 47). Included with the experimental data are results of recent theoretical calculations (Swanson et al. 1981a) employing the multiple-scattering model. The theoretical curves have been adjusted to correspond to the level ordering and the IP's listed in the last paragraph. Note that Swanson et al. (1981a) adopted the incorrect experimental ordering for the closely spaced 4e′ and 1e″ levels. The theoretical curves have not been folded with the finite instrumental resolution, but this does not affect the present comparison in any significant way. Each figure consists of three frames. The top frame presents the photoelectron asymmetry parameter, β, from the IP up to $h\nu = 30$ eV. The middle frame shows the photoelectron branching ratio for each channel. In the lower frame, the calculated partial cross section is displayed. Unfortunately, the total absorption cross section of BF_3 is not known in this wavelength range, so we are unable to convert the measured branching ratios to partial cross sections. A photoionization mass spectrometry measurement (Dibeler and Liston 1968) was made up to $h\nu \sim 20$ eV, but this wavelength range is too limited to be very helpful in the present discussion.

In the top frames of figs. 44–50, we see the degree to which the measured β's

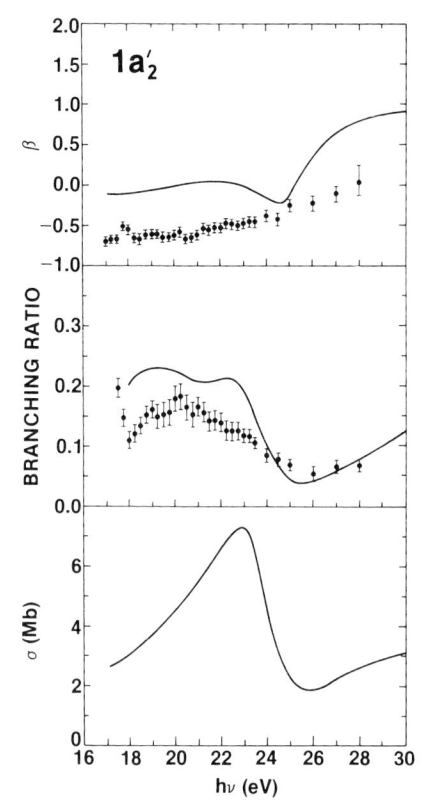

Fig. 44. Photoelectron asymmetry parameter, branching ratio, and partial cross section for photoionization of the $1a_2'$ orbital of BF_3. Solid dots: data from J.L. Dehmer et al. (1984); solid curves: theoretical results from Swanson et al. (1981a).

agree with the predictions of the multiple scattering calculation. The results for the $1e''$, $4e'$, and $1a_2''$ orbitals (figs. 45–48) show excellent agreement between experiment and theory. For the $1a_2'$ (fig. 44) and $3e'$ (fig. 49) orbitals, the experiment and theory agree fairly well in shape, but the magnitudes are different by ~0.5 β units on the average, a difference not uncommon even in much simpler molecules. The poorest agreement is found for the $4a_1'$ orbital (fig. 50), which is also by far the weakest channel. On the whole, the agreement is satisfactory, in view of the standards in the field, and it indicates that the theoretical results realistically reflect the gross photoionization dynamics of BF_3. In comparing the present results with those for SF_6 (section 5.5), it is interesting to note that the present β's tend to be rather anisotropic (ranging from $\beta < -0.5$ to $\beta \sim 1.5$), whereas those for the valence orbitals of SF_6 tended to gravitate strongly toward the isotropic value $\beta = 0$. This is not surprising, but it does show that the very simple isotropic pattern for F 2p derived orbitals in SF_6 is not in any sense typical of highly coordinated fluorides.

The branching ratios are shown in the middle frames of figs. 44–50. For the $1a_2'$

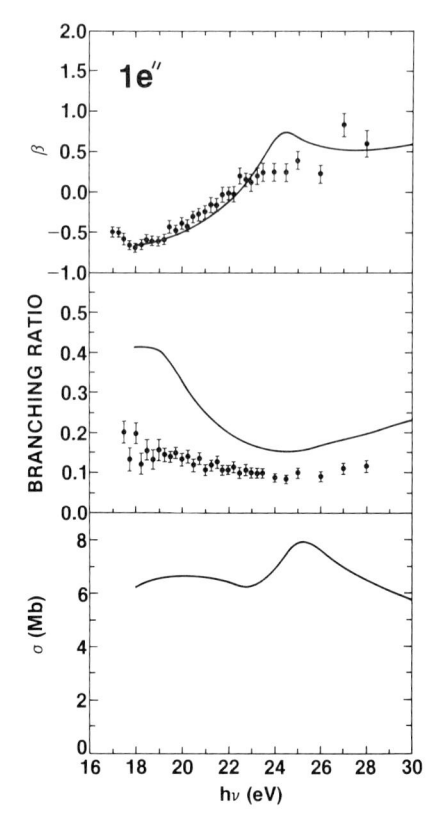

Fig. 45. Photoelectron asymmetry parameter, branching ratio, and partial cross section for photoionization of the 1e″ orbital of BF_3. Solid dots: data from J.L. Dehmer et al. (1984); solid curves: theoretical results from Swanson et al. (1981a).

(fig. 44) and 1a″$_2$ (fig. 48) cases, good agreement between theory and experiment is observed, both in shape and magnitude. For the overlapping 1e″ and 4e′ orbitals, two discrepancies between theory and experiment emerge. Most obvious, the bump at $h\nu \sim 25$ eV in the calculated branching ratio for the 4e′ orbital does not appear in the measured branching ratio. This will be discussed further below. The other discrepancy occurs on either side of this bump, where the measured branching ratios are lower than the theoretical curve for 1e″ and higher than the theoretical curve for the 4e′ orbital. The reason this is noteworthy is that this is consistent with the results of Haller et al. (1983) who predict that the 1e″ photoelectron band runs under the 4e′ band and that a sizable fraction of its intensity is thereby covered up by the 4e′ band. Our method of separating the intensity of the overlapping bands would have the effect of erroneously shifting intensity from the 1e″ peak to the 4e′ peak. In fact, when the two are summed in fig. 47, the agreement away from the $h\nu \sim 25$ eV bump is remarkably good, adding some support to the prediction by Haller et al. (1983). Note that the separately determined β's for these two channels should be much less sensitive to

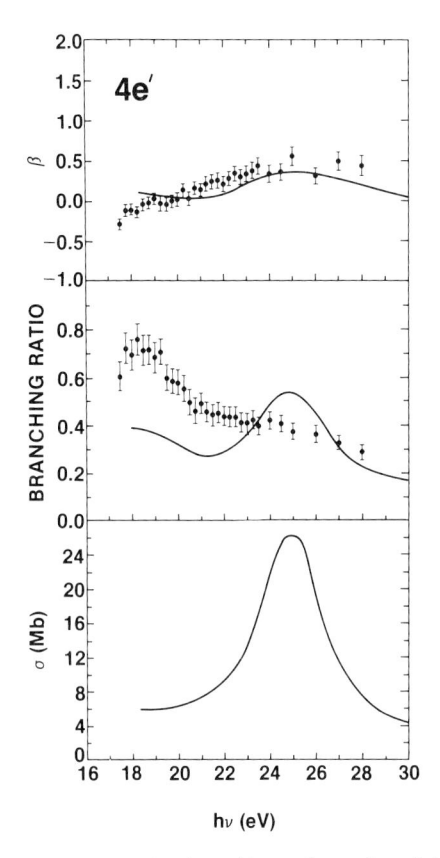

Fig. 46. Photoelectron asymmetry parameter, branching ratio, and partial cross section for photoionization of the 4e' orbital of BF$_3$. Solid dots: data from J.L. Dehmer et al. (1984); solid curves: theoretical results from Swanson et al. (1981a).

this issue. The remaining two channels reflect rather good agreement between theory and experiment. The 3e' branching ratio, in fact, reflects a maximum, similar in magnitude to, but slightly shifted from that in the theoretical curve. The 4a$_1'$ branching ratio agrees well in shape with the theoretical curve, both reflecting a sharp increase at high energy. The factor of two error in magnitude is not surprising in view of the very weak intensity in this channel just above its IP.

The bottom frames in figs. 44–50 contain the partial cross sections produced by the theoretical calculation (Swanson et al. 1981a). Comparison with experiment will require measurement of a total photoabsorption cross section, which, when multiplied by the present branching ratios, would yield experimental partial cross sections. Or, direct measurement by constant-ionic-state photoelectron spectroscopy would produce the needed experimental data. As neither is presently available, the theoretical curves are included for purposes of discussion, as partial cross sections and branching ratios present rather different views of the photoionization process. We anticipate ourselves by noting that the most definitive evidence

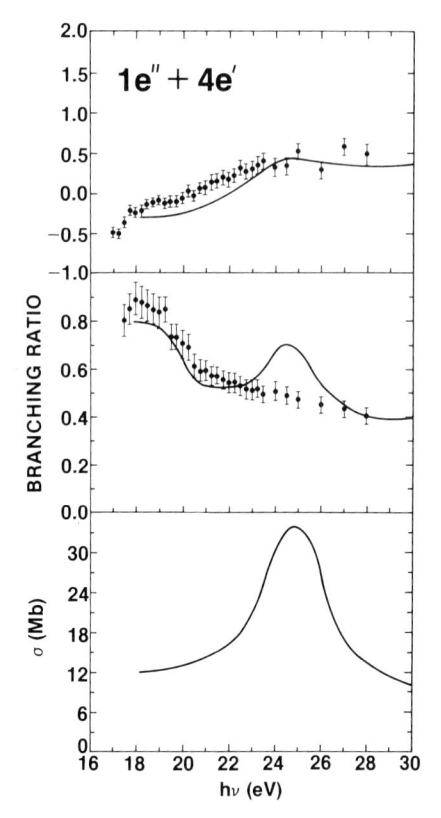

Fig. 47. Photoelectron asymmetry parameter, branching ratio, and partial cross section for photoionization of the 1e″ and 4e′ orbitals of BF_3. Solid dots: data from J.L. Dehmer et al. (1984); solid curves: theoretical results from Swanson et al. (1981a).

for the e′ shape resonance in valence-shell photoionization is likely to result from measurement of the 4e′ partial cross section.

The boron K-shell X-ray absorption spectrum (Fomichev 1967, Fomichev and Barinskii 1970, Hayes and Brown 1971, Dehmer 1972, Mazalov et al. 1974, Robin 1975, Swanson et al. 1981a, Ishiguro et al. 1982) displays two prominent features – an intense peak ~7 eV below the IP, and a broad (FWHM ~ 4 eV), intense peak centered at ~2.2 eV above the IP. Recent multiple-scattering calculations (Swanson et al. 1981a) show that these features can be understood at the independent-electron level and that they correspond to transitions to a $2a_2''$ bound state and an e′ shape resonance, respectively, in accordance with other interpretations (Fomichev 1967, Fomichev and Barinskii 1970, Hayes and Brown 1971, Dehmer 1972, Mazalov et al. 1974, Ishiguro et al. 1982). Since the e′ shape resonance is a final-state feature, it should also be accessed in symmetry-allowed transitions from the valence shells. One of the primary motivations of this work was to investigate the role of the e′ resonance in valence-shell photoionization dynamics of BF_3. Indeed, calculations (Swanson et al. 1981a) show that five of the

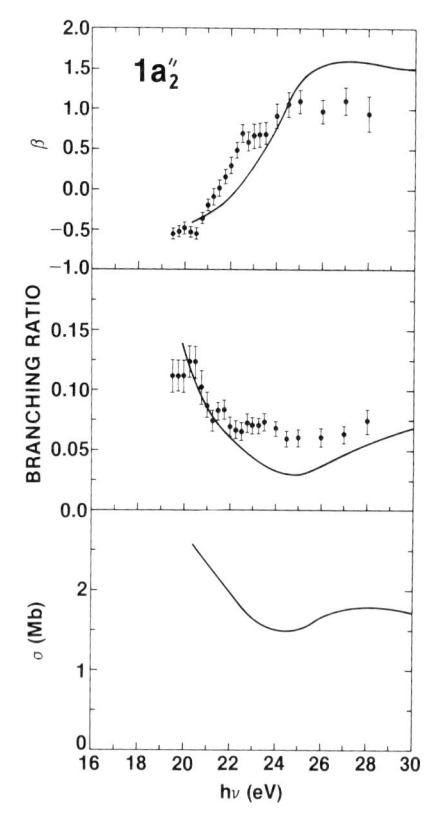

Fig. 48. Photoelectron asymmetry parameter, branching ratio, and partial cross section for photoionization of the $1a_2''$ orbital of BF_3. Solid dots: data from J.L. Dehmer et al. (1984); solid curves: theoretical results from Swanson et al. (1981a).

six valence orbitals of BF_3 (all but the $1a_a''$) are connected to the e' continuum by dipole selection rules and, further, that the e' is predicted to have clearly visible effects. As so often happens, however, the valence-shell properties do not follow the independent-electron predictions as clearly as do the inner-shell properties. Nevertheless, in this case, rather good, though indirect, evidence for the e' shape resonance is given by the β's, and the 3e' and $4a_1'$ branching ratios show direct evidence near the upper limit of the energy range. However, a predicted peak in the 4e' branching ratio is missing in the data, indicating the presence of interactions which are not adequately incorporated in the calculation.

Accordingly, it is important to have an independent way of estimating the location of the e' shape resonance in the valence-shell continua, so as to establish the presence or absence of such effects. Fortunately, this can be fairly reliably done based on the position of the e' shape resonance in the inner-shell spectra, plus a kinetic energy shift associated with differences in screening between a localized hole and a valence-shell hole. Examining well-characterized cases in N_2, CO, and SF_6, it is found that shape resonances experience a shift to higher kinetic

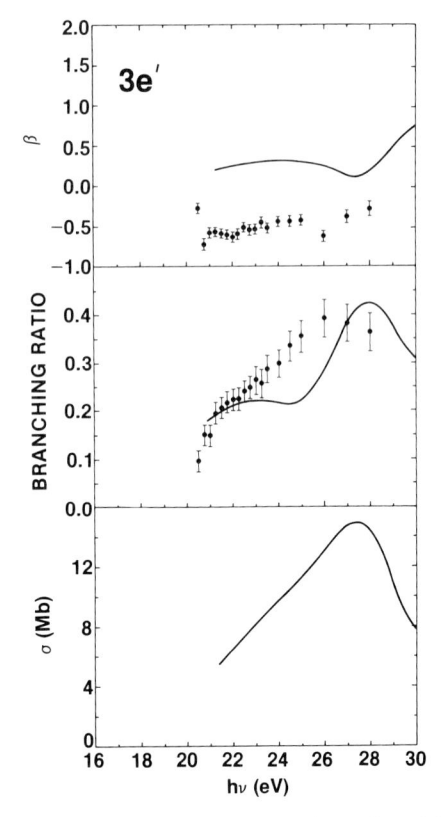

Fig. 49. Photoelectron asymmetry parameter, branching ratio, and partial cross section for photoionization of the 3e' orbital of BF_3. Solid dots: data from J.L. Dehmer et al. (1981a); solid curves: theoretical results from Swanson et al. (1981a).

energies when going from an inner-shell spectrum to a valence-shell spectrum. These shifts cluster around ~3 eV and always fall in the range 1–4 eV. Therefore, the e' shape resonance, which is centered at 2.2 eV kinetic energy in the boron K-shell spectrum, should fall in the 3–6 eV kinetic energy range in the valence-shell spectra. As the calculation quoted here placed the resonance at ~8 eV kinetic energy, the true resonance position should fall to the low-energy side of the predicted position. This means that the photon energy range studied in this work should suffice to investigate the role of the e' shape resonance in the six channels studied, though the resonance position in the 3e' and $4a'_1$ channels falls near the high-energy limit of the data reported here.

For completeness, we mention other states that will influence the valence-shell photoionization of BF_3 in this energy range. Multiple scattering model calculations (Swanson et al. 1981a) also predict a shape resonance in the a'_1 continuum, approximately 1–2 eV below the e' shape resonance. The a'_1 resonance derives from the trapping of p-waves on the fluorine sites, as does the e' resonance, but it is not dipole-allowed in boron K-shell photoexcitation and, hence, does not arise

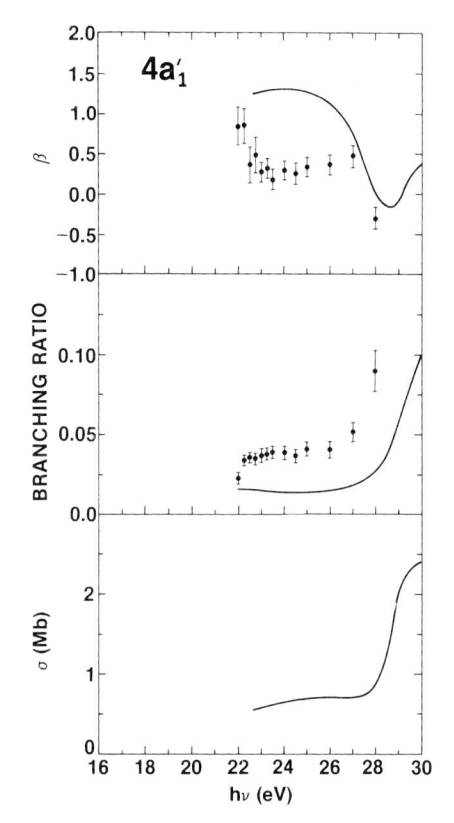

Fig. 50. Photoelectron asymmetry parameter, branching ratio, and partial cross section for photoionization of the $4a_1'$ orbital of BF_3. Solid dots: from J.L. Dehmer et al. (1984); solid curves: theoretical results from Swanson et al. (1981a).

in earlier discussions involving inner-shell processes. Among the valence shells, the a_1' continuum is dipole-allowed from the 4e′, $1a_2''$, and 3e′ initial states. According to the theoretical results (Swanson et al. 1981a), the a_1' resonance is masked by a much more intense e′ resonance in the 4e′ channel and is suppressed by a coincident zero in the dipole matrix element in the $1a_2''$ channel. However, in the 3e′ channel, it is equal in strength and shifted slightly from the e′ shape resonance. Therefore, the net resonant feature in the 3e′ channel must be considered a composite resonance with significant contributions from shape resonances in both the e′ and a_1' continua. However, as the e′ resonance is sharper, it still determines the peak position of the combined resonance feature in the 3e′ partial cross section. For this reason, we will continue to refer to the e′ resonance in what follows, although the likely contribution from the a_1' resonance should be recognized.

Another possible class of states to consider is autoionizing states converging to all but the lowest IP. Although we do not detect any narrow structure in the data that would indicate autoionizing structure, channel interaction with the Rydberg states converging to the 1e″ through $4a_1'$ thresholds may contribute to the failure to see features predicted from independent-electron calculations. Of particular interest is the possibility that transitions to the $2a_2''$ state (essentially a boron $2p_z$ orbital), so prominent in the boron K-shell spectra (Fomichev 1967, Fomichev and Barinskii 1970, Hayes and Brown 1971, Dehmer 1972, Mazalov et al. 1974, Robin 1975, Swanson et al. 1981a, Ishiguro et al. 1982), might affect the present data. In the K-shell spectra, this peak is ~ 7 eV below the IP. If we assume a 2–3 eV shift for the valence-shell spectra, this state should not affect the present range, except possibly in the $4a_1'$ channel where it would fall at $h\nu \sim 17$ eV, to a first approximation. However, as known (Levine and Soven 1983, 1984) from examples in both atoms and molecules, intrashell or intravalence transitions can undergo significant splitting and redistribution of oscillator strength to higher energies, relative to an independent-electron picture. Although these possibilities must be recognized in considering the experimental results, we have no particular reason to believe that they play an important role in this case.

We now examine the data in figs. 44–50 for the effects of the e′ shape resonance. In doing so, we examine separately the β's, branching ratios, and partial cross sections, as each reflects the photoionization dynamics in a different way. The β's differ from the other two in that they contain information on the relative phases of the continuum wavefunctions. However, as seen in an earlier study (Dehmer et al. 1982) of valence-shell photoionization of SF_6 (section 5.5), the branching ratios and partial cross sections also differ greatly in the way they display photoionization features. In the upper frames of figs. 44–50, the gross shapes of the measured curves agree reasonably well with theory, and the agreement is excellent, in shape and magnitude, for the 1e″, 4e′, and $1a_2''$ channels. The e′ shape resonance plays significant roles for all but the $1a_2''$ channel, so that one is tempted to consider this indirect evidence that the role of the e′ shape resonance in this spectral range is observed and reasonably accounted for by the calculations. Turning to the branching ratios, we note fair to good agreement between experiment and theory for all except the 4e′ channel. In particular, the broad maximum at $h\nu \sim 26$ eV in the 3e′ branching ratio and the rising branching ratio at high photon energy for $4a_1'$ represent direct evidence for the e′ shape resonance in those channels. Using the 3e′ branching ratio data, one can place the resonance position at $h\nu \sim 26$ eV, corresponding to a kinetic energy of ~ 6 eV, in accordance with expectations. The surprising aspect of these results is the absence of an e′-induced peak at $h\nu \sim 25$ eV in the 4e′ branching ratio. This will be discussed further below. The partial cross sections in the bottom frames are presently available from theory only. They show that the e′ shape resonance will emerge much more clearly when presented in this form. In particular, the peak at $h\nu \sim 25$ eV in the 4e′ partial cross section is preducted to have a much greater contrast ratio than the same feature in the branching ratio. Similarly, the e′ resonance will be displayed more clearly in the partial cross section than in the branching ratio for the 1e″ and the $1a_2'$ channels as well. For the 3e′ and $4a_1'$ channels, both parameters display the resonance equally clearly, and, in fact, these are the two cases in which the e′ shape resonance can be observed in the present branching ratio data. Recall that in the 3e′ channel, the a_1' shape resonance also contributes to the resonant feature at $h\nu \sim 26$ eV, as discussed earlier.

The failure to observe in valence-shell properties a final-state resonance that is well established in inner-shell spectra is not at all unprecedented. In SF_6, an e_g shape resonance causes an intense peak ~ 15 eV above the sulfur $L_{2,3}$ IP, but is absent from valence-shell partial cross sections where its presence is predicted by theory (Levinson et al. 1979, Wallace 1980). An even more subtle example is the $4\sigma_g$ ionization channel of CO_2 which was predicted (Grimm et al. 1980, Swanson et al. 1980, Padial et al. 1981a, Swanson et al. 1981b, Lucchese and McKoy 1982a,b) to access a strong σ_u shape resonance at ~ 20 eV kinetic energy. Nothing resembling the predicted resonance feature was observed in early partial cross section measurements (Brion and Tan 1978, Gustafsson et al. 1978a); however, a predicted dip in the β curve (Grimm et al. 1980, Swanson et al. 1981b, Lucchese and McKoy 1982a,b) at the resonance energy was subsequently observed experimentally (Carlson et al. 1981a), giving evidence that the "missing" resonance existed. More recently, the resonance has been observed (Roy et al. 1984) in the partial cross section, only shifted several eV and smeared out relative to predictions. Thus, its manifestation in the partial cross section has been drastically reduced and shifted by some as yet uncharacterized interaction(s). So we add to this list the 4e′ channel in BF_3 which has a β curve consistent with theory but fails to show the branching ratio feature resulting from the presence of the e′ shape resonance.

Without further evidence, we can only speculate as to possible causes for the missing e′ feature: First, it is well-known that multiple-scattering model calculations produce shape resonance profiles that are too narrow and too intense. Hence, effects such as intrachannel coupling (Levine and Soven 1983, 1984) and/or averaging over vibrational motion (Swanson et al. 1981b) will tend to smear out and diminish a shape resonant feature. Nevertheless, the feature may still be observable in the partial cross section, even if it is absent from the branching ratio. This is certainly possible in the present case since, if the total photoabsorption cross section peaks near $h\nu \sim 25$ eV, a flat branching ratio will produce a peak in the partial cross section. Second, interchannel coupling (either discrete–continuum or continuum–continuum) with other underlying valence channels can significantly alter the predictions based on an independent-electron theory. For instance, continuum–continuum coupling is known to have dramatic effects in $2\sigma_u$ photoionization in N_2 (Stephens and Dill 1985), and is believed to strongly influence valence-shell photoionization in SF_6 (Dehmer et al. 1982). Such channel interaction can be expected to be stronger among valence channels with their closely spaced IP's. The good agreement found for the β results tend to argue against this possibility. Note that in the N_2 case, the $2\sigma_u$ β is strongly affected (Stephens and Dill 1985), and in SF_6, the β results (Dehmer et al. 1982) showed very poor agreement between experiment and theory for the affected channels. Third, vibronic coupling has been shown (Haller et al. 1983) to play a very important role in photoelectron spectra of BF_3. This and other vibrational effects are excluded from the calculations (Swanson et al. 1981a) quoted here. We do not know how to assess the importance of such effects at this time. Other possibilities clearly exist, but these examples serve to indicate the types of mechanisms which may be causing the reduction of certain shape resonance effects in valence-shell spectra. Taken together, the examples quoted in CO_2, BF_3, and SF_6 pose a major challenge to our understanding of shape resonance phenomena.

We conclude by suggesting future work to help clarify the role of the e' shape resonance in BF_3 photoionization processes. Clearly it is very important to measure the partial photoionization cross sections for the valence shells to complete the comparisons begun in figs. 44–50. This would require the total photoabsorption cross section in order to convert the present branching ratios to partial cross sections, or that a constant-ionic-state photoelectron measurement be made on the valence shells of BF_3. New measurements at higher energy would also be very valuable, both to complete the study of the e' features at the high-energy limit of the present data, and to investigate the role of the e' resonance in the inner-valence $3a_1'$ and $2e'$ orbitals whose IP's are predicted to fall near $h\nu \sim 40$–43 eV.

6. Survey of related work

Molecular photoionization dynamics is a rich, multifaceted subject, requiring an eclectic approach to gain the greatest insight from the many complementary sources of information. In the main body of this chapter we have focussed on the use of triply differential photoelectron spectroscopy to probe the dynamics of two central resonant mechanisms – autoionization and shape resonances. We have further stressed the use of synchrotron radiation in six case studies to illustrate important topics of current interest. What we have failed so far to do is convey the diversity and extent of information from other techniques and for other physical circumstances. In this section we will now try to create a much broader perspective by noting very briefly various types of work which complement in an essential way what we have covered in detail. Even at this superficial level, we stress that this survey is not at all comprehensive, but, rather, is intended to give a general impression of the richness of the field.

(i) As indicated in the Introduction and in the shape resonance bibliography (see Appendix), the study of shape resonances has been very vigorous and productive over the last decade or so. Using a variety of probes (synchrotron radiation, electron energy loss spectroscopy, X-ray sources, electron spectroscopy, mass spectroscopy, fluorescence spectroscopy, etc.) shape resonances in well over fifty molecules have been studied. These include simple diatomics (N_2, O_2, CO, NO), triatomics (e.g., CO_2, CS_2, OCS, N_2O, HCN) and more highly-coordinated molecules and local molecular environments (e.g., SF_6, SO_4^{2-}, SF_5CF_3, SF_2O_2, SF_2O, BF_3, SiF_4, $SiCl_4$, SiF_6^{2-}, SiO_2, NF_3, CF_4, CCl_4, C_2H_2,

C_2N_2). There has also been extensive study of shape resonances in adsorbed molecules (Gustafsson et al. 1978b, Gustafsson 1980b, 1983, Stöhr and Jaeger 1982, Stöhr et al. 1983, 1984, Koestner et al. 1984) and molecular solids (Blechschmidt et al. 1972, Lau et al. 1982, Fock 1983, Fock et al. 1984, Fock and Koch 1984, 1985) which exploit the localized nature of shape resonances to probe the condensed state. Several important themes are developed in the literature on shape resonances which space did not permit covering here. These include electron optics of molecular fields (Dehmer and Dill 1979b), hole localization (Dill et al. 1978, 1979a) and relaxation effects (Lynch and McKoy 1984), Auger angular distribution anisotropies (Dill et al. 1980, Lindle et al. 1984, Truesdale et al. 1984), triply differential studies of deep inner-shell photoionization (Ferrett et al. 1986), chemical effects on shape resonances (many examples in the bibliography), as well as others highlighted below in other contexts. A somewhat related theme has involved the study of so-called "Cooper zeros" in molecular photoionization (Carlson et al. 1982b, 1983a,c, 1984a,c). These minima in the ionization cross section are well known in atomic physics (Fano and Cooper 1968, Starace 1979, Johnson and Cheng 1979, Kennedy and Manson 1972) and result from a cancellation in the dipole matrix element due to the passage of a final state node through the valence-shell part of the initial-state wavefunction. In several examples, e.g., valence p-shell photoionization in Ar, Kr, and Xe (Johnson and Cheng 1979, Kennedy and Manson 1972), the node of interest immediately follows the resonantly penetrating antinode which causes the shape resonance in those channels. In the molecular cases studied (Carlson et al. 1982b, 1983a,c, 1984a,c) so far, photoionization of halogen compounds have been treated and analyzed with respect to this closely related rare gas behavior.

(ii) As stated earlier, the literature on autoionization is extensive; however, detailed *ab initio* theoretical work was largely limited to prototypical work on H_2 until a few years ago. More recently, major progress has been made in larger molecules, e.g., N_2 (Lefebvre-Brion and Giusti-Suzor 1983, Raoult et al. 1983, Giusti-Suzor and Lefebvre-Brion 1984), O_2 (Morin et al. 1982a), NO (Collins and Schneider 1984, Giusti-Suzor 1982, Giusti-Suzor and Jungen 1984, Jungen 1984a), HI (Lefebvre-Brion et al. 1985), and C_2H_2 (Levine and Soven 1983, 1984). To date, various authors have treated rotational (Jungen and Dill 1980, Raoult et al. 1980), vibrational (Jungen and Dill 1980, Raoult et al. 1980, Raoult and Jungen 1981), electronic (Levine and Soven 1983, 1984, Collins and Schneider 1984, Raoult et al. 1983, Giusti-Suzor and Lefebvre-Brion 1984, Raseev 1985), spin–orbit (Lefebvre-Brion et al. 1985), and indirect autoionization mechanisms (or "complex resonances") (Giusti-Suzor and Lefebvre-Brion 1984), and the competition with predissociation has now been incorporated (Giusti-Suzor and Jungen 1984, Jungen 1984b, Mies 1984, Mies and Julienne 1984) in an MQDT treatment. On the experimental side, detailed triply differential photoelectron studies are now ripe for expansion with a few preliminary studies already complete, e.g., N_2 (Parr et al. 1981, West et al. 1981), CO (Ederer et al. 1981), CO_2 (Parr et al. 1982b, Hubin-Franskin et al. 1984), O_2 (Morin et al. 1980, Tabché-Fouhailé et al. 1981, Morin et al. 1982a,b, Codling et al. 1981), N_2O (Carlson et al. 1983b,

Truesdale et al. 1983b), C_2H_2 (Unwin et al. 1981, Keller et al. 1982, Parr et al. 1982a).

(iii) The interaction between shape resonances and autoionizing resonances in valence-shell spectra has been a most challenging problem. In the case of O_2, an analogous σ_u shape resonance to that discussed in connection with N_2 is expected to occur in the valence-shell spectra, but its identification in the photoionization spectrum has been complicated by the existence of extensive autoionization structure in the region of interest. Recent work (Morin et al. 1980, Tabché-Fouhailé et al. 1981, Morin et al. 1982a,b) using variable wavelength photoelectron measurements and an MQDT analysis of the principal autoionizing Rydberg series have sorted out this puzzle, with the result that the σ_u shape resonance was established to be approximately where expected, but was not at all clearly identifiable without the extensive analysis used in this case. Similar, but less well-analyzed examples occur in CO (see, e.g., Stockbauer et al. 1979, Stephens et al. 1981) and NO (see, e.g., Brion and Tan 1981, Delaney et al. 1982b, Southworth et al. 1982a, Wallace et al. 1982, Smith et al. 1983, Collins and Schneider 1984). Resolution of the joint shape resonance/autoionization dynamics in these and other cases is a current challenge.

(iv) A class of phenomena which appears to be very common in inner-valence-shell spectra is the breakdown of the single-particle model brought on by extensive vibronic coupling among the high density of states in the inner-valence region (Krummacher et al. 1980, 1983, Bagus and Viinikka 1977, Cederbaum and Domcke 1977, Schirmer et al. 1977, Cederbaum et al. 1977, 1978, 1980, Wendin 1981, Schirmer and Walter 1983). This breakdown manifests itself as a high density of satellites in the photoelectron spectrum (consisting of admixtures of single-hole states, two-hole one-particle states and certain higher-order combinations), to the extent that the main photoelectron line associated with a single inner-valence hole can even be difficult to recognize. This has been observed experimentally (Krummacher 1980, 1983) and treated successfully (Bagus and Viinikka 1977, Cederbaum and Domcke 1977, Schirmer et al. 1977, Cederbaum et al. 1977, 1978, 1980, Wendin 1981, Schirmer and Walter 1983), e.g., in the case of the "$2\sigma_g$" spectrum of N_2 by many-body Green's function techniques, and may be expected to be an important dynamical effect in the photoionization of molecular levels having IP's in the $h\nu \sim 30$ to 50 eV range.

(v) Multiphoton ionization is a recently developed technique which has great potential for expanding our understanding of photoionization dynamics in totally new directions. When used in conjunction with photoelectron detection, such as that described in section 4, vibrational intensities and photoelectron angular distributions can be measured for small molecules (see, e.g., Miller and Compton 1981a,b, Kimman et al. 1982, Miller et al. 1982, Miller and Compton 1982, Glownia et al. 1982, Achiba et al. 1982, White et al. 1982, Pratt et al. 1983a,b,c, Achiba et al. 1983, Anderson et al. 1984, Pratt et al. 1984a,b,c, White et al. 1984, Sato et al. 1984, Wilson et al. 1984, Müller-Dethlefs et al. 1984, Kimman 1984) which are windows onto the dynamics of the multiphoton process, just as they are in the single photon case. In addition, when the multiphoton process proceeds via resonances with excited neutral states, such as the excited valence states of N_2 in fig. 19, very high resolution ($\Delta\lambda < 0.05$ cm^{-1}) spectroscopy and dynamics of these excited molecular states can be examined. Although the laser sources used in multiphoton ionization are technically quite different from those used in the VUV and soft X-ray work emphasized in the body of this chapter, there is a strong scientific relationship which ties them together.

(vi) Another recently developed technique is polarization of fluorescence (Poliakoff et al. 1981, 1982, Guest et al. 1983, Greene and Zare 1983, Guest et al. 1984) from molecular ions formed by photoionization. This approach accesses information on the orientation of the molecular ion and the relative strengths of degenerate photoelectron channels in the photoionization process. To give a concrete example, photoionization of N_2 to form the B $^2\Sigma_u$ state of N_2^+ (fig. 19) leads to the rightmost photoelectron band in fig. 20. This band consists of electrons in degenerate $\varepsilon\sigma_g$ and $\varepsilon\pi_g$ ionization channels which cannot be separated by straight electron spectroscopy or by β measurements. Nevertheless, by observation of the polarization of the B–X fluorescence, information on the branching ratio for these degenerate photoelectron channels can be obtained (Poliakoff et al. 1981, Guest et al. 1983). This provides a unique test of dynamical information which exists but usually remains implicit in theoretical calculations. Another apparently dissimilar technique to examine relative strengths of degenerate photoionization channels is the measurement of angular distributions of photoions from dissociative photoionization (Dehmer and Dill 1978). The common link between these techniques is that they do not detect (and, hence integrate over the angular distribution of) the photoelectrons, thus eliminating the interference effects between the degenerate channels and isolating their relative strengths in the observed parameters.

(vii) An important extension of the triply differential studies discussed above is the extension to measuring the spin polarization of the photoelectrons. Spin polarization measurements access additional dynamical information described in detail elsewhere (Heinzmann 1980, Johnson et al. 1980). Full quadruply differential measurements have already been performed on atoms (Heinzmann 1980) and spin polarization studies have begun on molecules (Heinzmann et al. 1980, 1981, Cherepkov 1981a,b, Schäfers et al. 1983, Schönhense et al. 1984).

(viii) A variety of dissociative phenomena occur either in competition with or subsequent to molecular photoionization, and are an important part of the broader picture. For example, MQDT has been applied (Giusti-Suzor and Jungen 1984, Jungen 1984b, Mies 1984, Mies and Julienne 1984) to the competition between predissociation and autoionization in photoexcitation of H_2 and NO, representing the extension of MQDT analysis to dissociation channels and serving as a prototype for an extensive class of processes always present to some degree in molecular photoexcitation spectra. Another subject with a rich literature (see, e.g., Batten et al. 1978, Danby and Eland 1972, Brehm et al. 1973, Stockbauer 1973, Werner et al. 1974, Stockbauer and Inghram 1975, Peatman 1976, Mintz and Baer 1976, Batten et al. 1976, Stockbauer and Inghram 1976, Stockbauer 1977, Guyon et al. 1978, Peatman et al. 1978, Baer et al. 1979, Baer 1979,

Stockbauer 1980, Guyon et al. 1983) is fragmentation of molecular ions formed by photoionization. Usually studied by photoelectron–photoion coincidence techniques, this subfield focuses on the decay of excited molecular ions into alternative dissociation channels. In most work, a particular molecular ion state is selected by tuning the excitation energy so that a zero energy electron is produced, indicating that an ion state at an excitation energy equal to the photon energy is formed. Events producing zero electrons can be detected very efficiently either by an electrostatic zero-energy trap (Danby and Eland 1972, Brehm et al. 1973, Stockbauer 1973, Wernet et al. 1974, Stockbauer and Inghram 1975, Peatman 1976, Mintz and Baer 1976, Batten et al. 1976, Stockbauer and Inghram 1976, Stockbauer 1977, Guyon et al. 1978, Peatman et al. 1978, Baer et al. 1979, Baer 1979, Stockbauer 1980, Guyon et al. 1983) or by electron attachment to an electron scavenger (Chutjian and Ajello 1977, Ajello et al. 1980, Chutjian and Ajello 1980). In either case, measurement of the resulting fragment ions will indicate the fragmentation pattern of the parent ion. Alternatively, measurement of threshold electrons as a function of excitation energy (see, e.g., Kimura et al. 1981, Danby and Eland 1972, Brehm et al. 1973, Stockbauer 1973, Werner et al. 1974, Stockbauer and Inghram 1975, Peatman 1976, Mintz and Baer 1976, Batten et al. 1976, Stockbauer and Inghram 1976, Stockbauer 1977, Guyon et al. 1978, Peatman et al. 1978, Baer et al. 1979, Baer 1979, Stockbauer 1980, Guyon et al. 1983) map the spectroscopy of highly excited autoionizing states which decay to excited states of the ion by ejection of a thermal electron. This produces a source of information on the spectroscopy of highly vibrationally excited molecular ions which are otherwise difficult to observe; however, the detailed dynamics of these processes is often complicated and difficult to analyze.

7. Prospects for future progress

Discussing the future directions of a research field is an extremely limited exercise since it is inherently steeped in existing ideas and experience. Invariably, new ideas, insights, and/or techniques come along and redirect the field into unforeseen directions. Nevertheless, it is still useful to discuss prospects for future work if it is done briefly and not taken too seriously. We will do so here in terms of three themes: developing our prototypical concepts, new probes, and new systems.

Our present knowledge of shape and autoionizing resonances is still in the prototype stage of development. In the case of shape resonances, we have established deep insight for a broad range of phenomena in N_2, but attempts to extend this understanding to even closely related cases, e.g., CO, C_2H_2, O_2, NO, HCN, C_2N_2, has been met with new challenges, in many cases requiring extensive revision of our initial expectations. Moreover, serious study has only just begun on other important facets of shape resonant behavior: vibrational effects in polyatomic molecules, continuum–continuum coupling, interaction with autoionizing states, triply differential photoelectron studies of deep inner-shell spectra, and so on. In addition, the inventory of known shape resonances, though seemingly extensive, is only the tip of the iceberg and the depth of understanding in most cases is superficial. Exploration of shape resonances should remain vigorous, should explore new types of molecular environments, and should aim to establish the special dynamics in each case. A high order of challenge lies in the unification of different useful points of view and in understanding the fundamental similarities among different physical environments, excitation mechanisms, and observation channels.

The progress toward understanding autoionization phenomena is extremely impressive, yet it also is still in the development stage. Ultimate sophistication and insight has been achieved only for H_2, and attempts to extend this progress to other diatomics has made impressive progress, but has been arduous and has produced less complete results. Extension of theoretical work to larger molecules and a broader range of detection channels will be a high-priority and fruitful direction of work for many years to come. On the experimental side, only the most preliminary measurements have been carried out *within* autoionizing resonances at the triply differential level. Enhanced synchrotron light capabilities and more sophisticated instrumentation just coming on line will have a major impact on this fundamental type of measurement. Let us hope we will soon see a definitive experimental test of the MQDT predictions discussed in section 3.2.

Another widespread phenomena in need of further study is the breakdown of single particle behavior in inner-valence spectra. Prototype work succeeded in establishing the phenomena qualitatively; however, many theoretical predictions remain untested, and higher resolution is required to test the theoretical predictions in any detail.

These exemplars show clearly that expansion, refinement, and unification of molecular photoionization dynamics will provide a stimulating theme in the coming years.

New probes of molecular photoionization dynamics will invariably draw intense interest and stimulate new growth. At the time of this writing, several new approaches are in development and in preliminary use, but are not discussed in the literature to any significant extent. These include photoionization of laser excited molecular states (the atomic analog is much better established); photoionization of molecules oriented in a molecular beam (Kaesdorf et al. 1985); and the study of double ionization by measurement of both outgoing electrons in coincidence, either in the fast electron/slow electron case or in the Wannier limit of equal, low velocities. These are very exciting, new directions which will undoubtedly exhibit new dynamics. Their exploitation will benefit from new technology, e.g., undulators and free-electron lasers, and will stimulate new ideas and theoretical initiatives.

Nearly all work stressed in this chapter has dealt with standard molecular targets, partly owing to ease of handling and strong signal levels. With recent progress in synchrotron sources and the development of automated, more sensitive spectrometers, extension of these studies to more exotic species is ripe. It should now be straightforward to bring the broad spectral range and time structure of synchrotron radiation to bear on such species as clusters, free radicals, metastable states, and high-temperature molecules. There is strong interest in these species for both basic and applied reasons, and exploration of new species with a new experimental capability would seem to be a very enticing subject for future work.

In closing, we emphasize that, although this chapter is not at all comprehensive, we hope it does succeed in indicating the present richness of the field of molecular photoionization dynamics and the potential for significant progress in the future.

Appendix. A bibliography on shape resonances in molecular photoionization through early 1985

Reference	Molecule
1966	
LaVilla and Deslattes	SF_6
Zimkina and Fomichev	SF_6
1967	
Fomichev	BF_3
1969	
Nakamura et al.	N_2
1970	
Fomichev and Barinskii	BF_3, BCl_3
Nefedov	SF_6, $Cr(CO)_6$
1971	
Hayes and Brown	BF_3
Nakamura et al.	N_2, O_2, CO, SF_6
Zimkina and Vinogradov	review article
1972	
Blechschmidt et al.	SF_6
Cadioli et al.	BF_3
Dehmer	SF_6, BF_3, CS_2, SO_2, SO_4^{2-}, $SiCl_4$, SiO_2, SiF_4, SiF_6^{2-}, CF_3SF_5, SF_2O, SF_2O_2
El-Sherbini and van der Wiel	N_2, CO
Fano	review article
Gianturco et al.	SF_6
LaVilla	SF_6
van der Wiel and El-Sherbini	N_2, CO
Wight et al. (1972/73)	N_2, CO
1973	
Barinskii and Kulikova	SF_6, BF_3
LaVilla	fluoromethanes

Reference	Molecule
1974	
Dehmer	review article
Mazalov et al.	BF_3
Morioka et al.	NO
Sachenko et al.	SF_6
Tam and Brion	HCN
Vinogradov et al.	N_2
Wight and Brion (1974a)	CO_2, N_2O
Wight and Brion (1974b)	NO, O_2
Wight and Brion (1974c)	CF_4
Wight and Brion (1974d)	CS_2, COS
Wight and Brion (1974e)	$(CH_3)_2CO$
1975	
Dehmer and Dill	N_2
LaVilla	O_2, CO_2
Robin	CH_4, B_2H_6, BF_3, N_2, HCl, H_2S, PH_3, SiH_4, SiF_4
1976	
Davenport (1976a)	N_2, CO
Davenport (1976b)	N_2, CO, CO_2, H_2
Dehmer and Dill (1976a)	N_2
Dehmer and Dill (1976b)	review article
Dill et al.	N_2, CO
Eberhardt et al.	C_2H_2
Hamnett et al.	CO, N_2
Samson and Gardner	CO
Tronc et al.	CO, CH_4
Wight et al.	N_2, CO
1977	
Davenport	H_2, N_2, CO
Hitchcock and Brion	C_2H_2, C_2H_4, C_2H_6, C_6H_6
Kay et al.	N_2, CO
King et al.	N_2
Kondratenko et al.	N_2, CO
Langhoff	N_2, H_2CO
Langhoff et al.	H_2CO
Plummer et al.	N_2, CO
Rescigno and Langhoff	N_2
Samson et al. (1977a)	O_2
Samson et al. (1977b)	N_2
Schwarz et al.	NO_2

Reference	Molecule
1978	
Bianconi et al.	N_2, N_2O
Brion and Tan	N_2O, CO_2
Brown et al.	fluoromethanes
Dill et al.	N_2, CO
Gustafsson	SF_6
Gustafsson et al. (1978a)	CO_2
Hitchcock and Brion (1978a)	SF_6
Hitchcock and Brion (1978b)	chloromethanes
Hitchcock et al.	SF_6
Iwata et al.	N_2, CO, C_2H_2
Langhoff et al.	H_2CO
McCoy et al.	O_2
Padial et al.	CO
Rescigno et al.	N_2
Sasanuma et al.	SF_6
1979	
Barrus et al.	O_2, CO, CO_2, N_2O
Brion and Tan	N_2O, CO_2
Brion et al.	O_2
Dehmer and Dill (1979a)	review article
Dehmer and Dill (1979b)	review article
Dehmer et al.	N_2
Dill et al.	N_2, CO
Hitchcock and Brion (1979a)	HCN
Hitchcock and Brion (1979b)	HCN, C_2N_2
Hitchcock and van der Wiel	SF_6
Langhoff	review article
Langhoff et al.	O_2
Levinson et al.	SF_6
Marr et al.	N_2, CO
Stockbauer et al.	CO
Tronc et al.	CO, CH_4, CF_4, CO_2, COS, C_2H_2, C_2H_4
Wallace et al.	N_2, CO
1980	
Carlson et al.	N_2
Cole et al.	CO
Dehmer and Dill	review article
Dill et al.	N_2, CO
Friedrich et al.	SiF_4
Gerwer et al.	O_2

Reference	Molecule
1980 (cont'd)	
Grimm	CO, CO_2, COS
Grimm et al.	N_2, CO, CO_2, COS, CS_2
Gustafsson	O_2
Hitchcock and Brion (1980a)	CO_2, N_2, O_2
Hitchcock and Brion (1980b)	H_2CO, CH_3CHO, $(CH_3)_2CO$
Holmes and Marr	N_2, O_2, CO
Krummacher et al.	N_2
Langhoff et al.	review article
Morin et al.	O_2
Orel et al.	F_2
Raseev et al.	N_2
Ritchie and Tambe	CO
Swanson et al.	CO_2
Tronc et al.	N_2, NO, N_2O
van der Wiel	review article
Wallace	N_2, CO, NO, O_2, SF_6
West et al.	N_2
1981	
Brion and Tan	NO
Carlson et al. (1981a)	CO_2
Carlson et al. (1981b)	COS, CS_2
Carnovale et al.	CS_2
Grimm et al.	CO_2
Gustafsson and Levinson	NO
Hermann and Langhoff	H_2, N_2
Hitchcock and Brion	HF, F_2
Kreile et al.	C_2H_2
Langhoff et al. (1981a)	N_2CO
Langhoff et al. (1981b)	C_2H_2
Loomba et al.	N_2
Lucchese and McKoy (1981a)	CO_2
Lucchese and McKoy (1981b)	N_2
Ninomiya et al.	HCl, Cl_2
Padial et al. (1981a)	CO_2
Padial et al. (1981b)	O_3
Raseev et al.	O_2
Stephens et al.	CO
Swanson et al. (1981a)	BF_3
Swanson et al. (1981b)	CO_2
Tabché-Fouhailé et al.	O_2
Thiel	H_2, N_2, O_2, CO, CO_2

Reference	Molecule
1981 (cont'd)	
Unwin et al.	C_2H_2
White et al.	COS
Williams and Langhoff	N_2, CO
1982	
Carlson et al.	CS_2, COS
Dehmer et al.	SF_6
Delaney et al. (1982a)	O_2
Delaney et al. (1982b)	NO
Dittman et al.	O_2
Hayaishi et al.	C_2H_2
Hitchcock et al.	F_2
Ishiguro et al.	BF_3, BCl, BBr_3
Keller et al.	C_2H_2
Kreile et al.	HCN, C_2H_2
Lucchese and McKoy (1982a)	CO_2
Lucchese and McKoy (1982b)	CO_2
Lucchese et al.	N_2
Machado et al.	C_2H_2
Morin et al. (1982a)	O_2
Morin et al. (1982b)	O_2
Parr et al.	C_2H_2
Shaw et al.	N_2
Southworth et al.	NO
Thiel	CO_2, N_2
Wallace et al.	NO
1983	
Carlson et al.	N_2O
Dittman et al.	CO_2
Eberhardt et al. (1983a)	CO, $(CH_3)_2CO$
Eberhardt et al. (1983b)	N_2
Grimm	C_2H_4
Grimm and Carlson	N_2
Holland et al.	C_2N_2
Keller et al. (1983a)	chloromethanes
Keller et al. (1983b)	SiF_4, $Si(CH_3)_4$
Kreile et al.	C_2N_2
Krummacher et al.	CO
Levine and Soven	C_2H_2
Lucchese and McKoy	CO
McKoy et al.	N_2, CO, CO_2, C_2H_2
Morin	O_2, NO, N_2

Reference	Molecule
1983 (cont'd)	
Nenner	CO, O_2, N_2O
Schwarz et al.	BF_3, CF_4, KBF_4, $NaBF_4$, NH_4BF_4
Smith et al.	NO
Thiel	N_2, CO_2
Truesdale et al.	CO
1984	
Ågren and Arneberg	CO
Brion and Thomson (1984a)	data compilation – HF, HCl, HBr, O_2, NO, CO, N_2, H_2O, NH_3, CH_4, CO_2, COS, CS_2, N_2O
Brion and Thomson (1984b)	data compilation – H_2, CO, N_2, O_2, NO, HF, HCl, HBr, H_2O, NH_3, CH_4, N_2O, CO_2, COS, CS_2, SF_6
Carlson et al.	CF_4
Collins and Schneider	H_2, N_2, NO, CO_2
Dehmer	review article
Dehmer et al.	BF_3
Grimm et al.	C_3H_4
Hermann et al.	CO, H_2CO
Hitchcock et al.	1-butene, cis-2-butene, trans-2-butene, trans-1,3-butene, perfluoro-2-butene, review of other molecules with C–C bonds
Holland et al.	HCN
Kanamori et al.	BF_3
Keller et al. (1984a)	12 unsaturated organic molecules
Keller et al. (1984b)	H_2CO, CH_3OH
Kreile et al.	N_2, CO, C_2H_2, HCN, CO_2, N_2O, C_2N_2, C_4H_2, NC_3H
Langhoff	review article
Leal et al.	N_2
Levine and Soven	N_2, C_2H_2
Lindle et al.	N_2, NO
Lynch and McKoy	N_2
Lynch et al. (1984a)	C_2H_2
Lynch et al. (1984b)	review article
McKoy et al.	review article
Piancastelli et al.	CCl_4, $SiCl_4$, $GeCl_4$
Roy et al.	CO_2
Schneider and Collins	review article

Reference	Molecule
1984 (cont'd)	
Sette et al.	C_2H_2, C_2H_4, C_6H_6, CH_3HCO, $(CH_3)_2CO$, C_2H_6, HCN, C_2N_2, CH_3NH_2, N_2, CO, H_2CO, CH_3HCO, $(CH_3)_2CO$, BF_3, CH_3OCH_3, CH_3OH, NO, CF_4, CHF_3, CH_2F_2, CH_3F, O_2, NF_3
Shaw et al.	CO
Sodhi	NF_3, $Si(CH_3)_3$, PH_3, $P(CH_3)_3$, PF_3, PCl_3, PF_5, OPF_3, $OPCl_3$, methylamines, NH_3, $CH_2{=}C{=}CH_2$, t-1,3-butadiene, allene
Sodhi and Brion	SF_6, CO, N_2
Sodhi et al.	NF_3
Tossell and Davenport	CX_4, SiX_4 (X = H, F, Cl)
Truesdale et al.	CO, CO_2, CF_4, OCS
1985	
Dehmer et al.	review article
Ferrett et al. (1986)	SF_6
Kosman and Wallace	N_2
Sodhi and Brion (1985a)	methylamines, NH_3
Sodhi and Brion (1985b)	t-1,3-butadiene, allene
Sodhi and Brion (1985c)	PH_3, $P(CH_3)_3$, PCl_3, PF_3
Sodhi and Brion (1985d)	PF_5, OPF_3, $OPCl_3$
Sodhi et al.	$(CH_3)_4SI$
Stephens and Dill	N_2
Tossell	C_2H_2

References

Åberg, T., and J.L. Dehmer, 1973, J. Phys. C **6**, 1450.
Achiba, Y., K. Sato, K. Shobatake and K. Kimura, 1982, J. Chem. Phys. **77**, 2709.
Achiba, Y., K. Sato, K. Shobatake and K. Kumara, 1983, J. Chem. Phys. **78**, 5474.
Adam, M.Y., P. Morin, P. Lablanquie and I. Nenner, 1983, Int. Workshop on Atomic and Molecular Photoionization (Fritz-Haber-Institut der Max-Planck-Gesellschaft, Berlin).
Ågren, H., and R. Arneberg, 1984, Phys. Scr. **30**, 55.
Ågren, H., J. Nodgren, L. Selander, C. Nordling and K. Siegbahn, 1978, Phys. Scr. **18**, 499.
Ajello, J.M., A. Chutjian and R. Winchell, 1980, J. Electron Spectrosc. **19**, 197.
Anderson, S.L., G.D. Kubiak and R.N. Zare, 1984, Chem. Phys. Lett. **105**, 22.
Åsbrink, L., A. Svenson, W. von Niessen and G. Bieri, 1981, J. Electron Spectrosc. **24**, 293.
Atabek, O., D. Dill and Ch. Jungen, 1974, Phys. Rev. Lett. **33**, 123.
Azaroff, L.V., 1963, Rev. Mod. Phys. **35**, 1012.
Baer, T., 1979, in: Gas Phase Ion Chemistry, ed. M.T. Bowers (Academic Press, New York) ch. 5.

Baer, T., P.M. Guyon, I. Nenner, A. Tabché-Fouhailé, R. Botter, L.F.A. Ferreira and T.R. Govers, 1979, J. Chem. Phys. **70**, 1585.
Bagus, P.S., and H.F. Schaefer, 1972, J. Chem. Phys. **56**, 224.
Bagus, P.S., and E.K. Viinikka, 1977, Phys. Rev. A **15**, 1486.
Bahr, J.L., A.J. Blake, J.H. Carver, J.L. Gardner and V. Kumar, 1971a, J. Quant. Spectrosc. & Radiat. Transfer **11**, 1839.
Bahr, J.L., A.J. Blake, J.H. Carver, J.L. Gardner and V. Kumar, 1971b, J. Quant. Spectrosc. & Radiat. Transfer **11**, 1853.
Baker, C., and D.W. Turner, 1968, Proc. R. Soc. London Ser. A **308**, 19.
Bardsley, J.N., and F. Mandl, 1968, Rep. Prog. Phys. **31**, 472.
Barinskii, R.L., and I.M. Kulikova, 1973, Zh. Strukt. Khim. **14**, 372 [J. Struct. Chem. **14**, 335].
Barrus, D.M., R.L. Blake, A.J. Burek, K.C. Chambers and A.L. Pregenzer, 1979, Phys. Rev. A **20**, 1045.
Batten, C.F., J.A. Taylor and G.G. Meisels, 1976, J. Chem. Phys. **65**, 3316.
Batten, C.F., J.A. Taylor, B.P. Tsai and G.G. Meisels, 1978, J. Chem. Phys. **69**, 2547.
Benedict, M.G., and I. Gyemant, 1978, Int. J. Quantum Chem. **13**, 597.
Berkowitz, J., 1979, Photoabsorption, Photoionization and Photoelectron Spectroscopy (Academic Press, New York).
Berkowitz, J., and W.A. Chupka, 1969, J. Chem. Phys. **51**, 2341.
Berkowitz, J., and J.H.D. Eland, 1977, J. Chem. Phys. **67**, 2740.
Berry, R.S., and S.E. Nielsen, 1970a, Phys. Rev. A **1**, 383.
Berry, R.S., and S.E. Nielsen, 1970b, Phys. Rev. A **1**, 395.
Bianconi, A., H. Peterson, F.C. Brown and R.Z. Bachrach, 1978, Phys. Rev. A **17**, 1907.
Birtwistle, D.T., and A. Herzenberg, 1971, J. Phys. B **4**, 53.
Blake, A.J., J.L. Bahr, J.H. Carver and V. Kumar, 1970, Philos. Trans. Roy. Soc. London A **268**, 159.
Blechschmidt, D., R. Haensel. E.-E. Koch, U. Nielsen and T. Sagawa, 1972, Chem. Phys. Lett. **14**, 33.
Botter, R., V.H. Dibeler, J.A. Walker and H.M. Rosenstock, 1966, J. Chem. Phys. **44**, 1271.
Brehm, B., J.H.D. Eland, R. Frey and A. Kustler, 1973, Int. J. Mass. Spectrom. & Ion Phys. **12**, 197.
Brion, C.E., and K.H. Tan, 1978, Chem. Phys. **34**, 141.
Brion, C.E., and K.H. Tan, 1979, J. Electron Spectrosc. **15**, 241.
Brion, C.E., and K.H. Tan, 1981, J. Electron Spectrosc. **23**, 1.
Brion, C.E., and J.P. Thomson, 1984a, J. Electron Spectrosc. **33**, 287.
Brion, C.E., and J.P. Thomson, 1984b, J. Electron Spectrosc. **33**, 301.
Brion, C.E., K.H. Tan, M.J. van der Wiel and Ph.E. van der Leeuw, 1979, J. Electron Spectrosc. **17**, 101.
Brown, F.C., R.Z. Bachrach and A. Bianconi, 1978, Chem. Phys. Lett. **54**, 425.
Cadioli, B., U. Pincelli, E. Tosatti, U. Fano and J.L. Dehmer, 1972, Chem. Phys. Lett. **17**, 15.
Caprace, G., J. Delwiche, P. Natalis and J.E. Collin, 1976, Chem. Phys. **13**, 43.
Carlson, T.A., 1971, Chem. Phys. Lett. **9**, 23.
Carlson, T.A., and A.E. Jonas, 1971, J. Chem. Phys. **55**, 4913.
Carlson, T.A., and G.E. McGuire, 1972, J. Electron Spectrosc. **1**, 209.
Carlson, T.A., G.E. McGuire, A.E. Jonas, K.L. Cheng, C.P. Anderson, C.C. Lu and B.P. Pullen, 1972, in: Electron Spectroscopy, ed. D.A. Shirley (North-Holland, Amsterdam) p. 207.
Carlson, T.A., M.O. Krause, D. Mehaffy, J.W. Taylor, F.A. Grimm and J.D. Allen, 1980, J. Chem. Phys. **73**, 6056.
Carlson, T.A., M.O. Krause, F.A. Grimm, J.D. Allen, D. Mehaffy, P.R. Keller and J.W. Taylor, 1981a, Phys. Rev. A **23**, 3316.
Carlson, T.A., M.O. Krause, F.A. Grimm, J.D. Allen, D. Mehaffy, P.R. Keller and J.W. Taylor, 1981b, J. Chem. Phys. **75**, 3288.
Carlson, T.A., M.O. Krause and F.A. Grimm, 1982a, J. Chem. Phys. **77**, 1701.
Carlson, T.A., M.O. Krause, F.A. Grimm, P. Keller and J.W. Taylor, 1982b, J. Chem. Phys. **77**, 5340.

Carlson, T.A., M.O. Krause, F.A. Grimm and T.A. Whitley, 1983a, J. Chem. Phys. **78**, 638.
Carlson, T.A., P.R. Keller, J.W. Taylor, T. Whitley and F.A. Grimm, 1983b, J. Chem. Phys. **79**, 97.
Carlson, T.A., M.O. Krause, A. Fahlman, P.R. Keller, J.W. Taylor, T. Whitley and F.A. Grimm, 1983c, J. Chem. Phys. **79**, 2157.
Carlson, T.A., A. Fahlman, M.O. Krause, P.R. Keller, J.W. Taylor, T. Whitley and F.A. Grimm, 1984a, J. Chem. Phys. **80**, 3521.
Carlson, T.A., A. Fahlman, W.A. Svensson, M.O. Krause, T.A. Whitley, F.A. Grimm, M.N. Piancastelli and J.W. Taylor, 1984b, J. Chem. Phys. **81**, 3828.
Carlson, T.A., A. Fahlman, M.O. Krause, T.A. Whitley and F.A. Grimm, 1984c, J. Chem. Phys. **81**, 5389.
Carnovale, F., M.G. White and C.E. Brion, 1981, J. Electron Spectrosc. **24**, 63.
Carr, R.G., T.K. Sham and W.E. Eberhardt, 1985, Chem. Phys. Lett. **113**, 63.
Cederbaum, L.S., and W. Domcke, 1977, Adv. Chem. Phys. **36**, 205.
Cederbaum, L.S., J. Schirmer, W. Domcke and W. von Niessen, 1977, J. Phys. B **10**, L549.
Cederbaum, L.S., J. Schirmer, W. Domcke and W. von Niessen, 1978, J. Electron Spectrosc. **16**, 59.
Cederbaum, L.S., W. Domcke, J. Schirmer and W. von Niessen, 1980, Phys. Scr. **21**, 481.
Chandra, N., and A. Temkin, 1976, Phys. Rev. A **13**, 188.
Chase, D.M., 1956, Phys. Rev. **104**, 838.
Cherepkov, N.A., 1981a, J. Phys. B **14**, 2165.
Cherepkov, N.A., 1981b, J. Phys. B **14**, L73.
Child, M.S., 1974, Molecular Collision Theory (Academic Press, New York) p. 51.
Chupka, W.A., and J. Berkowitz, 1969, J. Chem. Phys. **51**, 4244.
Chutjian, A., and J.M. Ajello, 1977, J. Chem. Phys. **66**, 4544.
Chutjian, A., and J.M. Ajello, 1980, Chem. Phys. Lett. **72**, 504.
Codling, K., 1966, J. Chem. Phys. **44**, 4401.
Codling, K., A.C. Parr, D.L. Ederer, R. Stockbauer, J.B. West, B.E. Cole and J.L. Dehmer, 1981, J. Phys. B **14**, 657.
Cole, B.E., D.L. Ederer, R. Stockbauer, K. Codling, A.C. Parr, J.B. West, E.D. Poliakoff and J.L. Dehmer, 1980, J. Chem. Phys. **72**, 6308.
Collin, J.E., and J. Delwiche, 1967, Can. J. Chem. **45**, 1883.
Collin, J.E., and P.J. Natalis, 1969, Int. J. Mass Spectrom. & Ion Phys. **2**, 231.
Collin, J.E., J. Delwiche and P. Natalis, 1972, in: Electron Spectroscopy, ed. D.A. Shirley (North-Holland, Amsterdam) p. 401.
Collins, L.A., and B.I. Schneider, 1984, Phys. Rev. A **29**, 1695.
Connolly, J.W.D., and K.H. Johnson, 1971, Chem. Phys. Lett. **10**, 616.
Danby, C.J., and J.H.D. Eland, 1972, Int. J. Mass. Spectrom. & Ion Phys. **8**, 153.
Davenport, J.W., 1976a, Phys. Rev. Lett. **36**, 945.
Davenport, J.W., 1976b, Ph.D. Thesis (Univ. of Pennsylvania).
Davenport, J.W., 1977, Int. J. Quantum Chem. Quantum Chem. Symp. **11**, 89.
Davenport, J.W., W. Ho and J.R. Schrieffer, 1978, Phys. Rev. B **17**, 3115.
Dehmer, J.L., 1972, J. Chem. Phys. **56**, 4496.
Dehmer, J.L., 1974, Phys. Fenn. **9S**, 60.
Dehmer, J.L., 1984, in: Resonances in Electron–Molecule Scattering, van der Waals Complexes and Reactive Chemical Dynamics, ACS Symposium Series, No. 263, ed. D.G. Truhlar (American Chemical Society, Washington, DC) ch. 8, p. 139.
Dehmer, J.L., and D. Dill, 1975, Phys. Rev. Lett. **35**, 213.
Dehmer, J.L., and D. Dill, 1976a, J. Chem. Phys. **65**, 5327.
Dehmer, J.L., and D. Dill, 1976b, in: Proc. 2nd Int. Conf. on Inner-Shell Ionization Phenomena, eds W. Mehlhorn and R. Brenn (Fakultät für Physik, Universität Freiburg) p. 221.
Dehmer, J.L., and D. Dill, 1978, Phys. Rev. A **18**, 164.
Dehmer, J.L., and D. Dill, 1979a, in: Symp. on Electron–Molecule Collisions, eds I. Shimamura and M. Matsuzawa (Univ. of Tokyo Press, Tokyo) p. 95.
Dehmer, J.L., and D. Dill, 1979b, in: Electron–Molecule and Photon–Molecule Collisions, eds T. Rescigno, V. McKoy and B. Schneider (Plenum, New York) p. 225.

Dehmer, J.L., and D. Dill, 1980, in: Electronic and Atomic Collisions, eds N. Oda and K. Takayanagi (North-Holland, Amsterdam) p. 195.
Dehmer, J.L., A.F. Starace, U. Fano, J. Sugar and J.W. Cooper, 1971, Phys. Rev. Lett. **26**, 1521.
Dehmer, J.L., W.A. Chupka, J. Berkowitz and W.T. Jivery, 1975, Phys. Rev. A **12**, 1966.
Dehmer, J.L., J. Siegel and D. Dill, 1978, J. Chem. Phys. **69**, 5205.
Dehmer, J.L., D. Dill and S. Wallace, 1979, Phys. Rev. Lett. **43**, 1005.
Dehmer, J.L., J. Siegel, J. Welch and D. Dill, 1980, Phys. Rev. A **21**, 101.
Dehmer, J.L., A.C. Parr, S. Wallace and D. Dill, 1982, Phys. Rev. A **26**, 3283.
Dehmer, J.L., A.C. Parr, S.H. Southworth and D.M.P. Holland, 1984, Phys. Rev. A **30**, 1783.
Dehmer, J.L., D. Dill and A.C. Parr, 1985, in: Photophysics and Photochemistry in the Vacuum Ultraviolet, eds S. McGlynn, G. Findley and R. Huebner (Reidel, Dordrecht) p. 341.
Dehmer, P.M., and W.A. Chupka, 1975, J. Chem. Phys. **62**, 4525.
Dehmer, P.M., and W.A. Chupka, 1976, J. Chem. Phys. **65**, 2243.
Dehmer, P.M., and J.L. Dehmer, 1982, J. Electron Spectrosc. **28**, 145.
Dehmer, P.M., P.J. Miller and W.A. Chupka, 1984, J. Chem. Phys. **80**, 1030.
Delaney, J.J., I.H. Hillier and V.R. Saunders, 1982a, J. Phys. B **15**, L37.
Delaney, J.J., I.H. Hillier and V.R. Saunders, 1982b, J. Phys. B **15**, 1477.
Derenbach, H., R. Malutzki and V. Schmidt, 1983, Nucl. Instrum. Methods **208**, 845.
Dibeler, V.H., and S.K. Liston, 1968, Inorg. Chem. **7**, 1742.
Dibeler, V.H., and J.A. Walker, 1966, J. Chem. Phys. **44**, 4405.
Dibeler, V.H., and J.A. Walker, 1967, J. Opt. Soc. Am. **57**, 1007.
Dibeler, V.H., and J.A. Walker, 1973, Int. J. Mass Spectrom. & Ion Phys. **11**, 49.
Dill, D., 1972, Phys. Rev. A **6**, 160.
Dill, D., and J.L. Dehmer, 1974, J. Chem. Phys. **61**, 692.
Dill, D., and Ch. Jungen, 1980, J. Phys. Chem. **84**, 2116.
Dill, D., J. Siegel and J.L. Dehmer, 1976, J. Chem. Phys. **65**, 3158.
Dill, D., S. Wallace, J. Siegel and J.L. Dehmer, 1978, Phys. Rev. Lett. **41**, 1230.
Dill, D., S. Wallace, J. Siegel and J.L. Dehmer, 1979a, Phys. Rev. Lett. **42**, 411(E).
Dill, D., J. Welch, J.L. Dehmer and J. Siegel, 1979b, Phys. Rev. Lett. **43**, 1236.
Dill, D., R. Swanson, S. Wallace and J.L. Dehmer, 1980, Phys. Rev. Lett. **45**, 1393.
Dittman, P.M., D. Dill and J.L.Dehmer, 1982, J. Chem. Phys. **76**, 5703.
Dittman, P.M., D. Dill and J.L. Dehmer, 1983, Chem. Phys. **78**, 405.
Doolittle, P.H., and R.I. Schoen, 1965, Phys. Rev. Lett. **14**, 348.
Duzy, C., and R.S. Berry, 1976, J. Chem. Phys. **64**, 2431.
Eberhardt, W., R.P. Haelbich, M. Iwan, E.-E. Koch and C. Kunz, 1976, Chem. Phys. Lett. **40**, 180.
Eberhardt, W., T.K. Sham, R. Carr, S. Krummacher, M. Strongin, S.L. Wergand and D. Wesner, 1983a, Phys. Rev. Lett. **50**, 1038.
Eberhardt, W., J. Stöhr, J. Feldhaus, E.W. Plummer and F. Sette, 1983b, Phys. Rev. Lett. **51**, 2370.
Ederer, D.L., B.E. Cole and J.B. West, 1980, Nucl. Instrum. Methods **172**, 185.
Ederer, D.L., A.C. Parr, B.E. Cole, R. Stockbauer, J.L. Dehmer, J.B. West and K. Codling, 1981, Proc. R. Soc. London Ser. A **378**, 423.
El-Sherbini, Th.M., and M.J. van der Wiel, 1972, Physica **59**, 433.
Eland, J.H.D., 1980, J. Chem. Phys. **72**, 6015.
Fano, U., 1935, Nuovo Cimento **12**, 156.
Fano, U., 1961, Phys. Rev. **124**, 1866.
Fano, U., 1970, Phys. Rev. A **2**, 353.
Fano, U., 1972, Comments At. & Mol. Phys. **3**, 75.
Fano, U., 1975, J. Opt. Soc. Am. **65**, 979.
Fano, U., 1983, Rep. Prog. Phys. **46**, 97 and references cited therein.
Fano, U., and J.W. Cooper, 1968, Rev. Mod. Phys. **40**, 441.
Ferrett, T.A., D.W. Lindle, P.A. Heimann, H.G. Kerkhoff, V.E. Becker and D.A. Shirley, 1986, Phys. Rev. A **34**, 1916.
Fock, J.-H., 1983, Ph.D. Thesis (University of Hamburg).
Fock, J.-H., and E.-E. Koch, 1984, Chem. Phys. Lett. **105**, 38.

Fock, J.-H., and E.-E. Koch, 1985, Chem. Phys. **96**, 125.
Fock, J.-H., H.-J. Lau and E.-E. Koch, 1984, Chem. Phys. **83**, 377.
Fomichev, V.A., 1967, Fiz. Tverd. Tela **9**, 3167 [1968, Sov. Phys.-Solid State **9**, 2496].
Fomichev, V.A., and R.L. Barinskii, 1970, Zh. Strukt. Khim. **11**, 875 [J. Struct. Chem. **11**, 810].
Friedrich, H., B. Pittel, P. Rabe, W.H.E. Schwarz and B. Sonntag, 1980, J. Phys. B **13**, 25.
Gardner, J.L., and J.A.R. Samson, 1973, J. Electron Spectrosc. **2**, 153.
Gardner, J.L., and J.A.R. Samson, 1974a, J. Chem. Phys. **60**, 3711.
Gardner, J.L., and J.A.R. Samson, 1974b, Chem. Phys. Lett. **26**, 240.
Gardner, J.L., and J.A.R. Samson, 1978, J. Electron Spectrosc. **13**, 7.
Gelius, U., 1974, J. Electron Spectrosc. **5**, 985.
Gerwer, A., C. Asaro, B.V. McKoy and P.W. Langhoff, 1980, J. Chem. Phys. **72**, 713.
Gianturco, F.A., C. Guidotti, U. Lamanna and R. Moccia, 1971, Chem. Phys. Lett. **10**, 269.
Gianturco, F.A., C. Guidotti and U. Lamanna, 1972, J. Chem. Phys. **57**, 840.
Giardini-Guidoni, A., R. Fantoni, R. Tiribelli, D. Vinciguerra, R. Camilloni and G. Stefani, 1979, J. Chem. Phys. **71**, 3182.
Giusti-Suzor, A., 1982, in: Physics of Electronic and Atomic Collisions, ed. S. Datz (North-Holland, Amsterdam) p. 381.
Giusti-Suzor, A., and U. Fano, 1984a, J. Phys. B **17**, 215.
Giusti-Suzor, A., and U. Fano, 1984b, J. Phys. B **17**, 4267.
Giusti-Suzor, A., and Ch. Jungen, 1984, J. Chem. Phys. **80**, 986.
Giusti-Suzor, A., and H. Lefebvre-Brion, 1980, Chem. Phys. Lett. **76**, 132.
Giusti-Suzor, A., and H. Lefebvre-Brion, 1984, Phys. Rev. A **30**, 3057.
Glownia, J.H., S.J. Riley, S.D. Colson, J.C. Miller and R.N. Compton, 1982, J. Chem. Phys. **77**, 68.
Greene, C.H., and R.N. Zare, 1983, J. Chem. Phys. **78**, 6741.
Grimm, F.A., 1980, Chem. Phys. **53**, 71.
Grimm, F.A., 1983, Chem. Phys. **81**, 315.
Grimm, F.A., and T.A. Carlson, 1983, Chem. Phys. **80**, 389.
Grimm, F.A., T.A. Carlson, W.B. Dress, P. Agron, J.O. Thomson and J.W. Davenport, 1980, J. Chem. Phys. **72**, 3041.
Grimm, F.A., J.D. Allen, T.A. Carlson, M.O. Krause, D. Mehaffy, P.R. Keller and J.W. Taylor, 1981, J. Chem. Phys. **75**, 92.
Grimm, F.A., T.A. Whitley, P.R. Keller and J.W. Taylor, 1984, J. Electron Spectrosc. **33**, 361.
Guest, J.A., K.A. Jackson and R.N. Zare, 1983, Phys. Rev. A **28**, 2217.
Guest, J.A., M.A. O'Halloran and R.N. Zare, 1984, J. Chem. Phys. **81**, 2689.
Gurtler, P., V. Saile and E.-E. Koch, 1977, Chem. Phys. Lett. **48**, 245.
Gustafsson, T., 1978, Phys. Rev. A **18**, 1481.
Gustafsson, T., 1980a, Chem. Phys. Lett. **75**, 505.
Gustafsson, T., 1980b, Surf. Sci. **94**, 593.
Gustafsson, T., 1983, in: Atomic Physics 8, eds I. Lindgren, A. Rosen and S. Svanberg (Plenum, New York) p. 355.
Gustafsson, T., and H.J. Levinson, 1981, Chem. Phys. Lett. **78**, 28.
Gustafsson, T., E.W. Plummer, D.E. Eastman and W. Gudat, 1978a, Phys. Rev. A **17**, 175.
Gustafsson, T., E.W. Plummer, E.W. Liebsch, 1978b, in: Photoemission and the Electronic Properties of Surfaces, eds B. Feuerbacher, B. Fitton and R.F. Willis (Wiley, New York).
Guyon, P.-M., T. Baer, L.F.A. Ferreira, I. Nenner, A. Tabché-Fouhailé, R. Botter and T.R. Govers, 1978, J. Phys. B **11**, L141.
Guyon, P.-M., T. Baer and I. Nenner, 1983, J. Chem. Phys. **78**, 3665.
Haller, E., H. Köppel, L.S. Cederbaum, W. von Niessen and G. Bieri, 1983, J. Chem. Phys. **78**, 1359.
Hamnett, A., W. Stoll and C.E. Brion, 1976, J. Electron Spectrosc. **8**, 367.
Hancock, W.H., and J.A.R. Samson, 1976, J. Electron Spectrosc. **9**, 211.
Hara, S., 1985, J. Phys. B **18**, 3759.
Hara, S., and S. Ogata, 1985, J. Phys. B **18**, L59.
Harting, E., and F.H. Read, 1976, Electrostatic Lenses (North-Holland, Amsterdam).

Hay, P.J., 1977, J. Am. Chem. Soc. **99**, 1003.
Hayaishi, T., S. Iwata, M. Sasanuma, E. Ishiguro, Y. Morioka, Y. Iida and M. Nakamura, 1982, J. Phys. B **15**, 79.
Hayes, W., and F.C. Brown, 1971, J. Phys. B **4**, L85.
Heinzmann, U., 1980, Appl. Opt. **19**, 4087.
Heinzmann, U., F. Schäfers and B.A. Hess, 1980, Chem. Phys. Lett. **69**, 284.
Heinzmann, U., B. Osterheld, F. Schäfers and G. Schönhense, 1981, J. Phys. B **14**, L79.
Hermann, M.R., and P.W. Langhoff, 1981, Chem. Phys. Lett. **82**, 242.
Hermann, M.R., G.H.F. Diercksen, B.W. Fatyga and P.W. Langhoff, 1984, Int. J. Quantum Chem. **18**, 719.
Herzberg, G., and Ch. Jungen, 1972, J. Mol. Spectrosc. **41**, 425.
Herzenberg, A., and F. Mandl, 1962, Proc. R. Soc. London Ser. A **270**, 48.
Hitchcock, A.P., 1982, J. Electron. Spectrosc. **25**, 245.
Hitchcock, A.P., and C.E. Brion, 1977, J. Electron Spectrosc. **10**, 317.
Hitchcock, A.P., and C.E. Brion, 1978a, Chem. Phys. **33**, 55.
Hitchcock, A.P., and C.E. Brion, 1978b, J. Electron Spectrosc. **14**, 417.
Hitchcock, A.P., and C.E. Brion, 1979a, J. Electron Spectrosc. **15**, 201.
Hitchcock, A.P., and C.E. Brion, 1979b, Chem. Phys. **37**, 319.
Hitchcock, A.P., and C.E. Brion, 1980a, J. Electron Spectrosc. **18**, 1.
Hitchcock, A.P., and C.E. Brion, 1980b, J. Electron Spectrosc. **19**, 231.
Hitchcock, A.P., and C.E. Brion, 1981, J. Phys. B **13**, 4399.
Hitchcock, A.P., and M.J. van der Wiel, 1979, J. Phys. B **12**, 2153.
Hitchcock, A.P., C.E. Brion and M.J. van der Wiel, 1978, J. Phys. B **11**, 3245.
Hitchcock, A.P., C.E. Brion, G.R.J. Williams and P.W. Langhoff, 1982, Chem. Phys. **66**, 435.
Hitchcock, A.P., S. Beaulieu, T. Steel, J. Stöhr and F. Sette, 1984, J. Chem. Phys. **80**, 3927.
Holland, D.M.P., A.C. Parr, D.L. Ederer, J.L. Dehmer and J.B. West, 1982, Nucl. Instr. Methods **195**, 331.
Holland, D.M.P., A.C. Parr, D.L. Ederer, J.B. West and J.L. Dehmer, 1983, Int. J. Mass. Spectrom. & Ion Phys. **52**, 195.
Holland, D.M.P., A.C. Parr and J.L. Dehmer, 1984, J. Phys. B **17**, 1343.
Holmes, R.M., and G.V. Marr, 1980, J. Phys. B **13**, 945.
Hopfield, J.J., 1930a, Phys. Rev. **35**, 1133.
Hopfield, J.J., 1930b, Phys. Rev. **36**, 789.
Horton, V.G., E.T. Arakawa, R.N. Hamin and M.W. Williams, 1969, Appl. Opt. **8**, 667.
Hubin-Franskin, M.-J., J. Delwiche, P. Morin, M.Y. Adam, I. Nenner and P. Roy, 1984, J. Chem. Phys. **81**, 4246.
Ishiguro, E., S. Iwata, Y. Suzuki, A. Mikuni and T. Sasaki, 1982, J. Phys. B **15**, 1841.
Itikawa, Y., H. Takagi, H. Nakamura and H. Sato, 1983, Phys. Rev. A **27**, 1319.
Iwata, S., N. Kosugi and O. Nomura, 1978, J. Appl. Phys. Jpn. **17**(Suppl. 17-2), 109.
Johnson, W.R., and K.T. Cheng, 1979, Phys. Rev. A **20**, 978.
Johnson, W.R., K.T. Cheng, K.-N. Huang and M. LeDourneuf, 1980, Phys. Rev. A **22**, 989.
Jungen, Ch., 1980, J. Chim. Phys. **77**, 27.
Jungen, Ch., 1984a, Ann. Israel Phys. Soc. **6**, 491.
Jungen, Ch., 1984b, Phys. Rev. Lett. **53**, 2394.
Jungen, Ch., and O. Atabek, 1977, J. Chem. Phys. **66**, 5584.
Jungen, Ch., and D. Dill, 1980, J. Chem. Phys. **73**, 3338.
Jungen, Ch., and M. Raoult, 1981, Faraday Disc. Chem. Soc. **71**, 253.
Kaesdorf, S., G. Schönhense and U. Heinzmann, 1985, Phys. Rev. Lett. **54**, 885.
Kanamori, H., S. Iwata, A. Mikuni and T. Sasaki, 1984, J. Phys. B **17**, 3887.
Katsumata, S., Y. Achiba and K. Kimura, 1979, J. Electron Spectrosc. **17**, 229.
Kay, R.B., Ph.E. van der Leeuw and M.J. van der Wiel, 1977, J. Phys. B **10**, 2513.
Keller, P.R., D. Mehaffy, J.W. Taylor, F.A. Grimm and T.A. Carlson, 1982, J. Electron Spectrosc. **27**, 223.

Keller, P.R., J.W. Taylor, T.A. Carlson and F.A. Grimm, 1983a, Chem. Phys. **79**, 269.
Keller, P.R., J.W. Taylor, F.A. Grimm, P. Senn, T.A. Carlson and M.O. Krause, 1983b, Chem. Phys. **74**, 247.
Keller, P.R., J.W. Taylor, T.A. Carlson and F.A. Grimm, 1984a, J. Electron Spectrosc. **33**, 333.
Keller, P.R., J.W. Taylor, F.A. Grimm and T.A. Carlson, 1984b, Chem. Phys. **90**, 147.
Kennedy, D.J., and S.T. Manson, 1972, Phys. Rev. A **5**, 227.
Kennerly, R.E., 1980, Phys. Rev. A **21**, 1876.
Kennerly, R.E., R.A. Bonham and M. McMillan, 1979, J. Chem. Phys. **70**, 2039.
Kibel, M.H., F.J. Leng and G.L. Nyberg, 1979, J. Electron Spectrosc. **15**, 281.
Kimman, J., 1984, Ph.D. Thesis (FOM-Institute for Atomic and Molecular Physics, Amsterdam).
Kimman, J., P. Kruit and M.J. van der Wiel, 1982, Chem. Phys. Lett. **88**, 576.
Kimura, K., S. Katsumata, Y. Achiba, T. Yamazaki and S. Iwata, 1981, Handbook of HeI Photoelectron Spectra of Fundamental Organic Molecules (Japan Scientific Societies Press, Tokyo and Halsted Press, New York).
Kincaid, B.M., and P. Eisenberger, 1975, Phys. Rev. Lett. **34**, 1361.
King, G.C., F.H. Read and M. Tronc, 1977, Chem. Phys. Lett. **52**, 50.
Kleimenov, V.I., Yu.V. Chizhov and F.I. Vilesov, 1972, Opt. Spektrosk. **32**, 702 [Opt. Spectrosc. **32**, 371].
Koch, E.-E., ed., 1983, Handbook on Synchrotron Radiation, Vol. 1 (North-Holland, Amsterdam).
Koch, E.-E., and B.F. Sonntag, 1979, in: Synchrotron Radiation: Techniques and Applications, ed. C. Kunz (Springer, Berlin) p. 269.
Koestner, R.J., J. Stöhr, J.L. Gland and J.A. Horsley, 1984, Chem. Phys. Lett. **105**, 332.
Kondratenko, A.V., L.N. Mazalov, F.Kh. Gel'mukhanov, V.I. Avdeev and E.A. Saprykhina, 1977, Zh. Strukt. Khim. **18**, 546 [J. Struct. Chem. **18**, 437].
Kosman, W.M., and S. Wallace, 1985, J. Chem. Phys. **82**, 1385.
Krause, M.O., T.A. Carlson and P.R. Woodruff, 1981, Phys. Rev. A **24**, 1374.
Krauss, M., and F.H. Mies, 1970, Phys. Rev. A **1**, 1592.
Kreile, J., and A. Schweig, 1980, J. Electron Spectrosc. **20**, 191.
Kreile, J., A. Schweig and W. Thiel, 1981, Chem. Phys. Lett. **79**, 547.
Kreile, J., A. Schweig and W. Thiel, 1982, Chem. Phys. Lett. **87**, 473.
Kreile, J., A. Schweig and W. Thiel, 1983, Chem. Phys. Lett. **100**, 351.
Kreile, J., A. Schweig and W. Thiel, 1984, Chem. Phys. Lett. **108**, 259.
Kronig, R. de L., 1931, Z. Phys. **70**, 317.
Kronig, R. de L., 1932, Z. Phys. **75**, 191.
Krummacher, S., V. Schmidt and F. Wuilleumier, 1980, J. Phys. B **13**, 3993.
Krummacher, S., V. Schmidt, F.J. Wuilleumier, J.M. Bizau and D.L. Ederer, 1983, J. Phys. B **16**, 1733.
Kumar, V., and E. Krishnakumar, 1981, J. Electron Spectrosc. **22**, 109.
Kumar, V., and E. Krishnakumar, 1983, J. Chem. Phys. **78**, 46.
Kunz, C., ed., 1979, Synchrotron Radiation: Techniques and Applications (Springer, Berlin).
Lane, N.F., 1980, Rev. Mod. Phys. **52**, 29.
Langhoff, P.W., 1977, Int. J. Quantum Chem. Quantum Chem. Symp. **11**, 301.
Langhoff, P.W., 1979, in: Electron–Molecule and Photon–Molecule Collisions, eds T.N. Rescigno, V. McKoy and B. Schneider (Plenum, New York) p. 183.
Langhoff, P.W., 1984, in: Resonances in Electron–Molecule Scattering, van der Waals Complexes and Reactive Chemical Dynamics, ACS Symposium Series, No. 263, ed. D.G. Truhlar (Americal Chemical Society, Washington, DC) ch. 7, p. 113.
Langhoff, P.W., S.R. Langhoff and C.T. Corcoran, 1977, J. Chem. Phys. **67**, 1722.
Langhoff, P.W., A. Orel, T.N. Rescigno and B.V. McKoy, 1978, J. Chem. Phys. **69**, 4689.
Langhoff, P.W., A. Gerwer, C. Asaro and B.V. McKoy, 1979, Int. J. Quantum Chem. Quantum Chem. Symp. **13**, 645.
Langhoff, P.W., T.N. Rescigno, N. Padial, G. Csanak and B.V. McKoy, 1980, J. Chim. Phys. **77**, 590.
Langhoff, P.W., S.R. Langhoff, T.N. Rescigno, J. Schirmer, L.S. Cederbaum, W. Domcke and W. von Niessen, 1981a, Chem. Phys. **58**, 71.

Langhoff, P.W., B.V. McKoy, R. Unwin and A.M. Bradshaw, 1981b, Chem. Phys. Lett. **83**, 270.
Lau, H.-J., J.-H. Fock and E.-E. Koch, 1982, Chem. Phys. Lett. **89**, 281.
LaVilla, R.E., 1972, J. Chem. Phys. **57**, 899.
LaVilla, R.E., 1973, J. Chem. Phys. **58**, 3841.
LaVilla, R.E., 1975, J. Chem. Phys. **63**, 2733.
LaVilla, R.E., and R.D. Deslattes, 1966, J. Chem. Phys. **44**, 4399.
Leal, E.P., L.E. Machado and L. Mu-Tao, 1984, J. Phys. B **17**, L569.
Lee, L.C., E. Phillips and D.L. Judge, 1977, J. Chem. Phys. **67**, 1237.
Lee, P.A., P.H. Citrin, P. Eisenberger and B.M. Kincaid, 1981, Rev. Mod. Phys. **53**, 769.
Lefebvre-Brion, H., and A. Giusti-Suzor, 1983, in: Electron–Atom and Electron–Molecule Collisions, ed. J. Hinze (Plenum, New York) p. 215.
Lefebvre-Brion, H., A. Giusti-Suzor and G. Raseev, 1985, J. Chem. Phys. **83**, 1557.
Levine, Z.H., and P. Soven, 1983, Phys. Rev. Lett. **50**, 2074.
Levine, Z.H., and P. Soven, 1984, Phys. Rev. A **29**, 625.
Levinson, H., T. Gustafsson and P. Soven, 1979, Phys. Rev. A **19**, 1089.
Lindle, D.W., C.M. Truesdale, P.H. Kobrin, T.A. Ferrett, P.A. Heimann, U. Becker, H.G. Kerkhoff and D.A. Shirley, 1984, J. Chem. Phys. **81**, 5375.
Loomba, D., S. Wallace, D. Dill and J.L. Dehmer, 1981, J. Chem. Phys. **75**, 4546.
Lozes, R.L., O. Goscinski, U.I. Wahlgren, 1979, Chem. Phys. Lett. **63**, 77.
Lucchese, R.R., and V. McKoy, 1981a, J. Phys. Chem. **85**, 2166.
Lucchese, R.R., and V. McKoy, 1981b, J. Phys. B **14**, L629.
Lucchese, R.R., and V. McKoy, 1981c, Phys. Rev. A **24**, 770.
Lucchese, R.R., and V. McKoy, 1982a, Phys. Rev. A **26**, 1406.
Lucchese, R.R., and V. McKoy, 1982b, Phys. Rev. A **26**, 1992.
Lucchese, R.R., and V. McKoy, 1983, Phys. Rev. A **28**, 1382.
Lucchese, R.R., D.K. Watson and V. McKoy, 1980, Phys. Rev. A **22**, 421.
Lucchese, R.R., G. Raseev and V. McKoy, 1982, Phys. Rev. A **25**, 2572.
Lynch, D.L., and V. McKoy, 1984, Phys. Rev. A **30**, 1561.
Lynch, D.L., M.-T. Lee, R.R. Lucchese and V. McKoy, 1984a, J. Chem. Phys. **80**, 1907.
Lynch, D.L., V. McKoy and R.R. Lucchese, 1984b, 208.
Machado, L.E., E.P. Leal, G. Csanak, B.V. McKoy and P.W. Langhoff, 1982, J. Electron Spectrosc. **25**, 1.
Marr, G.V., and J.B. West, 1976, At. Data & Nucl. Data Tables **18**, 497.
Marr, G.V., and P.R. Woodruff, 1976, J. Phys. B **9**, L377.
Marr, G.V., J.M. Morton, R.M. Holmes and D.G. McCoy, 1979, J. Phys. B **12**, 43.
Mazalov, L.N., F.Kh. Gel'muskhanov and V.M. Chermoshentsev, 1974, Zh. Strukt. Khim. **15**, 1099 [J. Struct. Chem. **15**, 975].
McCoy, D.G., J.M. Morton and G.V. Marr, 1978, J. Phys. B **11**, L547.
McCulloh, K.E., 1973, J. Chem. Phys. **59**, 4250.
McKoy, V., D. Lynch and R.R. Lucchese, 1983, Int. J. Quantum Chem. Symp. **17**, 89.
McKoy, V., T.A. Carlson and R.R. Lucchese, 1984, J. Phys. Chem. **88**, 3188.
Mies, F.H., 1984, J. Chem. Phys. **80**, 2514.
Mies, F.H., and P.S. Julienne, 1984, J. Chem. Phys. **80**, 2526.
Miller, J.C., and R.N. Compton, 1981a, J. Chem. Phys. **75**, 22.
Miller, J.C., and R.N. Compton, 1981b, J. Chem. Phys. **75**, 2020.
Miller, J.C., and R.N. Compton, 1982, Chem. Phys. Lett. **93**, 453.
Miller, J.C., R.N. Compton, T.E. Carney and T. Baer, 1982, J. Chem. Phys. **76**, 5648.
Mintz, D.M., and T. Baer, 1976, J. Chem. Phys. **65**, 2407.
Mintz, D.M., and A. Kuppermann, 1978, J. Chem. Phys. **69**, 3953.
Morgenstern, R., A. Niehaus and M.W. Ruf, 1971, in: Electronic and Atomic Collisions, eds L. Branscomb, H. Ehrhardt, R. Geballe, F.J. de Heer, N.V. Fedorenko, J. Kistemaker, M. Barat, E.E. Nikitin and A.C.H. Smith (North-Holland, Amsterdam) p. 167.
Morin, P., 1983, Ph.D. Thesis (Université de Paris-Sud).
Morin, P., I. Nenner, P.-M. Guyon, O. Dutuit and K. Ito, 1980, J. Chim. Phys. **77**, 605.

Morin, P., I. Nenner, M.Y. Adam, M.-J. Hubin-Franskin, J. Delwiche, H. Lefebvre-Brion and A. Giusti-Suzor, 1982a, Chem. Phys. Lett. **92**, 609.
Morin, P., I. Nenner, P.-M. Guyon, L.F.A. Ferreira and K. Ito, 1982b, Chem. Phys. Lett. **92**, 103.
Morin, P., M.Y. Adam, I. Nenner, J. Delwiche, M.-J. Hubin-Franskin and P. Lablanquie, 1983, Nucl. Instrum. Methods **208**, 761.
Morin, P., I. Nenner, M.Y. Adam, P. Lablanquie, J. Delwiche and M.-J. Hubin-Franskin, 1987, to be published.
Morioka, Y., M. Nakamura, E. Ishiguro and M. Sasanuma, 1974, J. Chem. Phys. **61**, 1426.
Müller-Dethlefs, K., M. Sander and E.W. Schlag, 1984, Chem. Phys. Lett. **112**, 291.
Nakamura, M., M. Sasanuma, S. Sato, M. Watanabe, H. Yamashita, Y. Iguchi, A. Ejiri, S. Nakai, S. Yamaguchi, T. Sagawa, Y. Nakai and T. Oshio, 1969, Phys. Rev. **178**, 80.
Nakamura, M., Y. Morioka, T. Hayaishi, E. Ishiguro and M. Sasanuma, 1971, in: Conference Digest, Third Int. Conf. on Vacuum Ultraviolet Radiation Physics, ed. Y. Nakai (Physical Society of Japan, Tokyo) p. 1pA1-6.
Natalis, P., J. Delwiche, J.E. Collin, G. Caprace and M.-T. Praet, 1977, Chem. Phys. Lett. **49**, 177.
Nefedov, V.I., 1970, Zh. Strukt. Khim. **11**, 292 [J. Struct. Chem. **11**, 272].
Nenner, I., 1983, Laser Chem. **3**, 339.
Nenner, I., and A. Beswick, 1987, ch. 6, this volume.
Newton, R.G., 1966, Scattering Theory of Waves and Particles (McGraw-Hill, New york) p. 457.
Niehaus, A., and M.W. Ruf, 1972, Z. Phys. **252**, 84.
Ninomiya, K., E. Ishiguro, S. Iwata, A. Mikuni and T. Sasaki, 1981, J. Phys. B **14**, 1777.
Ogawa, M., and Y. Tanaka, 1962, Can. J. Phys. **40**, 1593.
Ono, Y., E.A. Osuch and C.Y. Ng, 1982, J. Chem. Phys. **76**, 3905.
Orel, A.E., T.N. Rescigno, B.V. McKoy and P.W. Langhoff, 1980, J. Chem. Phys. **72**, 1265.
Padial, N., G. Csanak, B.V. McKoy and P.W. Langhoff, 1978, J. Chem. Phys. **69**, 2992.
Padial, N., G. Csanak, B.V. McKoy and P.W. Langhoff, 1981a, Phys. Rev. A **23**, 218.
Padial, N., G. Csanak and P.W. Langhoff, 1981b, J. Chem. Phys. **74**, 4581.
Parr, A.C., R. Stockbauer, B.E. Cole, D.L. Ederer, J.L. Dehmer and J.B. West, 1980, Nucl. Instrum. Methods **172**, 357.
Parr, A.C., D.L. Ederer, B.E. Cole, J.B. West, R. Stockbauer, K. Codling and J.L. Dehmer, 1981, Phys. Rev. Lett. **46**, 22.
Parr, A.C., D.L. Ederer, J.B. West, D.M.P. Holland and J.L. Dehmer, 1982a, J. Chem. Phys. **76**, 4349.
Parr, A.C., D. Ederer, J.L. Dehmer and D.M.P. Holland, 1982b, J. Chem. Phys. **77**, 111.
Parr, A.C., S.H. Southworth, J.L. Dehmer and D.M.P. Holland, 1983, Nucl. Instrum. Methods **208**, 767.
Parr, A.C., S.H. Southworth, J.L. Dehmer and D.M.P. Holland, 1984, Nucl. Instrum. Methods **222**, 221.
Pavlovic, Z., M.J.W. Boness, A. Herzenberg and G.J. Schulz, 1972, Phys. Rev. A **6**, 676.
Peatman, W.B., 1976, J. Chem. Phys. **64**, 4368.
Peatman, W.B., B. Gotchev, P. Gürtler, E.-E. Koch and V. Saile, 1978, J. Chem. Phys. **69**, 2089.
Piancastelli, M.N., P.R. Keller, J.W. Taylor, F.A. Grimm, T.A. Carlson, M.O. Krause and D. Lichtenberger, 1984, J. Electron Spectrosc. **34**, 205.
Plummer, E.W., T. Gustafsson, W. Gudat and D.E. Eastman, 1977, Phys. Rev. A **15**, 2339.
Poliakoff, E.D., J.L. Dehmer, D. Dill, A.C. Parr, K.H. Jackson and R.N. Zare, 1981, Phys. Rev. Lett. **46**, 907.
Poliakoff, E.D., J.L. Dehmer, A.C. Parr and G.E. Leroi, 1982, J. Chem. Phys. **77**, 5243.
Potts, A.W., H.J. Lempka, D.G. Streets and W.C. Price, 1970, Philos. Trans. R. Soc. London A **268**, 59.
Pratt, S.T., E.D. Poliakoff, P.M. Dehmer and J.L. Dehmer, 1983a, J. Chem. Phys. **78**, 65.
Pratt, S.T., P.M. Dehmer and J.L. Dehmer, 1983b, J. Chem. Phys. **78**, 4315.
Pratt, S.T., P.M. Dehmer and J.L. Dehmer, 1983c, J. Chem. Phys. **79**, 3234.
Pratt, S.T., P.M. Dehmer and J.L. Dehmer, 1984a, Chem. Phys. Lett. **105**, 28.
Pratt, S.T., P.M. Dehmer and J.L. Dehmer, 1984b, J. Chem. Phys. **80**, 1706.

Pratt, S.T., P.M. Dehmer and J.L. Dehmer, 1984c, J. Chem. Phys. **81**, 3444.
Price, W.C., 1968, J. Mol. Spectrosc. **4**, 221.
Pulm, H., B. Marquardt, H.-J. Freund, R. Engelhardt, K. Seki, U. Karlsson, E.-E. Koch and W. von Niessen, 1985, Chem. Phys. **92**, 457.
Raoult, M., and Ch. Jungen, 1981, J. Chem. Phys. **74**, 3388.
Raoult, M., Ch. Jungen and D. Dill, 1980, J. Chim. Phys. **77**, 599.
Raoult, M., H. Le Rouzo, G. Raseev and H. Lefebvre-Brion, 1983, J. Phys. B **16**, 4601.
Raseev, G., 1985, J. Phys. B **18**, 423.
Raseev, G., and H. Le Rouzo, 1983, Phys. Rev. A **27**, 268.
Raseev, G., H. Le Rouzo and H. Lefebvre-Brion, 1980, J. Chem. Phys. **72**, 5701.
Raseev, G., H. Lefebvre-Brion, H. Le Rouzo and A.L. Roche, 1981, J. Chem. Phys. **74**, 6686.
Rescigno, T.N., and P.W. Langhoff, 1977, Chem. Phys. Lett. **51**, 65.
Rescigno, T.N., C.F. Bender, B.V. McKoy and P.W. Langhoff, 1978, J. Chem. Phys. **68**, 970.
Richards, J.A., and F.P. Larkins, 1984, J. Phys. B **17**, 1015.
Richards, J.A., and F.P. Larkins, 1986, J. Phys. B **19**, 1945.
Ritchie, B., and B.R. Tambe, 1980, J. Phys. B **13**, L225.
Robin, M.B., 1975, Chem. Phys. Lett. **31**, 140.
Rohr, K., 1979, J. Phys. B **12**, L185.
Roy, P., I. Nenner, M.Y. Adam, J. Delwiche, M.-J. Hubin-Franskin, P. Lablanquie and D. Roy, 1984, Chem. Phys. Lett. **109**, 607.
Rumble, J.R., D.G. Truhlar and M.A. Morrison, 1981, J. Phys. B **14**, L301.
Sachenko, V.P., V.E. Polozhentsev, A.P. Kovtun, Yu.F. Migal, R.V. Vedrinski and V.V. Kolesnikov, 1974, Phys. Lett. A **48**, 169.
Samson, J.A.R., and J.L. Gardner, 1976, J. Electron Spectrosc. **8**, 35.
Samson, J.A.R., and A.F. Starace, 1975, J. Phys. B **8**, 1806.
Samson, J.A.R., J.L. Gardner and G.N. Haddad, 1977a, J. Electron Spectrosc. **12**, 281.
Samson, J.A.R., G.N. Haddad and J.L. Gardner, 1977b, J. Phys. B **10**, 1749.
Sasanuma, M., E. Ishiguro, H. Masuko, Y. Morioka and M. Nakamura, 1978, J. Phys. B **11**, 3655.
Sasanuma, M., E. Ishiguro, T. Hayaisha, H. Masuko, Y. Morioka, T. Nakajima and M. Nakamura, 1979, J. Phys. B **12**, 4057.
Sato, K., Y. Achiba and K. Kimura, 1984, J. Chem. Phys. **81**, 57.
Schäfers, F., M.A. Baig and U. Heinzmann, 1983, J. Phys. B **16**, L1.
Schirmer, J., and O. Walter, 1983, Chem. Phys. **78**, 201.
Schirmer, J., L.S. Cederbaum, W. Domcke and W. von Niessen, 1977, Chem. Phys. **26**, 149.
Schneider, B.I., and L.A. Collins, 1984, in: Resonances in Electron–Molecule Scattering, van der Waals Complexes and Reactive Chemical Dynamics, ACS Symposium Series, No. 263, ed. D.G. Truhlar (American Chemical Society, Washington, DC) ch. 5, p. 65.
Schneider, B.I., M. LeDourneuf and Vo Ky Lan, 1979, Phys. Rev. Lett. **43**, 1926.
Schönhense, G., V. Dzidzonou, S. Kaesdorf and U. Heinzmann, 1984, Phys. Rev. Lett. **52**, 811.
Schulz, G.J., 1973, Rev. Mod. Phys. **45**, 422.
Schulz, G.J., 1976, in: Principles of Laser Plasmas, ed. G. Bekefi (Wiley, New York) p. 33.
Schwarz, W.H.E., T.C. Chang and J.P. Connerade, 1977, Chem. Phys. Lett. **49**, 207.
Schwarz, W.H.E., L. Mensching, K.H. Hallmeier and R. Szargan, 1983, Chem. Phys. **82**, 57.
Sell, J.A., and A. Kuppermann, 1978, Chem. Phys. **33**, 379.
Sell, J.A., A. Kuppermann and D.M. Mintz, 1979, J. Electron Spectrosc. **16**, 127.
Sette, F., J. Stöhr and A.P. Hitchcock, 1984, J. Chem. Phys. **81**, 4906.
Shaw, D.A., G.C. King, F.H. Read and D. Cvejanovic, 1982, J. Phys. B **15**, 1785.
Shaw, D.A., G.C. King, D. Cvejanovic and F.H. Read, 1984, J. Phys. B **17**, 2091.
Shimamura, I., and K. Takayanagi, eds, 1984, Electron-Molecule Collisions (Plenum, New York).
Simpson, J., C. Kuyatt and S. Mielczarek, 1966, J. Chem. Phys. **44**, 4403.
Smith, M.E., R.R. Lucchese and V. McKoy, 1983, J. Chem. Phys. **79**, 1360.
Sodhi, R.N.S., 1984, Ph.D. Thesis (Department of Chemistry, Univ. British Columbia).
Sodhi, R.N.S., and C.E. Brion, 1984, J. Electron Spectrosc. **34**, 363.
Sodhi, R.N.S., and C.E. Brion, 1985a, J. Electron Spectrosc. **36**, 187.

Sodhi, R.N.S., and C.E. Brion, 1985b, J. Electron Spectrosc. **37**, 1.
Sodhi, R.N.S., and C.E. Brion, 1985c, J. Electron Spectrosc. **37**, 97.
Sodhi, R.N.S., and C.E. Brion, 1985d, J. Electron Spectrosc. **37**, 125.
Sodhi, R.N.S., C.E. Brion and R.G. Cavell, 1984, J. Electron Spectrosc. **34**, 373.
Sodhi, R.N.S., S. Daviel, C.E. Brion and G.G.B. de Souza, 1985, J. Electron Spectrosc. **35**, 45.
Southworth, S., W.D. Brewer, C.M. Truesdale, P.H. Kobrin, D.W. Lindle and D.A. Shirley, 1982b, J. Electron Spectrosc. **26**, 43.
Southworth, S.H., C.M. Truesdale, P.H. Kobrin, D.W. Lindle, W.D. Brewer and D.A. Shirley, 1982a, J. Chem. Phys. **76**, 143.
Southworth, S.H., A.C. Parr, J.E. Hardis and J.L. Dehmer, 1986, Phys. Rev. A **33**, 1020.
Starace, A.F., 1979, in: Handbuch der Physik, Vol. 31, ed. W. Mehlhorn (Springer, Berlin) p. 1.
Stephens, J.A., and D. Dill, 1983, in: Abstracts of Papers, Proc. of the XIII Int. Conf. on the Physics of Electronics and Atomic Collisions, eds J. Eichler, W. Fritsch, I.V. Hertel, N. Stolterfoht and U. Wille (ICPEAC e.V., Berlin) p. 34.
Stephens, J.A., and D. Dill, 1985, Phys. Rev. A **31**, 1968.
Stephens, J.A., D. Dill and J.L. Dehmer, 1981, J. Phys. B **14**, 3911.
Stockbauer, R., 1973, J. Chem. Phys. **58**, 3800.
Stockbauer, R., 1977, Int. J. Mass Spectrom. & Ion Phys. **25**, 89.
Stockbauer, R., 1980, Adv. Mass Spectrom. **8**, 79.
Stockbauer, R., and M.G. Inghram, 1975, J. Chem. Phys. **62**, 4862.
Stockbauer, R., and M.G. Inghram, 1976, J. Chem. Phys. **65**, 4081.
Stockbauer, R., B.E. Cole, D.L. Ederer, J.B. West, A.C. Parr and J.L. Dehmer, 1979, Phys. Rev. Lett. **43**, 757.
Stöhr, J., and R. Jaeger, 1982, Phys. Rev. B **26**, 4111.
Stöhr, J., J.L. Gland, W. Eberhardt, D. Outka, R.J. Madix, F. Sette, R.J. Koestner and U. Doebler, 1983, Phys. Rev. Lett. **51**, 2414.
Stöhr, J., F. Sette and A.L. Johnson, 1984, Phys. Rev. Lett. **53**, 1684.
Swanson, J.R., D. Dill and J.L. Dehmer, 1980, J. Phys. B **13**, L231.
Swanson, J.R., D. Dill and J.L. Dehmer, 1981a, J. Chem. Phys. **75**, 619.
Swanson, J.R., D. Dill and J.L. Dehmer, 1981b, J. Phys. B **14**, L207.
Tabché-Fouhailé, A., I. Nenner, P.-M. Guyon and J. Delwiche, 1981, J. Chem. Phys. **75**, 1129.
Tam, W.-C., and C.E. Brion, 1974, J. Electron Spectrosc. **3**, 281.
Tanaka, K., and I. Tanaka, 1973, J. Chem. Phys. **59**, 5042.
Teo, B.K., and D.C. Joy, eds, 1981, EXAFS Spectroscopy, Techniques and Applications (Plenum, New York).
Thiel, W., 1981, Chem. Phys. **57**, 227.
Thiel, W., 1982, Chem. Phys. Lett. **87**, 249.
Thiel, W., 1983, J. Electron Spectrosc. **31**, 151.
Tossell, J.A., 1985, J. Phys. B **18**, 387.
Tossell, J.A., and J.W. Davenport, 1984, J. Chem. Phys. **80**, 813.
Trajmar, S., and A. Chutjian, 1977, J. Phys. B **10**, 2943.
Tronc, M., G.C. King, R.C. Bradford and F.H. Read, 1976, J. Phys. B **9**, L555.
Tronc, M., G.C. King and F.H. Read, 1979, J. Phys. B **12**, 137.
Tronc, M., G.C. King and F.H. Read, 1980, J. Phys. B **13**, 999.
Truesdale, C.M., S.H. Southworth, P.H. Kobrin, U. Becker, D.W. Lindle, H.G. Kerkhoff and D.A. Shirley, 1983a, Phys. Rev. Lett. **50**, 1265.
Truesdale, C.M., S.H. Southworth, P.H. Kobrin, D.W. Lindle and D.A. Shirley, 1983b, J. Chem. Phys. **78**, 7117.
Truesdale, C.M., D.W. Lindle, P.H. Kobrin, U.E. Becker, H.G. Kerkhoff, P.A. Heimann, T.A. Ferrett and D.A. Shirley, 1984, J. Chem. Phys. **80**, 2319.
Truhlar, D.G., S. Trajmar and W. Wjlliams, 1972, J. Chem. Phys. **57**, 3250.
Unwin, R., I. Khan, N.V. Richardson, A.M. Bradshaw, L.S. Cederbaum and W. Domcke, 1981, Chem. Phys. Lett. **77**, 242.

van der Wiel, M.J., 1980, in: Electronic and Atomic Collisions, eds N. Oda and K. Takayanagi (North-Holland, Amsterdam) p. 209.
van der Wiel, M.J., and Th.M. El-Sherbini, 1972, Physica **59**, 453.
Vinogradov, A.S., B. Shlarbaum and T.M. Zimkina, 1974, Opt. Spektrosk. **36**, 658 [Opt. Spectrosc. **36**, 383].
von Niessen, W., L.S. Cederbaum, G.H.F. Diercksen and G. Hohlneicher, 1975, Chem. Phys. **11**, 399.
von Niessen, W., P. Kraemer and G.H.F. Diercksen, 1979, Chem. Phys. Lett. **63**, 65.
Wallace, S., 1980, Ph.D. Thesis (Boston University).
Wallace, S., D. Dill and J.L. Dehmer, 1979, J. Phys. B **12**, L417.
Wallace, S., D. Dill and J.L. Dehmer, 1982, J. Chem. Phys. **76**, 1217.
Wendin, G., 1981, in: Structure and Bonding, Vol. 45 (Springer, Heidelberg).
Werner, A.S., B.P. Tsai and T. Baer, 1974, J. Chem. Phys. **60**, 3650.
West, J.B., A.C. Parr, B.E Cole, D.L. Ederer, R. Stockbauer and J.L. Dehmer, 1980, J. Phys. B **13**, L105.
West, J.B., K. Codling, A.C. Parr, D.L. Ederer, B.E. Cole, R. Stockbauer and J.L. Dehmer, 1981, J. Phys. B **14**, 1791.
White, M.G., R.A. Rosenberg, G. Gabor, E.D. Poliakoff, G. Thornton, S.H. Southworth and D.A. Shirley, 1979, Rev. Sci. Instrum. **50**, 1288.
White, M.G., K.T. Leung and C.E. Brion, 1981, J. Electron Spectrosc. **23**, 127.
White, M.G., M. Seaver, W.A. Chupka and S.D. Colson, 1982, Phys. Rev. Lett. **49**, 28.
White, M.G., W.A. Chupka, M. Seaver, A. Woodward and S.D. Colson, 1984, J. Chem. Phys. **80**, 678.
Wight, G.R., and C.E. Brion, 1974a, J. Electron Spectrosc. **3**, 191.
Wight, G.R., and C.E. Brion, 1974b, J. Electron Spectrosc. **4**, 313.
Wight, G.R., and C.E. Brion, 1974c, J. Electron Spectrosc. **4**, 327.
Wight, G.R., and C.E. Brion, 1974d, J. Electron Spectrosc. **4**, 335.
Wight, G.R., and C.E. Brion, 1974e, J. Electron Spectrosc. **4**, 347.
Wight, G.R., C.E. Brion and M.J. van der Wiel, 1972/73, J. Electron Spectrosc. **1**, 457.
Wight, G.R., M.J. van der Wiel and C.E. Brion, 1976, J. Phys. B **9**, 675.
Williams, G.R.J., and P.W. Langhoff, 1981, Chem. Phys. Lett. **78**, 21.
Wilson, W.G., K.S. Viswanathan, E. Sekreta and J.P. Reilly, 1984, J. Phys. Chem. **88**, 672.
Winick, H., and S. Doniach, eds, 1980, Synchrotron Radiation Research (Plenum, New york).
Woodruff, P.R., and G.V. Marr, 1977, Proc. R. Soc. London Ser. A **358**, 87.
Wu, C.Y.R., and C.Y. Ng, 1982, J. Chem. Phys. **76**, 4406.
Zimkina, T.M., and V.A. Fomichev, 1966, Dokl. Akad. Nauk SSSR **169**, 1304 [Sov. Phys.-Dokl. **11**, 726].
Zimkina, T.M., and A.S. Vinogradov, 1971, J. Phys. Colloq. (France) **32**, 3.

CHAPTER 6

MOLECULAR PHOTODISSOCIATION AND PHOTOIONIZATION

IRÈNE NENNER

Commissariat à l'Energie Atomique, IRDI,
Département de Physico-Chimie, Centre d'Etudes Nucléaires de Saclay,
91191 Gif sur Yvette, France

LURE, Centre National de la Recherche Scientifique et Université de Paris Sud,
91405 Orsay, France

J. ALBERTO BESWICK

LURE, Centre National de la Recherche Scientifique et Université de Paris Sud,
91405 Orsay, France

Contents

1. General introduction.	357
2. Experimental methods	361
2.1. Absorption and emission spectroscopy	364
2.1.1. Absorption	364
2.1.2. Fluorescence spectroscopy	365
2.1.3. Time-resolved fluorescence	366
2.2. Mass spectrometry	367
2.3. Photoelectron spectroscopy	368
2.3.1. Energy spectrum, angular distribution and constant ionic state spectrum	368
2.3.2. Spin polarization	373
2.3.3. Time of flight photoelectron spectroscopy	374
2.3.4. Threshold photoelectron spectroscopy	377
2.4. Coincidence techniques	377

Contents continued overleaf

Handbook on Synchrotron Radiation, Vol. 2, edited by G.V. Marr
© *Elsevier Science Publishers B.V., 1987*

Contents continued

        2.4.1. Threshold photoelectron–photoion coincidence 378
        2.4.2. Photoion–photoion coincidences. 380
        2.4.3. Other coincidence methods 381
3. Selected examples 382
    3.1. Photodissociation 382
        3.1.1. Electronic branching ratios and quantum yields 382
        3.1.2. Predissociation lineshapes 385
        3.1.3. Vibrational and rotational distributions of photofragments 390
    3.2. Valence photoionization. 396
        3.2.1. Spectroscopy of singly charged ions 396
        3.2.2. Dissociative ionization 402
        3.2.3. Photoionization dynamics: partial cross sections and branching ratios. 407
        3.2.4. Photoionization dynamics: angular distribution of photoelectrons . . 419
        3.2.5. Photoionization dynamics: spin polarization of photoelectrons . . . 427
    3.3. Competition between photodissociation into neutrals and photoionization . . 428
    3.4. Alignment and polarization of the fluorescence following photoionization and photodissociation 432
    3.5. Double ionization 434
    3.6. Inner-shell photoexcitation and photoionization 438
        3.6.1. Near-edge resonances 438
        3.6.2. Electronic relaxation near core edges 443
        3.6.3. Ionic relaxation 450
4. Miscellaneous and future trends. 455
References 458

1. General introduction

Photodissociation and photoionization in molecular systems are important processes in a wide range of photon energies, extending from a few to hundreds of eV. Interest in their study arises for several reasons. They are basic reactive processes which constitute the primary steps in a large number of photochemical phenomena. Knowledge of the nature of the products and their detailed internal energy content is needed in order to understand the reactions in which they may participate in subsequent collisions. This is of considerable importance in diverse fundamental fields like atmospheric chemistry, astrophysics and plasma physics, as well as in applications such as laser design, combustion, isotope separation and selective photochemistry.

Although many detailed experiments in this field can be performed using discharge lamps and lasers, synchrotron radiation provides tunability over a wide spectral range and in this respect supersedes all other presently available sources. It is the purpose of this chapter to present the basic gas phase molecular photoionization and photodissociation processes using synchrotron radiation. We shall not attempt to review all the experimental and theoretical work in this very rich area, but rather to select examples which illustrate the basic mechanisms of dissociation and ionization in isolated (low-density gas phase) simple molecules. We are particularly interested in sorting out the most important features as well as the dynamical information that can be extracted from the experiments. Another important aspect in this context is the relation between atomic and molecular processes. When the same process can occur in atoms and molecules we have attempted to show the molecular specificity and what new is learned.

In what follows we shall designate by *photofragmentation* [as was suggested by Greene and Zare (1982b)] the ensemble of processes corresponding to the ejection (after photon absorption by a stable system) of one or more particles, irrespective of their nature: electrons, ionic or neutral fragments. In fig. 1.1 we present a schematic view of the most relevant photofragmentation processes in isolated molecules together with typical rates.

Loosely speaking, one can classify three relevant energy regions for the study of photoionization and photofragmentation:

   (i) the low-energy domain (1–10 eV), i.e. the energy region below the first ionization potential where valence excitations are dominant;

   (ii) the intermediate-energy range (10–50 eV), where valence electrons are excited mainly to Rydberg and continuum orbitals (although excited valence states are also populated); in this region ionization as well as dissociation occurs;

   (iii) the high-energy domain (50–1000 eV), where in addition to the above, excitation of core electrons is also possible.

In the low-energy domain the basic processes occurring upon photon excitation in isolated molecules are:

(1) radiative decay (RD),
(2) electronic relaxation between different bound electronic states (ER),
(3) intramolecular vibrational relaxation or vibrational redistribution (IVR),
(4) predissociation (PD),
(5) direct photodissociation (DP).

We are particularly interested in describing the last two processes (4) and (5), but it should be realized that very often competition with processes (1) to (3) exists. In particular, for medium and large molecules, electronic relaxation between bound electronic states as well as intramolecular vibrational energy redistribution competes efficiently with fragmentation or alternatively provides indirect routes for this process to occur.

Figure 1.2 presents a schematic outline of these reactive and nonreactive decay channels. Thus predissociation can occur directly from the initial (doorway) state, or indirectly via IVR or ER. Several mechanisms are responsible for predissociation:

(a) electronic predissociation, induced by nonadiabatic (through the nuclear kinetic energy operator, for example), spin–orbit, spin rotation, gyroscopic, etc., interactions;

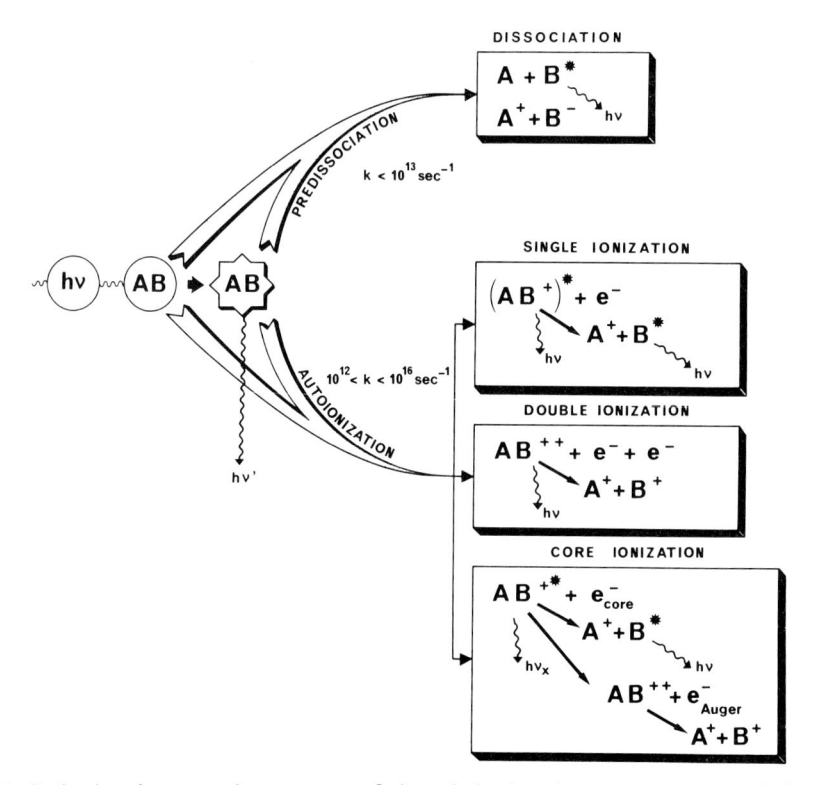

Fig. 1.1. Basic photofragmentation processes. Only typical values for rate constants are indicated. In particular, they exclude the Rydberg states with high principal quantum number, which are long lived.

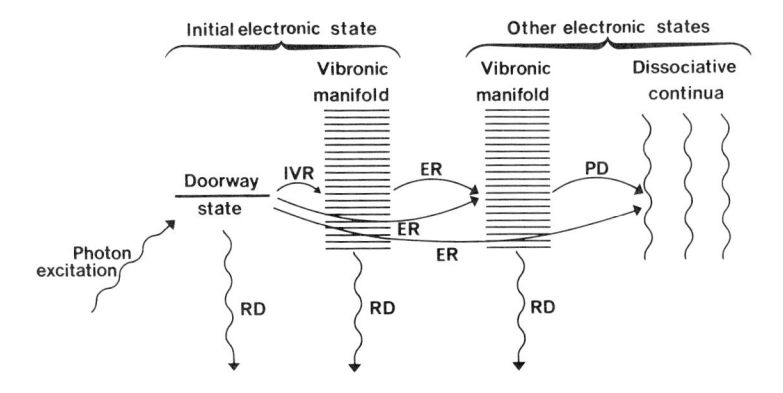

Fig. 1.2. Reactive and nonreactive decay channels in the low-energy domain.

IVR : Intramolecular Vibrational Relaxation
ER : Electronic Relaxation
PD : Predissociation
RD : Radiative decay

(b) vibrational predissociation and rotational predissociation, in which energy from internal vibrational and rotational degrees of freedom is transferred to the bond being broken;

(c) predissociation by rotation, which corresponds to tunneling through a centrifugal barrier.

In addition to predissociation, direct photodissociation (process 5) may also occur. In that case interference between these two processes will give rise to Fano–Beutler profiles (Fano 1961).

In the intermediate energy range (10–50 eV), ionization competes efficiently with the processes above. In complete analogy with the mechanisms in molecular fragmentation, there is direct ionization and autoionization (or preionization, as some prefer to call it to further stress the analogy with predissociation). Again, autoionization can be classified as:

(a) electronic autoionization, induced essentially by the electrostatic interaction between electrons, i.e. by the breakdown of the independent electron approximation, and by the spin–orbit interaction;

(b) vibrational and rotational autoionization,

(c) shape resonances, which are the analog of predissociation by rotation.

In contrast to the photodissociation case, interference between direct ionization and autoionization is fairly common, giving rise to the well-known Fano–Beutler profiles (Fano 1961).

We summarize in fig. 1.3 the different decay channels in this intermediate-energy range. The doorway state [often called "*the superexcited state*" after Platzman (1960)] can either autoionize, predissociate or fluoresce and competition between these processes has been observed in many cases. In this energy

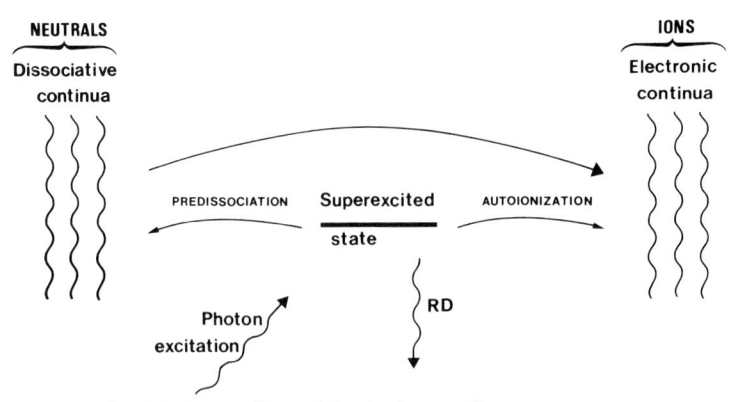

Fig. 1.3. Decay channels in the intermediate-energy range.

region, evidence for the departure from the independent-electron approximation is conspicuous: satellite lines, double ionization, etc.

Finally, the high-energy spectral region (50–1000 eV) corresponds to the excitation of inner-shell atomic-type orbitals (core electron excitation). The molecule is found to absorb in very specific energy regions concentrated around core ionization edges, each of them being very different from each other and characteristic of the atomic components of the molecule. In each core ionization region most features are found to be of pure molecular character. Excitation of a core electron into empty valence orbitals is a common event, followed by electronic and ionic relaxation processes showing strong molecular character because of the "delocalized" nature of these valence orbitals (see fig. 1.4). Above the core ionization threshold, the interaction of the outgoing electron with the molecular field may cause structures in the ionization cross sections. Interatomic Auger processes

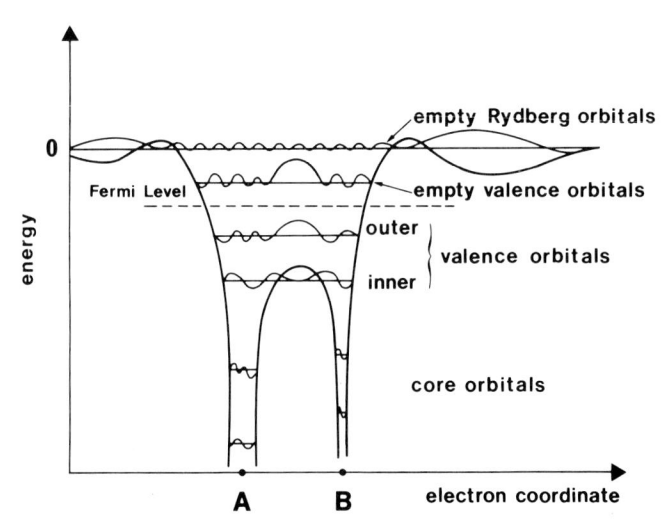

Fig. 1.4. Schematic view of the electronic potential in a diatomic molecule AB. The core, valence and Rydberg levels and wavefunctions are also shown.

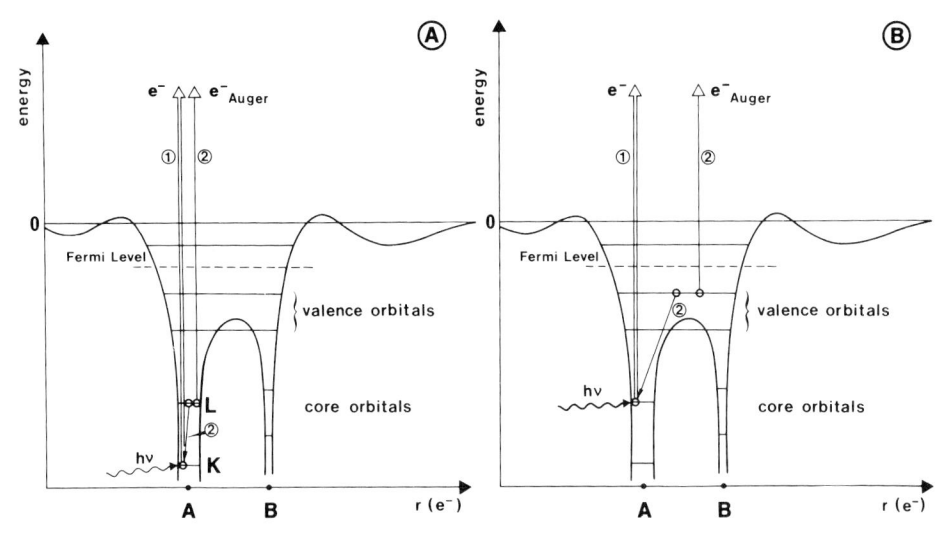

Fig. 1.5. Intra-atomic (a) and interatomic (b) Auger processes in a diatomic molecule AB.

(specific molecular effect) overcome intra-atomic Auger processes (atomic effect), because very often the outer energetically available orbitals, which may fill the core hole and eject the Auger electron, are valence orbitals and therefore delocalized over two or more atoms (see fig. 1.5). In addition to these specific molecular processes, many others exist which are very similar and/or exactly equivalent to those encountered in atoms: formation of a core hole followed by electronic relaxation leading to multiple excitation and ionization.

The basic information concerning the dynamics of the processes described above emerges from the following experimental observables:
– absorption cross sections,
– partial cross sections for the production of ions, electrons, photons, and neutrals,
– angular distributions of products,
– polarization of emitted photons and electrons,
– time-resolved photon counting and time-correlated measurements.

This chapter is organized as follows. Section 2 presents the experimental techniques. In section 3 we discuss selected examples of experimental work performed with synchrotron radiation, and their theoretical interpretation. We did not attempt to review all the work achieved in this rich field, but rather to present typical cases often chosen among those obtained at LURE (Orsay). Finally, section 4 is devoted to miscellaneous and future trends.

2. Experimental methods

Photodissociation and photoionization processes in molecules can be studied by a great variety of experimental techniques. In this section we shall first present the methods to produce monochromatized synchrotron radiation in the ultraviolet and soft X-ray regions. We shall consider almost exclusively the single-photon excitation of a neutral molecule. Note, however, that monochromatized synchrotron radiation can also be used in combination with another photon source (laser, discharge lamps) to perform two sequential photon experiments: excitation of a molecule and probing of the photodissociation products, for example. The general interest of those studies will be discussed in section 4.

Secondly, we shall present the detection techniques which allow the analysis of photon emission from the excited molecules, molecular ions and/or fragments, as well as those which detect ions (parents or fragments). It should be pointed out that the use of vacuum UV and soft X-ray radiation requires the detector to be under vacuum just as the beam transport system and the monochromator. A schematic view of the beam lines and monochromators installed on the ACO (LURE, Orsay) storage ring is shown in fig. 2.1. Also, the polarization properties of synchrotron radiation imposes additional geometrical arrangement constraints on the detection system. The polarization plane cannot be rotated at will (as with lasers). In addition, the source point in the experimental chamber is fixed. Consequently the detection system must be aligned or be eventually moveable with respect to the incident beam. This is usually tedious considering that the experimental set-up is under vacuum. On the other hand, an important property of synchrotron radiation sources is that they provide light pulses with a temporal width of the order of a nanosecond or less with a repetition rate in the MHz range. This is of particular importance for a UV experiment since not only it allows time-resolved fluorescence experiments to be performed in a new photon excitation region (see sections 2.2 and 2.3), but also it is very convenient for the analysis of the electron or ion energy spectrum by time of flight. Particularly important applications in this context are the detection of zero- or low-energy electrons with high efficiency and coincidence techniques (see section 2.4).

In addition, the wide tunability of synchrotron radiation sources in the far UV has revived and developed new techniques which could not be applied before. As seen in fig. 1.1, many fragmentation channels are usually in competition, i.e., they occur on the same time scale, and it is desirable to identify them independently. For example, when similar products are obtained, ion–ion or electron–electron coincidence techniques (see section 2.4) are important in order to distinguish double-ionization from single-ionization events.

The present intensity of synchrotron radiation sources produced by dipole magnets in the available storage rings is such that the number of photons per 10% bandwidth and per pulse ranges from 10^3 to 10^5 in the best cases. Due to the high repetition rates the average intensity ranges from 10^{10} to 10^{12} ph/10% bandwidth. These figures show clearly that synchrotron radiation sources have much lower average intensity per pulse than lasers. In addition, the resolving power reaches a maximum of 20 000 to 100 000 in the low-energy region ($\lambda \geqslant 400$ Å). On the other hand, synchrotron radiation sources and lasers do not cover the same spectral regions (except for the near UV) as lasers and, more importantly, synchrotron

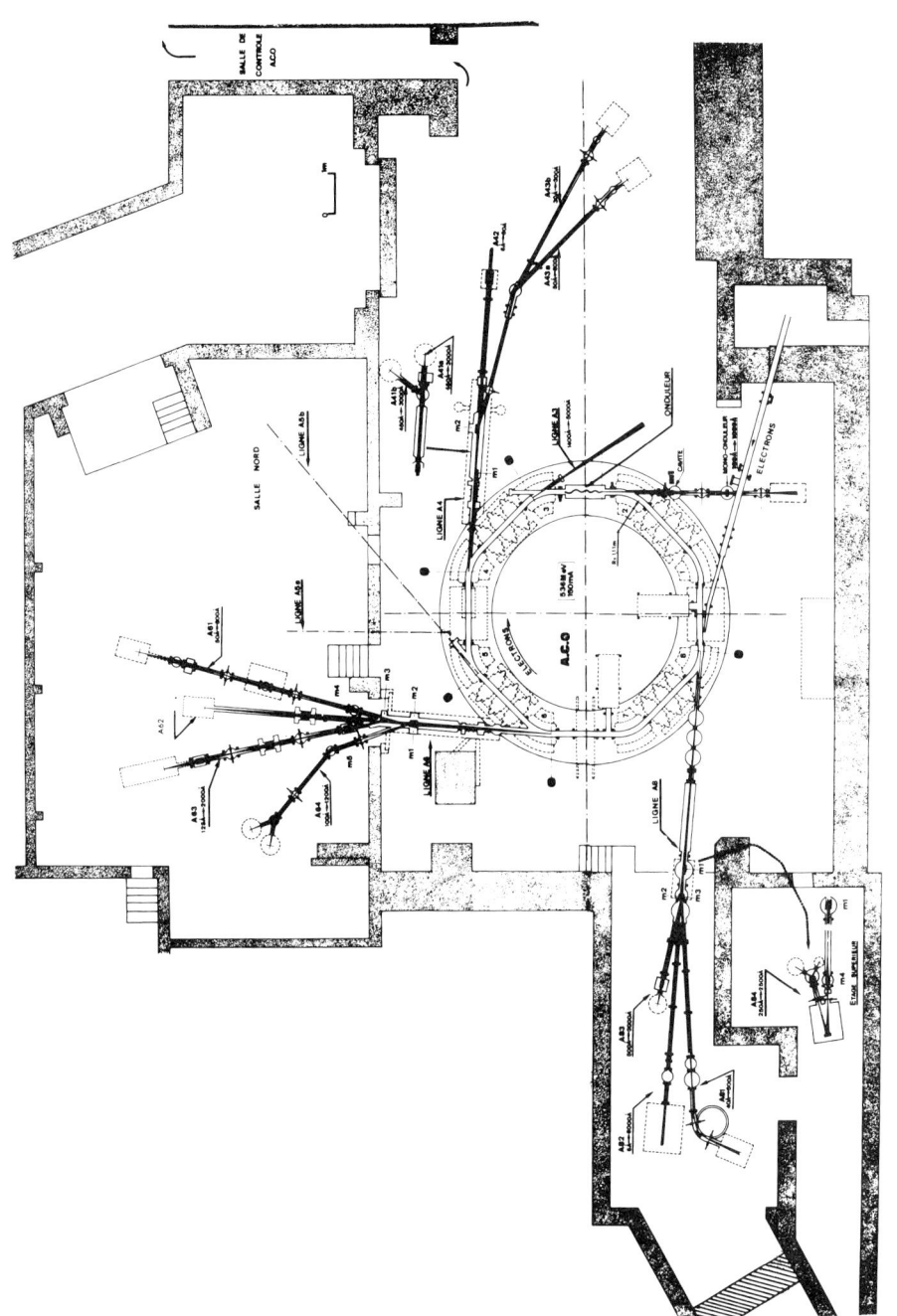

Fig. 2.1. Layout of the ACO (LURE, Orsay) storage ring for the UV and soft X-ray photon energy regions.

radiation sources have much larger tunability than lasers. [See Leach (1984) for a detailed comparison between lasers and synchrotron radiation sources.] Consequently, synchrotron radiation must be considered as a totally different source, which brings different information. At present, lasers are not routinely providing UV radiation below 1200 Å and they are not easily tunable. In addition, the low intensity per pulse and the high repetition rate of synchrotron radiation sources favors single and time-correlated detection methods, which are difficult to perform with lasers. In contrast, the high intensity and the spectral resolution of lasers provide information that synchrotron radiation sources obtained from dipole magnets will never provide. New developments in storage rings (existing or under construction), such as undulators and free-electron lasers, may provide in the near future (Petroff 1985 and references therein) a very significant increase in intensity (per pulse and on average) in the UV and soft X-ray regions as well as a decrease of the pulse width, allowing time-resolved experiments in the picosecond domain.

2.1. Absorption and emission spectroscopy

2.1.1. Absorption

Absorption spectroscopy combined with a tunable photon source in the low-, intermediate- and high-energy domains provides information on the first step of the excitation of the molecule (or atom) through allowed electric dipole transitions (resonances labeled AB* in fig. 1.1). Although absorption spectra do not give information on the subsequent individual decay channels (if there are several competing), they are important to "localize" interesting energy regions. The interest of an absorption spectrum is to provide a good measurement of the resonance width. This requires a suitable spectral resolution and rejection of high orders. Thus the quality of absorption spectra obtained with a photon source and a monochromator depends on the energy region.

Below 30 eV, which is the typical cut-off for normal incident grating monochromators, synchrotron radiation sources provide better quality absorption spectra than electron energy loss experiments (Sodhi and Brion 1984). Although rotational resolution is not generally achieved with synchrotron radiation sources [except for H_2, studied by Breton et al. (1980), and very recently in other diatomic and triatomic molecules, by Baig and Connerade (1984)], many resonances such as autoionization or shape resonances are short lived and are often resolved. In this energy domain various types of cells are used: windowless and traditional sealed cells equipped with suitable windows depending on the spectral range (Steele 1976, Gudat and Kunz 1978). Partial pressures cannot exceed 10^{-2} Torr in windowless cells because of the differential pumping efficiency and the vacuum requirements in the beam transport system and monochromator. The intensity of transmitted photons is measured with a combination of a scintillator and a visible photomultiplier, which provides a very good conversion efficiency from UV to visible radiation. Wavelength calibration is obtained with known resonances in standard gases.

At energies above 30 eV electron energy loss techniques (Daviel et al. 1984, King and Read 1985) provide better "absorption spectra". However, some recent developments on crystal monochromators have shown that a good signal to noise spectrum with a resolution of 6000 is feasible. As opposed to the low-energy domain, the cell is equipped with quite different materials. Polypropylene, for example, is 80% transparent at the sulfur edge (2500 eV) and only 40% at the fluorine edge. In addition, it must be thick enough (Steele 1976) to resist the pressure difference between the cell and the chamber. Other materials can be chosen according to the spectral region (see Gudat and Kunz 1978). Photon detectors are of various types. Flow proportional counters provide 40% efficiency, while H ionization chambers 80 to 100%. Channeltrons and multichannel plates are also used, but they have only ~2% efficiency. Energy calibration is more difficult than in the low-energy domain because resonances occur generally only in the vicinity of core edges. The electron energy loss method (Sodhi and Brion 1984, Daviel et al. 1984, King and Read 1985) has provided benchmarks with an accuracy of about 0.05 to 0.1 eV.

2.1.2. Fluorescence spectroscopy

Fluorescence is expected not only from excited neutrals but also from ions as indicated in fig. 1.1. When the emission spectral range of a well-defined species is known, the corresponding excitation spectrum is obtained by scanning the incident photon energy.

Generally, excitation spectra provide information on the energy threshold for the corresponding process as well as the relative probability as a function of the incident photon energy. Very often either narrow or wide resonant structures appear. Comparison with the absorption spectra measured with the same resolution provides valuable information on the decay channels.

When an excited molecular ion is produced by ejection of an electron, it can emit light by a transition to a lower electronic state. This fluorescence occurs when the emitting state is not (or only partially) predissociated.

For visible or near UV (i.e., $\lambda > 1200$ Å) a simple arrangement, schematically represented in fig. 2.2, can be used. Fluorescence is detected at right angles to the direction of the incident light by a photomultiplier and an appropriate optical filter. Since the optical filters are solid materials which can resist high pressures, the detector can be isolated from the cell at the atmospheric pressure of the laboratory. For emission in the far UV, the whole detection system must be at an appropriate low partial pressure and therefore included in the cell itself.

More refined measurements can be performed when the fluorescence is spectrally analysed using a secondary monochromator. After the pioneering experiments by Lavollée and Lopez-Delgado (1977) at ACO (Orsay), Monahan and Rehn (1978) at SPEAR (Stanford), and Brodmann et al. (1974) and Hahn et al. (1978) at DESY (Hamburg), a more sophisticated set-up (Wilcke et al. 1983,

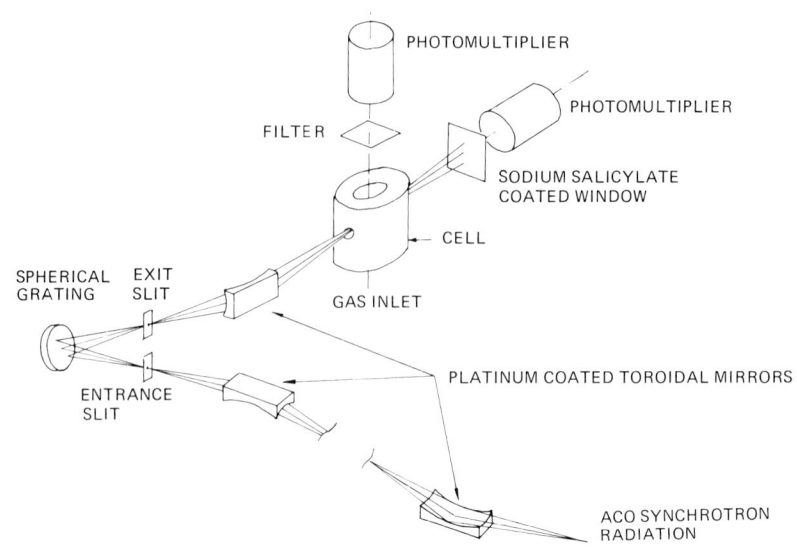

Fig. 2.2. Schematic diagram of the fluorescence experimental set-up operating on the A-84 beam line in ACO (LURE, Orsay).

Gürtler et al. 1983) was installed in HASYLAB at DESY and another at Brookhaven (Poliakoff et al. 1986a). The experimental station at DESY is called SUPERLUMI (see fig. 2.3). A combination of two monochromators allows the analysis of luminescence in an extremely large spectral range, from 50 to 1000 nm. The excitation VUV monochromator, specially designed with a large f-number, 1:2.8 in a working range of 50 to 300 nm, is associated with a secondary commercial monochromator.

Dispersed fluorescence spectra are a unique tool to analyse final states including regions of the potential energy surfaces located far outside the Franck–Condon region. This has been used successfully in various excimer studies (Zimmerer 1985 and references therein). More importantly, this technique combined with time-resolved measurements allows the study of transient processes in the sub-nanosecond time domain.

2.1.3. Time-resolved fluorescence

The pulsed character of synchrotron radiation has been applied to fluorescence lifetime measurements mainly from excited molecular states and from molecular ions. Generally the synchrotron light pulse is in the nanosecond domain (1.2 ns for ACO, Orsay, for example) or in the picosecond domain (50 ps in DESY, FRG), depending upon the size of the electron bunch. The repetition rate is on the order of a MHz, depending upon the size of the machine and the number of bunches. Up to now, these time characteristics have been found quite interesting for a number of molecular applications.

Since for neutral polyatomic molecules only low-lying excited states are expected to fluoresce, time-resolved experiments have been restricted to the low-

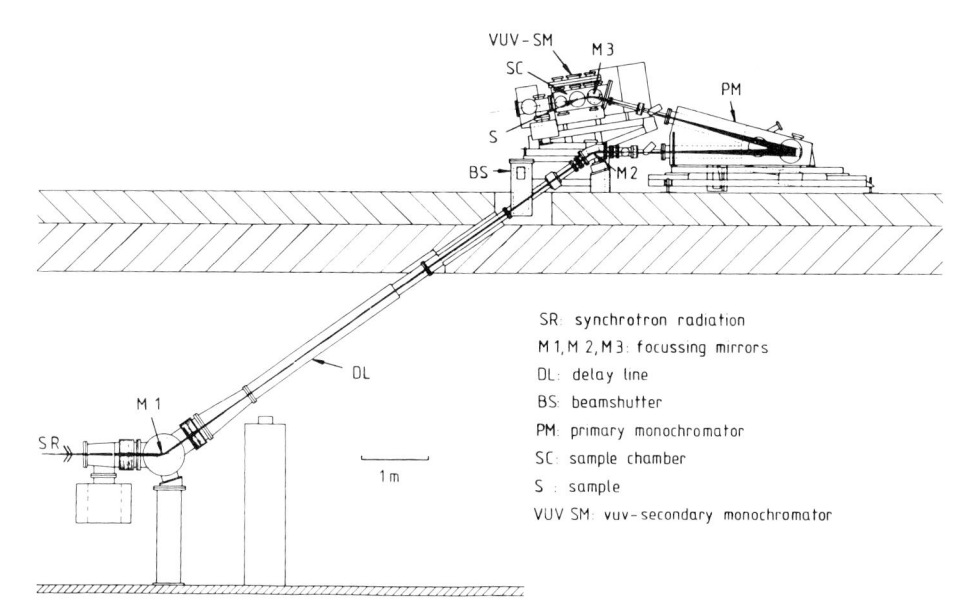

Fig. 2.3. Fluorescence experimental set-up SUPERLUMI, operating with the synchrotron radiation of the storage ring DORIS II at HASYLAB, DESY, Hamburg (from Wilcke et al. 1983).

energy domain, $E < 18$ eV. The selection of a neutral excited molecular or a neutral excited fragment state is obtained by choosing the photon energy and its fluorescence emission spectral range using an optical filter (see fig. 2.2) or a second monochromator (see fig. 2.3). The lifetime is obtained by measuring its decay with respect to the light pulse, using a standard single-photon counting technique. The time origin is actually provided by an electrode located in the storage ring, which senses the electron bunch.

With this method (see, for example, Zimmerer 1985), one can measure rise and decay curves with fixed excitation and fluorescence wavelengths, time-resolved fluorescence spectra with fixed excitation wavelength (within a time window at a fixed time delay) as a function of fluorescence wavelength, as well as time-resolved excitation spectra at fixed fluorescence wavelength (again within a time window) as a function of excitation wavelength.

2.2. Mass spectrometry

Above ~ 10 eV photon energy, the probability of ionizing a molecule is high and the corresponding quantum yield eventually reaches unity. Consequently, the detection of the total ion yield (parent and fragment ions) as a function of the incident photon wavelength is a very powerful method, which can be used as a complement to absorption and emission spectroscopy.

When the ionization quantum yield is unity and multiple ionization processes do not occur (or are negligible), the total ion yield is identical to the absorption cross section. The advantage in measuring the ion yield (or the total electron yield) is the great sensitivity of the method, allowing collision-free measurements. The method can also be very useful when absorption spectra are not available, because it can be used on standard mass spectrometers built for more sophisticated experiments.

When the ionization quantum yield is not unity, comparison with the absorption spectrum can give important information. The differences in resonance intensities or in intensities of broad continua reveal particular aspects of the competition between ionization and dissociation (see section 3.3).

Mass spectrometry provides, however, more detailed information on ionization pathways because of mass selection and ion kinetic energies measurements. Generally this method is applied to positive ions but also to negative ions. The interest of detecting the latter is in probing ion pair dissociative neutral states. This has been performed by Oertel et al. (1980).

Mass spectra combined with synchrotron radiation have been obtained with quadrupole (Beckmann et al. 1985) and more often with time of flight instruments, because of their high collection efficiency even for very energetic fragment ions, and also using coincidence techniques (see section 2.4).

Time of flight spectra require a time origin. This can be provided by different means:

(1) The synchrotron light pulse frequency may be used, but the very short period available between pulses considerably restricts the mass resolution.

(2) The total electron signal collected on a detector facing the time of flight spectrometer provides a time origin independently of the repetition rate of the photon source. It also allows the use of drift tubes of any length and therefore with any desirable mass resolution.

(3) The front edge of an extraction voltage can also provide a good time origin. This method is suitable when no electron detector is available or for some coincidence techniques (see section 2.4). The mass resolution depends upon the pulse voltage amplitude and the length of the drift region.

2.3. Photoelectron spectroscopy

2.3.1. Energy spectrum, angular distribution and constant ionic state spectrum

As opposed to mass spectrometry and total ionization techniques, which provide information on the various excitation processes leading to the formation of ions, PhotoElectron Spectroscopy (PES) provides a direct measurement of the state of excitation of the ions. In the ionization process leading to a singly charged ion,

$$M_{X,v=0} + h\nu \to M^+_{i,v} + e^-(\varepsilon\lambda, \varepsilon\lambda', \ldots), \tag{2.1}$$

the energy analysis of the outgoing electron gives the internal energy distribution of the residual ion M^+ with respect to the ground electronic and vibrational state of the initial neutral molecule, provided that the energy of the incident photon is known, that the kinetic energy of the molecule and the ion are negligible, and that the target molecule is in its ground vibrational state. Therefore, depending on the overall resolution of the experiment, i.e., the incident photon bandpass and the bandwidth of the electron analyser, a peak in the PES corresponds to a rotational or vibrational level of a particular electronic state of the ion. In practice, only vibrational analysis of diatomic and polyatomic ions has been obtained with synchrotron radiation, standard electrostatic analysers and effusive beam targets. Rotational analysis of the ion will be available in the future only if rovibrational cooling of the initial neutral molecule is obtained by supersonic molecular beam techniques and if major improvements on monochromator resolution, intensity of synchrotron radiation sources, as well as high resolution of electron analysers are achieved.

The analysis of the photoelectron spectrum intensity distribution is not as straightforward, although it is the most important quantity to measure. The intensity of the peak is related to the partial photoionization cross section, which represents the probability for producing an ion in a given state at a given excitation energy. However, this measurement requires the definition of the geometry of the detection system with respect to the direction as well as the degree of polarization of the incident light.

Consider a light beam propagating along the y axis, the electron detection being in the direction defined by the three angles θ_x, θ_y, θ_z. The differential partial cross section (assuming infinite angular resolution) is then given by

$$\frac{\partial \sigma_i}{\partial \Omega} = \frac{\sigma_i}{4\pi} \{1 - \tfrac{1}{2}\beta[P_2((\cos \theta_y) - \tfrac{3}{2}\mathcal{P}(\cos^2 \theta_x - \cos^2 \theta_z)]\} , \quad (2.2)$$

where

$$P_2(\cos \theta) = \tfrac{3}{2}\cos^2 \theta - \tfrac{1}{2} , \quad (2.3)$$

$$\mathcal{P} = (I_x - I_z)/(I_x + I_z) , \quad (2.4)$$

with I being the incident photon intensity and β the asymmetry parameter that characterizes the angular distribution of the photoelectrons.

If the light is unpolarized ($\mathcal{P} = 0$) one obtains

$$\frac{\partial \sigma_i}{\partial \Omega} = \frac{\sigma_i}{4\pi} [1 - \tfrac{1}{2}\beta P_2(\cos \theta_y)] , \quad (2.5)$$

and thus it is convenient to place the detector in the xy plane and to rotate the detector, $-\pi \leq \theta_y \leq \pi$ (see fig. 2.4 top). By choosing the "magic angle" $\theta_y = 54.7°$, which reduces the second Legendre polynomial $P_2(\cos \theta)$ to zero, the differential partial cross section becomes equal to the integral partial cross section.

If the light is polarized ($\mathcal{P} \neq 0$) as in the synchrotron radiation case, it is possible to use the above arrangement, but in addition, it is also possible to rotate the detector in a plane perpendicular to the direction of the incident light. In that

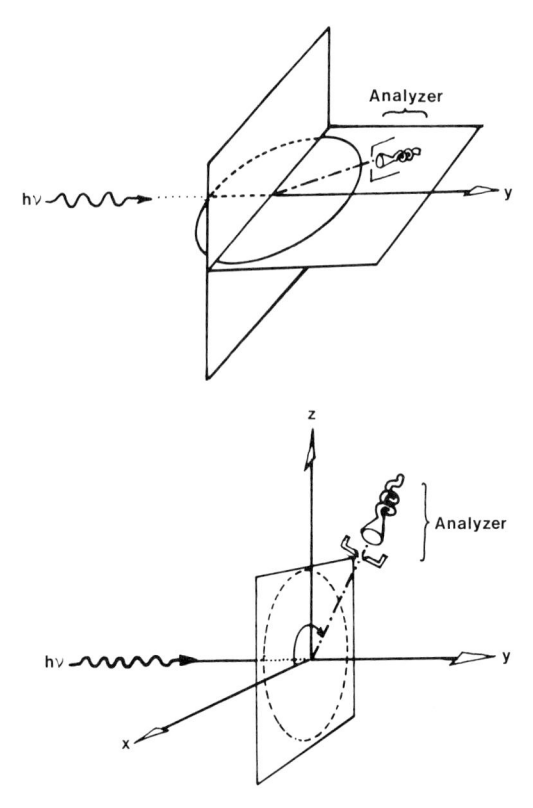

Fig. 2.4. Two possible geometrical arrangements for photoelectron angular distribution measurements. Top: The photoelectron analyser can rotate in a plane containing the propagation direction of the photon beam (xy). Bottom: The photoelectron analyser can rotate in a plane perpendicular to the propagation direction of the photon beam (xz).

case $\theta_y = 0$ and eq. (2.2) reduces to

$$\frac{\partial \sigma_i}{\partial \Omega} = \frac{\sigma_i}{4\pi} [1 + \tfrac{1}{4}\beta(3\mathcal{P}\cos 2\theta_x + 1)]. \tag{2.6}$$

With this arrangement the detector can be rotated between $0 \leq \theta_x \leq \pi$ (see fig. 2.4 bottom) and the differential partial cross section is equal to the integral partial cross section at $\theta_x = \tfrac{1}{2}\arccos(3\mathcal{P}/4)$, which is known as the "pseudo magic angle", since it is not constant but depends on the degree of polarization of the light, which in turn depends on the photon energy. This arrangement has been chosen by Morin (1983), Morin et al. (1983) (shown in fig. 2.5), Parr et al. (1984), Derenbach and Schmidt (1984), and White and Grover (1983).

The asymmetry parameter β is also a very important observable to measure. It depends primarily upon the symmetry of the initial orbital and it is related to the

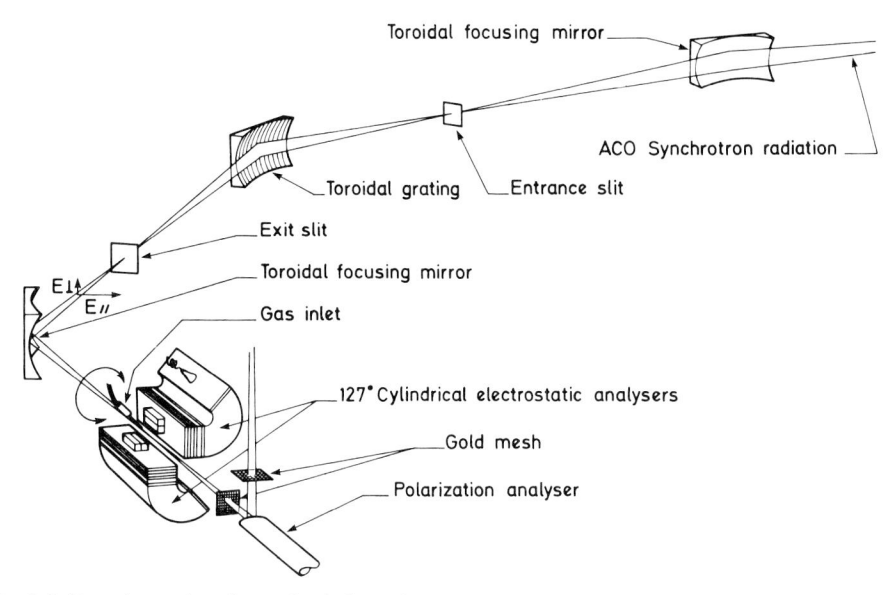

Fig. 2.5. Experimental angle-resolved photoelectron spectroscopy set-up, operating on the A-64 beam line in ACO (LURE, Orsay). From Morin et al. (1983).

dipole matrix elements in the various degenerate partial channels $\varepsilon\lambda$ (eq. 2.1), as well as to the cosine of the phase differences between the continuum wavefunctions of the outgoing electron. β can be obtained in two ways:

(1) By varying θ_x from 0 to π (fig. 2.4 bottom) or θ_y from 0 to π (fig. 2.4 top). The photon energy is fixed and the electron voltage is chosen to analyse a selected photoelectron band. Actually, measurements performed only at two angles are necessary (Krause et al. 1981) for obtaining β. However, more accurate measurements are obtained when the full angular dependence is measured (Derenbach and Schmidt 1984, Roy et al. 1984).

(2) By measuring two photoelectron spectra at two angles such as $\theta_x = 0$ and θ_{magic}. The photon energy is fixed and the ratio obtained as a function of binding energy is related to β by a relationship deduced from eq. (2.6),

$$\beta = 4(R-1)/(1+3P), \qquad (2.7)$$

with R being the ratio of the two PES measurements. This second method has the advantage of producing β(PES) and to allow comparison between the β of various bands.

The intensity of photoelectron bands can be measured continuously as a function of photon energy. If the ejection angle is chosen at the magic angle, the obtained spectrum is essentially proportional (after various corrections described below) to the partial cross section σ_i. The potential of the analyser (i.e. the analysed kinetic energy ε_k) and the photon energy are scanned simultaneously so that the binding energy E ($h\nu - \varepsilon_k$) remains constant. This method, known as the constant ionic state method (CIS), has been used by Plummer et al. (1977) and recently by Morin et al. (1983) and Morin (1983). It is very convenient to obtain in a short time a continuous measurement of a partial cross section and it is particularly powerful for studying autoionizing structures. However, it can only be applied to well-defined isolated photoelectron lines, for which the background is negligible or small. The CIS spectrum can be recorded at two angles, for example $\theta = 0$ and $\theta = \theta_{\text{magic}}$. From the ratio of the two curves, one can extract the continuous behavior of the β parameter.

The conversion of the intensity of a photoelectron peak obtained at a given photon wavelength into a differential partial cross section requires knowledge of the intensity of the incident photon beam, its polarization, the contribution of high diffraction orders of the monochromator, as well as the collection efficiency of the electron analyser as a function of electron energy. In addition, a comparison of a PES peak measured at two different photon wavelengths or two different angles requires the relative measurement of the gas density in the interaction region and the "source volume" correction, which may significantly affect the measurements (Morin 1983, Parr et al. 1984).

The incident photon intensity can be measured using a visible photomultiplier placed behind a scintillator such as a sodium salicylate coated window, which converts efficiently light into visible (Gudat and Kunz 1978). The photoelectron current measured on a gold mesh gives also a good measure, since the photoelectron yield of gold is known with great accuracy (Gudat and Kunz 1978). The determination of the degree of polarization can be done by this gold mesh method combined with a 45° mirror or a grazing-angle arrangement (Samson 1978). It is possible, however, to measure it with reasonable accuracy with a standard gas by measuring the ratio R of two PES recorded at two different angles (see eq. 2.7). Therefore, in a given ionization process such as 1s electron ejection in He, for which $\beta = 2$, one deduces the polarization at any photon energy above 24.6 eV. In addition it is highly desirable to have the highest polarization degree in order to keep good accuracy in the $3\mathscr{P}\cos 2\theta_x$ term in eq. (2.6).

The collection efficiency of any electron analyser varies with electron energy and must be determined accurately. Use of the tunability of the synchrotron radiation sources provides an efficient way to measure collection efficiencies when a rare gas is used, because their partial photoionization cross sections are known accurately in a very large range of electron energies. One chooses a rare gas (Holland et al. 1982, Morin 1983) or a small molecule (Gardner and Samson 1976), depending on the photon energy range covered by the monochromator.

The contribution of high orders to the incident light intensity can only be determined accurately by PES of standard gases. Finally, it is important to note that the overall alignment of the photon beam and the experimental chamber is critical, in particular when the detection plane is vertical. The difficulties arise from the requirement of good focusing conditions and position of the beam in the interaction region, because the light goes often through a capillary which is needed for a good differential pumping.

2.3.2. Spin polarization

A general property of photoionization processes is that electrons ejected from unpolarized molecules (or atoms) can be spin polarized, if the incident light is circularly polarized or unpolarized. Experimental evidence of spin polarization has been obtained by Heinzmann and collaborators (see Heinzmann 1980) in a variety of atoms and molecules using a Mott detector. This detector (see Heinzmann 1978) takes advantage of the spin-specific character of fast electron collisions with a gold foil. The interaction of the fast electrons with the nucleus makes the backscattered electrons to be preferentially ejected in well-defined directions depending on the original spin of the incident beam. Spin-resolved photoionization studies are particularly important because they provide information on the matrix elements of the dipole operator as well as on the difference between the phases of the continuum wavefunctions.

Circularly polarized light can be obtained with synchrotron radiation emitted out of the electron orbital plane. The spin polarization components A, $P_\perp$ and P_p (see fig. 2.6) are defined by the polar electron emission angle θ and azimuthal orientation angle ϕ. When $\phi = 0$, i.e. in the geometrical arrangement shown in fig. 2.6, the total photoelectron intensity measured at the angle θ is given by (see eq. 2.5)

$$I(\theta) \propto \sigma_i [1 - \tfrac{1}{2} \beta P_2(\cos \theta)], \qquad (2.8)$$

where β is the asymmetry parameter of the photoelectron angular distribution defined in the preceding section. The spin polarization components, on the other hand, are given by

$$P_\perp(\theta) = \frac{2\xi \sin \theta \cos \theta}{1 - \tfrac{1}{2} \beta P_2(\cos \theta)}, \qquad (2.9)$$

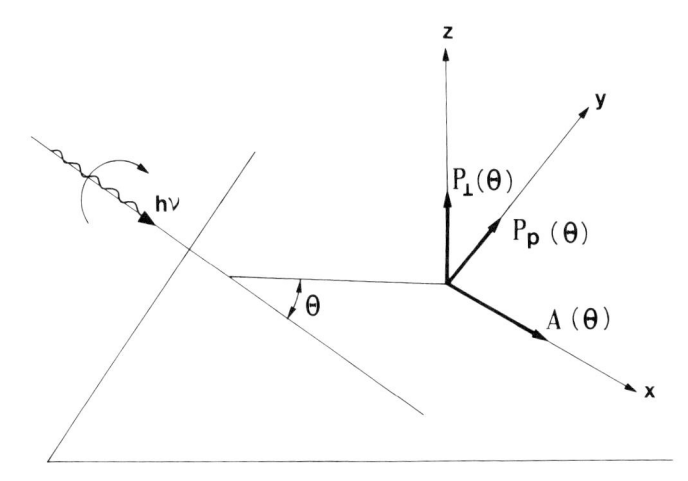

Fig. 2.6. A possible geometrical arrangement for spin polarization measurements.

$$A(\theta) = \pm \frac{A - \alpha P_2(\cos\theta)}{1 - \frac{1}{2}\beta P_2(\cos\theta)} , \tag{2.10}$$

$$P_p(\theta) = \pm \frac{\alpha \sin\theta \cos\theta}{1 - \frac{1}{2}\beta P_2(\cos\theta)} , \tag{2.11}$$

where the plus and minus sign correspond to right and left circularly polarized light, respectively.

Three new dynamical parameters appear in the above: ξ, A and α. The ξ parameter contains information on both the transition dipole moments and the phases of the continuum wavefunctions. For two continua ($\varepsilon d_{3/2}$ and $\varepsilon s_{1/2}$ after ionization of the $np_{1/2}$ subshell, for example),

$$\xi \propto \frac{D_s D_d}{D_s^2 + D_d^2} \sin(\delta_d - \delta_s) . \tag{2.12}$$

An interesting point to note is that β, the asymmetry parameter of the photoelectron angular distribution, also depends on those quantities, but it involves the cosine of the difference (see section 2.4), while ξ is directly proportional to the sine of this same difference. This makes the spin polarization measurements more sensitive to small differences in the phases.

Usually the spin measurements require the resolution of the photon energy, photon polarization, electron angular distribution, electron kinetic energy, and the electron spin. The experimental set-up used by Heinzmann and collaborators is shown in fig. 2.7. It contains a hemispherical analyser rotatable around the direction normal to the reaction plane. The electron beam is then deflected through 90° and impinges on the gold foil of the Mott detector. The low efficiency of the Mott detector (10^{-3}) requires a high photon flux. With the arrangement described above typical counting rates are a few counts per second.

2.3.3. Time of flight photoelectron spectroscopy

The pulsed nature of synchrotron radiation is ideal to develop Time Of Flight (TOF) spectrometers (especially for electrons) because of their very high transmission efficiency and their unique capability to analyse zero or very low kinetic energy particles. The repetition rate of synchrotron radiation defines the period of time during which a particle can be time analysed. The largest period leads to the largest time resolution, i.e. the largest energy resolution for electrons and the largest mass resolution for ions. The period between light pulses T is fixed by the circumference of the storage ring L and the number of bunches n by

$$T = L/nv , \tag{2.13}$$

where v is the speed of the electrons (or positrons) in the ring, which is close to the speed of light. The optimal storage ring will have the largest circumference and will operate with a very low number of bunches (ideally only one). Of course

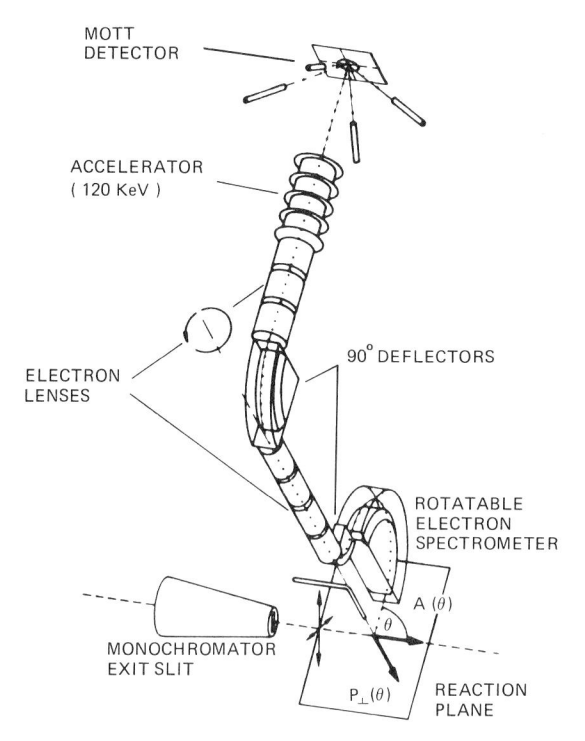

Fig. 2.7. Schematic diagram of the spin-resolved photoelectron spectroscopy set-up operating at the BESSY storage ring using circularly polarized light (from Heckenkamp et al. 1984).

the one-bunch operation mode also implies low total current and this may be a source of conflict with other user needs. There are two ways to overcome this problem. One is a multibunch operation mode combined with a mechanical rejection of bunches at an optimized frequency or pulsed electrical plates acting as a synchronized sweep in the time of flight spectrometer itself. The second is to operate at one or a few bunches combined with the use of wigglers or undulators to increase the intensity. Neither of these methods are used at present, but efforts are under way in connection with existing rings or with rings under construction. These methods could be also very interesting for other techniques, such as fluorescence (see section 2.1).

Time of flight electron spectrometers presently in operation with synchrotron radiation have been described by White et al. (1979) and Baer et al. (1979). In the LURE facilities at Orsay, for example, the ACO storage ring routinely provides a one-bunch operation mode, i.e. a 73 ns period between pulses. This period is too short to analyse electrons with kinetic energy higher than a few eV, because the energy analyser resolution deteriorates as the electron kinetic energy increases. However, this period is well suited for analysing very slow electrons down to zero kinetic energy.

Figure 2.8 shows a schematic diagram of the apparatus built at ACO (LURE,

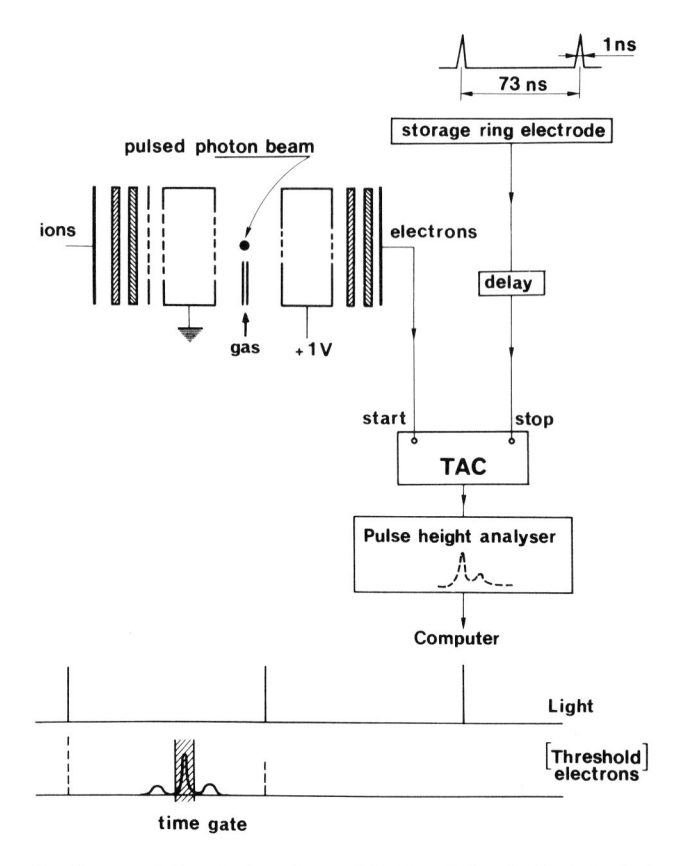

Fig. 2.8. Schematic diagram of the electron time of flight and threshold photoelectron spectroscopy spectrometer operating at ACO (LURE, Orsay), using the pulsed properties of the synchrotron radiation.

Orsay). The monochromatized photon beam crosses at right angles an effusive jet of gas. Ions and electrons are ejected from the source in opposite directions due to an electric field applied in the ionization region. The time of flight is measured between the signal given by an electrode located in the storage ring, which detects the passage of the electron bunch, and the photoelectron signal. The strength of the electric field is adapted to the period T between light pulses. So the time of flight of electrons corresponding to zero kinetic energy is less than T.

The angular discrimination provided by the finite opening on the electron side leads to a preferential collection of electrons with a velocity oriented along the detection axis (forward or backward). Energetic electrons are rejected more likely. The transmission function of such TOF analysers is then a strong function of the kinetic energy of the electrons. The advantage of this method compared to electrostatic analysers is that they provide a collection efficiency which can be of the order of 50% for zero kinetic energy electrons (Morin et al. 1980).

2.3.4. Threshold photoelectron spectroscopy

The first use of a time of flight spectrometer associated with synchroton radiation sources was for detecting threshold electrons, i.e. zero kinetic energy electrons. The selection of threshold electrons is done in two stages. The first is angular discrimination (diaphragms), which allows a maximum detection efficiency of threshold electrons and rejection of a large fraction of high-energy electrons (Peatman et al. 1969, Spohr et al. 1971, Peatman et al. 1975, Baer 1979). The time of flight is measured between the signal given by an electrode located in the storage ring, which detects the passage of the electron bunch, and the photoelectron signal. The second stage of selection is performed by the time window centered around zero kinetic energy electrons, schematically shown in the lower part of fig. 2.8.

The intensity of such electrons is recorded as a function of the incident photon energy. A Threshold PhotoElectron Spectrum (TPES) is obtained by taking the ratio of the electron signal and the incident photon intensity. One obtains a signal every time there is an ion state accessible. The most intense signals are given by ion states formed by direct ionization processes. This can be compared with the photoelectron spectrum obtained with line sources and conventional electrostatic analyzers. In addition, some weak structures are seen sometimes originating from resonant autoionization processes, i.e. involving an excited neutral molecular state which coincides in energy with an ionic state which cannot be reached by direction ionization. This autoionization process leads to threshold electrons associated with very high vibrationally excited states of the ion, which cannot be studied by standard techniques (see section 3.2.1).

2.4. Coincidence techniques

The detection in coincidence of two (or more) particles produced after photoabsorption by a molecule is of particular importance for studying photoionization processes. This is due to the fact that, in contrast to photoexcitation processes, photoionization produces at least two particles (see fig. 1.1). For example, in the simple event

$$h\nu + M \rightarrow M^+ + e^- \tag{2.14}$$

the photon energy is shared between the internal energy of the residual ion and the kinetic energy of the electron. As a consequence, for a given photon energy, the molecular ion can be formed in many different internal energy states, ranging from $h\nu$ to $(h\nu - IP)$, where IP denotes the first ionization potential. The selection of the internal state of the ion requires the measurement of both the photon and the electron energies. If a particular state of the ion decays via emission or by dissociation, the detection of the subsequent products has to be done in coincidence with the primary electron.

Consequently, the coincidence detection technique is essentially based on time-correlated events. The fastest particle (for example the photoelectron) gives the signal for an ionization event that has been produced some 20 to 100 ns earlier. Any other particle (fragment ion or photon) produced in the same event will arrive at the detector having a specific delay time with respect to the photoelectron signal. All uncorrelated particles will arrive at random time intervals. Actually this is true only if the light source is essentially continuous in time, or, in other words, if the time period between two successive light pulses is short compared to the time of flight of each one of the correlated particles under study. If the period of the light pulses is comparable to the time of flight of one of the particles, then the uncorrelated events, often called false coincidences, may produce a structured coincidence spectrum.

False coincidences can be drastically reduced if the time separation between two successive ionization events is much longer than the analysing time of the slowest particle. Thus the best signal to noise ratio is obtained by keeping the number of ionization events at a low level. This can be obtained by reducing the photon flux and/or the gas pressure in the source. However, additional important factors contribute to the coincidence count rate: the collection efficiency of both detectors. Thus high-transmission analysers such as the TOF are of particular importance for such experiments.

2.4.1. Threshold photoelectron–photoion coincidence

The Threshold PhotoElectron–PhotoIon COincidence (TPEPICO) method can be considered as a tool to study dissociative pathways of state-selected parent molecular ions. The measurement of kinetic energy distributions of mass-selected fragments provides the relative probabilities of dissociation. The principle is illustrated in fig. 2.9, where only two dissociative states of AB^+ can be produced

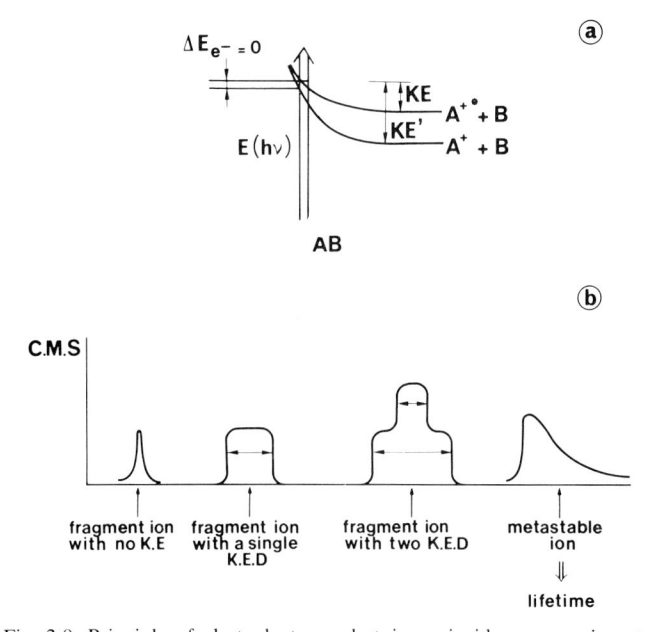

Fig. 2.9. Principle of photoelectron–photoion coincidence experiment.

at a given photon energy. One expects the fragment ion A^+ to have a bimodal kinetic energy distribution. Using TOF techniques one can extract, from the shape of the coincidence peak, the kinetic energy distribution (Baer 1979). Various situations are illustrated in fig. 2.9, i.e. no kinetic energy, single or double kinetic energy distribution. In addition, if the dissociation rate of the parent ion is such that its decay time is comparable to the time of flight scale of the spectrometer, one may expect an asymmetry of the Coincidence Mass Spectrometer (CMS) peak. This means that some parent ions are actually detected as their daughter ion. One may directly extract ion lifetimes, which, however, are limited to submicroseconds. Lastly, some asymmetry of the peaks may also arise from preferential angular directions of the outgoing fragments. Notice that, unlike photoelectron spectroscopy, with this method one can study the spectroscopy of repulsive ionic states.

Four instruments are presently in operation. The first is installed at LURE in Orsay and has been built by Guyon et al. (1978) (see also Nenner et al. 1980). The second is installed at BESSY in Berlin (Müller-Dethlefs et al. 1984), another has been installed at Daresbury by Frasinski et al. (1985), and more recently at UVSOR in Kyoto (Suzuki et al. 1986). We show in fig. 2.10 a schematic diagram

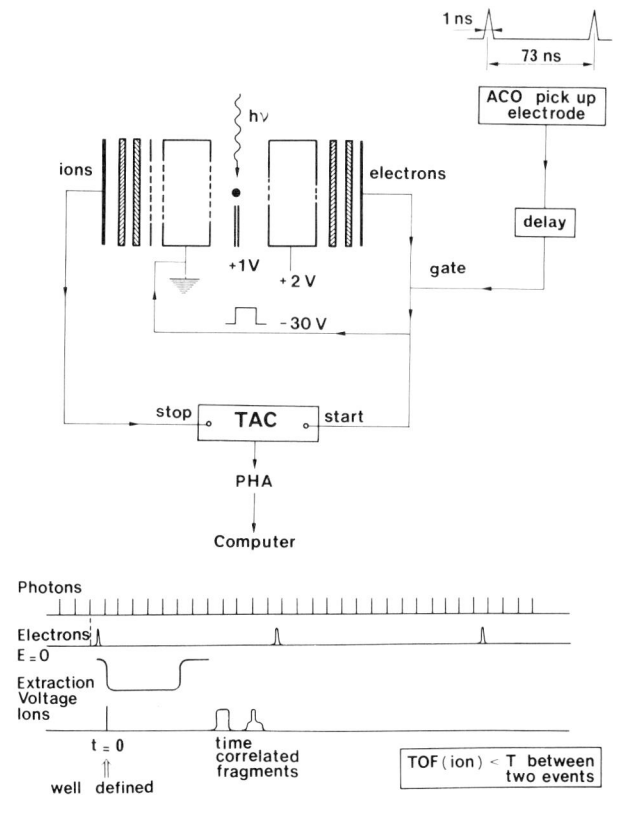

Fig. 2.10. Schematic diagram of the double time of flight spectrometer specially designed for the threshold photoelectron–photoion coincidence experiment.

of the TPEPICO set-up. The threshold electron signal detected above (see fig. 2.8) triggers a pulse voltage which extracts the ions at a well-defined time. The mass epectrum is then recorded by measuring the ion TOF with respect to the electron signal. More details of this technique can be found in the thesis of Richard-Viard (1982).

2.4.2. Photoion–photoion coincidences

The PhotoIon–PhotoIon COincidence (PIPICO) method is ideally suited to analyze any process producing a positive ion pair, i.e. soft double ionization and hard double ionization (Auger processes). For example (see fig. 1.1), $AB + h\nu$, giving AB^{++} plus two electrons with the subsequent decomposition of the doubly charged ion into $A^+ + B^+$ or $A^{++} + B$. AB^{++} and A^{++} can be detected by conventional mass spectrometry. The ion pair A^+ and B^+ can be detected by their time of flight difference, because they have different masses and/or different kinetic energies. This method, already known as autocorrelation (see Hitchcock and Van der Wiel 1979), is very interesting when coupled to synchrotron radiation (Dujardin et al. 1984, Lablanquie 1984, Lablanquie et al. 1985b) because of the tunability in the far UV up to the soft X-ray region. We present in fig. 2.11 a

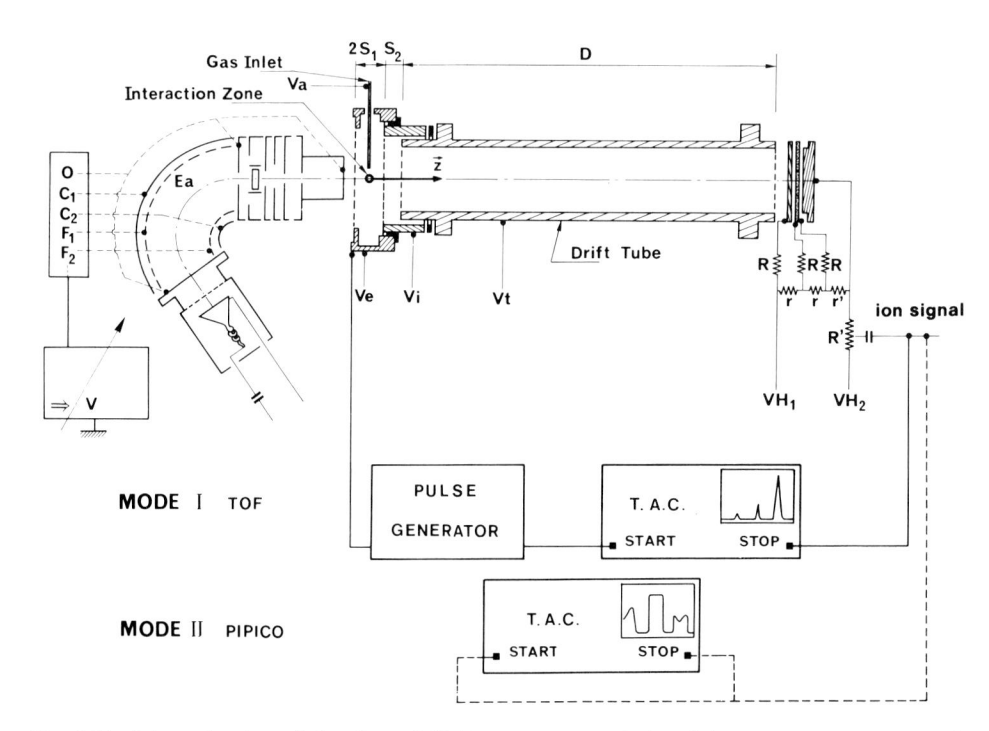

Fig. 2.11. Schematic view of the time of flight spectrometer designed for mass spectrometry and photoion–photoion coincidence measurements. The 127° electrostatic cylindrical analyser facing the TOF detector is designed for ion–electron and electron–electron coincidence experiments (from Lablanquie et al. 1985b, 1987).

schematic view of the time of flight spectrometer built by Lablanquie et al. (1985b), which can be used either in the conventional TOF mode with a pulsed voltage or in the PIPICO mode. In the latter, a constant voltage is applied in the ionization region and all ions are extracted continuously. A single detector provides both the start and the stop inputs of a time to amplitude converter. When a fragment ion A^+ is followed at a specific time by a fragment B^+ because they have been produced from the same dissociation event, the PIPICO spectrum shows well-defined peaks. From the mean time of flight difference one can extract the difference of masses and eventually the masses themselves.

From the width of the PIPICO peak, one can extract the kinetic energy released in the process. One obtains dynamical information on the dissociation process itself either for one, two or multiple bond breakings (Lablanquie 1984, Lablanquie et al. 1985b). The intensity of PIPICO peaks can be measured as a function of photon energy. From the onset of the curves one extracts information on the vertical transition into the AB^{++} potential curve (often repulsive). In the behavior of the curve, structures may reveal resonances, i.e. information on the dynamics of doubly charged ion formation.

2.4.3. Other coincidence methods

When a molecular ion fluoresces, one may measure its lifetime by a threshold electron–fluorescence coincidence technique. The threshold electron is measured by time of flight as discussed above. The luminescence is selected by an optical filter combined with a suitable photomultiplier. This method, known as TPEFCO, has been developed by Frey et al. (1978) and later by Dujardin et al. (1983). It can be applied to the study of radiationless transitions in the ion manifold states (Dujardin and Leach 1983).

In a double ionization process, the analysis of doubly charged ion states requires the energy analysis of both outgoing electrons in coincidence. In addition, the energy spectrum of the two electrons associated with the residual doubly charged ion (m^{2+}) can be obtained by photoelectron spectroscopy in coincidence with m^{2+}. This allows the rejection of all other electrons originating from single-ionization events. Both electron–electron and m^{2+}–electron coincidence measurements have been performed by Lablanquie et al. (1987) in argon. More sophisticated multicoincidence experiments (electron–ion–ion) have been achieved successfully by Frasinski et al. (1986) and Eland (1987). They have measured directly the partitioning of angular momenta in the fragments.

Triple ionization is also a rather efficient process at high energies, especially in the vicinity of core levels. Coulomb explosion of the molecule may lead to the formation of three singly charged fragments, which can be detected by a triple-coincidence experiment (Eland et al. 1986). When a doubly charged ion state (other than the ground state) is not predissociated, it may fluoresce. By detecting fluorescence photons in coincidence with m^{2+} ions selected by mass spectrometry, it is possible to trace out high lying bound m^{2+} states. This method known as Photoion–Fluorescence COincidence (PIFCO) has been introduced by Besnard et al. (1986). The schematic view of the apparatus is shown in fig. 2.12.

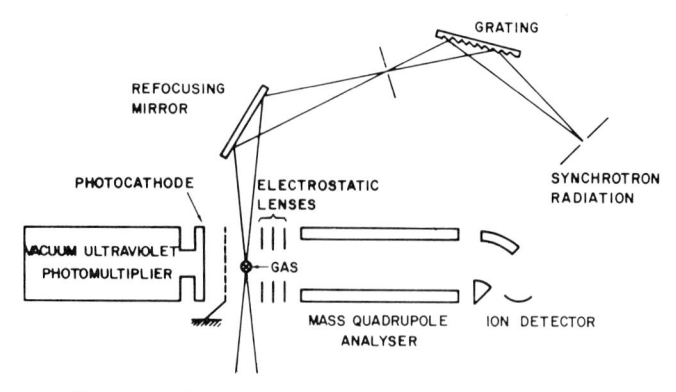

Fig. 2.12. Schematic diagram of the photoion–fluorescence photon coincidence (PIFCO) experiment (from Besnard et al. 1986).

3. Selected examples

3.1. Photodissociation

3.1.1. Electronic branching ratios and quantum yields

The fluorescence from photofragments has been measured in the VUV and in the extreme VUV regions using synchrotron radiation in several systems: N_2, O_2 (Lee et al. 1974), OCS (Lee and Chiang 1982, Tabché-Fouhailé et al. 1983), N_2O (Guyon et al. 1983), H_2O_2 (Suto and Lee 1983a), NH_3 (Suto and Lee 1983b), CF_3X, X = H, Cl, Br (Suto and Lee 1983c), H_2O (Lee et al. 1978, Wu and Judge 1981, Dutuit et al. 1985).

As we have already mentioned in the introduction, these measurements are important since the molecules studied play a central role in atmospheric chemistry and astrophysics. The branching ratios and quantum yields for photodissocation into different electronic states of the fragments are needed to understand the reactions in which they may participate in subsequent collisions. In addition, the fluorescence from the photofragments can be used as a means of detecting these molecules in various media.

We shall take H_2O as an example. Fluorescence excitation spectra of OH(A), OH(B), OH(C), Lyman α, Balmer α and Balmer β have been recorded in the 9–35 eV photon excitation region (Dutuit et al. 1985). Excitation of water below the first ionization potential (12.615 eV) leads to dissociation into several different fragmentation channels: $O(^1D) + H_2(X)$ above 6.947 eV, $O + H + H$ above 9.62 eV, OH(X) + H above 5.118 eV and OH(A) + H above 9.136 eV. Of all these channels OH(X) + H has been shown to be the dominant one. For example, at 10.2 eV the photodissociation of water vapor produces 10% of $O(^1D) + H_2$, 12% of O + H + H and 78% of H + OH (Slanger and Black 1982), 5 to 10% of which gives excited OH(A) (Lee 1980).

The OH(A) + H channel is of particular interest in atmospheric physics. The direct observation of OH(A) fluorescence can be used to measure the water concentration in the earth's atmosphere. The spectrum of fig. 3.1 represents the partial cross section for the production of OH(A). The first continuum, extending

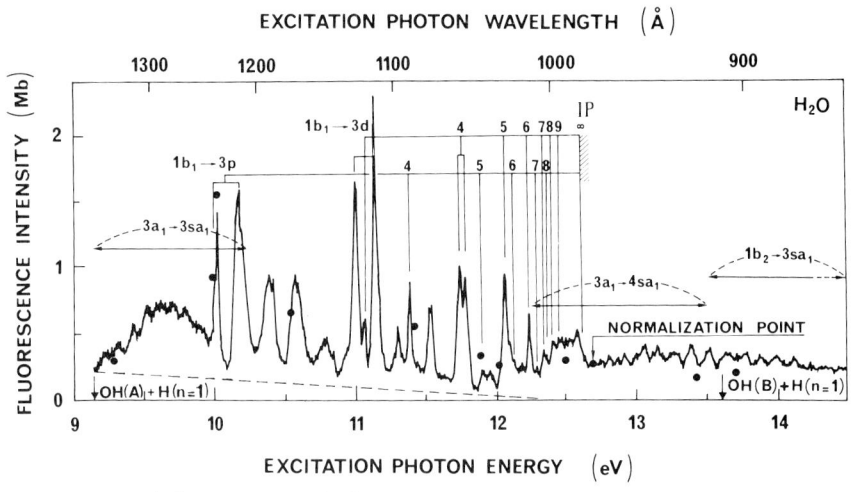

Fig. 3.1. 2800–6000 Å fluorescence excitation spectrum in the photodissociation of H_2O with photon excitation energies between 9 and 14 eV (from Dutuit et al. 1985).

from threshold to about 10 eV, shows diffuse structures. It has been assigned to the B state of H_2O with configuration

$$(1a_1)^2(2a_1)^2(1b_2)^2(3a_1)(1b_1)^2(3sa_1) .$$

Now, the ground state of H_2O has the configuration

$$(1a_1)^2(2a_1)^2(1b_2)^2(3a_1)^2(1b_1)^2 ,$$

and thus the B state corresponds to the excitation of a bonding electron in the $3a_1$ orbital into the Rydberg $3sa_1$ orbital, which has, however, some valence character coming from the contribution of the strong antibonding $4a_1$ valence orbital. This contribution causes the dissociative nature of the state.

Full three-dimensional quantum-mechanical calculations for photo excitation of the B state have been performed using the "artificial channel method" (Segev and Shapiro 1982). This method has been shown to be a powerful and very convenient technique to solve many problems in photofragmentation. The potential energy surface used was the one calculated by Flouquet and Horsley (1974). It has a minimum for the collinear configuration and a small barrier for dissocation in the bent configuration. In order to dissociate, the molecule has to change its geometry from the initially bent configuration of the ground state to a final more linear configuration. This induces a torque on the OH fragment and excites the bending motion at selected "resonance" energies. At these energies, the molecule can exist for a number of bending vibrations before it dissociates, thus greatly enhancing the photodissociation cross section, hence the appearance of the diffuse bands (Shapiro and Bersohn 1982). The torque exerted on the molecule during dissociation also explains the high degree of rotational excitation of the OH(A) fragments (Carrington 1964).

In the region from 10 eV up to 12.6 eV (the ionization threshold), the peaks observed in the fluorescence spectrum coincide with the position of the Rydberg series in the photoabsorption spectrum. They have been assigned to the series $1b_1 \to np$ and $1b_1 \to nd$, converging to the X state of H_2O^+. The fluorescence quantum yield has been measured. The value obtained for the peaks is different from that of the underlying continuum. Also, above the ionization threshold, the fluorescence quantum yield is approximately constant and equal to that of the underlying continuum below that threshold. This suggests that there is another purely repulsive state in the Franck–Condon region which contributes to dissociation into OH(A) + H. It has been tentatively assigned to the $^1B_2$ state calculated by Diercksen et al. (1982). It has a large oscillator strength with a broad continuum centered at 11.5 eV.

Just at the ionization potential, the OH(A) fluorescence quantum yield decreases significantly because the Rydberg states, converging to the first ionization limit, transform into direct ionization, while the other states, converging to higher ionic limits, autoionize. However, photodissociation of Rydberg states into neutral fragments remains an important channel even well above the ionization threshold. Two kinds of states can be distinguished: pure Rydberg states, which are bound, and the nsa_1 "Rydberg" states with strong valence character, which are repulsive. The latter dissociate fast while the pure Rydberg states are longer lived and so undergo both predissociation and autoionization.

Experiments have also been conducted for higher energies up to 35 eV. Fluorescence from OH(C → A) has been detected for excitation energies between 16 and 20 eV (see fig. 3.2) while OH(B → A) visible emission is detected between 13.6 and 14.5 eV. Just as for the A → X fluorescence, these emissions appear exactly at the thermodynamical thresholds. In the case of OH(A) this feature

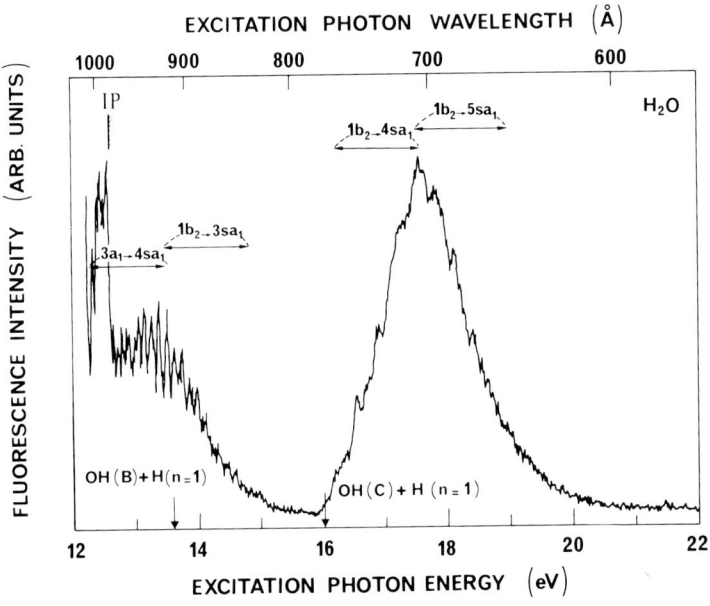

Fig. 3.2. 1800–3000 Å fluorescence excitation spectrum in the photodissociation of H_2O with photon excitation energies between 12 and 20 eV (from Dutuit et al. 1985).

could be explained as the result of direct dissociation occurring at HOH angles out of equilibrium due to the zero-point bending vibration in the initial ground state (Dutuit et al. 1985). In the ground state of H$_2$O the average deviation from equilibrium is 13°. At the equilibrium geometry the upper repulsive state can be vertically reached at 0.65 eV above the dissociation threshold, but for a deviation of 13° the potential energy surface of the excited state drops to about 0.25 eV above the threshold (Durand and Chapuisat 1985). Thus the fraction of the molecules which have a HOH angle larger than 20° out of equilibrium can reach directly the upper surface by photon absorption. This has been borne out by the three-dimensional quantum-mechanical calculations of Segev and Shapiro (1982). It may also be a common feature of the other excited states of H$_2$O.

3.1.2. Predissociation lineshapes

Predissociation of excited states of H$_2$ has been studied (Borrell et al. 1977, Mentall and Guyon 1977, Glass-Maujean et al. 1978, Guyon et al. 1979, Glass-Maujean et al. 1979, Breton et al. 1980) using synchrotron radiation from the ACO storage ring at Orsay in conjunction with fluorescence emission detection. The excitation wavelength was from 860 to 680 Å (i.e. from 14 to 18 eV) and the resolution bandwidth was typically of the order of 0.5 Å.

In fig. 3.3 we have represented some of the relevant potential energy curves in the region of the excitation. The states reached by photon absorption are of $^1\Sigma_u^+$ and $^1\Pi_u$ symmetries. The selection rule u $\not\leftrightarrow$ g for intramolecular perturbations and the small spin–orbit coupling in H$_2$ prevent those states to be effectively coupled to the ground X $^1\Sigma_g^+$ state or to the repulsive $^3\Sigma_u^+$ state, which dissociate both to H(1s) + H(1s) ground state hydrogen atoms. Thus excitation below 14 eV (see fig. 3.3) gives rise mainly to fluorescence from bound states of H$_2^*$, while excitation above that value may lead to dissociation into H(1s) + H(nl), $n \geqslant 2$, fragments. The excited H* fragments will then fluoresce, giving rise to the well-known Lyman α and Balmer α, β, γ emission.

Experimentally one observes, in addition to the atomic emission, the molecular H$_2$ fluorescence. Photon absorption in the region 14–18 eV gives rise to direct dissociation when the continuum of the B, B', C states are excited, but also to predissociation when the photon energy coincides with a transition to a bound level of the B", D, D', D", ... states (see fig. 3.3). Besides the B" and the D states, which have predissociation lifetimes of the order of picoseconds or less, the other states are much longer lived (of the order of nanoseconds). Thus molecular fluorescence competes with predissociation in their case.

One of the most interesting features of H$_2$ photofragmentation is that it provides an almost unique and clear example of Fano–Beutler profiles in predissociation. These profiles are well known in atomic autoionization (Fano 1961). They are the result of interference between direct and indirect dissociation (see fig. 3.4). The amplitude for the indirect process (predissociation) is given by

$$A_{\text{pred}} = \frac{\langle \Psi_\varepsilon | V | \Psi_d \rangle \langle \Psi_d | \mu | \Psi_g \rangle}{E - E_d + i\Gamma_d}, \qquad (3.1)$$

where $\langle \Psi_d | \mu | \Psi_g \rangle$ is the dipole matrix element for optical transition from the

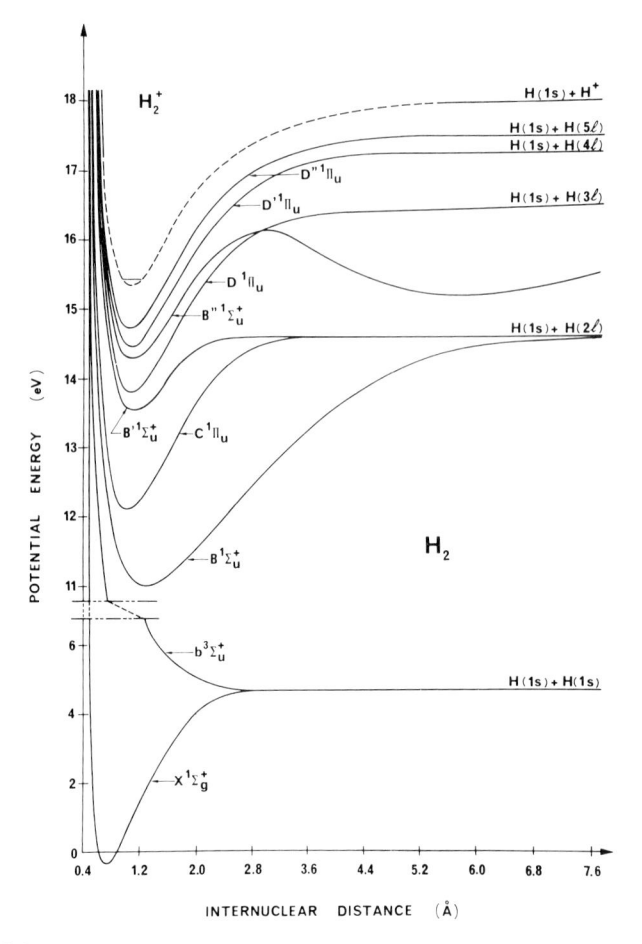

Fig. 3.3. Relevant potential energy curves for H_2 (adapted from Sharp 1971).

initial ground state $|\Psi_g\rangle$ to the discrete excited state (the doorway state) $|\Psi_d\rangle$, while $\langle\Psi_\varepsilon|V|\Psi_d\rangle$ is the intramolecular coupling between the discrete state and the final continuum state $|\Psi_\varepsilon\rangle$. Finally, Γ_d is the predissociation half-width at half-maximum (HWHM) of the $|\Psi_d\rangle$ state, which is given by the well-known golden-rule expression,

$$\Gamma_d = \pi|\langle\Psi_d|V|\Psi_\varepsilon\rangle|^2 . \tag{3.2}$$

In writing eq. (3.1), we have neglected off-energy shell contributions, which produce a slight modification of the resonant state $|\Psi_d\rangle$. For a complete and rigorous treatment the reader should refer to ch. 8, Vol. 1 of this Handbook (Almbladh and Hedin 1983). The square of eq. (3.1) provides the usual Lorentzian lineshape (with HWHM Γ_d and centered at E_d), which is expected in the case of pure predissociation. In the case where the system can also be excited to the final continuum state $|\Psi_\varepsilon\rangle$ by direct photoexcitation from the ground state (see

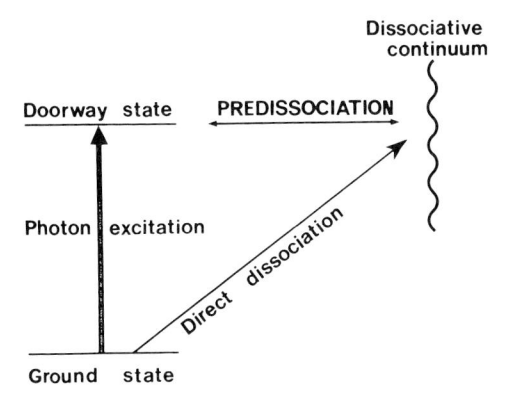

Fig. 3.4. Simple model for interference between direct dissociation and predissociation.

fig. 3.4), one should also consider the amplitude

$$A_{\text{direct}} = \langle \Psi_\varepsilon | \mu | \Psi_g \rangle \frac{E - E_d}{E - E_d + i\Gamma_d} . \tag{3.3}$$

The amplitudes (3.1) and (3.3) should be added for this case, and the photodissociation cross section will be given by

$$\sigma \propto |A_{\text{pred}} + A_{\text{direct}}|^2 . \tag{3.4}$$

Thus one obtains interference terms coming from the square of the sum. The fundamental reason for this interference is the existence of two different routes (direct and indirect) to reach the same final continuum state, a situation which is reminescent of the two-slit experiment. Using eqs. (3.1) and (3.3) into (3.4), we get

$$\sigma \propto |\langle \Psi_\varepsilon | \mu | \Psi_g \rangle|^2 \frac{(\varepsilon + q)^2}{\varepsilon^2 + 1} , \tag{3.5}$$

where $\varepsilon = (E - E_d)/\Gamma_d$ is the energy with respect to the doorway state (resonance) position measured in units of its half-width Γ_d, while

$$q = \frac{\langle \Psi_\varepsilon | V | \Psi_d \rangle \langle \Psi_d | \mu | \Psi_g \rangle}{\langle \Psi_\varepsilon | \mu | \Psi_g \rangle \Gamma_d} \tag{3.6}$$

is the asymmetry parameter. The latter can be easily shown to be real, irrespective of the phases of the wavefunctions.

Figures 3.5 and 3.6 present examples of predissociation profiles for the vibrational levels $v' = 3$ and $v' = 10$ of the D state (Glass-Maujean et al. 1979). The two peaks correspond to two different rotational transitions R(0) and R(1), i.e., they correspond to transitions N'' to $N' = N'' + 1$ with $N'' = 0$ and 1, respectively. The dissociative state responsible for this predissociation is the B' state (see fig.

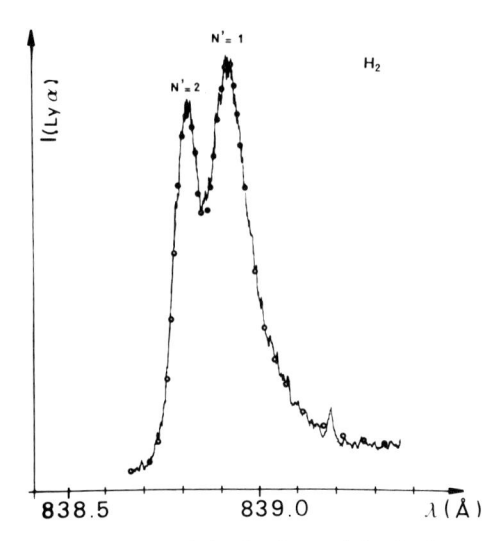

Fig. 3.5. R(0) and R(1) lines of X→D ($v' = 3$) observed in the Ly_α excitation spectrum (from Glass-Maujean et al. 1979). The points are obtained through a fit of the experimental profiles by the use of eq. (3.5).

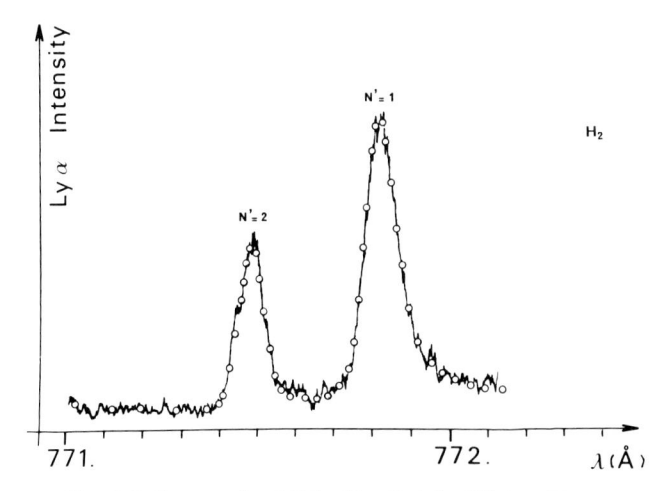

Fig. 3.6. Same as fig. 3.5 for X→D ($v' = 10$) transitions.

3.3) which is coupled to the D by coriolis coupling (*l*-uncoupling). The matrix elements of this interaction are proportional to $[N'(N'+1)]^{1/2}$. For R(N″) transitions the matrix elements of the electric dipole moment are given by

$$\langle D\Pi(N''+1)|\mu|X\Sigma N''\rangle \propto [(N''+2)/(2N''+1)]^{1/2},$$

$$\langle B'\Sigma(N''+1)|\mu|X\Sigma N''\rangle \propto [(N''+1)/(2N''+1)]^{1/2},$$

(3.7)

and thus one expects for the lineshape parameters q and Γ the following N' dependence:

$$q(N'+1) = q(N')N'/(N'+1), \qquad \Gamma(N'+1) = \Gamma(N')(N'+2)/N'. \qquad (3.8)$$

This dependence has been verified in the experiments. For the $v'=3$ level the best fit is obtained for $q(N'=2) = -9 \pm 1$ and $\Gamma(N'=2) = 14.5 \pm 0.5 \text{ cm}^{-1}$. This corresponds to a lifetime for predissociation of $\tau(N'=2) \simeq 0.4 \text{ ps}$.

The same type of fitting has been done for all other vibrational levels up to $v' = 13$ (Glass-Maujean et al. 1979). In fig. 3.7 the measured predissociation linewidths for the $N'=2$ levels are represented as a function of the vibrational quantum number v', together with two different theoretical calculations (Julienne 1971, Fiquet-Fayard and Gallais 1972). For low values of v' the agreement between theory and experiment is quite good: the widths decrease as v' increases, reflecting merely the behavior of the Franck–Condon overlaps between the bound levels of D and the continuum of B' (see fig. 3.3). A better agreement was obtained by Jungen (see Greene and Jungen 1985). On the other hand, for $v' > 9$ significant deviations are observed between the experimental and the calculated values. It has been suggested (Glass-Maujean et al. 1979) that this behavior may come from the interaction with another $^1\Sigma_u^+$ state, namely B". For large values of v', B" may interact strongly with D, perturbing the levels and inducing this peculiar behavior of the predissociation linewidths.

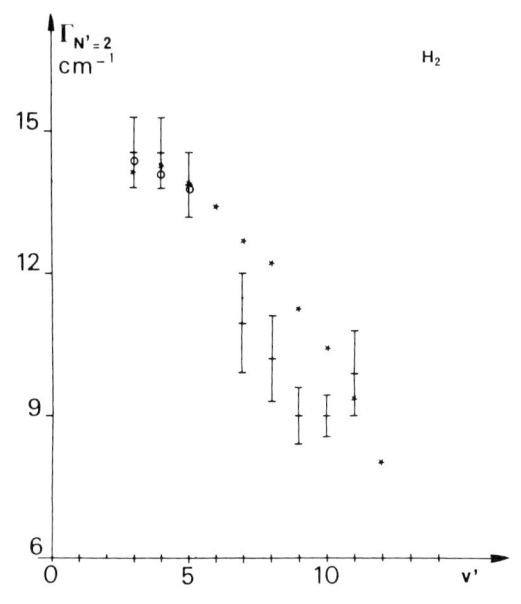

Fig. 3.7. Predissociation widths Γ for the $N'=2$ rotational levels of various vibrational levels of the D electronic state of H_2. The experimental values (Glass-Maujean et al. 1979) are indicated by error bars. The stars correspond to the theoretical values of Julienne (1971), while the open circles are the calculated values of Fiquet-Fayard and Gallais (1972).

Another important measurement in this context is the branching ratio H(2s, 2p). In the experiment described above the state B' is populated by rotational coupling with the D state at short internuclear distances. If the system dissociates adiabatically the fragments would be H(1s) + H(2s) and no H fragments in the 2p state would be produced. On the other hand, if some dissociation proceeds through the B state, then H(2p) fragments would be produced since the B state correlates with the H(1s) + H(2p) limit.

Using an electric field Mentall and Guyon (1977) have measured the branching ratio:

$$\Omega = \sigma(2s)/[\sigma(2s) + \sigma(2p)] = 0.57 . \quad (3.9)$$

The interpretation of this result was the following: the B' and B states interact strongly at large internuclear distances ($R \sim 5$ Å) where there is an avoided crossing between three states: B, B' and B'' (see fig. 3.3). Thus nonadiabatic radial couplings exist which induce transitions between the B' and B states. This explains that $\Omega \leq 1$.

Dynamical quantum-mechanical calculations have been performed by Borondo et al. (1982) and by Beswick and Glass-Maujean (1987). Using the adiabatic ab-initio potential energy curves and couplings they have obtained $\Omega \sim 0.7$–0.8 which is larger than the experimental value. It seems that further calculations of the non-adiabatic coupling are needed.

3.1.3. Vibrational and rotational distributions of photofragments

The analysis of fluorescence from photofragments provides a simple and direct determination of the internal (vibrational, rotational) energy content of the products formed in radiative electronic excited states. This addresses the very central question of how the excess energy is partitioned among the different degrees of freedom at a given photon wavelength. Synchrotron radiation provides in addition the possibility of measuring those final state distributions as a function of the photon wavelength in a very large domain.

The limitations are the following: (1) Because of the low photon flux the spectral resolution of the fluorescence is not high enough to resolve clearly the rotational transitions. In most cases only the rotational contours can be measured. (2) The study is limited to radiative excited states since, again, the number of photofragments is too low to perform laser-induced fluorescence experiments and to be able to study ground or metastable product states. Future high-intensity synchrotron radiation sources will probably allow us to overcome these limitations.

The resolved fluorescence spectra from NO fragments produced by photodissociation of a variety of NO-containing polyatomic molecules have been recorded. The following systems have been studied: CH_3ONO, CD_3ONO, C_2H_5ONO, i-C_3H_7ONO, CF_3NO, $ClNO$, NO_2 (Lahmani et al. 1980a,b, 1981a,b, 1982, 1983a,b, Lardeux 1983).

As an example, the absorption spectrum of CH_3ONO in the 1200–2600 Å region is presented in fig. 3.8, together with the relevant thresholds for dissociation. Up to 1700 Å photodissociation of CH_3ONO can only produce electronic ground state NO fragments. In the region 1300–1700 Å a series of Rydberg (A, C, D, E) as well as valence states (B, B') of NO are energetically accessible and fluorescence attributed to A→X, C→X and D→X transitions has been measured in the experiments. In the region of 1400–1700 Å mainly NO(A) is formed and the A→X fluorescence intensity reproduces very well the photoabsorption spectrum (see fig. 3.9). Thus NO(A) is produced in the primary process following photon absorption.

Analysis of the A→X fluorescence has revealed that NO(A) is formed in several vibrational levels ($v = 0, 1, 2, 3$). Their relative intensities as a function of the incident photon wavelength are represented in fig. 3.9. From these excitation spectra, after correction for the incident flux intensity and the emission probability of each observed band, one can deduce the relative vibrational population of the NO(A) fragments as a function of the excess energy.

In fig. 3.10 the ratio $P_{v=1}/P_{v=0}$ for the NO(A) fragments in the photodissociation of several RONO molecules is presented as a function of the excess energy E_{avl} in the region corresponding to 1300–1600 Å excitation wavelength. The behavior of this ratio is quite smooth in this region and one may expect that a simple statistical model may be applicable. According to the statistical assump-

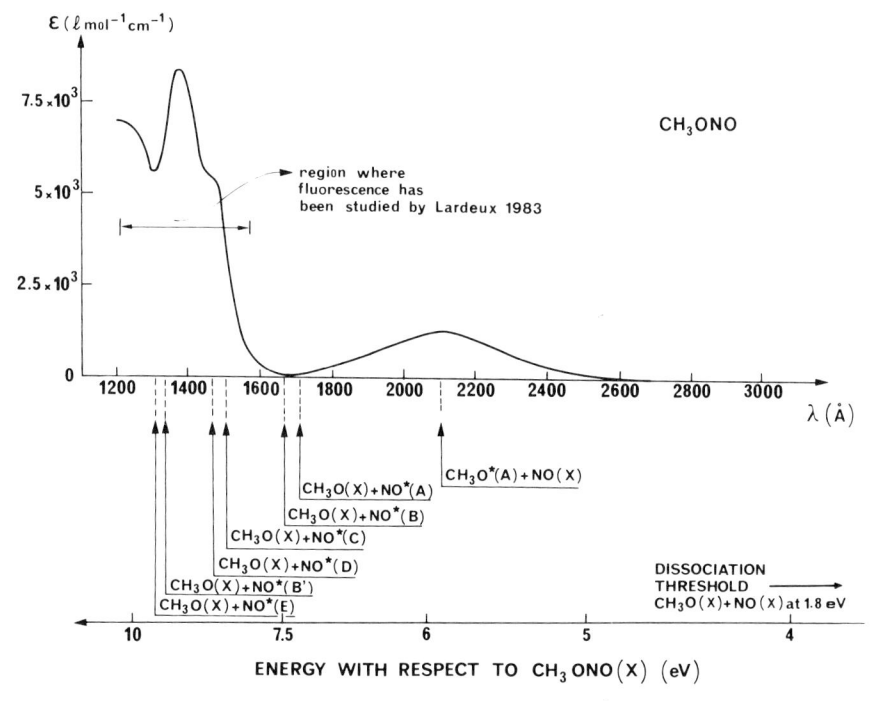

Fig. 3.8. UV absorption spectrum of CH_3ONO in the 1200–2600 Å region (adapted from Lardeux 1983).

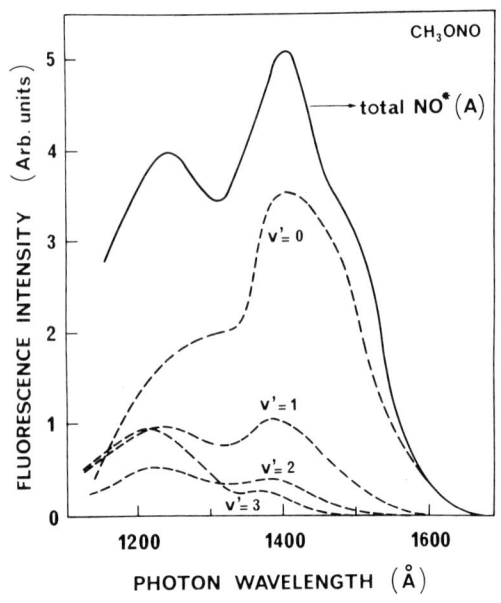

Fig. 3.9. Excitation spectra of vibronically excited NO(A) fragments from CH$_3$ONO photodissociation in vacuum UV (from Lahmani et al. 1980b).

tions for photodissociation of medium and large molecules, all the final quantum states allowed by the conservation of energy have equal probability to be produced. The ratio $P_{v=1}/P_{v=0}$ should therefore be related to the ratio of the density of states according to

$$P_{v=1}/P_{v=0} = \rho(E_{\text{avl}}, v=1)/\rho(E_{\text{avl}}, v=0), \qquad (3.10)$$

where E_{avl} is the total available energy i.e. the energy above the threshold corresponding to the particular electronic channel under consideration. The density of states can be evaluated according to (Kinsey 1971, Levine 1974)

$$\rho(E_{\text{avl}}, v) = \sum_{J=0}^{J_{\text{max}}} (2J+1)$$
$$\times \int_0^{E_{\text{avl}} - E_v^{\text{NO}} - E_J^{\text{NO}}} dE_I\, \rho_I(E_I)\rho_T(E_{\text{avl}} - E_v^{\text{NO}} - E_J^{\text{NO}} - E_I), \qquad (3.11)$$

where $\rho_T(\varepsilon) = \varepsilon^{1/2}$ is the translational density of states and $\rho_I(E_I)$ is the ro-vibrational internal density of states of the RO (R = CH$_3$, CD$_3$, C$_2$H$_5$, i-C$_3$H$_7$) fragment. In writing eq. (3.11) it was assumed that the internal density of states ρ_I of the RO fragments is high enough so that the sum over the discrete states has been replaced by an integration over E_I. This is justified for the examples considered here. The computed values of ρ_I for CH$_3$ONO show that it can be

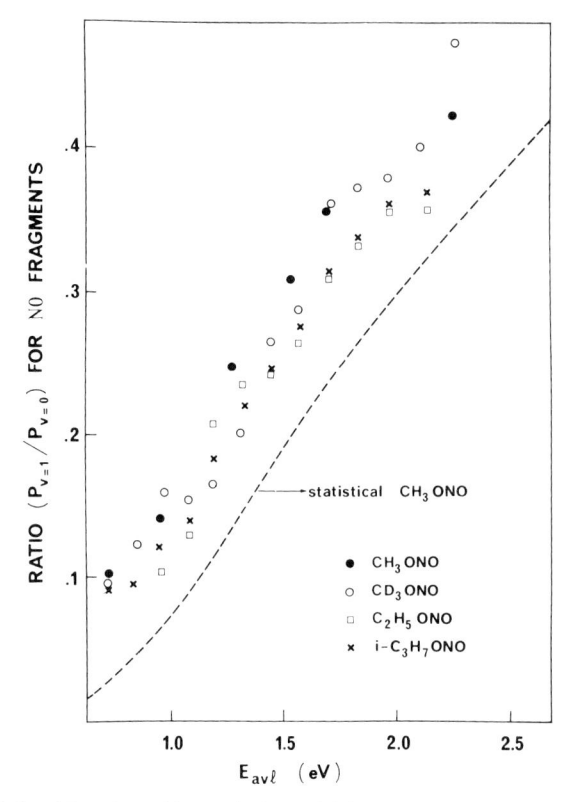

Fig. 3.10. Measured $P_{v=1}/P_{v=0}$ branching ratio for NO(A) fragments from vacuum UV photolysis of several nitrites (from Lahmani et al. 1981b), and comparison with the statistical expectation for CH_3ONO (see text).

approximated by (Lardeux 1983)

$$\rho_1(E_1) \sim (E_1)^{5.5} . \tag{3.12}$$

Using this expression in eq. (3.11) and replacing again the sum over J by an integral, obtains

$$\rho(E_{\text{avl}}, v) \propto (E_{\text{avl}} - E_v^{\text{NO}})^8 , \tag{3.13}$$

and thus

$$P_{v=1}/P_{v=0} = (1 - 0.28\,\text{eV}/E_{\text{avl}})^8 , \tag{3.14}$$

in the case of NO(A) fragments.

Equation (3.14) is represented in fig. 3.10. It is seen that the overall behavior agrees quite well with the experimental values, but a systematic deviation remains, the experimental points being higher than the statistical predictions. A better fit can be obtained by assuming that not all of the degrees of freedom in CH_3O are statistically populated, i.e., if one assumes only partial energy redistribution among the vibrational degrees of freedom. This will have the effect of decreasing the exponent of E_I in eq. (3.12). Assuming $\rho_I(E_I) \sim (E_I)^n$ then a reasonable fit is obtained with $n \sim 3-4$. This implies that only a few of the vibrational degrees of freedom in CH_3O are efficiently coupled during the photodissociation process. It is interesting to note at this point that the behavior for the other nitrites is quite the same. In all cases the vibrational ratio comes out to be essentially the same although the statistical model would predict a much lower ratio since n (the exponent of E_I in the $\rho_I(E_I)$ dependence) should be larger than for CH_3O. It was concluded that in the framework of the statistical model the vibrational degrees of freedom of the R radial in RONO photodissociation do not take part in the energy redistribution following photon absorption (Lardeux 1983).

In view of these very interesting results, it was tempting to study a small NO-containing molecule and to compare its behavior under photodissociation with that observed for the medium-size nitrites. The ClNO molecule has been chosen (Lahmani et al. 1982, Lardeux 1983). Between 1200 and 1700 Å ClNO was also shown to give rise to fluorescence from NO(A) fragments in different vibrational levels $v = 0, 1, 2$. In fig. 3.11 the ratio $P_{v=1}/P_{v=0}$ is represented as a function of the excess energy in the 1300–1600 Å excitation region together with the theoretical values expected on statistical grounds [in this case the density of states is given by $\rho(E_{avl}, v) \propto (E_{avl} - E_v^{NO})^{3/2}$]. As expected the experimental points deviate significantly from the statistical predictions.

Using a xenon resonance lamp ($\lambda = 1470$ Å), Lahmani et al. (1982) were able to study the rotational energy distribution of the NO fragments in detail. For NO(A, $v = 0$) in ClNO photodissociation they obtained a distribution with a maximum at $J \simeq 26$ and a width of $\Delta J \simeq 10$. The rotational level distribution expected on the basis of the statistical model would have a maximum at $J \simeq 50$ and a width of $\Delta J \simeq 50$. This result reinforces the conclusion that the process is highly nonstatistical. The dependence of the rotational energy content in NO fragments following ClNO photodissociation as a function of the available energy E_{avl}, has been estimated from the envelope of the emission bands obtained with synchrotron radiation.

In fig. 3.12 the rotational energy content E_{rot} of the NO(A) fragments, together with the vibrational energy E_{vib} and the total internal energy $E_I = E_{vib} + E_{rot}$, are plotted as a function of the available energy. They follow a quasi-linear dependence. This is consistent with an impulsive model for fragmentation. In this model it is assumed that the available energy is suddenly released along the Cl–N bond. From the conservation of energy and momentum it follows (Basco and Norrish 1962, Busch and Wilson 1972) that

$$E_{int} = E_{avl}(1 - \mu_{ClN}/\mu_{Cl,NO}) = 0.38 E_{avl}, \tag{3.15}$$

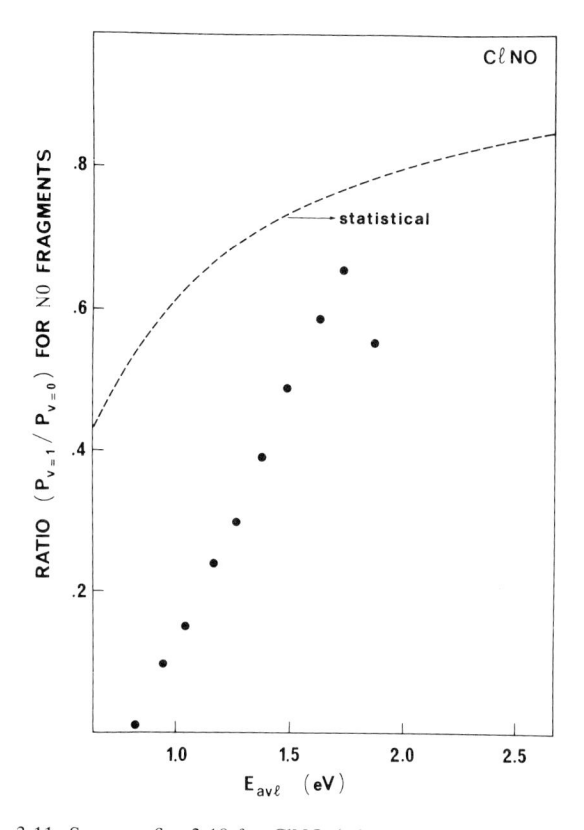

Fig. 3.11. Same as fig. 3.10 for ClNO (adapted from Lardeux 1983).

$$E_{\text{vib}} = E_{\text{int}} \cos^2 \chi , \qquad (3.16)$$

$$E_{\text{rot}} = E_{\text{int}} \sin^2 \chi , \qquad (3.17)$$

where χ is the dissociation bond angle. Comparison of the experimental results presented in fig. 3.12 with eq. (3.15) reveals that the slope of the total internal energy versus the available energy is indeed equal to 0.38 as predicted by the impulsive model. However, it does not extrapolate to zero; in other words, the experimental points obey the linear relation $E_{\text{int}} = 0.38(E_{\text{avl}} - 0.7 \text{ eV})$ instead of eq. (3.15). There is no simple explanation of this fact.

The partitioning of E_{int} into rotational and vibrational energy is consistent with the impulsive model if $\chi \simeq 130°$. In the ground state of ClNO the equilibrium bond angle is 113°. This suggests that in the excited surface the molecule needs to open its bond angle by about 15° to be able to dissociate. This situation is reminiscent of the case of $H_2O(A)$ dissociation (Shapiro and Bersohn 1982).

The impulsive model is, of course, too crude an approximation to expect fine

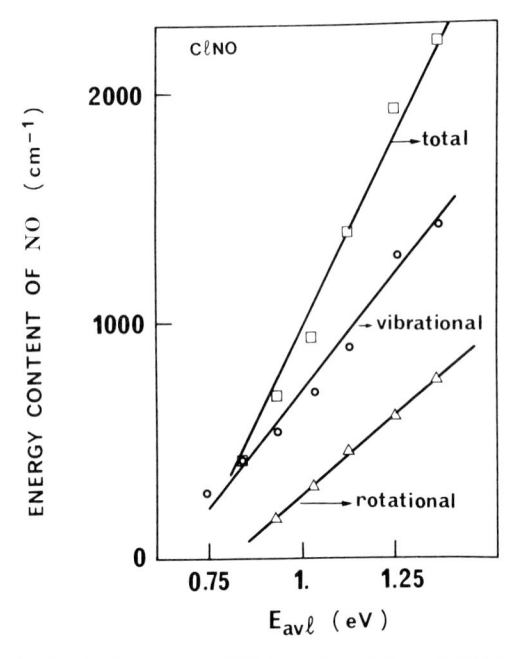

Fig. 3.12. Energy partitioning in the vacuum UV photodissociation of ClNO as a function of the total available energy (from Lahmani et al. 1982).

detailed analysis of the photodissociation process. It suggests, however, that the dynamics on the upper surface (interfragment interactions) is quite important for this system. The Franck–Condon models (Morse et al. 1979a,b, Morse and Freed 1981, Beswick and Gelbart 1980), based on the assumption that the dynamics on the upper surface has minor effects on the final vibrational and rotational distributions, should be inappropriate here. This has been shown by Lardeux (1983): the Franck–Condon model will predict, for example, that the rotational distribution in NO(A) fragments should peak at $J \sim 10$, in clear contradiction with the experimental value $J \sim 26$. It would be interesting to have a full dynamical calculation using a potential surface modeled by what the impulsive model suggests in order to confirm some of the assumptions made above.

3.2. Valence photoionization

3.2.1. Spectroscopy of singly charged ions

The binding energies associated with valence orbitals of common molecules were measured and identified many years ago by PhotoElectron Spectroscopy (PES) using laboratory sources such as discharge lamps. They have been reported in many textbooks (see, for example, Turner et al. 1970, Baker and Betteridge 1972, Rabalais 1977, Carlson 1978, Ballard 1978, Berkowitz 1979, Kimura et al. 1981, Eland 1984). In this section we shall present examples of spectroscopic studies of singly charged ions using synchrotron radiation, selecting those for which this technique has provided new information through its tunability over a wide range of energies extending over more than 20 eV. We shall restrict ourselves, however, to sufficiently low excitation energies so that only valence electrons are ejected. The spectroscopy of core electrons will be considered in section 3.6.

Excitation between 10 and 20 eV. In the 10 to 20 eV excitation energy region, ionization corresponds to the formation of singly charged ions produced by the ejection of an outer valence electron, i.e. from molecular orbitals (MO) built from atomic orbitals (AO) of the 2p (or 3p) type (for atoms of the second or third row of the periodic table). The PES spectrum shows various bands associated with the different valence orbitals. The energies, abundance and angular distributions of these electrons are characteristic of the individual molecular orbitals from which they originate. Thus photoelectron spectroscopy, in this energy range, allows the direct study of molecular orbitals.

Each electronic band is actually composed of peaks in the spectrum because of the possibility of vibrational excitation. The study of these vibrational progressions provides information on the geometry of the ionic state as compared to that of the ground state of the molecule. If a bonding electron is removed, the bond becomes weaker and the force constant smaller. Thus the vibrational frequency is lower and this is seen in the photoelectron spectrum through the spacings between vibrational peaks. The opposite is observed when an electron is removed from an antibonding orbital.

The intensity distribution within a vibronic band is also related to geometrical changes. The electron ejected is leaving the molecule in a time short compared to the vibrational periods. Thus, during the whole process of absorption of the photon and ejection of the electron, the internuclear distances remain constant. As a consequence, the intensity of the peaks will be directly proportional to the square of the overlaps (the so-called "Franck–Condon factors") between the initial vibrational wavefunction of the neutral and the final vibrational wavefunctions of the ion. When a change of the bond lengths and bond angles occurs, several vibrational levels of the ion will have appreciable overlaps with the initial state of the neutral and a series of peaks will appear in the photoelectron spectrum. If the electron removed is bonding, for example, an increase in bond length is expected, while the opposite is expected for an antibonding electron.

A He(I) PES spectrum usually shows empty regions, known as "Franck–Condon gaps", where other vibrational levels of the ion cannot be excited directly in a vertical transition because of vanishingly small Franck–Condon factors. However, in other experiments these states can be populated indirectly (i.e. via autoionization) if the photon energy is chosen properly. The tunability of synchrotron radiation provides here the opportunity to reveal these "hidden" ionic states. One method is to fix the photon energy corresponding to a transition into a Rydberg state converging to an electronically excited ionic state with an equilibrium internuclear distance different from both the initial state of the neutral and the final state of the ion. If the lifetime of the Rydberg state is long enough, the autoionization process (electronic autoionization in this case) can be described in two steps:

(1) $h\nu + M_i \rightarrow M_v^{**}$ (Rydberg),
(2) $M_v^{**} \rightarrow M_\alpha^+ + e^-$,

where i is the initial vibronic state of the molecule, v is the vibrational level of the Rydberg state, and α is the final ionic state. One can then understand why the vibrational distribution of the final ionic state can be different from the vibrational distribution obtained in a direct ionization process, since the Franck–Condon factors for the transition from the excited M_v^{**} state to the final state of the ion (step 2) are usually different from those for direct ionization.

Figure 3.13 presents the CO PES spectrum measured using synchrotron radiation at 21.58 eV and 17.33 eV (Morin et al. 1981). Three main bands are observed. They correspond to $CO^+(X\ ^2\Sigma^+)$, $(A\ ^2\Pi)$, and $(B\ ^2\Sigma^+)$ states resulting from the ejection of a 5σ, 1π, and 4σ valence electron from the ground CO molecule (configuration $\ldots 4\sigma^2 1\pi^4 5\sigma^2$). The Franck–Condon gap observed when the molecule is excited at 21.58 eV (see fig. 3.13a), is filled for 17.33 eV photon excitation (see fig. 3.13b). The X state of CO^+, which was produced only in the vibrational levels $v = 0$ and $v = 1$, is now formed up to $v = 8$. The photon energy has been chosen so that it corresponds to a transition to a Rydberg level $n = 3$, $v = 0$ (sharp series converging to the $CO^+(B\ ^2\Sigma^+)$ ionic state). A detailed analysis of the PES line intensity will be presented in section 3.2.3 for the case of O_2^+.

An alternative way of finding such "hidden" states was provided by Threshold PhotoElectron Spectroscopy (TPES) of H_2 (Peatman et al. 1983), N_2 (Peatman et al. 1978), O_2 (Guyon and Nenner 1980, Ferreira 1984), N_2O (Baer et al. 1979), OCS (Delwiche et al. 1981); references to earlier studies can be found in Koch and Sonntag (1978). The magnifying process comes from resonant autoionization, i.e., the Rydberg state matches in energy the final vibrational state of the ion and autoionizes into a zero kinetic energy electron. This experiment has been of particular importance for understanding the efficient production of fragment ions at their thermodynamical thresholds (see the review by Baer 1979). Figure 3.14 shows for N_2O a beautiful illustration of the use of TPES to reveal very high vibrational states of the $N_2O^+(X\ ^2\Pi)$ ion, which are totally absent in the He(I) PES spectra. Note that the density of rovibronic levels is such that they form a quasi-continuum. The intensity distribution of threshold electrons reproduces the total ion yield. This result is of particular importance because it reveals that resonant autoionization is the key mechanism for reaching specific parts of the excited N_2O^+ surface connected to both $O^+ + NO$ and $NO^+ + N$ dissociation channels (see section 3.2.2). This result must be associated with the dissociation of N_2O^{**} into excited neutral fragments observed in this region, which competes with autoionization (see section 3.2.3).

Excitation between 20 and 50 eV. Photon with 20 to 50 eV energy allow ionization of electrons from inner valence orbitals in most common molecules. The inner valence orbitals are built from 2s (or 3s) atomic orbitals (for atoms of the

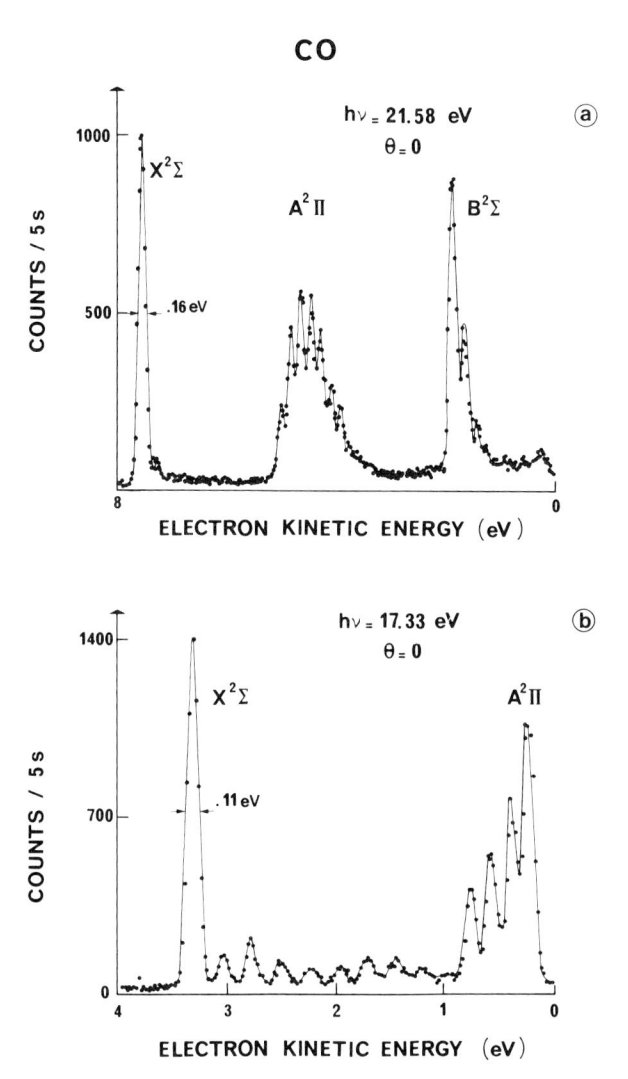

Fig. 3.13. Photoelectron spectra of CO measured at (a) 21.58 eV and (b) 17.33 eV photon energy (from Morin et al. 1981). The spectra are not corrected for transmission efficiency of the electron analyser.

second or third row of the periodic table). The peculiarity of these inner valence MO is that their photoelectron spectrum shows a great number of lines (satellites) instead of a limited number of well-defined bands as discussed above. This phenomenon reflects mainly the relaxation and reorganization of the electron cloud following the creation of an inner vacancy. Each of the satellites corresponds to an ionic state with multiple excited configurations, which are denoted as two holes–one electron, three holes–two electron, etc., depending on the number of outer valence electrons that are promoted into low-lying antibonding valence

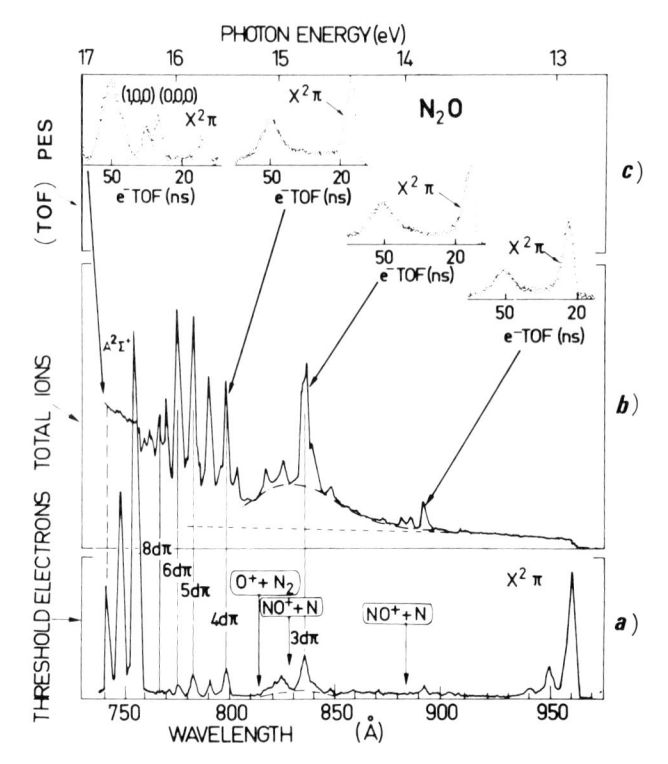

Fig. 3.14. (a) Threshold photoelectron and (b) total ionization spectra of N_2O as a function of photon energy from the X and A states of N_2O^+. (c) Four electron time of flight spectra taken at 742, 798, 835, and 893 Å (from Baer et al. 1979).

orbitals. Actually, these effects also exist for outer valence ionization but to a lesser extent. For the latter, a strong interaction occurs between the main line (one-hole configuration) and satellites having the same total symmetry. For the former, no real main line can be identified but rather a congestion of lines that originate from the interaction of multiply excited ion states. The theoretical analysis of the satellite configuration shows that there is no trace of the initial inner valence hole but rather multiply excited outer valence excitations (Cederbaum et al. 1986, Bagus and Viinikka 1977, Roy et al. 1986). This implies that if one can associate a well-defined physical meaning to the concept of inner valence orbital, this is not the case for the corresponding hole.

In the discussion above we have interpreted the satellites as resulting from a time-dependent process: excitation of one electron followed by relaxation of the excited ion. The interpretation in terms of the time-independent picture is the following: in the independent electron picture only states with one electron being excited can be coupled to the initial state via photon absorption. States corresponding to multiple excitations have zero oscillator strength for excitation from the initial state of the molecule but they are coupled to the one-electron inner valence excited state (the "main" line) by the electron–electron interactions (configuration mixing). Thus they borrow the oscillator strength from the main line through their mixing coefficients.

Experimentally, satellite lines have been studied for several molecules in photoelectron spectra by electron energy loss spectroscopy (e,2e dipole and binary), as reviewed by Brion and Hamnett (1981) and by Weigold and McCarthy (1978), and also by X-ray PES (see Siegbahn et al. 1969). More recently, synchrotron radiation has provided much higher resolution spectra in the energy

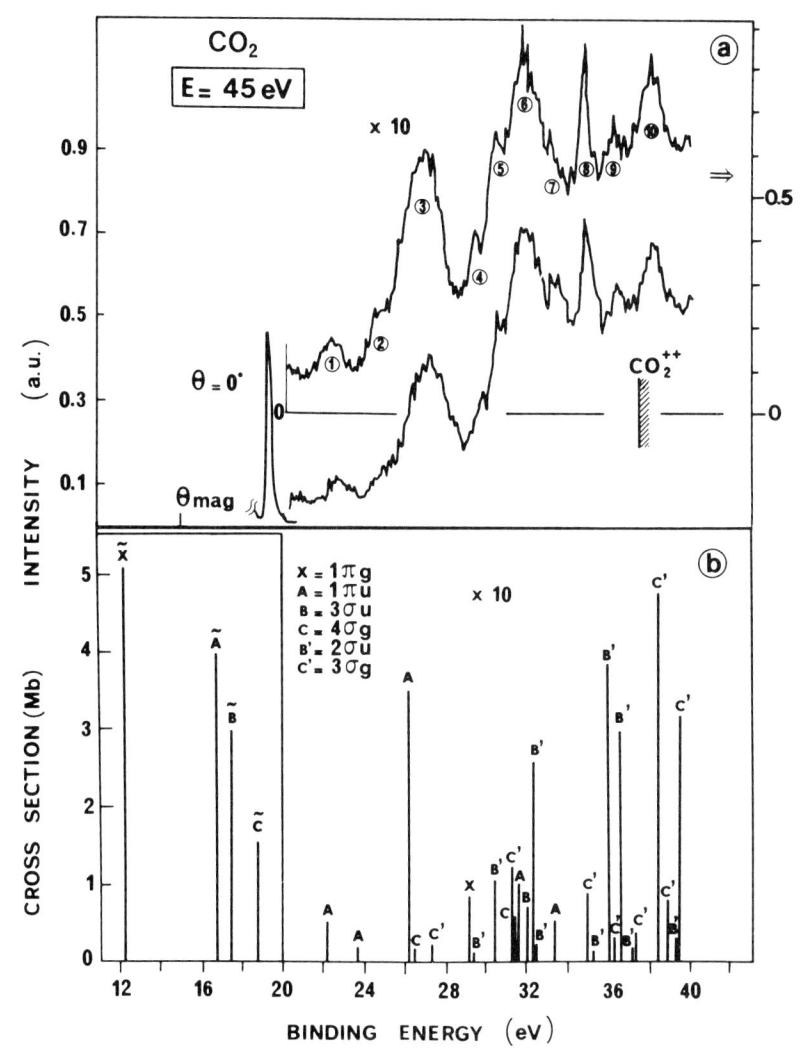

Fig. 3.15. Photoelectron spectrum of CO_2 at 45 eV incident photon energy (from Roy et al. 1986). (a) Experimental spectra. (b) Theoretical spectrum corrected for transition moments derived from experimental values.

range between 20 and 100 eV for the following systems: N_2 (Krummacher et al. 1980, Morin et al. 1983), CO (Krumacher et al. 1983), C_2H_2 (Bradshaw et al. 1980, Swensson et al. 1984), CS_2 (Carlson et al. 1982a), CO_2 (Roy et al. 1985), H_2S (Adam et al. 1985).

In fig. 3.15a the PES of CO_2 for excitation at 45 eV is presented. The measurements were performed for two different angles with respect to the polarization of the light. The theoretical spectrum reported in fig. 3.15b has been obtained by an ab-initio SCF-CI calculation with transition moments extracted from experimental partial cross sections for outer PES bands (Roy et al. 1986). These calculations have clearly shown that in order to reach reasonable quantitative agreement between theory and experiment, configurations involving three holes–two electrons have to be included.

3.2.2. Dissociative ionization
Ejection of a valence or a core electron from a molecule leads to an ion that is often predissociated or directly dissociated. Therefore the corresponding bands in the photoelectron spectrum are broadened (Eland 1984). This broadening can only be observed if it is larger than the total experimental resolution (photon plus electrons). For purely repulsive states, in particular those that are very steep in the Franck–Condon region, the width of the band may exceed 1 eV and it can be easily observed.

The Threshold Electron PhotoIon Coincidence technique (TEPICO) combined with a tunable photon source such as the helium continuum (Baer 1979, Rosenstock et al. 1981) or synchrotron radiation (Guyon et al. 1978, Schlag et al, 1977, Nenner et al. 1980, Richard-Viard 1982, Richard-Viard et al. 1985, Frasinski et al. 1985) has been shown to be a powerful tool to study the spectroscopy of such repulsive states as well as the predissociation mechanisms. For both, the threshold electron signal provides a selection of the internal energy state of the molecular ion. The subsequent analysis of mass and kinetic energy of the ionic fragments provides direct information on the dissociation limits. Notice that synchrotron radiation provides tunability above 20 eV photon energy as opposed to the helium continuum. It should also be stressed at this point that, although the coincidence methods cannot compete with laser photofragment spectroscopy (see, for example, Moseley et al. 1979) as far as the spectral and time resolution is concerned, they remain up to data tools for analysing dissociative ionization in a wide excitation energy range.

Diatomic molecules. The oxygen molecular ion provides an interesting case study for dissociative (direct and indirect) ionization and has been the subject of several investigations using synchrotron radiation (Guyon et al. 1978, Ferreira 1984, Guyon and Nenner 1980, Richard-Viard et al. 1985, Frasinski et al. 1985). Figure 3.16 presents a threshold photoelectron spectrum of O_2 in the region between 12 to 26 eV. At around 24 eV there is a clear example of broadening due to direct dissociation of O_2^+ (the $^2\Pi_u$ band).

Another very interesting problem in O_2 is the predissociation of the $c\,^4\Sigma_u^-$ state.

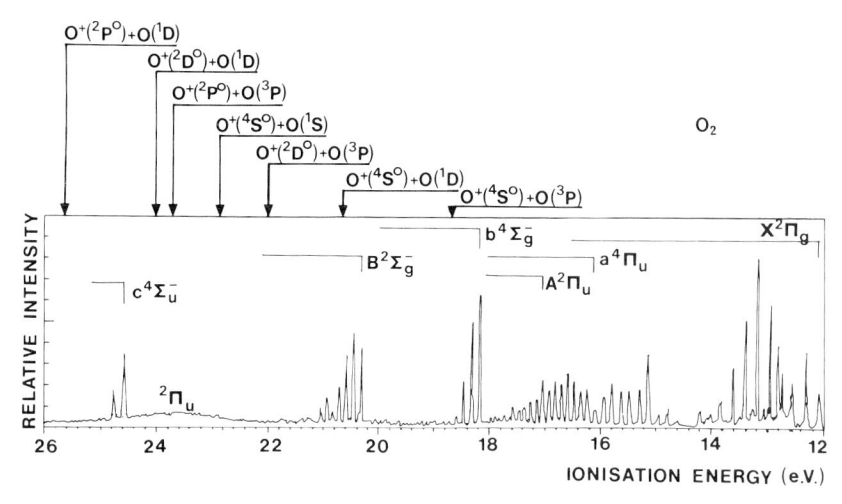

Fig. 3.16. Threshold photoelectron spectrum of O_2 for photon energies between 12 and 26 eV. The arrows on the top of the figure represent the different dissociation limits. From Guyon and Nenner (1980) and Ferreira (1984).

In fig. 3.17 the relevant potential curves (Beebe et al. 1976, Marian et al. 1982) are presented. The c state has a shallow well in the Franck–Condon region, which can accommodate only two vibrational levels ($v = 0$ and $v = 1$). Adiabatically it correlates to the $O^+(^4S) + O(^1D)$ limit at 20.7 eV (Tanaka and Yoshimine 1979).

The experimental TEPICO spectrum measured by Richard-Viard et al. (1985) is presented in fig. 3.18 with a corresponding Monte Carlo simulated spectrum. From this fit it was possible to conclude that two dissociation channels were populated: in addition to the above mentioned 20.7 eV limit leading to $O(^1D)$ atoms, there is also dissociation into the 18.73 eV limit leading to $O(^3P)$ atoms. The relative yields are 62% and 38%, respectively. The predissociation lifetime has been estimated to be ≤ 50 ns. The results show clearly that no dissociation into $O^+(^4S) + O(^1S)$, $O^+(^2D) + O(^3P)$ or $O^+(^2D) + O(^1D)$ is observed, although these limits are energetically accessible.

Recently, Frasinski et al. (1985) and Akakori et al. (1985) have also studied the decay channels of the c state using electron–ion coincidence techniques and synchrotron radiation. For the $v = 0$ level, predissociation to the two limits leading to $O(^1D)$ and $O(^3P)$ atoms is observed, in agreement with the experimental results of Richard-Viard et al. (1985). For $v = 1$, on the other hand, predissociation to $O(^1D)$ is predominant. From fig. 3.17 it is clear that tunneling through the barrier followed by dissociation along the $c\,^4\Sigma_u^-$ curve will produce essentially $O(^1D)$ atoms since the crossing probability to the $^6\Pi_u$ state by spin–orbit coupling will be very small (certainly much smaller than the experimental value of 38%). Thus one is tempted to conclude that the $v = 0$ level of the c state is predissociated by the $^{2,4,6}\Sigma$ states via spin–orbit interaction on a time scale comparable to its tunneling lifetime, while the $v = 1$ level has a much shorter tunneling lifetime and therefore leads to $O(^1D)$ atoms. However, this may very

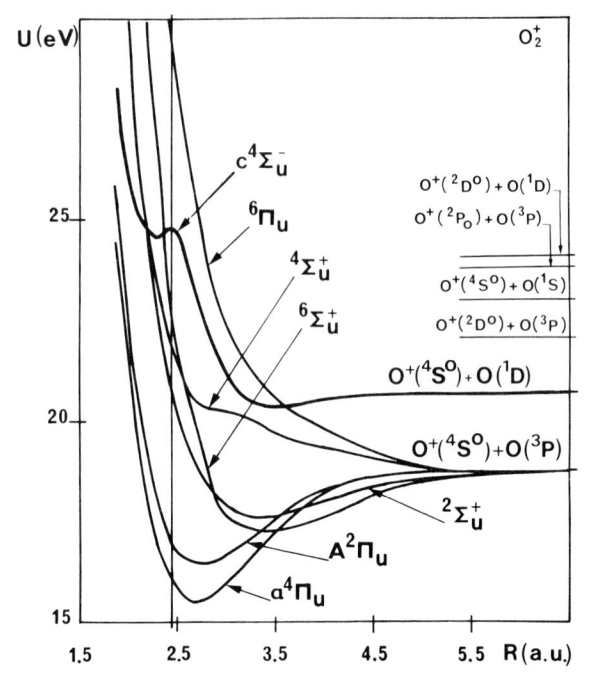

Fig. 3.17. Some of the potential energy curves of O_2^+ relevant for the predissociation of the c state. Adapted from Beebe et al. (1976) and Marian et al. (1982). The vertical line indicates the center of the Franck–Condon region.

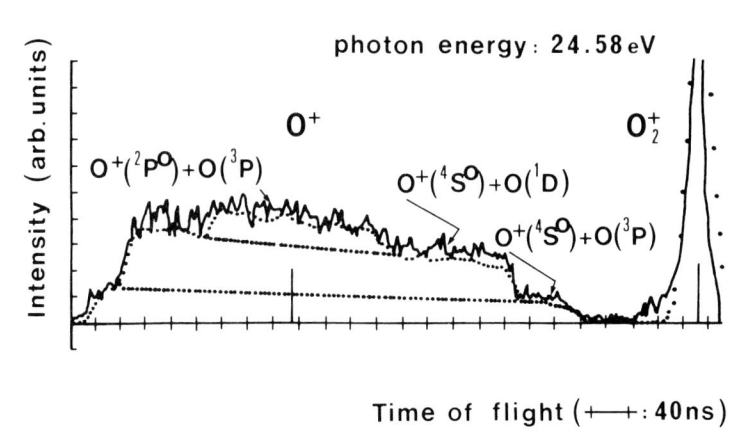

Fig. 3.18. TEPICO spectrum of O_2 at the energy of the $v=0$ level of the c state of O_2^+ (from Richard-Viard et al. 1985). The dotted lines are the Monte Carlo simulated spectrum.

well be an oversimplified picture. There are other states ($^4\Pi$ and $^4\Delta$), not represented in fig. 3.17, which cross the c state in the region of the well and which correlate with the $O^+(^4S) + O(^1D)$ limit. Thus predissociation may well be explained assuming that tunneling is very slow. Further experimental work, in particular for $^{18}O_2$ (Richard-Viard et al. 1985), as well as detailed calculations are needed to elucidate the predissociation mechanism of the c state in O_2^+.

Autoionization of excited states of neutral molecules may also lead to dissociative ionic states and enhance the production of ionic fragments at certain photon energies. Although this process will be discussed in more detail in section 3.2.3, we shall provide here some elements for that discussion, taking the case of H_2 as an example. Masuoka (1984) has measured the ratio H^+/H_2^+ between 18 and 35 eV, combining a time of flight spectrometer and synchrotron radiation. He has shown that proton production is dominated at threshold (18 eV) by the direct excitation of the repulsive part of the $^2\Sigma_g^+(1s\sigma_g)$ state of H_2^+. On the other hand, a wide continuum, extending from 25 to 35 eV with a maximum at 30 eV, is due to the contribution of the two-electron excited state $^1\Sigma_u^+(2p\sigma_u 2s\sigma_g)$ of H_2 in addition to direct ionization into the H_2^+ $^2\Sigma_u^+(2p\sigma_u)$ continuum.

Polyatomic molecules. The study of dissociative ionization of polyatomic molecules is of particular importance because it raises several new questions:

(1) What is the partitioning of the excess energy of the ion among the various degrees of freedom of the fragments, i.e., what is the energy content of the products? This question arises also, of course, in the case of photodissociation into neutral fragments (see section 3.1.3).

(2) Is there any selective dissociation pathway for certain photon energies? This general question arises not only for ions with a valence hole but also for those with a core hole (see section 3.6.3).

As opposed to diatomic molecules, there are experimental difficulties in measuring the energy content (electronic, vibrational and rotational) of the fragments. In addition, for polyatomic molecules several different fragmentation channels are possible, as for example

$$ABC^+ \rightarrow A^+ + BC,$$
$$\rightarrow AB^+ + C.$$

The electron–ion coincidence technique measures the mass and the kinetic energy of the ionic fragments. If for a given photon energy several dissociation limits are energetically allowed, it is impossible to distinguish the electronic from the vibrational energy content of the molecular fragment. Therefore this method has been applied successfully only to triatomic molecules and for a limited photon energy range above the first dissociation limit.

The dissociation of the N_2O^+ ion has been studied in detail by Nenner et al. (1980). This system provides an interesting case for comparison with theoretical calculations of the predissociated and repulsive states (Beswick and Horani 1981,

Komiha 1981, Miret-Artes et al. 1983, Gregory et al. 1984 and references therein). In addition, the photodissociation dynamics in this molecule is the "half-collision" analog of the well-known ion–molecule reaction

$$O^+ + N_2 \rightarrow NO^+ + N,$$

which plays a very important role in the earth's atmosphere.

An example of threshold electron coincidence spectra in N_2O^+ is shown in fig. 3.19. The photon energies have been chosen in the A–X Franck–Condon gap (15.54 eV) and at the energy corresponding to the $(1, 0, 0)$ level of the A state of N_2O^+ (16.55 eV). Both lie above the first two dissociation limits corresponding to $NO^+(^1\Sigma^+) + N(^4S_u)$ and $O^+(^4S_u) + N_2(^1\Sigma_g^+)$, but below all other possible ones. The results presented in fig. 3.19 show a quite different dissociation pattern as a function of the energy. For 15.54 eV, i.e. below the A state of the ion, the O^+ production dominates, while at the energy corresponding to the $(1, 0, 0)$ level of the A state, NO^+ is the main fragment and the production of O^+ vanishes.

This example illustrates how the dissociative pattern of a small ion depends critically on the preparation of the parent ion. In the A–X Franck–Condon gap the ion is prepared in the ground state with a large amount of vibrational energy. Those states are predissociated at large N–NO distances by a $^4A''$ state (Gregory et al. 1984), leading to the $O^+(^4S) + N_2(^1\Sigma_g^+)$ but also to $N(^4S) + NO^+(^1\Sigma)$. On the other hand, the $(1, 0, 0)$ level of the A state has only one quantum of vibration and because of unfavorable Franck–Condon factors it predissociates

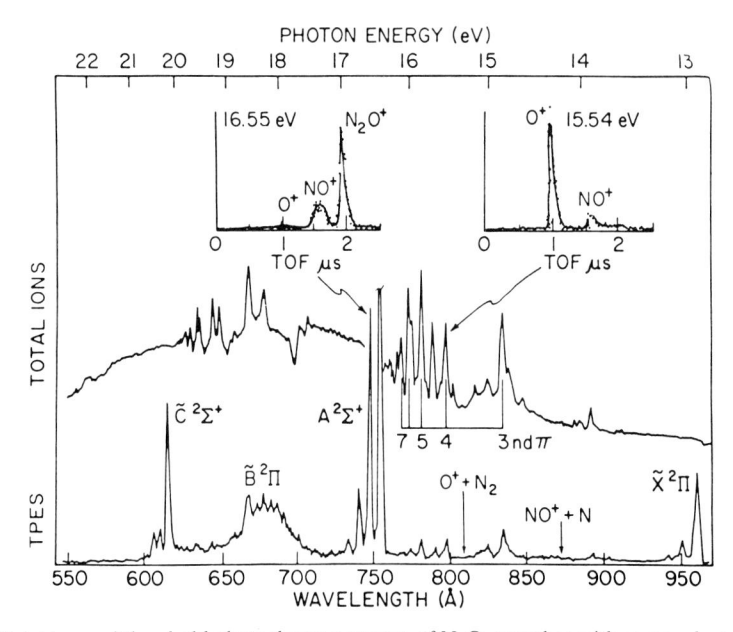

Fig. 3.19. Total ion and threshold photoelectron spectra of N_2O, together with some selected TEPICO spectra (from Nenner et al. 1980).

only to the N(^{4}S) + NO$^+$($^1\Sigma$) limit. Actually, this predissociation is quite slow and fluorescence from A to X can be detected. This explains the remaining N$_2$O$^+$ peak observed in the 16.55 eV coincidence spectrum shown in fig. 3.19.

Another very interesting aspect of the dissociation dynamics of N$_2$O$^+$ in this energy region has been brought about by the studies on the isotope-labelled ^{14}N^{15}NO molecule (Berkowitz and Eland 1977). Those experiments have shown that both ^{14}NO$^+$ and ^{15}NO$^+$ fragments are formed. This scrambling is more pronounced below the A state of the ion than above it. The theoretical study of Gregory et al. (1984) suggests that the ground state ion formed in the Franck–Condon gap between the X and the A states with a large amount of vibrational energy, samples an extended region of vibrational space (in particular highly bent configurations) before it is predissociated. Therefore, it can find its way into a cyclic configuration, in which the probability of scrambling is high.

Dissociative ionization into three fragments has been observed recently by Dujardin et al. (1985a) in SO$_2$. This case seems to be an example of simultaneous rather than sequential breaking of the two bonds. This is interesting because examples of simultaneous fragmentation into more than two fragments are scarce, not only in ions but also in neutral molecules.

3.2.3. Photoionization dynamics: partial cross sections and branching ratios
Electronically resolved partial cross sections have been measured in a number of molecules by Angular-Resolved PhotoElectron Spectroscopy (ARPES; see, for example, Plummer et al. 1977, Morin 1983, Lablanquie 1984, Lablanquie et al. 1985c, Woodruff and Marr 1977, Truesdale et al. 1983, Carlson et al. 1984a, and references therein), and by Fluorescence Excitation Spectroscopy (FES; see, for example, Hertz et al. 1974, Lee et al. 1976, Ito et al. 1985, and references therein). These methods complement nicely the pioneering work obtained by the dipole (e,2e) technique (Brion and Thomson 1984, Brion 1985), primarily because in the low-energy region (i.e. below 50 eV) PES are better resolved and have higher intensities than (e,2e) spectra.

The partial photoionization cross section for an electric dipole transition from an initial bound wavefunction $|\Psi_i\rangle$ to a final continuum wavefunction $|\Psi_{f\varepsilon}\rangle$, where ε is the kinetic energy of the electron, is given by (in mks units)

$$\sigma_{i,f\varepsilon} \propto \frac{4\pi^2(\hbar\omega)}{\hbar c} \left| \langle \Psi_i | \hat{n} \cdot q_e \sum_j r_j | \Psi_{f\varepsilon} \rangle \right|^2, \qquad (3.18)$$

where $\hbar\omega = I_{IP} + \varepsilon$ is the photon energy, $\hat{n}$ is the polarization of the incoming photon, q_e is the electronic charge, and the sum extends over all electrons. In the framework of the independent-particle one-electron picture, the energy dependence of the partial cross section will be essentially determined by the matrix element $D_{i,\varepsilon} = \langle \phi_i | \hat{n} \cdot r | \phi_{f\varepsilon} \rangle$, where ϕ_i and $\phi_{f\varepsilon}$ are electronic wavefunctions for the initial and final states of the outgoing electron. In the high-energy limit the continuum wavefunction $\phi_{f\varepsilon}$ can be described in first approximation by a plane wave. The radial integral in $D_{i,\varepsilon}$ will then be just the Fourier transform of the initial wavefunction ϕ_i (multiplied by r). Thus a localized initial wavefunction will have a partial photoionization cross section extending much further in the high-energy region than a delocalized one. For example, the ejection of an electron from a localized σ orbital will have a partial photoionization cross section at high energies much larger than one from a delocalized π orbital. This has been borne out in X-ray photoelectron spectroscopy. This fact is also well known in Compton scattering as well as in high-energy (e,2e) experiments.

Cooper minima are another manifestation of the same general (Fourier transform) behavior of the partial cross sections at high energies. A node in the radial initial wavefunction will be reflected as a minimum in the partial cross section. The Cooper minima are even more pronounced in the photon energy dependence of the angular distribution asymmetry parameter β. This will be discussed in section 3.2.4.

At low energies, partial photoionization cross sections are strongly dependent on the Coulomb interaction between the electron and the residual ion. The partial photoionization cross section has a finite value at zero kinetic energy and shows a step-like behavior. This has been evidenced by mass spectrometry only at the first ionization threshold (Berkowitz 1979). The total photoionization cross section is identical to the partial cross section because a single electronic continuum is open. For partial photoionization associated with ions produced through the ejection of inner electrons, only Fluorescence Excitation Spectroscopy (FES) has provided threshold measurements. As an example, fig. 3.20 presents the partial cross section for $N_2O^+(A\,^2\Sigma^+)$ production measured by FES (Guyon and Nenner 1980,

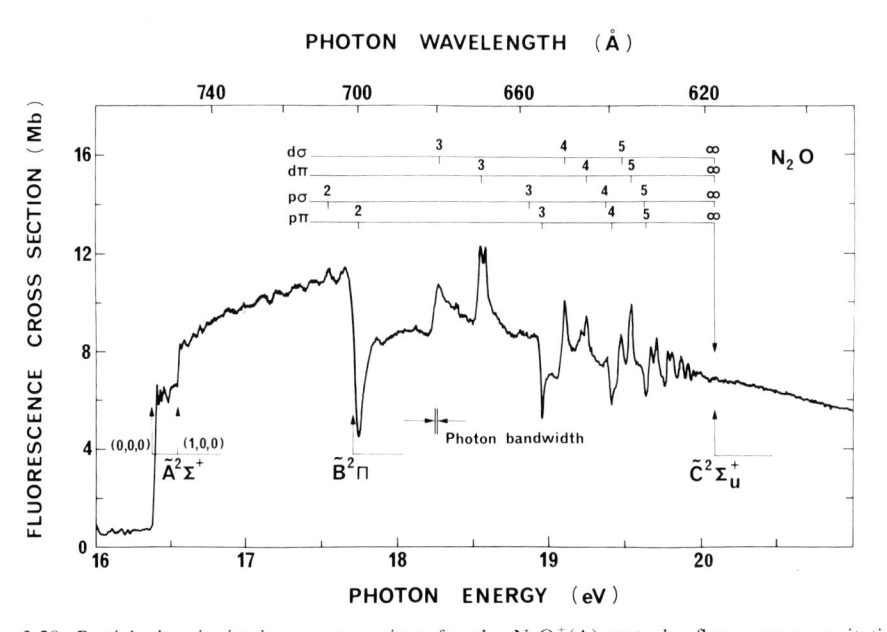

Fig. 3.20. Partial photoionization cross sections for the $N_2O^+(A)$ state by fluorescence excitation spectroscopy (from Guyon and Nenner 1980, Tabché-Fouhailé 1981).

Tabché-Fouhailé 1981). The ion results from the ejection of an electron from the second outer valence orbital 7σ of N_2O. At threshold one observes two steps associated with the formation of the $(0, 0, 0)$ and $(1, 0, 0)$ vibrational levels. The height of the steps are proportional to the vibrational Franck–Condon factors between the ground initial state of N_2O and the A state of the ion.

Resonances. In the low and intermediate electron energy regions (0 to ~20 eV) resonances are particularly conspicuous (Dehmer et al. 1987, Raseev et al. 1986, Keller and Lefebvre 1986). This is of considerable importance since resonances are often at the origin of the formation of ions in highly excited vibrational levels and/or the production of neutral and ionic fragments in specific electronic states. Those species are actually known to dominate the "chemistry" of the atmosphere (Fergusson 1979).

Resonances are also known to exist in atoms, of course, but in molecules there are many more since:

(1) molecular valence orbitals are much more numerous than in atoms and they are often close in energy,

(2) there is an extremely large number of final ionic states because each electronic continuum is actually composed of many rotational and vibrational continua,

(3) in addition to the expected resonant processes known in atomic physics, such as electronic and spin–orbit autoionization, specific molecular processes exist: vibronic, vibrational and dissociative autoionization as well as shape resonances induced by the angular anisotropy of the molecular potential.

An example of molecular resonances can be noticed in fig. 3.20. There are small fine structures superimposed on the two steps described above, which are likely due to Rydberg series converging to high vibrational levels of the $N_2O(A)$ state. Their presence on top of the steps reveals the existence of vibrational autoionization. The most intense features observed, showing asymmetric profiles, are four Rydberg series converging to the C state of N_2O^+. They are the result of a strong electronic autoionization process with the A continuum. In the following we shall consider electronic autoionization and shape resonances (which do not exist in the N_2O example presented in fig. 3.20).

Electronic autoionization and shape resonances. Electronic autoionization is certainly the most important process in molecular photon absorption between 10 to 20 eV. Experimentally it is evidenced by sharp structures superimposed on the direct ionization continuum. Total absorption spectra obtained with discharge lamps (helium continuum) and synchrotron radiation have demonstrated that for a number of molecules these structures can be very intense as compared to the direct ionization continuum.

From the theoretical point of view, electronic autoionization resonances are described in terms of Fano's configuration mixing (Fano 1961), which has already been discussed in section 3.1.2 for predissociation. The two processes are formally equivalent. The final wavefunction $\Psi_{f\varepsilon}$ in eq. (3.18) is written as a linear combination of a discrete (resonant) state, which in the case of electronic autoionization corresponds to a highly excited state of the neutral, and continuum wavefunctions corresponding to direct ionization. In the most common case, the coupling between the discrete state and the continuum is provided by the repulsion between electrons. Thus electronic autoionization is a manifestation of the breakdown of the independent-particle approximation (Raseev et al. 1986). From the electronic scattering point of view, electronic autoionization resonances correspond to an incoming electron being momentarily trapped by the ion with a fraction of its kinetic energy being transformed into internal excitation of the ion. In scattering theory they are known as Feshbach resonances (Newton 1966). An example of electronic autoionization is provided by the total photoionization spectrum of O_2 (Ferreira 1984, Guyon and Nenner 1980). The extremely low direct ionization probability above threshold (see fig. 3.21) has been explained qualitatively (Roche et al. 1981, Langhoff et al. 1980): the $1\pi_g$ orbital is very similar to an atomic 3d orbital. The 3d→ εf transition amplitude ($\Delta l = +1$), which usually dominates in atomic photoionization (as compared to 3d→ εp ($\Delta l = -1$), is small at threshold because of the centrifugal barrier preventing the f electron from being ejected. The intense peaks in the PhotoIonization Spectrum (PIS) presented in fig. 3.21 are Rydberg series converging to higher ionization limits, as indicated in the lower part of fig. 3.21, where the Threshold PhotoElectron Spectrum (TPES) is presented.

Shape resonances are also important in the valence ionization continuum from 0 to about 20 eV above threshold. They originate from the existence of potential barriers. In this respect shape resonances can be described by a single electronic configuration (at least to first order) and thus they can be understood in the framework of the independent-particle approximation (Dehmer et al. 1987). This is in contrast to the case of electronic autoionization. Also, shape resonances result in broad structures, which may extend up to a few eV, while autoionization resonances are usually sharp.

It is usually very difficult to observe shape resonances in total valence photoionization spectra because they are "washed out" by the contribution of the other ionization continua as well as autoionizing structures. A shape resonance is known to exist in O_2 around 21 eV photon energy, for instance, but cannot be seen in the PIS of fig. 3.21.

Two main properties of resonances will now be discussed:

(1) their electronic properties, i.e. the electronic configuration of the ionic core, the quantum numbers associated with the excited electron(s), the width (lifetime) and the transition moments;

(2) their nuclear properties, i.e. the potential curves which are involved (if they have a physical meaning) as a function of bond distances and angles.

Electronic properties of resonances. Electronically resolved partial cross sections in molecules are important observables in the analysis of some of the general properties of resonances since they provide information on the nature and the strength of the couplings with the ionization continua.

As we discussed above, shape resonances are essentially a one-channel proper-

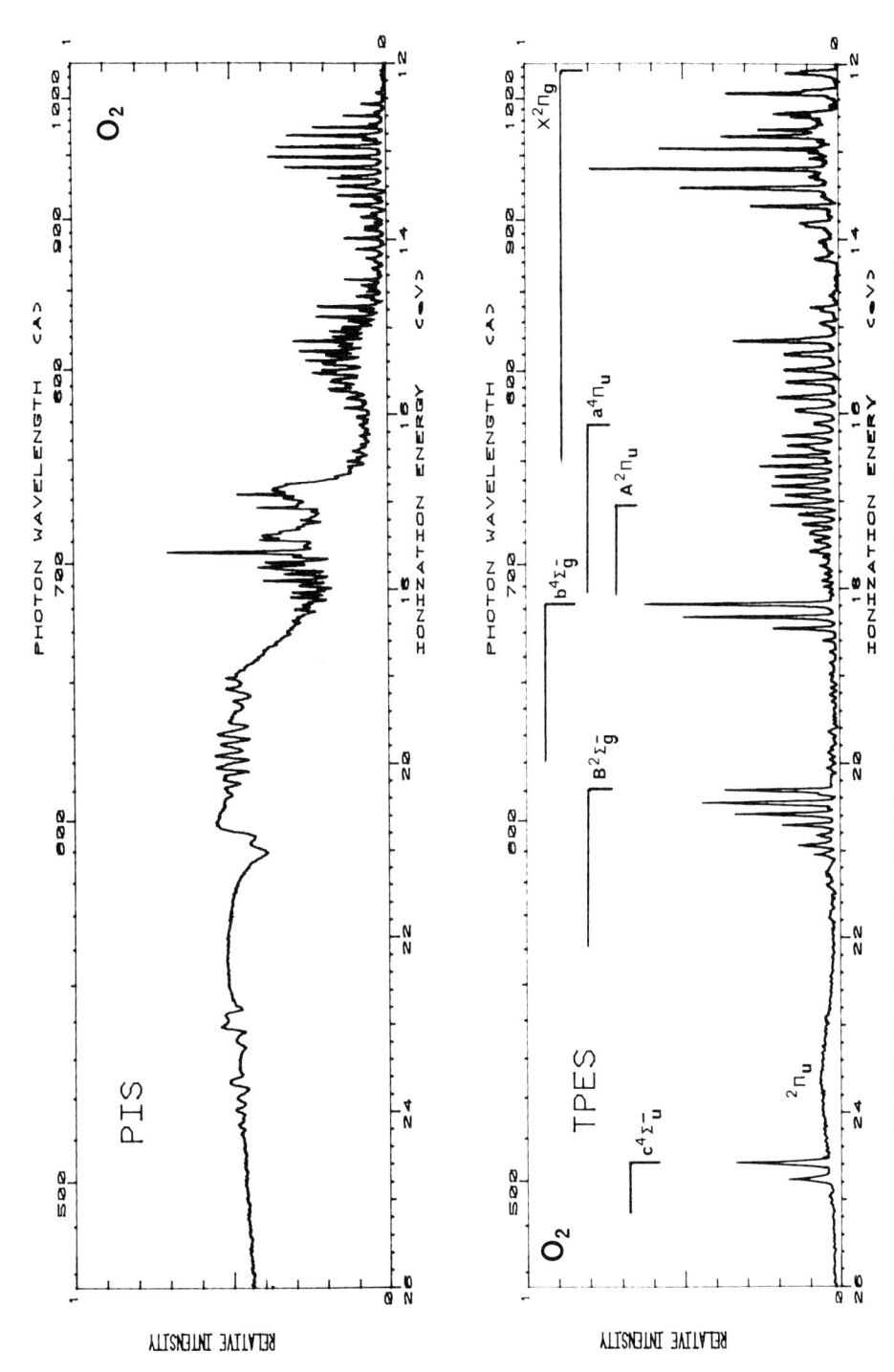

Fig. 3.21. Total ion and threshold photoelectron spectra in O_2 (from Guyon and Nenner 1980).

ty. Autoionization resonances, on the other hand, are generally coupled to several ionization continua. As an example let us consider the Hopfield series in N_2 (Dehmer and Chupka 1975). It corresponds to the excitation of a $2\sigma_u$ electron:

$$(2\sigma_u)^2(1\pi_u)^4(3\sigma_g)^2 N_2(^1\Sigma_g^+) \rightarrow (2\sigma_u)(1\pi_u)^4(3\sigma_g)^2 ns\sigma_g, nd\sigma_g, nd\pi_g, \ldots.$$

This series can autoionize in the continuum of $N_2^+(A\ ^2\Pi_u)[(2\sigma_u)^2(1\pi_u)^3(3\sigma_g)^2\varepsilon\lambda]$ as well as in the continuum of $N_2^+(X\ ^2\Sigma_g)$ with configuration $[(2\sigma_u)^2(1\pi_u)^4(3\sigma_g)\varepsilon\lambda]$. Figure 3.22 presents the partial photoionization cross sections associated with $N_2^+(A)$ and $N_2^+(X)$, as well as the total photoionization cross section measured by angular-resolved photoelectron spectroscopy, including the autoionization of the $n=3$ and $n=4$ members of the series (Morin 1983). The energy dependence of the partial cross sections has been studied by Raoult et al. (1983) by the use of the Multichannel Quantum Defect Theory (MQDT). The quantum defect parameters (47 in this case) have been determined by ab-initio calculations, neglecting rotational effects and the dependence on the internuclear distance.

From the comparison between the calculated values and the experiment (see fig. 3.22), it is possible to conclude that the main peak in the $N_2^+(X)$ partial cross section corresponds to a $nd\sigma_g$ resonant state, while the window-like feature in the $N_2^+(A)$ partial cross section corresponds to a $nd\pi_g$ state. The contribution of the $(n+1)s\sigma_g$ state is more difficult to sort out because of the superposition with the $nd\delta_g$ state. According to Raoult et al. (1983), the qualitative interpretation of

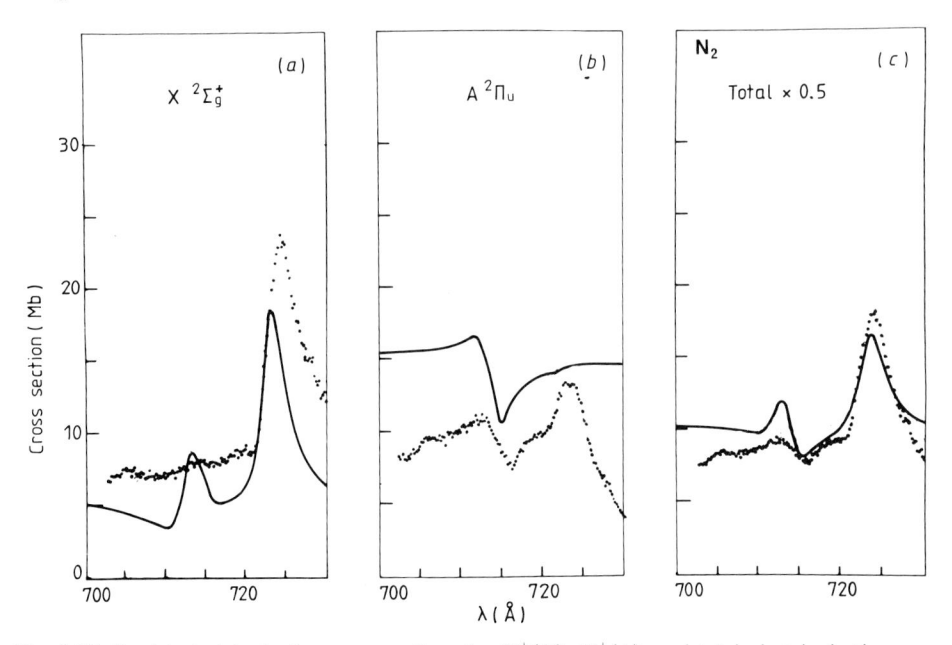

Fig. 3.22. Partial photoionization cross sections for $N_2^+(X)$, $N_2^+(A)$, and total photoionization cross section for N_2^+. From Morin (1983). The full lines are the theoretical results of Raoult et al. (1983).

both the profile and the branching ratio is the following. The $N_2^+(A)$ partial cross section is dominated by the $\varepsilon d\sigma_g$ continuum, which has a large transition moment from the initial ground state of N_2. This makes the profile asymmetry parameter q close to zero (see section 3.1.2). The $4s\sigma_g$ and $3d\sigma_g$ resonances interact weakly with the continuum and do not contribute significantly to the partial cross section. For the $N_2^+(X)$ partial cross section, on the other hand, the transition moment for the resonant $3d\sigma_g$ state is large compared to the direct transition to the continua. Therefore, the partial cross section for $N_2^+(X)$ presents a peak with large asymmetry parameter q (see eq. 3.6) and corresponds predominantly to autoionization into the $\varepsilon f\sigma_u$ continuum. The calculated electronic branching ratio (X/A) is almost 1, which compares reasonably well with the value of 0.85 measured by Morin (1983). According to Raoult et al. (1983), better agreement is obtained when final-state interactions (continuum–continuum couplings) are included in the calculations.

This example illustrates the importance of partial cross sections in order to identify the dominant photoionization channel associated with a given resonance. Although the peak and the window are seen in the total ionization spectra (see fig. 3.22c), a definite assignment of these features can only be made through the analysis of the partial cross sections and branching ratios.

We now consider an example showing the importance of partial cross section measurements for the case of a shape resonance. Theoretically, a σ_u shape resonance was expected in the 30 to 45 eV photon excitation range for photoionization of CO_2, giving $CO_2^+(\tilde{C})$. Because the total cross section is dominated in this energy region by photoionization into $CO_2^+(\tilde{X}, \tilde{A}$ and $\tilde{B})$ states, one needs to measure partial cross sections in order to be able to detect such a resonance. Roy et al. (1984) performed that experiment, using angular-resolved photoelectron spectroscopy. Their results are presented in fig. 3.23. A resonance is clearly seen.

Several theoretical calculations are also presented in fig. 3.23. The values obtained by Padial et al. (1981) are in good agreement with the experiment. Surprisingly enough, the calculations by Lucchese and McKoy (1982) and by Collins and Schneider (1984), which were doing extremely well for the $N_2(X)\varepsilon\sigma_u$ shape resonance, fail in this case. Several hypotheses have been advanced by Roy et al. (1984) to explain this discrepancy: final-state interactions and/or bent configuration of $CO_2^+(\tilde{C})$. New calculations are clearly needed to elucidate this point.

Nuclear properties of resonances. Important information on nuclear potential energy surfaces can be obtained from vibrationally resolved partial cross sections and branching ratios. Let us assume that the transition dipole moment is essentially constant as a function of nuclear distances and that the process can be separated in two: excitation of the Rydberg state and subsequent decay. Under these assumptions one can obtain information about the potential energy surface of the Rydberg state by analysing its decay into several different rovibronic continua. This can be done by measuring the photoelectron spectrum at selected photon energies. If the resonances are well separated (i.e. $\Gamma \ll \Delta E$), it is possible,

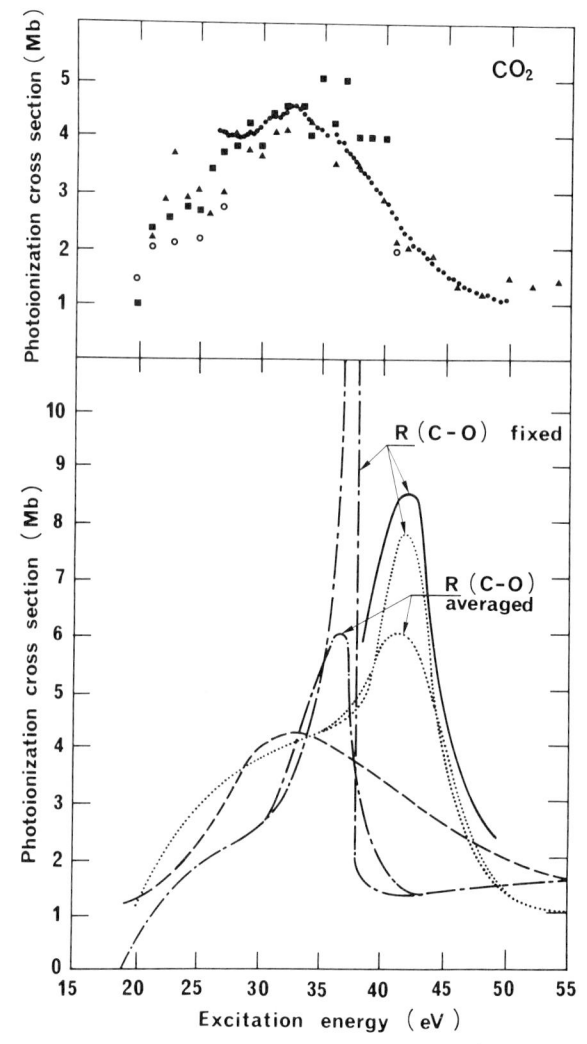

Fig. 3.23. Partial photoionization cross section for the C state of CO_2^+. From Roy et al. (1984). (a) Experimental data: triangles are from the electron impact experiment of Brion and Tan (1978), squares are from the PES experiment of Gustafson et al. (1978) using synchrotron radiation, open circles are the PES results of Samson and Gardner (1973) using a line source, and the solid circles are the results of Roy et al. (1984) using synchrotron radiation. (b) Calculations: $\cdots$ FCSE calculation of Lucchese and McKoy (1982), $-\cdot-\cdot-$ MSM calculations of Swanson et al. (1980, 1981), ——— STMT calculation of Padial et al. (1981), $---$ LA calculation of Collins and Schneider (1984).

using the measured vibronic branching ratios in PES or in dispersed fluorescence spectra, to extract the equilibrium internuclear distance of the Rydberg state in the case of diatomics. This can be done by using the theoretical analysis given by Smith (1970) and based on Fano's theory (Fano 1961). Experimentally this requires resolving the vibrational distribution of the residual ion in various PES or in the ion dispersed fluorscence recorded on and off resonance in order to extract the direct photoionization contribution.

Such measurements have been performed in O_2 (Ferreira 1984), in HCl (White and Grover 1985) as well as in N_2O (Poliakoff et al. 1985). In fig. 3.24 the relevant potential energy curves of O_2 in the 12–17 eV region are presented. Several Rydberg series have been studied by Dehmer and Chupka (1975), labelled H, H', M, I, I'. Time of flight electron spectroscopy has been used to measure PES at photon energies corresponding to the direct excitation of these Rydberg states. In fig. 3.24 the results of Ferreira (1984) for the I series are presented. The distribution of Franck–Condon factors is also shown. It is clear that many more higher vibrational levels of the $O_2^+(X)$ state are populated through electronic autoionization. Ferreira assumed that every autoionizing line could be considered as an isolated resonance. Since a single electronic ionic state can be energetically populated, the calculations of the vibrational branching ratio σ_α/σ_f measured at the energy of the resonance level ν, can be obtained using the

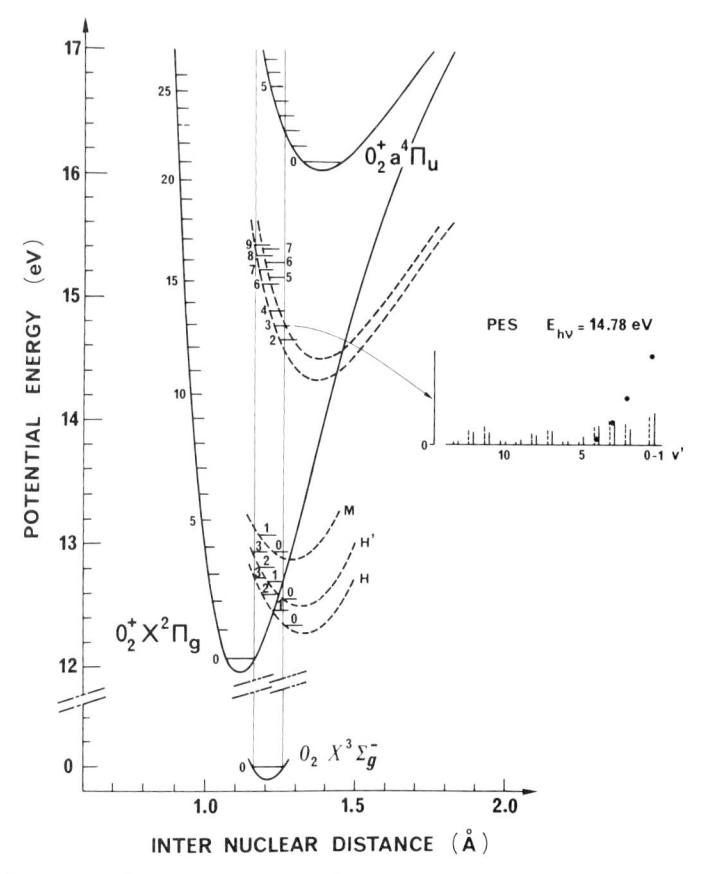

Fig. 3.24. Relevant potential energy curves of O_2^+ in the 12–17 eV region, together with an example of experimental and theoretical vibrational branching ratios obtained for excitation at 14.78 eV (from Ferreira 1984). The inset represents the photoelectron spectrum at 14.78 eV; full bars: experiment, dashed bars: calculation including autoionization of a Rydberg state with an equilibrium distance fixed at 1.37 Å, dots: calculation including direct ionization into the $O_2^+X_\nu$ continuum.

simplified equation (Smith 1970, Ferreira 1984)

$$\sigma_a/\sigma_f = F_{i,a}^2 + F_{i,v}^2 F_{i,v}^2 F_{v,a}^2 q^2 ,\qquad (3.19)$$

where $F_{i,a}^2$ is the Franck–Condon factor for the direct ionization transition, $F_{i,v}^2$ and $F_{v,a}^2$ are the Franck–Condon factors for the excitation step into the Rydberg level v and the autoionizing transition, respectively, while q is the Fano profile asymmetry parameter. Note that this formula is valid only if the photon energy is chosen at the resonance energy, if the experimental photon bandwidth is smaller than the autoionization width Γ and if the autoionization lifetime $\hbar/\Gamma$ is smaller than the rotational period but larger than a vibrational period.

Actually eq. (3.19) can be presented differently (Eland 1980):

$$\sigma_a/\sigma_f = F_{i,a}^2 + F_{v,a}^2 (\sigma_{v_{max}}/\sigma_{v_c} - 1) ,\qquad (3.20)$$

where $\sigma_{v_{max}}$ represents the photoionization cross section measured at the resonance maximum and σ_{v_c} is the photoionization cross section off resonance (i.e. in the continuum). Then, using the equilibrium distance of the Rydberg state as a parameter, Ferreira (1984) obtained the fit of fig. 3.24. The excellent agreement obtained for many different levels is impressive. Similar excellent agreement was obtained in HCl (White and Grover 1985).

A quite different situation occurs when the electronic transition moment depends on the internuclear distance. Well-known cases are shape resonances. Experimentally this is evidenced by a "shift" of the resonance maximum as a function of the final vibronic state of the residual ion as seen in the vibrationally resolved partial cross sections. An example is provided by NO. Ejection of a 5σ valence electron leads to the formation of two ionic states: b $^3\Pi$ and A $^1\Pi$. Vibrationally resolved partial cross sections for $v = 0$ and $v = 1$ levels have been measured by Morin (1983, 1985). They are shown in fig. 3.25. The broad maximum observed in each of the spectra corresponds to a shape resonance predicted by various theories. The position of the maximum is shifted towards higher energies when going from $v = 0$ to $v = 1$ in the b state. This can be interpreted as a departure from Franck–Condon transitions, arising from the change of the potential experienced by the outgoing electron as a function of the internuclear distance.

Another very interesting example is found in N_2. The partial cross section for $N_2^+(X)$ is well known (Plummer et al. 1977, Lablanquie 1984) to be dominated by two intense resonances, one around 23 eV and the other at around 30 eV. The one-electron picture was able to explain quantitatively the resonance at 30 eV in terms of a shape resonance in the σ_u continuum. Wendin (1979) suggested that the 23 eV resonance could be due to a doubly excited valence state, but no detailed calculations have been performed yet. If this prediction turns out to be correct, the resonance decay could be described as normal electronic autoionization, although the discrete state in this case is not a simple Rydberg one as for the Hopfield series discussed above.

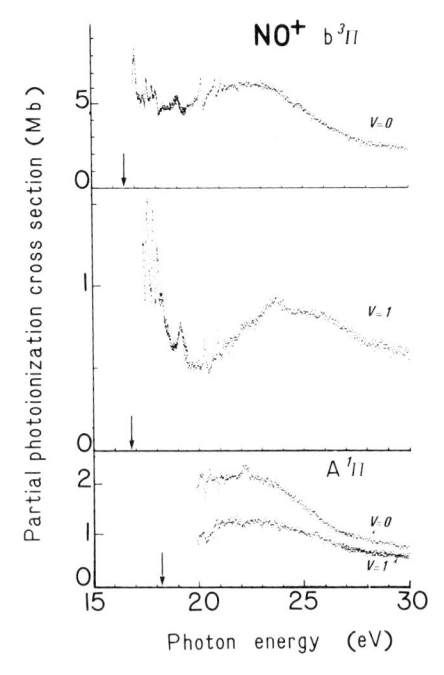

Fig. 3.25. Partial photoionization cross sections for NO leading to NO$^+$(b $^3\Pi$; $v = 0, 1$) and (A $^1\Pi$; $v = 0, 1$). From Morin (1985).

Let us now consider the vibrational branching ratio $P_{v=1}/P_{v=0}$ of $N_2^+(^2\Sigma_g^+)$ reported by West et al. (1980) and Lablanquie (1984) using angular-resolved partial photoelectron spectroscopy. The results are presented in fig. 3.26. The photon energy range covers the region of the two resonances. Around 23 eV the vibrational branching ratio is very close to the Franck–Condon value (9.3%), while for the shape resonance around 30 eV a large deviation is observed. Theoretical calculations (Dehmer et al. 1979, Lucchese and McKoy 1981) are also reported in the figure. They reproduce the experimental values very well.

The explanation for the behavior of the vibrational branching ratio in the region of the shape resonance is as follows. The barrier and, hence, the energy and width of the shape resonance are sensitive functions of the internuclear distance R (Dehmer et al. 1985). In particular for small values of R the resonance is pushed to higher energies and is weakened. The Franck–Condon factorization is no longer valid and the calculation has to include the proper matrix elements of the electronic amplitude between the initial and final vibrational wavefunctions. Since the $v = 1$ wavefunction has a larger amplitude in the small-R region than the $v = 0$ wavefunction, the partial photoionization cross section for $v = 1$ is shifted to higher energies.

Another usual assumption made in the analysis of autoionization for Rydberg series is to consider the potential curves to be independent of the principal quantum number and to be essentially parallel to the curve of the ionic core. This

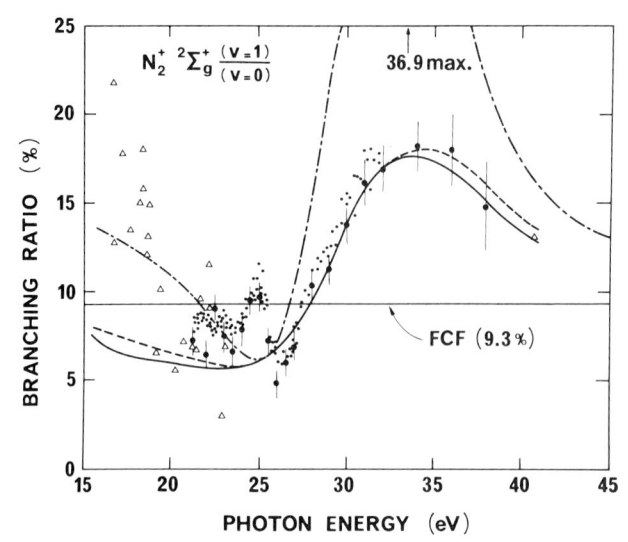

Fig. 3.26. Branching ratios for production of the $v = 0, 1$ levels of $N_2^+(X\ ^2\Sigma_g^+)$. Experiments: large dots: West et al. (1980); triangles: Gardner and Samson (1978); small dots: Morin (1983). Calculations: $-\cdot-$ Dehmer et al. (1979), ——— and $---$ Lucchese and McKoy (1981).

has been verified in many cases. However, several examples indicate that departures from this simple picture are not rare. There are several reasons to expect a distortion of the Rydberg potential curves, among which: (1) the variation of the molecular field with the internuclear distance, (2) configuration mixing with valence states, and (3) the change in nature of certain Rydberg orbitals, which, as R increases, transform to valence orbitals. The last has been described by Mulliken (1976) as "Rydbergization" and is illustrated by specific Rydberg series seen in the O_2 photoionization spectrum. The first has been illustrated by the theoretical analysis of the H_2 discrete absorption spectrum by Jungen using MQDT (see Greene and Jungen 1985, and references therein). He has shown that there are significant variations of the shape of the Rydberg potential energy curves when the principal quantum number increases due to the internuclear distance dependence of the quantum defect. The second point is illustrated by the Cl_2 (Moeller et al. 1983) and NO (Chergui et al. 1985a,b) molecules and will be discussed separately (see section 4).

Figure 3.27 presents vibrationally resolved partial cross sections associated with the formation of $O_2^+(b\ ^4\Sigma_g^-)$, $v = 0, 1, 2$ and 3, in the 18.5 to 21.5 eV photon energy region (Morin et al. 1982, 1983). The intense structures are Rydberg series converging to the $B\ ^2\Sigma_g^-$, $v = 0, 1, 2, \ldots$ of O_2^+ with $(np\sigma_u, nf\sigma_u)$ configurations. They have been shown to decay primarily into the $b\ ^4\Sigma_g^-$ continuum (Tabché-Fouhailé et al. 1981). Each member of the series corresponds to a particular final ionic vibrational state. A $\Delta v = 0$ propensity rule is observed and thus a strong vibrational selectivity in the autoionizing decay results. Also, this implies a similarity between the potential curves of the Rydberg and the ionic states of the same configuration. However, an unusual intensity and width (in particular for the first peaks a, b, c, d) is observed. This can be understood as originating from

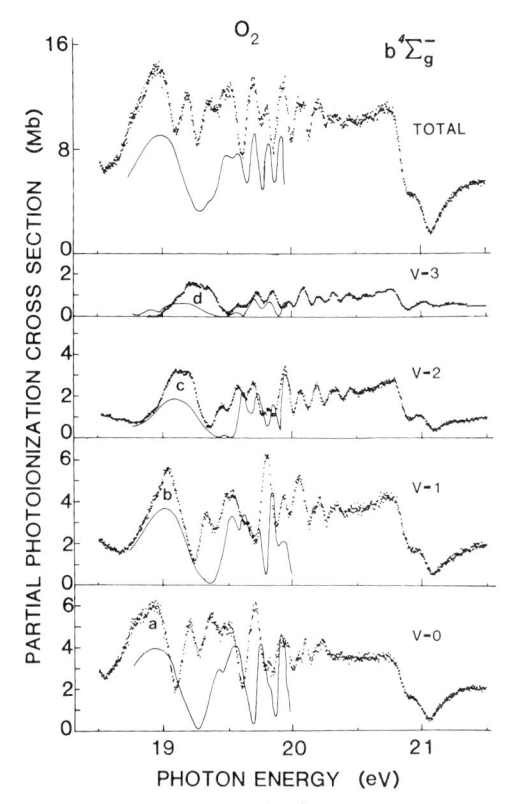

Fig. 3.27. Vibrational partial cross sections for $O_2^+(b\,^4\Sigma_g^-)$ in the 18.5–21.5 eV photon energy range (from Morin et al. 1982, 1983). Solid curves correspond to parametrized MQDT calculations (Raoult et al. 1983).

the σ_u valence orbital character of the $3p\sigma_u$ Rydberg state, which decreases as the principal quantum number n increases (see fig. 3.28). This effect can be described as the Rydbergization of the σ_u^* valence orbital, producing a slight distortion of the potential curves for the Rydberg series. In the $B\,^2\Sigma_g^-$ continuum, the σ_u valence orbital is actually responsible for the shape resonance of the same symmetry as calculated by Raseev et al. (1980) and observed experimentally by Morin et al. (1982). The MQDT calculations reproduced the main features of the spectrum (fig. 3.27) with the use of only three adjustable parameters.

Finally, it is worth mentioning that rotational distributions of molecular ions following resonance excitation, as successfully measured by Poliakoff et al. (1986b), shed new light on photoionization processes, namely that the partitioning of angular momenta between the photoelectron and the photoion depends on the photoejection dynamics.

3.2.4. Photoionization dynamics: angular distribution of photoelectrons

The polarization properties and tunability of synchrotron radiation have led to a large number of experimental measurements concerning the angular distribution of photoelectrons in molecules.

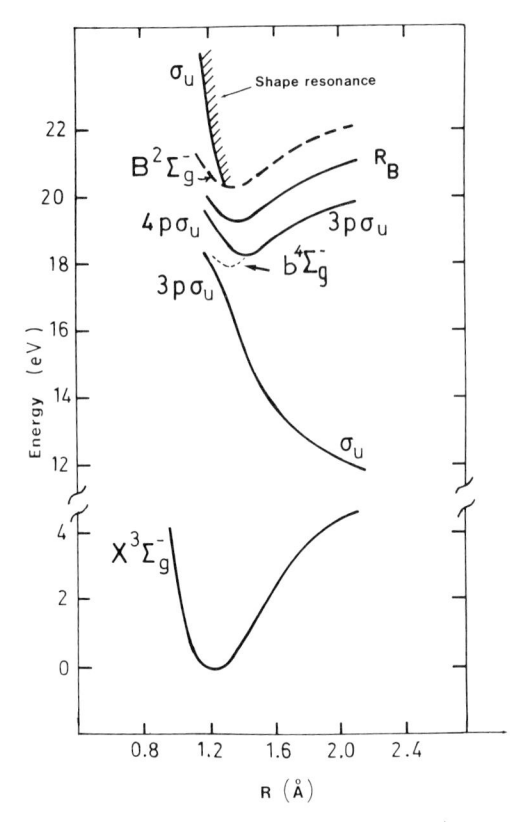

Fig. 3.28. Relevant potential curves for the Rydbergization in O_2^+. From Morin et al. (1982).

As already discussed in section 2.3 the angular distribution is governed by only one parameter: the asymmetry parameter β. Generally β gives important information on the photoionization process in particular from its dependence upon the kinetic energy of the outgoing electron [see, for example, the analysis given by Thiel (1982, 1983, 1984) for diatomic molecules]. Considering that the ejected photoelectron can take, according to the electric dipole selection rules, various asymptotic values for the orbital angular momentum l, the asymmetry parameter β actually is the sum of various terms, which depend on the transition dipole moments into the individual angular momentum channels, on the Coulomb phase shifts and on the intra- and interchannel phase shifts. On that basis one expects β to have specific values in particular circumstances and to show a variation with electron kinetic energy according to the variation with energy of those components.

For *low kinetic energy* ($E \leq 10$ eV) and nonresonant photoionization the first two partial waves usually dominate and the asymmetry parameter will be directly related to the Coulomb phase shift term $\cos(\eta_{l+2} - \eta_l)$. This term increases rapidly from -1 to $+1$ (Thiel 1983) and consequently so does β. This increase is faster with increasing l. This is verified for many outer valence direct ionization processes in linear molecules. One such example is the formation of $N_2^+(A\ ^2\Pi_u)$.

This process results from the ejection of a π_u outer valence electron from N_2,

$$\ldots (1\pi_u)^4(3\sigma_g)^2 \rightarrow (1\pi_u)^3(3\sigma_g)^2 + e^-(\varepsilon\sigma, \varepsilon\pi, \varepsilon\delta).$$

According to the analysis of Thiel (1982), the rapid increase of the asymmetry parameter associated with $N_2^+(A\,^2\Pi_u)$ is caused by the rapid variation of the Coulomb phase shifts ($l=0$ in this case).

At *intermediate kinetic energy*, the Coulomb phase shift differences are essentially constant and any variation of β is due to the behavior of the transition dipole moments and/or to the interchannel phase shifts.

In the case of a resonance (shape or autoionization), a single partial wave α dominates. Therefore the corresponding transition moment $D_{l\alpha}$ is maximum at a particular energy and β reduces to a purely geometrical factor, which can be calculated a priori since it depends only on the quantum numbers l and α [the values can be found in the table given by Thiel (1982, 1983)]. Actually β has often a local minimum as a function of energy around the resonance profile. The reason is the following: off resonance the interchannel phase shift terms give a positive contribution to the asymmetry parameter while they vanish on resonance. A beautiful example of this behavior is given by the autoionizing Hopfield series in N_2. As discussed already in the preceding section the Rydberg series converging to $B\,^2\Sigma_u^+$ state of N_2 autoionize in the $N_2^+(X\,^2\Sigma_g^+)$ and $N_2^+(A\,^2\Pi_u)$ continua as follows:

$$(2\sigma_u)^2(1\pi_u)^4(3\sigma_g)^2 N_2(X\,^1\Sigma_g^+) \rightarrow (2\sigma_u)(1\pi_u)^4(3\sigma_g)^2 ns\sigma_g, nd\sigma_g, nd\pi_g,$$

$$\rightarrow (2\sigma_u)^2(1\pi_u)^4(3\sigma_g) N_2^+(X\,^2\Sigma_g^+)\varepsilon\lambda,$$

$$\rightarrow (2\sigma_u)^2(1\pi_u)^3(3\sigma_g)^2 N_2^+(A\,^2\Pi_u)\varepsilon\lambda'.$$

Figure 3.29 presents the asymmetry parameter for each continuum as measured by West et al. (1981) and reproduced by Raoult et al. (1983). Around 724 Å there is a pronounced minimum where $\beta \approx 0.5$. This value is actually equal to the limiting value calculated by Thiel (1982) when the $\varepsilon f\sigma_u$ channel is taken to be the dominant one as calculated by ab-initio MQDT (Raoult et al. 1983). On the other hand, in the $A\,^2\Pi_u$ continuum β is essentially constant around 724 Å. This is due to the contributions of the $s\sigma$ and $d\delta$ channels, which are smooth functions of the energy in this region. Since there is no dominant channel in this case a full calculation such as the one performed by Raoult et al. (1983) is necessary. Actually a very slight increase of β, observed experimentally by West et al. (1981), arises from the Coulomb phase shift term between the $s\sigma$ and the $d\delta$ partial waves.

This example raises the problem of the possible variation of the asymmetry parameter with the internuclear distances. Up to now the analysis of various contributions to β has been made for a molecule at fixed geometry. As we have seen in section 3.2.3, the vibrational motion must be taken into account to properly describe transition moments at resonances. The electronic partial cross

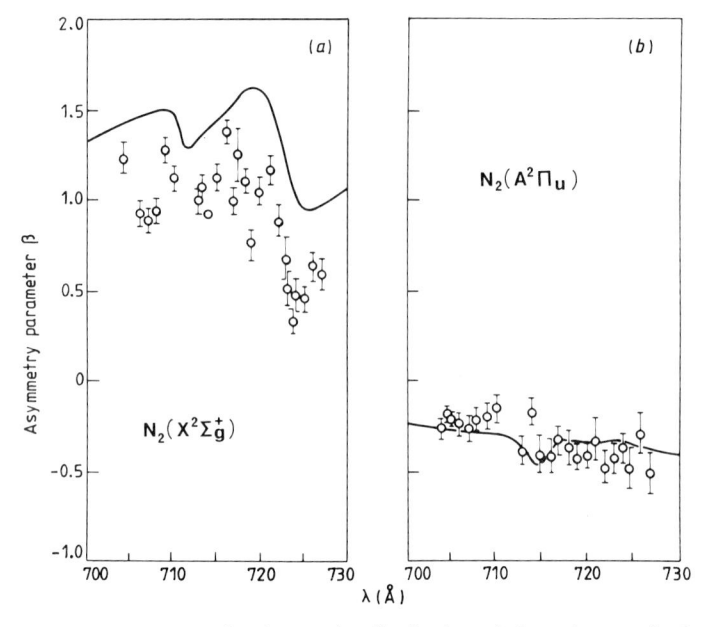

Fig. 3.29. Asymmetry parameter β for the angular distribution of photoelectrons in the ionization of N_2 for two different electronic states (X and A) of the residual ion. The solid curves correspond to the calculation of Raoult et al. (1983). The experimental points are from West et al. (1981).

section averaged over the vibrational wavefunctions and summed over all final vibrational levels gives a broader resonance peak as compared with the calculation at fixed internuclear coordinates. Since β depends also on transition moments, one expects β to show broader peaks in molecular processes (Dehmer et al. 1985, Raseev et al. 1980). This point is beautifully demonstrated by the H_2 calculations presented in fig. 3.30 (Raseev 1985). At fixed internuclear distance the asymmetry parameter shows three resonances as a function of the photon energy. However, the position and shape of the resonances change drastically with the internuclear distance. This should give rise to non-Franck–Condon effects in the vibrationally averaged results. One also expects the asymmetry parameter β for individual vibrational levels to vary slightly differently with energy from one vibrational level to another. The Hopfield series in N_2 studied experimentally by West et al. (1981) provides an example of this behavior. Around 720–700 Å the asymmetry parameter corresponding to the A state of N_2^+ is larger for $v = 1$ than for $v = 0$ (see also Carlson et al. 1980).

In a polyatomic molecule such as CO_2 this qualitative description of β may also apply. Figure 3.31 shows the energy dependence of β associated with the ejection of a $4\sigma_g$ electron from CO_2,

$$\ldots (4\sigma_g)^2(3\sigma_u)^2(1\pi_u)^4(1\pi_g)^4,$$

taken from the work of Roy et al. (1984).

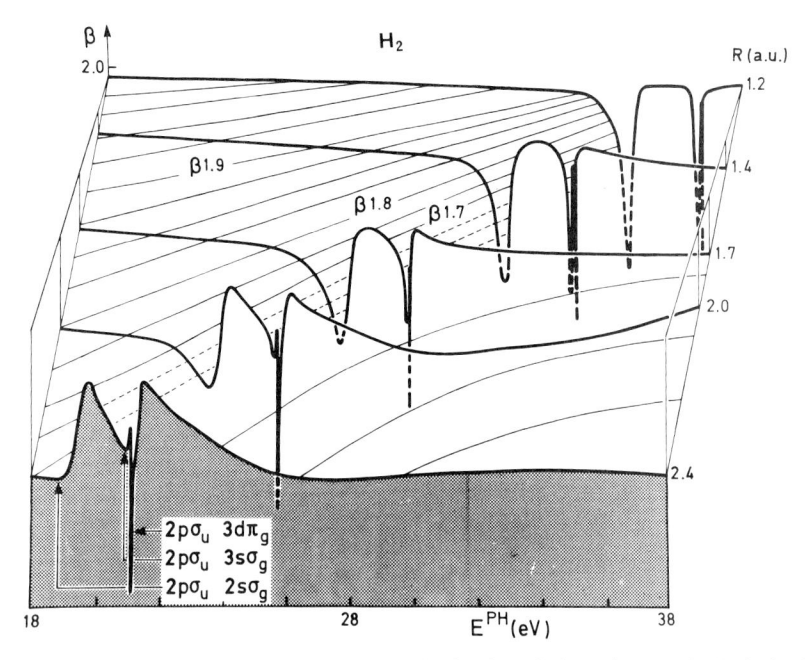

Fig. 3.30. Asymmetry parameter β for the angular distribution of photoelectrons in H_2 ionization as a function of photon energy (E^{PH}) and internuclear distance (R). From Raseev (1985).

The results show a well-defined minimum at 42 eV reproduced by several calculations (Swanson et al. 1980, 1981, Collins and Schneider 1984, Lucchese and McKoy 1982). It originates from the shape resonance in the σ_u continuum. According to Thiel (1982) this minimum is due to interchannel contributions which vanish at resonance (just as for the N_2 case examined above). The CO_2 case is, however, somewhat puzzling because the minimum in β does not match the maximum in σ, as can be seen in fig. 3.23. Moreover, the curve exhibits another shallow minimum around 28 eV (see fig. 3.31), which may reflect specific continuum–continuum interactions (Roy et al. 1984). Many different channels are possible candidates (particularly in a slightly bent geometry), but further theoretical work is needed to verify the suggestions of Roy et al. (1984).

At *high kinetic energy*, i.e. $E > 20$ eV, usually no resonances are found in the continuum and β is essentially related to the electronic angular momentum in the original orbital. However, in some cases the asymmetry parameter may vary due to the presence of a Cooper minimum. Just as for atoms, a node in the radial wavefunction of the initial orbital produces a change of sign of the relevant radial transition moments at a particular energy. This is expected in molecules for $n\sigma$ or $n\pi$ orbitals, with $n > 2$, i.e. valence orbitals. Many experimental examples of this have been found in molecules (Carlson et al. 1983, 1984b, 1986). As analysed by Thiel (1984), for diatomics the β parameter exhibits a minimum not only because some transition moments vanish but also because the positive interchannel

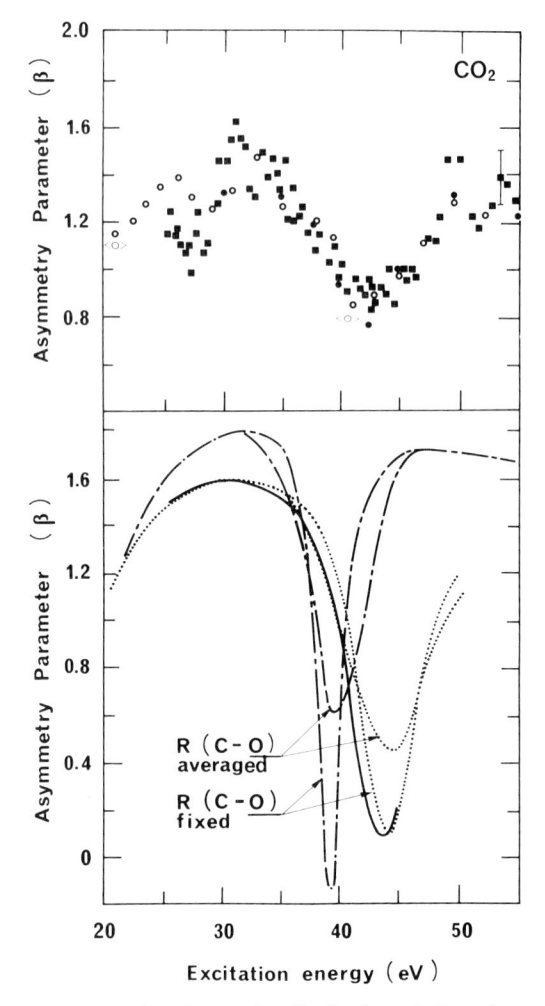

Fig. 3.31. Asymmetry parameter β for the angular distribution of photoelectrons in the ionization of CO_2^+ (from Roy et al. 1984). (a) Experimental data: the open circles are from Carlson et al. (1981) and Grimm et al. (1980, 1981), solid circles and squares are from Roy et al. (1986), the solid circles are the result of a full angular analysis while the squares were obtained in the CIS mode. (b) Calculations: $-\cdot-\cdot-$ MSM calculation of Swanson et al. (1980, 1981), $\cdots$ FCSE calculation of Lucchese and McKoy (1982) and ——— the LA calculation of Collins and Schneider (1984).

contributions to the phase shifts vanish and because the sσ wave dominates at the Cooper minimum.

An example of this has been provided by Adam et al. (1985) on H_2S (see fig. 3.32). The β parameter associated with the $2b_1$ outermost molecular orbital shows a pronounced Cooper minimum. Actually this beautiful example illustrates the strong atomic character of the nonbonding orbital, i.e. the 3p of sulfur, as revealed by the good agreement obtained theoretically by Roche et al. (1980) for

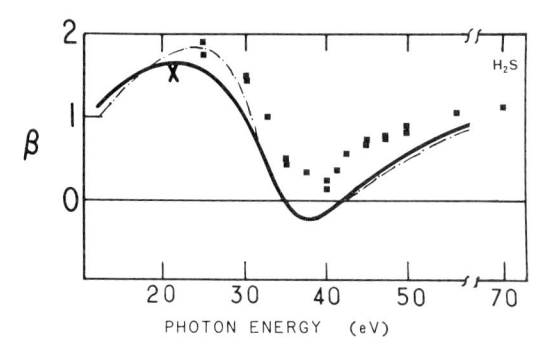

Fig. 3.32. Asymmetry parameter β for the angular distribution of photoelectrons in the ionization of H_2S. The squares are the experimental results of Adam et al. (1985). The cross is the experimental result of Brundle and Turner (1966). The solid and dot–dashed lines are the theoretical predictions of Roche et al. (1980), for S3p photoionization in H_2S and atomic sulfur.

both $H_2S(2b_1)$ and S(3p). Note that for bonding orbitals, the Cooper minimum is much shallower (see Adam et al. 1985, Carlson et al. 1982b, 1983, 1984a,b, 1986).

Another motivation for measuring β at high kinetic energy and/or to study the Cooper minimum, is to identify the symmetry of the final ionic state. This is particularly interesting for inner valence photoionization because of the presence of many "satellite" lines. It has been applied successfully to atoms such as argon (Adam et al. 1985). For molecules the situation is more complicated for the following reasons:

(1) outer valence satellite lines, which account for simultaneous ionization and excitation of one or several electrons into low-valence unoccupied orbitals, are expected in the ionization energy region as lines which originate from inner valence electron ejection;

(2) final ionic states are often predissociated or purely repulsive and the corresponding bands are broadened;

(3) for bound states, vibrational structures may enlarge the band in a photoelectron spectrum and may give rise to vibronic coupling (Köpel et al. 1984).

Angular distribution measurements for molecular satellites have been obtained, for example, in N_2 (Adam et al. 1983), H_2S (Adam et al. 1985) and CO_2 (Roy et al. 1986). In the case of the $2\sigma_u$ photoionization in N_2, the $N_2^+(B\ ^2\Sigma_u^+)$ (main band) and $N_2^+(C\ ^2\Sigma_u^+)$ (satellite) are produced. The B and C states can be described as the result of a hole in the $2\sigma_u$ orbital plus an additional excitation $1\pi_u \rightarrow 2\pi_u^*$. For both final ionic states, the ejected electron has the same distribution of partial waves since it is the same process. Experimentally, the energy dependence of β for the B and C states are quite different (see fig. 3.33). They reach the expected common value (Wallace et al. 1979, Lucchese et al. 1982) only at a kinetic energy of 80 eV. The differences seen at intermediate energies have been explained as the result of continuum–continuum interactions (Stephens and Dill 1985, see also Dehmer et al. 1987).

In fig. 3.15 the CO_2 PES spectrum in the inner valence photoionization region

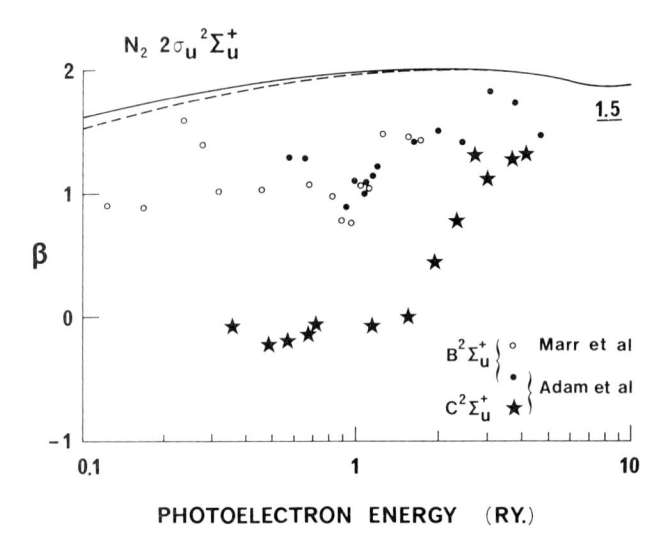

Fig. 3.33. Electron angular distribution asymmetry parameter β in the photoionization of N_2 in the $2\sigma_u$ channel. Full circles and stars are from Adam et al. (1983). The open circles correspond to the experimental results of Marr et al. (1979). Solid curve: theoretical results of Lucchese et al. (1982), dashed curve: calculations of Wallace et al. (1979).

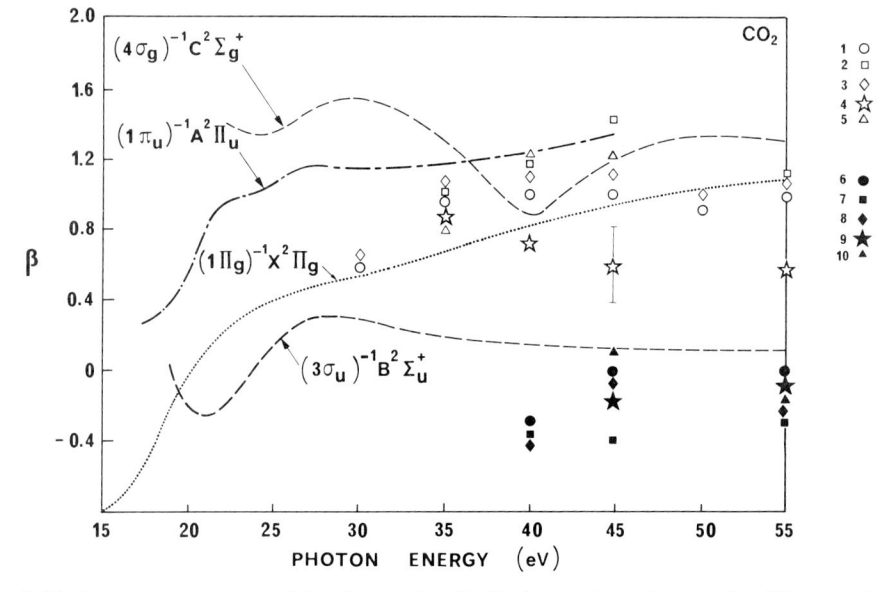

Fig. 3.34. Asymmetry parameter β for the angular distribution of photoelectrons for different valence satellite lines in CO_2 (from Roy et al. 1986), which are labeled on the right-hand side of the figure by numbers ranging from 1 to 10. The corresponding binding energies are those of fig. 3.15. The lines correspond to the experimental results of Grimm et al. (1980).

was presented together with some recent calculations. Although some preliminary conclusions can be drawn from this comparison, better insight is obtained by studying the energy dependence of β as presented in fig. 3.34 (Roy et al. 1986). Satellites 1, 2, 3 and 5 have quite similar values for the asymmetry parameter and different from the other group 6, 7, 8, 9 and 10. Satellite 4 shows a slightly different behavior. From the analysis of Roy et al. (1986) already presented in section 3.2.1, the first group of satellites is likely to be of π_u symmetry, while the second would result from $3\sigma_g$ and $2\sigma_u$ ionization. Actually, the asymptotic value for β may still not be reached at 30 eV. Some deviations from this value may occur due to interchannel couplings and a strong shape resonance amplitude in one of the channels (Roy et al. 1986). Recently, similar conclusions have been drawn in a similar study performed for CS_2 by Roy et al. (1987).

3.2.5. Photoionization dynamics: spin polarization of photoelectrons

For a complete understanding of the dynamics of photoionization processes in atoms or in molecules measurement of the spin polarization of photoelectrons is needed (Heinzmann 1980, Cherepkov 1980) in addition to the partial cross sections and angular distributions. Photoelectrons ejected from unpolarized atoms (or molecules) can be spin polarized, if the incident light is circularly polarized or completely unpolarized (Heinzmann 1980). This effect generally appears when spin–orbit interaction occurs in the ground, ionic or continuum state.

When the ionization threshold shows a spin–orbit splitting that can be resolved experimentally, then the spin polarization with circularly polarized light such as synchrotron radiation provides information that cannot be obtained by other experimental methods, i.e. the determination of the square of the transition matrix elements into the different energy degenerate subcontinua. This has been achieved for CO_2 and N_2O at their first ionization onset by Heinzmann et al. (1979). They found -7% for CO_2 and -18% for N_2O at threshold for the polarization of photoelectrons.

Using a simple formalism assuming only direct photoionization of a π_g electron into the σ, δ and π continua, Heinzmann et al. (1979) calculated -27% for the spin polarization ratio in CO_2 resulting from the following partial cross sections: 1.5 Mb (σ), 1 Mb (π) and 0.1 Mb (δ). The agreement with the experimental value is quite satisfactory considering the various simplifying assumptions made.

The complementarity of the spin polarization measurements and the angular distribution of the photoelectrons has been analysed theoretically by Cherepkov (1981a,b) and experimentally by Heinzmann et al. (1981). The parameters which are measured are A, ξ and β (see section 2.3). Unlike A, which is a sum of terms depending on the square of the transition moments without interference, ξ and β include phase shift terms $\sin(\delta_\alpha - \delta_\beta)$, and $\cos(\delta_\alpha - \delta_\beta)$, respectively, which come from interference. In the CH_3Br molecule (Heinzmann et al. 1981), A, ξ, and β have been measured at the ionization threshold ($^2E_{3/2}$) and below the ($^2E_{1/2}$) threshold. The three observables are consistent with phase shifts ($\delta_d - \delta_s$) ranging from 0.12π to 0.33π. The phase shifts are found to be quite different from the Coulomb phase shifts ($\eta_d - \eta_s = 0.56\pi$).

When autoionizing resonances occur in a well-defined spin–orbital continuum, the spin polarization measurements provide the relative importance of various partial channels. In CH_3Br the ejection of a 4π lone pair electron localized on the bromine atom leads to the $^2E_{3/2}$ and $^2E_{1/2}$ states of CH_3Br^+ with a large spin–orbit splitting of 0.32 eV. Several resonances have been resolved by Baig et al. (1982) just above the ionization onset. They are Rydberg states converging towards the $^2E_{1/2}$ threshold and they decay into the $^2E_{3/2}$ continuum by spin–orbit autoionization. The σ, π and δ channels are allowed by the process. Experimentally Schäfers et al. (1983) have measured rather large variations of spin polarization (using circularly polarized synchrotron radiation) when scanning on and off the resonances. They have found that the resonances have a strong δ character.

3.3. Competition between photodissociation into neutrals and photoionization

We have discussed in sections 3.1 and 3.2 photodissociation into neutrals and photoionization as if they were independent of each other. In some cases the two continua, one corresponding to ionization, the other to dissociation into neutrals, are decoupled and photoabsorption will lead to parallel decays. Usually, photoionization is fast compared with photodissociation (in particular in the case of predissociation) and the process can be treated as sequential. The opposite situation, namely photodissociation followed by ionization, can be found in particular for slow autoionization. This has been demonstrated for specific core excited states of HBr by Morin and Nenner (1986) and will be discussed in detail in section 3.6.2. In the valence excitation region similar fast dissociation decay was invoked by Cermak (1974) to explain the decay of some long lived Rydberg states. Another interesting situation arises when competition between dissociation and ionization occurs. This is the case we consider in this section. We shall take N_2O as an example. We have already discussed (see section 3.2.1) in some detail the photoionization of N_2O for energies between the X and the A state of the ion. Experimentally it was observed that highly vibrationally excited X state ions were formed with much higher efficiency than was predicted on the basis of Franck–Condon factors. This autoionization was accompanied by the ejection of electrons with very low kinetic energy (threshold electrons) and was termed "resonant autoionization" (Guyon et al. 1983). Resonant autoionization appears to be a common phenomenon in polyatomic molecules since it has been observed in several systems: OCS (Delwiche et al. 1981) and CO_2 (Stockbauer 1980).

In fig. 3.35, the absorption, the total fluorescence, the photoionization, and the threshold photoelectron spectrum (TPES) for N_2O are presented in the energy region from 14.5 to 16 eV. The lines which are seen in all these spectra were identified as members of autoionizing Rydberg series converging to the $A(0, 0, 0)$ and $A(1, 0, 0)$ levels of N_2O^+. The broad underlying bands have been assigned to dissociative valence states correlating to neutral fragments. The TPES spectrum (see fig. 3.35d) is superimposable on the fluorescence excitation spectrum. This indicates that the branching ratio between predissociation, producing fluorescence states, and the fraction of autoionization, yielding low-energy electrons, is constant over this energy range.

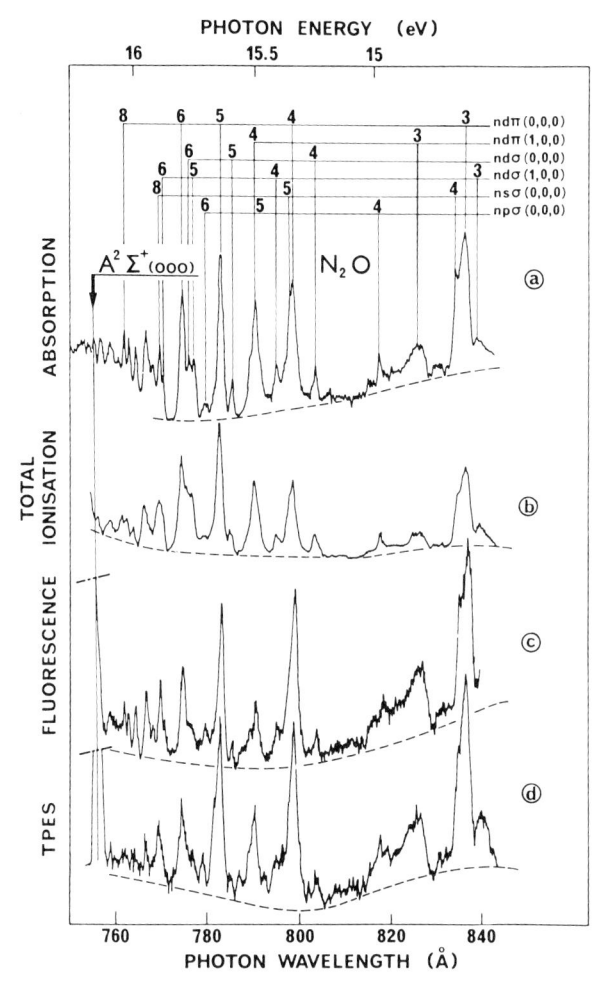

Fig. 3.35. Absorption, total ionization, fluorescence and threshold photoelectron spectrum for N_2O in the 760–840 Å photon wavelength region (from Guyon et al. 1983).

It should be noticed at this point, that threshold electrons represent only a small fraction of the total yield for photoionization. In fig. 3.36 the total photoelectron spectrum at 782 Å is presented. 75% of the photoelectrons are formed with 2.9 eV of kinetic energy. This corresponds to the formation of $X(0, 0, 0)$, which is expected since the two ground states for the neutral molecule and the ion are linear and the bond distances and frequencies do not change much. Hence the Franck–Condon principle will favor the $(0, 0, 0)$ to $(0, 0, 0)$ transition. Between this large peak and the zero electron energies, the distribution is essentially flat. Finally, there is a very narrow peak (<20 meV) at zero energies, which represents 10% of the total photoelectron yield. Although small this fraction is orders of magnitude larger than one expects on the basis of Franck–Condon factors.

The fact that there is a close relationship between the decay channels producing

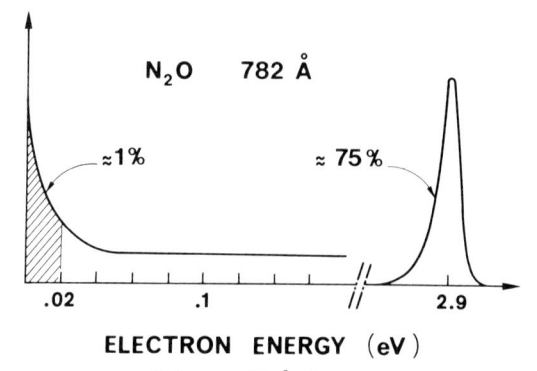

Fig. 3.36. Photoelectron spectrum of N_2O at 782 Å photon wavelength (from Guyon et al. 1983).

electronically excited neutral fragments and resonant autoionization suggests that valence dissociation states are the intermediate channels for the production of low- and intermediate-energy electrons (Guyon et al. 1983). In fig. 3.37 a schematic view of the relevant energy curves is presented (from Guyon et al. 1983). In addition to the Rydberg states (R_A) converging to the A state of the ion and those (R_X) converging to the ground state, the model considers a set of dissociative states, one member of which is shown in the figure as D. The initially excited R_A state can decay by autoionization to low vibrational levels of the ionic X state (Franck–Condon autoionization) or it can predissociate via D. While dissociating it can either autoionize or, if it is coupled to the R_X states, it can make a transition to highly vibrationally excited states of R_X, which may

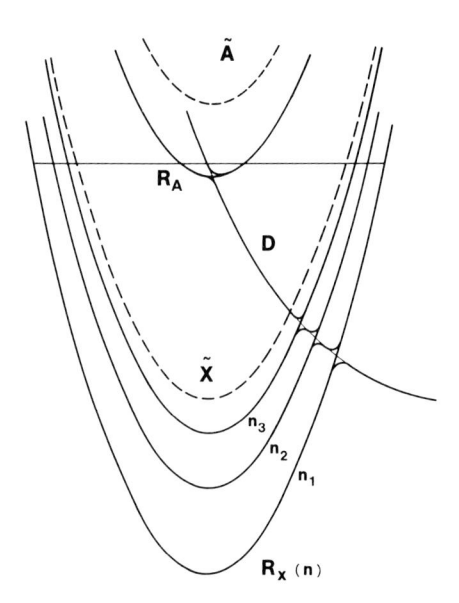

Fig. 3.37. Model for low-energy electron production and dissociation into neutral products (from Guyon et al. 1983).

autoionize by vibrational autoionization. This produces the low-energy electrons. If neither of these events occurs, the molecule will dissociate leading to excited neutral products.

A situation similar in nature to the above has been considered theoretically by Giusti-Suzor and Jungen (1984). The system studied was NO in the $^2\Pi$ excited states and the calculations were performed using MQDT theory. A schematic representation of the level scheme and couplings is presented in fig. 3.38. A discrete initial Rydberg state (n, v') interacts with an ionization continuum via vibronic couplings on the one hand, and to a dissociative continuum via electronic couplings on the other hand. If no other interactions are considered, the initial discrete state will decay independently (parallel decay) into both continua. The decay into the ionization continuum corresponds to vibrational autoionization and a $\Delta v = -1$ propensity rule is expected, while the decay into the dissociative continuum corresponds to electronic predissociation. The branching ratio will be given by the ratio of the independent golden-rule rates. In the case of NO, predissociation will be favored because of the strong Rydberg–valence interactions.

Actually the full MQDT calculation shows that the situation is more complex. The reason is the following. There is a coupling between the two continua (corresponding to electronic autoionization of a dissociative state). Therefore, even if the vibrational autoionization coupling vanishes, ionization can still occur indirectly through the continuum–continuum interaction. This has been clearly demonstrated in the MQDT calculations (Giusti-Suzor and Jungen 1984) by setting the quantum defect μ constant (in which case no direct vibrational autoionization can occur). Another very interesting outcome of these calculations is that the indirect mechanism induces $|\Delta v| > 1$ "vibrational" autoionization. This example clearly shows the importance of interactions between dissociative and ionization continua, a situation which may well be present in many other

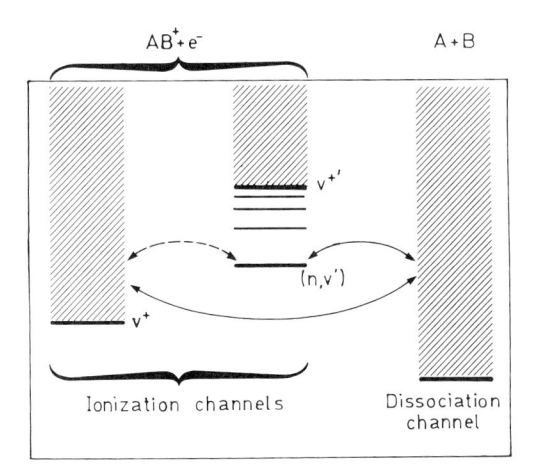

Fig. 3.38. Schematic representation of the couplings between two ionization channels (one open and one closed) and one dissociation channel. From Giusti-Suzor and Jungen (1984).

molecular systems. A unified treatment of this situation has been presented by Jungen (1984) and applied to molecular hydrogen.

3.4. Alignment and polarization of the fluorescence following photoionization and photodissociation

Although expected from theory, it is only very recently that polarization measurements on the fluorescence of the photofragments and/or photoions have been performed using synchrotron radiation (Poliakoff et al. 1981, 1982, Guest et al. 1984).

When a beam of light impinges on a molecular system, the absorption of a photon is known to be anisotropic and to produce an alignment and/or orientation of the excited molecule, i.e., different magnetic sublevels are not equally populated (Greene and Zare 1982a,b). If the excited molecule dissociates or ionizes, two consequences are expected: (i) the angular distribution of the fragments or electrons will also be anisotropic; (ii) the fragments or the ions may be aligned and/or oriented. If they are formed in an excited electronic state their alignment and/or orientation will be reflected in the degree of polarization of the emitted light [for a general presentation, see Greene and Zare (1982a,b)]. In the case of inner-shell ionization, the alignment of the ion may also be evidenced in the angular distribution of Auger electrons (see section 3.6.2).

By this means Poliakoff et al. (1981) first demonstrated molecular photoion alignment using synchrotron radiation. They photoionized N_2 by removal of an inner electron and this yields an electronically excited ion. The process may be described as

$$N_2(X\ ^1\Sigma_g^+) + h\nu \rightarrow N_2^+(B\ ^2\Sigma_u^+) + e^-(\varepsilon\sigma, \varepsilon\pi), \tag{3.21}$$

followed by

$$N_2^+(B\ ^2\Sigma_u^+) \rightarrow N_2^+(X\ ^2\Sigma_g^+) + h\nu'. \tag{3.22}$$

The absorption transition dipole is parallel (producing $\varepsilon\sigma$ electrons) or perpendicular (producing $\varepsilon\pi$ electrons) to the internuclear axis. The emission transition dipole is parallel to that axis ($\Sigma \rightarrow \Sigma$ transition). It follows (Poliakoff et al. 1981) that

$$P = \frac{I_\parallel - I_\perp}{I_\parallel + I_\perp} = \frac{1 - R}{7 + 13R}, \tag{3.23}$$

R being the ratio of the photoionization strengths for the $\varepsilon\sigma$ and $\varepsilon\pi$ channels. Thus the measurement of fluorescence polarization following photoionization yields the ratio of the dipole strengths for degenerate ionization channels. In photoelectron angular distribution studies this information is also present (in the anisotropy parameter β), but the channel ratios are folded together with the phase shift difference of the two outgoing waves. As a result dynamical information is somewhat more difficult to extract. Raoult et al. (1983) calculated P using ab-initio MQDT techniques (already mentioned in section 3.2.3) and found good agreement with experiment (Poliakoff et al. 1981, Guest et al. 1983). However, this was only possible through the introduction of continuum–continuum interactions.

It should be noticed that eq. (3.23) is obtained through the use of the classical expression

$$P = \frac{3\cos^2\langle\alpha\rangle - 1}{\cos^2\langle\alpha\rangle + 3}, \tag{3.24}$$

where $\langle\alpha\rangle$ is the average angle between the absorption and emission dipoles. This is valid in the limit where the fractional change in rotational angular momentum $\Delta j/j_i$ is small. In photoionization this condition is approached at high temperature (large j_i) or low electron kinetic energy (negligible angular momentum l removed by the photoelectron).

For low temperatures or high photoelectron energies, the problem must be treated quantum mechanically. In order to test some of the quantum-mechanical predictions, a cooled gas sample produced by a supersonic expansion may be used. However, one should keep in mind that fluorescence depolarization caused by electron and/or nuclear-spin angular momentum couplings in the photoion can be very important for low rotational quantum numbers. Unless one is interested in studying by this method the relevant couplings in the ion, the use of a cooled sample may not be an advantage for photoion alignment studies via this technique (Guest et al. 1983).

For the time being alignment measurements are limited to the study of processes that produce excited-state ions. It can also be probed by laser-induced fluorescence and thus it may be possible with future high-intensity machines to study ground state photoions by this technique.

Poliakoff et al. (1982) and also Guest et al. (1983) have studied fluorescence polarization and alignment in the case of autoionizing resonances. The system studies was CO_2. They have shown that this technique can be a significant tool in ascertaining the symmetry and the dynamical properties of autoionizing resonances which are superimposed on a nonresonant background of known symmetry. Along the same lines Kronast et al. (1984) measured the polarization of the fluorescence emitted from cadmium ionized in the 4d subshell by synchrotron radiation and they actually detected significant variations across some autoionizing resonances.

Finally, Poliakoff et al. (1979) have studied polarization of the fluorescence from photodissociation fragments in the case of ICN. Synchrotron radiation in the 1100–1700 Å region was used to photodissociate ICN and the polarization of the emission from the CN(B) fragment was analysed. They found that the degree of polarization varies between 0 and 8%. This yields information regarding the orientation of the diatomic rotational angular momentum, thus giving important insights into the dynamics of dissociation processes.

3.5. Double ionization

One of the most recent and spectacular applications of mass spectrometry, photoion–photoion and photoion–photon fluorescence coincidence techniques combined with the tunability of synchrotron radiation above 20 eV, is the single-photon double ionization of molecules, which corresponds to

$$h\nu + M \rightarrow M^{2+} + e^- + e^-,$$

i.e., a single photon excites a molecule M and ejects two valence electrons. The process requires the breakdown of the independent-electron model in the same way as the formation of excited ions showing up as satellites in PES (see section 3.2.1). In other words, the study of double-ionization processes is an ideal tool to elucidate electron correlations.

Several points relevant to double-ionization are of particular importance in molecules:

(1) the spectroscopy of molecular doubly charged ions, which includes the possibility of potential minima, and the dissociation lifetimes as well as fluorescence lifetimes;

(2) the double ionization dynamics, i.e. the intensity and relative importance of resonant processes as opposed to the direct ejection of the two electrons;

(3) the relaxation processes of the residual ion and the possibility of selective dissociation pathways.

If the two electrons are weakly bonding or nonbonding, the residual doubly charged ion M^{2+} may be metastable (i.e., it may be detected on the microsecond time scale in a mass spectrometer). More recently, Besnard et al. (1986) successfully applied the photoion–photon fluorescence coincidence method to detect high lying bound states of NO^{2+}. If, on the other hand, one or the two electrons are antibonding the corresponding M^{2+} state may have a shorter bond length. Finally, if the two electrons are bonding, the process will be accompanied by large changes in bond lengths and geometry. The corresponding potential curves will have minima outside the Franck–Condon region or they will be purely Coulomb repulsive. In that case dissociation of the doubly charged ion will lead to two singly charged ions. It is worth noting at this point that ions with two holes in the valence shell can also be produced by photoionization of a core shell followed by an Auger process. The physics of this phenomenon, which can be viewed as a "hard" double-ionization process, differs from the "soft" process considered in this section and will be discussed independently (see section 3.6.2).

From the experimental point of view, mass spectrometry combined with synchrotron radiation has been successfully used to detect metastable doubly charged molecular ions in several systems: CO_2 (Masuoka and Samson 1980, Lablanquie et al. 1985b), CS_2 (Lablanquie 1984, Lablanquie et al. 1985a,b). It is worth noting at this point that the results obtained with the dipole (e,ion) method for metastable CO_2^{2+} (Carnovale et al. 1982) compare favorably with the synchrotron radiation results. More importantly, the ion–ion coincidence technique is ideally suited to detect dissociative double-ionization processes producing ion pairs independently of other fragments issued from simple dissociative ionization

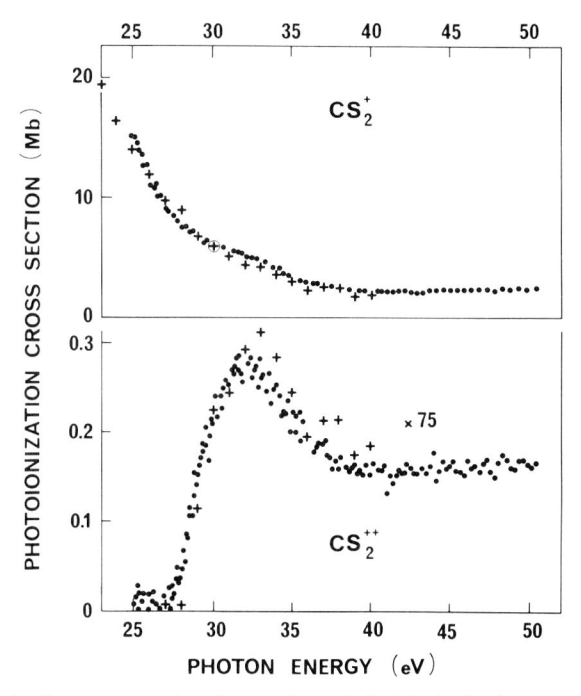

Fig. 3.41. Photoionization cross section for singly and doubly ionized parent ions in CS_2 (from Lablanquie et al. 1985b). The plusses correspond to the experimental results of Carnovale et al. (1982).

the dissociative nature of many of the doubly charged ion states relating to singly charged ion pairs explains most of the intensity of singly charged ions produced by absorption of high-energy photons in molecules (i.e. above 25 eV photon energy).

On the other hand, such excitation spectra show a quasi-linear energy dependence, which can be understood (neglecting the finite width of the Franck–Condon region) in terms of a direct double-ionization mechanism. According to Wannier (1953) one expects the corresponding cross section to rise as the 1.056th power of the energy above each threshold. This is very close to a linear slope. Many systems (Dujardin et al. 1984, Lablanquie 1984, Lablanquie et al. 1985a) show near threshold a more or less linear behavior. However, it is not clear at present whether one could attribute "breaks" in the excitation spectra (see fig. 3.40) as the occurrence of new CS_2^{2+} states or as resonances revealing indirect processes to double ionization, which would be analogous to autoionization in the simple one-electron ejection process.

Ion–ion coincidence spectra measured at a selected wavelength also provide energetic information. The peak shape is characteristic of the kinetic energy release during the process. Simulation of these peaks shows that in most cases (Dujardin et al. 1984, Lablanquie 1984, Lablanquie et al. 1985b) the available excess energy is transferred mainly into kinetic energy of the two outgoing ions and not into internal excitation of the fragments.

Further analysis of coincidence spectra such as that of CS_2 (see fig. 3.41) gives information on the dissociation mechanism of the doubly charged ion. For single-bond breaking processes, direct simulation of the peak provides the kinetic energy released and eventually the dissociation time (on the submicrosecond scale). One interesting observation in the dissociation of OCS^{2+} was the high probability of the $CO^+ + S^+$ channel as compared to the $CS^{2+}O^+$ channel (Lablanquie et al. 1985a). A very simple description of the difference has been provided in terms of the partitioning of charges for each of the calculated OCS^{2+} states using ab-initio SCF-CI techniques (Millié et al. 1986).

Multiple bond fragmentation is a quite general decay channel for doubly charged ions. One interesting question in this context is: do the bonds break sequentially or simultaneously like one would expect for a Coulomb explosion? In the case of CS_2^{2+}, simulation of PIPICO curves based on simple kinematics (Lablanquie et al. 1985b) shows that the $S^+ + C^+ + S$ channel proceeds via a two-step process whereas the production of $S^+ + C + S^+$ results from a direct three-body fragmentation. In polyatomic molecules such as benzene (Richardson et al. 1986) sequential decay seems to explain the multiple bond dissociation of the ion. Actually more sophisticated measurements involving multiple coincidences such as triple photoion (Eland et al. 1986) and photoelectron–photoion–photoion coincidence techniques (Frasinski et al. 1986, Eland 1987) are now providing more direct answers to those questions. Development of such techniques are therefore highly desirable in the future. Simultaneously, potential energy surface calculations and dynamical simulations of doubly charged molecular ions will be of great interest.

3.6. Inner-shell photoexcitation and photoionization

3.6.1. Near-edge resonances

Resonances observed in absorption spectra in the vicinity of core level excitation have been extensively studied in atoms and solids and on surfaces (Brown 1980, Smith and Himpsel 1983, Bianconi 1983). They are often known under the acronym XANES (X-ray Absorption Near-Edge Structures) and they sometimes interfere with early EXAFS (Extended X-ray Absorption Fine Structures) oscillations, which reflect the presence of neighboring atoms in condensed matter. Resonances have also been observed in gas phase molecules around the K or L core levels of their atomic constituents. However, for molecules having light atoms, i.e. those of the first two rows in the periodic table, photoabsorption data including those obtained with X-ray laboratory sources and synchrotron radiation [see the review by Koch and Sonntag (1978)] are very rarely of better quality and resolution than pseudo-absorption spectra obtained by Electron Energy Loss Spectroscopy (EELS) (see, for example, Brion and Hamnett 1981, Brion 1985, Sette et al. 1984, and references therein). The new monochromators which have become available recently (toroidal gratings and crystal monochromators) are providing a reasonable photon flux above 50 eV up to the X-ray region. This has allowed the study of the K and L edges of sulfur, fluorine, boron, carbon, chlorine, germanium, silicon, phosphorus, nitrogen and oxygen in a number of common molecules (Ishiguro et al. 1982, Ninomiya et al. 1981, Mazalov et al. 1974, Friedrich et al. 1979, Nakamura et al. 1971, Morioka et al. 1974, Bodeur and Esteva 1984, 1985, Chun 1969, Brown et al. 1978, Kincaid and Eisenberg 1975, 1976, Bianconi et al. 1978). Simultaneously the group of Eberhardt (Eberhardt et al. 1976) used the total electron yield method to analyse those structures in various hydrocarbons around the carbon K edge. More recently Morin et al. (1985a) used the total and partial ion yields to measure those structures in tetramethylsilane near the Si L edge. A bibliography of inner shell excitation studies of free atoms and molecules has been prepared by Hitchcock (1982) and updates of this are available.

Typical spectra obtained with synchrotron radiation (Bodeur and Esteva 1984, 1985) or EELS (Wight and Brion 1974) are presented in figs. 3.42 and 3.43. All features are excited neutral states, which can be classified in various categories:

(a) Some are "valence states", where the core electron is trapped into a low-lying unoccupied valence level. This is the case for the main features 1–4 in CS_2 (fig. 3.42) and the first two in SO_2 (see fig. 3.43). Note that they are intense and that they are located in the discrete part of the spectrum but also sometimes in the continuum (see below). They can also be described as shape resonances or Feshbach resonances. In general, the small spatial extent of core orbitals seems to favor transitions to more compact valence orbitals.

(b) Some others are "Rydberg states", but they have not been observed very often because of lack of resolution or because valence transitions dominate the spectra. Some examples of Rydberg states have been reported in EELS data on HF and F_2 (Hitchcock and Brion 1981) below the fluorine 1s threshold (F1s). In the SO_2 case (see fig. 3.43), the small features near the edge are probably Rydberg states.

(c) Broader and weaker features are often observed above the threshold. Some of them are explained in terms of shape resonances, i.e., the outgoing electron is trapped behind a centrifugal barrier (see Dehmer et al. 1987). Some of these shape resonances may also have a contribution from Feshbach resonances, i.e., the outgoing electron energy corresponds to that of an empty valence orbital. Others are not resonances associated with centrifugal barriers but rather with the potential itself [see, for example, the broad peak at 2495 eV for SO_2 in fig. 3.43 and the results for CF_4 (Brown et al. 1978)].

(d) Other features in the spectra are "doubly or multiply excited states", which correspond to states with a core hole and an additional valence excitation. An example is given by feature 6 in CS_2 (fig. 3.42) (Bodeur and Esteva 1984, 1985).

It is interesting to note at this point that some differences exist between pure photoabsorption spectra and total electron (or ion) yield spectra, as reported by Eberhardt et al. (1976) and by Morin et al. (1986) and Morin (1985). At this high energy Auger processes are expected to occur, leading to the ejection of secondary Auger electrons. Therefore, in some specific excitation regions, for a single event at least twice as many electrons as in simple photoionization processes are expected (Van der Wiel et al. 1970, Van der Wiel and El-Sherbini 1972, Hitchcock and Brion 1977). Auger relaxation processes lead also to two-hole molecular states (doubly charged ions), which are often unstable and

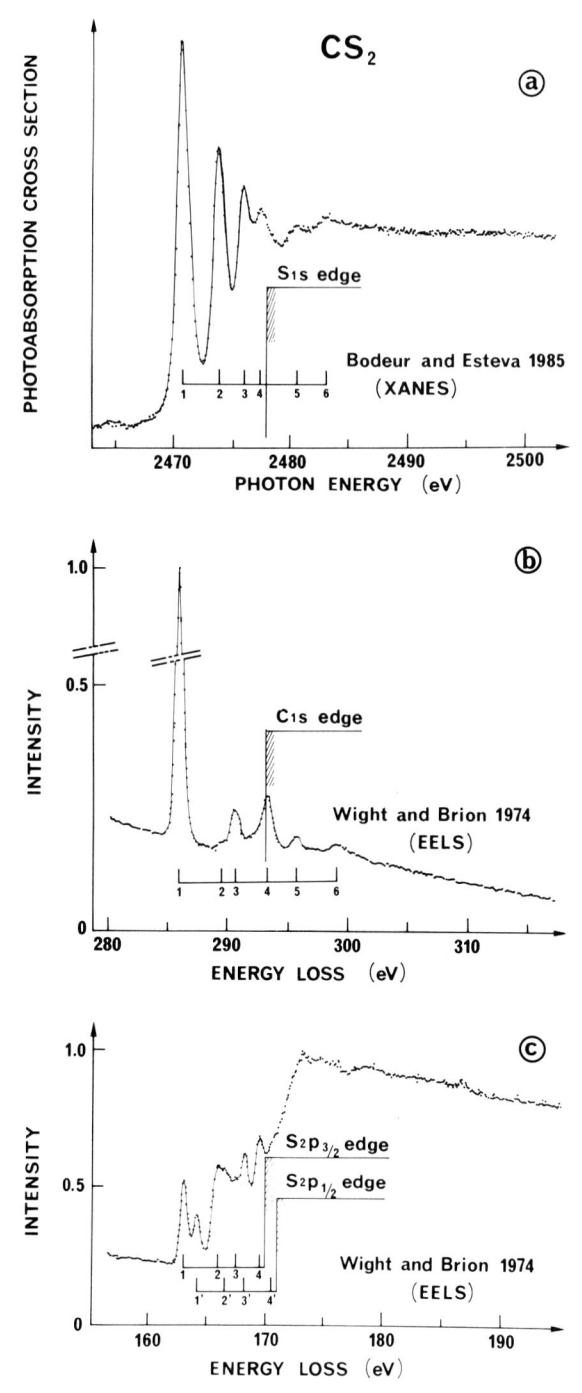

Fig. 3.42. Absorption cross section of CS_2 near the S1s edge (from Bodeur and Esteva 1984), and near the C1s and S2p edges (from Wight and Brion 1974).

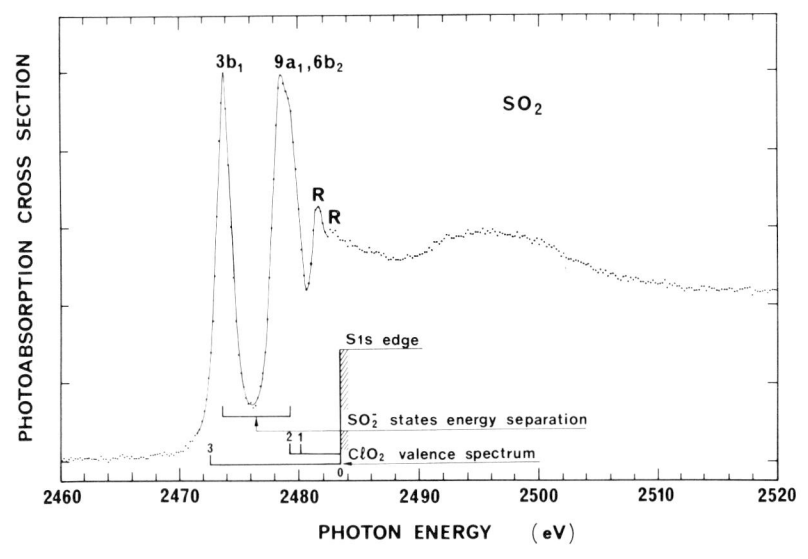

Fig. 3.43. Absorption cross section of SO_2 near the S1s edge (from Bodeur and Esteva 1985).

produce an ion pair. Consequently one expects an increase in the total electron (or ion) yield as compared to the optical absorption spectrum. This analysis is illustrated and borne out by refined measurements as discussed below.

On the basis of the many available electron energy loss (Hitchcock and Brion 1980, 1981, Brion 1985, Read 1978, King et al. 1977, Ishiguro et al. 1982) and photoabsorption data (Koch and Sonntag 1978, Bodeur and Esteva 1984, 1985, Ishiguro et al. 1982, Ninomiya et al. 1981) spectra, general trends of the position of the valence-type resonances with respect to threshold (term value) have been extracted in several molecules. Those resonances (at least the most prominent ones) result from the excitation of a core electron localized on a particular atom into a low-lying unoccupied valence orbital. In a very crude frozen orbital approximation any resonance energy which results from the excitation of an electron from a filled orbital (a) into an empty one (b), will be given by $E_r = |\varepsilon_a| - |\varepsilon_b| + k$, where ε_a and ε_b are the orbital energies measured with respect to the vacuum level and k includes the usual Coulomb and exchange integrals and the term corresponding to the difference between core relaxation and correlation effects in the ion and in the core excited neutral molecule. The k term may be considered as a constant for excited states of a molecule resulting from transitions originating from core shells of a given atom but not of different atoms. Consequently, if a resonance is characterized by its term value T, $T = |\varepsilon_a| - E_r$, then one can also say that T is related to the energy of the empty orbital: $T = \varepsilon_b - k$, and term values can be compared from one ionization region to another. Moreover, differences of term values ΔT can be compared directly to differences in energy of empty orbitals. It should be noted that the transferability of term values has been verified for excitation into Rydberg orbitals but not always for valence orbitals (Sodhi and Brion 1985). This simple description of core excited states may be rationalized in terms of the equivalent core approximation (see for example Schwarz 1975 and Kosugi and Kurada 1983). In this model the valence electron cloud "feels" the core as if it was replaced by the following atom in the periodic table. Then term values measured in a core photoabsorption spectrum are directly comparable with those of the equivalent core species.

Several consequences are then expected, which have been verified experimentally in many cases (note that the following does not take into account line intensities, which are governed primarily by electric dipole selection rules):

(1) The term values are constant from one core atomic edge to another. For example, in CS_2, fig. 3.42 shows the S1s (Bodeur and Esteva 1984), S2p (Wight and Brion 1974) core absorption (or EELS pseudo-absorption) regions when all IP have been aligned. The features 1 and 2 in figs. 3.42a,c have similar term values. Notice that the C1s spectrum (Wight and Brion 1974) is somewhat different from the S1s spectrum as one would expect. On the other hand, features 3 and 4 possibly have a different interpretation (Bodeur and Esteva 1984, Nenner et al. 1987a), which must still be clarified theoretically. Another example is provided by H_2S, where comparison can be made between the S1s and the S2p absorption regions (Bodeur and Esteva 1985).

(2) Isoelectronic molecules in their valence shell have similar core absorption spectra. For example, CS_2 and OCS have similar S1s absorption spectra (Bodeur and Esteva 1984, Nenner et al. 1987a). In contrast, $SiCl_4$ and SiF_4 have very different Si1s photoabsorption spectra (Bodeur et al. 1986).

(3) Along the lines of the equivalent core analogy model (Schwarz 1974), a molecule corresponding to the replacement of a core excited atom with another atom with nuclear charge one unit larger, shows a valence shell absorption spectrum similar to the core absorption spectrum of the original molecule. This has been verified in many cases for diatomic (Nakamura et al. 1969, Morioka et al. 1974, Wight et al. 1972), and for polyatomic molecules (Hitchcock and Brion 1981, Schwarz and Buenker 1976, Bodeur and Esteva 1985). For example, SO_2 has ClO_2 as an equivalent core model in the S1s edge region, where Cl is the $(Z+1)$ atom. SO_2 and ClO_2 have similar valence configurations except that ClO_2 has an extra electron in the $3b_1$ orbital. From the $3b_1$ valence excitation spectrum of ClO_2, Bodeur and Esteva extracted the corresponding term values. They are reported in fig. 3.43 and good correspondence has been found for the main peaks ($3b_1$ and $9a_1$ and $6b_2$).

(4) For a given molecule, the term value is related to the vertical electron affinity, as long as the same unoccupied orbital is concerned in both cases. Experimentally, low-energy electron impact measurements on a number of molecules have provided accurate energy values for low-lying unoccupied valence orbitals, by detecting temporarily negative ions, for instance (Schulz 1973, Nenner and Schulz 1975, Tronc et al. 1979).

As an example let us consider the SO_2^- ion (Bodeur and Esteva 1985). It is known that the two lowest SO_2^- states result in the attachment of the electron in the $3b_1$ and $6b_2$ empty orbitals. The associated electron affinities are 2.2 eV and -3.3 eV. Therefore the energy separation $3b_1$–$6b_2$ is 5.6 eV. Bodeur and Esteva (1985) found a very nice correspondence with the energy separation between the $3b_1$ and the $6b_2$ core resonances (see fig. 3.43).

Another general aspect of core resonances in diatomic as well as in polyatomic molecules has been described by Hitchcock and Brion (1981) and by Sette et al. (1984). When resonantly enhanced shape resonances of σ^* and π^* symmetry are observed, a very good correlation exists between the term value and the bond length as well as the sum of the atomic numbers of the heavy atoms in the molecule. Bianconi (1983) and Sette et al. (1984) suggest that the empirical relationship between $1s \rightarrow \sigma^*$ energies and bond lengths which has been verified for a number of light molecules, can be understood in terms of the molecular scattering potential, which depends upon the internuclear distances and the atomic number of the absorbing atoms and their neighbors. The existence of this correlation has been challenged by Piancastelli et al. (1987a) and continued to be discussed (Hitchcock and Stöhr 1987, Piancastelli et al. 1987b).

The general behavior of shape resonances in the continuum and the discrete region can be described in terms of the anisotropy of the molecular potential. For nonhydride molecules, the potential is clearly nonspherical (multicentral). As a result shape resonances are observed in the σ^* or π^* continua with rather large intensities. In contrast, the potential for hydride molecules such as HF, H_2O, NH_3, CH_4, etc. have a strong atomic character, since the hydrogens do not greatly perturb the heavy atom. Consequently, the shape resonances which may occur in the core ionization continua (or in the discrete region) of the heavy atom are much weaker or do not even exist in comparison with the nonhydride molecules. For example, the H_2S (Bodeur and Esteva 1985) and the HF (Hitchcock and Brion 1981) molecules show resonances only in the discrete region of the S1s and F1s spectrum respectively, and none in the continuum. The main features seen below the edge are described normally as valence transitions but do not result from the anisotropy of the potential.

Recently, oscillator strengths near core thresholds have been determined (Ishii and Hitchcock 1987). This allows a detailed investigation of the spatial distribution of the promoted electron in the core excited state (McLaren et al. 1987), Another interesting outcome of these studies is a comprehensive analysis of discrete resonances in complex molecules in terms of the bond strengths (Ishii et al. 1987).

3.6.2. Electronic relaxation near core edges
The formation of a core hole near the edge in a molecule is followed by a variety of electronic relaxation processes, which can be observed experimentally by angle-resolved photoelectron spectroscopy. Pioneering work in molecular systems has been performed by Siegbahn et al. (1969) and Carlson (1978), using PES with X-rays, i.e. at fixed photon energy. Since most common molecules are formed from light atoms (second or third row of the periodic table), the core energy edges correspond to soft X-rays. Recently, angle-resolved PES with synchrotron radiation has been used to study such processes in several molecules: Truesdale et al. (1984) studied CO, CO_2, CF_4 and OCS near the C1s edge, CO and CO_2 near the O1s edge and OCS near the S2p edge. De Souza et al. (1985) studied, using the same method, tetramethylsilane [TMS, $Si(CH_3)_4$] near the Si2p edge. Gerard (1984) studied Li_2 near the Li1s edge. Eberhardt et al. (1983a) investigated N_2 and CO near the N1s and C1s edges, respectively.

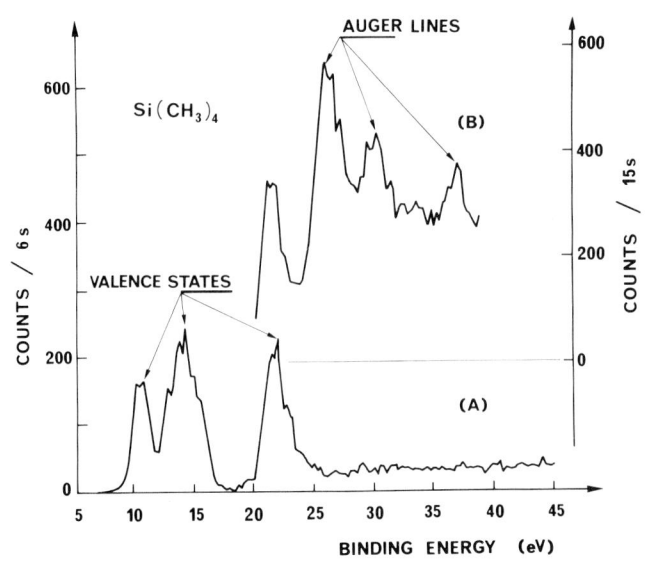

Fig. 3.44. Valence photoelectron spectra of TMS measured at (a) 100 eV, (b) 106.4 eV (from de Souza et al. 1985).

Auger processes. We consider the direct core ionization problems separately from the indirect ones, which proceed through resonances. When a core electron is ejected into the continuum, the residual ion has an internal energy high enough to eject another electron (or more than one). This is known as the Auger process. Schematically we have a two-step process (see fig. 1.5):

(i) $\quad h\nu + (\text{core})^2(\text{valence})^n \rightarrow (\text{core})^1(\text{valence})^n + e^-_{\text{core}}$,

(ii) $\qquad (\text{core})^1(\text{valence})^n \rightarrow (\text{core})^2(\text{valence})^{n-2} + e_{\text{Auger}}$.

The first point to be noted is that the Auger process is a two-electron problem very much like electronic autoionization discussed in section 3.2.3. In contrast to the primary photoelectron, whose energy depends upon the photon energy, the Auger electron has a fixed kinetic energy, corresponding to a particular transition in the ion. Generally speaking the Auger electron is produced because of the existence of a core hole, but it "ignores" the way the core hole was formed. Notice that more than one electron can be ejected. This is known as relaxation shake off where two electrons are ejected simultaneously (see, for example, Morin and Nenner 1987) or Auger cascades in which electrons are ejected sequentially. The residual ion is then formed with three or more positive charges (see, for example, the SF_6 study of Ferrett et al. 1986).

In fig. 3.44 the photoelectron spectrum of TMS is presented (de Souza et al. 1985). The photon energy is 106.4 eV and the molecule is excited just above the Si2p edge (binding energy 106.1 eV). In addition to valence bands and the core (Si2p) band (not shown in the figure), one observes three Auger bands on top of a large continuum. De Souza et al. (1985) identified them as Auger processes by using different photon energies.

As seen in step (ii) above, the residual ion is a doubly charged molecule. Its internal energy is given by the difference

$$E(M^{2+}) = I_c - \varepsilon_A ,$$

where I_c is the core ionization energy and ε_A the Auger electron energy. Generally, the Auger electron is labelled as KLL, KLV, LVV or LVVA, depending upon the nature of the initial core hole (K or L) and on the nature of the two final holes (LL, LV or VV). De Souza et al. (1985) observed three KVV lines (see fig. 3.44), which were identified as electronic states with different energies of TMS^{2+}. They have assigned them on the basis of the individual valence ionization energies of TMS, taking the simplified formula of Jennison (1980),

$$E(M^{2+}) = I_i + I_k + K ,$$

where I_i and I_k are the ionization energies for the i and k valence orbital and K is related to the hole–hole interaction, which includes electron correlation and relaxation effects. It turns out that the first Auger line in TMS corresponds to TMS^{2+} with two holes in the outermost valence shell.

The two others are TMS^{2+} states with holes in two different valence orbitals. The results of de Souza et al. (1985) raise the question of "localized" effects in Auger processes. In contrast to KLL Auger processes, which are localized on the initial atom, the KVV and LVV Auger processes have an extended effect all over the molecule because the two final holes are of the valence type. The KVV or LVV Auger processes may be described as more interatomic, as opposed to KLL, which has an indisputably intra-atomic character. Intra-atomic Auger is expected theoretically (Matthew and Komninos 1975) to dominate in intensity over interatomic by many orders of magnitude. However, because of the absence of electrons in outer orbitals localized on the initial atom, interatomic may overcome intra-atomic Auger. This situation is somewhat similar to that found in solids, where the outer electrons are completely delocalized in the conduction band (Bertel et al. 1984).

Just as for atoms, the angular distribution of the Auger electrons is characterized by the functional dependence of the intensity distribution

$$I(\theta) \propto 1 + \beta_A P_2(\cos \theta) . \tag{3.25}$$

Note that this equation has exactly the same form as the differential photoionization cross section described in section 3.2.4. We shall see that measurement of β_A provides information that is complementary to that obtained from the partial photoionization cross sections, angular distribution of the photoelectrons as well as polarization of the ion fluorescence.

In molecules, such measurements have been performed with synchrotron radiation only in a very limited number of cases. Truesdale et al. (1984) measured β_A for C(KVV) Auger lines in CO, CO_2, CF_4, and OCS, O(KVV) in CO and CO_2, and S(LVV) in OCS. Later de Souza et al. (1985) measured β_A for Si(LVV) Auger lines in tetramethylsilane and Ferrett et al. (1986) measured the β_A for S(LVV) Auger lines in SF_6.

We know from atomic photoionization studies that the use of a linearly polarized photon beam on randomly oriented target atoms leads to different populations of the magnetic sublevels JM, proportionally to the partial photoionization cross section $\sigma(JM)$. The angular distribution of Auger electrons, defined by the asymmetry parameter β_A, depends upon two factors:

$$\beta_A(\varepsilon) = A(\varepsilon)\alpha . \tag{3.26}$$

A is the alignment coefficient, which reflects the populations of the magnetic sublevels and depends upon the dipole transition matrix elements (and not upon phase shift terms). It is expected to vary with the photoelectron energy ε and a general dependence is expected for direct ionization processes. In addition, A may vary at resonances for particular photon energies. For K shell direct ionization the angular Auger electron distribution is expected to be isotropic. Truesdale et al. (1984) found a zero value for β_A of KVV lines in CO and CO_2 outside the resonance regions. Note that the similarity between atomic and molecular K shell ionization is simply due to the common "atomic" character of the core hole. In contrast to K shell ionization, the Auger electron angular distribution following L shell ionization is expected to be anisotropic. De Souza et al. (1985) actually measured $\beta_A = 0.5$ and 0.7 for Si(LVV) lines in TMS. However, it is not clear if this anisotropy comes from nonzero values of the angular momentum of the Si(2p) shell or from resonant processes. A much clearer case has been studied by Ferrett et al. (1986) who found in SF_6 that autoionization of discrete resonances near the S1s edge produces aligned ions.

Note that the alignment A can also be measured through the polarization of the fluorescence of the core-ionized ion when this fluorescence is detected. Such emission has been successfully measured by Kronast et al. (1984) in cadmium photoionized by synchrotron radiation; no equivalent observation exists for molecules.

The α coefficient in eq. (3.26) above depends upon the transition matrix elements of the Coulomb operator and on the phases of the different partial waves associated with the ejected Auger electron. In other words, α characterizes the final doubly charged ion. It varies among final M^{2+} states. An example of different α values for a given original hole was provided by de Souza et al. (1985) on the basis of β_A measurements for the LVV lines of TMS.

Resonances. Other important electronic relaxation processes are directly related to the near-edge resonances (see section 3.6.1). They are autoionizing and shape resonances. Shape resonances have been shown to exist near core edges in molecules by Truesdale et al. (1984) (in CO, CO_2, CF_4 and OCS), de Souza et al. (1985) (in TMS), and by Lindle et al. (1984) (in N_2 and NO), using partial cross sections and asymmetry parameter measurements associated with the electron ejected from the core orbital. Just as for the valence shells (see section 3.2.3), the outgoing electron experiences the anisotropy of the molecular field and can be trapped temporarily behind the centrifugal barrier (Dehmer et al. 1985, 1987).

When a shape resonance is observed in the continuum of a core electron, it is expected to be seen in the Auger partial cross section. Among other factors, the Auger transition moment varies with the transition moment of the primary core electron. This has been observed in CO_2, CO, CF_4 and OCS by Truesdale et al. (1984) and in TMS by de Souza et al. (1985). It should be noted that here the trapped electron is just a spectator with respect to the Auger process. We shall see below that this picture does not hold for other resonance processes in the discrete part of the spectrum. When the position of the shape resonance is very close to threshold, the primary core electron may be low enough that interference with the Auger electron may occur, giving rise to post-collision interaction effects. Although this is well known in atoms, such an effect has rarely been seen in molecules mainly because of the lack of resolution (both in the incident photon and in the analysed electrons). One PCI case reported is the ion yield measurements of CO by Kay et al. (1977).

Autoionization of resonances near core edges is also an important decay channel, in particular for those below threshold. Unlike most autoionization processes observed in the valence region, there is a wealth of energetically different electronic continua ranging from valence ionization continua (single-hole configuration) to excited valence ionization (satellites, two-hole–one-particle, three-hole–two-particle, etc. configurations) as well as other lower core ionization continua (in the case of K initial holes, see the extensive work on SF_6 near the S1s edge by Ferrett et al. 1986). However, the observation of such relaxation channels requires the analysis of photoelectrons on a very large kinetic energy scale ($KE \gg 50$ eV).

We shall consider TMS as an example. Sodhi et al. (1985) have evidenced by the use of electron energy loss experiments, a rather broad resonance extending below and above the Si2p edge (106.1 eV binding energy) namely from 103 to 109 eV.

In the preceding section evidence of a shape resonance in the Si2p continuum was mentioned (de Souza et al. 1985), which takes into account the upper part of this broad resonance. Below the edge, Morin (1985) and Morin et al. (1986), have observed a very specific autoionization pathway as illustrated in fig. 3.45. The various PES show a large enhancement of the 21.7 eV valence band (mostly Si3s) at 104.58 eV, while the other valence bands do not change in intensity (see also fig. 3.46). This observation illustrates nicely an autoionization process of a core-excited resonance into a valence ionization continuum. Here the excited electron may be ejected into the continuum while a valence electron may fill the core hole, or conversely the excited electron may fill the core hole while a valence electron is ejected into the continuum. This process has already been described as electronic autoionization for valence ionization processes (see section 3.2.3). However, this process is somewhat surprising considering the 80 eV energy separation between the initial and final states and the selectivity among the other valence bands.

Another important decay process is the singly and doubly resonant Auger interaction coupled to ionic continua and/or double- (or multiple-) ionization continua. Singly resonant Auger processes can be described as follows (see

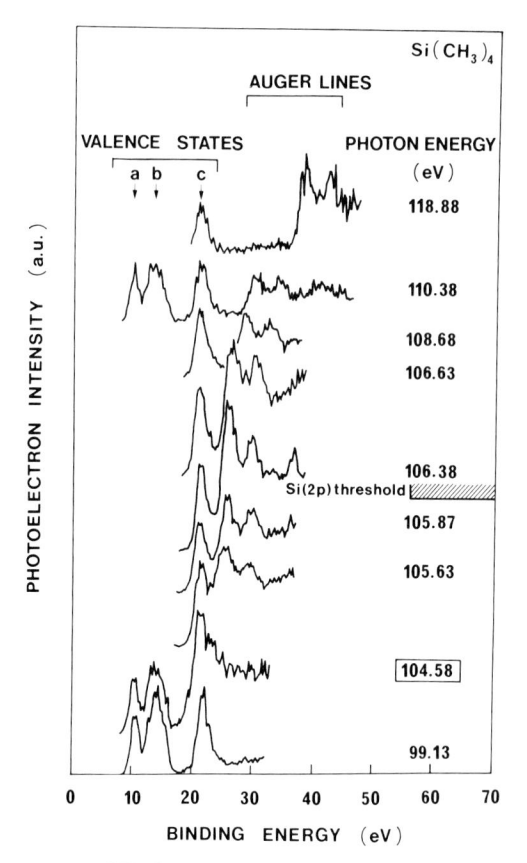

Fig. 3.45. Photoelectron spectra of TMS at selected photon energies in the vicinity of the Si2p edge (from Morin et al. 1986).

Combet-Farnoux 1982a,b). A core electron is excited into an unoccupied orbital and the excited molecule relaxes, ejecting one electron and leaving two holes in the valence shell:

$$h\nu + (\text{core})^2(\text{valence})^n(\text{unocc})^0 \to (\text{core})^1(\text{valence})^n(\text{unocc})^1,$$

followed by either

(i) $(\text{core})^1(\text{valence})^n(\text{unocc})^1 \to (\text{core})^2(\text{valence})^{n-2}(\text{unocc})^1 + e^-,$

which corresponds to the electron spectator model (the residual model is found with an excited configuration generally described as a valence satellite line), or

(ii) $(\text{core})^1(\text{valence})^n(\text{unocc})^1 \to (\text{core})^2(\text{valence})^{n-1}(\text{unocc})^0 + e^-,$

which corresponds to the nonspectator model and reduces to normal electronic autoionization. Here the residual ion is found with a single valence hole (main line in the PIS).

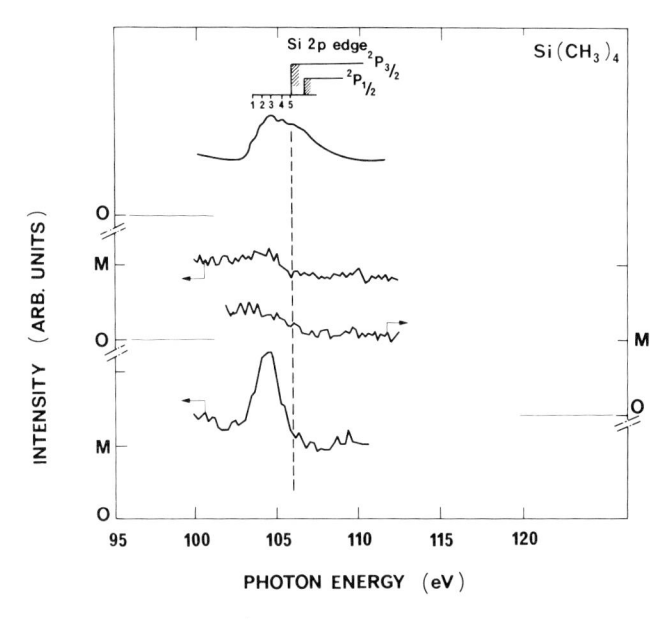

Fig. 3.46. Top curve: Electron energy loss spectrum of TMS (from Sodhi et al. 1985). Lower three curves: Partial photoionization cross sections associated with the first major valence bands in TMS in the vicinity of the Si2p edge (from Morin 1985, Morin et al. 1986).

The doubly resonant Auger process is an extension of the above, involving the ejection of two electrons. It is schematically represented by

(iii) $(\text{core})^1(\text{valence})^n(\text{unocc})^1 \rightarrow (\text{core})^2(\text{valence})^{n-3}(\text{unocc})^1 + 2e^-$,

the electron spectator model, and

(iv) $(\text{core})^1(\text{valence})^n(\text{unocc})^1 \rightarrow (\text{core})^2(\text{valence})^{n-2}(\text{unocc})^0 + 2e^-$, (a)

$\rightarrow (\text{core})^2(\text{valence})^{n-2}(\text{unocc})^1 + e^-$

$\hookrightarrow (\text{core})^2(\text{valence})^{n-2}(\text{unocc})^0 + e^-$

(b)

the electron nonspectator model, also described as cascade auto-ionization. The ejected electrons may share continuously the excess energy (processes iii and iva) or a finite kinetic energy (process ivb) and the residual ion is left with either a three-hole–one-particle or a two-hole configuration. Further ejection of a third valence electron may be energetically possible (Ferrett et al. 1986).

In the TMS case considered above, the significant enhancement above 25 eV is probably the result of processes (i), (iii) or (iv). The complicated valence orbital manifold of this molecule implies probably a congestion of satellite bands, which cannot be distinguished from a real double-ionization continuum. However, in other simpler molecules the enhancement of valence satellites near core resonances has been observed and illustrates the existence of the singly resonant

Auger process (i). Evidence of such spectator transcriptions has been found in molecules for the first time by Ungier and Thomas (1983) in CO (see also Haak et al. 1984, Ungier and Thomas 1985 and Becker et al. 1986). They have found that the satellite spectrum produced by core excitation is identical to the normal Auger spectrum, but shifted by an energy related to the Coulomb interaction between the core vacancy and the excited electron. In many cases the spectator channels are dominant. Evidence of a double-ionization continuum produced through a doubly resonant Auger process cannot be extracted easily from PES studies because it requires the detection of low kinetic energy electrons (process ivb) or the two electrons in coincidence, but rather from ionic relaxation studies as discussed in the following section.

All these electronic relaxation processes in resonances near core levels are described in terms of a frozen initial molecule at its equilibrium structure. In other words dissociation is assumed to be much slower than any electronic decay channel. This posture is too simplistic because some discrete resonances are purely dissociative states and may decay faster than the core vacancy relaxation. Such processes have been established for the first time by Morin and Nenner (1986) in the HBr molecule excited near the bromine 3d edge at the σ^* resonance. The PES spectrum (see fig. 3.47) shows numerous additional lines which reflect the electronic relaxation pathways of the excited bromine fragment and not the excited molecule. The mechanism is described in fig. 3.48 by a very fast dissociation on femtosecond time scale followed by autoionization. Similar conclusions have been drawn from SiH_4 PES studies performed at the σ^* resonance below the Si2p edge. Competition between the two processes is found when the departing fragment originally bonded to the excited atom becomes heavier. A review of such cases has been completed by Morin and Nenner (1987). More generally, when the time scales for Auger decay and vibrational motion are comparable the frequently made assumption that the excitation and decay can be treated separately, may be invalid (Carroll et al. 1987).

3.6.3. Ionic relaxation

Various core excitation and ionization processes in molecules lead to a variety of singly, doubly or multiply charged molecular ions with a specific electronic configuration and internal energy. Actually, before the electron–ion coincidence spectroscopy work of Hitchcock et al. (1978, 1979) and photoionization mass resolved spectrometry results of Eberhardt et al. (1983), followed by Müller-Dethlefs et al. (1984), Morin (1985) and Morin et al. (1986), no specificity in dissociative ionization was expected by most photochemists. The main reason for this expectation was that the extremely large amount of energy stored in the molecule, compared with the dissociation binding energy, was supposed to literally produce an explosion of the molecule. Actually this was supported by the X-ray mass spectrometry measurements of Carlson and White (1966), which have shown full atomization in the case of CH_3I. In the particular energy region near the inner-shell edges, where resonances are dominant, the situation can be drastically different. The atomic character of the excited (or ionized) electron

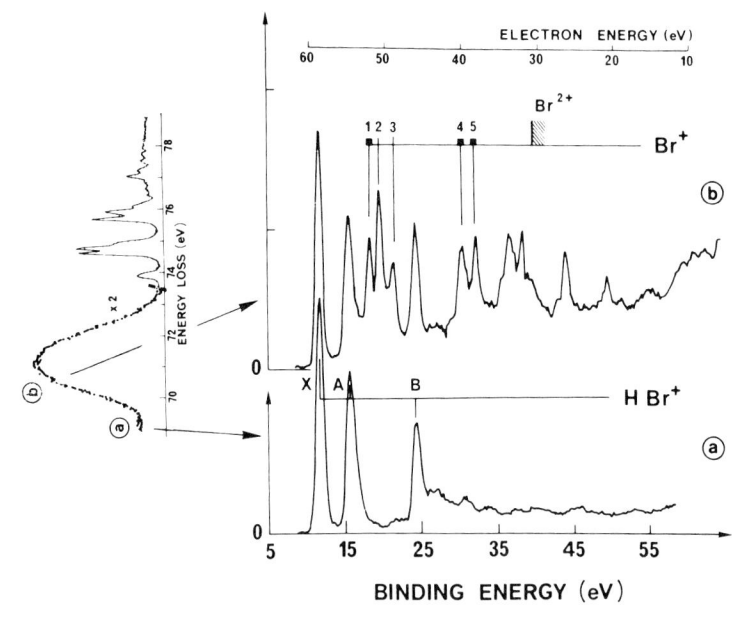

Fig. 3.47. Left part: Electron energy loss spectrum by Shaw et al. (1984). Right part: PES recorded at (a) "off" ($h\nu = 68.2$ eV) and at (b) "on" the resonance ($h\nu = 70.6$ eV). The Br autoionization lines are positioned on a kinetic-energy scale.

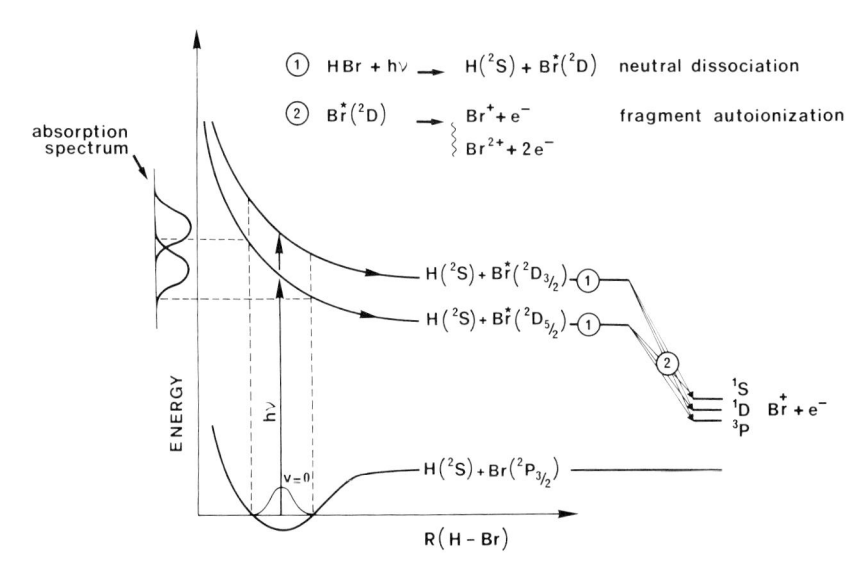

Fig. 3.48. Schematic potential energy curves of HBr describing the two-step relaxation process.

leads to a localization of the process on a single atom, but the configuration of the residual ion with holes in the valence shells (localized or delocalized) gives molecular properties to the relaxation.

The main processes in ionic relaxation are:
(1) dissociative ionization processes such as those presented in section 3.2.2,

$$h\nu + AB \rightarrow AB^*(\text{core excited}) \rightarrow AB^+ + e^-,$$

where AB^+ is an excited molecule with a valence hole or in a multiply excited configuration, followed by dissociation into $A^+ + B$ and possibly also $A^{++} + B^-$ for the case of multiply excited AB^+;

(2) soft dissociative double ionization,

$$h\nu + AB \rightarrow AB^{++} + e^-(\text{valence}) + e^-(\text{valence}),$$

followed by dissociation into $A^+ + B^+$ or $A^{++} + B$;

(2) hard dissociative double ionization,

$$h\nu + AB \rightarrow AB^{+*} + e^-(\text{core}),$$

followed by the Auger process

$$AB^{+*} \rightarrow AB^{++} + e^-(\text{valence}),$$

and the dissociation of AB^{2+} into $A^+ + B^+$ or $A^{2+} + B$.

A very clear example of some of these main processes has been provided recently by Morin (1985) and by Morin et al. (1986), in the TMS molecule excited by synchrotron radiation near the Si2p edge, i.e. in the 102–112 eV energy range. The singly charged fragment yields in TMS, obtained by mass spectrometry (Morin 1985), show (see fig. 3.49) a quite different behavior as compared to the absorption spectrum (see top curve of fig. 3.46) and a specific variation with photon energy (see fig. 3.49a). Observation of specific ion pairs in TMS using the photoion–photoion coincidence technique already described in section 2.4 (see fig. 3.50a), shows dramatic changes when the excitation is on the resonance and off the resonance below or above the threshold (Morin et al. 1985). Continuous ion pair intensity measurements as a function of photon energy are reported in fig. 3.50b (upper part). Above the edge, the ion pair yield [$Si(CH_3)_3^+ + CH_3^+$, for example] increases drastically, showing the formation of unstable TMS^{2+} after an Auger process (see section 3.6.2). Note that triply (or multiply) charged TMS^{n+} ions may also be formed because the other observed pairs refer to small fragments associated either with a third neutral or other charged fragments. Below the edge (see fig. 3.50) one observes a significant ion pair yield, which can be understood in two ways:

(1) Soft double-ionization and, more important, double resonant Auger processes (see section 3.6.2) explain the formation of doubly charged ions. Triple

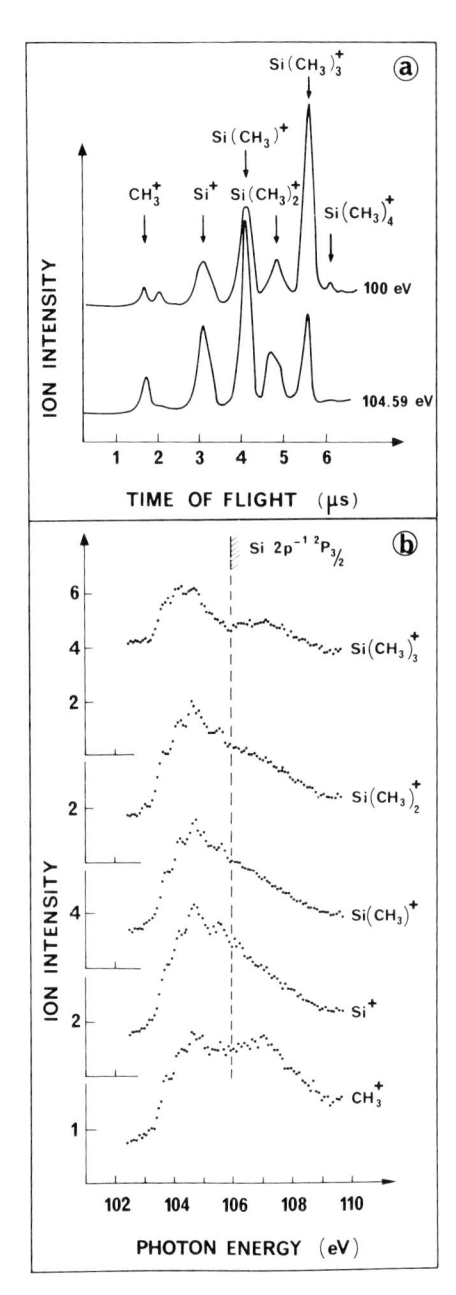

Fig. 3.49. (a) Typical mass spectra of TMS measured below the Si2p edge, off resonance (100 eV) and on resonance (104.59 eV). (b) Ion yield for the most abundant fragments in the 102–110 eV photon energy range (from Morin 1985).

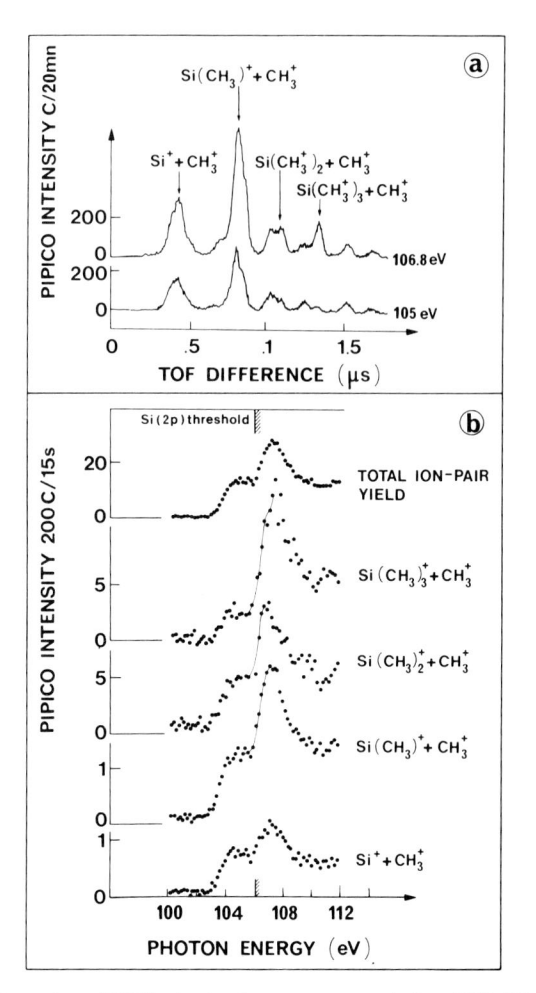

Fig. 3.50. (a) PIPICO spectra of TMS obtained on resonances below (105 eV) and above (106.8 eV) the Si2p edge (from Morin et al. 1985). (b) Photon energy dependence of the intensity of the most important positive ion pairs, in the 100–112 eV region (from Morin et al. 1986).

ionization (soft or hard) may also be present because of the uncertainty in the charge of the third fragment.

(2) The ion pair observed is actually associated with a negative ion fragment. In that case charge conservation is compatible with the formation of excited ions (satellites) through a simple resonant Auger process (see section 3.6.2). This would require, however, a rather strong anisotropy of the charge distribution in the final state.

It is clear that theoretical calculations are badly needed in order to test the various hypotheses already discussed in section 3.6.2. However, the experimental results obtained with both mass spectrometry and ion–ion coincidence techniques show that the origin of the selectivity in the formation of singly charged fragments near core resonances is linked with the formation of doubly (or multiply) charged ions as illustrated in fig. 3.49b, formed via specific intra- or interatomic Auger processes. In addition, such doubly charged ionic states must be highly repulsive to provide dissociation at a high enough rate to compete with internal energy redistribution in the molecule. Recent results on a polyatomic model molecule Br–CF_2–CF_2–I (Nenner et al. 1987b) show that selective dissociation pathways of the Br–C bond and the C–I bond are favored when one ionizes selectively either the bromine 3d continuum or the iodine 4d continuum.

A related question concerning specificity in dissociative inner-shell ionization is whether or not continuum shape resonances can influence ionic relaxation. This has been addressed with an (e,e + ion) coincidence simulation of the photoionization spectra of SF_6 in the S2p region (Hitchcock et al. 1978). Although changes in the fragmentation between the discrete and the continuum occur there is no detectable change in the ion branching ratios between the strong S2p continuum resonances and the non-resonant continuum. This has been explained by the short resonance lifetime and the rapid decay into the "normal" S2p continuum. However, complex electronic decay of continuum resonances in SF_6 has been observed by Ferrett (1986) near the S2p edge. Related studies of the electronic decay and ionic fragmentation of inner-shell ionization (Kay et al. 1977, Reimer et al. 1986, Hitchcock et al. 1987) have shown that the above simple picture is not always valid.

4. Miscellaneous and future trends

Several studies using synchrotron radiation have been conducted on the formation and decay of excimer molecules like rare gas dimers, rare gas halides and rare gas hydrides (see Zimmerer 1985). The gas mixture is irradiated in the energy region corresponding to an excited state of one of the monomers. The excimers are formed in collisions of the electronically excited atoms or molecules with ground state species. The fluorescence excitation spectra provide detailed information on the potential energy curves and relaxation of these species. This has been demonstrated very clearly for the rare gas dimers, where oscillatory structures in the bound–free fluorescence spectra following state-selected pulsed synchrotron radiation excitation have been observed (Dutuit et al. 1980, Moeller et al. 1985).

In the case of rare gas halides and hydrides, synchrotron radiation is used to excite a mixture of rare gas atoms (R) and diatomic molecules (Cl_2, for example). For this system, Castex et al. (1980), LeCalvé et al. (1985) and Jordan et al. (1985) have demonstrated clearly that the reaction proceeds through

$$R^* + Cl_2 \rightarrow (RCl)^* + Cl,$$

but also through

$$R + Cl_2^* \rightarrow (RCl)^* + Cl.$$

The mechanism for this reaction is the following: the collision $R^* + Cl_2$ or $R + Cl_2^*$ gives rise, by an electron jump, to an intermediate "ionic" or charge-transfer state, which predissociates giving $(RCl)^* + Cl$.

Another, complementary, way to study those processes is by the use of supersonic expansions. In a supersonic expansion it is possible to form van der Waals complexes, $R \cdots Cl_2$ for instance, in the ground electronic state. It is then possible to excite the system in the intermediate states $(R \cdots Cl_2)^*$ and to study their predissociation (Castex et al. 1985). It is the half-collision analog of the studies discussed above. This new class of photochemistry studies via photoexcitation of van der Waals complexes is a very promising area of research. The combination of a supersonic molecular beam with synchrotron radiation is particularly attractive for the study of ion–molecule reactions.

It should be noted at this point that for systems with higher van der Waals bonding energy, the preparation of the complex in a supersonic expansion is not absolute necessary. The XeF_2 molecule, for instance, has been studied at room temperature using synchrotron radiation by Black et al. (1981).

Other experiments on van der Waals molecules and clusters using synchrotron radiation have been performed (Trevor et al. 1984, Ding et al. 1985, Kamke et al. 1986). Neon dimers, for instance, have been studied from 560 to 620 Å (Trevor et al. 1984). The dimers are produced in a supersonic expansion at 77 K and photoionization is monitored by mass spectrometric techniques. The photoionization was observed to be dominated by autoionization near the threshold. Careful consideration of the autoionization mechanisms, absorption spectra, and electronic configurations of the molecular Rydberg states led to several assignments in the photoionization efficiency. The molecular Rydberg states with $B\ ^2\pi_g$ core configuration were found to make the most important contribution to the photoionization cross section near threshold. This is most probably due to the fact that they have internuclear equilibrium distances intermediate between the large internuclear distance in the van der Waals dimer and the small one in the molecular ion. They are the most efficient indirect route for ionization since they provide the best Franck–Condon overlaps.

A method has been developed for the measurement of the dissociation energies of neutral dimers using photoionization spectra of mass-selected ions from molecular beams. The dissociation energy for $C_4H_8 \cdots SO_2$, $(C_4H_8)_2$, $C_6H_6 \cdots SO_2$ and $C_6H_6 \cdots HCl$ has been determined (Grover et al. 1985, Walters et al. 1985).

Mercury clusters have also been studied using synchrotron radiation (Bréchignac et al. 1985a,b). The changes in electronic structure when going from the isolated atom to the bulk are important for an understanding of the evolution of metallic character in metal clusters. The behavior of autoionizing lines as a function of cluster size has been studied as a means to address that question. The mercury clusters were generated in a free jet expansion, ionized and mass selected by a quadrupole. The autoionization peaks are red shifted following an inverse $1/n$ law, where n is the size of the cluster. The mercury clusters with n smaller than 8 have not yet a metallic character. Double-ionization experiments have also been performed on those clusters (Bréchignac et al. 1985b).

An interesting aspect of photoexcitation of molecules in the range of 15 to 30 eV concerns ion pair formation. The photoionization efficiency curves for O^- and O^+ from molecular oxygen, for O^- and N^+ from nitric oxide and for O^- and C^+ and O^+ and C^- from carbon monoxide, have been measured using synchrotron radiation (Oertel et al. 1979). Because the mass-spectrometric determination of negative photoions exhibits a very low noise level, the investigation of ion pair formation processes provides considerable and very clear information on weak Rydberg series in the VUV energy region.

We have discussed in section 3.2.5 the photoelectron spin polarization measurements. The investigation of the angular distribution of photoelectrons with defined spin polarization provides the most complete information on photoionization dynamics. In particular, Cherepkov (1982, 1983) has shown theoretically that a measurement of the circular dichroism ($I_{+1} - I_{-1}$, where I_{+1} and I_{-1} are the intensities for left and right circularly polarized light) may give very detailed information on the photoionization dynamics in the case of oriented molecules or for unoriented optically active (chiral) molecules. In the latter case, this effect appears due to the lack of inversion symmetry and a plane of symmetry of the molecule. Therefore, angle- and spin-resolved photoelectron spectroscopy may provide, in the near future, important information on the structure and dynamics of chiral molecules.

Another very interesting type of experiments are photoionization studies in external electric or magnetic fields. In particular, electric fields may provide important insight into photoionization dynamics because they allow one to modify the external field experienced by the electron without changing the internal core field (see, for example, Fano 1981). Recently Poliakoff et al. (1984) have shown that fluorescence excitation spectroscopy can be used to measure partial photoionization cross sections of molecules in external electric fields. The system studies was N_2 and the threshold for the production of $N_2^+(B\,^2\Sigma_u^+)$ was found to shift linearly with the square root of the applied field. Although this result is expected from classical considerations, the slope of the line is smaller than that predicted for a pure Coulomb plus Stark potential. This is most probably due to the deviation of the molecular potential from the pure Coulomb potential.

In this review we have concentrated on gas phase studies. An extremely important class of experiments are those performed in matrices and surfaces. Chergui et al. (1985a,b, 1986) have studied the Rydberg state spectroscopy and intramolecular relaxation of NO trapped in rare gas matrices, using synchrotron radiation at DESY on the SUPERLUMI set-up (see section 2.1). Their results raise the question of the validity of the Wannier exciton model (see, for example, Schwentner et al. 1985) in the case of matrix-isolated molecules. In addition the Rydberg bands have broad multiphonon sidebands. They represent the participation by the cavity breathing modes in the electronic transition and they provide information on the dynamics of the Rydberg state in the matrix.

Finally, photoemission experiments have been performed in molecular solids (Fock and Koch 1985). Solid films of SF_6 and CCl_4 have been excited with synchrotron radiation in the 10 to 40 eV photon energy range. Shape resonances have been observed.

In the future, double-excitation experiments (laser plus synchrotron radiation) in molecular systems will become possible. This opens the possibility of detecting ground state fragments by laser-induced fluorescence or to study photoionization and photodissociation from initially laser-excited molecules. Photoionization of laser-excited atoms has already been done and similar studies in molecules should be feasible.

Acknowledgments

We are very grateful to C.R. Botter, E. Brion, A. Giusti, M. Glass-Maujean, P.M. Guyon, U. Heinzmann, A. Hitchcock, S. Leach, T. Lebrun, H. Lefebvre-Brion, P. Millié and P. Morin for very stimulating discussions and for a critical reading of the manuscript. We thank M.Y. Adam and S. Bodeur for making their results available to us prior to publication. We also thank all the synchrotron radiation users from LURE, who, through their publications, contributed largely to the material presented in this chapter.

References

Adam, M.Y., P. Morin, P. Lablanquie and I. Nenner, 1983, Int. Workshop Atomic and Molecular Photoionization (Berlin), unpublished.
Adam, M.Y., P. Morin, C. Cauletti and M. Piancastelli, 1985, J. Electron Spectrosc. **36**, 377.
Akahori, T., Y. Morioka, M. Watanabe, T. Hayaishi, K. Ito and M. Nakamura, 1985, J. Phys. B **18**, 2219.
Almbladh, C.-O., and L. Hedin, 1983, Beyond the One-Electron Model: Many-Body Effects in Atoms, Molecules, and Solids, in: Handbook on Synchrotron Radiation, Vol. 1, ed. E.-E. Koch (North-Holland, Amsterdam) ch. 8.
Baer, T., 1979, in: Gas Phase Ion Chemistry, Vol. 1, ed. M.T. Bowers (Academic Press, New York) ch. 5, p. 153.
Baer, T., P.M. Guyon, I. Nenner, A. Tabché-Fouhailé, R. Botter, L. Ferreira and T. Govers, 1979, J. Chem. Phys. **70**, 1585.
Bagus, P., and W. Viinikka, 1977, Phys. Rev. A **15**, 1486.
Baig, M.A., and J.P. Connerade, 1984, J. Phys. B **16**, L757.
Baig, M.A., J.P. Connerade and J. Hormes, 1982, J. Phys. B **14**, L25.
Baker, A., and D. Betteridge, 1972, Photoelectron Spectroscopy (Pergamon Press, Oxford).
Ballard, R., 1978, Photoelectron Spectroscopy and Molecular Orbital Theory (Hilger, Bristol).
Basco, N., and R. Norrish, 1962, Proc. R. Soc. London Ser. A **268**, 291.
Becker, U., H. Hölzel, H.G. Kerkhoff, B. Langer, D. Szostak and R. Wehlirz, 1986, Phys. Rev. Lett. **56**, 1455.
Beckmann, O., W. Baun, H.-W. Jochims, E. Rühl and H. Baumgärtel, 1985, Chem. Phys. Lett. **121**, 499.
Beebe, N., E. Thulstrup and A. Anderson, 1976, J. Chem. Phys. **64**, 2080.
Benfatto, M., C. Natoli, A. Bianconi, J. Garcia, A. Marcelli, M. Fantoni and I. Davoli, 1987, Phys. Rev. B, in press.
Berkowitz, J., 1979, Photoionization and Photoelectron Spectroscopy (Academic Press, New York).
Berkowitz, J., and J. Eland, 1977, J. Chem. Phys. **67**, 2740.
Bertel, E., R. Stockbauer and T. Madey, 1984, Surf. Sci. **141**, 355.
Besnard, M., L. Hellner, Y. Malinovich and G. Dujardin, 1986, J. Chem. Phys. **85**, 1316.

Beswick, J.A., and W. Gelbart, 1980, J. Phys. Chem. **84**, 3148.
Beswick, J.A., and M. Glass-Maujean, 1987, Phys. Rev. A **35**, 3339.
Beswick, J.A., and M. Horani, 1981, Chem. Phys. Lett. **78**, 4.
Bianconi, A., 1983, in: Proc. 1st Int. Conf. on EXAFS and XANES, Frascati, 1982, Springer Series on Chemical Physics, Vol. 27, eds A. Bianconi, L. Incoccia and S. Stipcich (Springer, New York) p. 118.
Bianconi, A., H. Petersen, F. Brown and R. Bachrach, 1978, Phys. Rev. A **17**, 1907.
Black, G., R. Sharpless, D. Lorents, D. Huestic, R. Gutcheck, T. Bonifield, D. Helms and G. Walters, 1981, J. Chem. Phys. **75**, 4840.
Bodeur, S., and J. Esteva, 1984, Proc. of 16th EGAS, London, Europhysics Conf. Abstr., eds A.P. Thorn, K. Burnett, J.P. Connerade and R.C.M. Learner, pp. A4–41.
Bodeur, S., and J. Esteva, 1985, Chem. Phys. **100**, 415.
Bodeur, S., I. Nenner and P. Millié, 1986, Phys. Rev. A **34**, 2986.
Borondo, F., L. Eguiagaray and A. Riera, 1982, J. Phys. B **15**, 899.
Borrell, P., P.M. Guyon and M. Glass-Maujean, 1977, J. Chem. Phys. **66**, 818.
Bradshaw, A., W. Eberhardt, H. Levinson, W. Domcke and L. Cederbaum, 1980, Chem. Phys. Lett. **70**, 36.
Bréchignac, C., M. Broyer, Ph. Cahuzac, G. Delacretaz, P. Labastie and L. Wöste, 1985a, Chem. Phys. Lett. **118**, 174.
Bréchignac, C., M. Broyer, Ph. Cahuzac, G. Delacretaz, P. Labastie and L. Wöste, 1985b, Chem. Phys. Lett. **120**, 559.
Breton, J., P.M. Guyon and M. Glass-Maujean, 1980, Phys. Rev. A **21**, 1909.
Brion, C., 1985, Comments At. & Mol. Phys. **16**, 249.
Brion, C., and A. Hamnett, 1981, Adv. Chem. Phys. **45**, 1.
Brion, C., and K. Tan, 1978, Chem. Phys. **34**, 141.
Brion, C., and J. Thomson, 1984, J. Electron Spectrosc. **33**, 287, 301.
Brodmann, R.R., R. Haensel, U. Hahn, U. Nielsen and G. Zimmerer, 1974, Chem. Phys. Lett. **29**, 250.
Brown, F., 1980, in: Synchrotron Radiation Research, eds H. Winick and S. Doniach (Plenum Press, New York) p. 61.
Brown, F., R. Bachrach and A. Bianconi, 1978, Chem. Phys. Lett. **54**, 425.
Brundle, C., and D. Turner, 1966, Proc. R. Soc. London Ser. A **307**, 27.
Busch, G., and K. Wilson, 1972, J. Chem. Phys. **56**, 3655.
Carlson, T., 1971, Chem. Phys. Lett. **9**, 23.
Carlson, T., 1978, Photoelectron and Auger Spectroscopy (Plenum, New York).
Carlson, T., and R. White, 1966, J. Chem. Phys. **44**, 4510.
Carlson, T., M. Krause, D. Mehaffy, J. Taylor, F. Grimm and J. Allen, 1980, J. Chem. Phys. **73**, 6056.
Carlson, T., M. Krause, F. Grimm, J. Allen Jr, D. Mehaffy, P. Keller and J. Taylor, 1981, Phys. Rev. A **23**, 3316.
Carlson, T., M. Krause and F. Grimm, 1982a, J. Chem. Phys. **77**, 1701.
Carlson, T., M. Krause, F. Grimm, P. Keller and J. Taylor, 1982b, J. Chem. Phys. **77**, 5340.
Carlson, T., M. Krause, A. Fahlman, P. Keller, J. Taylor, T. Whitley and F. Grimm, 1983, J. Chem. Phys. **79**, 2157.
Carlson, T., A. Fahlman, M. Krause, P. Keller, J. Taylor, T. Whitley and F. Grimm, 1984a, J. Chem. Phys. **80**, 3521.
Carlson, T., A. Fahlman, W. Svensson, M. Krause, T. Whitley, F. Grimm, M. Piancastelli and J. Taylor, 1984b, J. Chem. Phys. **81**, 3828.
Carlson, T., M. Krause, W. Svensson, P. Gerard, F. Grimm, T. Whitley and B. Pullen, 1986, Z. Phys. D **2**, 309.
Carnovale, C., A. Hitchcock, J. Cook and C. Brion, 1982, Chem. Phys. **66**, 249.
Carrington, T., 1964, J. Chem. Phys. **41**, 2012.
Carroll, T., S. Anderson, L. Ungier and T. Thomas, 1987, Phys. Rev. Lett. **58**, 867.
Castex, M.C., J. Le Calvé, D. Haaks, B. Jordan and G. Zimmerer, 1980, Chem. Phys. Lett. **70**, 106.

Castex, M.C., I. Dimicoli, J. Le Calvé, F. Piuzzi and A. Tramer, 1985, in: Photophysics and Photochemistry above 6 eV, ed. F. Lahmani (Elsevier, Amsterdam) p. 661.
Cederbaum, L., W. Domcke, J. Schirmer and W. Von Niessen, 1986, Adv. Chem. Phys. LXV, ed. S.A. Rice, p. 115.
Cermak, V., 1974, J. Electron Spectrosc. **3**, 329.
Cherepkov, N., 1980, Opt. Spectrosc. **49**, 1067.
Cherepkov, N., 1981a, J. Phys. B **14**, L73.
Cherepkov, N., 1981b, J. Phys. (France) **14**, 2165.
Cherepkov, N., 1982, Chem. Phys. Lett. **87**, 344.
Cherepkov, N., 1983, J. Phys. B **16**, 1543.
Chergui, M., V. Chandrasekharan, N. Schwentner and H. Kühle, 1985a, in: Photophysics and Photochemistry above 6 eV, ed. F. Lahmani (Elsevier, Amsterdam) p. 433.
Chergui, M., N. Schwentner, W. Böhmer and R. Haensel, 1985b, Phys. Rev. A **31**, 527.
Chergui, M., N. Schwentner and W. Böhmer, 1986, J. Chem. Phys. **85**, 2472.
Chun, M., 1969, Phys. Lett. A **30**, 445.
Collins, L., and B. Schneider, 1984, Phys. Rev. A **29**, 1695.
Combet-Farnoux, F., 1982a, Phys. Rev. A **25**, 285.
Combet-Farnoux, F., 1982b, Abstracts 8th Int. Conf. on Atomic Physics (Göteborg, Sweden).
Curtis, D., and J. Eland, 1985, J. Mass. Spect. Ion. Phys. **63**, 241.
Daviel, S., C. Brion and A. Hitchcock, 1984, Rev. Sci. Instrum. **55**, 182.
De Souza, G., P. Morin and I. Nenner, 1985, J. Chem. Phys. **83**, 492.
Dehmer, J.L., D. Dill and S. Wallace, 1979, Phys. Rev. Lett. **43**, 1005.
Dehmer, J.L., P. Miller and W. Chupka, 1984, J. Chem. Phys. **80**, 1030.
Dehmer, J.L., D. Dill and A. Parr, 1985, in: Photophysics and Photochemistry in the Vacuum Ultraviolet, eds S. McGlynn, G. Findley and R. Huebner (Reidel, Dordrecht) p. 341.
Dehmer, J.L., A.C. Parr and S.H. Southworth, 1987, this volume.
Dehmer, P.M., and W. Chupka, 1975, J. Chem. Phys. **62**, 4525.
Delwiche, J., M.J. Hubin-Franskin, P.M. Guyon and I. Nenner, 1981, J. Chem. Phys. **74**, 4219.
Derenbach, H., and V. Schmidt, 1984, J. Phys. B **17**, 83.
Diercksen, G., W. Kraemer, T. Rescigno, C. Bender, B.V. McKoy, S. Langhoff and P. Langhoff, 1982, J. Chem. Phys. **76**, 1043.
Ding, A., R. Cassidy, L. Cordis and F. Lampe, 1985, J. Chem. Phys. **83**, 3426.
Dujardin, G., and S. Leach, 1983, J. Chem. Phys. **79**, 658.
Dujardin, G., and D. Winkoun, 1985, J. Chem. Phys. **83**, 6222.
Dujardin, G., S. Leach, O. Dutuit, T. Govers and P.M. Guyon, 1983, J. Chem. Phys. **79**, 644.
Dujardin, G., S. Leach, O. Dutuit, P.M. Guyon and M. Richard-Viard, 1984, Chem. Phys. **88**, 339.
Dujardin, G., T. Govers, S. Leach and D. Winkoun, 1985a, in: Photophysics and Photochemistry above 6 eV, ed. F. Lahmani (Elsevier, Amsterdam) p. 163.
Dujardin, G., D. Winkoun and S. Leach, 1985b, Phys. Rev. A **31**, 3027.
Durand, G., and X. Chapuisat, 1985, Chem. Phys. **96**, 381.
Dutuit, O., M.C. Castex, J. Le Calvé and M. Lavollée, 1980, J. Chem. Phys. **73**, 3107.
Dutuit, O., A. Tabché-Fouhailé, H. Fröhlich, P.M. Guyon and I. Nenner, 1985, J. Chem. Phys. **83**, 584.
Eberhardt, W., R. Haelbich, M. Iwan, E.-E. Koch and C. Kunz, 1976, Chem. Phys. Lett. **40**, 180.
Eberhardt, W., J. Stöhr, J. Feldhaus, E. Plummer and F. Sette, 1983a, Phys. Rev. Lett. **51**, 2370.
Eberhardt, W., T. Sham, R. Carr, S. Krummacher, M. Strongin, S. Weng and D. Wesner, 1983b, Phys. Rev. Lett. **50**, 1038.
Eland, J., 1980, J. Chem. Phys. **72**, 6015.
Eland, J., 1984, Photoelectron Spectroscopy (Butterworths, London).
Eland, J., 1987, Mol. Phys., in press.
Eland, J., S. Wort, P. Lablanquie and I. Nenner, 1986, Z. Phys. D **4**, 31.
Fano, U., 1961, Phys. Rev. **124**, 1866.
Fano, U., 1981, Phys. Rev. A **24**, 619.
Fergusson, E., 1979, Kinetic of Ion Molecule Reactions (Plenum Press, New York).

Ferreira, L., 1984, Thèse (Université d'Orsay).
Ferrett, T.A., 1986, Ph.D. Thesis, LBL Rep. 22418 (Lawrence Berkeley Laboratory, University of California).
Ferrett, T.A., D. Lindle, P. Heimann, H.G. Kerkhoff, U. Becker and D.A. Shirley, 1986, Phys. Rev. A **34**, 1916.
Fiquet-Fayard, F., and O. Gallais, 1972, Chem. Phys. Lett. **16**, 18.
Flouquet, F., and J. Horsley, 1974, J. Chem. Phys. **60**, 3767.
Fock, J.-H., and E.-E. Koch, 1985, Chem. Phys. **96**, 125.
Frasinski, L., K. Randall and K. Codling, 1985, J. Phys. B **18**, L129.
Frasinski, L., M. Stankiewicz, K. Randall, P. Hatherby and K. Codling, 1986, J. Phys. B **19**, L819.
Frey, R., B. Gotchev, W. Peatman, H. Pollak and E. Schlag, 1978, Chem. Phys. Lett. **54**, 411.
Friedrich, H., B. Sonntag, P. Rabe, W. Butscher and W. Schwarz, 1979, Chem. Phys. Lett. **64**, 360.
Gardner, J., and J. Samson, 1976, J. Electron Spectrosc. & Relat. Phenom. **8**, 469.
Gardner, J., and J. Samson, 1978, J. Electron Spectrosc. & Relat. Phenom. **13**, 7.
Gerard, P., 1984, Thèse (Université d'Orsay).
Giusti-Suzor, A., and Ch. Jungen, 1984, J. Chem. Phys. **80**, 986.
Glass-Maujean, M., J. Breton and P.M. Guyon, 1978, Phys. Rev. Lett. **40**, 181.
Glass-Maujean, M., J. Breton and P.M. Guyon, 1979, Chem. Phys. Lett. **63**, 591.
Greene, Ch., and Ch. Jungen, 1985, Adv. At. & Mol. Phys. **21**, 51.
Greene, Ch., and R. Zare, 1982a, Phys. Rev. A **25**, 2031.
Greene, Ch., and R. Zare, 1982b, Ann. Rev. Phys. Chem. **33**, 119.
Gregory, A., K. Sunil and K. Jordan, 1984, Chem. Phys. Lett. **108**, 439.
Grimm, F., T. Carlson, W. Press, P. Agron, J. Thomson and J. Davenport, 1980, J. Chem. Phys. **72**, 3041.
Grimm, F., J. Allen Jr, T. Carlson, M. Krause, D. Mehaffy, P. Keller and J. Taylor, 1981, J. Chem. Phys. **75**, 92.
Grover, J., E. Walters, J. Newman and M. White, 1985, J. Am. Chem. Soc. **107**, 7329.
Gudat, W., and C. Kunz, 1978, in: Synchrotron Radiation, ed. C. Kunz, Topics in Current Physics (Springer, Heidelberg) ch. 3.
Guest, J., K. Jackson and R. Zare, 1983, Phys. Rev. A **28**, 2217.
Guest, J., M. O'Halloran and R. Zare, 1984, J. Chem. Phys. **81**, 2689.
Gürtler, P., E. Roick, G. Zimmerer and M. Pouey, 1983, Nucl. Instrum. Methods **208**, 835.
Gustafsson, T., E. Plummer, D. Eastman and W. Gudat, 1978, Phys. Rev. A **17**, 175.
Guyon, P.M., and I. Nenner, 1980, Appl. Opt. **19**, 4068.
Guyon, P.M., T. Baer, L. Ferreira, I. Nenner, A. Tabché-Fouhailé, R. Botter and T. Govers, 1978, J. Phys. B **11**, L141.
Guyon, P.M., J. Breton and M. Glass-Maujean, 1979, Chem. Phys. Lett. **68**, 314.
Guyon, P.M., T. Baer and I. Nenner, 1983, J. Chem. Phys. **78**, 3665.
Haak, H., G.A. Sawatzky, L. Ungier, J.K. Gimzewski and T.D. Thomas, 1984, Rev. Sci. Instrum. **55**, 696.
Hahn, U., N. Schwentner and G. Zimmerer, 1978, Nucl. Instrum. Methods **152**, 261.
Heckenkamp, Ch., F. Schäfers, G. Schönhense and U. Heinzmann, 1984, Phys. Rev. Lett. **52**, 421.
Heinzmann, U., 1978, J. Phys. B **11**, 399.
Heinzmann, U., 1980, Appl. Opt. **19**, 4087.
Heinzmann, U., F. Schäfers, K. Thim, A. Wolcke and J. Kessler, 1979, J. Phys. B **12**, L679.
Heinzmann, U., B. Osterheld, F. Schäfers and G. Schönhense, 1981, J. Phys. B **14**, L79.
Hertz, H., H. Jochims and W. Sroka, 1974, J. Phys. B **7**, L548.
Hitchcock, A.P., 1982, J. Electron Spectrosc. **25**, 245.
Hitchcock, A.P., and C.E. Brion, 1977, J. Electron Spectrosc. **10**, 317.
Hitchcock, A.P., and C.E. Brion, 1980, J. Electron Spectrosc. **18**, 1.
Hitchcock, A.P., and C. Brion, 1981, J. Phys. **14**, 4399.
Hitchcock, A.P., and J. Stöhr, 1987, J. Chem. Phys., in press.
Hitchcock, A.P., and M. Van der Wiel, 1979, Phys. B **12**, 2153.
Hitchcock, A.P., C.E. Brion and M.J. van der Wiel, 1978, J. Phys. B **11**, 3245.

Hitchcock, A.P., C.E. Brion and M.J. van der Wiel, 1979, Chem. Phys. Lett. **66**, 213.
Hitchcock, A.P., P. Lablanquie, P. Morin, E. Lizon, A. Lergrin, P. Thiry and I. Nenner, 1987, unpublished.
Holland, D., A. Parr, D. Ederer, J.L. Dehmer and J.B. West, 1982, Nucl. Instrum. Methods **195**, 331.
Ishiguro, E., S. Iwata, Y. Susuki, A. Mikuni and T. Sasaki, 1982, J. Phys. B **15**, 1841.
Ishii, I., and A.P. Hitchcock, 1987, J. Chem. Phys., in press.
Ishii, I., R. McLaren, A.P. Hitchcock and M. Robin, 1987, J. Chem. Phys., in press.
Ito, K., A. Tabché-Fouhailé, H. Fröhlich, P.M. Guyon and I. Nenner, 1985, J. Chem. Phys. **82**, 1231.
Jennison, D., 1980, Chem. Phys. Lett. **69**, 435.
Jordan, B., T. Moeller, G. Zimmerer, D. Haaks, J. Le Calvé and M.C. Castex, 1985, to be published.
Julienne, P., 1971, Chem. Phys. Lett. **8**, 27.
Kamke, W., B. Kamke, H.V. Kiefl and I.V. Hertel, 1986, J. Chem. Phys. **84**, 1325.
Kay, R.B., P. van der Leeuw and M.J. van der Wiel, 1977, J. Phys. B **10**, 2521.
Keller, F., and H. Lefebvre, 1986, Z. Phys. D **4**, 15.
Kimura, K., S. Katsumata, Y. Achiba, T. Yamazaki and S. Iwata, 1981, Handbook of He(I) Photoelectron Spectroscopy of Fundamental Organic Molecules (Halsted, New York).
Kincaid, B., and P. Eisenberger, 1975, Phys. Rev. Lett. **34**, 1361.
Kincaid, B., and P. Eisenberger, 1976, Phys. Rev. Lett. **36**, 1346.
King, G., and F. Read, 1985, in: Atomic Inner Shell Physics, ed. B. Crasemann (Plenum, New York) ch. 8, p. 317.
King, G., M. Tronc, F. Read and R. Bradford, 1977, J. Phys. B **10**, 2479.
Kinsey, L., 1971, J. Chem. Phys. **54**, 1206.
Koch, E.-E., and B. Sonntag, 1978, Topics in Current Physics, ed. C. Kunz (Springer, Heidelberg) ch. 6.
Komiha, N., 1981, Thèse (Université de Paris).
Köpel, H., W. Domcke and L. Cederbaum, 1984, Adv. Chem. Phys. **57**, 59.
Kosugi, N., and H. Kurada, 1983, Chem. Phys. Lett. **94**, 377.
Krause, M., T. Carlson and P. Woodruff, 1981, Phys. Rev. A **24**, 1374.
Kronast, W., R. Huster and W. Melhorn, 1984, J. Phys. B **17**, L51.
Krummacher, S., V. Schmidt and F. Wuilleumier, 1980, J. Phys. B **13**, 3993.
Krummacher, S., V. Schmidt, F. Wuilleumier, J.M. Bizau and D. Ederer, 1983, J. Phys. B **16**, 1733.
Lablanquie, P., 1984, Thèse (Université d'Orsay).
Lablanquie, P., I. Nenner, J. Eland, J. Delwiche, M.J. Hubin-Franskin and P. Morin, 1985a, in: Photophysics and Photochemistry above 6 eV, ed. F. Lahmani (Elsevier, Amsterdam) p. 53.
Lablanquie, P., I. Nenner, P. Millié, P. Morin, J. Eland, M.J. Hubin-Franskin and J. Delwiche, 1985b, J. Chem. Phys. **82**, 2951.
Lablanquie, P., P. Morin, Y. Adam and I. Nenner, 1985c, unpublished.
Lablanquie, P., J. Eland, I. Nenner, P. Morin, J. Delwiche and M.J. Hubin-Franskin, 1987, Phys. Rev. Lett. **58**, 992.
Lahmani, F., C. Lardeux, M. Lavollée and D. Solgadi, 1980a, J. Chem. Phys. **73**, 1187.
Lahmani, F., C. Lardeux and D. Solgadi, 1980b, J. Chem. Phys. **73**, 4433.
Lahmani, F., C. Lardeux and D. Solgadi, 1981a, Il Nuovo Cimento **63**, 233.
Lahmani, F., C. Lardeux and D. Solgadi, 1981b, J. Photochem. **15**, 37.
Lahmani, F., C. Lardeux and D. Solgadi, 1982, J. Chem. Phys. **77**, 275.
Lahmani, F., C. Lardeux and D. Solgadi, 1983a, Laser Chem. **3**, 97.
Lahmani, F., C. Lardeux and D. Solgadi, 1983b, J. Chim. Phys. & Phys.-Chim. Biol. **80**, 705.
Langhoff, P., A. Gerwer, C. Asaro and B.V. McKoy, 1980, J. Chem. Phys. **72**, 713.
Lardeux, C., 1983, Thèse (Université d'Orsay).
Lavollée, M., and R. Lopez Delgado, 1977, Rev. Sci. Instrum. **48**, 816.
Le Calvé, J., M.C. Castex, B. Jordan, G. Zimmerer, T. Moeller and D. Haaks, 1985, in: Photophysics and Photochemistry above 6 eV, ed. F. Lahamani (Elsevier, Amsterdam) p. 639.
Leach, S., 1984, in: Laser Applications in Chemistry, eds K. Kompa and J. Wanner (Plenum, New York) p. 35.
Lee, L.C., 1980, J. Chem. Phys. **72**, 6642.

Lee, L.C., and C.C. Chiang, 1982, Chem. Phys. Lett. **92**, 425.
Lee, L.C., R. Carlson, D. Judge and M. Ogawa, 1974, J. Chem. Phys. **61**, 3261.
Lee, L.C., R. Carlson and D. Judge, 1976, J. Phys. B **9**, 855.
Lee, L.C., L. Oren, E. Phillips and D. Judge, 1978, J. Phys. B **11**, 47.
Levine, R., 1974, Ber. Bunsenges. Phys. Chem. **78**, 111.
Lindle, D., C. Truesdale, P. Kobrin, T. Ferrett, P. Heimann, U. Becker, H. Kerkhoff and D.A. Shirley, 1984, J. Chem. Phys. **81**, 5375.
Lucchese, R., and B.V. McKoy, 1981, J. Phys. B **14**, L629.
Lucchese, R., and B.V. McKoy, 1982, Phys. Rev. A **26**, 1406, 1992.
Lucchese, R., G. Raseev and B.V. McKoy, 1982, J. Phys. B **12**, L417.
Marian, C., R. Marian, S. Peyerimhoff, B. Hess, R. Buenker and G. Seger, 1982, Mol. Phys. **46**, 779.
Marr, G., J. Morton, R. Holmes and D. McCoy, 1979, J. Phys. B **12**, 43.
Masuoka, T., 1984, J. Chem. Phys. **81**, 2652.
Masuoka, T., and J. Samson, 1980, J. Chim. Phys. & Phys.-Chim. Biol. **77**, 623.
Matthew, J., and Y. Komninos, 1975, Surf. Sci. **53**, 716.
Mazalov, L., A. Sadovskii, E. Gluskin, G. Dolenko and A. Krasnoperova, 1974, J. Struct. Chem. **15**, 705.
McLaren, R., S. Clark, I. Ishii and A.P. Hitchcock, 1987, Phys. Rev. A, in press.
Mentall, J., and P.M. Guyon, 1977, J. Chem. Phys. **67**, 3845.
Millié, Ph., I. Nenner, P. Archirel, P. Lablanquie, P. Fournier and J. Eland, 1986, J. Chem. Phys. **84**, 1259.
Miret-Artes, S., G. Delgado-Barrio, O. Atabek and J.A. Beswick, 1983, Chem. Phys. Lett. **98**, 554.
Moeller, T., B. Jordan, P. Gürtler, G. Zimmerer, D. Haaks, J. Le Calvé and M.C. Castex, 1983, Chem. Phys. **76**, 295.
Moeller, T., J. Stapelfeldt, M. Beland and G. Zimmerer, 1985, Chem. Phys. Lett. **117**, 301.
Monahan, K., and V. Rehn, 1978, Nucl. Instrum. Methods **152**, 255.
Morin, P., 1983, Thése (Université d'Orsay).
Morin, P., 1985, in: Photophysics and Photochemistry above 6 eV, ed. F. Lahmani (Elsevier, Amsterdam) p. 1.
Morin, P., and I. Nenner, 1986, Phys. Rev. Lett. **56**, 1913.
Morin, P., and I. Nenner, 1987, Phys. Scr., in press.
Morin, P., I. Nenner, P.M. Guyon, O. Dutuit and K. Ito, 1980, J. Chim. Phys. & Phys.-Chim. Biol. **77**, 605.
Morin, P., M.Y. Adam, I. Nenner, J. Delwiche and M.J. Hubin-Franskin, 1981, unpublished.
Morin, P., I. Nenner, M.Y. Adam, M.J. Hubin-Franskin, J. Delwiche, H. Lefebvre-Brion and A. Giusti-Suzor, 1982, Chem. Phys. Lett. **92**, 609.
Morin, P., M.Y. Adam, I. Nenner, J. Delwiche, M.J. Hubin-Franskin and P. Lablanquie, 1983, Nucl. Instrum. Methods **208**, 761.
Morin, P., G.G.B. de Souza, I. Nenner and P. Lablanquie, 1986, Phys. Rev. Lett. **56**, 131.
Morioka, Y., N. Nakamura, E. Ishiguro and M. Sasanuma, 1974, J. Chem. Phys. **61**, 1426.
Morse, M., and K. Freed, 1981, J. Chem. Phys. **74**, 4395.
Morse, M., K. Freed and Y. Band, 1979a, J. Chem. Phys. **70**, 3604.
Morse, M., K. Freed and Y. Band, 1979b, J. Chem. Phys. **70**, 3620.
Moseley, J., P. Cosby, J. Ozenne and J. Durup, 1979, J. Chem. Phys. **70**, 1474.
Müller-Dethlefs, K., M. Sander, L. Chewter and E. Schlag, 1984, J. Chem. Phys. **88**, 6098.
Mulliken, R., 1976, Acc. Chem. Res. **9**, 7.
Nakamura, M., M. Sasanuma, S. Sato, M. Watanabe, H. Yamashita, Y. Iguchi, A. Ejiri, S. Nakai, S. Yamaguchi, T. Sagawa, Y. Nakai and T. Oshio, 1969, Phys. Rev. **178**, 80.
Nakamura, N., Y. Morioka, T. Hayaiski, E. Ishiguro and M. Sasanuma, 1971, 3rd. Int. Conf. on VUV, ed. Y. Nakai (Physical Society of Japan, Tokyo).
Nenner, I., and G. Schulz, 1975, J. Chem. Phys. **62**, 1747.
Nenner, I., P.M. Guyon, T. Baer and T. Govers, 1980, J. Chem. Phys. **72**, 6587.
Nenner, I., P. Morin, M. Hubin-Franskin, J. Delwiche and S. Bodeur, 1987a, J. Molec. Struct., to be published.
Nenner, I., P. Lablanquie, M. Simon, P. Morin and A. Zewail, 1987b, to be published.

Newton, R., 1966, Scattering Theory of Waves and Particles (McGraw-Hill, New York).
Ninomiya, K., E. Ishiguro, S. Iwata, A. Mikuni and T. Sasaki, 1981, J. Phys. B **14**, 1777.
Oertel, H., H. Schenk and H. Baumgärtel, 1979, DESY Rep. SR-79/28.
Oertel, H., H. Schenk and H. Baumgärtel, 1980, Chem. Phys. **46**, 251.
Padial, N., G. Csanak, B.V. McKoy and P. Langhoff, 1981, Phys. Rev. A **23**, 218.
Parr, A., S. Soutworth, J.L. Dehmer and J. Holland, 1984, Nucl. Instrum. Methods **222**, 221.
Peatman, W., T. Borne and E. Schlag, 1969, Chem. Phys. Lett. **3**, 492.
Peatman, W., G. Kasting and D. Wilson, 1975, J. Electron Spectrosc. **7**, 233.
Peatman, W., B. Gotchev, P. Gürtler, E.-E. Koch and V. Saile, 1978, J. Chem. Phys. **69**, 2089.
Peatman, W., F. Wolf and R. Unwin, 1983, Chem. Phys. Lett. **95**, 453.
Petroff, Y., 1985, in: Atomic and Molecular Collisions in a Laser Field, eds J. Picqué, G. Spiess and F. Wuilleumier, (Les Éditions de Physique, France) p. C1-147 and references cited therein.
Piancastelli, M., D. Lindle, T.A. Ferrett and D.A. Shirley, 1987a, J. Chem. Phys. **86**, 2765.
Piancastelli, M., D. Lindle, T.A. Ferrett and D.A. Shirley, 1987b, J. Chem. Phys., in press.
Platzman, R.L., 1960, J. Phys. Radium **21**, 853.
Plummer, E., T. Gustafsson, W. Gudat and D. Eastman, 1977, Phys. Rev. A **15**, 2339.
Poliakoff, E., S. Soutworth, D.A. Shirley, K. Jackson and R. Zare, 1979, Chem. Phys. Lett. **65**, 407.
Poliakoff, E., J.L. Dehmer, D. Dill, A. Parr, K. Jackson and R. Zare, 1981, Phys. Rev. Lett. **46**, 907.
Poliakoff, E., J.L. Dehmer, A. Parr and G. Leroi, 1982, J. Chem. Phys. **77**, 5243.
Poliakoff, E., J.L. Dehmer, A. Parr and G. Leroi, 1984, Chem. Phys. Lett. **111**, 128.
Poliakoff, E., M. Hang Ho, G. Leroi and M. White, 1985, J. Chem. Phys. **85**, 5529.
Poliakoff, E., M. Hang Ho, G. Leroi and M. White, 1986a, J. Chem. Phys. **84**, 4779.
Poliakoff, E., J. Chan and M. White, 1986b, J. Chem. Phys. **85**, 6232.
Rabalais, J., 1977, Principles of Ultraviolet Photoelectron Spectroscopy (Wiley, New York).
Raoult, M., H. Le Rouzo, G. Raseev and H. Lefebvre-Brion, 1983, J. Phys. B **16**, 4601.
Raseev, G., 1985, in: Photophysics and Photochemistry above 6 eV, ed. F. Lahmani (Elsevier, Amsterdam) p. 47.
Raseev, G., H. Le Rouzo and H. Lefebvre-Brion, 1980, J. Chem. Phys. **72**, 570.
Raseev, G., B. Leyh and H. Lefebvre-Brion, 1986, Z. Phys. D **2**, 319.
Read, F., 1978, J. Phys. (France) **39**, C1 82.
Reimer, A., J. Schirmer, J. Feldhaus, A. Bradshaw, U. Becker, H.G. Kerkhoff, B. Langer, D. Szostak, K. Wehlirz and W. Braun, 1986, Phys. Rev. Lett. **57**, 1707.
Richard-Viard, M., 1982, Thèse (Université d'Orsay).
Richard-Viard, M., O. Dutuit, M. Lavollée, T. Govers, P.M. Guyon and J. Durup, 1985, J. Chem. Phys. **82**, 4054.
Richardson, P., J. Eland and P. Lablanquie, 1986, Org. Mass. Spectrom. **21**, 289.
Roche, A., K. Kirby, S. Guberman and A. Dalgarno, 1981, J. Electron Spectrosc. **22**, 223.
Roche, M., D. Salahub and R. Messmer, 1980, J. Electron Spectrosc. **19**, 273.
Rosenstock, H., R. Stockbauer and A. Parr, 1981, Int. J. of Mass Spectrosc. & Ion Phys. **38**, 323.
Roy, P., I. Nenner, M.Y. Adam, J. Delwiche, M.J. Hubin-Franskin, P. Lablanquie and D. Roy, 1984, Chem. Phys. Lett. **109**, 607.
Roy, P., I. Nenner, P. Millié, P. Morin and D. Roy, 1986, J. Chem. Phys. **84**, 2050.
Roy, P., I. Nenner, P. Millié, P. Morin and D. Roy, 1987, J. Chem. Phys., in press.
Samson, J., 1978, Nucl. Instrum. Methods **152**, 225.
Samson, J., and J. Gardner, 1973, J. Electron Spectrosc. **2**, 259.
Schäfers, F., M.A. Baig and U. Heinzmann, 1983, J. Phys. B **16**, L1.
Schlag, E., R. Frey, B. Gotchev, W. Peatman and H. Pollak, 1977, Chem. Phys. Lett. **51**, 406.
Schönhense, G., V. Dzidzonou, S. Kaesdorf and U. Heinzmann, 1984, Phys. Rev. Lett. **52**, 811.
Schulz, G., 1973, Rev. Mod. Phys. **45**, 423.
Schwarz, W., 1974, Angew. Chemie Int. Ed. **13**, 454.
Schwarz, W., 1975, Chem. Phys. **11**, 217.
Schwarz, W., and R. Buenker, 1976, Chem. Phys. **13**, 153.
Schwentner, N., E.-E Koch and J. Jortner, 1985, Electronic Excitations in Condensed Rare Gases, Springer Tracts in Modern Physics, Vol. 107 (Springer, Berlin).

Segev, E., and M. Shapiro, 1982, J. Chem. Phys. **77**, 5604.
Sette, F., J. Stöhr and A. Hitchcock, 1984, J. Chem. Phys. **81**, 4906.
Shapiro, M., and R. Bersohn, 1982, Ann. Rev. Phys. Chem. **33**, 409.
Sharp, T., 1971, At. Data **2**, 119.
Shaw, D., D. Cvejanovic, G.C. King and F.H. Reed, 1984, J. Phys. B **17**, 1173.
Siegbahn, K., C. Nordling, G. Johansson, J. Hedman, P. Heden, K, Hamrin, U. Gelius, T. Bergmark, L. Werme, R. Manne and Y. Baer, 1969, ESCA Applied to Free Molecules (North-Holland, Amsterdam).
Slanger, T., and G. Black, 1982, J. Chem. Phys. **77**, 2432.
Smith, A., 1970, J. Quant. Spectrosc. & Radiat. Transfer **10**, 1129.
Smith, N., and F. Himpsel, 1983, Photoelectron Spectroscopy, in: Handbook on Synchrotron Radiation, Vol. 1, ed. E.-E. Koch (North-Holland, Amsterdam) ch. 9.
Sodhi, R., and C. Brion, 1984, J. Electron Spectrosc. & Relat. Phenom. **34**, 363.
Sodhi, R., and C. Brion, 1985, J. Electron Spectrosc. & Relat. Phenom. **35**, 45.
Sodhi, R., S. Daviel, C. Brion and G. De Souza, 1985, J. Electron Spectrosc. **35**, 45.
Southworth, S.H., A.C. Parr, J. Hardis and J.L. Dehmer, 1986, Phys. Rev. A **33**, 1020.
Spohr, R., P.M. Guyon, W. Chupka and J. Berkowitz, 1971, Rev. Sci. Instrum. **42**, 1872.
Steele, G., 1976, Opt. Spectra, p. 37.
Stephens, A., and D. Dill, 1985, Phys. Rev. A **31**, 1968.
Stockbauer, R., 1980, Adv. Mass. Spectrom. **8**, 79.
Susuki, S., S. Nagaoka, I. Koyana, K. Tanaka and T. Kato, 1986, Z. Phys. D **4**, 111.
Suto, M., and L.C. Lee, 1983a, Chem. Phys. Lett. **98**, 152.
Suto, M., and L.C. Lee, 1983b, J. Chem. Phys. **78**, 4515.
Suto, M., and L.C. Lee, 1983c, J. Chem. Phys. **79**, 1127.
Swanson, J., D. Dill and J.L. Dehmer, 1980, J. Phys. B **13**, 1231.
Swanson, J., D. Dill and J.L. Dehmer, 1981, J. Phys. B **14**, 1207.
Swensson, S., P. Malmquist, M.Y. Adam, P. Lablanquie, P. Morin and I. Nenner, 1984, Chem. Phys. Lett. **111**, 574.
Tabché-Fouhailé, A., 1981, Thése (Université d'Orsay).
Tabché-Fouhailé, A., I. Nenner, P.M. Guyon and J. Delwiche, 1981, J. Chem. Phys. **75**, 1129.
Tabché-Fouhailé, A., M.J. Hubin-Franskin, J. Delwiche, H. Fröhlich, K. Ito, P.M. Guyon and I. Nenner, 1983, J. Chem. Phys. **79**, 5894.
Tanaka, K., and M. Yoshimine, 1979, J. Chem. Phys. **70**, 1626.
Thiel, W., 1982, Chem. Phys. Lett. **87**, 249.
Thiel, W., 1983, Chem. Phys. **77**, 103.
Thiel, W., 1984, J. Electron Spectrosc. **34**, 399.
Trevor, D., J. Pollard, W. Brewer, S. Southworth, C. Truesdale, D.A. Shirley and Y. Lee, 1984, J. Chem. Phys. **80**, 6083.
Tronc, M., R. Azria and Y. Le Coat, 1979, J. Phys. B **13**, 2327.
Truesdale, C., S. Southworth, P. Kobrin, D. Lindle and D.A. Shirley, 1983, J. Chem. Phys. **78**, 7117.
Truesdale, C., D. Lindle, P. Kobrin, U. Becker, H. Kerkhoff, P. Heimann, T. Ferrett and D.A. Shirley, 1984, J. Chem. Phys. **80**, 2319.
Tsai, B., and J. Eland, 1980, Int. J. Mass & Spectrosc. Ion Phys. **36**, 143.
Turner, D., A. Baker, C. Baker and C. Brundle, 1970, Molecular Photoelectron Spectroscopy (Wiley, London).
Ungier, L., and T.D. Thomas, 1983, Chem. Phys. Lett. **96**, 247.
Ungier, L., and T.D. Thomas, 1985, J. Chem. Phys. **82**, 3146.
Van der Wiel, M., and Th. El-Sherbini, 1972, Physica **59**, 453.
Van der Wiel, M., Th. El-Sherbini and C. Brion, 1970, Chem. Phys. Lett. **7**, 161.
Wallace, S., D. Dill and J.L. Dehmer, 1979, J. Phys. B **12**, L417.
Walters, E., J. Grover, M. White and E. Hui, 1985, J. Phys. Chem. **89**, 3814.
Wannier, G., 1953, Phys. Rev. **90**, 817.
Weigold, E., and I. McCarthy, 1978, Adv. At. & Mol. Phys. **14**, 127.
Wendin, G., 1979, Int. J. Quantum Chem. **13**, 659.

West, J.B., A. Parr, B. Cole, D. Ederer, R. Stockbauer and J.L. Dehmer, 1980, J. Phys. B **13**, L105.
West, J.B., K. Codling, A. Parr, D. Ederer, B. Cole, R. Stockbaer and J.L. Dehmer, 1981, J. Phys. B **14**, 1791.
White, M., and J. Grover, 1983, J. Chem. Phys. **79**, 4124.
White, M., and J. Grover, 1985, Proc. Conf. on Autoionization of Atoms and Small Molecules, ed. J. Berkowitz (Argonne National Laboratory, Argonne, IL).
White, M., R. Rosenberg, G. Gabor, E. Poliakoff, G. Thornton, S. Southworth and D.A. Shirley, 1979, Rev, Sci. Instrum. **50**, 1288.
Wight, G., and C. Brion, 1974, J. Electron Spectrosc. **4**, 335.
Wight, G., M. Van der Wiel and C. Brion, 1972, J. Electron Spectrosc. **1**, 454.
Wilcke, H., W. Böhmer, R. Haensel and N. Schwentner, 1983, Nucl. Instrum. Methods **208**, 59.
Woodruff, P., and G.V. Marr, 1977, Proc. R. Soc. London Ser. A **358**, 87.
Wu, C., and D. Judge, 1981, J. Chem. Phys. **75**, 172.
Zimmerer, G., 1985, in: Photophysics and Photochemistry above 6 eV, ed. F. Lahmani (Elsevier, Amsterdam) p. 357.

CHAPTER 7

SURFACE SCIENCE WITH SYNCHROTRON RADIATION

I.T. McGOVERN
Department of Pure and Applied Physics, Trinity College, Dublin, Republic of Ireland

D. NORMAN
SERC Daresbury Laboratory, Warrington, UK

R.H. WILLIAMS
Physics Department, University College, Cardiff, UK

Contents

1. Introduction	469
2. Techniques	469
2.1. Photoemission	469
2.2. SEXAFS and XANES	471
2.3. Photon-stimulated ion desorption	473
2.4. X-ray methods	473
3. Clean surfaces	476
3.1. Bulk electronic structure	476
3.1.1. Photoemission cross-section effects	476
3.1.2. Mapping of bulk bandstructure	478
3.1.3. Polarisation-dependent selection rules	478
3.2. Special topics	480
3.2.1. Magnetism	480
3.2.2. Mixed valence	483
3.3. Surface states on metals	484
3.3.1. Itinerant surface states	485
3.3.2. Surface core shifts	487

Contents continued overleaf

Handbook on Synchrotron Radiation, Vol. 2, edited by G.V. Marr
© *Elsevier Science Publishers B.V., 1987*

Contents continued

- 3.4. Surface states on semiconductors ... 489
  - 3.4.1. Itinerant surface states on Ge(111)(2 × 1) ... 490
  - 3.4.2. Itinerant surface states on Ge(001) ... 490
  - 3.4.3. Surface core shifts on GaAs(110) ... 492
- 3.5. Clean surface geometry ... 494
- 3.6. Small metal clusters ... 498
4. Surfaces with adsorbates ... 500
  - 4.1. Chemisorption ... 500
    - 4.1.1. Hydrogen ... 500
    - 4.1.2. Oxygen ... 502
    - 4.1.3. Chlorine ... 503
    - 4.1.4. Carbon monoxide ... 504
    - 4.1.5. Water vapour ... 506
    - 4.1.6. Sulphur ... 508
  - 4.2. Physisorption ... 511
  - 4.3. Geometry of adsorbate structures ... 513
    - 4.3.1. Chemical effects on Pauling radii ... 513
    - 4.3.2. Non-standard bonding sites ... 514
    - 4.3.3. Adsorbate-induced changes in substrate geometry ... 515
    - 4.3.4. Geometry of molecular adsorbates ... 515
  - 4.4. Reactions ... 519
5. Interfaces ... 521
  - 5.1. Metal–metal interfaces ... 521
  - 5.2. Metal–semiconductor interfaces ... 522
    - 5.2.1. Abrupt interfaces ... 522
    - 5.2.2. Non-abrupt interfaces ... 525
    - 5.2.3. Schottky barrier height measurements ... 526
  - 5.3. Semiconductor–semiconductor interfaces ... 527
    - 5.3.1. Interface band offsets ... 528
    - 5.3.2. Interface reactivity ... 530
  - 5.4. Insulator–semiconductor interfaces ... 533
6. Conclusions ... 533
References ... 533

1. Introduction

The study of solid surfaces is one of the most prominent areas of research at almost all of the world's synchrotron radiation laboratories. However, it is evident to the non-specialist that surface science is a particularly difficult, time-consuming and expensive discipline, and that to combine such a subject with the problems of working with synchrotron radiation must only lessen the likelihood of successful experiments. Why, then, is the topic of surface science with synchrotron radiation pursued so widely and so arduously? The answer lies both in the great importance of surface science and in the advantages which use of synchrotron radiation can bring to investigations of surfaces. Many modern technological processes, such as catalysis, semiconductor device fabrication and inhibition of corrosion, are fundamentally controlled by surface properties. There is now widespread acceptance that full understanding of a surface can be achieved only by a combination of experimental (and theoretical) approaches, and synchrotron radiation has considerably widened the range of techniques which can be brought to bear on surface problems.

This chapter discusses the application of synchrotron radiation techniques to the study of solid surfaces and interfaces. It does so largely through illustrative examples of specific studies. These examples have been taken from an extensive and rapidly expanding literature; it is emphasised that this is not a review of that literature, nor is it a judgemental exercise! What is intended is a fairly simple overview of a complex subject and one which is directed at a more general readership than the synchrotron radiation surface science community.

The layout of this chapter is as follows. The next section outlines the major techniques, with brief illustrations of their usefulness. The following three sections deal with applications to the three broad areas of modern surface science: section 3 covers clean surfaces; section 4 deals with surfaces with adsorbates; section 5 treats general interfaces.

2. Techniques

In this section synchrotron radiation techniques with surface science applications are discussed in a general way. Details of the experimental methods can be obtained through the references in sections 3 to 5. Four areas are considered, viz., photoemission (section 2.1), SEXAFS and XANES (section 2.2), photon-stimulated ion desorption (section 2.3) and X-ray methods (section 2.4).

2.1. Photoemission

The basic physics of the photoemission process is simple: When monochromatic light is shone upon a sample, some of the photons are absorbed and give their energy to electrons, which may then be emitted from the specimen. Analysis of the kinetic energy and angles of emission of the electrons then gives information on the binding energy and momentum of electron states within the sample. Mean free paths for electrons in solids are rather short – typically of the order of two atomic layers for electrons with kinetic energies in the range 20 to 100 eV (see fig. 1) – and so photoexcited electrons can emerge without loss of energy only if they were created near to the surface. Photoemission from solids is thus inherently a surface-sensitive technique.

Although experiments can be accomplished with conventional light sources [where the subject has traditionally been divided, according to the type of photon source, into ultra-violet photoelectron spectroscopy (UPS) and X-ray photoelectron spectroscopy (XPS)], the use of synchrotron radiation has revolutionised the application of the photoemission technique to surface science. Continuous tunability across a wide photon energy range allows matching of binding energy and photon energy so that cross section and/or surface sensitivity may be enhanced; it also makes possible different ways of conducting the experiment such as the constant final-state (CFS) or constant initial-state mode (CIS), both of which call for scanning of the photon energy. The natural polarisation of synchrotron radiation is invaluable in deducing the symmetry and orbital character of electron states via the application of selection rules. Finally, and probably of greatest importance, the high intensity of synchrotron radiation allows high resolution of the energy, the angle of emission or the spin polarisation of photoelectrons.

The subject of photoelectron spectroscopy using synchrotron radiation has been well covered by Smith and Himpsel in Vol. 1 of this Handbook (Smith and Himpsel 1983). In consequence, further details are not reproduced here. Other general reviews of photoemission include Plummer and Eberhardt (1982), Feuer-

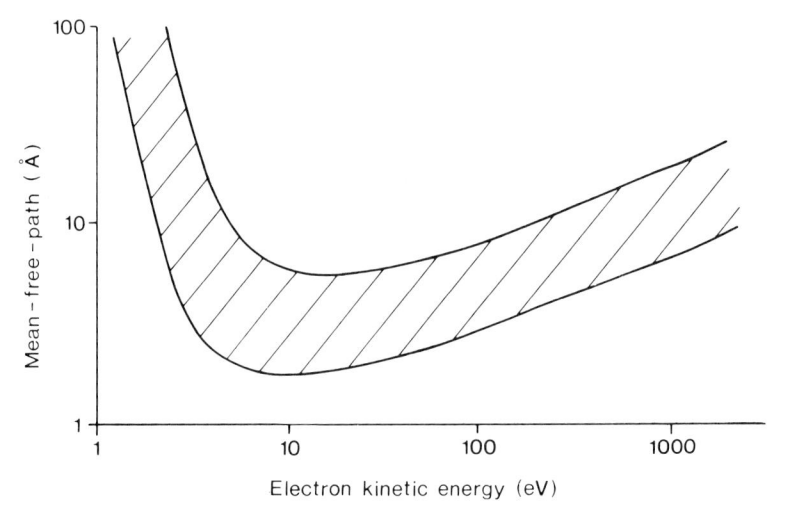

Fig. 1. Mean free path range of electrons in solids as a function of energy.

bacher et al. (1978) and Williams et al. (1980). For discussion of spin-polarised photoemission, the reader is referred to the recent review by Kisker (1983). Photoemission continues to be the most important single technique in the investigation of surfaces using synchrotron radiation.

2.2. SEXAFS and XANES

X-ray absorption spectroscopy has for a long time been used in the study of solids; its application to surfaces is, however, a fairly recent phenomenon, and one which has been directly spurred by the availability of synchrotron radiation. This subject has recently been extensively reviewed by Norman (1986) and Stöhr (1987), and the reader is referred to these articles for details of the technique, its advantages and pitfalls, and examples of its range of application.

The basic physical processes involved in X-ray absorption are depicted schematically in fig. 2. The energies of the discrete core electron levels are uniquely determined by the atom type, so tuning the photon energy to a particular core level immediately gives an *atom-specific* probe. When its photon energy is greater than the binding energy of the core level, a photon may be absorbed, giving its energy to a photoelectron. In a molecule or solid, part of the photoelectron wave may be backscattered from neighbouring atoms, with consequent modulation of the matrix element for the absorption. The backscattered wave will interfere with the outgoing wave, and the interference may be constructive or destructive, depending on the wavevector and distance. The modulated part of the absorption coefficient thus contains information on the distance, number and type of atoms surrounding the absorbing atom. Optical (dipole)

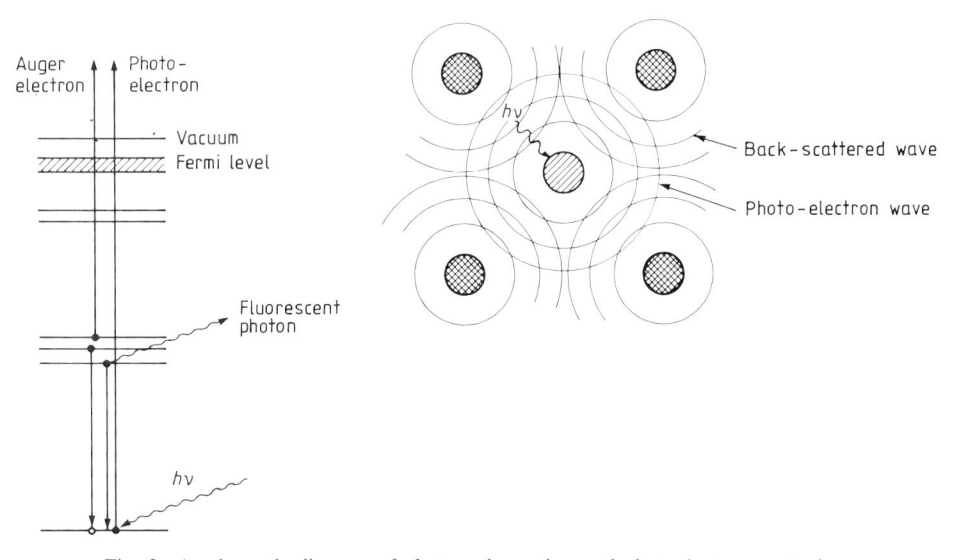

Fig. 2. A schematic diagram of photon absorption and photoelectron scattering.

selection rules apply to the photon absorption, so that varying the angle of incidence of plane-polarised light onto a surface can give particularly valuable information on adsorption sites.

Far above the absorption edge, in the EXAFS (Extended X-ray Absorption Fine Structure) regime with electron kinetic energies of ca. 50 eV or more, a single scattering analysis is usually applicable and similar methods to that of bulk EXAFS [see article by Stern and Heald (1983) in Vol. 1 of this Handbook] can be used. In the XANES (X-ray Absorption Near-Edge Structure) region, electron mean free paths are longer and scattering is stronger, making multiple scattering more important and complicating the interpretation. It has been shown for Ni(100)$c(2 \times 2)$O (Norman et al. 1983) and for Ni(110)$c(2 \times 2)$S (Norman et al. 1986) that the XANES spectra of *atomic* adsorbates are not dominated by just the nearest-neighbour (nn) substrate atoms but rather that the scattering from as many as 30 neighbour atoms has to be included in the calculation before convergence is achieved. This is in contrast to the near-edge structure of *molecular* adsorbates, where the spectra are dominated by intramolecular scattering with only small or negligible scattering contributions from the surface substrate atoms. Deduction of the orientation and bonding of molecules on surfaces can thus be quite straightforward (Stöhr and Jaeger 1982a).

The primary requirement for surface X-ray absorption studies is a photon beam which is intense ($>10^{10}$ photons/s), monochromatic (ca. 0.5–5 eV resolution), linearly polarised, tunable over a wide range (ca. 250–10 000 eV), and preferably in ultra high vacuum to enable experiments on well-characterised surfaces. The only source with these properties is synchrotron radiation with a suitable monochromator. Rather few facilities for X-ray absorption measurements at surfaces are currently available, and details are in the literature (e.g. Stöhr et al. 1980, Hussain et al. 1982, MacDowell et al. 1986).

The experiment consists of measuring the intensity of photons incident on the sample, and the proportion of them which is absorbed, as a function of photon energy. To enhance surface specificity, most SEXAFS experiments have involved detection of the absorption coefficient by indirect means. The core hole created is subsequently annihilated, either by fluorescence, with emission of another photon, or in the Auger process, with emission of an Auger electron. Measurement of the yield of Auger electrons or fluorescent photons thus gives the absorption coefficient. Most Auger electrons lose energy on their way out of the solid, creating a cascade of lower energy electrons, whose cross section is still proportional to the number of core holes produced; detection of these secondary electrons is thus a valid way of measuring SEXAFS or surface XANES spectra.

The key to the usefulness of SEXAFS is that it is a short-range probe, requiring no long-range order. Thus a reaction may be followed through its various stages (e.g. Norman et al. 1981), for instance starting from initial chemisorption of isolated adatoms on a single crystal, through the formation of ordered overlayers, and then growth of a bulk compound on the surface. No other technique can follow the detailed geometry through all these phases.

2.3. Photon-stimulated ion desorption

It has long been known that ions could be desorbed from a surface by an electron beam in the process known as electron-stimulated desorption (ESD), but the existence of direct photon-stimulated desorption (PSD) of ions – as opposed to ESD by photoexcited electrons – was in dispute until unequivocal evidence was obtained using synchrotron radiation. Interest in the technique, and in ESD, was revived by a new suggestion for the process initiating the desorption (Knotek and Feibelman 1978). The absorption of a photon (or scattering of an electron) at a core level produces a hole which is then filled by Auger decay; by this mechanism two or more holes may be localised in the valence levels of the surface atom, reversing its charge and leading to its desorption as a positive ion following the graphically named "Coulomb explosion". As originally introduced, this mechanism applied only to ionically bonded solids and to adsorbates in an ionic state on a substrate. However, it has since been empirically determined (see review by Knotek 1984) that some covalently bonded species (e.g. H on Si) exhibit ion desorption behaviour similarly correlated with core level thresholds. Since the Auger filling of the core hole may be either intra-atomic or interatomic, adsorbed atoms may be desorbed following excitation of a core level in either the adsorbate itself, or in a substrate atom to which the absorbate is bonded. PSD thus gives a means, via selective excitation of substrate species, of determining the type of atom to which an adsorbate is bonded (in a compound or heterogeneous catalyst for example).

The rate of production of ions (γ) above threshold is given by $(1-f)P\mu$, where P is the probability of Auger transitions leading to a repulsive final state, f is the probability of reneutralisation for the ion and μ is the photon absorption coefficient. P and f are independent of photon energy, and so $\gamma \propto \mu$; thus the yield of ions can be used as a measure of the absorption coefficient, and may even be used to monitor SEXAFS. However, it should be noted that f depends strongly on the overlap of wavefunctions of the repelled ion and its neighbours and so the ion yield can be highly site specific. In particular, $f \sim 1$ for bulk atoms, making ion yield very surface sensitive, and $f \to 1$ for many covalent bonds. Thus we should be aware that ion desorption may pick out a minority surface species and particularly an ionic one.

Synchrotron radiation is well suited to study PSD: the strong edge effects exploit the tunability; the intensity is important for the low cross section of desorption of the order of 10^{-6} ions/photon or much less; the time structure of the photon flux facilitates "time of flight" ion mass spectrometers (Owen et al. 1986).

2.4. X-ray methods

Scattering of X-rays has been for many years the standard method of determining the crystallography of bulk samples, but it is only recently that the same techniques have been applied to surface structures. X-rays have bulk attenuation lengths of the order of microns, so that Bragg scattering involves thousands of atoms; since the peak intensity of a diffracted beam is proportional to the square of the number of atoms involved, the peak intensity of X-rays diffracted from a monolayer is lower than from the bulk by at least a factor of 10^7. It is therefore not surprising that X-ray diffraction has not found routine use as a surface structural tool. However, the surface sensitivity can be greatly enhanced by making the X-ray beam incident on the sample at a very grazing angle and this is the key to the experimental approaches now being adopted. Indeed, below a certain critical angle, X-rays will undergo total external reflection, limiting the photon penetration of the surface to their extinction length, typically 20 Å. Working at very small grazing angles of course presents severe experimental problems, particularly with respect to the beam size and the flatness and the necessarily precise alignment of the sample. The use of intense synchrotron radiation sources, with the natural high collimation of the photon beam, facilitates such measurements.

The potential advantages of grazing-incidence X-ray scattering lie in two areas. Theoretically, it should be relatively easy to evaluate surface structures, since – unlike electrons – X-rays are only weakly scattered and analysis should be possible via direct Fourier inversion of diffracted intensities. Experimentally, X-ray scattering is sensitive to long-range order across the surface, and, most importantly, may be used to study clean surfaces, or, since the photons will penetrate a less dense overlayer, solid–solid interfaces (such as semiconductor heterojunctions) or solid–liquid interfaces.

One example serves to illustrate the usefulness of the technique. Marra and co-workers (Marra et al. 1982, Brennan 1985) have studied the structure of layers of Pb on Cu(110), in particular the thermodynamics of the surface melting of the lead overlayer. A second-order phase transition shows a continuous change in scattered intensity as a function of temperature, as is seen in fig. 3a for a monolayer coverage of Pb. By contrast, fig. 3b shows the scattering from a surface covered with several layers of lead, where a change in temperature from 245°C to 247°C is sufficient to reduce the peak scattered intensity from 3×10^5 counts per second to less than 1 count per second; a rather dramatic example of a first-order, discontinuous, phase transition. Finally, mention should be made of the exciting possibility of performing time-resolved surface X-ray scattering studies – if the photon source is bright enough – as has been demonstrated by Larson et al. (1982) in their study of the pulsed laser annealing of silicon.

A further application of X-ray techniques to the determination of surface structures is via the use of interference to produce X-ray standing waves parallel to a crystal surface (Materlik 1985). As the incident beam angle is scanned through the region of total reflectivity, the standing waves outside the crystal surface will move, and as a node (or antinode) passes through the position of an atomic overlayer of impurity atoms, the characteristic fluorescence signal from these overlayer atoms will go through a minimum (or maximum). The technique is experimentally elegant but is limited to substrates of sufficiently perfect

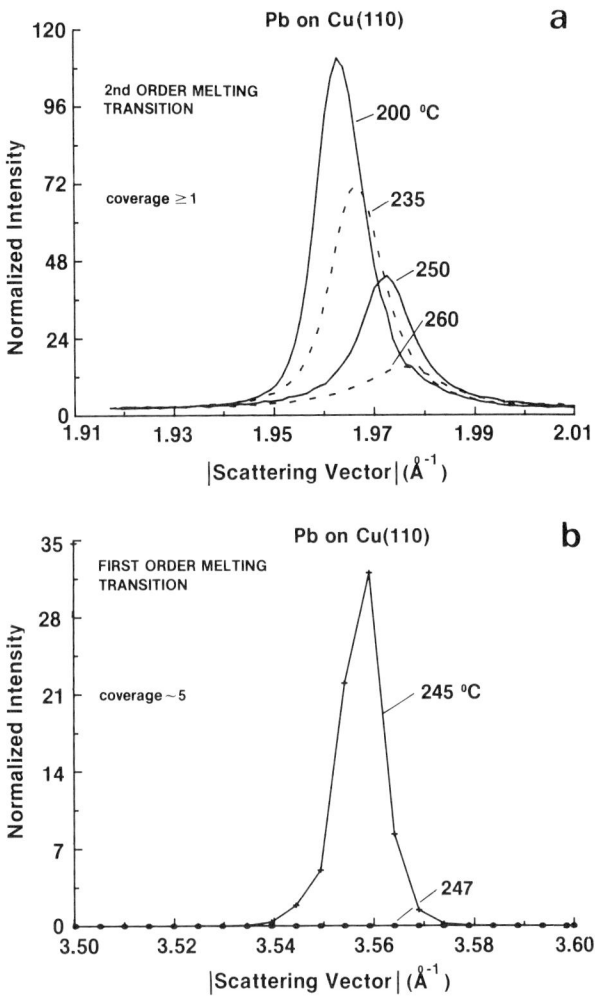

Fig. 3. (a) Radial scans of the monolayer commensurate phase of Pb on Cu(100) as a function of temperature as the structure undergoes a second-order reversible melting transition. The shift in peak position corresponds to a decrease in the interplanar distance of 0.7%. These planes are perpendicular to the (110) direction. (After Brennan 1985.) (b) Radial scans of a multilayer coverage of Pb for two temperatures, 245 and 247°C. The curves represent scattering from bulk-like lead planes at 3.5 Å. The circles at the bottom of the figure represent the scattering from these same planes after melting and the intensity is less than a count per second, indicating a first-order reversible phase transition has occurred. These planes are 30° off the (110) direction. (After Brennan 1985.)

crystallinity to enable their use as a component of an X-ray interferometer. We note in passing that this technique actually measures the spacing of the overlayer atoms from a continuation of the perfect bulk lattice rather than from the real surface layer, which may well be relaxed or reconstructed in some way. Thus the use of X-ray standing waves in conjunction with some other technique might allow deductions on the presence and the nature of such a surface reconstruction.

3. Clean surfaces

Surface science begins with the clean surface. Usually this implies an ordered, low-index face of a single crystal, stoichiometric if a compound and containing less than 1% of adsorbed impurities. In themselves such surfaces present formidable experimental and theoretical challenges. The key areas of interest are surface electronic states (sections 3.3 and 3.4) and surface geometry (section 3.5). However, through the agency of photoemission, surface scientists have become major investigators of *bulk* electronic properties; indeed, bulk and surface are intimately related in experiment and theory. Accordingly, this section begins with a *brief* discussion of photoemission spectroscopy of bulk electronic structure (section 3.1) and of the special topics of magnetism and mixed valence (section 3.2).

3.1. Bulk electronic structure

Photoemission spectroscopy can be performed in two collection modes, "angle-integrated" and "angle-resolved". Whereas the latter selects emission in a narrow cone (half-angle of order one degree), the former is liberally applied to the wide range of collection geometries which are not "angle-resolved". In valence photoemission, the angle-integrated spectra broadly reflect electronic density of states, while the angle-resolved spectra carry the details of band dispersion. That continuous photon tunability enhances both regimes is discussed in this section; differential photoemission cross-section effects in angle-integrated spectra can reveal partial densities of states (section 3.1.1), the variable photon energy is the third parameter necessary for three-dimensional bandstructure determination (section 3.1.2). Polarisation properties facilitate the application of selection rules in angle-resolved photoemission (section 3.1.3).

3.1.1. Photoemission cross-section effects
Three main cross-section effects have been exploited to distinguish valence orbital contributions. These are the Cooper minimum (Cooper 1964), the delayed onset and resonant photoemission (Williams et al. 1980). At the Cooper minimum the theoretical atomic cross section goes to zero due to cancellation of positive and negative contributions to the matrix element. The number of Cooper minima is determined by the $n - (l + 1)$ nodes in the radial part of the nl wavefunction. The layered solid MoS_2 provides a ready example of this effect. The lowest binding energy valence feature is Mo-4d derived, while the remaining features are largely S-3p derived (McGovern et al. 1981); the relative intensity of the valence band edge feature is markedly reduced at a photon energy of 110 eV. That this atomic effect can be strongly modified in the solid state is shown in fig. 4a. The lower panel shows the characteristic Cooper minimum for Mo-4d in MoS_2, while the upper panel shows that obtained for Mo-4d in polycrystalline molybdenum. There is an obvious quenching of the minimum, which has been qualitatively related to the different bonding in the two systems (Abbati et al. 1985).

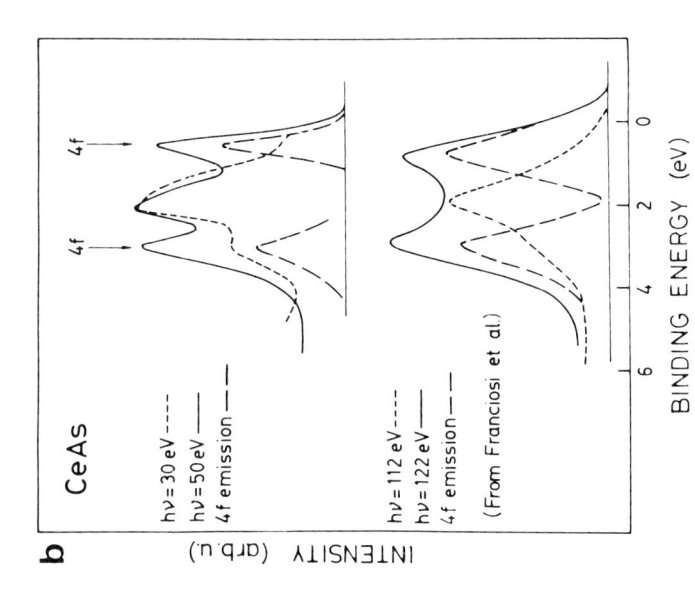

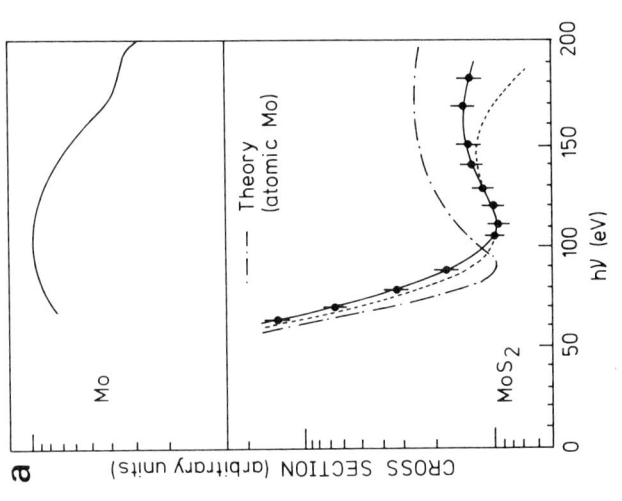

Fig. 4. (a) Logarithmic plot in arbitrary units of photoionization cross sections σ. Lower panel: experimental σ for Mo-4d_{z^2} orbitals in 2H-MoS$_2$, corrected for diffractive effects (solid line) and uncorrected (dotted line), and theoretical σ for isolated Mo atoms (dot–dashed line). Upper panel: measured σ for polycrystalline bcc Mo. The error bars give the statistical errors; the measured σ could have systematic distortions due to inaccuracies of photon flux evaluations; this could produce slowly varying distortions ($\pm 10\%$) over the whole $h\nu$ range. The theoretical σ includes a small $h\nu$ dependence due to the asymmetry parameter in the geometry of the experiment. (After Abbati et al. 1985.) (b) Synchrotron radiation excited spectra for CeAs, which demonstrate two different ways of delineating the 4f spectral intensity. (After Mårtensson 1985.)

The centrifugal barrier terms in the potential results in reduced cross section near threshold for higher l orbitals. This "delayed onset" effect has been exploited by Franciosi et al. (1981) to distinguish 4f and 5d valence contributions in CeAs. As shown in fig. 4b (upper panel), the 4f is reduced relative to the 5d as the photon energy is lowered from 50 to 30 eV. Delayed onset of 4f combines with a higher 5d cross section in this region to distinguish the two orbitals. The 4f contribution is resonantly enhanced at 122 eV, the threshold for 4d excitation; this excitation can decay via a super Coster–Kronig process to the same final state as the direct 4f emission. As shown in fig. 4b (lower panel), the separate 4f and 5d contributions to CeAs are further confirmed.

Photoemission cross-section effects are widely used to distinguish overlapping adsorbate and substrate valence emission. Resonant photoemission is discussed in more detail in ch. 10 of this Volume (Jugnet et al. 1987).

3.1.2. Mapping of bulk bandstructure

The measurement of photoelectron momentum can determine the electronic bandstructure of a solid. Fundamental difficulties with momentum conservation perpendicular to the surface $(k_\perp)$ can be overcome by using a calculated final-state bandstructure or by employing a free-electron final state over an appropriate photon energy range. In a classic example of the latter approach Chiang et al. (1979) probed normal emission $(k_\parallel = 0)$ for GaAs(110) over the photon energy range 25 to 100 eV. The dispersion of initial-state features was transformed to the bandstructure shown in fig. 5a via the relation

$$\hbar k_\perp = [2m(E_k - V_0)]^{1/2},$$

where E_k is the kinetic energy of emission relative to the vacuum level and V_0 is the inner potential. There is excellent agreement with the pseudopotential calculation of Pandey and Phillips (1974), as shown.

Similar methods have been employed to determine bandstructures of metals, semiconductors and semi-insulators. The one-electron approach assumed here may not always be adequate; for narrow-band metals, hole–hole correlation effects must be considered (Liebsch 1981); other many-body effects have been investigated (Plummer 1985) but in general, the one-electron approach is reasonably accurate. A bandstructure determination requires only a sample that does not charge and for which a clean ordered surface can be prepared (Cerrina et al. 1984).

3.1.3. Polarisation-dependent selection rules

The provision of linearly polarised light in the plane of the orbiting electrons allows state symmetry to be probed via selection rules (Eberhardt and Himpsel 1980). Of particular importance are the mirror plane emission and normal emission regimes, where the even symmetry of the final state requires the parity of the initial state to match that of the operator (Hermanson 1977). The operator

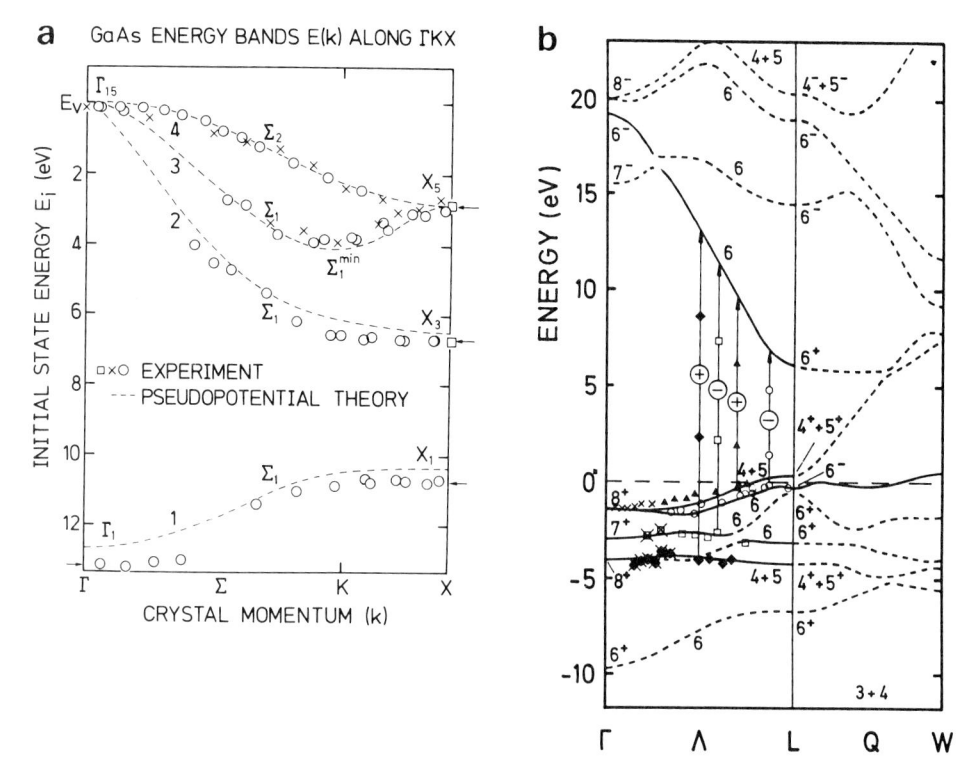

Fig. 5. (a) Band dispersion $E(k)$ of GaAs along the [110] direction. Circles are experimental points for normal emission, crosses and squares are determined from off-normal data, and the dashed curves are theoretical dispersion curves for valence bands. The symmetry characters of the bands and critical points (horizontal arrows) are labelled, $\Gamma KK = 1.57 \text{ Å}^{-1}$. (After Chiang et al. 1979.) (b) Symmetry-resolved band mapping of Pt in comparison with the calculated bandstructure; the band mapping procedure used the calculated final band No. 7 for photon energies <20 eV and a free-electron parabola (not shown) with $V_0 = -1.85$ eV and $m^* = 1.10 m_e$ for $h\nu > 20$ eV, which yields the mapping points without and with crosses, respectively. (After Eyers et al. 1984.)

$(A \cdot P)$ has even (odd) parity when A is parallel (perpendicular) to the mirror plane. These methods have been extensively applied to bulk, surface and adsorbate band emission (Dietz and Himpsel 1979, Plummer and Eberhardt 1982).

Spin-polarised photoemission provides further selection rules which are appropriate to non-magnetic spin–orbit interacting systems. These relativistic selection rules exploit the circularly polarised properties of the synchrotron radiation flux emitted out of the orbit plane. A novel monochromator/spectrometer facility at BESSY has been used to probe the symmetry of the bulk bandstructure of platinum(111). The development of spin-resolved contributions to normal emission with increasing photon energy is transformed into the symmetry-resolved band map shown in fig. 5b (Eyers et al. 1984).

3.2. Special topics

Magnetism and mixed valence have become special topics in the photoemission spectroscopy of the bulk electronic structure. For magnetism the coupling of precise bandstructure measurements with spin-polarisation analysis has been extremely beneficial; for mixed valence it is the combination of high resolution, surface sensitivity and cross-section effects.

3.2.1. Magnetism

The electronic bandstructure of a 3d ferromagnetic metal (viz, iron, cobalt and nickel) is exchange split; in effect, there are two bandstructures, one for majority-spin electrons and another for minority-spin electrons. Near the Fermi level the shape of the bands and low hole lifetime broadening allows this energy splitting to be directly observed (Eastman et al. 1980a, Turner et al. 1984). Figure 6 (Eastman et al. 1980a) illustrates this approach, from which critical point values of the exchange splitting for iron(111), cobalt(0001) and nickel(111) were deter-

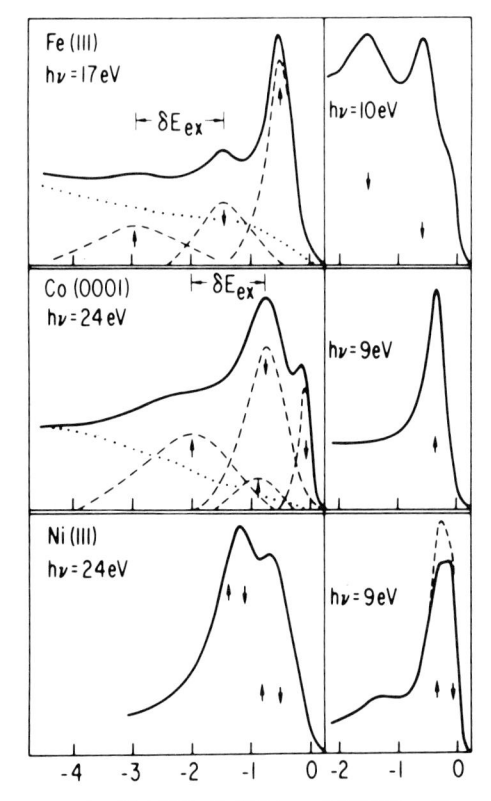

Normal-emission spectra for Fe(111), Co(001), and Ni(111) with use of s-polarized radiation. ↕ indicate spectral peak locations, while exchange splittings for Fe and Co are indicated. (After Eastman et al. 1980a.)

mined as 1.5 eV at P, 1.1 eV at Γ and 0.3 eV near L, respectively. The more direct spin-polarisation measurement considerably enhances this analysis. Sample spin- and energy-resolved normal emission spectra for Fe(100) (Kisker et al. 1984) are shown in fig. 7. The minority-spin peak near E_F and the majority-spin peak $\sim$3 eV below E_F are the exchange-split states of Γ'_{25} symmetry, yielding an exchange splitting of 2.2 eV. It is necessary to include self-energy corrections to these one-electron bandstructures, particularly in the case of nickel (Liebsch 1981).

Exhange splitting is fundamental to a Stoner-type model of these ferromagnets. However, a major difficulty is encountered in the temperature dependence of exchange splitting. In a Stoner-type model, exchange splitting should vanish at T_C, the Curie temperature. The temperature dependence of photoemission and spin-resolved photoemission indicates that this does not happen. Rather, the long-range magnetic order is replaced by short-range magnetic order, in which microscopic moments fluctuate in direction while maintaining magnitude. Spin-polarised measurements for Fe(100) near T_C are also shown in fig. 7. Not only is the exchange splitting maintained, but a new structure is seen in the minority-spin spectrum at the energy of the majority-spin feature (Kisker et al. 1984). This is

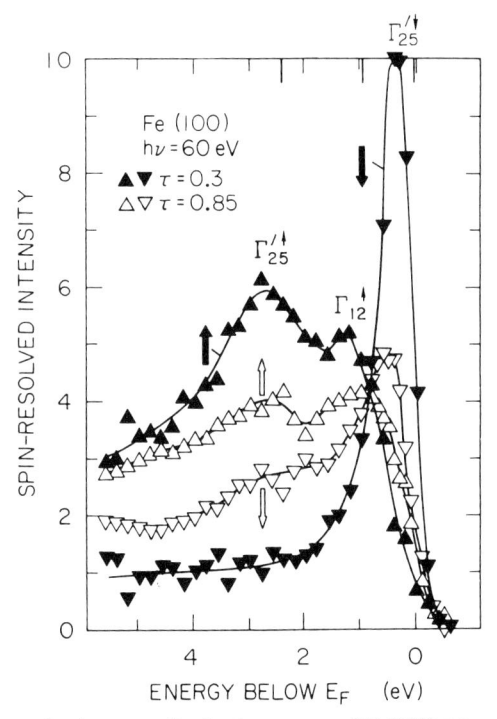

Fig. 7. Spin- and angle-resolved energy distribution curves of Fe(100) taken at 60 eV photon energy for two different temperatures, $\tau = T/T_c = 0.3$ and 0.85, for normal emission and s-polarized light (unsmoothed data). The arrows labelling the curves refer to the direction of the spontaneous magnetization. (After Kisker et al. 1984.)

taken as evidence of regions which are magnetised in a different direction from the macroscopic low-temperature direction of spontaneous magnetisation.

These measurements have generated considerable debate as to the spatial extent of the microscopic regions. The "local band" (Korenman et al. 1977) and "disordered local moment" (Pindor et al. 1983) models represent extremes of substantial and zero correlations. Comparisons with calculations have indicated short-range magnetic order of at least 4 Å in iron (Haines et al. 1985) and a substantially larger 20 Å in nickel (Korenman and Prange 1984). That the study of itinerant magnetism has entered an exciting phase is due in no small measure to these synchrotron radiation based techniques. For further details of temperature effects see the review by Kisker et al. (1985).

A direct surface issue is that of magnetic "dead layers". The question has been posed whether the surface of a magnetic material is magnetic or non-magnetic (Liebermann et al. 1969). This has a theoretical basis in Landau's (1937) postulation of the inability of an infinite two-dimensional system to support long-range magnetic order at non-zero temperature. Surface-sensitive photoemission and spin-resolved photoemission of single crystals do not show much evidence for this effect. Indeed, exchange-split surface states have been identified for nickel (Eberhardt et al. 1980) and for iron (Turner and Erskine 1983). It is argued that these surface layers can interact with the magnetic bulk, rendering Landau's principle inoperative. An alternative scheme is to investigate a monolayer of the metal of interest deposited on a non-magnetic substrate. For example, Miranda et al. (1982) have determined the photoemission exchange splitting for monolayer cobalt on copper(111) to be the same as that of bulk cobalt. On the other hand, Hezaveh et al. (1986) observed no exchange splitting for monolayer iron on copper(001). This is in agreement with predicted loss of long-range magnetic order in the largely fcc structure of the film, compared with the bcc structure of bulk iron. Binns et al. (1985) have compared monolayer iron on Pd(111) and Ag(111) substrates. That the former appears to be a "dead layer" is attributed to coupling between the adlayer and substrate. These latter measurements probe the magnetism through exchange (3s–3d) splitting of the 3s core emission. Surface enhancement of the magnetic properties of chromium has also been studied in this way (Klebanoff and Shirley 1986). It has recently been suggested that this exchange effect offers the possibility of local source, spin-resolved diffraction studies (Sinkovic et al. 1985).

Similar techniques have been applied to magnetic systems other than 3d ferromagnets. Johansson et al. (1980) have followed the antiferromagnetic/paramagnetic phase transition in chromium via photoemission. Temperature-dependent measurements of the antiferromagnetic compound uranium nitride have been reported (Reihl et al. 1983). The surface magnetism of 4f-ferromagnetic gadolinium(0001) has been probed by spin-polarised photoemission through the 4f surface core shift (see section 3.3 below); the observed spin polarisation is best fitted by assuming that there is antiferromagnetic coupling between surface and bulk spins. This surface is also found to exhibit enhanced magnetic order (Weller et al. 1985).

3.2.2. Mixed valence

Rare earth compounds exhibit the phenomenon of mixed or intermediate valence; that is, a ground state in which the rare earth atom has on average a non-integral number of 4f electrons. For example homogeneously mixed-valence $YbAl_2$ has a mean valence of ~2.4 resulting from a slow (relative to photoemission timescale) fluctuation between two configurations $4f^{14}V^2$ (divalent) and $4f^{13}V^3$ (trivalent); in effect, a 4f electron is being "promoted" to a valence level V. In a photoemission spectrum each configuration contributes a multiplet structure. The lower-valence structure will be at the Fermi level (or nearby due to final-state effects) with the higher-valence structure at a higher binding energy. The mean valence is simply obtained from the relative intensities of these spectral features. The mixed-valence state is associated with special magnetic, electrical and structural properties (Campagna et al. 1978, Mårtensson 1985).

Surface science enters this spectroscopy directly in that the bulk valence is often modified at the surface. This can lead to an incorrect estimation of bulk valence by photoemission. Early XPS studies of samarium (Wertheim and Crecelius 1978) indicated a surface origin for the observed divalent signal through angle of emission escape depth variability. However, as discussed in section 3.3 below, surface core emission is energy shifted from that of the bulk. With synchrotron radiation it is then possible to distinguish bulk and surface valence on the basis of binding energy. The example of $YbAl_2$ is illustrated in fig. 8. The strong divalent emission near E_F comprises bulk and surface doublets (final-state $4f^{13}$); the weaker bulk trivalent emission located between 6 and 11 eV binding energy is a

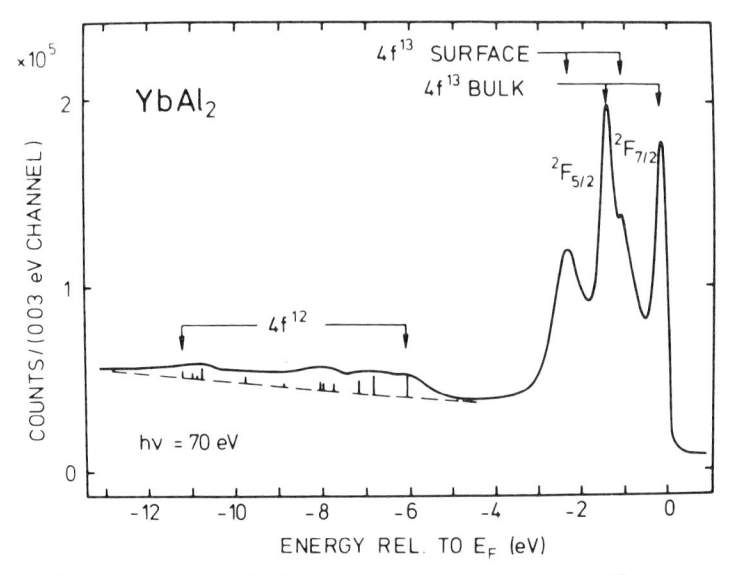

Fig. 8. Photoemission spectrum of $YbAl_2$ at $h\nu = 70$ eV. Two final-state $4f^{13}$ doublets are seen, a narrow bulk doublet and a much wider surface doublet shifted to higher binding energy. Weaker bulk final-state $4f^{12}$ multiplet lines are also seen. (After Kaindl et al. 1982.)

multiplet (final-state $4f^{12}$) structure. In fact, the divalent spectrum is best fitted by three doublets, one bulk and two surface corresponding to the topmost and underlying surface layers (Kaindl et al. 1982). This analysis yields a mean bulk valence in agreement with other measurements.

Cerium and cerium compounds present a closely related field of study. The compounds can be classified as γ-like (magnetic) or α-like (non-magnetic, contracted), reflecting the properties of the room-temperature (γ) and low-temperature (α) phases of cerium metal. This phase transition has a wide significance. γ-phase cerium is known to be trivalent so a promotional model anticipates a valence change of the type $4f^1V^3$ to $4f^0V^4$ at the phase transition. When the 4f content is small, it has proved useful to enhance this emission selectively by suitable choice of photon energy. The example of 4f identification in CeAs has already been cited in section 3.1 above. Similar methods have been applied to the γ–α phase transition in cerium (Mårtensson et al. 1982, Wieliczka et al. 1982). Again, the absence of major 4f intensity changes invalidates the promotional model. The separate 4f features at E_F and 2 eV binding energy are attributed to screened and unscreened excitations. The spectral changes observed during the phase transitions are attributed to small increases in 4f hybridisation (Wieliczka et al. 1982).

Specific studies of the valence of surface layers have been made. Whereas bulk-trivalent Sm becomes divalent at the surface, bulk-trivalent Tm ordinarily does not. This is understood in terms of different cohesive energy reductions at the surfaces of the two metals. Domke et al. (1986) have illustrated this connection in an examination of the valence of rough surfaces of Tm. These surfaces have further reduced cohesive energy, which is reflected in a small divalent component in the photoemission spectrum. Adsorption-induced valence changes have also been investigated. The oxidation of europium (Barth et al. 1983) results in photoemission spectra characteristic of both divalent and trivalent Eu ions, with a close correspondence to mixed-valence Eu_3O_4. In this "inhomogeneous" system, the two valences are associated with different lattice sites.

3.3. Surface states on metals

In general, termination of the crystalline lattice at a surface results in electronic states localised at the surface, whose characteristics differ from those of the bulk. Historically, the term "surface states" has been reserved for the outer electron valence or conduction band states. Having band-like character parallel to the surface, these states may be termed itinerant surface states. More recently (and largely through synchrotron radiation photoemission), attention has been focussed on inner electron core states, which also see a modified potential at the surface. As the experimental result is a relatively simply binding energy change, the more usual nomenclature of these surface states is that of "surface core shifts" (SCS).

3.3.1. Itinerant surface states

Synchrotron radiation photoemission has contributed significantly to the study of filled itinerant surface states. Photon energy–momentum selectivity provides the important litmus test for surface character of absence of dispersion normal to the surface. In fig. 9 are shown normal emission spectra for Zn(0001) from Himpsel et al. (1981). The tic marks track the dispersion of bulk features with $h\nu$, while the feature at -3.6 eV showing no dispersion is identified as a Λ_1 surface state. The historically older test of surface character, namely sensitivity to adsorption, is not always unambiguous, as discussed by Smith and Himpsel (1983). A third test, based on momentum selectivity, is that the energy and parallel momentum of the state are such that it lies within a gap of the bulk bandstructure projected onto the surface Brillouin zone. Otherwise the state is more correctly termed a surface resonance. Surface states on metals may be identified as either Shockley-like or Tamm-like. In broad terms Shockley states can exist only in a hybridisation gap, while Tamm states have their origin in a potential change on the surface atom;

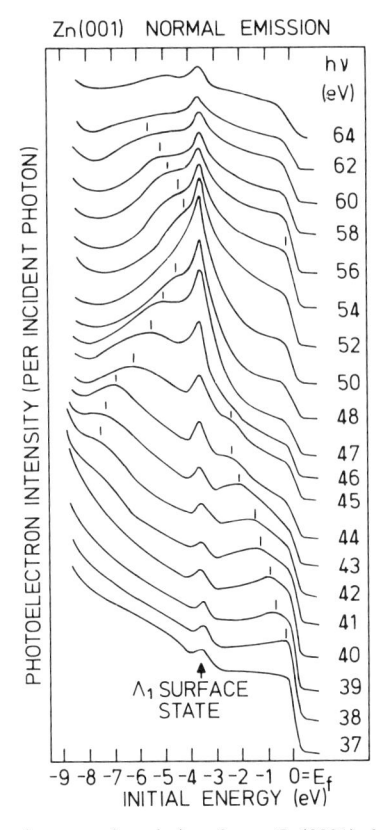

Fig. 9. Photoelectron spectra in normal emission from Zn(0001) for various photon energies $h\nu$. Interband transitions from the bulk s, p band are shown with tic marks. The s, p-like Λ_1 surface state is seen at -3.6 eV. (After Himpsel et al. 1981.)

more precise existence criteria have been described (Goodwin 1939, Shockley 1939). Tamm states and metal surface core shifts are related effects.

Occupied surface states on low-index faces of most metals have been catalogued by angle-resolved photoemission techniques (Smith and Himpsel 1983, Inglesfield 1982, 1985). Dispersion parallel to the surface is determined by the relation $\hbar k_\| = (2mE_k)^{1/2} \sin \theta$. In many cases there is semiquantitative agreement with calculations. Copper is a good example of the complexity that may arise, having both s–p and d bands in the bulk. Kevan et al. (1985a) have tabulated some nine surface states (both Tamm and Shockley) on the three low-index faces of copper, for which there are experimental and theoretical correlations. Considerably more states have been predicted than have been observed. A topic of current interest is how many surface states may exist in a gap (Bartynski and Gustafsson 1986).

Photon energy variability can contribute to surface state spectroscopy beyond simply illustrating the absence of dispersion perpendicular to the surface. In normal emission ($k_\| = 0$), surface states in s–p gaps exhibit periodic intensity oscillations as a function of photon energy (Louie et al. 1980, Kevan et al. 1985b). This may mean that they are not readily observed at certain photon energies. More intrinsically this phenomenon has been linked to the decay length of the surface state, the distance it penetrates the bulk. In effect the state has a preferred value of $k_\perp$ associated with a bulk band edge. The intensity enhancement results for direct transitions which approximately conserve this momentum. An early example, again from copper (Louie et al. 1980) is shown in fig. 10. The enhancement of the two surface states S_1 and S_3 at ~70 eV photon energy translates into a value of $k_\perp$ equivalent to the L critical point. The depth to which a surface state extends into the bulk is inversely related to both its distance from the band edge and the gap size. Extremes of long and short decay lengths are observed for Al(111) (Kevan et al. 1985c) and Be(0001) (Bartynski et al. 1985), respectively. Similar behaviour for a Tamm state (normally highly localised) on Cu(001) has been attributed to spin–orbit-induced symmetry-breaking effects (Kevan et al. 1985b).

At lower photon energies surface states on Al(100) (Levinson et al. 1980) and Be(0001) (Bartynski et al. 1985) exhibit intensity variations which have their origin in a surface photoeffect. There are rapid variations in the effective vector potential at the metal surfae as the photon energy passes through the threshold for plasmon creation. The very similar lower photon energy (i.e. photofield) and different higher photon energy (i.e. decay length) behaviours of surface states on these two metals are illustrated in fig. 11 (Bartynski et al. 1985).

Surface state spectroscopy can bear strongly on surface reconstruction effects (see section 3.5 below) and surface magnetism (see section 3.2 above). Exchange-split surface states have been observed on 3d ferromagnetic metals (Turner and Erskine 1983, Eberhardt et al. 1980). On Ni(100) a surface state is observed where there are gaps in only one spin band, i.e. a magnetic surface state (Plummer and Eberhardt 1979). Vacant surface states have been probed by absorption techniques such as "partial yield" or "constant final-state" spectros-

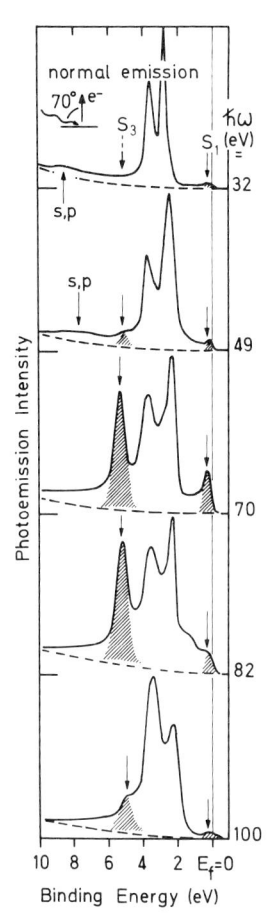

Fig. 10. Normal-emission ($k_\parallel = 0$) angle-resolved photoemission distribution curves from Cu obtained at various photon energies. S_1 and S_3 are surface states. The light is p-polarized. (After Louie et al. 1980.)

copies. However, these methods have no momentum selectivity and have been largely superseded by inverse photoemission methods. For metals, these experiments also reveal "image potential" states. These states are determined by the potential in the vacuum, whereas Shockley and Tamm states are determined by the potential associated with the bulk and surface, respectively (Kevan et al. 1985b).

3.3.2. Surface core shifts
The field of surface core shifts is almost exclusively synchrotron radiation based; the combination of high monochromaticity at reasonable flux and $h\nu$-dependent surface sensitivity allows core photoemission spectra to be resolved into bulk and energy-shifted surface components. A large variation in binding energy shift has been observed for metals ranging from -0.7 eV to $+0.6$ eV. The positive sign

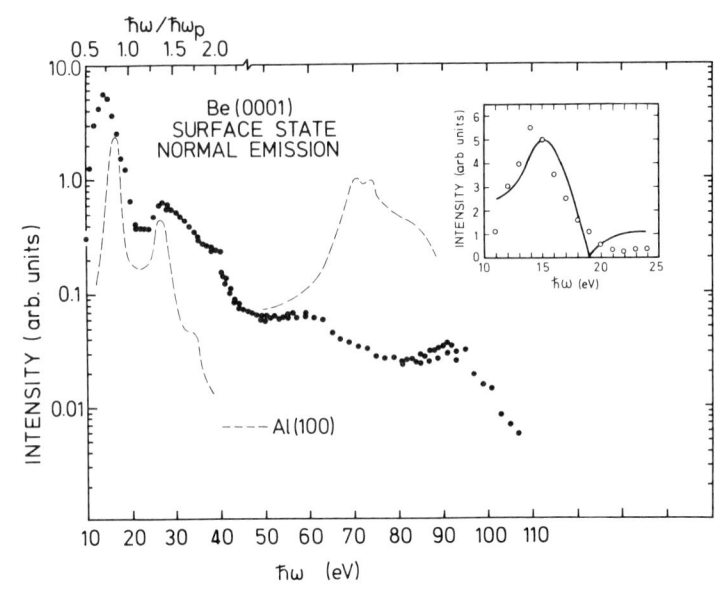

Fig. 11. Semilogarithmic plot of the surface state intensity on Be(0001) at Γ as a function of $\hbar\omega$ (dots). The angle of incidence is 45°. The dashed curve shows the intensity profile for the Al(100) surface state. The low-energy Al data are plotted as a function of $\hbar\omega/\hbar\omega_p$ (the upper horizontal axis). The high-energy Al data are plotted on the same horizontal scale as the Be data (the lower horizontal axis). The Be data may of course be referenced to either scale. The inset shows the low-energy data for Be on a linear intensity scale. The solid line is the calculation scaled to account for the difference in plasmon energy between Be and Al. (After Bartynski et al. 1985.)

indicates an increase in binding energy. In general (see review by Eastman et al. 1982) shifts are large for transition and rare earth metals due to their valence d electrons, and much smaller for metals with sp valence electrons. There is also a marked dependence on the number of d electrons, resulting in a polarity reversal. For Ta the shift is positive, while for W it is negative (van der Veen et al. 1981).

In a simple tight binding model, assuming layer-wise neutrality for a metal, the shift follows the shift of the valence d band centre of gravity (Desjonquères et al. 1980). At the surface the reduced atom coordination leads to a narrowing of the d band, which for a less than half-filled symmetric d band means a downward shift. Conversely, for a more than half-filled symmetric d band the shift is upward. For real, i.e. non-symmetric, d bands this model correctly predicts the observed sign reversal between Ta and W. It also explains the observed dependence on surface crystallography (Guillot et al. 1985). These general features are also reproduced by a semi-empirical thermodynamical description. The surface core shift of a metal of atomic number Z is approximately the difference in surface energies of Z and $(Z + 1)$ atomic species. This model also reveals the correspondence of surface core shifts and heats of segregation (Rosengren and Johansson 1981).

The spectroscopy of surface core shifts is an important addition to the range of surface techniques. The sensitivity to surface crystallography has allowed surface reconstruction to be examined (see section 3.5 below). As with itinerant surface states, there is a sensitivity to adsorption (see section 4 below). Surface core shifts have allowed surface magnetism and surface valence to be probed (see section 3.2 above). For further details see the recent review by Spanjaard et al. (1985).

3.4. Surface states on semiconductors

Surface states on semiconductors are of great importance because in certain circumstances they can appreciably influence the macroscopic electrical properties of the bulk material via a surface space charge layer. As illustrated in fig. 12, if electrons from the n-type bulk can be trapped in mid-gap surface states, a positive space charge layer of width W is established in the semiconductor. Simple electrostatics yield that $W = (2\varepsilon\varepsilon_0 V_s/eN_D)^{1/2}$, where ε is the dielectric constant, N_D is the donor density and V_s is the magnitude of the band bending. Depending on N_D, W may vary from nanometres to microns. This depletion of charge from the bulk can severely influence the electrical conductivity of thin films or small devices whose sizes are comparable to W. A surface charge density as low as 10^{16} electrons/m^2 can cause large effects; the modulation of surface charge (and hence the bulk conductivity) by adsorbed gases has been exploited in solid state gas sensors. Space charge layers are also important at metal–semiconductor interfaces. Here, related interface states may influence contact and Schottky barrier formation [see section 5 below and the following chapter (Brillson 1987)].

Surface states on semiconductors are generally described in terms of broken

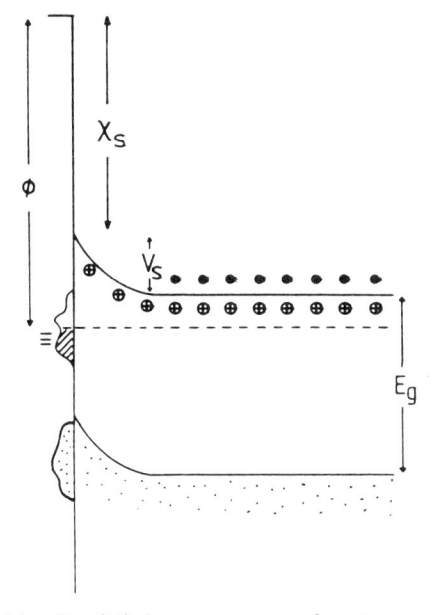

Fig. 12. Illustration of band bending (V_s) due to acceptor surface states on an n-type semiconductor. ϕ is the thermionic work function, χ_s the electron affinity and E_g the bandgap.

("dangling") bonds. Energy minimisation usually results in substantial rearrangement of the surface atoms (see section 3.5 below), and the surface electronic structure is a direct monitor of this surface structure. This approach is highlighted in the examples presented here: the dispersion of itinerant surface states on Ge(111) and Si(111) confirm the "π-bonded chain" model of the (2×1) reconstruction (section 3.4.1); the reconstruction phase transition on Ge(001) is followed through the "metallic" surface state (section 3.4.2); surface core shifts reflect the relaxation geometry of the GaAs(110) surface (section 3.4.3).

3.4.1. Itinerant surface states on Ge(111)(2 × 1)

Until recently the (2×1) reconstruction on cleaved Si(111) and Ge(111) surfaces was viewed as the up and down displacement of alternate rows of atoms (Haneman 1961). Calculations of surface state dispersion based on this model and its variations (Chadi 1980, Schlüter et al. 1975) differed markedly from that measured by angle-resolved photoemission using synchrotron radiation and discharge lamp sources (Uhrberg et al. 1982, Nicholls et al. 1983, 1984). Synchrotron radiation spectra for Ge(111) are shown in fig. 13a. The lowest binding energy feature (A) is the surface state; the $k_\parallel$ dispersion of this feature along $\bar{\Gamma}\bar{J}$ is essentially the same as that shown in fig. 13b (Nicholls et al. 1984). This composite diagram (Nicholls et al. 1985) also shows a calculated dispersion (Northrup and Cohen 1983) based on a radically different model of (2×1) reconstruction due to Pandey (1981). This is known as the π-bonded chain model, and a ball and stick representation is given in fig. 14. Although the binding energies do not concur, the calculated dispersion closely follows the experimental points.

Further support for this model derives from an elegant probe of the *vacant* surface state. By using highly doped n-type germanium, this state can be partially populated (Nicholls et al. 1985). Its dispersion in the region $\bar{\Gamma}\bar{K}$, (determined by discharge lamp photoemission) is shown as B in fig. 13b. The magnitude of the experimental surface state gap (A–B separation) is in good agreement with optical absorption measurements (Nannarone et al. 1980). This understanding of Ge(111) and Si(111) surfaces is complemented by recent measurements on Si(111) (Uhrberg et al. 1984) which reassign features from back-bond surface states to bulk transitions.

3.4.2. Itinerant surface states on Ge(001)

The ideal (001) surface of group IV semiconductors has two dangling bonds per surface atom. As shown schematically in fig. 15a, a short-range, nominal (2×1) reconstruction is achieved by the formation of dimers, leaving one dangling bond per surface atom. If the dimers are asymmetric, larger surface reconstructions are possible, viz. $c(2 \times 2)$ and $p(2 \times 2)$ and the $c(4 \times 2)$, which is shown in fig. 15a. For Ge(001) a phase transition between room-temperature (2×1) and low-temperature $c(4 \times 2)$ has been observed in LEED (ref. [9] in Kevan 1985). This

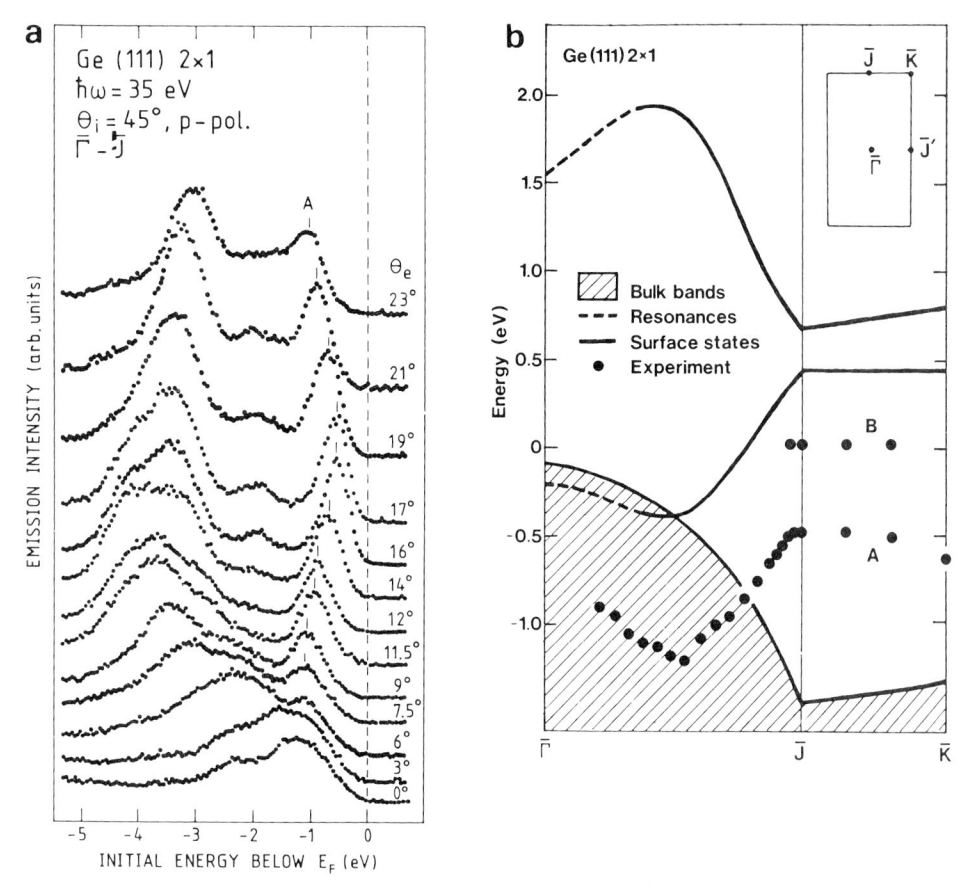

Fig. 13. (a) Photoemission spectra recorded at room temperature for various angles of emission (θ_e) along the $\bar{\Gamma}$–$\bar{J}$ symmetry line in the (2 × 1) surface Brillouin zone of the Ge(111) surface. The peak marked A corresponds to a surface state. Photon energy = 35 eV. (After Nicholls et al. 1984.) (b) Initial-state energy dispersion for Ge(111)(2 × 1) surface bonding (A) and antibonding (B) surface states, along the $\bar{\Gamma}$–$\bar{J}$–$\bar{K}$ lines, and the corresponding calculated bands for the π-bonded chain model. (After Nicholls et al. 1985.)

transition has recently been followed through high angle and energy resolution photoemission of a surface state (Kevan 1985).

The surface state is located at the Fermi level in normal emission, giving the surface a metallic nature. It is very narrow both in energy ($\Delta E < 100$ meV) and in angle ($\Delta\theta \sim 3.5°$), which probably explains why it has not previously been observed. The temperature dependence of this surface state and of LEED spot profiles are shown in fig. 15b. As the temperature is lowered from 500 K, the metallic surface state intensity decreases while the profile sharpens for the LEED feature due to ordered $c(4 \times 2)$. In this ordered phase the electrons from the metallic surface state are relocated in a higher binding energy shoulder believed to be characteristic of a $c(4 \times 2)$ surface state.

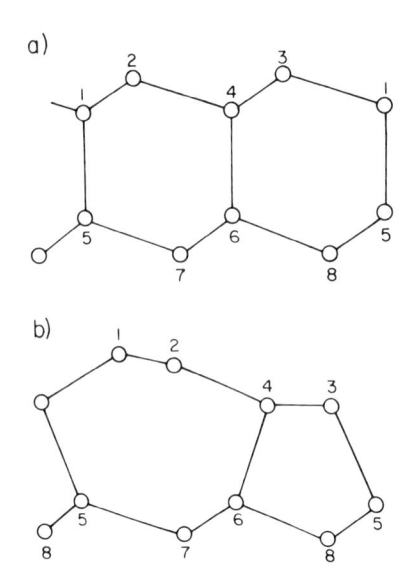

Fig. 14. (a) Side view of the ideal Ge(111) surface. (b) Side view of the Ge(111) surface in the π-bonded chain model.

These photoemission measurements have been analysed in terms of possible driving mechanisms of the phase transition. In particular, a long-range, Fermi surface instability mechanism can be ruled out by the absence of a suitable dispersive band crossing the Fermi level. A short-range coupling between the dangling bonds of the dimers is consistent with the observations. It is suggested that the metallic surface state is produced by flipping one dimer to promote a dangling bond electron state into the bandgap. An appropriate spatial extent for this state is deduced from the observed angular width of the feature. It is not yet understood why similar effects are not observed on Si(001).

3.4.3. Surface core shifts on GaAs(110)

The (110) cleaved surface of III–V compound semiconductors undergoes a relaxation in which the surface anions move outwards and become negatively charged while the surface cations move inwards and become positively charged. For most members of this group, the relaxation results in the exclusion of itinerant surface states from the bandgap (Spicer et al. 1979, Chadi 1979). In contrast to the situation of layer-wise neutrality in metals, surface core shifts in semiconductors largely reflect the charge localised on the surface atoms. As shown in fig. 16, the Ga-3d and As-3d levels exhibit surface components which are shifted in binding energy by $+0.28$ eV and -0.37 eV, respectively (Eastman et al. 1980b). These opposite polarities are consistent with the charge transfer from Ga to As associated with the relaxation. However, to quantify the charge transfer would require an estimation of final-state effects such as reduced screening of the hole. Indeed, Mönch (1986) has suggested that the observed shift is due to the

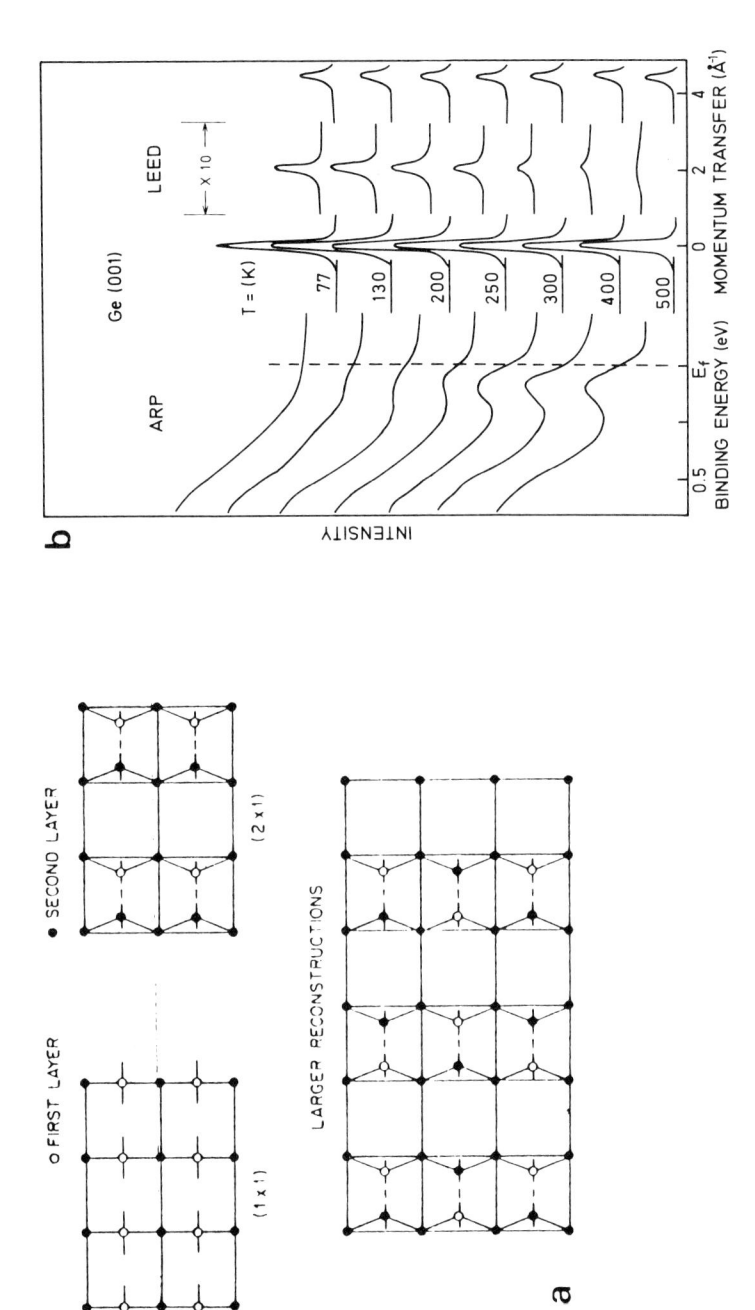

Fig. 15. (a) Schematic representation of the Ge(001) surface. Simple bulk termination (1×1) yields two dangling bonds per surface atom; asymmetric dimerization of nearest-neighbour surface atoms forms a (2×1) unit cell, satisfying one dangling bond per surface atom. Open circles indicate surface layer atoms buckled inwards, while solid circles indicate atoms buckled outwards. The larger $c(4 \times 2)$ unit cell is formed by alternating the orientation of nearest-neighbour dimers. (After Kevan 1985.) (b) Temperature-dependent results on Ge(001). Left panel: EDCs of the Fermi level region at the temperatures shown. The metallic peak disappears as the temperature is lowered. Right panel: Corresponding LEED profiles extending from the $(0, 0)$ beam to the $(1, \frac{1}{2})$ beam at the same temperatures. The $c(4 \times 2)$ beam at momentum transfer $2.2 \, \text{Å}^{-1}$ is seen to sharpen dramatically as the temperature is lowered. (After Kevan 1985.)

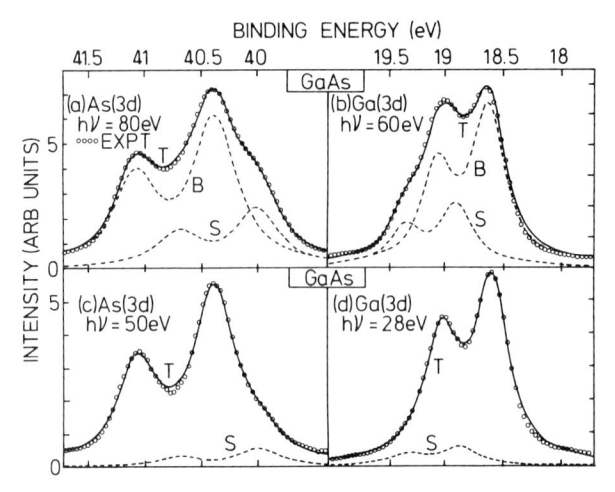

Fig. 16. Photoemission spectra for Ga- and As-3d core levels in GaAs(110). Spectra for low photon energies (c), (d) (~10 eV above threshold) show mainly bulk emission (B), while spectra for higher photon energies (a), (b) (~40 eV above threshold, small escape depth) show additional surface core level emission (S). (After Eastman et al. 1980b.)

difference in surface and bulk Madelung energies, and that no additional surface charge transfer is required. Similar techniques have been applied to silicon surfaces (Himpsel et al. 1980).

3.5. Clean surface geometry

Many of the determinations of the crystallography of clean surfaces have been made with LEED (Jona 1978), ion scattering spectroscopy (van der Veen 1985) and other techniques, with synchrotron radiation based studies playing mainly a supporting role. The subject is advancing fast, and a review article is likely to become rapidly out of date; in consequence this section attempts to bring out general principles rather than specific details. There are some significant differences between metals, semiconductors and insulators, caused by factors such as directional (covalent) bonding, and Coulomb (electrostatic) interactions, and examples are chosen for each of these types of solid.

The surfaces of most solids show a different structure from that expected in a straightforward termination of the bulk. The top layer of atoms usually moves inwards (relaxation) and there may be a change in the two-dimensional periodicity of the surface layer (reconstruction). Sub-surface layers sometimes show relaxation as well. The usual notation for a surface net is in the form $(m \times n)RA°$, indicating that the two-dimensional unit cell has lattice vectors m and n times that of the underlying bulk, with the principal direction rotated by an angle A.

The key to all surface relaxations and reconstructions is the change in periodicity of the electronic potential near the surface. The microscopic forces acting on the near-surface atoms differ from those in the bulk, and the atoms move to minimise the total potential energy. This can result in some very complicated atomic rearrangements, of which probably the most famous is the silicon(111) (7 × 7) reconstruction. Viewed along a (111) plane, the structure consists of successive double layers. This structure has 98 atoms in the outer double layer and 49 broken bonds localised on the top layer atoms. Si(111) has defied solution for many years, but a good understanding of its geometry has now been obtained, using a combination of techniques including transmission electron microscopy (Takayanagi et al. 1985a), scanning tunnelling microscopy (Binnig et al. 1983, Hamers et al. 1986), grazing-angle X-ray diffraction (Robinson et al. 1986) and medium-energy ion scattering (Tromp and van Loenen 1985). It appears that, in the energy balance between forming bonds and strain forces, the surface stretches by removing eight surface atoms per unit cell, the missing atoms being accommodated at occasional step sites. The model of Takayanagi et al. (1985b) (fig. 17) has three key ingredients – dimers, adatoms and a stacking fault. The stacking fault, a partial dislocation caused by a changed orientation of the distorted hexagons in successive double layers, is clearly visible. Atoms dimerise along the stacking fault and each boundary of the unit cell, further minimising energy. Finally, there are 12 adatoms (the top layer silicons) per unit cell, filling broken bonds from alternate trios of the second-layer atoms. This is illustrated to demonstrate the great complexity of some reconstructions resulting from the balance between microscopic forces at a surface.

X-ray diffraction using extreme grazing-angle geometry promises to aid the solution of various surface structures. Because scattering is weak, an intense photon source is essential, at least for light atoms, and this technique has

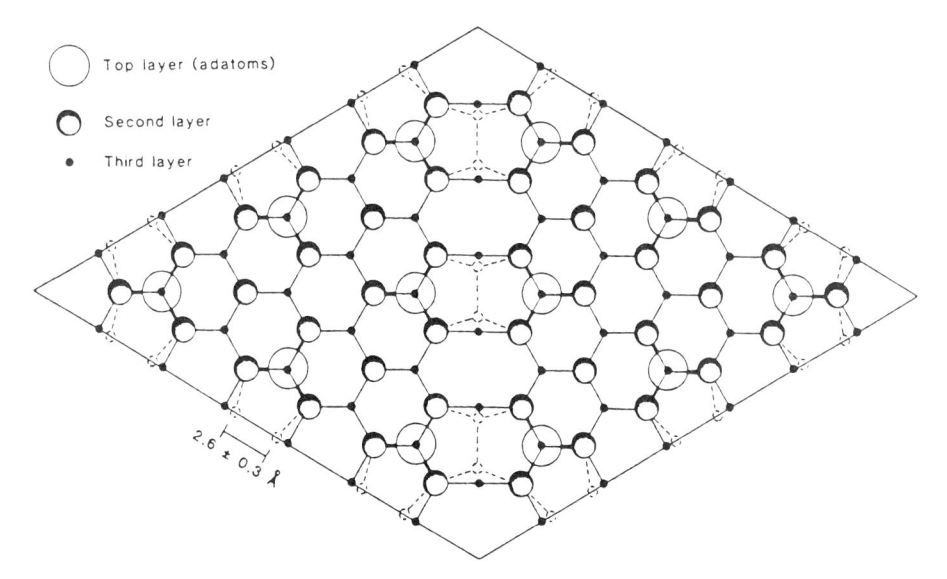

Fig. 17. The 7 × 7 reconstructed Si(111) surface. Top view of the unit cell of the dimer/adatom/stacking fault model, showing three layers of atoms, the adatom and underlying double layer.

burgeoned since the advent of bright synchrotron radiation sources. Most of the momentum transfer in the scattering is parallel to the surface plane, giving highly accurate in-plane atomic positions, whereas LEED is rather better at deriving perpendicular layer spacings. By contrast with LEED, a kinematical approach to X-ray diffraction is valid, since photon scattering is much weaker than low-energy electron scattering. As in bulk X-ray diffraction, amplitude information is collected but the phase term is lost; the electron density cannot be directly calculated. However, the Patterson (pair correlation) function gives the interatomic vectors and such Fourier methods can yield structures without any *a priori* model. Several clean surface reconstructions have been studied including Ge(100)(2×1) (Eisenberger and Marra 1981) and Au(110)(2×1) (Robinson 1983).

The kind of structural information which may be obtained from surface X-ray diffraction is illustrated in fig. 18a for InSb(111)(2×2), which shows a difference Fourier map of one unit cell of the 2×2 structure (Bohr et al. 1985). These data were combined with LEED information (Tong et al. 1984) on the analogous GaAs(111)(2×2) structure to yield the projected atomic positions depicted in fig. 18b. The Sb atoms are displaced outwards by a radial distance of 0.45 ± 0.04 Å, and the inward radial displacement of the In atoms is 0.23 ± 0.05 Å. The top layer atoms in effect rotate alternately in and out of the surface plane, keeping the In–Sb bond length very close to its bulk value.

The temperature-induced reconstruction of the clean W(001) surface is an interesting example of the many metal surface reconstructions, which have recently been reviewed by Inglesfield (1985). On cooling below room temperature, the surface undergoes a continuous phase transition from a 1×1 to a centred 2×2 net (Felter et al. 1977, Debe and King 1977a). The structure of the low-temperature phase was deduced from the symmetry of the LEED pattern (Debe and King 1977b), with the surface layer having zig-zag chains in $\langle 11 \rangle$ directions. Photoemission studies of the W $4f_{7/2}$ core level show a broad peak, which may be deconvolved into three constituents, corresponding to atoms in the bulk, atoms in the surface layer, and sub-surface atoms. These each have slightly different binding energies according to their different local coordinations. On reducing the temperature through the phase transition, there is found to be a small increase in binding energy of the surface peak, towards the bulk value (Guillot et al. 1984), confirming the increased coordination of the surface atoms on reconstruction. In this regard there has been both experimental and theoretical investigation of the W(001) itinerant surface states; the suggestion is that the surface states may drive the phase transition by splitting and hence reducing the one-electron energy. Polarised synchrotron radiation has been used to derive the symmetry of the surface states (Holmes and Gustafsson 1981) and there is clear evidence of a surface bandgap opening up on going through the reconstruction (Campuzano et al. 1981). There is still disagreement over the nature of the phase transition, particularly whether it is displacive or of the order–disorder type. Similar studies have been conducted of the reconstructions of a large number of

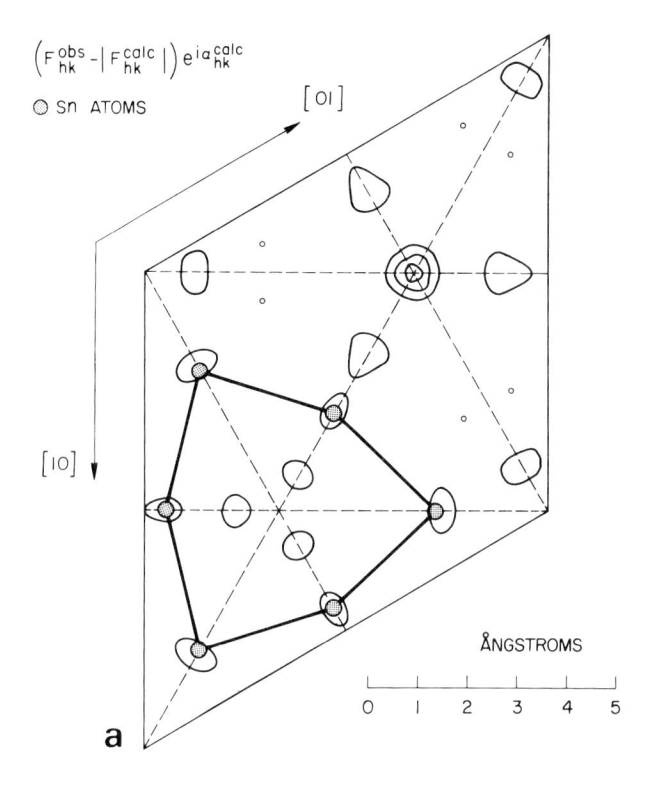

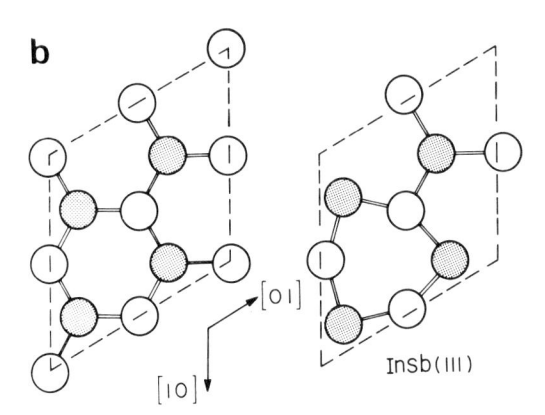

Fig. 18. (a) Difference Fourier map of one InSb(111) 2×2 unit cell. Positive contours above zero are shown. (b) Scale drawing of the projected atomic positions in (left) an unreconstructed III–V(111) surface, and (right) the InSb(111)(2×2) surface. Group V atoms are shaded. (After Bohr et al. 1985.)

clean metal surfaces, with synchrotron radiation photoemission studies of core levels and surface states playing a part in helping to unravel the strong interaction between crystallographic and electronic structure.

It has long been predicted that the (100) surface of ionic binary compounds would exhibit differential relaxation or "rumpling", owing to the different electronic polarisabilities of the anion and cation. MgO has been extensively examined by electron diffraction techniques. It seems from recent LEED data that the cleaved (100) surface is extremely smooth, with only a slight rumpling of 0.04 Å with the oxygen atoms above the surface plane (Welton–Cook and Berndt 1982). This class of materials has not received much study by synchrotron radiation techniques, although this could be particularly worthwhile in view of their susceptibility to electron beam damage.

It should be mentioned that SEXAFS, whilst providing precise lengths for adsorbate–substrate bonds, has not yielded any credible high-precision data for clean surfaces. This is because there is no clear way of distinguishing the X-ray absorption of an atom at the surface from the same type of atom in the bulk. Attempts to enhance surface specificity have involved detection of electrons (emitted following photoabsorption) at grazing angles or having kinetic energies giving short mean free paths. Comin et al. (1985) applied a nice analysis to ion-bombarded silicon, using the difference in escape depth between Si KLL (1600 eV) and LVV (90 eV) Auger electrons to discriminate surface atoms, and showed that the top layers of silicon recrystallise following ion bombardment. This depth-selective Auger probe SEXAFS technique is quite generally applicable, but a short data range is enforced by the relatively low binding energies of the core levels necessarily used to achieve really short mean free paths; this hinders attempts to analyse data with the EXAFS single-scattering formalism, and limits the precision of the results.

3.6. Small metal clusters

Small metallic clusters have received much experimental attention recently, for two main reasons. First, they are expected to represent a bridge between atomic/molecular and solid behaviour. Secondly, many commercial catalysts are in the form of small particles dispersed on a substrate, and a basic understanding of the structural and electronic properties of clusters may lead to advances in catalytic science. Evaporation of metals onto some substrates, such as amorphous carbon, produces isolated clusters, rather than a continuous overlayer. Many of the measurements on clusters have used synchrotron radiation techniques, particularly photoemission and EXAFS.

Most studies of clusters of many metals have shown qualitatively similar results: as the mean cluster size is decreased, and the average coordination of the metal atoms is reduced, near-neighbour distances shrink (Apai et al. 1979, Balerna et al. 1985). The typical effects on the electronic structure are illustrated in fig. 19, namely, the narrower, more atomic-like spectrum of small Pd clusters, and the evolution of a metallic valence band as the cluster volume increases. It has also

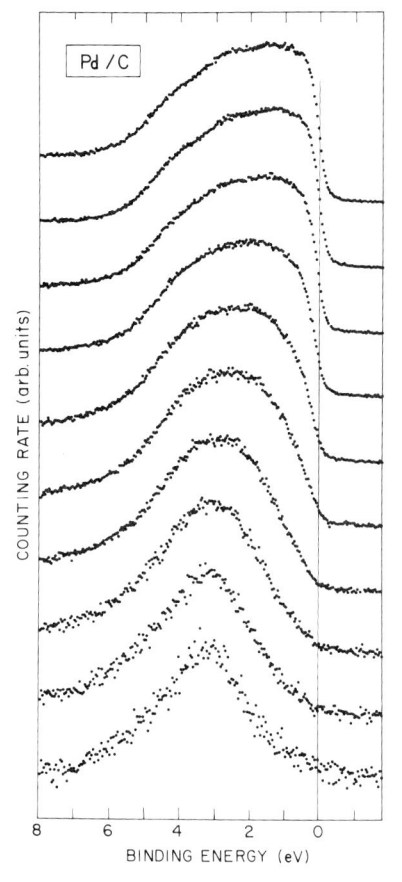

Fig. 19. Valence band spectra of Pd clusters, with cluster volume increasing by factors of 2 between adjacent spectra. The top spectrum corresponds closely to that of bulk palladium. (After DiCenzo and Wertheim 1985.)

been shown, using synchrotron radiation photoemission, that there is no photon energy dependence of valence band features at small cluster sizes, indicating that k is not a good quantum number. It appears that clusters of 100–150 atoms (for silver or gold) are needed before metallic behaviour is exhibited (Lee et al. 1981, Apai et al. 1981).

Core level effects are more puzzling. Small clusters of several metals (e.g. Sn, Pd, Au, Pt) have been observed to exhibit a core level shift to higher binding energy, which appears to conflict with the negative core level shift reported for surface atoms on single crystals [e.g. for Au and Pt (Citrin et al. 1983a)]. The surface core shifts arise from band narrowing due to the reduced coordination number at the surface (see section 3.3 above). The sign of the surface core level shift depends on the occupation of the valence (d) band, being negative for metals with more than half-filled bands (e.g. W, Au, Pt) and positive when the band is less than half-filled (e.g. Ta). Since the mean coordination number of atoms in clusters decreases with decreasing cluster volume (a higher proportion of surface atoms), a similar behaviour of core level shifts might be expected, but in fact the opposite is observed. It has been suggested (Wertheim et al. 1985, DiCenzo and Wertheim 1985) that this apparent paradox is in fact due to the differences in screening between a metallic surface and a single isolated cluster: the cluster retains a net charge, due to the emission of a photoelectron, for a time sufficiently long (in comparison to the photoemission event) to affect the apparent binding energy, whereas the charge is quickly screened on a metal surface. It appears that this final-state effect has a larger magnitude than the initial-state effect of band narrowing.

4. Surfaces with adsorbates

The next stage in the study of surfaces is the interaction of an adsorbate with a clean surface. In the experimental tradition of replacing the vacuum ambient with another, the adsorbate is either a gas or an element which is deposited by cracking of a gas; this tradition reflects the "catalysis" connection in surface science. The more recent "electronic devices" connection demands a broader range of adsorbates, usually furnished by molecular beam deposition techniques. In general, these adsorbates tend to form continuous overlayers, and it is convenient (albeit somewhat arbitrary) to discuss these adsorbate complexes as interfaces (see section 5 below). In this section, then, the discussion is restricted to the traditional gas-related adsorbates. The topic is further subdivided into chemisorption (section 4.1), physisorption (section 4.2), geometry of adsorbate structures (section 4.3) and reactions (section 4.4).

4.1. Chemisorption

The chemisorption regime is characterised by a strong chemical-bonding interaction between adsorbate and substrate atoms. In keeping with the general illustrative theme of this chapter, selected chemisorption studies are presented so as to highlight a particular approach or technique. Material is presented under the heading of chemisorbing species, that is, the archetypal gases hydrogen (section 4.1.1), oxygen (section 4.1.2), chlorine (section 4.1.3), carbon monoxide (section 4.1.4) and water vapour (section 4.1.5), and the archetypal chalcogen sulphur (section 4.1.6).

4.1.1. Hydrogen
Chemisorption of hydrogen on metals has attracted attention for two reasons. First, the simple nature of the adsorbate makes for a prototype chemisorption system and secondly, hydrogen uptake in the bulk has potential for energy storage. A supplementary aspect is that itinerant surface state contributions to complex angle-resolved photoemission spectra may be identified in part by their

quenching by hydrogen. These aspects are illustrated in a recent photoemission study of hydrogen on Pd(111) (Eberhardt et al. 1983).

Normal emission spectra of Pd(111), clean and hydrogen covered at 100 K, are shown in fig. 20a for three photon energies. Well removed from the metal d bands, there is a hydrogen-induced state near -8 eV at $\bar{\Gamma}$. The two-dimensional dispersion of this state confirms the 1×1 symmetry of the overlayer and allows the application of difference curves (fig. 20b) to reveal the induced changes in the d bands. Negative features are assigned to two surface states (-0.3 and -2.2 eV) – in conjunction with photon energy dependence (section 3.3) – while positive features are assigned to three hydrogen-induced states (-1.2, -3.1 and -7.9 eV). Comparison with a self-consistent pseudopotential calculation confirms the nature and energy of the two surface states. It also indicates that the two lower binding energy, hydrogen-induced features are these same states shifted to higher binding energy by a hydrogen-induced change in surface potential. A third surface state predicted to lie at -4.1 eV is too weak to be observed. This state combines with the H-1s orbital to form the bonding adsorbate state near -8 eV. There is good overall agreement between the observed spectra and that calculated for a three-fold chemisorption site, with a H–Pd bond length of 1.69 Å.

Upon warming this adsorbate system to near room temperature (or exposing at room temperature), the hydrogen-induced features vanish and the d-like intrinsic

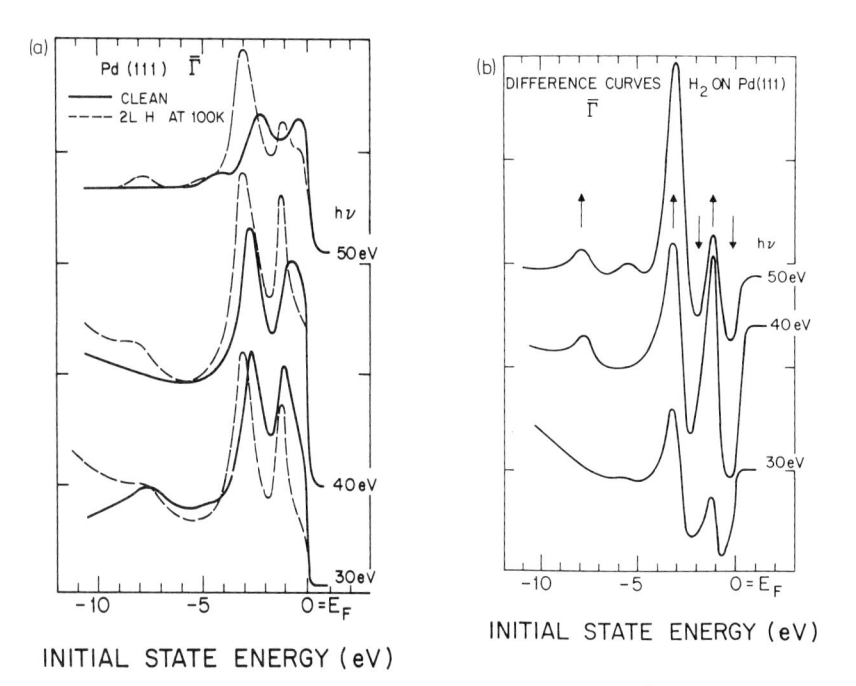

Fig. 20. (a) Normal-emission spectra from Pd(111) and following 4×10^{-6} Torr s exposure of H_2 at 100 K. (b) Difference curves for the three sets of spectra shown in (a). Arrows pointing up are H-induced states, arrows pointing down are intrinsic surface states or resonances. (After Eberhardt et al. 1983.)

surface states are restored to clean surface condition. The hydrogen is essentially invisible and yet other measurements indicate the presence of approximately half a monolayer. This is interpreted as hydrogen taking up a sub-surface site such as would be occupied by a hydride. However, comparison with hydride spectra suggests that this bonding configuration is different from that of both hydride and low-temperature overlayer. For further discussion see Greuter et al. (1986).

4.1.2. Oxygen

The influence of extrinsic experimental conditions on the chemisorption of oxygen, particularly on semiconductor surfaces, is now well known. The concept of "excited oxygen" (produced by hot filaments, electron beams etc.), explaining order of magnitude variations in oxidation of GaAs, emerged from synchrotron radiation soft X-ray photoemission spectroscopy of oxygen-induced core level satellites (Spicer et al. 1979). This technique is illustrated here through high-resolution measurements of O/Si and O/GaAs chemisorption systems. Enhanced oxide formation on semiconductors is further discussed in section 4.4.

Interest in the chemisorption of oxygen on silicon stems from the need to understand the Si/SiO_2 interface, which is important in semiconductor device technology. Many diverse models of the bonding configuration have been proposed. Using a display-type analyser coupled with synchrotron radiation, Hollinger and Himpsel (1983) have examined the oxygen-induced silicon-2p core satellites with high-resolution (~ 0.2 eV) photoemission. Four discrete core satellites are observed (at -0.9, -1.8, -2.6 and -3.6 eV). These are attributed to silicon atoms bonded to one, two, three and four oxygen atoms, respectively. It is apparent that beyond 100 L exposure, SiO_2 is already being formed.

The same experimental apparatus has been employed to re-examine the oxidation of GaAs under high resolution (Miller and Chiang 1984, Landgren et al. 1984). The early photoemission data on this system (Spicer et al. 1979) indicated that some 10^7 L exposure was required to produce a distinct 3 eV shift of the As core as against a slight broadening of the Ga core. The more recent high-resolution, high-sensitivity data probe the small oxygen-induced shifts which occur before the regime dominated by the 3 eV As shift. The widths of the satellites are interpreted as a distribution of shifts arising from disorder and/or contributions from deeper layers (Miller and Chiang 1984). The data underpin the roles of *both* the As and Ga species in the oxidation from the initial stages. Landgren et al. (1984) present evidence that at 10^4 L only the Ga is involved, but that both Ga and As are involved in sub-surface oxide formation beyond this exposure. These authors suggest that the oxidation nucleates and proceeds at extended, cleavage-induced defect sites.

In general, chemisorption information from the adsorbate core level has been less forthcoming. The oxygen-1s core is within the range of the new generation of soft X-ray monochromators, but to date the measurements have been largely those requiring only moderate resolution such as photoelectron diffraction and surface EXAFS (see sections 4.1.6 and 4.3). Hughes et al. (1985) have reported moderate-resolution (0.7 eV) oxygen 1s photoemission measurements of the oxidation of GaAs. These studies parallel the higher-exposure (10^6–10^{14} L) Ga- and As-3d spectra of Landgren et al. (1984), and support the notion of a sub-surface oxidation which is strongly cleavage dependent.

4.1.3. Chlorine
The saturated chemisorption of chlorine on silicon and germanium has long been an important prototype system for testing experimental and theoretical methods. Studies of Cl/Si(111) provided early indications of the potential of synchrotron radiation photoemission in mapping adsorbate bandstructure, revealing orbital character and discriminating between bonding geometries via polarisation selection rules (Larsen et al. 1978). An early example of the latter due to Schlüter et al. (1976) compared detailed pseudopotential calculations of electronic densities of states for "one-fold covalent" and "three-fold ionic" chemisorption sites with polarisation-dependent angle-integrated photoemission spectra; the behaviour of a σ-p_z feature indicated the "one-fold covalent" site for Cl/Si(111) but the "three-fold ionic" site for Cl/Ge(111). The latter was a surprising result since both surfaces, having the same half-filled dangling bond, would be expected to support the same (one-fold covalent) chemisorption site. Recently, these systems have been re-examined using SEXAFS and NEXAFS by Citrin et al. (1983b). From the SEXAFS data it is deduced that, contrary to the earlier photoemission study, the chlorine does indeed occupy the "one-fold covalent" site on Ge(111). Moreover, the NEXAFS data (which probe the vacant states above E_F) indicate that the electronic differences between Cl/Si and Cl/Ge observed in photoemission are due to matrix element effects. A detailed photoemission study has been reported by Schnell et al. (1985).

The chemisorption of the other halogens has been extensively examined both by photoemission (Dowben et al. 1985) and by SEXAFS techniques (Jones et al. 1985). They have also been used as prototypes of the novel X-ray standing wave analysis. As described in section 2.4, this technique employs adsorbate fluorescence yield to follow the standing wave antinode and hence locate the position of the adsorbate relative to the bulk d spacing of the substrate. Measurements on a chemically prepared Br/Ge(111) overlayer system by Bedzyk and Materlik (1985) are shown in fig. 21; the Br fluorescence goes through a minimum near the centre of the rocking curve. In effect, the Br atoms are close to the bulk d spacing position ($\sim 0.8d$), which is consistent with the "one-fold covalent" chemisorption site discussed above (with Ge atoms in ideal bulk-like positions). Similar standing wave measurements on chemically prepared Br/Si(111) indicate the same on-top adsorption site (Materlik et al. 1984).

However, more recent measurements on UHV prepared Br/Si(111) provide a rather different picture (Funke and Materlik 1985). For this system the fluorescence profile indicates that the Br atoms occupy a position of $0.64d$. In this case, the Br atoms appear to prefer the "three-fold ionic" site or sub-surface three-fold equivalents. It is suggested that sample preparation may cause a silicon surface relaxation, or there may be coverage-dependent effects. For further details, see the recent review by Materlik (1985).

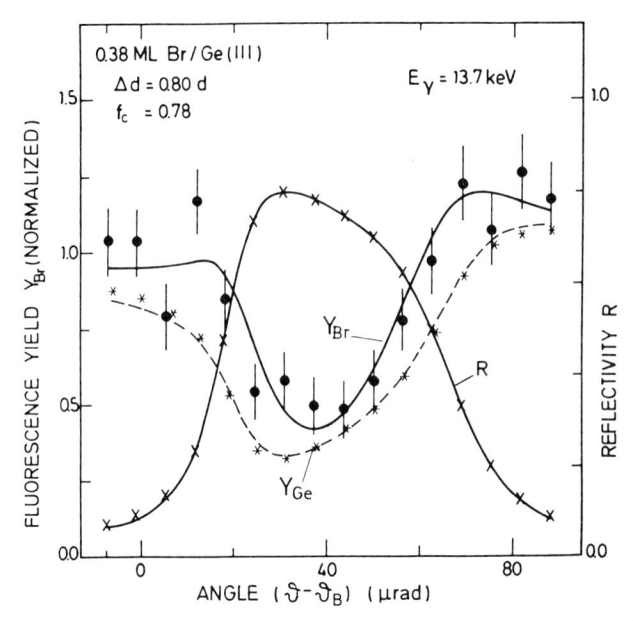

Fig. 21. Experimental data and theoretical curves for the Br Kα fluorescence yield (Y_{Br}), the Ge Kα fluorescence yield (Y_{Ge}) and the Ge(111) reflectively (R) versus Bragg reflection angle. (After Bedzyk and Materlik 1985.)

4.1.4. Carbon monoxide

The chemisorption of carbon monoxide onto transition metal surfaces is a classic case of non-dissociative adsorption of a molecule as an ordered overlayer. As such, it presents interesting questions concerning the spectral correspondence with the free molecule and the polar orientation of the adsorbate molecule. An early synchrotron radiation angle-resolved photoemission study of CO/Ni(001) by Smith et al. (1976) successfully used polarisation selection rules coupled with theoretical calculations to demonstrate that CO chemisorbed perpendicular to the surface, carbon end down.

A more recent example, that of CO/Co(0001) is illustrated in fig. 22. The familiar assignment of molecular orbitals 4σ and $5\sigma/1\pi$ to the features at -10.7 and -7.9 eV, respectively, derives from comparison with gas phase data; the 5σ (with amplitude largely on the carbon end) is deemed to undergo a shift of ~ 3 eV towards higher binding energy due to formation of the CO–substrate bond. The orientation of the molecule is determined by the allowed/forbidden geometry selection rule for emission in a mirror plane; the requirement is whether the A vector of normal incidence light is parallel or perpendicular to the collection plane. The almost complete disappearance of the even-symmetry 4σ level in the forbidden geometry shows that the CO prefers to bond perpendicular to the surface (Greuter et al. 1983).

Further evidence for this bonding configuration comes from photoemission cross-section resonances, which carry over from the gas phase. The 4σ shape

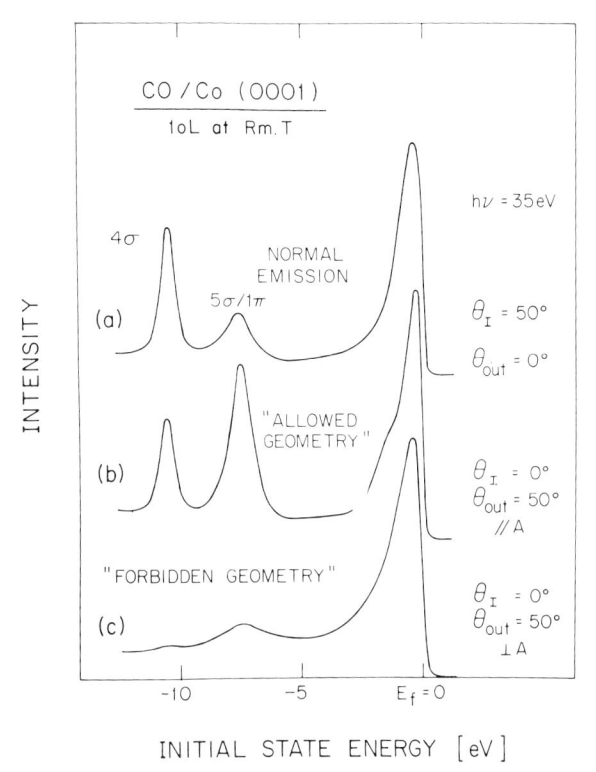

Fig. 22. Typical spectra for saturation coverage of CO on Co(0001) at room temperature (RT) for p-polarized light and normal emission (a) and s-polarized light with the polarization vector A parallel (b) or perpendicular (c) to the collection plane, which was the $\bar{\Gamma}\bar{K}$ mirror plane. (After Greuter et al. 1983.)

resonance, centred at ~35 eV and strongly directed out the oxygen end of the free molecule, is also observed for the adsorbate. A similar resonance exists for the 5σ orbital but directed out the carbon end of the free molecule. The involvement of this level in the bonding to the surface reduces the likelihood of simple carry-over of the resonance from the gas phase. The effect is to make the 5σ resonance also appear to come from the oxygen end. A number of mechanisms have been suggested to account for an increased 5σ resonant emission upon adsorption (Greuter et al. 1983). The bonding geometry is also confirmed by NEXAFS measurements (Stöhr et al. 1981) as discussed in section 4.3 below.

Carbon monoxide exhibits satellite structure on both core and valence levels in photoemission. In the free molecule, two prominent satellites (~8 and 15 eV binding energy) on C-1s and O-1s core levels are attributed to multi-electron shake-up originating from $1\pi \rightarrow 2\pi^*$ excitation. Upon chemisorption a third satellite may be observed [at ~5 eV binding energy for CO on Co, for example (Freund et al. 1983a)]. In a general view, this satellite is described as a poorly screened version of the fully screened main line, where the metal-induced screening operates via charge transfer into the 2π molecular level. In detail, the satellite has been assigned to shake-up between bonding and antibonding metal–2π combinations (Freund et al. 1983a). Clearly, such satellites carry information on the chemisorption process (Umbach 1982).

These multi-electron satellites are also associated with valence features, where they may confuse the one-electron valence electron analysis. For most CO chemisorption systems, the chemisorption satellite intensity is small; according to Freund and Plummer (1981) the relative weight of the satellite grows (and its binding energy decreases) as the CO–metal interaction weakens. The case of CO/Cu is intermediate with comparable intensity in "satellite" and "main line". Mariani et al. (1982) have used the photon energy dependence to monitor valence satellites in this system. A further influence of these satellites is to be seen in angle-resolved photoemission of ordered overlayers of CO on Cu (Freund et al. 1983b). Whereas there is a simple relation between the observed dispersion and the CO–CO spacing in general, the presence of a strong satellite causes a reduction in the measured band dispersion. Resonance of valence satellites has also been observed [for CO/Pt(111)], associated with the carbon-1s (strong) and platinum-4f (weak) thresholds (Loubriel et al. 1982, Grider et al. 1983).

Although CO chemisorbed on transition metal surfaces presents a simple geometry, the details of chemisorption bonding are not fully understood. In the synergistic model of Blyholder (1964) there is charge donation from the 5σ level to the metal and back donation from the metal to the empty 2π level. However, it appears that the magnitude of the 5σ binding energy shift is not directly related to the bond strength, and a more recent model views the bonding largely in terms of metal to 2π donation (Bagus et al. 1983). The experimental range is being expanded by core absorption (Jugnet et al. 1984) and Auger (Umbach and Hussain 1984) spectroscopies, which complement the one-hole photoemission. For further details, see recent reviews by Messmer (1985) and Plummer et al. (1985).

4.1.5. Water vapour
The chemisorption of water vapour may be regarded as an archetype of more complex molecules which are likely to dissociate on adsorption. Valence photoemission spectroscopy is employed to determine whether dissociation has occurred but, as discussed below, the interpretation of the data may not be unambiguous. Photon-stimulated desorption (PSD) (especially of H^+) is an important monitor of H_2O chemisorption and as this technique may employ the same analyser as photoemission, these are often combined studies.

Owen et al. (1986) have compared photoemission spectra of H_2O adsorbed at different temperatures on $SrTiO_3(100)$ with the H_2O gas phase spectra. As shown in fig. 23, difference curves for (a) 300 K and (b) 150 K are markedly different. The latter corresponds closely to the gas phase spectrum and indicates undissociated chemisorption. Dissociative chemisorption at 300 K is characterised by the absence of the $1b_2$ molecular feature. Similar data on $H_2O/Si(100)$ (Schmeisser et al. 1983a) show features (at 6.2, 7.2 and 11.5 eV binding energy) which are

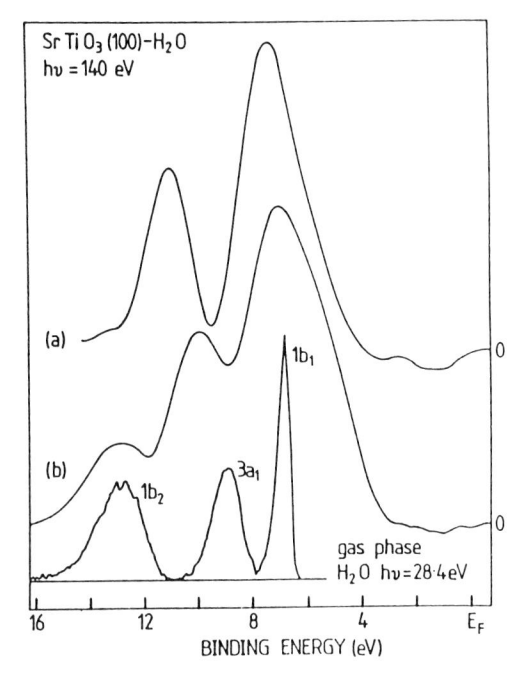

Fig. 23. UPS difference spectra (H_2O dosed-clean) for (a) 10 L H_2O on a $SrTiO_3(100)$ surface adsorbed at 300 K; (b) 0.5 L H_2O on $SrTiO_3(100)$ at 150 K, compared with (c) the gas phase photoelectron spectrum of H_2O (from Truesdale et al. 1982), aligned at the $1b_2$ energy. (After Owen et al. 1986.)

closer to the dissociated case in fig. 23. By contrast, the preferred interpretation here is that of undissociated chemisorption, along the lines discussed for CO chemisorption in section 4.1.4. The highest binding energy feature is assigned to $1b_2$. The lowest binding energy feature (assigned to the oxygen lone pair orbital b_1) is shifted downward, indicating bonding of the oxygen end to the surface, while the remaining (a_1) OH orbital is shifted upwards due to a weakening of the OH bonds by the extra Si–O bond. In these studies, it is necessary to invoke a wider range of experimental evidence to determine whether the water vapour has dissociated upon adsorption. That the $H_2O/Si(100)$ analysis is not in agreement with electron energy loss measurements is the subject of some debate (see Schmeisser and Demuth 1986).

Similar measurements of H_2O adsorption on stepped Ti(100) surfaces have been analysed in terms of dissociative adsorption (Stockbauer et al. 1982). This combination study also reported that photostimulated desorption gives a strong H^+ signal with little or no OH^+ or O^+. Separate experiments involving the adsorption of hydrogen and the co-adsorption of oxygen and hydrogen indicate that the strong H^+ signal is associated with the presence of OH on the surface. Thresholds at ~25 eV and ~45 eV are tentatively assigned to O-2s and Ti-3p edges, leading to a discussion of possible bonding configurations. In principle, more direct geometry information can be obtained by using the ion yield to monitor the SEXAFS modulation of the photoabsorption. Employing a "time of flight" technique based on the time structure of the synchrotron radiation flux, Owen et al. (1986) have investigated the PSD SEXAFS of the $H_2O/SrTiO_3$ system. It was found that in this system the desorption is dominated by an electron-stimulated mechanism, so that the EXAFS is bulk dominated. A recent study of the $H_2O/Si(100)$ system is more encouraging (McGrath et al. 1986). A photon-stimulated mechanism appears to dominate, implying that the EXAFS reflects the silicon substrate dimer geometry. Further studies of this potential technique are required.

4.1.6. Sulphur
The chemisorption of sulphur on metal surfaces has implications for such diverse industrial problems as steel embrittlement and catalytic poisoning. It also provides a well-characterised system with which the details of photoemission spectroscopy of chemisorption may be investigated. This and related chalcogen/transition metal chemisorption systems have also been the testing ground of photoelectron diffraction techniques.

The $c(2 \times 2)S/Fe(100)$ chemisorption system has been extensively investigated. As discussed in section 4.1.1, the 2D bandstructure of adsorbate levels is readily obtained from the *energy positions* of features in ARPES data. Application of polarisation selection rules (of the type described in section 3.1) labels the symmetry of these bands as odd or even with respect to the surface mirror plane. The results of a more detailed investigation of these sulphur 3p bands is shown in fig. 24 (Didio et al. 1984a). Here attention is focussed on the *lineshapes* of the adsorbate features. Whereas the odd band (not shown), which is the in-plane sulphur–sulphur interaction, shows no variation in line width across the zone, the two even bands show marked variation in width. This variation is plotted against the projected even Fe bands (shaded regions). It is apparent that the broadening is largest when a band simultaneously has p_z character and overlaps the Fe sp bands; that is, at Γ for the lower band and approximately one third of the way along ΓM for the upper band. This hybridisation broadening [which was predicted by Liebsch (1978) for oxygen] is analysed to show that an important component of the S–Fe bond is the $S(3p_z)$–Fe(p) interaction, in contrast to nickel where the $S(3p)$–$Ni(3d)$ interaction is considered to be more important (Didio et al. 1984b). These high-resolution studies also noted an analogous dispersion with photon energy in normal emission of the adsorbate p_z level (see surface state identification in section 3.3). This effect is the result of differential cross-section variation of the separate components of S-p_z and Fe-s and -p character resulting from the hybridisation (Didio et al. 1984a).

Initial experimental observations of normal emission photoelectron diffraction (NPD) from an adsorbate core level employed the Se/Ni(100) system (Kevan et

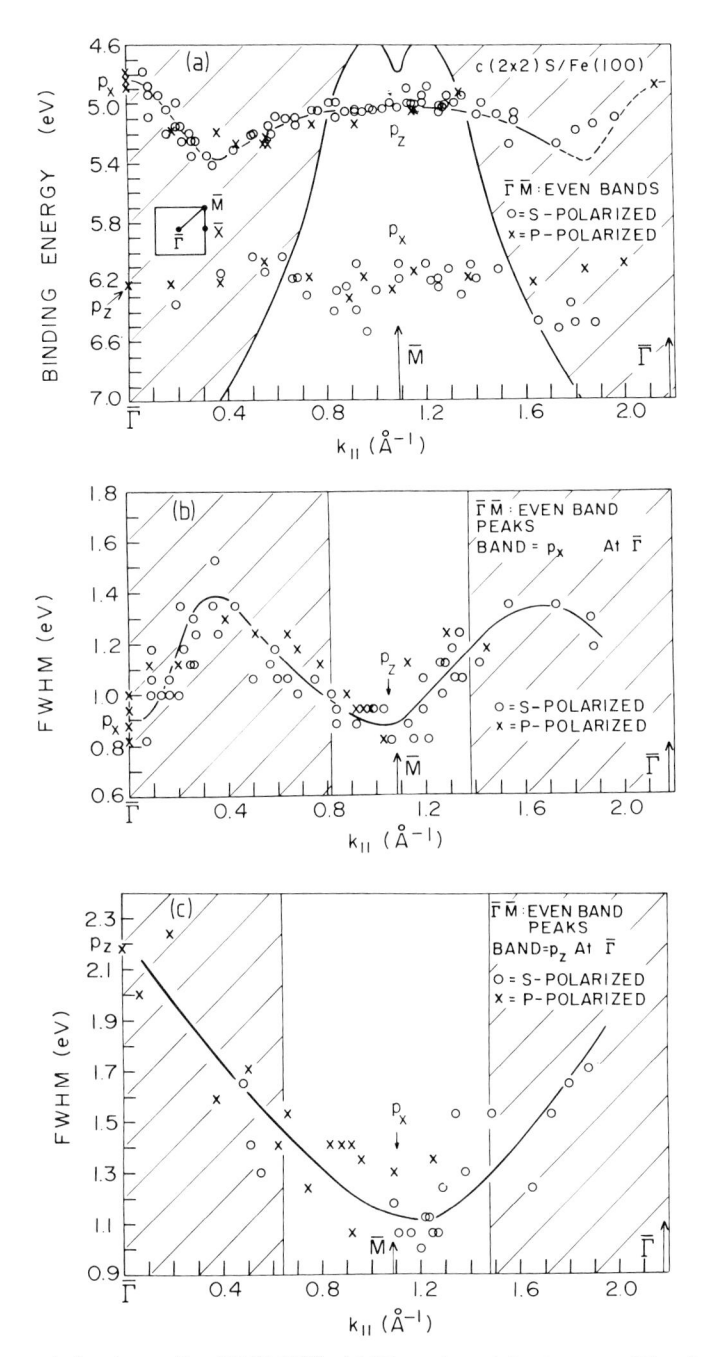

Fig. 24. Photoemission from $c(2 \times 2)$S/Fe(100). (a) Dispersion of the two even S bands. The shaded region is the projection of the bulk bands. (b) Hybridization widths along the upper band in (a). (c) Hybridization widths along the lower band in (a). (After Didio et al. 1984a.)

al. 1978). As discussed by Smith and Himpsel (1983), a comparison of experimental and theoretical NPD curves (core intensity versus electron kinetic energy) yields an accurate determination of the spacing ($d_\perp$) between an adsorbate layer and the substrate. More recent data on the same system (Rosenblatt et al. 1982) followed the progression of *off-normal* PD curves as the polar angle of emission (θ) is varied; the structure in the profiles becomes less pronounced off normal due to overlapping peaks, so that these curves are more useful as a self-consistency check of the NPD results. However, two different adsorption sites may yield an accidental coincidence in peak positions in the theoretical NPD curves, which will be lifted in off-normal curves. The authors comment that the close similarity between experimental PD curves with $\theta \leq 10°$ in two different azimuths presents the possibility that NPD data could be taken into a much larger solid angle of emission (resulting in a dramatic reduction in collection time) without significantly degrading the structural accuracy of NPD.

The potential of NPD for structural analysis is limited on two fronts. First, like LEED, it depends critically on complex theoretical analysis, and secondly, it emphasises the single parameter $d_\perp$. In attempting to address this second point by off-normal PD, Barton et al. (1983, 1984) have demonstrated a novel variation which they have called ARPEFS (Angle-Resolved PhotoEmission Fine Structure). In the 100–500 eV kinetic energy range the modulations appeared to be related to single rather than multiple scattering. The data were therefore amenable to simple Fourier analysis. Moreover, the atom–atom scattering path lengths could be directly probed rather than $d_\perp$. The technique compares favourably with SEXAFS and its lower signals are compensated by much higher modulations in the profile.

The first results of ARPEFS, shown in fig. 25, are of the $c(2 \times 2)S/Ni(100)$ system. The kinetic energy profile of angle-resolved emission from the S-1s core shows pronounced modulations, which have a cosine dependence on wave vector. An autoregressive Fourier transform analysis reveals nearest-neighbour scattering lengths and a unique structure of a four-fold hollow site with S–Ni bond length of 2.23(3) Å. While the Fourier transform method has been questioned (Orders and Fadley 1983), single-scattering calculations appear to be successful in simulating

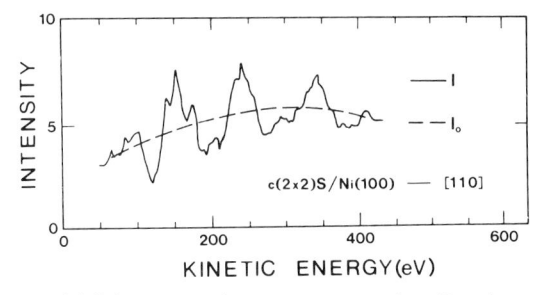

Fig. 25. Photoemission partial S-1s cross-section measurements for $c(2 \times 2)S/Ni(100)$. The solid curve I is the S-1s photoemission intensity with the dotted curve I_0 the estimate for a free atom. (After Barton et al. 1984.)

the experimental data. A single-scattering model is known to be appropriate at higher energies (~1 keV), while multiple scattering is considered still important in LEED up to several hundred eV. In an examination of the correlations between LEED, EXAFS and PD, Woodruff (1986) suggests that, given the "local source" nature of PD, many of the EXAFS simplifications are indeed appropriate to diffraction of photoelectrons in this energy range. In recent studies of the CO/Cu(100) structure, a LEED-style model calculation approach based on single scattering is preferred to the Fourier transform approach (McConville et al. 1986). This technique, together with azimuthal (Sinkovic et al. 1984) and polar (Daimon et al. 1984) photoelectron diffraction methods, seems likely to find increasing application in adsorbate structure determination.

4.2. Physisorption

The physisorption regime is characterised by low heats of adsorption, of the order of 0.1 eV. While classic physisorption can be ascribed to the inert nature of the adsorbate (as for a rare gas) or of the substrate (as for a layered solid), a wider range of adsorbate/substrate combinations can be described as physisorption systems. There are two broad areas of interest in these systems. First, for gases other than rare gases a useful comparison can be made with the stronger chemisorption of that gas under other conditions. Secondly, the weak interaction between adlayer and substrate offers the opportunity to investigate electronic and structural properties of simple, two-dimensional systems.

Whereas the chemisorption of water vapour presents a classic dissociative/non-dissociative problem (see section 4.1.5 above), physisorbed water vapour retains its molecular nature. Schmeisser et al. (1983b) have examined the valence photoemission of water physisorbed on different metal surfaces. Marked differences are observed which are dependent on substrate temperature and film thickness, but which are independent of the substrate. The data are used to distinguish three regimes of adsorption, namely, monomeric adlayer, dimeric or small cluster and multilayer. As shown in fig. 26, a splitting of the three molecular orbitals is observed at temperatures below that at which the monomer is seen (160 K). The data are analysed in terms of hydrogen bonding dimers, resulting in proton acceptor and proton donor contributions to the spectra. The splitting disappears in the multilayer case when most of the molecules are equivalent.

The chemisorption of CO on transition metal surfaces exhibits a bonding configuration where the molecular axis is perpendicular to the surface (see section 4.1.4 above). In a weakly interacting physisorption system, a bonding configuration with the molecular axis parallel to the surface is more likely. Schmeisser et al. (1985) have made an angle-resolved photoemission study of ordered, physisorbed CO on Ag(111). The spectra exhibit a momentum-dependent splitting of the 5σ and 4σ features. By comparing the 5σ dispersion with that observed for physisorbed nitrogen on graphite, the same "herringbone" in-plane structure of the latter is deduced for CO/Ag(111). The splitting arises from the presence of two molecules per unit cell in the "herringbone" structure.

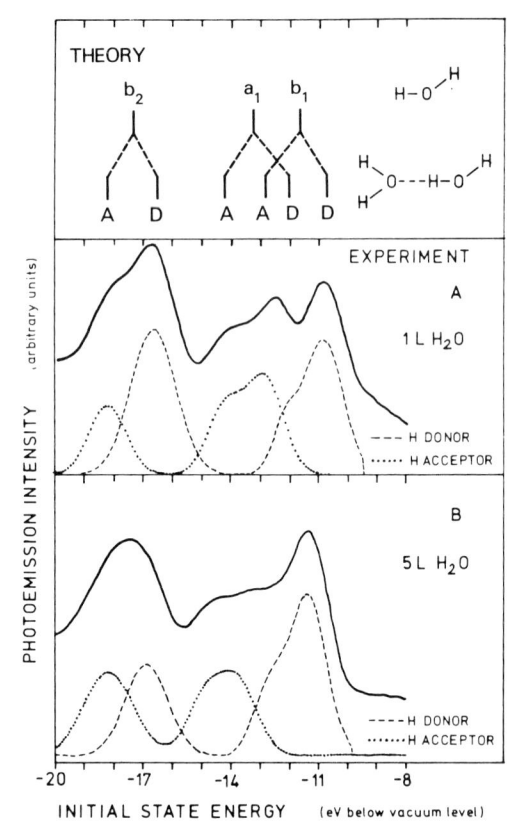

Fig. 26. Analysis of the photoemission spectrum of a monolayer (1 L exposure) of physisorbed water at 41 eV photon energy. The observed splitting of the three molecular orbitals is explained by comparison with the calculated binding energies of the proton acceptor (A) and proton donor (D) levels for the linear water dimer (schematic sketch). For comparison, the spectrum of a multilayer is shown in the lower panel, where the splitting disappears since most of the molecules are equivalent (such as in ice). (After Schmeisser et al. 1983b.)

Rare gases are physisorbed onto surfaces in a layer-wise fashion. This is reflected in layer-dependent shifts of photoelectron and Auger electron energies. Chiang et al. (1986) have examined these effects for a wide range of rare gas/metal surface combinations. The shifts in Auger electron energy are approximately three times the shifts in photoelectron energy. The shifts are explained in terms of hole screening, both by the distance-dependent, image-charge potential on the metal and by dielectric screening via polarisation of the neighbouring rare gas atoms. A jellium model of the substrate and a dielectric continuum model of the adsorbate yields screening energies in good agreement with experimental shifts. Similar measurements on a graphite substrate confirm this model, with reduced shifts appropriate to the smaller electron density in graphite (Mandel et al. 1985a). These layer-dependent shifts have allowed the investigation of the layer-by-layer bandstructure of physisorbed Xe/Al(100) (Mandel et al. 1985b).

Rare gas/graphite physisorption also provides important model systems for studying two-dimensional melting by X-ray diffraction techniques (Dimon et al. 1985).

4.3. Geometry of adsorbate structures

It is now well established, mainly from LEED studies, that in most cases atoms adsorbed on single-crystal surfaces occupy the sites which would have been taken by the next layer of substrate atoms, and assume adsorbate–substrate distances which are close to the sum of radii taken from compilations such as those of Pauling (1960). This result is exemplified by studies of Ni(001)$c(2 \times 2)$S, which has been used as the archetypal system for testing out new surface structural techniques. LEED (Demuth et al. 1974), photoelectron diffraction (Rosenblatt et al. 1981) and SEXAFS (Brennan et al. 1981) all agree that the sulphur adatoms sit in the four-fold hollows of the unreconstructed nickel substrate, at a S–Ni distance of 2.23 Å.

Since there is now a considerable bank of data for systems which fit this typical adsorption pattern, interest has shifted somewhat from the determination of straightforward adsorption geometries towards the chemically important small differences from the Pauling radii (section 4.3.1), systems where the "standard" adsorption site is not adopted (section 4.3.2), changes to the substrate geometry induced by adsorption (section 4.3.3), and molecular adsorbates (section 4.3.4).

4.3.1. Chemical effects on Pauling radii

Unfortunately, in most cases, LEED is not sufficiently accurate to reveal the fine differences between adsorbate radii which are chemically significant (Mitchell 1985), and much of the work in this area is now being carried out using SEXAFS. A striking example of a large change in adsorbate distance is given by Cs adsorbed on Ag(111) (Lamble et al. 1986b), where the nearest-neighbour (nn) Cs–Ag distance is found to be 3.20 Å for a coverage of 0.15 monolayers of Cs and 3.50 Å for a higher coverage of 0.3 monolayers of Cs (this high-coverage phase represents almost a full close-packed overlayer of the large caesium atoms, with coverages expressed in terms of the silver surface atoms). With all random and systematic errors included, a change in nn distance of 0.3 ± 0.06 Å is quoted between the two phases, attributed to a change in the nature of the caesium adatom, from an ionic state at low coverage towards a covalent state at higher coverage.

This contrasts markedly with studies of chlorine adsorption on the same Ag(111) surface, where an identical Cl–Ag nn distance is seen at different Cl coverages (Lamble et al. 1986a). The reason for this difference in behaviour is postulated to be related to the change in potential induced at the surface by the adatom. Around an adsorbed chlorine atom, the total charge distribution one lattice vector away, along the surface, is little different from that of the clean metal and thus another chlorine atom may be adsorbed in a site with a nn substrate distance the same as that of the first Cl atom. However, the situation is dramatically different for adsorption of Cs atoms: the charge distribution on the surface near to an adsorbed Cs atom is considerably different from the potential of the clean metal surface. This is another way of expressing the fact that a relatively low coverage of alkali metal atoms gives a marked reduction in the metal work function. Thus the adsorption of further Cs atoms near to one already on the surface will not be at the same adsorbate–substrate distance. Presumably the detailed energy balance will lead to an equilibrium distribution of adatoms at a particular adsorbate–substrate distance rather than a spread of dissimilar nn distances. The determination of distances at the accuracy obtainable from SEXAFS, ±0.01 or 0.02 Å in favourable cases, should allow deductions about detailed chemistry at surfaces, in the same way as X-ray crystallographic data have for bulk crystals.

4.3.2. Non-standard bonding sites
Rather few adsorbate systems adopt an unconventional bonding site on metal surfaces. The open (110) face of fcc metals offers several possible adsorption sites, and on Cu(110) (Döbler et al. 1984) and on Ag(110) (Puschmann and Haase 1984) oxygen has been shown by SEXAFS to occupy the two-fold bridge in the (100) direction. The situation for semiconductor surfaces is somewhat different. On metals bonding is largely non-directional, but the covalent character and consequent directional bonding of semiconductors suggests that a local orbital approach based on saturation of dangling bonds might be successful in predicting their surface adsorption behaviour. Many experiments (e.g. Ludeke and Koma 1975, Pianetta et al. 1975) have been performed on the surface structure of reconstructed semiconductors using photoemission, electron energy loss spectroscopy (EELS) and LEED, but none of these studies has been able reliably to test this idea. This is in part because no technique other than SEXAFS provides the necessary accuracy, but also because all other measurements require long-range order within the overlayer, which may be impossible to achieve. Experiments on the adsorption of Te and I on clean Si(111)(7 × 7) and Ge(111)(2 × 8) surfaces (Citrin et al. 1982) showed that SEXAFS, not being hampered by either of those limitations, could enable successful testing of the bond saturation arguments.

The data were analysed to give nearest-neighbour and second nearest-neighbour distances and also effective surface atom coordination numbers, with the polarisation dependence of this latter parameter proving particularly useful in distinguishing between possible adsorption sites. From these measurements iodine is found to adsorb on both Si(111) and Ge(111) in the atop site, which is what would be anticipated for a monovalent atom saturating the semiconductor's dangling bond. The case of tellurium adsorption, however, is distinctly different. It is not too surprising to find a difference between the adsorption behaviour on Si and Ge in view of the change in anion coordinations in the analogous bulk compounds (SiI_4 versus GeI_2 and $SiTe_2$ versus GeTe). However, this would suggest a different site for iodine adsorption on the two surfaces, which does not occur. The data obtained lead to the conclusion that a two-fold bridge site is occupied for Te on Si(111) but that, on Ge(111), the Te atoms occupy the "hcp" three-fold hollow, in alternate rows.

It is understandable *a posteriori* that divalent Te should favour a two-fold bridging site on Si(111), with one electron per dangling bond, and this again fits in with the directional bond saturation arguments advanced earlier. Te on Ge is more complex, and indeed the model suggested does not explain all of the observations relating to longer-range order, for instance the (2×2) symmetry of the annealed surface. However, it is clear that the three-fold hollow site above a substrate atom is the only one of the possible high-symmetry sites which fits the short-range SEXAFS data. Reconstructions involving several layers of Ge atoms may have to be invoked to explain the long-range structure: such an analysis could well be beyond the capabilities of SEXAFS, which is essentially a short-range, single-scattering phenomenon. Finally it is worth noting that, out of the four high-symmetry sites possible for the (111) surface, three have been identified with Te or I on Si and Ge, and none of them corresponds to the fcc three-fold site most commonly observed for adsorption on (111) metal surfaces.

4.3.3. Adsorbate-induced changes in substrate geometry
In general, the structures of the more close-packed faces of crystals are perturbed little by adsorption. However, more open faces may show significant effects. The (110) surface of nickel has been shown by ion scattering (van der Veen et al. 1979) and LEED (Baudoing et al. 1985) to exhibit "differential relaxation", with the clean surface showing a contraction between the top two layers, and an expansion between the second and third layer, relative to the bulk lattice spacing. The same studies showed that this is reversed by adsorption of sulphur in the half-monolayer $c(2 \times 2)$ coverage, which causes an expansion of the top layer nickel spacing.

The nitrogen-induced reconstruction of W(100) is more dramatic. The $c(2 \times 2)$ LEED pattern with split half-order spots has been interpreted as showing a large-scale movement (a displacive phase transition) of top layer tungsten atoms into "rafts" containing about 49 atoms (Griffiths et al. 1981) (fig. 27). Photoemission core level shift measurements (Jupille et al. 1986) show that the centre of gravity of the tungsten $4f_{7/2}$ peak changes continuously as a function of nitrogen coverage, with a discontinuity at around 0.4 monolayers, where the reconstruction is complete. This is attributed to a continuous contraction of the surface layer into domains, giving a higher average coordination for the surface tungsten atoms.

4.3.4. Geometry of molecular adsorbates
Adsorption of molecular species is of particular interest for its relevance to chemical reactions and catalysis. Some studies have been performed using LEED, but this is a tricky area since many molecules are liable to "crack" in an electron beam. The orientation and bonding of molecular adsorbates may be determined in many cases by photoemission, but the analysis and interpretation is often complex. However, the study of molecular adsorption has been revolutionised

○ Tungsten atom
● Nitrogen atom

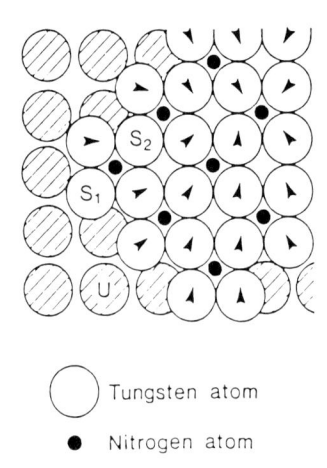

Fig. 27. A contracted domain of the nitrogen-induced reconstructed surface of W(100).

since the pioneering synchrotron radiation X-ray absorption experiments of Stöhr and Jaeger. Figure 28 shows the absorption spectra near the C K-edge (NEXAFS) for a saturation coverage of CO on Ni(100), measured at different angles to the photon beam (Stöhr et al. 1981, Stöhr and Jaeger 1982a). The spectra are dominated by two peaks having opposite polarisation dependence. A π resonance is found near threshold, arising from transitions from the C-1s state to the unfilled π^* bound state (antisymmetric) molecular orbital. The second, broader peak arises from a transition from the C-1s state to a resonance of σ symmetry in the continuum, a so-called "shape resonance" (Dehmer and Dill 1975). A σ shape resonance in CO adsorbed on Ni has been observed in valence photoemission studies (Allyn et al. 1977), but the π resonance, lying below the continuum K-shell ionisation threshold, cannot be seen by photoemission.

The variation of cross section of the resonances with polarisation angle gives the molecular orientation. The transitions are governed by dipole selection rules so that, with absorption at a K absorption edge, a π resonance has maximum amplitude when the polarisation vector E is perpendicular to the intramolecular bond and zero intensity when E lies parallel to this axis; the σ shape resonance has the opposite behaviour, being zero when E is perpendicular to the interatomic axis and a maximum if E lies along this bond. It is then easy to see from fig. 28 that the CO molecule stands upright on the Ni(100) surface. However, this experiment alone gives only the orientation with respect to the top plane of nickel atoms, not the detailed arrangement of the bonding to the individual surface atoms, nor does it indicate which end of the molecule is next to the surface. These facts may be determined from SEXAFS measurements on the carbon or oxygen atoms. These findings may also be applied to more complex molecular adsorbates. It is readily determined that benzene (C_6H_6) lies down on Pt(111) where pyridine (C_5H_5N) stands up, bonding to the surface through the nitrogen lone pair orbitals (Johnson et al. 1983).

It has further been shown that the energy of the σ shape resonance above

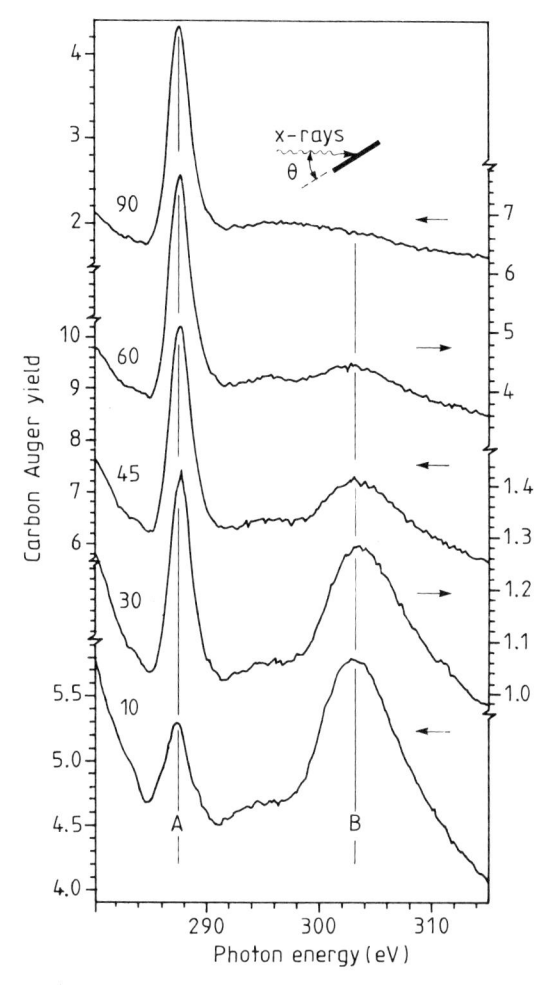

Fig. 28. NEXAFS spectra above the C K-edge for CO on Ni(100) at $T = 180$ K as a function of incidence angle θ. (After Stöhr et al. 1981.)

threshold is inversely correlated with the intramolecular bond length, with a long bond giving a shape resonance close to threshold and a shorter bond showing a peak at higher energy. This effect is nicely demonstrated in fig. 29, which shows the NEXAFS above the O K-edge (Stöhr et al. 1983) for three molecules that have carbon–oxygen bonds and are chemisorbed on Cu(100). In each case the angle of the E vector to the surface has been chosen to maximise the intensity of the σ shape resonance. The three molecular species are CO, with a short (1.13 Å) triple C–O bond, formate (HCO_2), with a quasi-double C–O bond, and methoxy (CH_3O), whose single C–O bond is long (1.43 Å). Close to the K-edge threshold, which coincides with the O 1s binding energy as determined by photoemission (marked "XPS"), a π resonance peak is seen for CO and formate, corresponding

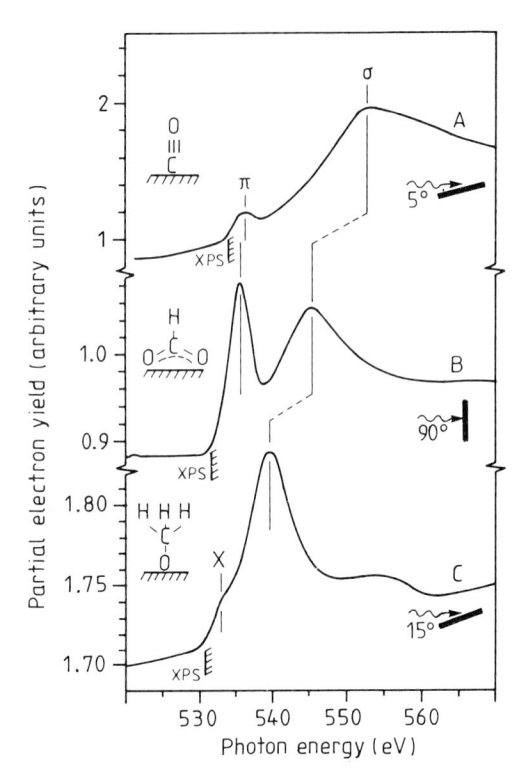

Fig. 29. NEXAFS spectra of the O K-edge region for CO, formic acid (HCOOH), and methanol (CH$_3$OH) on Cu(100). HCOOH and CH$_3$OH were dissociated to formate, HCO$_2$, and methoxy, CH$_3$O, respectively, under controlled experimental conditions. π and σ are two final-state intramolecular resonances. The photon incidence angle was chosen to maximise the intensity of the σ feature in each spectrum. The O-1s binding energies determined by photoemission are labelled "XPS". (After Stöhr et al. 1983).

to transitions from the O 1s state into the partially unoccupied π^* antibonding orbital in these molecules. Note that the π peak is minimised for CO with the depicted experimental geometry, as for CO on Ni(100). There is no such peak for methoxy, since there is no π character in the single C–O bond. A σ shape resonance is seen for all three molecules, as marked in fig. 29, and its position does indeed decrease in energy with increasing C–O bond lengths. With care, such spectra may be used to estimate intramolecular bond lengths in adsorbed molecules.

The short-range nature of SEXAFS uniquely gives the possibility of examining adsorption on other than single-crystal substrates, provided that a well-defined local order is adopted. This has recently been demonstrated in studies of the barium/oxygen/tungsten complex on the surfaces of real thermionic cathodes (Norman et al. 1987). SEXAFS oscillations were analysed to give the distances from barium to its near-neighbour oxygen and to the substrate metal; these show that the barium sits on top of the oxygen atoms, which like to lie in hollows of the more open planes of tungsten present at the polycrystalline surface. This orientation is consistent with that expected for lowering of work function by formation of a favourable surface dipole, but such surface structural information had not previously been obtained, and probably could not be directly found from any other technique.

4.4. Reactions

Given appropriate conditions of exposure and substrate temperature, an adsorbate will proceed from the relatively simple chemisorption phase to a complex reacted phase. Typically, there is simultaneous loss of surface specificity and long-range order, which limits the surface techniques that can be applied to these systems. The reaction regime is illustrated here with examples of semiconductor oxide formation, co-adsorption of CO and the dissociative adsorption of complex molecules. These "surface" systems can be related to the wider effort in EXAFS of thin oxide films and metal catalyst systems.

In the past decade, the technique of soft X-ray photoemission has been extensively applied to study oxide formation at semiconductor surfaces (Spicer et al. 1976, 1977, Bauer et al. 1979, Landgren et al. 1984). A topic of current interest is the enhancement or control of oxide formation by predeposition of thin metal overlayers. For example, chromium enhances the oxidation of silicon by several orders of magnitude (Franciosi et al. 1985). As shown in fig. 30, there is a chromium coverage threshold for the effect, which is related to the onset of chromium–silicon intermixing. Similar measurements have been reported for other metals on silicon (Abbati et al. 1982, Rossi et al. 1982) and GaAs (Su et al. 1979, Chang et al. 1986). Suggested mechanisms for the effect include catalysis of the oxygen by the intermixed phase (Franciosi et al. 1985) or by the metal (Su et al. 1979), or the provision of less strongly bound silicon at the surface (Abbati et al. 1982). The observation of enhancement has been used as an indirect indicator of metal–semiconductor intermixing (Rossi et al. 1982). The extensive field of direct spectroscopy of metal–semiconductor reactions is discussed in section 5.1 below.

The bonding of CO to clean metal surfaces is effectively probed by angle-resolved photoemission techniques (see section 4.1.4 above). The promoting ability of alkali metals in catalytic reactions prompts the many investigations of the co-adsorption of CO and alkali metals on metal surfaces using the same techniques. For example, Heskett et al. (1985) have investigated the co-adsorption of CO and potassium on Cu(100). The CO/Cu(100) system is relatively weakly bonding and the effects of the potassium are easier to follow. Dramatic reduction in the satellite intensity (see section 4.1.4 above) is observed and a hybridisation of the CO 5σ and 1π levels is reported. This latter interpretation means that an observed lowering of symmetry does not necessarily imply a tilting of CO axis relative to the surface normal. Building on the model of Blyholder (1964), a model is proposed in which the bonding becomes predominantly π-like,

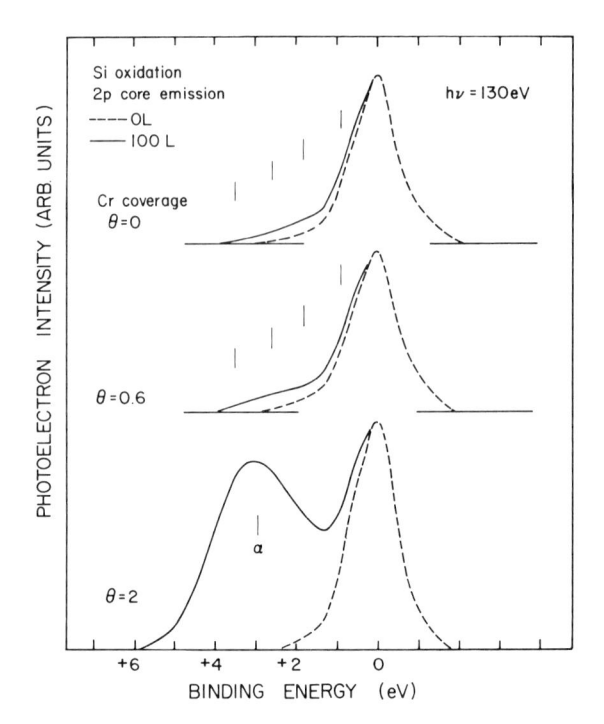

Fig. 30. Si-2p core emission from cleaved Si(111)(2 × 1) surfaces before (dashed line) and after exposure (solid line) to 100 L of activated oxygen. The vertical bars mark the position of Si-2p oxide features associated by Hollinger and Himpsel (1983) with silicon atoms coordinated with one, two, three, and four oxygen atoms. Top: A freshly cleaved surface. Midsection: A 0.6 Å Cr overlayer deposited on a freshly cleaved surface. Bottom: Effect of a 2 Å Cr overlayer on the Si(111) surface oxidation. After exposure to 100 L of activated oxygen (solid line) a major oxide band emerges. The vertical bar marks the position of a major Si-2p oxide feature identified during oxidation of amorphous silicon. (After Franciosi et al. 1985.)

involving both charge transfer into the CO 2π level and direct interaction between the potassium and the CO 1π level.

In favourable circumstances, these photoemission techniques can be extended to the adsorption of more complex molecules on surfaces. For example, Albert et al. (1982) have employed "forbidden geometry" methods (see section 4.1.4 above) to deduce the structures of low-temperature and high-temperature phases of acetylene and ethylene adsorbed on Pt(111). NEXAFS and SEXAFS techniques are more generally applicable. Puschmann et al. (1985) have applied the NEXAFS π/σ ratio method (see section 4.3 above) to determine the orientation of formate (HCOO) on Cu(110); SEXAFS methods reveal an adsorption site in which the oxygen atoms almost bridge a pair of top-layer ("ridge") copper atoms. The thermal decomposition of the thiophene (C_4H_4S) molecule on the Pt(111) surface has also been investigated using NEXAFS (Stöhr et al. 1984), in conjunction with thermal desorption, XPS and EELS measurements; the mechanism for breaking the sulphur out of the thiophene ring is deduced. Studies of thiophene on Ni(100) (Stöhr et al. 1985) show that sulphur is broken out of the thiophene ring by bonding to four-fold hollow nickel sites, and this dissociation can be blocked by pre-adsorption of oxygen atoms, which occupy the active sites on the nickel surface. The hydrodesulphurisation process, involving the breaking of C–S bonds in polyatomic organic molecules, is of great industrial importance. It seems likely that these techniques will profitably contribute to our understanding of promotion and poisoning in catalytic systems.

5. Interfaces

The clean surface and the adsorbate-covered surface may be viewed as vacuum–solid and ambient–solid interfaces, respectively; the solid–solid interface is the third area of study within surface science. In greater part the motivation for this area derives from microelectronics. All devices require electrical contacts, most commonly at metal–semiconductor interfaces. Some devices are based on Schottky barriers at such interfaces; others depend on heterojunctions at semiconductor–semiconductor interfaces. The move towards 3D stacking of devices requires a greater understanding of insulator–semiconductor interfaces.

With a few possible exceptions (e.g. X-ray "total external reflection" techniques), surface science techniques are not applicable to interfaces *per se*. Application is restricted to test systems in which the interface is of the order of an escape depth (1 nm) below the free surface. Fortuitously, these test interfaces exhibit much the same electronic properties as more realistic interface systems. In any event, the shrinking of the dimension of electronic devices normal to the interface is creating a closer matching of system to techniques.

This section begins with a brief discussion of metal–metal interfaces (section 5.1), where the interest is rather more traditional surface than interface. Then, reflecting their relative weights in device-related research, there are subsections on metal–semiconductor (section 5.2), semiconductor–semiconductor (section 5.3) and insulator–semiconductor (section 5.4) interfaces.

5.1. Metal–metal interfaces

The primary interest in these structures is in the adlayer metal film. Monolayer films can exhibit distinct magnetic (see section 3.2 above), electronic, chemical and structural properties. Preparation of clean, ordered films presents experimental difficulties so that noble metals are generally preferred. For example, angle-resolved valence photoemission techniques have been applied to Cu/Ag(001), and the resultant 2D bandstructure compared with slab calculations (Stoffel et al. 1985). Similar measurements on the inverse system Ag/Cu(001) have been reported (Tobin et al. 1986).

The structural properties have also attracted attention. For some overlayer systems the dimensionality is further reduced to quasi-1D. Binns et al. (1986) have employed core and valence photoemission techniques together with LEED/

Auger methods to investigate thallium chains on silver(100). The chain structure has the potential for Peierls distortion, as was shown for thallium chains on copper(100) (Binns et al. 1984). No equivalent distortion is found for Tl/Ag(100), which is attributed to the presence of first-order commensurability in this system.

The "atomic layer engineering" of thin structures afforded by molecular beam epitaxy (MBE) techniques has been coupled with short escape depths in photoemission to probe distance-dependent effects. In one experiment a Au(111) monolayer deposited onto Ag(111) was buried beneath a growing Ag(111) overlayer (Hsieh et al. 1986). Measurement of the Au-4f core binding energy as a function of Ag overlayer thickness showed the effective spatial range of surface effects to be about three atomic layers. This implies that significant sub-surface core shifts can occur for close-packed metal surfaces. A similar approach has been used to probe the spatial extent of the Ag(111) surface state (Hsieh et al. 1985).

5.2. Metal–semiconductor interfaces

In the past decade a considerable effort has been expended on Schottky barrier formation at metal–semiconductor interfaces; the general approach is the controlled deposition of sub-monolayer metal onto clean semiconductor surfaces. A detailed review of this subject is to be found in the following chapter of this volume (Brillson 1987). To complete this surface science survey, however, major pointers are also presented here. Perhaps the greatest single contribution of these studies has been the demonstration that few metal–semiconductor interfaces are abrupt. Examples of abrupt interface studies (which are really chemisorption systems, see section 4.1 above) are presented in section 5.2.1. Non-abrupt interfaces (more akin to reactions, see section 4.4 above) are treated in section 5.2.2. The spectroscopic determination of Schottky barrier height by photoemission is discussed in section 5.2.3.

5.2.1. Abrupt interfaces

The techniques of angle-resolved valence photoemission have been applied to the ordered chemisorption of several metals on semiconductor surfaces. A recent example due to Uhrberg et al. (1986) is the As/Si(100) system. This surface shows the two-domain, (2 × 1) LEED pattern also characteristic of clean Si(100), but with a lower background intensity. This invites comparison with the cleansurface structure. Experimental surface state $k_{\parallel}$ dispersions for the [001] and [0$\bar{1}$1] directions are shown in fig. 31a. Corresponding theoretical dispersions shown in fig. 31b are the results of energy minimisation calculations of an As–As dimer structure, in which the As atoms have replaced the silicon atoms involved in dimer formation on the clean surface; each As atom is bonded to two Si atoms and to the partner As atom in a symmetric dimer.

The agreement between experiment and calculation is critically dependent on bond length and dimer buckling. Along [0$\bar{1}$1] the experiment samples two domains, resulting in the two surface state bands S_{α} and S_{β}. In the theory each

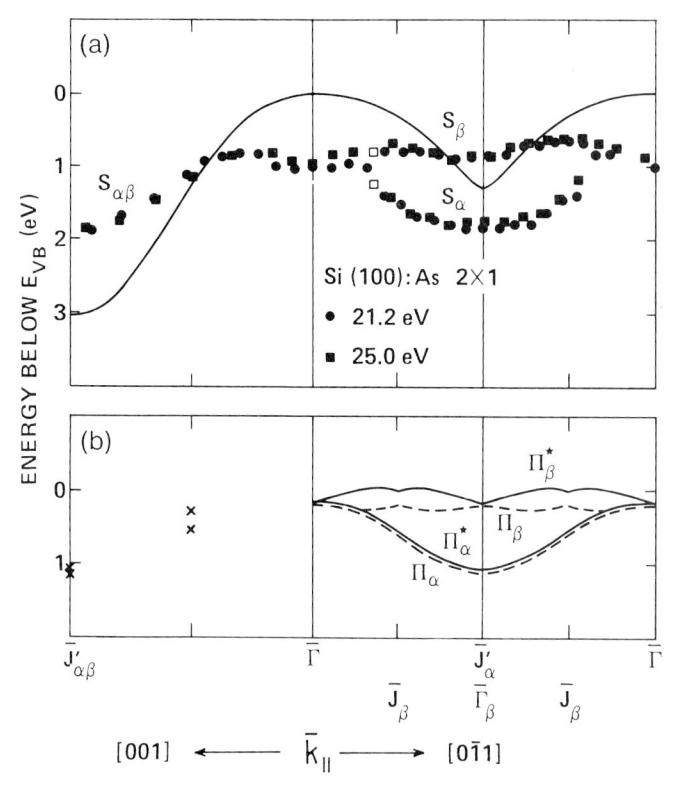

Fig. 31. (a) Experimental surface state dispersion for Si(100)As(2 × 1) in the [001] and [0$\bar{1}$1] directions. The full curve shows the edge of the projected bulk band structure. Full (empty) symbols correspond to strong (weaker) spectral features. (b) Corresponding theoretical surface state dispersions for the optimised symmetric As–As dimer model along the $\bar{\Gamma} \to \bar{J}'_\alpha$, $\bar{\Gamma} \to \bar{J}_\beta$, and $\bar{\Gamma} \to \bar{J}'_{\alpha\beta}$ lines. (After Uhrberg et al. 1986.)

band is split into π and π^* states, which are the symmetric (bonding) and antisymmetric (antibonding) combinations of the dangling hybrids on the As atoms [cf. dangling bonds on clean Si(100)]. The $\pi-\pi^*$ splitting depends on the As–As dimer bond length d. Theory predicts $d = 2.55$ Å giving a maximum splitting of 0.23 eV, which would not be resolved in these measurements. Reduction of the bond length increases the splitting dramatically, as does buckling of the dimer. The observation of single states constrains the geometry to that described above. This method of structure deduction by surface electronic properties parallels that discussed in section 3.4 above for clean surfaces.

The local geometry of metal atoms adsorbed on a semiconductor surface has been probed by SEXAFS. Stöhr and Jaeger (1982b) have recorded Ag L_2-edge SEXAFS for different silver coverages on Si(111)(7 × 7). Figure 32 shows corresponding Fourier transforms for thick (50 Å) layer and 2.5 monolayer (ML) room temperature deposits, together with polarisation-dependent data for a $\sqrt{3} \times$

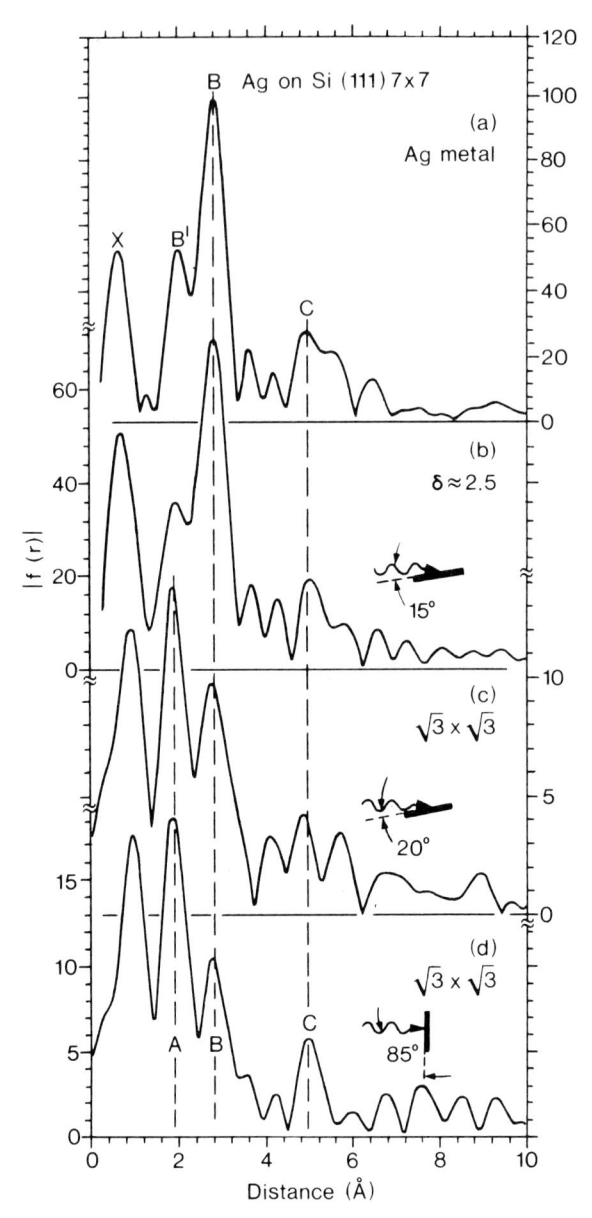

Fig. 32. Absolute value of Fourier transform of the SEXAFS signal for (a) Ag metal, (b) $\delta = 2.5$ monolayers Ag on Si(111)(7×7), (c) $\delta = \frac{1}{3}$ monolayers Ag on Si(111)(7×7) heated to 500°C ($\sqrt{3} \times \sqrt{3}$) at $\theta = 20°$, (d) same as (c) for $\theta = 85°$. (After Stöhr and Jaeger 1982b.)

√3 Ag/Si(111) surface obtained by annealing a $\frac{1}{3}$ ML deposit to 500°C. (The peak X can be ignored for reasons discussed in the reference.) Peak B is the first nearest-neighbour distance in silver metal, peak B' is an associated satellite and peak C is the third nearest-neighbour distance. It is apparent that the 2.5 ML data are identical to those of the bulk metal, indicating an abrupt interface.

The Fourier transforms of the $\sqrt{3} \times \sqrt{3}$ Ag overlayer are markedly different from that of the metal. The strong feature A corresponds to the Ag–Si nearest-neighbour distance; appropriate phase shifts result in a value of 2.50 ± 0.05 Å. The relative strength of A and B is interpreted as the co-existence of silver in the $\sqrt{3} \times \sqrt{3}$ chemisorption sites and in clusters. The variation in the A:B ratio with angle of incidence implies that the Ag adsorption site is close to (or possibly underneath) the plane of silicon atoms. The Ag–Si distance then requires a lateral displacement of surface silicon atoms, leading to a honeycomb structure and a silver coverage of two-thirds of a monolayer. This analysis supports the model suggested by low-energy ion scattering (Saitoh et al. 1980) and disagrees with the trimer model suggested by LEED/Auger (Wehking et al. 1978) and angle-resolved photoemission (Hansson et al. 1981).

5.2.2. Non-abrupt interfaces

Strong chemical reactions commonly occur when metals are deposited on clean semiconductor surfaces. The primary probe of these interactions is soft X-ray core photoemission spectroscopy. For example, the Al/InP(110) interface provides the spectra shown in fig. 33 (Zhao et al. 1983). With increasing coverage of Al, the

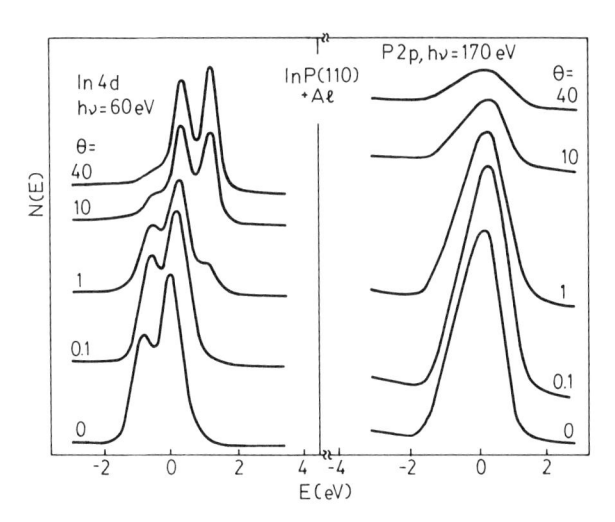

Fig. 33. Evolution of the In-4d and P-2p peaks for increasing Al coverage of InP(110). The horizontal scales are referred to the high-coverage positions of the In-4d$_{5/2}$ peak and of the P-2p peak. The spectra were *not* corrected for bandbending change effects, and the low-coverage shifts reflect the shifts of the Fermi level in the gap. The appearance of a free In, low binding energy component given by the exchange reaction is very clear in the In-4d spectra above 0.1 monolayer coverage. (After Zhao et al. 1983.)

P-2p core emission is strongly attenuated while the In-4d core emission progressively converts to a lower binding energy satellite. This "chemical shift" feature is attributed to phase-segregated indium, which has been liberated from the InP lattice by an exchange reaction. That this reaction is thermodynamically favourable can be shown from the relative heats of formation of InP and AlP (McGilp and McGovern 1985). In general, heats of metal–cation alloying have also to be considered.

The variation of the adlayer metal core emission also carries information on the reaction (although it is usually more difficult to interpret). In this Al/InP study the Al-2p binding energy for a thick Al overlayer was found to be about 1 eV lower than that for a sub-monolayer (0.05 ML) deposit. Between these two extremes there were significant continuous changes in the Al-2p profile. The general interpretation of these spectra is in terms of cluster formation. It has been suggested that critical size clustering triggers the exchange reaction (Zunger 1981, McKinley et al. 1982). Cluster formation has also been connected to Fermi level pinning (Daniels et al. 1982) – see section 5.2.3 below. Further information on the adlayer growth mode can be deduced from the rate of attenuation of the substrate core emission (Williams 1985).

The local geometry of non-abrupt interfaces has been probed by SEXAFS. For example, Stöhr and Jaeger (1982b) have examined the Pd/Si(111) interface. Unlike the abrupt Ag/Si(111) interface discussed in section 5.2.1 above, the SEXAFS spectrum of a 1.5 monolayer Pd room temperature deposit closely resembles that of bulk Pd_2Si. Silicide formation is an important reactive interface phenomenon, which can produce an abrupt silicide–silicon interface.

5.2.3. Schottky barrier height measurements

In most photoemission spectrometers binding energy is referenced to the Fermi level of the sample. As illustrated in fig. 34, the introduction of bandbending (see

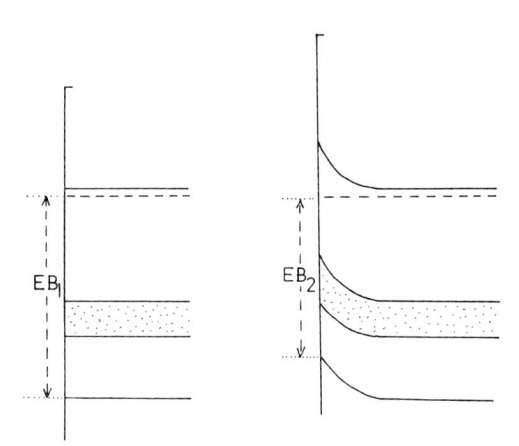

Fig. 34. Illustration of the way bandbending leads to a change in binding energy of a core level at the surface with respect to the Fermi level.

also section 3.4 above) can cause a rigid shift of measured binding energy compared to "flat-band" emission; this assumes that the escape depth is much less than the bandbending depth, which is usually the case. If the clean surface is in the flat-band condition, the difference in the core binding energy $EB_1 - EB_2$, before and after metal deposition, is equal to the bandbending contribution to the induced barrier height. These "bandbending" or "Fermi-level" shifts occur along with the chemical shifts, and sometimes it can be difficult to distinguish the two. In the spectra shown in fig. 33, a bandbending shift of ~0.3 eV occurs, most of it for the first deposition shown. In principle, the bandbending information can also be determined from valence band emission, but here there are potential problems of overlapping metal emission and metal-induced scattering of substrate emission.

At an abrupt interface, the estimation of induced bandbending is easier. For example, Ludeke et al. (1983) have followed the development of the Ag/GaAs(110) interface bandbending. The Ga- and As-3d core emissions were carefully deconvolved into surface and bulk components in order to rule out additional contributions indicative of a metal–semiconductor interaction. A composite of the progression of the Fermi level within the surface bandgap of n- and p-type GaAs with Ag coverage is shown in fig. 35. There is steady movement over four decades of thickness. In general, the barrier is well established by the monolayer stage. Barrier formation at this particular interface has been analysed in terms of screened silver-derived interface states. Other models of Schottky barrier formation more appropriate to non-abrupt interfaces are discussed in detail by Brillson (1987).

5.3. Semiconductor–semiconductor interfaces

The junction between two dissimilar semiconductors (the heterojunction) is a potentially more flexible device unit than either the metal–semiconductor junc-

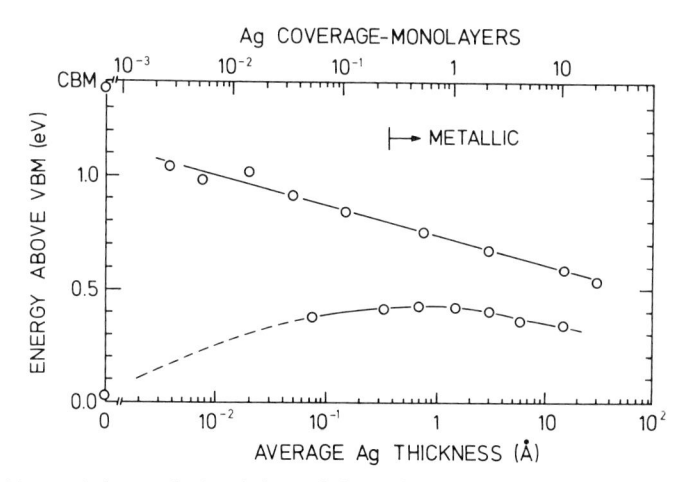

Fig. 35. Positions relative to the bandedges of the surface Fermi level for n-type (top) and p-type (bottom) GaAs(110) with Ag coverage. (After Ludeke et al. 1983.)

tion or the semiconductor homojunction; it is also more complex to understand. A programme of study of these interfaces has developed alongside that on Schottky barrier formation, employing the same techniques and, indeed, similar microscopic modelling. The measurement of interface band offsets (section 5.3.1) parallels the barrier measurements discussed in section 5.2.3 above. The spectroscopy of interface reactivity (section 5.3.2) is essentially that discussed in section 5.2.2 above.

5.3.1. Interface band offsets

A general energy band scheme for a heterojunction is shown in fig. 36a. The dissimilar bandgaps of the two semiconductors result in band offsets in the valence (ΔE_v) and conduction (ΔE_c) band edges; the band gap difference is $\Delta E_g = \Delta E_c + \Delta E_v$. In addition to these barriers, there may be bandbending barriers extending on both sides of the interface. The general approach of working with a thin overlayer of one semiconductor deposited onto another is illustrated in fig. 36b.

In the spectroscopy of states below the vacuum level, photoemission can probe only the filled states, in this case the valence band offset ΔE_v. Two methods have been described by Katnani and Margaritondo (1983). The first, which is only applicable when the valence offset is large, is illustrated in fig. 37. Following the deposition of 4 Å of Si onto clean GaP(110), the Si band edge is clearly visible, as is a distinct GaP feature at higher binding energy. Alignment of this latter feature

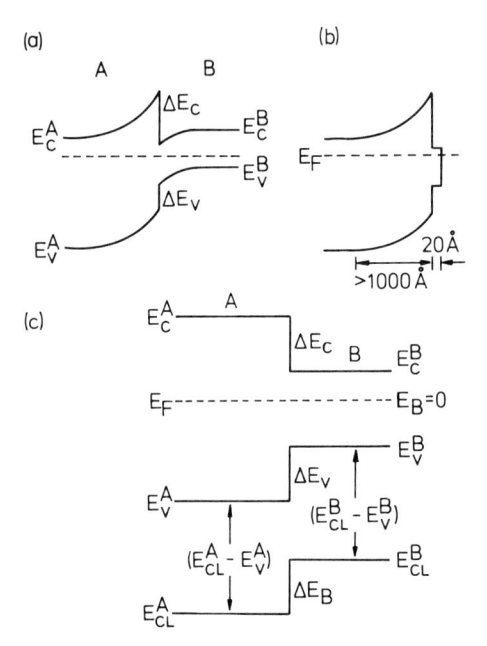

Fig. 36. (a) Typical "buried" heterojunction, (b) "exposed" thin heterojunction accessible to XPS analysis, and (c) schematic energy band diagram of thin, abrupt heterojunction interface. (After Waldrop et al. 1985.)

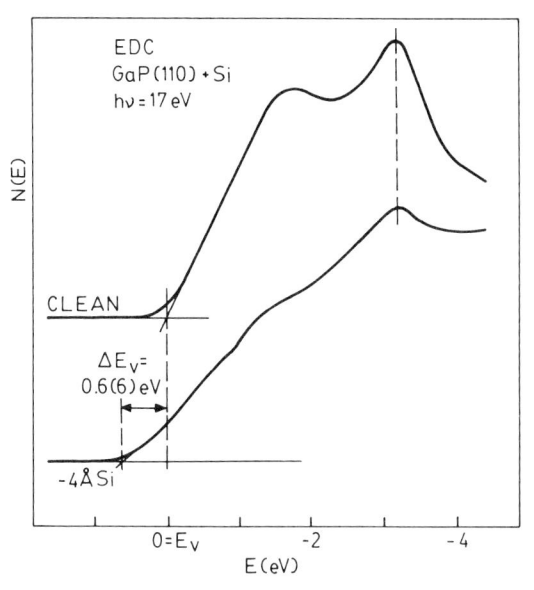

Fig. 37. Energy distribution curves taken at a photon energy of 17 eV on cleaved, clean, and Si-covered GaP. The estimated valence band discontinuity from these curves is 0.6 ± 0.1 eV. (After Perfetti et al. 1984.)

in the two spectra allows a direct measurement of $\Delta E_v = 0.6 \pm 0.1$ eV at this interface (Perfetti et al. 1984). The second method is less direct, employing a combination of valence edge and core binding energy measurements. As shown in fig. 36c, the core to valence edge separation in each semiconductor ($E_{cl} - E_v$), together with the separation of the two cores (ΔE_B) at the interface, gives the valence offset via the relation ($\Delta E_v(A-B) = \Delta E_B(A-B) + (E_{cl}^B - E_v^B) - (E_{cl}^A - E_v^A)$) (Waldrop et al. 1985). Core chemical shifts can be problematical in this general method. Bandbending can be monitored as discussed above for metal–semiconductor interfaces.

The conduction band offset can be inferred from the valence band offset and the band gap difference; it may also be probed directly by the partial yield technique. The example shown in fig. 38 is again that of Si/GaP(110). The yield spectrum at the Ga-3d core absorption edge reveals the two conduction band edges, and their separation can be directly measured. Possible sources of inaccuracy here are differential excitonic effects, chemically shifted core edges and, as shown in fig. 38, vacant surface states (Perfetti et al. 1984).

Many heterojunction systems have been studied with these techniques. Margaritondo (1986) has recently tabulated valence band offsets for Ge and Si interfaces with III–V semiconductors. The values compare well with the predictions of a general LCAO calculation (Tersoff 1984). The agreement is of the order of 0.1 eV, which is also the general experimental accuracy. However, it is important to get below this limiting value and this goal is being actively pursued. As with metal–semiconductor junctions, the interface electronic states present a

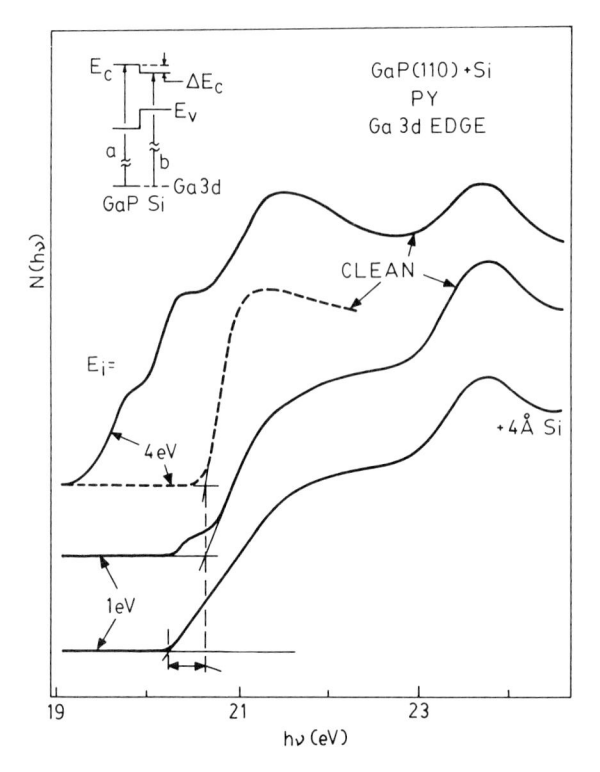

Fig. 38. Partial yield spectra for clean and Si-covered GaP, showing the Ga-3d optical absorption edge. The curve taken at a kinetic energy $E_k = 4$ eV is more surface sensitive than those taken at $E_k = 1$ eV, and it exhibits optical transitions to surface states. The dashed line shows the same spectrum after correction for those surface contributions. It was argued that the Si-induced shift in the $E_k = 1$ eV PY spectral edge reflects the conduction band discontinuity. (After Perfetti et al. 1984.)

formidable spectroscopic challenge. A discussion of future directions is given by Margaritondo (1985).

5.3.2. Interface reactivity

Interface reactivity is best studied by soft X-ray core photoemission. Two examples of this approach are presented here. In the first Katnani et al. (1984) have studied the epitaxial growth of germanium on different reconstructed surfaces [$c(4 \times 4)$, $c(8 \times 2)$ and 4×6] of GaAs(100). The starting surface differences are reflected in different As-3d core level shifts. However, following deposition of 16 Å of Ge onto the surfaces held at 340°C, each interface exhibited essentially the same As-3d core emission. In fact, the distinct starting surfaces converge to the common interface by 8 Å coverage; the final LEED pattern has the same two-domain (2×1) structure in each case. This uniformity is also reflected in the band offset measurements, which yield a common value of 0.46 ± 0.05 eV. That ΔE_v appears to be independent of surface conditions is somewhat supportive of models based on bulk parameters.

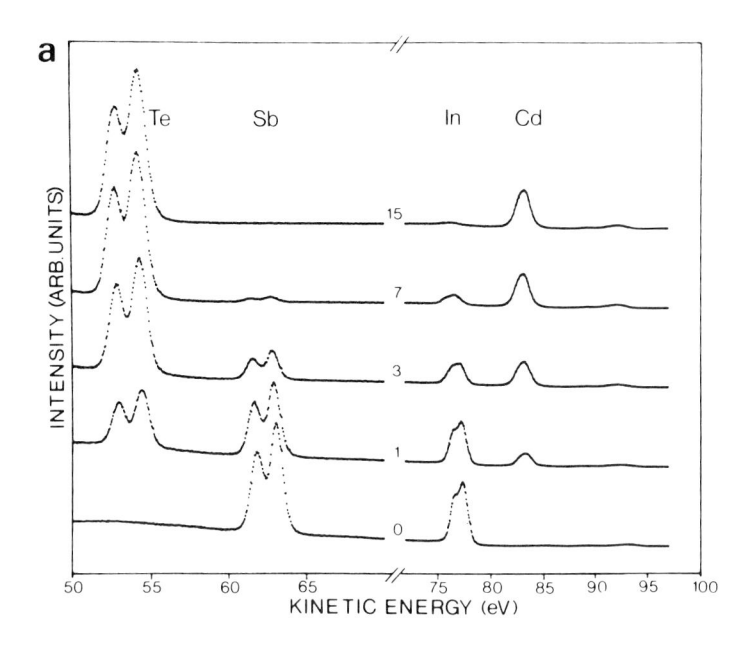

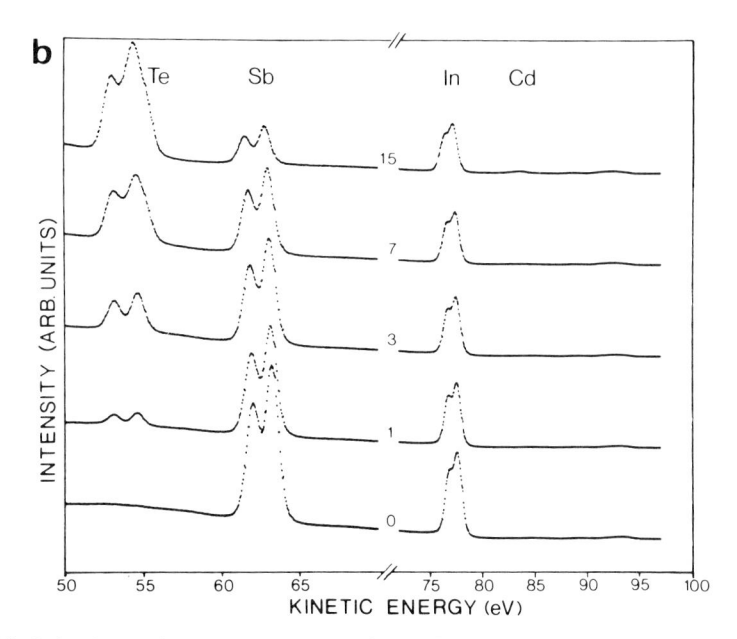

Fig. 39. (a) Soft X-ray photoelectron spectra ($h\nu = 100$ eV) of clean InSb(100) and following deposition of CdTe layers onto a room temperature substrate, showing core levels of Te, Sb, In and Cd with the valence bands just visible. Each set is labelled with the CdTe layer thickness in Å. (b) Photoelectron spectra ($h\nu = 100$ eV) of clean InSb(100) and following the deposition of CdTe layers onto a substrate at 500 K. Each set is labelled with the CdTe layer thickness in Å. (After Mackey et al. 1986.)

The second example presents a rather different picture. Mackey et al. (1986) have examined the CdTe/InSb interface, prepared by evaporation of CdTe onto clean InSb(100) surfaces. This is a II–VI/III–V heterojunction and it is also an epitaxial system. The progression of 4d cores of Cd, In, Sb and Te with room temperature deposition is shown in fig. 39a; the overlayer is stoichiometric and the interface appears to be quite abrupt. The valence band offset is measured to be 0.87 ± 0.1 eV, which may be compared with theoretical predictions of 0.96 eV (Anderson 1962) and 0.84 eV (Tersoff 1986). This room temperature deposit is not ordered and epitaxial growth requires the substrate to be held at an elevated temperature. Equivalent spectra for deposition under these conditions (500 K) are markedly different (fig. 39b). The In signal does not decrease with the Sb signal, the Te signal is broadened and the Cd signal is barely visible. The interface is neither abrupt nor stoichiometric. Further measurements indicate that the interfacial layer (largely indium telluride) can be several ångströms thick.

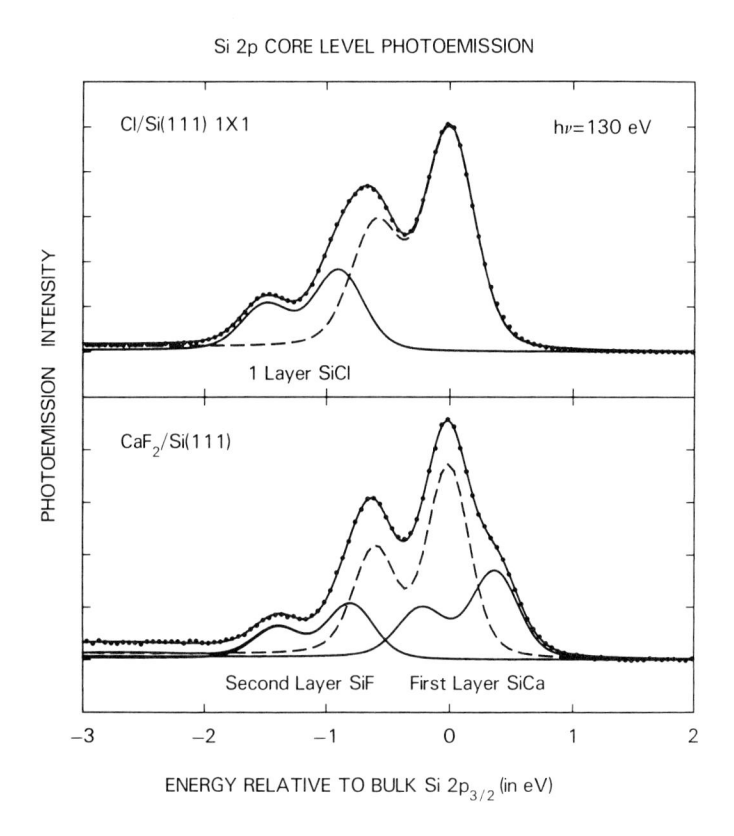

Fig. 40. Core level photoelectron spectra from the Si-$2p_{1/2,3/2}$ doublet (dotted) and their decomposition into several chemically shifted components. The Cl-terminated Si(111)(1 × 1) surface is used as an intensity standard to determine the number of interface Si atoms at the CaF_2/Si(111) interface (8 Å CaF_2 grown by MBE at 750°C). (After Himpsel et al. 1986.)

5.4. Insulator–semiconductor interfaces

One of the attractive features of silicon as a device material is the insulating properties of its oxide SiO_2. Other advantages of this oxide are its relative ease of preparation and mask etching. One disadvantage of SiO_2 is that it is amorphous, which inhibits further epitaxial growth on top of an insulating layer. There is a need for alternative insulators which can be deposited in ordered structures to facilitate stacking of layers.

One such system, CaF_2/Si, has recently been investigated by photoemission techniques (Himpsel et al. 1986). The close lattice matching permits ordered deposition, both by MBE at elevated substrate temperature and at room temperature followed by annealing. Si-2p core emission spectra are shown in fig. 40. The upper spectrum is that of the chlorine/silicon chemisorption system, which is employed to calibrate core satellite intensities. The lower spectrum is that of the CaF_2/Si interface. Here two chemical-shift satellites are observed, one on either side of the bulk silicon feature. The higher binding energy satellite is attributed to Si–F interaction, which transfers charge from the silicon; the lower binding energy feature is attributed to Si–Ca interaction, which transfers charge to the silicon. This is in agreement with the relative electronegativity of these species. The calibration indicates that the satellite intensities are consistent with models of the interface in which Ca bonds to the first layer Si atoms and F bonds to the second layer Si atoms. The interface is abrupt and independent of overlayer thickness. These measurements have been used to discriminate between possible structural models of the interface (Himpsel et al. 1986). Other insulators of this type, viz. BaF_2, SrF_2 and mixtures, have potential as lattice-matched insulators on other semiconductors.

6. Conclusions

Synchrotron radiation has, over the last ten years, revolutionised the study of the electronic and geometric structure of surfaces. This selective review has summarised the advances made in a variety of areas. New techniques, and novel applications of old ones, are appearing so frequently that it would be unwise to attempt to predict future directions in this ever-growing subject. We feel sure, however, that synchrotron radiation will continue to have a major role in surface science.

References

Abbati, I., G. Rossi, L. Calliari, L. Braicovich, I. Lindau and W.E. Spicer, 1982, J. Vac. Sci. & Technol. **21**, 409.

Abbati, I., L. Braicovich, C. Carbone, J. Nogami, J.J. Yeh, I. Lindau and U. del Pennino, 1985, Phys. Rev. B **32**, 5459.

Albert, M.R., G. Sneddon, W. Eberhardt, F. Greuter, T. Gustafsson and E.W. Plummer, 1982, Surf. Sci. **120**, 19.
Allyn, C.L., T. Gustafsson and E.W. Plummer, 1977, Chem. Phys. Lett. **47**, 127.
Anderson, R.L., 1962, Solid-State Electron. **5**, 341.
Apai, G., J.F. Hamilton, J. Stöhr and A. Thompson, 1979, Phys. Rev. Lett. **43**, 165.
Apai, G., S.-T. Lee and M.G. Mason, 1981, Solid State Commun. **37**, 213.
Bagus, P.S., C.J. Nelin and C.W. Bauschlicher, 1983, Phys. Rev. B **28**, 5423.
Balerna, A., E. Bernieri, P. Picozzi, A. Reale, S. Santucci, E. Burattini and S. Mobilio, 1985, Phys. Rev. B **31**, 5058.
Barth, J., F. Gerken, J. Schmidt-May, A. Flodström and L.I. Johansson, 1983, Chem. Phys. Lett. **96**, 532.
Barton, J.J., C.C. Bahr, Z. Hussain, S.W. Robey, J.G. Tobin, L.E. Klebanoff and D.A. Shirley, 1983, Phys. Rev. Lett. **51**, 272.
Barton, J.J., C.C. Bahr, Z. Hussain, S.W. Robey, L.E. Klebanoff and D.A. Shirley, 1984, J. Vac. Sci. & Technol. A **2**, 847.
Bartynski, R.A., and T. Gustafsson, 1986, Phys. Rev. B **33**, 6588.
Bartynski, R.A., E. Jensen, T. Gustafsson and E.W. Plummer, 1985, Phys. Rev. B **32**, 1921.
Baudoing, R., Y. Gauthier and Y. Joly, 1985, J. Phys. C **18**, 4061.
Bauer, R.S., J.C. McMenamin, R.Z. Bachrach, A. Bianconi, L. Johansson and H. Petersen, 1979, Inst. Phys. Conf. Ser. **43**, 797.
Bedzyk, M., and G. Materlik, 1985, Surf. Sci. **152/153**, 10.
Binnig, G., H. Rohrer, Ch. Gerber and E. Weibel, 1983, Phys. Rev. Lett. **50**, 120.
Binns, C., M.G. Barthès-Labrousse and C. Norris, 1984, J. Phys. C **17**, 1465.
Binns, C., P.C. Stephenson, C. Norris, G.C. Smith, H.A. Padmore, G.P. Williams and M.G. Barthès-Labrousse, 1985, Surf. Sci. **152/153**, 237.
Binns, C., D.A. Newstead, C. Norris, M.G. Barthès-Labrousse and P.C. Stephenson, 1986, J. Phys. C **19**, 829.
Blyholder, G., 1964, J. Chem. Phys. **68**, 2772.
Bohr, J., R. Feidenhans'l, M. Neilsen, M. Toney, R.L. Johnson and I.K. Robinson, 1985, Phys. Rev. Lett. **54**, 1275.
Brennan, S., 1985, Surf. Sci. **152/153**, 1.
Brennan, S., J. Stöhr and R. Jaeger, 1981, Phys. Rev. B **24**, 4871.
Brillson, L.J., 1987, ch. 8, this volume.
Campagna, M., G.K. Wertheim and Y. Baer, 1978, in: Photoemission in Solids II, Vol.27, eds L. Ley and M. Cardona, in Topics in Applied Physics (Springer, Berlin).
Campuzano, J.-C., J.E. Inglesfield, D.A. King and C. Somerton, 1981, J. Phys. C **14**, 3099.
Cerrina, F., J.R. Myron and G.J. Lapeyre, 1984, Phys. Rev. B **29**, 1798.
Chadi, D.J., 1979, Phys. Rev. B **19**, 2074.
Chadi, D.J., 1980, Surf. Sci. **99**, 1.
Chang, S., A. Rizzi, C. Caprile, P. Phillip, A. Wall and A. Franciosi, 1986, J. Vac. Sci. & Technol. A **4**, 799.
Chiang, T.-C., J.A. Knapp, D.E. Eastman and M. Aono, 1979, Solid State Commun. **31**, 917.
Chiang, T.-C., G. Kaindl and T. Mandel, 1986, Phys. Rev. B **33**, 695.
Citrin, P.H., P. Eisenberger and J.E. Rowe, 1982, Phys. Rev. Lett. **48**, 802.
Citrin, P.H., G.K. Wertheim and Y. Baer, 1983a, Phys. Rev. B **27**, 3160.
Citrin, P.H., J.E. Rowe and P. Eisenberger, 1983b, Phys. Rev. B **28**, 2299.
Comin, F., L. Incoccia, P. Lagarde, G. Rossi and P.H. Citrin, 1985, Phys. Rev. Lett. **54**, 122.
Cooper, J.W., 1964, Phys. Rev. Lett. **13**, 762.
Daimon, H., H. Ito, S. Shin and Y. Murata, 1984, J. Phys. Soc. Jpn. **53**, 3488.
Daniels, R.R., A.D. Katnani, T.-X. Zhao, G. Margaritondo and A. Zunger, 1982, Phys. Rev. Lett. **49**, 895.
Debe, M.K., and D.A. King, 1977a, J. Phys. C **10**, L303.
Debe, M.K., and D.A. King, 1977b, Phys. Rev. Lett. **39**, 708.
Dehmer, J.L., and D. Dill, 1975, Phys. Rev. Lett. **35**, 213.

Demuth, J.E., D.W. Jepsen and P.M. Marcus, 1974, Phys. Rev. Lett. 32, 1182.
Desjonquères, M.C., D. Spanjaard, Y. Lassailly and C. Guillot, 1980, Solid State Commun. 34, 807.
DiCenzo, S.B., and G.K. Wertheim, 1985, Comments Solid State Phys. 11, 203.
Didio, R.A., E.W. Plummer and W.R. Graham, 1984a, Phys. Rev. Lett. 52, 683.
Didio, R.A., E.W. Plummer and W.R. Graham, 1984b, J. Vac. Sci. & Technol. A 2, 983.
Dietz, E., and F.J. Himpsel, 1979, Solid State Commun. 30, 235.
Dimon, P., P.M. Horn, M. Sutton, R.J. Birgeneau and D.E. Moncton, 1985, Phys. Rev. B 31, 437.
Döbler, U., K. Baberschke, J. Haase and A. Puschmann, 1984, Phys. Rev. Lett. 52, 1437.
Domke, M., C. Laubschat, M. Prietsch, T. Mandel, G. Kaindl and W.D. Schneider, 1986, Phys. Rev. Lett. 56, 1287.
Dowben, P.A., Y. Sakisaka and T.N. Rhodin, 1985, J. Vac. Sci. & Technol. A 3, 1855.
Eastman, D.E., F.J. Himpsel and J.A. Knapp, 1980a, Phys. Rev. Lett. 44, 95.
Eastman, D.E., T.-C. Chiang, P. Heimann and F.J. Himpsel, 1980b, Phys. Rev. Lett. 45, 656.
Eastman, D.E., F.J. Himpsel and J.F. van der Veen, 1982, J. Vac. Sci. & Technol. 20, 609.
Eberhardt, W., and F.J. Himpsel, 1980, Phys. Rev. B 21, 5572.
Eberhardt, W., E.W. Plummer, K. Horn and J. Erskine, 1980, Phys. Rev. Lett. 45, 273.
Eberhardt, W., S.G. Louie and E.W. Plummer, 1983, Phys. Rev. B 28, 465.
Eisenberger, P., and W.C. Marra, 1981, Phys. Rev. Lett. 46, 1081.
Eyers, A., F. Schäfers, G. Schönhense, U. Heinzmann, H.D. Oepen, K. Hünlich, J. Kirschner and G. Borstel, 1984, Phys. Rev. Lett. 52, 1559.
Felter, T.E., R.A. Barker and P. Estrup, 1977, Phys. Rev. Lett. 38, 1138.
Feuerbacher, B., B. Fitton and R.F. Willis, 1978, Photoemission and Electronic Properties of Surfaces (Wiley, New York).
Franciosi, A., J.H. Weaver, N. Mårtensson and M. Croft, 1981, Phys. Rev. B 24, 3651.
Franciosi, A., S. Chang, P. Phillip, C. Caprile and J. Joyce, 1985, J. Vac. Sci. & Technol. A 3, 933.
Freund, H.-J., and E.W. Plummer, 1981, Phys. Rev. B 23, 4859.
Freund, H.-J., F. Greuter, D. Heskett and E.W. Plummer, 1983a, Phys. Rev. B 28, 1727.
Freund, H.-J., W. Eberhardt, D. Heskett and E.W. Plummer, 1983b, Phys. Rev. Lett. 50, 768.
Funke, P., and G. Materlik, 1985, Solid State Commun. 54, 921.
Goodwin, E.T., 1939, Proc. Cambridge Philos Soc. 35, 221.
Greuter, F., D. Heskett, E.W. Plummer and H.-J. Freund, 1983, Phys. Rev. B 27, 7117.
Greuter, F., I. Strathy and E.W. Plummer, 1986, Phys. Rev. B 33, 736.
Grider, D.E., K.G. Purcell and N.V. Richardson, 1983, Chem. Phys. Lett. 100, 320.
Griffiths, K., C. Kendon and D.A. King, 1981, Phys. Rev. Lett. 46, 1585.
Guillot, C., M.C. Desjonquères, D. Chauveau, G. Treglia, J. Lecante, D. Spanjaard and Tran Minh Duc, 1984, Solid State Commun. 50, 393.
Guillot, C., D. Chaveau, P. Roubin, J. Lecante, M.C. Desjonquères, G. Treglia and D. Spanjaard, 1985, Surf. Sci. 162, 46.
Haines, E., R. Clauberg and R. Feder, 1985, Phys. Rev. Lett. 54, 932.
Hamers, R.J., R.M. Tromp and J.E. Demuth, 1986, Phys. Rev. Lett. 56, 1972.
Haneman, D., 1961, Phys. Rev. 121, 1093.
Hansson, G.V., R.Z. Bachrach, R.S. Bauer and P. Chiaradia, 1981, Phys. Rev. Lett. 46, 1033.
Hermanson, J., 1977, Solid State Commun. 22, 9.
Heskett, D., I. Strathy, E.W. Plummer and R.A. de Paola, 1985, Phys. Rev. B 32, 6222.
Hezaveh, A.A., G. Jennings, D. Pescia, R.F. Willis, K. Prince, M. Surman and A. Bradshaw, 1986, Solid State Commun. 57, 329.
Himpsel, F.J., P. Heimann, T.-C. Chiang and D.E. Eastman, 1980, Phys. Rev. Lett. 45, 1112.
Himpsel, F.J., D.E. Eastman and E.-E. Koch, 1981, Phys. Rev. B 24, 1687.
Himpsel, F.J., F.U. Hillebrecht, G. Hughes, J.L. Jordan, U.O. Karlsson, F.R. McFeely, J.F. Morar and D. Rieger, 1985, Appl. Phys. Lett. 48, 596.
Hollinger, G., and F.J. Himpsel, 1983, Phys. Rev. B 28, 3651.
Holmes, M.I., and T. Gustafsson, 1981, Phys. Rev. Lett. 47, 443.
Hsieh, T.C., T. Miller and T.-C. Chiang, 1985, Phys. Rev. Lett. 55, 2483.
Hsieh, T.C., T. Miller and T.-C. Chiang, 1986, Phys. Rev. B 33, 2865.

Hughes, G., R. Ludeke, J.F. Morar and J.L. Jordan, 1985, J. Vac. Sci. & Technol. B **3**, 1079.
Hussain, Z., E. Umbach, D.A. Shirley, J. Stöhr and J. Feldhaus, 1982, Nucl. Instrum. Methods **195**, 115.
Inglesfield, J.E., 1982, Rep. Prog. Phys. **45**, 223.
Inglesfield, J.E., 1985, Prog. Surf. Sci. **20**, 105.
Johansson, L.I., L.-G. Petersson, K.-F. Berggren and J.W. Allen, 1980, Phys. Rev. B **22**, 3294.
Johnson, A.L., E.L. Muetterties and J. Stöhr, 1983, J. Am. Chem. Soc. **105**, 7183.
Jona, F., 1978, J. Phys. C **11**, 4271.
Jones, R.G., S. Ainsworth, M.D. Crapper, C. Somerton, D.P. Woodruff, R.S. Brooks, J.C. Campuzano, D.A. King, G.M. Lamble and M. Prutton, 1985, Surf. Sci. **152/153**, 443.
Jugnet, Y., F.J. Himpsel, Ph. Avouris and E.-E. Koch, 1984, Phys. Rev. Lett. **53**, 198.
Jugnet, Y., G. Grenet and Tran Minh Duc, 1987, ch. 10, this volume.
Jupille, J., K.G. Purcell and D.A. King, 1986, Solid State Commun. **58**, 529.
Kaindl, G., B. Reihl, D.E. Eastman, R.A. Pollack, N. Mårtensson, B. Barbara, T. Penney and T.S. Plaskett, 1982, Solid State Commun. **41**, 157.
Katnani, A.D., and G. Margaritondo, 1983, Phys. Rev. B **28**, 1944.
Katnani, A.D., P. Chiaradia, H.W. Sang Jr and R.S. Bauer, 1984, J. Vac. Sci. & Technol. B **2**, 471.
Kevan, S.D., 1985, Phys. Rev. B **32**, 2344.
Kevan, S.D., D.H. Rosenblatt, D. Denley, B.-C. Lu and D.A. Shirley, 1978, Phys. Rev. Lett. **41**, 1565.
Kevan, S.D., N.G. Stoffel and N.V. Smith, 1985a, Phys. Rev. B **31**, 3348.
Kevan, S.D., N.G. Stoffel and N.V. Smith, 1985b, Phys. Rev. B **32**, 4956.
Kevan, S.D., N.G. Stoffel and N.V. Smith, 1985c, Phys. Rev. B **31**, 1788.
Kisker, E., 1983, J. Phys. Chem. **87**, 3597.
Kisker, E., K. Schröder, M. Campagna and W. Gudat, 1984, Phys. Rev. Lett. **52**, 2285.
Kisker, E., R. Clauberg and W. Gudat, 1985, Z. Phys. B **61**, 453.
Klebanoff, L.E., and D.A. Shirley, 1986, Phys. Rev. B **33**, 5301.
Knotek, M.L., 1984, Rep. Prog. Phys. **47**, 1499.
Knotek, M.L., and P.J. Feibelman, 1978, Phys. Rev. Lett. **40**, 964.
Korenman, V., and R.E. Prange, 1984, Phys. Rev. Lett. **53**, 186.
Korenman, V., J.L. Murray and R.E. Prange, 1977, Phys. Rev. B **16**, 4032.
Lamble, G.M., R. Brooks, S. Ferrer, D.A. King and D. Norman, 1986a, Phys. Rev. B **34**, 2975.
Lamble, G.M., D.J. Holmes, D.A. King and D. Norman, 1986b, J. Phys. Colloq. C **8**, 509.
Landau, L.D., 1937, Z. Physik Sovjetunion **11**, 26.
Landgren, G., R. Ludeke, J.F. Morar, Y. Jugnet and F.J. Himpsel, 1984, Phys. Rev. B **30**, 4839.
Larsen, P.K., N.V. Smith, M. Schlüter, H.H. Farrell, K.M. Ho and M.L. Cohen, 1978, Phys. Rev. B **17**, 2612.
Larson, B.C., C.W. White, T.S. Noggle and D. Mills, 1982, Phys. Rev. Lett. **48**, 337.
Lee, S.-T., G. Apai, M.G. Mason, R. Benbow and Z. Hurych, 1981, Phys. Rev. B **23**, 505.
Levinson, H.J., E.W. Plummer and P. Feibelman, 1980, J. Vac. Sci. & Technol. **17**, 216.
Liebermann, L.N., D.R. Fredkin and H.B. Shore, 1969, Phys. Rev. Lett. **22**, 539.
Liebsch, A., 1978, Phys. Rev. B **17**, 1653.
Liebsch, A., 1981, Phys. Rev. B **23**, 5203.
Loubriel, G., T. Gustafsson, L.I. Johansson and S.J. Oh, 1982, Phys. Rev. Lett. **49**, 571.
Louie, S.G., P. Thiry, R. Pinchaux, Y. Petroff, D. Chandesris and J. Lecante, 1980, Phys. Rev. Lett. **44**, 549.
Ludeke, R., and A. Koma, 1975, Phys. Rev. Lett. **34**, 1170.
Ludeke, R., T.-C. Chiang and T. Miller, 1983, J. Vac. Sci. & Technol. B **1**, 581.
MacDowell, A.A., D. Norman and J.B. West, 1986, Rev. Sci. Instrum. **57**, 2667.
Mackey, K.J., P.M.G. Allen, W.G. Herrenden-Harker, R.H. Williams, C.R. Whitehouse and G.M. Williams, 1986, Appl. Phys. Lett. **49**, 354.
Mandel, T., M. Domke, G. Kaindl, C. Laubschat, M. Prietsch, U. Middelmann and K. Horn, 1985a, Surf. Sci. **162**, 453.

Mandel, T., G. Kaindl, M. Domke, W. Fischer and W.D. Schneider, 1985b, Phys. Rev. Lett. **55**, 1638.
Margaritondo, G., 1985, Z. Phys. B **61**, 447.
Margaritondo, G., 1986, Surf. Sci. **168**, 439.
Mariani, C., H.-U. Middelmann, M. Iwan and K. Horn, 1982, Chem. Phys. Lett. **93**, 308.
Marra, W.C., P.H. Fuoss and P. Eisenberger, 1982, Phys. Rev. Lett. **49**, 1169.
Mårtensson, N., 1985, Z. Phys. B **61**, 457.
Mårtensson, N., B. Reihl and R.D. Parks, 1982, Solid State Commun. **41**, 573.
Materlik, G., 1985, Z. Phys. B **61**, 405.
Materlik, G., A. Frahm and M.J. Bedzyk, 1984, Phys. Rev. Lett. **52**, 441.
McConville, C.F., D.P. Woodruff, K.C. Prince, G. Paolucci, V. Chab, M. Surman and A.M. Bradshaw, 1986, Surf. Sci. **166**, 221.
McGilp, J.F., and I.T. McGovern, 1985, J. Vac. Sci. & Technol. B **3**, 1641.
McGovern, I.T., K.D. Childs, H.M. Clearfield and R.H. Williams, 1981, J. Phys. C **14**, L243.
McGrath, R., I.T. McGovern, D.R. Warburton, G. Thornton and D. Norman, 1986, Surf. Sci. **178**, 101.
McKinley, A., G.J. Hughes and R.H. Williams, 1982, J. Phys. C **15**, 7049.
Messmer, R.P., 1985, Surf. Sci. **158**, 40.
Miller, T., and T.-C. Chiang, 1984, Phys. Rev. B **29**, 7034.
Miranda, R., F. Yndurain, D. Chandesris, J. Lecante and Y. Petroff, 1982, Phys. Rev. B **25**, 527.
Mitchell, K.A.R., 1985, Surf. Sci. **149**, 93.
Mönch, W., 1986, Solid State Commun. **58**, 215.
Nannarone, S., P. Chiaradia, F. Ciccacci, R. Memeo, P. Sassaroli, S. Selci and G. Chiarotti, 1980, Solid State Commun. **33**, 593.
Nicholls, J.M., G.V. Hansson, R.I.G. Uhrberg and S.A. Flodström, 1983, Phys. Rev. B **27**, 2594.
Nicholls, J.M., G.V. Hansson, U.O. Karlsson, R.I.G. Uhrberg, R. Engelhardt, K. Seki, S.A. Flodström and E.-E. Koch, 1984, Phys. Rev. Lett. **52**, 1555.
Nicholls, J.M., P. Mårtensson and G.V. Hansson, 1985, Phys. Rev. Lett. **54**, 2363.
Norman, D., 1986, J. Phys. C **19**, 3273.
Norman, D., S. Brennan, R. Jaeger and J. Stöhr, 1981, Surf. Sci. **105**, L297.
Norman, D., J. Stöhr, R. Jaeger, P.J. Durham and J.B. Pendry, 1983, Phys. Rev. Lett. **51**, 2052.
Norman, D., C.H. Richardson, D.R. Warburton, G. Thornton, R. McGrath and F. Sette, 1986, J. Phys. Colloq. C **8**, 525.
Norman, D., R.A. Tuck, H.B. Skinner, P.J. Wadsworth, T.M. Gardiner, I.W. Owen, C.H. Richardson and G. Thornton, 1987, Phys. Rev. Lett. **58**, 519.
Northrup, J.E., and M.L. Cohen, 1983, Phys. Rev. B **27**, 6553.
Orders, P.J., and C.S. Fadley, 1983, Phys. Rev. B **27**, 781.
Owen, I.W., N.B. Brookes, C.H. Richardson, D.R. Warburton, F.M. Quinn, D. Norman and G. Thornton, 1986, Surf. Sci. **178**, 897.
Pandey, K.C., 1981, Phys. Rev. Lett. **47**, 1913.
Pandey, K.C., and J.C. Phillips, 1974, Phys. Rev. B **9**, 1552.
Pauling, L., 1960, The Nature of the Chemical Bond (Cornell Univ. Press, Ithaca, NY).
Perfetti, P., F. Patella, F. Sette, C. Quaresima, C. Capasso, A. Savoia and G. Margaritondo, 1984, Phys. Rev. B **29**, 5941.
Pianetta, P., I. Lindau, C.M. Garner and W.E. Spicer, 1975, Phys. Rev. Lett. **35**, 1356.
Pindor, A.J., J. Staunton, G.M. Stocks and H. Winter, 1983, J. Phys. F **13**, 979.
Plummer, E.W., 1985, Surf. Sci. **152/153**, 162.
Plummer, E.W., and W. Eberhardt, 1979, Phys. Rev. B **20**, 1444.
Plummer, E.W., and W. Eberhardt, 1982, Advance Chem. Phys. **49**, 533.
Plummer, E.W., C.T. Chen, W.K. Ford, W. Eberhardt, R.P. Messmer and H.-J. Freund, 1985, Surf. Sci. **158**, 58.
Puschmann, A., and J. Haase, 1984, Surf. Sci. **144**, 559.
Puschmann, A., J. Haase, M.D. Crapper, C.E. Riley and D.P. Woodruff, 1985, Phys. Rev. Lett. **54**, 2250.

Reihl, B., G. Hollinger and F.J. Himpsel, 1983, Phys. Rev. B **28**, 1490.
Robinson, I.K., 1983, Phys. Rev. Lett. **50**, 1045.
Robinson, I.K., W.K. Waskiewicz, P.H. Fuoss, J.B. Stark and P.A. Bennett, 1986, Phys. Rev. B **33**, 7013.
Rosenblatt, D.H., J.G. Tobin, M.G. Mason, R.F. Davis, S.D. Kevan, D.A. Shirley, C.H. Li and S.Y. Tong, 1981, Phys. Rev. B **24**, 3828.
Rosenblatt, D.H., S.D. Kevan, J.G. Tobin, R.F. Davis, M.G. Mason, D.A. Shirley, J.C. Tang and S.Y. Tong, 1982, Phys. Rev. B **26**, 3181.
Rosengren, A., and B. Johansson, 1981, Phys. Rev. B **23**, 3852.
Rossi, G., L. Calliari, I. Abbati, L. Braicovich, I. Lindau and W.E. Spicer, 1982, Surf. Sci. **116**, L202.
Saitoh, M., F. Shoji, K. Oura and T. Hanawa, 1980, J. Appl. Phys. (Japan) **19**, L421.
Schlüter, M., J.R. Chelikowsky, S.G. Louie and M.L. Cohen, 1975, Phys. Rev. B **12**, 4200.
Schlüter, M., J.E. Rowe, G. Margaritondo, K.M. Ho and M.L. Cohen, 1976, Phys. Rev. Lett. **37**, 1632.
Schmeisser, D., and J.E. Demuth, 1986, Phys. Rev. B **33**, 4233.
Schmeisser, D., F.J. Himpsel and G. Hollinger, 1983a, Phys. Rev. B **27**, 7813.
Schmeisser, D., F.J. Himpsel, G. Hollinger, B. Reihl and K. Jacobi, 1983b, Phys. Rev. B **27**, 3279.
Schmeisser, D., F. Greuter, E.W. Plummer and H.-J. Freund, 1985, Phys. Rev. Lett. **54**, 2095.
Schnell, R.D., D. Rieger, A. Bogen, F.J. Himpsel, K. Wandelt and W. Steinmann, 1985, Phys. Rev. B **32**, 8057.
Shih, H.D., F. Jona, D.W. Jepsen and P.M. Marcus, 1976, Phys. Rev. Lett. **36**, 798.
Shockley, W., 1939, Phys. Rev. **56**, 317.
Sinkovic, B., P.J. Orders, C.S. Fadley, R. Trehan, Z. Hussain and J. Lecante, 1984, Phys. Rev. B **30**, 1833.
Sinkovic, B., B. Hermsmeier and C.S. Fadley, 1985, Phys. Rev. Lett. **55**, 1227.
Smith, N.V., and F.J. Himpsel, 1983, in: Handbook on Synchrotron Radiation, Vol. 1, ed. E.-E. Koch (North-Holland, Amsterdam) p. 905.
Smith, R.J., J. Anderson and G.J. Lapeyre, 1976, Phys. Rev. Lett. **37**, 1081.
Spanjaard, D., C. Guillot, M.C. Desjonquères, G. Treglia and J. Lecante, 1985, Surf. Sci. Repts **5**, 1.
Spicer, W.E., I. Lindau, P.E. Gregory, C.M. Garner, P. Pianetta and P.W. Chye, 1976, J. Vac. Sci. & Technol. **13**, 780.
Spicer, W.E., P. Pianetta, I. Lindau and P.W. Chye, 1977, J. Vac. Sci. & Technol. **14**, 885.
Spicer, W.E., P.W. Chye, C.M. Garner, I. Lindau and P. Pianetta, 1979, Surf. Sci. **86**, 763.
Stern, E.A., and S.M. Heald, 1983, in: Handbook on Synchrotron Radiation, Vol. 1, ed. E.-E. Koch (North-Holland, Amsterdam) p. 955.
Stockbauer, R., D.M. Hanson, S.A. Flodström and T.E. Madey, 1982, Phys. Rev. B **26**, 1885.
Stoffel, N.G., S.D. Kevan and N.V. Smith, 1985, Phys. Rev. B **32**, 5038.
Stöhr, J., 1987, in: X-Ray Absorption, Principles, Applications, Techniques of EXAFS, SEXAFS and XANES, eds R. Prins and D.C. Koningsberger (Wiley, New York).
Stöhr, J., and R. Jaeger, 1982a, Phys. Rev. B **26**, 4111.
Stöhr, J., and R. Jaeger, 1982b, J. Vac. Sci. & Technol. **21**, 619.
Stöhr, J., R. Jaeger, J. Feldhaus, S. Brennan, D. Norman and G. Apai, 1980, Appl. Opt. **19**, 3911.
Stöhr, J., K. Baberschke, R. Jaeger, R. Treichler and S. Brennan, 1981, Phys. Rev. Lett. **47**, 381.
Stöhr, J., J.L. Gland, W. Eberhardt, D. Outka, R.J. Madix, F. Sette, R.J. Koestner and U. Döbler, 1983, Phys. Rev. Lett. **51**, 2414.
Stöhr, J., J.L. Gland, E.B. Kollin, R.J. Koestner, A.L. Johnson, E.L. Muetterties and F. Sette, 1984, Phys. Rev. Lett. **53**, 2161.
Stöhr, J., E.B. Kollin, D.A. Fischer, J.B. Hastings, F. Zaera and F. Sette, 1985, Phys. Rev. Lett. **55**, 1468.
Su, C.Y., P.W. Chye, P. Pianetta, I. Lindau and W.E. Spicer, 1979, Surf. Sci. **86**, 894.
Takayanagi, K., Y. Tanishiro, S. Takahashi and M. Takahashi, 1985a, Surf. Sci. **164**, 367.
Takayanagi, K., Y. Tanishiro, S. Takahashi and M. Takahashi, 1985b, J. Vac. Sci. & Technol. A **3**, 1502.
Tersoff, J., 1984, Phys. Rev. B **30**, 4874.

Tersoff, J., 1986, J. Vac. Sci. & Technol. B **4**, 1066.
Tobin, J.G., S.W. Robey and D.A. Shirley, 1986, Phys. Rev. B **33**, 2270.
Tong, S.Y., G. Xu and W.N. Mei, 1984, Phys. Rev. Lett. **52**, 1911.
Tromp, R.M., and E.J. Loenen, 1985, Surf. Sci. **155**, 441.
Truesdale, C.M., S.H. Southworth, P.H. Kobrin, D.W. Lindle, G. Thornton and D.A. Shirley, 1982, J. Chem. Phys. **75**, 860.
Turner, A.M., and J.L. Erskine, 1983, Phys. Rev. B **28**, 5628.
Turner, A.M., A.W. Donoho and J.L. Erskine, 1984, Phys. Rev. B **29**, 2986.
Uhrberg, R.I.G., G.V. Hansson, J.M. Nicholls and S.A. Flodström, 1982, Phys. Rev. Lett. **48**, 1032.
Uhrberg, R.I.G., G.V. Hansson, U.O. Karlsson, J.M. Nicholls, P.E.S. Persson, S.A. Flodström, R. Engdhardt and E.-E. Koch, 1984, Phys. Rev. Lett. **52**, 2265.
Uhrberg, R.I.G., R.D. Bringans, R.Z. Bachrach and J.E. Northrup, 1986, Phys. Rev. Lett. **56**, 520.
Umbach, E., 1982, Surf. Sci. **117**, 482.
Umbach, E., and Z. Hussain, 1984, Phys. Rev. Lett. **52**, 457.
van der Veen, J.F., 1985, Surf. Sci. Rep. **5**, 199.
van der Veen, J.F., R.M. Tromp, R.G. Smeenk and F.W. Saris, 1979, Surf. Sci. **82**, 468.
van der Veen, J.F., P. Heimann, F.J. Himpsel and D.E. Eastman, 1981, Solid State Commun. **37**, 555.
Waldrop, J.R., R.W. Grant, S.P. Kowalczyk and E.A. Kraut, 1985, J. Vac. Sci. & Technol. A **3**, 835.
Wehking, F., H. Beckermann and R. Niedermayer, 1978, Surf. Sci. **71**, 364.
Weller, D., S.J. Alvarado, W. Gudat, K. Schröder and M. Campagna, 1985, Phys. Rev. Lett. **54**, 1555.
Welton-Cook, M.R., and W. Berndt, 1982, J. Phys. C **15**, 5691.
Wertheim, G.K., and G. Crecelius, 1978, Phys. Rev. Lett. **40**, 813.
Wertheim, G.K., S.B. DiCenzo, D.N.E. Buchanan and P.A. Bennett, 1985, Solid State Commun. **53**, 377.
Wieliczka, D., J.H. Weaver, D.W. Lynch and C.G. Olson, 1982, Phys. Rev. B **26**, 7056.
Williams, R.H., 1985, in: The Physics and Chemistry of III–V Compound Semiconductor Interfaces, ed. C.W. Wilmsen (Plenum, New York) ch. 1.
Williams, R.H., G.P. Srivastava and I.T. McGovern, 1980, Rep. Prog. Phys. **43**, 1357.
Woodruff, D.P., 1986, Surf. Sci. **166**, 377.
Zhao, T.-X., R.R. Daniels, A.D. Katnani, G. Margaritondo and A. Zunger, 1983, J. Vac. Sci. & Technol. B **1**, 610.
Zunger, A., 1981, Phys. Rev. B **24**, 4372.

CHAPTER 8

METAL–SEMICONDUCTOR INTERFACE STUDIES BY SYNCHROTRON RADIATION TECHNIQUES

L.J. BRILLSON

Xerox Webster Research Center, 800 Phillips Road W114, Webster, NY 14580, USA

Contents

1. Introduction 543
2. Metal–semiconductor interactions and electronic properties 544
   2.1. Historical perspective 544
   2.2. The nonideal metal–semiconductor interface 545
   2.3. Role of metal–semiconductor interactions in Schottky barrier formation . . . 546
3. Soft X-ray photoemission spectroscopy for microscopic interface characterization . 547
   3.1. The soft X-ray photoemission spectroscopy technique 547
   3.2. Ultrahigh vacuum fabrication and analysis of metal–semiconductor interfaces . 548
4. Chemical bonding at metal–semiconductor interfaces 549
   4.1. Evolution of metal–semiconductor bonding with interface formation 549
      4.1.1. Silicon–metal interfaces 550
      4.1.2. III–V compound semiconductor–metal interfaces 557
      4.1.3. II–VI compound semiconductor–metal interfaces 562
      4.1.4. A III–VI compound semiconductor–metal interface 565
      4.1.5. Additional metal–semiconductor interfaces 567
   4.2. Chemical bond variations with depth 569
   4.3. Metal island formation 570
5. Metal–semiconductor interdiffusion 576
   5.1. Interface diffusion near room temperature 576
   5.2. Dynamics of semiconductor outdiffusion 577
   5.3. Marker identification of diffusing species 581

Contents continued overleaf

Handbook on Synchrotron Radiation, Vol. 2, edited by G.V. Marr
© *Elsevier Science Publishers B.V., 1987*

Contents continued

    6. Fermi level pinning and semiconductor band bending. 582
        6.1. Fermi level movement with metal deposition 582
        6.2. Semiconductor band bending with metal deposition. 587
        6.3. Comparison of spectroscopic and electrical barrier heights 588
        6.4. Models of Schottky barrier formation 589
    7. Other synchrotron radiation techniques for interface analysis 594
        7.1. Surface extended X-ray absorption fine structure analysis 594
        7.2. Angle-resolved techniques 596
        7.3. Additional techniques. 598
    8. Conclusions and future directions 600
References 603

1. Introduction

Over the past few years, considerable progress has been made in understanding the structure and properties of metal–semiconductor interfaces. These rapid changes in interface science reflect a large body of work, which has only recently been reviewed in a number of articles, monographs, and books (Brillson 1982a, 1983a, Margaritondo 1983, Williams 1982, Bachrach 1984, Henisch 1984, Bauer et al. 1983). Synchrotron radiation techniques have been responsible for many developments in this field, notwithstanding the many other experimental techniques which researchers have marshalled to probe solid-state junctions. Soft X-ray photoemission spectroscopy (SXPS) in particular has led to a number of advances in understanding the metal–semiconductor contact. SXPS reveals that metal–semiconductor interfaces are far from the abrupt junctions commonly envisioned and that the detailed chemical structure on an atomic scale influences and in some cases dominates the interface electronic properties. In this chapter, we will review the synchrotron radiation techniques which have been used to characterize metal–semiconductor interface properties. Of primary importance will be the analytical methods associated with SXPS, which has of all surface science techniques provided the widest variety of critical information in this field. To the extent that other synchrotron radiation techniques can contribute to fundamental analysis of interface properties, they are reviewed here as well.

In section 2, we present a brief historical review of developments in modeling metal–semiconductor interfaces and relating their electronic properties to Schottky barrier formation. Based on developments involving a number of experimental techniques, but especially SXPS, we introduce a general picture of the metal–semiconductor interface which includes a wide range of possible chemical and electronic features. We then indicate how these features can influence the measured electrical properties related to Schottky barrier formation.

In section 3, we review soft X-ray photoemission methods and describe how SXPS is used to characterize interfaces as they are created, layer by atomic layer.

In section 4, we illustrate the various aspects of chemical bonding which can be involved in the formation of metal–semiconductor contacts. From the SXPS energies and lineshapes of electrons photoemitted from atomic core levels, we extract information on the evolution of chemical bonding at each stage of interface formation. Such evolution may involve strong chemical reactions, interdiffusion of species without stable compound formation, and spatial variations in bonding both laterally and normal to the initial interface plane. The salient features of chemical bonding vary between different classes of semiconductors. Hence we include representative examples of each under separate subheadings to emphasize these qualitative differences.

In section 5, we describe how SXPS core level spectra permit the identification of diffusing species across the metal–semiconductor boundary. Such detailed identification of diffusing species can be linked with strong electronic changes near the semiconductor surface. Furthermore, soft X-ray techniques permit us to identify key factors determining the kinetics of the semiconductor diffusion.

In section 6, we discuss the use of spectral energy shifts in determining Fermi level (E_F) movements within the semiconductor as a function of overlayer metal deposition. These E_F movements can be related to chemical changes which occur in the same initial stages of metal deposition as well as to E_F positions in the band gap corresponding to the barrier heights measured electrically. On this basis, we are able to assess various models of Schottky barrier formation.

In section 7, we describe other techniques which can provide information on the chemical, geometrical, and electronic structure of metal–semiconductor interfaces. Techniques such as surface extended X-ray fine structure analysis (SEXAFS), angle-resolved SXPS, and X-ray diffraction have been applied to interface characterization in only a limited way, yet they have already yielded valuable results.

In section 8, we summarize the relationships between interface chemical and electronic features elucidated by synchrotron radiation techniques and we suggest new areas of opportunity for which these powerful techniques are likely to contribute to our understanding of electronically active interfaces.

2. Metal–semiconductor interactions and electronic properties

2.1. Historical perspective

The study of metal–semiconductor interfaces is one of the most active areas in condensed matter physics today. In large part this is due to the issue of Schottky barrier formation, i.e. the contact rectification or band bending within the semiconductor surface space charge region (Sze 1981), which is of both technological and scientific importance.

Rectification effects were the subject of investigation as early as the 1870's (Braun 1874), but it was not until the refinement of semiconductor materials in the 1930's and 1940's that the role of the semiconductor interior could be separated from semiconductor interface effects. For a historical review of these developments, see Teal (1976). The central issue in Schottky barrier studies has been the failure of the classical relation between contact potential difference and barrier height:

$$\varphi_{SB} = \varphi_M - \chi_{SC}, \tag{1}$$

where φ_{SB}, φ_M and χ_{SC} are the Schottky barrier height, metal work function, and semiconductor electron affinity, respectively. These parameters are illustrated in fig. 1, along with E_C, E_V, E_{VAC}, and E_F^{SC}, the conduction band minimum, valence band maximum, vacuum level and semiconductor Fermi level, respectively. A considerable body of research has focussed on why metals of different work

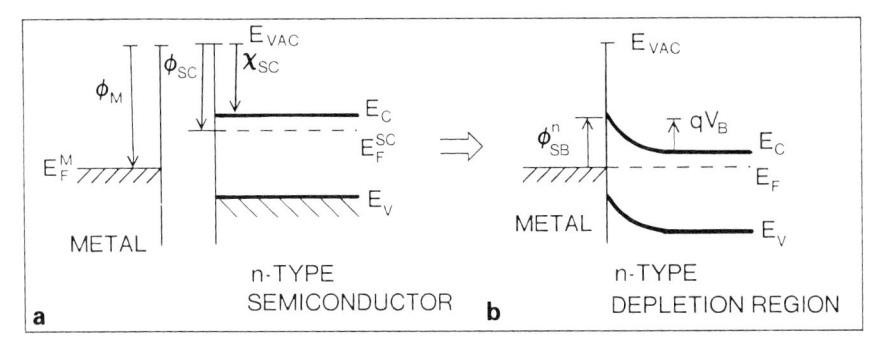

Fig. 1. Schematic diagrams of Schottky barrier formation (a) before and (b) after metal–semiconductor contact for a high work function metal and an n-type semiconductor. Band bending is upward. For a low work function metal and a p-type semiconductor, the sign of band bending is reversed (downward).

function do not produce a proportional range of barrier heights and why applied voltages to such junctions produce changes in band bending which are smaller than expected (Milnes and Feucht 1972, Sze 1981).

The pioneering work of John Bardeen (1947) described the importance of localized charge states near the interface in determining rectification properties (Bardeen 1947). Such interfacial charge can accommodate much of the contact potential difference between metal and semiconductor. If localized to a single monolayer at the interface, such charge layers permit tunneling [for a general review, see Duke (1969)] and hence may not contribute to the macroscopic barrier height. Indeed, early field effect experiments demonstrated the existence of such trapped charge at the metal–semiconductor contact (Schockley and Pearson 1948).

2.2. The nonideal metal–semiconductor interface

With the advent of surface science techniques in the 1960's, efforts increased to characterize localized interface states in detail. Up until the early 1970's, such efforts focussed on identifying intrinsic surface states of semiconductors, that is, states associated with the clean, ordered semiconductor crystal face prepared under ultrahigh vacuum (UHV) conditions. Early attempts to identify intrinsic surface states were hampered by difficulties in eliminating extrinsic electronic effects. Factors such as surface contamination (Spicer et al. 1976, Brillson 1975a,b, Goodwin and Mark 1972), lattice damage (see, e.g., Henzler 1970), cleavage steps (van Laar and Scheer 1967, Huijser and van Laar 1975, van Laar and Huijser 1976) and bulk impurity levels (see, e.g., Milnes 1973) can all contribute to states within the band gap. While intrinsic surface states were found, their energy positions were in general outside the energy band gap (Brillson 1975a, 1977a, van Laar and Scheer 1967, Williams et al. 1977) or near the forbidden band edges (Eastman and Grobman 1972, Wagner and Spicer 1972).

Furthermore, in many cases, metal overlayers caused these states to disappear (Rowe et al. 1975). Hence, intrinsic surface states did not provide a straightforward explanation of Schottky barrier formation.

By the early 1970's, evidence began to appear suggesting that extrinsic effects, i.e. due to interactions between metal and semiconductor, could be influencing barrier formation. A number of experimental techniques, mostly notably Rutherford backscattering spectrometry (RBS), provided evidence for semiconductor diffusion into thick metal overlayers (Hiraki et al. 1971, 1977) as well as changes in chemical and electronic properties as a function of high-temperature annealing treatments (Sinha and Poate 1973, 1978, Robinson 1975). Theorists and experimentalists alike established correlations between measured Schottky barrier heights and parameters of the interface which reflected metal–semiconductor interactions (Andrews and Phillips 1975, Ottaviani et al. 1980, Freeouf 1980, 1981, Brillson 1978a,b). Experimentalists using surface science techniques discovered that large atomic and charge rearrangements occurred at the initial metal–semiconductor interface and that the semiconductor E_F moved from its initial bulk position to that of the macroscopic barrier with deposition of only a few monolayers or less of metal (Brillson 1982a, 1983a, Margaritondo 1983, Williams 1982, Bachrach 1984, Henisch 1984, Bauer et al. 1983). Furthermore, theorists showed that even small changes in the arrangement of local charge and atomic cores at the interface could alter critically the band bending at the metal–semiconductor junction (Bennett and Duke 1967, Duke 1967, Louie and Cohen 1975, 1976, Chelikowsky 1977, Zhang and Schlüter 1978).

2.3. Role of metal–semiconductor interactions in Schottky barrier formation

Surface science techniques applied to the determination of interface atomic and electronic structure have led to a new picture of the metal–semiconductor interface, which differs significantly from fig. 1b. In addition to new electronic states produced by bonding changes and lattice distortion at the outer monolayer in contact with the metal, substantial chemical reaction and/or interdiffusion can occur. As a result, there can be formation of new semiconducting and metallic alloys, dipole layers, native defects, impurities, and their defect complexes. Figure 2 illustrates schematically the effect of such phenomena on the energy band diagram of the metal–semiconductor interface (Brillson et al. 1981a, Brillson 1982a,b). Formation of native defects, impurities, and their complexes can contribute new charge states within the semiconductor, which can alter the band bending within the surface space charge region. Such changes need not be uniform and can lead to nonparabolic band bending. New compound formation can produce new semiconducting layers or dipoles of thickness sufficient to alter the electrical properties of the junction. Metallic alloy overlayers formed by interdiffusion may contribute to a new contact potential difference between overlayer and substrate by a classical work function change. Such overlayers need be only a few monolayers thick or less to have a pronounced electrical effect and need not even be uniform across the interface plane. Figure 2 represents a

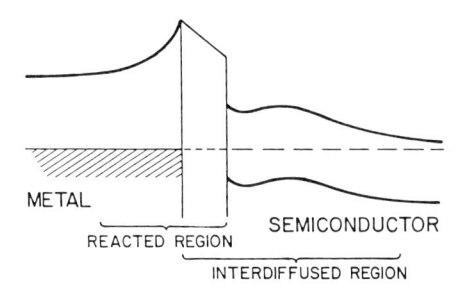

Fig. 2. Schematic band diagram of the nonideal, extended metal–semiconductor interface. Reacted and interdiffused regions are determined by the strength and nature of interfacial chemical bonding. The reacted region can have new dielectric properties and built-in potential gradients. The interdiffused region can have nonparabolic band bending determined by the movement and activation of the various atomic constituents. After Brillson et al. (1981a) and Brillson (1982a,b).

junction which extends from a few to hundreds of monolayers, depending upon the kinetics of the metal–semiconductor interaction. With soft X-ray techniques, electronic and chemical changes can be monitored together as a function of metal deposition with a precision of fractions of a monolayer. Hence, detailed spatial variations normal to the interface can be monitored, which are extremely difficult to obtain by other, more conventional interface characterization tools.

3. Soft X-ray photoemission spectroscopy for microscopic interface characterization

3.1. The soft X-ray photoemission spectroscopy technique

To date, soft X-ray photoemission spectroscopy (SXPS) has been the most successful solid state technique for characterizing metal–semiconductor interfaces on a microscopic scale. The basis for the extremely fine depth resolution of SXPS is the short and variable scattering length of electrons in a solid by inelastic collisions. Figure 3 illustrates that this scattering length λ_e exhibits a well-defined minimum in the energy range between roughly 50 and 100 eV kinetic energy relative to the Fermi level of the solid (Seah and Dench 1979). At energies near the minimum, electrons photoemitted from atomic cores only 4–6 Å below the free surface can escape from the solid and be detected. The photoelectron kinetic energy is given by

$$E_k = h\nu - E_B - (E_{VAC} - E_F), \qquad (2)$$

where $h\nu$ is the incident photon energy and E_B is the binding energy relative to E_F. E_{VAC} is the vacuum level, which the electron must exceed in energy to escape from the solid. Since

$$E - E_F = E_k + (E_{VAC} - E_F) = h\nu - E_B, \qquad (3)$$

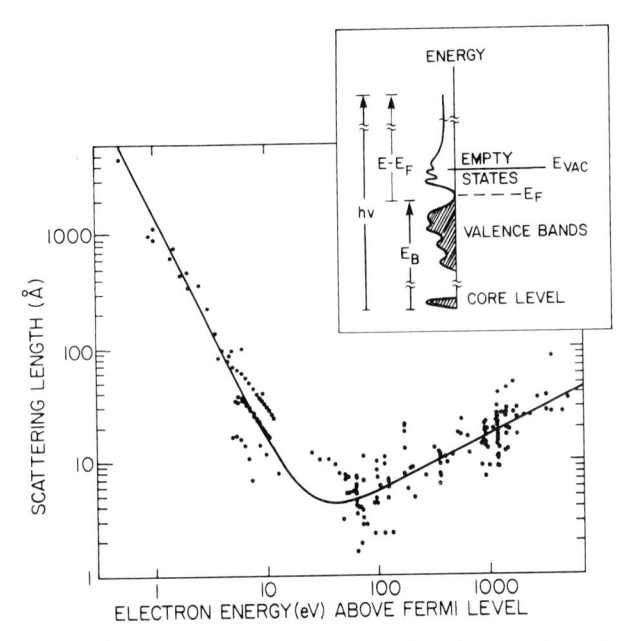

Fig. 3. Scattering length in Å of electrons in a solid as a function of energy above the Fermi level. The solid line represents an empirical least squares fit over the entire energy range for elemental solids. After Seah and Dench (1979). Inset shows schematically the photoexcitation of electrons from an atomic core level to an empty state at an energy $E - E_F$ above the Fermi level by a photon of energy $h\nu$. Valence electrons are photoexcited to correspondingly higher energy.

$\lambda_e(E)$ can be adjusted by selecting an appropriate $h\nu$. The inelastic mean free path increases significantly at much higher or lower energies, so that, by varying the incident photon energy, one can tune the escape depth away from the minimum and probe several atomic layers or more to the surface. For example, with the "grasshopper" (Brown et al. 1974) monochromator at the Tantalus ring of the University of Wisconsin, photoelectrons with energies ranging from 40 eV to 200 eV can be monochromatized and directed into a UHV chamber and to a particular surface or interface understudy (Weaver and Margaritondo 1979, Rowe 1981). An electron energy analyzer, such as a cylindrical mirror analyzer (CMA), counts the photoemitted electrons (Palmberg et al. 1969). Depending upon the photoexcited material and the core level probed, electron escape depth ranges of up to 50–100 Å can be achieved, particularly at low kinetic energies.

3.2. Ultrahigh vacuum fabrication and analysis of metal–semiconductor interfaces

In practice, soft X-ray radiation from an electron storage ring is passed through an appropriate monochromator to a UHV chamber, in which clean or controllably treated semiconductor surfaces can be prepared. Metal films can then be deposited by evaporation on these surfaces such that electronic and chemical properties can be analyzed as an interface is built up layer by atomic layer in stepwise fashion. In order to avoid contamination, which could easily perturb the interfacial properties, one performs these experiments at typical pressures of 10^{-10} Torr or less, rising to 10^{-9} Torr or less during evaporation. For a surface of sticking coefficient $s = 1$ (that is, every incident atom sticks to the surface), an exposure of 10^{-6} Torr pressure for 1 s produces a monolayer of contamination. Hence a pressure of 10^{-10} Torr prevents ambient contamination of metals ($s \leqslant 1$) for tens of minutes and of semiconductors ($s \ll 1$) for several hours or more.

Clean semiconductor surfaces can be obtained by crystal cleavage or by ion bombardment and annealing within the UHV chamber. Cleaved surfaces must be relatively free of steps or other imperfections in order to avoid extrinsic sites with electrical activity. Likewise, ion bombardment may produce electrically active sites by introducing lattice disorder or nonstoichiometry, which may not be removed completely by annealing. Metal atoms can be deposited onto the semiconductor surfaces controlled with a precision of fractions of a monolayer, using a quartz crystal oscillator to monitor the thickness of condensed atoms near the position of the exposed semiconductor substrate.

To complement the SXPS measurements, the UHV chamber can incorporate a variety of other surface science techniques. Most commonly found are Auger electron spectroscopy (AES) for monitoring chemical contamination and low-energy electron diffraction (LEED) for determining surface atomic geometry. (Such electron techniques should be employed cautiously since the incident beam can perturb the surface composition and bonding.) A vibrating Kelvin probe (Brillson 1975b) can be used for surface work function measurements and, when compared to ionization thresholds extracted from SXPS valence band spectra, can yield semiconductor electron affinities as well (Brillson 1978c). A host of other techniques can be coupled with SXPS to relate the chemical and electronic features evolving at the initial interface to more conventional measurements. In addition, molecular beam epitaxy (MBE) attachments to the UHV test chamber provide new methods of preparing ordered metal–semiconductor as well as semiconductor–semiconductor junctions for in situ SXPS analysis (Bachrach and Bauer 1979, Bauer et al. 1981).

4. Chemical bonding at metal–semiconductor interfaces

4.1. Evolution of metal–semiconductor bonding with interface formation

Metal atoms deposited on semiconductor surfaces can (1) bond weakly to the substrate as uniform adlayers, (2) bond weakly as three-dimensional clusters, (3) chemisorb as two-dimensional reacted layers, (4) chemisorb as three-dimensional clusters which induce a chemical reaction at a critical stage or, (5) diffuse into the substrate either from randomly distributed surface atoms or from three-dimensional surface clusters. At higher metal coverages, metals which react may (a) form new bonded phases, which limit further reaction at a particular thickness and temperature, or (b) propagate into the semiconductor by reactive diffusion, producing extended reacted products both normal to and across the interface plane (Ludeke 1983). At higher metal coverages, less reactive metals may induce semiconductor outdiffusion, metal indiffusion, and surface segregation, whose extent, composition, and homogeneity depend upon overlayer thickness, temperature, solid solubility, surface morphology and lattice disorder. These interactions manifest themselves in characteristic features of core level and valence band spectra, which vary substantially from one semiconductor to another.

In this section we illustrate the use of SXPS to analyze these effects in terms of core level lineshapes on a monolayer by monolayer scale. We have organized results into classes of semiconductors in order to emphasize characteristic similarities within and differences between the features of each class. While only a representative sampling of work appears in each subsection, the intent is to provide some scope to the large body of research spanning many diverse types of semiconductor–metal systems.

4.1.1. Silicon–metal interfaces
In general, reactions at metal–Si interfaces lead to a variety of interfacial layers, which depend upon the specific type and amount of metal involved, its epitaxial relationship on the free Si surface, and the temperature at which the interface forms. Depending upon these parameters, the interfacial region between metal and Si can extend from monolayers to microns. SXPS studies focus on: (a) the composition of the reacted silicide phases formed by metal deposition on Si, (b) the density-of-states features associated with the interfacial reaction, which can account for the Fermi level stabilization, (c) the bonding energies and lineshapes of core levels, which provide information on charge transfer, bond ionicity, electron–hole excitations across the Fermi surface and correlation effects, and (d) the analysis of bonding between metal and semiconductor in terms of orbital character, through the use of cross section effects.

Figure 4 shows the evolution of interface chemical bonding with metal deposition for two common metals, Au and Al, on Si(111) surfaces (Brillson et al. 1984, Brillson 1984). A photon energy $h\nu = 130$ eV was used to excite Si 2p core electrons from UHV-cleaved surfaces overlaid with successive deposits of metal. The choice of $h\nu$ and core binding energy results in extreme surface sensitivity. Au and Al on the same Si surface lead to quite different spectral features. The clean Si surface exhibits a spin–orbit-split Si 2p feature, which changes substantially with Au deposition. With an initial deposition of 1 Å Au ($=5.77 \times 10^{14}$ atoms/cm^2 = 0.74 monolayer equivalent), a second Si 2p component appears, which is shifted to higher binding energy. This shift is consistent with a charge transfer from Si to Au due to the higher electronegativity of Au as compared to Si. With increasing Au coverage, the Si 2p component at higher binding energy dominates the lower binding energy (substrate) component. At a coverage of 20 Å Au, the spectrum consists almost entirely of the higher binding component, and its spin–orbit splitting is now clearly resolved. In earlier studies, these features were not discernable due to lower energy resolution (Braicovitch et al. 1979, 1980).

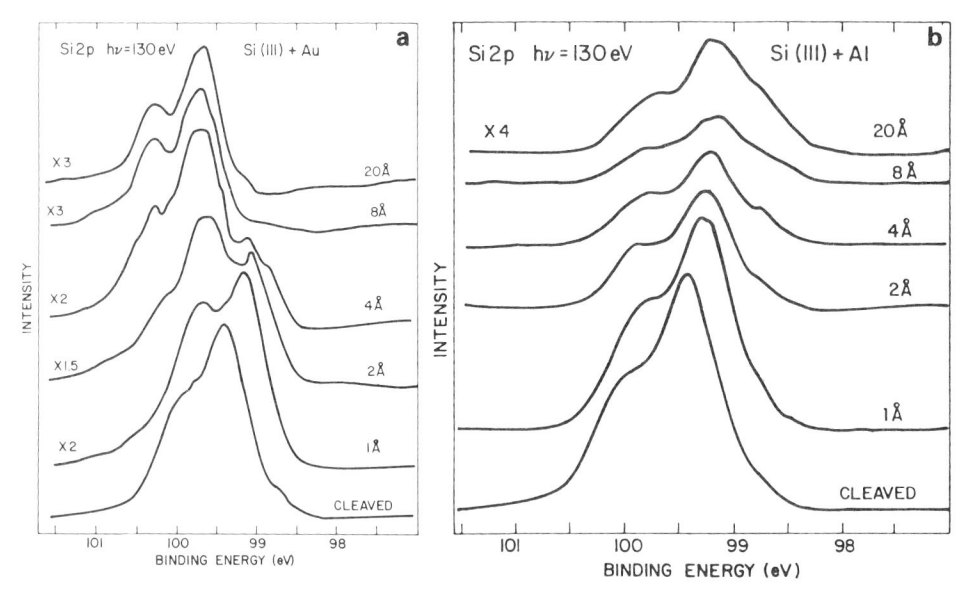

Fig. 4. SXPS Si 2p core level spectra as a function of increasing metal coverage for (a) Au and (b) Al on Si(111) with incident energy $h\nu = 130$ eV. After Brillson et al. (1984) and Brillson (1984).

The slow decrease in the Si 2p peak intensity and the dramatic change in lineshape with Au coverage indicates that substantial diffusion takes place. While the intensity decrease for the first 8 Å Au deposition is consistent with a photoelectron escape depth of 4–6 Å, no significant decrease in Si 2p peak intensity occurs for higher coverages. Significantly, the substrate component of the spectra of fig. 4a does decrease exponentially, corresponding to a 3–4 Å escape depth, which suggests that Au on Si forms a uniform overlayer rather than islands. This laterally uniform layer must contain Si in a bonding environment different from that of the substrate.

SXPS studies of Al on similar UHV-cleaved Si(111) surfaces provide a striking contrast to the Au-Si(111) results (Brillson et al. 1984, Brillson 1984). Figure 4b illustrates Si 2p core level spectra as a function of Al overlayer thickness. In contrast to fig. 4a, the Si 2p intensity decreases rapidly at all coverages and only minor changes in core lineshape are evident. The only evidence for diffusion is the low binding energy shoulder at higher Al coverages, which corresponds to a slight segregation of dissociated Si to the free Al surface. Hence, Al on Si(111) near room temperature exhibits no strong interdiffusion or reaction, as opposed to the Au–Si(111) case. Depth-dependent SXPS studies to be described below confirm these different pictures.

Despite the fact that interface interactions are often modeled by uniform overlayers on substrates, surface diffusion and chemical reactions can lead to metal–semiconductor junctions which are heterogeneous in the plane of the interface. Such heterogeneity complicates the interpretation of photoemission spectra. Nevertheless, by deconvoluting core lineshapes and analyzing rates of intensity decrease with equivalent deposition, it is possible to model various types of heterogeneous film growth on semiconductors. Figures 5 and 6 illustrate such an analysis for Ce on Si(111). The left side of fig. 5 shows Si 2p core level spectra for increasing equivalent coverages of Ce (Grioni et al. 1984a). The clean surface spectrum reveals only the characteristic Si 2p doublet broadened by surface-shifted components (Eberhardt et al. 1978, Himpsel et al. 1980a). At an equiva-

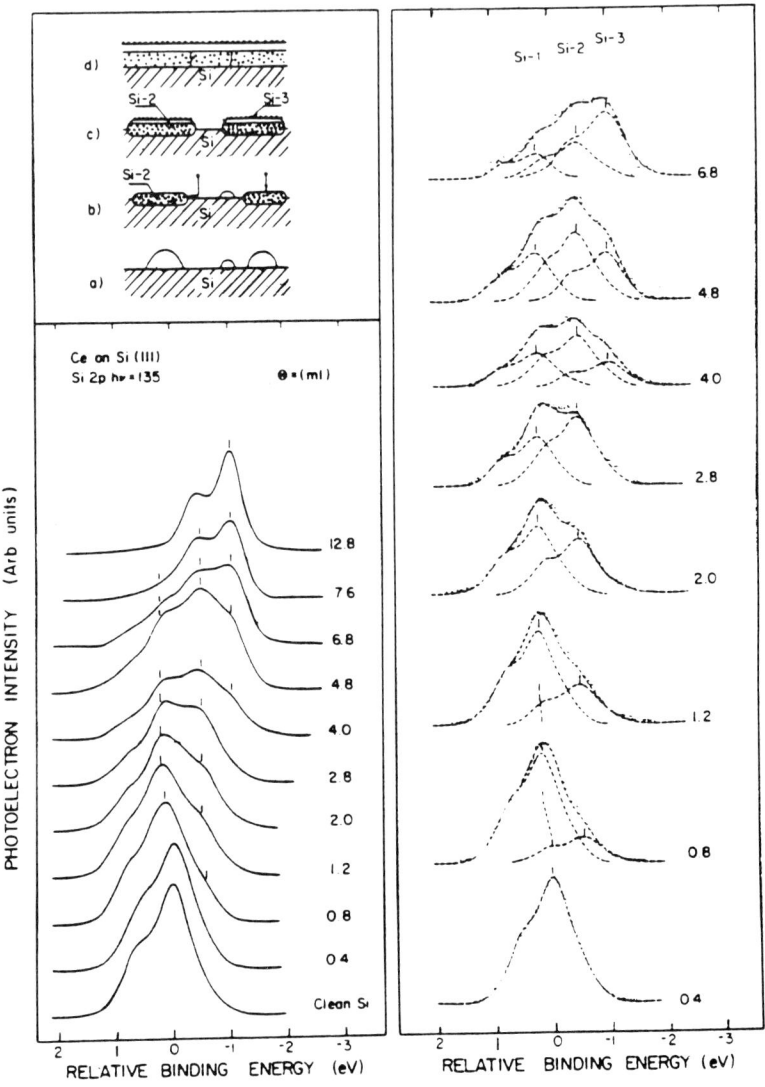

Fig. 5. Evolution of Si 2p core level spectra taken at $h\nu = 135$ eV for Si(111) with increasing Ce deposition. Spectra on the right are deconvoluted for representative coverages. Spin–orbit-split components attributed to clean Si (Si-1), a reacted phase (Si-2), and surface-segregated Si (Si-3) comprise the overall spectra. Inset shows a model for the heterogeneous interface. After Grioni et al. (1984a).

lent coverage of 0.6 monolayer, a reacted component appears and grows relative to the original doublet. At multilayer coverages, a third component appears at lower binding energy, which persists to high coverages and which corresponds to segregated Si atoms.

Deconvolution of these spectra at representative coverages into equivalent spin–orbit-split pairs with appropriate energy splittings and branching ratios of intensity reveals three distinct local bonding configurations of Si. At any coverage, the results can be fit with components corresponding to the clean Si substrate, a reacted phase, and a surface-segregated Si phase denoted by Si-1, Si-2, and Si-3, respectively. As shown on the right, only three unique chemical configurations (three doublets) are needed for all spectra.

Figure 6 shows the attenuation of the total Si 2p emission (solid line on top) and the attenuation of each of the three components (Grioni et al. 1984a). The substrate attenuation is rapid and monotonic but not necessarily proof of two-dimensional growth. Indeed, SXPS valence band, LEED, and angle-resolved AES (Chambers and Swanson 1983) studies indicate the presence of weakly interacting Ce clusters. Beyond a critical coverage of 0.6 monolayer, the reacted phase Si-2 increases abruptly, suggesting that the cluster stage is a precursor to chemical reaction. This critical coverage for reaction coupled with the large, observed chemical shifts reflect substantial charge transfer in the reacted phase. Besides the Ce/Si system, Sm/Si (Franciosi et al. 1984), Eu/Si (Grioni et al.

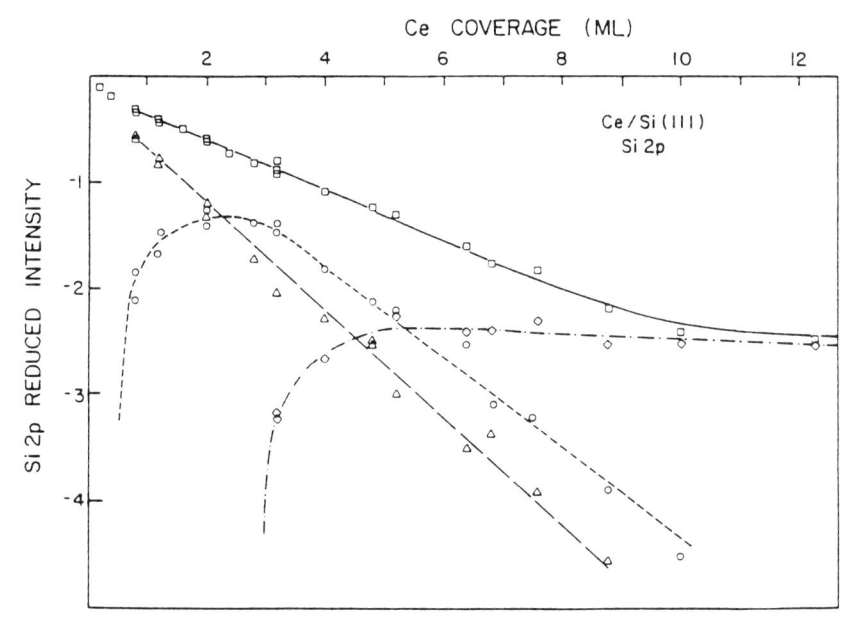

Fig. 6. Attenuation of the total Si 2p emission (solid line on top) and each of its three components. Substrate attenuation (long dashed line) is rapid and monotonic. The reacted component (short dashed line) grows to a maximum at 2.5 monolayers and then decreases. Above this coverage, a surface-segregated component (dot–dashed line) appears, which remains constant to high coverage. After Grioni et al. (1984b).

1984b) and Yb/Si (Rossi et al. 1983) exhibit similar phenomena as well. A model for the evolving Ce/Si interface, including lateral growth of the silicide, overgrowth of metal, and surface segregation of Si at the free metal surface appears as an inset in fig. 5.

Another method of analyzing the chemical bonding at a metal–semiconductor interface exploits the energy dependence of the photoionization cross section, which can exhibit large variations for different orbitals of the metal relative to the semiconductor. For example, the energy dependence of the 4d and 5d electron photoionization cross sections exhibit a deep minimum [termed the Cooper minimum (Cooper 1962)] in the range from 130 to 180 eV (Johannson et al. 1980). This decrease is associated with the presence of a node in the radial wavefunction of 4d and 5d orbitals and the resultant cancellation of matrix elements for transitions to particular final states. Solid state effects are evident in cases when the initial state wave function is strongly modified. By varying $h\nu$, one can probe the valence states of interacting metal–silicon interfaces and preferentially modulate the contribution of metal d states. Hence, sp contributions to the total density of states can be separated out which would otherwise be overshadowed by emission from metal states and states involved in bonds with Si. In this way one can characterize the detailed nature of Si/d-metal chemical bonding.

A prime example is the interface between the reactive transition metal Pt on Si. Figure 7 contains a comparison of valence band spectra at $h\nu = 80$ eV and $h\nu = 130$ eV (Cooper minimum) for three representative coverages of Pt on Si(111) (Rossi et al. 1982b). At $h\nu = 80$ eV, the Pt 5d cross section is much larger than that of Si sp orbitals (Abbati et al. 1980a), whereas at the Cooper minimum for Pt, the Pt emission is reduced such that the evolution of Si valence states is discernable. Below one monolayer, a dramatic change occurs in the relative intensities of Si valence peak features. This suggests that a small number of Pt atoms has a strong effect on Si surface rearrangement and that mixed orbitals form between Pt 5d and Si sp states. At intermediate coverage (1–10 monolayers), three features are evident: (a) mixed-orbital contributions from Si p and metal d electrons with antibonding character, (b) Si-p/Pt-d hybrid bonding contributions typical of the silicide bond and relatively stable over a relatively wide range of coverage (indicative of a wide silicide interface), and (c) residual features of the Si–Pt chemical bond, whose d character at deeper binding energy is not affected as much as shallow states at the Cooper minimum (Rossi et al. 1982b). Corresponding labels denote these features in fig. 7, which demonstrates how key features of the Si bonding can be uncovered by appropriate choices of synchrotron photon energies.

Another technique to identify the bonding nature of valence electronic states involves resonant enhancement of the photoelectron cross section (Lenth et al. 1978). Typically, such enhancement is associated with the optical absorption threshold of an atomic core level. Here the excitation of the core electron leaves behind a hole in the corresponding atom, which can be filled by an Auger-type process (e.g., a second, higher lying electron fills the hole and a third electron leaves the solid with the overall energy gain). Since this process competes with

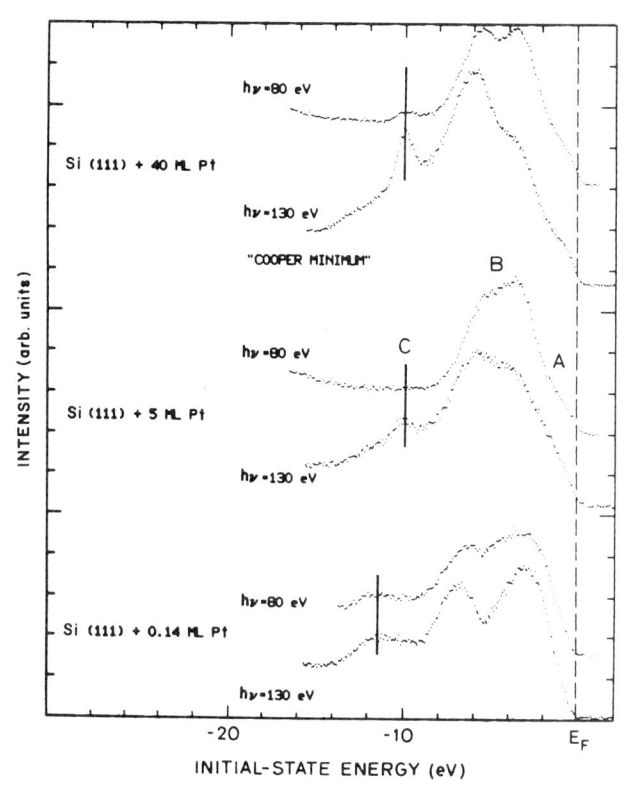

Fig. 7. Comparison of valence band spectra at $h\nu = 80$ eV and 130 eV (Cooper minimum) for cleaved Si and with low, intermediate and high coverages of Pt. After Rossi et al. (1982b).

direct photoemission in ejecting electrons from the solid, the two excitation channels interfere with each other quantum mechanically, yielding a resonance in the $h\nu$ dependence of the emission probability.

Figure 8 shows a representative valence spectrum for VSi_2 at $h\nu = 37$ eV in the inset (Weaver et al. 1984). Arrows point to spectral features whose photoemission cross sections appear in the main figure. Based on self-consistent, augmented-spherical-wave calculations of the density of states for VSi_2 (Weaver et al. 1984) and other transition metal silicides, the character of initial states is nonbonding d near E_F, p–d hybrid near 4 eV, p-like near 6 eV and s-like near 10 eV. Several of these features exhibit pronounced resonance effects associated with the competition between the direct valence band emission $3p^6(4sd)^n + h\nu \rightarrow 3p^6(4sd)^{n-1} + e$ and a process $3p^6(4sd)^n + h\nu \rightarrow 3p^5(4sd)^{n+1} \rightarrow 3p^6(4sd)^{n-1} + e$. Here the strength of the enhancement is a measure of the overlap of the p core hole and the d valence state. Thus the enhancement decreases as the d states become hybridized and the overlap is reduced. Figure 8 shows a clear enhancement of all five of these features at the V 3p core threshold (37.4 eV). For these states near E_F, there is a threefold enhancement between 38 and 50 eV. For states in the p–d bonding region, there is weaker enhancement for $h\nu > 38$ eV but similar in shape. States

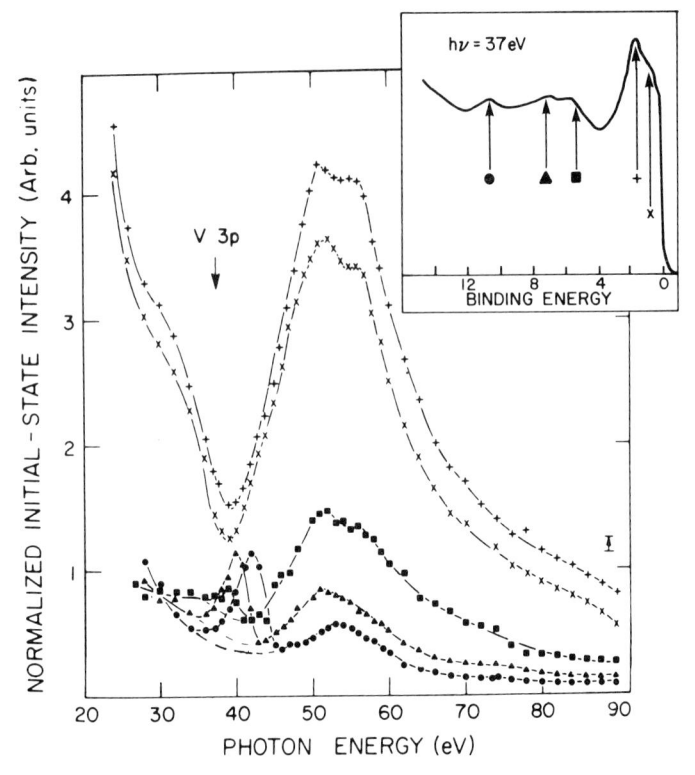

Fig. 8. Photoionization cross sections for VSi_2 corresponding to initial states marked by arrows and labels in the inset. Nonbonding d states near E_F exhibit a threefold enhancement due to the 3p–3d resonance for the p–d hybrid states near 4 eV; the enhancement depends on the degree of hybridization, decreasing with increasing Si p contribution (near 6 eV). States near 10 eV have s-like character. After Weaver et al. (1984).

with increased Si-derived character exhibit much less enhancement. The relative enhancements support the theoretical analysis and demonstrate that resonant photoemission can highlight d states in the photon energy range of the p core level excitation energy. Other silicides such as $MoSi_2$ (Weaver et al. 1984) and Pd_2Si (Franciosi and Weaver 1983a) show similar but smaller enhancements. Resonant enhancement complements the Cooper minimum technique for 4d and 5d metals. As shown for VSi_2, it has the additional advantage of being applicable to 3d systems, which have no Cooper minimum.

For the noble metals Au (Abbati et al. 1980b), Ag (Rossi et al. 1982c) and Cu (Abbati and Grioni 1981), as well as the near-noble metals Pt and Pd (Abbati et al. 1980c), metal–semiconductor interactions can occur even at liquid nitrogen temperature. For the near-noble metals Ni and Pd, the growth of epitaxial layers can be studied by a combination of SXPS, surface-extended X-ray fine structure analysis (SEXAFS, to be discussed in section 7), and transmission electron microscopy (TEM) to describe the atomic ordering and the complex stages of reaction on a layer-by-layer basis. Also at issue are the stoichiometry at the intimate interface, the structural defects which may be electrically active, and any interface-specific electronic states which arise from the transitional bonding. For many other metals which form silicides, the interactions are stronger and more difficult to characterize. Franciosi and Weaver have laid important groundwork for identifying the products of interaction between these metals and Si. Their extensive characterization of bulk transition metal silicides of Ti, V, Nb, Cr, Fe, Co, and Ni cleaved in UHV and probed at a wide range of photon energies provides a valuable data base for comparison with interfacial species (Franciosi et al. 1982a,b, Franciosi and Weaver 1983a,b,c, Weaver et al. 1981, 1984). However, theoretical calculations are not yet in close agreement with the SXPS results. To further complicate matters, many workers have found large concentration gradients at the microscopic silicide interfaces (Braicovitch et al. 1979, Rossi et al. 1983). Deviations in silicide stoichiometry at the outer Si layer [for example, Si-rich PdSi (Franciosi and Weaver 1983b)] can produce new electronic states localized at the interface. Furthermore, there are apparently differences between chemical interactions at metal–Si versus metal–Ge interfaces (Perfetti et al. 1982a,b, 1983, Rossi et al. 1981, 1982a, LeLay 1983a) [e.g. the stoichiometric Ge–Au interface phase versus the graded Si–Au junction (Perfetti et al. 1982a,b)].

For an extensive review of chemical bonding and electronic structure at transition metal/Si interfaces, with particular attention to SXPS results, the reader is directed to the article by Calandra et al. (1985). Also valuable are LeLay's review of noble metals on Si (LeLay 1983b) and Rubloff's review of silicide interfaces (Rubloff 1983). The latter addresses results obtained by a wider range of techniques and focusses on the role of interfacial chemistry and microscopic mechanisms which determine the Schottky barrier formation.

4.1.2. III–V compound semiconductor–metal interfaces
The interface between metals and III–V compound semiconductors constitutes another system studied extensively using SXPS techniques. Indeed, much current insight into mechanisms of chemical interaction at interfaces and their effect of Schottky barrier formation derives from such studies. Perhaps the best example in this regard is the Al–GaAs(110) interface, for which the spectra of SXPS core level shifts have stimulated considerable controversy and have led to new models of surface chemical activity. Figure 9 illustrates photoemission spectra for Al 2p and Ga 3d core levels as a function of Al coverage on the cleaved (110) surface (Bachrach and Bauer 1979). For the initial coverage of 0.53 Å Al, Bachrach, Bauer, and coworkers found the Al 2p core level to be chemically shifted by 0.7 eV to higher binding energy, while the Ga 3d peak exhibited a shoulder at 0.95 eV lower binding energy (Bachrach et al. Bachrach 1978, Bachrach and Bauer 1979, Brillson et al. 1979). The As 3d level simply decreased in intensity with only a 0.1 eV shift to higher binding energy. From straightforward escape depth arguments, the numbers of chemically shifted Ga atoms and deposited Al atoms are equal. Bachrach's core level results were the first evidence for a

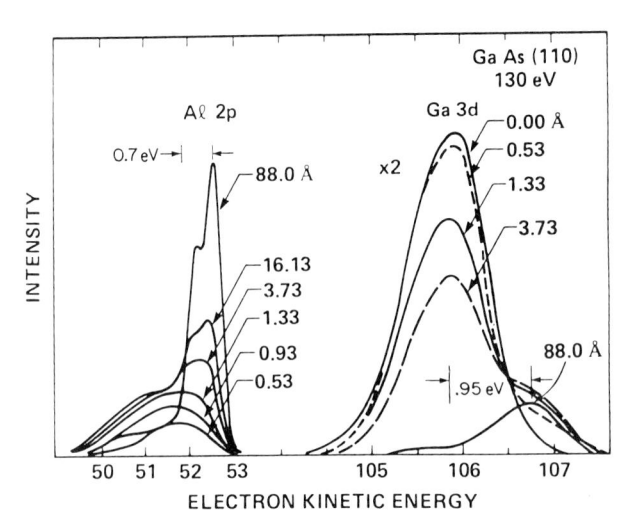

Fig. 9. Photoemission spectra with $h\nu = 130$ eV of chemically shifted Al 2p and Ga 3d core levels for various Al overlayer thicknesses on GaAs(110). After Bachrach and Bauer (1979).

metal–semiconductor exchange reaction, whereby the Al replaces the Ga in the top layer and nearby free Ga resides outside the surface (Bachrach et al. 1978, Bachrach and Bauer 1979).

From other SXPS and LEED measurements, however, the room temperature deposition of Al does not appreciably modify the surface structure of clean GaAs(110). Instead, dramatic changes in core level features occur for thin Al layers on GaAs after annealing to relatively low temperatures (Kahn et al. 1981a,b,c, Duke et al. 1981, Kahn 1983). Figure 10 demonstrates the SXPS core level changes upon heating one monolayer of Al on GaAs(110) (Kahn et al. 1981a). The Al 2p core level exhibits a well-defined shift to higher binding energy beginning with a 15 min anneal at 200°C, corresponding to a surprisingly low activation barrier to reaction. For a 400°C, 15 min anneal, the Al 2p core level shifts by 0.5–0.65 eV and decreases with respect to Ga and As peaks. The ratio of Ga to As peak intensities decreases for most thin coverages as well.

All of these results suggest the following picture for the UHV-cleaved GaAs surface. At room temperature, the interaction is either a partial and randomized Al–Ga replacement reaction, which does not modify the LEED diffraction patterns substantially (Bachrach and Bauer 1979), or a mild interaction between Al and substrate, dominated by strong Al–Al interaction on the surface (Huijser et al. 1981, Swartz et al. 1980), which results in formation of two-dimensional islands (Skeath et al. 1980a,b), similar to the inset of fig. 5. Hence the Al 2p core level in fig. 9 corresponds to an average of several possible Al states chemisorbed or reacted on or within the top GaAs layer. After annealing, the decrease in Al 2p peak intensity indicates that the Al penetrates the lattice below the original absorption/reaction site. The shift to higher binding energy is consistent with the formation of AlAs, as Al is driven into the lattice at sites of maximum coordina-

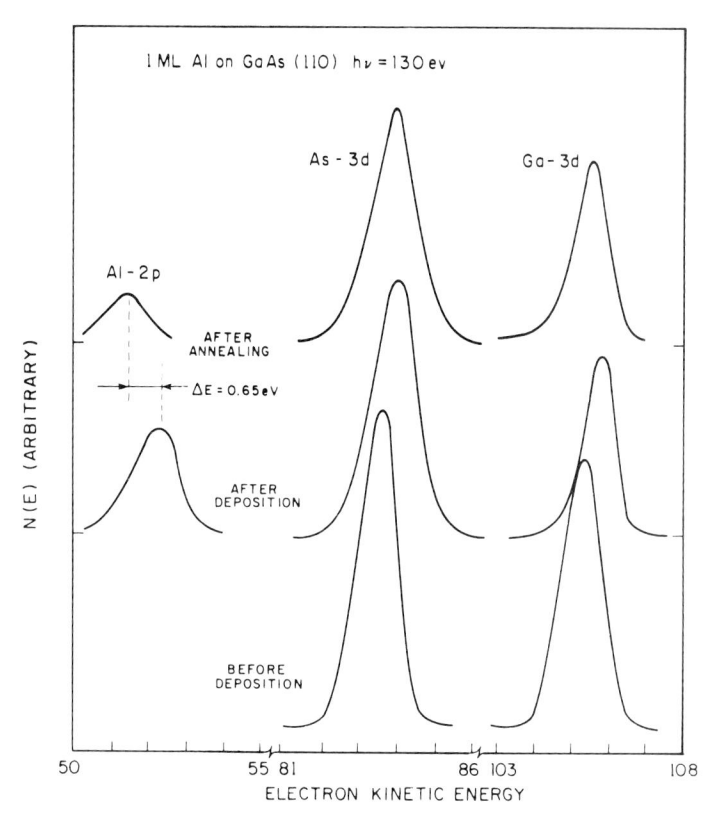

Fig. 10. Photoemission spectra with $h\nu = 130$ eV of Al 2p, As 3d, and Ga 3d core levels for the cleaved GaAs(110) surface before Al deposition, after deposition of one Al monolayer and after heat treatment of the Al–GaAs interface. After Kahn et al. (1981a).

tion with As (Kahn et al. 1981a,b,c, Duke et al. 1981, Huijser et al. 1981, Kahn 1983). Dynamical analysis of the LEED spot intensities indicates that Al penetrates to the second and deeper atomic layers and that several initial monolayers are needed to saturate the layers close to the GaAs surface after annealing (Duke et al. 1981, Kahn et al. 1981b,c, Kahn 1983). The thermal treatment appears to produce a depletion of Ga near the interface, whereas other room temperature experiments suggest that Ga or In segregates to the free Al surface on GaAs (Bachrach and Bauer 1979, Brillson et al. 1980d,e, 1981d, Huijser et al. 1981, Shapira et al. 1984) or InP (Brillson et al. 1981a, Shapira and Brillson 1983, Kendelewicz et al. 1983b, 1984a), respectively.

Mild interaction between Al with polished and annealed GaAs substrates (perhaps a Ga–Al exchange interaction) indicates that the kinetic barrier to surface reconstruction is highly sensitive to the degree of perfection of the GaAs(110) surface (Skeath et al. 1980a, Duke et al. 1981, Kahn et al. 1981b, Kahn 1983). In addition, the progression from weakly interacting to exchange reactions with coverage has suggested the role of cluster formation in initiating reactions, to be discussed in section 4.4. Interpretation of many room temperature results involving SXPS, LEED, AES, and electron loss spectroscopy (ELS) on GaAs(110) and (100) are not yet clear, in large part because of the strong sensitivity of the electron spectroscopies to surface perfection, orientation, and temperature.

Kendelewicz et al. have used high-resolution SXPS measurements to show that exchange reactions are more pronounced at the Al–InP(110) interface, due to the larger heat of reaction involving Al (Kendelewicz et al. 1983b, 1984a). Along with McKinley et al. (1982) and Zhao et al. (1983), they find evidence for exchange reactions between Al and InP, starting at coverages below a monolayer, to form AlP and to promote In segregation at the free Al surface. As with Al on GaAs, the reactions appear to be preceded by formation of clusters weakly bonded to the substrate. Nevertheless, up to coverages of 5 Å, they observe no photoemission from metallic Al, indicating that all metal atoms are coordinated into covalent bonds. This result can be seen more clearly for InP than GaAs because of the larger Al 2p core level shifts observed.

Richter et al. (1984) have used pulsed laser annealing to "step through" the Al–InP reaction and have shown graphically how the Al 2p lineshape evolves with the extent of interface reaction. For the data in fig. 11, a 5 ns, 308 nm XeCl excimer laser was used to promote the chemical reaction between Al and InP at elevated temperatures without significant diffusion. AES depth profiling studies confirm the shallow depth of such treated layers. Without the anneal, the Al 2p core level in fig. 11 displays a wing at higher binding energy consistent with a limited reaction near room temperature. At the lowest energy density of $0.1 \, J/cm^2$, there is little or no change in the Al 2p (as well as In 4d and P 2p) core level, indicating that this density is too low to promote a chemical reaction during the short annealing time. Between 0.1 and $0.17 \, J/cm^2$ more than half the Al atoms change from Al–Al bonding to a stronger chemical bond. This trend continues until, at $0.3 \, J/cm^2$, only a small "metallic" Al contribution remains. A chemical shift of 1.5 eV between "metallic" and strongly bonded Al core level positions is apparent. At higher energy densities, significant disruption of the surface occurs. Consistent with SXPS studies of conventionally annealed Al–InP(110) interfaces (Kendelewicz et al. 1983b), Richter et al. observe In segregation to the free surface and no significant change in P 2p lineshape or position. Thus SXPS provides a unique diagnosis of chemical bonding at discrete stages of reaction in microscopically thin interfacial layers.

Up until the 1980's, it was thought that Schottky barrier formation could be understand in general from the behavior of weakly interacting metal–semiconductor systems. However, SXPS, ELS, and AES studies have made clear that strong contact interactions are more the rule than the exception and that chemical phenomena play a first-order role in the rectification process (Brillson 1978b, 1979, 1982b,c,d, 1983b). Many metals (e.g. Al, Ti, and Ni) can react with III–V compound semiconductors to form stable interfacial compounds with the substrate anion. Even metals with no strong compound formation (e.g. Au and Cu) can promote extensive interdiffusion of both anion and cation near room tem-

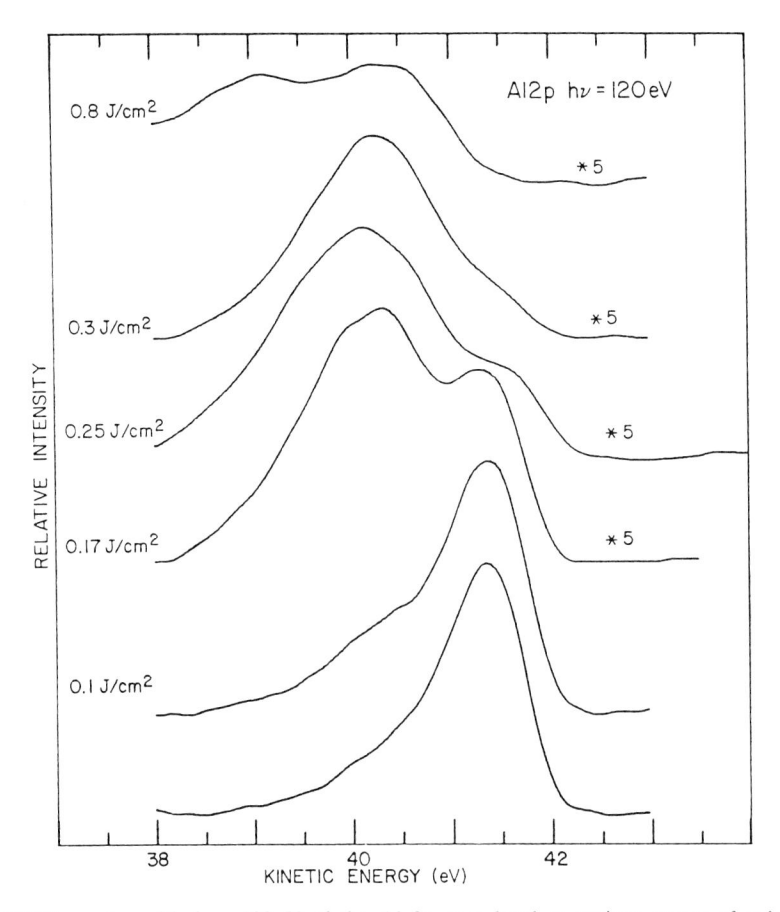

Fig. 11. SXPS spectra with $h\nu = 120$ eV of the Al 2p core level at various stages of pulsed laser annealing for 20 Å Al on UHV-cleaved InP(110). After Richter et al. (1984).

perature. This atomic redistribution depends heavily on the strength and character of chemical bonding, a point now recognized by several groups (Brillson 1979, 1982b,c,d, 1983b, Williams 1979, Brillson et al. 1981b, Margaritondo 1983, Kendelewicz et al. 1984b, 1985). In the same cases, the metal can interact with both anion and cation. Indeed, by taking into account the free energy of all parts of the interaction (McGilp 1984), one can parametrize the strength of chemical interaction thermodynamically (Kubaschewski et al. 1967, Wagman et al. 1968/1971, Mills 1974) and relate this parameter to observed Schottky barrier heights (Brillson 1978a).

More recent work shows that different metals on InP and GaAs produce complex chemical interactions, with phase segregation, island formation, spatial redistribution, and gradients in composition and bonding. Extensive high-resolution SXPS work by Kendelewicz et al. for InP interfaces with Pd (Kendelewicz et al. 1983d, 1984b,c,d), Cu (Kendelewicz et al. 1983a,c, 1984b,d, Petro et al. 1984),

Ag (Kendelewicz et al. 1984b,d, Babalola et al. 1984), Au (Babalola et al. 1983, Kendelewicz et al. 1984b,d, Petro et al. 1984), Al (Kendelewicz et al. 1983b, 1984, 1987), Ni (Kendelewicz et al. 1984d, 1987), Ti (Kendelewicz et al. 1985), Tl (Kendelewicz et al. 1985), Cr (Kendelewicz et al. 1985, List et al. 1985) and Mn (Kendelewicz et al. 1985, List et al. 1985) overlayers reveal that such complexities exhibit corresponding differences in the evolution of the Schottky barrier, in agreement with an earlier medium-resolution analysis (Brillson et al. 1982). In addition, Kendelewicz et al. (1984b) have shown that, for InP and GaAs, reactions with noble and near-noble metals follow a similar pattern, although the GaAs–metal reactions appear to be less pronounced. This is consistent with the weaker bonds between As and metals versus P and metals. SXPS studies of Williams et al. (1984) for Ni, Cu, Sb, Sn and Ga on InP(110) show significant reaction or diffusion in all cases. Likewise, Weaver et al. found strong chemical reactions for the rare earth metals Ce and Sm on GaAs(110) (Grioni et al. 1986). For the refractory metals V and Cr on the same surface, however, they found both metal–anion and metal–cation bonding as well as complex interface morphologies analogous to fig. 5 (Weaver et al. 1985, Grioni et al. 1986). Other III–V compound semiconductors for which strong contact interactions appear include GaSb, GaP, InAs, and InSb (Brillson et al. 1980a, 1981b, Brillson 1982a,d). Notable for their absence of chemical activity are Ag on MBE-grown GaAs surfaces (Ludeke et al. 1982, 1983) and Pb on GaAs (van der Veen et al. 1982). In addition, the formation of metallic clusters can occur for both strongly and weakly interacting interfaces and will be discussed in section 4.4.

A number of current reviews deals with chemical interactions at metal interfaces with III–V compound semiconductors. Brillson's monograph (Brillson 1982a) discusses the microscopic bonding in the context of both SXPS measurements near room temperature and more conventional RBS and AES studies at elevated temepratures. Here the strength and nature of chemical bonding is presented as the dominant influence in Schottky barrier formation. Bachrach (1984) emphasizes the influence of atomic ordering and surface reconstruction of the initial metal layers on the interface electronic structure. Williams' review (Williams 1982) describes surface science results of a similar nature with an emphasis on the role of semiconductor defects in the Fermi level pinning. Spicer and Eglash (1985) amplify this point of view in a review of their defect model and its relation to GaAs integrated circuits. Margaritondo and Franciosi's review articles have incorporated both of these perspectives and extended them to the semiconductor heterojunction interface (Margaritondo 1983, Margaritondo and Franciosi 1984).

4.1.3. II–VI compound semiconductor–metal interfaces
Chemical reactions for II–VI compound semiconductors with metals are more spatially extended near room temperature than for III–V compounds (Brucker and Brillson 1981b,c, 1982, Stoffel et al. 1982, Brillson 1983b, Brillson et al. 1983). SXPS studies of metals with CdS, CdSe, CdTe, ZnS, ZnSe, and ZnTe reveal semiconductor anions and cations in new bonding configurations in interfacial regions extending tens to hundreds of Å. Interface atomic redistribution

appears to be qualitatively different from III–V compound semiconductors as well, with evidence for both reactive outdiffusion of anions and localization of dissociated cations in many II–VI systems (Brillson 1982a, Brillson et al. 1983).

The interface between Al and a UHV-cleaved CdS($10\bar{1}0$) surface is a representative case. Figure 12 illustrates the evolution of Al 2p and Cd 4d core level spectra with metal deposition (Brillson et al. 1980b). Here photon energies of 140 eV and 85 eV were used to excite Al 2p and Cd 4d core electrons, respectively, with high surface sensitivity. With the initial deposition of about one monolayer, the Al 2p spectrum exhibits only a single peak, characteristic of Al bonded to the substrate. With additional Al coverage, a second peak appears which is shifted to lower binding energy, characteristic of metallic Al. The appearance of only a single peak at monolayer coverage indicates that no islands are forming. At thick Al coverages, the metallic Al feature dominates the spectrum. The bonding of Al with S atoms promotes the dissociation of surface Cd, as evident from the valence band spectra of fig. 12b. The Cd 4d feature broadens and gradually produces a peak, chemically shifted to 0.6 eV lower binding energy. Comparison of S and Cd attenuation with Al coverage (Brillson 1982a) reveals that the dissociated Cd is spatially localized at the microscopic interface. The chemical picture arising from these SXPS figures is of a strong reaction between Al and CdS to form an Al chalcogenide with displacement of Cd. Interface dielectric measurements (Brillson 1977b) using ELS support this model.

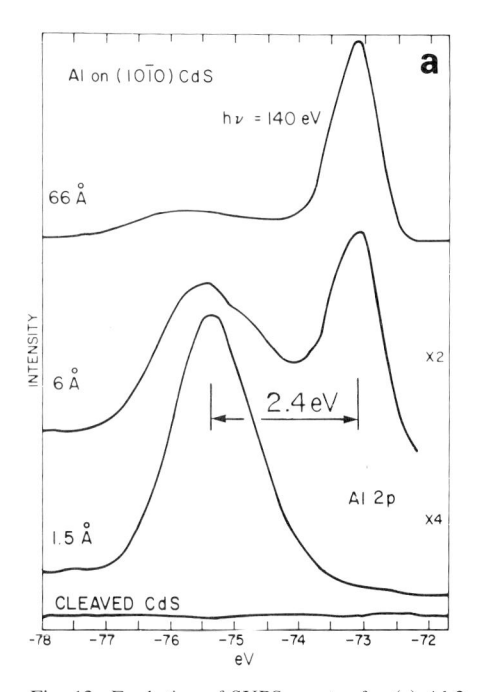

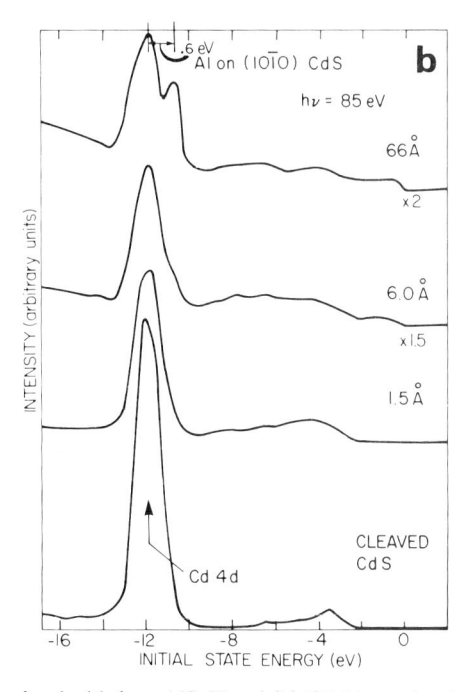

Fig. 12. Evolution of SXPS spectra for (a) Al 2p core level with $h\nu$ = 140 eV and (b) Cd 4d core level with $h\nu$ = 85 eV, as a function of Al overlayer coverage on UHV-cleaved CdS($10\bar{1}0$). After Brillson et al. (1980b).

Chemical bonding changes can be even more dramatic for metals on ternary compound semiconductors for which the two cation constituents can exhibit qualitatively different behavior. In the case of $Hg_{0.72}Cd_{0.28}Te$, monolayer deposits of both Al and In were found to deplete the surface of much of its surface Hg. Figure 13 provides SXPS features for the Al 2p, Te 4d, Cd 4d, and Hg 5d core levels at surface-sensitive energies (Davis et al. 1983, 1985). With increasing deposition of Al, the valence band spectra exhibit a pronounced decrease in Hg 5d versus Cd 4d intensity. Both intensities decrease with respect to the Te signal, which diminishes very slowly.

Broadening of the Te 4d feature reflects the presence of at least two chemical states for Te in the lattice. Similarly, the Al 2p core level feature exhibits a chemical shift between the initial chemisorption layers and metallic coverages, indicative of nonmetallic bonding near the intimate junction. Figure 13 provides clear evidence for the presence of dissociated Hg but not necessarily dissociated Cd. The relative increase in Te signal with Al coverage reflects a greater fraction of Te in Al_2Te_3 than in CdTe. Davis et al. (1985) have constructed a three-dimensional diagram of surface behavior to illustrate that Al replaces Hg in the

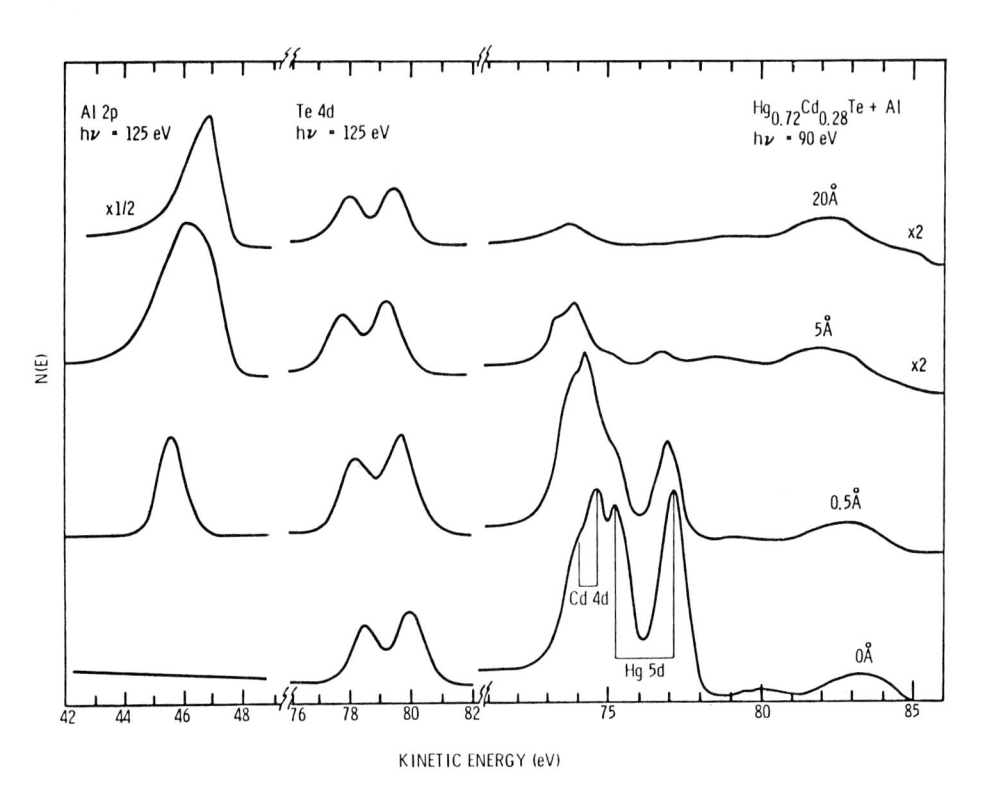

Fig. 13. SXPS features for the Al 2p, Te 4d, Cd 4d and Hg 5d core levels at $h\nu = 125$, 125, 90, and 90 eV, respectively, with increasing deposits of Al on the UHV-cleaved $Hg_{0.72}Cd_{0.28}Te$ surface. After Davis et al. (1985).

HgCdTe lattice with no significant attenuation of the CdTe portion of the lattice. With the appearance of metallic Al, Te begins to diffuse out of the lattice as well.

Similar evolution occurs for In on HgCdTe, albeit without chemical shifts of the metallic In overlayer (Davis et al. 1985). Again, there is a strong Hg depletion and a slow Te attenuation, suggesting the formation of an In telluride. For Au on HgCdTe, the Hg to Cd ratio is constant at low coverages, but Hg decreases preferentially for Au deposition above a few monolayers (Davis et al. 1984). Significantly, the Fermi level movement obtained from rigid shifts of all the core levels (to be discussed in section 6) is in the opposite direction from that induced by Al and In. For Al or In on HgCdTe, the Hg freed during the interfacial process can form an n^+ surface space charge region. Au diffusion into HgCdTe is believed to produce the opposite (p^+) effect. In addition to a Hg-depleted subsurface, Cr on $Hg_{1-x}Cd_xTe$ produces a surface layer of elemental Te as well (Philip et al. 1985, Peterman and Franciosi 1984). Franciosi et al. (1985) report analogous phenomena occurring for metals on the magnetic ternary semiconductor HgMnTe.

The pronounced decrease in Hg concentration near the ternary semiconductor surfaces with metal deposition reflects the weak Hg–Te bond. The stability of the HgCdTe surface appears to depend sensitively upon bulk specimen preparation (Silberman et al. 1982, Spicer et al. 1982a), suggesting that the weak Hg–Te bond promotes rapid communication between surface and bulk, which dominates the interface chemical behavior (Spicer et al. 1983).

No analogous SXPS studies are available yet for other classes of ternary compound semiconductor. For example, whether or not the III–V ternaries GaAlAs and GaInAs exhibit preferential dissociation would be of considerable relevance to the control of contacts to MBE-grown devices.

4.1.4. A III–VI compound semiconductor–metal interface

The III–VI compound GaSe possesses a peculiar lamellar structure, which is illustrated as an inset in fig. 14b. Unlike the semiconductor classes discussed thus far, one may achieve a high perfection of crystal cleavage in such materials since only weak van der Waals bonding exists between the flat layers. The resultant surfaces can be free of native defects, steps, or other imperfections, which can contribute to the chemical activity. Williams and coworkers (Hughes et al. 1982, Williams et al. 1982) have emphasized the desirability of such surfaces for Schottky barrier studies, since there are no broken chemical bonds or dangling bonds to introduce states within the semiconductor band gap which can influence the Fermi level position. We discuss this unique Fermi level behavior in section 6.

The core level spectra in fig. 14a show that the deposition of Au on UHV-cleaved GaSe merely shifts both Ga and Se peaks uniformly, indicative of a weak interaction with only Fermi level movement with respect to the band edges and core levels. See the inset of fig. 3. Ag and Ni exhibit weak chemical interactions with GaSe as well, and all three show a slow attenuation of the core level emission with metal deposition (fig. 14b), which is consistent with either metal

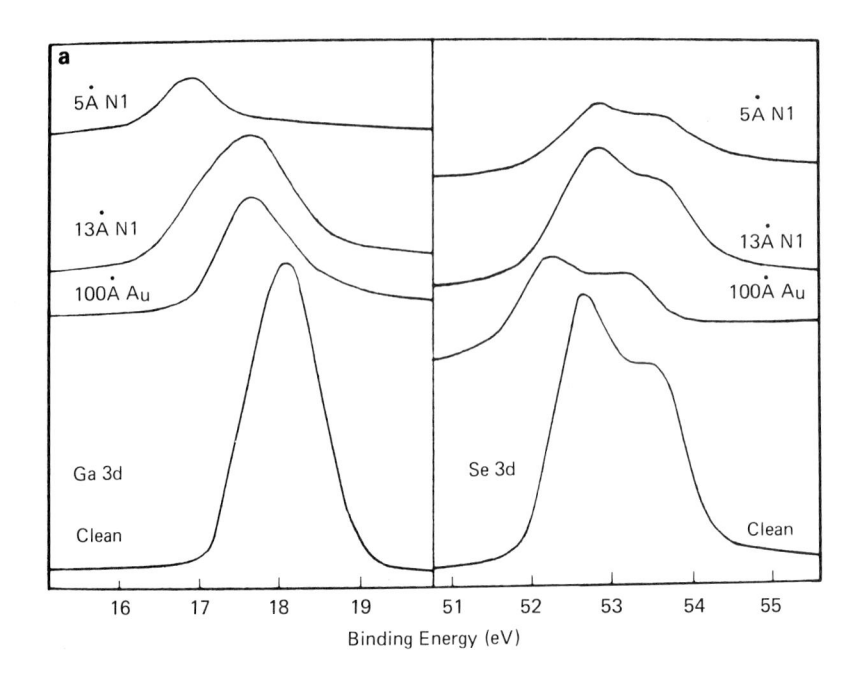

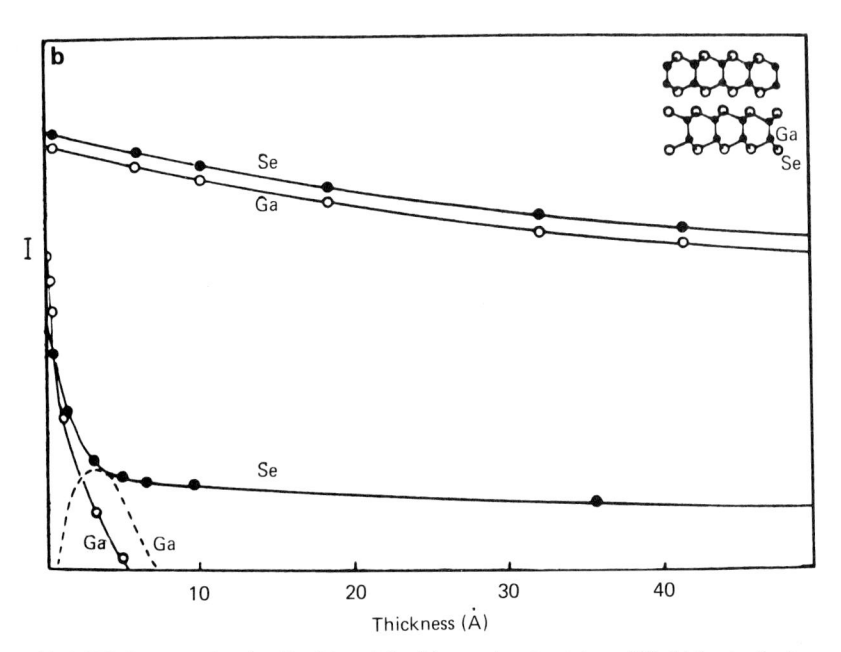

Fig. 14. (a) SXPS features for the Ga 3d and Se 3d core levels at $h\nu = 100$ eV for both Au and Ni deposition on the UHV-cleaved GaSe(0001) surface. (b) Attenuation of the Ga 3d and Se 3d core level intensities for $h\nu = 100$ eV with Ag and Ni deposition. Dotted line is the chemically shifted dissociated component of the Ga emission. Structure of GaSe appears as an inset. After Williams et al. (1982).

indiffusion or island formation. Transmission electron microscopy results demonstrate the latter for Ag and GaSe (Williams et al. 1982).

Figure 14a illustrates the case of Ni, a more reactive metal, on GaSe. Here the Ga 3d peak splits into two components, while the Se 3d features shift to higher binding energies, analogous to the Al–CdS (Brillson et al. 1980b) or CdSe system (Brucker and Brillson 1981b,c, Brucker et al. 1982). Hence the behavior of Ni on GaSe resembles that of Al, Cu and In and contrasts with that of Sn, Au, and Ag (Williams et al. 1982, 1984, Hughes et al. 1982). These two types of chemical behavior can be related to striking differences in Schottky barrier properties, to be discussed in section 6.

4.1.5. Additional metal–semiconductor interfaces

The SXPS technique has been applied to three additional classes of semiconductor with metal overlayers. The least ionic of these is the IV–VI compound PbTe. Similar to the covalent III–V compounds described above, deposition of Ge, Al and Au all produce exchange reactions with the PbTe surfaces, resulting in a significant outdiffusion of dissociated Pb. The latter manifests itself as a shift to lower binding energy of the Pb 5d level in the valence band. Cerrina et al. (1983a,b) have measured changes in E_F position with coverage and observed the formation of inversion layers, the degree of which depends upon the extent of Pb outdiffusion – the larger the amount of Pb released, the higher the Schottky barrier height. This effect is consistent with the stoichiometric doping properties of this material, namely Pb (Te) vacancies produce two extra holes (electrons). Hence the Pb excess produces a heavily n-type (inverted) surface on the normally p-type bulk material. Cerrina's results demonstrate that for a variety of metals on this covalent semiconductor chemical interactions are strong and electronic effects are large.

The more ionic insulator (i.e. wide band gap semiconductor) SiO_2 exhibits characteristic differences in its chemical activity with metals of different reactivity. Figure 15 shows Si 2p spectra obtained with high surface sensitivity before and after deposition of 1.5 Å equivalent thickness of Al or Au on "clean" oxidized Si (Bauer et al. 1980). Here the surface SiO_{2x} ($x < 1$) is representative of thermally oxidized Si since the SiO_2–Si transition observed spectrally during oxidation is identical to that reported for conventional, etched Si (Helms et al. 1978, Aspnes and Theeten 1979). In fig. 15a, the relative increases in the unoxidized Si photoemission connote the presence of free Si at the Al–SiO_2 interface. Correspondingly, the Al 2p spectra (not shown) exhibit an evolution in time to a higher binding energy indicative of Al_2O_3 (Flodstrom et al. 1976, Johannson and Stöhr 1979) in which three Al atoms cluster in a closed pack configuration around each surface O atom of the outer SiO_2 layer. Thus for the Al–SiO_2 interface, Al clusters first about each surface oxygen atom, then forms Al_2O_3 by reducing the SiO_{2x} ($x < 1$). Au deposited under identical conditions causes quite different changes in spectra, which are indicative of metal bonding to Si rather than to O atoms and the formation of SiO_x at the now strained Au–SiO_2 interface (Bauer et al. 1980).

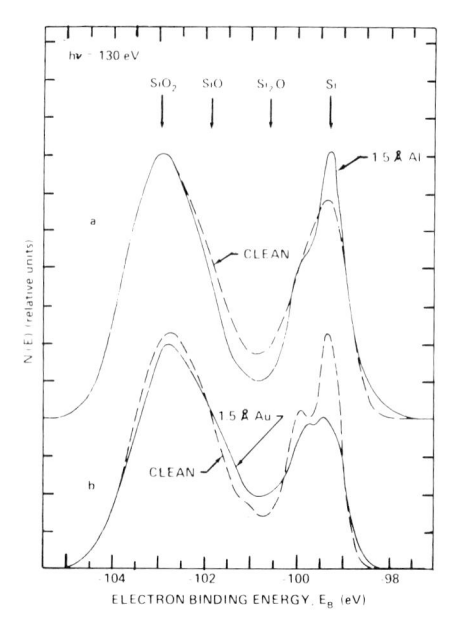

Fig. 15. SXPS features for Si 2p core level spectra with $h\nu = 130$ eV for SiO_{2x} ($x < 1$) films on Si before and after deposition of 1.5 Å equivalent thicknesses of (a) Al and (b) Au overlayers. Peak heights are normalized to constant Si 2p total yield. The arrows indicate various oxidation states of Si. The dashed curves correspond to the oxidized Si surface. After Bauer et al. (1980).

In contrast to SiO_2, wide band gap semiconducting diamond shows no dramatic reaction or diffusion of dissociated species with metal deposition near room temperature. For example, Al deposited on a clean, ordered (111) surface of diamond produced no observable interface reaction (Himpsel et al. 1980b). Only after an 800°C anneal does the Al 2p core level broaden with the appearance of a second, more strongly bonded chemical species. Himpsel et al (1980b) find no evidence for chemical interaction with the less reactive metal Au on diamond. Hence diamond resembles a highly ionic compound in terms of its activation barriers to chemical interaction. That such activation barriers are high is not surprising since the strength of lattice bonding scales with band gap and ionicity (Pauling 1939).

The trends in chemical bonding from one semiconductor class to another can be summarized as follows: With increasing semiconductor ionicity, the bonding of metals to the semiconductor anion becomes more pronounced (especially for "reactive" metals such as Al, Ni, and Ti). As a result, the more covalent group IV and III–V semiconductors have a greater tendency to form both metal–anion and metal–cation phases (especially for relatively "unreactive" metals such as Au). Because the interaction strengths between metal, cation, and anion are comparable, reaction and diffusion may be relatively complex and sensitive to conditions of surface preparation (e.g. stoichiometry, substrate orientation and cleavage step density), metal deposition rate, temperature and crystalline perfection. On the other hand, the more ionic II–VI and insulating compounds form metal–anion reaction products and localized concentrations of dissociated cations near the interface. The II–VI ternary HgCdTe appears to resemble its binary counterpart, with the relative strength of Hg–Te and Cd–Te bonds as an additional factor in the atomic redistribution. Interestingly, for the IV–VI compound PbTe the cation dissociation and outdiffusion resembles that of a covalent compound, while its reactive anion outdiffusion resembles that of a more ionic II–VI compound.

4.2. Chemical bond variations with depth

The tunable nature of synchrotron radiation over a selected energy range permits the nondestructive analysis of bond variations at the metal–semiconductor interface. As indicated in fig. 3, photoexcitation of core electrons with kinetic energies in the range of 50–100 eV provides a surface sensitivity of only one or two monolayers, while photoexcitation of the same core electrons with just a few eV above the excitation threshold energy allows a depth of tens of monolayers to be probed.

Figure 16 provides examples for which this variation in escape depth reveals dramatic changes in interface chemical bonding. Figure 16a illustrates SXPS Si 2p core level spectra for a 4 Å Au deposit on a Si(111) surface (Brillson et al. 1984,

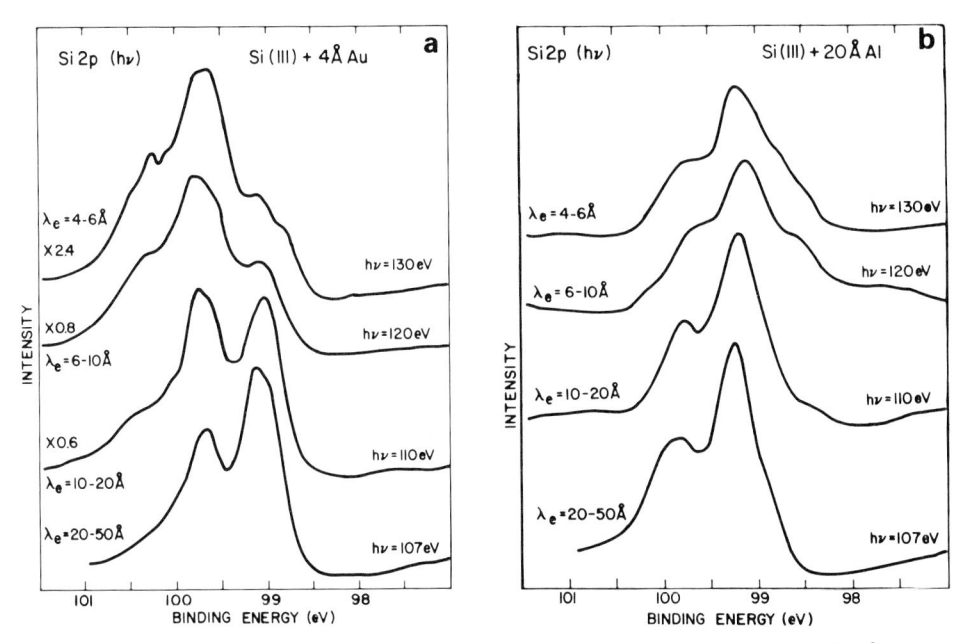

Fig. 16. SXPS Si 2p core level spectra as a function of incident photon energy for (a) 4 Å Au on Si(111) and (b) 20 Å Al on Si(111). The photoelectron escape depth λ_e corresponding to the incident $h\nu$ appears for each curve. After Brillson et al. (1984) and Brillson (1984).

Brillson 1984). This system represents one stage of the deposition given in fig. 4a. By varying $h\nu$ with only a 4 Å Au overlayer, one observes a clean separation of Si phases with depth. Figure 4a shows that at $h\nu = 130$ eV, the surface Au–Si phase dominates the Si 2p spectrum, while at $h\nu = 107$ eV, only the bulk Si phase appears to be present. Significantly, at $h\nu = 107$ eV, the spin–orbit splitting of the Si $2p_{3/2,1/2}$ doublet appears to be well resolved, due to the absence of contributions from chemically shifted surface core levels. The latter produce an effective broadening, which is evident in the cleaved spectra of figs 4a and 4b.

Figure 16b illustrates similar spectra for 20 Å Al deposited on Si(111), corresponding to one stage of the deposition sequence in fig. 4b (Brillson et al. 1984, Brillson 1984). Here the variation in $h\nu$ reveals that the lower binding energy shoulder is due to dissociated Si at the free Al surface. This shoulder is most prominent for $h\nu = 130$ eV, the most surface-sensitive excitation energy, and disappears at more bulk-sensitive energies. In contrast, the variation of $h\nu$ for 20 Å Au on Si(111) (not shown) produces no significant variation of peak features, indicating a more uniform distribution of Au with Si above the free Si surface. Thus, the $h\nu$-dependent features in fig. 16 confirm the weak Al–Si interaction and its contrast with the strong Au–Si interdiffusion.

Similarly, variation of $h\nu$ has been used for Al on CdS (see fig. 12a) to demonstrate the relative enhancement of the subsurface reacted Al–S phase relative to a metallic Al overlayer at lower surface sensitivity (Brillson et al. 1980b). The increase of the sublayer intensity at both high and low energies relative to the escape depth minimum (e.g. 50–100 eV) eliminates any steep cross section variations with $h\nu$ as a possible contributing factor to these intensity changes.

4.3. Metal island formation

An important consideration in the study of metal overlayers on semiconductor substrates is whether the metals possess lateral inhomogeneity, i.e., whether the metals form uniform two-dimensional layers, three-dimensional clusters, or combinations of layers plus clusters (Stranski–Krastanov growth). Here the growth modes are to be distinguised from any subsequent metal–semiconductor interaction. The latter can lead to a wide variety of interface inhomogeneity: morphological, crystallographic, and compositional. Ludeke (1983, 1984) has addressed this variety of interface inhomogeneity in considerable detail.

Early surface studies of metals on semiconductors concentrated on systems believed to be laterally homogeneous. Nevertheless considerable evidence for island growth in such systems has appeared in recent years, particularly for cases of weak metal–substrate bonding, for which surface mobility and the probability of clustering are high. In this section, we examine techniques for determining the presence of such islands. We describe trends among different semiconductor subclasses and for metals of different chemical reactivity. Finally, we discuss how cluster formation may act as a precursor to chemical reaction and Schottky barrier formation.

The Ag interface with GaAs(110) provides a clear example of metal island formation. Here exchange reactions and interdiffusion are not observed and the substrate signal is attenuated too slowly for any uniform layer-by-layer growth. Furthermore, substrate LEED features persist for coverages above an equivalent monolayer. SXPS studies of Ag clustering are particularly suitable for two reasons: (1) the shape of the Ag 4d level is very sensitive to cluster size, increasing in width with cluster size due to band formation (Baetzold 1978, 1981, Head and Mitchell 1978, Basch 1981) and (2) the Ag 4d level has a large photoionization cross section so that clusters can be investigated at very low coverages.

Figure 17 illustrates the subtractive technique for obtaining Ag 4d peak features at ultralow coverages and the evolution of the peak shape with Ag deposition (Ludeke et al. 1982, Ludeke 1984). At the lowest coverage, the spectra are believed to be representative of monoatomic or diatomic clusters (Ludeke et al. 1983). This figure makes evident the direct correlation between spectra width of the Ag 4d level and onset of Ag cluster formation during epitaxial growth (Ludeke 1983, 1984). Such an onset occurs at 0.1 monolayer for the GaAs(100) surface versus a coverage nearly two orders of magnitude lower for the GaAs(110) surface. Thus Ag atoms on (110) surfaces have a dramatically higher surface mobility, indicating a low reactivity between absorbate and surface. Ludeke has used the shape of substrate attentuation curves to model the three-dimensional morphology of Ag films (Ludeke et al. 1982). In general the

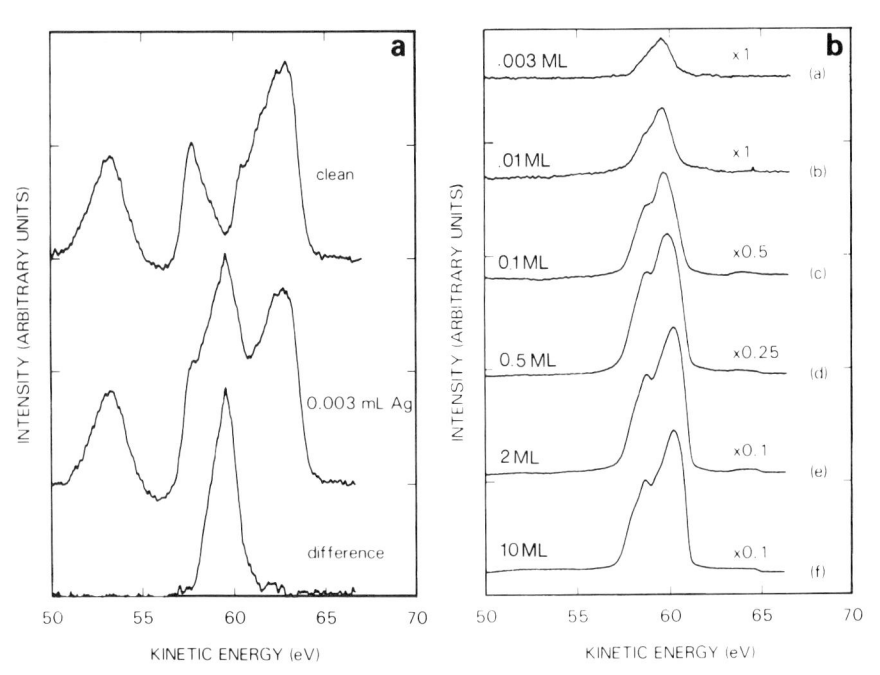

Fig. 17. Valence band spectra of clean GaAs(110) surface, (a) before and after deposition of 0.003 monolayer Ag and their difference spectrum, and (b) the difference spectra for increasing Ag deposition. After Ludeke et al. (1983).

growth sequence from nucleation to thick film may be modeled in terms of two or more growth modes, each of which is characterized by a crystalline aspect ratio and capture cross section for migrating surface atoms.

Significantly, the number of surface electronic charges associated with the substrate band bending at the lowest coverage in fig. 17 coincides almost exactly with the number of Ag atoms. Hence, interface state associated with these individual Ag atoms could play a role in Schottky barrier formation.

For epitaxial Al growth on GaAs(100), Ludeke (1984) found that nucleation of three-dimensional Al clusters preceded the initial two-dimensional coverage and that the thickness of Al necessary for nucleation is proportional to the As coverage of the bare GaAs surface. Al forms metallic bonds to the Ga surface atoms, which are weaker and less directional than a covalent bond between Al and As atoms. Hence the more As-rich surfaces lower the surface mobility of Al atoms, suppressing cluster formation, and they impart a directional constraint to any resultant nuclei. Cluster formation at the Ga-rich (100) surfaces appears linked to the onset of an exchange reaction as well, providing experimental confirmation of the triggering mechanism proposed by Zunger (1981).

For Au on III–V compounds, anomalously low Au 5d splittings appear in the valence band, which have been interpreted both as uniformly dispersed atoms (Liang et al. 1986, Chye et al. 1977) and as metal clusters (Ludeke 1983). While Au clusters also produce low Au 5d splittings, their valence band features are qualitatively different from those ascribed to dispersed atoms (Lee et al. 1981). Furthermore, the valence band spectra for Au on GaAs, GaSb, and InP exhibit no Au Fermi edge for Au coverages below approximately 2 monolayers and no metallic behavior until 4–5 Au monolayers, in contrast to the coverage dependence of Au clusters (Chye et al. 1978).

Group III, IV, and V atoms show different absorption behavior on the same InP surface. In fig. 18, the slow attenuation of the In 4d and P 2p core level peaks of the UHV-cleaved InP(110) surface with increasing Ga deposition contrasts dramatically with the attenuation upon Sb deposition (Williams et al. 1984). The Ga–InP interaction is the weakest of these three adsorbates. On the other hand, Sb adsorbs on GaAs(110) in an ordered layer-by-layer fashion at room temperature (Skeath et al. 1981) and the exponential attenuation corresponding to a 4 Å escape depth bears this out for InP. Williams (1983) finds similar evidence for cluster formation with Ag, Au, Cu, and Al deposition. Figures 6 and 14 contain evidence for island formation on Si and GaSe, respectively.

Stoffel et al. (1983, 1984) have derived evidence for island formation on GaAs from angle-resolved photoemission studies. Essentially, they compare photoemission at grazing and normal emission angles to see differences in attenuation of the substrate core levels. These studies for both (100) and (110) surfaces show slower attenuation of the substrate As signal at normal incidence, suggesting that the substrate is not completely covered at multilayer coverage. However, the interpretation favoring either island formation or simply the longer escape distance for Al electrons exiting the solid at grazing angles depends sensitively upon the choice of scattering length in a narrow range, e.g. 2–6 Å (Seah and Dench 1979). What

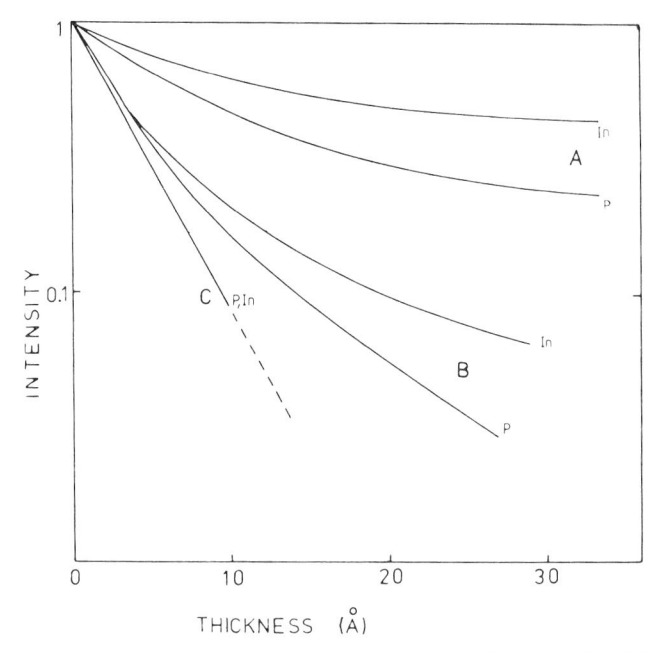

Fig. 18. Attenuation of In 4d and P 2p emission peaks of the UHV-cleaved InP(110) surface as a function of (a) Ga deposition, (b) Sn deposition, and (c) Sb deposition. Photon energies are 100 eV and 170 eV for the In 4d and P 2p peaks, respectively. The dashed line indicates the attenuation for a mean free path of 4 Å, assuming uniform two-dimensional growth. After Williams et al. (1984).

is perhaps less subject to interpretation is a strikingly slow attenuation of Ga versus As core level intensities at grazing emission angles, indicative of Ga segregation to the free metal surface. In support of island formation, Stoffel et al. (1984) observe a much more rapid attenuation in normal-incidence As signal for Al on GaAs(100) at 200 K, where surface migration of Al should be greatly reduced. Hence Al coverage at 200 K on GaAs(110) appears to be more uniform, consistent with the LEED Auger work of Bonapace et al. (1984). Interestingly, the free Ga signal at 200 K decreases by a much lower factor than that of normal-incidence As, indicating that some exchange reaction continues to occur. Similarly, McKinley et al. (1982) observe evidence for exchange reactions between Al and InP(110) at 100 K. Such low-temperature data suggest that exchange reactions may not depend as sensitively on cluster formation as originally believed (Zunger 1981).

Angle-resolved photoemission has been used to identify the persistence of bare-surface features with multilayers of metal deposition. Figure 19 illustrates the variation in valence band spectra obtained with $h\nu = 21.2$ eV for UHV-cleaved InP at increasing coverages of Cu and at various polar emission angles along the $\Gamma X'$ direction (see inset) (Hughes et al. 1983). Photoemission from the anion-derived surface state is denoted by S and is clearly visible at polar angles of 20–30°. This state is very sensitive to surface chemisorption. Thus its peak

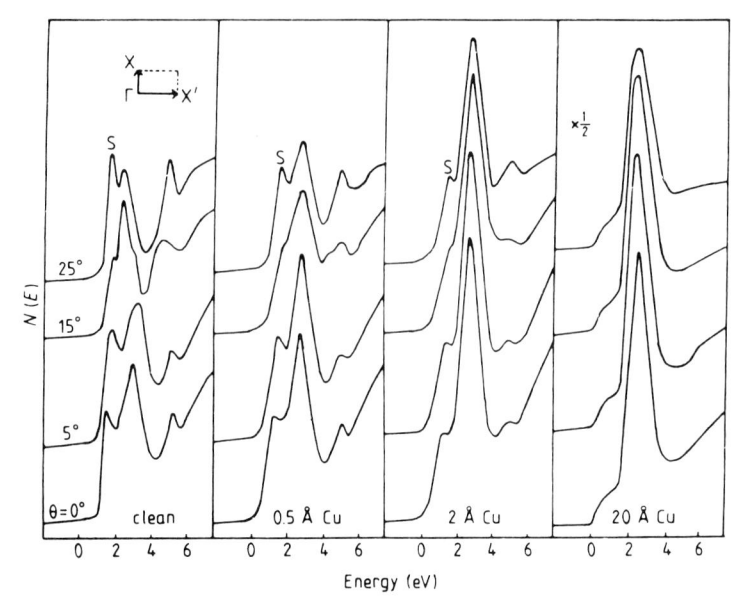

Fig. 19. Angle-resolved valence band spectra with $h\nu = 21.2$ eV for various deposited thicknesses of Cu on UHV-cleaved InP(110). The surface state feature is denoted by S at an angle of 25° along $\Gamma X'$ (shown in inset) with respect to the surface normal. After Hughes et al. (1983).

amplitude decreases rapidly when the InP surface is exposed to gases such as hydrogen or water vapor as well as to reactive metals such as Al. However, when Cu is deposited on this surface, emission remains strong for coverages up to 2 Å. Similar behavior occurs for Ag in InP (McKinley et al. 1980). Hence the persistence of emission from the dangling-bond, anion-derived surface states argues for metal-free regions between metal clusters (or an extremely weak interaction between Cu and this state, which is less likely).

For the II–VI compound CdS, Stoffel et al. (1982) have shown that Cu islands form on Cu-chemisorbed UHV-cleaved CdS(11$\bar{2}$0). The first 4 Å Cu deposited appear to be laminar, while subsequent deposition may cause Stranski–Krastanov growth on Cu and/or Cu diffusion into the CdS. The former growth mode is favored since emission from an Al marker on the CdS surface falls off at least as slowly as the CdS substrate signals. Three-dimensional growth is less likely for more reactive metals on II–VI compounds [i.e. Al on CdS or CdSe (Brillson 1977b, 1978c)].

In addition to the above techniques, the observation of chemical shifts at ultralow coverages can provide evidence for cluster formation. Figure 20a illustrates three different states of Al adsorption for submonolayer Al coverages above and below ~0.1 monolayer. The 0.3–0.55 eV shift of the Al 2p peak to lower binding energies as the coverage increases from 0.05 monolayer to 0.2–1 monolayer is consistent with a ground state predicted by Zunger (1981). At submonolayer coverages and low temperature, this state involves molecular clusters of Al atoms which interact only weakly with the substrate rather than a strongly chemisorbed, ordered overlayer (Daniels et al. 1982, 1984). Above 0.4 monolayer Al coverage, the Ga 3d peak displays a high-energy component due to free Ga atoms released by replacement reactions between the adatoms and the substrate, as in fig. 9. These results point to cluster formation as the precursor stage of the chemical reaction between Al and GaAs as well as the dominant process occurring as the Fermi level shifts (see section 6). Thus the states in fig. 20a represents chemisorption of isolated Al atoms, cluster formation along with the onset of an exchange reaction, and metal overlayer growth.

By way of contrast, fig. 20b illustrates a case in which only a gradual formation of clusters takes place (Daniels et al. 1984). Here photoemission and LEED data indicate that the In overlayer growth is not layer-by-layer, yet there appears to be no well-defined shift in metal core level features. This gradual formation of clusters promotes no exchange reaction, consistent with the energetically unfavorable bulk heat of reaction (Brillson 1978a) for In on GaAs. In section 6, we address the qualitative difference in Fermi level pinning for these two cases.

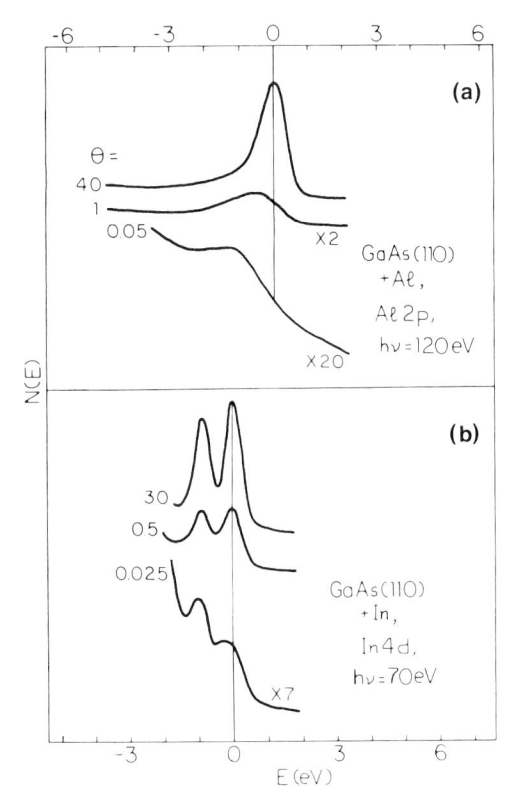

Fig. 20. SXPS spectra of the (a) Al 2p core level as a function of Al deposition with $h\nu = 120$ eV and (b) In 4d core level as a function of In deposition with $h\nu = 70$ eV. The peak positions are shifted to compensate for changes in band bending and are referenced to the thick overlayer position of the Al 2p and In $4d_{5/2}$ peaks. After Daniels et al. (1982, 1984).

Similar evidence for the above three states of Al adsorption is observed for chemisorption of Al on InP(110) along with a new bonded state at intermediate coverage (Zhao et al. 1983).

To recapitulate, SXPS provides a wide variety of techniques with which to characterize the structure of metal overlayers on semiconductor substrates as well as the chemical reactions at their interface. In the following section, we will present SXPS techniques for elucidating the mechanisms and character of atomic redistribution at the metal–semiconductor interface.

5. Metal–semiconductor interdiffusion

5.1. Interface diffusion near room temperature

SXPS and a variety of other surface science techniques reveal that substantial interdiffusion can occur at metal–semiconductor junctions. Since the atomic structure at a surface or interface is known to affect the associated electronic structure, the movement of metal and semiconductor atoms at their interface is of considerable interest. Early RBS and AES work showed that such interdiffusion could take place even near room temperature (Hiraki et al. 1971, 1977). Initial results were obtained from chemically etched and air-exposed surfaces. Even more pronounced effects appear for clean surfaces prepared in UHV (Chye et al.

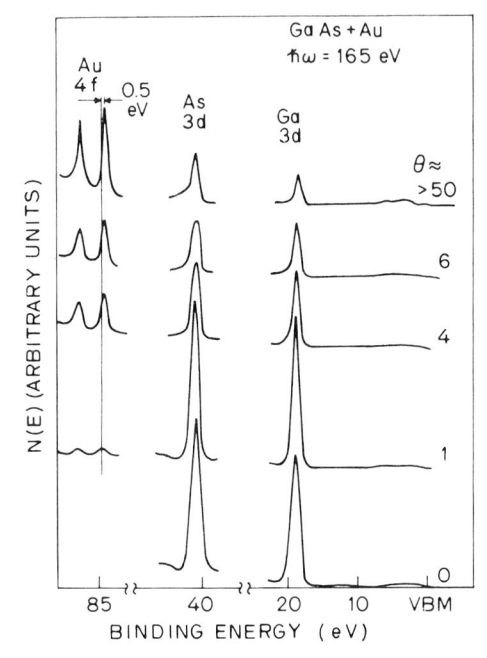

Fig. 21. SXPS features obtained with $h\nu = 165$ eV excitation for UHV-cleaved GaAs(110) with increasing thickness of deposited Au. After Chye et al. (1978).

1978) and especially at elevated temperatures (Sinha and Poate 1973, Robinson 1975). Thus Chye et al. (1978) and Lindau et al. (1978) found similar dissociation and semiconductor outdiffusion for III–V compound semiconductors using SXPS. As shown in fig. 21 for increasing Au coverages, both the Ga 3d and As 3d core level intensities decrease exponentially to residual levels, beyond which further decreases are slow. Similar behavior is observed for Au on other III–V compounds as well. Compositional depth profiles of the Au overlayers on these semiconductors show that semiconductor outdiffusion is continuous within the metal overlayer. Such diffusion can extend smoothly over several hundred Å, although for overlayers thicker than 20–50 Å, such redistribution may be complex, involving for example, surface segregation or species localized near the interface. As already mentioned in section 4.3, Au deposited on the cleavage faces of III–V compound semiconductors is uniformly distributed since the Au 5d splitting at coverages below several monolayers is anomalously low, characteristic of Au atoms dispersed as a dilute alloy, and the valence band features of chemisorbed Au are unlike those of Au clusters. Here we present SXPS results obtained for Au and other metals on semiconductors, which reveal the factors governing semiconductor outdiffusion, the systematic differences between atomic redistribution on III–V and II–VI compound semiconductors and the diffusion of metal atoms into the semiconductor lattice.

5.2. Dynamics of semiconductor outdiffusion

For metal films on UHV-cleaved surfaces, the outdiffusion phenomena represented by fig. 22 are quite representative and systematic. Characteristic measures of this phenomenon are the concentration of dissociated anions and cations with migrate to the free surface of a uniform, 20 Å thick Au overlayer. (Brillson 1982a,d). Up to coverages of at least 20–40 Å, anion and cation intensities decrease monotonically, as illustrated schematically by the inset of fig. 22. Here I/I_0 is the integrated cation-to-anion peak area at the characteristic 20 Å Au coverage relative to that of the UHV-cleaved surface. For both III–V and II–VI compound semiconductors, the extent of dissociation and outdiffusion decreases with increasing heat of formation H_F, a measure of semiconductor stability (Kubaschewski et al. 1967, Wagman et al. 1968/1971, Mills 1974). For II–VI compounds, both anion and cation exhibit a strong correlation with H_F over a wide range. Such a relationship is particularly significant, given that II–VI compounds with both wurzite and zincblende structure, n- and p-type doped material, and both Cd and Zn cation families lie on the same curves. Hence fig. 22 demonstrates that a rate-limiting step to the semiconductor outdiffusion is the breaking of anion–cation bonds and that the less stable the bulk compound, the greater the outdiffusion (Brillson 1982a,d). These conclusions are not affected by the uniformity of the Au overlayers. Indeed, if islands form, one expects opposite behavior: the more stable semiconductors would interact less with adsorbates, leading to higher surface mobilities and more clustering at intermediate cover-

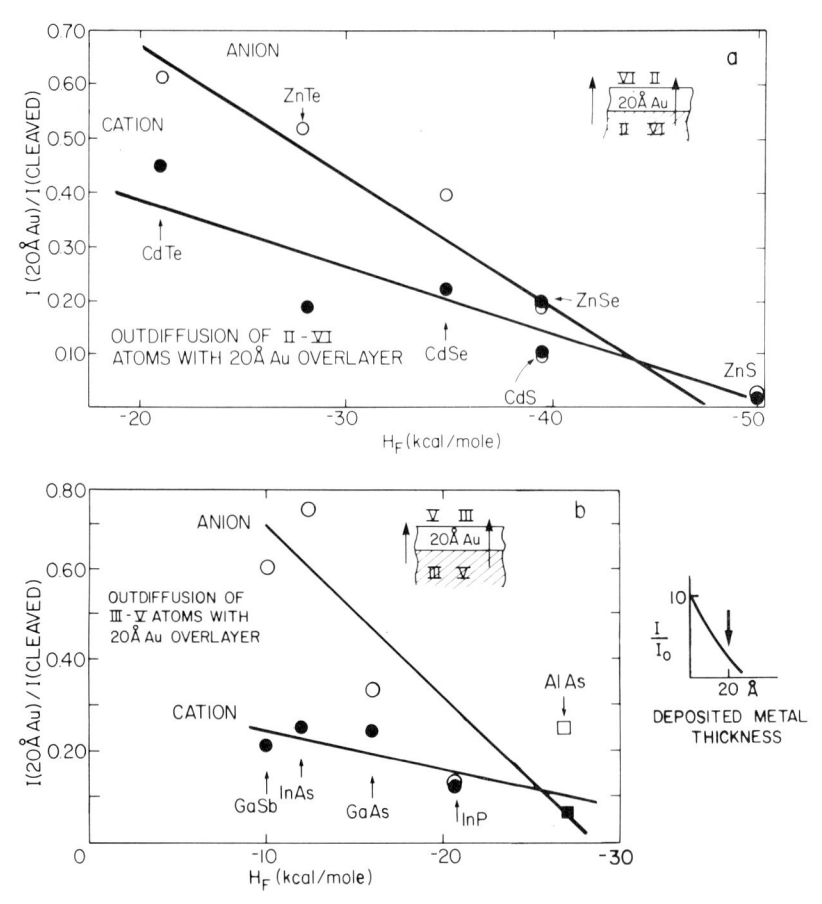

Fig. 22. Normalized intensity I/I_0 of outdiffused anions and cations with a 20 Å Au overlayer versus the semiconductor heat (Wagman et al. 1968/1971, Mills 1974). Intensities are normalized with respect to the UHV-cleaved surface. After Brillson (1982a,d). Data points for AlAs are after Bauer et al. (1981).

ages. Since more of the surface then remain uncovered, the anion and cation signals would decrease more slowly, contrary to what is observed.

A second rate-limiting step for interdiffusion is the movement of semiconductor atoms into the metal. Bonapace et al. (1984) showed that "reactive" metals (i.e. metals which can form a stable metal-anion compound) (Kubaschewski et al. 1967, Wagman et al. 1968/1971, Mills 1974) cause a preferential attenuation of anion atoms due to metal–anion bonding at the intimate semiconductor–metal interface. Figure 23 illustrates how reactive Al interlayers between a GaAs substrate and a Au overlayer (see geometry in the inset of fig. 23) attenuate the As relative to Ga outdiffusion preferentially (Brillson et al. 1980c). Indeed, Al coverages of one monolayer or less can affect the diffusion process measurably and only 10 Å can cause an order-of-magnitude change. Any cluster formation

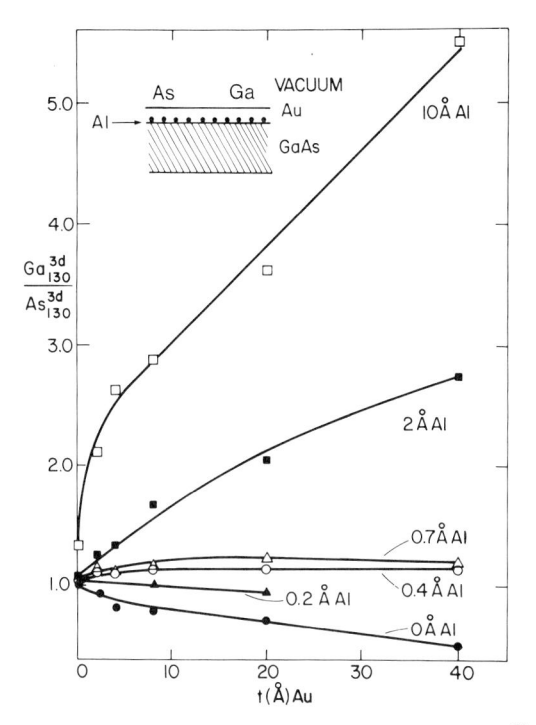

Fig. 23. SXPS surface cation/anion core level intensities at $h\nu = 130$ eV, $Ga^{3d}_{130}/As^{3d}_{130}$, relative to the UHV-cleaved GaAs surface and as a function of Au overlayer thickness. Each curve represents a different initial Al coverage, as indicated. Inset shows interlayer configuration schematically. After Brillson et al. (1980c).

serves only to decrease the magnitude of such interlayer effects. Consistent with a "chemical trapping" of the anion, a constant (e.g. 10 Å) thickness of interlayer with metals of different reactivity demonstrates that the more "reactive" the interface and the stronger the metal–anion bond (Brillson 1978a), the stronger the interlayer effect (Brillson et al. 1980a). Here one measures anion/cation variations in stoichiometry of nearly two orders of magnitude.

Franciosi et al. (1983) have observed analogous effects for Cr interlayers at Au–Si interfaces. A large reduction of interdiffusion occurs for a critical thickness of Cr, which corresponds to a fully reacted Cr–Si interface.

The anion attenuation by reactive metal overlayers suggests that the metal–semiconductor interface has a "width" characterized by the extent of metal–anion bonding, which decreases with increasing metal–semiconductor reactivity (Brillson et al. 1981b). Notable exceptions to such a trend are metals which have no strong chemical interaction with the substrate [e.g. Ag on GaAs (Ludeke et al. 1982, 1983) and yet form atomically abrupt interfaces. The fact that such systems are also epitaxial suggests that the polycrystalline or disordered structure of most metal overlayers provides channels of reduced activation energy for diffusion, channels which are not present for continuous, ordered metal overlayers.

SXPS results indicate that outdiffusion into metals from II–VI compounds is substantially different from that of III–V compounds. While appreciable dissociation and outdiffusion takes place in both, metal deposition (especially "reactive" metals) on II–VI compounds produces more rapid attenuation of the cation versus anion intensity. This behavior suggests that the free cations formed are localized at the intimate metal–semiconductor interface (see fig. 12, for example) and that anions move into the metal overlayer by "reactive diffusion" (Brucker and Brillson 1981b,c, Brillson et al. 1983).

The outdiffusion of II–VI anions and cations can vary substantially, depending upon the semiconductor ionicity. Figure 24 illustrates the differences between CdS, an ionic II–VI compound, and ZnSe, which is much more covalent (Brillson 1983b, Brillson et al. 1983). Here, integrated SXPS core level intensities normalized to cleaved surface values provide a gauge of semiconductor outdiffusion for different reactive interlayer thicknesses of Al plus Au overlayers. The insets in (a)–(d) illustrate the geometry and the particular elements measured. The progression in elemental intensity with interlayer thickness demonstrates that the Al interlayer promotes anion outdiffusion for CdS but anion "trapping" for ZnSe. For both semiconductors, cation outdiffusion is reduced. Similar studies for ZnS, ZnTe, CdSe, and CdTe reveal that chemical trapping versus cation localization and reactive anion outdiffusion have a systematic dependence on semiconductor ionicity (Brillson 1982, 1983b). These chemical differences vary monotonically in the semiconductor [Pauling ionicity (Pauling 1939)] sequence of CdS (0.59), ZnS

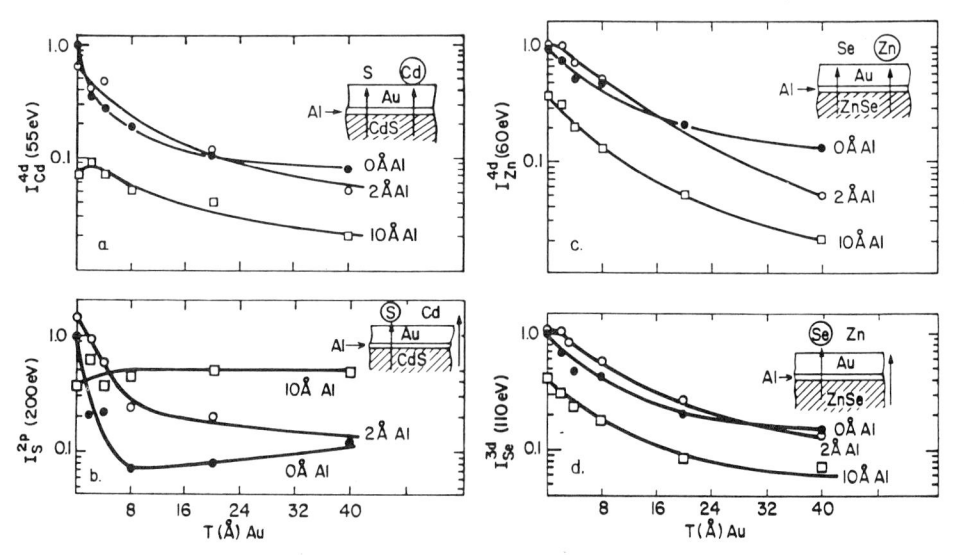

Fig. 24. Integrated SXPS peak areas for (a) Cd 4d, (b) S 2p, (c) Zn 4d, and (d) Se 3d core levels as a function of Au coverage for different Al interlayer thicknesses. Intensities are normalized to the cleaved surface. Insets show corresponding diffusion of anions and cations through the metal. Al interlayers act to increase anion outdiffusion for CdS, decrease anion outdiffusion for ZnSe and decrease cation outdiffusion for both semiconductors. After Brillson et al. (1983) and Brillson (1983b).

(0.59), CdSe (0.58), ZnSe (0.57), CdTe (0.52), and ZnTe (0.53). Scaling by Phillips ionicity (Phillips and Van Vechten 1969, 1970) reinforces this sequence. Thus the more ionic the semiconductor, the more pronounced are its chemical differences with respect to III–V compounds. The systematic behavior exhibited in figs. 22–24 demonstrates that bulk thermodynamic parameters can be useful in gauging the different stages of metal–semiconductor interaction. Of course, the thin film structures typically studied impose additional constraints on the kinetics and thermodynamics of interaction. The fact that thermodynamic trends can be extracted from the microscopic redistribution of interface atoms near room temperature attests to the value of monatomic depth resolution and nondestructive analysis available with the SXPS technique.

5.3. Marker identification of diffusing species

The relative movement of metal and semiconductor atoms at their interface is of central importance in identifying the basis for interface electronic states. Specifically, the movement of metal atoms into the semiconductor or the diffusion of semiconductor atoms out of their lattice can lead to impurity and defect states in the semiconductor band gap which influence the Fermi level position. In order to distinguish between these processes, one requires not only fine depth resolution but also a marker at the original interface. Such a marker must not interfere significantly with the diffusion process and thus must be only a monolayer or less in thickness. Furthermore, this marker must be strongly chemisorbed to the semiconductor surface in order not to drift away from the initial interface during the interdiffusion process.

SXPS combined with the marker technique was used first to characterize interdiffusion at Au–GaAs interfaces (Brillson et al. 1980d, 1981d). Here Ga and As core level intensities decreased relative to an Al marker signal of the Au–Al–GaAs interface. This indicated Au diffusion past the interface into the GaAs, effectively diluting or screening the GaAs subsurface. Further Au deposition caused an increase in both Ga and As intensities relative to Al, indicating an outdiffusion of GaAs into the Au overlayer. Brillson (1983b) observed analogous diffusion of Cu into CdS and InP.

Figure 25 illustrates results obtained for the Au–Si and Al–Si interfaces, already described in figs. 4 and 16, but now using one monolayer of Ni as a marker layer (Davis et al. 1983, 1984). As shown, initial deposition of up to 4 Å leads to a net decrease of the Si to Ni intensity ratio, corresponding to Au indiffusion. At higher coverage, this ratio increases, corresponding to Si outdiffusion. In contrast, Si 2p (130 eV)/Ni 3d (130 eV) increases for all Al coverages on Si, indicating just Si outdiffusion. These conclusions are consistent with the spectral features of figs. 4 and 16.

The SXPS marker experiments have an intrinsic difficulty in that the marker layer must be extremely thin in order to minimize effects on the diffusion process. Thus SXPS marker intensities are weak and are attenuated even further by the metal overlayer which promotes the interdiffusion.

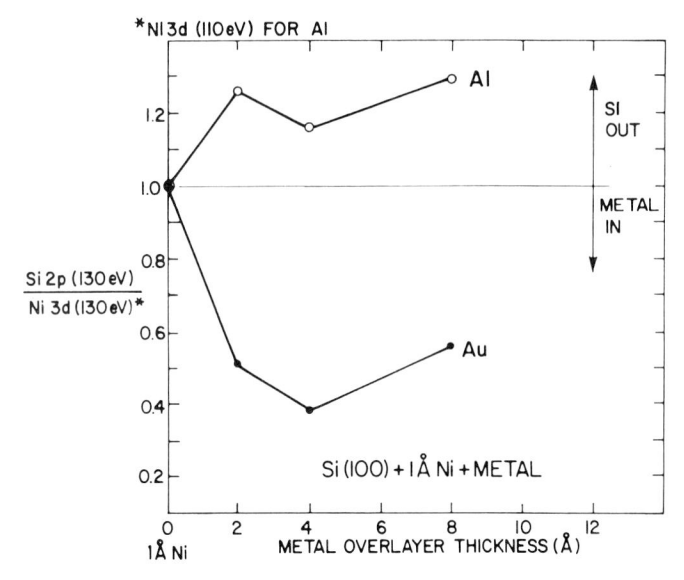

Fig. 25. SXPS intensity ratios at $h\nu = 130$ eV of Si relative to Ni marker layer, Si 2p (130 eV)/Ni 3d (130 eV), with increasing overlayer thickness of Al or Au on the clean Si(100) surface. Intensity ratios are normalized to unity at zero overlayer coverage. After Brillson et al. (1984) and Brillson (1984).

The SXPS techniques described in this section have been used to uncover considerable atomic redistribution at the metal–semiconductor interface even near room temperature, where conventional diffusion extrapolated from elevated temperatures is expected to be orders-of-magnitude smaller. That such diffusion occurs and is systematic in simple chemical terms can provide new clues in unraveling the basis for Schottky barrier formation.

6. Fermi level pinning and semiconductor band bending

6.1. Fermi level movement with metal deposition

An important application of SXPS to metal–semiconductor interface studies is the measurement of the position of E_F with respect to the band edges. This technique involves measurement of energy shifts of core level and/or valence band features as a function of metal deposition on the clean semiconductor surface. It is based on the fact that, as illustrated in fig. 3, a change in E_F with respect to the semiconductor band edges causes a shift of all of the photoemitted core level and valence band features by the same energy. An energy analyzer for photoemitted electrons detects the same shift since it shares a common E_F with the specimen. Since E_F remains fixed with respect to the bulk semiconductor band edges, the E_F changes with surface treatment are equivalent to a change in band bending within the surface space charge region. For example, a rigid shift of semiconductor features to higher kinetic energy with metal deposition signifies an increase in n-type (upward) band bending and vice versa for lower kinetic energy. Hence, SXPS provides a gauge of semiconductor band bending as the Schottky barrier evolves with metal deposition, layer by atomic layer.

Using this approach, Rowe (1976) and Margaritondo et al. (1976) demonstrated that band bending at metal–Si interfaces can evolve completely over only a few monolayers of metal. Furthermore, the final position of E_F in the Si band gap corresponded to the Schottky barrier heights extracted from conventional techniques (Turner and Rhoderick 1968, Thanailakis 1975). Spicer et al. (1980a) (see also Spicer and Eglash 1985) showed that such E_F movements were even more rapid for simple metals such as Ga and Al on III–V compounds such as GaAs. Figure 26 illustrates the rapid E_F movement at submonolayer coverages for Al deposition on UHV-cleaved GaAs(110) and the asymptotic E_F limit (e.g. the "pinning" position) reached at approximately monolayer Al coverage (Spicer et al. 1980a). Spicer and coworkers (Lindau et al. 1978, Spicer et al. 1979a,b, 1980a,b, Spicer and Eglash 1985) as well as Williams (1981, 1982) observed that most metal overlayers as well as oxygen on UHV-cleaved GaAs led to the same E_F position, as illustrated in fig. 27. Here pinning levels for n- and p-type GaAs are separated by 0.2 eV. Similar behavior was proposed for InP and GaSb (Lindau et al. 1978, Spicer et al. 1979a,b, 1980a,b, 1982b), although InP appears to have a

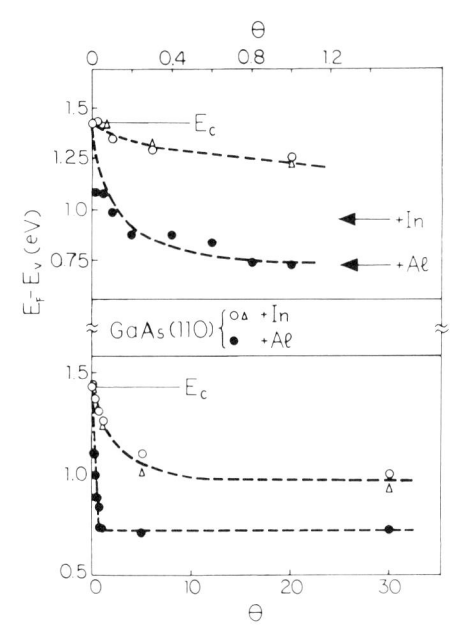

Fig. 26. Shifts in Fermi level position with equivalent metal coverage for Al and In and UHV-cleaved GaAs(110) as deduced from photoemitted core electron energy shifts. Energies are with respect to the valence band edge. Top plot shows the evolution of E_F at very low coverages and emphasizes the different pinning position and rate of pinning of the GaAs–Al and GaAs–In interfaces. The horizontal arrows indicate the final pinning positions of E_F for the two overlayers as extracted from the lower plot. After Daniels et al. (1984).

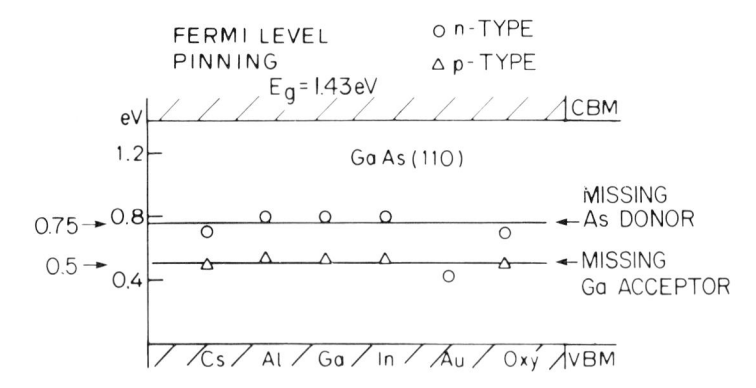

Fig. 27. Pinning positions within the band gap of GaAs (as derived from SXPS data) produced by a monolayer or less of various absorbates. Circles and triangles indicate energy positions for n- and p-type semiconductors, respectively, to an accuracy of ±0.1 eV or better. See text for n-type In point. Labels for various energy levels indicate their proposed electrical activity and chemical nature. After Spicer et al. (1980a).

wider spread of E_F positions and the narrow gap of GaSb constricts E_F movement naturally. Slower rates of E_F movement occur for deposited metals with different morphologies on GaAs. Thus fig. 26 illustrates the E_F changes for In deposition on clean GaAs(110) as well (Daniels et al. 1984). In contrast to Al, In deposition shifts E_F much more slowly, reaching an asymptotic "pinning" value at an order of magnitude higher coverage. Daniels et al. (1984) interpret this difference in terms of a much more gradual growth of clusters and the absence of an exchange reaction (see fig. 20). Furthermore, careful and repeated measurements show that the ultimate pinning position for In is significantly different than that of Al and the other metals in fig. 27. Such results imply that the kinetics of interface formation can influence the final interface parameters. The reasons for the discrepancy between the In value consistently obtained in fig. 26 and that of fig. 27 are not yet clear. Kendelewicz et al. (1984e) observe a similarly slow evolution of E_F with deposition for Tl on UHV-cleaved GaAs. In this case, however, a value consistent with fig. 27 was obtained. Ludeke et al. (1983) have observed a slow evolution of E_F position for Ag on GaAs(110) and reported significant differences between n- and p-type GaAs. As shown in fig. 28, Ag deposition on n-type GaAs causes a steady decrease in the E_F position within the band gap over a coverage range spanning four decades. For p-type GaAs, E_F increases then decreases with Ag deposition. Au on GaAs appears to show similar behavior (Petro et al. 1982). The pinning values shown in fig. 28 differ considerably from the values for n- and p-type GaAs shown in fig. 27.

E_F movement for metals on other semiconductors can exhibit a variety of pinning positions. Figure 29 illustrates E_F movements for various metals on InP(110) (Brillson et al. 1982, Kendelewicz et al. 1985). Each set of data points denotes E_F shifts extracted from rigid shifts of both In 4d and P 2p core levels. One may conclude from fig. 29 that different metals on InP produce a spectrum of

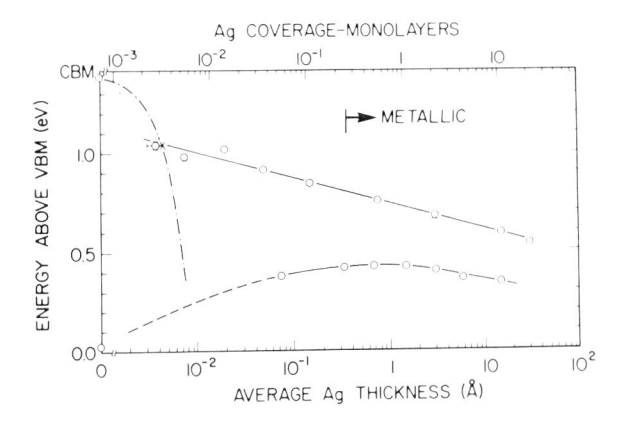

Fig. 28. Shifts in Fermi level position with equivalent Ag coverage on GaAs(110) as deduced from surface- and bulk-sensitive shifts of core level peaks. Energies are with respect to the valence band edge. Top and bottom plots are for n- and p-type GaAs(110), respectively. After Ludeke et al. (1983).

E_F behavior – both in terms of the pinning positions at thick coverages and their rate of change at monolayer coverages.

A number of factors may complicate the analysis of E_F position within the band gap. First, the starting position of E_F for the clean semiconductor surface must be established. In general, clean cleaved semiconductor surfaces exhibit no intrinsic surface states so that there is no band bending and the E_F position is that of the bulk (Brillson 1982a). However, the initial band bending is sensitive to the quality

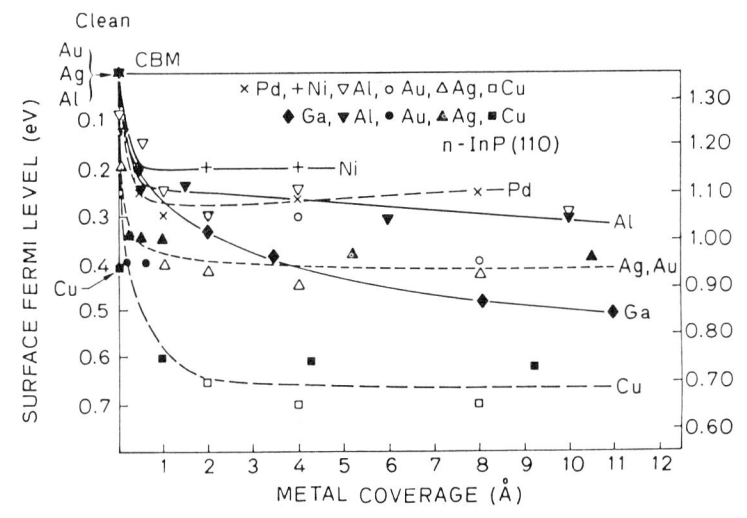

Fig. 29. Shifts in Fermi level position with equivalent metal coverage for Pd, Ni, Al, Au, Ag, Cu, and Ga on UHV-cleaved InP(110), as deduced from photoemitted core electron energy shifts. Energies are with respect to the valence band edge. After Kendelewicz et al. (1985), and Brillson et al. (1982).

of cleavage, e.g. the density of cleavage steps and subsequent damage (van Laar and Scheer 1967, Huijser and van Laar 1975, van Laar and Huijser 1976). In order to establish the initial E_F position, one determines the energies corresponding to the conduction band and valence band edges E_c and E_V, respectively. This can be accomplished by comparison of spectra features for n^+- and p^+-type material. For well-cleaved surfaces, corresponding spectral features should differ in energy by just the band gap, since the n- (p-)type E_F position is at the conduction (valence) band edge. Alternatively, one may obtain the absolute E_F energy from a sharp valence band edge of a metal in good electrical contact with the semiconductor. For the bulk doping, one obtains $E_F - E_V$ in the bulk. Comparison of E_V^{bulk} with E_V^{surface} obtained by photoemission yields the band bending.

Another complication of the analysis is that chemical effects also produce core level shifts with changes in metal deposition. Typically, chemical reaction, dissociation and diffusion shift the core levels of the constituent atoms by different amounts and in different directions, as evident from figs. 9 and 12–14. Unless the energies of all constituent cores shift rigidly, assignment of an E_F shift is problematic. In some cases, the valence band edge may be used as an indicator. However, chemical interactions may complicate this analysis as well, especially above monolayer coverages.

Multiple chemical states of an atomic core may also distort the SXPS peak lineshape, especially when the energy splittings are smaller than the spectral resolution. Furthermore, core levels of the clean surface exhibit an additonal shift due to the termination of the lattice and the reduced dielectric screening (Eberhardt et al. 1978, Himpsel et al. 1980a). Two procedures may be employed to minimize these complications. First, one can use photon energies away from the escape depth minimum in fig. 3 in order to maximize the core level contribution from the substrate. Second, one can deconvolute the spectral peaks into their constituents (i.e. as in fig. 6) in order to identify the substrate contribution. In both cases, one attempts to measure energy shifts of subsurface core levels, free of any complication due to altered chemical bonding or screening. Unless the semiconductor is highly degenerate (see section 5.2) the probe depth far from the escape depth minimum is still a small fraction of the width of the surface space charge region. Thus the "bulk" determination is still an accurate measure of the surface E_F position.

Landgren and Ludeke (1984) have investigated E_F movements for several metal/GaAs interfaces using deconvolution techniques and both surface-sensitive and bulk SXPS excitation energies. Their comparison of surface-sensitive and bulk core level spectra reveal significant differences in rate of change and energy of E_F position. While such deviations are minor for simple metals such as Al, they appear to be significant for transition metals such as Ti, where screening effects in the metal introduce large excitonic shifts for substrate atoms dispersed in the overlayer (Ludeke and Landgren 1986).

A final complication is that the metal overlayer may not be uniform, so that the photoemission signal may be an average of different phases weighted by their relative surface areas. However, such cases should exhibit anomalous broadening of the constituent core levels.

6.2. Semiconductor band bending with metal deposition

In principle, the band bending within the surface space charge region should scale linearly with the E_F position within the band gap. However, SXPS studies reveal that this band bending need not be parabolic, as conventionally assumed. Instead, new phases near the interface may lead to rapid band bending over distances which are narrow compared with the width of the surface space charge region of

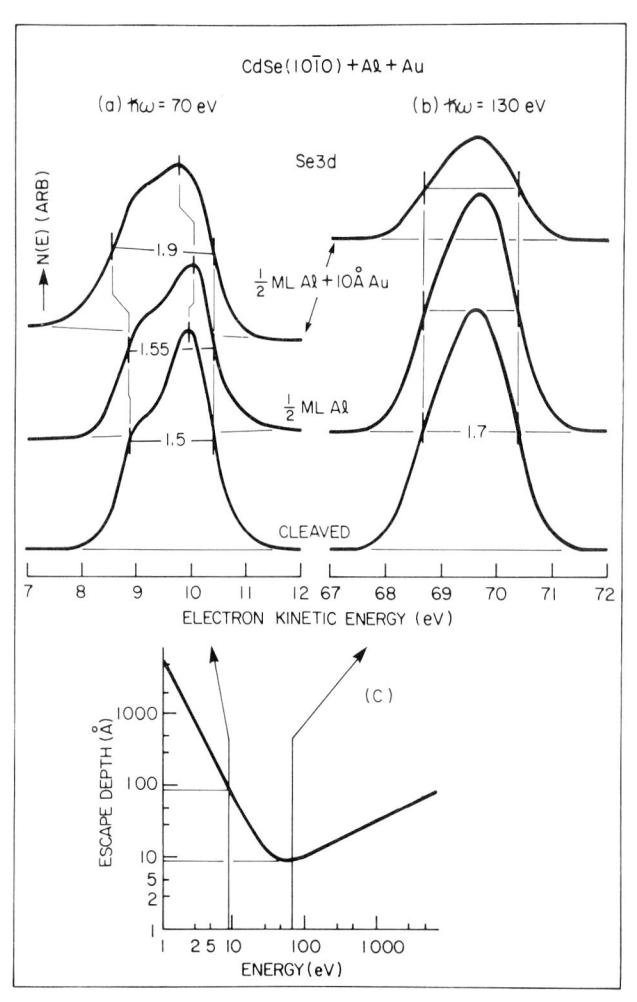

Fig. 30. SXPS Se 3d core level spectra under (a) bulk-sensitive ($h\nu = 70$ eV) and (b) surface-sensitive ($h\nu = 130$ eV) excitation conditions for Au and Al deposited on UHV-cleaved CdSe($10\bar{1}0$). FWHM values for each peak are given in eV. Corresponding escape depths are shown in (c). After Brucker et al. (1982).

the bulk semiconductor. A case in point is the Al–CdS junction, at which cations accumulate at the interface (see fig. 12). Such a cation excess can produce degenerate n-type doping of the CdS just below its free surface and an effective barrier height reduced by tunneling (Brucker and Brillson 1981a, 1982, Brucker et al. 1982). This is evident from a comparison of the Au–CdS(10$\bar{1}$0) interface with and without a 2 Å interlayer of Al. For the Au–CdS interface, E_F lies 0.8 eV below E_C, consistent with a capacitance–voltage (C–V) barrier of 0.76 eV. On the other hand, the Au–2 Å Al–CdS interface has a similar E_F position but an "ohmic" contact. These results are reconciled by a rapid band bending at the latter interface, which permits extensive tunneling through the thin surface barrier.

One can use SXPS to determine the presence of such a rapid band bending at the metal–semiconductor interface. By varying the escape depth of core level photoelectrons at Au–Al–CdSe(10$\bar{1}$0) interfaces, Brucker and Brillson (1982) and Brucker et al. (1982) demonstrated an anomalous broadening of the Se 3d core level when bulk versus surface regions are probed, as shown in fig. 30. For $h\nu = 130$ eV, only the surface (escape depth $\lambda_e = 5$–10 Å) region was probed, whereas for $h\nu = 70$ eV, both the surface and bulk ($\lambda_e = 50$–100 Å) were examined. Surface core level shifts would be expected to broaden the surface-sensitive spectra (Eberhardt et al. 1978, Himpsel et al. 1980a). Instead, the surface-sensitive spectra exhibit a constant width, while the bulk-sensitive spectra broaden by ~30%. Al on CdSe or CdS produces no significant band bending or broadening. The broadening in fig. 30 becomes apparent only after the Au deposition induces the E_F movement.

6.3. Comparison of spectroscopic and electrical barrier heights

Several photoemission studies of E_F movement with metal deposition have shown a correspondence with barrier heights measured by conventional techniques on interfaces prepared in non-UHV environments. Such comparisons are available for Si (Rowe 1976, Margaritondo et al. 1976), GaAs (Newman et al. 1985a,b,c), InP (List et al. 1985), CdS and CdSe (Brucker and Brillson 1981, 1982, Brucker et al. 1982), for example. Recently Newman et al. (1985a,b,c) have investigated the relationship between E_F position determined by SXPS, and barrier height determined by I–V, C–V, and internal photoemission measurements, for interfaces prepared under essentially identical UHV conditions. Such a comparison was motivated in part (Spicer et al. 1984) by the fact that the Fermi level stabilization measured by SXPS requires a charge density of only 10^{12} cm^{-2}, whereas one to two orders of magnitude larger charge density is required to pin E_F for a thick film with metallic screening of these states (Zur et al. 1983a,b). Furthermore, the barrier heights measured for interfaces prepared under non-UHV conditions could be shifted due to the presence of even a thin contamination layer. Conventional measurements must be used to ascertain the E_F position, because photoemission spectroscopy is not capable of measuring barriers for thick metal coverages. Newman et al. (1985c) show a close agreement between thin and thick metal coverages on UHV-cleaved GaAs with near-unity ideality factors for the metals Au, Ag, and Cu. On the other hand, they report a significant discrepancy for the Al/n-GaAs interface, where device measurements display a barrier 0.2 eV higher than that derived from photoemission spectroscopy. They attributed this discrepancy to either an atomic dipole or the formation of a reacted GaAlAs interfacial layer (Brillson 1979, Okamoto et al. 1981). For annealed Au–GaAs interfaces, Newman et al. (1985a,b) find similar shifts in E_F between SXPS measurements of 0.2 and 15 monolayers and device measurements of thick overlayers. They attribute the thermal degradation observed to formation of Au–Ga eutectic formed by interdiffusion. Hence such measurements confirm the general utility of SXPS for monitoring the evolution of Schottky barrier formation, but suggest that chemical effects can occur at multilayer coverages which can influence the macroscopic barrier height.

6.4. Models of Schottky barrier formation

The analysis of metal–semiconductor interfaces by SXPS and other surface science techniques (Brillson 1982a) has led to a number of new models for describing various aspects of Schottky barrier formation. These models reflect the interactions which can occur between metals and semiconductors and incorporate native defects, chemically reacted interfacial layers, alloy overlayers, and interface-specific charge exchange as factors which can determine the Fermi level pinning. Detailed discussions of each of these models requires considerably more space than is available or appropriate here. Nevertheless, we provide here a brief overview of the main, currently active models in the context of SXPS results described in earlier sections.

The model of Schottky barrier formation which has attracted most attention in recent years is based on the formation of native defects near the metal–semiconductor interface. This defect model has been developed primarily by Spicer et al. (1979a,b, 1980a,b, 1982b), along with Lindau et al. (1978), Wieder (1978), and Williams (1981, 1982). It is motivated by several characteristics of E_F behavior for adsorption on III–V compound semiconductors. First, the UHV-cleaved surfaces of III–V compounds exhibit little or no band bending, consistent with the absence of intrinsic surface states (van Laar and Scheer 1967, Huijser and van Laar 1975, van Laar and Huijser 1976, Brillson 1982a). Second, E_F varies rapidly with submonolayer adsorption of metal or oxygen atoms, indicative of the strong effect which the initial interfacial layer has on the E_F position. Third, the E_F position asymptotes to a constant value within a few monolayers coverage or less and these pinning positions are quite close in energy to each other and to the macroscopic Schottky barrier heights, despite the different chemical nature of the chemisorbed species. Based on the last observation, Spicer et al. (1979a,b, 1980a,b, 1982b) concluded that electronically active defects are created which "pin" E_F at the same position within the semiconductor band gap. This picture appears in fig. 27. Here, the constant pinning levels for n- and p-type GaAs are separated by 0.2 eV. Furthermore, these states are created in an indirect way, which does not depend on the chemical characteristics of the atom [i.e. by energy release of the heat of condensation to the substrate upon metal deposition (Lindau et al. 1978)]. The observation that oxides and metals induce the same pinning within GaAs and possibly other semiconductors is valuable, since most contaminated semiconductor surfaces and device structures have such oxide layers. Williams (1981, 1982) reached similar conclusions regarding defect formation in III–V compounds, especially InP, on the basis of electrical, AES, and UPS measurements.

The defect model has undergone modifications to account for a number of apparent discrepancies. Thus the density of defects and the occupancy of a given defect level can change with the deposited atom. In this way, one can account for the different E_F positions for different metals on n-InP(110) as well as, to a lesser extent, GaAs(100) and (110) surfaces. Recently, the pinning of noble metals at higher than expected barrier values on GaAs (e.g. Au in fig. 27) has been explained in terms of interface charge transfer due to electronegativity differences (Pauling 1939) and compensating centers due to metal substitution on the semiconductor lattice (Newman et al. 1985c). A reacted interface layer of GaAlAs has been included to account of the higher barrier value of Al on n-type GaAs(110) (Newman et al. 1985c).

A serious difficulty of the defect model is that it is not easy to identify which of many types of defects (Milnes 1973, 1983) is responsible for the E_F behavior, thereby limiting the model's predictive value. For example, Allen and Dow (1981, 1982) and Dow and Allen (1982) have argued that antisite defects are responsible for E_F pinning in III–V compounds, as opposed to a vacancy defect mechanism described by Daw and Smith (1980a,b). Both Dow and Allen and Daw and Smith have used their different defect pictures to account, for the same trends in barrier heights for Au Schottky barrier contacts and metal–insulator–semiconductor structures in the $Ga_{1-x}Al_xAs$ (Kajiyama et al. 1973, Best 1979, Wieder 1981) and $In_{1-x}Ga_xAs$ alloy series. Comparison of calculated or experimentally observed energies alone is not sufficient to rule out pinning by one or the other defect.

Other researchers have found evidence for a single defect providing the pinning mechanism. From the temperature dependence of E_F position within the GaAs band for Ge deposition on GaAs(110), Mönch and Grant (1982) deduce acceptor- and donor-like surface states with equal density and therefore a common origin. Grant et al. (1981) have also postulated a single-defect model for GaAs(100) and (110) surfaces based on the XPS observation that the difference in E_F position is constant for n- and p-type GaAs exposed to identical surface treatments. They interpret variations in E_F pinning position as due to changes in local defect environment, whereas the constant energy difference between n- and p-type GaAs is due to multiple charge states of the same defect.

The defect model does not appear to extend to III–V compound semiconductor interfaces with other semiconductors. SXPS studies at the GaAs–Ge(100) heterojunction grown by MBE show that, by varying GaAs surface As and ambient As_4 during growth, the E_F position can vary by 0.35 eV at the interface and its position is uncorrelated with the constant valence band discontinuity (Chiaradia et al. 1984, Katnani et al. 1984).

To summarize, defect models provide a straightforward explanation of the nearly constant (± 0.2 eV) pinning position of E_F at metal–GaAs(110) and possibly other III–V compound semiconductor interfaces. While they supply a clue to the slow variations in barrier height with different metals, these models provide few if any predictions based on a given semiconductor and metal prepared under specific conditions.

A second perspective on the Schottky barrier problem has emerged from the observations of pronounced chemical reactions and interdiffusion at metal–semiconductor interfaces. For III–V compounds, this chemical perspective appears to be complementary to defect models, while for II–VI compounds, it provides a qualitatively different chemical basis for Schottky barrier variations. For example, Brillson et al. (1981a,c) have demonstrated a correlation between Schottky barrier heights for a variety of metals on UHV-cleaved InP and the stoichiometry of In versus P outdiffusion measured by SXPS. Figure 31 demonstrates that metals such as Au, Cu, Pd, and Ag, which produce anion-rich outdiffusion, also produce high barrier heights, as measured by I–V and C–V techniques and pictured in the inset (Williams et al. 1978). Likewise, more reactive metals such as Al, Ni, and Ti cause cation-rich outdiffusion and lower barrier heights. The division in both chemical and electronic behavior pictured in fig. 31 provides a chemical basis for Schottky barrier formation at InP–metal interfaces, suggesting that the nature of outdiffusion and the resulting changes in the semiconductor lattice below the surface determines the type of electrically active sites formed. In turn, these defect and/or impurity states can stabilize E_F at the different positions within the band gap. Metals on UHV-cleaved GaAs exhibit a similar trend, as discussed in section 5.2, although the splitting between high and low E_F positions is smaller than for InP (Mead 1966, Newman et al. 1985c).

A key attribute of the chemical framework is that it recognizes qualitative differences in interface chemistry for metals with different classes of semiconductors (Brillson 1982a, 1983b). These changes in chemical systematics provide an explanation for the substantial difference in interface behavior (i.e. the sensitivity of a semiconductor barrier height to different metals) between III–V and II–VI compound semiconductors. Specifically, different physical phenomena may contribute to formation of the electrical barrier, depending upon the class of semiconductor. Thus in section 6.2, localization of dissociated cations at reactive metal interfaces with II–VI compounds leads to lowering of the barrier due to quantum mechanical tunneling (Brucker and Brillson 1981a, 1982, Brucker et al. 1982). Such effects are not apparent for III–V compounds in general (Brillson 1983b). On the other hand, the reversal in stoichiometry observed in III–V compounds is not seen for II–VI/metal interfaces (Brillson 1983b). The chemically interactive interface model of Schottky barrier formation pictured in fig. 2 also includes the possibility that new dielectric layers can form between metal overlayer and semiconductor substrate (Brillson 1977b, 1978c). II–VI/metal interfaces tend to

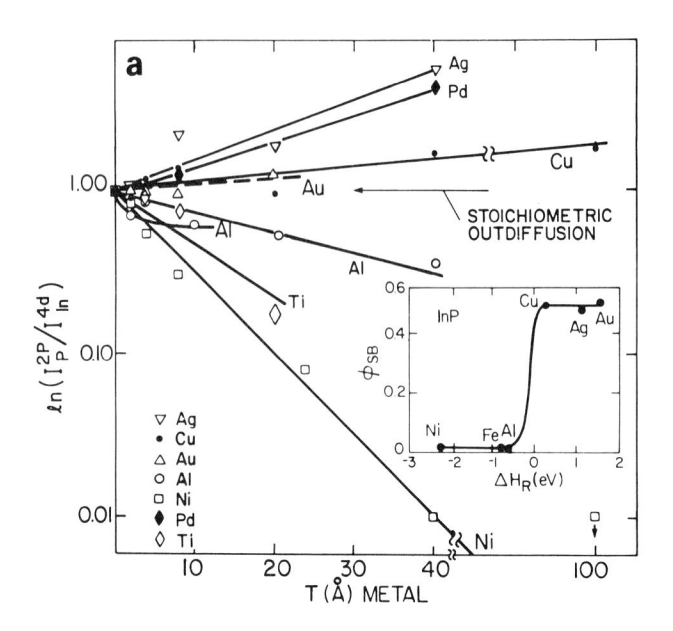

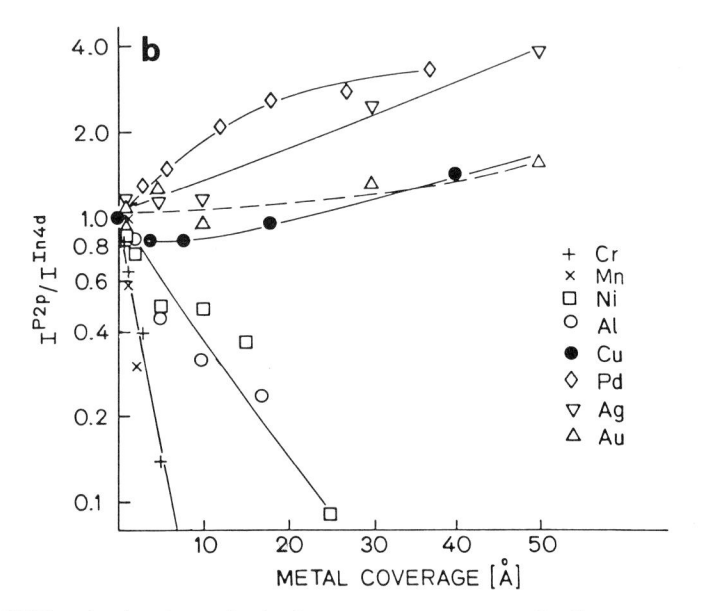

Fig. 31. (a) SXPS ratio of surface anion/cation core level intensities I_P^{2p}/I_{In}^{4d} versus metal coverage on InP(110) relative to the UHV-cleaved surface ratio for Ag, Pd, Cu, Au, Al, Ti, and NI. The inset contains a barrier height φ_{SB} versus ΔH_R plot [after Williams et al. (1978)] and emphasizes the correlation between φ_{SB} and the stoichiometry of outdiffusion. After Brillson et al. (1981c). (b) Similar data for Ni, Al, Cu, Pd, Ag, Au as well as for Cr and Mn. After Kendelewicz et al. (1985).

be more extended spatially than III–V/metal interfaces and, to the extent that they form nonmetallic interface layers, these interfaces will contribute more heavily to the electrical measurement (Brillson 1982a). All these differences lead to a wider range of electrical barriers for II–VI versus III–V compounds, as is reported for device measurements.

Another possibility pictured in fig. 2 is that chemical interactions at the metal–semiconductor interface can form a metallic alloy layer which represents a new effective work function to the semiconductor surface (Freeouf 1980, 1981, Freeouf and Woodall 1981, Woodall and Freeouf 1982). For metals on GaAs, Freeouf and Woodall (1981) and Woodall and Freeouf (1982) have proposed that As precipitates from such an interfacial layer, whose effective work function is relatively constant. Hence, the classical charge transfer between metal and GaAs changes only slightly, so that the semiconductor band bending exhibits only smaller variations with different metal overlayers. Freeouf (1980, 1981) has advanced similar arguments for the relatively weak dependence of Si barrier heights on the type of metal overlayer. Here the formation of a Si-rich interfacial layer dilutes the effect of different metals on the band bending.

In support of the effective work function model, Schwartz et al. (1979) have observed interfacial layers of As at a significant number of metal–GaAs interfaces. Also, Freeouf (1980, 1981) has presented a strong correlation between Schottky barrier height expected from the Schottky relation $\varphi_{SB^n} = \varphi_I - \chi_{SC}$ (n-type Si) and $\varphi_I = (\varphi_{Si}^4 \varphi_M)^{1/5}$, the weighted average of metal and Si work functions for an interface layer of MSi_4 stoichiometry.

Recently, Woodall et al. (1983) have found evidence that another artifact of the metal–semiconductor interface – line dislocations – can influence E_F position at strained III–V overlayers grown by MBE. Indeed, they demonstrated that the degree of Fermi level pinning scaled proportionally with the density of line dislocations induced within the pseudomorphic layer.

Another model which can account for E_F pinning involves the formation of interfacial states within the band gap by chemisorption and charge transfer involving metal atoms and clusters. The resultant charged clusters will be screened due to their proximity with the semiconductor dielectric medium. With such a dielectric screening model, Ludeke et al. (1983) have successfully accounted for the general trends in surface E_F position with Ag coverage on GaAs, pictured in fig. 28. In addition, this figure indicates that these screened, Ag-derived interface states can be distinguished from defect states, which are calculated to have a much more rapid movement of E_F as indicated by the dashed–dotted line. Thus the metal-induced interface state model proposed by Ludeke provides an alternative mechanism for the rapid E_F movement in III–V compounds, which is particularly applicable to metal–semiconductor junctions at which no chemical reaction or diffusion is apparent.

Another model for localized charge states at the interface involves quantum mechanical tunneling of the metal's wavefunction into the semiconductor at energies within the band gap (Louie and Cohen 1975, 1976, Chelikowsky 1977, Zhang and Schlüter 1978, Tersoff 1984). However, such "metal-induced gap states (MIGS) appear to be ruled out by SXPS experiments. Results of this section show that E_F stabilization occurs at metal coverages well below those for which metallic valence and core level features are observed. Thus metallic wave functions are not involved at the critical stages of Schottky barrier formation.

It is particularly appropriate that almost all of the viable models discussed here have been motivated largely by SXPS studies. This fact should not be too surprising given the ability of SXPS to analyze chemical and electronic structure within the same experiment and with submonolayer precision. However, SXPS and other surface-sensitive techniques have not yet provided an understanding of the physical mechanisms during Schottky barrier formation in sufficient detail such that one can *predict* electronic behavior for a given metal–semiconductor interface formed under a given set of conditions.

7. Other synchrotron radiation techniques for interface analysis

Besides SXPS, a number of other synchrotron radiation techniques can provide valuable information in the analysis of metal–semiconductor interfaces. These methods involve absorption, diffraction, and multi-electron excitation within the near-surface region of the surface or interface under investigation, and they rely heavily upon the tunable X-ray characteristics of synchrotron radiation.

7.1. Surface extended X-ray absorption fine structure analysis

An aspect of metal–semiconductor interfaces not yet addressed in this chapter is the atomic structure of the junction. Since the electronic structure (e.g. local charge rearrangement, band bending, dipole formation, band structure) depends sensitively on atomic positions, any detailed theoretical analysis of such phenomena requires an accurate modeling of interface chemical structure.

Among the synchrotron radiation techniques which can provide structural information for surfaces and interfaces, surface extended X-ray absorption fine structure (SEXAFS) analysis is perhaps the foremost. For a metal deposited on a semiconductor, one monitors the absorption spectrum at energies just above one of the metal's core level absorption thresholds. Absorption of X-rays in this region creates a core hole, which initiates an Auger process leading to ejection of an electron from the solid. Selection of particular Auger transitions rather than the total electron yield or the X-ray absorption is required to achieve high surface sensitivity. The choice of absorption edge and corresponding Auger transitions is determined by the particular metal deposited and by the relative intensities of the Auger signals. Essentially, the interference between the outgoing photoelectron wave and the component backscattered from nearest neighbors causes oscillations in absorption at energies above the threshold for a particular atomic core level. This fine structure exhibits a period of oscillation corresponding to the distance between the atom whose core electrons are excited and nearby atoms. Variations in absorption appear as changes in Auger electron emission. In contrast to SXPS, the SEXAFS method involves monitoring electron emission at a constant Auger energy while scanning the incident photon energy across the absorption threshold and above. SEXAFS is a particularly powerful technique because it provides direct structural analysis including extremely accurate determination of bond lengths without the need for an ordered overlayer. However, the technique is difficult to apply to complicated structures for which many similar lengths involving a chemisorbed species could be present which SEXAFS could not resolve.

Stöhr and Jaeger (1982) have applied the SEXAFS technique to metals deposited on semiconductors and have found, for example, qualitatively different results for two different metals on Si(111). For Pd deposition at room temperature and 1.5 monolayer coverage, they determined a local structure around the Pd atoms which resembles that of thick Pd_2Si grown on Si(111). In contrast, spectra for 2.5 monolayers of Ag on Si(111) indicated a local structure around the deposited Ag which resembled bulk Ag. At 500°C, polarization-dependent SEXAFS revealed Ag atoms in a threefold site.

Comin et al. (1983) have applied SEXAFS to the structure and nucleation of Ni silicide on Si(111). They monitored absorption from the Ni:K edge using Auger electrons from a KLL ^1D Auger transition. SEXAFS absorption data were obtained as a function of Ni deposition and compared with bulk EXAFS data obtained from NiSi with the same detection scheme and from $NiSi_2$ and NiSi by detection of the transmission. This absorption data showed a strong similarity with $NiSi_2$ at low Ni coverage, indicating a single first-neighbor coordination shell of only Si atoms with a bond length close to the Ni–Si distance in bulk $NiSi_2$. Furthermore, they determined the amplitude of EXAFS oscillators, proportional to the number of atoms in the coordination shell, to be 7.4 ± 0.9, compared with 8 in $NiSi_2$. The real space Fourier transformed data [amplitude of EXAFS oscillators versus k (Å^{-1})] yields a K-edge amplitude function which, coupled with the coordination number and bond length, allows the chemisorption site to be determined unambiguously. Comin et al. (1983) found only the six-fold hollow between first and second layers to be consistent with the transformed data.

At higher coverage (1–5 Ni monolayers), similar analysis indicates Ni substitution into some of the Si sites in a $NiSi_2$ structure. Thus the SEXAFS data indicate a new model for the growth of nickel silicide as shown in fig. 32: (a) the first Ni–Si–Ni layer forms with Ni in six-fold sites, which weakens Si–Si bonds between first and second layers, (b) the surface atoms expand outward and form a $NiSi_2$-like structure, (c) further diffusion weakens Si–Si bonds in the third and fourth layers and expansion of the third layer leads to either (d) two silicide layers with expansion of the overlayer or (e) a shear glide of the silicide layer and (f) a three-dimensional type-B $NiSi_2$ structure. This model is consistent with many of the photoemission, ion scattering and transmission electron microscopy results for the Ni–Si(111) system and is indicative of the detailed level of information obtainable with the SEXAFS technique.

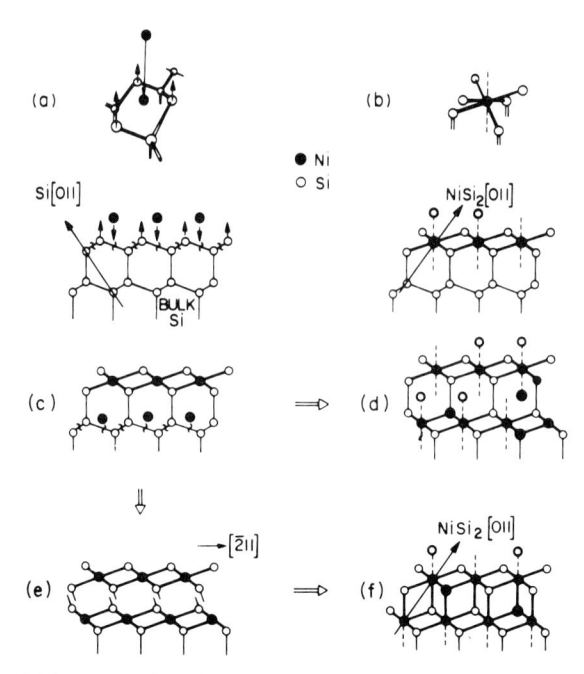

Fig. 32. Atomic model for nucleation of NiSi$_2$ on Si(111) as projected on the (001) plane. (a) and (b) present three-dimensional views as well. Stages of nucleation (a)–(f) are described in the text. After Comin et al. (1983).

7.2. Angle-resolved techniques

Angle-resolved techniques provide a method of structural analysis beyond the morphological studies described in section 4.3. Angle-resolved photoemission spectroscopy (ARPES) utilizes surface photoemission to derive symmetry information about atomic bonding within the surface plane. By collecting electrons emitted into a small solid angle as well as a narrow energy range, one can determine the component of momentum parallel to the surface. Since only $k_\parallel$, the momentum parallel to the surface, is conserved (Kane 1964),

$$k_\parallel = (\sin\theta)(2mE_k)^{1/2}/\hbar, \qquad (4)$$

where E_k is the kinetic energy, m the mass, and θ the angle with respect to the surface normal of photoemitted electrons. Thus angle-resolved data can be analyzed along particular lines of the Brillouin zone rather than over the entire volume.

Figure 33 illustrates how ARPES can reveal significant changes in valence band features upon deposition of various metals on Si(111) surfaces (Hansson et al. 1981a,b). Figure 33a shows that angle-resolved spectra of the Si(111) valence band exhibit pronounced and characteristic changes with monolayer coverages of metal. Here the normal emission at $h\nu = 22$ eV exhibits a local maximum in the

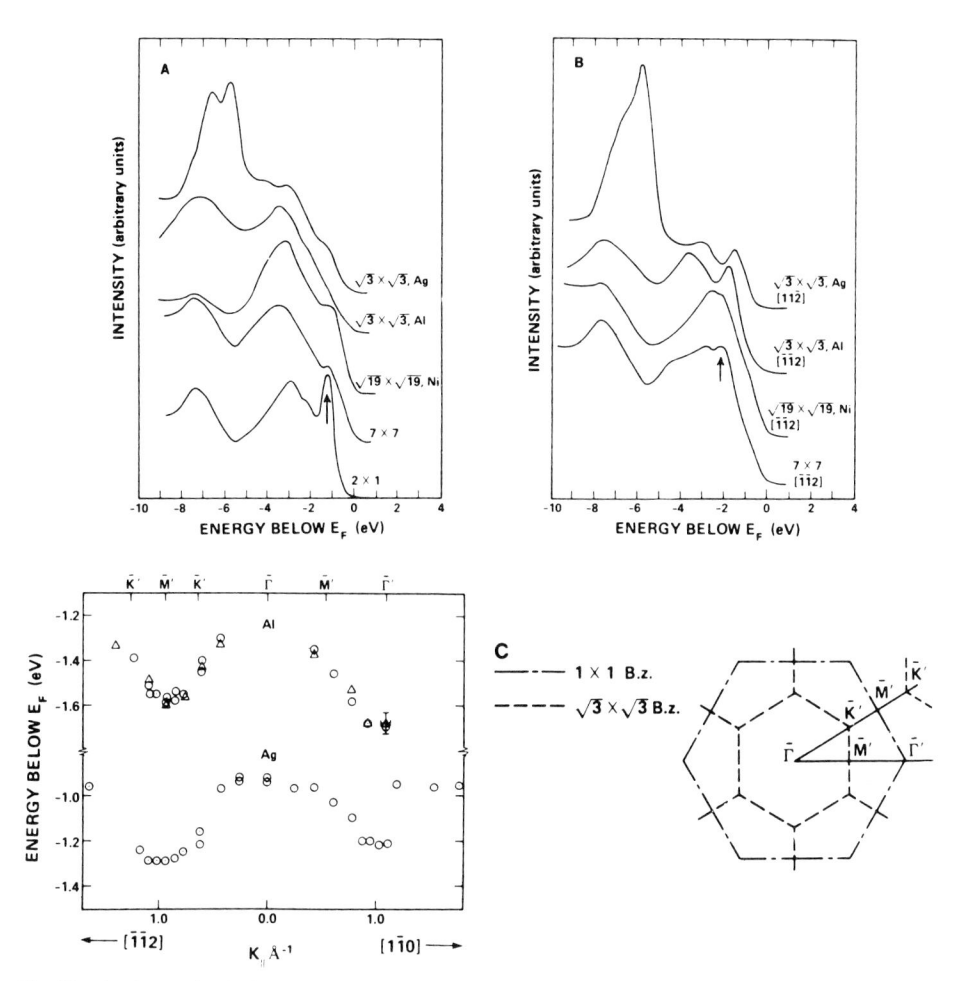

Fig. 33. Angle-resolved photoemission spectra obtained at (a) normal and (b) off-normal direction for Si(111) surface with monolayer or lower coverages of various metals on the 7 × 7 surface. 2 × 1 surface shown for comparison. $h\nu = 22$ eV. Arrows indicate the dangling bond emission. Dispersion for the metal-induced state emission appears in (c) for Ag and Al surfaces, derived from the full range of spectra as a function of angle. After Hansson et al. (1981a).

dangling bond surface bonds indicated by the arrows. The metals Ni, Ag, and Al form different ordered overlayers on clean ordered Si(111). The designations $\sqrt{3} \times \sqrt{3}$ Ag, $\sqrt{3} \times \sqrt{3}$ Al, $\sqrt{19} \times \sqrt{19}$ Ni, 7 × 7 Si, and 2 × 1 Si correspond to the size of the geometry of the LEED pattern obtained for each surface. The strong similarity between electronic states of 7 × 7 and $\sqrt{19} \times \sqrt{19}$ surfaces in both normal (fig. 33a) and off-normal (fig. 33b) directions suggests a similarity between atomic reconstructions of the two surfaces. On the other hand, monolayer coverages of Al and Ag on the same surface yield qualitatively different electronic features. Valence band features also exhibit dependence on polar and azimuthal angles, as shown in fig. 33c and the inset. The states replacing the dangling bond features in fig. 30b have a three-fold symmetry, peaking toward the [112] direction for Al on Si and toward [$\bar{1}\bar{1}2$] for Ag on Si. These metal-induced surface states exhibit the characteristic E versus k dispersions shown in fig. 30c. Based on these observations, Hansson et al. (1981a,b) have proposed structural models for the various metal-induced surface reconstructions.

Angle-resolved photoemission measurements can provide structural information analogous to that of the SEXAFS technique. Barton et al. (1983) have demonstrated that core level angle-resolved photoemission intensity oscillates sinosoidally with increasing electron momentum. This effect is due to interference between direct and scattered photoemission. Angle-resolved photoemission extended fine structure (ARPEFS) can be analyzed to yield a Fourier spectrum with peaks for individual scattering atoms. The peak positions provide a scattering path length and the peak area gives a scattering power, from which one can extract structural information. As with SEXAFS, one requires wide kinetic and photon energy ranges, typically 500 eV, Fourier transform methods, as well as an analytic single-scattering theory for interpretation (Orders and Fadley 1983). The ARPEFS technique has a distinct advantage relative to SEXAFS, which provides accurate bond lengths to nearest neighbors without requiring long-range surface order. ARPEFS provides path length differences along chosen directions which involve more than just nearest neighbors. Hence, ARPEFS is more suitable than SEXAFS for complex structural analysis, i.e. multi-atom adsorbates, multiple site adsorption, and sites which are distinguished by second rather than first nearest neighbors. Disadvantages of ARPEFS include reliance on a number of assumptions which may not be universally valid: (a) scattering potential of each atom concentrated to a point and no overlap with other potentials or the scattering wave variations, (b) single, elastic scattering, and (c) charge state of the scattering atom, exchange, randomized atomic positions and excitonic effects ignored. While the ARPEFS method has not yet been applied to metals on semiconductors, Barton et al. (1984) have demonstrated its potential for extracting surface geometric information with simple adsorbates on metal surfaces.

7.3. Additional techniques

X-ray diffraction techniques can now provide structural analyses of surfaces and interfaces. Marra et al. (1979) and Eisenberger and Marra (1981) have used X-rays at glancing incidence from a synchrotron source or high-power anode in a reflection geometry and have limited detection depths to only hundreds or thousands of Å. Thus detection of an adsorbate on an interfacial layer is enhanced substantially relative to X-ray transmission experiments. Intensities of Bragg peak measured can be interpreted in a straightforward way. Furthermore, interface structures below the surface can be examined with this technique. Marra and coworkers have applied the X-ray reflection–Bragg diffraction method to Ge reconstruction (Eisenberger and Marra 1981) and to buried Al–GaAs interface structures (Marra et al. 1979).

X-ray standing wave interference techniques provide positions of ordered metal overlayers on semiconductor surfaces without the use of glancing incidence. Thus by rocking a Si(111) substrate through a Bragg reflection, Durbin et al. (1985) have determined Au adatom positions with respect to the Si(111) planes. These results confirm SXPS results which show that Au diffuses into the top Si surface layers. Alternatively, one can use high-energy synchrotron radiation for X-ray standing wave interference spectrometry (SWXIS), in which $h\nu$ is scanned for a particular orientation of incident excitation. In addition to measurement of the adatom location normal to the crystal surface plane, one can obtain atom positions relative to planes inclined to the surface without crystal rocking. Illustrative of this technique, Dev et al. (1985) have reported SWXIS results for Se chemisorption on Si.

The continuous nature of synchrotron radiation provides the basis for two other spectroscopies, constant final state spectroscopy (CFS) and constant initial state spectroscopy (CIS). With CFS (also termed partial yield spectroscopy), one measures photoemission yield within some energy interval E to $E + \Delta E$ above the photoemission threshold as $h\nu$ is varied (Gudat and Kunz 1972, Eastman and Freeouf 1974, Lapeyre and Anderson 1975). After the primary photoabsorption, recombination with the excited core hole occurs, largely via Auger emission for low-energy ($E < 10^3$ eV) core holes. Thus an optical transition takes place from the core hole to an unoccupied final state, the core hole recombining with a valence electron with the emission of a second valence electron. The resultant electron emission spectrum is a measure of the probability for creating a core hole as a function of $h\nu$. Essentially, it represents a photoabsorption spectrum for optical transitions from a core level to empty states above E_F. The empty state's energy equals the core level binding energy minus the CFS transition energy, corrected for excitonic effects.

Lindau et al. (1978) have used the CFS method to determine the role of empty states of the semiconductor surface in Schottky barrier formation. For Au on GaSb(110), they monitored Ga 3d transitions into the first empty states above the Fermi level as a function of metal coverage. Suppression of these transitions required monolayer Au coverage, whereas E_F stabilization took place at much smaller coverages. Hence the CFS data ruled out the involvement of any intrinsic semiconductor surface states with the Fermi level pinning.

Stoffel et al. (1982) have used CFS to monitor the unoccupied conduction bands of CdS as a function of Cu deposition. From S 2p core level transitions to empty states, they observed the development of a new absorption peak corresponding to a state $\sim$1 eV above the conduction band minimum. This state provides additional evidence for a chemical reaction between Cu and CdS at monolayer metal coverages.

The CIS technique involves measurement of the photoemission yield within some energy interval E to $E + \Delta E$ at a fixed initial state energy, $E_i = E - h\nu$ (Lapeyre et al. 1974a,b). In this technique, optical monochromator and electron energy analyzer are scanned synchronously so that $E - h\nu$ is kept constant. By selecting E_i such that valence band emission is minimized and Auger emission to the final state is maximized, one can enhance the signal due to core level to empty state transitions. Gudat and Eastman (1976) have used CIS to identify transitions from III–V compound core levels to empty states induced by Au adsorption which correlate with E_F pinning positions. However, absolute determination of these energies is complicated by excitonic effects (Gudat and Eastman 1976).

8. Conclusions and future directions

In this chapter, we have examined the synchrotron radiation techniques capable of characterizing interface chemical and electronic structure. These techniques have been instrumental in changing the conventional view of metal–semiconductor interface structure – from an abrupt, noninteracting contact to the extended junction pictured in section 2. SXPS studies have revealed that reacted and/or diffused regions can exist above and below the initial semiconductor surface, which can extend over tens or hundreds of Å. With its extreme surface sensitivity, SXPS readily detects such phenomena, which occur even at room temperature and under ideal conditions of surface purity and crystalline perfection.

Using the UHV tools described in section 3, one can now create ideal semiconductor surfaces free of contamination or lattice disorder, maintain them for reasonable periods of time, and modify their surface chemistry in a controllable and reproducible way. With this prescription, SXPS studies can provide the correlated changes in surface electronic and chemical structure which take place as the interface grows, layer by atomic layer.

Electronic and chemical phenomena at the microscopic interface can be quite varied, depending upon the particular metal–semiconductor combination, and in some cases quite complex. Thus section 4 provides examples of striking differences in chemical interaction between metals on the same semiconductor surface. The deconvolution of core level lineshapes reveals multiple phases evolving with increasing metal deposition. Different metals on Si exhibit a rich variety of materials changes for noble, near-noble, and transition metal junctions. Photoionization cross section dependences on $h\nu$ provide several effective techniques for identifying the bond character of their microscopic interfacial layers. Compound semiconductors exhibit a multiplicity of chemical behavior as well. III–V compound semiconductors display exchange reactions for metals which bond strongly with semiconductor anions. Such reactions can proceed in stages, with metals chemisorbing, then propagating into the semiconductor bulk at multilayer coverage or elevated temperature to form new reacted compound phases. New excitation methods such as pulsed laser annealing can promote such reactions over discrete, microscopic thicknesses. Qualitatively different bonding and diffusion are evident for metals on II–VI compounds, where metal–anion reactions are more spatially extended. Ternary II–VI compound interface behavior exhibits the additional complication of two cations reacting differently with the metal overlayer. III–V, IV–VI and more insulating semiconductor materials exhibit characteristic diffusion as well, with the more ionic semiconductors showing the most pronounced metal–anion bonding.

The variable depth sensitivity of SXPS permits nondestructive analysis of chemical phases spatially separated in depth by only a few monolayers. Such analysis can reveal evidence of segregation to the thin overlayer film surface, another facet of metal–semiconductor interactions.

SXPS also provides characterization of surface morphology for monolayer metal deposits on semiconductor surfaces. A number of techniques – valence band photoemission, substrate intensity attenuation, and comparative changes in angle-resolved photoemission features with metal deposition – demonstrate that clusters of metal atoms can form for many metal–semiconductor systems at initial submonolayer depositions. Here the strength of metal–substrate bonding, modified by surface orientational and composition differences, plays a dominant role.

The strength of chemical bonding plays a central role in each stage of the interdiffusion which can occur at the metal–semiconductor junction. As discussed in section 5, SXPS identifies at least two rate-limiting steps: (1) dissociation and diffusion out of the semiconductor lattice, and (2) diffusion through the metal overlayer. The latter stage involves "chemical trapping" at III–V compounds, "reactive outdiffusion" at ionic II–VI compounds, and a continuous variation in chemical activity as a function of ionicity between these two extremes. Similarly, metal diffusion into the semiconductor can be identified and distinguished from semiconductor outdiffusion by marker techniques.

The chemical activity revealed by SXPS core level spectra is directly related to electronic changes in the interface region. In section 6, E_F movements within the semiconductor band gap and band bending changes were identified from energy shifts of spectral features, suggestive of a number of mechanisms which may underly Schottky barrier formation. Such mechanisms include E_F pinning by native defects, interfacial dielectric and metallic alloy layers, tunneling through highly doped surface layers induced by chemical reactions, as well as weakly interacting metal chemisorbates. While our understanding of Schottky barrier formation has not advanced to a stage at which Schottky barrier heights can be predicted, we now realize the overriding significance of extrinsic surface phenomena. Indeed, the nature and strength of chemical bonding determines how important a role each of these mechanisms will play as well as the ultimate electrical properties of the macroscopic junction for a given metal–semiconductor pair.

Although SXPS has surpassed all other synchrotron radiation techniques in its application to metal–semiconductor interfaces, a number of new methods has demonstrated considerable promise. Section 7 described how soft X-ray absorption, interference, and diffraction can provide new approaches for structural analysis during the initial stages of inference growth. Such techniques may bring a new level of refinement to our understanding of interfaces on an atomic scale.

Synchrotron radiation methods offer a number of potentially fruitful directions to advance our knowledge of metal–semiconductor interactions. In order to explore further the initial stages of metal–semiconductor interaction and interface growth beyond the initial deposited monolayer, more structural analysis using SEXAFS and X-ray diffraction techniques are needed. These measurements can be compared with results of other structural techniques such as LEED, electron microscopy, RBS, and photoemission features in order to obtain unique interface geometries. In this regard, considerably more energy- and angle-resolved SXPS analysis of bulk compounds, possibly formed by interfacial reaction (e.g. $NiSi_2$ and NiSi for Ni–Si studies), is needed for comparison with observed interfacial layers. Studies of epitaxial, ordered interfaces offer particular promise for examining growth stages. In addition, ordered overlayers can provide a measure of the effect which controlled disorder and controlled strain have on the diffusion and Schottky barrier height.

In future, interface morphology may be examined with fine-focus SXPS probes. With such a technique, currently available with XPS, photoemission from regions as small as a micron or less can provide a measure of surface inhomogeneity in chemical bonding, chemical composition and band bending. Coupled with in-situ electron microscopy, this approach can supply new correlations between islands, steps, and other morphological features with the spectral energies of valence and core level features.

Temperature-dependent measurements will provide a measure of activation barriers to surface chemical interactions. At low temperatures, thermally – versus electronically – driven chemical reactions will decrease, and surface atom mobilities will decrease as well, reducing any cluster formation. Comparison of the rate of E_F pinning movement with that of cluster formation should indicate how strong a role clusters play in Schottky barrier formation. At elevated temperatures, SXPS can monitor changes in E_F with the evolution of chemical reactions and diffusion. Such studies will be particularly valuable in thin film studies, for which the supply of one interface constituent is limited so that the extent of chemical interaction can be controlled. SXPS can not investigate which parameters of the contact formation may affect the initial interface morphology. Factors which may affect interface morphology and chemical activity include: deposition rate, substrate temperature, density of surface imperfections (steps, dislocations, etc.), chemical composition, and crystal substrate orientation. Likewise, SXPS and structural probes can determine how substrate processing by gas exposure, addition of solid interlayer, and preannealing alter the contact morphology.

With the introduction of wigglers and undulators and with the high energy stored beams now becoming available at many synchrotron facilities, interface researchers will be able to capitalize on higher photon fluxes in many ways. Higher photon flux provides proportionally faster characterization of reactive systems such as Ti- or Cr-deposited semiconductors, which are rapidly contaminated even under UHV conditions. SEXAFS statistics will also improve while suppressing the likelihood of contamination during the otherwise lengthy data acquisition stages. Very high resolution photoemission spectra with reasonable signal levels will be possible, permitting more unambiguous deconvolution of spectral features for identifying chemical species and E_F movement. Higher photon fluxes will also enhance the use of the synchrotron source for infrared absorption experiments, so that electronic and vibrational structure can be probed with the same UHV chamber. With the high-energy stored beams now available, signal intensities at the deep C 1s and O 1s core levels will be significantly higher, permitting much more effective analysis of oxide– and organic–semiconductor interfaces.

Besides further investigation of Fermi level pinning mechanisms and characterization of metal–semiconductor interface chemistry, synchrotron radiation techniques can contribute insight into the control of interface chemical properties. SXPS analysis of atomic redistribution for interfaces processed in several ways may lead to new procedures for controlling thermal degradation of thin film structures, improving adhesion, and reducing surface charge recombination. For example, SXPS represents an ideal tool for probing highly localized interfacial layers promoted by laser annealing, permitting the characterization of ultrathin metastable species and, by virtue of the pulsed nature of annealing, various stages of the high-temperature chemical interaction. The coupling of reaction chambers for plasma film growth, etching, deposition or even wet chemical treatments will provide new opportunities for high-resolution analysis of chemical species left at the semiconductor surface as well as the electronic effects on subsequent barrier formation.

Synchrotron radiation techniques have provided key observations of the metal–semiconductor interface, which have altered substantially our understanding of their electronic and chemical properties. As this variety of chemical, electronic and structural tools is extended to new materials under new preparation and processing conditions, the prospects appear bright for continued advances in the understanding of metal–semiconductor interfaces.

Acknowledgement

The author gratefully acknowledges Ms. Cathryn Albright for her expert typing of the manuscript.

References

Abbati, I., and M. Grioni, 1981, J. Vac. Sci. & Technol. **19**, 631.
Abbati, I., L. Braicovitch, B. de Michelis, O. Bisi, C. Calandra, V. del Pennino and S. Valeri, 1980a, in: Proc. 15th Int. Conf. on the Physics of Semiconductors, Kyoto, Japan, J. Phys. Soc. Jpn. **49**, 1071.
Abbati, I., L. Braicovitch, A. Franciosi, I. Lindau, P.R. Skeath, C.Y. Su and W.E. Spicer, 1980b, J. Vac. Sci. & Technol. **17**, 930.
Abbati, I., L. Braicovitch, B. Michelis, U. del Pennino and S. Valeri, 1980c, Solid State Commun. **35**, 917.
Allen, R.E., and J.D. Dow, 1981, J. Vac. Sci. & Technol. **19**, 383.
Allen, R.E., and J.D. Dow, 1982, Phys. Rev. B **25**, 1423.
Andrews, J.M., and J.C. Phillips, 1975, Phys. Rev. Lett. **35**, 56.
Aspnes, D.E., and J.B. Theeten, 1979, Phys. Rev. Lett. **43**, 1046.

Babalola, I.A., W.G. Petro, T. Kendelewicz, I. Lindau and W.E. Spicer, 1983, J. Vac. Sci. & Technol. A **1** 762.
Babalola, I.A., W.G. Petro, T. Kendelewicz, I. Lindau and W.E. Spicer, 1984, Phys. Rev. B **29**, 6614.
Bachrach, R.Z., 1978, J. Vac. Sci. & Technol. **15**, 1340.
Bachrach, R.Z., 1984, in: Metal–Semiconductor Schottky Barrier Junctions and Their Applications (Plenum, New York) pp. 61–112.
Bachrach, R.Z., and R.S. Bauer, 1979, J. Vac. Sci. & Technol. **16**, 1149.
Bachrach, R.Z., R.S. Bauer and J.C. McMenamin, 1978, in: Proc. 14th Int. Conf. on the Physics of Semiconductors, Edinburgh, Inst. Conf. Ser. 43, ed. B.L.H. Wilson (Institute of Physics, Bristol) p. 1073.
Baetzold, R.C., 1978, J. Chem. Phys. **68**, 555.
Baetzold, R.C., 1981, Surf. Sci. **106**, 243.
Bardeen, J., 1947, Phys. Rev. **71**, 717.
Barton, J.J., C.C. Bahr, Z. Hussain, S.W. Robey, J.G. Tobin, L.E. Klebanoff and D.A. Shirley, 1983, Phys. Rev. Lett. **51**, 272.
Barton, J.J., C.C. Bahr, Z. Hussain, S.W. Robey, L.E. Klebanoff and D.A. Shirley, 1984, in: Proc. Brookhaven Conf. on Advances in Soft X-Ray Science and Technology, eds. F.J. Himpsel and R.W. Klaffky (SPIE, Bellingham, WA) p. 82.
Basch, H., 1981, J. Am. Chem. Soc. **103**, 4657.
Bauer, R.S., R.Z. Bachrach and L.J. Brillson, 1980, Appl. Phys. Lett. **37**, 1006.
Bauer, R.S., R.Z. Bachrach, G.V. Hansson and P. Chiaradia, 1981, J. Vac. Sci. & Technol. **19**, 674.
Bauer, R.S., ed, 1983, Proc. of the 2nd IUPAP Semiconductor Symp. on Surfaces and Interfaces, Surf. Sci. Vol. 132, and articles therein.
Bennett, A.J., and C.B. Duke, 1967, Phys. Rev. B **162**, 578.
Best, J.S., 1979, Appl. Phys. Lett. **34**, 522.
Bonapace, C.R., K. Li and A. Kahn, 1984, J. Vac. Sci. & Technol. A **2**, 566.
Braicovitch, L., C.M. Garner, P.R. Skeath, C.Y. Su, P.W. Chye, I. Lindau and W.E. Spicer, 1979, Phys. Rev. B **20**, 5131.
Braicovitch, L., I. Abbati, J.N. Miller, I. Lindau, S. Schwarz, P.R. Skeath and W.E. Spicer, 1980, J. Vac. Sci. & Technol. **17**, 1005 and references cited therein.
Braun, F., 1874, Pagg. Ann. **153**, 556.
Brillson, L.J., 1975a, J. Vac. Sci. & Technol. **12**, 76.
Brillson, L.J., 1975b, Surf. Sci. **51**, 45.
Brillson, L.J., 1977a, Surf. Sci. **69**, 62.
Brillson, L.J., 1977b, Phys. Rev. Lett. **38**, 245.
Brillson, L.J., 1978a, Phys. Rev. Lett. **40**, 260.
Brillson, L.J., 1978b, J. Vac. Sci. & Technol. **15**, 1378.
Brillson, L.J., 1978c, Phys. Rev. **18**, 2341.
Brillson, L.J., 1979, J. Vac. Sci. & Technol. **16**, 1137.
Brillson, L.J., 1982a, The Structure and Properties of Metal–Semiconductor Interfaces, Surf. Sci. Rep. **2**, 123 (and references cited therein).
Brillson, L.J., 1982b, Appl. Surf. Sci. **11/12**, 249.
Brillson, L.J., 1982c, J. Vac. Sci. & Technol. **20**, 652.
Brillson, L.J., 1982d, Thin Solid Films **89**, 461.
Brillson, L.J., 1983a, Int. J. Phys. Chem. Solids **44**, 703.
Brillson, L.J., 1983b, Surf. Sci. **132**, 212.
Brillson, L.J., 1984, in: Proc. Brookhaven Conf. on Advances in Soft X-Ray Science and Technology, eds F.J. Himpsel and R.W. Klaffky (SPIE, Bellingham, WA) p. 89.
Brillson, L.J., R.Z. Bachrach, R.S. Bauer and J. McMenamin, 1979, Phys. Rev. Lett. **42**, 397.
Brillson, L.J., C.F. Brucker, G. Margaritondo, J. Slowik and N.G. Stoffel, 1980a, Proc. 15th Int. Conf. on the Physics of Semiconductors, Kyoto, Japan, J. Phys. Soc. Jpn. **49**, 1089.
Brillson, L.J., R.S. Bauer, R.Z. Bachrach and M.C. McMenamin, 1980b, J. Vac. Sci. & Technol. **17**, 476.
Brillson, L.J., G. Margaritondo and N.G. Stoffel, 1980c, Phys. Rev. Lett. **44**, 667.

Brillson, L.J., R.S. Bauer, R.Z. Bachrach and G. Hansson, 1980d, Appl. Phys. Lett. **36**, 326.
Brillson, L.J., G. Margaritondo, N.G. Stoffel, R.S. Bauer, R.Z. Bachrach and G. Hansson, 1980e, J. Vac. Sci. & Technol. **17**, 880.
Brillson, L.J., C.F. Brucker, A.D. Katnani, N.G. Stoffel and G. Margaritondo, 1981a, J. Vac. Sci. & Technol. **19**, 661.
Brillson, L.J., C.F. Brucker, A.D. Katnani, N.G. Stoffel and G. Margaritondo, 1981b, Phys. Rev. Lett. **46**, 838.
Brillson, L.J., C.F. Brucker, A.D. Katnani, N.G. Stoffel and G. Margaritondo, 1981c, Appl. Phys. Lett. **38**, 784.
Brillson, L.J., R.S. Bauer, R.Z. Bachrach and G. Hansson, 1981d, Phys. Rev. B **23**, 6204.
Brillson, L.J., C.F. Brucker, A.D. Katnani, N.G. Stoffel, R. Daniels and G. Margaritondo, 1982, J. Vac. Sci. & Technol. **21**, 564.
Brillson, L.J., C.F. Brucker, N.G. Stoffel, A.D. Katnani, R. Daniels and G. Margaritondo, 1983, Proc. 16th Int. Conf. on the Physics of Semiconductors, Montpellier, France, Physica B&C **117**, 848.
Brillson, L.J., A.D. Katnani, M. Kelly and G. Margaritondo, 1984, J. Vac. Sci. & Technol. A **2**, 551.
Brown, F.C., R.Z. Bachrach, S.B.M. Hagström, N. Lien and C.H. Pruett, 1974, in: Proc. 4th. Int. Conf. on Vacuum Ultraviolet Radiation Physics, Hamburg, eds E.E. Koch, R. Haensel and C. Kunz (Pergamon–Vieweg, Braunschweig, FRG).
Brucker, C.F., and L.J. Brillson, 1981a, Appl. Phys. Lett. **39**, 67.
Brucker, C.F., and L.J. Brillson, 1981b, J. Vac. Sci. & Technol. **19**, 617.
Brucker, C.F., and L.J. Brillson, 1981c, J. Vac. Sci. & Technol. **19**, 787.
Brucker, C.F., and L.J. Brillson, 1982, Thin Solid Films **89**, 67.
Brucker, C.F., L.J. Brillson, A.D. Katnani, N.G. Stoffel and G. Margaritondo, 1982, J. Vac. Sci. & Technol. **21**, 590.
Calandra, C., O. Bisi and G. Ottaviani, 1985, Surf. Sci. Rept. **4**, 271.
Cerrina, F., R.R. Daniels, T.-X. Zhao and V. Fano, 1983a, J. Vac. Sci. & Technol. B **1**, 570.
Cerrina, F., R.R. Daniels and V. Fano, 1983b, Appl. Phys. Lett. **43**, 182.
Chambers, S.A., and L.W. Swanson, 1983, Surf. Sci. **131**, .
Chelikowsky, J.R., 1977, Phys. Rev. B **16**, 3618.
Chiaradia, P., A.D. Katnani, H.W. Sang Jr and R.S. Bauer, 1984, Phys. Rev. Lett. **52**, 1246.
Chye, P.W., I. Lindau, P. Pianetta, C.M. Garner and W.E. Spicer, 1977, Phys. Lett. A **63**, 387.
Chye, P.W., I. Lindau, P. Pianetta, C.M. Garner, C.Y. Su and W.E. Spicer, 1978, Phys. Rev. B **18**, 5545.
Comin, F., J.E. Rowe and P.H. Citrin, 1983, Phys. Rev. Lett. **51**, 2401.
Cooper, J.W., 1962, Phys. Rev. **128**, 681.
Daniels, R.R., A.D. Katnani, T.-X. Zhao, G. Margaritondo and A. Zunger, 1982, Phys. Rev. Lett. **49**, 895.
Daniels, R.R., T.-X. Zhao and G. Margaritondo, 1984, J. Vac. Sci. & Technol. A **2**, 831.
Davis, G.D., N.E. Byer, R.R. Daniels and G. Margaritondo, 1983, J. Vac. Sci. & Technol. B **1**, 1726.
Davis, G.D., W.A. Beck, N.E. Byer, R.R. Daniels and G. Margaritondo, 1984, J. Vac. Sci. & Technol. A **2**, 546.
Davis, G.D., N.E. Byer, R.A. Riedel and G. Margaritondo, 1985, J. Appl. Phys. **57**, 1919.
Daw, M.S., and D.L. Smith, 1980a, Appl. Phys. Lett. **8**, 690.
Daw, M.S., and D.L. Smith, 1980b, J. Vac. Sci. & Technol. **17**, 1028.
Dev, B.N., T. Thundat and W.M. Gibson, 1985, J. Vac. Sci. & Technol. A **3**, 946.
Dow, J.D., and R.E. Allen, 1982, J. Vac. Sci. & Technol. **20**, 659.
Duke, C.B., 1967, J. Vac. Sci. & Technol. **6**, 152.
Duke, C.B., 1969, Tunneling in Solids (Academic Press, New York).
Duke, C.B., A. Paton, R.J. Meyer, L.J. Brillson, A. Kahn, D. Kanani, J. Carelli, J.L. Yeh, G. Margaritondo and A.D. Katnani, 1981, Phys. Rev. Lett. **46**, 440.
Durbin, S.M., L.E. Berman and B.W. Batterman, 1985, J. Vac. Sci. & Technol. A **3**, 913.
Eastman, D.E., and J.L. Freeouf, 1974, Phys. Rev. Lett. **33**, 1601.
Eastman, D.E., and W.D. Grobman, 1972, Phys. Rev. Lett. **28**, 1376.

Eberhardt, W., G. Kalkoffen, C. Kunz, D. Aspnes and M. Cardona, 1978, Phys. Status Solidi B **88**, 135.
Eisenberger, P., and W.C. Marra, 1981, Phys. Rev. Lett. **46**, 1081.
Flodstrom, S.A., R.Z. Bachrach, R.S. Bauer and S.B. Hagstrom, 1976, Phys. Rev. Lett. **37**, 1282.
Franciosi, A., and J.H. Weaver, 1983a, Phys. Rev. B **27**, 3554.
Franciosi, A., and J.H. Weaver, 1983b, Surf. Sci. **132**, 324.
Franciosi, A., and J.H. Weaver, 1983c, Physica B **117/118**, 846.
Franciosi, A., J.H. Weaver and F.A. Schmidt, 1982a, Phys. Rev. B **26**, 546.
Franciosi, A., J.H. Weaver, D.G. O'Neill, Y. Chabal, J.E. Rowe, J.M. Poate, O. Bisi and C. Calandra, 1982b, J. Vac. Sci. & Technol. **21**, 624.
Franciosi, A., D.G. O'Neill and J.H. Weaver, 1983, J. Vac. Sci. & Technol. B **1**, 524.
Franciosi, A., P. Perfetti, A.D. Katnani, J.H. Weaver and G. Margaritondo, 1984, Phys. Rev. B **29**, 5611.
Franciosi, A., R. Reifenberger and J.K. Furdyna, 1985, J. Vac. Sci. & Technol. A **3**, 124.
Freeouf, J.L., 1980, Solid State Commun. **33**, 1059.
Freeouf, J.L., 1981, J. Vac. Sci. & Technol. **18**, 910.
Freeouf, J.L., and J.M. Woodall, 1981, Appl. Phys. Lett. **39**, 727.
Goodwin, T.A., and P. Mark, 1972, in: Progress in Surface Science, Vol. I (Pergamon, New York) p. 1.
Grant, R.W., J.R. Waldrop, S.P. Kowalczyk and E.A. Kraut, 1981, J. Vac. Sci. & Technol. **19**, 477.
Grioni, M., J. Joyce, M. del Guidice, D.G. O'Neill and J.H. Weaver, 1984a, Phys. Rev. B **30**, 7370.
Grioni, M., J. Joyce, S.A. Chambers, D.G. O'Neill, M. del Guidice and J.H. Weaver, 1984b, Phys. Rev. Lett. **53**, 2331.
Grioni, M., J. Joyce and J.H. Weaver, 1986, J. Vac. Sci. & Technol. A **4**, 965.
Gudat, W., and D.E. Eastman, 1976, J. Vac. Sci. & Technol. **13**, 831.
Gudat, W., and C. Kunz, 1972, Phys. Rev. Lett. **29**, 169.
Hansson, G.V., R.Z. Bachrach, R.S. Bauer and P. Chiaradia, 1981a, Phys. Rev. Lett. **46**, 1033.
Hansson, G.V., R.Z. Bachrach, R.S. Bauer and P. Chiaradia, 1981b, J. Vac. Sci. & Technol. **18**, 550.
Head, J.D., and K.A.R. Mitchell, 1978, Mol. Phys. **35**, 1681.
Helms, C.R., W.E. Spicer and N.M. Johnson, 1978, Solid State Commun. **25**, 673.
Henisch, H.K., 1984, Semiconductor Contacts: An Approach to Ideas & Models (Oxford Univ. Press, Oxford).
Henzler, M., 1970, Surf. Sci. **22**, 12.
Himpsel, F.J., P. Heimann, T.-C. Chiang and D.E. Eastman, 1980a, Phys. Rev. Lett. **45**, 1112.
Himpsel, F.J., P. Heimann and D.E. Eastman, 1980b, Solid State Commun. **36**, 631.
Hiraki, A., M.A. Nicolet and J.W. Mayer, 1971, Appl. Phys. Lett. **18**, 178.
Hiraki, A., K. Shuto, F. Kim, W. Kammura and W. Iwami, 1977, Appl. Phys. Lett. **31**, 611.
Hughes, G.J., A. McKinley, R.H. Williams and I.T. McGovern, 1982, J. Phys. C **15**, L159.
Hughes, G.J., A. McKinley and R.H. Williams, 1983, J. Phys. C **16**, 2391.
Huijser, A., and J. van Laar, 1975, Surf. Sci. **52**, 202.
Huijser, A., J. van Laar and T.L. Rooy, 1981, Surf. Sci. **102**, 264.
Johannson, L.I., and J. Stöhr, 1979, Phys. Rev. Lett. **43**, 1882.
Johannson, L.I., I. Lindau, M.H. Hecht and E. Kallne, 1980, Solid State Commun. **34**, 83 and references cited therein.
Kahn, A., 1983, Semiconductor Surface Structures, Surf. Sci. Rep. **3**, 193, and references cited therein.
Kahn, A., L.J. Brillson, G. Margaritondo and A.D. Katnani, 1981a, Solid State Commun. **38**, 1269.
Kahn, A., D. Kanani, J. Carelli, J.L. Yeh, C.B. Duke, R.J. Meyer, A. Paton and L.J. Brillson, 1981b, J. Vac. Sci. & Technol. **18**, 792.
Kahn, A., J. Carelli, D. Kanani, C.B. Duke, A. Paton and L.J. Brillson, 1981c, J. Vac. Sci. & Technol. **19**, 331.
Kajiyama, K., Y. Mizushima and S. Sakata, 1973, Appl. Phys. Lett. **23**, 458.
Kane, E.O., 1964, Phys. Rev. Lett. **12**, 97.

Katnani, A.D., P. Chiaradia, H.W. Sang Jr and R.S. Bauer, 1984, J. Vac. Sci. & Technol. B **2**, 471.
Kendelewicz, T., W.G. Petro, I.A. Babalola, I. Lindau and W.E. Spicer, 1983a, J. Vac. Sci. & Technol. B **1**, 564.
Kendelewicz, T., W.G. Petro, I.A. Babalola, J.A. Silberman, I. Lindau and W.E. Spicer, 1983b, J. Vac. Sci. & Technol. B **1**, 623.
Kendelewicz, T., W.G. Petro, I.A. Babalola, J.A. Silberman, I. Lindau and W.E. Spicer, 1983c, Phys. Rev. B **27**, 3366.
Kendelewicz, T., W.G. Petro, I. Lindau and W.E. Spicer, 1983d, Phys. Rev. B **28**, 3618.
Kendelewicz, T., W.G. Petro, I. Lindau and W.E. Spicer, 1984a, Phys. Rev. B **30**, 5800.
Kendelewicz, T., W.G. Petro, I. Lindau and W.E. Spicer, 1984b, J. Vac. Sci. & Technol. B **2**, 453.
Kendelewicz, T., W.G. Petro, I. Lindau and W.E. Spicer, 1984c, J. Vac. Sci. & Technol. A **2**, 542.
Kendelewicz, T., W.G. Petro, I. Lindau and W.E. Spicer, 1984d, Appl. Phys. Lett. **44**, 1066.
Kendelewicz, T., W.G. Petro, I. Lindau and W.E. Spicer, 1984e, J. Vac. Sci. & Technol. B **2**, 582.
Kendelewicz, T., N. Newman, R.S. List, I. Lindau and W.E. Spicer, 1985, J. Vac. Sci. & Technol. B **3**, 1206.
Kubaschewski, O., E.Ll. Evans and C.B. Alcock, 1967, Metallurgical Thermochemistry (Pergamon, Oxford).
Landgren, G., and R. Ludeke, 1984, Bull. Am. Phys. Soc. **29**, 552.
Lapeyre, G.J., and J. Anderson, 1975, Phys. Rev. Lett. **35**, 117.
Lapeyre, G.J., J. Anderson, P.L. Gobby and J.A. Knapp, 1974a, Phys. Rev. Lett. **33**, 1290.
Lapeyre, G.J., A.D. Baer, J. Anderson, J.C. Hermannson, J.A. Knapp and P.L. Goppy, 1974b, Solid State Commun. **15**, 1601.
Lee, S.T., G. Apai, M.G. Mason, R. Benhow and Z. Hurych, 1981, Phys. Rev. B **23**, 505.
LeLay, G., 1983a, J. Vac. Sci. & Technol. B **1**, 354.
LeLay, G., 1983b, Surf. Sci. **132**, 169.
Lenth, W., F. Lutz, J. Barth, G. Kalkoffen and C. Kunz, 1978, Phys. Rev. Lett. **41**, 1185.
Liang, K.S., W.R. Salaneck and I.A. Aksay, 1976, Solid State Commun. **19**, 329 and references cited therein.
Lindau, I., P.W. Chye, C.M. Garner, P. Pianetta, C.Y. Su and W.E. Spicer, 1978, J. Vac. Sci. & Technol. **15**, 1332.
List, R.S., T. Kendelewicz, M.D. Williams, I. Lindau and W.E. Spicer, 1985, J. Vac. Sci. & Technol. A **3**, 1002.
Louie, S.G., and M.L. Cohen, 1975, Phys. Rev. Lett. **35**, 866.
Louie, S.G., and M.L. Cohen, 1976, Phys. Rev. B **13**, 2461.
Ludeke, R., 1983, Surf. Sci. **132**, 143.
Ludeke, R., 1984, J. Vac. Sci. & Technol. **52**, 400.
Ludeke, R., and G. Landgren, 1986, Phys. Rev. B **33**, 5526.
Ludeke, R., T.-C. Chiang and D.E. Eastman, 1982, J. Vac. Sci. & Technol. **21**, 599.
Ludeke, R., T.-C. Chiang and T. Miller, 1983, J. Vac. Sci. & Technol. B **1**, 581.
Margaritondo, G., 1983, Solid-State Electron. **26**, 499.
Margaritondo, G., and A. Franciosi, 1984, Ann. Rev. Mater. Sci. **14**, 67.
Margaritondo, G., J.E. Rowe and S.B. Christman, 1976, Phys. Rev. B **14**, 5396.
Marra, W.C., P. Eisenberger and A.Y. Cho, 1979, J. Appl. Phys. **50**, 6927.
McGilp, J., 1984, J. Phys. C **17**, 2249.
McKinley, A., A. Park and R.H. Williams, 1980, J. Phys. C **13**, 6723.
McKinley, A., G.J. Hughes and R.H. Williams, 1982, J. Phys. C **15**, 7049.
Mead, C.A., 1966, Solid State Electron. **9**, 1023 and references cited therein.
Mills, K.C., 1974, Thermodynamic Data for Inorganic Sulphides, Selenides and Tellurides (Butterworths, London).
Milnes, A.G., 1973, Deep Impurities in Semiconductors (Wiley–Interscience, New York).
Milnes, A.G., 1983, Advance Electron. & Electron. Phys., **61**, 63.
Milnes, A.G., and D.L. Feucht, 1972, Heterojunctions and Metal–Semiconductor Junctions (Academic Press, New York) pp. 156–200.

Mönch, W., and H. Grant, 1982, Phys. Rev. Lett. **48**, 512.
Newman, N., W.G. Petro, T. Kendelewicz, S.H. Pan, S.J. Eglash and W.E. Spicer, 1985a, J. Appl. Phys. **57**, 1247.
Newman, N., K.K. Chin, W.G. Petro, T. Kendelewicz, M.D. Williams, C.E. McCants and W.E. Spicer, 1985b, J. Vac. Sci. & Technol. A **3**, 996.
Newman, N., T. Kendelewicz, D. Thomson, S.H. Pan, S.J. Eglash and W.E. Spicer, 1985c, Solid State Electron. **28**, 307.
Okamoto, K., C.E.C. Wood and L.F. Eastman, 1981, Appl. Phys. Lett. **38**, 636.
Orders, P.J., and C.S. Fadley, 1983, Phys. Rev. B **27**, 781.
Ottaviani, G., K.N. Tu and J.W. Mayer, 1980, Phys. Rev. Lett. **44**, 284.
Palmberg, P.W., G.K. Bohn and J.C. Tracy, 1969, Appl. Phys. Lett. **15**, 254.
Pauling, L., 1939, The Nature of the Chemical Bond (Cornell Univ. Press, Ithaca, NY).
Perfetti, P., A.D. Katnani, T.-X. Zhao, G. Margaritondo, O. Bisi and C. Calandra, 1982a, J. Vac. Sci. & Technol. **21**, 628.
Perfetti, P., A.D. Katnani, R.R. Daniels, T.-X. Zhao and G. Margaritondo, 1982b, Solid State Commun. **41**, 213.
Perfetti, P., G. Rossi, I. Lindau and O. Bisi, 1983, Surf. Sci. **124**, L19.
Peterman, D.J., and A. Franciosi, 1984, Appl. Phys. Lett. **42**, 1305.
Petro, W.G., I.A. Babalola, P. Skeath, C.Y. Su, I. Hino, I. Lindau and W.E. Spicer, 1982, J. Vac. Sci. & Technol. **21**, 585.
Petro, W.G., T. Kendelewicz, I.A. Babalola, I. Lindau and W.E. Spicer, 1984, J. Vac. Sci. & Technol. A **2** 835.
Philip, P., A. Franciosi and D.J. Peterman, 1985, J. Vac. Sci. & Technol. A **3**, 1007.
Phillips, J.C., and J.A. Van Vechten, 1969, Phys. Rev. Lett. **23**, 115.
Phillips, J.C., and J.A. Van Vechten, 1970, Phys. Rev. B **2**, 2147.
Richter, H.W., L.J. Brillson, M.K. Kelly, R.R. Daniels and G. Margaritondo, 1984, J. Vac. Sci. & Technol. B **2**, 591.
Robinson, G.Y., 1975, Solid State Electron. **18**, 331.
Rossi, G., I. Abbati, L. Braicovitch, I. Lindau and W.E. Spicer, 1981, Surf. Sci. Lett. **112**, L765.
Rossi, G., I. Abbati, L. Braicovitch, I. Lindau and W.E. Spicer, 1982a, Phys. Rev. B **25**, 3619.
Rossi, G., I. Abbati, L. Braicovitch, I. Lindau and W.E. Spicer, 1982b, Phys. Rev. B **25**, 3627.
Rossi, G., I. Abbati, I. Lindau and W.E. Spicer, 1982c, Appl. Surf. Sci. **11/12**, 348.
Rossi, G., J. Nogami, I. Lindau, L. Braicovitch, I. Abbati, U. del Pennino and S. Nannarone, 1983, J. Vac. Sci. & Technol. A **1**, 781.
Rowe, E.M., 1981, Physics Today **34**, 28.
Rowe, J.E., 1976, J. Vac. Sci. & Technol. **13**, 798.
Rowe, J.E., S.B. Christman and G. Margaritondo, 1975, Phys. Rev. Lett. **35**, 1471.
Rubloff, G.W., 1983, Surf. Sci. **132**, 268.
Schockley, W., and G.L. Pearson, 1948, Phys. Rev. **74**, 232.
Schwartz, G.P., J.E. Griffiths and B. Schwartz, 1979, J. Vac. Sci. & Technol. **16**, 1383.
Seah, M.P., and W.A. Dench, 1979, Surf. & Interface Anal. **1**, 2.
Shapira, Y., and L.J. Brillson, 1983, J. Vac. Sci. & Technol. B **1**, 618.
Shapira, Y., L.J. Brillson, A.D. Katnani and G. Margaritondo, 1984, Phys. Rev. B **30**, 4586.
Silberman, J.A., P. Morgan, I. Lindau, W.E. Spicer and J.A. Wilson, 1982, J. Vac. Sci. & Technol. **21**, 154.
Sinha, A.K., and J.M. Poate, 1973, Appl. Phys. Lett. **23**, 666.
Sinha, A.K., and J.M. Poate, 1978, in: Thin Films – Interdiffusion and Reactions, eds J.M. Poate, K.N. Tu and J.W. Mayer (Wiley–Interscience, New York) p. 407 and references cited therein.
Skeath, P., I. Lindau, C.Y. Su, P.W. Chye and W.E. Spicer, 1980a, J. Vac. Sci. & Technol. **17**, 511.
Skeath, P., I. Lindau, C.Y. Su, P.W. Chye and W.E. Spicer, 1980b, J. Vac. Sci. & Technol. **17**, 874.
Skeath, P., I. Lindau, C.Y. Su and W.E. Spicer, 1981, J. Vac. Sci. & Technol. **19**, 556.
Spicer, W.E., and S.J. Eglash, 1985, in: Surface and Interface Effects in VLSI, Vol. 10, ed. R.S. Bauer (Academic Press, New York) p. 79.
Spicer, W.E., I. Lindau, P.E. Gregory, C.M. Garner, P. Pianetta and P.W. Chye, 1976, J. Vac. Sci. & Technol. **13**, 780.

Spicer, W.E., P.W. Chye, P.R. Skeath, C.Y. Su and I. Lindau, 1979a, J. Vac. Sci. & Technol. 16, 1422.
Spicer, W.E., P.W. Chye, C.M. Garner, I. Lindau and P. Pianetta, 1979b, Surf. Sci. 86, 763.
Spicer, W.E., I. Lindau, P. Skeath and C.Y. Su, 1980a, J. Vac. Sci. & Technol. 17, 1019.
Spicer, W.E., I. Lindau, P. Skeath and C.Y. Su, 1980b, Phys. Rev. Lett. 44, 420.
Spicer, W.E., J.A. Silberman, P. Morgan, I. Lindau and J.A. Wilson, 1982a, J. Vac. Sci. & Technol. 21, 149.
Spicer, W.E., S. Eglash, I. Lindau, C.Y. Su and P.R. Skeath, 1982b, Thin Solid Films 89, 447.
Spicer, W.E., J.A. Silberman, I. Lindau, A.-B. Chen, A. Sher and J.A. Wilson, 1983, J. Vac. Sci. & Technol. A 1, 1735.
Spicer, W.E., S. Pan, D. Mo, N. Newman, P. Mahowald, T. Kendelewicz and S.J. Eglash, 1984, J. Vac. Sci. & Technol. B 2, 476.
Stoffel, N.G., R.R. Daniels, G. Margaritondo, C.F. Brucker and L.J. Brillson, 1982, J. Vac. Sci. & Technol. 20, 701.
Stoffel, N.G., M.K. Kelly and G. Margaritondo, 1983, Phys. Rev. B 27, 6561.
Stoffel, N.G., M. Turowski and G. Margaritondo, 1984, Phys. Rev. B 30, 3294.
Stöhr, J., and R. Jaeger, 1982, J. Vac. Sci. & Technol. 21, 619.
Swartz, C.A., J.J. Barton, W.A. Goddard and T.C. McGill, 1980, J. Vac. Sci. & Technol. 17,. 869.
Sze, S.M., 1981, Physics of Semiconductor Devices, 2nd Ed. (Wiley, New York) ch. 5.
Teal, G.K., 1976, IEEE Trans. Electron Devices ED-23, 621 and references cited therein.
Tersoff, J., 1984, Phys. Rev. Lett. 57, 465.
Thanailakis, A., 1975, J. Phys. C 8, 655.
Turner, M.J., and E.H. Rhoderick, 1968, Solid State Electron. 11, 291.
van der Veen, J.F., L. Smit, P.K. Larsen, J.H. Neave and B.A. Joyce, 1982, J. Vac. Sci. & Technol. 21, 375.
van Laar, J., and A. Huijser, 1976, J. Vac. Sci. & Technol. 13, 769.
van Laar, J., and J.J. Scheer, 1967, Surf. Sci. 8, 342.
Wagman, D.D., W.H. Evans, V.B. Parker, I. Halow, S.M. Bailey and R.H. Schumm, 1968/1971, Natl. Bur. Std. Technical Notes 270 3 270 7 (U.S. Government Printing Office, Washington, DC).
Wagner, L.F., and W.E. Spicer, 1972, Phys. Rev. Lett. 28, 1381.
Weaver, J.H., and G. Margaritondo, 1979, Science 206, 151.
Weaver, J.H., V.L. Moruzzi and F.A. Schmidt, 1981, Phys. Rev. B 23, 2916.
Weaver, J.H., A. Franciosi and V.L. Moruzzi, 1984, Phys. Rev. B 29, 3293.
Weaver, J.H., M. Grioni, J.J. Joyce and M. del Guidice, 1985, Phys. Rev. B 31, 5290.
Wieder, H.H., 1978, J. Vac. Sci. & Technol. 15, 1498.
Wieder, H.H., 1981, Appl. Phys. Lett. 38, 170.
Williams, R.H., 1981, J. Vac. Sci. & Technol. 18, 929.
Williams, R.H., 1982, Contemp. Phys. 23, 329.
Williams, R.H., 1983, Surf. Sci. 132, 122.
Williams, R.H., R.R. Varma and A. McKinley, 1977, J. Phys. C 10, 4545.
Williams, R.H., V. Montgomery and R.R. Varma, 1978, J. Phys. C 11, L735.
Williams, R.H., R.R. Varma and V. Montgomery, 1979, J. Vac. Sci. & Technol. 16, 1418.
Williams, R.H., A. McKinley, G.J. Hughes, V. Montgomery and I.T. McGovern, 1982, J. Vac. Sci. & Technol. 21, 594.
Williams, R.H., A. McKinley, G.J. Hughes, T.P. Humphreys and C. Maani, 1984, J. Vac. Sci. & Technol. B 2, 561.
Woodall, J.M., and J.L. Freeouf, 1982, J. Vac. Sci. & Technol. 21, 574.
Woodall, J.M., G.D. Pettit, T.N. Jackdon, C. Lanza, K.L. Kavanaugh and J.W. Mayer, 1983, Phys. Rev. Lett. 31, 1783.
Zhang, H.I., and M. Schlüter, 1978, Phys. Rev. B 15, 1923.
Zhao, T.-X., R.R. Daniels, A.D. Katnani, G. Margaritondo and A. Zunger, 1983, J. Vac. Sci. & Technol. B 1, 610.
Zunger, A., 1981, Phys. Rev. B 24, 4372.
Zur, A., T.C. McGill and D.L. Smith, 1983a, J. Vac. Sci. & Technol. A 1, 608.
Zur, A., T.C. McGill and D.L. Smith, 1983b, Phys. Rev. B 28, 2060.

CHAPTER 9

INNER SHELL PHOTOELECTRON PROCESS IN SOLIDS

AKIO KOTANI

Department of Physics, Faculty of Science, Tohoku University,
Sendai 980, Japan

Contents

1. Introduction 613
2. Fundamental theory of many-body effects in inner shell photoelectron process . . 614
   2.1. General description 614
   2.2 Generating function and dielectric response 616
   2.3. Photoelectron spectrum and its limiting forms 617
      2.3.1. Slow modulation limit 618
      2.3.2. Rapid modulation limit 619
   2.4. Typical examples of photoelectron spectra 619
      2.4.1. Lattice relaxation effect 619
      2.4.2. Plasmon satellite 620
      2.4.3. Orthogonality catastrophe 621
      2.4.4. Shake-up satellite 621
      2.4.5. Lifetime effect 622
3. Simple metals 623
4. Rare earth metals and compounds 627
   4.1. Exchange splitting 627
   4.2. Satellite in La metal and its compounds 629
   4.3. Mixed valence state in Ce compounds 635
   4.4. Other rare earth systems 639
5. Transition metals and compounds 639
   5.1. Transition metals 639
   5.2. Transition metal compounds 644
6. Related topics 650
   6.1. Satellite in 3d photoemission of Ni 650

Contents continued overleaf

Handbook on Synchrotron Radiation, Vol. 2, edited by G.V. Marr
© *Elsevier Science Publishers B.V., 1987*

Contents continued

    6.2. Resonant photoemission 652
7. Concluding remarks 656
Appendix: Derivation of basic photoemission formulae 657
References 660

1. Introduction

In the inner shell photoelectron spectroscopy in solids, a core electron is excited by an incident photon, and the excited electron is taken out of the solid as a photoelectron. In the first approximation, the photoemission process is described as a simple *one-electron process*, and *many-body effects* give the correction to the one-electron approximation. Historically, the inner shell photoelectron spectroscopy has been developed from the study of the one-electron process, but now the photoelectron spectroscopy has proved to be a very powerful technique in the study of the many-body effects.

Within the one-electron approximation, the kinetic energy ε of the photoelectron is expressed, from the energy conservation law, as

$$\varepsilon = \hbar\omega_i + \varepsilon_c, \tag{1}$$

where $\hbar\omega_i$ is the incident photon energy, and ε_c is the one-electron energy of the relevant core state measured from the vacuum level. Theoretically, ε_c is obtained by the Hartree–Fock approximation, and eq. (1) represents the *Koopman's theorem*. On the other hand, experimentally ε is observed with a known value of $\hbar\omega_i$, and the core level ε_c can be determined from eq. (1). Since the value of ε_c in solids is not very different from the corresponding free atom value, which is characteristic in each element, the inner shell photoelectron spectroscopy can be used for the analysis of elements. Furthermore, the small deviation of ε_c from its free atom value, which is denoted by the *chemical shift*, provides us with the information on the chemical bonding of valence electrons. The application of the chemical shift to the chemical analysis of molecules and solids was extensively made by Siegbahn and his group (Siegbahn et al. 1967), and their study played an important role in the development of the inner shell photoelectron spectroscopy.

However, Koopman's theorem does not generally hold because of the many-body effect. Owing to the recent progress in the experimental technique, the many-body effect in the photoelectron spectrum can be observed with sufficient accuracy as an *asymmetry* of spectral shape and as a *satellite* structure in various materials. The many-body effect occurs from the interaction between the outer electrons (i.e., the valence electrons) and the core hole left behind in the final state of the photoemission. Let us denote the outer electron system (including the lattice system) as "medium". The medium is *polarized* by the core hole charge and *screens* it. Thus, the core hole plays a role of "test charge" which induces the many-body dielectric response of the medium. The core hole has not only a charge but also a spin, so that it can also play a role of "test spin" to induce the magnetic response of the medium. The dynamics of the many-body response of the medium to the core hole are directly reflected in the observed inner shell photoelectron spectrum.

A well-known example, which demonstrates the importance of the many-body response of the medium, is the singularity in the inner shell photoelectron spectrum of simple metals. As shown by Anderson (1967) and by Nozières and De Dominicis (1969), the screening effect of conduction electrons around the core hole in simple metals gives rise to an asymmetric photoemission line shape diverging at the threshold, due to the "orthogonality catastrophe". From this fact, the inner shell photoelectron spectrum is found to be very sensitive to the many-body effect of outer electrons. More interesting is the recent study of many-body effects in *magnetic materials*, which include the incomplete 3d and 4f states. In these materials the electrons in the incomplete shells couple very strongly with the core hole, resulting in the characteristic splitting of inner shell photoelectron spectra, from which we can obtain important information on the mixed valence property of 4f electrons, the electron correlation effect in the 3d band, and so on.

In this chapter, we describe the many-body effects in the inner shell photoelectron process. Starting from a general and fundamental description of the photoelectron process including many-body effects, we explain how the photoelectron spectra observed in various materials reflect various many-body processes, emphasizing the most essential physical pictures. Main space is devoted to the theoretical aspects, but typical examples of experimental data are also introduced.

In section 2 we give general descriptions of the inner shell photoelectron spectrum and their relationship with the many-body response of the medium, without specifying the material. Typical examples of the spectral shape are also given. The analysis of the photoelectron spectra in various materials are described in sections 3–5; sections 3, 4 and 5 are devoted to simple metals, rare earth metals and compounds, and transition metals and compounds, respectively. A satellite of 3d photoelectron spectrum in Ni and its resonance enhancement effect are discussed in section 6, since they are the topics intimately related with the inner shell photoemission discussed in section 5. Brief concluding remarks are given in section 7.

As references closely related with this article, we refer the two volumes of "Photoemission in Solids I and II" (Cardona and Ley 1978, Ley and Cardona 1979) and the review articles by Kotani and Toyozawa (1979) and by Almbladh and Hedin (1983). An outline of the present article has also been published as a brief review paper (Kotani 1983).

2. *Fundamental theory of many-body effects in inner shell photoelectron process*

2.1. *General description*

We show here, from a general viewpoint, how the response of outer electrons to the core hole is reflected in the inner shell photoelectron spectrum. The system of outer electrons (including photons) is denoted as "medium", separated from the core electrons and the photoelectron. The Hamiltonian of the medium in the initial state of the photoemission is formally written as H_0, and its ground state is written as $|0\rangle$ with energy E_0, i.e.,

$$H_0|0\rangle = E_0|0\rangle . \tag{2}$$

In the final state of the photoemission a core hole is left behind, so that the Hamiltonian of the medium should be written as

$$H = H_0 + U , \tag{3}$$

where U represents any disturbance (charge interaction or spin interaction) caused by the core hole. We write the eigenstates of H as $|f\rangle$ with energy E_f, i.e.,

$$H|f\rangle = E_f|f\rangle . \tag{4}$$

The inner shell photoelectron spectrum is expressed (see Appendix) in the form

$$F = \sum_f |\langle 0|f\rangle|^2 \, \delta(\varepsilon - \varepsilon_c + E_f - E_0 - \hbar\omega_i) . \tag{5}$$

This equation is rewritten in the following integral form:

$$F = \langle 0|\delta(\varepsilon - \varepsilon_c + H - E_0 - \hbar\omega_i)|0\rangle$$

$$= \frac{1}{\pi\hbar} \, \mathrm{Re} \int_0^\infty dt \, \exp\!\left(i\,\frac{E_B}{\hbar}\,t\right) g(t) , \tag{6}$$

where

$$E_B = \hbar\omega_i - \varepsilon + \varepsilon_c , \tag{7}$$

$$g(t) = \langle 0|f(t)\rangle , \tag{8}$$

$$|f(t)\rangle = \exp\!\left[-\frac{i}{\hbar}(H - E_0)t\right]|0\rangle . \tag{9}$$

For $t \geq 0$, $|f(t)\rangle$ represents the time variation of the medium caused by the interaction with the core hole, which is suddenly switched on at $t = 0$. Thus, the function $g(t)$ reflects directly the *many-body response of the medium* to the core hole, as the overlap integral between $|f(t)\rangle$ and $|0\rangle$, and the inner shell photoelectron spectrum F is just the Fourier transform of $g(t)$. We denote $g(t)$ as the *generating function*.

In the following, we put $\langle 0|U|0\rangle = 0$ by assuming that $\langle 0|U|0\rangle$ is already included in ε_c. Then, U describes the relaxation effect (or the redistribution effect) of the medium caused by the core hole. When we disregard the relaxation effect by putting $U = 0$, the generating function $g(t)$ is always unity, so that the photoelectron spectrum F reduces to $\delta(\varepsilon - \varepsilon_c - \hbar\omega_i)$ and eq. (1) is restored. When $U \neq 0$, the time variation of $g(t)$ is represented formally, from eqs. (8) and (9), in the form

$$g(t) = \langle 0|T \exp\left[-\frac{i}{\hbar}\int_0^t dt' \, U(t')\right]|0\rangle . \tag{10}$$

In eq. (10), T is the time ordering operator and $U(t)$ is defined by

$$U(t) = \exp\left(i\frac{H_0}{\hbar}t\right) U \exp\left(-i\frac{H_0}{\hbar}t\right) . \tag{11}$$

2.2. Generating function and dielectric response

By using the second order cumulant approximation in eq. (10), the generating function $g(t)$ is expressed as

$$g(t) = \exp\left[-\frac{1}{\hbar^2}\int_0^t dt_1 \int_0^{t_1} dt_2 \, C(t_1 - t_2)\right], \tag{12}$$

where $C(t_1 - t_2)$ is the *correlation function* of U defined by

$$C(t_1 - t_2) = \langle 0|U(t_1) \, U(t_2)|0\rangle . \tag{13}$$

In many cases this approximation is considerably good and in some special cases (see section 2.4.1) eq. (12) holds exactly.

We apply this formula to the *dielectric response* of the medium. When we take account of the Coulomb potential of the core hole charge, U is given by

$$U = \frac{1}{V}\sum_q \rho_q \phi_{-q} , \tag{14}$$

where V is the volume of the system, ρ_q is the q-component of the polarization charge of the medium and

$$\phi_q = 4\pi e/q^2 , \quad (e > 0) . \tag{15}$$

Hereafter we choose the volume of the system as unity for simplicity of description. Substituting eq. (14) into eq. (13), we have

$$C(t_1 - t_2) = \sum_q \phi_q^2 \langle 0|\rho_q(t_1) \, \rho_{-q}(t_2)|0\rangle . \tag{16}$$

In order to extend the theory to the finite temperature case, the ground state expectation value in eq. (16) is replaced by the thermal average $\langle \rho_q(t_1) \rho_{-q}(t_2) \rangle$. We use the following relation between the charge correlation function and the dielectric function $\varepsilon(q, \omega)$:

$$\langle \rho_q(t_1) \rho_{-q}(t_2) \rangle = -\frac{\hbar q^2}{4\pi^2} \int_{-\infty}^{\infty} d\omega \, \frac{e^{-i\omega(t_1-t_2)}}{1-e^{-\beta\hbar\omega}} \, \text{Im}\left[\frac{1}{\varepsilon(q,\omega)}\right], \tag{17}$$

where $\beta = 1/k_B T$. The generating function is finally expressed in the form

$$g(t) = \exp\left\{\frac{1}{4\pi^2 \hbar} \sum_q q^2 \phi_q^2 \int_{-\infty}^{\infty} d\omega \, \frac{-i\omega t + 1 - e^{-i\omega t}}{\omega^2(1-e^{-\beta\hbar\omega})} \, \text{Im}\left[\frac{1}{\varepsilon(q,\omega)}\right]\right\}. \tag{18}$$

2.3. Photoelectron spectrum and its limiting forms

We define here several basic quantities which characterize the features of the photoelectron spectrum. First, a *spectral function* $J(\omega)$ is defined by the Fourier transform of the correlation function $C(t)$:

$$J(\omega) = \frac{1}{2\pi} \int_{-\infty}^{\infty} dt \, C(t) \, e^{i\omega t}. \tag{19}$$

Then, eq. (12) is written as

$$g(t) = \exp\left[-\frac{1}{\hbar^2} \int_{-\infty}^{\infty} d\omega \, J(\omega) \, \frac{-i\omega t + 1 - e^{-i\omega t}}{\omega^2}\right], \tag{20}$$

so that the photoelectron spectrum is expressed as

$$F = \frac{1}{\pi\hbar} \text{Re} \int_0^{\infty} dt \, \exp\left[i\frac{E_B - \Delta}{\hbar} t - S + \int_{-\infty}^{\infty} d\omega \, \frac{J(\omega)}{(\hbar\omega)^2} e^{-i\omega t}\right]$$

$$= \sum_{n=0}^{\infty} F_n(E_B), \tag{21}$$

where

$$F_n(E_B) = \begin{cases} e^{-S} \delta(E_B - \Delta), & \text{for } n=0, \\ \frac{1}{n!} e^{-S} \int_{-\infty}^{\infty} d\omega_1 \sim d\omega_n \, \frac{J(\omega_1) \cdots J(\omega_n)}{(\hbar\omega_1)^2 \cdots (\hbar\omega_n)^2} \delta(E_B - \Delta - \hbar\omega_1 - \cdots - \hbar\omega_n), & \text{for } n \geq 1. \end{cases} \tag{22}$$

Here, E_B represents the binding energy which is measured from the core electron

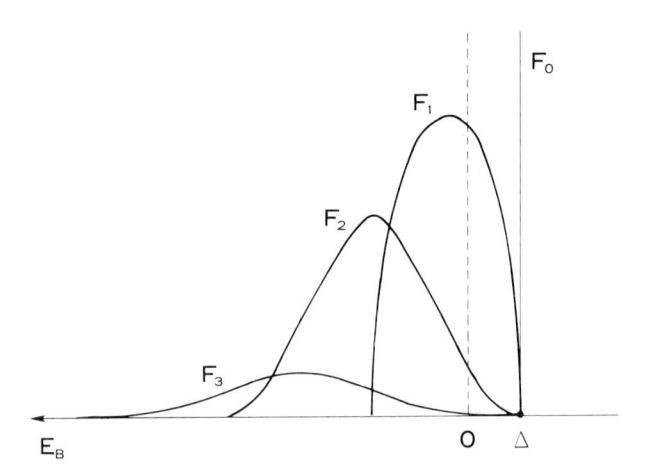

Fig. 1. Schematic shape of decomposed photoelectron spectra $F_0 \sim F_3$ as a function of the binding energy.

binding energy in the case of $U = 0$ (see the definition of E_B in eq. (7)). In eqs. (21) and (22), S and Δ are defined by

$$S = \int_{-\infty}^{\infty} d\omega \, \frac{J(\omega)}{(\hbar\omega)^2}, \tag{23}$$

$$\Delta = -\int_{-\infty}^{\infty} d\omega \, \frac{J(\omega)}{\hbar\omega}. \tag{24}$$

We remark that $\exp(-S)$ is the intensity of "zero line" and Δ represents the *relaxation energy* of the medium due to the interaction U. The behavior of $F_n(E_B)$ is shown schematically in fig. 1. In the case of $U = 0$, we have $F = \delta(E_B)$ as shown with the dashed line. When $U \neq 0$, the discrete line F_0 (the zero line) shifts by Δ with the intensity reduced to $\exp(-S)$, and the sideband structures F_n occur. The integrated intensity of F_n obeys the *Poisson distribution*:

$$\int_{-\infty}^{\infty} dE_B \, F_n(E_B) = e^{-S} \frac{S^n}{n!}. \tag{25}$$

In the following we give two limiting forms of $F(E_B)$, in the *slow and rapid modulation limits*. To this end, we define two characteristic times. One is the decay time τ_c of the correlation function $C(t)$ and the other is the decay time τ_g of the generating function $g(t)$.

2.3.1. Slow modulation limit
When $\tau_c \gg \tau_g$, $C(t)$ changes scarcely in the time interval $0 \leq t \leq \tau_g$, so that we can put $C(t) \simeq C(0)$ in the integrand of eq. (12). Then, $g(t)$ can be approximated by the *Gaussian form*,

$$g(t) = \exp\left(-\frac{D^2}{2\hbar^2}t^2\right), \tag{26}$$

so that $F(E_B)$ is also given by the *Gaussian function*,

$$F(E_B) = \frac{1}{\sqrt{2\pi D^2}} \exp\left(-\frac{E_B^2}{2D^2}\right), \tag{27}$$

where

$$D^2 = C(0) = \int_{-\infty}^{\infty} d\omega\, J(\omega). \tag{28}$$

Since τ_g is estimated as $\tau_g \simeq \hbar/D$, the condition $\tau_c \gg \tau_g$ is equivalent to $D\tau_c/\hbar \gg 1$.

2.3.2. Rapid modulation limit
When $\tau_c \ll \tau_g$, $C(t)$ decays so rapidly that we can approximate $J(\omega)$ as a constant in the integrand of eq. (20). We define

$$J(\omega) = \hbar\Gamma/\pi, \tag{29}$$

where

$$\Gamma = D^2\tau_c/\hbar, \tag{30}$$

then $g(t)$ decays exponentially as

$$g(t) = \exp\left(-\frac{\Gamma}{\hbar}t\right). \tag{31}$$

Therefore, the Fourier transform of eq. (31) gives the photoelectron spectrum in the *Lorentzian form*,

$$F(E_B) = \frac{\Gamma/\pi}{E_B^2 + \Gamma^2}. \tag{32}$$

2.4. Typical examples of photoelectron spectra

2.4.1. Lattice relaxation effect
In ionic crystals, the core hole charge is screened by the lattice polarization in the final state of the photoemission. Let us assume that the medium is a diatomic ionic crystal, and take account of the lattice relaxation due to the optical phonon. The dielectric function of the medium is given by

$$\varepsilon(\omega) = \frac{\varepsilon_\infty(\omega_\ell^2 - \omega^2)}{\omega_t^2 - \omega^2}, \tag{33}$$

so that we obtain

$$\text{Im}\left[\frac{1}{\varepsilon(\omega)}\right] = -\frac{\pi}{2}\omega_\ell\left(\frac{1}{\varepsilon_\infty} - \frac{1}{\varepsilon_0}\right)[\delta(\omega - \omega_\ell) - \delta(\omega + \omega_\ell)]. \tag{34}$$

Here, ε_∞ and ε_0 are, respectively, the high frequency and static dielectric constants, and ω_ℓ and ω_t are, respectively, frequencies of longitudinal and transverse optical phonons, whose dispersion is disregarded. In this system, the correlation time τ_c is estimated to be of the order of $1/\omega_\ell$, and if we assume $\beta\hbar\omega_\ell \ll 1$, the condition, $\tau_c \gg \tau_g$, of the *slow modulation limit* is usually satisfied. Then, the photoelectron spectrum $F(E_B)$ is given by the *Gaussian function* of eq. (27), where D^2 is obtained from eqs. (16), (17), (28) and (34) as

$$D^2 = \frac{1}{4\pi\beta}\left(\frac{1}{\varepsilon_\infty} - \frac{1}{\varepsilon_0}\right)\sum_q q^2 \phi_q^2. \tag{35}$$

Some remarks are given in the following.

(i) The slow modulation limit corresponds to the classical limit, where the Gaussian spectrum is easily obtained by the *Franck–Condon principle* with the configuration coordinate model.

(ii) The above result (eqs. 27 and 35) is also obtained by taking the interaction U as the usual *Fröhlich-type* interaction. Expression (12) holds exactly when U is linear with respect to the phonon (creation and annihilation) operator (e.g., Fröhlich-type interaction, deformation potential-type interaction, etc.).

(iii) When the condition $\tau_c \gg \tau_g$ is satisfied, the Gaussian broadening of the photoelectron spectrum occurs irrespective of the type of the medium (insulators or metals) and the type of the relevant phonon (optical or acoustic phonon).

2.4.2. Plasmon satellite

In the case that the medium is an electron gas, the dielectric function is expressed, by the RPA approximation, as

$$\varepsilon(q, \omega) = 1 + \frac{4\pi e^2}{q^2}\sum_k \frac{f(\varepsilon_k) - f(\varepsilon_{k+q})}{\varepsilon_{k+q} - \varepsilon_k - \hbar\omega}, \tag{36}$$

where $\varepsilon_k = (\hbar k)^2/2m$, and $f(\varepsilon_k)$ is the Fermi distribution function. Here we consider the contribution from the collective mode, i.e., the *plasmon*. By disregarding the dispersion of the plasmon, we obtain from eq. (36)

$$\varepsilon(q, \omega) \sim \varepsilon(0, \omega) = 1 - \left(\frac{\omega_p}{\omega}\right)^2, \tag{37}$$

where ω_p is the plasma frequency given by

$$\omega_p = (4\pi n e^2/m)^{1/2} \tag{38}$$

with the electron density n. Therefore, we have

$$J(\omega) = -\frac{\hbar}{4\pi^2} \sum_q q^2 \phi_q^2 \frac{1}{1-e^{-\beta\hbar\omega}} \text{Im}\left[\frac{1}{\varepsilon(\mathbf{q},\omega)}\right] \sim \frac{\hbar\omega_p}{8\pi} \sum_q q^2 \phi_q^2 \delta(\omega - \omega_p) \tag{39}$$

by assuming $\beta\hbar\omega_p \gg 1$, and from eqs. (21) and (22) we obtain

$$F(E_B) = e^{-S} \sum_{n=0}^{\infty} \frac{S^n}{n!} \delta(E_B - \Delta - n\hbar\omega_p), \tag{40}$$

where

$$S = \frac{1}{8\pi\hbar\omega_p} \sum_q q^2 \phi_q^2, \qquad \Delta = -\hbar\omega_p S.$$

Thus, the simultaneous excitation of the plasmon is found to give the satellite (at $E_B = \Delta + n\hbar\omega_p$) in the photoelectron spectrum.

2.4.3. Orthogonality catastrophe

The low-lying excitation mode of the electron gas is the individual electron–hole pair excitation, which screens the core hole charge. From eq. (36), we see

$$\text{Im}\left[\frac{1}{\varepsilon(\mathbf{q},\omega)}\right] \propto \omega, \quad \text{for} \quad \omega \to 0,$$

so that from eq. (18) it is found that $g(t)$ is expressed for large t in the asymptotic form

$$g(t) \propto t^{-\gamma} \tag{41}$$

with a positive constant γ. Then, the photoelectron spectrum $F(E_B)$ diverges at the threshold in the inverse power form with respect to the binding energy measured from the threshold. This spectral anomaly is called *orthogonality catastrophe*, since it originates from that the ground state $|0\rangle$ of H_0 is orthogonal with the ground state (say, $|\hat{0}\rangle$) of H (Anderson 1967), as seen from eq. (41):

$$\langle 0|\hat{0}\rangle = \lim_{t \to \infty} g(t) = 0.$$

More details on the orthogonality catastrophe will be given in section 3.

2.4.4. Shake-up satellite

The electron–hole pair excitation in metals starts from $\omega = 0$, but in semiconductors and insulators the energy of electronic excitation modes (exciton and band-to-band excitation) is finite. When we express the energy of the *longitudinal excitation mode* with wave vector $\mathbf{q}$ as $\varepsilon_{ex}(\mathbf{q})$, we have

$$\text{Im}\left[\frac{1}{\varepsilon(\mathbf{q},\omega)}\right] \propto \delta(\omega - \varepsilon_{ex}(\mathbf{q})),$$

so that the simultaneous excitation of the longitudinal exciton or band-to-band electron excitation gives a satellite of the photoelectron spectrum at E_B separated by ε_{ex} from the main line. This is denoted by the *shake-up satellite*. The intra-atomic or intra-molecular electronic excitation also gives the shake-up satellite, although the formalism with the dielectric function is not appropriate in this case. For the system consisting of molecular orbitals, a mechanism of "photo-induced covalency", which will be discussed in section 5.2, describes the shake-up satellite.

2.4.5. Lifetime effect

(a) *Auger transition*

The core hole has a finite lifetime due to the *Auger transition*, where two occupied electrons with energies ε_i and ε_j are scattered by the Coulomb interaction to the core level ε_c and a state with ε_k above the ionization threshold. The Auger transition is decribed by the interaction

$$U' = \sum_{i,j,k} v_A(ij; ck) a_k^+ a_c^+ a_i a_j + \text{h.c.} , \tag{42}$$

where a_l^+ and a_l ($l = i, j, c$ and k) are the creation and annihilation operators of the state l, respectively, and

$$v_A(ij; ck) = \int d\mathbf{r}\, d\mathbf{r}'\, \phi_k^*(\mathbf{r})\, \phi_c^*(\mathbf{r}')\, \frac{e^2}{|\mathbf{r} - \mathbf{r}'|}\, \phi_j(\mathbf{r}')\, \phi_i(\mathbf{r})$$

$$- \int d\mathbf{r}\, d\mathbf{r}'\, \phi_k^*(\mathbf{r})\, \phi_c^*(\mathbf{r}')\, \frac{e^2}{|\mathbf{r} - \mathbf{r}'|}\, \phi_i(\mathbf{r}')\, \phi_j(\mathbf{r}) , \tag{43}$$

with the wave functions ϕ_l of the states l. Since U' includes explicitly the creation operator of the core state c, we apply our formalism to this case after including $\varepsilon_c a_c^+ a_c$ in H_0 and H of eqs. (2) and (3). Then the spectral function is obtained as

$$J(\omega) = \frac{1}{2\pi} \int_{-\infty}^{\infty} dt \langle 0|a_c^+ U'(t_1) U'(t_2) a_c|0\rangle\, e^{i\omega(t_1 - t_2)}$$

$$= \sum_{i,j,k} |v_A(ij; ck)|^2\, \delta\!\left(\omega - \frac{\varepsilon_k + \varepsilon_c - \varepsilon_i - \varepsilon_j}{\hbar}\right). \tag{44}$$

It is to be remarked that the Auger transition corresponds usually to the *rapid modulation* case with a very small τ_c, because the excited electron with ε_k has a large velocity and moves very rapidly away from the relevant core electron site. Therefore, the photoelectron spectrum is expressed by the *Lorentzian function* of eq. (32) with

$$\Gamma_A = \frac{\pi}{\hbar} J(0) = \pi \sum_{i,j,k} |v_A(ij; ck)|^2\, \delta(\varepsilon_i + \varepsilon_j - \varepsilon_k - \varepsilon_c) . \tag{45}$$

(b) *Radiative transition*
The core hole has also a lifetime due to the radiative transition of any occupied electrons i to the core state c. The radiative interaction is expressed by the dipole approximation as

$$U' = \sum_{i,q,\lambda} v_R(i; cq\lambda) a_c^+ a_i b_{q\lambda}^+ + \text{h.c.} , \qquad (46)$$

where

$$v_R(i; cq\lambda) = \frac{e}{m} \sqrt{\frac{2\pi\hbar}{cq}} \langle \phi_c | \boldsymbol{p} \cdot \boldsymbol{\eta}_{q\lambda} | \phi_i \rangle . \qquad (47)$$

Here, $b_{q\lambda}^+$ is the creation operator of a photon with wave vector $\boldsymbol{q}$ and polarization λ ($\boldsymbol{\eta}_{q\lambda}$ being the unit vector in the polarization direction). The radiative transition is also a very rapid process, and from the calculation similar to that of the Auger transition the photoelectron spectrum is given by the Lorentzian function with

$$\Gamma_R = \pi \sum_{i,q,\lambda} |v_R(i; cq\lambda)|^2 \delta(\varepsilon_i - \varepsilon_c - \hbar cq) . \qquad (48)$$

The radiative transition is less important in determining the core hole lifetime than the Auger transition, except for some deep core holes. (For K-shell holes, the radiative process becomes more important when the atomic number Z becomes $Z \geqslant 31$.)

Before closing this section, we give a comment on the magnetic materials, where the medium contains the incomplete d or f electrons. In the final state of photoemission, the exchange coupling between the d or f electron spin and the core hole spin, as well as the screening of the core hole charge by d or f electrons, give rise to important many-body effects, which will be discussed in detail in sections 4 and 5.

3. Simple metals

The conduction band of simple metals can be described with the nearly free-electron model. The orthogonality catastrophe and the plasmon satellite, which were mentioned in section 2.4, are actually observed in the inner shell photoelectron spectra of simple metals. In order to gain further insight into the orthogonality catastrophe, let us consider a simple model where the conduction electrons are noninteracting spinless electrons described by the Hamiltonian

$$H_0 = \sum_k \varepsilon_k a_k^+ a_k , \qquad (49)$$

and the core hole potential is a weak short-range potential

$$U = v \sum_{k,k'} a_k^+ a_{k'} . \qquad (50)$$

Then the spectral function $J(\omega)$ is calculated as

$$J(\omega) = \frac{1}{2\pi} \int_{-\infty}^{\infty} dt \, \langle 0|U(t) \, U|0\rangle \, e^{i\omega t}$$

$$= v^2 \sum_{k>k_F} \sum_{k'<k_F} \delta\left(\omega - \frac{\varepsilon_k - \varepsilon_{k'}}{\hbar}\right)$$

$$= (\rho v)^2 \hbar^2 \omega , \quad \text{for} \quad \hbar\omega \ll D , \qquad (51)$$

where ρ is the density of states (at the Fermi level) of the conduction band, k_F is the Fermi wave number, and D is a cut-off energy of the order of the Fermi energy. We note that $J(\omega)$ is proportional to the excitation spectrum of an electron–hole pair in the conduction band, and the singularity described below comes from the fact that $J(\omega) \propto \omega$ for $\omega \to 0$. By substituting eq. (51) into eq. (20), the generating function is obtained as

$$g(t) = \exp\left[-\frac{i\Delta t}{\hbar} - \alpha \log\left(\frac{iDt}{\hbar}\right)\right], \qquad (52)$$

and the photoelectron spectrum is given by

$$F(E_B) = \begin{cases} \dfrac{1}{D\Gamma(\alpha)\left(\dfrac{E_B - \Delta}{D}\right)^{1-\alpha}}, & \text{for} \quad E_B \geq \Delta , \\ 0, & \text{for} \quad E_B < \Delta , \end{cases} \qquad (53)$$

where

$$\alpha = (\rho v)^2 , \qquad (54)$$

and $\Gamma(\alpha)$ is the gamma function. Equation (52) is the asymptotic expression for $t \to \infty$, so that eq. (53) is also the asymptotic form for $E_B - \Delta \to 0$.

When we expand the right-hand side of eq. (52) with respect to α, the n-th order term gives the contribution from the excitation of n electron–hole pairs, from which $F_n(E_B)$ of eq. (22) results. Let us consider how many electron–hole pairs are excited, on an average, with the photoemission of one core electron. Bearing in mind that the integrated intensity of $F_n(E_B)$ obeys the Poisson distribution (see eq. 25), we obtain the average value of n as

$$\langle n \rangle = \sum_n n \, e^{-S} \frac{S^n}{n!} = S .$$

Further, from eqs. (23) and (51) we obtain

$$S \propto \lim_{\omega \to 0} (-\log \omega) = \infty.$$

Thus, $\langle n \rangle$ diverges *logarithmically*, corresponding to an infinite number of electron–hole pair excitations with vanishingly small excitation energy.

What we have found is as follows: When a core electron is removed in the photoemission, the conduction electrons screen the core hole potential. The screening process is equivalent to the excitation process of electron–hole pairs by the core hole potential U, as shown in fig. 2. The excitation of electron–hole pairs causes the relaxation of the conduction electron system around the core hole, the relaxation energy $|\Delta|$ being of the order of $(\rho v)^2 D$. At the same time, the excitation of electron–hole pairs manifests itself in the photoelectron spectrum as the sideband structure $F_n(E_B)$, where the excitation energy corresponds to $E_{ex} = E_B - \Delta$. Since the number of electron–hole pairs diverges logarithmically with vanishingly small excitation energy E_{ex}, the intensity of the photoelectron spectrum diverges as E_B tends to Δ. This is the mechanism of the orthogonality catastrophe. We have shown that $\langle n \rangle \to \infty$ comes from $S \to \infty$, which is equivalent to the orthogonality between $|0\rangle$ and $|\hat{0}\rangle$ (see also the discussion in section 2.4.3), because the intensity of the zero line F_0 is expressed as

$$|\langle 0|\hat{0}\rangle|^2 = e^{-S}.$$

When we extend the model by taking account of the electron spin and the core hole potential with arbitrary strength and finite force range, α in eqs. (52) and (53) is replaced by

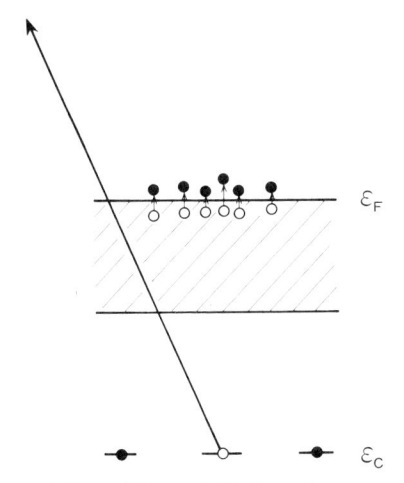

$$\alpha = 2 \sum_l (2l+1)[\delta_l(\varepsilon_F)/\pi]^2, \tag{55}$$

Fig. 2. Schematic representation of inner shell photoelectron process in simple metals.

where $\delta_l(\varepsilon_F)$ is the phase shift of the conduction electron (partial wave with angular momentum l) at the Fermi energy due to the scattering by the core hole potential. The singular photoelectron spectrum (eqs. 53 with 55), as well as the Fermi edge singularity in the inner shell photoabsorption spectrum, were extensively studied from theoretical side by Mahan (1967), Anderson (1967), Nozières and De Dominicis (1969), Hopfield (1969), Friedel (1969), Doniach and Šunjić (1970) and others, and gave a strong impact to the theoretical and experimental studies of inner shell spectroscopy. With this as a start, the study of many-body response of the outer electrons to the core hole in the inner shell photoemission has been developed in various materials, as will be shown in sections 4–6.

As an example of experimental data of inner shell photoemission in simple metals, we show in fig. 3 the 2p photoelectron spectrum of Na, which was observed at 300 K by Citrin et al. (1977). The spectral shape near the 2p threshold is analysed by a convolution among the singular spectrum (53), the Gaussian spectrum with D_{phonon} due to the lattice relaxation (section 2.4.1) and the Lorentzian spectrum with Γ_{hole} due to the core hole lifetime (section 2.4.5). By further taking account of the Gaussian broadening due to the instrumental resolution, as well as the effect of the spin–orbit splitting of the 2p level, Citrin et al. (1977) obtained the following values: $\alpha = 0.198 \pm 0.015$, $2\Gamma_{hole} = 0.02 \pm 0.02$ eV, and $2.35 D_{phonon}$ (= FWHM value of Gaussian) = 0.18 ± 0.03 eV.

In fig. 3, a plasmon satellite is also observed. When we take account of the plasmon energy dispersion quadratic with respect to the wave number, $J(\omega)$ has a ω-dependence of the form

$$J(\omega) \sim \frac{\omega_p^2}{\omega} \sqrt{\frac{\omega_p}{\omega - \omega_p}}, \quad \text{for} \quad \omega > \omega_p.$$

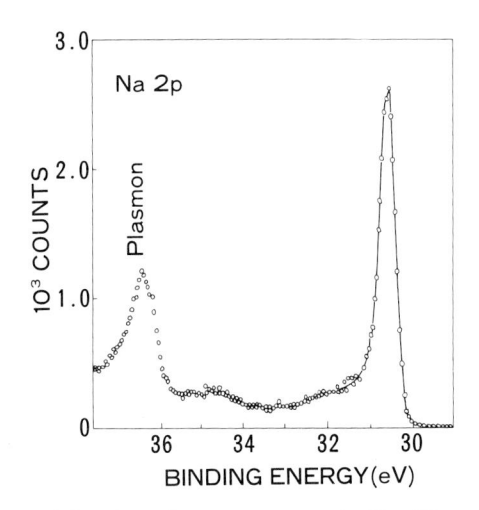

Fig. 3. Experimental data of 2p photoelectron spectrum in Na. The solid curve is a result of least-squares analysis (from Citrin et al. 1977).

Therefore, the one plasmon satellite is expressed as

$$e^{-s}\frac{J(\omega)}{(\hbar\omega)^2}, \quad \text{with} \quad \omega = \frac{E_B - \Delta}{\hbar},$$

which can explain the experimental asymmetric spectral shape with a tail in the high-energy side, as shown in fig. 3.

4. Rare earth metals and compounds

In magnetic materials, such as rare earth metals, transition metals, actinide metals and their compounds, the medium (i.e., the outer electron system) includes incompletely filled f or d shells. The electron wave function of incomplete shells is much more localized than that of the conduction electrons in simple metals, so that the f or d electron couples very strongly with the well-localized core hole in the final state of the photoemission. Reflecting this fact, remarkable spectral splittings are often observed experimentally in the inner shell photoelectron spectra of magnetic materials. In this section, we discuss those spectral splittings in rare earth metals and their compounds.

4.1. Exchange splitting

A well-known origin of the above mentioned splitting is the *exchange coupling* between the core hole spin s and the 4f electron spin S at the core hole site. In this case, the core hole plays a role of "test spin" to induce the magnetic response of the medium. As an example, we show in fig. 4 the 4s photoelectron spectra in GdF_3, TbF_3, DyF_3 and HoF_3 (Cohen et al. 1972). Since the 4f electrons of rare earths have a well-localized spin S ($S = \frac{7}{2}, \frac{6}{2}, \frac{5}{2}$ and $\frac{4}{2}$ for Gd, Tb, Dy and Ho, respectively), the exchange interaction U is now written in the form

$$U = -2J\mathbf{s} \cdot \mathbf{S}. \tag{56}$$

We disregard the coupling between the 4f spins at different atomic sites. Then, the states S_z with $S_z = -S \sim S$ are degenerate in the initial state of the photoemission. In the final state, we have the multiplets with the total spin $S_{tot} = |\mathbf{s} + \mathbf{S}| = S + \frac{1}{2}$ and $S - \frac{1}{2}$, whose energies are $-JS$ and $J(S+1)$, respectively. The photoelectron spectrum (5) is given by

$$F(E_B) = \frac{S+1}{2S+1}\delta(E_B + JS) + \frac{S}{2S+1}\delta[E_B - J(S+1)]. \tag{57}$$

Thus the 4s photoelectron spectrum splits into two peaks with the intensity ratio $I_R = (S+1)/S$ and the energy separation $\Delta E_B = J(2S+1)$. The observed I_R and ΔE_B in $GdF_3 \sim HoF_3$ changes with S consistently with these expressions, as found from fig. 4. For the absolute value of ΔE_B, however, the Hartree–Fock theory is

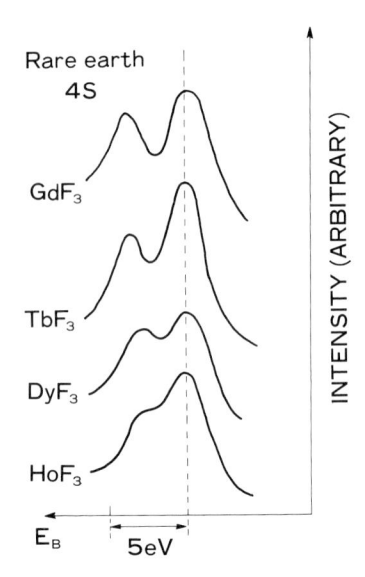

Fig. 4. Schematic drawing of 4s photoelectron spectra in rare earth compounds (after Cohen et al. 1972).

not accurate enough to explain the experimental results. In the Hartree–Fock approximation, J is given by $J_{HF} = G^3(4s, 4f)/7$ with the Slater exchange integral $G^3(4s, 4f)$, but the experimental value of ΔE_B is well reproduced by $J(2S+1)$ with $J = 0.55 J_{HF}$ (McFeely et al. 1972). This discrepancy can be explained by the *electron correlation effect*, because the 4s and 4f electrons have a strong correlation due to the same principal quantum number. For the 5s photoelectron spectra, the observed ΔE_B is in good agreement with $J_{HF}(2S+1)$.

We give some remarks.

(i) Both the 4f and core electrons have well-localized atomic-like wave functions, so that the exchange splitting reflects essentially the atomic character. Thus, in various materials including a certain rare earth element (irrespective of metals and insulators), almost the same exchange splitting is observed. An exceptional case is the *mixed valence materials* (see section 4.3), where the valence number of rare earth ions is not an integer and depends on materials even when they contain an identical rare earth element, so that the exchange splitting also depends on materials.

(ii) The exchange splitting is also observed for photoelectrons from core states with the p or d symmetry. In this case, the splitting becomes more complicated than that of the s-symmetric core state, because the splitting depends on the orbital angular momentum of both the 4f and core states. The splitting in this case is denoted by "multiplet splitting".

(iii) The exchange splitting (or the multiplet splitting) decreases as the core level becomes deeper. For the 3d core level (or still deeper levels) the splitting is usually too small to be observed as separate photoemission peaks, and the exchange coupling contributes only to the broadening of the photoelectron spectrum.

4.2. Satellite in La metal and its compounds

In the 3d photoelectron spectra of La metal and its compounds, as well as those of Ce, Sm and Eu metals and their compounds, spectral splittings of the order of 5–10 eV are often observed (Nagakura et al. 1972, Wertheim and Campagna 1978, Crecelius et al. 1978, Fuggle et al. 1983b). This energy separation is too large to be attributed to the multiplet splitting. Furthermore, in the case of La metal and its compounds, there is *no 4f occupation* in the ground state (i.e., $4f^0$ configuration), so that the splitting can *never* be explained by the multiplet splitting mechanism. We show in fig. 5a the experimental $3d_{5/2}$ photoelectron spectra of La and $LaPd_3$ (Fuggle et al. 1983b). We find a satellite peak in the lower binding energy side of the main line. The intensity of the satellite is very weak in La, but considerably strong in $LaPd_3$.

In order to interpret this satellite structure, we consider a model proposed by Kotani and Toyozawa (1973a,b, 1974a). Let us consider a system consisting of a conduction band (denoted by the s band), well-localized 4f shells (denoted by f states) and core states. We assume a hybridization V between s and f states. In the initial state of the photoemmission, the f level ε_f^0 is well above the Fermi level ε_F, so that the f level is empty. However, in the final state of the photoemission, as shown in fig. 6, the f level at the core hole site is assumed to be pulled down below ε_F due to the attractive potential of the core hole. Then, we expect to have *two types* of final states: In one type (type A), an s electron near ε_F *jumps* into the f level ε_f through the s–f hybridization V, and in the other type (type B) the f level is still *empty* even after pulled down below ε_F. Then the photoelectron spectrum splits into two peaks, corresponding to the two types of the final states. In order to obtain the photoelectron spectrum by this mechanism through a simple explicit calculation, we assume that (i) the level ε_f^0 is infinitely high, (ii) electrons are spinless, and (iii) the orbital degeneracy of the f state is disregarded.

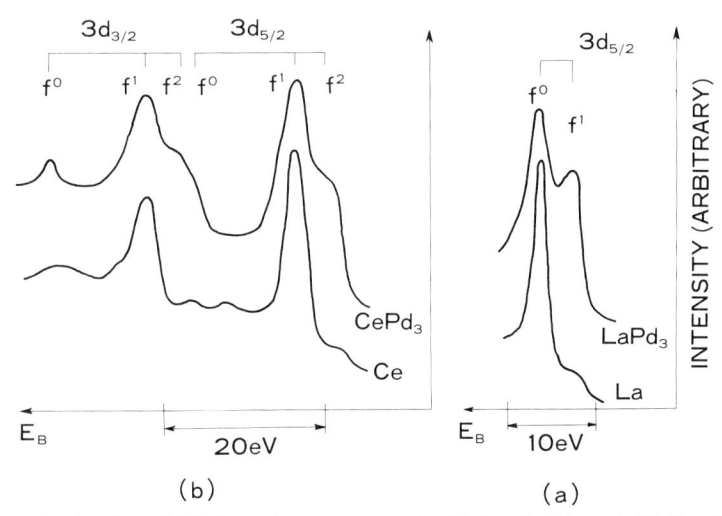

Fig. 5. Schematic drawing of 3d photoelectron spectra in (a) La, $LaPd_3$ and (b) Ce, $CePd_3$ (after Fuggle et al. 1983b).

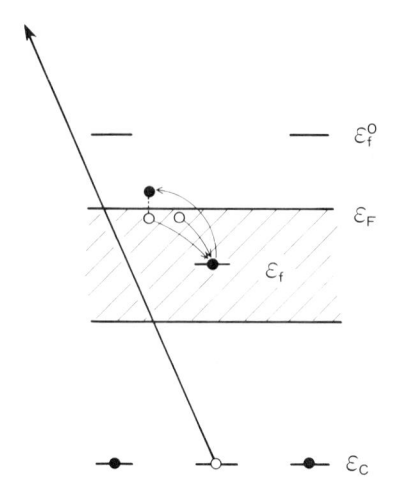

Fig. 6. Model of inner shell photoemission in La.

Then the Hamiltonian of the "medium" is written as

$$H_0 = \sum_k \varepsilon_k a_k^+ a_k \tag{58}$$

in the initial state, and as

$$H = H_0 + \varepsilon_f a_f^+ a_f + V \sum_k (a_k^+ a_f + a_f^+ a_k) \tag{59}$$

in the final state. Here a_k^+ and a_f^+ are the creation operators of the s and f electrons, respectively. The photoelectron spectrum (5) is rewritten as

$$F(E_B) = -\frac{1}{\pi} \operatorname{Im} \langle 0 | \frac{1}{z-H} | 0 \rangle, \tag{60}$$

where

$$z = E_B + E_0 + i\eta, \quad \eta \to +0. \tag{61}$$

In the following, we calculate $F(E_B)$ for the two types of final states, separately.

(a) *Final state of type (A)*

The f state is occupied in the final state of type (A). It is convenient to introduce a new state $|f\rangle = a_f^+ |0\rangle$, and we rewrite $\langle 0 | 1/(z-H) | 0 \rangle$ as follows:

$$\langle 0 | \frac{1}{z-H} | 0 \rangle = \langle f | a_f^+ \frac{1}{z - H_{0f} - H'} a_f | f \rangle$$

$$= \langle f | a_f^+ \left[\frac{1}{z - H_{0f}} + \frac{1}{z - H_{0f}} H' \frac{1}{z - H_{0f}} + \frac{1}{z - H_{0f}} H' \frac{1}{z - H} H' \frac{1}{z - H_{0f}} \right] a_f | f \rangle$$

$$= G_f^0 + (V G_f^0)^2 G, \tag{62}$$

where

$$H_{0f} = H_0 + \varepsilon_f a_f^+ a_f, \tag{63}$$

$$H' = V \sum_k (a_k^+ a_f + a_f^+ a_k), \tag{64}$$

$$G_f^0 = \langle f| a_f^+ \frac{1}{z - H_{0f}} a_f |f\rangle = \frac{1}{E_B + i\eta}, \tag{65}$$

$$G = \sum_{k<k_F} \sum_{k'<k_F} \langle f| a_k^+ \frac{1}{z - H} a_{k'} |f\rangle. \tag{66}$$

When we define a "generating function" $\hat{g}(t)$ by

$$-\frac{1}{\pi} \operatorname{Im} G = \frac{1}{2\pi\hbar} \int_{-\infty}^{\infty} dt \exp\left(i \frac{E_B}{\hbar} t\right) \hat{g}(t), \tag{67}$$

then $\hat{g}(t)$ is written by using the *linked-cluster theorem* as

$$\hat{g}(t) = \sum_{k<k_F} \sum_{k'<k_F} \langle f| a_k^+ \exp\left[-\frac{i}{\hbar}(H - E_0)t\right] a_{k'} |f\rangle$$

$$= L_2(t) \exp[L_1(t)], \tag{68}$$

where

$$\exp[L_1(t)] = \langle f| S(t) |f\rangle, \tag{69}$$

$$L_2(t) = \sum_{k<k_F} \sum_{k'<k_F} \exp\left[-\frac{i}{\hbar}(\varepsilon_f - \varepsilon_k)t\right] [\langle f| a_k^+ S(t) a_{k'} |f\rangle]_c. \tag{70}$$

Here $S(t)$ is the S-matrix defined by

$$S(t) = \exp\left(i \frac{H_{0f}}{\hbar} t\right) \exp\left(-i \frac{H}{\hbar} t\right),$$

and $[\ ]_c$ means the contribution from connected diagrams. By the *most divergent approximation*, L_1 and L_2 are obtained as

$$L_1(t) \sim \int_0^t d\tau_1 \int_0^{\tau_1} d\tau_2 \int_0^{\tau_2} d\tau_3 \int_0^{\tau_3} d\tau_4 \langle f| H'(\tau_1) H'(\tau_2) H'(\tau_3) H'(\tau_4) |f\rangle$$

$$\sim -i \frac{\Delta_f}{\hbar} t - (\rho v_{\text{eff}})^2 \log\left(\frac{iDt}{\hbar}\right), \tag{71}$$

$$L_2(t) \sim \frac{\rho}{it} \exp\left[\frac{i}{\hbar}(\varepsilon_F - \varepsilon_f)t\right]\left(\frac{iDt}{\hbar}\right)^{-2\rho v_{\text{eff}}}, \tag{72}$$

where Δ_f represents an appropriate energy shift and v_{eff} is defined by

$$v_{\text{eff}} = \frac{-V^2}{\varepsilon_F - \varepsilon_f}. \tag{73}$$

Substituting eqs. (68), (71) and (72) into eq. (67), we obtain

$$-\frac{1}{\pi}\operatorname{Im} G = \begin{cases} \dfrac{\rho}{\Gamma(1-2g+g^2)\left(\dfrac{E_B + \varepsilon_F - \tilde{\varepsilon}_f}{D}\right)^{2g-g^2}} & \text{for } E_B \geq -(\varepsilon_F - \tilde{\varepsilon}_f), \\ 0 & \text{for } E_B < -(\varepsilon_F - \tilde{\varepsilon}_f), \end{cases} \tag{74}$$

where

$$g = -\rho v_{\text{eff}}, \qquad \tilde{\varepsilon}_f = \varepsilon_f + \Delta_f.$$

The photoelectron spectrum near the threshold $E_B = -(\varepsilon_F - \tilde{\varepsilon}_f)$ is expressed, by using eqs. (60), (62) and (65), as

$$F(E_B) = \frac{\rho V^2}{(\varepsilon_F - \tilde{\varepsilon}_f)^2}\left(-\frac{1}{\pi\rho}\operatorname{Im} G\right). \tag{75}$$

Thus $F(E_B)$ is found, from eqs. (74) and (75), to diverge at the threshold.

(b) *Final state of type (B)*
In the final state (B), the f level is empty, so that this final state can be obtained from $|0\rangle$ by including the self-energy correction due to the interaction H'. Thus, eq. (60) is written as

$$F(E_B) = -\frac{1}{\pi}\operatorname{Im}\frac{1}{z - E_0 - \Sigma_0(z)}, \tag{76}$$

where the self-energy Σ_0 is given, up to the second order in H', by

$$\Sigma_0(z) = \langle 0|H'\frac{1}{z - H_{0f}}H'|0\rangle$$

$$= V^2 \sum_{k<k_F}\frac{1}{z + \varepsilon_k - \varepsilon_f - E_0}$$

$$\sim \Delta_0 - i\pi\rho V^2. \tag{77}$$

The photoelectron spectrum is given by the *Lorentzian function* with a maximum at $E_B = \Delta_0$:

$$F(E_B) = \frac{\rho V^2}{(E_B - \Delta_0)^2 + (\pi \rho V)^2}. \tag{78}$$

The schematic shape of the photoelectron spectrum is shown in fig. 7. In the final state of type (A), which corresponds to the peak (A) in fig. 7, the core hole charge is *screened* by the f electron due to the charge transfer from s to f states. On the other hand, in the final state of type (B) (corresponding to the peak (B) in fig. 7) the core hole charge is *not screened* by the f electron. The spectrum (A) diverges due to the *orthogonality catastrophe* similar to that in section 3, because we have the simultaneous excitation of electron–hole pairs by the repetition of the process where an electron in ε_f jumps out above ε_F through V and another s electron below ε_F jumps into ε_f (see fig. 6). In other words, the scattering of s electrons by the core hole occurs due to the second order process of V via the intermediate f state, and the scattering is described by the effective potential v_{eff} of eq. (73). The peak (B) has a Lorentzian broadening because of a finite lifetime $\tau = \hbar/(\pi \rho V^2)$ of the *f hole* (the empty f state below ε_F) due to its resonant transfer to the s hole through V. This lifetime effect is somewhat similar to that discussed in section 2.4.5. The final states of (A) and (B), corresponding to $4f^1$ and $4f^0$ configurations, are denoted by "well-screened state" and "poorly-screened state", respectively. Furthermore, it is to be remarked that in the final state (B) we have two holes – one in the core level and another in the f level – and the two holes are bound in a single atomic site. Therefore, we also denote the final state (B) by "two-hole bound state".

In the La metal, the intensity of the well-screened peak is very small because of the small prefactor $\rho V^2/(\varepsilon_F - \tilde{\varepsilon}_f)^2$ in eq. (75). However, in $LaPd_3$ this factor is not very small because of the increase of ρ and V due to the existence of 4d states of Pd. A similar satellite structure is also observed in the insulating compounds of La, such as LaF_3, $LaCl_3$, $LaBr_3$, La_2O_3 and so on (Signorelli and Hayes 1973, Suzuki et al. 1974). Our model system for the photoemission of La metal can be applied to the insulating La compounds only by relacing the conduction band (s band) by a completely filled valence band. Even in the insulators, both of the

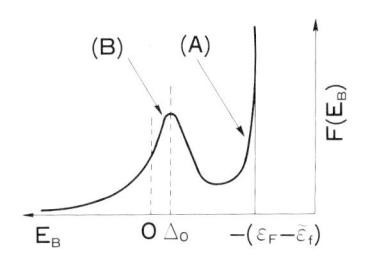

Fig. 7. Schematic shape of the photoelectron spectrum obtained with the model of fig. 6.

well-screened and poorly-screened final states are possible to occur due to the charge transfer from the valence band to the 4f state (although the orthogonality catastrophe does not occur). This problem has also been analysed with a cluster model by Fujimori (1983a).

In the rest of this subsection we discuss briefly the effect of finite ε_f^0 in the metallic system. In fig. 8 we show the photoelectron spectra calculated for various values of ε_f^0 by taking account of the finite ε_f^0 only at the core hole site (Kotani and Toyozawa 1974a). The calculation is made by exactly diagonalizing H_0 and H for a finite system where the conduction band consists of 50 discrete levels, and the photoelectron spectrum (5) is computed numerically. In this figure we fix $\rho V^2 = 0.05$ and $\varepsilon_f - \varepsilon_F = -0.5$, where the half width of the s band is taken as energy units. It is found that when ε_f^0 is lowered, the intensity of the well-screened $4f^1$ peak increases. This is understood as follows: When ε_f^0 is finite, the $4f^1$ configuration is already mixed in the ground state $|0\rangle$ with a finite weight, although the weight is very small in the case of $\varepsilon_f^0 - \varepsilon_F > \rho V^2$. Since the photoemission intensity for the final state $|f\rangle$ is given by the overlap integral $|\langle f|0\rangle|^2$, the initial mixing of the $4f^1$ state enhances the intensity of the $4f^1$ final

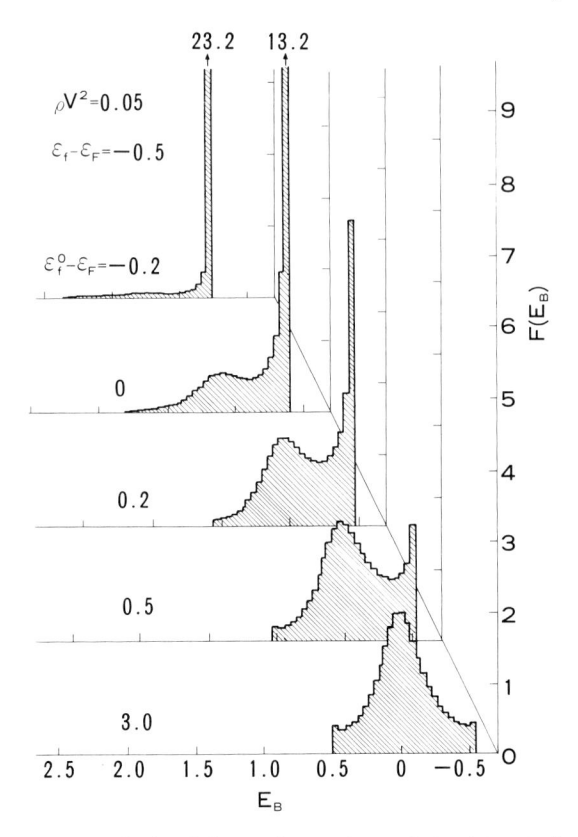

Fig. 8. Photoelectron spectra calculated for various values of the f level ε_f^0 (from Kotani and Toyozawa 1974a).

state. When $\varepsilon_f^0 - \varepsilon_F \sim \rho V^2$, the weight of the 4f^1 configuration in the initial state becomes comparable with that of the 4f^0 configuration, and this special state is denoted by the *mixed valence state*, which occurs in some Ce compounds and will be discussed in the next subsection.

Calculations of absorption (emission) and photoemission spectra with models similar to that presented here were also made, from somewhat different viewpoints, by Combescot and Nozières (1971) and Schönhammer and Gunnarsson (1977, 1978).

4.3. Mixed valence state in Ce compounds

When the ground state is in the 4f^0 configuration, as in the case of La, it has been shown that there appear two types of the photoemission final states, i.e., the well-screened 4f^1 and the poorly-screened 4f^0 states. In a similar way, when the ground state is in the 4f^1 configuration, as in the case of the trivalent Ce, we expect to have two types of the final states corresponding to the well-screened 4f^2 and the poorly-screened 4f^1 states. Actually, in various materials including the trivalent Ce, such as CeAl$_2$, CeCu$_2$Si$_2$, CeSe and so on, the splitting of the inner shell photoelectron spectrum is observed, originating from the 4f^1 and 4f^2 final states (Lässer et al. 1980, Fuggle et al. 1983b).

Very interesting is the case where the ground state is in the mixed valence state between 4f^0 and 4f^1 configurations. In the mixed valence system, the final states associated with both the 4f^0 ground state (as in the case of La) and 4f^1 ground state (as in the trivalent Ce) are expected to coexist, so that the photoelectron spectrum will split into *three peaks* corresponding to 4f^2, 4f^1 and 4f^0 configurations. Actually, such a three-peak structure is observed in various mixed valence Ce compounds (Lässer et al. 1980, Krill et al. 1981, Hillebrecht et al. 1982, Fuggle et al. 1983b). As an example, we show in fig. 5b the 3d photoelectron spectrum of CePd$_3$, where both of the 3d$_{3/2}$ and 3d$_{5/2}$ peaks are found to split into three peaks.

In order to investigate theoretically the photoelectron spectrum of the mixed valence system, it is necessary to extend our model by taking account of the electron spin, as well as the finite value of ε_f^0. Then the Hamiltonians (58) and (59) are replaced by

$$H_0 = \sum_{k,\sigma} \varepsilon_k a_{k\sigma}^+ a_{k\sigma} + \sum_{\sigma} \varepsilon_f^0 a_{f\sigma}^+ a_{f\sigma} + \sum_{k,\sigma} (V_{kf} a_{k\sigma}^+ a_{f\sigma} + V_{kf}^* a_{f\sigma}^+ a_{k\sigma})$$

$$+ \frac{U_{ff}}{2} \sum_{\sigma} a_{f\sigma}^+ a_{f\sigma} a_{f-\sigma}^+ a_{f-\sigma} \tag{79}$$

and

$$H = H_0 + \sum_{\sigma} (\varepsilon_f - \varepsilon_f^0) a_{f\sigma}^+ a_{f\sigma}. \tag{80}$$

In eq. (79), U_{ff} is the Coulomb interaction between the f electrons, and the

k-dependence of V is explicitly written. An approximate calculation of the photoelectron spectrum in this system was first made by Oh and Doniach (1982) with a Green's function decoupling method. Their calculation is valid for small s–f hybridization, but with increasing hybridization the decoupling scheme breaks down, resulting in the occurrence of a region of negative photoelectron spectrum.

More recently Gunnarsson and Schönhammer (1983) proposed a new method, where they paid their attention to the importance of the *orbital degeneracy* of the 4f state. They calculated the photoelectron spectrum which is *exact* in the limit of $N_f \to \infty$, where N_f is the degeneracy of the 4f level including both the spin and orbital components. According to their method, the initial and final states of the photoemission are limited to those which are inside the space spanned by the following four types of basis functions:

$$|F\rangle, \tag{81}$$

$$|\varepsilon\rangle = \frac{1}{\sqrt{N_f}} \sum_\nu a^+_{f\nu} a_{\varepsilon\nu} |F\rangle, \tag{82}$$

$$|E, \varepsilon\rangle = \frac{1}{\sqrt{N_f}} \sum_\nu a^+_{E\nu} a_{\varepsilon\nu} |F\rangle, \tag{83}$$

$$|\varepsilon, \varepsilon'\rangle = \frac{1}{\sqrt{N_f(N_f-1)}} \sum_{\nu,\nu'} a^+_{f\nu} a_{\varepsilon\nu} a^+_{f\nu'} a_{\varepsilon'\nu'} |F\rangle. \tag{84}$$

Here $|F\rangle$ is the Fermi vacuum, in which all the s electrons below ε_F are occupied. The index ν represents the spin and orbital states (σ, m), and the operator $a^+_{\varepsilon\nu}$ is defined by

$$a^+_{\varepsilon\nu} = V(\varepsilon)^{-1} \sum_k V^*_{km} \delta(\varepsilon - \varepsilon_k) a^+_{k\sigma} \tag{85}$$

with $V(\varepsilon)$ expressed as

$$\sum_k V^*_{km} V_{km'} \delta(\varepsilon - \varepsilon_k) = |V(\varepsilon)|^2 \delta_{mm'}. \tag{86}$$

The ground state $|0\rangle$ of H_0 is expressed as a linear combination of eqs. (81)–(84), and the ground state energy E_0, as well as the coefficients of the linear combination, are determined by the minimization of $\langle 0|H_0|0\rangle$. The photoelectron spectrum (60) is then calculated from

$$F(E_B) = -\frac{1}{\pi} \sum_{i,j} \mathrm{Im}\, \langle 0|i\rangle \langle i| \frac{1}{z-H} |j\rangle \langle j|0\rangle, \tag{87}$$

where for $|i\rangle$ and $|j\rangle$, all the states of the types (81)–(84) are taken into account.

For small N_f, this approximation is not very good, but for $N_f \geq 6$ it gives sufficiently accurate results. In fig. 9, we show, as an example, a calculated

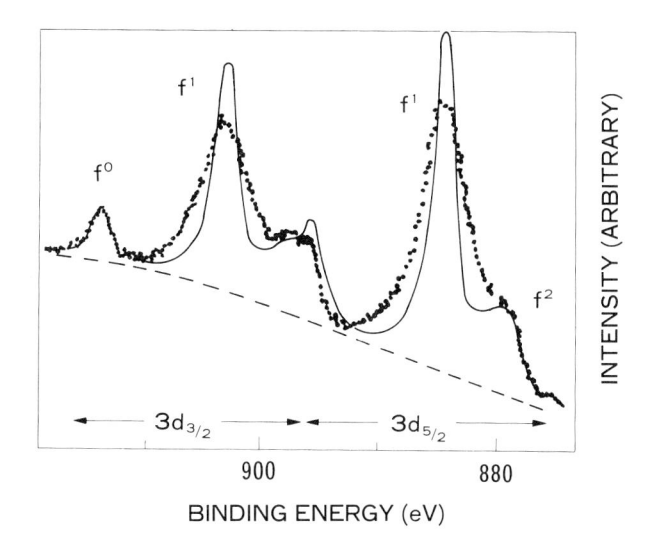

Fig. 9. Comparison of theoretical (solid curve) and experimental (dots) results of the 3d photoelectron spectrum in CeNi$_2$ (from Gunnarsson and Schönhammer 1983).

spectrum (solid line), as well as the experimental one (dots), for the 3d photoemission of CeNi$_2$. In this calculation $|V(\varepsilon)|^2$ is assumed in the form

$$|V(\varepsilon)|^2 = \frac{2V^2[D^2 - (\varepsilon - \varepsilon_0)^2]^{1/2}}{\pi D^2}$$

with $\varepsilon_0 - \varepsilon_F = -1.995$ eV, $D = 2.005$ eV and $\Delta \equiv 2V^2/D = 0.13$ eV. The other parameters are taken as $\varepsilon_f^0 - \varepsilon_F = -1.3$ eV, $\varepsilon_f^0 - \varepsilon_f = 10.3$ eV, $U_{ff} = 5.5$ eV and $N_f = 14$. In order to describe the lifetime broadening and instrumental resolution, η in eq. (61) is taken as a finite value, $\eta = 0.9$ eV. Two theoretical spectra $F(E_B)$ are superimposed with the weights 0.4 and 0.6 to simulate the 3d spin–orbit splitting. The fractional intensity of 4f^2 photoemission peak is found to be sensitive to the value of Δ described above, whereas that of 4f^0 peak is sensitive to the weight, $w(4f^0)$, of the 4f^0 configuration in the ground state. In this analysis, the value of $w(4f^0)$ is obtained as $w(4f^0) = 0.19$, and the averaged 4f electron number in the ground state is $n_f \simeq 1 - w(4f^0) = 0.81$ because $w(4f^2)$ is negligibly small. From the systematic analysis of the photoelectron spectra, the values of n_f and Δ for various Ce compounds have been obtained (Gunnarsson and Schönhammer 1983, Fuggle et al. 1983). Before such an analysis has been made, the value of n_f was estimated from the data of lattice constant and susceptibility, and then the value of n_f distributed between 0 and 1, depending on the material. However, n_f which is estimated from photoemission analysis is found to be always larger than about 0.7, ranging mainly between 0.8 and 1. Even for the materials which were traditionally considered to have $n_f = 0$, such as the intermetallic Ce compounds with Ni, Ru, Rh and so on, the photoemission analysis gives n_f around 0.8, as already described in the case of CeNi$_2$. It is to be emphasized that the inner shell photoemission is a powerful technique in the study of the mixed valence state, because it detects directly the *microscopic* and *local* properties of the 4f state. The results of the photoemission analysis are also consistent with the other spectroscopic data, such as the X-ray absorption and the bremsstrahlung isochromat spectra (Gunnarsson and Schönhammer 1983, Fuggle et al. 1983a).

Next we discuss the photoelectron spectrum of the insulating compound, CeO_2. Traditionally, CeO_2 was considered to be a typical ionic crystal consisting of the Ce^{4+} and O^{2-} ions (so that $n_f = 0$). However, in the 3d photoelectron spectrum of CeO_2, as shown in the inset of fig. 10, the conspicuous three-peak structure is observed (Burroughs et al. 1976, Beaurepaire 1983, Wuilloud et al. 1984), very similarly to that of mixed valence Ce compounds. The analysis of this photoelectron spectrum has been made by using a cluster model (Fujimori 1983b), as well as a band model (Kotani and Parlebas 1985, Kotani et al. 1985). In the band model analysis, the model of Gunnarsson and Schönhammer (1983) is used only by replacing the conduction band as the filled valence band with

$$|V(\varepsilon)|^2 = V^2/2d \quad \text{for} \quad \varepsilon_0 - D \le \varepsilon \le \varepsilon_0 + D.$$

The width of the valence band is taken as $2D = 3$ eV, according to the result of energy band calculation (Koelling et al. 1983). So as to reproduce the experimental spectrum, the parameter values are taken as $\Delta \equiv \pi V^2/2D = 0.59$ eV, $\varepsilon_f^0 - \varepsilon_0 = 1.6$ eV, $\varepsilon_f^0 - \varepsilon_f = 12.5$ eV, $U_{ff} = 10.5$ eV, $\eta = 1$ eV and $N_f = 14$. From this analysis, we found that the 4f state is hybridized strongly with the valence band in the

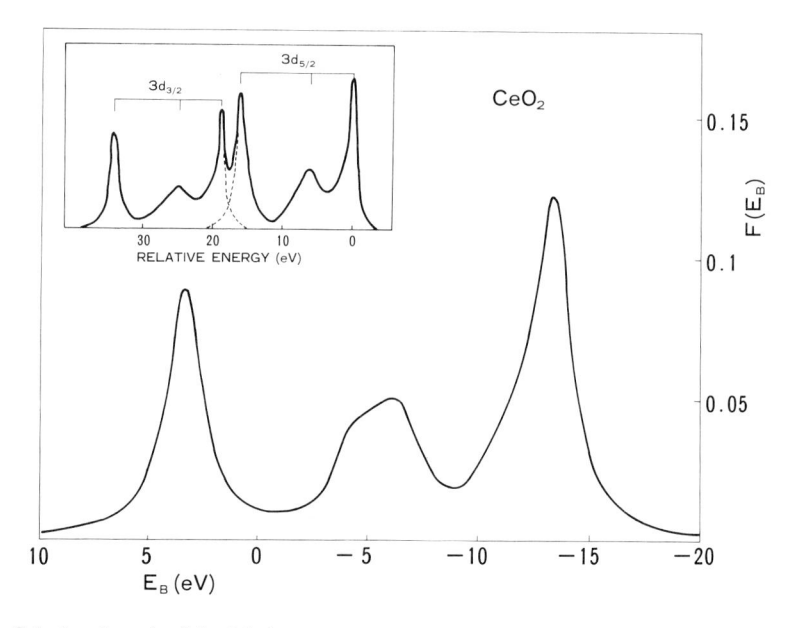

Fig. 10. Calculated result of the 3d photoelectron spectrum in CeO_2. Experimental result by Wuilloud et al. (1984) is shown in the inset.

ground state. Thus the ground state is the strong mixture between $4f^0$ and $4f^1$ configurations with $n_f \sim 0.5$. For the final state, the peak with the highest binding energy corresponds to an almost pure $4f^0$ configuration, but its fractional intensity is only about 60% of $w(4f^0)$ because of the strong final state interaction. The other two photoemission peaks correspond to the final states in which $4f^1$ and $4f^2$ configurations are also mixed strongly. Comparison between the 3d photoemission and the 2p absorption edge of CeO_2 are also made, both experimentally (Beaurepaire 1983) and theoretically (Jo and Kotani 1985).

4.4. Other rare earth systems

In general, for the rare earth element with $4f^n$ configuration in the ground state, two types of final states are expected to occur in the inner shell photoemission: the well-screened $4f^{n+1}$ state and the poorly-screened $4f^n$ state. However, in order that the well-screened state is observable as a photoemission satellite peak, the hybridization V should be reasonably large. As a general trend, the hybridization is largest in La and decreases with increasing atomic number (i.e., with increasing n), because the spatial extension of the 4f wave function decreases with the atomic number. Hillebrecht and Fuggle (1982) compared experimentally the satellite intensity of the 3d photoemission in various compounds RPd_3 (R = La, Ce, Pr, Nd and Sm), and found that the intensity of the $4f^{n+1}$ peak is appreciable for R = La, Ce, Pr and Nd, and decreases in going from La to Nd, but it is not detectable in $SmPd_3$. Therefore, we can roughly say that the effect of the final state interaction is negligible for Sm and heavier elements, although the situation is to some extent different for different types of materials (see Herbst et al. 1980, Schneider et al. 1981). On the other hand, the photoemission splitting due to the mixing of 4f configurations in the ground state (i.e., due to the mixed valence effect) may be observed in some compounds of Ce, Sm, Eu, Tm and Yb. The spectral structure is most complicated in the case of Ce, where the mixing effect both in the initial and final states is important, as we have shown in section 4.3.

Finally, it is to be remarked that a similar spectral splitting also occurs in some actinide metals and their compounds (Fuggle et al. 1980, Schneider and Laubschat 1981).

5. Transition metals and compounds

5.1. Transition metals

In the 2p photoelectron spectrum of Ni, a satellite structure is observed (Hüfner and Wertheim 1975a, Kemeney and Schevchik 1975), as shown in fig. 11a. A very similar satellite also occurs in the 3d band photoemission of Ni as shown in fig. 11b, but we defer until section 6 the discussion on the 3d band photoemission. Since the exchange interaction between 2p and 3d spins is too weak to give rise to this satellite, the effect of the core hole potential, somewhat similar to that in the

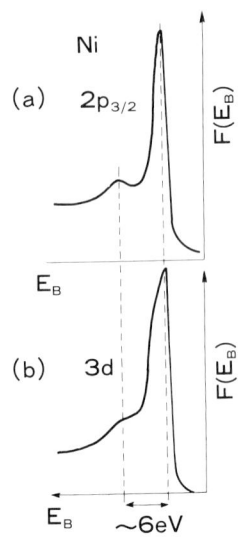

Fig. 11. Schematic drawing of (a) 2p and (b) 3d photoelectron spectra in Ni (after Hüfner and Wertheim 1975).

3d photoemission of rare earth metals, is considered to be responsible for it. However, the 3d incomplete shell state of transition metals is not so well-localized as the 4f state of rare earths, but forms a *narrow energy band*, which overlaps with a wide 4s band. In the ferromagnetic state of Ni, the 3d band has an exchange splitting Δ_{ex} (here the term "exchange splitting" does not mean the splitting of the inner shell photoemission spectrum discussed in section 4.1, but means the splitting of the 3d bands with majority and minority spins due to the ferromagnetic order). As a result, the 3d majority spin band (denoted by 3d ↑ band) is completely filled and the minority spin band (3d ↓ band) has empty states of about 0.6 electrons per atom, as shown schematically in fig. 12. Even in this case, when the 2p core hole potential is strong enough, a well-localized state (or a virtual bound state) consisting mainly of the 3d wave function is expected to occur (at ε_{d} in fig. 12) below the 3d band. Then, two types of the final states, where the level ε_{d} is filled or empty, will cause the splitting of the photoelectron spectrum.

When we take account of the d band effect, a simplest extension of the model Hamiltonians H_0 and H from those used for La metal, will be

$$H_0 = \sum_{k,\sigma} \varepsilon_{sk} a^+_{sk\sigma} a_{sk\sigma} + \sum_{R,\sigma} \left(\varepsilon^0_d \pm \frac{\Delta_{\text{ex}}}{2} \right) a^+_{dR\sigma} a_{dR\sigma}$$

$$+ \sum_{R,R'} t(R - R') a^+_{dR\sigma} a_{dR'\sigma} + \sum_{k,R,\sigma} (V_k e^{ik \cdot R} a^+_{sk\sigma} a_{dR\sigma} + V^*_k e^{-ik \cdot R} a^+_{dR\sigma} a_{sk\sigma})$$

(88)

and

$$H = H_0 - (\varepsilon^0_d - \varepsilon_d) \sum_\sigma a^+_{dR_0\sigma} a_{dR_0\sigma}.$$

(89)

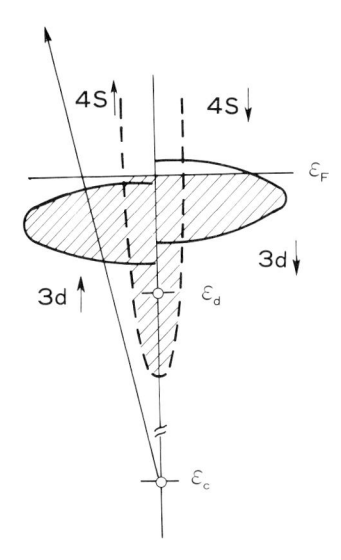

Fig. 12. Schematic representation of inner shell photoelectron process in Ni.

where $a^+_{sk\sigma}$ is the creation operator of the s electron with wave vector k and spin σ, $a^+_{dR\sigma}$ is that of the d electron at the site R and with spin σ, $t(R - R')$ represents the transfer of the d electron between the sites R and R', V_k describes the hybridization between s and d electrons, and R_0 is the core hole site. Photoelectron spectra $F(E_B)$ calculated with this model system are shown in fig. 13, where $F(E_B)$ is plotted as a function of E_B for various values of the core hole potential strength $\varepsilon^0_d - \varepsilon_d$ (Kotani 1979). In this calculation, the s and d bands (before the hybridization) are assumed to be simple one-dimensional bands with linear dispersion, and the parameters are fixed as $\rho_s V^2 = 0.05$, $\varepsilon^0_d - \varepsilon_F = 0$ and $D_d = 0.4$ with the half-width of the s band as energy units, where ρ_s is the density of states of the s band, D_d is the half-width of the d band, V is the hybridization (the k-dependence of V_k being disregarded), and the system is assumed to be in the paramagnetic state (so that $\Delta_{ex} = 0$). It is found that when the core hole potential $\varepsilon^0_d - \varepsilon_d$ is sufficiently strong the satellite occurs, and its intensity increases with increasing $\varepsilon^0_d - \varepsilon_d$. From a similar calculation, the intensity of the satellite is also found to increase with the increase of the unfilled states in the d band.

For more realistic calculations, the model should be extended further by including the orbital degeneracy of the 3d states, the exchange splitting Δ_{ex}, and the correlation effect of the 3d electrons. Some of these extensions have been made in the calculations by Tersoff et al. (1979) and by Feldkamp and Davis (1980). After all, we obtain a picture of the 2p photoelectron process in Ni as illustrated schematically in fig. 14 in the real space representation. In the initial state of the photoemission, the 3d↓ band has empty states of 0.6 electrons per atom, so that by using the hole picture, there exist the 3d↓ band holes, as shown in (A) of fig. 14. These 3d↓ holes are spatially moving with strong correlation, and fig. 14 corresponds to their snapshot. In the final state, we have a core hole

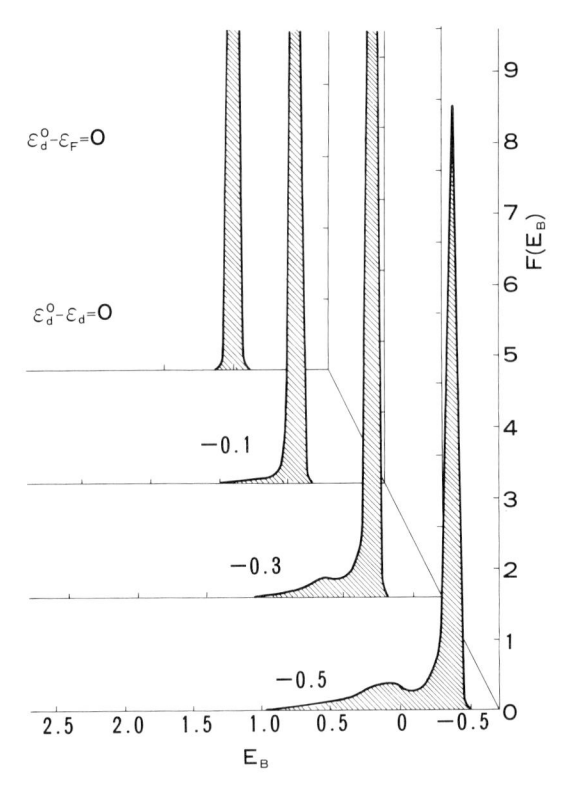

Fig. 13. Photoelectron spectra calculated for various values of the core hole potential strength $\varepsilon_d^0 - \varepsilon_d$ (after Kotani 1979).

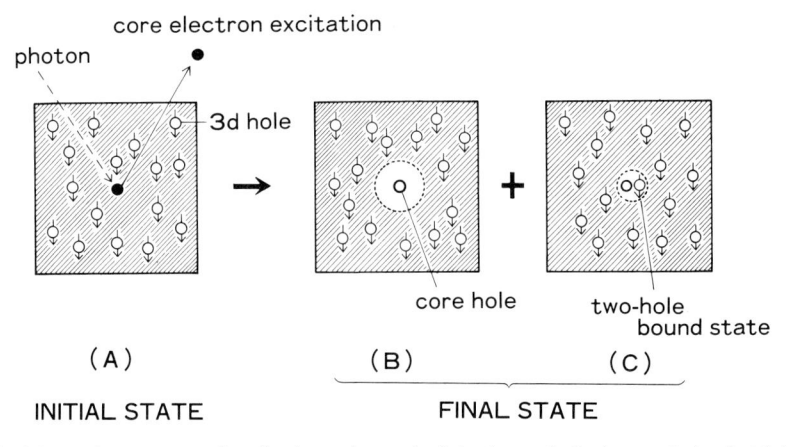

Fig. 14. Schematic representation (in the real space) of the inner shell photoemission in Ni (see text).

which couples with the 3d ↓ holes through the repulsive Coulomb interaction. As a result, there occur two types of the final states, (B) and (C). In (B), the 3d ↓ holes do not approach the core hole (corresponding to the case where the level ε_d is occupied in fig. 12), and in (C) one 3d ↓ hole is bound at the core hole site (corresponding to the case where ε_d is empty). The final state of (C) is denoted by "two-hole bound state". The energy of the two-hole bound state is higher than that of the final state of (B) due to the Coulomb repulsion between the bound pair of the 3d and core holes. Thus the two-hole bound state gives the *satellite* of the photoemission, while the final state (B) gives the *main peak*. It is to be reminded that the two-hole bound state consisting of a core hole and a 4f hole also causes the splitting of the 3d photoemission of La metal. However, in the case of La the two-hole bound state corresponds to the *main line* (stronger peak), whereas in Ni the two-hole bound state corresponds to the *satellite* (weaker peak). This difference comes from the following two facts: In the initial state of the photoemission in Ni, the unoccupied 3d states (i.e., the 3d holes) are limited, whereas in La all of the 4f states are unoccupied. Furthermore, in the final state of the photoemission in Ni, the core hole charge is screened by 3d electrons through the strong 3d electron transfer $t(\boldsymbol{R} - \boldsymbol{R}')$, in contrast to the screening by the 4f electrons through the weak s–f hybridization in La.

Extensive experimental studies of the 2p photoemission in various intermetallic Ni compounds were made by Hillebrecht et al. (1983). They observed the change of the satellite intensity as a function of stoichiometry in alloys with electropositive metals, and found that the satellite intensity decreases as the 3d band filling increases, in agreement with the mechanism of the satellite mentioned above. The satellite is also observed in the 2p photoemission of other transition metals, Pd, Fe and so on, but their intensities are much lower than that of Ni (Richardson and Hisscott 1976, Hillebrecht et al. 1983). For the main line of the 2p photoemission, on the other hand, the spectral shape is expected to be asymmetric (as shown in fig. 13). The asymmetry can be expressed as an inverse-power form such as eq. (53) or eq. (74). Systematics of the asymmetry in various transition metals were studied experimentally by Hüfner and Wertheim (1975b,c).

In the 3s and 3p photoemission of transition metals the spectral splitting due to the exchange interaction between the 3d and core hole spins are also observed. For instance, Fadley and Shirley (1970) observed the splitting of the 3s photoelectron spectrum in Fe with the intensity ratio 2.6:1 and the energy separation about 4 eV. The splitting was found to be almost unchanged at temperatures below and above the Curie temperature T_C. This result was analysed by approximating the 3d spin as a well-localized spin S and by using the model just the same as that mentioned in section 4.1. Then, the intensity ratio is given by $S+1:S$, which is rather consistent with the experimental result by using $S=1$ or $\frac{3}{2}$, and the exchange coupling constant J is estimated from the energy separation $J(2S+1)$. The local spin S is considered to persist above T_C, even though the long-range magnetic order vanishes, so that the splitting does not change at T_C. However, this picture is too much simplified, because the localized spin model is only valid in the limit of the vanishing electron transfer $t(\boldsymbol{R} - \boldsymbol{R}')$ (i.e., in the *atomic limit*).

When the effect of the electron transfer becomes important, the exchange-split spectrum will be motionally narrowed, and in the limit of the large electron transfer (*free-electron limit*) it reduces to a single peak with singularity due to the orthogonality catastrophe mentioned in section 3. Transition metals are considered to be in the intermediate situation between the atomic and free-electron limits. Another problem is that the effect of the core hole charge, as in the case of 2p photoemission, may also contribute to the 3s photoemission, coexisting with the effect of the exchange interaction. Some theoretical attempts to resolve these problems have been made by Kaga et al. (1976), Kakehashi et al. (1984) and Kakehashi and Kotani (1984), but these theories are not yet in a position to be compared quantitatively with the experimental results.

5.2. Transition metal compounds

In insulating transition metal compounds, the satellite peak of the 2p photoemission is often observed more clearly than that of transition metals. We show in fig. 15 the experimental data in $MnF_2 \sim ZnF_2$ observed by Rosencwaig et al. (1971). The intensity of the satellite is found to increase with increasing 3d electron number (from Mn^{2+} to Cu^{2+}), but it vanishes in the case of Zn^{2+}. This satellite was first considered to originate from the shake-up transition between the metal 3d and 4s orbitals, but now it is interpreted more reasonably as originating from the charge transfer between the 3d and ligand orbitals (or between the molecular orbitals consisting of the 3d and ligand orbitals) by using a cluster model (Kim and Davis 1972, Kim 1975, Larsson 1975, 1976, Asada and Sugano 1976, Van der Laan et al. 1981). Let us consider a cluster consisting of a transition metal ion and its neighboring anions. For simplicity, we consider only two electronic states $|l\rangle$ and $|d\rangle$ with energies ε_l and $\varepsilon_d^0 (\varepsilon_d^0 > \varepsilon_l)$. In the state $|l\rangle$, 3d electrons are in the lowest configuration $3d^n$ and an electron is occupied in a ligand orbital (we consider only one ligand orbital), while in $|d\rangle$ the ligand electron is transferred to the 3d state, so that the 3d electrons are in the $3d^{n+1}$ configuration. The Hamiltonian in the initial state of the photoemission is written as

$$H_0 = \varepsilon_l |l\rangle\langle l| + \varepsilon_d^0 |d\rangle\langle d| + V(|l\rangle\langle d| + |d\rangle\langle l|) \tag{90}$$

with the electron transfer energy V (we assume V is real and positive). The eigenstates of H_0 are given by

$$|1\rangle = \cos\theta_0 |d\rangle + \sin\theta_0 |l\rangle, \tag{91}$$

$$|0\rangle = -\sin\theta_0 |d\rangle + \cos\theta_0 |l\rangle, \tag{92}$$

where

$$\tan\theta_0 = \frac{\varepsilon_l - E_0}{V}, \tag{93}$$

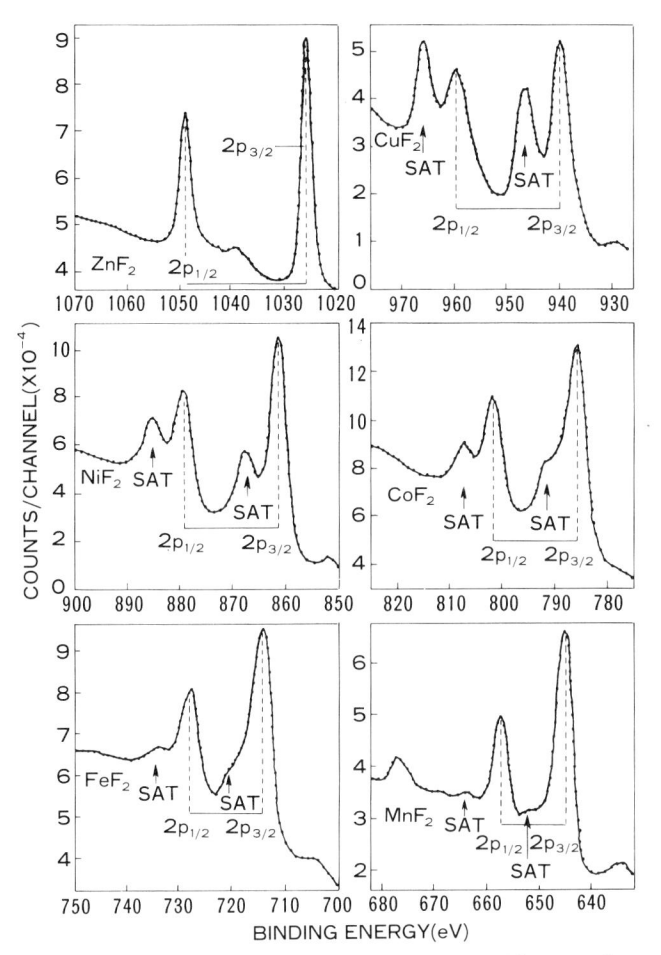

Fig. 15. Experimental data of 2p photoelectron spectra in transition metal compounds (from Rosencwaig et al. 1971).

$$E_0 = \frac{\varepsilon_d^0 + \varepsilon_l}{2} - \sqrt{\left(\frac{\varepsilon_d^0 - \varepsilon_l}{2}\right)^2 + V^2}. \tag{94}$$

The initial state of the photoemission is the ground state $|0\rangle$ with the energy E_0. As a parameter representing the strength of the covalent bond between $|l\rangle$ and $|d\rangle$, we define

$$\gamma_0 = \tan \theta_0 \tag{95}$$

and denote it by "covalency parameter". In the limit of $\varepsilon_d^0 \to \infty$, we have no covalency and $\gamma_0 = 0$. With decreasing $\varepsilon_d^0 - \varepsilon_l$, γ_0 increases and takes the maximum value $\gamma_0 = 1$ in the limit of $\varepsilon_d^0 \to \varepsilon_l$, corresponding to the maximum covalency.

In the final state of the photoemission, we assume that the Hamiltonian is expressed as

$$H = H_0 + (\varepsilon_d - \varepsilon_d^0)|d\rangle\langle d| \tag{96}$$

due to the attractive potential of the core hole of the 3d state. The Hamiltonians (90) and (96) are just the cluster version of the Hamiltonians (88) and (89), respectively. By the diagonalization of H, we obtain

$$H|\pm\rangle = E_\pm|\pm\rangle, \tag{97}$$

where

$$|+\rangle = \cos\theta|d\rangle + \sin\theta|l\rangle, \tag{98}$$

$$|-\rangle = -\sin\theta|d\rangle + \cos\theta|l\rangle, \tag{99}$$

$$\tan\theta = \frac{\varepsilon_l - E_-}{V}, \tag{100}$$

$$E_\pm = \frac{\varepsilon_d + \varepsilon_l}{2} \pm \sqrt{\left(\frac{\varepsilon_d - \varepsilon_l}{2}\right)^2 + V^2}. \tag{101}$$

Therefore, the photoelectron spectrum splits into two peaks corresponding to the final states $|\pm\rangle$ with intensities $I_\pm = |\langle 0|\pm\rangle|^2$. Note that the final state $|+\rangle$ gives the photoelectron peak with *higher* binding energy. The energy separation of the two peaks is given by

$$\Delta E_B = E_+ - E_- = \sqrt{(\varepsilon_d - \varepsilon_l)^2 + 4V^2}, \tag{102}$$

and their intensity ratio is expressed as

$$\frac{I_+}{I_-} = \tan^2(\theta - \theta_0). \tag{103}$$

We discuss the photoelectron spectrum in the following two cases.

(A) *Case of* $\varepsilon_d > \varepsilon_l$
In this case, the final states $|+\rangle$ and $|-\rangle$ correspond mainly to the $3d^{n+1}$ and $3d^n$ configurations, respectively. We define the covalency parameter *in the final state* by

$$\gamma = \tan\theta, \tag{104}$$

so that eq. (103) is rewritten as

$$\frac{I_+}{I_-} = \left(\frac{\gamma - \gamma_0}{1 + \gamma\gamma_0}\right)^2. \qquad (105)$$

Since $\varepsilon_d < \varepsilon_d^0$, the covalency increases in the final state of the photoemission (i.e., $\gamma > \gamma_0$), and this effect is denoted by "photoinduced covalency" after Asada and Sugano (1976). Due to the photoinduced covalency, as found from eq. (105), a *shake-up satellite* with the intensity I_+ occurs in the photoemission. In the case of Mn^{2+}, both of $\varepsilon_d^0 - \varepsilon_l$ and $\varepsilon_d - \varepsilon_l$ are much larger than V, so that γ_0 and γ are small and the satellite intensity is also small. When we go from Mn^{2+} to Fe^{2+}, ε_d^0 and ε_d approach ε_l, and the satellite intensity increases mainly due to the increase of γ. A systematic analysis of the satellite in $Mn^{2+} \sim Ni^{2+}$ was made by Asada and Sugano (1976).

(B) *Case of $\varepsilon_d < \varepsilon_l$*
In this case, the final states $|+\rangle$ and $|-\rangle$ correspond mainly to $3d^n$ and $3d^{n+1}$ configurations, respectively. The covalency parameter is now defined by

$$\gamma = \cot\theta, \qquad (106)$$

so that γ takes the maximum value $\gamma = 1$ at $\varepsilon_d = \varepsilon_l$, and decreases with increasing $\varepsilon_l - \varepsilon_d$, corresponding to the decrease of the covalency. Equation (103) is rewritten as

$$\frac{I_+}{I_-} = \left(\frac{1 - \gamma\gamma_0}{\gamma + \gamma_0}\right)^2. \qquad (107)$$

When both of γ_0 and γ are much smaller than unity, the intensity I_+ is much larger than I_-, so that the final state $|+\rangle$ corresponds to the main line, rather than the satellite. This situation is similar to that of La metal discussed in section 4.2. On the other hand, when both γ_0 and γ are close to unity, the state $|+\rangle$ gives the satellite whose intensity is very small, similarly to the situation of Ni metal discussed in section 5.1.

On the analogy of the satellite in the Ni metal, the insulating Cu^{2+} compounds are considered to be in the case of $\varepsilon_d < \varepsilon_l$, where the satellite (with higher E_B) and the main line (with lower E_B) correspond to the $3d^9$ and $3d^{10}$ configurations, respectively. This fact was suggested by Kotani and Toyozawa (1974b) and Larsson (1975, 1976), and it was confirmed more recently by Van der Laan et al. (1981) and Hüfner (1983). In $CuCl_2$ and $CuBr_2$, the experimental value of the optical absorption gap E_g, which corresponds to the energy difference between $|1\rangle$ and $|0\rangle$, is smaller than the experimental value of ΔE_B. Since E_g is expressed as

$$E_g = \sqrt{(\varepsilon_d^0 - \varepsilon_l)^2 + 4V^2}, \qquad (108)$$

the inequality $E_g < \Delta E_B$ cannot be satisfied when $\varepsilon_d > \varepsilon_l$. The observed E_g

decreases with decreasing electronegativity of the anion (in the order of F$^-$, Cl$^-$ and Br$^-$), consistently with the decrease of $\varepsilon_d^0 - \varepsilon_l$. On the other hand, the observed ΔE_B increases with the decrease of the electronegativity, and this is interpreted reasonably only when $\varepsilon_d < \varepsilon_l$ because $\varepsilon_d^0 - \varepsilon_d$ is not much affected by the electronegativity of the anion. The behavior of ΔE_B was analysed more quantitatively by Hüfner (1983) with essentially the same model system, by using the experimental value of E_g and the quantities

$$V_0 = \langle 1|H|0 \rangle \sim 3 \text{ eV},$$

$$X_c = E_g - \langle 1|H|1 \rangle + \langle 0|H|0 \rangle \sim 12 \text{ eV}$$

as fixed parameters.

In this way, it is clear that $\varepsilon_d < \varepsilon_l$ in Cu^{2+}, whereas $\varepsilon_d > \varepsilon_l$ in Mn^{2+}. However, for Ni^{2+}, Co^{2+} and Fe^{2+} the situation is not so obvious, and further investigations are required.* In the case of Zn^{2+}, the 3d states are completely filled (i.e., in the 3d^{10} configuration) in the ground state, so that the charge transfer from the ligand orbital to the 3d state cannot occur in the final state of the photoemission. This is the reason that the satellite is absent in the photoemission of Zn^{2+}.

Although the cluster model so far considered is very much simplified, the most essential features are unchanged in more realistic cluster calculations (Asada and Sugano 1976), where the degeneracy of 3d and ligand orbitals, the crystalline field effect, and various Coulomb and exchange integrals concerning the 3d, ligand and core hole states are taken into account. If we extend the model by taking account of the energy band effect of anion electronic states, the Hamiltonian H_0 is written, by replacing the ligand orbital in eq. (90) by the filled valence band, as

$$H_0 = \sum_k \varepsilon_k |k\rangle\langle k| + \varepsilon_d^0 |d\rangle\langle d| + V \sum_k (|k\rangle\langle d| + |d\rangle\langle k|) . \tag{109}$$

By using the final state Hamiltonian (96) and by assuming the density of states of the valence band as

$$(2\rho/\pi)\sqrt{1 - (\varepsilon - \varepsilon_l)^2} , \tag{110}$$

the fractional intensity of the satellite is given by

---

*After the manuscript of the present article was completed and submitted to the editor in February 1985, Sawatzky and co-workers published their detailed analysis for the 2p photoemission of nickel dihalides (Zaanen, Westra and Sawatzky 1986). According to their analysis, NiF$_2$ corresponds to the case of $\varepsilon_d < \varepsilon_l$ (in our notation). Taking account of three different configurations 3d^8, 3d^9 and 3d^{10}, they showed that the energies of these configurations in the final state are classified into the following cases: $E_{d^{10}} > E_{d^8} > E_{d^9}$ for NiF$_2$, $E_{d^8} > E_{d^{10}} > E_{d^9}$ for NiCl$_2$ and NiBr$_2$, and $E_{d^8} > E_{d^9} > E_{d^{10}}$ for NiI$_2$. They also studied the energy band effect of ligand orbitals, in addition to their cluster model calculation.

$$\frac{I_+}{I_{\text{tot}}} = \frac{\{1 + \rho V^2[f(\varepsilon_d^0 - E_0) - f(\varepsilon_d - \omega - E_0)]/(\varepsilon_d - \omega - \varepsilon_d^0)\}^2}{\left[1 + 2\rho V^2\left(\dfrac{\omega + E_0 - \varepsilon_d}{\sqrt{(\varepsilon_d - \omega - E_0)^2 - 1}} - 1\right)\right]\left[1 + 2\rho V^2\left(\dfrac{\varepsilon_d^0 - E_0}{\sqrt{(\varepsilon_d^0 - E_0)^2 - 1}} - 1\right)\right]},$$

(111)

where

$$E_0 = -2\rho V^2[\varepsilon_d^0 - \sqrt{(\varepsilon_d^0)^2 - 1 + 4\rho V^2}]/(1 - 4\rho V^2),$$

$$\omega = -E_0 - 2\rho V^2[\varepsilon_d + \sqrt{(\varepsilon_d)^2 - 1 + 4\rho V^2}]/(1 - 4\rho V^2),$$

$$f(\varepsilon) = \text{Re } 2(z - \sqrt{z^2 - 1}), \quad (z \equiv \varepsilon - i\eta, \eta \to +0).$$

Kotani and Toyozawa (1974b) showed that by assuming $\varepsilon_d^0 = 3$, $\varepsilon_d = -3$ and $\rho V^2 = 3$ (with the half-width of the valence band as units), eq. (111) gives $I_+/I_{\text{tot}} \sim 0.42$ in agreement with the experimental result of CuF_2 and also in consistence with the cluster model calculation. However, their analysis is not valid for other transition metal ions because the change of ε_d^0 is not taken into account. The effect of the 3d electron itinerancy was also studied by Asada and Sugano (1978) by using a model system of four clusters.

Next we discuss the multiplet splitting of the photoemission in transition metal compounds. Fadley and Shirley (1970) observed the multiplet splitting of 3s and 3p photoemission in various compounds of Mn and Fe, and analysed them by the Hartree–Fock theory. For instance, the 3s photoelectron spectrum of Mn^{2+} splits into two peaks corresponding to the multiplets ^7S and ^5S. The observed energy separation ΔE_B (about 6.5 eV) and the intensity ratio I_R (about 2.0) of the two peaks were not in agreement with the Hartree–Fock calculation, which gives $\Delta E_B \sim 13$ eV and $I_R = \frac{7}{5}$. More systematic observations of the multiplet splitting in various transition metal compounds including Mn^{2+} have also been made by Carver et al. (1972), Wertheim et al. (1973) and Shirley (1975). The discrepancy between the experiment and the Hartree–Fock theory in Mn^{2+} mentioned above was found to originate from the electron correlation effect, similarly to the 4s photoemission of rare earth compounds mentioned in section 4.1. According to the theory by Bagus et al. (1973), the ^7S final state is not much affected by the electron correlation effect, but the ^5S final state couples strongly with the configuration where two 3p electrons in ^5S are transferred to the 3s and 3d states. Due to this confiruration interaction, the energy of the ^5S multiplet is lowered, resulting in the decrease of the energy difference between ^5S and ^7S multiplets. Furthermore, the intensity of the ^5S photoemission peak decreases by this configuration interaction. Bagus et al. predicted that the decrease of the ^5S intensity should be compensated by the occurrence of new satellite peaks in the higher energy side of the ^5S peak. This prediction was confirmed experimentally by Kowalczyk et al. (1973), who observed two weak satellites in the higher energy side of the ^5S peak of the 3s photoemission of MnF_2. In the 2s photoemission, on the other hand, the electron correlation effect is not important. It is also to be remarked that in the 3s and 3p photoemission of Cu^{2+} and Ni^{2+} both the multiplet coupling and the charge transfer between the 3d and ligand orbitals will be essentially important.

6. Related topics

In the 3d valence band photoemission of Ni, a satellite structure very similar to that of the 2p inner shell photoemission of Ni is observed, as shown in fig. 11b. The satellite is found about 6 eV below the Fermi energy, so that it is well known as "6 eV satellite". Since the mechanism of the 6 eV satellite is closely related with that of the 2p photoemission satellite discussed in section 5, we describe in this section some topics on the 3d photoemission in Ni (see also review articles by Eastman et al. (1979) and by Kotani (1985)).

6.1. Satellite in 3d photoemission of Ni

The existence of the 6 eV satellite in the 3d photoemission of Ni was already reported in 1970 (Baer et al. 1970), but in those days the satellite was considered to originate from the surface plasmon loss mechanism. Kemeney and Schevchik (1975) denied the possibility of the plasmon loss mechanism by observing the 3d photoemission of NiZn alloys, where they found that the Ni 3d photoemission, is accompanied by the satellite but the Zn 3d photoemission does not exhibit it. Furthermore, Kemeney and Schevchik (1975), as well as Hüfner and Wertheim (1975a), discovered the similarity of the satellites in the 2p and 3d photoelectron spectra, and proposed that the mechanism of the satellite is also similar in the two cases.

We have shown in fig. 14 the mechanism of the satellite in the 2p photoemission. The mechanism of the 6 eV satellite in the 3d photoemission can also be explained by using fig. 14. Let us assume that a 3d↑ electron is excited by the incident photon. Then the core hole in (B) and (C) of fig. 14 should be replaced by a 3d hole in the ↑ band (denoted by the 3d↑ hole). The 3d↑ hole moves throughout the crystal in contrast to the localized core hole, but even in this case we have the two types of final states similar to (B) and (C). For the final state (B), although the 3d↑ hole moves, its motion is strongly restricted such that it cannot approach the 3d↓ holes as a result of the repulsive Coulomb interaction between 3d↑ and 3d↓ holes (so that the motion is the strongly correlated one). In the final state (C), the two-hole bound state is now constructed by the two 3d holes with ↑ and ↓ spins. We denote it as "two-d-hole bound state".

The process of the two-d-hole bound state formation is schematically shown in fig. 16a. The photocreated 3d hole in the $(k, ↑)$ state interacts with another 3d hole of $(k', ↓)$ through the intra-atomic Coulomb interaction U_{dd}, and the 3d

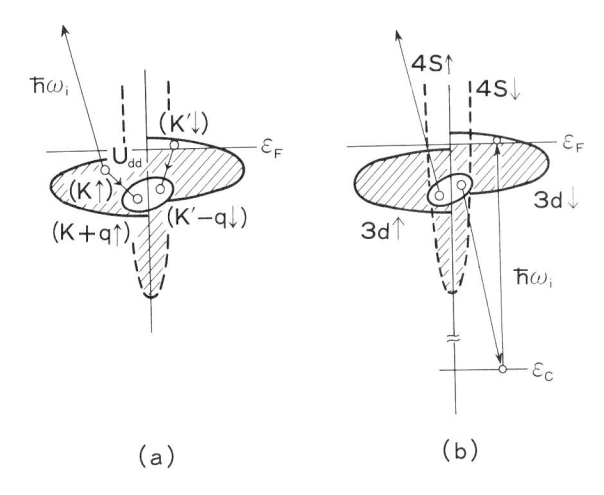

Fig. 16. Schematic representation of (a) 3d photoemission and (b) resonant photoemission processes in Ni.

hole pair is scattered to another 3d hole pair $(k+q, \downarrow)$ and $(k'-q, \uparrow)$, which further repeats the scattering. As a result of this multiple scattering, the two-d-hole bound state is formed. The first theoretical calculation of the two-d-hold bound state was carried out by Penn (1979), who treated the multiple scattering by Kanamori's t-matrix method (Kanamori 1963). The theory has further been improved by Liebsch (1979, 1981), Davis and Feldkamp (1979, 1980a), Tréglia et al. (1980, 1982) and Igarashi (1983, 1985). When we use the single-band Hubbard model for the 3d band, the two-d-hole bound state is a singlet spin state, as shown in fig. 16a, but when we take account of the orbital degeneracy of 3d states, the triplet bound state also occurs.

We have shown that the 6 eV satellite originates from the two-d-hole bound state, while the main band of the 3d photoemission corresponds to the final state where 3d holes move with the strong correlation. This correlation effect in the main band is also reflected in *spin-polarized* photoelectron spectra, as well as in *angle-resolved* photoelectron spectra as mentioned below. The spin polarization of photoelectrons from the 3d band of Ni was first observed by Bänninger et al. (1970). They observed the positive spin polarization of the photoelectrons excited from just below the Fermi level ε_F, where the positive spin polarization means that the spin polarization is parallel to the ferromagnetic spin direction. However, according to the energy band theory, only the 3d$\downarrow$ band has large density of states at ε_F (see the schematical illustration in fig. 12), so that the electron excited from near ε_F should have negative spin polarization, contrary to the experimental result. In order to resolve the discrepancy between experiment and theory, many experimental and theoretical investigations have been performed. The spin polarization P of the 3d photoemission was measured more precisely by Eib and Alvarado (1976) as a function of the incident photon energy $\hbar\omega_i$. They found that when $\hbar\omega_i$ is sufficiently close to the work function (so that the photoelectron is excited from the state sufficiently close to ε_F) the observed P is negative, in agreement with the energy band theory. As $\hbar\omega_i$ increases, however, P changes its sign at a value of $\hbar\omega_i$ which is much smaller than that estimated from the band theory. It was shown by Moore and Pendry (1978) that in order to reproduce the experimental P versus $\hbar\omega_i$ relation the exchange splitting Δ_{ex} of the 3d band should be about 0.3 eV, which is only half of Δ_{ex} obtained from self-consistent energy band calculations (see, for instance, Moruzzi et al. 1978).

Himpsel et al. (1979) observed the angle-resolved photoelectron spectra from the valence band of Ni, and succeeded to obtain experimentally the energy dispersion of 3d and 4s electrons along several high-symmetry axes in the wave vector space. They estimated the exchange splitting Δ_{ex} from their experimental energy separation of a pair of exchange-split branches near the L_3 point and obtained $\Delta_{ex} \sim 0.3$ eV, in good agreement with the Δ_{ex} given by Moore and Pendry. Furthermore, Himpsel et al. (1979) revealed that the experimental value of the occupied 3d band width (denoted by W) is also much smaller than that of the energy band calculation. For instance, the experimental value of W at the L point is about 3.4 eV, whereas the theoretical value is about 4.7 eV (Moruzzi et al. 1978).

The narrowings of experimental values Δ_{ex} and W suggest the importance of the electron correlation effect, which is not taken into account sufficiently in the energy band theories. Indeed, the discrepancy between the experiment and the band theory can be explained by the electron correlation effect mentioned before. It is to be noted that the spin-polarized photoemission and the angle-resolved photoemission mentioned so far were measured for the *main band* of the 3d photoemission. In the final state of the main band, the 3d holes are kept away from each other with the strong correlation, so that the space for the motion of the 3d holes is relatively decreased, resulting in the narrowing of the band width W. Furthermore, the effective exchange interaction responsible for the exchange splitting Δ_{ex} is also weakened by this correlation effect, resulting in the decrease of Δ_{ex}. This interpretation is confirmed by recent theoretical calculations (Liebsch 1981, Igarashi 1983, and others) not only qualitatively but also semi-quantitatively. In order to further improve the theory, it will be necessary to take into account explicitly the screening effect of 4s electrons, as pointed out by Kanamori (1981).

6.2. Resonant photoemission

Guillot et al. (1977) discovered that the intensity of the 6 eV satellite of the 3d photoemission in Ni is resonantly enhanced when the incident photon energy $\hbar\omega_i$ approaches the 3p core electron excitation threshold $\hbar\omega_0 \sim 67$ eV, as shown schematically with the solid curves in fig. 17. this phenomenon is denoted by the "resonant photoemission". When $\hbar\omega_i$ becomes larger than $\hbar\omega_0$, an Auger electron peak appears, but the resonant enhancement of the satellite also persists.

The mechanism of the resonant photoemission is considered to be the following second order process (as shown in fig. 16b), arising from the interplay between the core electron excitation and the valence electron excitation: (i) By absorbing

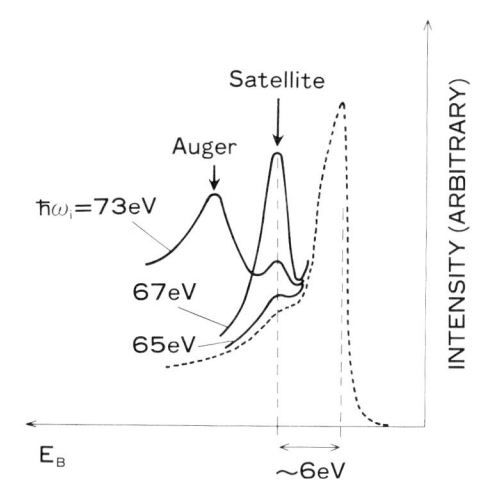

Fig. 17. Schematic drawing of the resonant photoemission spectra in Ni (after Guillot et al. 1977).

the incident photon, a 3p core electron is excited to the 3d (or 4s) band. (ii) Then, the super Coster–Kronig transition occurs, where one 3d electron makes a transition to the 3p level and another 3d electron is excited to become a photoelectron. Therefore, in the final state of this second order process, two extra 3d holes are created and they form the two-d-hole bound state through the multiple scattering by the interaction U_{dd}. We note that the final state of fig. 16b is the same as that of fig. 16a, which contributes to the satellite of the 3d photoemission. In this way, it is understood that the satellite intensity is enhanced when $\hbar\omega_i$ is close to the 3p threshold. More precisely, when the processes (i) and (ii) occur *coherently*, we obtain the resonant enhancement of the *satellite*, but when they occur *independently*, we obtain the *Auger electron* emission. For instance, when $\hbar\omega_i$ is sufficiently larger than $\hbar\omega_0$, the 3p electron is excited by the incident photon to the 4s band (or higher conduction bands). Then the excited electron moves away from the core hole site before the super Coster–Kronig transition occurs, so that the processes (i) and (ii) are independent and we have the Auger electron emission. However, when $\hbar\omega_i$ is near $\hbar\omega_0$, the excited 3d or 4s electron does not move away so rapidly, and contributes essentially to the final state of the super Coster–Kronig transition. When the 3p electron is photoexcited to the 3d band, this process contributes mainly to the satellite, as mentioned above. In this case, since the final states of (a) and (b) of fig. 16 are the same, the two processes *interfere* with each other. When the 3p electron is excited to the 4s band near ε_F, this process contributes partly to the satellite because the 4s electron relaxes around the two-d-hole bound state to screen its attractive potential.

Theoretical calculations of resonant photoemission were made by Penn (1979) and by Davis and Feldkamp (1981), but they disregarded the existence of the 4s band. Recently, Jo et al. (1983) have calculated the resonant photoemission spectrum by taking account of both the 3d and 4s bands with s–d hybridization. The formation of the two-d-hole bound state is treated self-consistently through the interaction U_{dd}, where the two-d-hole bound state is assumed to be well-localized. The 4s electron screening effect is also taken into account by introducing the potential U_{sd} of the two-d-hole bound state. A result is shown in fig. 18, where the intensities of the satellite and Auger peaks are plotted as a function of the incident photon energy $\hbar\omega_i - \hbar\omega_0$. The parameter α represents the ratio of the photoexcitation amplitudes of the 3d and 3p electrons. When $\alpha = 0$, we have only the second order process (fig. 16b), and the satellite intensity is enhanced symmetrically around $\hbar\omega_i = \hbar\omega_0$. When $\alpha \neq 0$, the satellite intensity becomes asymmetric around $\hbar\omega_i = \hbar\omega_0$ (decreased for $\hbar\omega_i < \hbar\omega_0$ and increased for $\hbar\omega_i > \hbar\omega_0$) due to the *interference effect* between the processes of fig. 16a and 16b. Such an interference is well known as the *Fano effect* (Fano 1961). According to the experiments by Guillot et al. (1977) and Barth et al. (1979), the satellite intensity is asymmetric as shown in the inset of fig. 18, suggesting the importance of the interference effect. On the other hand, the Auger intensity is not much influenced by the interference.

Instead of the electron correlation mechanism mentioned so far, another mechanism based on a simple one-electron picture was also proposed to explain

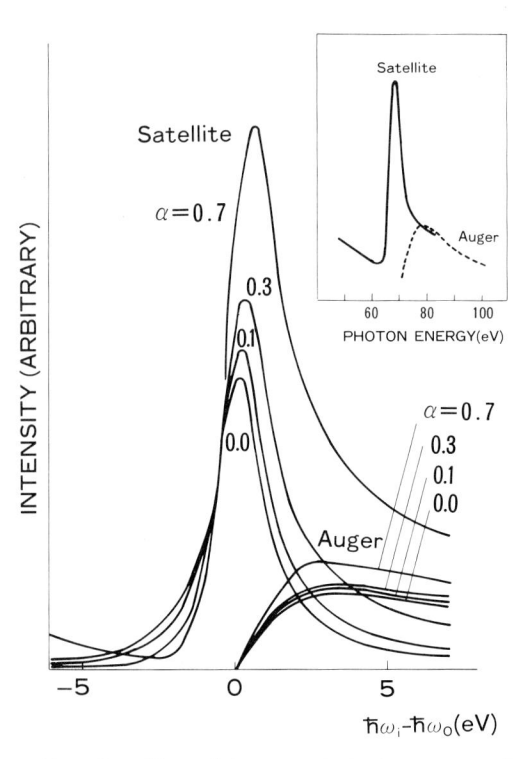

Fig. 18. Calculated intensities of satellite and Auger peaks in the resonant photoemission of Ni (after Jo et al. 1983). The inset shows the experimental result by Barth et al. (1979).

the occurrence of the satellite and its resonant enhancement. Kanski et al. (1980) calculated the 3d photoemission spectra by assuming a simple one-electron transition between the 3d band and the high-energy photoelectron band. They obtained a satellite peak at about 5 eV below the Fermi energy, originating from the flat top region of the lowest Δ_1 energy band. Furthermore, the intensity of this peak was found to be strongly enhanced for $\hbar\omega_i \sim 66$ eV, as a result of the increase of the joint density of states.

In order to study experimentally which is the more essential of the two mechanisms (i.e., the electron correlation mechanism and the one-electron mechanism), it is most effective to observe the photoelectron *spin polarization of the satellite* for $\hbar\omega_i \sim \hbar\omega_0$. If the electron correlation mechanism is more essential, the photoelectrons of the satellite should have a large positive spin polarization by the following reason (which was first pointed out by Feldkamp and Davis (1979) with an atomic model): At the 3p threshold, only the 3p ↓ electron can be excited strongly to the 3d ↓ band, so that only the 3d ↑ electron can become the photoelectron, as shown in fig. 16b, if the two-d-hole bound state is in the singlet state (the contribution from the triplet spin state is actually minor). On the other hand, if the one-electron mechanism is more essential, the photoelectrons of the satellite should not have such a large spin polarization.

Clauberg et al. (1981) observed the photoelectron spin polarization P of the satellite and obtained a large positive spin polarization of about 50% just at $\hbar\omega_i = \hbar\omega_0$, as shown in the inset of fig. 19. This result supports strongly the importance of the electron correlation effects. They also found that P of the satellite decreases when $\hbar\omega_i$ is away from $\hbar\omega_0$, where the decrease of P is more rapid for $\hbar\omega_i < \hbar\omega_0$ than for $\hbar\omega_i > \hbar\omega_0$. Also they reported that P of the Auger peak is much smaller than that of the satellite. These results can also be explained

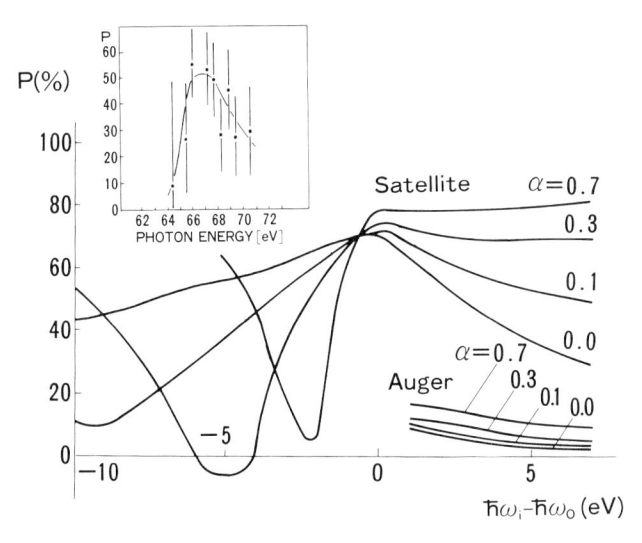

Fig. 19. Calculated spin polarization of satellite and Auger peaks in the resonant photoemission of Ni (after Jo et al. 1983). The inset shows the experimental result by Clauberg et al. (1981).

by the theory of Jo et al. (1983). The calculated result of the spin polarization of satellite and Auger peaks is shown in fig. 19 as a function of $\hbar\omega_i - \hbar\omega_0$. It is found that the satellite has a large positive value of P at $\hbar\omega_i = \hbar\omega_0$, and the P versus $\hbar\omega_i$ relation depends strongly on α. In order to understand this behavior of P, we note that there exist four independent processes in the resonant second order process, i.e., $3p\sigma$–$3d\sigma$ and $3p\sigma$–$4s\sigma$ photoexcitations followed by the emission of $-\sigma$ spin photoelectron ($\sigma = \uparrow, \downarrow$) due to the super Coster–Kronig transition. The intensity of the $3p\downarrow$–$3d\downarrow$ process is always much stronger than that of $3p\uparrow$–$3d\uparrow$, which is only weakly allowed due to the s–d hybridization, while the $3p\downarrow$–$4s\downarrow$ and $3p\uparrow$–$4s\uparrow$ processes occur with the comparable intensities. When $\hbar\omega_i \sim \hbar\omega_0$, the $3p\downarrow$–$3d\downarrow$ process is dominant of the four processes, so that the satellite has a large positive P, as mentioned before. When $\hbar\omega_i$ goes away from $\hbar\omega_0$, the intensity of the $3p\downarrow$–$3d\downarrow$ process decreases and becomes comparable with $3p\sigma$–$4s\sigma$ processes. Therefore, P of the satellite decreases with increasing $|\hbar\omega_i - \hbar\omega_0|$. For $\alpha = 0$, as shown in fig. 19, the decrease of P for $\hbar\omega_i > \hbar\omega_0$ is more rapid than that for $\hbar\omega_i < \hbar\omega_0$.

Let us consider the case of $\alpha \neq 0$. Then the $3p\downarrow$–$3d\downarrow$ process interferes with the direct photoexcitation process of the $3d\uparrow$ electron, whereas the $3p\sigma$–$4s\sigma$ processes do not interfere with it. As a result, the intensity of the $3p\downarrow$–$3d\downarrow$ process, which contributes to the positive P, decreases for $\hbar\omega_i < \hbar\omega_0$ and increases for $\hbar\omega_i > \hbar\omega_0$, so that P also follows this trend, as shown in fig. 19. Thus, by taking account of the interference effect the calculated P versus $\hbar\omega_i$ relation becomes consistent with the experimental result. Further, the calculated P of the Auger peak is much smaller than that of the satellite, which also agrees with the experiment. The reason of the small P of the Auger peak is that the Auger electron emission comes mainly from the $3p\sigma - 4s\sigma$ processes.

Recently, the experimental observations of resonant photoemission spectra have been carried out for various materials including other transition metals (Iwan et al. 1979, Chandesris et al. 1983, Sugawara et al. 1984), transition metal compounds (Thuler et al. 1983, Kakizaki et al. 1983, Ishii 1983), rare earth compounds (Allen et al. 1982, Peterman et al. 1983) and semiconductors (Kobayashi et al. 1983, Taniguchi et al. 1984). The theoretical analysis of resonant photoemission has also been made for Cu metal by Davis and Feldkamp (1980b) and Parlebas et al. (1982), for transition metal compounds by Davis (1982), and for semiconducting black phosphorus by Kotani and Nakano (1984).

7. Concluding remarks

In this chapter, we have described the fundamental theory of inner shell photoelectron processes, and presented typical experimental data and their interpretations. Large space is devoted to the many-body effects in the photoemission of magnetic materials, such as rare earth metals, transition metals, and their compounds. The competition between the local and itinerant properties of f and d electrons in these materials is one of the most interesting many-body problems in solid state physics, and we have shown that the photoelectron spectroscopy provides us with very important information on this problem. For the interpretation of various photoemission data, emphasis is placed on the understanding of the most essential mechanism with the use of the model system as simple as possible. To this end, we have omitted the description on some realistic but complicated aspects of photoelectron processes; for instance, the energy loss of photoelectrons (i.e., the effect of the interaction V in the Appendix), the spectrum of secondary electrons, the effect of solid surfaces and so on. We also omitted the subjects on the time-resolved spectrum, as well as the spin-polarized spectrum, of inner shell photoelectrons, although they will be interesting subjects in future investigations.

In the final state of the inner shell photoemission, a core hole is left behind, so that this state is, at the same time, the initial state of the Auger transition and the optical emission. When these secondary processes occur coherently with the primary core excitation process, we obtain the resonant photoemission (as briefly discussed in this chapter) and the resonant X-ray scattering. The investigation of these second order processes will be much developed in future, with the development of more powerful sources of synchrotron radiation.

Acknowledgements

The author would like to express his sincere thanks to Professor J. Kanamori and Professor Y. Toyozawa for their encouragement and enlightening discussion. He thanks Dr. T. Jo, Dr. J. Igarashi, Dr. J.C. Parlebas, Mr. T. Nakano and Mr. H. Mizuta for their stimulating interest and support. Thanks are also due to Mrs. A. Egusa for her help in preparing the manuscript.

Appendix. Derivation of basic photoemission formulae

Let us consider an inner shell photoelectron process, where a photon with wave vector q and polarization λ is incident on a material sample. A photoelectron which is emitted from the sample is observed at a position R, well apart from the sample. We write the Hamiltonian of the material system (i.e., the sample) as H_m, and denote its ground state as $|g\rangle$ (with energy E_g). Then the initial state of the photoemission is expressed as

$$|\psi_0\rangle = |q\lambda\rangle|g\rangle.$$

After switching on the electron–photon interaction, by which the incident photon is absorbed and a core electron c is excited to a photoelectron state, the eigenstate of the total system is expressed, by using the scattering theory, as

$$|\psi\rangle = |\psi_0\rangle + \frac{1}{\hbar\omega_i + E_g - H_m + i\eta} M_c^+ |g\rangle, \quad \eta \to +0,$$

where $\hbar\omega_i$ ($\equiv \hbar cq$) is the incident photon energy, M_c^+ represents the photo-excitation of the core electron

$$M_c^+ = \sum_i v_R^*(i; cq\lambda) a_i^+ a_c$$

with v_R expressed as eq. (47) in the text and with an arbitrary one-electron complete set i.

We divide the material system into two sub-systems: the excited photoelectron (described by Hamiltonian h) and the remaining material system (described by Hamiltonian $\tilde{H}_m$). Therefore, H_m is written as

$$H_m = h + \tilde{H}_m + V,$$

where V is the interaction between the two sub-systems. The Hamiltonian h is further divided as follows:

$$h = h_0 + v,$$

where h_0 is the Hamiltonian of a free electron and v is the periodic potential of crystalline lattice (v vanishes outside the sample). We write the eigenstates of h_0 and $\tilde{H}_m$ as $|k\rangle$ and $|\tilde{m}\rangle$, respectively, so that

$$h_0|k\rangle = \varepsilon_k|k\rangle, \quad \varepsilon_k = \hbar^2 k^2/2m \quad \text{and} \quad \tilde{H}_m|\tilde{m}\rangle = \tilde{E}_m|\tilde{m}\rangle.$$

Then the overlap integral between $|\psi\rangle$ and $|k, \tilde{m}\rangle$ ($\equiv |k\rangle|\tilde{m}\rangle$) is given by

$$\langle k, \tilde{m}|\psi\rangle = \frac{1}{\hbar\omega_i + E_g - \tilde{E}_m - \varepsilon_k + i\eta} \langle k, \tilde{m}|(1 + TG)M_c^+|g\rangle,$$

where we have used the following resolvent expansion formula (with respect to the interaction $v + V$):

$$\frac{1}{\hbar\omega_i + E_g - H_m + i\eta} = G(1 + TG)$$

with

$$G = \frac{1}{\hbar\omega_i + E_g - \tilde{H}_m - h_0 + i\eta},$$

$$T = (v + V)(1 + GT).$$

In order to obtain the probability amplitude that the photoelectron is detected at R, we use the real-space representation of the photoelectron state,

$$|r\rangle = \sum_k \frac{1}{(2\pi)^{3/2}} e^{i k \cdot r} |k\rangle .$$

The asymptotic form of $\langle R, \tilde{m}|\psi\rangle$ for large R is obtained as

$$\langle R, \tilde{m}|\psi\rangle = -\frac{2m}{\hbar^2} \frac{\exp(i k_f R)}{4\pi R} \langle k_f, \tilde{m}|(1 + TG) M_c^+|g\rangle ,$$

where

$$k_f = \sqrt{2m(\hbar\omega_i + E_g - \tilde{E}_m)}/\hbar ,$$

and the direction of k_f is parallel to R. Thus, we obtain the photocurrent per unit solid angle at R as follows:

$$I = \sum_m \frac{m k_f}{4\pi^2 \hbar^3} |\langle k_f, \tilde{m}|(1 + TG)|g\rangle|^2 \delta(\varepsilon_{k_f} + \tilde{E}_m - E_g - \hbar\omega_i) .$$

Next, we impose several simplifying assumptions. We first disregard the effect of V, then the photocurrent is expressed as

$$I = \sum_m A(k_f) |\langle \tilde{m}|a_c|g\rangle|^2 \delta(\varepsilon_{k_f} + \tilde{E}_m - E_g - \hbar\omega_i) ,$$

where

$$A(k_f) = \frac{e^2 k_f}{2\pi \hbar^2 m \omega_i} |\langle \phi_{k_f}|(1 + tg) p \cdot \eta_{q\lambda}|\phi_c\rangle|^2 ,$$

$$g = 1/(\varepsilon_{k_f} - h_0 + i\eta) ,$$

$$t = v(1 + gt) .$$

We further assume $A(k_f)$ to be a constant after averaging over the angle of k_f (this assumption is not very poor when ε_{k_f} is sufficiently large). Then we define the photoelectron spectrum F, which is normalized so as to be unity when integrated over the photoelectron kinetic energy $\varepsilon(=\varepsilon_{k_f})$:

$$F = \sum_m |\langle \tilde{m}|a_c|g\rangle|^2 \delta(\varepsilon + \tilde{E}_m - E_g - \hbar\omega_i) .$$

Finally, we decompose the system into the core electrons and outer electrons (including the lattice of ions), and eliminate the operator a_c and the core states from the expression of F. We denote the outer electron system as "medium". In the initial state of the photoemission, the Hamiltonian of the medium is expressed as H_0, and in the final state of the photoemission it is written as $H = H_0 + U$, where U represents the disturbance caused by the core hole. By expressing the ground state of H_0 as $|0\rangle$ (with energy E_0) and eigenstates of H as $|f\rangle$ (with energies E_f), we obtain

$$F = \sum_f |\langle 0|f\rangle|^2 \, \delta(\varepsilon - \varepsilon_c + E_f - E_0 - \hbar\omega_i) \, ,$$

where ε_c is the one-electron energy of the core state.

References

Allen, J.W., S.-J. Oh, I. Lindau, M.B. Maple, J.F. Suassuna and S.B. Hagström, 1982, Phys. Rev. B **26**, 445.
Almbladh, C.-O., and L. Hedin, 1983, Beyond the One-Electron Model, in: Handbook on Synchrotron Radiation, Vol. 1, ed. E.-E. Koch (North-Holland, Amsterdam) p. 607.
Anderson, P.W., 1967, Phys. Rev. Lett. **18**, 1049.
Asada, S., and S. Sugano, 1976, J. Phys. Soc. Jpn. **41**, 1291.
Asada, S., and S. Sugano, 1978, J. Phys. C **11**, 3911.
Baer, Y., P. Hedén, J. Hedman, M. Klasson, C. Nordling and K. Siegbahn, 1970, Phys. Scr. **1**, 55.
Bagus, P.S., A.J. Freeman and F. Sasaki, 1973, Phys. Rev. Lett. **30**, 850.
Bänninger, U., G. Busch, M. Campagna and H.C. Siegmann, 1970, Phys. Rev. Lett. **25**, 383.
Barth, J., G. Kalkoffen and C. Kunz, 1979, Phys. Rev. A **74**, 360.
Beaurepaire, E., 1983, Thesis (Institut Polytechnique, Nancy).
Burroughs, P., A. Hamnett, A.F. Orchard and G. Thornton, 1976, J. Chem. Soc. Dalton Trans. **17**, 1686.
Cardona, M., and L. Ley, eds, 1978, Photoemission in Solids I. General Principles (Springer, Berlin).
Carver, J.C., G.K. Schweitzer and T.A. Carlson, 1972, J. Chem. Phys. **57**, 973.
Chandesris, D., J. Lecante and Y. Pétroff, 1983, Phys. Rev. B **27**, 2630.
Citrin, P.H., G.K. Wertheim and Y. Baer, 1977, Phys. Rev. B **16**, 4256.
Clauberg, R., W. Gudat, E. Kisker, E. Kuhlmann and G.M. Rothberg, 1981, Phys. Rev. Lett. **47**, 1314.
Cohen, R.L., G.K. Wertheim, A. Rosencwaig and H.J. Guggenheim, 1972, Phys. Rev. B **5**, 1037.
Combescot, M., and P. Nozières, 1971, J. Phys. (France) **32**, 913.
Crecelius, G., G.K. Wertheim and D.N.E. Buchanan, 1978, Phys. Rev. B **18**, 6519.
Davis, L.C., 1982, Phys. Rev. B **25**, 2912.
Davis, L.C., and L.A. Feldkamp, 1979, J. Appl. Phys. **50**, 1944.
Davis, L.C., and L.A. Feldkamp, 1980a, Solid State Commun. **34**, 141.
Davis, L.C., and L.A. Feldkamp, 1980b, Phys. Rev. Lett. **44**, 673.
Davis, L.C., and L.A. Feldkamp, 1981, Phys. Rev. B **23**, 6239.
Doniach, S., and M.J. Šunjić, 1970, J. Phys. C **3**, 285.
Eastman, D.E., J.F. Janak, A.R. Williams, R.V. Coleman and G. Wendin, 1979, J. Appl. Phys. **50**, 7423.
Eib, W., and S.F. Alvarado, 1976, Phys. Rev. Lett. **37**, 444.
Fadley, C.S., and D.A. Shirley, 1970, Phys. Rev. A **2**, 1109.
Fano, U., 1961, Phys. Rev. **124**, 1866.
Feldkamp, L.A., and L.C. Davis, 1979, Phys. Rev. Lett. **43**, 151.
Feldkamp, L.A., and L.C. Davis, 1980, Phys. Rev. B **22**, 3644.
Friedel, J., 1969, Comments Solid State Phys. **2**, 21.
Fuggle, J.C., M. Campagna, Z. Zolnierek, R. Lässer and A. Platau, 1980, Phys. Rev. Lett. **45**, 1597.
Fuggle, J.C., F.U. Hillebrecht, J.-M. Esteva, R.C. Karnatak, O. Gunnarsson and K. Schönhammer, 1983a, Phys. Rev. B **27**, 4637.
Fuggle, J.C., F.U. Hillebrecht, Z. Zolnierek, R. Lässer, Ch. Freiburg, O. Gunnarsson and K. Schönhammer, 1983b, Phys. Rev. B **27**, 7330.

Fujimori, A., 1983a, Phys. Rev. B **27**, 3992.
Fujimori, A., 1983b, Phys. Rev. B **28**, 2281.
Guillot, C., Y. Ballu, J. Paigné, J. Lecante, K.P. Jain, P. Thiry, R. Pinchaux, Y. Pétroff and L.M. Falicov, 1977, Phys. Rev. Lett. **39**, 1632.
Gunnarsson, O., and K. Schönhammer, 1983, Phys. Rev. B **28**, 4315.
Herbst, J.F., J.M. Burkstrand and J.W. Wilkins, 1980, Phys. Rev. B **22**, 531.
Hillebrecht, F.U., and J.C. Fuggle, 1982, Phys. Rev. B **25**, 3550.
Hillebrecht, F.U., J.C. Fuggle, P.A. Bennett, Z. Zolnierek and Ch. Freiburg, 1983, Phys. Rev. B **27**, 2179.
Himpsel, F.J., J.A. Knapp and D.E. Eastman, 1979, Phys. Rev. B **19**, 2919.
Hopfield, J.J., 1969, Comments Solid State Phys. **2**, 40.
Hüfner, S., 1983, Solid State Commun. **47**, 943.
Hüfner, S., and G.K. Wertheim, 1975a, Phys. Lett. A **51**, 299.
Hüfner, S., and G.K. Wertheim, 1975b, Phys. Lett. A **51**, 301.
Hüfner, S., and G.K. Wertheim, 1975c, Phys. Rev. B **11**, 678.
Igarashi, J., 1983, J. Phys. Soc. Jpn. **52**, 2827.
Igarashi, J., 1985, J. Phys. Soc. Jpn. **54**, 260.
Ishii, T., 1983, Ann. Israel Phys. Soc. **6**, 587.
Iwan, M., F.J. Himpsel and D.E. Eastman, 1979, Phys. Rev. Lett. **43**, 1829.
Jo, T., and A. Kotani, 1985, Solid State Commun. **54**, 451.
Jo, T., A. Kotani, J.C. Parlebas and J. Kanamori, 1983, J. Phys. Soc. Jpn. **52**, 2581.
Kaga, H., A. Kotani and Y. Toyozawa, 1976, J. Phys. Soc. Jpn. **41**, 1851.
Kakchashi, Y., and A. Kotani, 1984, Phys. Rev. B **29**, 4292.
Kakehashi, Y., K. Becker and P. Fulde, 1984, Phys. Rev. B **29**, 16.
Kakizaki, A., K. Sugeno, T. Ishii, H. Sugawara and I. Nagakura, 1983, Phys. Rev. B **28**, 1026.
Kanamori, J., 1963, Prog. Theor. Phys. **30**, 275.
Kanamori, J., 1981, in: Electron Correlation and Magnetism in Narrow-Band Systems, ed. T. Moriya (Springer, Berlin) p. 102.
Kanski, J., P.O. Nilsson and C.G. Larsson, 1980, Solid State Commun. **35**, 397.
Kemeney, P.C., and N.Y. Schevchik, 1975, Solid State Commun. **17**, 255.
Kim, K.S., 1975, Phys. Rev. B **11**, 2177.
Kim, K.S., and R.E. Davis, 1972, J. Electron Spectrosc. & Relat. Phenom. **1**, 251.
Kobayashi, K.L.I., H. Daimon and Y. Murata, 1983, Phys. Rev. Lett. **50**, 1701.
Koelling, D.D., A.M. Boring and J.H. Wood, 1983, Solid State Commun. **47**, 227.
Kotani, A., 1979, J. Phys. Soc. Jpn. **46**, 488.
Kotani, A., 1983, Ann. Israel Phys. Soc. **6**, 539.
Kotani, A., 1985, J. Appl. Phys. **57**, 3632.
Kotani, A., and T. Nakano, 1984, Solid State Commun. **51**, 97.
Kotani, A., and J.C. Parlebas, 1985, J. Phys. (France) **46**, 77.
Kotani, A., and Y. Toyozawa, 1973a, J. Phys. Soc. Jpn. **35**, 1073.
Kotani, A., and Y. Toyozawa, 1973b, J. Phys. Soc. Jpn. **35**, 1082.
Kotani, A., and Y. Toyozawa, 1974a, J. Phys. Soc. Jpn. **37**, 912.
Kotani, A., and Y. Toyozawa, 1974b, J. Phys. Soc. Jpn. **37**, 563.
Kotani, A., and Y. Toyozawa, 1979, Theoretical Aspects of Inner-Level Spectroscopy, in: Synchrotron Radiation, ed. C. Kunz (Springer, Berlin) p. 169.
Kotani, A., H. Mizuta, T. Jo and J.C. Parlebas, 1985, Solid State Commun. **53**, 805.
Kowalczyk, S.P., L. Ley, R.A. Pollak, F.R. McFeely and D.A. Shirley, 1973, Phys. Rev. B **7**, 4009.
Krill, G., J.P. Kappler, A. Meyer, L. Abadli and M.F. Ravet, 1981, J. Phys. F **11**, 1713.
Larsson, S., 1975, Chem. Phys. Lett. **32**, 401.
Larsson, S., 1976, Chem. Phys. Lett. **40**, 362.
Lässer, R., J.C. Fuggle, M. Beyss, M. Campagna, F. Steglich and F. Hullinger, 1980, Physica B **102**, 360.
Ley, L., and M. Cardona, eds, 1979, Photoemission in Solids II. Case Studies (Springer, Berlin).
Liebsch, A., 1979, Phys. Rev. Lett. **43**, 1431.

Liebsch, A., 1981, Phys. Rev. B **23**, 5203.
Mahan, G.D., 1967, Phys. Rev. **163**, 612.
McFeeley, F.R., S.P. Kowalczyk, L. Ley and D.A. Shirley, 1972, Phys. Lett. A **49**, 301.
Moore, J.D., and J.B. Pendry, 1978, J. Phys. C **11**, 4615.
Moruzzi, V.L., J.F. Janak and A.R. Williams, 1978, Calculated Electronic Properties of Metals (Pergamon, New York).
Nagakura, I., T. Ishii and T. Sagawa, 1972, J. Phys. Soc. Jpn. **33**, 754.
Nozières, P., and C.T. De Dominicis, 1969, Phys. Rev. **178**, 1097.
Oh, S.-J., and S. Doniach, 1982, Phys. Rev. B **26**, 2085.
Parlebas, J.C., A. Kotani and J. Kanamori, 1982, J. Phys. Soc. Jpn. **51**, 124.
Penn, D.R., 1979, Phys. Rev. Lett. **42**, 921.
Peterman, D.J., J.H. Weaver, M. Croft and D.T. Peterson, 1983, Phys. Rev. B **27**, 808.
Richardson, D., and L.A. Hisscott, 1976, J. Phys. F **6**, L127.
Rosencwaig, A., G.K. Wertheim and H.J. Guggenheim, 1971, Phys. Rev. Lett. **27**, 479.
Schneider, W.-D., and C. Laubschat, 1981, Phys. Rev. Lett. **46**, 1023.
Schneider, W.-D., C. Laubschat, I. Nowik and G. Kaindl, 1981, Phys. Rev. B **24**, 5422.
Schönhammer, K., and O. Gunnarsson, 1977, Solid State Commun. **23**, 691.
Schönhammer, K., and O. Gunnarsson, 1978, Z. Phys. B **30**, 297.
Shirley, D.A., 1975, Phys. Scr. **11**, 177.
Siegbahn, K., C. Nordling, A. Fahlman, R. Nordberg, K. Hamrin, J. Hedman, G. Johansson, T. Bergmark, S.-E. Karlsson, I. Lindgren and B. Lindberg, 1967, in: ESCA-Atomic, Molecular and Solid State Structure Studied by Means of Electron Spectroscopy (Almqvist and Wicksells, Uppsala, Sweden).
Signorelli, A.J., and R.G. Hayes, 1973, Phys. Rev. B **8**, 81.
Sugawara, H., K. Naito, T. Miya, A. Kakizaki, I. Nagakura and T. Ishii, 1984, J. Phys. Soc. Jpn. **53**, 279.
Suzuki, S., T. Ishii and T. Sagawa, 1974, J. Phys. Soc. Jpn. **37**, 1334.
Taniguchi, M., S. Suga, M. Seki, H. Sakamoto, H. Kanzaki, Y. Akahama, S. Endo, S. Terada and S. Narita, 1984, Solid State Commun. **49**, 867.
Tersoff, J., L.M. Falicov and D.R. Penn, 1979, Solid State Commun. **32**, 1045.
Thuler, M.R., R.L. Benbow and Z. Hurych, 1983, Phys. Rev. **27**, 2082.
Tréglia, G., F. Ducastelle and D. Spanjaard, 1980, Phys. Rev. B **21**, 3729.
Tréglia, G., F. Ducastelle and D. Spanjaard, 1982, J. Phys. (France) **43**, 341.
Van der Laan, G., C. Westra, C. Haas and G.A. Sawatzky, 1981, Phys. Rev. B **23**, 4369.
Wertheim, G.K., and M. Campagna, 1978, Solid State Commun. **26**, 553.
Wertheim, G.K., S. Hüfner and H.J. Guggenheim, 1973, Phys. Rev. B **7**, 556.
Wuilloud, E., B. Delley, W.-D. Schneider and Y. Bear, 1984, Phys. Rev. Lett. **53**, 202.
Zaanen, J., C. Westra and G.A. Sawatzky, 1986, Phys. Rev. B **33**, 8060.

CHAPTER 10

SURFACE CORE LEVEL SHIFT

Y. JUGNET, G. GRENET and TRAN MINH DUC

*Institut des Sciences de la Matière
and Institut de Physique Nucléaire de Lyon (and I N2 P3),
Université Claude Bernard Lyon-I,
43 Bd. du 11 Novembre 1918, F69622 Villeurbanne Cédex, France*

Contents

1. Introduction	665
2. Surface–bulk core level shift for clean surfaces	666
2.1. Core level binding energy shift	667
2.2. Surface–bulk core level shift (SCS)	669
2.3. Surface sensitivity enhancement	670
2.3.1. Tunability of photon energy	670
2.3.2. Detection angle	671
2.3.3. Light polarization	672
2.3.4. Photoelectron backscattering and diffraction effects	672
2.4. Review of results: SCS for clean surfaces	673
2.5. General trends	673
2.5.1. 5d transition metals	673
2.5.2. Rare earths	685
2.5.3. Surface core level shifts with sp electron states	688
2.5.3.1. sp free-electron-like metals	688
2.5.3.2. Semi conductors	690
3. Applications	693
3.1. Stepped surfaces	693
3.2. Surface reconstruction and relaxation	695
3.3. Chemisorption	697
3.4. Alloys and compounds; correlation between the observed shift and thermodynamical properties	703
3.5. Quantitative aspects	709
3.6. Surface crystallography versus photoelectron diffraction	710
4. Conclusion	718
References	719

*Handbook of Synchrotron Radiation, Vol. 2, edited by G.V. Marr
© Elsevier Science Publishers B.V., 1987*

1. Introduction

Electron spectroscopies, and more specifically, photoelectron spectroscopy, are by now well recognized to be quite valuable for surface studies. Angle resolved photoemission has been proved during this last decade to be the most powerful tool for probing the intrinsic surface states for clean crystals and for the chemisorption bond between adsorbate species and substrate as well. Firstly, photons interact very weakly with matter and thus the incident beam is feebly destructive regarding clean surface and/or adsorbate species. Secondly, the electrons produced in photoemission experiments usually have a very short mean free path (from a few tenths of a nanometer up to ~5 nm, depending on the photoelectron kinetic energy), and the probed depth is limited to ~10 nm, an upper limit under the most general analysis conditions. The latter can even be reduced to a few tenths of a nanometer under appropriate conditions as we shall describe later on. In a case like this, one is in a position to probe the very first atomic surface layer, the layer which participates in a definitive way in most of the reactions involving the different fields of surface chemistry and physics.

Up to very recently, clean surfaces have been investigated exclusively through their electronic band structure in the UPS (Ultraviolet Photoelectron Spectroscopy) regime, while the characterization of the core levels of the uppermost layer of these clean surfaces by XPS (X-ray Photoelectron Spectroscopy) was failing.

Surface atoms appear to be intermediate between free atoms and bulk atoms as far as the coordination number is concerned, due to the fact that the chemical environment of a surface atom is different from that of a bulk atom. All this implies a different distribution of the valence and conduction electron states. For example, the presence of broken bonds at the surface leads to a narrowing of the local density of states (Cyrot Lackmann 1967, 1968). The core levels are most sensitive to the electronic environment of the atom (Siegbahn et al. 1967) as demonstrated by the pioneering work of K. Siegbahn using ESCA (Electron Spectroscopy for Chemical Analysis, another acronym for XPS) experiments. Surprisingly, no real XPS identification of surface atoms had been achieved until very recent measurements were reported on surface core level shifts (SCS) relative to the bulk.

The first successful attempt to measure a surface core level shift was made by Citrin et al. (1978) with a conventional monochromatized XPS experiment. They demonstrated that by increasing the surface sensitivity, using photoelectron grazing detection, the 4f core levels of a polycrystalline gold sample produced a shoulder on the low binding energy side of the main bulk peak, which since then has been attributed to surface emission.

However, it is only through the utilization of synchrotron radiation (SR) that SCS can be identified as a tool for investigating surface properties. The first SR experiment on SCS (Tran Minh Duc et al. 1979) has proved that the W $4f_{7/2}$ core level spectrum of the W(110) crystalline face shows a surface atom core level peak clearly resolved from the bulk counterpart. Following this work, a wealth of data appears on SCS covering transition metals, free-electron-like metals, rare earth metals, alloys and semiconductors, all of which were clean or covered with an adsorbate. Several review articles have been published recently on some particular systems (Eastman et al. 1982, Tran Minh Duc et al. 1982, Citrin and Wertheim 1983, Spanjaard et al. 1985).

The primary purpose of this chapter is to present a rather broad overview of investigated SCS, and to attempt to provide comprehensive tables summarizing the extent of SCS over the entire periodic table, as well as their applications in surface crystallography.

In section 2 the experimental requirement for the observation of a surface–bulk core level shift will be described, and will be illustrated by the W(110) case. The parameters contributing to surface sensitivity enhancement, such as polarization effect, photon energy, detection angle and photoelectron diffraction effects, will be summarized. A compilation of the data will be given for clean surfaces, including free-electron-like metals, transition metals, rare earth metals and semiconductors. It is intended that this discussion will be from an experimental rather than a theoretical point of view, since while the qualitative understanding of SCS is well established, a quantitative theoretical model is still lacking. The two principal theoretical approaches – thermodynamic and microscopic – which have been reviewed recently (Johansson and Mårtensson 1980, Rosengren and Johansson 1980, Desjonquères et al. 1980), will be briefly described.

Section 3 will be devoted to the applications of such a shift for investigating the atomic structure of surfaces. SCS being dependent on crystalline surface orientation and closely related to the detailed surface atomic arrangement, these shifts can be used for monitoring surface reconstruction, as well as relaxation. Another field of investigation is the study of chemisorption, the SCS being specific to the atomic nature and to the crystallographic geometry of the adsorption site. In the cases of alloys or metal overlayers on substrates, the SCS have been related to the thermodynamical properties of these compounds, such as the heat of adsorption or the heat of segregation.

Finally, we shall describe an aspect of the technique opening up a new field of investigation: the measurement at high angular resolution of the photoelectron diffraction effects on surface core levels, which appears to be quite promising in the field of surface crystallography.

2. Surface–bulk core level shift for clean surfaces

The basic principles underlining photoemission techniques, including XPS or ESCA and UPS, are now well documented (Siegbahn et al. 1967, Baker and Brundle 1977, Martin and Shirley 1977, Pendry 1978, Feuerbacher et al. 1978, Fadley 1978, Ley et al. 1979, Smith and Himpsel 1983) and need not be considered in detail in this chapter. We shall first present some general considerations on the core level binding energy shift. Then, we shall limit ourselves to the surface–bulk binding energy (BE) shift, without discussing at any length other effects on the core levels due to the presence of the surface, such as line shape or width modifications, since these topics are poorly documented and need more investigation (Wertheim and Citrin 1984, Spanjaard et al. 1985).

2.1. Core level binding energy shift

The shift of the binding energy of a core level, as measured by XPS, is of considerable interest for probing the electronic structure of solids in general, and more particularly of their surfaces, as far as we are concerned here with SCS. Formally, XPS binding energy is defined as the total energy difference between the final core hole state, with $(N-1)$ electrons, and the unperturbed initial N electron state:

$$E_B = E_T(N-1) - E_T(N). \tag{1}$$

Hence, in the magnitude of core level binding energy shift, we may recognize two main contributions – initial ground state and final state core hole effects:

$$\Delta E_B = \Delta_{init} + \Delta_{relax}. \tag{2}$$

The most interesting initial state contribution reflects the change of the electrostatic potential in the atomic core region of the atom under consideration, which characterizes the electronic structure and the chemical environment of this atom. All this constitutes the information we wish to obtain, and in turn has to be partitioned into several additional terms:

$$\Delta_{init} = \Delta_{config} + \Delta_{charg} + \Delta_{Madelung}. \tag{3}$$

Δ_{config} is the change in the core level BE due to a change in the valence electronic configuration of the considered atom, i.e. going from free atoms to surface and/or solid state. The spatial distribution of the different valence orbitals, and the corresponding core-valence Coulomb interaction are quite different according to their quantum numbers. As a matter of fact, nd valence electron orbitals are expanded much less than their $(n+1)s$ and/or p counterparts, and their Coulomb repulsion from a core orbital can be twice as large. Therefore, an s→d conversion within the valence shell of an atom may decrease the (positive) BE of the core level of this atom. On the other hand, an s⇄p conversion has little effect in this respect. When going from a bulk site to a surface site, sp⇄d rehybridization effects probably take place, and can contribute to modify the core level BE. Following the same order of ideas, the renormalization at the surface of valence charge within the Wigner–Seitz cell may result in a compression (expansion) of valence charge into a negative (positive) core level BE shift.

We define Δ_{charg} as the effect associated with a modification in the valence charge density surrounding an atom when this atom is included in a chemical bond. As is widely admitted, large charge transfers (often correlated with semiconductor surface reconstruction) occur at the surface of a solid – metal or semiconductor, and these constitute the main factor considered for explaining SCS. $\Delta_{\text{config}} + \Delta_{\text{charg}}$ are generally interplayed and difficult to separate, and usually may be linked together in a general term: the intra-atomic chemical effect, Δ_{chem}. This effect can be expressed by the very simple potential model as follows:

$$\Delta_{\text{chem}} = \Delta \sum_i \left\langle \frac{q_i^2}{r_i} \right\rangle + \text{exchange term} . \tag{4}$$

q_i is the charge residing on the valence orbitals, with a mean radius $\langle r_i \rangle$ of the considered atom. $\langle 1/r_i \rangle$ is then the Coulomb potential, approximated as a potential induced in a hollow sphere of radius $\langle r_i \rangle$, which delimits the core atomic region, and sampled by the core electron to be photo-ejected. In this simplest scheme for interpreting Δ_{chem}, exchange terms are ignored. Δ_{chem} can, in principle, be calculated at different higher levels of approximation from the tight-binding model up to the self-consistent Hartree–Fock calculations.

If Δ_{chem} includes the intra-atomic Coulomb potential, the last initial state component takes into account the other charges sitting on the surrounding atomic sites of the crystal lattice where the considered atom is embodied. This term is referred to as the interionic Madelung summation:

$$\Delta_{\text{Madelung}} = \sum_{j, \text{ all latt}} \frac{e_j^2}{r_{ij}} , \tag{5}$$

e_j being the total valence charge density on atom j, i being the atom from which the core electron is ejected, and r_{ij} being the distance between atoms i and j.

The Madelung potential acts in a reverse way on the intra-atomic term. In other words, the Madelung term algebraically reduces (increases) the core level BE of a cation (anion). Then the reduction of Δ_{Madelung} shifts positively (negatively) the cation (anion) core level BE.

Turning now to the final state core hole relaxation energy Δ_{relax}, this term is taken to be zero at the lowest level of the fully independent electron approximation. This implies that when ejecting one electron from an N electron system, the $(N-1)$ remaining electrons are essentially unaffected by the photoevent. This "frozen" orbital approximation is also known as Koopmans' theorem, which states that the core level BE is opposite in magnitude to the one-electron energy of this orbital:

$$E_{\text{B}} = -\varepsilon_l . \tag{6}$$

By delineating this zero level approximation, the creation of the photohole shifts this core level BE by the quantity Δ_{relax}, which can be understood to be the screening energy of the final hole by the electrons from the medium surrounding the photoionized atom. This screening, polarization or relaxation energy always appears upon reduction to the core level BE. Δ_{relax} is obviously also connected with the electronic structure and properties of the medium: energy released by a screening d-like charge is smaller than the energy from more mobile s or p screening. Relaxation energy should diminish when going from the bulk to the surface simply by considering that a core hole localized at the surface experiences only semi-infinite dielectric screening, as compared with the full infinite dielectric screening in the bulk. In this case, Δ_{relax} should contribute positively to the SCS.

2.2. Surface–bulk core level shift (SCS)

We start here with a general presentation of features demonstrated by a surface core level shift when measured with synchrotron radiation (SR).

Figure 1 shows the high resolution soft X-ray photoelectron spectra obtained

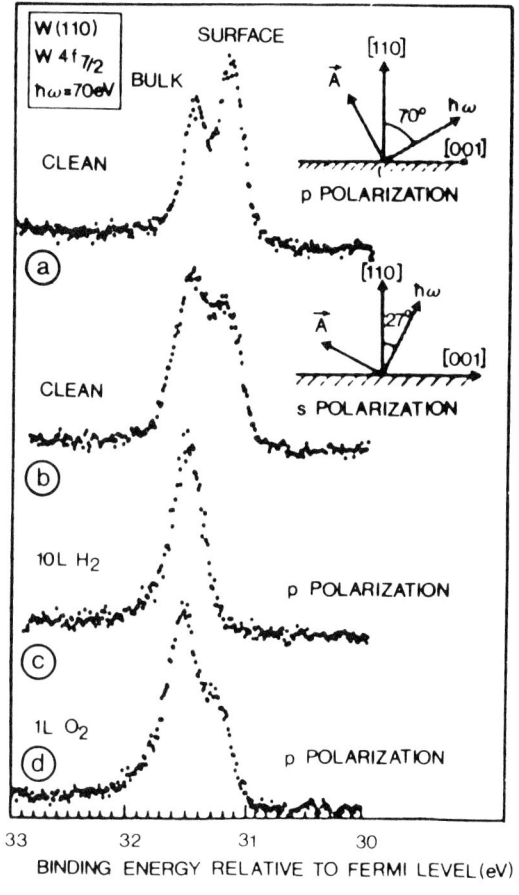

Fig. 1. W4f$_{7/2}$ core levels of clean W(110) excited by (a) p polarized light, (b) s polarized light, and after (c) 10 L H$_2$ and (d) 1 L O$_2$ exposures with p polarization. (From Tran Minh Duc et al. 1979.)

with the synchrotron radiation delivered by ACO at LURE on a W(110) oriented single crystal (Tran Minh Duc et al. 1979). The photon energy is set at 70 eV, to achieve maximum surface sensitivity, and the measured level is the localized $4f_{7/2}$ core level. The $\frac{5}{2}j$ component is out of the spectrum scale, the spin orbit splitting being 2.2 eV. The $4f_{7/2}$ peak is clearly resolved into two components at 31.50 eV and 31.20 eV binding energy, referred to the Fermi level, with a full width at half maximum of 0.25 eV and 0.22 eV, respectively. The energy splitting between the two peaks is constant when photon energy is varied. The low binding energy peak is quite intense when p polarization is used for excitation, i.e., when the potential vector of the light, A, is nearly perpendicular to the sample surface. The intensity of this peak decreases considerably when s polarization is used (A almost parallel to the surface). In addition, this peak vanishes when the surface is exposed to some contaminant gas. These criteria are specific for surface states and lead us to assign the lower binding energy component to the first layer surface atoms, and the higher binding energy component to the bulk atoms. Thus, this experiment provides us with irrefutable evidence that surface atoms have a binding energy different from that of the bulk atoms. This shift has been readily expected due to the change of surface and bulk coordination number. It is worthwhile to point out that a SCS is small, <-0.9 eV, one order of magnitude smaller than the usual ESCA chemical shift, and that such an observation needs a very high experimental energy resolution, namely 0.15 eV. Such a resolution is achieved with high counting rate, which is dependent on the availability of intense sources of SR, and high performance monochromators. Before probing further into the interpretation of these data, it is important to comment on the selection of physical parameters for the experiment in order to achieve the best enhancement of the surface peak intensity.

2.3. Surface sensitivity enhancement

Surface core level shift had been researched unsuccessfully with conventional XPS for quite some time. The key points were that previous experiments did not achieve the surface sensitivity required for focusing on the outermost surface layer.

There are at least three parameters which can be dealt with in order to increase the surface sensitivity of the technique as far as synchrotron radiation experiments are concerned. These include the photon energy tunability, the choice of the light polarization, and the variation of the detection angle. A fourth parameter, which has not been identified up to now for increasing surface sensitivity, is photoelectron backscattering from diffraction processes.

2.3.1. Tunability of photon energy
The tunability of photon energy represents the major advantage of the SR over the fixed-energy laboratory X-ray sources. One way to increase the surface sensitivity is to decrease the depth under the surface probed by photoelectron spectroscopy. This depth is primarily determined by the electron mean free path,

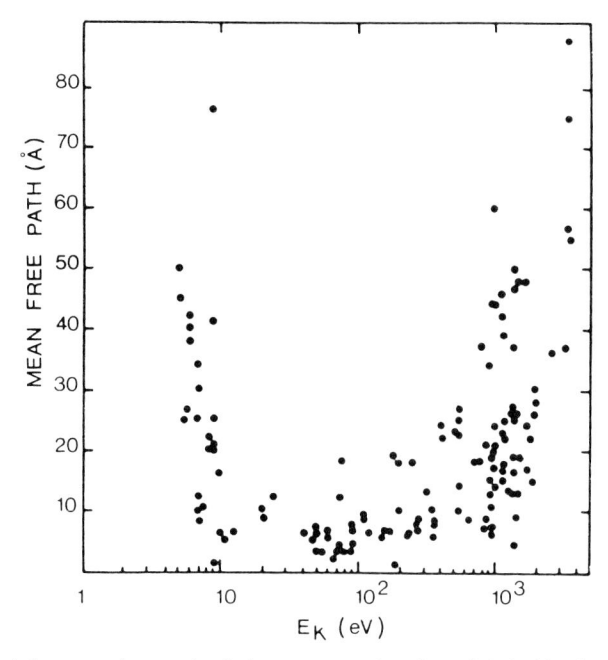

Fig. 2. Variation of the mean free path of electrons as a function of their kinetic energy E_K. (From Pijolat 1979 and references cited therein.)

$\lambda(E_K)$. The general dependence of λ on E_K is reported in fig. 2 (Pijolat 1979). The most important point in this curve is the minimum observed for a kinetic energy of about 50 eV, λ having a value close to 0.5 nm, i.e., one or two atomic layers. On both sides of this minimum, λ is increasing steadily up to 2.5 nm at 1 keV, which is the usual photoelectron kinetic energy in conventional XPS experiments. Thus, by tuning to a suitable photon energy, it is possible for a given electronic level of binding energy, E_B, to eject photoelectrons with a kinetic energy E_K, corresponding to the minimal mean free path, ($\hbar\omega \sim E_K + E_B - \Phi$), Φ being the work function of the sample. For example, in the case of the W $4f_{7/2}$ level, the suitable choice for the photon energy is $\hbar\omega = 70$ eV, the binding energy of the $4f_{7/2}$ level being 31.5 eV.

2.3.2. Detection angle

Another way of shortening the inspection depth is to change the detection angle. The analyzed depth varies as $\sim 3\,\lambda(E_K)\cos\theta$, θ being defined relative to the line perpendicular to the surface. Thus, by using grazing detection, the surface effect is greatly enhanced; the analyzed depth can be decreased to as little as 2–3 atomic layers when θ is on the order of 80°. Simple quantitative models, as well as the complicating effects which occur with grazing detection, including instrument response function, surface roughness effects, elastic electron scattering and electron refraction at the surface, have been extensively described by Fadley (1984).

2.3.3. Light polarization

The radiation delivered from storage rings is linearly polarized parallel to the light potential vector, A, in the plane of the ring, and perpendicular to the propagation direction of the light. By convention, we can define two extreme cases of polarization depending on the orientation of A relative to the normal of the sample surface. In real p polarization A is normal to the surface, whereas in s polarization A is parallel to the surface.

In the independent particle approximation, the photoelectric current is estimated using the Fermi golden rule,

$$\frac{d^2 j}{dE_f \, d\Omega} \propto E_f \sum_i |\langle f | A \cdot p + p \cdot A | i \rangle|^2 \, \delta(E_f - E_i - \hbar\omega) , \qquad (7)$$

where the matrix element corresponds to the photoemission transition between an initial state, $|i\rangle$, and the one-hole final states, $|f\rangle$, A and p being the light potential vector and the moment operator of the electron, respectively. The δ function insures energy conservation. The interaction between the incoming photon and the photoelectrons is described by the familiar dipolar term $A \cdot p + p \cdot A$, at least for the UPS regime. The optical transition element, M_{if}, which is proportional to $\langle f | A \cdot p + p \cdot A | i \rangle$, may be written as (Feuerbacher and Willis 1976)

$$M_{if} \propto -\frac{i}{\omega} [\langle f | A \cdot \nabla_B | i \rangle + \langle f | A \cdot \nabla_S | i \rangle] - i\hbar \left\langle f \left| \frac{\partial A}{\partial Z} \right| i \right\rangle \qquad (8)$$

where the total contribution is subdivided into a bulk contribution, $A \cdot \nabla_B$, and a surface contribution, $A \cdot \nabla_S$. Thus, the three terms in the previous equation describe, respectively, the bulk effect, the potential surface effect, and the surface field effect arising from the spatial dependence of A near the surface (Feibelman 1975, Kliever 1976, Kliever 1977). When the A vector is parallel to the surface, the $\partial A / \partial Z$ component equals zero. On the other hand, the $\partial A / \partial Z$ component is maximal as soon as A and the perpendicular to the surface are colinear. Thus in the case of p polarization, surface emission is expected to be at a maximum.

2.3.4. Photoelectron backscattering and diffraction effects

Most of the investigations thus far on SCS have been conducted with angle integrated spectroscopy. If such measurements are useful for deriving the atom number in the surface layer (see section 3.5), the surface peak always remains at a lower intensity compared to the bulk peak. With higher angular resolution for specific conditions of angle emission and photoelectron kinetic energy, the surface peak surpasses the bulk peak in intensity (Tran Minh Duc et al. 1979, Horn 1985). Such results are explained by photoelectron diffraction effects, revealed only when the k vector of the photoelectrons is resolved. The wave initially emitted from surface overlayer atoms toward the bulk can be backscattered toward the detector by the underlying atoms, and then, interfere constructively

2.4. Review of results: SCS for clean surfaces

Table 1 gives an overview of measured data on SCS. In this table a negative surface bulk core level shift means that the surface binding energy is less than the bulk binding energy, and vice versa. When no orientation is mentioned, the sample is polycrystalline. Finally, when two values are reported for the same surface bulk core level, they correspond to the first and the second surface planes, both of which can be detected on widely open surfaces.

2.5. General trends

Since the first experimental observation of a surface bulk core level shift a wealth of results appeared, as well as different theoretical models which tried to explain this shift. Figure 3 shows the list of the elements across the periodic table studied in this respect. A compilation of results has been included earlier in this chapter, but we shall now try to describe the general trends as a function of crystallographic face, and the number of neighbors, making a distinction between 5d metals, sp metals, semiconductors and rare earths.

2.5.1. 5d transition metals

Core level shift in solids includes two principal contributions related to the initial state (chemical shift induced by a modification of the electrostatic potential at the core), and to the final state (relaxation due to the reorganization of orbitals after the ejection of a photoelectron). In order to explain the surface bulk core level shift, Desjonquères et al. (1980) proposed a model restricted to transition metals with a d-localized band, in which the bulk surface splitting echoes the shift of the center of gravity of the local density of states of surface valence levels. In their interpretation, they consider surface and bulk atoms as being affected in the same way by relaxation energy. This hypothesis seems to be justified as far as intra-atomic relaxation is concerned. The extra-atomic relaxation, mainly due to well-localized d-valence electrons, is expected to not be too sensitive to the existence of the surface (Williams and Lang 1978). Thus, in a first approximation, all the modifications induced by the surface can be interpreted in terms of chemical shift (initial state effects). Such an interpretation, i.e., in terms of initial states, has also been widely supported by Citrin and Wertheim (1983).

The core electrons of surface atoms experience an electrostatic potential different from those in the bulk since the coordination number is reduced on the surface as compared to in the bulk. This reduced coordination number results in a narrowing of the d valence band. In a tight-binding energy scheme, the width of the d local density of states is proportional to N_s/d^5, where N_s and d are the coordination number and the nearest-neighbor distance, respectively (Desjon-

Table 1
List of measured SCS on clean surfaces published prior to 1986.

Element	Face	Core level	SCS (eV)	References
Be	0001	1s	−0.5	Nyholm et al. (1985)
C, diamond	111	1s	−0.8	Morar et al. (1986)
Na		2p	+0.22	Kammerer et al. (1982)
Mg		2p	+0.14	Kammerer et al. (1982)
Al	100	2p	−0.12	Kammerer et al. (1982)
			−0.57	Chiang and Eastman (1981)
Si	100 (2 × 1)	2p	−0.52	Himpsel et al. (1980)
	111 (7 × 7)		−0.7	Himpsel et al. (1980)
	111 (2 × 1)		−0.37	Himpsel et al. (1980)
			$\begin{cases} -0.59 \\ +0.3 \end{cases}$	Brennan et al. (1980)
	111 H(1 × 1)		$\begin{cases} +0.26 \\ +0.15 \end{cases}$	Himpsel et al. (1980)
Ca		3p	0.0	Erbudak et al. (1983)
Sc		3p	+0.5	Erbudak et al. (1983)
Cu		2p	−0.24	Citrin et al. (1983)
	100		$\begin{cases} -0.3 \\ -0.7 \end{cases}$	Webber and Morris (1984)
Ga, GaAs	110	3d	+0.28	Eastman et al. (1980)
				Miller and Chiang (1984)
	100 c(4 × 4)		+0.49	Ludeke et al. (1983)
	100 c(2 × 8)		+0.30	Ludeke et al. (1983)
	100 c(4 × 6)		$\begin{cases} -0.21 \\ +0.4 \end{cases}$	Ludeke et al. (1983)
GaSb	110	3d	+0.30	Eastman et al. (1980)
GaP	110	3d	+0.28	Eastman et al. (1980)
Ge	100 (2 × 1)	3d	−0.41	Miller et al. (1983)
	100 c(4 × 2)		−0.43	Schnell et al. (1985)
	111 c(2 × 8)		$\begin{cases} -0.26 \\ -0.76 \end{cases}$	Schnell et al. (1985)
As, GaAs	110	3d	−0.37	Eastman et al. (1980)
			−0.38	Miller and Chiang (1984)
	100 c(4 × 4)		$\begin{cases} -0.28 \\ +0.55 \end{cases}$	Ludeke et al. (1983)
	100 c(2 × 8)		$\begin{cases} -0.25 \\ +0.10 \end{cases}$	Ludeke et al. (1983)
	100 c(4 × 6)		−0.61	Ludeke et al. (1983)
Rh		3d	−0.58	Erbudak et al. (1983)
Pd		3d	+0.23	Erbudak et al. (1983)
Ag		3d	−0.08	Citrin et al. (1983)
In, InSb	110	4d	+0.22	Taniguchi et al. (1983)
			+0.23	Horn (1985)
Sb, InSb	110	4d	−0.29	Taniguchi et al. (1983)
			−0.29	Horn (1985)
Hf		4f	+0.42	Nyholm and Schmidt-May (1984)
Ta	111	4f	$\begin{cases} +0.40 \\ +0.19 \end{cases}$	Van der Veen et al. (1981c)
	100		$\begin{cases} +0.14 \\ +0.74 \end{cases}$	Guillot et al. (1984b)

Table 1 (*Continued*)

Element	Face	Core level	SCS (eV)	References
W	110	4f	−0.3	Tran Minh Duc et al. (1979)
	100		−0.16 −0.40	Guillot et al. (1982)
			−0.19 −0.39	Jupille et al. (1986)
			−0.35	Van der Veen et al. (1982)
	100 Hc(2 × 2)		−0.09 −0.25	Guillot et al. (1982)
			−0.13 −0.35	Van der Veen et al. (1981b)
	temperature 100 c(2 × 2)		−0.16 −0.35	Guillot et al. (1982)
	111		−0.43 −0.10	Van der Veen et al. (1982)
	6 (110) (1$\bar{1}$0)		−0.58 −0.3 −0.44 −0.18	Chauveau et al. (1984)
Re	0001	4f	−(0.15, 0.20)	Ducros and Fusy (1986)
Ir	111	4f	−0.5	Van der Veen et al. (1980)
	100		−0.68	Van der Veen et al. (1980)
	100 (5 × 1)		−0.49	Van der Veen et al. (1980)
	332		−0.48 −0.75	Van der Veen et al. (1981a)
Pt	111	4f	−0.40	Baetzold et al. (1982)
			−0.35	Bertolini et al. (1985)
	110 (1 × 2)		−0.21 −0.55	Baetzold et al. (1982)
	6 (111) (100)	4f	−0.30 −0.60	Apai et al. (1983)
	3 (111) (111)		−0.30 −0.57	Apai et al. (1983)
Au		4f	−0.4	Citrin et al. (1978)
			−0.4	Citrin et al. (1983)
			−0.39	Erbudak et al. (1983)
	111	4f	−0.35	Heimann et al. (1981)
	100		−0.38	Heimann et al. (1981)
	100 (5 × 20)		−0.28	Heimann et al. (1981)
	110 (2 × 1)		−0.35	Heimann et al. (1981)
La		5p	+0.48	Nilsson et al. (1985)
Ce		4f	+0.3	Parks et al. (1982)
Pr		4f	+0.4	Gerken et al. (1985)
Nd		4f	+0.5	Gerken et al. (1985)
Eu		4f	+0.63	Kammerer et al. (1982)
			+0.63	Gerken et al. (1985)
			+0.60	Kaindl et al. (1982)
Gd		4f	+0.48	Kammerer et al. (1982)
			+0.50	Gerken et al. (1985)
Tb		4f	+0.55	Gerken et al. (1985)
Dy		4f	+0.55	Gerken et al. (1985)

Table 1 (*Continued*)

Element	Face	Core level	SCS (eV)	References
Ho		4f	+0.63	Gerken et al. (1985)
Er		4f	+0.65	Gerken et al. (1985)
Tm		4f	+0.70	Gerken et al. (1985)
TmSe	100		+0.32	Kaindl et al. (1982)
TmTe	100		+0.41	Gerken et al. (1985)
Yb		4f	+0.62	Kaindl et al. (1983)
		5p	+0.60	
		4f	+0.60	Alvarado et al. (1980)
			+0.60	Gerken et al. (1985)
		4f	+0.56	Nilsson et al. (1985)
		5p	+0.56	
Lu		4f	+0.77	Gerken et al. (1985)
		4f	+0.83	Kaindl et al. (1983)
		5p	+0.74	Kaindl et al. (1983)

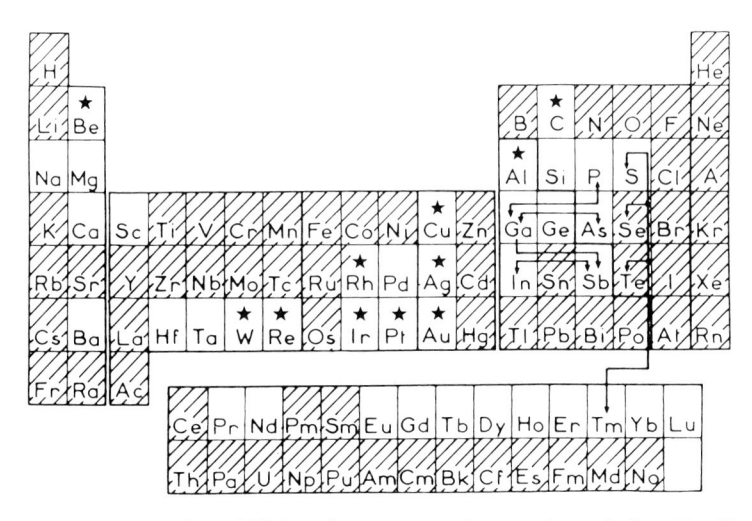

Fig. 3. List of the elements whose SCS have been measured across the periodic table. Hatched parts have not yet been measured. The symbol ★ means that the SCS is negative i.e., the surface binding energy is less than the bulk one. The arrows indicate that the elements have been measured in the resulting binary semiconductors.

quères and Cyrot Lackmann 1975a). In order to insure a charge neutrality in the surface plane, the surface d band has to be shifted with respect to the bulk d band, as shown in fig. 4. Both bulk and surface (narrower) bands are illustrated. In fig. 4a we assume that the d band is less than half-filled; the surface band lacks electrons in comparison with the bulk band. Charge neutrality requires a shift of the surface band towards higher binding energy. In fig. 4b where the d band is

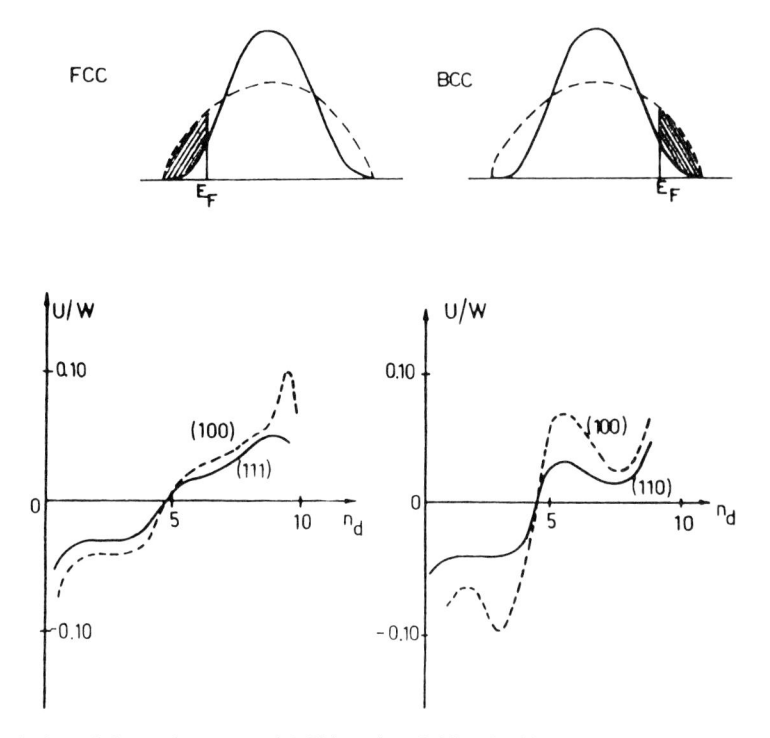

Fig. 4. Evolution of the surface potential U in units of d band with W as a function of N_d-filling for different faces of FCC (left) and BCC (right) transition metals with hypothetical surface (---) and bulk (——) density of states. Hatched areas visualize the excess FCC or lack BCC of electrons in bulk DOS as compared to the surface one. (From Desjonquères et al. 1980.)

more than half-filled, an excess of electrons appears in the surface band, which means a consequent shift towards lower binding energy. This charge rearrangement modifies the potential of surface atoms, U, which has been calculated using a tight-binding formalism by Desjonquères and Cyrot Lackmann (1975a). The shift of the surface band corresponds to the surface potential, U. Figure 4 reports the calculated variation of U expressed in units of d-band width, W, as a function of the d-band filling n_d. The same trends are evident for (100) and (111) faces of FCC metals, and for (100) and (110) faces of BCC metals. The potential takes on a value of zero and changes sign for a nearly half-filled band. In the illustrations of fig. 4 the bands appear to be symmetric, which is not realistic. It is the natural asymmetry of the actual d band which leads to a reverse of the sign of U, and therefore, of the surface–bulk core level shift for an n_d value of less than 5 electrons. Furthermore, U increases with the number of broken bonds. Desjonquères et al. (1980) have shown that the core levels rigidly follow the shift of the d band, although the shift is slightly larger for core levels by a factor of $\sim 10\%$.

In the tight-binding approximation, the second moment which describes the bandwidth is proportional to $\sqrt{Z}/d^5$, where Z is the atom coordination, and d is the nearest-neighbor distance. Thus for a given element, the surface–bulk core level shift can be roughly estimated by $\Delta E_{SCS} \propto \sqrt{Z_B} - \sqrt{Z_S}$, where Z_B and Z_S are the bulk and the surface coordination numbers, respectively. Knowing the surface core level shift, $\Delta E_{SCS,A}$, of a given face A and assuming that there is no surface relaxation, it is possible to estimate the shift $\Delta E_{SCS,B}$ expected on face B by the following relation:

$$\frac{\Delta E_{SCS,B}}{\Delta E_{SCS,A}} = \frac{1 - \sqrt{Z_S^B}/\sqrt{Z_B}}{1 - \sqrt{Z_S^A}/\sqrt{Z_B}}, \tag{9}$$

if there is no surface reconstruction.

To summarize for transition metals, the surface core level shift is expected to change sign for a d band filled with slightly less than five electrons. For a single element, the more open the surface plane, the larger the surface core level shift.

A similar formulation in terms of initial state effects has been put forth by Citrin and Wertheim (1983) to explain the surface–bulk core level shift observed in noble metals and transition metals. Assuming that the properties of the surface atoms are intermediate between those of the free and bulk atoms, they consider the total energy involved in going from the free-atom state to the bulk, i.e., the cohesive energy. This cohesive energy is partitioned among five terms: (1) the reconfiguration energy, Δ_{config}; (2) the renormalization energy, Δ_{norm}; (3) the conduction band energy, $\Delta_{s\,band}$, gained by allowing the free-electron-like s,p charge to broaden into a band; (4) the d band energy, $\Delta_{d\,band}$, gained when d electrons form a band; and (5) the sd hybridization energy, $\Delta_{sd\,hybrid}$, gained by mixing s and d bands. The reduced coordination number on the surface leads to more atomic-like d bands, and to a reduced hybridization as compared to the bulk. For noble metals, the delocalized s and d bulk states of bonding character are reduced for the surface atoms, and surface atoms are expected to be less bound than the bulk. With free-electron-like metals (s, p), where a large number of unfilled, highly delocalized states are available for hybridization, the surface–bulk core level shift is predicted to be small.

The above models do not consider final ionized state effects, i.e., core hole screening energy. On the other hand, the thermodynamic approach, as proposed by Johansson and coworkers (Johansson and Mårtensson 1980), calculates bulk and surface core level binding energies by including these energies in thermodynamic cycles. By doing such, the core level binding energy is actually written as the total energy difference between the $(N-1)$ electron core-hole final state and the N electron initial state, according to eq. (1).

Thus, by definition, core hole screening energy is explicitly considered in this equation, and with this thermodynamic approach, final state effects are treated on an equivalent basis with chemical initial state effects.

In addition, the metal work function is straightforwardly included in the cycle, allowing for values of the core level binding energy, relative to the Fermi level (and not to the vacuum level), to be derived. These values are then directly comparable to the experimental values measured relative to the sample Fermi level.

In the thermodynamic model for surface core level shift, two basic assumptions are made: (i) the final core hole is considered to be fully screened; (ii) the equivalent $(Z + 1)$ core approximation is assumed to be valid. If they are fulfilled, these assumptions allow one to utilize available standard thermochemical data in SCS calculations, instead of the available data for highly excited core hole states. If in the final state the core hole is fully screened by the conduction electrons of the metal, the valence electron distribution of the atom with a core hole, i.e., with a core of $(Z + 1)$ positive charges, is stated to be a completely relaxed configuration. This configuration can be described in turn as the ground state electron configuration of the succeeding $(Z + 1)$ element in the periodic table.

Within these assumptions, the ionized core atom can be treated as a $(Z + 1)$ substitutional impurity in the metallic matrix. Interesting information on important thermochemical data can be extracted, such as surface segregation heat.

Thermodynamic theory for surface core level shift has been reviewed recently (Johansson and Mårtensson 1983). Thus, in this chapter, only the main results from this model will be recalled. In their formulation, initially proposed for the calculation of the core level binding energy shift between a free atom and a metal, the combination of the complete screening and of the $(Z + 1)$ or equivalent core approximation, makes it possible to introduce a Born–Haber cycle, which connects the initial state with the final states of the core ionization process. This cycle is shown in detail in fig. 5.

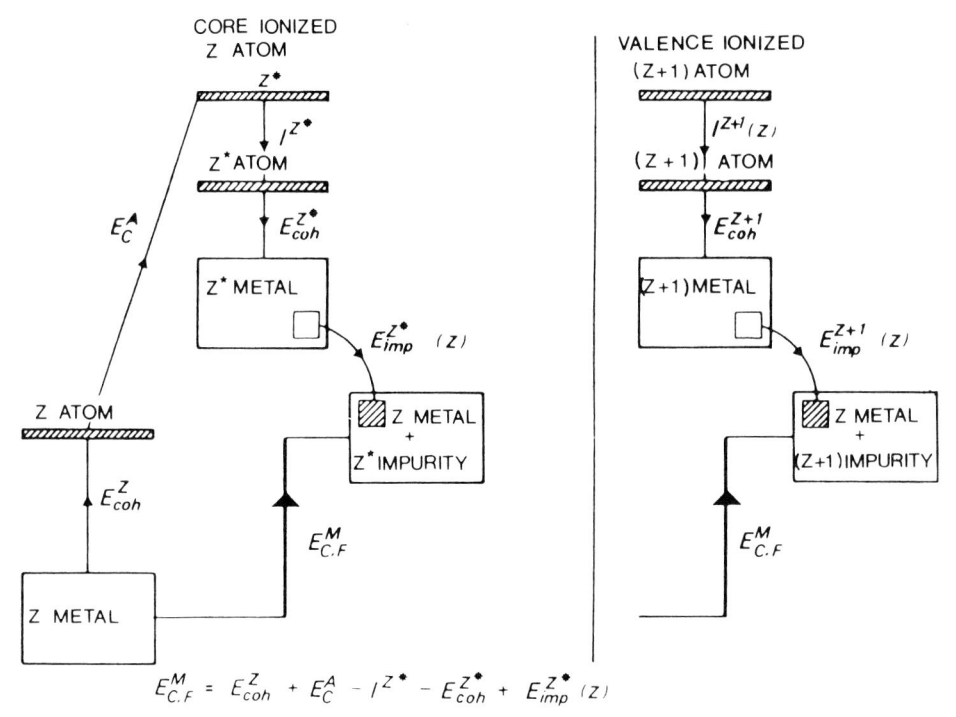

$$E^M_{C.F} = E^Z_{coh} + E^A_C - I^{Z*} - E^{Z*}_{coh} + E^{Z*}_{imp}(Z)$$

Fig. 5. Scheme of the thermodynamical model used for the estimation of SCS. (From Johansson and Mårtensson 1980.)

Let us consider a Z metal from which we first isolate a Z atom. We then eject a core electron. The energy involved in this process is the sum of the cohesive energy, $E_{\text{coh}}^{(Z)}$ of the Z metal and of the binding energy of the excited core level, E_c^A. Let Z^* be this ionized atom. By screening, a valence electron neutralizes the Z^* atom, which releases the ionization energy $I^{Z^*} \sim I^{Z+1}$ in the equivalent core approximation. The adjunction of a macroscopic number of such atoms to make a solid requires the cohesive energy, $E_{\text{coh}}^{(Z^*)}$, which is equal to $E_{\text{coh}}^{(Z+1)}$. Finally, this solid has to be diluted into the A metal with atomic number Z, the solution energy involved being $E_{\text{imp}}^{(Z+1,A)}$. In this scheme, the energy which allows for the closing of the cycle is nothing other than the excited core level binding energy as referred to the metal Fermi level, $E_{\text{C,F}}^M$. The energy balance of this process can be written as

$$E_{\text{C,F}}^M = E_{\text{coh}}^{(Z)} + E_c^A - I^{Z+1} - E_{\text{coh}}^{(Z+1)} + E_{\text{imp}}^{(Z+1,A)} . \tag{10}$$

Now let us make a distinction between bulk atom core level photoemission and surface atom core level photoemission. The only terms, within the Born–Haber cycle, to be affected by such a distinction are the cohesive energy and the solution energy, presenting different values for the bulk and for the surface: E_{coh}^B and E_{imp}^B against E_{coh}^S and E_{imp}^S. The difference between these two values describes the surface–bulk core level shift, ΔE_{SCS}, which can be expressed as

$$\Delta E_{\text{SCS}} = E_{\text{C,F}}^{M,S} - E_{\text{C,F}}^{M,B} , \tag{11}$$

$$\Delta E_{\text{SCS}} = [E_{\text{coh}}^B(Z+1) - E_{\text{coh}}^B(Z) - E_{\text{imp}}^B(Z+1)]$$
$$- [E_{\text{coh}}^S(Z+1) - E_{\text{coh}}^S(Z) - E_{\text{imp}}^S(Z+1)] . \tag{12}$$

This result can be visualized in a straightforward manner by a much simpler Born–Haber cycle (fig. 6) summing up the difference between the two previous, more detailed cycles devised for bulk and surface photoemission (Rosengren and Johansson 1981, Egelhoff 1983). From this figure, ΔE_{SCS} appears to be nothing other than the difference between $E_{\text{coh}}^B(Z+1, A)$, the energy of a system consisting of a Z metal containing a $(Z+1)$ impurity in a bulk site, and $E_{\text{coh}}^S(Z+1, A)$, the corresponding energy in a surface site. These energies are exactly

$$E_{\text{coh}}^{B(S)}(Z+1, A) = E_{\text{coh}}^{B(S)}(Z+1) - E_{\text{coh}}^{B(S)}(Z) - E_{\text{imp}}^{B(S)}(Z+1) . \tag{13}$$

By taking the difference (Kumar et al. 1981)

$$\Delta E_{\text{SCS}} = E_{\text{coh}}^B(Z+1, A) - E_{\text{coh}}^S(Z+1, A) \tag{14}$$

we are left with eq. (12).

Figure 6 has additional merit in demonstrating with an illuminating simplicity

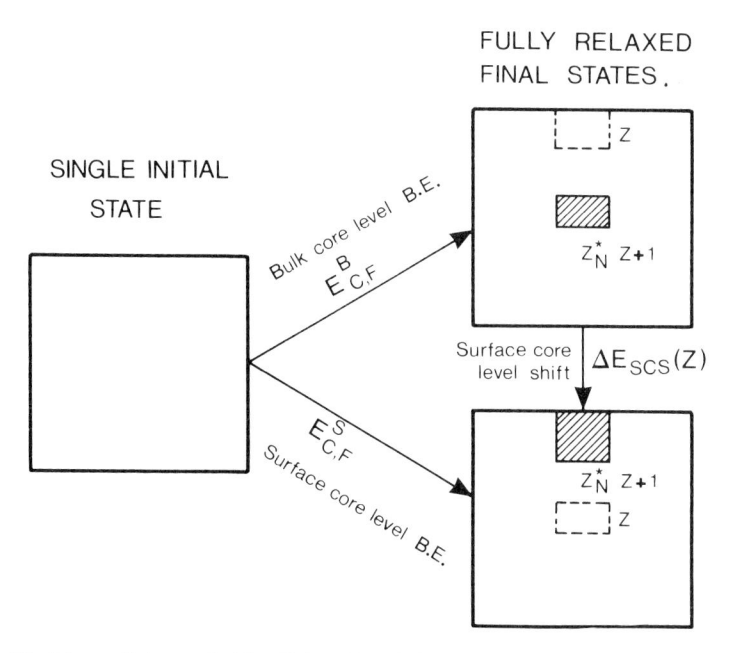

Fig. 6. Simplified Born–Haber cycle identifying $\Delta E_{SCS}(Z)$ as the heat of surface segregation due to the $(Z + 1)$ impurity in the metal Z [or an alloy between two neighboring Z and $(Z + 1)$ atoms]. From the same neutral initial state, bulk emission creates a Z^* fully relaxed core ionized state in the bulk, while surface emission creates the same on the surface. By equivalent core approximation, Z^* is set equivalent to $(Z + 1)$ and then $\Delta E_{SCS}(Z)$ is identified to the energy needed for exchanging a Z atom on the surface with a $(Z + 1)$ atom in the bulk.

the important physical meaning of surface core level shift in a Z metal to be exactly the heat of surface segregation of a $(Z + 1)$ impurity in the Z metallic matrix (Rosengren and Johansson 1981, Kumar et al. 1981 and Feibelman 1983).

Equation (12) can be rewritten by considering the physical definitions of the cohesive and dissolution energies within the framework of the quasichemical theory of solids and alloys (Miedema 1978). Cohesive energy is related to the bonding of an atom to its identical neighbors within the solid, and the $E_{imp}^{(Z+1)}$ dissolution energy is the bonding energy of the $(Z + 1)$ impurity to the surrounding Z metal atoms. The reduced coordination on the surface attenuates these interaction energies from simple geometrical consideration. The reduction factor α can be reasonably estimated to be the same for both the cohesive and dissolution energies. The numerical value of 0.8 is generally considered an acceptable value for α.

$$E_{coh}^S = \alpha E_{coh}^B \quad \text{and} \quad E_{imp}^S(Z+1) = \alpha E_{imp}^B . \tag{15}$$

Then, SCS can be expressed as

$$\Delta S_{SCS} = (1 - \alpha)[E_{coh}^B(Z+1) - E_{coh}^B(Z) - E_{imp}(Z+1, A)] \tag{14'}$$

or

$$\Delta E_{SCS} = E_S(Z+1) - E_S(Z) - (1-\alpha)E_{imp}(Z+1). \tag{14''}$$

$E_S(Z)$ and $E_S(Z+1)$ are the surface energies of pure Z and $Z+1$ metals, respectively, and are defined as

$$E_S(Z) = E_{coh}^B(Z) - E_{coh}^S(Z). \tag{16}$$

Values for cohesive energy in the transition metals series are available. Figure 7 shows the variation of E_{coh}^B across the 3d, 4d and 5d series (Gschneidner 1964). Since the s(p) configuration remains identical along the transition metal series, d bonding is fundamental in these metals. The elements in the first half of the series correspond to the filling of the bonding d orbitals, whereas after the middle of the series, with the appearance of the heaviest elements, antibonding orbitals are filled. Correspondingly, the cohesive energy increases with d occupancy for the first half of the series, and decreases after a maximum at the midpoint in the series (fig. 7).

From this data, SCS can be evaluated following the empirical expression

$$\Delta E_{SCS} = 0.2[E_{coh}^B(Z+1) - E_{coh}^B(Z)], \tag{14'''}$$

where $\alpha = 0.8$ and $E_{imp}(Z+1)$ is considered to be a negligibly small contribution to the shift. Results calculated as such for surface core level shifts are reported in fig. 8 for 3d, 4d and 5d transition metals.

Figure 7 shows that the cohesive energy of the final core hole state is larger

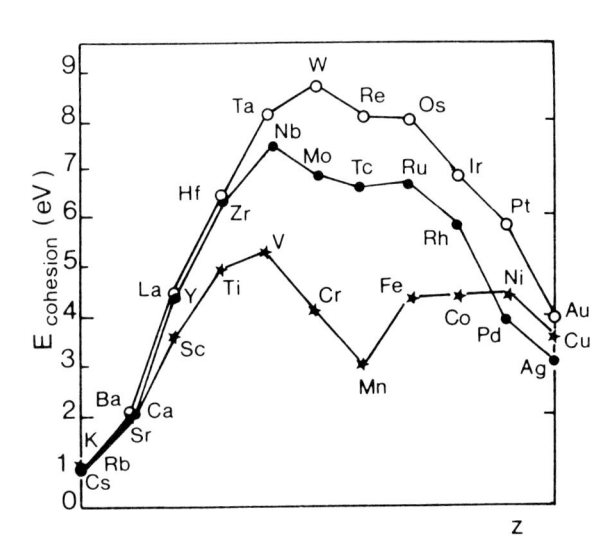

Fig. 7. Variation of the cohesion energy as a function of the atomic number Z. (From Gschneidner 1964.)

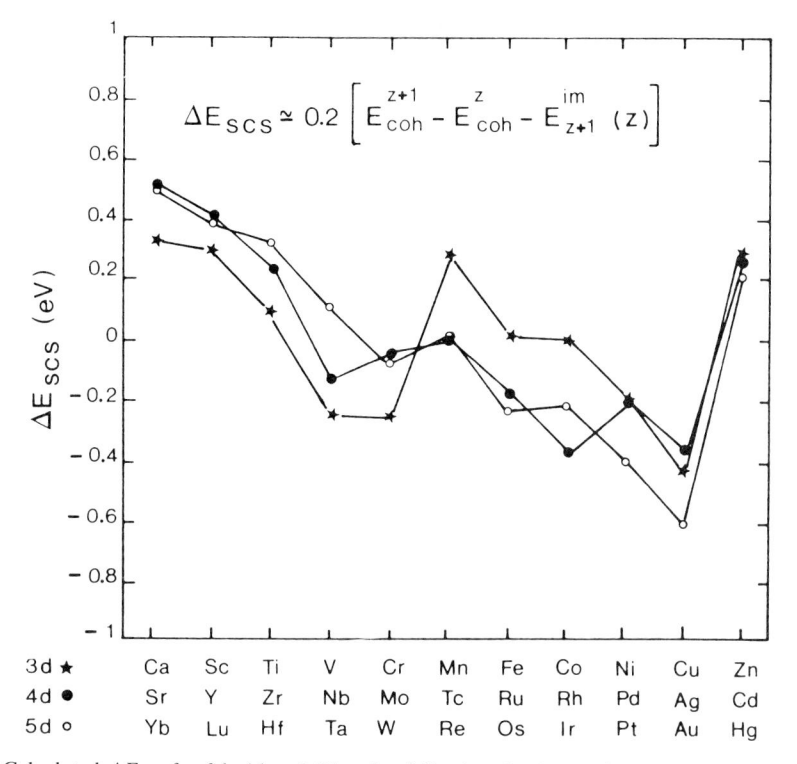

Fig. 8. Calculated ΔE_{SCS} for 3d, 4d and 5d series following the inserted formula. (From Citrin and Wertheim 1983, and references cited therein.)

Formula in figure: $\Delta E_{SCS} \simeq 0.2 \left[E_{coh}^{z+1} - E_{coh}^{z} - E_{z+1}^{im}(z) \right]$

than that of the initial state at the beginning of the series of transition metals, the reverse being true for the second half. Consequently, surface core level shift is positive for the first elements of the series, and surface peaks possess higher binding energy relative to the bulk peaks. In addition, the shift is the largest with n_d occupancy equal to one. Conversely, ΔE_{SCS} is negative for the second half of the series and its absolute magnitude is largest at the end of the series. This reversal in the sign of the surface core level shift at the midpoint of the series is in satisfactory agreement with experimental data.

Experimentally, surface core level shifts demonstrate a detailed relation on the crystallographic orientation of the surfaces (see table 1). Then to reproduce this crystallographic orientation dependence, relation (16) cannot be accommodated of crudely estimated heats of segregation, instead one must use refined values of surface energies specific to the different single-crystal surface orientations. The crystallographic anisotropy of surface energies was explicitly taken into account within the tight-binding model for FCC and BCC structures by Desjonquères and Cyrot Lackmann (1975a,b) as

$$E_S(Z) = \int^{E_F(Z)} E \sum_i \Delta n_i(E, U_0) \, dE - Z_M U_0 - \tfrac{1}{2}(Z_S - Z_M)U_0, \qquad (17)$$

where U_0 describes the surface potential. Δ_{ni} represents the variation of the local density of states on the nth plane, and Z_M and Z_S are, respectively, the d electron number in the bulk and in the surface. Values thus derived for surface energies can then be used for calculating surface core level shifts for different low index surfaces of 5d transition metal (Rosengren and Johansson 1980) and are reported in fig. 9 in comparison with experimental data. The correspondence between model and experiment is compelling. The results are intuitively easy to understand: the more open the surface, the larger the surface shift. Closer theoretical values should be obtained if some bond length relaxation can be included (Johansson and Mårtensson 1980). The same trends predicted by Desjonquères and Cyrot Lackmann (1975b) are observed as a function of n_d filling as well as a function of crystallographic orientation.

This is not so surprising since from a Taylor series expansion of ΔE_{SCS}, the surface–bulk core level shift can be written as

$$\Delta E_{SCS} = \frac{\partial E}{\partial Z} + \frac{1}{2}\frac{\partial^2 E_S}{\partial Z^2} + \cdots, \tag{18}$$

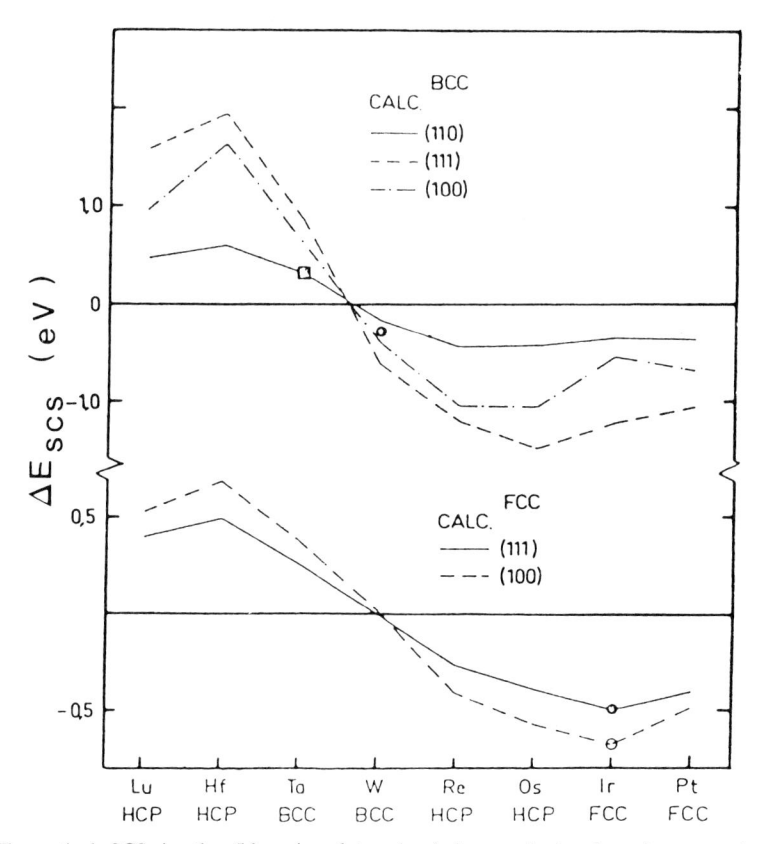

Fig. 9. Theoretical SCS in the 5d series determined from calculated surface energies. (From Rosengren and Johansson 1980.)

where the first term corresponds to the variation of the chemical shift at the surface, and the other terms being identified with relaxation effects (Spanjaard et al. 1985). The identification of surface core level shift with heat of segregation has been formally generalized to the case of all transition metals by Feibelman (1983). In this study he showed that surface core level shifts evaluated as the difference in the one-electron ground state energies of the outermost layer and a central one calculated self-consistently within the local exchange correlation approximation for a system of five-layer metal slabs, are in close agreement with the heat of surface segregation of $(Z + 1)$ impurity in the host Z metal, obtained with the Miedema formulation (Miedema 1978). From both theories, band structure calculations and thermodynamical approach leading to quite similar results, we can conclude that relaxation effects are quite small. This is confirmed by a self-consistent calculation of a core electron binding energy at a surface in the case of a 3s orbital of Cu(100) (Smith et al. 1982). Although both initial and final state effects participate importantly in the surface binding energy shift, short metal screening lengths make surface relaxation very similar to bulk relaxation. It has been concluded thus far that relaxation shifts may be safely ignored in estimating surface core level shift. The thermodynamical approach has been extended to the case of 3d and 4d elements (Johansson and Mårtensson 1980), to the case of rare earths implanted in the noble metals (Johansson and Mårtensson 1980), and to the lanthanide metals (Gerken et al. 1985). This approach has also been used for interpreting SCS in adsorption systems (Fiebelman 1983, Tomanek et al. 1982). Finally, Mårtensson et al. (1984) extended this model to Auger energy shifts, which are slightly more complicated since the final states, in this case, are two core-hole final states. As for the single core hole case, the surface shifts change sign as we proceed through a transition series due to the bonding–antibonding division of the d band.

2.5.2. Rare earths

In some respects, SCS for RE metals presents many gross similarities with the 5d transition metal series: rare earths have localized 4f levels which give shallow binding energy (BE) and sharp structures in their photoemission spectra. From the experimental standpoint, this situation is most favorable for resolving the surface core peak from the bulk. It is interesting to note that the deeper 5p doublet may also be used for detecting SCS due to the large value of the latter (Kaindl et al. 1983, Nilsson et al. 1985). For La, with an unfilled 4f level, this alternative is fundamental. In addition, the valence electrons in the lanthanide series are mainly of 5d character, with sp mixing to some extent, and therefore, rare earth metals are expected to exhibit SCS as large as 5d metals (i.e. <1 eV). The theoretical scheme set up to explain SCS in the 5d transition series should hold for rare earths as well.

However, some specificities and complications arise from the gradual filling of the 4f level along the lanthanide series. Since the initial $4f^n$ configuration is incomplete, the photoionisation of a 4f electron leads to a $4f^{n-1}$ final state spin multiplet manifold, reflecting complex structures in the 4f photoemission spectrum. In order to separate surface structures from the bulk in the 4f core level spectra, one has to know the energy separation and the fractional parentage intensity of the final $4f^{n-1}$ multiplet structure. This can be achieved with a theoretical and/or a bulk-sensitive 4f XPS spectrum (Gerken et al. 1985). The degeneracy of the 4f core levels with the spd valence states comes as a second complication because of the resulting mixed valence configuration. A well-known example is the Sm metal, which is trivalent in the bulk and divalent on the surface because of the decrease of cohesive energy due to the surface-reduced coordination number. Detailed modification within the actual (spd) valence configuration markedly influences the SCS across the series.

Numerous experimental articles have reported on SCS for the rare earths, covering all the elements of the lanthanide series from La to Lu, except for Sm and Pm. Lack of data on Sm must be imputed to the valence change on the surface of this metal. When measured on both $5p_{3/2}$ and $4f_{7/2}$ core levels,, SCS yields essentially identical shifts (Kaindl et al. 1983). Thus $5p_{3/2}$ SCS on La can be relevantly compared to the 4f SCS for the other lanthanides. The first SCS obtained for the rare earths have been seen very early on Yb (Alvarado et al. 1980). The different experimental results are summarized in table 1. Of particular interest are the results of the very comprehensive investigation on the lanthanide series by Gerken et al. (1985). The agreement between the different groups is quite satisfying. Except for Ce, where some ambiguity occurs, a 0.3 eV shift was reported by Parks et al. (1982), while Gerken et al. (1985) concluded that a definite value was impossible to obtain. All of the published data have been obtained with polycrystalline film samples. Results with single crystal surfaces are now becoming highly desirable for deepening our understanding of SCS for RE which, will now be seen, is not as clear as was previously expected when given a more detailed inspection.

From table 1 and more specifically from fig. 10, reproduced from the work by Gerken et al. (1985), and displaying the SCS values as a function of the 4f filling across the lanthanide series, it can be seen, firstly, that values of SCS are positive, i.e., surface peaks are shifted to higher BE relative to the bulk, and are markedly large by 0.3 to 0.8 eV. Secondly, there is a clear trend in the series of trivalent lanthanides with the 4f filling in that SCS increases monotonously.

Large positive SCS are easily expected for rare earths if we consider, within the tight binding model for SCS, the formal $(5d6s)^2$ and $(5d6s)^3$ initial state valence configuration for, respectively, the divalent and trivalent lanthanides. Since 5d electrons play an important role in the bonding of the lanthanides, the situation of the SCS for these elements is quite similar with that for the 5d transition metals. In the first half of the 5d band filling, SCS are large. More specifically, the smaller the d occupation number, the larger the SCS. From fig. 10, SCS are found to be larger than the average for the divalent lanthanides Sm and Yb than for the trivalent ones. In this figure, Gerken et al. (1985) have also presented the values they calculated within the thermodynamical model. Since the surface cohesive energies, E_S^{coh}, are not available either experimentally or theoretically, they approximate the surface energy, $E_S(Z)$, by $E_S(Z) = 0.2 E_{coh}^B(Z)$, and then use relation (15).

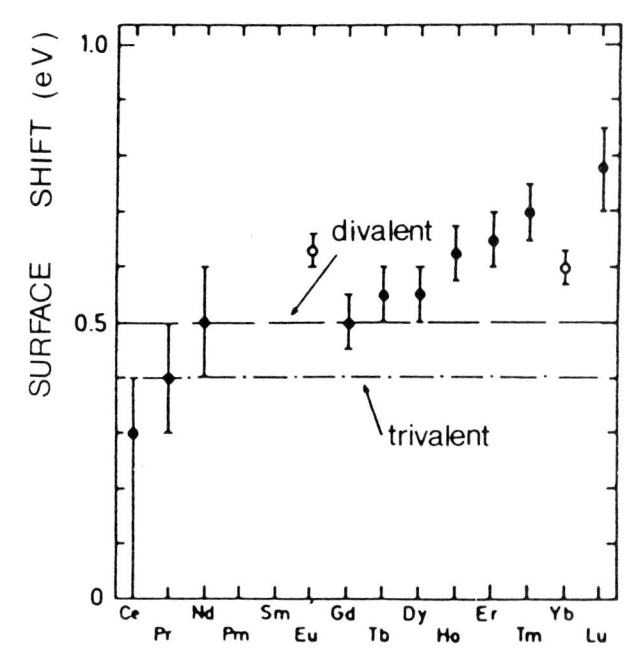

Fig. 10. Experimental SCS of the lanthanide series. Estimated SCS for the divalent and trivalent lanthanides within the thermodynamical model are reported. (From Gerken et al. 1985.)

The values for the bulk cohesive energy for the $(spd)^2$ divalent, $(spd)^3$ trivalent and $(spd)^4$ tetravalent metallic states of the lanthanides, to be included in this relation, were obtained from an interpolation procedure.

Except for Eu and Yb, which are divalent, and Ce, Gd and Lu, which are trivalent in both the atomic and solid states, most lanthanides experience a configuration renormalization $4f^{n+1}s^2 \rightarrow 4f^n (spd)^3$ upon condensation. Therefore, the cohesive energies for the divalent metallic lanthanides are to be interpolated between the values for Eu and Yb, and comparably, the cohesive energies for the trivalent metallic lanthanides are to be interpolated between the values for Ce, Gd and Lu. No lanthanide exists as tetravalent, but an element from the following series, such as Hg, is a good approximation. An alternative equivalent method consists of adding the $4f^{n+1}s^2 \rightarrow 4f^n$ transition energy to the cohesive energy of the lanthanides in their atomic states.

To sum up, a value of $\Delta E_{SCS} = 0.5$ eV is obtained for the divalent lanthanides (Eu, Sm and Yb), and 0.4 eV for the trivalent lanthanides (Ce, Pr, Nd, Gd, Tb, Dy, Ho, Er, Tm) (Gerken et al. 1985). If we consider all lanthanides to have the same formal valence configuration, both tight-binding and thermodynamical theoretical approaches obviously predict constant SCS within each lanthanide valence state. Thus, they do not reproduce the monotonic experimental increase of SCS across the trivalent lanthanides. This variation is probably due, as stated by Gerken et al. (1985), to the actual variation of the d occupation which is known from different physical parameters to decrease from La to Lu. Thus, two different band calculations give for the d occupation numbers of La and Lu, respectively, $n_d = 2.5$ and 1.5, and $n_d = 2.0$ and 1.4 (Duthie and Pettifor 1977, Gerken et al. 1985). Because a systematic decrease takes place in the d occupancy across the series from the lightest to the heaviest trivalent lanthanide, the tight binding model can be used to qualitatively explain the corresponding increase of SCS. A similar trend is seen with an increase of cohesive energy along this series following the sp→d transfer, and according to eq. (14‴) with an increase of ΔE_{SCS}. Further investigation with single-crystal surfaces is needed to clarify this point. On the other hand, surface peaks measured for rare earth polycrystalline samples are systematically broader than their bulk counterparts, with the explanation arising most probably from the existence of different single-crystal grains in the film, thus allowing for different SCS (Johansson and Mårtensson 1983, Schneider et al. 1983 and Gerken et al. 1985). A second-layer SCS may also be suspected. With all of this evidence, an investigation of the dependence of rare earth SCS on crystalline orientation is quite justified.

2.5.3. Surface core level shifts with sp electron states

A fundamental point at the origin of SCS in transition and rare earth metals arises from the presence of localized tight-binding d states, and the subsequent surface band narrowing occurs upon lower surface coordination. However, the phenomenon thus far appears to be much more general. SCS have been recorded for classes of solids other than d band metals; such classes being characterized by non-localized sp electrons states: free-electron-like metals, sp covalent bonded semiconductors, and wide gap insulators. As a consequence, different physical properties should act to induce SCS in these materials, and there is no reason for disregarding these possible contributions to d-band-like metal SCS.

2.5.3.1. sp free-electron-like metals. The case of sp metals represents a very interesting challenge for future theoretical and experimental research in the field of SCS. Indeed, the SCS present in free-electron-like metals is not relevant to the tight-binding approach, since for these solids the cohesive energy is essentially insured by the delocalized sp binding electrons. Thus, only the thermodynamic model can be used to calculate the SCS occurring in the sp metals, provided that sufficiently accurate values are available for the relevant surface cohesive energies. This implies that a microscopic theory is not yet established. Experimentally, SCS have been measured with success for Be, Na, Mg and Al, all samples being prepared as either polycrystalline films or single crystals. Very small values were measured for Al: the Al(100) oriented face shows a $\Delta E_{SCS} = -0.057$ eV, as found by partial yield spectroscopy on L edges (Chiang and Eastman 1981), and a $\Delta E_{SCS} = -0.120$ eV, as found with the photoelectron spectrum of the 2p doublet, while the Al(111) face does not show any detectable SCS or surface broadening at all (Kammerer et al. 1982). On the other hand, much larger values were recorded on the 2p core levels for polycrystalline Na: $\Delta E_{SCS} = +0.22$ eV, polycrystalline Mg: $\Delta E_{SCS} = +0.14$ eV (Kammerer et al. 1982), and on the 1s peak for Be(0001): $\Delta E_{SCS} = +0.50$ eV (Nyholm et al. 1985).

When predicting the SCS for Na, Mg and Al, using the thermodynamic relation $\Delta E_{SCS} = E_S(Z+1) - E_S(Z)$, surface energies must be known for the metals Na, Mg, Al and Si. By using values obtained theoretically as well as experimentally, Kammerer et al. (1982) calculated the following values for Na, Mg and Al, respectively: $\Delta E_{SCS} = +0.199$ eV, $+0.105$ eV, and -0.106 eV. Agreement between theory and experiment is acceptable for Na and Mg, and the Al case will be briefly reviewed later.

The calculation of SCS for Be is less straightforward, the estimation of surface energies being more difficult, especially because the replacement of the final core-ionized state of Be within the equivalent $(Z+1)$ core approximation by the B atom is highly questionable since B is by no means a metal. Following the procedure adopted by Nyholm et al. (1985), the bulk cohesive energy of the initial $(sp)^2$ configuration for metallic Be is set equal to the bulk energy of the atomic s^2 state, augmented by the atomic $s^2 \rightarrow sp$ optical energy = 6.05 eV/atom. For the final $(sp)^3$ metallic configuration of the ionized Be they simply used the bulk energy of metallic $(sp)^3$ Al = 3.39 eV/atom. Then, by using the familiar approximation $E_S(Z) = 0.2\, E_{coh}^B(Z)$ they found $\Delta E_{SCS} = -0.53$ eV, which is in close agreement with experiment, justifying this procedure per se.

In addition to the large BE shift for surface atoms, the results reported for Be(0001) present many interesting, specific features: a very strong surface enhancement, a sensitive dependence on the polarization of the exciting light, and a noticeable broadening of the surface core level (Nyholm et al. 1985). These results are intriguing, even more so since the data were obtained with no angular resolution. An enhancement of the surface peak intensity relative to the bulk by 20% is measured for all photon energies when using p polarization instead of s polarization, together with an overall increase of photoemission intensity by an average factor of about 2. These results are similar to angle-resolved results on W(110) (Tran Minh Duc et al. 1979). The most surprising result is the surface peak intensity which is about 6 times more intense than the bulk peak intensity for surface sensitive spectra excited by photons within the energy range $\hbar\omega = 130$ to 150 eV. This result can be coupled with the broadening of the Be 1s peak from 0.24 eV to 0.60 eV upon going from the bulk to surface atoms. For explaining this broadening, Nyholm et al. (1985) disregarded the possible presence of a second surface layer peak, although they recognized that the fit using two surface peaks was of better quality for reproducing their data. Instead, they suggested an interpretation which relies on enhanced phonon broadening at the surface. Clearly, this system deserves more investigation with better angle resolution, and on different orientation faces.

The discussion of the Al(100) results is complicated by the fact that a surface broadening attributed to crystal field splitting (Eberhardt et al. 1979) may be as important as SCS (Wimmer et al. 1981, Kammerer et al. 1982). By using a jellium model for Al(100), Chiang and Eastman (1981) related the SCS to the 0.120 eV variation of the electrostatic potential in the conduction electron sea at the jellium surface. This value is in agreement with the -0.106 eV value derived from surface energies. Using a full potential self-consistent linearized augmented plane wave calculation, Wimmer et al. (1981) also found a -0.120 eV shift for the first layer, and a -0.050 eV shift for the consequent layers, together with a 0.038 eV broadening. By themselves, both SCS and width broadening can separately reproduce the Al(100) results (Kammerer et al. 1982). Thus, no clear conclusion can be derived. However, for the Al(111) face, which has a smaller predicted SCS than Al(100), no broadening is observed (Chiang and Eastman 1981). Thus, it seems unlikely that broadening would occur only on Al(100) and not on Al(111).

2.5.3.2. Semiconductors. In contrast with the small SCS (<0.1 eV) presented by the sp band metals, sp covalent semiconductor surfaces exhibit core level shifts relative to the bulk as large as d transition and rare earth metals, viz. <1 eV. A very interesting feature with semiconductor single crystals is that SCS can be measured for monoatomic systems as well as binary systems. In principle, the difference between the SCS for the two binary systems may provide valuable clues for separating the different contributions in SCS. Contrary to the situation for metals, the ground state screening lengths in semiconductors are much larger. This, therefore, removes the layerwise charge neutrality condition at the semiconductor surfaces which, in fact, happens to be rather polar. In actuality, most semiconductors exhibit structural reconstruction accompanied by a large charge transfer, in the case of monoatomic crystals. When considering an initial state, these charge transfers may induce large core level shifts. If we examine the case of ionic binary semiconductors, the intersite Madelung potential has important effects on core level XPS BE; for surface planes, the truncation in the Madelung summation may be quite different, resulting in important consequences for SCS.

From a theoretical aspect, our understanding of these different initial state contributions – intra-atomic charge and interatomic Madelung potential – and their relative importance is rather unclear and controversial. The body of published experimental investigations is surprisingly somewhat meager in comparison to the technological interest in semiconductor surfaces. Furthermore, no general agreement can be ruled out from these studies. Obviously, more detailed measurements at higher energy and angle resolution are desirable in order to clarify the understanding of the semiconductor SCS.

Monoatomic semiconductors
The two monoatomic semiconductors investigated thus far are Si and Ge. Different clean Si surface orientations have been reported: Si(111) (2×1) (Himpsel et al. 1980, Brennan et al. 1980), Si(111) (7×7) (Himpsel et al. 1980), Si(100) (2×1) (Himpsel et al. 1980). All of these Si surfaces have been reported to furnish two or more surface core level peaks. For example, the Si 2p spin–orbit doublet spectrum of the freshly cleaved Si(111) (2×1) reconstructed surface reported by Brennan et al. (1980) exhibits two surface components, S1 and S2, located respectively at -0.6 eV (lower BE) and 0.3 eV (higher BE) relative to the bulk. In addition, the two surface $2p_{3/2}$ peaks present different widths upon comparison, as well as different widths from the bulk peaks: FWHM for S1 = 0.67 eV, for S2 = 0.41 eV and for B = 0.5 eV. In contrast with the above work,

Himpsel et al. (1980) reported for this Si(111) face three surface peaks at -0.37 eV, -0.14 eV and $+0.16$ eV BE relative to the bulk. This discrepancy probably arises from the spectrum resolution procedure, and should be negligible for higher resolution measurements. Nevertheless, from these works, the Si(111) (2×1) reconstructed surface presents certainly at least two SCS, one toward lower BE and the other one toward higher BE. This same conclusion, which remains to be confirmed, should hold equally for the annealed Si(111) (7×7) and Si(100) (2×1) reconstructed surfaces. However, in contrast to Si, Ge(100) (2×1) is found to present a single SCS $(-0.41$ eV) tending toward lower BE relative to the bulk (Miller et al. 1983).

Multiple SCS are expected as well from the various asymmetric structural models proposed for the different reconstructions generally experienced by the polar semiconductor surfaces. In the Si(110) (2×1) reconstruction, the outermost atomic plane undergoes an inward relaxation of 0.04 Å, with the formation of dimers between the atoms of adjacent rows. The reconstruction is achieved by an additional buckling of the dimer atoms, which raises and lowers alternate rows. From a total energy minimization approach, a charge transfer of 0.25–0.30 electrons is found to take place from the inward to the outward relaxed atoms of the asymmetric dimers (Brennan et al. 1980). This charge transfer results in a partly ionic dimer with the electropositive Si atom pointing outward from the surface plane.

Up to now, SCS in Si surfaces have been qualitatively interpreted solely on the basis of the variation of the intra-atomic Coulomb potential following the above charge transfer. The consideration of final state relaxation energy is particularly difficult for semiconductors where the screening of the core hole is incomplete due to final ionized states described by excitonic states. As already discussed, final state effects should be reduced on the surface relative to the bulk, and should affect equally the two (or more) surface core levels by raising their BE. Within the initial state framework, Madelung potential energy also has to be considered, since the surface plane is ionic, and long range terms have to be summed over all the surface anionic and cationic sites. The surface Madelung energy has been ignored. By considering only the intra-atomic electrostatic potential on the core orbitals due to the charge transfer, evaluated to be on the order of -2.2 eV per transferred electron, the difference between the BE of the inward and outward buckled Si atoms can be crudely estimated to be 0.55–0.66 eV, on the same order of magnitude as the experimentally observed 0.9 eV. This conclusion is confirmed by the intensity of the angle integrated measurements, which yield the same intensity for both core levels, and link each of them to half-monolayer coverage. It is clear, however, that self-consistent calculations of SCS for reconstructed Si surfaces are sadly lacking to clarify our understanding. Some attempts at tight-binding and self-consistent Hartree–Fock calculations were reported to give much larger theoretical values (2–3 eV) for the Si(111) (2×1) reconstruction.

By extending the empirical considerations made above for the Si(111) (2×1) surface to other systems, the same understanding is achieved for Si(111) (7×7) and Si(110) (2×1) reconstructions. However, the Ge(100) (2×1) orientation, which should be described by the same asymmetric buckling dimer model as its Si counterpart presents, in contrast with the latter, only one single SCS at −0.41 eV, toward lower BE relative to the bulk. Consequently, the surface intensity corresponds to the number of atoms equal to 0.94–1.1 of the coverage of the (100) atomic layer (Miller et al. 1983). With these results, the conclusion can be reached that the two asymmetric buckled surface atoms present the same SCS, in contradiction with the Si(100) (2 × 1) and with the charge transfer arguments previously used for the analysis of Si results. Clearly, it is necessary to reinvestigate these systems under better experimental conditions.

Binary semiconductors
SCS have been reported for several III–V semiconductor surfaces: GaAs(110) (Eastman et al. 1980, Miller and Chiang 1984), different reconstructed GaAs(100) (Ludeke et al. 1983), GaP(110) and GaSb(110) (Eastman et al. 1980), and InSb (Taniguchi et al. 1983, Horn et al. 1985). Measured on 3d core levels, the SCS are single and quite large (<0.7 eV) (see table 1), and generally, both elements have their core level shifted. A general statement, confirmed to date by all published investigations, seems to indicate that the SCS moves in opposite directions for the two ions: for cations the shift is positive, and conversely for anions, it is negative. Let us present as a typical example the well-known GaAs(110) surface (Eastman et al. 1980): only one SCS is observed for the two elements, at $\Delta E_B^{Ga} = +0.28$ eV for the Ga 3d level, and at $\Delta E_B^{As} = -0.37$ eV for the As 3d level, respectively. Thus, the BE difference between the two ions increases: $\Delta |E_B^{Ga} - E_B^{As}| = 0.65$ eV, by going from the bulk to the surface, whereas their mean value remains nearly constant: $\Delta E_B^{Ga} + \Delta E_B^{As} = -0.09$ eV.

From this last experimental observation, final state core hole relaxation, which affects the two SCS by an equal and positive amount, has been concluded to be of negligible effect and/or compensated by other contributions, such as the building of a dipole normal to the surface from the reconstruction (Tong et al. 1978, Kahn et al. 1978). Therefore, the SCS for these binary surfaces have been interpreted in terms of initial state effects. Since these compounds are ionic, both contributions – intra-atomic charge transfer and Madelung potential – should contribute equal weight. Since the BE difference between the two component core levels increases at the surface, we can conclude that the charge transfer between the two elements increases at the surface, and/or the Madelung interionic potential is reduced. These two interpretations have been proposed separately at both empirical and semi-empirical levels: Eastman et al. (1980) used purely charge transfer terms, and Davenport et al. (1981) proposed only Madelung potential reduction. A pure intra-atomic interpretation implies that at the surface the cation charge is more positive and the anion charge is more negative. This conclusion is reported by Davenport et al. (1981), arguing that from the reduction of coordination of the surface atoms the surface is expected to be less ionic. However, from the buckling reconstruction model, recognized to be valid for GaAs(11) reconstruction, charge may most likely be transferred from the inward buckled Ga atoms to the outward buckled As atoms. Theoretical tight-binding and ab initio cluster (Barton et al. 1979) calculations, however, conclude that, on the contrary, the relaxation reduces the surface charge transfer when compared to the ideal truncated (110) surface. On the other hand, a self-consistent calculation of the GaAs(110) SCS predicts the correct sign for the Ga and As shift, but not their magnitudes. Obviously, full consistent calculations for GaAs relaxed surfaces are required to extract quantitative information on the important chemical and structural properties of these surfaces from SCS.

Molecular beam epitaxy (MBE) has been extensively developed for elaborating III–V compounds in the design of solid state electronic devices. Understanding and control of the MBE-grown semiconductor surfaces are of fundamental and technological interest. Ludeke et al. (1983) have measured surface core level spectra for GaAs and GaSb (110) surfaces grown by MBE. These surfaces are polar and highly non-stoichiometric according to the growing conditions. In general, more than one surface core level peak has been found on these surfaces. Ludeke et al. (1983) have assigned these SCS to inequivalent surface stoichiometry of several GaAs surfaces. This work clearly demonstrates the interest of surface core level spectroscopy in the physics and chemistry of MBE surfaces and interfaces.

3. Applications

3.1. Stepped surfaces

Generally speaking, special interest is devoted to stepped surfaces because of their role in chemical reactions on metals. For example, the chemisorption characteristics of vicinal platinum surfaces are very different than those with low Miller indexes; atomic steps play an important role in dissociating hydrogen or oxygen molecules from Pt surfaces (Estrup and Anderson 1968).

Now considering these surfaces in terms of energy, it is well known that surface tensions, τ, are different between step- and terrace-atoms since τ is proportional to the average number, q_S, of broken bonds per unit area:

$$\tau \sim \tfrac{1}{2} q_S L , \qquad (19)$$

the energy, L, of a broken bond being related to the cohesive energy per atom by

$$E_S \sim \tfrac{1}{2} q L , \qquad (20)$$

where q is the average number of bonds of each atom in the solid. τ being directly related to the average number of broken bonds, it is expected to have approximately the same value for base surface as well as terrace atoms, and to increase upon going from terraces to step edges.

On the basis of the thermodynamical model described previously in terms of cohesive energy, we can then expect a different core level shift for surface steps and terraces. In this respect, different experiments have been run on Ir(332) (Van der Veen et al. 1981a), on Pt(557) or Pt⟨6(111)(100)⟩ and Pt(331) or Pt⟨3(111)(111)⟩ (Apai et al. 1983), and on W⟨6(110)(110)⟩ (Chauveau et al. 1984). Figure 11 compiles the results obtained on Ir(332) by Van der Veen et al. (1981a).

Ir $4f_{7/2}$ core levels excited by 100 eV photon energy are displayed for clean Ir(111) (fig. 11b) and Ir(332) surfaces (fig. 11c), and after hydrogen adsorption on Ir(332) surfaces (figs. 11a and 11d). On clean Ir(111), the surface–bulk core level shift is -0.53 eV, the surface component remaining on the low binding energy side. The $4f_{7/2}$ level of the (332) stepped surface exhibits three components obtained with a least-squares fitting procedure, attributed to bulk, terraces, and steps (fig. 11c). The most intense surface level is shifted by 0.48 eV relative to the bulk, i.e., almost degenerate with the surface level for Ir(111), and is attributed to terrace atoms. The component at lower binding energy and with a lower intensity is assigned to step edge atoms. The measured shift between bulk and

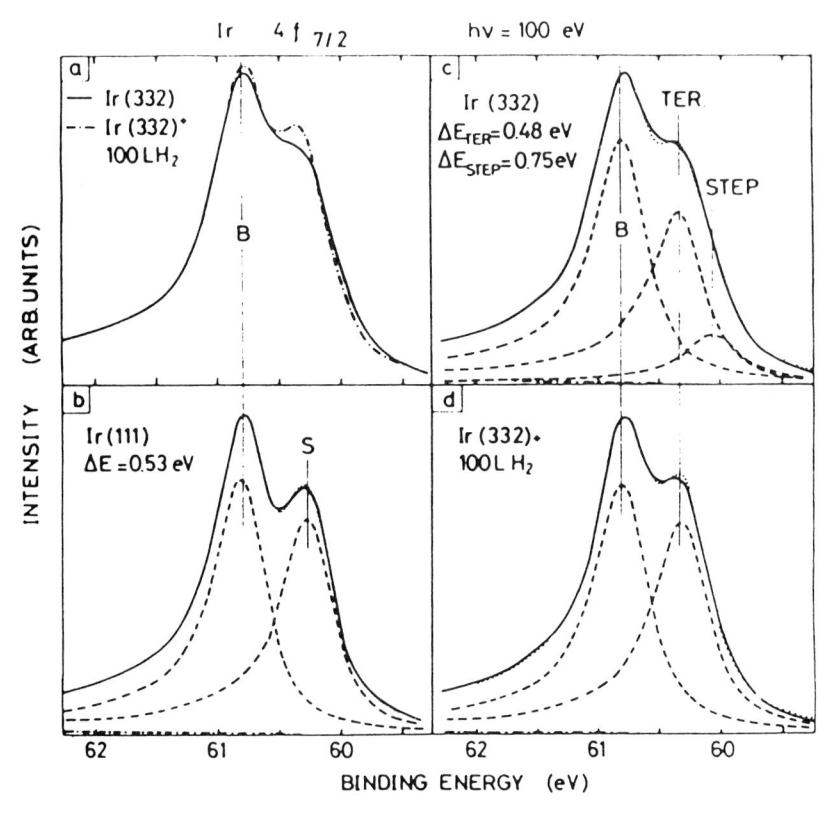

Fig. 11. Ir $4f_{7/2}$ levels of various surfaces of Ir after background subtraction: (a) Ir(332) clean and after 100 L H_2 exposure, (b) clean Ir(111) after decomposition into two Doniach–Sunjic lines attributed to bulk and surface, (c) clean Ir(332) resolved into three components due to bulk, terrace and steps, and (d) Ir(332) + 100 L H_2. (From Van der Veen et al. 1981a.)

step edge atoms is 0.75 eV. This larger shift is consistent with the reduced coordination number on step edge atoms, as compared to terrace atoms.

Upon hydrogen adsorption (figs. 11a and 11d), a shift of the step atoms towards higher binding energy is observed, whereas the terrace atoms binding energy is almost unaffected, which indicates that no hydrogen is adsorbed on terrace surfaces.

3.2. Surface reconstruction and relaxation

A large number of transition metal or semiconductor surfaces reconstruct and show a temperature or chemisorption-induced phase transition. (100) surfaces of FCC metals like Au, Ir and Pt reconstruct with a (5×1) or (5×20) structure; these reconstructed surfaces generally being more reactive than unreconstructed ones. W(110) $c(2 \times 2)$ is also an example of reconstruction, induced either by hydrogen chemisorption or by temperature. It is also well known that Si(111) undergoes a (2×1) reconstruction if cleaved, and a (7×7) reconstruction if heated in the 200–400°C temperature range. These are just a few examples. Several of them have been studied by high resolution photoemission in order to detect a modification of the surface bulk core level shift resulting from a change of coordination number (table 1). In fact, such a reconstruction can be monitored by surface core levels.

Let us consider the W(100) reconstruction. It is well known that this (100) face undergoes a reconstruction induced either by a cooling down to <300 K, or by hydrogen adsorption. In both cases, surface atoms are reorganized into a $c(2 \times 2)$ superstructure. Barker and Estrup (1978, 1981) have shown that under hydrogen, reconstruction proceeds by atom displacement along the (10) direction; surface atoms regroup into pairs along this direction. Thermal reconstruction is different, and results in the formation of zig-zag chains along the (11) direction. Figure 12 reports the data obtained for the H_2 reconstruction (Guillot et al. 1982).

The clean surface shows three components assigned to bulk (B), first (S_1), and second (S_2) surface layers. S_1 and S_2 are, respectively, shifted by -0.4 and -0.16 eV relative to B. After exposure to hydrogen (0.3 L), a $c(2 \times 2)$ structure is observed by low energy electron diffraction (LEED) and the surface peaks S_1 and S_2 are shifted toward the bulk, from which they are now separated by 0.25 eV (S_1') and 0.09 eV (S_2'). By pursuing the hydrogen exposure, a $p(1 \times 1)$ LEED pattern is observed, indicating that the surface atoms may recover their unreconstructed positions. This interpretation is confirmed for the 4f surface peaks which are, in fact, now shifting conversely back to 0.32 eV (S_1'') and 0.15 eV (S_2''), away from the bulk. These new values do not recover their original unreconstructed surface values, S_1 and S_2, because of an additional chemical shift due to hydrogen bonding.

A similar experiment has been run by Van der Veen et al. (1981b). Observed shifts on a clean surface are analogous (table 1). However, some disagreement appears in the interpretation of the two components, S_1 and S_2, in the reconstructed phase. On clean surfaces, S_1 and S_2 are attributed to the first and second

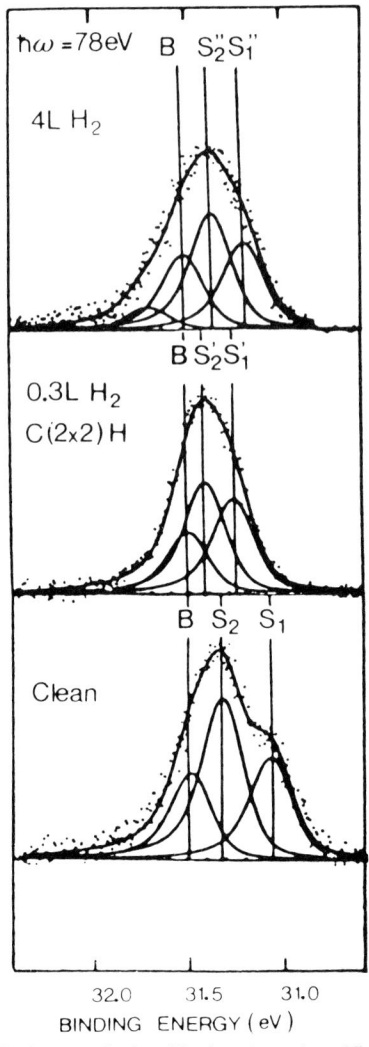

Fig. 12. $4f_{7/2}$ spectra of W(100) clean and after H_2 chemisorption. These spectra have been resolved into three components: B, S_1, and S_2, corresponding respectively to bulk, first and second surface layers. (From Guillot et al. 1982.)

surface atomic layers. Upon adsorption of a small amount of hydrogen, the intensity of S_2 grows dramatically at the cost of S_1, while the binding energies are much less affected. This is taken as evidence that hydrogen triggers the growth of reconstructed domains until reaching a 0.2 coverage whereupon all surface atoms become paired, which corresponds to the most stable configuration. S_1 originates from atoms on "normal" lattice sites, while S_2, nearly degenerate with the bulk level, originates from atoms that have some contact with each other in the surface plane.

At this level of interpretation, based on peak decomposition, it almost seems impossible to decide upon one model rather than another since the convolution of one peak at a fixed energy by another, either moving in energy or changing in intensity, can lead to the same result. However, one of the major trends resulting from this experiment is that the reconstruction leads to a more bulk-like character for the surface atoms. This has been confirmed for other metals, like Au (Heimann et al. 1981) and Ir (Van der Veen et al. 1980) (table 1).

In the case of Au and Ir experiments, Heimann et al. and Van der Veen et al. use the intensity variation of the surface and bulk components to interpret the reconstruction. For the Au(110) (2 × 1) surface, their data are consistent with a relaxed missing-row model, whereas the Au(100) (5 × 20) reconstructed surface gives surface core level shifts and intensity ratios which are consistent with a buckled hexagonal overlayer model with surface relaxation (Heimann et al. 1981). The Ir(100) (5 × 1) face is described by a compressed hexagonal monolayer (Van der Veen et al. 1980).

Experimental results on thermal reconstruction of W(100) (Guillot et al. 1984a) are different from those obtained with hydrogen reconstruction. The surface peaks, S_1 and S_2, undergo a quite small, almost indiscernible shift, as compared to the shift measured on an H_2 reconstructed surface. This is in contradiction with the theoretical predictions based on lateral reconstruction along the (11) direction. A calculation, including a reconstruction along the normal to the surface, would lead to better agreement with surface–bulk core level shift. However, this last model seems to be discredited by other experiments (Guillot et al. 1984a and references cited therein).

Before concluding this subsection let us note very recent results concerning nitrogen-induced reconstruction on W(100) (Jupille et al. 1986). These authors present the evidence for the occurrence of N_2-induced reconstruction of W(100) from the SCS study of the W $4f_{7/2}$ level. From the shift of the center of gravity of the spectrum as a function of the N_2 coverage, they infer that at low coverage N adatoms cause a local displacement of surface W atoms with formation of small contracted islands. Removal of surface reconstruction is observed between 0.4 and 0.5 monolayers.

To summarize, surface reconstruction can be confirmed by surface–bulk core level shift, and in the case of angular integrated experiments, by variation in the surface–bulk intensity ratio. However, to go further in the interpretation and modeling of reconstruction, this technique is certainly not the best choice. Other developing experiments, such as scanning tunneling microscopy (STM), or the use of photoelectron diffraction by angular resolved experiments, along with the help of multiple scattering calculations (see section 3.5) make prospects in this field quite promising.

3.3. Chemisorption

Photoemission has been recognized for a long time now as a valuable technique for adsorption studies. During adsorption, electron transfer to (or from) the adsorbate from (or to) the substrate occurs so that the electronic structure of both the adsorbate and substrate is modified. UPS, which mainly probes the valence bands, appeared as a fingerprinting technique: the spectrum of the adsorbed gas is superposed on the metal valence band. Then, from the comparison of the adsorbate–substrate system data with the gas phase spectrum of the adsorbate, we obtain direct information about the nature of the chemisorption process, whether it be molecular or dissociative. Until now, XPS was used mainly to probe the adsorbate core levels, since the substrate shift induced by chemisorption was expected to be small and the energetic resolution was large. During the past few years however, a fresh interest appeared in this field since, as shown previously, it is now possible to detect very small shifts and to be much more sensitive to the surface layer. The appearance of new high resolution monochromators in the 100 eV–1 keV range opens a new field of investigation in the surface science domain since the K-edges of C, N, O, i.e., the constituants of most of the adsorbates: CO, N_2, NH_3, O_2 and hydrocarbides, can be reached, and this way, it becomes possible to run experiments at photon energies close to the edge. This results in, among other things, an enhanced photoionisation cross section.

Consider now a substrate–adsorbate system. What we are interested in includes the site of adsorption – linear, bridge or multibonded – and the charge transfer, if any (amplitude, sign, etc.). We shall limit ourselves in the following to the strong adsorption case, i.e., chemisorption with real bonds between adsorbate and substrate.

Although we are not directly concerned with the shift measured on the adsorbate, but rather with the shift induced by the adsorbate on the substrate, we will mention just a few results relative to the measured adsorbate core level shift, which can lead to valuable information. In fact, we shall demonstrate how the core level shifts can be predicted, once thermodynamic properties of the adsorbate are known or assumed in advance. Actually, Grunze et al. (1982), and then Tomanek et al. (1983) and Feibelman (1983), calculated core level binding energies for adsorbates on metal surfaces using the equivalent core approximation in combination with a Born–Haber cycle. The scheme used is shown briefly in fig. 13.

Generally speaking, the adsorbed molecule can be expressed as $X(Z)Y_n$ (CO, NH_3, O_2, N_2, ...), Z being the atomic number for the X atoms, and n being the number of Y atoms. The initial state is described by the $X(Z)Y_n$ molecule on the metal, M. The binding energy of the core level, i.e. 1s, of the X atom in the $X(Z)Y_n$ molecule, E_b^F (X 1s, $X(Z)Y_n$/M), can be expressed as

$$E_b^F[X(Z)Y_n/M] = E_{chem}[X(Z)Y_n/M] + D[X(Z)Y_n] + E_B^V[X(Z)] - I^{X(Z+1)}$$
$$- D[X(Z+1)Y_n] - E_{chem}[X(Z+1)Y_n/M], \qquad (21)$$

where

$E_{chem}[X(Z)Y_n/M]$	is the adsorption energy of $X(Z)Y_n$;
$D[X(Z)Y_n]$	is the dissociation energy needed to obtain isolated atoms;

$E_B^V[X(Z)]$ is the core level ionization of X atoms;

$I^{X(Z+1)}$ is the ionization energy released to neutralize the $[X(Z+1)]^+$ ion;

$D[X(Z+1)Y_n]$ is the dissociation energy of the $X(Z+1)Y_n$ molecule; and

$E_{chem}[X(Z+1)Y_n/M]$ is the adsorption energy of the $X(Z+1)Y_n$ molecule.

The decomposition of the core level binding energy into a sum of thermodynamic quantities helps to provide quantitative predictions. Following this formulation, Grunze et al. (1982) predicted N 1s binding energies for different nitrogen-containing species on a W(110) surface (NH_3, NH_2, NH, N). Similar work has been done by Tomanek et al. (1983) for CO, NO, N_2 on various metal surfaces, and for Br, Br_2, I and I_2 on a Fe(110) surface. Reasonable agreement between theoretical and experimental values is achieved with adsorbates for which the adsorption energies are accurately known, and complete final state screening of the adsorbate is expected. More recently, Egelhoff (1984), with similar arguments, demonstrates how the core level binding energy shift analysis of adsorption and dissocation provides important thermodynamic quantities, such as heats of adsorption and dissociation, that often cannot be measured by any other method. He also provides a new approach to interpreting XPS peaks of adsorbates when the peak assignment is not immediately clear. Thus, from the N 1s XPS spectrum of N_2/Ni (100), he can determine the heats of adsorption of NO molecules adsorbed either nitrogen end down or oxygen end down (fig. 14).

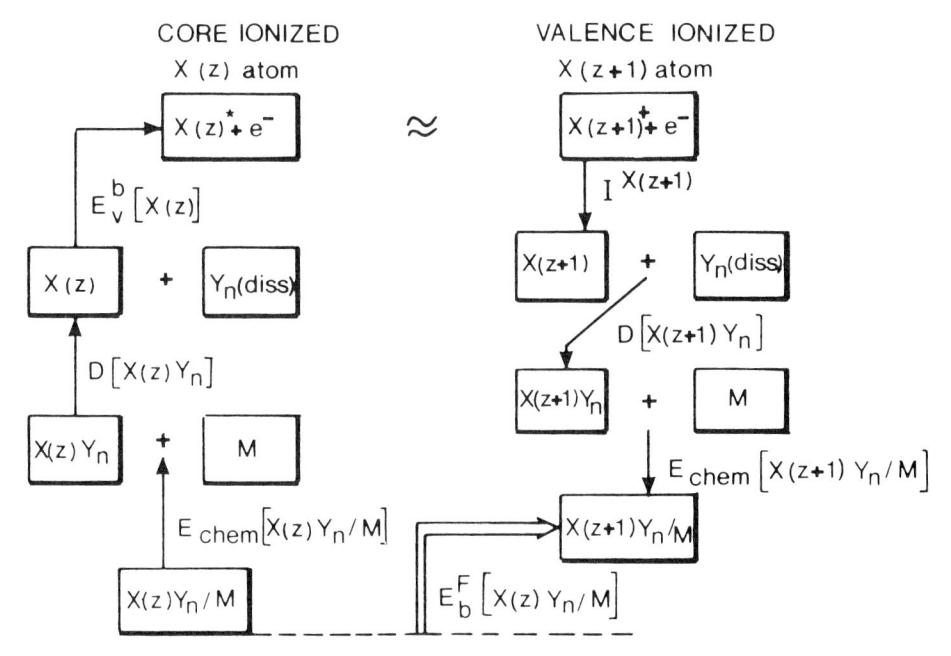

Fig. 13. Scheme of the thermodynamical model used for the estimation of core level shift induced by chemisorption.

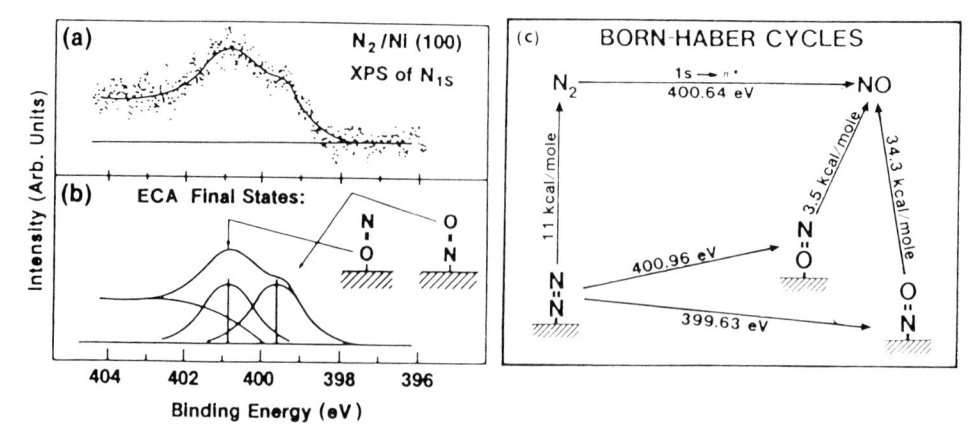

Fig. 14. N 1s X-ray photoelectron spectra of N_2: (a) adsorbed on Ni(100), and (b) after resolution into two components. Also shown in (b) are the equivalent core approximation final states; in (c) the Born–Haber cycle used for interpreting fig. 14a is reported. (From Egelhoff 1984.)

The deduced values from the Born–Haber cycle for the heats of adsorption of NO/Ni (100), 34.3 and 3.5 kcal/mol for NO bonded nitrogen end down and oxygen end down, respectively, indicate, as expected, that NO is strongly chemisorbed nitrogen end down and weakly physisorbed oxygen end down.

Let us return now to the substrate core level shift. In the presence of an adsorbate, we expect the core level shifts of the surface atoms to be modified mostly for those surface atoms directly bound to the adsorbate. For example, in the case of O_2 adsorption on transition metal surfaces, there is a small electron transfer toward oxygen which leads to a displacement of the d band to larger binding energy.

In the following, with the help of a few examples, we shall try to illustrate how the measurement of chemical shifts for substrate atom core levels can provide useful information on the nature of the chemical bond between adsorbate and substrate atoms. Such information can include the number of surface atoms bound to an adsorbate atom (site-specific technique), charge transfer, and the determination in a compound of the element principally involved during gas–solid interaction.

Site-specific technique: $O_2/W(110)$

The interaction of oxygen with W(110) proceeds via several steps depending on the oxygen coverage (Engel et al. 1975, Bauer and Engel 1978). A chemisorption phase corresponds to oxygen coverage less than 0.5 monolayer. As a function of coverage, LEED experiments first exhibit a $p(2 \times 1)$ structure, then a $p(2 \times 2)$ structure, and finally a more complicated structure resulting from the formation of a bidimensional oxide. Figure 15 reports the W 4f level spectra for different oxygen coverages (Tréglia et al. 1981, Jugnet 1981). Measurements have been run at a photon energy of 70 eV, corresponding to the maximal surface sensitivity.

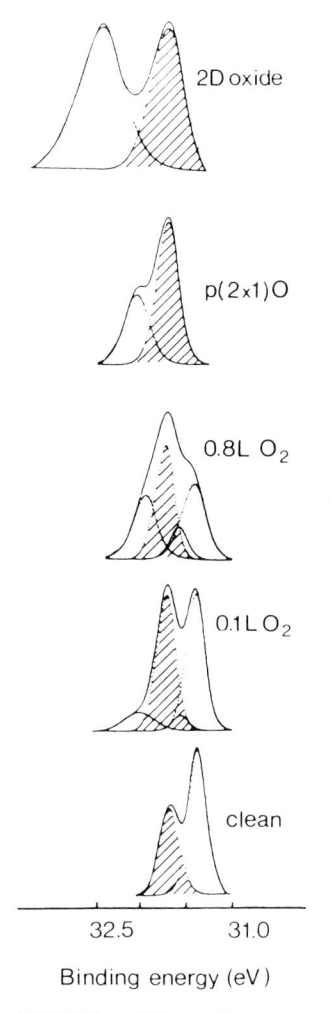

Fig. 15. W $4f_{7/2}$ core level spectra of W(110) at different O_2 exposures. A 70 eV photon energy with p polarization has been used to excite these levels. The bulk component is hatched. (From Jugnet 1981.)

After adsorption of 0.1 Langmuir (L; 1 L corresponds to an exposure of 10^{-6} Torr/s) of O_2, the surface component at 31.2 eV begins to vanish whereas at the same time, two new components appear at 31.4 and 31.8 eV. This trend is observed until the p(2 × 1) is formed (~3 L O_2). At this time, just one surface component remains at 31.8 eV. At higher oxygen coverages (10 min at 5×10^{-7} Torr) and after heating to ~800°C, a new component appears at 32.2 eV. For the sake of comparison, bulk oxides WO_2 or WO_3 lead to chemical shifts of 1.2 and 4.2, respectively, relative to the bulk (Hollinger 1979). From a theoretical model in which the binding energy change for core levels with W bound to oxygen is due to a small charge transfer from these atoms to oxygen, the different components are assigned in the following manner:

bulk	31.5 eV
surface (clean)	31.2 eV
surface + isolated O atoms	31.4 eV
surface + O(p(2 × 1))	31.8 eV

These results show that islands of p(2 × 1) exist from the beginning of chemisorption.

O_2/GaAs (110)

The oxidation of GaAs(110), although widely studied, has been reexamined by high resolution photoemission on the 3d core levels of both Ga and As (Landgren et al. 1984). Atomic rather than molecular adsorption is now well accepted. As to the question of oxidation mechanism, however, a lot of discrepancies occurs. Thus, the spectral evolution of As 3d and Ga 3d core levels upon oxygen exposure taken with maximal surface sensitivity has been examined from the beginning of exposure to oxygen. 3d spectra show a multicomponent substructure which increases in complexity with oxygen exposure over the range 10^6–10^{14} L. Spectral changes are already evident for Ga at 10^4 L and for As near 10^6 L, all of which indicate a preferential Ga oxidation and that separate Ga and As oxide phases form. Thus, by high resolution measurement of core levels, it has been shown that, contrary to previous proposals, the oxidation of a cleaved GaAs surface is not a homogeneous surface process involving predominantly the surface As. Rather, the oxidation is both spatially and chemically inhomogeneous, and the subsurface oxidation is the dominant process beyond an initial low coverage chemisorption stage. It is clear that obtaining such data requires very high resolution experiments, the measured shifts being quite small, at least at the beginning of the reaction.

Donor versus acceptor chemisorption on transition metals

The influence of chemisorption on the surface core level had been explored until now, through relatively simple systems: mainly O_2 and H_2 on the different faces of W and Ta (Tréglia et al. 1981, Guillot et al. 1982, Van der Veen et al. 1982, Guillot et al. 1984a,b). These configurations are considered as similar in that O_2 and H_2 are both electron acceptor species. In both these cases, surface core levels shift towards higher binding energy, which is explained in terms of charge transfer from the metal to the adsorbate. However, there exist some cases where adsorbates are electron donor species, such as K (often used as a methanation promoter), NH_3, Cs, etc. (Bonzel and Krebs 1981, Apai et al. 1983, Soukiassian et al. 1985). In such cases where electron transfer proceeds via a donation to the substrate, the surface core levels are expected to shift towards lower binding energy. To our knowledge, only two experiments of this type have been attempted: one with CO, K and NH_3 on different faces of Pt (Apai et al. 1983) and one with Cs on W(100) and Ta(100) (Soukiassian et al. 1985). These results are reported in table 2. In the case of donor species chemisorbed on Pt, the surface

Table 2
Surface core level shifts induced by donor species chemisorption.

Element (face)	CO	NH$_3$	K	Cs	Ref.[a]
Pt(111)	+1.3	–	+0.6	–	[1]
Pt(331)	+1.3	+0.75	–	–	[1]
W(100)	–	–	–	−0.06	[2]
Ta(100)	–	–	–	−0.1	[2]

[a] References:
[1] Apai et al. (1983).
[2] Soukiassian et al. (1985).

core level shifts toward higher binding energy. The authors interpret this positive shift in terms of a metal rehybridization, leading to a population decrease in the metal d valence band as a result of bonding with unoccupied adsorbate orbitals. Using an extended Huckel version of the tight-binding calculation, involving both the rehybridization induced by the presence of an adsorbate near the surface and a possible charge transfer, they calculated the charge transfer in the case of K atoms in a threefold site on a Pt(111) surface. They found that when K has a charge of $0.15\,e^-$, the metal atom surface has its sp population enhanced by $0.15\,e^-$ and its d population decreased by $0.11\,e^-$. They conclude that the main effect of the chemisorption is the rehybridization of the transition metal surface.

The only case of a negative shift (toward lower binding energy) observed as yet is for Cs adsorbed over (110) faces of W and Ta (Soukiassian et al. 1985).

3.4. Alloys and compounds; correlation between the observed shift and thermodynamical properties

Surface core level shifts in alloys and compounds clearly require a more complex theoretical evaluation than for pure elements. Thus, only a macroscopic approach, such as with the thermodynamic model, can then be applied. Furthermore, a general binary alloy cannot always be numerically treated since some of the thermochemical data for a ternary system formed by two initial elements plus the $(Z+1)$ element created by the photoevent, are unavailable. Hence, the model has to be restricted to alloys formed by two neighboring Z and $Z+1$ metals. By extending the Born–Haber cycle, as shown in fig. 6, to this type of alloy, Rosengren and Johansson (1981), have shown that the SCS can be directly connected to the heat of surface segregation for the $(Z+1)$ atom in the alloy. This result comes out directly from the now familiar and basic assumption of a fully screened core hole Z^* state due to the electron configuration of the equivalent core $(Z+1)$ atom.

If A denotes the Z atom, and B the $(Z+1)$ atom in an AB alloy, the surface shift experienced by the A atom core levels is expressed as

$$\Delta E_{\text{SCS}}[Z, \text{AB}] = [E_{\text{S}}(Z+1) + E^{\text{B}}_{\text{alloy}}(Z+1, \text{AB}) - E^{\text{S}}_{\text{alloy}}(Z+1, \text{AB})]$$
$$- [E_{\text{S}}(Z) + E^{\text{B}}_{\text{alloy}}(Z, \text{AB}) - E^{\text{S}}_{\text{alloy}}(Z, \text{AB})], \qquad (22)$$

where $E_S(Z)$ and $E_S(Z+1)$ are the surface energies of the pure metals A and B (relation 16), and $E^B_{\text{alloy}}[Z, AB]$ and $E^B_{\text{alloy}}[Z+1, AB]$ are the heats of solution of the metallic A and B atoms, respectively, in the bulk alloy. $E^S_{\text{alloy}}(Z, AB)$ and $E^S_{\text{alloy}}(Z+1, AB)$ are the comparable heats of solution, but for the surface of the alloy.

If we modify fig. 6 for the case of an alloy AB, the result is a surface core level shift for element A (the Z atom) in alloy AB ($\Delta E_{\text{SCS}}[Z, AB]$), exactly equal to the heat of surface segregation of element B (the $(Z+1)$ atom) in alloy AB (Rosengren and Johansson 1981, Kumar et al. 1981). As a matter of fact, if we join in eq. (22) the quantities related to Z and $Z+1$, respectively, into a single term:

$$E^S(Z, AB) = E^B_{\text{coh}}(Z) - E^S_{\text{coh}}(Z) + E^B_{\text{alloy}}(Z, AB) - E^S_{\text{alloy}}(Z, AB) \tag{23a}$$

and

$$E^S(Z+1, AB) = E^B_{\text{coh}}(Z+1) - E^S_{\text{coh}}(Z+1) + E^B_{\text{alloy}}(Z+1, AB) - E^S_{\text{alloy}}(Z+1, AB) \tag{23b}$$

we obtain (Kumar et al. 1981)

$$\Delta E_{\text{SCS}} = E^S(Z+1, AB) - E^S(Z, AB), \tag{24}$$

where, physically, $E^S(Z+1, AB)$ represents the energy required for moving an atom B from the bulk of the alloy to the surface, whereas $[-E^S(Z, AB)]$ is the energy gained by transferring the atom B from the surface back into the bulk. $\Delta E_S(Z, AB)$ corresponds, therefore, to the heat involved in the reaction of exchanging a B atom in the bulk with an A atom on the surface.

From the point of view of the applications of surface core level shifts in alloys, a positive $\Delta E_{\text{SCS}}(Z, AB)$ predicts an enrichment of constituent A on the alloy surface. Conversely, if $\Delta E_{\text{SCS}}(Z, AB)$ is negative, suggesting lower binding energy, we expect a surface enrichment of the B constituent, since according to eq. (24), this segregation is exothermic.

It is worthwhile to note that eq. (24) holds true for any general AC alloy, C being any element, but the energies $E^S(Z, AC)$ and $E^S(Z+1, AC)$ now have ternary values. If $\Delta E_S(Z, AC)$ is no longer subject to the simple physical definition given above for a $Z/Z+1$ alloy, the surface core level shift, in the case of dilute alloy of Z in AC, is identified to be the difference in the heat of segregation between the Z and $(Z+1)$ atom in the alloy AC.

A more detailed insight into the understanding of the properties of a surface core level shift in alloys can be obtained by relating $\Delta E_{\text{SCS}}(Z, AB)$ with $\Delta E_{\text{SCS}}(Z, A)$, the surface core level shift in the pure metal, and $\Delta E^B_C(Z, AB)$, the bulk chemical shift of Z core levels, in other words, by going from pure metal A to alloy AB.

According to Johansson and Mårtensson (1980), the alloy bulk shift is given by

$$\Delta E_C^B(Z, AB) = E_C^B(Z, AB) - E_C^B(Z, A)$$
$$= E_{alloy}^B(Z, AB) - E_{alloy}^B(Z+1, AB) + E_{imp}^B(Z+1, A). \quad (25)$$

We may then express $\Delta E_{SCS}(Z, AB)$ by rewriting eq. (22) as

$$\Delta E_{SCS}(Z, AB) = E_S(Z+1) - E_S(Z)$$
$$+ (1-\alpha)[E_{alloy}^B(Z+1, AB) - E_{alloy}^B(Z, AB)], \quad (26)$$

where we have made the reasonable assumption that $E_{alloy}^S(Z, AB)$ and $E_{alloy}^S(Z+1, AB)$ are reduced from the corresponding bulk energies by the same α factor as with the cohesive energies:

$$E_{alloy}^S(Z, AB) = \alpha E_{alloy}^B(Z, AB) \quad (27a)$$

and

$$E_{alloy}^S(Z+1, AB) = \alpha E_{alloy}^B(Z+1, AB). \quad (27b)$$

Keeping $\Delta E_{SCS}(Z, AB)$ as it was expressed in eqs. (26), and from eqs. (25) and (14″), we obtain

$$\Delta E_{SCS}(Z, AB) = \Delta E_{SCS}(Z, A) - (1-\alpha)[\Delta E_C^B(Z, AB) + 2E_{imp}(Z+1, A)]. \quad (28)$$

From eq. (28), if we ignore the small quantity $2E_{imp}(Z+1, A)$, we are led to the conclusion that the surface core level shift magnitude is larger (smaller) in an alloy than in pure metals if $\Delta E_{SCS}(Z, AB)$ (and/or $\Delta E_{SCS}(Z, A)$) and ΔE_C^B are of opposite (same) signs. The increase or decrease is given approximately by $(1-\alpha)\Delta E_C^B(Z, AB) \sim 0.2\, \Delta E_C^B(Z, AB)$. This is illustrated more clearly in fig. 16. In this figure, it is clearly illustrated that our conclusion comes from the fact that for alloys, surface core levels are shifting in the same direction as bulk core levels, but with a reduced magnitude. As a matter of fact, this result can be easily derived using eq. (28) by simply disregarding the negligible $2E_{imp}(Z+1, A)$ term:

$$\Delta E_C^S(Z, AB) = E_C^S(Z, AB) - E_C^S(Z, A) = \alpha \Delta E_C^B(Z, AB). \quad (29)$$

Until now, experimental results for SCS have been reported for an extremely scarce number of alloys. Furthermore, only one alloy consisting of two neighboring Z and $Z+1$ elements has been investigated thus far, i.e., the Pt–Au alloy, by Hörnström et al. (1985), who reported SCS for both constituents. For Au, the SCS relative to the bulk is towards lower BE by -0.26 and -0.19 eV, respectively for Au–Pt alloys with 15 wt% and 2 wt% Au concentration. In pure Au, the SCS

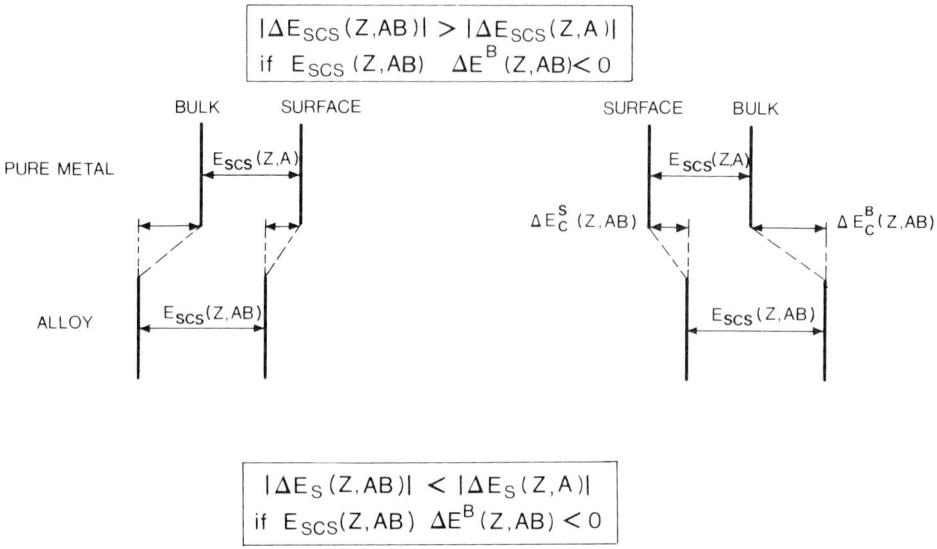

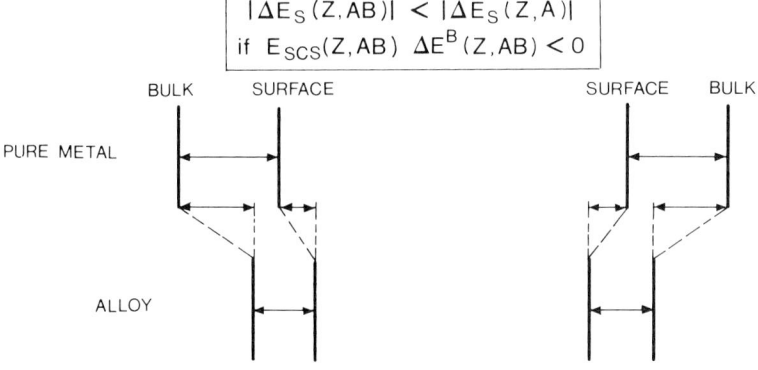

Fig. 16. Illustration of the codification of the variation of surface core level shift between pure metals and alloys.

is -0.34 eV. For Pt, the SCS is found to be negative as well at -0.30, -0.30 and -0.29 eV, respectively for pure Pt, Pt–2 wt% Au and Pt–15 wt% Au alloys.

According to the thermodynamical model for SCS in alloys, since the SCS is negative for Pt, the Z element of the model, the $Z+1$ constituent Au is expected to segregate to the surface of these alloys. This is indeed the case as demonstrated in the measurements of Hörnström et al. (1985) by the Au to Pt 4f intensity ratio measured as 0.7 for the Pt–2 wt% Au sample, and 5.3 for the Pt–15 wt% Au sample. The strong enrichment in Au at the alloy surface is more clearly indicated by the bulk-to-surface ratios, which are decreasing for Au from 0.80–0.90 in pure Au, to 0.50–0.56 for Pt–2 wt% Au and 0.08–0.10 for Pt–15 wt% Au, while they appear to slightly increase for Pt from 0.90–1.08 in pure Pt to 1.00–1.40 for Pt–2 wt% Au and 1.00–1.30 for Pt–15 wt% Au.

SCS for Au, as measured by Hörnström et al. (1985), are reduced in magnitude upon alloying with Pt. The trend for Pt SCS is less clear, the values remaining somewhat constant. The behavior of Au SCS is consistent with the empirical conclusions drawn from fig. 16 and eq. (29) by considering the chemical shift of the bulk core levels in these alloys. The bulk shifts for Au were measured at -0.19 and -0.39 eV, respectively for Pt–2 wt% Au and Pt–15 wt% Au relative to the pure metal. Since the chemical shifts in the alloys are of the same sign as the SCS, a reduction in magnitude of the latter is expected (see fig. 16). However, no conclusive physical meaning should be, for the moment, extracted from such an analysis since, for example, eq. (29) was derived through very crude approximations. The usefulness of relation (29), predicting the SCS in alloys from the bulk chemical shift in these alloys, may nevertheless be tested. From relation (29), Au SCS is evaluated to be $-0.34 - 0.2 \times (-0.39) = -0.26$ eV and $-0.34 - 0.2 \times (-0.19) = -0.30$ eV, for respectively the Pt–15 wt% Au sample and the Pt–2 wt% Au sample. These estimated values are somewhat larger in magnitude than the experimental values of -0.19 and -0.26 eV reported by Hörnström et al. (1985). For Pt, the estimated values for SCS are $-0.30 - 0.2 \times (-0.18) = -0.26$ eV and $-0.30 - 0.2 \times (-0.01) = -0.30$ eV, as compared to the experimental values of -0.30 and -0.29 eV.

Relation (29) has been used as well by Nilsson et al. (1985) for the evaluation of Yb 4f SCS, with Yb alloyed with both Au and Ag. From a value of SCS of $+0.56$ eV in pure Yb, and chemical shifts of bulk Yb 4f peaks of -0.9 eV in Yb–Au alloys and -0.45 eV in Yb–Ag alloys, they found, using relation (29), a Yb SCS of $0.56 + 0.24 \times (0.9) = 0.78$ eV in Yb–Au and a Yb SCS of $0.56 + 0.24 \times (0.45) = 0.67$ eV in Yb–Ag. Their experimental values are respectively $+0.75$ and $+0.70$ eV. From these few examples, the consistency of relation (29) with experiment is gratifying enough to deserve more investigation.

A Yb SCS of $+0.75$ eV was also measured by Johansson et al. (1982), for Yb–Au. This group has also investigated the chemical shifts of both surface and bulk 4f levels of Eu and Yb when these elements are alloyed with Au with several Au concentrations up to 50%. As already stated, SCS for lanthanides are always towards higher BE. On alloying with Au, the Eu and Yb bulk core peaks are shifted towards lower BE. As a consequence of relation (29) and fig. 16, Johansson et al. (1982) found an increase in the SCS for Eu from 0.60 eV in pure Eu to 0.70 for Eu–Au alloys with high Au concentrations. In the same manner, the Yb SCS increases from 0.63 to 0.75 eV when going from pure Yb to Yb–Au. Concerning the surface segregation in these alloys, the surface-to-bulk peak intensity ratios indicate a strong segregation of Eu and Yb towards the surface, which is in agreement with their lower heat of vaporization as compared to Au.

SCS have been reported as well for a number of intermetallic rare earth compounds (Mårtenson et al. 1982, Kaindl et al. 1982a,b, 1983). The general trends in these compounds have been reviewed by Kaindl et al. (1983) in fig. 17. SCS for rare earths are consistently larger for the intermetallic compounds as compared to the pure metals. In addition, this magnification, by a factor of 2, for the divalent and mixed-valent lanthanides – Eu and Yb – is of particular importance in contrast with the trivalent lanthanides, where the increase of the SCS is

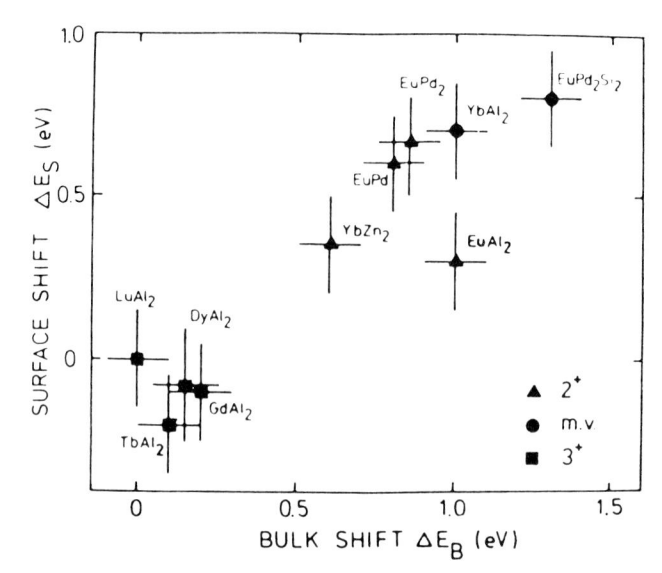

Fig. 17. Correlation of SCS with bulk shifts for various rare earth compounds. (From Kaindl et al. 1983.)

of a much smaller magnitude (see fig. 17). This enhancement of the SCS can be explained by the bulk chemical shift in these compounds toward lower BE relative to the metal, in the opposite direction of SCS. Further, the difference in SCS magnification between the divalent and the trivalent rare earth compounds is explained by the much larger chemical bulk shifts for the divalent compounds, of about 1 eV, as compared to those of trivalent compounds, which are practically zero. The correlation of the SCS enhancement with the bulk chemical shift of fig. 17, qualitatively follows the predictions of relation (29).

The lower BE shifts of the bulk 4f core levels with respect to the pure metal for the divalent rare earth compounds are unexpected when considering the positive atomic charge in the (+2) valence state, in a purely initial state interpretation. By including the terms due to the screening of the final state hole, Kaindl et al. (1983) relate this negative bulk shift in the divalent compounds to the difference in the heat of formation between the rare earth compounds in the initial (+2) state and the fully relaxed final (+3) ionized state. This statement implies that the heats of formation of the trivalent compounds are higher than those for the divalent compounds. This result is quite in agreement with the conclusions drawn from the thermodynamical approach for the positive values of SCS in these rare earth compounds.

As a result of the above examples, the surface and chemical bulk shifts of the core levels for the rare earth elements and their compounds or alloys are closely related to their detailed electronic structure and to the thermochemistry within these compounds. The thermodynamical model is a most valuable tool for interpreting these shifts, but still awaits a more quantitative level.

3.5. Quantitative aspects

In the following, we shall concentrate on peak intensities rather than their energies, and show that the measurement of surface and bulk intensity can be used to determine the mean free path of the electron, λ, as a function of their kinetic energy, E_K. Actually, the experimentalist is left with a universal curve giving approximate values of λ as a function of E_K (fig. 2). This curve results from a compilation of experimental data (Pijolat 1979, and references cited therein), and can be approximately fitted by a E^{-2} curve when E_K is less than ~15 eV, and by a $E^{1/2}$ curve when $E_K > 75$ eV. Some empirical relations have been proposed to describe $\lambda(E_K)$ variation, such as

$$\lambda(E_K) = \frac{2170}{E_K^2} + 0.72(aE_K)^{1/2}, \tag{30}$$

where a is the thickness of the monolayer (in nm), this relation being valid for $E_K > 150$ eV (Seah and Dench 1979).

In fact, the knowledge of this parameter is quite important as soon as we become interested in quantitative results, such as profile, thickness of overlayers, etc.

Different quantitative models have been described recently by Fadley (1984) for various ESCA experiments: semi-infinite specimens, specimens of thickness t, and semi-infinite substrates with uniform layer. Here we shall limit ourselves to the case of a surface plane over an infinite bulk.

Let N_B and N_S be the surface and the bulk layer densities, respectively, expressed in atoms/cm². I_S and I_B are the measured intensities of surface and bulk components of a core level k, respectively. σ is the cross section for inelastic scattering per atom. Furthermore, we consider here a layer model for electron escape and a perfectly smooth surface, a quite utopist case. Lastly, we are reminded that such a model cannot be applied directly to the case of angular resolved experiments.

In a first case, we consider that electrons coming from surface atoms are autoabsorbed in the surface layer. Thus, the intensity coming from the surface layer, I_S, can be expressed as

$$I_S \propto KN_S(1 - e^{-d/\lambda_S}), \tag{31}$$

where K is an instrumental factor, and λ is the mean free path. The intensity coming from the successive bulk underlayers can be written as

$$I_{B_1} \propto KN_B(1 - e^{-d/\lambda_B}) e^{-d/\lambda_S}, \tag{32a}$$

$$I_{B_n} \propto KN_B(1 - e^{-d/\lambda_B})(e^{-d/\lambda_S})^{n-1}, \tag{32b}$$

$$I_B \propto KN_B(1 - e^{-d/\lambda_B}) e^{-\lambda_S} \left[1 + \sum_{i=1}^{\infty} (e^{-d/\lambda_S})^{i-1} \right] \sim N_B e^{-d/\lambda_S}. \tag{32c}$$

From these expressions, we can deduce the surface–bulk ratio I_S/I_B to be

$$\frac{I_S}{I_B} = \frac{N_S}{N_B}(e^{d/\lambda_S} - 1). \tag{33}$$

If, in a second case, we neglect the autoabsorption, we are left with the following expression:

$$\frac{I_S}{I_B} = \frac{N_S}{N_B} e^{d/\lambda_S}(1 - e^{-d/\lambda_B}). \tag{34}$$

λ is primarily dependent on the kinetic energy of the electrons, and thus, in a first approximation, $\lambda_S = \lambda_B$. This is not really true since λ depends as well on the atomic concentration. Regardless, such a description can be used in principle either to determine the mean free path, λ, as a function of the kinetic energy by simply tuning the photon energy, or to indicate a change of atomic density in the surface layer, which might have resulted from a surface reconstruction. In this case, the mean free path has to be available from other experiments.

At this point, it is worthwhile to mention a severe limitation to the use of this formula in the case of monocrystalline samples. As a matter of fact, photodiffraction effects which can play an important role in angular resolved experiments, can completely change the intensity ratio between bulk and surface components, and this application becomes invalid.

3.6. Surface crystallography versus photoelectron diffraction

Up to this point, we have mainly discussed the SCS in terms of energy. Indeed two pieces of information can be extracted from a photoemission measurement: the energy term, which reflects the electronic structure of an atom in a solid, and the intensity term, which is more directly sensitive to solid state effects. We shall now focus on this last point, the surface and bulk peak intensities.

When the experiment is not angle resolved, the ratio of surface over bulk components leads directly to the number of surface atoms compared to bulk atoms (see section 3.5). This is not true for angle resolved experiments because of complicating but useful effects due to photoelectron diffraction.

In order to clarify the following, let us just remind ourselves of a few characteristics of an angle resolved experiment. Compared to an integrated experiment where electron detection is only dependent on their kinetic energy, an angular resolved experiment leads to the knowledge of a new parameter: the propagation direction, **OD**, of those electrons, defined by two angles, θ and φ, as shown in fig. 18. The surface plane is defined by the **OX** and **OY** axes, **OZ** being normal to the surface. α defines the incidence angle, and **A** the polarization vector of the light (**A** is perpendicular to the direction of light propagation and is in the plane of the ring). Electrons are detected in a small solid angle, typically $\sim 2°$ along **OD**. The detection geometry is defined by the polar angle, θ, and the

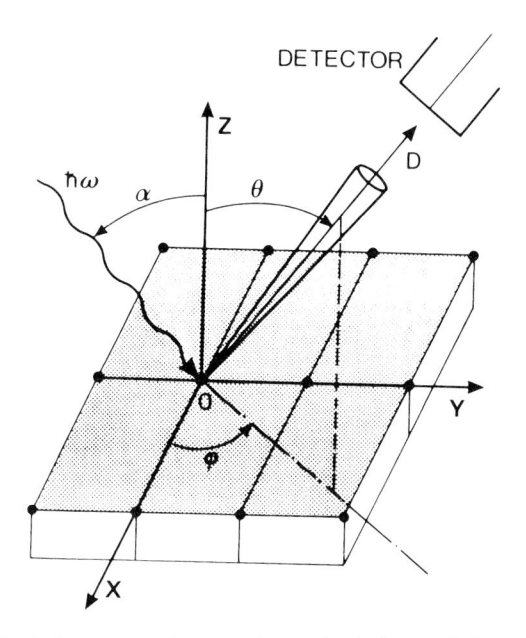

Fig. 18. Analysis geometry in an angle resolved photoemission experiment.

azimuthal angle, φ, the angle between the projection of **OD** on the surface plane and the axis **OX**. Generally A, **OZ** and **OD** are in the same plane. The kinetic energy, E_{ext}, of the emitted photoelectrons is related to their wave vector, K, through the relation

$$E_{ext} = \frac{\hbar^2 K_{ext}^2}{2m} \tag{35}$$

with

$$K_{ext}^X = \left(\frac{2mE_{ext}}{\hbar^2}\right)^{1/2} \sin\theta \cos\varphi ,$$

$$K_{ext}^Y = \left(\frac{2mE_{ext}}{\hbar^2}\right)^{1/2} \sin\theta \sin\varphi ,$$

$$K_{ext}^Z = \left(\frac{2mE_{ext}}{\hbar^2}\right)^{1/2} \cos\theta .$$

Thus, the knowledge of θ and φ leads to the determination of the wave vector of the photoelectron. θ is defined by an appropriate alignment of the light with the sample, and φ has to be determined from a LEED diagram.

Let us consider now what happens when light impinges on a monocrystalline sample under such experimental conditions. Core level photoelectrons are emitted from localized and incoherent atomic sources (fig. 19). The detected current

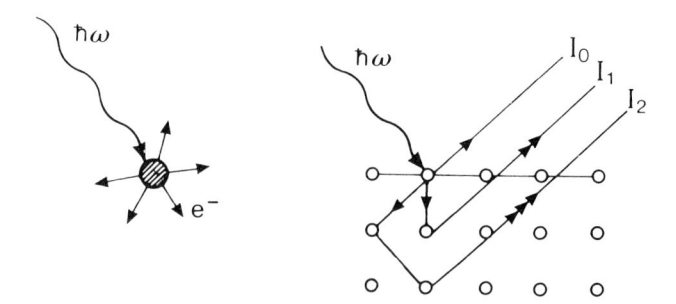

Fig. 19. Schematic representation of photoelectron diffraction effects. The detected photocurrent appears as the interference between the directly emitted wave I_s and the waves emitted after scattering on one atom I_1 or several atoms I_2.

can be pictured as the coherent sum of the direct photoelectron wave (I_0) from the emitting atom toward the detector, and all of the indirect waves (I_1 and I_2) emitted in other directions and backscattered by the atomic sites surrounding the emitter toward the detector. The interference between these different waves leads to the photodiffraction effect.

Depending on the kinetic energy of the emitted electrons, the scattering can take place on just one atom (single scattering), or on a sequence of several atoms (multiple scattering). There is no definite limit between these two models; however, beyond a kinetic energy of approximately 200 eV, we have to consider a multiple scattering formalism, which leads to quite complicated theoretical treatments (dynamical theory). For a kinetic energy larger than about 200 eV, the problem is easier since scattering can be treated in the simpler formalism of single scattering or kinematic theory. This will be discussed later.

Experimentally, a photodiffraction experiment consists of measuring the intensity modulations as a function of the wave vector **K**. There are several possible experiments which vary the wave vector **K** of the emitted electrons. Indeed, the diffraction effect may be observed either via angle variations of the light incidence and/or detection, or via kinetic energy variation of the photoelectron (in fact, light wave length variation). However, energy distribution experiments, normally easier to perform than angle resolved experiments when synchrotron radiation is available, make further theoretical interpretations more difficult since numerous quantities depend on energy.

As an example, let us summarize here the principal results obtained by Tran Minh Duc et al. (1979) using synchrotron radiation. In this pioneer work on a W(110) single crystal, the $4f_{7/2}$ core state diffraction pattern is studied. Radiation wave length ($\hbar\omega = 70$ eV) and polarization have been optimized to emphasize surface contributions accordingly to the above discussed considerations. Two basic types of measurements have been made on photoelectron intensity distributions: (i) as a function of polar angle of detection at $\varphi = 0°$ and as a function of azimuthal angle of detection for $\theta = 45°$ (fig. 20), and (ii) as a function of excitation light wave length at $\theta = 0$ between $\hbar\omega = 50$ eV and $\hbar\omega = 140$ eV (fig.

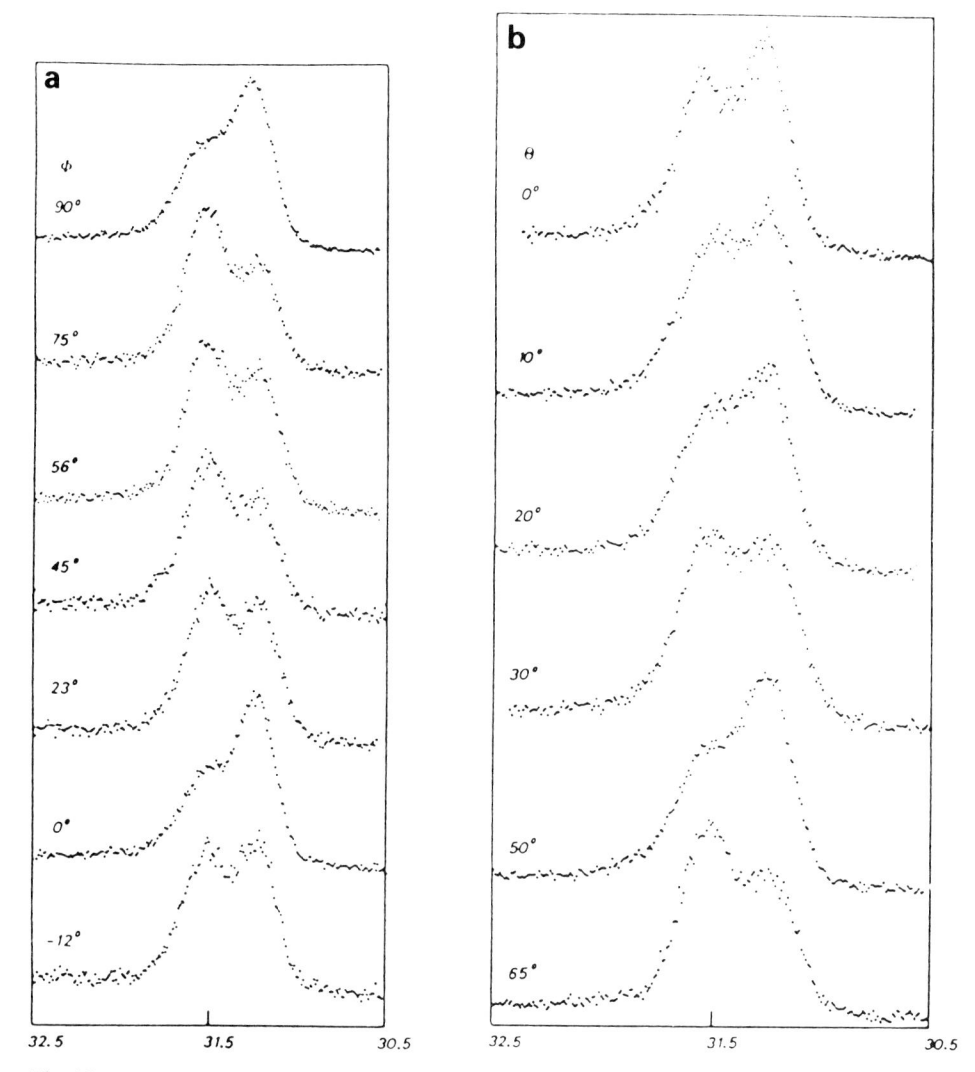

Fig. 20. W $4f_{7/2}$ core level spectra of clean W(110) measured at a 70 eV photon energy as a function of (a) polar angle θ, and (b) azimuthal angle φ. These raw data show the evolution of the surface (31.2 eV) over bulk (31.5 eV) components due to diffraction effects. (From Tran Minh Duc et al. 1980.)

21). In both cases, as expected, resolution and surface shift were sufficient (figs. 1, 20 and 21) to clearly resolve a surface contribution (due to photoemission from surface atoms), and bulk contributions (due to the mean emission from bulk atoms). Furthermore, a different anisotropic pattern between these two contributions has been found. It is on the origin of these anisotropies that we focus this section.

Theoretically (Spicer 1958, Mahan 1978, Fadley 1978, 1984), the entire

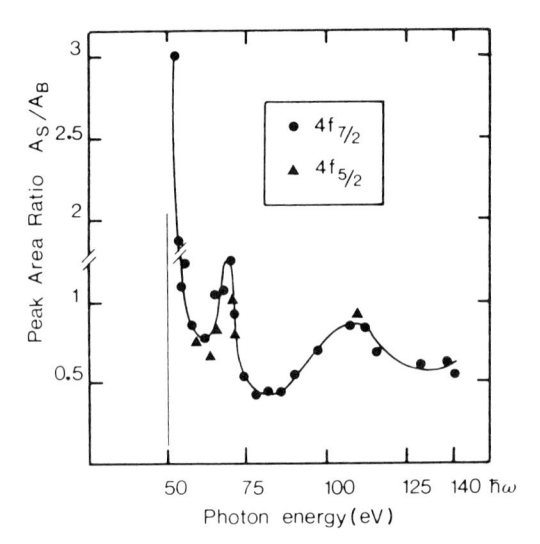

Fig. 21. Variation of the ratio of surface over bulk components of W $4f_{7/2}$ and W $4f_{5/2}$ core levels of clean W(110) as a function of photon energy, for detection along the normale to the surface. (From Tran Minh Duc et al. 1979.)

phenomenon may be explained as a photoionization process (step 1) followed by a transport of these photoionized electrons to and through (steps 2 and 3) the surface. Note that this distinction in steps is an artifice, since in accurate calculations individual step amplitudes, rather than intensities, must be combined to take into account the interdependence (induced selection rules and phase shifts) within the three steps. In such a model, the two first steps, viz. photoionization and transport, intertwine and therefore compete to give rise to the observed anisotropy.

When a hole is created by the photoionization of an electron from an inner shell nl, the other electrons may be assumed not to participate in the transition (frozen orbitals model). If use is made of the standard decoupling formalism (Cox 1975, Schirmer et al. 1980, Gupta and Sen 1974), this approximation enables us to concentrate our attention on the one-electron problem. As already mentioned, the one-electron partial transition amplitude is dependent on the matrix elements, $\langle \varepsilon l'm' | \boldsymbol{A} \cdot \boldsymbol{p} + \boldsymbol{p} \cdot \boldsymbol{A} | nlm \rangle$, where $|nlm\rangle$ is the initial discrete (core level) state, and $|\varepsilon l'm'\rangle$ is the final continuum state labelled by its kinetic energy, ε, and by its angular and magnetic momentum, $l'm'$. The diamagnetic term $|A|^2$ is always small and thus may be neglected. The transition is mediated by $\boldsymbol{A} \cdot \boldsymbol{p} + \boldsymbol{p} \cdot \boldsymbol{A}$, which describes the photon–electron interaction. On the other hand, this interaction may be written as

$$\boldsymbol{A} \cdot \boldsymbol{p} + \boldsymbol{p} \cdot \boldsymbol{A} = 2\boldsymbol{A} \cdot \boldsymbol{p} - i\hbar \boldsymbol{\nabla} \cdot \boldsymbol{A}, \tag{36}$$

using the commutation relation. The term $\boldsymbol{\nabla} \cdot \boldsymbol{A}$ is usually assumed to be zero, according to the Coulomb gauge. In particular, if the potential vector $\boldsymbol{A}$ is a plane wave, this assumption is automatically fulfilled since it is a transverse wave of type

$$A = A_0 \exp(-i\omega t + i\mathbf{k} \cdot \mathbf{r}). \tag{37}$$

Note that in bulk medium, the Coulomb gauge may be assumed, but that this assumption must be (at least) relaxed at the surface. It is clear that such a lack of allowance for the Coulomb gauge is a very serious theoretical problem which is commonly overcome by the formulation:

$$i\hbar(\mathbf{A} \cdot \nabla_{\mathrm{B}} + \mathbf{A} \cdot \nabla_{\mathrm{S}}) - i\hbar \frac{\partial}{\partial Z} \mathbf{A}, \tag{38}$$

where ∇_{B} and ∇_{S} denote the gradient for the bulk and the surface, respectively. At present, for low photon energies (below 200 eV), the dipole approximation ($\mathbf{k} \cdot \mathbf{r} \ll 1$) appears to be acceptable, and the matrix element becomes

$$\langle \varepsilon l'm'|\mathbf{A} \cdot \mathbf{p} + \mathbf{p} \cdot \mathbf{A}|nlm\rangle \propto 2A_0[\langle \varepsilon l'm'|\nabla_{\mathrm{B}}|nlm\rangle + \langle \varepsilon l'm'|\nabla_{\mathrm{S}}|nlm\rangle]$$

$$- i\hbar \left\langle \varepsilon l'm' \left| \frac{\partial \mathbf{A}}{\partial Z} \right| nlm \right\rangle, \tag{39}$$

where $\mathbf{A}$ is the polarization vector. It follows that assuming quasi-sphericity of the potential (a reasonable approximation for core level state), both initial and final states are separated into a radial part, R_{nl} and $R_{\varepsilon l'}$ respectively, and an angular part Y_{lm}, the spherical harmonic. If the gradient formula is used:

$$\nabla R_{nl} Y_{lm} = -\left(\frac{l+1}{2l+1}\right)^{1/2} \left(\frac{\mathrm{d}}{\mathrm{d}r} - \frac{l}{r}\right) R_{nl} Y_{l\,l+1\,m}$$

$$+ \left(\frac{l}{2l+1}\right)^{1/2} \left(\frac{\mathrm{d}}{\mathrm{d}r} + \frac{l+1}{r}\right) R_{nl} Y_{l\,l-1\,m}, \tag{40}$$

where $Y_{l\,l\pm1\,m}$ denotes the vector spherical harmonics, and is combined with the dipole approximation; the only pertinent quantities needed are:

(i) The radial parts R_{nl}, $R_{\varepsilon l+1}$ and $R_{\varepsilon l-1}$. The usual way to handle this first step consists of a more or less sophisticated self-consistent calculation (Hartree–Fock–Slater, Dirac–Slater, or Relativistic Random Phase Approximation).

(ii) An evaluation of the one-electron dipole matrix elements:

$$I_{l+1} = \int_0^\infty R_{\varepsilon l+1} \left(\frac{\mathrm{d}}{\mathrm{d}r} - \frac{l}{r}\right) R_{nl} r^2 \, \mathrm{d}r \tag{41}$$

and

$$I_{l-1} = \int_0^\infty R_{\varepsilon l-1} \left(\frac{\mathrm{d}}{\mathrm{d}r} + \frac{l+1}{r}\right) R_{nl} r^2 \, \mathrm{d}r. \tag{42}$$

(iii) The corresponding Coulomb plus non-Coulomb phase shifts, ξ_{l+1} and ξ_{l-1}.

However, actual calculations (Goldberg et al. 1981, Manson 1985) are not conducted as straightforwardly as they might appear at first glance. If the initial state radial part may be readily taken from tabulations or standard programs (Hermann and Skillman 1963), the final state vector partial wave decomposition must be obtained directly by solving the one-electron Schrödinger equation for the central field potential assumed. In a first approach, it may be a standard Hartree potential plus Slater free-electron exchange potential. The ξ_{l+1} and ξ_{l-1} phase shifts (Coulomb plus non-Coulomb) are consequences of the matching of the so obtained numerical solutions (inside the muffin-tin spheres) with the analytical asymptotic solutions (outside the muffin-tin spheres).

With a view to getting around the difficulty inherent in this type of calculation, one may disconnect steps 1 and 2 by neglecting possible interferences between them, and thus be able to use the numerous published calculations of the parameter $\beta(\varepsilon)$, known as the asymmetry parameter, which describes the emission anisotropic dependence on kinetic energy, ε. This parameter is defined as

$$\beta(\varepsilon) = \frac{l(l+1)I_{l-1}^2 + (l+1)(l+2)I_{l+1}^2 - 6l(l+1)I_{l-1}I_{l+1}\cos(\xi_{l+1} - \xi_{l-1})}{(2l+1)[lI_{l-1}^2 + (l+1)I_{l+1}^2]} \tag{43}$$

and thus takes into account interferences between the two outgoing $l+1$ and $l-1$ partial waves. The asymmetry parameter is directly connected to the differential photoionization cross section $d\sigma/d\Omega$,

$$\frac{d\sigma}{d\Omega} = \frac{\sigma}{4\pi}\left[1 - \frac{\beta(\varepsilon)}{4}(3\cos^2\omega - 1)\right], \tag{44a}$$

for unpolarized light, ω being the angle between the incident light and the photoejection direction, and

$$\frac{d\sigma}{d\Omega} = \frac{\sigma}{4\pi}\left[1 + \frac{\beta(\varepsilon)}{2}(3\cos^2\omega - 1)\right] \tag{44b}$$

for polarized light, ω being the angle between A and the photoejection direction.

The allowed range of the β parameter is $[-1, 2]$ to ensure cross-section positivity. Let us recall that the two limiting cases $\beta = -1$ and $\beta = 2$ (ns electron case) correspond to the photoionization cross section angularly peaked, respectively along and perpendicular to the potential vector of light. On the other hand, the case $\beta = 0$ implies no anisotropy at all (an alternative to the "magic" angle $\omega = 54.74°$). This parameter, because it reflects interferences between the two partial waves involved in the final state vector, shows rapid variations (related to Cooper minima) near the threshold, and then varies slowly to reach its asymptotic value. Such a situation explains part of the angular sensitivity with energy of the electron emission distribution near the threshold.

The second step, namely, the diffraction process, has been investigated through

LEED-type formalism using single or multiple scattering (Tong et al. 1978, 1985, Pendry 1974, 1975, 1978, 1980) and/or EXAFS-type formalism (Fadley 1984), depending on the electron mean free path and the kinetic energy. To be more precise, at high kinetic energy, the final state scattering is dominated by elastic single scattering in the forward direction, instead of multiple back scattering which prevails at low kinetic energy. In both cases, the problem turns out to be a determination of the phase shifts which result from the electron diffusion on a single atom of the crystal (using a formalism similar to those used in step 1). Then, assembling atoms in layers (see Pendry's discussion 1974) or in clusters (see Fadley's discussion 1984), the diffraction pattern is built. The photoionization final state determined in step 1 is expressed as a linear combination of (Hankel, Bessel or Coulomb) spherical waves, and then re-injected in the chosen diffraction formalism. The diffraction pattern depends on characteristics of phase shifts and diffusion amplitudes of each individual atomic diffusion, but also on crystal geometry (layer lattice and displacements between layers in LEED type formalism, or geometrical path length differences for cluster-type formalism). Very early in the development of angle resolved electron spectroscopies, it was recognized that the two-dimensional reciprocal lattice of layers for a given face governs the angular positions of the pattern extrema, especially in the XPS regime. Indeed, the periodicity supposed perfect of the lattice implies Bloch theorem. This situation suggests a possible use in surface analysis: surfaces with overlayers, relaxed or reconstructed surfaces, and step surfaces do not present exactly the same two-dimensional lattice as the bulk does. However, in the special case of low-energy angle resolved photoemission data, the observed structures may depend drastically on the emission distribution and phase shifts.

The last step, the crossing of the solid–vacuum interface, is commonly treated as simple refraction except for special surface effects (Pendry 1974, Fadley 1984). The external kinetic energy, E_{ext}, decreases from its internal value, E_{int}, by an amount equal to the inner potential, V_0 (typically 10–15 eV):

$$E_{ext} = E_{int} - V_0 . \tag{45}$$

The momentum conservation laws give rise to the well-known refraction relation,

$$\left(\frac{2mE_{ext}}{\hbar^2}\right)^{1/2} \sin\theta_{ext} = \left(\frac{2mE_{int}}{\hbar^2}\right)^{1/2} \sin\theta_{int} , \tag{46}$$

where θ_{ext} and θ_{int} are respectively the external and internal polar angles of the wave vector, $\mathbf{K}$. The internal angle, θ_{int}, is always smaller than the external angle, θ_{ext}. The transmission and reflection coefficients are thus given by

$$T = 4\left[\left(\frac{2mE_{int}}{\hbar^2}\right)^{1/2}\left(\frac{2mE_{ext}}{\hbar^2}\right)^{1/2} \cos\theta_{int} \cos\theta_{ext}\right]\Big/\Delta , \tag{47}$$

$$R = \left[\left(\frac{2mE_{int}}{\hbar^2}\right)^{1/2} \cos\theta_{int} - \left(\frac{2mE_{ext}}{\hbar^2}\right) \cos\theta_{ext}\right]^2\Big/\Delta , \tag{48}$$

with

$$\Delta = \left[\left(\frac{2mE_{\text{int}}}{\hbar^2}\right)^{1/2} \cos\theta_{\text{int}} + \left(\frac{2mE_{\text{ext}}}{\hbar^2}\right)^{1/2} \cos\theta_{\text{ext}}\right]^2. \qquad (49)$$

In the XPS regime, internal and external angles are very similar, and this refraction effect does not significantly affect the detected intensity. However, in the UPS regime, its dependence on the polar angle has to be taken into account, especially when grazing angles are approached.

4. Conclusion

Core level shift of surface atoms relative to the bulk has been measured by XPS for a large variety of solids including sp-metals, d-transition metals, rare earths, monoatomic and binary semiconductors, alloys and compounds. With very few exceptions, the quasi-majority of these experiments benefited from the development of synchrotron radiation as a powerful excitation source. Although in some cases the shift in BE can be as large as 1 eV, SCS commonly are in the range of a few tenths of an electronvolt (<0.3 eV). Therefore very high energy resolution experiments are required together with extreme surface sensitivity, obtained by tuning photon energy to lowest photoelectron mean free path. These capabilities are achieved in the soft X-ray range ($h\nu < 200$ eV), suitable for exciting the shallow and narrow 4f, 3d and 4d levels. Thus the subject is well documented for 5d transition and lanthanide series; however, further work should be continued for other solids such as semiconductors, free-electron metals and 3d-transition series, requiring an extension to higher photon energy, viz, 1 keV.

From the theoretical standpoint a good semi-quantitative understanding of SCS is now available from two models using respectively a tight binding approach and a thermodynamical approach. For example, the algebraic variation of SCS along the 5d series with the reversal of the sign is well reproduced by both models. However, the dependence of SCS on the surface crystallographic orientation and on the detailed surface electronic structure remains to be solved. The thermodynamical model is of more general application than the tight-binding model, confined to only d band metals, and allows to extract from SCS valuable information on thermochemical data such as heats of surface segregation in metals and alloys.

The capability of SCS to provide information on surface crystallography relaxation, chemisorption geometry, has to be confirmed. Further work should be continued particularly regarding the photoelectron diffraction process. Since low energy photoelectrons are to be used, the theoretical models necessary for using the diffraction patterns should be developed to include multiple scattering and refraction effects.

Acknowledgements

The authors would like to acknowledge J. Lecante, C. Guillot and Y. Lassailly for their participation to the experimental work, performed in the Laboratoire d'Utilisation du Rayonnement Electromagnétique, and M.C. Desjonquères, D. Spanjaard and G. Tréglia for their contribution to the interpretation of data.

We would like to thank members of the Laboratoire de l'Accélérateur Linéaire d'Orsay for their help in operating the storage ring.

References

Alvarado, S.F., M. Campagna and W.J. Gudat, 1980, J. Electron Spectrosc. & Relat. Phenom. **18**, 43.
Apai, G., R.C. Baetzold, P.J. Jupiter, A.J. Viescas and I. Lindau, 1983, Surf. Sci. **134**, 122.
Baetzold, R.C., G. Apai, E. Shustorovich and R. Jaeger, 1982, Phys. Rev. B **26**, 4022.
Baker, A.D., and C.R. Brundle, 1977, in: Electron Spectroscopy, Vol. 1, eds C.R. Brundle and A.D. Baker (Academic Press, London) p. 2.
Barker, J.A., and P.J. Estrup, 1978, Phys. Rev. Lett. **41**, 1307.
Barker, J.A., and P.J. Estrup, 1981, J. Chem. Phys. **74**, 1442.
Barton, J.J., W.A. Goddard III and T.C. Mc Gill, 1979, J. Vac. Sci. & Technol. **16**, 1178.
Bauer, E., and T. Engel, 1978, Surf. Sci. **71**, 695.
Bertolini, J.C., J. Massardier, Y. Jugnet, G. Grenet and J. Lecante, 1985, unpublished results.
Bonzel, H.P., and H.J. Krebs, 1981, Surf. Sci. **109**, L527.
Brennan, S., J. Stöhr, R. Jaeger and J.E. Rowe, 1980, Phys. Rev. Lett. **45**, 1414.
Chauveau, D., P. Roubin, C. Guillot, J. Lecante, G. Tréglia, M.C. Desjonquères and D. Spanjaard, 1984, Solid State Commun. **52**, 635.
Chiang, T.C., and D.E. Eastman, 1981, Phys. Rev. B **23**, 6836.
Citrin, P.H., and G.K. Wertheim, 1983, Phys. Rev. B **27**, 3176.
Citrin, P.H., G.K. Wertheim and Y. Baer, 1978, Phys. Rev. Lett. **41**, 1425.
Citrin, P.H., G.K. Wertheim and Y. Baer, 1983, Phys. Rev. B **27**, 3160.
Cox, P.A., 1975, Structure and Bonding **24**, 59.
Cyrot Lackmann, F., 1967, Adv. Phys. **16**, 393.
Cyrot Lackmann, F., 1968, J. Phys. Chem. Solids **29**, 1235.
Davenport, J.W., R.E. Watson, M.L. Perlman and T.K. Sham, 1981, Solid State Commun. **40**, 999.
Desjonquères, M.C., and F. Cyrot Lackmann, 1975a, J. Phys. (France) **36**, L45.
Desjonquères, M.C., and F. Cyrot Lackmann, 1975b, Surf. Sci. **50**, 257.
Desjonquères, M.C., D. Spanjaard, Y. Lassailly and C. Guillot, 1980, Solid State Commun. **34**, 907.
Ducros, R., and J. Fusy, 1986, unpublished results.
Duthie, J.C., and D.G. Pettifor, 1977, Phys. Rev. Lett. **38**, 564.
Eastman, D.E., T.C. Chiang, P. Heimann and F.J. Himpsel, 1980, Phys. Rev. Lett. **45**, 656.
Eastman, D.E., F.J. Himpsel and J.F. Van der Veen, 1982, J. Vac. Sci. & Technol. **20**, 609.
Eberhardt, W., G. Kalkoffen and C. Kunz, 1979, Solid State Commun. **32**, 901.
Egelhoff, W.F., 1983, Phys. Rev. Lett. **50**, 587.
Egelhoff, W.F., 1984, Phys. Rev. B **29**, 3681.
Engel, T., H. Niehus and E. Bauer, 1975, Surf. Sci. **52**, 237.
Erbudak, M., P. Kalt, L. Schalpbach and K. Bennemann, 1983, Surf. Sci. **126**, 101.
Estrup, P.J., and J. Anderson, 1968, J. Chem. Phys. **49**, 523.
Fadley, C.S., 1978, Electron Spectroscopy, Vol. 2, eds C.R. Brundle and A.D. Baker (Academic Press, London) p. 2.
Fadley, C.S., 1984, in: Progress in Surface Science, Vol. 16, ed. S. Davison (Pergamon, New York) p. 275.

Feibelman, P.J., 1975, Phys. Rev. Lett. **12**, 1092.
Feibelman, P.J., 1983, Phys. Rev. B **27**, 2531.
Feuerbacher, B., and R.F. Willis, 1976, J. Phys. C **9**, 169.
Feuerbacher, B., B. Fitton and R.F. Willis, 1978, in: Photoemission and the Electronic Properties of Surfaces, eds B. Feuerbacher, B. Fitton and R.F. Willis (Wiley, New York) p. 12.
Gerken, F., A.S. Flodstrom, J. Barth, L.I. Johansson and C. Kunz, 1985, Phys. Scr. **32**, 43.
Goldberg, S.M., C.S. Fadley and S. Kono, 1981, J. Electron Spectrosc. & Relat. Phenom. **21**, 285.
Grunze, M., C.R. Brundle and D. Tomanek, 1982, Surf. Sci. **119**, 133.
Gschneidner, K.R., 1964, Solid State Phys. **16**, 275.
Guillot, C., C. Thuault, Y. Jugnet, D. Chauveau, R. Hoogewijs, J. Lecante, Tran Minh Duc, G. Tréglia, M.C. Desjonquères and D. Spanjaard, 1982, J. Phys. C **15**, 4023.
Guillot, C., M.C. Desjonquères, D. Chauveau, G. Tréglia, J. Lecante, D. Spanjaard and Tran Minh Duc, 1984a, Solid State Commun. **50**, 393.
Guillot, C., P. Roubin, J. Lecante, M.C. Desjonquères, G. Tréglia, D. Spanjaard and Y. Jugnet, 1984b, Phys. Rev. B **30**, 5487.
Gupta, R.P., and S.K. Sen, 1974, Phys. Rev. B **10**, 71.
Heimann, P., J.F. Van der Veen and D.E. Eastman, 1981, Solid State Commun. **38**, 595.
Herman, F., and S. Skillman, 1963, Atomic Structure Calculations (Prentice-Hall, Englewood Cliffs, NJ).
Himpsel, F.J., P. Heimann, T.-C. Chiang and D.E. Eastman, 1980, Phys. Rev. Lett. **45**, 1112.
Hollinger, G., 1979, Ph.D. Thesis (Université Lyon, France).
Horn, K., 1985, personal communication.
Hörnström, S.E., L.I. Johansson, A. Flodström, R. Nyholm and J. Schmidt-May, 1985, Surf. Sci. **160**, 561.
Johansson, B., and N. Mårtensson, 1980, Phys. Rev. B **21**, 4427.
Johansson, B., and N. Mårtensson, 1983, Acta Phys. Helv. **56**, 405.
Johansson, L.I., A. Flodström, S.E. Hörnström, B. Johansson, J. Barth and F. Gerken, 1982, Solid State Commun. **41**, 427.
Jugnet, Y., 1981, Ph.D. Thesis (Université Lyon, France).
Jupille, J., K.G. Purcell and D.A. King, 1986, submitted to Solid State Commun.
Kahn, A., E. So, P. Mark and C.B. Duke, 1978, J. Vac. Sci. & Technol. **15**, 580.
Kaindl, G., C. Laubschat, B. Reihl, R.A. Pollak, N. Mårtensson, F. Holtzberg and D.E. Eastman, 1982a, Phys. Rev. B **26**, 1713.
Kaindl, G., B. Reihl, D.E. Eastman, R.A. Pollak, N. Mårtensson, B. Barbara, T. Penney and T.S. Plaskett, 1982b, Solid State Commun. **41**, 157.
Kaindl, G., W.D. Schneider, C. Laubschat, B. Reihl, and N. Mårtensson, 1983, Surf. Sci. **126**, 105.
Kammerer, R., J. Barth, F. Gerken, C. Kunz, S.A. Flodström and L.I. Johansson, 1982, Phys. Rev. B **26**, 3491.
Kliever, K.L., 1976, Phys. Rev. B **14**, 1412.
Kliever, K.L., 1977, Phys. Rev. B **15**, 3759.
Kumar, V., D. Tomanek and K.H. Bennemann, 1981, Solid State Commun. **39**, 987.
Landgren, G., R. Ludeke, Y. Jugnet, J.F. Morar and F.J. Himpsel, 1984, J. Vac. Sci. & Technol. B **2**, 351.
Ley, L., M. Cardona and R.A. Pollak, 1979, Photoemission in Solids II, Vol. 27, eds. L. Ley and M. Cardona (Springer, Berlin) p. 11.
Lindau, I., and W.E. Spicer, 1974, J. Electron Spectrosc. & Relat. Phenom. **3**, 409.
Ludeke, R., T.-C Chiang and D.E. Eastman, 1983, Physica B **117/118**, 819.
Mahan, G.D., 1978, Phys. Rev. B **2**, 4334.
Manson, S.T., 1985, J. Electron Spectrosc. & Relat. Phenom. **37**, 37.
Mårtensson, N., B. Reihl, W.D. Schneider, V. Murgai, L.C. Gupta and R.D. Parks, 1982, Phys. Rev. B **25**, 1446.
Mårtensson, N., P. Hedegard and B. Johansson, 1984, Phys. Scr. **29**, 154.

Martin, R.L., and D.A. Shirley, 1977, Electron Spectroscopy, Vol. 1, eds C.R. Brundle and A.D. Baker (Academic Press, London) p. 76.
Miedema, A.R., 1978, Z. Metallkd. **69**, 455 and references therein.
Miller, T., and T.C. Chiang, 1984, Phys. Rev. B **29**, 7034.
Miller, T., E. Rosenwinkel and T.C. Chiang, 1983, Solid State Commun. **47**, 935.
Morar, J.F., F.J. Himpsel, G. Hollinger, J.L. Jordan, G. Hughes and F.R. McFeely, 1986, Phys. Rev. B **33**, 1340.
Nilsson, A., N. Mårtensson, J. Hedman, B. Eriksson, R. Bergman and U. Gelius, 1985, Surf. Sci. **162**, 51.
Nyholm, R., and J. Schmidt-May, 1984, J. Phys. C **17**, L113.
Nyholm, R., A.S. Flodström, L.I. Johansson, S.E. Hörnström and J. Schmidt-May, 1985, Surf. Sci. **149**, 449.
Parks, R.D., N. Mårtensson and B. Reihl, 1982, in: Proc. Int. Conf. on Valence Instabilities, Zürich, April 13–16, eds P. Wachter and H. Boppart (North-Holland, Amsterdam) p. 239.
Pendry, J.B., 1974, Low Energy Electron Diffraction (Academic Press, London).
Pendry, J.B., 1975, J. Phys. C **8**, 2413.
Pendry, J.B., 1978, Photoemission and the Electronic Properties of Surfaces, eds B. Feuerbacher, B. Fitton and R.F. Willis (Wiley, New York) ch. 4.
Pendry, J.B., 1980, J. Phys. C **13**, 937.
Pijolat, M., 1979, 3rd cycle Thesis (Université Lyon, France) and references cited therein.
Powell, C.J., 1974, Surf. Sci. **44**, 29.
Rosengren, A., and B. Johansson, 1980, Phys. Rev. B **22**, 3706.
Rosengren, A., and B. Johansson, 1981, Phys. Rev. B **23**, 3852.
Schirmer, J., L.S. Cederbaum and J. Kiessling, 1980, Phys. Rev. B **22**, 2696.
Schneider, W.D., C. Laubschat and B. Reihl, 1983, Phys. Rev. B **27**, 10.
Schnell, R.D., F.J. Himpsel, A. Bogen, D. Rieger and W. Steinmann, 1985, Phys. Rev. B **32**, 8052.
Seah, M.P., and N.A. Dench, 1979, Surf. & Interface Anal. **1**, 2.
Siegbahn, K., C. Nordling, A. Fehlman, R. Nordberg, K. Hamrin, J. Hedman, G. Johansson, T. Bergmark, S.E. Karlsson and I. Lindgren, 1967, ESCA, Atomic, Molecular and Solid State Structure studied by means of Electron Spectroscopy (Almquist and Wiksell, Stockholm).
Smith, J.R., F.J. Arlinghaus and J.G. Gay, 1982, Phys. Rev. B **26**, 1071.
Smith, N.V., and F.J. Himpsel, 1983, Handbook on Synchrotron Radiation, ed E.E. Koch (North-Holland, Amsterdam) Vol. 1, p. 905.
Soukiassian, P., R. Riwan, J. Cousty, J. Lecante and C. Guillot, 1985, Surf. Sci. **152 & 153**, 290.
Spanjaard, D., C. Guillot, M.C. Desjonquères, G. Tréglia and J. Lecante, 1985, Surf. Sci. Reports **5**, 1.
Spicer, W., 1958, Phys. Rev. **112**, 114.
Taniguchi, M., S. Suga, M. Seki, S. Shin, K.L.I. Kobayashi and H. Kanzaki, 1983, J. Phys. C **16**, L45.
Tomanek, D., V. Kumar, S. Holloway and K.H. Bennemann, 1982, Solid State Commun. **41**, 273.
Tomanek, D., P.A. Dowben and M. Grunze, 1983, Surf. Sci. **126**, 112.
Tong, S.Y., A.R. Lubinsky, B.J. Mr Stik and M.A. Van Hove, 1978, Phys. Rev. B **17**, 3303.
Tong, S.Y., H.C. Poon and D.R. Snider, 1985, Phys. Rev. B **32**, 2096.
Tran Minh Duc, C. Guillot, Y. Lassailly, J. Lecante, Y. Jugnet and J.C. Vedrine, 1979, Phys. Rev. Lett. **43**, 789.
Tran Minh Duc, Y. Jugnet, J. Lecante and C. Guillot, 1980, unpublished results.
Tran Minh Duc, C. Guillot, Y. Lassailly, J. Lecante, Y. Jugnet, M.C. Desjonquères, D. Spanjaard and G. Tréglia, 1982, Nucl. Instrum. Methods **194**, 633.
Tréglia, G., M.C. Desjonquères, D. Spanjaard, Y. Lassailly, C. Guillot, Y. Jugnet, Tran Minh Duc and J. Lecante, 1981, J. Phys. C **14**, 3463.
Van der Veen, J.F., F.J. Himpsel and D.E. Eastman, 1980, Phys. Rev. Lett. **44**, 189.
Van der Veen, J.F., D.E. Eastman, A.M. Bradshaw and S. Holloway, 1981a, Solid State Commun. **39**, 1301.

Van der Veen, J.F., F.J. Himpsel and D.E. Eastman, 1981b, Solid State Commun. **40**, 57.
Van der Veen, J.F., P. Heimann, F.J. Himpsel and D.E. Eastman, 1981c, Solid State Commun. **37**, 555.
Van der Veen, J.F., F.J. Himpsel and D.E. Eastman, 1982, Phys. Rev. B **25**, 7388.
Webber, P.R., and M.A. Morris, 1984, Phys. Rev. B **29**, 5957.
Wertheim, G.K., and P.H. Citrin, 1984, Phys. Rev. B **30**, 57.
Wertheim, G.K., P.H. Citrin and J.F. Van der Veen, 1984, Phys. Rev. B **30**, 4343.
Williams, A.R., and N.D. Lang, 1978, Phys. Rev. Lett. B **40**, 954.
Wimmer, E., M. Weinert, A.J. Freeman and H. Krakauer, 1981, Phys. Rev. B **24**, 2292.

CHAPTER 11

OPTICAL CONSTANTS

DAVID W. LYNCH

Department of Physics and Ames Laboratory-USDOE*, Iowa State University, Ames, IA 50011, USA

Contents

1. Introduction 725
2. Definitions 726
3. Sample characteristics 738
4. Measurement methods 742
    4.1. Photometric methods 742
    4.2. Kramers–Kronig methods 751
    4.3. Photoyield methods 755
    4.4. Ellipsometry 758
    4.5. Electron energy loss measurements 759
    4.6. X-ray methods 760
5. Summary 762
6. Some examples 764
7. Comments on data collections and recent literature 770
Appendix. Recent literature reference to optical data for $E \geq 6$ eV . . 772
References 775

---

* Ames Laboratory is operated for the U.S. Department of Energy by Iowa State University under contract no. W-7405-Eng-82. This work was supported by the Director for Energy Research, Office of Basic Energy Sciences.

Handbook on Synchrotron Radiation, Vol. 2, edited by G.V. Marr
© *Elsevier Science Publishers B.V., 1987*

1. Introduction

This chapter discusses the optical properties of solids with a view toward their use in synchrotron radiation research, and their determination by the use of synchrotron radiation. The optical properties of materials are of interest for several reasons. In the first place, reliable data are needed for the design of optical instrumentation to be used at synchrotron radiation facilities. Such instrumentation is also used with conventional laboratory sources and in astrophysical instrumentation, as well as in instrumentation for plasma diagnostics. On a second level, optical data are sometimes needed to interpret an experiment on a material. A common example is photoemission from a solid, in which the optical properties are needed to evaluate the electric field in the sub-surface region as a function of photon energy, angle of incidence, and polarization. This field distribution is needed to calculate the photoelectron production rate. At a deeper level, the study of the optical properties leads to a partial understanding of the electronic structure of the material. As has atomic spectroscopy, solid-state spectroscopy has contributed greatly to our understanding of the electronic structure of matter. For both atoms and molecules, photoelectron spectroscopy now augments vacuum ultraviolet and soft X-ray optical spectroscopy, for it gives information on initial and final electronic states separately, rather than jointly, but a significant role remains for optical spectroscopy. It often has higher resolution, and in some cases, reveals electronic excitations difficult or impossible to study by photoelectron spectroscopy. The same is true of the study for condensed matter. Photoelectron spectroscopy now is more widely practiced than vacuum ultraviolet and soft X-ray optical spectroscopy. Although optical spectra, in a sense, form a background for photoelectron spectra, it now often happens that photoelectron spectra and their interpretation exist for materials whose optical spectra are unknown.

Volume 1a of this Handbook contains a chapter on the interaction of radiation with condensed matter (Bassani and Altarelli 1983). In that chapter, the authors define the optical response functions and their interrelations. We repeat some of those definitions below, but with less discussion. They then show how these response functions, "the optical properties", are related to microscopic properties of the solid, i.e., to the electronic structure. Finally, they illustrate several optical phenomena and their interpretation with examples, usually prototypical solids. There are many other reviews and books on the interpretation of optical properties of solids (Stern 1963, Greenaway and Harbeke 1968, Wooten 1972, Abelès 1972, Greenaway 1972, Nilsson 1974, Bassani and Pastori Parravicini 1975). We do not devote space to the interpretation of optical properties in this chapter. Instead, we review the more common methods of determining the optical response of a solid and discuss the errors that often arise with such measurements. We also discuss problems with the samples used for such measurements, and how they affect the results and their interpretation. We illustrate the measured optical properties of a few solids, with particular attention paid to the agreement, or lack thereof, of the data taken by different investigators. It is possible to learn a great deal about a material from optical spectra in arbitrary units, e.g., un-normalized reflectance or absorption spectra. The peaks in such spectra are significant, even though the magnitudes may not be. Still more information is available if the spectra are measured quantitatively. We emphasize measurement techniques which give quantitative data. There are now several collections of optical data, but only for certain classes of solids. There are also several reviews of optical properties which contain references to optical data on many more solids. We close with a description of these collections, and a list of recent references to bring them up to date.

Since synchrotron radiation covers an extremely wide spectral region, from the far infrared to hard X-rays, this entire region should be discussed. In this region the optical properties of solids are determined by lattice vibrations and impurities (in non-metals), the valence electrons, and the core electrons. One should discuss the interpretation of the optical properties of individual materials and classes of materials in terms of their electronic and vibrational properties. Because this has been done in numerous reviews of the optical properties of solids in the visible and near-ultraviolet regions (see previously-mentioned references), in several reviews of optical properties in the vacuum ultraviolet and soft X-ray regions (Harbeke 1972, Priol and Robin 1974, Damany 1974, Brown 1974, Kunz 1975, Koch et al. 1977, Lynch 1979, Brown 1980), and in several reviews of the optical properties of metals in the infrared (Stern 1963, Sokolov 1967, Abelès 1972, Nilsson 1974), we do not discuss these to any great extent in this chapter. We also neglect the infrared optical properties arising from lattice vibrations in non-metals.

2. Definitions

Cubic crystals, and polycrystalline samples of non-cubic materials are optically iostropic. Their optical properties can be described by two scalar functions, with the exceptions described below. As discussed by Bassani and Altarelli (1983), these functions can be the real and imaginary parts of a complex dielectric function $\tilde{\varepsilon}$ or of a complex conductivity $\tilde{\sigma}$,

$$\tilde{\varepsilon} = \varepsilon_1 + i\varepsilon_2, \tag{1}$$

$$\tilde{\sigma} = \sigma_1 + i\sigma_2, \tag{2}$$

with

$$\tilde{\varepsilon} = 1 + 4\pi i \tilde{\sigma}/\omega \quad \text{(esu)}. \tag{3}$$

These are functions of wavelength λ, angular frequency ω, or photon energy $h\nu$.

The real and imaginary parts are related by dispersion, or Kramers–Kronig, relations, integral relations of the form

$$\varepsilon_1(\omega) - 1 = \frac{2}{\pi} P \int_0^\infty \frac{\varepsilon_2(\omega') - 4\pi\sigma(0)/\omega'}{\omega'^2 - \omega^2} \omega' \, d\omega', \tag{4}$$

$$\varepsilon_2(\omega) - \frac{4\pi}{\omega}\sigma(0) = -\frac{2\omega}{\pi} P \int_0^\infty \frac{\varepsilon_1(\omega') - 1}{\omega'^2 - \omega^2} \, d\omega', \tag{5}$$

where the P denotes principal value. These are further discussed by Bassani and Altarelli (1983). It is also useful to introduce a complex refractive index

$$\tilde{N} = n + ik = \sqrt{\tilde{\varepsilon}}. \tag{6}$$

From this

$$\varepsilon_1 = n^2 - k^2, \tag{7}$$

$$\varepsilon_2 = 2nk, \tag{8}$$

$$2n^2 = \varepsilon_1 + \sqrt{\varepsilon_1^2 + \varepsilon_2^2}, \tag{9}$$

$$2k^2 = -\varepsilon_1 + \sqrt{\varepsilon_1^2 + \varepsilon_2^2}. \tag{10}$$

Relations between these and other functions used to describe optical properties are given by Bell et al. (1985). The goal of optical studies on solids is to obtain these functions for a given material over as wide a frequency range as is possible.

If the material is a crystal of symmetry lower than cubic, the optical properties are anisotropic and the dielectric function becomes a tensor, each component of which has a real and imaginary part (Nye 1957, Smith 1958). Using axes fixed in the crystallographic unit cell, the dielectric function tensor is diagonal for crystals of the hexagonal, tetragonal, trigonal, and orthorhombic classes, with two, two, two, and three independent components, respectively. These are uniaxial crystals. Monoclinic, and triclinic crystals have three unequal diagonal components and one and three independent off-diagonal components, respectively. (There is twice this number of off-diagonal components, but they are related as $\tilde{\varepsilon}_{ij} = \tilde{\varepsilon}_{ji}$.) These are biaxial crystals. As with anisotropic transparent materials, the dielectric function tensor can be diagonalized, but for biaxial crystals the directions of the principal axes with respect to the crystallographic axes will change with frequency. Moreover, the real and imaginary parts of the tensor need not have the same principal axes. Thus, it is common to use axes fixed in the unit cell. The complete specification of the optical properties of a triclinic crystal calls for knowledge of six complex spectra, twelve functions of frequency in all.

There are a number of cases in which the optical properties are inadequately described by a frequency-dependent dielectric function tensor. These are cases in which the optical properties are non-local. This means that the constitutive relation for a cubic crystal, $D = \tilde{\varepsilon}E$, which actually means $D(r, t) = \tilde{\varepsilon}E(r, t)$, is invalid. D is then not a local function of r, but the response of the material at a point r depends on the electric field in a region around that point, making the dielectric function depend on r and r', the latter in a region around r. Thus, $D(r) = \int \tilde{\varepsilon}(r, r') \, d^3r'$. For periodic systems, the spatial dependence simplifies to that of $r - r'$, and it can be Fourier transformed, leading to a dielectric function which depends on both frequency and wavevector. When a non-local description must be introduced, even cubic or isotropic materials require two complex dielectric functions, one for longitudinal excitations, e.g., electrons, and one for transverse excitations, e.g., photons (Stern 1963, Wooten 1972). Examples are the anomalous skin effect for metals in the infrared (Stern 1963, Sokolov 1967), and the effects of exciton dispersion in non-metals, the latter discussed by Bassani and Altarelli (1983). We do not pursue non-local optics further here.

The dielectric function is not directly measurable at optical frequencies. It must be obtained from measurements of other optical quantities, reflectances of surfaces, changes in the state of polarization of reflected radiation, or transmission measurements. This necessarily introduces the interface between the material under study and an ambient medium, nearly always air or vacuum. We assume for now that these interfaces are mathematical planes, perfectly flat and sharp, and that there are no surface layers such as oxides. The ratios of reflected to incident electric fields are given by the Fresnel relations. (In addition to previously-mentioned references, see also Wolter 1956 and Born and Wolf 1959.) These are

$$\tilde{r} = r\, e^{i\theta} = \frac{\tilde{E}_r}{\tilde{E}_i} = \frac{\tilde{g} - g_0}{\tilde{g} + g_0}, \tag{11}$$

with

$$\tilde{g}_s = \tilde{N} \cos \tilde{\phi}, \tag{12}$$

$$\tilde{g}_p = (\cos \tilde{\phi})/\tilde{N}, \tag{13}$$

$$g_{0s} = n_0 \cos \phi_0, \tag{14}$$

$$g_{0p} = (\cos \phi_0)/n_0, \tag{15}$$

$$n_0 \sin \phi_0 = \tilde{N} \sin \tilde{\phi}, \tag{16}$$

in which the radiation is incident on the interface at angle ϕ_0, measured from the interface normal, from a non-absorbing ambient medium of refractive index n_0 (often $n_0 = 1$, for air or vacuum), toward the material of refractive index $\tilde{N}$. The polarization subscripts denote s- and p-linearly polarized radiation with the electric vector perpendicular to, or in, the plane of incidence, respectively.

Equation (11) gives $\tilde{r}_p = -\tilde{r}_s$ for normal incidence. This is the result of the choice of directions for the electric fields for the two polarizations. A different choice leads to the more satisfying $\tilde{r}_p = \tilde{r}_s$. In either case $R_p = R_s$ at normal incidence. The above expressions assume cubic or isotropic media. Since the ratios are complex and not equal for the two polarizations, the reflected electric field is phase shifted differently with respect to the incident field. Incident plane-polarized radiation which is not purely s or p polarized is reflected at non-normal incidence as elliptically polarized radiation.

These complex reflectance ratios for the electric fields cannot be measured directly at optical frequencies. The ratio of reflected power or flux to that incident, an "intensity" ratio, is measurable. The measured quantity is then a reflectance R, which depends on the angle of incidence and on the polarization of the incident radiation:

$$\tilde{R} = |\tilde{r}|^2 = r^2 . \tag{17}$$

One can also measure transmission through a slab, if the slab is thin enough. Within the slab, the intensity varies as $\exp(-\mu z)$, where μ is the absorption coefficient. It is related to the complex refractive index as

$$\mu = 4\pi k/\lambda . \tag{18}$$

The absorption coefficient for a system of non-interacting atoms or molecules is related to the atomic or molecular photoabsorption cross section Σ by $\mu = N\Sigma$, where N is the number density of atoms or molecules. In condensed matter, the assignment of absorption to individual atoms or molecules is not justified, although at high enough energies, where essentially only core electrons are involved, it is better justified and widely used. (Even in the X-ray region, an atomic cross section loses its meaning at subshell thresholds and in the regions above, where the absorption coefficient depends on the surroundings of the atom whose core electron is excited.)

In a transmission measurement, there is partial reflection at the two slab interfaces with the ambient, and the multiply-reflected beams must be summed. If the multiply-reflected beams are coherent, as would be the case for thin slabs and well-collimated, narrow bandpass radiation, there will be interference effects in the transmission spectrum. The result for a film of thickness d with vacuum on each side is, for normal incidence,

$$T = |\tilde{t}|^2 , \tag{19}$$

$$= \frac{4\tilde{N}}{(\tilde{N}+1)^2 \exp(\tilde{\rho}d) - (\tilde{N}-1)^2 \exp(-\tilde{\rho}d)} , \tag{20}$$

$$\tilde{\rho} = 2\pi i \tilde{N}/\lambda . \tag{21}$$

(For non-normal incidence, see Heavens 1955, Wolter 1956, or Born and Wolf 1959.) When the multiply-reflected fields do not add coherently, an average over a cycle of $\mathrm{Im}(\tilde{\rho}d)$ leads to the same result that addition of multiply-reflected beam intensities does:

$$T = \frac{[(1-R)^2 + 4R\sin^2\Psi]\exp(-\mu d)}{1 - R^2\exp(-2\mu d)}, \tag{22}$$

$$R = \left|\frac{\tilde{N}-1}{\tilde{N}+1}\right|^2, \tag{23}$$

$$\tan\Psi = \frac{2k}{n^2 + k^2 - 1}, \tag{24}$$

where R is the reflectance at normal incidence for a semi-infinite slab (Kessler 1963). The last term in the numerator is often dropped, either because $n \ll k$ (at the fundamental absorption edge of a semiconductor or insulator), or because $R \ll 1$ (at high enough photon energies).

These expressions can be generalized in a number of ways. Consider first the semi-infinite sample. For an isotropic medium, an isotropic overlayer, or several, can be placed on the surface, and the attendant multiple reflections treated either coherently or incoherently (Wolter 1956, Born and Wolf 1959, McIntyre and Aspnes 1971, McIntyre 1973, 1976). An overlayer only a monolayer thick can produce a measurable change in the reflectance in some circumstances. This can be seen by using the approximate expression (McIntyre and Aspnes 1971) for the change in normal-incidence reflectance of a semi-infinite material of dielectric function $\tilde{\varepsilon}_\mathrm{b}$ with a layer of thickness d of a material of dielectric function $\tilde{\varepsilon}_0$ placed on top. The medium from which the radiation is incident has a (real) dielectric function ε_a. (The only approximation is $d \ll \lambda$.)

$$\frac{\Delta R}{R} = -\frac{8\pi d}{\lambda}\,\mathrm{Im}\!\left(\frac{\tilde{\varepsilon}_\mathrm{b} - \tilde{\varepsilon}_0}{\tilde{\varepsilon}_\mathrm{b} - \varepsilon_\mathrm{a}}\right). \tag{25}$$

Using values appropriate for an overlayer of aluminum oxide on bulk aluminum in vacuum with a photon energy of 12 eV, this leads to $\Delta R/R = -0.065\, d/\text{Å}$. Thus a 1 Å overlayer, to the extent that it can be viewed as a continuum, leads to a reflectance decrease of 6.5% of the original value of the reflectance. Such an overlayer may be placed on an anisotropic medium. An anisotropic overlayer may be placed on an isotropic medium. All these situations may be treated by a suitable application of the Fresnel relations, along with a description of the propagation of the electromagnetic wave inside the various media. The most complex case, an anisotropic layer over an anisotropic medium, has been worked out by Yeh (1979, 1980), with the optical properties of the media treated as known. If both dielectric functions are known, the optical properties of such a system can be calculated, but the possibility of learning something about one component of such a system from a series of measurements seems remote. The case of an overlayer on an optically isotropic medium deserves more comment. Even if the layer is of a material which is optically isotropic in bulk, it should be treated as anisotropic (uniaxial) when it is in the form of a thin layer (Dignam et al. 1971, Dignam and Moskovitz 1973, Dignam and Fedyk 1977). The reason is that the response of the thin layer to an electric field will be different for the components of the field normal to the plane of the layer and within the plane. As the layer grows thicker, it begins to resemble the bulk more, and a single dielectric function component usually is appropriate for thicknesses of the order of a wavelength. The expressions for the transmission of a slab may also be generalized to allow for non-normal incidence, for overlayers on the slab, for thick, transparent substrates, and for anisotropy.

There are many optical spectra for a material: the directly measured quantities, R and μ, the spectra of ε_1 and ε_2, which may be calculated from microscopic models and are thus the spectra usually interpreted, and the spectra of n and k, often useful for determining the properties of a material in an optical system. Over the wide region of the synchrotron radiation spectrum the magnitudes of these spectra change by orders of magnitude. This is illustrated by the optical properties of two common materials, aluminum and aluminum oxide, Al_2O_3. The optical data for aluminum from many sources have been thoroughly analyzed by Shiles et at. (1980), and by Smith (1985a). We use their data. Al_2O_3 exists in three phases. Optical data for any of the phases are far less extensive than those of aluminum, and, as will be discussed later, are far less accurate as well. Here we use the data collected and evaluated by Hagemann et al. (1974) for amorphous Al_2O_3. Figures 1–4 show the spectra of ε_1 and ε_2, of n and k, of R, and of μ. Figure 5 shows the reflectance as a function of angle of incidence for both these materials for photon energies in three characteristic regions.

In the infrared region aluminum is characterized by large values of n and k, both greater than unity, and both increasing with increasing wavelength. Such behavior is characteristic of all metals in the infrared, although for some other metals the rise might not be as structure-free as it appears to be in fig. 2. Such values of n and k give rise to reflectances very close to unity for all angles of incidence and both types of polarization, again a characteristic of all metals. The absorption coefficient is also very large. ε_1 is large and negative, while ε_2 is large and positive. This region can be qualitatively described by a free-electron model for the metal. Aluminum oxide is transparent in the infrared, visible, and ultraviolet. Thus at 1 eV, $k = \mu = \varepsilon_2 = 0$, $\varepsilon_1 = n^2$, and the reflectance is that of a dielectric. The reflectance for p polarization goes to zero at the Brewster angle, ϕ_B, given by $\tan \phi_B = n$. Note that aluminum shows a minimum in R. This occurs at the pseudo-Brewster angle, which is slightly greater than arctan n because of the non-zero value of k, which also causes the minimum in R_p to be greater than zero. The angle of this minimum is the pseudo-Brewster angle, and it is close to 90° in the infrared, moving to somewhat smaller angles at higher energies. Humphreys-Owen (1961) gives a cubic equation from which the pseudo-Brewster angle may be calculated for an absorbing medium. (Further in the infrared the spectra for Al_2O_3 show some structure due to the photoexcitation of lattice

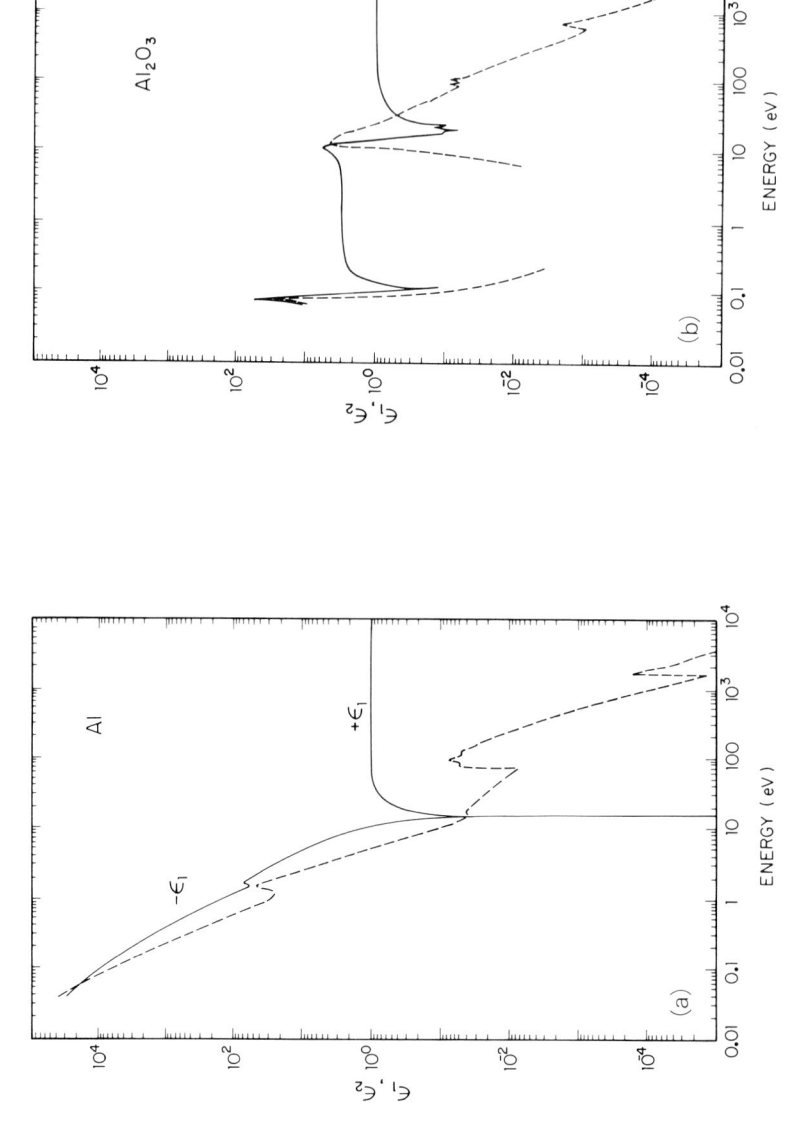

Fig. 1. Real (solid) and imaginary (dashed) parts of the dielectric function of (a) aluminum and (b) amorphous Al_2O_3. The data for aluminum are from Shiles et al. (1980), those for Al_2O_3 from Hagemann et al. (1974). For Al_2O_3 the infrared data are for crystalline α-Al_2O_3. In the region between 0.2 and 6 eV, ε falls to very low, sample-dependent values.

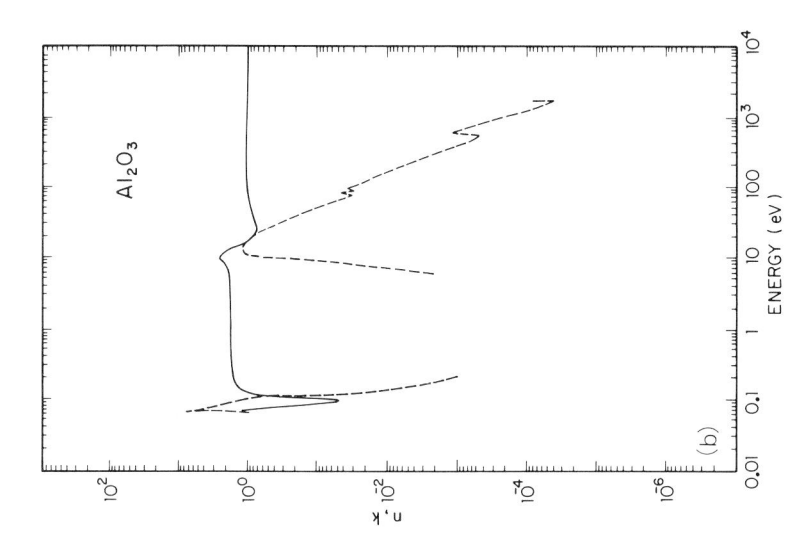

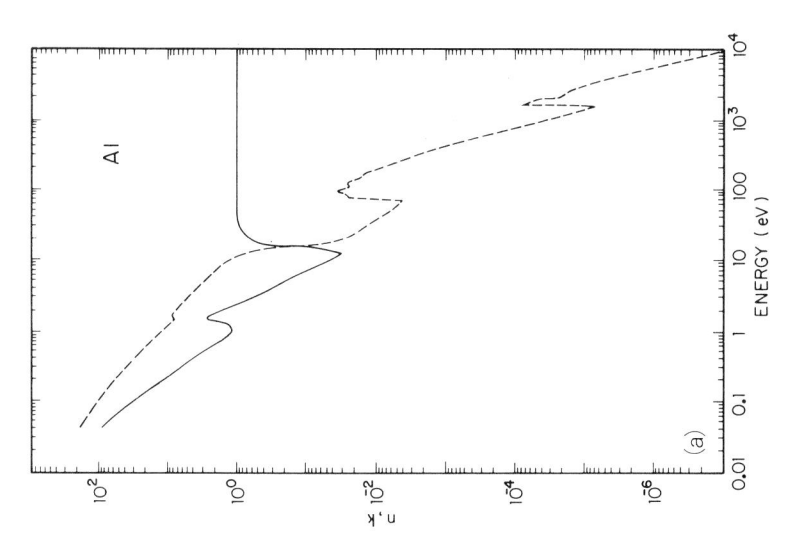

Fig. 2. Real (solid) and imaginary (dashed) parts of the refractive index of (a) aluminum and (b) amorphous Al_2O_3. The data for aluminum are from Shiles et al. (1980), those for Al_2O_3 from Hagemann et al. (1974). For Al_2O_3, the infrared data are for crystalline α-Al_2O_3. In the region between 0.2 and 6 eV, k falls to very low, sample-dependent values.

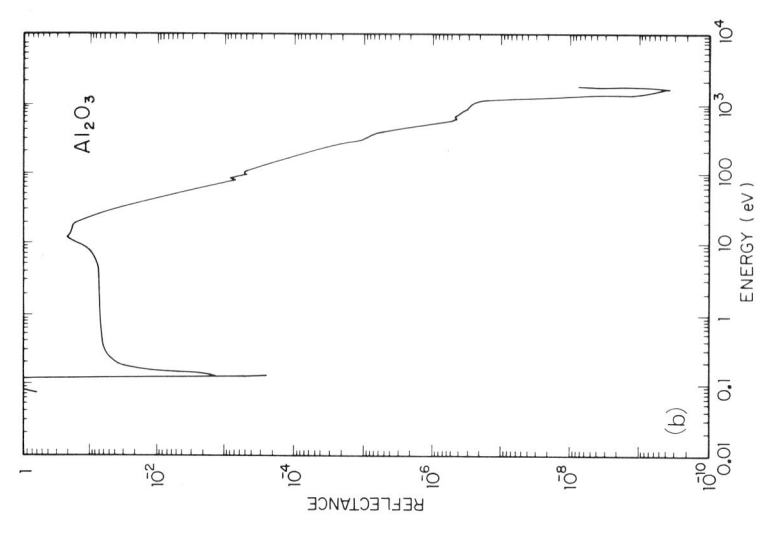

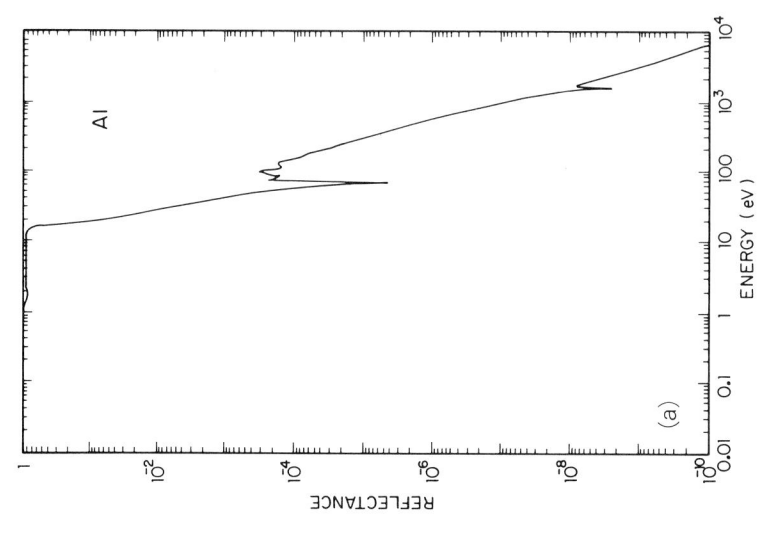

Fig. 3. Normal-incidence reflectance of (a) aluminum and (b) amorphous Al_2O_3. The data for aluminum are from Shiles et al. (1980), those for Al_2O_3 from Hagemann et al. (1974). For Al_2O_3 the infrared data are for crystalline α-Al_2O_3.

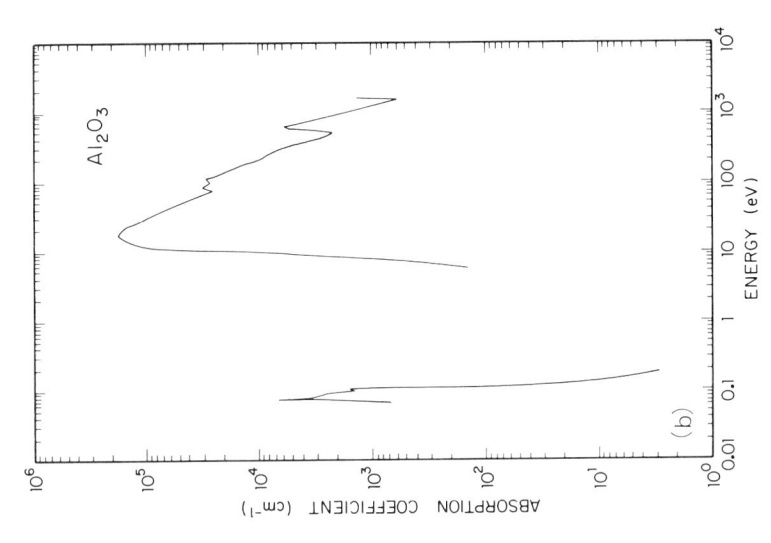

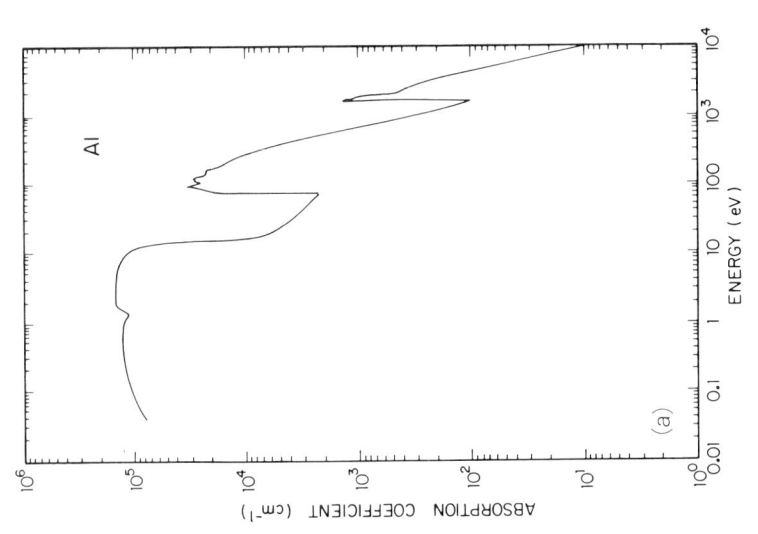

Fig. 4. Absorption coefficient of (a) aluminum and (b) amorphous Al_2O_3. The data for aluminum are from Shiles et al. (1980), those for Al_2O_3 from Hagemann et al. (1974). For Al_2O_3 the infrared data are for crystalline α-Al_2O_3. In the region between 0.2 and 6 eV, μ falls to very low, sample-dependent values.

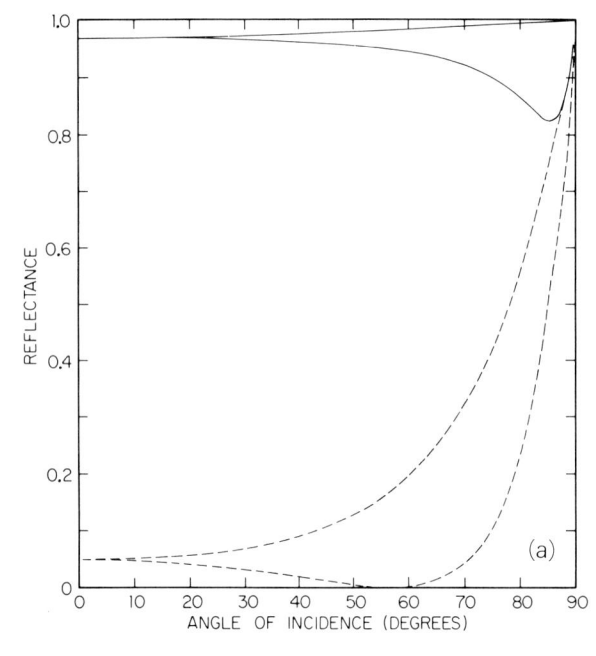

Fig. 5(a)

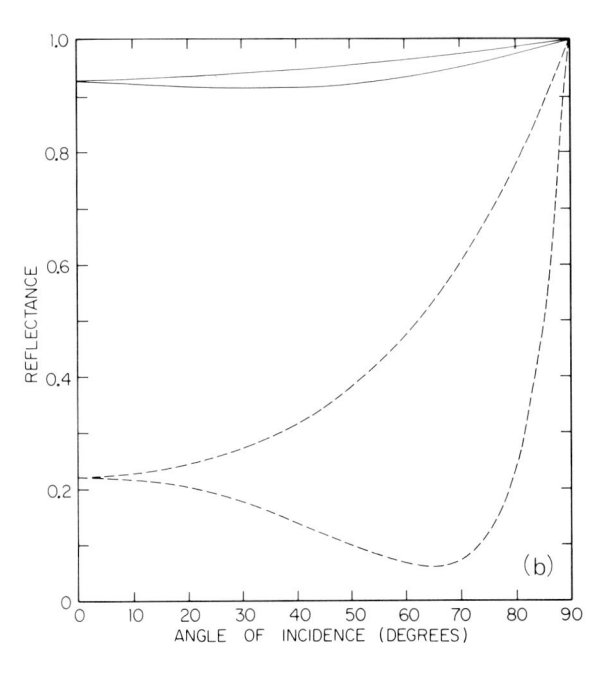

Fig. 5(b)

Optical constants 737

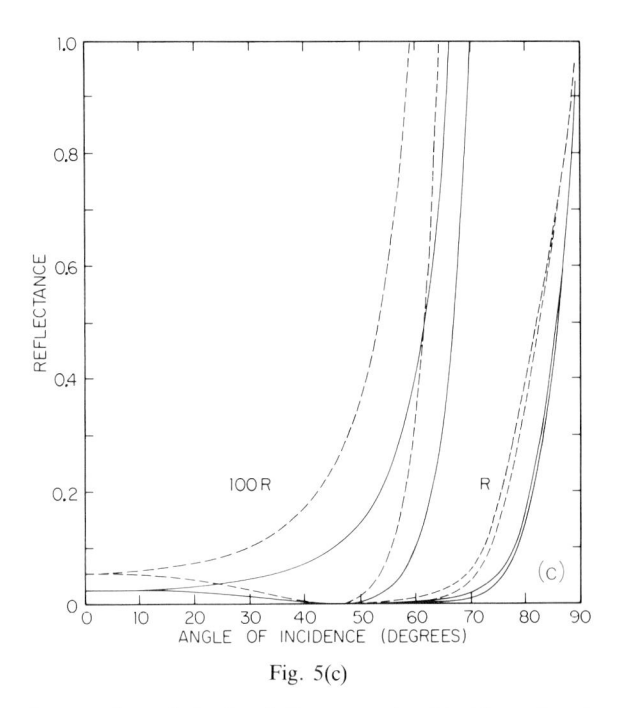

Fig. 5(c)

Fig. 5. Reflectance for s- and p-polarized radiation vs angle of incidence for aluminum (solid) and amorphous Al_2O_3 (dashed) at several energies characteristic of different spectral regions: (a) 1 eV, (b) 12 eV, (c) 100 eV. The complex refractive indices used for aluminum are $1.2188 + 12.464i$, $0.0328 + 0.7907i$, and $0.9906 + 0.2992i$ for 1, 12, and 100 eV, respectively, while those for Al_2O_3 are 1.75, $1.87 + 1.17i$, and $0.966 + 0.0332i$.

vibrations; however, at still longer wavelengths the material behaves as a dielectric again, but with a larger value of ε_1 than in the visible.)

The visible and near-ultraviolet regions usually are characterized by structured n and k spectra, often more structured than the spectra for aluminum, with magnitudes roughly in the range 0.1 to 10. All materials, not just metals, have n and k values in this range of values in this spectral region, provided that the energy is above the band gap energy in the case of insulators. The corresponding ε_1 and ε_2 spectra exhibit structures on a monotonic background, with ε_1 tending toward 1 from below, and ε_2 tending toward zero from above (both with increasing photon energy). The reflectance at normal incidence usually is significantly smaller than unity for metals: aluminum being unusual in that the reflectance is still high. The absorption coefficient is smaller than in the infrared, but still large. This is the region of interband absorption, superimposed on a free-electron background for metals, and a detailed description of the electronic structure is needed to interpret the spectra. Note that at 12 eV, the pseudo-Brewster angle for Al is no longer near 90°, a consequence of the values of n and k in this spectral region. At higher photon energies, k falls rapidly, with structures occurring at core electron absorption edges. n approaches unity from below, with small structures near the core thresholds. Far above any edge, the spectra for all materials resemble those of a free-electron gas with a density corresponding to the number of electrons in the atoms with binding energies less than that of the nearest edge at lower energy. ε_1 and n approach unity from below, and ε_2 and k approach zero. The reflectances become very small, except for angles very near grazing. The absorption coefficient rises at each core excitation threshold, then falls as the photon energy increases, first slowly, then more rapidly above the last absorption edge. This behavior is characteristic of all materials as the photon energy increases. Note the qualitative similarities in all of these spectra for metallic aluminum and insulating Al_2O_3 above the band gap energy of about 8.5 eV.

Anisotropic materials have similar spectra. (The data for Al_2O_3 neglect anisotropy, which is, in fact, present for crystalline Al_2O_3, but not for the amorphous form.) The anisotropy in k or μ is usually largest at absorption thresholds for valence electrons, and even core electrons. The anisotropy in n and ε_1 is also largest in these regions. (The anisotropy of the refractive index when $k \approx 0$ is actually very small, although it is adequate to make polarizing optical elements.)

3. Sample characteristics

Real surfaces are not mathematical planes. The simple termination of a three-dimensional crystal leads to surface-specific phenomena, even for a defect-free, unreconstructed surface. These effects usually fall in the category of non-local optics, although they may include surface electronic states in any material, in addition to the effects on the conduction electrons in metals. We do not consider effects of this type here. Actual surfaces may depart from ideality in a number of ways. They may have small-scale (smaller than a wavelength) geometrical irregularities, generally called roughness. They may have larger-scale geometrical non-planarities, sometimes called waviness. Finally, the surfaces may have an overlying layer, usually an oxide, whose effect on the reflectance was described in section 2. Samples in vacuum, unless outgassed, also will have at least a monolayer of adsorbed molecules which can effect the optical properties measurably in regions where the adsorbate is absorbing. There is an additional problem with thin-film samples. They may consist of many small grains which do not pack well. The intergrain material may have a significantly lower density, which, although not a surface effect, will have an effect on the measured spectra and their interpretation (Hunderi 1973, Nagel and Schnatterly 1974, 1975, Aspnes et al. 1979, 1980, Nestell et al. 1980). Finally, evaporated films often have densities lower than bulk densities, which will make measured absorption coefficients too low. (See, e.g., Chopra 1969.) In fact, measurements of the optical constants have been used to study the inhomogeneity of films (Borgogno et al. 1984).

These departures from ideality will have an effect on the optical properties that depends on the type of measurement used to obtain them. In principle, one might eliminate several, or all, of these by proper sample preparation, but that has rarely been done. For example, one can remove an oxide overlayer in an ultrahigh vacuum system by bombarding the surface with energetic Ar^+ ions. This induces surface damage, usually extensive enough that the surface may be characterized as amorphous. The damage can be reduced or removed by annealing. The annealing usually will roughen any polycrystalline sample. It is necessary to bear in mind the effects of non-ideal samples whenever one uses literature values for optical constants, for there are many inconsistencies in the published spectra. Some of these can be understood on the basis of sample preparation and measurement technique, sometimes enabling one to select the data set most appropriate to use. This set may not always be the one most representative of an ideal sample, for one may be interested in, for example, film samples with grain boundaries constituting a significant fraction of the volume, or with an equilibrium oxide layer on the surface. In such cases, the use of the optical properties which were directly measured in the literature may be appropriate, but the use of other optical spectra derived from them may lead to errors.

Small-scale roughness on the surface scatters radiation out of any specularly reflected beam. In a reflectance measurement this leads to a measured value that is too low, even if all the radiation scattered above the surface is collected. (Some is scattered into the subsurface region and some may be lost in the creation of surface plasmons.) In the simplest model, a scalar model, the surface roughness is characterized by a Gaussian distribution of surface heights above and below the mean surface plane, with an rms height h, assumed to be much smaller than a wavelength. This is the only parameter characterizing the surface in this scalar model for loss of normally-incident, specularly-reflected radiation due to scattering by a rough metallic surface. The model also gives the angular distribution of the diffusely-scattered radiation, but it requires another parameter describing the surface, the autocorrelation length of the roughness, a, measured in the plane of the surface. It predicts that the reflectance of a rough metallic surface at normal incidence will fall below that of the flat surface by a factor of (Bennett and Porteus 1961, Porteus 1963)

$$\frac{R}{R_0} = \exp\left[-\left(\frac{4\pi h}{\lambda}\right)^2\right] + \left\{1 - \exp\left[-\left(\frac{4\pi h}{\lambda}\right)^2\right]\right\}$$
$$\times \left\{1 - \exp\left[-\left(\frac{\sqrt{2}\,\pi h \Delta\theta}{m\lambda}\right)^2\right]\right\}. \qquad (26)$$

The first term represents radiation lost from the beam and the second represents the diffusely-scattered radiation which is collected by the detector subtending solid angle $\Delta\theta$. Here m is the rms value of the slope of the rough surface. It is related to h and a by $a = \sqrt{2}\,h/m$. A very flat superpolished quartz surface might have a value for h of about 10 Å. The first term in eq. (26) gives a reduction of 0.1% in the normal incidence reflectance at a wavelength of 1000 Å. At 330 Å, the same term causes a 1.4% reduction of the measured value for the same surface. Most measurements in this spectral region have not used surfaces this smooth.

At normal incidence the s and p reflectances are identical, but polarization effects may occur in the angular distribution of the diffusely-scattered radiation. For non-normal incidence the situation is more complicated, for polarization effects are important, both in the reflectance and in the nature of the scattered radiation. The more complicated vector theory of scattering by rough surfaces must be used (Elson and Ritchie 1974, Maradudin and Mills 1975, Mills and Maradudin 1975, Marvin et al. 1975, Celli et al. 1975, Elson 1975, 1977, Toigo et al. 1977, Elson and Bennett 1979, Raether 1982, Rehn et al. 1980). It requires a description of the surface roughness, as does the scalar theory, and it also assumes that $h \ll \lambda$. We do not discuss the theory here. A first comparison of the theory with experiment in the soft X-ray region can be found in the report by Rehn et al. (1980).

Needless to say, transmission measurements on thin films are similarly affected by roughness of both the film–air interface and the film–substrate interface. Since these are usually carried out at near-normal incidence, the scalar theory may be used, unless the wavelength of the measuring radiation becomes too short. It often is stated that ellipsometric measurements are less sensitive to scattering by surface roughness than are photometric measurements, because the intensity of the reflected beam is not measured. This is so for metals only insofar as the s- and p-electric field reflection coefficients are reduced by the same factor by scattering. This is seen not to be the case in the work by Maradudin and Mills (1975). For aluminum at 7 eV (where ellipsometry is not normally carried out) a larger fraction of p-polarized radiation than of s-polarized radiation is lost by roughness-induced scattering and absorption at large angles of incidence. This should lead to errors in ellipsometric measurements, but at lower energies, where such measurements are normally carried out, the effects of rough surfaces are much smaller (Burge and Bennet 1964, Berremann 1970b).

At higher photon energies, the condition that the wavelength of the radiation is much larger than the rms height of the surface roughness is not met. Moreover, at such energies, reflection is usually carried out at grazing incidence, so the vector scattering theory must be used. In the X-ray region, nearly all materials studied by reflection or used for their reflective properties are used at angles greater than the critical angle. We return to this, and the role of scattering, in section 4.5.

The majority of optical data have been taken at room temperature. Some spectra have been obtained at cryogenic temperatures, but only a little work has been done at temperatures above 300 K. The temperature dependence of the optical properties are of fundamental and practical interest.

An increase in temperature causes an increase in the phonon population and a change, generally, but not always, an increase, in the interatomic spacing. The phonon increase shortens the electron–phonon scattering time and increases the free-electron absorption. Thus for metals in the infrared, ε_2 will increase with temperature, and there will be a concomitant change in ε_1. Higher phonon populations also broaden interband transitions, making all structures in the optical spectra broader, and, from sum rules, less prominent. Interband structures have contributions to their widths other than electron–phonon scattering, however, so the broadening may not increase as rapidly as expected.

Although the increased phonon populations lead to thermal expansion, it is convenient to separate the effects of temperature on electron energy eigenvalues into two terms, one depending on volume alone, and one depending on temperature at fixed volume. (This separation can be carried out experimentally by studying optical properties as a function of pressure to obtain the first term, and as a function of temperature to obtain the sum.) A change in interatomic spacing leads to a change in crystal potential, hence, to a change in the eigenvalue spectrum for the electrons. Thus the electronic structure will change. Since eigenvalues may shift in either direction upon a volume change, band gaps important for optical properties may increase or decrease with decreasing volume. The Fermi level may shift from the motion of the bands below it. There are two effects of temperature at constant volume. The increased lattice vibration amplitudes weaken the Fourier coefficients of the crystal potential that are most important in determining the electronic structure. This may be treated by a factor like the Debye–Waller factor, which weakens the potential as the temperature increases (Antoncik 1955). This makes gaps decrease as the temperature increases. The electron–phonon interaction produces the other effect, a broadening of the electron energies (which may be viewed as a self-energy), which also lowers gaps (Fan 1950, 1951). When all these terms are considered together, the result is that band gaps may shift with temperature by an amount of the order of $n \times 10^{-4}$ eV/K, where n is of the order of 1–10, and may be of either sign. (Negative is more common.) This estimate is valid at temperatures above about liquid nitrogen temperature, except for materials with a very large Debye temperature.

Measurements at high temperature are complicated by two factors. If the vacuum environment is not good, the surface may oxidize during measurements. The sample may also recrystallize, thereby becoming rougher. The other problem is that the samples emit radiation at high temperatures which must be subtracted, but this is not a problem in the ultraviolet and beyond. Nonetheless, few measurements have been made at high temperatures outside the visible and near infrared. Many of the high-temperature measurements made to date have been by ellipsometry. Laser calorimetry has also been used at several discrete energies in the visible and near-infrared. Modulation techniques have been used to obtain qualitative temperature dependences of optical spectra for many materials, including some work in the vacuum ultraviolet and soft X-ray regions (Aspnes et al. 1976, Olson and Lynch 1979, 1980, Olson et al. 1980). The amplitude of modulation is not known accurately, so the spectra of the temperature derivatives of the optical spectra remain in arbitrary units unless they can be calibrated at one energy with data from another measurement. These spectra do show, however, that the temperature derivatives of the optical constants are not simply constants; sign reversals are common, even at photon energies above 50 eV.

4. Measurement methods

An extremely large number of techniques exists for determining the optical constants of solids. All have limitations, and all become relatively poor for certain spectral regions, either because of the values of the optical constants in those regions, or because of limitations of optical components, including sources and detectors. Some of these technqiues are applicable only to special cases, e.g., transparent materials, while others are more general. We consider only the latter here. We divide them into four general categories: (1) photometric methods, (2) Kramers–Kronig methods (actually a special type of photometric method), (3) ellipsometry, and (4) electron energy loss measurements. This classification scheme does not include all known methods, e.g., the determination of infrared optical constants from the emissivity spectrum, but it does cover nearly all the methods widely used.

4.1. Photometric methods

Photometric measurements involve measuring photon fluxes, e.g., those incident on a sample and those reflected from it or transmitted through it. The accuracy with which one can carry out such a measurement depends on many things, and the way measurement errors propagate to the resultant spectrum of the complex refractive index or dielectric function depends on the technique adopted and on the values of the optical constants themselves. All methods of flux measurement have several features in common. The flux is assumed to be sufficiently monochromatic that spectral features will not be broadened. Monochromator resolution requirements therefore depend on the spectrum to be measured. Resolution is not the only monochromator characteristic to be controlled, however. With a diffraction grating dispersing a continuous spectrum there is always second- and higher-order radiation present to some degree. Some gratings produce more than others. The amount of higher-order radiation must be known. If possible, it should be eliminated by suitable filters. Diffraction gratings and other optical elements also produce scattered radiation, essentially white radiation, usually having components with wavelengths longer than that being diffracted in first order. This is much more difficult to eliminate, and corrections to the data are usually uncertain. Scattered radiation is especially severe when continuum sources, e.g., synchrotron radiation, are being used. At the high energy region of a scan, the grating efficiency falls rapidly, and the ratio of scattered radiation to first-order diffracted radiation can exceed unity. Higher-order radiation can introduce false structures in a spectrum, and both higher-order and scattered radiation can cause the magnitudes of measured spectra to be incorrect. Finally, gratings and other optical elements can alter the state of polarization of any beam. It must be assumed that the radiation incident on a sample is partly polarized. In some cases the state of polarization must be known in order to avoid additional errors.

The detector also provides sources of error common to all measuring strategies.

Wavelength-dependent sensitivity of the detector is usually not a problem, for one often deals with the ratio of two measurements made at any one wavelength. Much more troublesome is the fact that the sensitivity of a photocathode depends on the part of the cathode struck by the photon, the result of inhomogeneities in the cathode material itself, and in the collection efficiency for the photoelectrons from the cathode. More significant is the fact that these inhomogeneities may depend on wavelength and polarization. If the incident beam is not homogeneous, significant errors can arise from such sources, especially if the incident beam becomes left–right inverted upon reflection. Such effects can be reduced by working only with reflected beams and by using photocathodes or phosphor converters resembling integrating spheres.

The sample itself may introduce measurement problems in the vacuum ultraviolet if it fluoresces, as most insulators do, especially if they contain impurities at even rather low concentrations. If the luminescence cannot be removed from the detector by suitable filters, it may at least be identified by a signal detected in an off-specular direction. Its role may be reduced by separating the sample further from the detector. It may also be possible to reduce its effect by gating the detector on only while the measuring radiation pulse is present, if synchrotron radiation is used, for often the luminescence lifetime is longer than the duration of synchrotron radiation pulses.

The simplest reflectance measurement is at near-normal incidence, for which knowledge of the degree of polarization is not necessary. However, only one measurement is insufficient to obtain the real and imaginary parts of the dielectric function. A common method for obtaining optical constants is to measure the reflectance at two angles of incidence, then use the Fresnel relations to obtain the dielectric function, using graphical or numerical methods. For application of the Fresnel equations, the state of polarization of the incident beam must be known. The simplest cases are complete linear polarization, or completely unpolarized radiation, the latter often being harder to achieve than the former. For incomplete polarization, the state of polarization must either be measured, or treated as an unknown, to be found in the ensuing analysis, which then requires three measurements. We first assume that nearly-linearly polarized radiation is available and that a correction can be applied for imperfect polarization. Then one must choose which polarization, s or p, and which angles of incidence to use. The choice is determined primarily by the sensitivity of the resultant dielectric function to errors in the measured reflectances. An early discussion of such errors was given by Humphreys-Owen (1961). Reflectances are calculated from the Fresnel relations for a series of n and k values, and curves of constant n and k are constructed in the plane of the two measured variables. Figure 6 shows such a plot for measurements of R_p/R_s at 60° and 80° angles of incidence. Assuming n and k have the values 1.8 and 1.4, and that the measured quantities have errors of $\pm 3\%$ each, the experimental errors in the measured quantities lead to errors in n and k of ± 0.04 and ± 0.03, respectively. From fig. 6 it is clear that the errors will depend on the optical constants themselves, and the choice of the best method, or even of a good one, will depend on the values of the optical constants to be measured.

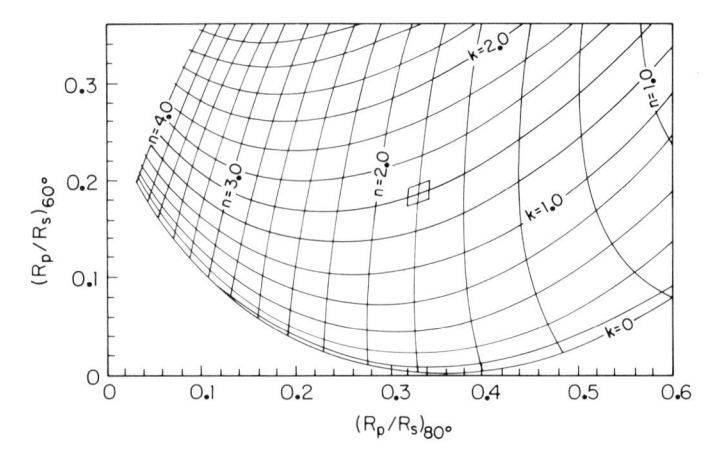

Fig. 6. Curves of constant n and k plotted in the space of two measured parameters, the ratio of p- to s-polarized reflectances at angles of incidence of 20° and 80°. (From Humphreys-Owen 1961.)

For example, in fig. 6 the curves of constant n and k become more closely spaced for larger values of n, making errors in n and k larger for equal errors in the measurement of the reflectance ratios.

Hunter (1965) introduced a different plot to analyze sensitivity to errors. For measurements of R_p at two angles of incidence, he first calculated R_p versus the angle of incidence for many pairs of n and k. Two such plots are shown in fig. 7a. From many such plots, curves of constant R_p for each of several angles can be made. These are plotted in the $n-k$ plane, as shown in fig. 7b. Any one curve in this plot gives all pairs, n, k, which will yield the same value of R_p at a constant angle. One of the pairs is the anticipated true value; in the figure the two n, k pairs, 0.8, 1.3, and 0.8, 3.3 are used as examples. A second curve is a plot of constant R_p at another angle of incidence. The two curves should intersect at a point, giving the correct n and k values, as can be seen in fig. 7b. If the values of R are not known precisely, the intersection is somewhere within a region in this plot, shown expanded in fig. 7c. The extent of this region indicates uncertainty in n and k resulting from uncertainties in the two measured quantities. For a reflectance uncertainty of $\pm 1\%$ in measuring R_p at 20° and 70° angles of incidence, the uncertainty in n and k is very small when n and k have the values 0.8 and 1.3, but quite large when they have the values 0.8 and 3.3.

One can also measure at several angles of incidence and use the overdetermined data set to get a best value for n and k. Klucker et al. (1974) used this method on the basal plane of graphite. They measured the (un-normalized) reflectance at nine angles of incidence. Using one measurement for normalization gave them eight measured reflectance ratios from which to extract five parameters: the degree of polarization and the real and imaginary parts of each of two dielectric function tensor components. In making a least-squares fit to the data, they gave each reflectance a weight according to the sum of the expected sensitivities of each reflectance R to the four dielectric function components:

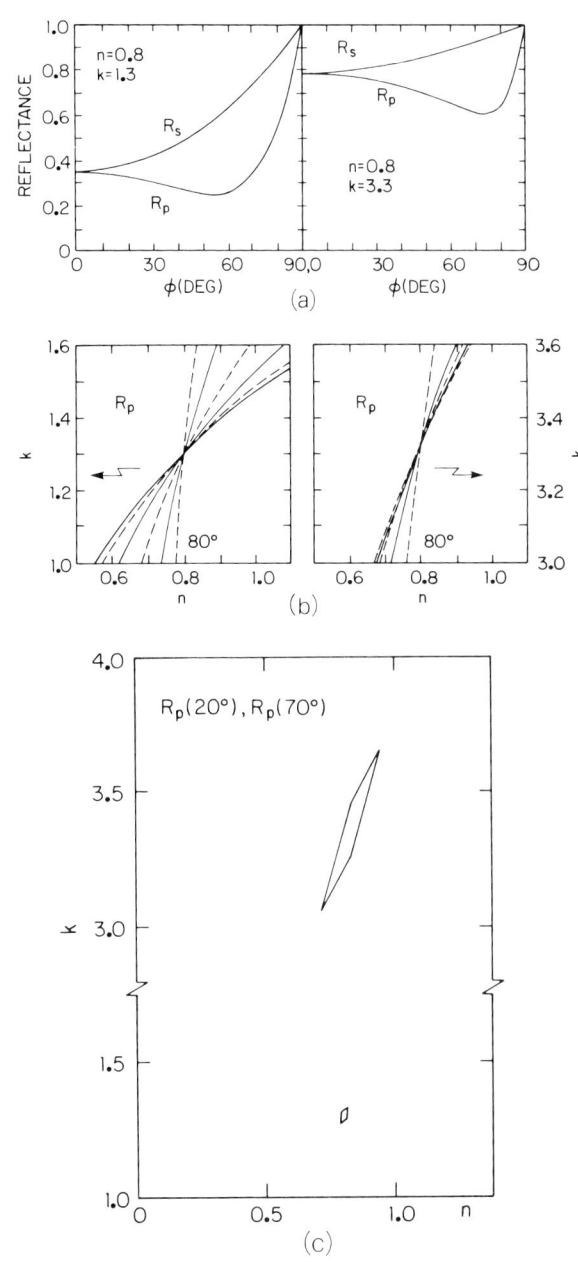

Fig. 7. (a) Reflectances for polarized radiation vs angle of incidence for materials with refractive indices of $0.8 + 1.3i$ and $0.8 + 3.3i$. (b) Isoreflectance curves (p polarization) in $n-k$ space. One curve is labelled 80° (angle of incidence). Adjacent curves are for angles smaller in 10° steps, until at small angles, the curves overlap. The values of the reflectances are those corresponding to the complex refractive indices used in (a), hence, all curves intersect at those values. (c) Expected errors in n and k from errors in measured values of R_p at 20° and 70°. The center of each "parallelogram" is the point of intersection of the curves in (b). The boundaries are sections of the curves for R_p values 1% larger and 1% smaller than the true values. (From Hunter 1965.)

$$\delta_{ij} = \left| \frac{\delta \ln R_i}{\delta \ln X_j} \right|, \quad X_j = \varepsilon_{1\parallel}, \varepsilon_{2\parallel}, \varepsilon_{1\perp}, \varepsilon_{2\perp}. \tag{27}$$

This weighting is not often used, but it appears to be an improvement in the method.

Miller et al. (1970) analyzed measurements of R_s and R_p at a fixed angle. First they determined the regions in the R_s–R_p plane in which solutions for n and k could be found for each of many angles of incidence. They assumed that, in general, the greatest sensitivity in n and k to the measured quantities occurs when the area of this region is maximized. This occurs at angles of incidence of 16° and 74°. (There is symmetry in these areas about 45°.) Plots like those of fig. 6 showed that for 16° the greatest sensitivity was for values of n and k both less than unity, while for 74°, the greatest sensitivity was for n in the range 1–2 and k less than about 1.5, a much more useful range. A similar analysis was made by Armaly et al. (1972).

These plots are useful in selecting a method, and in evaluating the propagation of errors, but they require an estimate in the errors in the measured quantities. Such an estimate is difficult to make. Statistical errors due to low count rates can usually be made negligible. Hunter (1967) has discussed errors in the angles of incidence, and how to reduce them in the design of a reflectometer. Remaining errors result from the degree of collimation of the beam of radiation, imperfect polarization, the inhomogeneous response of the cathode, and the sample inhomogeneity itself, for by increasing the angle of incidence on the sample, one is increasing the area sampled.

In addition to reflectances, one can also measure such quantities as the pseudo-Brewster angle, at which R_p is a minimum, the derivative of the reflectance with respect to angle, and, for some materials, the critical angle for total external reflectance. The latter measurement, actually of the reflectance as a function of angle for angles near the critical angle, is very useful when n is less than unity, or less than n of a transparent substrate (limiting measurements to energies below about 11 eV) and k rather small (Hunter 1964). As shown in fig. 8, if k is zero, the reflectance (s and p) is total for angles of incidence larger than the critical angle $\phi_c = \arccos n$. A plot of reflectance versus angle of incidence has an infinite slope as the critical angle is approached from below. As k increases, the slope becomes finite, but there is an inflection point for k less than $0.63(1-n)$ (Rehn 1981). In the example of fig. 8, this region is for $k < 0.18$. Reflectance versus angle measurements can be fitted to the Fresnel relations with n and k as fitting parameters. As k becomes larger, the curves become closer together for equal increments in k, and the accuracy of this method diminishes. This method is widely applicable in the X-ray region, but it is also appropriate for a few metals in limited spectral regions in the vacuum ultraviolet and soft X-ray regions (Rehn 1981). For other materials or spectral regions, k is too large or $n > 1$. In the visible region, a variant of this method is widely used to obtain optical data, where it is called attenuated total reflectance. In such applications, the radiation reflects from the interface between a transparent solid, e.g., fused quartz, being

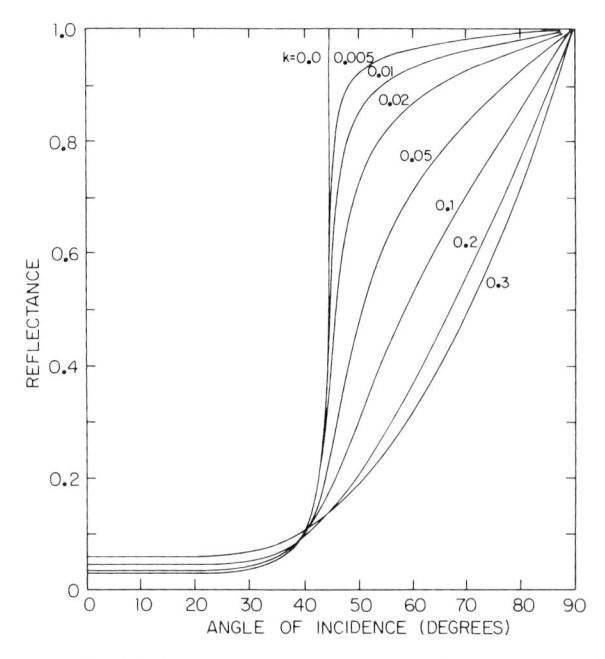

Fig. 8. Reflectance vs angle of incidence for materials with $n = 0.707$, and several values of k. The critical angle is 45°, and reflectance is less than total for $\phi < 90°$ for any $k > 0$. As k increases, the inflection point is lost and curves for different k become more closely spaced.

incident from within the medium at an angle above the critical angle. The addition of a film of absorbing material spoils the total reflection. Usually the reflectance versus angle of incidence is measured. (See, e.g. Otto 1976.)

The result of analyses of the propagation of errors like those just illustrated is that for most absorbing media measurements of R_p at two angles, one of which is large, near the pseudo-Brewster angle, is a good method of wide applicability. Hunter (1982) has discussed this method extensively, especially with regard to measurements in the vacuum ultraviolet and soft X-ray region. He also discusses some other photometric measurements. Additional discussion of these methods will be found in the work by Damany (1974).

Uzan (1968) has re-evaluated a number of photometric methods with consideration of the degree of polarization. He analyzed errors in determining the degree of polarization from measurements of the reflectances for nominal s- and p-polarized radiation at an angle of incidence of 45°. (For a clean surface and complete linear polarization, the Fresnel equations yield $R_p = R_s^2$ at 45° angle of incidence, independent of the optical constants.) The errors in the measurement depend on the value of the reflectance of the material used. He then considered the effect of incompletely polarized radiation on a number of photometric methods used to determine optical constants. The use of radiation which is not highly polarized leads to additional errors in several of the methods, making some of them unsuitable.

The foregoing methods can, in some cases, be applied to anisotropic materials. Uniaxial materials are easier. If it is possible to obtain single crystal samples with the optic axis in the plane of a large flat face, as well as those of the more common orientation, with the optic axis normal to the sample face, many of the previous methods can be adapted. However, one must note that different faces of an anisotropic crystal may be mechanically and chemically anisotropic. (This is true of cubic crystals, too.) This means that the different faces may have different reflectances because of different morphology or overlayers. Many anisotropic materials, however, grow only as thin plates or they cleave too easily to permit preparation of a sample with the optic axis in the face, the usual situation being that the optical axis is normal to the cleavage planes. In such cases, one has only one surface with which to work, and the need to determine the dielectric function for radiation polarized parallel and perpendicular to the optic axis. By measuring R_s at several angles of incidence, one can obtain the dielectric function for $E \perp c$, c denoting the optic axis. This can be done exactly as for isotropic materials. Measuring R_p at several angles of incidence gives the rest of the needed data. As the angle of incidence increases from zero, R_p depends increasingly on the $E \parallel c$ component of the dielectric function, and less on the $E \perp c$ component. This is described in detail in the work by Mosteller and Wooten (1968), Graves and Lenham (1968), Berman et al. (1970), and Wooten (1984). This method is difficult to apply for highly anisotropic materials, as the history of measurements of the optical anisotropy of graphite shows, for one must know the dielectric function for $E \perp c$ with great accuracy in order for it not to introduce errors in determining the other component from the measurements of R_p. The graphite measurements were carried out in the visible and ultraviolet by Greenaway et al. (1969), and in the vacuum ultraviolet by Klucker et al. (1974). The latter measurements should be compared with those of Zeppenfeld (1971), taken by electron energy loss spectroscopy. There is significant disagreement in the spectra for $E \parallel c$, both in magnitudes and in the presence or absence of a peak.

Biaxial crystals present a more difficult problem, but they may be studied in some cases, subject to some approximations. If the anisotropic crystal can be prepared with high-quality surfaces of several different crystallographic planes then reflectance methods can be applied. Koch et al. (1974) have dealt with a monoclinic crystal, which has two (equal) off-diagonal components in the dielectric function, and three unequal diagonal components. If two suitable crystallographic planes can be prepared for a reflectance measurement, they show that three normal-incidence measurements of the reflectance of one face with polarized radiation and a polarization-sensitive detector can give three uncoupled equations for two diagonal components and the off-diagonal component of the dielectric function, but each involves both the real and imaginary parts. A similar reflectance measurement on the other face gives a reflectance which depends only on the third diagonal component of the dielectric function. To separate real and imaginary parts, Kramers–Kronig techniques, as described in section 4.2, must be used. (Also see Pavinich and Belousov 1978, Belousov and Pavinich 1978, and

Emslie and Aronson 1983.) Other methods are discussed in section 4.2 on Kramers–Kronig methods. The more usual case is that in which only one crystallographic plane, usually a close-packed one, can give a good surface. If the symmetry in this plane is fairly high, one may assume that the off-diagonal component of the dielectric tensor for a monoclinic crystal is zero, which makes the biaxial crystal appear uniaxial. This is probably a reasonable approximation for highly anisotropic layered materials. This approximation was made by Leveque et al. (1984) for orthorhombic GaTe. Eight reflectance spectra were measured, using two angles of incidence and the four possible orientations of the electric vector and one crystallographic axis. Six dielectric tensor spectra and the polarization spectrum were extracted by fitting the reflectance spectra (Koch et al. 1974), using both a weighting function (Klucker et al. 1974), and a Kramers–Kronig analysis during the fitting to keep the real and imaginary parts of a dielectric function component consistent.

At higher photon energies, usually above 30 eV, reflectances at near-normal incidence become very small and measurement is difficult. Working at large angles of incidence extends the range somewhat, but large samples are needed and eventually the reflectance is too low. Thus at higher energies, almost all data are taken by transmission measurements on thin films. The films cannot be mounted on monolithic substrates for energies above 11 eV. They must be either unsupported films, films evaporated on a parting agent, floated free in water or another solvent, and mounted over a mesh or an aperture (Hunter 1977), or films evaporated in situ onto another thin film. The latter technique is necessary for studying any reactive material, although it is possible to make a composite film with the layer of reactive material sandwiched between two protective layers. The supporting thin film often is carbon or a hydrocarbon such as formvar, both of which have very little absorption structure of their own, except at the carbon K-edge. Films evaporated this way may have densities lower than that of a monolithic sample (Chopra 1969), leading to lower magnitudes for the absorption coefficient. Often they are not evaporated in ultrahigh vacuum. This leads to the incorporation of oxygen in the films, for some materials, which leads to absorption by the oxygen 2p electrons. The latter can be assessed by the relative increase in the absorption edge for one of the sample elements, if one has an edge in the 50–150 eV region, because the oxygen 2p electrons contribute essentially the same absorption coefficient just below and just above the edge, thus lowering the jump.

The transmission is given by eq. (22) (with the denominator set equal to unity and usually with $\Psi = 0$), which is valid for the incoherent superposition of multiply-reflected beams, the usual case. The unknowns are the absorption coefficient and R, the reflectance for a single interface. The former depends on k, and the latter on k and n. For large photon energies R is very small because $n \approx 1$ and $k \ll 1$, and can be neglected in zeroth order. The transmission spectrum then gives the absorption coefficient, hence k. The spectrum of k can be Kramers–Kronig analyzed to give n. This n spectrum and the original k spectrum can then be used to calculate R. The data are reanalyzed until self-consistency is achieved.

Another approach is to use two films with different thicknesses. The ratio of the two transmission spectra does not contain R.

Errors in transmission measurements can be rather high. Besides the aforementioned problems with samples, there are more. Pinholes obviously cause problems, as does inhomogeneous thickness. The accurate measurement of thickness is difficult, and the use of two samples to eliminate the need for R exacerbates this, for it requires the difference of two sample thicknesses. Finally, a large range of sample thicknesses is required for good accuracy, with a typical thickness being of the order of $2.557/\mu$, the ideal thickness for minimizing errors from shot noise in the detector. Samples much thicker than $2.557/\mu$ give more severe troubles with pinholes and detector nonlinearity, while thinner samples require better knowledge of R. If μ changes much throughout the spectrum, different samples should be used. With such a range, one can plot the logarithm of the transmission versus thickness and obtain the absorption coefficient from the slope (Haensel et al. 1968).

At lower photon energies, R is not so small, but the method can be modified and extended. At such energies, below 11 eV, the film may be on a flat transparent substrate. The transmission may be measured as above, and the reflectance measured as well, often both at near-normal incidence. These two quantities can give the optical constants. Nilsson (1968) evaluated the use of normal-incidence reflectance and transmission measurements on a supported film (see also Paulick 1986). A variation on this is to treat the film thickness as unknown and to make a third measurement. This can be another reflectance or transmission measurement at an angle not near normal incidence (Abelès and Theye 1966, Nestell and Christy 1972). In solving for the three unknowns at each wavelength, the same value of the thickness should result. There are many possible measurements, using different angles of incidence, polarization, and measuring with the radiation incident first on the film or on its substrate. Nestell and Christy (1972) carried out an error analysis similar to that of Hunter (1965). They found that in the visible and near ultraviolet, measurements of the normal-incidence transmission and reflectance, and the transmission for p-polarized radiation incident at around 60° will give good results.

Recently, transmission gratings have been made of unsupported thin films. The diffraction patterns they produce depend on the optical constants of the material, for the normally-opaque regions of the grating are, in fact, semi-transparent. This offers yet another way to determine the optical properties of materials, one currently being tested in the soft X-ray region. In this method, one illuminates the grating at normal incidence with monochromatic radiation and measures the ratio of the first-order diffracted intensity to the zero-order intensity. By doing this with two gratings with different geometrical parameters (grating constant, width and thickness of the metallic regions) one can extract the two optical constants. This method has been used once for gold in the 280–640 eV region. It appears to give good results in a relatively unexplored region of the spectrum (Tatchyn et al. 1980a,b, 1984).

4.2. Kramers–Kronig methods

Kramers–Kronig (KK) methods have been the most popular methods for obtaining the optical spectra of solids in the past twenty years. One first measures the reflectance over as wide a frequency range as possible. Then one produces an extrapolation of the reflectance to zero frequency and to very high frequencies, then carries out the integral given below. The disadvantages of the technique are that the dielectric function is not available until this procedure has been carried out, often days after the measurement has been made, and that the extrapolation procedures may cause errors in addition to the errors in the reflectance measurement itself, making the accuracy poorer than that of previously described methods. Nevertheless, the fact that only one quantity needs to be measured apparently has been viewed as a great advantage for the technique. Most of its practitioners have been interested in the shapes of the dielectric function spectra, not in the magnitudes, and the KK method usually provides spectra with the correct shapes, even when the magnitudes may be in error.

The method relies on a dispersion integral (the KK integral) to obtain the phase shift of the reflected electric field from the spectrum of the reflected field amplitude, $|\tilde{r}| = R$ (Velicky 1961, Stern 1963, Bassani and Altarelli 1983, Smith 1985a). The phase is given by

$$\theta(\omega) = -\frac{\omega}{\pi} P \int_0^\infty \frac{\ln R(\omega')}{\omega'^2 - \omega^2} \, d\omega', \tag{28}$$

$$= \frac{1}{2\pi} P \int_0^\infty \frac{d \ln R(\omega')}{d\omega'} \ln\left|\frac{\omega' - \omega}{\omega' + \omega}\right| d\omega', \tag{29}$$

where the second form is the result of a partial integration of the first. Often R is normal-incidence reflectance, but the reflectance for s- or p-polarized radiation may be used at any angle of incidence, although problems can arise at large angles with p-polarization, and with anisotropic crystals and p-polarization (Roessler 1965b, Berremann 1967, Hale et al. 1973, Querry and Holland 1973, Makarova and Morozov 1976, Smith 1977, and Young 1977). The accuracy in carrying out the numerical integration can be better than that of the reflectance measurement (Roessler 1965a, Kopilevich and Makarova 1981). Once θ is obtained, the real and imaginary parts of the complex refractive index are obtained from

$$n = \frac{1 - R}{1 + R - 2\sqrt{R} \cos \theta}, \tag{30}$$

$$k = \frac{2R \sin \theta}{1 + R - 2\sqrt{R} \cos \theta}, \tag{31}$$

for the case of normal incidence.

The integral is broken into three integrals, one for the low-frequency extrapolation, one for the measured data, which may contain data from other measurements to augment the range of one's own, and one for the high-energy extrapolation. The extrapolations have been the subject of many publications (Rimmer and Dexter 1960, Miloslavskii 1961, Roessler 1965a, 1966, White and Straley 1968, Scouler 1969, Tulvinskii and Terentex 1969, Leveque 1977, Jezierski 1984, 1986). The low-frequency extrapolation is the easier of the two to deal with. For a metal, one expects free-electron-like behavior at sufficiently low energy, and a free-electron-gas model may be used with parameters taken from the dc conductivity. Not many metals are truly free-electron-like, and the reflectance near the low-energy limit of the measurements may not fit very well to a free-electron model with reasonable parameters. In such a case one may construct a model with one or more Lorentzian oscillators near the low-energy limit of the data (on both sides of the limit) to represent the interband contributions to the optical properties in the low-energy region (Veal et al. 1974, Belousov and Pogarev 1977). It is very difficult to assess errors introduced by such a technique. For semiconductors or insulators, the low-energy extrapolation is easier if the measurements extend below the band gap energy. One then can use the reflectance calculated from the refractive index for all lower energies, ignoring the contribution of the lattice vibrations to the optical properties as long as one is interested in the region above the band gap.

The region above the high-energy limit of the data is more difficult to deal with. There are many approaches, which we list. All of them can be made quite accurate if any parameters in them can be adjusted to force the resultant dielectric function to agree with an accurately known value at one particular or a few frequencies, either the result of an ellipsometric measurement in the visible region or the fact that $k = 0$ below the band gap in a semiconductor or insulator (Wu and Andermann 1968, Ahrenkiel 1971). These approaches include the following:

(1) The free-electron model gives a reflectance falling as E^{-4} at high energy. This has been used, despite the fact that it might not fit smoothly the measured data at the highest energy. This model is valid strictly only above the K-edge of all of the atoms in a material. A variation of this is to use a power law for the high-energy reflectance, and adjust the exponent to fit some criterion, either the one listed above, or one to be given below. The extrapolation need not appear as a table of reflectances to high energy. For a power law, one can evaluate the integral and obtain a simple polynomial for the contribution to the phase from the high-energy extrapolation (Miloslavskii 1961, Roessler 1965a, MacRae et al. 1967, Jezierski 1986).

(2) One can use not only literature values for optical data, but absorption cross sections that have been calculated for each atom in the periodic chart (McMaster et al. 1969, Band et al. 1979, Henke et al. 1982), weighting them according to the contribution of each atom to the composition of the material. Measured absorption coefficients may be used, as well. In calculating the high energy reflectance from these, the real part of the refractive index often is assumed to be unity, but it is more accurate to use a refractive index spectrum obtained from a KK analysis of the absorption coefficient.

In addition to testing the result of the KK analysis by a comparison with accurate data in a limited spectral region, when available, there are other checks of the result which are useful when there are no other optical data. The sum rule on ε_2 must be valid. Since partial sum rules may not always be valid, the total sum rule is the more reliable (Klucker and Nielsen 1973, Ellis and Stevenson 1975, King 1977, Smith and Shiles 1978). If it is not nearly satisfied, then, if the experimental data are not in error, the extrapolations need to be altered, or possibly, the numerical integration has not been carried out accurately enough in the high energy region. In addition to the sum rules on ε_2 there are also "super-convergence" sum rules which the data should obey. Since these have different weightings for the data, they form a valuable additional test (Altarelli et al. 1972, Altarelli and Smith 1974, Furuya et al. 1977, Inagaki et al. 1979, Smith and Manogue 1981, Kimel 1982, Grundler 1983, Smith 1985). Leveque (1986) has introduced augmented partial sum rules which converge toward the $\omega \to \infty$ limit from above rather than from below, as do super-convergence sum rules making the use of both sets of rules attractive.

One can also test the method by repeating the analysis using alternate extrapolations, and even data with a reflectance measurement error simulated by multiplying the reflectance by a factor close to one. The result of such exercises is usually a spectrum that resembles the best one, but with magnitudes that are different. Peak positions are reproduced quite faithfully, but relative peak heights may change significantly.

Kramers–Kronig integrals exist for the transmission of a thin film. The analogous integral to that used for reflectance relates the phase shift of the transmitted beam to an integral of the transmission. The phase shift and the transmission then give the two optical constants at each energy (Nilsson 1968, Neufeld and Andermann 1972, Palmer and Williams 1985).

There have been a number of discussions of errors in the Kramers–Kronig method of determining optical constants. An early assessment was made by Bowlden and Wilmshurst (1963) for one and two rather sharp absorption peaks. Hinson and Stevenson (1967) measured reflectances at several angles and obtained the dielectric function by two methods, the use of two reflectances, as in the preceding section, and a Kramers–Kronig analysis of one reflectance spectrum. They concluded that the two-angle method was more accurate for their data, provided the degree of polarization was known. Since that time the extrapolation problem has been studied more, and better extrapolations can be made. Nilsson and Munkby (1969) used exact reflectance data, data from an oscillator model, to study the effect of extrapolations (primarily to high energy) and of assumed errors in the reflectance on the optical constants obtained by the Kramers–Kronig method. They showed that poor high-energy extrapolations produced dielectric function spectra with the right shapes, i.e., spectral features, but with magnitudes that were very poor. (Such incorrect magnitudes then would produce poor spectra for $\text{Im}(-1/\tilde{\varepsilon})$.) By fitting several parameters in the extrapolation to force the dielectric function to agree with a few accurate values, the effect of the extrapolation on the magnitudes of the dielectric function spectra could be reduced to below that expected from errors in the reflectance measurement, well under 1%. Errors in reflectance measurements were simulated by subtracting a constant from R. The resultant errors in $\tilde{\varepsilon}$ were not all of the same sign. They depended on wavelength, and the error $\Delta\tilde{\varepsilon}/\tilde{\varepsilon}$ could be larger than $\Delta R/R$, but by less than an order of magnitude. For the materials with a dielectric function similar to that simulated, e.g., many metals, if the reflectance could be measured to within 0.1%, the errors in the dielectric function due to these errors and the extrapolation would be comparable. Leveque (1977) presented spectra for GaTe, in which there is an absorption edge above the highest energy for which reflectance data were available. He used three extrapolations, a free-electron E^{-4} extrapolation, a Lorentz-oscillator representation of the (unmeasured) edge absorption, and actual absorption coefficient data. The latter method combined reflectance data over one energy range with absorption coefficient data, over another, but overlapping, range. The analysis required that the end result was a dielectric function over the combined range that was consistent with both measured spectra. The method can also be applied if there is no overlap of the ranges of the measured spectra, and it may be used to generate a low-energy extrapolation as well. The effects of the three higher-energy extrapolations were changes in the magnitudes of the various optical spectra, with a shift in the energy of the peak in $Im(-1/\tilde{\varepsilon})$ at the plasmon energy. The E^{-4} extrapolation was poorer than the other two.

It is clear that errors in the reflectance data will lead to errors in the derived optical functions. A common occurrence in early applications of the technique was the appearance of a region of unphysical, negative values of k, hence, of μ, in the region just below the band gap of an insulator or semiconductor. Since the energy region was often quite limited in early reflectance studies, this problem was attributed to the extrapolation. Bauer et al. (1974) showed that in the case of AgCl, the problem could be traced to an error in the reflectance right in the region of the gap, i.e., very close to the region of negative k. Correcting this error resulted in a more physical spectrum near the gap, with little change elsewhere. However, this seems not to be a universal conclusion. Jezierski and Misiewicz (1984) had the same problem with Zn_3P_2. They could remove the region of negative k by a local change in the reflectance, but without being able to assign the errors to a known problem in the instrumentation, as were Bauer et al. (1974). They showed that an alternate problem could have been the cause. They corrected their reflectance data for scattering by surface roughness (Porteus 1963), using the rms height irregularity and slope as free parameters. Physically reasonable values of these parameters led to a reflectance, the KK transformation of which gave no region of negative k.

A method with some similarities to the Kramers–Kronig treatment of a single reflectance or transmission spectrum was described by Verleur (1968). In it, one fits the reflectance spectrum by the reflectance of a series of damped harmonic oscillators, with one having zero resonance frequency to represent the conduction electrons in a metal. The calculated spectrum is fit in a least-squares sense by adjusting the resonance frequencies, oscillator strengths, and damping parameters of each oscillator. These oscillator parameters need not have any direct physical significance. The number of oscillators is not unique. A relatively small number appears adequate to produce a fit to the reflectance to within experimental errors for the reflectance. A similar treatment of transmission data can also yield a dielectric function. Rivory (1970) has applied this method to reflectance and transmission data on films. In common with the Kramers–Kronig method, there is a problem with the end regions of the reflectance spectrum. In Verleur's (1968) method some of the oscillators may have to be located outside the region of the measured spectrum. Even then, sometimes a good fit is hard to achieve. In such cases, as with the Kramers–Kronig method, one can obtain a better fit by forcing the oscillators to give one or more accurately-known values of n and k (from another measurement) at some point in the spectrum.

There are yet other ways to analyze a single reflectance spectrum to obtain the complex dielectric function. Peterson and Knight (1973) used the Fourier transform of the reflectance spectrum for their analysis, finding that this gave increased computational speed. King (1977) represented the phase by a Fourier series, the terms of which could be determined from the measured reflectance. Neither of these two methods has been widely employed.

4.3. Photoyield methods

Measurements of photoyield have been used frequently in the past ten years to obtain qualitative absorption spectra in the photon energy range above about 20 eV. Photoexcited electrons are produced in a depth of the order of μ^{-1} below the surface. Since the energy-dependent mean free path for inelastic scattering is much shorter than μ^{-1}, the majority of these electrons are not emitted. However, those which scatter only a few times still have enough energy to escape, and they form a nearly structure-free background under the primary photoelectron spectrum, constituting a dominant majority of all the electrons emitted. These electrons come from a depth of the order of 100 Å. They may be measured by collecting only low-energy electrons (partial yield) or all electrons (total yield). In either case, the spectrum has the structure of the absorption spectrum, but it is a qualitative spectrum on a background (see Gudat and Kunz 1972, and Kunz 1979).

Recently Birken et al. (1986) developed photoyield measurements into a quantitative method of obtaining optical constants, and applied it in the 40–600 eV range to a film of Al about 800 Å thick. They measured the reflectance versus angle of incidence for a large number of angles, and fitted the Fresnel relations to obtain optical constants. This was done for both the glass substrate and the Al-covered substrate. They then measured photoyield at over a hundred angles for each photon energy. The analysis of the yield in terms of the optical constants of the film and substrate, the film thickness, and the escape length had been made by Pepper (1970). It was used in a least-squares fit to obtain the optical constants of the Al film, the film thickness, and the electron escape length at each photon energy, the optical constants of the substrate being known from multiple-angle reflectance measurements. (The escape length here is one averaged over a range of electron kinetic energies. It varies from 10 to 30 Å with photon energy, larger than the mean free path for inelastic scattering, but smaller than expected.) Agreement of the resultant ε_1 and ε_2 spectra, the latter ranging over two decades in magnitude, was excellent. This method seems very promising, requiring primarily a better understanding of the resultant mean free path, which may come about through a reanalysis of the derivation of Pepper's expressions for the yield.

4.4. Ellipsometry

Ellipsometry is now widely practiced in the visible, near infrared, and near ultraviolet (Aspnes 1976, 1985, Azzam and Bashara 1977). Recent results from this technique have provided the most accurate optical properties of a number of solids in the applicable spectral region. Despite there being essentially no ellipsometrically-obtained data at short wavelengths, we discuss the technique here because of the high quality of recent ellipsometric data in the 1–6 eV region, data which should match well with any shorter wavelength data obtained in the future.

Linearly polarized radiation at non-normal incidence is reflected from a material interface as elliptically polarized radiation, unless the incident radiation is pure s or p polarized. The ratio of the reflected electric fields is

$$\tilde{r}_p/\tilde{r}_s = \rho\, e^{i\Delta} = \tan \Psi\, e^{i\Delta}. \tag{32}$$

The measured quantities in ellipsometry are $\tan \Psi$ and Δ.

There are many ways of carrying out ellipsometric measurements. The original methods involved creating elliptically polarized light which fell on the sample at non-normal incidence. The elliptical polarization was adjusted with a compensator so that when it was reflected from the sample, its state of polarization was converted to plane polarization. This could be nulled by rotating a polarizer just before the detector, often the eye.

This null ellipsometry has been supplanted by photometric ellipsometry in which either the polarizer before the sample or the analyzer after the sample is rotated (Cahan and Spanier 1969, Hauge and Dill 1973, Aspnes 1974, Aspnes and Studna 1975). As an example, we consider the former. The polarizer is rotated mechanically. Radiation incident on the sample is linearly polarized, with the plane of polarization rotating at angular frequency Ω. After reflection from the sample, the radiation at any instant of time is, in general, elliptically polarized. The parameters of the ellipse change with time at frequency Ω. This radiation passes through a fixed polarizing element, the analyzer, and then is detected. The signal from the detector is, in general (Budde 1962),

$$I = A + B \sin(\Omega t + \delta), \tag{33}$$

$$= A + B \cos \delta \sin \Omega t + B \sin \delta \cos \Omega t, \tag{34}$$

where the phase shift δ is with respect to a reference orientation of the rotating polarizer. Fourier analysis of the signal yields A, B, and δ. The dielectric function of the sample can be obtained from the quantities B/A and δ. When using a computer, one can scan wavelengths and obtain the dielectric function, not just A/B and δ, essentially in real time. The method, and variations of it, are now widely used, and can yield data of extremely high precision and accuracy. The precision is limited by counting statistics and the accuracy by the quality of the components. At the moment, for work in the visible and near-ultraviolet, precisions of a few parts in 10^5 and accuracies of a few parts in 10^3 have been achieved. Other forms of automatic ellipsometers exist, relying on modulation of the state of polarization of the beam by stress-induced anisotropy in a plate of cubic or isotropic material (Kemp 1969, Jasperson and Schnatterly 1969, Mollenauer et al. 1969, Treu et al. 1973, O'Handley 1973, Bermudez and Ritz 1978).

Ellipsometry has been carried out in the vacuum ultraviolet region at least once (Schledermann and Skibowski 1971) and another attempt is being made (Johnson 1984). The problem is the quality of components. At the minimum, two polarizers must be used. These are prisms made of birefringent materials for use in the visible and near-ultraviolet. In the vacuum ultraviolet, mica has been used, but its range of transmission is small. LiF pile-of-plates polarizers have been used, but their efficiency is poor. The best polarizers are combinations of three or four metallic mirrors (Hamm et al. 1965, Hunter 1978). For optimum polarization efficiency, the angles of incidence on the mirrors should be adjusted as the wavelength changes, but a reasonably wide range of near-optimum efficiency can be had. Such polarizers were used in the only vacuum ultraviolet ellipsometry to date. (It is possible to use the natural polarization of synchrotron radiation and dispense with one of the polarizers, but the problem of the second polarizer remains. Moreover, the degree of polarization of the radiation as it impinges on the sample must be known quite accurately to assess accuracy, and this, too, requires a good polarizer.)

As mentioned in section 3, ellipsometry is considered to be relatively free from errors caused by rough surfaces (Berremann 1970), the explanation being that lost intensity plays no role, because the intensity is not measured. However, when the rms roughness, h, becomes as large as or larger than a wavelength, surface roughness definitely influences ellipsometric measurements (Vorburger and Ludema 1980). Such rough surfaces rarely would be used for the determination of optical constants. Because phase differences upon reflection are measured, ellipsometry is very sensitive to overlayers (Burge and Bennett 1964). Indeed, it is often employed to study the growth of overlayers on surfaces.

4.5. Electron energy loss measurements

When energetic electrons pass through matter they lose energy by exciting electrons in the system. Measurement of the spectrum of energies lost, i.e., energy analysis of the exiting electron beam, constitutes electron energy loss spectroscopy (Raether 1965, 1980, Daniels et al. 1969, Schnatterly 1979). As we shall outline below, electron energy loss spectroscopy can lead to a description of the optical properties of a material. Moreover, it is particularly convenient for work on anisotropic materials which can be prepared with large surfaces bounded by only one type of crystallographic plane. The acronym EELS is now commonly used for electron energy loss spectroscopy, but the majority of recent EELS has used the electron beam reflected from the sample. This gives a spectrum resembling the optical spectrum, but the complete optical spectrum, including proper magnitudes, is difficult to derive from such a spectrum. Moreover, in some cases, the incident electrons are not energetic enough, and the spectrum obtained contains non-dipole transitions, of great interest, but not the same spectrum of electric dipole transitions that is the optical spectrum. We emphasize only transmission energy loss spectroscopy here.

An electron of energy E_0, sufficient that the Born approximation can be applied (in practice, at least several keV, depending on the material), has a differential probability per unit path length for inelastic scattering given by (Hubbard 1955, Ritchie 1957)

$$\frac{W}{d\Omega\, dE} = \left(\frac{e}{\pi h v}\right)^2 \text{Im}\left(\frac{-1}{\tilde{\varepsilon}}\right) \frac{1}{\theta_E^2 + \theta^2}. \tag{35}$$

$$\theta_E = \frac{E}{2E_0}, \tag{36}$$

$$v = \sqrt{2mE_0}. \tag{37}$$

The energy lost by the electron as it scatters into differential solid angle $d\Omega$ located at angle θ from the original direction is E. The momentum transferred to the electron is q, and it has components $q_\parallel$ parallel to the original momentum, and $q_\perp$ perpendicular to it. $\tilde{\varepsilon}(E, q)$ is the longitudinal dielectric function (a cubic or isotropic medium is assumed), for the electric field of the probe electron is longitudinal. At small values of q the longitudinal dielectric function is the same as the transverse dielectric function which describes the optical response (Wooten 1972). The differential cross section can be related to the measured current of electrons at energy $E_0 - E$, when the detector intercepts a known solid angle, and, in principle, $\text{Im}(-1/\tilde{\varepsilon}(E, q))$ may be determined. A Kramers–Kronig analysis of this function gives the real part, so that $\tilde{\varepsilon}(E, q)$ may be determined. For small q this agrees with the spectrum determined by optical measurements, and for some regions of the spectrum and for some materials, this method may be superior to optical methods.

The above expression neglects a number of complications which must be addressed before a dielectric function can be obtained (Daniels et al. 1969). It ignores multiple scattering. As the thickness of the sample grows, the probability of multiple scattering grows. If there is a weak spectral feature at energy E, a strong feature at energy $\tfrac{1}{2}E$ could appear at energy E as a result of double scattering and thereby obscure it. Techniques for removing the multiple scattering contribution to the spectrum appear to be quite successful (Misell 1970, Wehenkel

1975, Feldkamp et al. 1977). Measurements on samples of different thicknesses are an aid to identifying such effects. There are also contributions to the loss spectrum not given by the above expression, contributions arising from the surfaces of the slab sample (Ritchie 1957, Geiger and Witmaak 1966). These contributions can be identified by their independence of sample thickness, by changes upon surface alteration, e.g., oxidation, and by their different dependence on q. They, too, must be subtracted from the spectrum before further analysis. Finally, the thickness of the sample must be known if an absolute measurement is to be made. Since the ratio of the scattered spectrum to the unscattered current may be small, detector linearity over a wide range is needed for an absolute determination of $\text{Im}(-1/\tilde{\varepsilon})$. A more common method is to measure the loss spectrum in arbitrary units so that d does not have to be known accurately (but knowledge of d is needed to remove surface and multiple scattering effects), then fit the amplitude to a reliable value of $\text{Im}(-1/\tilde{\varepsilon})$ from an optical measurement, e.g., the result of an ellipsometric measurement in the visible or near-ultraviolet region.

The electron energy loss technique nicely complements reflectance measurements. Reflectance techniques and ellipsometry sample the surface region, typically a distance of an inverse absorption coefficient, several hundred angstrom, while the energy loss technique, when used with 100 keV electrons, samples the entire depth of a sample, 1000 Å or more. The energy loss function has a peak at the zero of the dielectric function, where the reflectance exhibits a sharp drop to small values with increasing energy. Thus, the loss function is most easily measurable in a region of the spectrum where the reflectance becomes small and difficult to measure.

The loss function,

$$\text{Im}\left(\frac{-1}{\tilde{\varepsilon}}\right) = \frac{\varepsilon_2}{\varepsilon_1^2 + \varepsilon_2^2}, \tag{38}$$

does not resemble either the real or imaginary part of the dielectric function in general. At energies below the plasmon energy, it has been shown (Chiarello et al. 1984) that the negative of the second derivative of the loss function exhibits peaks at peaks in ε_2. [This aspect of the loss function makes reflection EELS usefully quantitative, especially since corrections for surface effects can be made (Henrich et al. 1980).] The largest peaks in the loss function occur at zeroes of ε_1, i.e., at the energies of longitudinal excitations in the system, e.g., plasmons and longitudinal excitons. At higher energies ε_1 approaches unity and ε_2 becomes very small. The loss function then approaches ε_2, giving a direct picture of the optical absorption spectrum.

Electron energy loss measurements can give the wave-vector dependence of the dielectric function, by making measurements of the spectrum at several scattering angles, i.e., at several momentum transfers (Zeppenfeld 1971). Moreover, one can work with anisotropic crystals and get several components of the dielectric function tensor, and do so without the need for different samples. This can be done because the loss function for uniaxial crystals is proportional to Im(1/$(\tilde{\varepsilon}_{xx}q_x^2 + \tilde{\varepsilon}_{yy}q_y^2 + \tilde{\varepsilon}_{zz}q_z^2)$) (Tosatti 1969, Chen and Silcox 1979), and for biaxial crystals this is a good approximation, provided the axes are principal axes of the dielectric tensor, and that they do not disperse much in direction (Venghaus 1970). By using a single crystal at non-normal electron incidence and collecting electrons at several scattering angles, one can obtain the individual components of the dielectric tensor. By going to larger scattering angles, one obtains the wavevector dependence of the dielectric function components.

4.6. X-ray methods

Above about 3 keV photon energy, the optical properties and measurement techniques are somewhat different. We will call this region the X-ray region. The real part of the refractive index is usually less than unity, but by an amount rarely larger than 0.001, usually much smaller, and the imaginary part is small, of the same order of magnitude. Thus $\tilde{N} = 1 - \delta + i\beta$, with δ and β each very small. As the energy increases above an absorption edge, both of these quantities become yet smaller, and the absorption coefficient falls as well. Because edges are usually well separated in energy, this rapid falloff leads to the relative transparency of all materials for hard X-rays. Because of the smallness of δ it is necessary to measure δ itself, rather than $1 - \delta$. Sometimes δ and β are replaced by the atomic scattering factor $\tilde{f}$, or its departure from the number of electrons on the atoms, $\Delta \tilde{f}$, defined as

$$1 - i\delta + i\beta = 1 - \frac{r_0 \lambda^2}{2\pi} \Sigma_j N_j \tilde{f}_j , \qquad (39)$$

$$= 1 - \frac{r_0 \lambda^2}{2\pi} \Sigma_j N_j (Z_j + \Delta \tilde{f}_j) , \qquad (40)$$

$$\tilde{f}_j = f'_j - i f''_j , \qquad (41)$$

$$\Delta \tilde{f}_j = \Delta f'_j - i \Delta f''_j . \qquad (42)$$

N_j is the number of atoms of type j (with Z_j electrons) per unit volume and r_0 is the classical radius of the electron, $r_0 = e^2/(mc^2)$. Note that this form considers contributions to the refractive index from individual atoms. Near absorption edges, and for a few hundred electron volt above them, the edge and EXAFS regions, the absorption from an atom will depend on its surroundings, and f'_j and f''_j become material parameters, rather than atomic parameters. Nevertheless, out of necessity, tabulated atomic parameters are often used.

There are three general measurement methods in use: absorption coefficient measurements, interferometer measurements, and reflectance measurements at grazing incidence. These have been reviewed extensively by Creagh (1985).

We have already discussed absorption measurements. They are carried out in the X-ray range in the same way as in other regions, except that the source and detector may be different. In most cases, they are carried out in air or a helium atmosphere. Since the absorption coefficient is smaller than in other spectral regions, sample thicknesses may be larger, and for metals, rolled foils may be used. The problem of oxides on the sample surface still remains, but for thicker samples it is less serious. Too-low sample densities are still a problem. The absorption coefficient data give β directly, since the correction for reflection at the interfaces is negligible. δ is found either from a Kramers–Kronig analysis of β, or a fit of the absorption coefficient to a model for the core level transitions. The latter method was formerly used frequently (Compton and Allison 1935), but it ignores effects such as EXAFS, which can have an effect of up to 10% or so in a region many hundred electron volt above any absorption edge.

X-ray interferometers are similar to optical interferometers (Bonse and Materlik 1976, Cusatis and Hart 1977, Hart 1980, Bonse et al. 1980, Hart and Siddons 1981, Siddons 1981). They use Bragg diffraction for beam splitting and recombination. The interferometer is made of three plates of high-quality silicon cut from one large single crystal. The fringe pattern shifts if a thin slab of semitransparent material is inserted in one beam. Measurements of this shift give a value for δ. The measured spectrum of δ can be Kramers–Kronig analyzed to give a spectrum of β, or β may be determined from the fringe visibility in the interferometer, or by an absorption coefficient measurement.

The reflectance of a bulk sample is extremely small except near grazing incidence. Since the real part of $\tilde{N}$ is less than unity, "total" external reflection can occur, provided β is small enough. The critical angle, ϕ_c, measured from grazing, is $\phi_c = \cos^{-1} n \approx \sqrt{2} \beta$, provided $\beta \leq 0.63 \delta$. Larger values of β lead to a plot of reflectance versus angle that lacks an inflection point (Rehn 1981). ϕ_c is usually less than 1 or 2 degrees. The reflectance can be measured as a function of the grazing angle and the data fitted to the Fresnel relations to obtain δ and β. One may also use just a few angles above the critical angle and fit the reflectance, as is done in other spectral regions. The absolute reflectance is difficult to measure, so one often must normalize the reflectances to that at one angle (André et al. 1982). All measurements made at angles more grazing than the critical angle sample only a small depth below the surface. This depth is approximately $\lambda/4\pi\sqrt{\beta}$, which amounts to only 10–100 Å, making such measurements dependent on the quality of the surface. Measurements made at angles less grazing than the critical angle are affected by surface roughness (Bilderback 1982, Bilderback and Hubbard 1982).

There are other methods, some no longer used (Compton and Allison 1935), and some still in use. For polar crystals the intensity ratio of a pair of reflections $\overline{hkl}$ and hkl is equal to the ratio of structure factors. For a polar crystal this allows δ to be obtained, provided β is known, usually from an absorption measurement (Fukamachi and Mosoya 1975). Recently, Deutsch and Hart (1984) used the deviation of an X-ray beam by a wedge-shaped sample to obtain δ. Instead of measuring the angle of beam deviation directly, a historical method, they used a diffractometer, two parallel plates of Si. Inserting the wedge between them deviated the beam, reducing the transmitted intensity, which was restored by rotating the second Si crystal by an amount equal to the deviation. δ could be determined from two such measurements at each wavelength, one for each of two different Bragg planes. Fontaine et al. (1985) also used refraction of monochromated X-ray by a prism to obtain δ at an absorption edge. A very recent method makes use of synthetic microstructures, a periodic array of alternating layers of two materials (see the chapter by Spiller in Volume 1 of this Handbook, Spiller 1983). These can diffract X-rays, and the shape of the reflectance versus angle of incidence curve can be used to determine δ, β being assumed known from absorption coefficient measurements. One could also fix the angle of incidence and scan the energy. The first results by this technique are just emerging (Barbee et al. 1984). Since the period of such structures is limited, at present, to at least 20 Å or so, this method cannot be used at extremely short wavelengths.

5. Summary

The number of ways to measure the optical properties of materials is large, especially in the visible and near-visible regions. Despite many years of measurements and analyses of errors, there does not yet seem to be a single best method. Probably the reason for this is that sample preparation is not standardized, and even when a single method of preparation is adopted, the sample is not ideal, and the departures from ideality affect the results of different methods in different ways. Probably the most study has been done on transparent materials used in the optical coating industry. Here the number of methods is even larger than it is for absorbing media, and lasers make a number of methods particularly precise. Two recent studies point to the fact that sample inhomogeneities are important. A study of evaporated TiO_2 films was made by Borgogno et al. (1984). They measured the refractive index by photometric measurements in the visible and near-infrared, both in vacuo during deposition, and in the atmosphere after deposition. The resultant refractive index spectra depended strongly on deposition conditions, with high and low spectra differing by up to 20%. The measurements made during deposition were analyzed to show the refractive index changes as additional material was deposited. There were decreases of up to 20% in the refractive index between the material deposited first and that deposited last, for films about 700 Å thick. Upon admitting air to the chamber, there were changes in n of about 5%, attributed to adsorption of water by the films. A different kind of study was carried out by Arndt et al. (1984). Many samples of Sc_2O_3 and Rh films were made at the same time and sent to a number of laboratories. Each laboratory measured n and k at eight wavelengths on each of two samples of each material, the samples differing in thickness by about a factor of two. Widely different methods were used, including ellipsometry and several photometric methods. In a few cases, the optical method gave the film thickness as well, although it had been measured by mechanical profiling. The results were better for the transparent Sc_2O_3 than for the opaque Rh. For the former, the mean variance in the refractive index was 0.01 for a refractive index ranging from 1.8 to

1.9 in the spectral range. Inhomogeneities of up to 0.2 in n were found (difference in n for the innermost and outermost regions). The results for Rh were more disappointing. The variances in n and k were 13% and 8%, respectively, far more than expected. The different results were attributed to the different effects on sample inhomogeneities and ageing effects (oxidation) in each measurement procedure.

The results of the measurements on Rh were later clarified by Aspnes and Craighead (1986). The large scatter in n and k was ascribed to the errors in the film thickness. Analysis showed that a film thickness error of about 1 Å would be consonant with the accuracy of both photometric and ellipsometric measurements, but thickness measurements reported by Arndt et al. (1984) varied by up to 60 Å. They also made ellipsometric measurements on two of the same films used in the previous study. The end result was a pseudo-dielectric function spectrum for the film material which was consistent with the previous direct optical measurements, but not those of thickness. The pseudo-dielectric function was interpreted with an effective medium model, the results of which were in agreement with scanning electron microscopic determinations of the films' microstructure, and which gave a dielectric function characteristic of bulk Rh.

It is clear from the foregoing that discrepancies in the results of different measurements of the optical constants of a material are more likely to be due to differences in samples than to errors in the measured quantities and their propagation to the final result. This has been discussed thoroughly by Aspnes et al. (1980). They measured the dielectric function in the 1.5–5.8 eV range of variously-prepared gold films. They also made comparisons with some of the 28 cited reports on the optical properties of gold in this region of the spectrum. They found that discrepancies between data sets were larger than could be attributed to measurement errors, and arose from differences in sample parameters, often traceable to the method of preparation. In the region below about 1.8 eV, where intraband absorption by the conduction electrons dominates, the differences were attributed to differences in electron scattering times. Small-grained samples had shorter scattering times, hence larger values of ε_2. Thus large-grained samples better represent bulk gold for optical measurements in this spectral region. Surface roughness produced by annealing (e.g., to obtain larger grains) also reduced the scattering times. In the interband region, above 1.8 eV for gold, the results were quite different. Here, a higher value of ε_2 was more representative of bulk gold, and lower values usually could be attributed to a lower density of gold in the region of the surface sampled by the radiation. The lower density arises from a non-negligible volume fraction of lower-density intergranular gold in fine-grained samples, fractions of 0.2 to 0.3 being possible. Annealing reduces this fraction, and produces data close to those of bulk gold. Surface roughness, which increases upon annealing, has a larger effect on ε_1 than on ε_2 in this region. The data from any sample could be described quantitatively with an effective medium model (Aspnes et al. 1979), with bulk gold and vacuum ("voids") as the two constituents. This model can be used for the bulk, and for an overlayer representing surface roughness. Note that for the two spectral regions, the criteria for selecting the best set of data, lowest ε_2 or highest ε_2, are different, and the preparation method for the best (film) sample to use are not necessarily the same in both regions. The results of this study are probably qualitatively applicable to all materials in the region of their valence band interband absorption, including non-metals, and to metals in the high-energy part of the region of free-electron absorption, for those metals where interband absorption is not strong at low energies. Additional evidence for the effect of voids on the interband dielectric function is supplied by the work of Nestell et al. (1980) on evaporated films of V, Cr, Nb, Mo, and W which, depending on the evaporation and condensation conditions, exhibit bulk optical properties or show the effects of low-density grain boundary material.

A similar study of silicon (and seven other semiconductors) was carried out by Aspnes and Studna (1983) in essentially the same spectral region. The new ellipsometric data were compared with a large number of earlier measurements, some at only one energy, and comments made on the discrepancies. Such extensive analyses of optical data, considering the effects of surface overlayers, void content, and some experimental errors, do not exist for other materials. The extensive data on aluminum have been discussed by Shiles et al. (1980) and by Smith (1985), using Kramers–Kronig consistency and satisfaction of sum rules as criteria of quality, but attempts to assign reasons for discordant data were not made.

6. Some examples

We consider here a few other examples, all far less exhaustively studied than aluminum or gold. They are chosen to illustrate an even wider variety of measurement techniques, problems with surface oxidation, and anisotropy, and the lower accuracy of most measurements made at energies above 5 or 6 eV.

Molybdenum is a bcc metal whose optical spectrum is prototypical of that of many bcc transition metals. It has been measured many times by a variety of techniques. We refer the reader to data compilations for complete references to all of the data, and excerpt only a few measurements here. Veal and Paulikas (1974) and Weaver et al. (1974) measured the reflectance, the former in the 0.5–6 eV region, the latter in the 4.9–35 eV region, supplemented by measurements of the absorbance ($A = 1 - R$) in the 0.1–4.9 eV range. The samples were electropolished bulk metal, both single crystal and polycrystal. Both used a Kramers–Kronig analysis, but with different extrapolations and different criteria for the choice of parameters. However, as can be seen in fig. 9 the resultant dielectric functions are in rather good agreement in the spectral regions where they overlap. Structures in the spectra are identical, but the magnitudes differ, especially in ε_1. Note the additional zero-crossing near 3 eV in the ε_1 spectrum of Weaver et al. (1974). Manzke (1980) used high-energy transmission electron energy loss spectroscopy to derive the dielectric function. His samples were films 500–600 Å thick. Data in the 1.5–38 eV region were Kramers–Kronig analyzed.

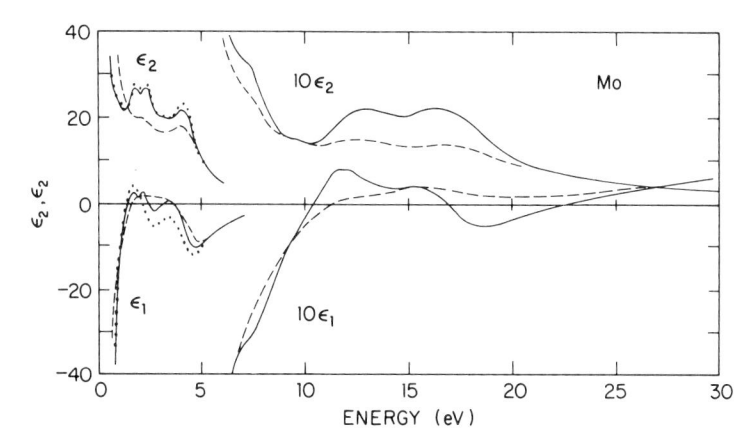

Fig. 9. Dielectric function of Mo. Solid line: from optical data of Weaver et al. (1974), dots: optical data from Veal and Paulikas (1974), dashed line: from electron energy loss data from Manzke (1980).

Rather than rely on the film thickness measurement to obtain the magnitudes of the $1/\tilde{\varepsilon}$ spectra, the data were normalized to make $\text{Re}(-1/\tilde{\varepsilon})$ vanish in the limit of zero energy. The resulting dielectric functions are shown in fig. 9. There is less structure in the 2–3 eV region, compared with the optical data. This is probably due to the low resolution, about 0.5 eV, in the EEL measurement. In the entire spectral region, the magnitudes are not always in close agreement with those from the optical spectra. Above 6 eV, where the effects of an oxide layer on the sample should be large for reflectance measurements, but small for transmission EEL measurements, the discrepancies in magnitude are not much worse than they are at lower energy. However, where ε_2 is small and ε_1 approaches zero, the discrepancies may be small in absolute units, but large in relative units. Such discrepancies are more easily seen in the loss function, $\text{Im}(-1/\tilde{\varepsilon})$, the dominant structure of which is the peak near the zero of ε_1. In this region, the reflectance is small, thus difficult to measure accurately, and it is sensitive to the effects of thin, even monolayer, surface oxides, which are absorbing in this spectral region. This is also the spectral region in which EEL measurements are most sensitive. The more prominent 10–20 eV structure in the dielectric functions derived from reflectance may be the result of an oxide overlayer (see below). Since interband structure is expected in this spectral region, one cannot assign all of the structure to oxygen on the surface or incorporated in the films, even though the oxygen 2p electrons have a large photoabsorption cross section here.

For more reactive materials, agreement is worse, and not only are the magnitudes of ε_1 and ε_2 different in different measurements, the presence or absence of some peaks is determined by an oxide layer. We illustrate this with data on Ti. Ti is quite reactive, reactive enough to be used as a getter, but in one relatively recent study in the vacuum ultraviolet (Lynch et al. 1975), it was hoped that the large-grained polycrystal used would not oxidize as rapidly as polycrystalline films. The sample was electropolished and exposed to air before insertion in the vacuum chamber, which was not an ultrahigh-vacuum chamber. The samples, as did the Mo samples cited above, definitely had an oxide layer on the surface at least 10–20 Å thick, possibly thicker. A later study by Wall et al. (1980) used polycrystalline bulk Ti which had been polished and etched. Its reflectance was measured as introduced into the ultrahigh-vacuum chamber, after cleaning by Ar^+ bombardment and annealing (with LEED and Auger spectra to monitor surface quality), and after additional oxidation. The reflectance measurements were accompanied by Auger spectra which gave the amount of oxygen coverage. The samples were of unknown morphology, as were those of Lynch et al. (1975). However, the effects of oxygen were far greater than any expected effects of anisotropy. Figure 10 shows the reflectance for ex-situ prepared surfaces of Ti before any treatment, after cleaning, and after surface oxidation. It is clear that the large reflectance peak at 10 eV, and possibly the sharp structure just above 5 eV in the as-prepared sample, and in the earlier measurements by Lynch et al. (1975), are the result of the presence of oxygen on the surface. Girouard and Truong (1984) measured the reflectance (6–20 eV) at several angles of incidence of 250 Å-thick films of Ti. They derived dielectric function spectra from which the normal-incidence reflectance was calculated. In this work, the reflectance spectrum (not shown) resembled that of Wall et al. (1980) in shape, but the magnitude was lower by about a factor of two over most of the spectral range. It appears that their samples were relatively free of oxygen, for the extra structure in the reflectance was absent, even though the films were prepared in a vacuum of 2×10^{-7} Torr, where Ti films are used as pumps. Their low reflectance may be the result of low density of the film or scattering by surface roughness.

There have been very few studies of the optical properties of materials in the region above 6 eV in which the samples were truly clean, for very few reflectome-

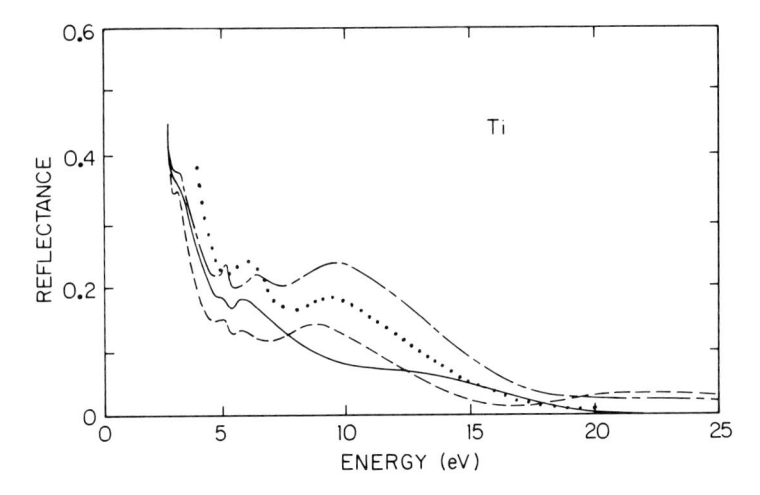

Fig. 10. Normal-incidence reflectance of Ti. Dashed line: as prepared, solid line: after cleaning in vacuo, dot–dashed line: after surface oxidation (from Wall et al. 1980). Dots: from Lynch et al. (1975).

ters have been built which operate in ultrahigh vacuum. The work reported by Wall et al. (1980) was carried out in such a chamber, but that chamber is no longer in use. Hogrefe et al. (1983) describe a new ultrahigh vacuum chamber for reflectance measurements in the spectral region above 6 eV. It was used in the work of Birken et al. (1986).

As a last illustration of the agreement, or lack thereof, in optical spectra of solids, we take aluminum oxide. This material should not be sensitive to oxidation, being fully oxidized, but it can react with water to some extent. Unfortunately, there are several allotropes of Al_2O_3. The two important ones are α-Al_2O_3, a rhombohedral, hence uniaxial, crystal, called corundum, sapphire, or ruby (when Cr-doped), and amorphous Al_2O_3, which is the oxide that forms on metallic aluminum. It can also be produced by electrolytic oxidation of aluminum. γ-Al_2O_3 is a cubic form. Of the numerous studies of the optical properties of Al_2O_3 above 10 eV, we cite only a few. Transmission measurements were made on films of amorphous Al_2O_3 by Hagemann et al. (1974). These, when augmented by other data, were Kramers–Kronig transformed to obtain the optical functions. Arakawa and Williams (1968) measured the reflectance of a single crystal of α-corundum. They used partially polarized light, and measured the reflectance at several angles of incidence throughout the 8.5–28.5 eV region. (The optical axis was in the plane of the surface, but its orientation with respect to the plane of incidence was not stated.) They attempted to obtain the optical constants from the reflectance at two angles, but did not obtain consistent results over most of the spectral region. They then Kramers–Kronig analyzed the reflectance to obtain the optical constants. Later measurements on single-crystal α-corundum by Osmun (1972) and by Olson (1975) with nearly completely polarized radiation showed the anisotropy in the spectra, and suggest that the data of Arakawa and Williams are primarily for $E \parallel c$. Finally, Tews and Gründler (1982) made transmission electron energy loss measurements on films of amorphous, polycrystalline α- and polycrystalline γ-Al_2O_3. Figure 11 shows Osmun's reflectance of single-crystal α-Al_2O_3 and that of a sample of fused Al_2O_3 (not the same as amorphous Al_2O_3). First, note the anisotropy. There is more structure for $E \perp c$ than for $E \parallel c$. The unequal magnitudes of the two spectra cannot be interpreted unambiguously as optical anisotropy, for the samples were polished with diamond, but not subsequently etched. The mechanical anisotropy of Al_2O_3 could lead to anisotropic surface morphology and attendant anisotropic surface scattering. That this might be so can be judged from the fact that the reflectance of the fused sample is considerably higher than that of the crystal. The fused sample presumably was not polished, and was microscopically smoother. Additional evidence for this comes from the data by Olson (1975), whose reflectance spectrum for $E \parallel c$ agrees with that of Osmun (1972), but whose spectrum for $E \perp c$ is of lower magnitude, and missing the peak at 15 eV. These differences arise most probably from differences in surface quality. Figure 12 shows the reflectance measured by Arakawa and Williams on a single crystal of α-Al_2O_3, and that derived from EEL measurements on polycrystalline α-Al_2O_3 by Tews and Gründler. The spectral features of the single-crystal reflectance here match those

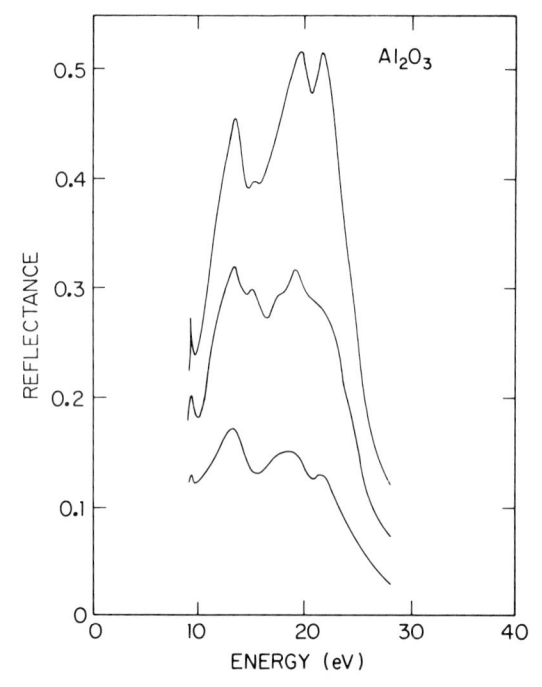

Fig. 11. Normal-incidence reflectance of Al_2O_3. The upper curve is for fused Al_2O_3. The lower curves are for a single crystal of α-Al_2O_3. The lowest is for $E \parallel c$, the middle curve for $E \perp c$. (From Osmun 1972.)

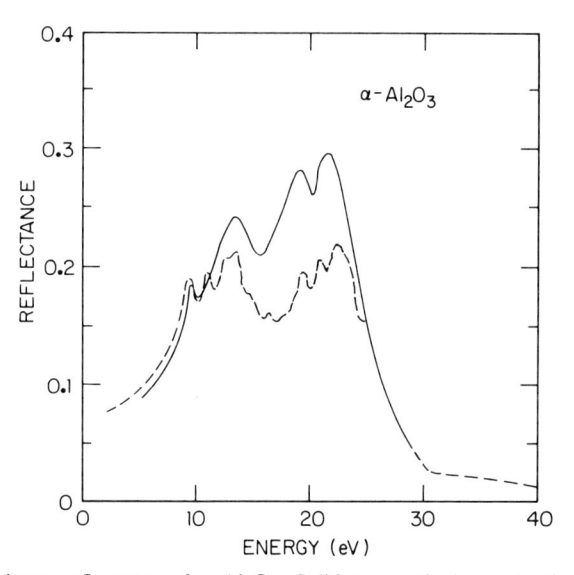

Fig. 12. Normal-incidence reflectance of α-Al_2O_3. Solid curve: single crystal (probably for $E \parallel c$). (From Arakawa and Williams 1968.) Dashed curve: polycrystalline film. (From electron energy loss measurements of Tews and Gründler 1982.)

of Osmun's spectrum for $E \parallel c$, but the relative magnitudes of the peak heights differ considerably. Arakawa and Williams (1968) tested their data with several sum rules, from which one can conclude only that their reflectance and ε_2 spectra are not too high. These sum rules allow higher values of ε_2, but they do not require them. If the crystal reflectances were as high as those shown for fused Al_2O_3 in fig. 11, the sum rules would not be violated. The crystal reflectances may be too low as a result of polishing with diamond paste without subsequent etching. The reflectance spectrum derived from the EEL data also is shown in fig. 12. It has a lower overall magnitude than the single crystal data, possibly due to a low density in the film. However, the structures are different. One can simulate the polycrystal data from the single crystal data by averaging the two single-crystal spectra, weighting the $E \perp c$ spectrum twice as much as the $E \parallel c$, but this can be done only in the $\tilde{\varepsilon}$ spectra, not in the reflectance. The average reflectance spectrum then can be calculated from the average $\tilde{\varepsilon}$ spectrum. However, in this spectral region the R and ε_2 spectra resemble each other, so such an average of reflectances may be meaningful. Such an average reflectance (not illustrated) does not show all the structure present in the spectrum for the polycrystal. The reason for this is not clear. Figure 13 shows ε_2 from the EEL measurements on polycrystalline films of α-, γ-, and amorphous Al_2O_3, along with that of Arakawa and Williams (1968) for single-crystal α-Al_2O_3. Since different data sets on α-Al_2O_3 alone cannot be reconciled, it is not useful to comment on the effect of crystal structure on the optical properties. Al_2O_3 is a reasonably simple material, and its occurrence as an oxide on metallic aluminum makes it an important optical material, but we do not yet have optical spectra above 8.5 eV in which we have much confidence.

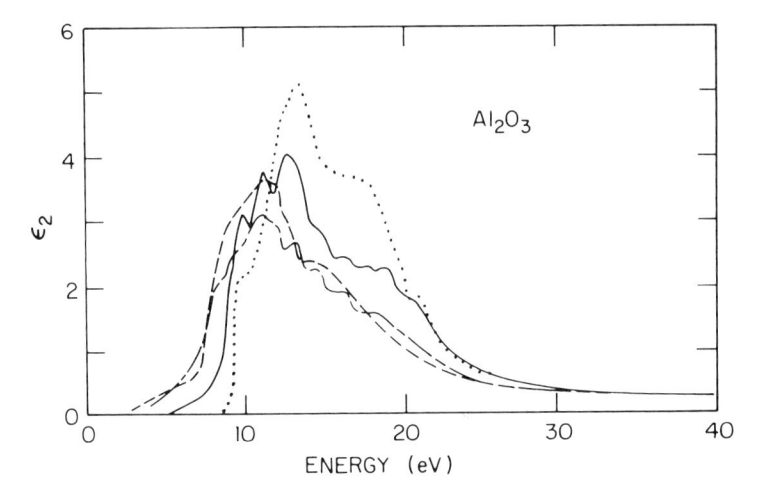

Fig. 13. Imaginary part of the dielectric function for several allotropes of Al_2O_3. Dots: single crystal α-Al_2O_3. (From Arakawa and Williams 1968.) Solid line: polycrystalline α-Al_2O_3, dashed line: polycrystalline γ-Al_2O_3, dot-dashed line: amorphous Al_2O_3. (From Tews and Gründler 1980.)

7. Comments on data collections and recent literature

There have been several reviews of the optical properties of materials in recent years, written primarily to summarize the current state of understanding of the spectra. These include those of Abelès (1972), Brown (1974), Nilsson (1974), Priol and Robin (1974), Stephan and Robin (1974), Kunz (1975), Koch et al. (1977), Lynch (1979), Brown (1980), and Bassani and Altarelli (1983). Many of these include extensive references. Those of Priol and Robin, and Stephan and Robin include references to measurements organized by material. A recent book edited by Palik (1985), contains a number of chapters on the interpretation of the optical properties of solids. Many of these reviews contain extensive lists of references to original publications on optical properties of particular materials. In addition to these, there are a number of compilations of optical data. We describe these in some detail.

McMaster et al. (1969) tabulated free-atom X-ray absorption cross sections in the 1 keV–1 MeV region of the elements from hydrogen through plutonium. They are presented as tables, log–log plots, and as a set of coefficients for each element for writing $\ln \Sigma$ as a cubic polynomial in $\ln E$. The total attenuation cross sections were obtained from a variety of literature sources (119 of them) reporting measured values, with the data from some sources assigned a higher weight than those from other sources. Interpolation was used for gaps in the spectra of several elements. Coherent (Rayleigh) and incoherent (Compton) scattering cross sections were calculated. These involved some extrapolation and interpolation. Both of these cross sections were subtracted from the total cross section to obtain the photoelectric cross section. [We discussed only the photoelectric cross section in this chapter, as have Bassani and Altarelli (1983). For energies below 10 keV it completely dominates the total absorption.] The tables, plots, and coefficients are for each contribution to the cross section, as well as for the total. These tables, as well as the other tables for "X-ray use" that follow, do not take account of any structure at or above the absorption edge: white lines, NEXAFS, EXAFS, for these arise from the surroundings of the atom and the details of chemical bonding.

Storm and Israel (1970) present calculated cross sections (coherent, incoherent, photoelectric and pair-production, as well as their sum) in the 1 keV–100 MeV energy range for the first 100 elements. These are presented as tables.

Hubbell (1971) reports measured values for the total cross section in the 10 eV–100 GeV region for the elements up to atomic number 94. Two hundred and ninety references are organized by element. Data are not tabulated, and plots are given for only 17 elements. These log–log plots compare the experimental data with those of McMaster et al. (1969).

Veigele (1973) calculated the scattering cross sections, and subtracted them from the total cross sections to get the photoelectric cross sections. These were fitted with theory to obtain a best set, then the scattering cross sections added to give a total cross section. Log–log plots and tables are given in the energy range 100 eV–1 MeV for the elements up to $Z = 94$. Above 10 keV, the results agree within about 5% with those of Hubbell (1971), of McMaster et al. (1969), and of Storm and Israel (1970). Further discussion of the agreement can be found in the article by Pratt et al. (1973). Below 10 keV, agreement with the data set of McMaster et al. is poorer, by up to 20%, with Veigele's (1973) data being consistently lower. The cause is assigned to the use of hydrogenic wavefunctions by McMaster et al. (1969).

Hubbell et al. (1974) combine data from several previous compilations and report total cross section values for the elements at 24 discrete energies, those of characteristic radiation from common targets, between 4.509 and 24.942 keV.

Band et al. (1979) calculated relativistically the photoionization cross-section, subshell by subshell, for free atoms ($Z = 1$ through 100) at 14 discrete energies between 132.3 and 4509 eV, energies of line sources of radiation. They were not adjusted to fit measured values. A comparison at 4509 eV with the tables of McMaster et al. (1969), using the polynomial fit of the latter, shows differences of up to 20% between the two sets of cross sections, for the few elements checked. The differences are not all in the same direction.

Smith (1985b), has shown that some of the inconsistencies between data sets arise from the use of different forms of the dispersion integrals relating f'_j and f''_j when the input data are inaccurate. For f''_j data not satisfying the sum rule, different f'_j spectra will result from each form. He also points out that in the relativistic calculations of f'_i from f''_j often cited (Cromer and Liberman 1970), a term was omitted which can cause errors of the order of 10% in f'_i in the soft X-ray region for heavy elements.

Hagemann et al. (1974) compiled complete optical spectra for Mg, Al, Cu, Ag, Au, Bi, glassy C, and amorphous Al_2O_3. They used literature values, where available, and described the samples and methods used to obtain them. They measured the absorption coefficient for all of these materials in the 20–150 eV region, using thin-film samples. They combined their data with selected literature values. (This involved occasional interpolation), and subjected the extended data set to a Kramers–Kronig analysis, with partial sum-rule tests. The results of this are tabulated and plotted for the 0.005 eV–50 keV region, although there are few data points at the extremes of this range. There was not an unambiguous best data set for Au, so two sets are presented.

Haelbig et al. (1977) collected optical data on all the rare gas solids and alkali halides. For each material they tabulate the energy range, method of measurement, and sample characteristics. The data are displayed graphically, not numerically, in the form of absorption coefficient or cross section, sometimes in arbitrary units if quantitative data were not available. They cover the energy region above the band gap, extending, with breaks in the data, to regions of absorption edges at energies up to 5 keV. Over 250 references are given.

Weaver et al. (1981) followed the example set by Haelbig et al. (1977) and compiled optical data on many elemental metals: all of the transition metals, rare earths, actinides, and Al, Cu, Ag, and Au. In addition to graphical presentation, and the use exclusively of quantitative data, they present one data set for each metal in tabular form. The plots of the optical spectra contain the results of many investigators, so the agreement, or lack thereof, is manifest. They do not comment directly on the individual data sets, nor do they select a "best set". (The numerical tabulations were selected, in part, for reasons of convenience.) The energy range covered is usually 0.1–30 eV for the dielectric functions and 25 to several hundred electronvolt for the absorption coefficient, although not all metals tabulated have been measured over these entire ranges. Over 250 references are given.

Henke et al. (1982) present real and imaginary parts of the scattering factor for the first 94 elements at 50 discrete energies in the 100–10 000 eV region. They began with experimental absorption coefficient data over a wider energy range than this, using suitable interpolations and extrapolations. These gave f'', the spectra of which were Kramers–Kronig analyzed to obtain f'. Tables are presented, along with log–log plots and data derived from the scattering factors, e.g., reflectances of eight selected mirror materials as a function of angle of incidence.

A recent book, edited by Palik (1985), discusses methods for measuring optical constants, interpretation of spectra, and presents a collection of data for 11 elemental metals, 14 elemental and compound semiconductors, and 12 insulating compounds. In general, the materials were selected for their application in optical devices, and because they are representative of a class of materials. For the latter reason, the data encompass a wide spectral region, far beyond that needed for a given optical device. The data for each material are presented in tables and log–log plots. An attempt was made to select the best data for each material, and the sample characteristics and method of measurement are given for each data set presented. The energy range covered varies with the material, but generally it extends from the infrared through the vacuum ultraviolet region, sometimes going to 2 keV.

Lynch (1985) presented data for 21 materials (10 are metals), 6 of which were not covered by Palik (1985). Log–log plots of n, k, R, and A are presented for the wavelength range 0.01–10 μm. The data selected are not discussed extensively, nor is there a complete set of references to all data for each material.

The foregoing reviews and data compilations contain little data published after 1979. To supplement these, a list of recent (1979–1985) references on the optical properties of the solids in the spectral region above 6 eV has been added as an appendix. This list is not complete, but it should contain the majority of quantitative dielectric function or reflectance spectra published in the time interval covered.

Appendix. Recent literature reference to optical data for $E \geqslant 6$ eV

Reference	Material
1977	
Inagaki et al.	polystyrene
1979	
Abramov et al.	Al_2O_3
Ambrazevivius et al.	$CdSiP_2$, $ZnSiP_2$

Reference	Material
1979 (cont'd)	
Amitin et al.	$CsPbCl_3$
Beal and Hughes	$2H-MoS_2$, $2H-MoSe_2$, $2H-MoTe_2$
Bertrand et al.	$SnSe_2$, SnS_2
Bruhn et al.	Fe, Co, Ni, Cu
Feldkamp et al. (1979a)	TiS_2
Feldkamp et al. (1979b)	Ni
Galuza et al.	Fe_2O_3
Hunter et al.	Pt
Mamy and Couget	InSe
Mitani et al.	$(SN)_x$
Nicklaus	$BaCl_2$, $BaBr_2$, BaI_2, $SrBr_2$, BaFCl, BaFBr, BaFI, ScFCl, SrFBr
Nosenzo and Reguzzoni	Cs-halides
Sato et al.	Na, K, Rb, Cs
Schlegel et al.	Fe_2O_3
Sprüssel et al.	KF, KCl, KBr, KI
Zhuze et al.	La_2S_3, Nd_2S_3, Dy_2S_3
1980	
Amitin et al.	$CsPbCl_3$, $PbCl_2$, $CsCaCl_3$
Cukier et al..	Sc, Sc_2O_3, Y, U_2O_3, Gd, Gd_2O_3, Dy, LaF_3, GdF_3, DyF_3, Th, ThF_4, U, UF_4
Fäldt and Nilsson	Th
Gross et al. (1980a)	CuCl
Gross et al. (1980b)	CuCl, CuBr, CuI
Gurin et al.	EuB_6, LaB_6
Khumalo and Hughes (1980a)	TiS_3, ZrS_3, $ZrSe_3$, $ZrTe_3$, HfS_3, $HfSe_3$, $HfTe_3$
Khumalo and Hughes (1980b)	$NaCrS_2$
Kinno and Onaka	$MgCl_2$
Lynch et al.	TiC_x ($0.64 \leq x \leq 0.90$)
Martzen and Walker	CuI, $CuBr_2$
Miyahara et al.	K–Rb, K–Cs alloys
Ritsko and Mele	$FeCl_3$-intercalated graphite
Vishnevskii et al.	$ZrCl_2$
Zhuze et al.	LaS, LaSe, LaTe, GdS
1981	
Arakawa et al.	$(C_{22}H_{10}N_2O_5)_n$ (Kapton H)
Batlogg	TmS, TmSe, Tm_xSe
Boudriot et al.	Zn/SiP_2
Galuza et al.	FeO
Goel et al.	RuO_2, IrO_2

Reference	Material
1981 (cont'd)	
Heidrich et al.	$CsPbCl_3$, $CsPbBr_3$
Kaneko et al.	CaS
Kendelewicz	$Cd_{1-x}Mn_xTe$
Khumalo	$NaCrS_2$
Khumalo and Hughes	$NaCrS_2$, $FePS_3$, $NiPS_3$, $MnPS_3$, $FePS_3$-type layered compounds
Kinno and Onaka	$MgBr_2$
Kubier et al.	$CdGeP_2$
Manzke et al.	NbS_2
Quemerais et al.	Gd, Dy
Ritsko et al.	K-intercalated graphite
Sobolev et al.	ZnTe–CdTe
Trepakov et al.	$PbMg_{1/3}Nb_{2/3}O_3$
Trykozko and Huffman	$CdIn_2Se_4$
1982	
Bartlett and Olson	Ni
Kondo and Matsumoto	$CdCl_2$, $CdBr_2$
Laporte and Subtil	LiF
Meixner et al.	Nb
Mitani et al.	$(SN)_x$, brominated $(SN)_x$
Miyahara et al.	La, Ce, Sm, Eu, Gd
Onuki and Rife	KMF_3(M = Mn, Fe, Co, Ni, Cu)
Ritsko (1982a)	K-intercalated graphite, KC_8, KC_{24}
Ritsko (1982b)	polyacetylene
Rossinelli and Bosch	SiO_2
Suga et al.	$ZnCr_2Se_4$ (spinel)
Zhuze et al.	ScS, ScSe, ScTe
1983	
Abramov et al.	Sc_2O_3, Y_2O_3
Batson and Silcox	Al
Bedford et al.	BeF_2, BeF_2 (glassy)
Farberovich et al.	SmB_6
Filatova et al.	Si
Gluskin et al.	Ru, Os, PdIn
Kaneko et al.	CaO, CaS, SrO, SrS, SrSe, BaO, BaSe
Leveque et al.	Ag
Pollini et al.	$NiBr_2$, $NiCl_2$
Suga et al.	FeS_2, CoS_2, NiS_2, NiO
Valiukonis et al.	GeSe
Vinogradov et al.	SiO_2, SiO_x, Si_3N_4, SiC

Reference	Material
1984	
Bagdasarov et al.	Y_2O_3, Ho_2O_3, Er_2O_3, Tm_2O_3, Yb_2O_3, Lu_2O_3
Deutsch and Hart (1984a)	Si, LiF
Deutsch and Hart (1984b)	LiF
Girouard and Truong	Ti, Mn
Hanyu et al.	Ba, Sm, Tb, Ho, Er, Yb
Hoffman et al.	BN
Pajasova et al.	a-GeS
Pfluger et al.	TiC_x, TiN_x, VC_x, VN_x
Pollini et al. (1984a)	$FeCl_2$, $CoCl_2$, $NiCl_2$, $FeBr_2$, $CoBr_2$, $NiBr_2$
Pollini et al. (1984b)	$NiCl_2$, $NiBr_2$, NiI_2
Pollini et al. (1984c)	FeI_2, CoI_2, NiI_2
Preil et al.	$KHgC_4$, $KHgC_8$
Schoenes	CeS, CeN, US, UAs, NSb, $U_xY_{1-x}Sb$
Tokura et al.	polydiacetylene (PTS, HPU, TCPU)
1985	
Arakawa et al.	C
Bauliss and Liang	TiS_2, $TiSe_2$, ZrS_2, $ZrSe_2$, HfS_2, $HfSe_2$
Filatova et al. (1985a)	SiC, Si_3N_4
Filatova et al. (1985b)	SiO_2, SiO_x
Filatova et al. (1985c)	Si compounds
Fisher et al.	KC_8, LiC_6, $KHgC_4$
Katoh et al.	Ge
Pfluger et al.	ZrN, NbC, NbN

References

Abelès, F., 1972, Optical Properties of Metals, in: Optical Properties of Solids, ed. F. Abelès (North-Holland, Amsterdam) pp. 93–162.

Abelès, F., and T. Lopez-Rios, 1980, Surf. Sci. **96**, 32.

Abelès, F., and M.L. Theye, 1966, Surf. Sci. **5**, 325.

Abramov, V.N., M.G. Karin, A.I. Kuznetsov and K.K. Sidorin, 1979, Fiz. Tverd. Tela **21**, 80 [Sov. Phys.-Solid State **21**, 47].

Abramov, V.N., A.N. Ermoshkin and A.I. Kuznetsov, 1983, Fiz. Tverd. Tela **25**, 1703 [Sov. Phys.-Solid State **25**, 981].

Ahrenkiel, R.F., 1971, J. Opt. Soc. Am. **61**, 1651.

Altarelli, M., and D.Y. Smith, 1974, Phys. Rev. B **9**, 1290.

Altarelli, M., D.L. Dexter, H.M. Nussenzweig and D.Y. Smith, 1972, Phys. Rev. B **6**, 4502.
Ambrazevivius, G., G. Babonas and A. Vileika, 1979, Phys. Status Solidi b **95**, 643.
Amitin, L.N., A.T. Anistratov and A.I. Kuznetsov, 1979, Fiz. Tverd. Tela **21**, 3535 [Sov. Phys.-Solid State **21**, 2041].
Amitin, L.N., A.T. Anistratov and A.I. Kuznetsov, 1980, Phys. Status Solidi b **101**, K65.
André, J.M., A. Maquet and R. Barchewitz, 1982, Phys. Rev. B **25**, 5671.
Antoncik, E., 1955, Czech. J. Phys. **5**, 449.
Arakawa, E.T., and M.W. Williams, 1968, J. Phys. Chem. Solids **29**, 735.
Arakawa, E.T., M.W. Williams, J.C. Ashley and L.R. Painter, 1981, J. Appl. Phys. **52**, 3579.
Arakawa, E.T., S.M. Dolfini, J.C. Ashley and M.W. Williams, 1985, Phys. Rev. B **31**, 8097.
Armaly, B.F., J.G. Ochoa and D.C. Look, 1972, Appl. Opt. **11**, 2907.
Arndt, D.P., R.M.A. Azzam, J.M. Bennett, J.P. Borgogno, C.K. Carniglia, W.E. Case, J.A. Dobrowlski, U.J. Gibson, T. Tuttle Hart, F.C. Ho, V.A. Hodgkin, M.P. Klapp, M.A. Macleod, E. Pelletier, M.K. Purvis, D.M. Quinn, D.H. Strome, R. Swenson, P.A. Temple and T.F. Thonn, 1984, Appl. Opt. **23**, 3571.
Aspnes, D.E., 1974, J. Opt. Soc. Am. **64**, 639.
Aspnes, D.E., 1976a, J. Opt. Soc. Am. **64**, 639.
Aspnes, D.E., 1976b, Spectroscopic Ellipsometry of Solids, in: Optical Properties of Solids–New Developments, ed. B.O. Seraphin (North-Holland, Amsterdam) p. 799–846.
Aspnes, D.E., 1985, The Accurate Determination of Optical Properties by Ellipsometry, in: Handbook of Optical Constants of Solids, ed. E.D. Palik (Academic, Orlando) ch. 5.
Aspnes, D.E., and H.G. Craighead, 1986, Appl. Opt. **25**, 1299.
Aspnes, D.E., and A.A. Studna, 1975, Appl. Opt. **14**, 220.
Aspnes, D.E., and A.A. Studna, 1983, Phys. Rev. B **27**, 985.
Aspnes, D.E., C.G. Olson and D.W. Lynch, 1976, Phys. Rev. Lett. **36**, 1563.
Aspnes, D.E., J.B. Theeten and F. Hottier, 1979, Phys. Rev. B **20**, 3292.
Aspnes, D.E., E. Kinsbron and D.D. Bacon, 1980, Phys. Rev. B **21**, 3290.
Azzam, R.M.A., and N.M. Bashara, 1977, Ellipsometry and Polarized Light (North-Holland, Amsterdam).
Bagdasarov, K.S., V.P. Zhuze, M.G. Karin, K.K. Sidorin and A.I. Shelykh, 1984, Fiz. Tverd. Tela **26**, 1134 [Sov. Phys.-Solid State **26**, 687].
Band, I.M., Yu.I. Kharitonov and M.B. Trzhasleovskaya, 1979, At. Data & Nucl. Data Tables **23**, 443.
Barbee, T.W., W.K. Warburton and J.H. Underwood, 1984, J. Opt. Soc. Am. B **1**, 691.
Bartlett, R.J., and C.G. Olson, 1982, Phys. Status Solidi b **111**, K33.
Bassani, F., and M. Altarelli, 1983, Interaction of Radiation with Condensed Matter, in: Handbook on Synchrotron Radiation, Vol. 1, ed. E.-E. Koch (North-Holland, Amsterdam) pp. 463–605.
Bassani, F., and G. Pastori Parravicini, 1975, Electronic States and Optical Transitions in Solids (Pergamon Press, Oxford).
Batlogg, B., 1981, Phys. Rev. B **23**, 1827.
Batson, P.E., and J. Silcox, 1983, Phys. Rev. B **27**, 5224.
Bauer, R.S., W.E. Spicer and J.J. White III, 1974, J. Opt. Soc. Am. **64**, 830.
Bauliss, S.C., and W.Y. Liang, 1985, J. Phys. C **18**, 3327.
Beal, A.R., and H.P. Hughes, 1979, J. Phys. C **12**, 881.
Bedford, K.L., R.T. Williams, W.R. Hunter, J.C. Rife, M. Lif, J. Weber, D.D. Kingman and C.F. Cline, 1983a, Phys. Rev. B **27**, 2446.
Bedford, K.L., R.T. Williams, W.R. Hunter, J.C. Rife, M.J. Weber, D.D. Kingman and C.F. Cline, 1983b, Phys. Rev. B **27**, 2446.
Bell, R.J., M.A. Ordal and R.W. Alexander, 1985, Appl. Opt. **24**, 3680.
Belousov, M.V., and V.F. Pavinich, 1978, Opt. Spectrosc. **45**, 771.
Belousov, M.V., and D.E. Pogarev, 1977, Opt. Spectrosc. **43**, 223.
Bennett, H.E., and J.O. Porteus, 1961, J. Opt. Soc. Am. **51**, 123.
Berman, M., M.R. Kerchner and S. Ergun, 1970, J. Opt. Soc. Am. **60**, 646.
Bermudez, V.M., and V.H. Ritz, 1978, Appl. Opt. **17**, 542.

Berremann, D.W., 1967, Appl. Opt. **6**, 1519.
Berremann, D.W., 1970a, Phys. Rev. B **1**, 381.
Berremann, D.W., 1970b, J. Opt. Soc. Am. **60**, 499.
Bertrand, Y., G. Leveque, C. Raisin and F. Levy, 1979, J. Phys. C **12**, 2907.
Bilderback, D.M., 1982, in: Reflecting Optics for Synchrotron Radiation, Proc. SPIE, Vol. 315, ed. M.R. Howells (SPIE, Bellingham, WA) p. 90.
Bilderback, D.M., and S. Hubbard, 1982, Nucl. Instrum. Methods **195**, 85, 91.
Birken, H.-G., W. Jark, C. Kunz and R. Wolf, 1986, Nucl. Instrum. Methods A **253**, 166.
Bonse, U., and G. Materlik, 1976, Z. Physik B **24**, 189.
Bonse, U., P. Spieker, J.-T. Hein and G. Materlik, 1980, Nucl. Instrum. Methods **172**, 223.
Borgogno, J.P., F. Flory, P. Roche, B. Schmitt, G. Albrand, E. Pelletier and H.A. Macleod, 1984, Appl. Opt. **23**, 3567.
Born, M., and E. Wolf, 1959, Principles of Optics (Pergamon Press, Oxford).
Boudriot, H., R. Gründler, B. Kubier, K. Deus and H.A. Schneider, 1981, Phys. Status Solidi b **108**, K129.
Bowlden, H.J., and J.K. Wilmshurst, 1963, J. Opt. Soc. Am. **53**, 1073.
Brown, F.C., 1974, Solid State Physics **29**, 1.
Brown, F.C., 1980, Inner-Shell Threshold Spectra, in: Synchrotron Radiation Research, eds H. Winick and S. Doniach (Plenum Press, New York) pp. 61–100.
Bruhn, R., B. Sonntag and H.W. Wolff, 1979, J. Phys. B **12**, 203.
Budde, W., 1962, Appl. Opt. **1**, 201.
Burge, D.K., and H.E. Bennett, 1964, J. Opt. Soc. Am. **54**, 1428.
Cahan, B.D., and R.F. Spanier, 1969, Surf. Sci. **16**, 166.
Carroll, J.J., S.T. Ceyer and A.J. Melmed, 1982, J. Opt. Soc. Am. **72**, 668.
Celli, V., A. Marvin and F. Toigo, 1975, Phys. Rev. B **11**, 1779.
Ceyer, S.T., A.J. Melmed, J.J. Carroll and W.R. Graham, 1984, Surf. Sci. **144**, L444.
Chen, C.H., and J. Silcox, 1979, Phys. Rev. B **20**, 3605.
Chiarello, G., E. Colavita, M. Decrescenzi and S. Nannarone, 1984, Phys. Rev. B **29**, 4878.
Chopra, K.L., 1969, Thin Film Phenomena (McGraw-Hill, New York) p. 189.
Compton, A.M., and S.K. Allison, 1935, X-Rays in Theory and Experiment, 2nd Ed. (Van Nostrand, New York) ch. IV.
Creagh, D., 1985, Austr. J. Phys. **38**, 271.
Cromer, D.T., and D. Libermann, 1970, J. Chem. Phys. **53**, 1891.
Cukier, M., B. Gauthe and C. Wehenkel, 1980, J. Phys. (France) **41**, 603.
Cusatis, C., and M. Hart, 1977, Proc. R. Soc. London Ser. A **354**, 291.
Damany, H., 1974, Phenomenological Description of Optical Properties of Solids and Methods of Determination of Optical Constants in the Vacuum Ultraviolet, in: Some Aspects of Vacuum Ultraviolet Radiation Physics, eds J. Romand and B. Vodar (Pergamon Press, Oxford).
Daniels, J., C. von Festenberg, H. Raether and K. Zeppenfeld, 1969, Springer Tracts Mod. Phys. **54**, 78.
Deutsch, M., and M. Hart, 1984a, Phys. Rev. B **30**, 640.
Deutsch, M., and M. Hart, 1984b, Phys. Rev. B **30**, 643.
Dignam, M.J., and J. Fedyk, 1977, J. Phys. (France) **38**, C5-57.
Dignam, M.J., and M. Moskovitz, 1973, J. Chem. Soc. Faraday Trans. II **69**, 56.
Dignam, M.J., M. Moskovitz and R.W. Stobie, 1971, Trans. Faraday Soc. **67**, 3306.
Ellis, H.W., and J.R. Stevenson, 1975, J. Appl. Phys. **46**, 3066.
Elson, J.M., 1975, Phys. Rev. B **12**, 2541.
Elson, J.M., 1977, Appl. Opt. **16**, 2872.
Elson, J.M., and J.M. Bennett, 1979, J. Opt. Soc. Am. **69**, 31.
Elson, J.M., and R.M. Ritchie, 1974, Phys. Status Solidi b **62**, 461.
Emslie, A.G., and J.R. Aronson, 1983, J. Opt. Soc. Am. **73**, 916.
Fäldt, A., and P.O. Nilsson, 1980, Phys. Rev. B **22**, 1740.
Fan, H.Y., 1950, Phys. Rev. **78**, 808.

Fan, H.Y., 1951, Phys. Rev. **82**, 900.
Farberovich, O.V., S.I. Kurganskii, K.K. Sidorin, M.G. Karin, V.N. Bobrikov, G.P. Nizhnikova, A.I. Shelykh, M.M. Korsukova and V.N. Gurin, 1983, Fiz. Tverd. Tela **25**, 708 [Sov. Phys.-Solid State **25**, 404].
Feldkamp, L.A., L.C. Davis and M.B. Stearns, 1977, Phys. Rev. B **15**, 5535.
Feldkamp, L.A., S.S. Shinozaki, C.A. Kukkonen and S.P. Faile, 1979a, Phys. Rev. B **19**, 2291.
Feldkamp, L.A., M.B. Stearns and S.S. Shinozaki, 1979b, Phys. Rev. B **20**, 1310.
Filatova, E.O., A.S. Vinogradov, I.A. Sorokin and T.M. Zimkina, 1983, Fiz. Tverd. Tela **25**, 1280 [Sov. Phys.-Solid State **25**, 736].
Filatova, E.O., A.S. Vinogradov, T.M. Zimkina and I.A. Sorokin, 1985a, Fiz. Tverd. Tela **27**, 678 [Sov. Phys.-Solid State **27**, 419].
Filatova, E.O., A.S. Vinogradov, T.M. Zimkina and I.A. Sorokin, 1985b, Fiz. Tverd. Tela **28**, 991 [Sov. Phys.-Solid State **27**, 603].
Filatova, E.O., A.S. Vinogradov and T.M. Zimkina, 1985c, Fiz. Tverd. Tela **28**, 997 [Sov. Phys.-Solid State **28**, 606].
Fisher, J.E., J.M. Block, C.C. Shieh, M.E. Preil and K. Jeiley, 1985, Phys. Rev. B **31**, 4773.
Fontaine, A., W.K. Warburton and K.F. Ludwig, 1985, Phys. Rev. B **31**, 3599.
Fukamachi, T., and S. Mosoya, 1975, Acta Crystallogr. A **31**, 215.
Furuya, K., A. Villani and A.M. Zimmerman, 1977, J. Phys. C **10**, 3189.
Galuza, A., V.V. Eremenko and A.P. Kirichenko, 1979, Fiz. Tverd. Tela **21**, 1125 [Sov. Phys.-Solid State **21**,654].
Galuza, A., V.V. Eremenko, A.P. Kirichenko and V.A. Konstantinov, 1981, Fiz. Tverd. Tela **23**, 251 [Sov. Phys.-Solid State **23**, 140].
Geiger, J., and K. Wittmaack, 1966, Z. Physik **195**, 44.
Girouard, F.E., and V.-V. Truong, 1984, J. Opt. Soc. Am. B **1**, 76.
Gluskin, E.S., A.V. Druzhinin, M.M. Kirillova, V.I. Kochubei, L.V. Nomerovannaya and V.M. Maevskii, 1983, Opt. Spektrosk. **55**, 891 [Opt. Spectrosc. **55**, 537].
Goel, A.K., G. Skorinko and F.H. Pollak, 1981, Phys. Rev. B **24**, 7342.
Graves, R.H.W., and A.P. Lenham, 1968, J. Opt. Soc. Am. **58**, 884.
Greenaway, D.L., and G. Harbeke, 1968, Optical Properties and Band Structure of Semiconductors (Pergamon Press, Oxford).
Greenaway, D.L., G. Harbeke, F. Bassani and E. Tosatti, 1969, Phys. Rev. B **17**, 1340.
Gross, J.G., S. Lewonczak, M.A. Khan, R. Pinchaux and J. Rigeissen, 1980a, Solid State Commun. **35**, 445.
Gross, J.G., S. Lewonczak, M.A. Khan and J. Rigeissen, 1980b, Solid State Phys. **36**, 907.
Gründler, R., 1983, Phys. Status Solidi b **115**, K147.
Gudat, W., and C. Kunz, 1972, Phys. Rev. Lett. **29**, 169.
Gurin, V.N., M.M. Korsukova, M.G. Karin, K.K. Sidorin, I.A. Smirnov and A.I. Shelykh, 1980, Fiz. Tverd. Tela **22**, 715 [Sov. Phys.-Solid State **22**, 418].
Haelbig, R.-P., M. Iwan and E.-E. Koch, 1977, Optical Properties of Some Insulators in the Vacuum Ultraviolet Region, Vol. 8 (Fachinformationszentrum Energie, Physik, Mathematik, Karlsruhe).
Haensel, R., C. Kunz, T. Sasaki and B. Sonntag, 1968, Appl. Opt. **7**, 301.
Hagemann, H.J., W. Gudat and C. Kunz, 1974, DESY Report SR-74/17 (Deutsches Elektronensynchrotron, Hamburg).
Hagemann, H.J., W. Gudat and C. Kunz, 1975, J. Opt. Soc. Am. **65**, 742.
Hale, G.M., W.E. Holland and M.R. Querry, 1973, Appl. Opt. **12**, 48.
Hamm, R.M., R.A. MacRea and E.T. Arakawa, 1965, J. Opt. Soc. Am. **55**, 1460.
Hanyu, T., T. Miyahara, T. Kamada, K. Asada, H. Ohkuma, H. Ishii, K. Naito, H. Kato and S. Yamaguchi, 1984, J. Phys. Soc. Jpn. **53**, 3667.
Harbeke, G., 1972, Optical Properties of Semiconductors, in: Optical Properties of Solids, ed. F. Abelès (North-Holland, Amsterdam) pp. 21–92.
Hart, M., 1980, Nucl. Instrum. Methods **172**, 209.

Hart, M., and D.P. Siddons, 1981, Proc. R. Soc. London Ser. A **376**, 465.
Hauge, P.S., and R.H. Dill, 1973, IBM. J. Res. Dev. **17**, 472.
Heavens, O.S., 1955, Optical Properties of Thin Films (Butterworths, London).
Heidrich, K., W. Schafer, M. Schreiber, J. Sochtig, G. Trendel, J.H. Treusch, T. Grandke and H.J. Stolz, 1981, Phys. Rev. B **24**, 5642.
Henke, B.L., P. Lee, T.J. Tanaka, R.Shimabukuro and B.K. Fujikawa, 1982, At. Data & Nucl. Data Tables **27**, 1.
Henrich, V.E., G. Dresselhaus and M.J. Zeiger, 1980, Phys. Rev. B **22**, 4764.
Hinson, D.C., and J.R. Stevenson, 1967, Phys. Rev. **159**, 711.
Hoffman, D.M., G.L. Doll and P.C. Eklund, 1984, Phys. Rev. B **30**, 6051.
Hogrefe, H., M.D. Giesenberg, R.-P. Haelbig and C. Kunz, 1983, Nucl. Instrum. Methods **208**, 415.
Hubbard, J., 1955, Proc. Phys. Soc. London A **68**, 976.
Hubbell, J.R., W.H. McMaster, N. Kerr Del Grande and J.M. Mallett, 1974, in: International Tables for Crystallography, Vol. IV, eds J.A. Ibers and W.C. Hamilton (Kynoch Press, Birmingham).
Hubell, J.M., 1971, At. Data & Nucl. Data Tables **3**, 241.
Hulthen, R., 1982, J. Opt. Soc. Am. **72**, 794.
Humphreys-Owen, S.P.F., 1961, Proc. Phys. Soc. London **77**, 949.
Hunderi, O., 1973, Phys. Rev. B **7**, 3419.
Hunter, W.R., 1964, J. Opt. Soc. Am. **54**, 15.
Hunter, W.R., 1965, J. Opt. Soc. Am. **55**, 1197.
Hunter, W.R., 1967, Appl. Opt. **6**, 2140.
Hunter, W.R., 1977, Thin Solid Films **8**, 43.
Hunter, W.R., 1978, Appl. Opt. **17**, 1259.
Hunter, W.R., 1982, Appl. Opt. **21**, 2103.
Hunter, W.R., D.W. Angel and G. Hass, 1979, J. Opt. Soc. Am. **69**, 1695.
Inagaki, T., E.T. Arakawa, R.N. Hamm and M.W. Williams, 1977, Phys. Rev. B **15**, 3243.
Inagaki, T., H. Kuwata and A. Ueda, 1979, Phys. Rev. B **19**, 2400.
Jasperson, S.N., and S.E. Schnatterly, 1969, Rev. Sci. Instrum. **40**, 761.
Jezierski, K., 1984, J. Phys. C **17**, 475.
Jezierski, K., 1986, J. Phys. C **19**, 2103.
Jezierski, K., and J. Misiewicz, 1984, J. Opt. Soc. Am. B **1**, 850.
Johnson, R.L., 1984, private communication.
Kaneko, Y., K. Morimoto and T. Koda, 1981, J. Phys. Soc. Jpn. **50**, 1047.
Kaneko, Y., K. Morimoto and T. Koda, 1983, J. Phys. Soc. Jpn. **52**, 4358.
Katoh, H., H. Shimakura, T. Ogawa, S. Hattori, Y. Kobayashi, K. Umezawa, T. Ishikawa and K. Ishida, 1985, J. Phys. Soc. Jpn. **564**, 881.
Kemp, J.C., 1969, J. Opt. Soc. Am. **59**, 950.
Kendelewicz, T., 1981, J. Phys. C **14**, L407.
Kessler, F.R., 1963, in: Festkörperprobleme, Bd II, ed. F. Sauter (Vieweg, Braunschweig) p. 1.
Khumalo, F., 1981, Phys. Rev. B **24**, 4919.
Khumalo, F.S., and H.P. Hughes, 1980a, Phys. Rev. B **22**, 2078.
Khumalo, F.S., and H.P. Hughes, 1980b, Phys. Rev. B **22**, 4066.
Khumalo, F.S., and H.P. Hughes, 1981, Phys. Rev. B **23**, 5375.
Kimel, I., 1982, Phys. Rev. B **25**, 6561.
King, F.W., 1977, J. Phys. C **10**, 3199.
Kinno, S., and R. Onaka, 1980, J. Phys. Soc. Jpn. **49**, 1379.
Kinno, S., and R. Onaka, 1981, J. Phys. Soc. Jpn. **50**, 2073.
Klucker, R., and U. Nielsen, 1973, Comput. Phys. Commun. **6**, 189.
Klucker, R., M. Skibowski and W. Steinmann, 1974, Phys. Status Solidi b **65**, 703.
Koch, E.-E., A. Otto and K.L. Kliewer, 1974, Chem. Phys. **3**, 362.
Koch, E.-E., C. Kunz and B. Sonntag, 1977, Phys. Rep. C **28**, 154.
Kondo, D., and H. Matsumoto, 1982, J. Phys. Soc. Jpn. **51**, 1441.

Kopilevich, Y.I., and E.G. Makarova, 1981, Opt. Spectrosc. **51**, 591.
Kubier, B., H. Boudriot, K. Deus, R. Gründler, H. Zscheile and E. Bührig, 1981, Phys. Status Solidi b **106**, K103.
Kunz, C., 1975, Soft X-Ray Excitation of Core Electrons in Metals and Alloys, in: Optical Properties of Solids – New Developments, ed. B.O. Seraphin (North-Holland, Amsterdam) pp. 473–553.
Kunz, C., 1979, Synchrotron Radiation, in: Photoemission in Solids II. Case Studies, eds L. Ley and M. Cardona (Springer, Berlin) pp. 299–348, especially sect. 4, 6, 7, p. 322 ff.
Laporte, P., and J.L. Subtil, 1982, J. Opt. Soc. Am. **72**, 1558.
Leveque, G., 1977, J. Phys. C **10**, 4877.
Leveque, G., 1986, Phys. Rev. B **34**, 5070.
Leveque, G., C.G. Olson and D.W. Lynch, 1983, Phys. Rev. B **27**, 4654.
Leveque, G., C.G. Olson and D.W. Lynch, 1984, J. Phys. (France) **45**, 1699.
Lynch, D.W., 1979, Solid State Spectroscopy, in: Synchrotron Radiation: Techniques and Application, Topics in Current Physics, Vol. 10, ed. C. Kunz (Springer, Berlin).
Lynch, D.W., 1985, Mirror and Reflector Materials, in: Handbook of Laser Science and Technology, Vol. III, Part A, ed. M.J. Weber (Chemical Rubber Co. Press, Boca Raton) sect. 1.3.
Lynch, D.W., C.G. Olson and J.H. Weaver, 1975, Phys. Rev. B **11**, 3617.
Lynch, D.W., C.G. Olson, D.-J. Peterman and J.H. Weaver, 1980, Phys. Rev. B **22**, 3991.
MacRae, R.A., E.T. Arakawa and M.W. Williams, 1967, Phys. Rev. **162**, 615.
Makarova, E.G., and V.N. Morozov, 1976, Opt. Spectrosc. **40**, 138.
Mamy, R., and A. Couget, 1979, Solid State Commun. **32**, 1124.
Manzke, R., 1980, Phys. Status Solidi b **97**, 157.
Manzke, R., G. Crecelius, J. Fink and R. Schollhorn, 1981, Solid State Commun. **40**, 103.
Maradudin, A.A., and D.L. Mills, 1975, Phys. Rev. B **11**, 1392.
Martzen, P.D., and W.C. Walker, 1980, Phys. Rev. B **21**, 2562.
Marvin, A., F. Toigo and V. Celli, 1975, Phys. Rev. B **11**, 2777.
McIntyre, J.D.E., 1973, Specular Reflection Spectroscopy of the Electrode–Solution Interphase, in: Advances in Electrochemistry and Electrochemical Engineering, Vol. 9, eds F. Delahay and C. Tobias (Wiley–Interscience, New York) pp. 61–166.
McIntyre, J.D.E., 1976, Optical Reflection Spectroscopy of Chemisorbed Monolayers, in: Optical Properties of Solids: New Developments, ed. B.O. Seraphin (North-Holland, Amsterdam) pp. 555–630.
McIntyre, J.D.E., and D.E. Aspnes, 1971, Surf. Sci. **24**, 417.
McMaster, W.H., N. Kerr Del Grande, J.M. Mallett and J.M. Hubbell, 1969, Compilation of X-Ray Cross Sections, Rep. UCRL-50174, Sect. II, Rev. 1 (Lawrence Radiation Laboratory, Livermore, CA). See also Sect. I, issued in 1970.
Meixner, A.E., C.H. Chen. J. Geerk and P.H. Schmidt, 1982, Phys. Rev. B **25**, 5032.
Miller, R.F., A.J. Taylor and L.S. Julien, 1970, J. Phys. D **3**, 1957.
Mills, D.L., and A.A. Maradudin, 1975, Phys. Rev. B **12**, 2943.
Miloslavskii, V.K., 1961, Opt. Spectrosc. **21**, 193.
Misell, D.L., 1970, Z. Physik **235**, 353.
Mitani, T., H. Mori, S. Suga, T. Koda, S. Shin, K. Inoue, I. Nakada and H. Kanzaki, 1979, J. Phys. Soc. Jpn. **47**, 679.
Mitani, T., K. Koyama, H. Mori, T. Koda and I. Nakada, 1982, J. Phys. Soc. Jpn. **51**, 3197.
Miyahara, T., S. Sato, T. Hanyu, A. Kakizaki, S. Yamaguchi and T. Ishii, 1980, J. Phys. Soc. Jpn. **49**, 194.
Miyahara, T., H. Ishii, T. Hanyu, H. Ohkuma and S. Yamaguchi, 1982, J. Phys. Soc. Jpn. **51**, 1834.
Mollenauer, L.F., D. Downie, H. Engstrom and M.B. Grant, 1969, Appl. Opt. **8**, 661.
Mosteller, L.P., and F. Wooten, 1968, J. Opt. Soc. Am. **58**, 511.
Nagel, S.R., and S.E. Schnatterly, 1974, Phys. Rev. B **9**, 1229.
Nagel, S.R., and S.E. Schnatterly, 1975, Phys. Rev. B **12**, 6002.
Nestell, J.E., and R.W. Christy, 1972, Appl. Opt. **11**, 643.
Nestell, J.E., R.W. Christy, M.H. Cohen and G.C. Ruben, 1980, J. Appl. Phys. **51**, 655.
Neufeld, J.D., and G. Andermann, 1972, J. Opt. Soc. Am. **62**, 1156.
Nicklaus, E., 1979, Phys. Status Solidi a **53**, 217.

Nilsson, P.O., 1968, Appl. Opt. **7**, 435.
Nilsson, P.O., 1974, Solid State Physics **29**, 139.
Nilsson, P.O., and L. Munkby, 1969, Phys. Kondens. Mater. **6**, 187.
Nosenzo, L., and E. Reguzzoni, 1979, Phys. Rev. B **19**, 2314.
Nye, J.F., 1957, Physical Properties of Crystals (Oxford Univ. Press, London).
O'Handley, R.C., 1973, J. Opt. Soc. Am. **63**, 523.
Olson, C.G., 1975, unpublished.
Olson, C.G., and D.W. Lynch, 1979, Solid State Commun. **31**, 601.
Olson, C.G., and D.W. Lynch, 1980, Solid State Comm. **33**, 849.
Olson, C.G., D.W. Lynch and R. Rosei, 1980, Phys. Rev. B **22**, 543.
Onuki, H., and J.C. Rife, 1982, Phys. Rev. B **26**, 654.
Osmun, J.W., 1972, unpublished.
Otto, A., 1976, Spectroscopy of Surface Polaritons by Attenuated Total Reflection, in: Optical Properties of Solids – New Developments, ed. B.O. Seraphin (North-Holland, Amsterdam) pp. 677–729.
Pajasova, L., P. Pahas, O.A. Makarov and V.M. Zakharov, 1984, Phys. Status Solidi b **121**, 293.
Palik, E.D., ed, 1985, Handbook of Optical Constants of Materials (Academic Press, Orlando).
Palmer, K.F., and M.Z. Williams, 1985, Appl. Opt. **24**, 1788.
Paulick, T.C., 1986, Appl. Opt. **25**, 1562.
Pavinich, V.F., and M.V. Belousov, 1978, Opt. Spectrosc. **45**, 881.
Pepper, S.V., 1970, J. Opt. Soc. Am. **60**, 805.
Peterson, C.M., and B.W. Knight, 1973, J. Opt. Soc. Am. **63**, 1238.
Pflüger, J., J. Fink, W. Weber, K.-P. Bohnen, G. Crecelius, 1984, Phys. Rev. B **30**, 1155.
Pflüger, J., J. Fink, W. Weber, K.-P. Bohnen and G. Crecelius, 1985, Phys. Rev. B **31**, 1244.
Pollini, I., J. Thomas, G. Jezequel, J.C. Lemonnier and R. Mamy, 1983, Phys. Rev. B **27**, 1303.
Pollini, I., G. Benedek and J. Thomas, 1984a, Phys. Rev. B **29**, 3617.
Pollini, I., J. Thomas and A. Lenselink, 1984b, Phys. Rev. B **30**, 2140.
Pollini, I., J. Thomas, G. Jezequel, J.C. Lemonnier and A. Lenselink, 1984c, Phys. Rev. B **29**, 4716.
Porteus, L.O., 1963, J. Opt. Soc. Am. **53**, 1394.
Pratt, R.M., A. Ron and M.K. Tseng, 1973, Rev. Mod. Phys. **45**, 273.
Preil, M.E., L.A. Grunes, J.J. Ritsko and J.E. Fisher, 1984, Phys. Rev. B **30**, 5852.
Priol, M., and S. Robin, 1974, Optical Properties of Metals, in: Some Aspects of Vacuum Ultraviolet Radiation Physics, eds J.H. Romand and B. Vodar (Pergamon Press, Oxford).
Quemerais, A., B. Loisel, G. Jezequel, J. Thomas and J.C Lemonnier, 1981, J. Phys. F **11**, 293.
Querry, M.R., and W.E. Holland, 1973, Appl. Opt. **13**, 595.
Raether, H., 1965, Springer Tracts Mod. Phys. **38**, 84.
Raether, H., 1980, Springer Tracts Mod. Phys. **88**, 1.
Raether, H., 1982, Surface Plasmons and Roughness, in: Surface Polaritons, eds V.M. Agranovich and D.L. Mills (North-Holland, Amsterdam) pp. 331–403.
Rehn, V., 1981, in: Low Energy X-Ray Diagnostics–1981, eds D.T. Attwood and B.L. Menke (American Institute of Physics, New York) p. 162.
Rehn, V., V.O. Jones, J.M. Elson and J.M. Bennett, 1980, Nucl. Instrum. Methods **172**, 307.
Rimmer, M.P., and D.L. Dexter, 1960, J. Appl. Phys. **31**, 775.
Ritchie, R.M., 1957, Phys. Rev. **106**, 874.
Ritsko, J.J., 1982a, Phys. Rev. B **26**, 2192.
Ritsko, J.J., 1982b, Phys. Rev. B **25**, 6452.
Ritsko, J.J., and E.J. Mele, 1980, Phys. Rev. B **21**, 730.
Ritsko, J.J., E.J. Mele and I.P. Gates, 1981, B **24**, 6114.
Rivory, J., 1970, Opt. Commun. **1**, 334.
Roessler, D.M., 1965a, Br. J. Appl. Phys. **16**, 1119.
Roessler, D.M., 1965b, Br. J. Appl. Phys. **16**, 1359.
Roessler, D.M., 1966, Br. J. Appl. Phys. **17**, 1313.
Rossinelli, M., and M.A. Bosch, 1982, Phys. Rev. B **25**, 6482.
Sato, S., T. Miyahara, T. Hanyu, S. Yamaguchi and T. Ishii, 1979, J. Phys. Soc. Jpn. **47**, 836.

Schledermann, M., and M. Skibowski, 1971, Appl. Opt. **10**, 321.
Schlegel, A., S.F. Alvarado and P. Wachter, 1979, J. Phys. C **12**, 1157.
Schnatterly, S.E., 1979, Solid State Phys. **34**, 275.
Schoenes, J., 1984, in: Moment Formation in Solids, ed. W.J.L. Buyers (Plenum Press, New York) p. 237.
Scouler, W.J., 1969, Phys. Rev. **178**, 1353.
Shiles, E., T. Sasaki, M. Inokuti and D.Y. Smith, 1980, Phys. Rev. B **22**, 1612.
Shin, S., K. Inoue, I. Nakada and H. Kanzaki, 1979, J. Phys. Soc. Jpn. **47**, 679.
Siddons, D.P., 1981, in: Low Energy X-Ray Diagnostics – 1981, eds D.T. Attwood and B.L. Menke (American Institute of Physics, New York) p. 236.
Smith, C.S., 1958, Solid State Phys. **6**, 175.
Smith, D.Y., 1977, J. Opt. Soc. Am. **67**, 570.
Smith, D.Y., 1985a, Dispersion Theory, Sum Rules and their Application to the Analysis of Optical Data, in: Handbook of Optical Constants of Solids, ed. E.D. Palik (Academic, Orlando).
Smith, D.Y., 1985b, private communication.
Smith, D.Y., and C.A. Manogue, 1981, J. Opt. Soc. Am. **71**, 935.
Smith, D.Y., and E. Shiles, 1978, Phys. Rev. B **17**, 4689.
Sobolev, V.V., O.G. Maksimova and S.G. Kroitoru, 1981, Phys. Status Solidi b **103**, 499.
Sokolov, A.V., 1967, Optical Properties of Metals (Blackie, London).
Spiller, E., 1983, Soft X-Ray Optics and Microscopy, in: Handbook on Synchrotron Radiation, Vol. 1, ed. E.-E. Koch (North-Holland, Amsterdam) pp. 1091–1129.
Sprüssel, G., M. Skibowski and V. Saile, 1979, Solid State Commun. **32**, 1091.
Stephan, G., and S. Robin, 1974, Optical Properties of Ionic Insulators, in: Some Aspects of Vacuum Ultraviolet Radiation Physics, eds J. Romand and B. Vodar (Pergamon Press, Oxford).
Stern, F., 1963, Solid State Phys. **15**, 299.
Storm, E., and H.I. Israel, 1970, Nucl. Data Tables A **7**, 565.
Suga, S., S. Shin, M. Taniguchi, K. Inoue, M. Seki, I. Nakada, S. Shibuya and T. Yamaguchi, 1982, Phys. Rev. B **25**, 5486.
Suga, S., K. Inoue, M. Tanuguchi, S. Shin, S. Seki, K. Sato and T. Teranishi, 1983, J. Phys. Soc. Jpn. **52**, 1848.
Tatchyn, R., I. Lindau and E. Källne, 1980a, Nucl. Instrum. Methods **172**, 315.
Tatchyn, R., I. Lindau, E. Källne, M. Hecht, E. Spiller, R. Bartlett, J. Källne, J.M. Dijkstra, A. Mawryluk and R.Z. Bachrach, 1980b, Nucl. Instrum. Methods **195**, 423.
Tatchyn, R., I. Lindau, E. Källne and E. Spiller, 1984, Phys. Rev. Lett. **53**, 1264.
Tews, W., and R. Gründler, 1982, Phys. Status Solidi b **109**, 255.
Toigo, F., A. Marvin, V. Celli and N.R. Hill, 1977, Phys. Rev. B **15**, 5618.
Tokura, Y., Y. Oowaki, Y. Kaneko, T. Koda and T. Mitani, 1984, J. Phys. Soc. Jpn. **53**, 4054.
Tosatti, E., 1969a, Il Nuovo Cimento B **63**, 54.
Tosatti, E., 1969b, Il Nuovo Cimento B **65**, 280.
Trepakov, V.A., L.N. Amitin, A.I. Kuznetsov and A.L. Kholkin, 1981, Phys. Status Solidi a **69**, 425.
Treu, J.I., A.B. Callender and S.E. Schnatterly, 1973, Rev. Sci. Instrum. **44**, 793.
Trykozko, R., and D.R. Huffman, 1981, J. Appl. Phys. **52**, 5283.
Tulvinskii, V.B., and N.I. Terentex, 1969, Opt. and Speckt. **36**, 484.
Uzan, E., 1968, Opt. Acta **15**, 237.
Valiukonis, G., F.M. Gashimzade, D.A. Guseinova, G. Krivaite, A.M. Kulibekov, G.S. Orudzhev and A. Sileika, 1983, Phys. Status Solidi b **117**, 81.
Veal, B.W., and A.P. Paulikas, 1974, Phys. Rev. B **10**, 1280.
Veigele, W.J., 1973, At. Data & Nucl. Data Tables **5**, 51.
Velicky, B., 1961, Czech. J. Phys. B **11**, 787.
Venghaus, M., 1970, Z. Physik **239**, 289.
Verleur, H.M., 1968, J. Opt. Soc. Am. **58**, 1356.
Vinogradov, A.S., E.O. Filatova and T.M. Zimkina, 1983, Fiz. Tverd. Tela **25**, 1120 [Sov. Phys.-Solid State **25**, 643].

Vishnevskii, V.N., I.P. Pashuk, M.S. Pidzyrailo and M.V. Tokarivskii, 1980, Fiz. Tverd. Tela **22**, 3159 [Sov. Phys.-Solid State **22**, 1847].
Vorburger, T.V., and K.C. Ludema, 1980, Appl. Opt. **19**, 561.
Wall, W.E., M.W. Ribarsky and J.R. Stevenson, 1980, J. Appl. Phys. **51**, 661.
Weaver, J.H., D.W. Lynch and C.G. Olson, 1974, Phys. Rev. B **10**, 501.
Weaver, J.H., C. Krafka, D.W. Lynch and E.-E. Koch, 1981, Optical Properties of Metals, Parts I, II, Vols. 18-1, 18-2 (Fachinformationszentrum Energie, Physik, Mathematik, Karlsruhe).
Wehenkel, C., 1975, J. Phys. (France) **36**, 199.
White, J.J., and J.W. Straley, 1968, J. Opt. Soc. Am. **58**, 759.
Wolter, H., 1956, Optik Dünner Schichten, in: Handbuch der Physik, Vol. 24, ed. S. Flügge (Springer-Verlag, Berlin) pp. 461–645.
Wooten, F., 1972, Optical Properties of Solids (Academic Press, New York).
Wooten, F., 1984, Appl. Opt. **23**, 4226.
Wu, C.-K., and G. Andermann, 1968, J. Opt. Soc. Am. **58**, 519.
Yeh, P., 1979, J. Opt. Soc. Am. **69**, 742.
Yeh, P., 1980, Surf. Sci. **96**, 41.
Young, R.H., 1977, J. Opt. Soc. **67**, 520.
Zeppenfeld, K., 1971, Z. Phys. **243**, 229.
Zhuze, V.P., A.A. Kamarzin, M.G. Karin, K.K. Sidorin and A.L. Shelvkh, 1979, Fiz. Tverd. Tela **21**, 3410 [Sov. Phys. Solid state **21**, 1968].
Zhuze, V.P., M.G. Karin, D.P. Lukirskii, V.M. Sergeeva and A.L. Shelykh, 1980, Fiz. Tverd. Tela **22**, 2669 [Sov. Phys.-Solid State **22**, 1558].
Zhuze, V.P., A.V. Golubkov, M.G. Karin, K.K. Sidorin and A.I. Shelykh, 1982, Fiz. Tverd. Tela **24**, 1017 [Sov. Phys.-Solid State **24**, 577].

AUTHOR INDEX

Abadli, L., see Krill, G. 635
Abbati, I. 476, 477, 519, 554, 556
Abbati, I., see Braicovich, L. 550
Abbati, I., see Rossi, G. 519, 554–557
Abelès, F. 45, 276, 725, 750, 770
Åberg, T. 246
Abramov, V.N. 772, 774
Abramowitz, M. 30
Achiba, Y. 332
Achiba, Y., see Katsumata, S. 245, 264
Achiba, Y., see Kimura, K. 320, 334, 396
Achiba, Y., see Sato, K. 332
Adam, M.Y. 295, 296, 402, 424–426
Adam, M.Y., see Hubin-Franskin, M.-J. 244, 264, 331
Adam, M.Y., see Lablanquie, P. 407
Adam, M.Y., see Morin, P. 244, 263, 264, 275, 292, 331, 332, 340, 370–372, 398, 399, 402, 418–420
Adam, M.Y., see Roy, P. 329, 341, 371, 413, 414, 422–424
Adam, M.Y., see Swensson, S. 402
Ågren, H. 305, 309, 341
Agron, P., see Grimm, F.A. 329, 339, 424, 426
Ahituv, N. 164
Ahrenkiel, R.F. 752
Ainsworth, S., see Jones, R.G. 503
Ajello, J.M. 334
Ajello, J.M., see Chutjian, A. 334
Akahama, Y., see Taniguchi, M. 656
Akahori, T. 403
Aksay, I.A., see Liang, K.S. 572
Albert, M.R. 520
Albrand, G., see Borgogno, J.P. 738, 762
Alcock, C.B., see Kubaschewski, O. 561, 577, 578
Alexa, B., see Garton, W.R.S. 230
Alexander, J.R. 144
Alexander, R.W., see Bell, R.J. 727
Allen, J.D., see Carlson, T.A. 280, 284, 286, 287, 329, 338, 339, 422, 424
Allen, J.D., see Grimm, F.A. 339, 424
Allen, J.W. 656
Allen, J.W., see Johansson, L.I. 482
Allen, P.M.G., see Mackey, K.J. 531, 532

Allen, R.E. 590
Allen, R.E., see Dow, J.D. 590
Allison, S.K., see Compton, A.M. 761
Allyn, C.L. 516
Almbladh, C.-O. 386, 614
Altarelli, M. 753
Altarelli, M., see Bassani, F. 725–728, 751, 770
Alvarado, S.F. 676, 686
Alvarado, S.F., see Eib, W. 651
Alvarado, S.F., see Schlegel, A. 773
Alvarado, S.J., see Weller, D. 482
Ambrazevivius, G. 772
Amitin, L.N. 773
Amitin, L.N., see Trepakov, V.A. 774
Andermann, G., see Neufeld, J.D. 753
Andermann, G., see Wu, C.-K. 752
Anderson, A., see Beebe, N. 403, 404
Anderson, C.P., see Carlson, T.A. 264
Anderson, J., see Estrup, P.J. 693
Anderson, J., see Lapeyre, G.J. 599
Anderson, J., see Smith, R.J. 504
Anderson, P.W. 614, 621, 626
Anderson, R.L. 532
Anderson, S.L. 332
Anderson, S.L., see Carroll, T. 450
André, J.M. 761
Andrew, K.L., see Griffin, D.C. 201
Andrews, J.M. 546
Angel, D.W., see Hunter, W.R. 773
Anistratov, A.T., see Amitin, L.N. 773
Anokhin, B.S., see Smirnov, L.A. 52
Antoncik, E. 741
Aono, M., see Chiang, T.-C. 478, 479
Apai, G. 498, 499, 675, 694, 702, 703
Apai, G., see Baetzold, R.C. 675
Apai, G., see Lee, S.T. 499, 572
Apai, G., see Stöhr, J. 472
Arakawa, E.T. 767–769, 773, 775
Arakawa, E.T., see Hamm, R.N. 45, 757
Arakawa, E.T., see Horton, V.G. 278
Arakawa, E.T., see Inagaki, T. 772
Arakawa, E.T., see MacRae, R.A. 752
Archirel, P., see Millié, Ph. 436, 438
Arfken, G. 30
Arlinghaus, F.J., see Smith, J.R. 685

Armaly, B.F. 746
Arndt, D.P. 762, 763
Arneberg, R., see Ågren, H. 341
Aronson, J.R., see Emslie, A.G. 749
Asada, K., see Hanyu, T. 775
Asada, S. 644, 647–649
Asaro, C., see Gerwer, A. 338
Asaro, C., see Langhoff, P.W. 338, 410
Åsbrink, L. 320
Ashfold, M.N.R. 220
Ashley, J.C., see Arakawa, E.T. 773, 775
Aspnes, D.E. 567, 738, 741, 756, 763, 764
Aspnes, D.E., see Eberhardt, W. 552, 586, 588
Aspnes, D.E., see McIntyre, J.D.E. 730
Atabek, O. 244, 264, 268
Atabek, O., see Jungen, Ch. 244, 264, 268
Atabek, O., see Miret-Artes, S. 406
Avan, P., see Cohen-Tannoudji, C. 193
Avdeev, V.I., see Kondratenko, A.V. 337
Avouris, Ph., see Jugnet, Y. 506
Aymar, M., see Robaux, O. 185
Azaroff, L.V. 253
Azria, R., see Tronc, M. 442
Azzam, R.M.A. 756
Azzam, R.M.A., see Arndt, D.P. 762, 763

Babalola, I.A. 562
Babalola, I.A., see Kendelewicz, T. 559–562
Babalola, I.A., see Petro, W.G. 561, 562, 584
Baberschke, K., see Döbler, U. 514
Baberschke, K., see Stöhr, J. 505, 516, 517
Babonas, G., see Ambrazevivius, G. 772
Bachrach, R.Z. 543, 546, 549, 557–559, 562
Bachrach, R.Z., see Bauer, R.S. 519, 549, 567, 568, 578
Bachrach, R.Z., see Bianconi, A. 338, 439
Bachrach, R.Z., see Brillson, L.J. 557, 559, 563, 567, 570, 581
Bachrach, R.Z., see Brown, F.C. 338, 439, 548
Bachrach, R.Z., see Cerino, J. 94, 137
Bachrach, R.Z., see Flodström, S.A. 567
Bachrach, R.Z., see Hansson, G.V. 525, 596–598
Bachrach, R.Z., see Tatchyn, R. 750
Bachrach, R.Z., see Uhrberg, R.I.G. 522, 523
Bacon, D.D., see Aspnes, D.E. 738, 763
Baer, A.D., see Lapeyre, G.J. 599
Baer, A.D., see Rehn, V. 106
Baer, T. 244, 264, 333, 334, 375, 377, 379, 398, 400, 402
Baer, T., see Guyon, P.M. 333, 334, 379, 382, 402, 428–430

Baer, T., see Miller, J.C. 332
Baer, T., see Mintz, D.M. 333, 334
Baer, T., see Nenner, I. 379, 402, 405, 406
Baer, T., see Werner, A.S. 333, 334
Baer, Y. 650
Baer, Y., see Campagna, M. 483
Baer, Y., see Citrin, P.H. 499, 626, 665, 674, 675
Baer, Y., see Siegbahn, K. 401, 443
Baer, Y., see Wuilloud, E. 638
Baetzold, R.C. 571, 675
Baetzold, R.C., see Apai, G. 675, 694, 702, 703
Bagdasarov, K.S. 775
Bagus, P.S. 251, 263, 332, 400, 506, 649
Bahr, C.C., see Barton, J.J. 510, 598
Bahr, J.L. 244, 264
Bahr, J.L., see Blake, A.J. 244, 264
Baig, M.A. 181, 182, 186, 189, 191, 194, 201, 202, 204, 205, 214, 219, 226, 228, 230, 364, 428
Baig, M.A., see Connerade, J.P. 181, 193, 194, 209, 220, 226, 235
Baig, M.A., see Dagata, J.A. 227
Baig, M.A., see Garton, W.R.S. 230
Baig, M.A., see Mayhew, C. 226
Baig, M.A., see Schäfers, F. 236, 333, 428
Baig, M.A., see Sommer, K. 200
Bailey, J.M., see Ashfold, M.N.R. 220
Bailey, S.M., see Wagman, D.D. 561, 577, 578
Bailey, W.P., see Franks, A. 106
Baker, A. 396, 666
Baker, C. 298, 299
Balerna, A. 498
Ballard, R. 396
Ballu, Y., see Guillot, C. 652–654
Band, I.M. 211, 752, 771
Band, Y., see Morse, M. 396
Bänninger, U. 651
Barbara, B., see Kaindl, G. 483, 484, 675, 676, 707
Barbee, T.W. 762
Barchewitz, R., see André, J.M. 761
Bardeen, J. 545
Bardsley, J.N. 245, 246, 252, 261, 262, 285
Barinskii, R.L. 336
Barinskii, R.L., see Fomichev, V.A. 319, 324, 328, 336
Barker, J.A. 695
Barker, R.A., see Felter, T.E. 496
Barrus, D.M. 338
Barth, J. 484, 654
Barth, J., see Gerken, F. 675, 676, 685–688

Barth, J., see Johansson, L.I. 707
Barth, J., see Kammerer, R. 674, 675, 688–690
Barth, J., see Lenth, W. 554
Barthès-Labrousse, M.G., see Binns, C. 482, 521, 522
Bartlett, R.J. 49, 774
Bartlett, R.J., see Tatchyn, R. 750
Barton, J.J. 510, 598, 693
Barton, J.J., see Swartz, C.A. 558
Bartynski, R.A. 486, 488
Basch, H. 571
Basco, N. 394
Bashara, N.M., see Azzam, R.M.A. 756
Bassani, F. 725–728, 751, 770
Bassani, F., see Greenaway, D.L. 748
Batlogg, B. 773
Batson, P.E. 774
Batten, C.F. 320, 333, 334
Batterman, B.W., see Durbin, S.M. 599
Baudoing, R. 515
Bauer, E. 700
Bauer, E., see Engel, T. 700
Bauer, R.S. 519, 543, 546, 549, 567, 568, 578, 754
Bauer, R.S., see Bachrach, R.Z. 549, 557–559
Bauer, R.S., see Brillson, L.J. 557, 559, 563, 567, 570, 581
Bauer, R.S., see Chiaradia, P. 591
Bauer, R.S., see Flodström, S.A. 567
Bauer, R.S., see Hansson, G.V. 525, 596–598
Bauer, R.S., see Katnani, A.D. 530, 591
Bauliss, S.C. 775
Baumann, R. 153
Baumgärtel, H., see Beckmann, O. 368
Baumgärtel, H., see Oertel, H. 368, 457
Bauschlicher, C.W., see Bagus, P.S. 506
Beal, A.R. 773
Beaulieu, S., see Hitchcock, A.P. 341
Beaurepaire, E. 638, 639
Beck, W.A., see Davis, G.D. 565, 581
Becker, K., see Kakehashi, Y. 644
Becker, U.E. 450
Becker, U.E., see Ferrett, T.A. 306, 331, 342, 444–447, 455
Becker, U.E., see Lindle, D.W. 331, 341, 446
Becker, U.E., see Reimer, A. 455
Becker, U.E., see Truesdale, C.M. 331, 340, 342, 443, 445–447
Beckermann, H., see Wehking, F. 525
Beckmann, O. 368
Beckmann, P. 50
Bedford, K.L. 774

Bedwell, M.O. 131
Bedzyk, M. 503, 504
Bedzyk, M., see Materlik, G. 503
Beebe, N. 403, 404
Beland, M., see Moeller, T. 455
Bell, R.J. 727
Belousov, M.V. 748, 752
Belousov, M.V., see Pavinich, V.F. 748
Benbow, R.L., see Lee, S.T. 499
Benbow, R.L., see Thuler, M.R. 656
Bender, C.F., see Diercksen, G. 384
Bender, C.F., see Rescigno, T.N. 338
Benedek, G., see Pollini, I. 775
Benedict, M.G. 305
Benhow, R., see Lee, S.T. 572
Bennemann, K.H., see Erbudak, M. 674, 675
Bennemann, K.H., see Kumar, V. 680, 681, 704
Bennemann, K.H., see Tomanek, D. 685
Bennett, A.J. 546
Bennett, H.E. 50, 739
Bennett, H.E., see Burge, D.K. 740, 757
Bennett, J.M. 83
Bennett, J.M., see Arndt, D.P. 762, 763
Bennett, J.M., see Elson, J.M. 740
Bennett, J.M., see Franks, A. 83, 105
Bennett, J.M., see Rehn, V. 105, 740
Bennett, P.A., see Hillebrecht, F.U. 635, 643
Bennett, P.A., see Robinson, I.K. 495
Bennett, P.A., see Wertheim, G.K. 500
Berggren, K.-F., see Johansson, L.I. 482
Bergman, R., see Nilsson, A. 675, 676, 685, 707
Bergmark, T., see Siegbahn, K. 401, 443, 613, 665, 666
Berkowitz, J. 244, 263, 264, 298, 396, 407, 408
Berkowitz, J., see Chupka, W.A. 263
Berkowitz, J., see Dehmer, J.L. 98, 279
Berkowitz, J., see Ruščić, B. 202
Berkowitz, J., see Spohr, R. 377
Berman, L.E., see Durbin, S.M. 599
Berman, M. 748
Bermudez, V.M. 757
Berndt, W., see Welton-Cook, M.R. 498
Bernieri, E., see Balerna, A. 498
Berremann, D.W. 740, 751, 757
Berry, R.S. 264, 274, 291
Berry, R.S., see Duzy, C. 264
Bersohn, R., see Shapiro, M. 383, 395
Bertel, E. 445
Bertolini, J.C. 675
Bertrand, Y. 773
Besnard, M. 381, 382, 434

Best, J.S. 590
Beswick, J.A. 369, 390, 405
Beswick, J.A., see Miret-Artes, S. 406
Beswick, J.A., see Nenner, I. 292
Betteridge, D., see Baker, A. 396
Beutler, H.G. 64, 204
Beyss, M., see Lässer, R. 635
Bianconi, A. 338, 438, 439, 443
Bianconi, A., see Bauer, R.S. 519
Bianconi, A., see Brown, F.C. 338, 439
Bieri, G., see Åsbrink, L. 320
Bieri, G., see Haller, E. 320, 321, 329
Bilderback, D.H. 48, 51, 52, 761
Binnig, G. 495
Binns, C. 482, 521, 522
Birgeneau, R.J., see Dimon, P. 513
Birken, H.-G. 755, 767
Birtwistle, D.T. 285
Bisi, O., see Abbati, I. 554
Bisi, O., see Calandra, C. 557
Bisi, O., see Franciosi, A. 557
Bisi, O., see Perfetti, P. 557
Bizau, J.M., see Krummacher, S. 263, 332, 340, 402
Black, G. 456
Black, G., see Slanger, T. 382
Blake, A.J. 244, 264
Blake, A.J., see Bahr, J.L. 244, 264
Blake, R.L., see Barrus, D.M. 338
Blechschmidt, D. 246–248, 305, 331, 336
Block, J.M., see Fisher, J.E. 775
Blyholder, G. 506, 519
Bobrikov, V.N., see Farberovich, O.V. 774
Bodeur, S. 439–443
Bodeur, S., see Nenner, I. 442, 455
Bogen, A., see Schnell, R.D. 503, 674
Böhmer, W., see Chergui, M. 418, 547
Böhmer, W., see Wilcke, H. 365, 367
Bohn, G.K., see Palmberg, P.W. 548
Bohnen, K.-P., see Pflüger, J. 775
Bohr, J. 496, 497
Boller, K. 108
Bonapace, C.R. 573, 578
Bondybey, V.E., see Katayama, D.H. 181
Boness, M.J.W., see Pavlovic, Z. 262
Bonham, R.A., see Kennerly, R.E. 261, 306
Bonifield, T., see Black, G. 456
Bonse, U. 761
Bonzel, H.P. 702
Borgogno, J.P. 738, 762
Borgogno, J.P., see Arndt, D.P. 762, 763
Boring, A.M., see Koelling, D.D. 638
Born, M. 728, 730
Borne, T., see Peatman, W. 377

Borondo, F. 390
Borrell, P. 385
Borstel, G., see Eyers, A. 479
Bosch, M.A., see Rossinelli, M. 774
Botter, R. 298
Botter, R., see Baer, T. 244, 264, 333, 334, 375, 398, 400, 402
Botter, R., see Guyon, P.M. 333, 334, 379, 402
Boudriot, H. 773
Boudriot, H., see Kubier, B. 774
Bowlden, H.J. 753
Bradford, R.C., see King, G. 441
Bradford, R.C., see Tronc, M. 337
Bradshaw, A. 402
Bradshaw, A., see Hezaveh, A.A. 482
Bradshaw, A., see Reimer, A. 455
Bradshaw, A.M., see Dietz, E. 78
Bradshaw, A.M., see Langhoff, P.W. 297, 298, 303, 339
Bradshaw, A.M., see McConville, C.F. 511
Bradshaw, A.M., see Unwin, R. 264, 297, 298, 303, 332, 340
Bradshaw, A.M., see van der Veen, J.F. 675, 694
Braicovich, L. 550, 557
Braicovich, L., see Abbati, I. 476, 477, 519, 554, 556
Braicovich, L., see Rossi, G. 519, 554, 555, 557
Braun, F. 544
Braun, W., see Beckmann, O. 368
Braun, W., see Dietz, E. 78
Braun, W., see Reimer, A. 455
Bréchignac, C. 456
Brehm, B. 333, 334
Brennan, S. 474, 475, 513, 674, 690, 691
Brennan, S., see Norman, D. 472
Brennan, S., see Stöhr, J. 472, 505, 516, 517
Breton, J. 364, 385
Breton, J., see Glass-Maujean, M. 385, 387–389
Breton, J., see Guyon, P.M. 385
Brewer, W.D., see Southworth, S.H. 274, 275, 332, 340
Brewer, W.D., see Trevor, D. 456
Brillson, L.J. 489, 522, 527, 543, 545–547, 549–551, 557, 559–563, 567, 569, 570, 574, 575, 577–582, 584, 585, 589, 591–593
Brillson, L.J., see Bauer, R.S. 567, 568
Brillson, L.J., see Brucker, C.F. 562, 567, 580, 587, 588, 591
Brillson, L.J., see Duke, C.B. 558, 559
Brillson, L.J., see Kahn, A. 558, 559

Brillson, L.J., see Richter, H.W. 560, 561
Brillson, L.J., see Shapira, Y. 559
Brillson, L.J., see Stoffel, N.G. 562, 574, 599
Bringans, R.D., see Uhrberg, R.I.G. 522, 523
Brion, C.E. 332, 338, 339, 341, 401, 407, 414, 438, 441
Brion, C.E., see Carnovale, F. 339, 434, 437
Brion, C.E., see Daviel, S. 365
Brion, C.E., see Hamnett, A. 337
Brion, C.E., see Hitchcock, A.P. 251, 262, 306, 337–340, 439, 441–443, 450, 455
Brion, C.E., see Sodhi, R.N.S. 342, 364, 365, 441, 447, 449
Brion, C.E., see Tam, W.-C. 337
Brion, C.E., see van der Wiel, M.J. 439
Brion, C.E., see White, M.G. 340
Brion, C.E., see Wight, G.R. 251, 262, 336, 337, 439, 440, 442
Brodmann, R.R. 365
Brookes, N.B., see Owen, I.W. 473, 506, 507
Brooks, R., see Jones, R.G. 503
Brooks, R., see Lamble, G.M. 513
Brookshier, W.K. 125
Brown, C.M., see Ginter, M.L. 177, 214
Brown, F.C. 338, 438, 439, 548, 726, 770
Brown, F.C., see Bianconi, A. 338, 439
Brown, F.C., see Hayes, W. 319, 324, 328, 336
Brown, G. 116
Broyer, M., see Bréchignac, C. 456
Brucker, C.F. 562, 567, 580, 587, 588, 591
Brucker, C.F., see Brillson, L.J. 546, 547, 559, 561–563, 579, 580, 591, 592
Brucker, C.F., see Stoffel, N.G. 562, 574, 599
Bruhn, R. 773
Brundle, C.R. 425
Brundle, C.R., see Baker, A.D. 666
Brundle, C.R., see Grunze, M. 698, 699
Buchanan, D.N.E., see Crecelius, G. 629
Buchanan, D.N.E., see Wertheim, G.K. 500
Budde, W. 756
Buenker, R., see Marian, C. 403, 404
Buenker, R., see Schwarz, W. 442
Buff, R., see Butler, J.J. 134
Bührig, E., see Kubier, B. 774
Buras, B. 5
Burattini, E., see Balerna, A. 498
Burek, A.J., see Barrus, D.M. 338
Burge, D.K. 740, 757
Burkstrand, J.M., see Herbst, J.F. 639
Burroughs, P. 638
Busch, G. 394
Busch, G., see Bänninger, U. 651
Butler, J.J. 134

Butscher, W., see Friedrich, H. 439
Byer, N.E., see Davis, G.D. 564, 565, 581

Cadioli, B. 245, 336
Cahan, B.D. 756
Cahuzac, Ph., see Bréchignac, C. 456
Calandra, C. 557
Calandra, C., see Abbati, I. 554
Calandra, C., see Franciosi, A. 557
Calandra, C., see Perfetti, P. 557
Callender, A.B., see Treu, J.I. 757
Calliari, L., see Abbati, I. 519
Calliari, L., see Rossi, G. 519
Camilloni, R., see Giardini-Guidoni, A. 306
Campagna, M. 483
Campagna, M., see Alvarado, S.F. 676, 686
Campagna, M., see Bänninger, U. 651
Campagna, M., see Fuggle, J.C. 639
Campagna, M., see Kisker, E. 481
Campagna, M., see Lässer, R. 635
Campagna, M., see Weller, D. 482
Campagna, M., see Wertheim, G.K. 629
Campuzano, J.-C. 496
Campuzano, J.-C., see Jones, R.G. 503
Canfield, L.R. 108
Canfield, L.R., see Jacobus, G.F. 108
Canfield, L.R., see Rabinovitch, K. 45
Capasso, C., see Perfetti, P. 529, 530
Caprace, G. 245, 264
Caprace, G., see Natalis, P. 245, 264
Caprile, C., see Chang, S. 519
Caprile, C., see Franciosi, A. 519, 520
Carbone, C., see Abbati, I. 476, 477
Cardona, M. 614
Cardona, M., see Eberhardt, W. 552, 586, 588
Cardona, M., see Ley, L. 614, 666
Carelli, J., see Duke, C.B. 558, 559
Carelli, J., see Kahn, A. 558, 559
Carlson, R., see Lee, L.C. 382, 407
Carlson, T.A. 244, 245, 264, 280, 284, 286, 287, 329, 331, 338–341, 396, 402, 407, 422–425, 443, 450
Carlson, T.A., see Carver, J.C. 649
Carlson, T.A., see Grimm, F.A. 329, 339, 340, 424, 426
Carlson, T.A., see Keller, P.R. 264, 297, 298, 303, 305, 332, 340, 341
Carlson, T.A., see Krause, M.O. 244, 275, 371
Carlson, T.A., see McKoy, V. 341
Carlson, T.A., see Piancastelli, M.N. 341
Carney, T.E., see Miller, J.C. 332
Carniglia, C.K., see Arndt, D.P. 762, 763

Carnovale, F. 339, 434, 437
Carr, R. 246, 254
Carr, R., see Eberhardt, W. 340, 450
Carrington, T. 383
Carroll, T. 450
Carver, J.C. 649
Carver, J.H., see Bahr, J.L. 244, 264
Carver, J.H., see Blake, A.J. 244, 264
Case, W.E., see Arndt, D.P. 762, 763
Cassidy, R., see Ding, A. 456
Castex, M.C. 455, 456
Castex, M.C., see Dutuit, O. 455
Castex, M.C., see Jordan, B. 455
Castex, M.C., see Le Calvé, J. 455
Castex, M.C., see Moeller, T. 418
Cauletti, C., see Adam, M.Y. 402, 424, 425
Cavell, R.G., see Sodhi, R.N.S. 342
Cederbaum, L.S. 263, 332, 400
Cederbaum, L.S., see Bradshaw, A. 402
Cederbaum, L.S., see Haller, E. 320, 321, 329
Cederbaum, L.S., see Köpel, H. 425
Cederbaum, L.S., see Langhoff, P.W. 263, 339
Cederbaum, L.S., see Schirmer, J. 263, 332, 714
Cederbaum, L.S., see Unwin, R. 264, 297, 298, 303, 332, 340
Cederbaum, L.S., see von Niessen, W. 305, 308, 309
Celli, V. 740
Celli, V., see Marvin, A. 740
Celli, V., see Toigo, F. 740
Cerino, J. 94, 137
Cermak, V. 428
Cerrina, F. 478, 567
Chab, V., see McConville, C.F. 511
Chabal, Y., see Franciosi, A. 557
Chaban, E.E., see Weeks, S.P. 137
Chadi, D.J. 490, 492
Chambers, F.B. 162
Chambers, K.C., see Barrus, D.M. 338
Chambers, S.A. 553
Chambers, S.A., see Grioni, M. 553, 554
Chan, J., see Poliakoff, E. 419
Chandesris, D. 656
Chandesris, D., see Louie, S.G. 486, 487
Chandesris, D., see Miranda, R. 482
Chandra, N. 285
Chandrasekharan, V., see Chergui, M. 418, 457
Chang, S. 519
Chang, S., see Franciosi, A. 519, 520
Chang, T.C., see Schwarz, W.H.E. 337
Chapuisat, X., see Durand, G. 385
Chase, D.M. 282

Chauveau, D. 675, 694
Chauveau, D., see Guillot, C. 488, 496, 675, 695–697, 702
Chelikowsky, J.R. 546, 593
Chelikowsky, J.R., see Schlüter, M. 490
Chen, A.-B., see Spicer, W.E. 565
Chen, C.H. 760
Chen, C.H., see Meixner, A.E. 774
Chen, C.T. 78
Chen, C.T., see Plummer, E.W. 506
Cheng, K.L., see Carlson, T.A. 264
Cheng, K.T., see Hill, W.T. 227
Cheng, K.T., see Johnson, W.R. 295, 331, 333
Cherepkov, N.A. 333, 427, 457
Chergui, M. 418, 457, 547
Chermoshentsev, V.M., see Mazalov, L.N. 319, 324, 328
Chewter, L., see Müller-Dethlefs, K. 379, 450
Chiang, C.C., see Lee, L.C. 382
Chiang, T.-C. 478, 479, 512, 674, 688–690
Chiang, T.-C., see Eastman, D.E. 492, 494, 674, 692
Chiang, T.-C., see Himpsel, F.J. 494, 552, 586, 588, 674, 690, 691
Chiang, T.-C., see Hsieh, T.C. 522
Chiang, T.-C., see Ludeke, R. 527, 562, 571, 579, 593, 674, 692, 693
Chiang, T.-C., see Miller, T. 502, 674, 691, 692
Chiaradia, P. 591
Chiaradia, P., see Bauer, R.S. 549, 578
Chiaradia, P., see Hansson, G.V. 525, 596–598
Chiaradia, P., see Katnani, A.D. 530, 591
Chiaradia, P., see Nannarone, S. 490
Chiarello, G. 759
Chiarotti, G., see Nannarone, S. 490
Child, M.S. 248, 249
Childs, K.D., see McGovern, I.T. 476
Chin, K.K., see Newman, N. 588, 589
Chizhov, Yu.V., see Kleimenov, V.I. 245, 264
Cho, A.Y., see Marra, W.C. 598
Chopra, K.L. 738, 749
Choyke, W.J., see Rehn, V. 106
Christman, S.B., see Margaritondo, G. 583, 588
Christman, S.B., see Rowe, J.E. 546
Christman, S.B., see Weeks, S.P. 137
Christy, R.W., see Nestell, J.E. 738, 750, 764
Chun, M. 439
Chupka, W.A. 263
Chupka, W.A., see Berkowitz, J. 244, 264
Chupka, W.A., see Dehmer, J.L. 279
Chupka, W.A., see Dehmer, P.M. 263, 266–272, 341, 412, 415

Chupka, W.A., see Spohr, R. 377
Chupka, W.A., see White, M.G. 332
Chutjian, A. 334
Chutjian, A., see Ajello, J.M. 334
Chutjian, A., see Trajmar, S. 306
Chye, P.W. 572, 576, 577
Chye, P.W., see Braicovich, L. 550, 557
Chye, P.W., see Lindau, I. 577, 583, 589, 590, 599
Chye, P.W., see Skeath, P. 558, 559
Chye, P.W., see Spicer, W.E. 493, 502, 519, 545, 583, 589
Chye, P.W., see Su, C.Y. 519
Ciccacci, F., see Nannarone, S. 490
Citrin, P.H. 499, 503, 514, 626, 665, 666, 673–675, 678, 683
Citrin, P.H., see Comin, F. 498, 595, 596
Citrin, P.H., see Lee, P.A. 253
Citrin, P.H., see Wertheim, G.K. 667
Clark, S., see McLaren, R. 443
Clauberg, R. 655
Clauberg, R., see Haines, E. 482
Clauberg, R., see Kisker, E. 482
Clearfield, H.M., see McGovern, I.T. 476
Cline, C.F., see Bedford, K.L. 774
Clout, P.N. 138–140, 142, 146, 149, 155, 157
Codling, K. 28, 244, 264, 305, 307, 316, 331
Codling, K., see Cole, B.E. 338
Codling, K., see Ederer, D.L. 244, 264, 331
Codling, K., see Frasinski, L. 379, 381, 402, 403, 438
Codling, K., see Madden, R.P. 194, 200, 201
Codling, K., see Parr, A.C. 264, 268, 287, 290, 291, 331
Codling, K., see West, J.B. 61, 264, 268, 287, 292, 293, 331, 421, 422
Cohen, M.H., see Nestell, J.E. 738, 764
Cohen, M.L., see Larsen, P.K. 503
Cohen, M.L., see Louie, S.G. 546, 593
Cohen, M.L., see Northrup, J.E. 490
Cohen, M.L., see Schlüter, M. 490, 503
Cohen, R.L. 627, 628
Cohen-Tannoudji, C. 193
Colavita, E., see Chiarello, G. 759
Cole, B.E. 338
Cole, B.E., see Codling, K. 244, 264, 331
Cole, B.E., see Ederer, D.L. 85, 244, 264, 276, 331
Cole, B.E., see Parr, A.C. 129, 244, 264, 268, 275, 276, 278, 279, 287, 290, 291, 331
Cole, B.E., see Stockbauer, R. 332, 338
Cole, B.E., see West, J.B. 264, 268, 280, 285–287, 292, 293, 331, 339, 417, 418, 421, 422

Coleman, R.V., see Eastman, D.E. 650
Collin, J.E. 244, 245, 264, 298
Collin, J.E., see Caprace, G. 245, 264
Collin, J.E., see Natalis, P. 245, 264
Collins, L.A. 243, 246, 263, 264, 331, 332, 341, 413, 414, 423, 424
Collins, L.A., see Schneider, B.I. 243, 246, 251, 341
Colson, S.D., see Glownia, J.H. 332
Colson, S.D., see White, M.G. 332
Combescot, M. 635
Combet-Farnoux, F. 448
Comin, F. 498, 595, 596
Compton, A.M. 761
Compton, R.N., see Glownia, J.H. 332
Compton, R.N., see Miller, J.C. 332
Connerade, J.P. 181, 186, 190, 191, 193, 194, 200, 201, 204, 205, 208–210, 212, 220, 226, 231, 233, 235
Connerade, J.P., see Baig, M.A. 181, 182, 186, 189, 191, 194, 201, 202, 204, 205, 214, 219, 226, 228, 230, 364, 428
Connerade, J.P., see Dagata, J.A. 227
Connerade, J.P., see Garton, W.R.S. 230
Connerade, J.P., see Mayhew, C. 226
Connerade, J.P., see Rose, S.J. 205, 208
Connerade, J.P., see Schwarz, W.H.E. 337
Connolly, J.W.D. 305, 308
Cook, J.M., see Carnovale, F. 434, 437
Cook, J.M., see Katayama, D.H. 181
Cooke, W.E., see Rinneberg, H. 193
Cooper, J.W. 476, 554
Cooper, J.W., see Dehmer, J.L. 304
Cooper, J.W., see Fano, U. 248, 263, 331
Corcoran, C.T., see Langhoff, P.W. 337
Cordis, L., see Ding, A. 456
Cosby, P., see Moseley, J. 402
Couget, A., see Mamy, R. 773
Coulson, C.A. 222
Cousty, J., see Soukiassian, P. 667, 702, 703
Cowan, P.L., see Golovchenko, J.A. 49
Cowan, R.D., see Griffin, D.C. 201
Cox, J.T. 108
Cox, P.A. 714
Craighead, H.G., see Aspnes, D.E. 763
Crapper, M.D., see Jones, R.G. 503
Crapper, M.D., see Puschmann, A. 520
Creagh, D. 760
Crecelius, G. 629
Crecelius, G., see Manzke, R. 764, 765, 774
Crecelius, G., see Pflüger, J. 775
Crecelius, G., see Wertheim, G.K. 483
Croce, P., see Névot, L. 51, 52
Croft, M., see Franciosi, A. 478

Croft, M., *see* Peterman, D.J.   656
Cromer, D.T.   771
Csanak, G., *see* Langhoff, P.W.   339
Csanak, G., *see* Machado, L.E.   297, 298, 303, 340
Csanak, G., *see* Padial, N.   329, 338, 339, 413, 414
Cukier, M.   773
Curtis, D.   435
Cusatis, C.   761
Cvejanovic, D., *see* Shaw, D.A.   340, 342, 450
Cyrot Lackmann, F.   665
Cyrot Lackmann, F., *see* Desjonquères, M.C.   673, 676, 677, 683, 684

Dabrowsky, I.   213
Dagata, J.A.   227, 228
Dagata, J.A., *see* Baig, M.A.   214, 226, 228
Daimon, H.   511
Daimon, H., *see* Kobayashi, K.L.I.   656
Dalgarno, A., *see* Roche, A.   410
Damany, H.   44, 726, 747
Danby, C.J.   333, 334
Daniels, J.   757, 758
Daniels, R.R.   526, 575, 583, 584
Daniels, R.R., *see* Brillson, L.J.   562, 563, 580
Daniels, R.R., *see* Cerrina, F.   567
Daniels, R.R., *see* Davis, G.D.   564, 565, 581
Daniels, R.R., *see* Perfetti, P.   557
Daniels, R.R., *see* Richter, H.W.   560, 561
Daniels, R.R., *see* Stoffel, N.G.   562, 574, 599
Daniels, R.R., *see* Zhao, T.-X.   525, 560, 576
Davenport, J.W.   246, 337, 692
Davenport, J.W., *see* Grimm, F.A.   329, 339, 424, 426
Davenport, J.W., *see* Tossell, J.A.   342
Daviel, S.   365
Daviel, S., *see* Sodhi, R.N.S.   342, 447, 449
Davies, D.G.T., *see* Popper, P.   106
Davies, H.   156, 157
Davis, G.D.   564, 565, 581
Davis, L.C.   651, 653, 656
Davis, L.C., *see* Feldkamp, L.A.   641, 655, 759
Davis, R.E., *see* Kim, K.S.   644
Davis, R.F., *see* Rosenblatt, D.H.   510, 513
Daw, M.S.   590
Day, J.D.   162
Day, R.H., *see* Bartlett, R.J.   49
De Dominicis, C.T., *see* Nozières, P.   614, 626
de Laat, C.T.A.M.   156
de Michelis, B., *see* Abbati, I.   554
de Paola, R.A., *see* Heskett, D.   519
de Souza, G.G.B.   443–447

de Souza, G.G.B., *see* Morin, P.   450, 452, 454
De Souza, G.G.B., *see* Sodhi, R.N.S.   447, 449
de Souza, G.G.B., *see* Sodhi, R.N.S.   342
Debe, M.K.   496
Decrescenzi, M., *see* Chiarello, G.   759
Dehmer, J.L.   98, 243, 245, 246, 248, 250, 252, 256, 257, 259, 261–263, 279, 280, 282, 285–289, 295, 304–306, 308–315, 318, 319, 321–329, 331, 333, 336–338, 340–342, 409, 410, 417, 418, 422, 425, 439, 446, 516
Dehmer, J.L., *see* Åberg, T.   246
Dehmer, J.L., *see* Cadioli, B.   245, 336
Dehmer, J.L., *see* Codling, K.   244, 264, 331
Dehmer, J.L., *see* Cole, B.E.   338
Dehmer, J.L., *see* Dehmer, P.M.   299
Dehmer, J.L., *see* Dill, D.   243, 245, 246, 256, 257, 261, 331, 337, 338
Dehmer, J.L., *see* Dittman, P.M.   263, 340
Dehmer, J.L., *see* Ederer, D.L.   244, 264, 331
Dehmer, J.L., *see* Holland, D.M.P.   279, 340, 341, 372
Dehmer, J.L., *see* Loomba, D.   255, 258, 259, 339
Dehmer, J.L., *see* Parr, A.C.   129, 244, 264, 268, 275–279, 287, 290, 291, 297–299, 303–305, 331, 332, 340, 370, 372
Dehmer, J.L., *see* Poliakoff, E.D.   333, 432, 433, 457
Dehmer, J.L., *see* Pratt, S.T.   332
Dehmer, J.L., *see* Southworth, S.H.   294, 296, 297
Dehmer, J.L., *see* Stephens, J.A.   332, 339
Dehmer, J.L., *see* Stockbauer, R.   332, 338
Dehmer, J.L., *see* Swanson, J.R.   254, 255, 318–330, 339, 414, 423, 424
Dehmer, J.L., *see* Wallace, S.   295, 296, 303, 332, 334, 339, 425, 426
Dehmer, J.L., *see* West, J.B.   264, 268, 280, 285–287, 292, 293, 331, 339, 417, 418, 421, 422
Dehmer, P.M.   263, 266–272, 299, 341, 412, 415
Dehmer, P.M., *see* Pratt, S.T.   332
del Guidice, M., *see* Grioni, M.   552–554
del Guidice, M., *see* Weaver, J.H.   562
del Pennino, U., *see* Abbati, I.   476, 477, 554, 556
del Pennino, U., *see* Rossi, G.   554, 557
Delacretaz, G., *see* Bréchignac, C.   456
Delaney, J.J.   332, 340
Delgado-Barrio, G., *see* Miret-Artes, S.   406
Delley, B., *see* Wuilloud, E.   638
Delwiche, J.   398, 428

Delwiche, J., see Caprace, G. 245, 264
Delwiche, J., see Collin, J.E. 245, 264, 298
Delwiche, J., see Hubin-Franskin, M.-J. 244, 264, 331
Delwiche, J., see Lablanquie, P. 380, 381, 434–438
Delwiche, J., see Morin, P. 244, 263, 264, 275, 292, 331, 332, 340, 370–372, 398, 399, 402, 418–420
Delwiche, J., see Natalis, P. 245, 264
Delwiche, J., see Nenner, I. 442, 455
Delwiche, J., see Roy, P. 329, 341, 371, 413, 414, 422–424
Delwiche, J., see Tabché-Fouhailé, A. 264, 331, 332, 339, 382, 418
Demuth, J.E. 513
Demuth, J.E., see Hamers, R.J. 495
Demuth, J.E., see Schmeisser, D. 507
Dench, W.A., see Seah, M.P. 547, 548, 572, 709
Denley, D., see Kevan, S.D. 508, 510
Derenbach, H. 244, 275, 370, 371
Desjonquères, M.C. 488, 666, 673, 676, 677, 683, 684
Desjonquères, M.C., see Chauveau, D. 675, 694
Desjonquères, M.C., see Guillot, C. 488, 496, 674, 675, 695–697, 702
Desjonquères, M.C., see Spanjaard, D. 489, 666, 685
Desjonquères, M.C., see Tran Minh Duc 666
Desjonquères, M.C., see Tréglia, G. 700, 702
Deslattes, R.D., see LaVilla, R.E. 246, 248, 305, 336
Deus, K., see Boudriot, H. 773
Deus, K., see Kubier, B. 774
Deutsch, M. 761, 775
Dev, B.N. 599
Dexter, D.L., see Altarelli, M. 753
Dexter, D.L., see Rimmer, M.P. 752
Dibeler, V.H. 263, 303, 306, 320
Dibeler, V.H., see Botter, R. 298
DiCenzo, S.B. 499, 500
DiCenzo, S.B., see Wertheim, G.K. 500
Didio, R.A. 508, 509
Diercksen, G.H.F. 384
Diercksen, G.H.F., see Hermann, M.R. 341
Diercksen, G.H.F., see von Niessen, W. 305, 308, 309
Dietrich, H. 60
Dietz, E. 78, 479
Dietz, E., see Rehfeld, N. 89, 90
Dignam, M.J. 731
Dijkstra, J.M., see Tatchyn, R. 750

Dill, D. 243–246, 256, 257, 261, 264, 268, 331, 337, 338
Dill, D., see Atabek, O. 244, 264, 268
Dill, D., see Dehmer, J.L. 243, 245, 246, 250, 252, 256, 257, 259, 261–263, 280, 282, 285–287, 289, 295, 305, 306, 308–315, 318, 329, 331, 333, 337, 338, 340, 342, 417, 418, 422, 446, 516
Dill, D., see Dittman, P.M. 263, 340
Dill, D., see Jungen, Ch. 244, 268–271, 331
Dill, D., see Loomba, D. 255, 258, 259, 339
Dill, D., see Poliakoff, E.D. 333, 432, 433
Dill, D., see Raoult, M. 244, 264, 268, 331
Dill, D., see Stephens, J.A. 263, 294–297, 310, 318, 329, 332, 339, 342, 425
Dill, D., see Swanson, J.R. 254, 255, 318–330, 339, 414, 423, 424
Dill, D., see Wallace, S. 295, 296, 303, 332, 334, 339, 425, 426
Dill, R.H., see Hauge, P.S. 756
Dimicoli, I., see Castex, M.C. 456
Dimon, P. 513
Ding, A. 456
Dittman, P.M. 263, 340
Dixon, R.N., see Ashfold, M.N.R. 220
Dobinson, R.W. 151
Döbler, U. 514
Döbler, U., see Stöhr, J. 246, 254, 331, 517, 518
Dobrowlski, J.A., see Arndt, D.P. 762, 763
Dolenko, G., see Mazalov, L. 439
Dolfini, S.M., see Arakawa, E.T. 775
Doll, G.L., see Hoffman, D.M. 775
Domcke, W., see Bradshaw, A. 402
Domcke, W., see Cederbaum, L.S. 263, 332, 400
Domcke, W., see Köpel, H. 425
Domcke, W., see Langhoff, P.W. 263, 339
Domcke, W., see Schirmer, J. 263, 332
Domcke, W., see Unwin, R. 264, 297, 298, 303, 332, 340
Domke, M. 484
Domke, M., see Mandel, T. 512
Doniach, S. 626
Doniach, S., see Oh, S.-J. 636
Doniach, S., see Winick, H. 244, 253
Donoho, A.W., see Turner, A.M. 480
Doolittle, P.H. 244, 264
Dow, J.D. 590
Dow, J.D., see Allen, R.E. 590
Dowben, P.A. 503
Dowben, P.A., see Tomanek, D. 698, 699
Downie, D., see Mollenauer, L.F. 757
Dress, W.B., see Grimm, F.A. 329, 339

Dresselhaus, G., *see* Henrich, V.E. 759
Druzhinin, A.V., *see* Gluskin, E.S. 774
Dubau, J. 186
Ducastelle, F., *see* Tréglia, G. 651, 6651
Duce, D.A., *see* Chambers, F.B. 162
Duce, D.A., *see* Hopgood, F.R.A. 160
Ducros, R. 675
Dujardin, G. 380, 381, 407, 427, 435, 436
Dujardin, G., *see* Besnard, M. 381, 382, 434
Duke, C.B. 545, 546, 558, 559
Duke, C.B., *see* Bennett, A.J. 546
Duke, C.B., *see* Kahn, A. 558, 559, 692
Durand, G. 385
Durbin, S.M. 599
Durham, P.J., *see* Norman, D. 472
Durup, J., *see* Moseley, J. 402
Durup, J., *see* Richard-Viard, M. 380, 402–405
Duthie, J.C. 688
Dutuit, O. 382–385, 455
Dutuit, O., *see* Dujardin, G. 380, 381, 427, 435, 436
Dutuit, O., *see* Morin, P. 263, 331, 332, 339, 376
Dutuit, O., *see* Richard-Viard, M. 380, 402–405
Duzy, C. 264
Dzidzonou, V., *see* Schönhense, G. 333

Eastman, D.E. 480, 488, 492, 494, 545, 599, 650, 666, 674, 692
Eastman, D.E., *see* Chiang, T.-C. 478, 479, 674, 688–690
Eastman, D.E., *see* Gudat, W. 600
Eastman, D.E., *see* Gustafsson, T. 329, 338, 414
Eastman, D.E., *see* Heimann, P. 675, 697
Eastman, D.E., *see* Himpsel, F.J. 485, 494, 552, 568, 586, 588, 652, 674, 690, 691
Eastman, D.E., *see* Iwan, M. 656
Eastman, D.E., *see* Kaindl, G. 483, 484, 675, 676, 707
Eastman, D.E., *see* Koch, E.-E. 3
Eastman, D.E., *see* Ludeke, R. 562, 571, 579, 674, 692, 693
Eastman, D.E., *see* Plummer, E.W. 337, 372, 407, 416
Eastman, D.E., *see* van der Veen, J.F. 488, 674, 675, 694, 695, 697, 702
Eastman, L.F., *see* Okamoto, K. 589
Eberhardt, W. 337, 340, 439, 443, 450, 478, 482, 483, 501, 552, 586, 588, 689
Eberhardt, W., *see* Albert, M.R. 520
Eberhardt, W., *see* Bradshaw, A. 402
Eberhardt, W., *see* Carr, R.G. 246, 254
Eberhardt, W., *see* Freund, H.-J. 506
Eberhardt, W., *see* Plummer, E.W. 470, 471, 486, 506
Eberhardt, W., *see* Stöhr, J. 246, 254, 331, 517, 518
Ederer, D.L. 23, 85, 244, 264, 276, 331
Ederer, D.L., *see* Codling, K. 244, 264, 331
Ederer, D.L., *see* Cole, B.E. 338
Ederer, D.L., *see* Holland, D.M.P. 279, 340, 372
Ederer, D.L., *see* Krummacher, S. 263, 332, 340, 402
Ederer, D.L., *see* Madden, R.P. 66, 70, 194
Ederer, D.L., *see* Parr, A.C. 129, 244, 264, 268, 275, 276, 278, 279, 287, 290, 291, 297–299, 303–305, 331, 332, 340
Ederer, D.L., *see* Stockbauer, R. 332, 338
Ederer, D.L., *see* West, J.B. 264, 268, 280, 285–287, 292, 293, 331, 339, 417, 418, 421, 422
Egelhoff, W.F. 680, 699, 700
Eglash, S.J., *see* Newman, N. 588–591
Eglash, S.J., *see* Spicer, W.E. 562, 583, 588, 589
Eguiagaray, L., *see* Borondo, F. 390
Eib, W. 651
Eisenberger, P. 496, 598
Eisenberger, P., *see* Citrin, P.H. 503, 514
Eisenberger, P., *see* Kincaid, B.M. 253, 439
Eisenberger, P., *see* Lee, P.A. 253
Eisenberger, P., *see* Marra, W.C. 474, 598
Ejiri, A., *see* Nakamura, M. 336, 442
Eklund, P.C., *see* Hoffman, D.M. 775
El-Sherbini, Th.M. 336
El-Sherbini, Th.M., *see* van der Wiel, M.J. 336, 439
Eland, J.H.D. 245, 264, 381, 396, 402, 416, 438
Eland, J.H.D., *see* Berkowitz, J. 263, 407
Eland, J.H.D., *see* Brehm, B. 333, 334
Eland, J.H.D., *see* Curtis, D. 435
Eland, J.H.D., *see* Danby, C.J. 333, 334
Eland, J.H.D., *see* Lablanquie, P. 380, 381, 434–438
Eland, J.H.D., *see* Millié, Ph. 436, 438
Eland, J.H.D., *see* Richardson, P. 438
Eland, J.H.D., *see* Tsai, B.P. 435
Ellis, H.W. 753
Elson, J.M. 740
Elson, J.M., *see* Rehn, V. 105, 740
Emslie, A.G. 749
Enderby, M.J. 134
Endo, S., *see* Taniguchi, M. 656

Engel, T. 700
Engel, T., see Bauer, E. 700
Engelhardt, R., see Nicholls, J.M. 490, 491
Engelhardt, R., see Pulm, H. 246
Engelhardt, R., see Uhrberg, R.I.G. 490
Engstrom, H., see Mollenauer, L.F. 757
Erbudak, M. 674, 675
Eremenko, V.V., see Galuza, A. 773
Ergun, S., see Berman, M. 748
Eriksson, B., see Nilsson, A. 675, 676, 685, 707
Ermoshkin, A.N., see Abramov, V.N. 774
Erskine, J.L., see Eberhardt, W. 482, 483
Erskine, J.L., see Turner, A.M. 480, 482, 486
Esherick, P. 184, 185
Esteva, J.-M., see Bodeur, S. 439–443
Esteva, J.-M., see Fuggle, J.C. 637, 638
Estrup, P.J. 693
Estrup, P.J., see Barker, J.A. 695
Estrup, P.J., see Felter, T.E. 496
Evans, E.Ll., see Kubaschewski, O. 561, 577, 578
Evans, W.H., see Wagman, D.D. 561, 577, 578
Eyers, A. 35, 479

Fadley, C.S. 643, 649, 666, 671, 709, 713, 717
Fadley, C.S., see Goldberg, S.M. 716
Fadley, C.S., see Orders, P.J. 510, 598
Fadley, C.S., see Sinkovic, B. 482, 511
Fahlman, A., see Carlson, T.A. 331, 340, 341, 407, 423, 425
Fahlman, A., see Siegbahn, K. 613
Faile, S.P., see Feldkamp, L.A. 773
Fäldt, A. 773
Falicov, L.M., see Guillot, C. 652–654
Falicov, L.M., see Tersoff, J. 641
Fan, H.Y. 741
Fano, U. 186, 190, 191, 194, 244, 248, 263, 264, 268, 331, 336, 359, 385, 409, 414, 457, 654
Fano, U., see Cadioli, B. 245, 336
Fano, U., see Cerrina, F. 567
Fano, U., see Dehmer, J.L. 304
Fano, U., see Giusti-Suzor, A. 244, 264, 268
Fano, U., see Lu, K.T. 184, 227
Fantoni, R., see Giardini-Guidoni, A. 306
Farberovich, O.V. 774
Farge, Y., see Koch, E.-E. 3
Farrell, H.H., see Larsen, P.K. 503
Fatyga, B.W., see Hermann, M.R. 341
Feder, R., see Haines, E. 482
Fedyk, J., see Dignam, M.J. 731
Fehlman, A., see Siegbahn, K. 665, 666

Feibelman, P.J. 672, 681, 685, 698
Feibelman, P.J., see Knotek, M.L. 473
Feibelman, P.J., see Levinson, H.J. 486
Feidenhans'l, R., see Bohr, J. 496, 497
Feldhaus, J., see Eberhardt, W. 340, 443, 450
Feldhaus, J., see Hussain, Z. 472
Feldhaus, J., see Reimer, A. 455
Feldhaus, J., see Stöhr, J. 472
Feldkamp, L.A. 641, 655, 759, 773
Feldkamp, L.A., see Davis, L.C. 651, 653, 656
Felter, T.E. 496
Fergusson, E. 409
Ferreira, L.F.A. 398, 402, 403, 408, 410, 415, 416
Ferreira, L.F.A., see Baer, T. 244, 264, 333, 334, 375, 398, 400, 402
Ferreira, L.F.A., see Guyon, P.M. 333, 334, 379, 402
Ferreira, L.F.A., see Morin, P. 244, 264, 331, 332, 340
Ferrer, S., see Lamble, G.M. 513
Ferrett, T.A. 306, 331, 342, 444–447, 455
Ferrett, T.A., see Lindle, D.W. 331, 341, 446
Ferrett, T.A., see Piancastelli, M.N. 443
Ferrett, T.A., see Truesdale, C.M. 331, 342, 443, 445–447
Feucht, D.L., see Milnes, A.G. 545
Feuerbacher, B. 470, 471, 666, 672
Filatova, E.O. 774, 775
Filatova, E.O., see Vinogradov, A.S. 774
Findley, G.L., see Dagata, J.A. 227
Fink, J., see Manzke, R. 764, 765, 774
Fink, J., see Pflüger, J. 775
Fiquet-Fayard, F. 389
Fischer, D.A., see Stöhr, J. 521
Fischer, W., see Mandel, T. 512
Fisher, J.E. 775
Fisher, J.E., see Preil, M.E. 775
Fitton, B., see Feuerbacher, B. 470, 471, 666
Flamand, J. 82
Flamand, J., see Neviere, M. 82, 83
Flodström, A., see Barth, J. 484
Flodström, A., see Gerken, F. 675, 676, 685–688
Flodström, A., see Hörnström, S.E. 705–707
Flodström, A., see Johansson, L.I. 707
Flodström, A., see Nyholm, R. 674, 688, 689
Flodström, S.A. 567
Flodström, S.A., see Kammerer, R. 674, 675, 688–690
Flodström, S.A., see Nicholls, J.M. 490, 491
Flodström, S.A., see Stockbauer, R. 507
Flodström, S.A., see Uhrberg, R.I.G. 490

Flory, F., see Borgogno, J.P. 738, 762
Flouquet, F. 383
Fock, J.-H. 246, 262, 316, 331, 457
Fock, J.-H., see Lau, H.-J. 246, 331
Fomichev, V.A. 319, 324, 328, 336
Fomichev, V.A., see Zimkina, T.M. 246, 305, 336
Fontaine, A. 762
Ford, W.K., see Plummer, E.W. 506
Fournier, P., see Millié, Ph. 436, 438
Frahm, A., see Materlik, G. 503
Franciosi, A. 478, 519, 520, 553, 556, 557, 565, 579
Franciosi, A., see Abbati, I. 556
Franciosi, A., see Chang, S. 519
Franciosi, A., see Margaritondo, G. 562
Franciosi, A., see Peterman, D.J. 565
Franciosi, A., see Phillip, P. 565
Franciosi, A., see Weaver, J.H. 555–557
Franks, A. 83, 105, 106
Franz, L.S. 164
Frasinski, L. 379, 381, 402, 403, 438
Fredkin, D.R., see Liebermann, L.N. 482
Freed, K., see Morse, M. 396
Freeman, A.J., see Bagus, P.S. 649
Freeman, A.J., see Wimmer, E. 689, 690
Freeouf, J.L. 546, 593
Freeouf, J.L., see Eastman, D.E. 599
Freeouf, J.L., see Woodall, J.M. 593
Freiburg, Ch., see Fuggle, J.C. 629, 635, 637
Freiburg, Ch., see Hillebrecht, F.U. 635, 643
Freund, H.-J. 505, 506
Freund, H.-J., see Greuter, F. 504, 505
Freund, H.-J., see Plummer, E.W. 506
Freund, H.-J., see Pulm, H. 246
Freund, H.-J., see Schmeisser, D. 511
Frey, R. 381
Frey, R., see Brehm, B. 333, 334
Frey, R., see Schlag, E. 402
Friedel, J. 626
Friedrich, H. 338, 439
Froese-Fischer, C. 204
Fröhlich, H., see Dutuit, O. 382–385
Fröhlich, H., see Ito, K. 407
Fröhlich, H., see Tabché-Fouhailé, A. 382
Fuggle, J.C. 629, 635, 637–639
Fuggle, J.C., see Hillebrecht, F.U. 635, 639, 643
Fuggle, J.C., see Lässer, R. 635
Fujikawa, B.K., see Henke, B.L. 42, 47, 49, 752, 772
Fujimori, A. 634, 638
Fujimori, A., see Koide, T. 108
Fukamachi, T. 761
Fulde, P., see Kakehashi, Y. 644

Funke, P. 503
Fuoss, P.H., see Marra, W.C. 474
Fuoss, P.H., see Robinson, I.K. 495
Furdyna, J.K., see Franciosi, A. 565
Furuya, K. 753
Fusy, J., see Ducros, R. 675

Gabor, G., see White, M. 375
Gabor, G., see White, M.G. 132, 244, 275
Gale, B., see Franks, A. 106
Gallais, O., see Fiquet-Fayard, F. 389
Gallop, J.R., see Hopgood, F.R.A. 160
Galuza, A. 773
Galwey, R.K. 138
Ganz, J., see Lewandowski, B. 208
Gardiner, T.M., see Norman, D. 518
Gardner, J.L. 245, 264, 286, 287, 372, 418
Gardner, J.L., see Bahr, J.L. 244, 264
Gardner, J.L., see Samson, J.A.R. 337, 414
Garner, C.M., see Braicovich, L. 550, 557
Garner, C.M., see Chye, P.W. 572, 576, 577
Garner, C.M., see Lindau, I. 577, 583, 589, 590, 599
Garner, C.M., see Pianetta, P. 514
Garner, C.M., see Spicer, W.E. 493, 502, 519, 545, 583, 589
Garton, W.R.S. 230
Garton, W.R.S., see Baig, M.A. 214, 226, 228
Garton, W.R.S., see Connerade, J.P. 220, 226
Gashimzade, F.M., see Valiukonis, G. 774
Gates, I.P., see Ritsko, J.J. 774
Gauthe, B., see Cukier, M. 773
Gauthier, Y., see Baudoing, R. 515
Gay, J.C., see Connerade, J.P. 226, 233
Gay, J.G., see Smith, J.R. 685
Geerk, J., see Meixner, A.E. 774
Geiger, J. 759
Gelbart, W., see Beswick, J.A. 369
Gelius, U. 305–309
Gelius, U., see Nilsson, A. 675, 676, 685, 707
Gelius, U., see Siegbahn, K. 401, 443
Gel'mukhanov, F.Kh., see Kondratenko, A.V. 337
Gel'mukhanov, F.Kh., see Mazalov, L.N. 319, 324, 328
Gerard, P. 443
Gerard, P., see Carlson, T.A. 423, 425
Gerber, Ch., see Binnig, G. 495
Gerhardt, U., see Rehfeld, N. 89, 90
Gerken, F. 675, 676, 685–688
Gerken, F., see Barth, J. 484
Gerken, F., see Johansson, L.I. 707
Gerken, F., see Kammerer, R. 674, 675, 688–690

Gerwer, A. 338
Gerwer, A., see Langhoff, P.W. 338, 410
Gianturco, F.A. 255, 305, 336
Giardini-Guidoni, A. 306
Gibson, U.J., see Arndt, D.P. 762, 763
Gibson, W.M., see Dev, B.N. 599
Giesenberg, M.D., see Hogrefe, H. 767
Gimzewski, J.K., see Haak, H. 450
Ginter, D.S., see Ginter, M.L. 177, 214
Ginter, M.L. 177, 214
Girouard, F.E. 766, 775
Giusti-Suzor, A. 227, 244, 264, 268, 331, 333, 431
Giusti-Suzor, A., see Lefebvre-Brion, H. 244, 264, 268, 331
Giusti-Suzor, A., see Morin, P. 244, 263, 264, 331, 332, 340, 418-420
Gland, J.L., see Koestner, R.J. 246, 254, 331
Gland, J.L., see Stöhr, J. 246, 254, 331, 517, 518, 520
Glass-Maujean, M. 385, 387-389
Glass-Maujean, M., see Beswick, J.A. 390
Glass-Maujean, M., see Borrell, P. 385
Glass-Maujean, M., see Breton, J. 364, 385
Glass-Maujean, M., see Guyon, P.M. 385
Glownia, J.H. 332
Gluskin, E.S. 49, 774
Gluskin, E.S., see Mazalov, L. 439
Gobby, P.L., see Lapeyre, G.J. 599
Goddard, W.A., see Barton, J.J. 693
Goddard, W.A., see Swartz, C.A. 558
Goel, A.K. 773
Goldberg, S.M. 716
Golovchenko, J.A. 49
Golubkov, A.V., see Zhuze, V.P. 774
Goodwin, E.T. 486
Goodwin, T.A. 545
Goppy, P.L., see Lapeyre, G.J. 599
Goscinski, O., see Lozes, R.L. 251
Gotchev, B., see Frey, R. 381
Gotchev, B., see Peatman, W.B. 333, 334, 398
Gotchev, B., see Schlag, E. 402
Goto, K., see Namioka, T. 54, 55
Govers, T.R., see Baer, T. 244, 264, 333, 334, 375, 398, 400, 402
Govers, T.R., see Dujardin, G. 381, 407, 435, 436
Govers, T.R., see Guyon, P.M. 333, 334, 379, 402
Govers, T.R., see Nenner, I. 379, 402, 405, 406
Govers, T.R., see Richard-Viard, M. 380, 402-405
Graham, W.R., see Didio, R.A. 508, 509
Grandke, T., see Heidrich, K. 774

Granneman, E.H.A. 125, 133, 134
Grant, H., see Mönch, W. 590
Grant, I.P., see Connerade, J.P. 208
Grant, I.P., see Rose, S.J. 205, 208
Grant, M.B., see Mollenauer, L.F. 757
Grant, R.W. 590
Grant, R.W., see Waldrop, J.R. 528, 529
Graves, R.H.W. 748
Green, G.K. 26, 28, 31
Greenaway, D.L. 725, 748
Greene, C.H. 333, 357, 389, 418, 432
Greene, J.P., see Ruščić, B. 202
Gregory, A. 406, 407
Gregory, P.E., see Spicer, W.E. 519, 545
Grenet, G., see Bertolini, J.C. 675
Grenet, G., see Jugnet, Y. 478
Greuter, F. 502, 504, 505
Greuter, F., see Albert, M.R. 520
Greuter, F., see Freund, H.-J. 505, 506
Greuter, F., see Schmeisser, D. 511
Grider, D.E. 506
Griffin, D.C. 201
Griffiths, J.E., see Schwartz, G.P. 593
Griffiths, K. 515
Grimm, F.A. 329, 339-341, 424, 426
Grimm, F.A., see Carlson, T.A. 244, 264, 280, 284, 286, 287, 329, 331, 338-341, 402, 407, 422-425
Grimm, F.A., see Keller, P.R. 264, 297, 298, 303, 305, 332, 340, 341
Grimm, F.A., see Piancastelli, M.N. 341
Grioni, M. 552-554, 562
Grioni, M., see Abbati, I. 556
Grioni, M., see Weaver, J.H. 562
Grobman, W.D., see Eastman, D.E. 545
Gross, J.G. 773
Grover, J. 456
Grover, J., see Walters, E. 456
Grover, J., see White, M.G. 370, 415, 416
Gründler, R. 753
Gründler, R., see Boudriot, H. 773
Gründler, R., see Kubier, B. 774
Gründler, R., see Tews, W. 767-769
Grunes, L.A., see Preil, M.E. 775
Grunze, M. 698, 699
Grunze, M., see Tomanek, D. 698, 699
Gschneidner, K.R. 682
Guberman, S., see Roche, A. 410
Gudat, W. 364, 365, 372, 599, 600, 755
Gudat, W., see Alvarado, S.F. 676, 686
Gudat, W., see Clauberg, R. 655
Gudat, W., see Gustafsson, T. 329, 338, 414
Gudat, W., see Hagemann, H.J. 44-46, 731-735, 767, 771
Gudat, W., see Kisker, E. 481, 482

Gudat, W., see Plummer, E.W. 337, 372, 407, 416
Gudat, W., see Weller, D. 482
Guest, J.A. 333, 432, 433
Guggenheim, H.J., see Cohen, R.L. 627, 628
Guggenheim, H.J., see Rosencwaig, A. 644, 645
Guggenheim, H.J., see Wertheim, G.K. 649
Guggenheimer, K., see Beutler, H. 204
Guidotti, C., see Gianturco, F.A. 255, 305, 336
Guillot, C. 488, 496, 652–654, 674, 675, 695–697, 702
Guillot, C., see Chauveau, D. 675, 694
Guillot, C., see Desjonquères, M.C. 488, 666, 673, 677
Guillot, C., see Soukiassian, P. 667, 702, 703
Guillot, C., see Spanjaard, D. 489, 666, 685
Guillot, C., see Tran Minh Duc 665, 666, 669, 670, 672, 675, 689, 712–714
Guillot, C., see Tréglia, G. 700, 702
Gunnarsson, O. 636–638
Gunnarsson, O., see Fuggle, J.C. 629, 635, 637, 638
Gunnarsson, O., see Schönhammer, K. 635
Gupta, L.C., see Mårtensson, N. 707
Gupta, R.P. 714
Gurin, V.N. 773
Gurin, V.N., see Farberovich, O.V. 774
Gürtler, P. 263, 366
Gürtler, P., see Moeller, T. 418
Gürtler, P., see Peatman, W.B. 333, 334, 398
Guseinova, D.A., see Valiukonis, G. 774
Gustafsson, T. 246, 254, 305–307, 309, 310, 317, 318, 329, 331, 338, 339, 414
Gustafsson, T., see Albert, M.R. 520
Gustafsson, T., see Allyn, C.L. 516
Gustafsson, T., see Bartynski, R.A. 486, 488
Gustafsson, T., see Holmes, M.I. 496
Gustafsson, T., see Levinson, H. 305, 309, 317, 329, 338
Gustafsson, T., see Loubriel, G. 506
Gustafsson, T., see Plummer, E.W. 337, 372, 407, 416
Gutcheck, R., see Black, G. 456
Guyon, P.M. 333, 334, 379, 382, 385, 398, 402, 403, 408, 410, 411, 428–430
Guyon, P.M., see Baer, T. 244, 264, 333, 334, 375, 398, 400, 402
Guyon, P.M., see Borrell, P. 385
Guyon, P.M., see Breton, J. 364, 385
Guyon, P.M., see Delwiche, J. 398, 428
Guyon, P.M., see Dujardin, G. 380, 381, 427, 435, 436

Guyon, P.M., see Dutuit, O. 382–385
Guyon, P.M., see Glass-Maujean, M. 385, 387–389
Guyon, P.M., see Ito, K. 407
Guyon, P.M., see Mentall, J. 385, 390
Guyon, P.M., see Morin, P. 244, 263, 264, 331, 332, 339, 340, 376
Guyon, P.M., see Nenner, I. 379, 402, 405, 406
Guyon, P.M., see Richard-Viard, M. 380, 402–405
Guyon, P.M., see Spohr, R. 377
Guyon, P.M., see Tabché-Fouhailé, A. 264, 331, 332, 339, 382, 418
Gyemant, I., see Benedict, M.G. 305

Haak, H. 450
Haaks, D., see Castex, M.C. 455
Haaks, D., see Jordan, B. 455
Haaks, D., see Le Calvé, J. 455
Haaks, D., see Moeller, T. 418
Haas, C., see Van der Laan, G. 644, 647
Haase, J., see Döbler, U. 514
Haase, J., see Puschmann, A. 514, 520
Haber, H. 64, 68
Haddad, G.N., see Samson, J.A.R. 337
Haelbich, R.-P. 50, 771
Haelbich, R.-P., see Boller, K. 108
Haelbich, R.-P., see Eberhardt, W. 337, 439
Haelbich, R.-P., see Hogrefe, H. 767
Haensel, R. 750
Haensel, R., see Blechschmidt, D. 246–248, 305, 331, 336
Haensel, R., see Brodmann, R.R. 365
Haensel, R., see Chergui, M. 418, 547
Haensel, R., see Wilcke, H. 365, 367
Hagemann, H.J. 44–46, 731–735, 767, 771
Hagström, S.B.M., see Allen, J.W. 656
Hagström, S.B.M., see Brown, F.C. 548
Hagström, S.B.M., see Flodström, S.A. 567
Hahn, U. 365
Hahn, U., see Brodmann, R.R. 365
Haines, E. 482
Halbach, K., see Brown, G. 116
Hale, G.M. 751
Haller, E. 320, 321, 329
Hallmeier, K.H., see Schwarz, W.H.E. 341
Halow, I., see Wagman, D.D. 561, 577, 578
Hamers, R.J. 495
Hamilton, J.F., see Apai, G. 498
Hamin, R.N., see Horton, V.G. 278
Hamm, R.N. 45, 757
Hamm, R.N., see Inagaki, T. 772
Hamnett, A. 337

Hamnett, A., see Brion, C.E. 401, 438
Hamnett, A., see Burroughs, P. 638
Hamrin, K., see Siegbahn, K. 401, 443, 613, 665, 666
Hanawa, T., see Saitoh, M. 525
Hancock, W.H. 245, 264
Haneman, D. 490
Hang Ho, M., see Poliakoff, E.D. 366, 415
Hansen, J.E. 208
Hanson, D.M., see Stockbauer, R. 507
Hansson, G.V. 525, 596–598
Hansson, G.V., see Bauer, R.S. 549, 578
Hansson, G.V., see Brillson, L.J. 559, 581
Hansson, G.V., see Nicholls, J.M. 490, 491
Hansson, G.V., see Uhrberg, R.I.G. 490
Hanyu, T. 775
Hanyu, T., see Miyahara, T. 773, 774
Hanyu, T., see Sato, S. 773
Hara, S. 275
Harbeke, G. 725, 726
Harbeke, G., see Greenaway, D.L. 725, 748
Hardis, J.E., see Southworth, S.H. 294, 296, 297
Harris, J., see Brown, G. 116
Hart, M. 761
Hart, M., see Cusatis, C. 761
Hart, M., see Deutsch, M. 761, 775
Harting, E. 278
Hartman, P.L., see Tomboulian, D.H. 28
Hass, G. 45, 108
Hass, G., see Canfield, L.R. 108
Hass, G., see Cox, J.T. 108
Hass, G., see Hunter, W.R. 108, 773
Hastings, C. 29
Hastings, J.B. 131
Hastings, J.B., see Stöhr, J. 521
Hatherby, P., see Frasinski, L. 381, 438
Hattori, S., see Katoh, H. 775
Hauge, P.S. 756
Hay, P.J. 305, 308, 309
Hayaishi, T. 263, 264, 297, 298, 303, 340
Hayaishi, T., see Akahori, T. 403
Hayaishi, T., see Nakamura, M. 248, 305, 336, 439
Hayaishi, T., see Sasanuma, M. 306
Hayes, R.G., see Signorelli, A.J. 633
Hayes, W. 319, 324, 328, 336
Head, J.D. 571
Heald, S.M., see Stern, E.A. 472
Heavens, O.S. 730
Hecht, M., see Johannson, L.I. 554
Hecht, M., see Tatchyn, R. 750
Heckenkamp, Ch. 35, 375
Heckenkamp, Ch., see Eyers, A. 35

Hedegard, P., see Mårtensson, N. 685
Hedén, P., see Baer, Y. 650
Heden, P., see Siegbahn, K. 401, 443
Hedin, L., see Almbladh, C.-O. 386, 614
Hedman, J., see Baer, Y. 650
Hedman, J., see Nilsson, A. 675, 676, 685, 707
Hedman, J., see Siegbahn, K. 401, 443, 613, 665, 666
Heidrich, K. 774
Heimann, P.A. 675, 697
Heimann, P.A., see Eastman, D.E. 492, 494, 674, 692
Heimann, P.A., see Ferrett, T.A. 306, 331, 342, 444–447, 455
Heimann, P.A., see Himpsel, F.J. 494, 552, 568, 586, 588, 674, 690, 691
Heimann, P.A., see Lindle, D.W. 331, 341, 446
Heimann, P.A., see Truesdale, C.M. 331, 342, 443, 445–447
Heimann, P.A., see van der Veen, J.F. 488, 674
Hein, J.-T., see Bonse, U. 761
Heinzmann, U. 230, 235, 236, 333, 373, 427
Heinzmann, U., see Eyers, A. 35, 479
Heinzmann, U., see Heckenkamp, Ch. 35, 375
Heinzmann, U., see Kaesdorf, S. 335
Heinzmann, U., see Schäfers, F. 236, 333, 428
Heinzmann, U., see Schönhense, G. 333
Hellner, L., see Besnard, M. 381, 382, 434
Hellwege, K.H. 82, 83
Helms, C.R. 567
Helms, D., see Black, G. 456
Henisch, H.K. 543, 546
Henke, B.L. 42, 47, 49, 752, 772
Henrich, V.E. 759
Henzler, M. 545
Herbst, J.F. 639
Herbst, L.J. 126
Herman, F. 716
Hermann, M.R. 256, 339, 341
Hermanson, J. 478
Hermanson, J., see Lapeyre, G.J. 599
Hermsmeier, B., see Sinkovic, B. 482
Herrenden-Harker, W.G., see Mackey, K.J. 531, 532
Hertel, I.V., see Kamke, W. 456
Hertz, H. 407
Herzberg, G. 182, 212, 213, 223, 226, 244, 264, 268, 272
Herzberg, G., see Dabrowsky, I. 213
Herzberg, G., see Huber, K.P. 213

Herzenberg, A. 285
Herzenberg, A., see Birtwistle, D.T. 285
Herzenberg, A., see Pavlovic, Z. 262
Heskett, D. 519
Heskett, D., see Freund, H.-J. 505, 506
Heskett, D., see Greuter, F. 504, 505
Hess, B., see Marian, C. 403, 404
Hess, B.A., see Heinzmann, U. 333
Hezaveh, A.A. 482
Hieronymus, H., see Rinneberg, H. 193
Hill, N.R., see Toigo, F. 740
Hill, W.T. 227
Hillebrecht, F.U. 635, 639, 643
Hillebrecht, F.U., see Fuggle, J.C. 629, 635, 637, 638
Hillebrecht, F.U., see Himpsel, F.J. 532, 533
Hillier, I.H., see Delaney, J.J. 332, 340
Hillier, I.H., see MacDowell, A.A. 134, 135
Himpsel, F.J. 485, 494, 532, 533, 552, 568, 586, 588, 652, 674, 690, 691
Himpsel, F.J., see Dietz, E. 479
Himpsel, F.J., see Eastman, D.E. 480, 488, 492, 494, 666, 674, 692
Himpsel, F.J., see Eberhardt, W. 478
Himpsel, F.J., see Hollinger, G. 502, 520
Himpsel, F.J., see Iwan, M. 656
Himpsel, F.J., see Jugnet, Y. 506
Himpsel, F.J., see Landgren, G. 502, 503, 519, 702
Himpsel, F.J., see Morar, J.F. 674
Himpsel, F.J., see Reihl, B. 482
Himpsel, F.J., see Schmeisser, D. 506, 511, 512
Himpsel, F.J., see Schnell, R.D. 503, 674
Himpsel, F.J., see Smith, N.V. 438, 470, 485, 486, 510, 666
Himpsel, F.J., see van der Veen, J.F. 488, 674, 675, 695, 697, 702
Hines, D.P., see Tasker, R. 168
Hino, I., see Petro, W.G. 584
Hinson, D.C. 753
Hiraki, A. 546, 576
Hisscott, L.A., see Richardson, D. 643
Hitchcock, A.P. 243, 251, 262, 306, 307, 337–341, 380, 439, 441–443, 450, 455
Hitchcock, A.P., see Carnovale, F. 434, 437
Hitchcock, A.P., see Daviel, S. 365
Hitchcock, A.P., see Ishii, I. 443
Hitchcock, A.P., see McLaren, R. 443
Hitchcock, A.P., see Sette, F. 342, 438, 443
Ho, F.C., see Arndt, D.P. 762, 763
Ho, K.M., see Larsen, P.K. 503
Ho, K.M., see Schlüter, M. 503
Ho, W., see Davenport, J.W. 246

Hodgkin, V.A., see Arndt, D.P. 762, 763
Hoffman, A. 17
Hoffman, D.M. 775
Hogrefe, H. 82, 767
Hogrefe, H., see Boller, K. 108
Hohlneicher, G., see von Niessen, W. 305, 308, 309
Holland, D.M.P. 279, 340, 341, 372
Holland, D.M.P., see Butler, J.J. 134
Holland, D.M.P., see Dehmer, J.L. 263, 287, 288, 318, 319, 321–328, 341
Holland, D.M.P., see Enderby, M.J. 134
Holland, D.M.P., see Parr, A.C. 244, 264, 275–278, 297–299, 303–305, 331, 332, 340
Holland, J., see Parr, A.C. 370, 372
Holland, W.E., see Hale, G.M. 751
Holland, W.E., see Querry, M.R. 751
Hollinger, G. 502, 520, 701
Hollinger, G., see Morar, J.F. 674
Hollinger, G., see Reihl, B. 482
Hollinger, G., see Schmeisser, D. 506, 511, 512
Holloway, S., see Tomanek, D. 685
Holloway, S., see van der Veen, J.F. 675, 694
Holmes, D.J., see Lamble, G.M. 513
Holmes, M.I. 496
Holmes, R.M. 303, 339
Holmes, R.M., see Marr, G.V. 244, 275, 284, 295, 296, 338, 426
Holtzberg, F., see Kaindl, G. 675, 676, 707
Hölzel, H., see Becker, U.E. 450
Hoogewijs, R., see Guillot, C. 675, 695, 696, 702
Hopfield, J.J. 287, 626
Hopgood, F.R.A. 160
Horani, M., see Beswick, J.A. 405
Horelick, D. 146
Hormes, J. 137, 230
Hormes, J., see Baig, M.A. 214, 226, 228, 230, 428
Hormes, J., see Garton, W.R.S. 230
Hormes, J., see Sommer, K. 200
Horn, K. 672, 674, 692
Horn, K., see Eberhardt, W. 482, 483
Horn, K., see Mandel, T. 512
Horn, K., see Mariani, C. 506
Horn, P.M., see Dimon, P. 513
Hörnström, S.E. 705–707
Hörnström, S.E., see Johansson, L.I. 707
Hörnström, S.E., see Nyholm, R. 674, 688, 689
Horsley, J.A., see Flouquet, F. 383
Horsley, J.A., see Koestner, R.J. 246, 254, 331

Horton, V.G. 278
Hotop, H. 208
Hotop, H., see Lewandowski, B. 208
Hottier, F., see Aspnes, D.E. 738, 763
Howe, L.L., see Herzberg, G. 213
Howells, M.R. 61, 66–69, 98
Howells, M.R., see Chen, C.T. 78
Howells, M.R., see Hogrefe, H. 82
Howells, M.R., see McKinney, W.R. 68
Howells, M.R., see Williams, G.P. 87, 111
Hower, N., see Cerino, J. 94, 137
Howson, J.M., see Alexander, J.R. 144
Hoyer, E., see Hogrefe, H. 82
Hsieh, T.C. 522
Huang, K.-N., see Johnson, W.R. 333
Hubbard, D.J. 73, 99
Hubbard, J. 758
Hubbard, S., see Bilderback, D.H. 48, 761
Hubbell, J.M. 770, 771
Hubbell, J.M., see McMaster, W.H. 752, 770, 771
Huber, K.P. 213
Hubin-Franskin, M.-J. 244, 264, 331
Hubin-Franskin, M.-J., see Delwiche, J. 398, 428
Hubin-Franskin, M.-J., see Lablanquie, P. 380, 381, 434–438
Hubin-Franskin, M.-J., see Morin, P. 244, 263, 264, 275, 292, 331, 332, 340, 370–372, 398, 399, 402, 418–420
Hubin-Franskin, M.-J., see Nenner, I. 442, 455
Hubin-Franskin, M.-J., see Roy, P. 329, 341, 371, 413, 414, 422–424
Hubin-Franskin, M.-J., see Tabché-Fouhailé, A. 382
Huestic, D., see Black, G. 456
Huffman, D.R., see Trykozko, R. 774
Hüfner, S. 639, 640, 643, 647, 648, 650
Hüfner, S., see Wertheim, G.K. 649
Hughes, G. 502
Hughes, G., see Himpsel, F.J. 532, 533
Hughes, G., see Morar, J.F. 674
Hughes, G.J. 565, 567, 573, 574
Hughes, G.J., see McKinley, A. 526, 560, 573
Hughes, G.J., see Williams, R.H. 562, 565–567, 572, 573
Hughes, H.P., see Beal, A.R. 773
Hughes, H.P., see Khumalo, F.S. 773, 774
Hui, E., see Walters, E. 456
Huijser, A. 545, 558, 559, 586, 589
Huijser, A., see van Laar, J. 545, 586, 589
Hullinger, F., see Lässer, R. 635

Humphreys, T.P., see Williams, R.H. 562, 567, 572, 573
Humphreys-Owen, S.P.F. 44, 731, 743, 744
Hunderi, O. 738
Hünlich, K., see Eyers, A. 479
Hunt, D.J., see Franks, A. 83, 105
Hunter, W.R. 45, 60, 61, 108, 744–747, 749, 750, 757, 773
Hunter, W.R., see Bedford, K.L. 774
Hunter, W.R., see Canfield, L.R. 108
Hunter, W.R., see Cox, J.T. 108
Hunter, W.R., see Hass, G. 45, 108
Hupp, J.A., see Shoch, J.F. 168
Hurych, Z., see Lee, S.T. 499, 572
Hurych, Z., see Thuler, M.R. 656
Hussain, Z. 472
Hussain, Z., see Barton, J.J. 510, 598
Hussain, Z., see Sinkovic, B. 511
Hussain, Z., see Umbach, E. 506
Huster, R., see Kronast, W. 433, 446
Hutchison, D. 166, 168

Igarashi, J. 651, 652
Iguchi, Y., see Nakamura, M. 336, 442
Iida, Y., see Hayaishi, T. 263, 264, 297, 298, 303, 340
Inagaki, T. 753, 772
Incoccia, L., see Comin, F. 498
Inghram, M.G., see Stockbauer, R. 333, 334
Inglesfield, J.E. 486, 496
Inglesfield, J.E., see Campuzano, J.-C. 496
Inokuti, M., see Shiles, E. 731–735, 764
Inoue, K., see Mitani, T. 773
Inoue, K., see Suga, S. 774
Ishida, K., see Katoh, H. 775
Ishiguro, E. 319, 324, 328, 340, 439, 441
Ishiguro, E., see Hayaishi, T. 263, 264, 297, 298, 303, 340
Ishiguro, E., see Morioka, Y. 337, 439, 442
Ishiguro, E., see Nakamura, M. 248, 305, 336, 439
Ishiguro, E., see Ninomiya, K. 339, 439, 441
Ishiguro, E., see Sasanuma, M. 305, 306, 338
Ishii, H., see Hanyu, T. 775
Ishii, H., see Miyahara, T. 774
Ishii, I. 443
Ishii, I., see McLaren, R. 443
Ishii, T. 656
Ishii, T., see Kakizaki, A. 656
Ishii, T., see Miyahara, T. 773
Ishii, T., see Nagakura, I. 629
Ishii, T., see Sato, S. 113, 773
Ishii, T., see Sugawara, H. 656
Ishii, T., see Suzuki, S. 633

Ishikawa, T., see Katoh, H. 775
Israel, H.I., see Storm, E. 770, 771
Itikawa, Y. 275
Ito, H., see Daimon, H. 511
Ito, K. 407
Ito, K., see Akahori, T. 403
Ito, K., see Morin, P. 244, 263, 264, 331, 332, 339, 340, 376
Ito, K., see Tabché-Fouhailé, A. 382
Iwami, W., see Hiraki, A. 546, 576
Iwan, M. 656
Iwan, M., see Eberhardt, W. 337, 439
Iwan, M., see Haelbig, R.-P. 771
Iwan, M., see Mariani, C. 506
Iwata, S. 338
Iwata, S., see Hayaishi, T. 263, 264, 297, 298, 303, 340
Iwata, S., see Ishiguro, E. 319, 324, 328, 340, 439, 441
Iwata, S., see Kanamori, H. 341
Iwata, S., see Kimura, K. 320, 334, 396
Iwata, S., see Ninomiya, K. 339, 439, 441

Jackdon, T.N., see Woodall, J.M. 593
Jackson, J.D. 3
Jackson, K.H., see Guest, J.A. 333, 433
Jackson, K.H., see Poliakoff, E.D. 333, 432, 433
Jacobi, K., see Schmeisser, D. 511, 512
Jacobus, G.F. 108
Jacobus, G.F., see Hass, G. 108
Jaeger, R., see Baetzold, R.C. 675
Jaeger, R., see Brennan, S. 513, 674, 690, 691
Jaeger, R., see Norman, D. 472
Jaeger, R., see Stöhr, J. 246, 254, 331, 472, 505, 516, 517, 523, 524, 526, 595
Jain, K.P., see Guillot, C. 652–654
James, R.W. 42
Janak, J.F., see Eastman, D.E. 650
Janak, J.F., see Moruzzi, V.L. 652
Jark, W. 50, 82
Jark, W., see Birken, H.-G. 755, 767
Jark, W., see Boller, K. 108
Jasperson, S.N. 757
Jeiley, K., see Fisher, J.E. 775
Jennings, G., see Hezaveh, A.A. 482
Jennison, D. 445
Jensen, E., see Bartynski, R.A. 486, 488
Jepsen, D.W., see Demuth, J.E. 513
Jezequel, G., see Pollini, I. 774, 775
Jezequel, G., see Quemerais, A. 774
Jezierski, K. 752, 754
Jivery, W.T., see Dehmer, J.L. 279
Jo, T. 639, 653–656

Jo, T., see Kotani, A. 638
Jochims, H., see Beckmann, O. 368
Jochims, H., see Hertz, H. 407
Johansson, B. 666, 678, 679, 684, 685, 688, 705
Johansson, B., see Johansson, L.I. 707
Johansson, B., see Mårtensson, N. 685
Johansson, B., see Rosengren, A. 488, 666, 680, 681, 684, 703, 704
Johansson, G., see Siegbahn, K. 401, 443, 613, 665, 666
Johansson, L., see Bauer, R.S. 519
Johansson, L.I. 482, 554, 567, 707
Johansson, L.I., see Barth, J. 484
Johansson, L.I., see Gerken, F. 675, 676, 685–688
Johansson, L.I., see Hörnström, S.E. 705–707
Johansson, L.I., see Kammerer, R. 674, 675, 688–690
Johansson, L.I., see Loubriel, G. 506
Johansson, L.I., see Nyholm, R. 674, 688, 689
Johns, J.W.C. 200, 221
Johnson, A.L. 516
Johnson, A.L., see Stöhr, J. 254, 331, 520
Johnson, K.H., see Connolly, J.W.D. 305, 308
Johnson, N.M., see Helms, C.R. 567
Johnson, P.N., see Clout, P.N. 138–140
Johnson, R.L. 23, 83, 757
Johnson, R.L., see Bohr, J. 496, 497
Johnson, R.L., see Dietz, E. 78
Johnson, W.R. 295, 331, 333
Johnson, W.R., see Hill, W.T. 227
Joly, Y., see Baudoing, R. 515
Jona, F. 494
Jonas, A.E., see Carlson, T.A. 245, 264, 284
Jones, G.P., see Chambers, F.B. 162
Jones, R.G. 503
Jones, V.O., see Rehn, V. 105, 106, 740
Jonsson, G., see Rinneberg, H. 193
Jordan, B. 455
Jordan, B., see Castex, M.C. 455
Jordan, B., see Le Calvé, J. 455
Jordan, B., see Moeller, T. 418
Jordan, J.L., see Himpsel, F.J. 532, 533
Jordan, J.L., see Hughes, G. 502
Jordan, J.L., see Morar, J.F. 674
Jordan, K., see Gregory, A. 406, 407
Jortner, J., see Schwentner, N. 457
Joy, D.C., see Teo, B.K. 253
Joyce, B.A., see van der Veen, J.F. 562
Joyce, J., see Franciosi, A. 519, 520
Joyce, J., see Grioni, M. 552–554, 562
Joyce, J., see Weaver, J.H. 562
Judge, D., see Lee, L.C. 305, 307, 382, 407

Judge, D., see Wu, C. 382
Jugnet, Y. 478, 506, 700, 701
Jugnet, Y., see Bertolini, J.C. 675
Jugnet, Y., see Guillot, C. 674, 675, 695, 696, 702
Jugnet, Y., see Landgren, G. 502, 503, 519, 702
Jugnet, Y., see Tran Minh Duc 665, 666, 669, 670, 672, 675, 689, 712–714
Jugnet, Y., see Tréglia, G. 700, 702
Julien, L.S., see Miller, R.F. 746
Julienne, P.S. 389
Julienne, P.S., see Mies, F.H. 244, 264, 268, 331, 333
Jungen, Ch. 227, 244, 264, 268–271, 331, 333
Jungen, Ch., see Atabek, O. 244, 264, 268
Jungen, Ch., see Dill, D. 244, 264, 268
Jungen, Ch., see Giusti-Suzor, A. 244, 264, 268, 331, 333, 431
Jungen, Ch., see Greene, C.H. 389, 418
Jungen, Ch., see Herzberg, G. 244, 264, 268, 272
Jungen, Ch., see Raoult, M. 244, 264, 265, 268, 271–273, 291, 331
Jupille, J. 515, 675, 697
Jupiter, P.J., see Apai, G. 675, 694, 702, 703

Kabler, M.N., see Hunter, W.R. 60, 61
Kaesdorf, S. 335
Kaesdorf, S., see Schönhense, G. 333
Kaga, H. 644
Kahn, A. 558, 559, 692
Kahn, A., see Bonapace, C.R. 573, 578
Kahn, A., see Duke, C.B. 558, 559
Kaindl, G. 483, 484, 675, 676, 685, 686, 707, 708
Kaindl, G., see Chiang, T.-C. 512
Kaindl, G., see Domke, M. 484
Kaindl, G., see Mandel, T. 512
Kaindl, G., see Schneider, W.-D. 639
Kajiyama, K. 590
Kakehashi, Y. 644
Kakizaki, A. 656
Kakizaki, A., see Miyahara, T. 773
Kakizaki, A., see Sugawara, H. 656
Kalkoffen, G., see Barth, J. 654
Kalkoffen, G., see Eberhardt, W. 552, 586, 588, 689
Kalkoffen, G., see Lenth, W. 554
Källne, E., see Bartlett, R.J. 49
Källne, E., see Johansson, L.I. 554
Källne, E., see Tatchyn, R. 750
Kalt, P., see Erbudak, M. 674, 675
Kamada, T., see Hanyu, T. 775

Kamarzin, A.A., see Zhuze, V.P. 773
Kamimura, W., see Hiraki, A. 546, 576
Kamke, B., see Kamke, W. 456
Kamke, W. 456
Kammerer, R. 674, 675, 688–690
Kanamori, H. 341
Kanamori, J. 651, 652
Kanamori, J., see Jo, T. 653–656
Kanamori, J., see Parlebas, J.C. 656
Kanani, D., see Duke, C.B. 558, 559
Kanani, D., see Kahn, A. 558, 559
Kane, E.O. 596
Kaneko, Y. 774
Kaneko, Y., see Tokura, Y. 775
Kania, D.R., see Bartlett, R.J. 49
Kanski, J. 655
Kanzaki, H., see Mitani, T. 773
Kanzaki, H., see Taniguchi, M. 656, 674, 692
Kappler, J.P., see Krill, G. 635
Karin, M.G., see Abramov, V.N. 772
Karin, M.G., see Bagdasarov, K.S. 775
Karin, M.G., see Farberovich, O.V. 774
Karin, M.G., see Gurin, V.N. 773
Karin, M.G., see Zhuze, V.P. 773, 774
Karlsson, S.-E., see Siegbahn, K. 613, 665, 666
Karlsson, U., see Pulm, H. 246
Karlsson, U.O., see Himpsel, F.J. 532, 533
Karlsson, U.O., see Nicholls, J.M. 490, 491
Karlsson, U.O., see Uhrberg, R.I.G. 490
Karnatak, R.C., see Fuggle, J.C. 637, 638
Kasting, G., see Peatman, W. 377
Katayama, D.H. 181
Katayama, T., see Namioka, T. 54, 55
Katnani, A.D. 528, 530, 591
Katnani, A.D., see Brillson, L.J. 546, 547, 550, 551, 559, 561–563, 569, 570, 579, 580, 582, 591, 592
Katnani, A.D., see Brucker, C.F. 567, 587, 588, 591
Katnani, A.D., see Chiaradia, P. 591
Katnani, A.D., see Daniels, R.R. 526, 575
Katnani, A.D., see Duke, C.B. 558, 559
Katnani, A.D., see Franciosi, A. 553
Katnani, A.D., see Kahn, A. 558, 559
Katnani, A.D., see Perfetti, P. 557
Katnani, A.D., see Shapira, Y. 559
Katnani, A.D., see Zhao, T.-X. 525, 560, 576
Kato, H., see Hanyu, T. 775
Kato, R., see Miyake, K.P. 60, 61
Kato, T., see Susuki, S. 379
Katoh, H. 775
Katsumata, S. 245, 264
Katsumata, S., see Kimura, K. 320, 334, 396

Kavanaugh, K.L., see Woodall, J.M.   593
Kay, R.B.   251, 262, 337, 447, 455
Keller, F.   409
Keller, P.R.   264, 297, 298, 303, 305, 332, 340, 341
Keller, P.R., see Carlson, T.A.   244, 264, 329, 331, 339–341, 407, 423–425
Keller, P.R., see Grimm, F.A.   339, 341, 424
Keller, P.R., see Piancastelli, M.N.   341
Kelly, M., see Brillson, L.J.   550, 551, 569, 570, 582
Kelly, M.K., see Richter, H.W.   560, 561
Kelly, M.K., see Stoffel, N.G.   572
Kelly, M.M.   106
Kemeney, P.C.   639, 650
Kemp, J.C.   757
Kendelewicz, T.   559–562, 584, 585, 592, 774
Kendelewicz, T., see Babalola, I.A.   562
Kendelewicz, T., see List, R.S.   562, 588
Kendelewicz, T., see Newman, N.   588–591
Kendelewicz, T., see Petro, W.G.   561, 562
Kendelewicz, T., see Spicer, W.E.   588
Kendon, C., see Griffiths, K.   515
Kennedy, D.J.   331
Kennerly, R.E.   261, 262, 306
Kerchner, M.R., see Berman, M.   748
Kerkhoff, H.G., see Becker, U.E.   450
Kerkhoff, H.G., see Ferrett, T.A.   306, 331, 342, 444–447, 455
Kerkhoff, H.G., see Lindle, D.W.   331, 341, 446
Kerkhoff, H.G., see Reimer, A.   455
Kerkhoff, H.G., see Truesdale, C.M.   331, 340, 342, 443, 445–447
Kerr Del Grande, N., see Hubbell, J.M.   771
Kerr Del Grande, N., see McMaster, W.H.   752, 770, 771
Kessler, F.R.   730
Kessler, J., see Heinzmann, U.   236, 427
Kevan, S.D.   486, 487, 490, 491, 493, 508, 510
Kevan, S.D., see Rosenblatt, D.H.   510, 513
Kevan, S.D., see Smith, N.V.   125
Kevan, S.D., see Stoffel, N.G.   521
Khan, I., see Unwin, R.   264, 297, 298, 303, 332, 340
Khan, M.A., see Gross, J.G.   773
Kharitonov, Yu.I., see Band, I.M.   752, 771
Kholkin, A.L., see Trepakov, V.A.   774
Khumalo, F.S.   773, 774
Kibel, M.H.   245, 264
Kiefl, H.V., see Kamke, W.   456
Kiessling, J., see Schirmer, J.   714
Kim, F., see Hiraki, A.   546, 576
Kim, K.S.   644
Kimel, I.   753

Kimman, J.   332
Kimura, K.   320, 334, 396
Kimura, K., see Achiba, Y.   332
Kimura, K., see Katsumata, S.   245, 264
Kimura, K., see Sato, K.   332
Kincaid, B.M.   4, 253, 439
Kincaid, B.M., see Lee, P.A.   253
King, D.A., see Campuzano, J.-C.   496
King, D.A., see Debe, M.K.   496
King, D.A., see Griffiths, K.   515
King, D.A., see Jones, R.G.   503
King, D.A., see Jupille, J.   515, 675, 697
King, D.A., see Lamble, G.M.   513
King, F.W.   753, 755
King, G.C.   337, 365, 441
King, G.C., see Shaw, D.A.   340, 342, 450
King, G.C., see Tronc, M.   337–339
King, J.L.   164, 165
Kingman, D.D., see Bedford, K.L.   774
Kinno, S.   773, 774
Kinsbron, E., see Aspnes, D.E.   738, 763
Kinsey, L.   392
Kirby, K., see Roche, A.   410
Kirichenko, A.P., see Galuza, A.   773
Kirillova, M.M., see Gluskin, E.S.   774
Kirkland, J.P., see Hunter, W.R.   60, 61
Kirschner, J., see Eyers, A.   479
Kisker, E.   471, 481, 482
Kisker, E., see Clauberg, R.   655
Kitagawa, K.   235
Kitamura, H., see Maezawa, H.   117
Kjato, H., see Koide, T.   108
Klapp, M.P., see Arndt, D.P.   762, 763
Klasson, M., see Baer, Y.   650
Klebanoff, L.E.   482
Klebanoff, L.E., see Barton, J.J.   510, 598
Kleimenov, V.I.   245, 264
Klein, A., see Hormes, J.   137, 230
Kliewer, K.L.   672
Kliewer, K.L., see Koch, E.-E.   748, 749
Klucker, R.   744, 748, 749, 753
Knapp, J.A., see Chiang, T.-C.   478, 479
Knapp, J.A., see Eastman, D.E.   480
Knapp, J.A., see Himpsel, F.J.   652
Knapp, J.A., see Lapeyre, G.J.   599
Knight, B.W., see Peterson, C.M.   755
Knotek, M.L.   473
Kobayashi, K.L.I.   656
Kobayashi, K.L.I., see Taniguchi, M.   674, 692
Kobayashi, Y., see Katoh, H.   775
Kobrin, P.H., see Lindle, D.W.   331, 341, 446
Kobrin, P.H., see Southworth, S.H.   274, 275, 332, 340
Kobrin, P.H., see Truesdale, C.M.   244, 264,

331, 332, 340, 342, 407, 443, 445–447, 507
Koch, E.-E. 3, 243, 244, 253, 263, 398, 438, 441, 726, 748, 749, 770
Koch, E.-E., see Blechschmidt, D. 246–248, 305, 331, 336
Koch, E.-E., see Eberhardt, W. 337, 439
Koch, E.-E., see Fock, J.-H. 246, 262, 316, 331, 457
Koch, E.-E., see Gurtler, P. 263
Koch, E.-E., see Haelbich, R.-P. 771
Koch, E.-E., see Himpsel, F.J. 485
Koch, E.-E., see Jugnet, Y. 506
Koch, E.-E., see Lau, H.-J. 246, 331
Koch, E.-E., see Nicholls, J.M. 490, 491
Koch, E.-E., see Peatman, W.B. 333, 334, 398
Koch, E.-E., see Pulm, H. 246
Koch, E.-E., see Schwentner, N. 457
Koch, E.-E., see Uhrberg, R.I.G. 490
Koch, E.-E., see Weaver, J.H. 771
Kochubei, V.I., see Gluskin, E.S. 774
Koda, T., see Kaneko, Y. 774
Koda, T., see Mitani, T. 773, 774
Koda, T., see Tokura, Y. 775
Koelling, D.D. 638
Koestner, R.J. 246, 254, 331
Koestner, R.J., see Stöhr, J. 246, 254, 331, 517, 518, 520
Koide, T. 108
Koide, T., see MacDowell, A.A. 102, 106, 107
Koide, T., see Sato, S. 113
Kolesnikov, V.V., see Sachenko, V.P. 305, 337
Kollin, E.B., see Stöhr, J. 520, 521
Koma, A., see Ludeke, R. 514
Komiha, N. 406
Komninos, Y., see Matthew, J. 445
Kondo, D. 774
Kondratenko, A.V. 337
Konig, G., see Rinneberg, H. 193
Kono, S., see Goldberg, S.M. 716
Konstantinov, V.A., see Galuza, A. 773
Kopilevich, Y.I. 751
Köppel, H. 425
Köppel, H., see Haller, E. 320, 321, 329
Korenman, V. 482
Korsukova, M.M., see Farberovich, O.V. 774
Korsukova, M.M., see Gurin, V.N. 773
Kosman, W.M. 342
Kostroun, V.O. 30
Kostroun, V.O., see Hastings, J.B. 131
Kosugi, N. 442
Kosugi, N., see Iwata, S. 338
Kotani, A. 614, 629, 634, 638, 641, 642, 647, 649, 650, 656
Kotani, A., see Jo, T. 639, 653–656

Kotani, A., see Kaga, H. 644
Kotani, A., see Kakehashi, Y. 644
Kotani, A., see Parlebas, J.C. 656
Kovacs, J. 214
Kovtun, A.P., see Sachenko, V.P. 305, 337
Kowalczyk, S.P. 650
Kowalczyk, S.P., see Grant, R.W. 590
Kowalczyk, S.P., see McFeeley, F.R. 628
Kowalczyk, S.P., see Waldrop, J.R. 528, 529
Kowalski, E. 126, 128
Koyama, K., see Mitani, T. 774
Koyana, I., see Susuki, S. 379
Kraemer, P., see von Niessen, W. 305, 308, 309
Kraemer, W., see Diercksen, G. 384
Krafka, C., see Weaver, J.H. 771
Krakauer, H., see Wimmer, E. 689, 690
Krasnoperova, A., see Mazalov, L. 439
Krause, M.O. 244, 275, 371
Krause, M.O., see Carlson, T.A. 280, 284, 286, 287, 329, 331, 338–341, 402, 407, 422–425
Krause, M.O., see Grimm, F.A. 339, 424
Krause, M.O., see Keller, P.R. 340
Krause, M.O., see Piancastelli, M.N. 341
Krauss, M. 252
Kraut, E.A., see Grant, R.W. 590
Kraut, E.A., see Waldrop, J.R. 528, 529
Krebs, H.J., see Bonzel, H.P. 702
Krebs, W., see Hormes, J. 137, 230
Kreile, J. 245, 264, 279, 297, 298, 303, 339–341
Krill, G. 635
Krinsky, S. 3, 6, 10
Krishnakumar, E., see Kumar, V. 245, 264
Krivaite, G., see Valiukonis, G. 774
Kroitoru, S.G., see Sobolev, V.V. 774
Kromme, J.G., see de Laat, C.T.A.M. 156
Kronast, W. 433, 446
Kronig, R. de L. 253
Kruit, P., see Kimman, J. 332
Krummacher, S. 263, 332, 339, 340, 402
Krummacher, S., see Eberhardt, W. 340, 450
Kubaschewski, O. 561, 577, 578
Kubiak, G.D., see Anderson, S.L. 332
Kubier, B. 774
Kubier, B., see Boudriot, H. 773
Kühle, H., see Chergui, M. 418, 457
Kuhlmann, E., see Clauberg, R. 655
Kukkonen, C.A., see Feldkamp, L.A. 773
Kulibekov, A.M., see Valiukonis, G. 774
Kulikova, I.M., see Barinskii, R.L. 336
Kumar, V. 245, 264, 680, 681, 704
Kumar, V., see Bahr, J.L. 244, 264
Kumar, V., see Blake, A.J. 244, 264

Kumar, V., see Tomanek, D. 685
Kumara, K., see Achiba, Y. 332
Kummer, P.S., see Tasker, R. 168
Kunz, C. 244, 253, 726, 755, 770
Kunz, C., see Barth, J. 654
Kunz, C., see Birken, H.-G. 755, 767
Kunz, C., see Boller, K. 108
Kunz, C., see Dietrich, H. 60
Kunz, C., see Eberhardt, W. 337, 439, 552, 586, 588, 689
Kunz, C., see Gerken, F. 675, 676, 685–688
Kunz, C., see Gudat, W. 364, 365, 372, 599, 755
Kunz, C., see Haensel, R. 750
Kunz, C., see Hagemann, H.J. 44–46, 731–735, 767, 771
Kunz, C., see Hogrefe, H. 767
Kunz, C., see Jark, W. 50
Kunz, C., see Kammerer, R. 674, 675, 688–690
Kunz, C., see Koch, E.-E. 726, 770
Kunz, C., see Lenth, W. 554
Kuo, B.C. 140
Kuppermann, A., see Mintz, D.M. 245, 264
Kuppermann, A., see Sell, J.A. 245, 264, 305–307, 309
Kurada, H., see Kosugi, N. 442
Kurganskii, S.I., see Farberovich, O.V. 774
Kustler, A., see Brehm, B. 333, 334
Kuwata, H., see Inagaki, T. 753
Kuyatt, C., see Simpson, J. 306
Kuznetsov, A.I., see Abramov, V.N. 772, 774
Kuznetsov, A.I., see Amitin, L.N. 773
Kuznetsov, A.I., see Trepakov, V.A. 774

Laaser, W., see Hormes, J. 137, 230
Labastie, P., see Bréchignac, C. 456
Lablanquie, P. 380, 381, 407, 416, 417, 434–438
Lablanquie, P., see Adam, M.Y. 295, 296, 425, 426
Lablanquie, P., see Eland, J. 381, 438
Lablanquie, P., see Hitchcock, A.P. 455
Lablanquie, P., see Millié, Ph. 436, 438
Lablanquie, P., see Morin, P. 244, 275, 292, 340, 370–372, 402, 418, 419, 450, 452, 454
Lablanquie, P., see Nenner, I. 442, 455
Lablanquie, P., see Richardson, P. 438
Lablanquie, P., see Roy, P. 329, 341, 371, 413, 414, 422–424
Lablanquie, P., see Swensson, S. 402
Lagarde, P., see Comin, F. 498
Lagerqvist, A. 214
Lahmani, F. 390, 392–394, 396

Lamanna, U., see Gianturco, F.A. 255, 305, 336
Lamble, G.M. 513
Lamble, G.M., see Jones, R.G. 503
Lampe, F., see Ding, A. 456
Landau, L.D. 482
Landgren, G. 502, 503, 519, 586, 702
Landgren, G., see Ludeke, R. 586
Lane, A.M. 193
Lane, A.M., see Connerade, J.P. 193, 194, 209
Lane, N.F. 245, 246, 252, 261, 262, 285
Lang, N.D., see Williams, A.R. 673
Langer, B., see Becker, U. 450
Langer, B., see Reimer, A. 455
Langhoff, P.W. 243, 246, 251, 255, 263, 297, 298, 303, 337–339, 341, 410
Langhoff, P.W., see Diercksen, G. 384
Langhoff, P.W., see Gerwer, A. 338
Langhoff, P.W., see Hermann, M.R. 256, 339, 341
Langhoff, P.W., see Hitchcock, A.P. 340
Langhoff, P.W., see Machado, L.E. 297, 298, 303, 340
Langhoff, P.W., see Orel, A.E. 339
Langhoff, P.W., see Padial, N. 329, 338, 339, 413, 414
Langhoff, P.W., see Rescigno, T.N. 337, 338
Langhoff, P.W., see Williams, G.R.J. 340
Langhoff, S.R., see Diercksen, G. 384
Langhoff, S.R., see Langhoff, P.W. 263, 337, 339
Lanza, C., see Woodall, J.M. 593
Lapeyre, G.J. 599
Lapeyre, G.J., see Cerrina, F. 478
Lapeyre, G.J., see Smith, R.J. 504
Laporte, P. 774
Lardeux, C. 390, 391, 393–396
Lardeux, C., see Lahmani, F. 390, 392–394, 396
Larkins, F.P., see Richards, J.A. 243, 246, 275
Larsen, P.K. 503
Larsen, P.K., see van der Veen, J.F. 562
Larson, B.C. 474
Larsson, C.G., see Kanski, J. 655
Larsson, S. 644, 647
Lassailly, Y., see Desjonquères, M.C. 488, 666, 673, 677
Lassailly, Y., see Tran Minh Duc 665, 666, 669, 670, 672, 675, 689, 712, 714
Lassailly, Y., see Tréglia, G. 700, 702
Lässer, R. 635
Lässer, R., see Fuggle, J.C. 629, 635, 637, 639

Lau, H.-J. 246, 331
Lau, H.-J., see Fock, J.-H. 246, 331
Laubschat, C., see Domke, M. 484
Laubschat, C., see Kaindl, G. 675, 676, 685, 686, 707, 708
Laubschat, C., see Mandel, T. 512
Laubschat, C., see Schneider, W.D. 639, 688
LaVilla, R.E. 246–248, 305, 309, 317, 336, 337
Lavollée, M. 365
Lavollée, M., see Dutuit, O. 455
Lavollée, M., see Lahmani, F. 390
Lavollée, M., see Richard-Viard, M. 380, 402–405
Le Calvé, J. 455
Le Calvé, J., see Castex, M.C. 455, 456
Le Calvé, J., see Dutuit, O. 455
Le Calvé, J., see Jordan, B. 455
Le Calvé, J., see Moeller, T. 418
Le Coat, Y., see Tronc, M. 442
Le Rouzo, H., see Raoult, M. 230, 244, 264, 268, 290, 292–294, 331, 412, 413, 419, 421, 422, 433
Le Rouzo, H., see Rascev, G. 243, 244, 246, 264, 268, 280, 285, 331, 338, 339, 419, 422
Leach, S. 364
Leach, S., see Dujardin, G. 380, 381, 407, 427, 435, 436
Leal, E.P. 280, 285, 341
Leal, E.P., see Machado, L.E. 297, 298, 303, 340
Lecante, J., see Bertolini, J.C. 675
Lecante, J., see Chandesris, D. 656
Lecante, J., see Chauveau, D. 675, 694
Lecante, J., see Guillot, C. 488, 496, 652–654, 674, 675, 695–697, 702
Lecante, J., see Louie, S.G. 486, 487
Lecante, J., see Miranda, R. 482
Lecante, J., see Sinkovic, B. 511
Lecante, J., see Soukiassian, P. 667, 702, 703
Lecante, J., see Spanjaard, D. 489, 666, 685
Lecante, J., see Tran Minh Duc 665, 666, 669, 670, 672, 675, 689, 712–714
Lecante, J., see Tréglia, G. 700, 702
LeDourneuf, M., see Johnson, W.R. 333
LeDourneuf, M., see Schneider, B.I. 285
Lee, G. 155
Lee, L.C. 305, 307, 382, 407
Lee, L.C., see Suto, M. 382
Lee, M.-T., see Lynch, D.L. 297, 298, 303, 304, 341
Lee, P., see Henke, B.L. 42, 47, 49, 752, 772
Lee, P.A. 253
Lee, S.T. 499, 572

Lee, S.T., see Apai, G. 499
Lee, S.T., see Rosenberg, R.A. 208
Lee, S.T., see Suzer, S. 205
Lee, Y., see Trevor, D. 456
Lefebvre, H., see Keller, F. 409
Lefebvre-Brion, H. 244, 264, 268, 331
Lefebvre-Brion, H., see Giusti-Suzor, A. 227, 244, 264, 268, 331
Lefebvre-Brion, H., see Morin, P. 244, 263, 264, 331, 332, 340, 418–420
Lefebvre-Brion, H., see Raoult, M. 230, 244, 264, 268, 290, 292–294, 331, 412, 413, 419, 421, 422, 433
Lefebvre-Brion, H., see Raseev, G. 243, 246, 280, 285, 331, 338, 339, 409, 410, 419, 422
LeLay, G. 557
Lemonnier, J.C., see Pollini, I. 774, 775
Lemonnier, J.C., see Quemcrais, A. 774
Lempka, H.J., see Potts, A.W. 305–307, 309, 320
Leng, F.J., see Kibel, M.H. 245, 264
Lenham, A.P., see Graves, R.H.W. 748
Lenselink, A., see Pollini, I. 775
Lenth, W. 554
Lepère, D. 68
Lergrin, A., see Hitchcock, A.P. 455
Lerner, J.M., see Neviere, M. 82, 83
Leroi, G.E., see Poliakoff, E.D. 333, 366, 415, 432, 433, 457
Leung, K.T., see White, M.G. 340
Leveque, G. 749, 752–754, 774
Leveque, G., see Bertrand, Y. 773
Levesque, R.A., see Golovchenko, J.A. 49
Levine, R. 392
Levine, Z.H. 243, 246, 264, 297, 298, 303–305, 328, 329, 331, 340, 341
Levinson, H.J. 305, 309, 317, 329, 338, 486
Levinson, H.J., see Bradshaw, A. 402
Levinson, H.J., see Gustafsson, T. 339
Levy, F., see Bertrand, Y. 773
Lewandowski, B. 208
Lewonczak, S., see Gross, J.G. 773
Ley, L. 614, 666
Ley, L., see Cardona, M. 614
Ley, L., see Kowalczyk, S.P. 650
Ley, L., see McFeeley, F.R. 628
Leyh, B., see Raseev, G. 409, 410
Li, C.H., see Rosenblatt, D.H. 513
Li, K., see Bonapace, C.R. 573, 578
Liang, K.S. 572
Liang, W.Y., see Bauliss, S.C. 775
Liberman, S., see Connerade, J.P. 226, 233
Libermann, D., see Cromer, D.T. 771
Lichtenberger, D., see Piancastelli, M.N. 341

Liebermann, L.N. 482
Liebsch, A. 478, 481, 508, 651, 652
Liebsch, E.W., see Gustafsson, T. 246, 254, 331
Lien, N., see Brown, F.C. 548
Lienard, A. 3
Lif, M., see Bedford, K.L. 774
Lindau, I. 577, 583, 589, 590, 599
Lindau, I., see Abbati, I. 476, 477, 519, 556
Lindau, I., see Allen, J.W. 656
Lindau, I., see Apai, G. 675, 694, 702, 703
Lindau, I., see Babalola, I.A. 562
Lindau, I., see Braicovich, L. 550, 557
Lindau, I., see Chye, P.W. 572, 576, 577
Lindau, I., see Johannson, L.I. 554
Lindau, I., see Kendelewicz, T. 559–562, 584, 585, 592
Lindau, I., see List, R.S. 562, 588
Lindau, I., see Perfetti, P. 557
Lindau, I., see Petro, W.G. 561, 562, 584
Lindau, I., see Pianetta, P. 514
Lindau, I., see Rossi, G. 519, 554–557
Lindau, I., see Silberman, J.A. 565
Lindau, I., see Skeath, P.R. 558, 559, 572
Lindau, I., see Spicer, W.E. 493, 502, 519, 545, 565, 583, 584, 589
Lindau, I., see Su, C.Y. 519
Lindau, I., see Tatchyn, R. 750
Lindberg, B., see Siegbahn, K. 613
Lindgren, I., see Siegbahn, K. 613, 665, 666
Lindle, D.W. 331, 341, 446
Lindle, D.W., see Ferrett, T.A. 306, 331, 342, 444–447, 455
Lindle, D.W., see Piancastelli, M.N. 443
Lindle, D.W., see Southworth, S.H. 274, 275, 332, 340
Lindle, D.W., see Truesdale, C.M. 244, 264, 331, 332, 340, 342, 407, 443, 445–447, 507
Lindsey, K. 106
Lindsey, K., see Franks, A. 83, 105, 106
Lippiatt, A.G. 151
List, R.S. 562, 588
List, R.S., see Kendelewicz, T. 561, 562, 584, 585, 592
Lister, A.M. 152, 157
Liston, S.K., see Dibeler, V.H. 263, 320
Lizon, E., see Hitchcock, A.P. 455
Lloyd, D.E., see Kelly, M.M. 106
Loenen, E.J., see Tromp, R.M. 495
Loisel, B., see Quemerais, A. 774
Look, D.C., see Armaly, B.F. 746
Loomba, D. 255, 258, 259, 339
Lopez Delgado, R., see Lavollée, M. 365
Lorents, D., see Black, G. 456

Lorin, H. 162
Loubriel, G. 506
Louie, S.G. 486, 487, 546, 593
Louie, S.G., see Eberhardt, W. 501
Louie, S.G., see Schlüter, M. 490
Lozes, R.L. 251
Lu, B.-C., see Kevan, S.D. 508, 510
Lu, C.C., see Carlson, T.A. 264
Lu, K.T. 184, 227
Lubinsky, A.R., see Tong, S.Y. 692, 717
Lucatorto, T.B., see Hill, W.T. 227
Lucchese, R.R. 243, 246, 280, 285–287, 295, 296, 318, 329, 339, 340, 413, 414, 417, 418, 423–426
Lucchese, R.R., see Lynch, D.L. 243, 246, 251, 297, 298, 303, 304, 341
Lucchese, R.R., see McKoy, V. 340, 341
Lucchese, R.R., see Smith, M.E. 332, 341
Ludeke, R. 514, 527, 550, 562, 570–572, 579, 584–586, 593, 674, 692, 693
Ludeke, R., see Hughes, G. 502
Ludeke, R., see Landgren, G. 502, 503, 519, 586, 702
Ludema, K.C., see Vorburger, T.V. 757
Ludwig, K.F., see Fontaine, A. 762
Lukirskii, D.P., see Zhuze, V.P. 773
Lund, P.A., see Snyder, P.A. 230
Lutz, F., see Lenth, W. 554
Lynch, D.L. 243, 246, 251, 297, 298, 303, 304, 331, 341
Lynch, D.L., see McKoy, V. 340
Lynch, D.W. 726, 765, 766, 770, 772, 773
Lynch, D.W., see Aspnes, D.E. 741
Lynch, D.W., see Leveque, G. 749, 774
Lynch, D.W., see Olson, C.G. 741
Lynch, D.W., see Weaver, J.H. 764, 765, 771
Lynch, D.W., see Wieliczka, D. 484

Maani, C., see Williams, R.H. 562, 567, 572, 573
MacDowell, A.A. 97, 102, 106, 107, 134, 135, 472
Machado, L.E. 297, 298, 303, 340
Machado, L.E., see Leal, E.P. 280, 285, 341
Mack, R.A. 28
Mackey, K.J. 531, 532
Macleod, H.A., see Borgogno, J.P. 738, 762
Macleod, M.A., see Arndt, D.P. 762, 763
MacRae, R.A. 752
MacRae, R.A., see Hamm, R.N. 45, 757
Madden, R.P. 66, 70, 194, 200, 201
Madden, R.P., see Jacobus, G.F. 108
Madden, R.P., see Rabinovitch, K. 45
Madden, R.P., see Timothy, J.G. 125, 134

Madey, T.E., see Bertel, E. 445
Madey, T.E., see Stockbauer, R. 507
Madix, R.J., see Stöhr, J. 246, 254, 331, 517, 518
Maevskii, V.M., see Gluskin, E.S. 774
Maezawa, H. 54, 117
Mahan, A.I. 42
Mahan, G.D. 626, 713
Mahowald, P., see Spicer, W.E. 588
Mahr, D., see Hotop, H. 208
Makarov, O.A., see Pajasova, L. 775
Makarova, E.G. 751
Makarova, E.G., see Kopilevich, Y.I. 751
Maksimova, O.G., see Sobolev, V.V. 774
Malinovich, Y., see Besnard, M. 381, 382, 434
Mallett, J.M., see Hubbell, J.R. 771
Mallett, J.M., see McMaster, W.H. 752, 770, 771
Malmquist, P., see Swensson, S. 402
Malutzki, R., see Derenbach, H. 244, 275
Mamy, R. 773
Mamy, R., see Pollini, I. 774
Mandel, T. 512
Mandel, T., see Chiang, T.-C. 512
Mandel, T., see Domke, M. 484
Mandl, F., see Bardsley, J.N. 245, 246, 252, 261, 262, 285
Mandl, F., see Herzenberg, A. 285
Manne, R., see Siegbahn, K. 401, 443
Manogue, C.A., see Smith, D.Y. 753
Mansfield, M.W.D., see Connerade, J.P. 235
Manson, S.T. 716
Manson, S.T., see Kennedy, D.J. 331
Manzke, R. 764, 765, 774
Maple, M.B., see Allen, J.W. 656
Maquet, A., see André, J.M. 761
Maradudin, A.A. 740
Maradudin, A.A., see Mills, D.L. 740
Marcus, P.M., see Demuth, J.E. 513
Margaritondo, G. 529, 530, 543, 546, 561, 562, 583, 588
Margaritondo, G., see Brillson, L.J. 546, 547, 550, 551, 559, 561–563, 569, 570, 578–580, 582, 591, 592
Margaritondo, G., see Brucker, C.F. 567, 587, 588, 591
Margaritondo, G., see Daniels, R.R. 526, 575, 583, 584
Margaritondo, G., see Davis, G.D. 564, 565, 581
Margaritondo, G., see Duke, C.B. 558, 559
Margaritondo, G., see Franciosi, A. 553
Margaritondo, G., see Kahn, A. 558, 559
Margaritondo, G., see Katnani, A.D. 528

Margaritondo, G., see Perfetti, P. 529, 530, 557
Margaritondo, G., see Richter, H.W. 560, 561
Margaritondo, G., see Rowe, J.E. 546
Margaritondo, G., see Schlüter, M. 503
Margaritondo, G., see Shapira, Y. 559
Margaritondo, G., see Stoffel, N.G. 562, 572–574, 599
Margaritondo, G., see Weaver, J.H. 548
Margaritondo, G., see Zhao, T.-X. 525, 560, 576
Marian, C. 403, 404
Marian, R., see Marian, C. 403, 404
Mariani, C. 506
Mark, P., see Goodwin, T.A. 545
Mark, P., see Kahn, A. 692
Marquardt, B., see Pulm, H. 246
Marr, G.V. 244, 264, 275, 279, 284, 290, 295, 296, 338, 426
Marr, G.V., see Holmes, R.M. 303, 339
Marr, G.V., see McCoy, D.G. 338
Marr, G.V., see West, J.B. 61
Marr, G.V., see Woodruff, P.R. 244, 264, 290, 407
Marra, W.C. 474, 598
Marra, W.C., see Eisenberger, P. 496, 598
Mårtensson, N. 477, 483, 484, 685, 707
Mårtensson, N., see Franciosi, A. 478
Mårtensson, N., see Johansson, B. 666, 678, 679, 684, 685, 688, 705
Mårtensson, N., see Kaindl, G. 483, 484, 675, 676, 685, 686, 707, 708
Mårtensson, N., see Nilsson, A. 675, 676, 685, 707
Mårtensson, N., see Parks, R.D. 675, 686
Mårtensson, P., see Nicholls, J.M. 490, 491
Martin, M.A.P., see Connerade, J.P. 208
Martin, R.L. 666
Martin, W.C. 181
Martzen, P.D. 773
Marvin, A. 740
Marvin, A., see Celli, V. 740
Marvin, A., see Toigo, F. 740
Mason, M.G., see Apai, G. 499
Mason, M.G., see Lee, S.T. 499, 572
Mason, M.G., see Rosenblatt, D.H. 510, 513
Massardier, J., see Bertolini, J.C. 675
Masuko, H., see Sasanuma, M. 305, 306, 338
Masuoka, T. 405, 434
Materlik, G. 474, 503
Materlik, G., see Bedzyk, M. 503, 504
Materlik, G., see Bonse, U. 761
Materlik, G., see Funke, P. 503
Matsumoto, H., see Kondo, D. 774

Matthew, J. 445
Mawryluk, A., see Tatchyn, R. 750
Mayer, J.W., see Hiraki, A. 546, 576
Mayer, J.W., see Ottaviani, G. 546
Mayer, J.W., see Woodall, J.M. 593
Mayhew, C. 226
Mayhew, C., see Baig, M.A. 181, 182, 201, 205
Mazalov, L.N. 319, 324, 328, 439
Mazalov, L.N., see Kondratenko, A.V. 337
McCants, C.E., see Newman, N. 588, 589
McCarthy, I., see Weigold, E. 401
McConville, C.F. 511
McCoy, D.G. 338
McCoy, D.G., see Marr, G.V. 244, 275, 284, 295, 296, 338, 426
McCulloh, K.E. 263, 303
McFeeley, F.R. 628
McFeeley, F.R., see Himpsel, F.J. 532, 533
McFeeley, F.R., see Kowalczyk, S.P. 650
McFeeley, F.R., see Morar, J.F. 674
McGill, T.C., see Barton, J.J. 693
McGill, T.C., see Swartz, C.A. 558
McGill, T.C., see Zur, A. 588
McGilp, J.F. 526, 561
McGlynn, S.P., see Baig, M.A. 214, 226, 228
McGlynn, S.P., see Connerade, J.P. 220, 226
McGlynn, S.P., see Dagata, J.A. 227
McGovern, I.T. 476
McGovern, I.T., see Hughes, G.J. 565, 567
McGovern, I.T., see McGilp, J.F. 526
McGovern, I.T., see McGrath, R. 508
McGovern, I.T., see Williams, R.H. 471, 476, 565–567
McGrath, R. 508
McGrath, R., see Norman, D. 472
McGuire, G.E., see Carlson, T.A. 245, 264
McIlrath, T.J., see Hill, W.T. 227
McIntyre, J.D.E. 730
McKinley, A. 526, 560, 573
McKinley, A., see Hughes, G.J. 565, 567, 573, 574
McKinley, A., see Williams, R.H. 545, 562, 565–567, 572, 573
McKinney, W.R. 68
McKoy, B.V., see Diercksen, G. 384
McKoy, B.V., see Gerwer, A. 338
McKoy, B.V., see Langhoff, P.W. 297, 298, 303, 338, 339, 410
McKoy, B.V., see Lucchese, R.R. 413, 414, 417, 418, 423–426
McKoy, B.V., see Machado, L.E. 297, 298, 303, 340
McKoy, B.V., see Orel, A.E. 339

McKoy, B.V., see Padial, N. 329, 338, 339, 413, 414
McKoy, B.V., see Rescigno, T.N. 338
McKoy, V. 340, 341
McKoy, V., see Lucchese, R.R. 243, 246, 280, 285–287, 295, 296, 318, 329, 339, 340
McKoy, V., see Lynch, D.L. 243, 246, 251, 297, 298, 303, 304, 331, 341
McKoy, V., see Smith, M.E. 332, 341
McLaren, R. 443
McLaren, R., see Ishii, I. 443
McMaster, W.H. 752, 770, 771
McMaster, W.H., see Hubbell, J.R. 771
McMenamin, J.C., see Bachrach, R.Z. 557, 558
McMenamin, J.C., see Bauer, R.S. 519
McMenamin, J.C., see Brillson, L.J. 557, 563, 567, 570
McMillan, M., see Kennerly, R.E. 261, 306
Mead, C.A. 591
Mehaffy, D., see Carlson, T.A. 280, 284, 286, 287, 329, 338, 339, 422, 424
Mehaffy, D., see Grimm, F.A. 339, 424
Mehaffy, D., see Keller, P.R. 264, 297, 298, 303, 305, 332, 340
Mei, W.N., see Tong, S.Y. 496
Meisels, G.G., see Batten, C.F. 320, 333, 334
Meixner, A.E. 774
Mele, E.J., see Ritsko, J.J. 773, 774
Melhorn, W., see Kronast, W. 433, 446
Memeo, R., see Nannarone, S. 490
Mensching, L., see Schwarz, W.H.E. 341
Mentall, J. 385, 390
Merritt, R. 146
Messmer, R.P. 506
Messmer, R.P., see Plummer, E.W. 506
Messmer, R.P., see Roche, M. 424, 425
Metcalf, M. 161
Meyer, A., see Krill, G. 635
Meyer, R.J., see Duke, C.B. 558, 559
Meyer, R.J., see Kahn, A. 558, 559
Michelis, B., see Abbati, I. 556
Middelmann, H.-U., see Mariani, C. 506
Middelmann, U., see Mandel, T. 512
Miedema, A.R. 681, 685
Mielczarek, S., see Simpson, J. 306
Mies, F.H. 244, 264, 268, 331, 333
Mies, F.H., see Krauss, M. 252
Migal, Yu.F., see Sachenko, V.P. 305, 337
Mikuni, A., see Ishiguro, E. 319, 324, 328, 340, 439, 441
Mikuni, A., see Kanamori, H. 341
Mikuni, A., see Koide, T. 108
Mikuni, A., see Ninomiya, K. 339, 439, 441

Miller, J.C. 332
Miller, J.C., see Glownia, J.H. 332
Miller, J.N., see Braicovitch, L. 550
Miller, P.J., see Dehmer, P.M. 341
Miller, R.F. 746
Miller, T. 502, 674, 691, 692
Miller, T., see Hsieh, T.C. 522
Miller, T., see Ludeke, R. 527, 562, 571, 579, 593
Miller, T.A., see Katayama, D.H. 181
Millié, P. 436, 438
Millié, P., see Bodeur, S. 442
Millié, P., see Lablanquie, P. 380, 381, 434–438
Millié, P., see Roy, P. 400–402, 424–427
Mills, D., see Larson, B.C. 474
Mills, D.L. 740
Mills, D.L., see Maradudin, A.A. 740
Mills, K.C. 577, 578
Milnes, A.G. 545, 590
Miloslavskii, V.K. 752
Mintz, D.M. 245, 264, 333, 334
Mintz, D.M., see Sell, J.A. 245, 264
Miranda, R. 482
Miret-Artes, S. 406
Misell, D.L. 758
Misiewicz, J., see Jezierski, K. 754
Mitami, S., see Maezawa, H. 54, 117
Mitani, T. 773, 774
Mitani, T., see Tokura, Y. 775
Mitchell, A.C.G. 232
Mitchell, K.A.R. 513
Mitchell, K.A.R., see Head, J.D. 571
Miya, T., see Sugawara, H. 656
Miyahara, T. 773, 774
Miyahara, T., see Hanyu, T. 775
Miyahara, T., see Koide, T. 108
Miyahara, T., see Sato, S. 773
Miyake, K.P. 60, 61
Mizushima, Y., see Kajiyama, K. 590
Mizuta, H., see Kotani, A. 638
Mo, D., see Spicer, W.E. 588
Mobilio, S., see Balerna, A. 498
Moccia, R., see Gianturco, F.A. 305
Moeller, T. 418, 455
Moeller, T., see Jordan, B. 455
Moeller, T., see Le Calvé, J. 455
Mollenauer, L.F. 757
Monahan, K. 365
Mönch, W. 492, 590
Moncton, D.E., see Dimon, P. 513
Montgomery, V., see Williams, R.H. 561, 565–567, 591, 592
Moore, C.E. 202

Moore, J.D. 652
Morar, J.F. 674
Morar, J.F., see Himpsel, F.J. 532, 533
Morar, J.F., see Hughes, G. 502
Morar, J.F., see Landgren, G. 502, 503, 519, 702
Morgan, P., see Silberman, J.A. 565
Morgan, P., see Spicer, W.E. 565
Morgenstern, R. 245, 264
Mori, H., see Mitani, T. 773, 774
Morimoto, K., see Kaneko, Y. 774
Morin, P. 244, 263, 264, 275, 292, 331, 332, 339, 340, 370–372, 376, 398, 399, 402, 407, 412, 413, 416–420, 428, 439, 444, 447, 448, 450, 452–454
Morin, P., see Adam, M.Y. 295, 296, 402, 424–426
Morin, P., see De Souza, G. 443–447
Morin, P., see Hitchcock, A.P. 455
Morin, P., see Hubin-Franskin, M.-J. 244, 264, 331
Morin, P., see Lablanquie, P. 380, 381, 407, 434–438
Morin, P., see Nenner, I. 442, 455
Morin, P., see Roy, P. 400–402, 424–427
Morin, P., see Swensson, S. 402
Morioka, Y. 337, 439, 442
Morioka, Y., see Akahori, T. 403
Morioka, Y., see Hayaishi, T. 263, 264, 297, 298, 303, 340
Morioka, Y., see Nakamura, M. 248, 305, 336, 439
Morioka, Y., see Sasanuma, M. 305, 306, 338
Morioka, Y., see Sato, S. 113
Morozov, V.N., see Makarova, E.G. 751
Morrell, R., see Lindsey, K. 106
Morris, M.A., see Webber, P.R. 674
Morris, R.J.T., see Wang, Y.-T. 164
Morrison, M.A., see Rumble, J.R. 263
Morrison, R. 125
Morse, M. 396
Morton, J.M., see Marr, G.V. 244, 275, 284, 295, 296, 338, 426
Morton, J.M., see McCoy, D.G. 338
Moruzzi, V.L. 652
Moruzzi, V.L., see Weaver, J.H. 555–557
Moseley, J. 402
Moskovitz, M., see Dignam, M.J. 731
Mosoya, S., see Fukamachi, T. 761
Mosteller, L.P. 748
Mu-Tao, L., see Leal, E.P. 280, 285, 341
Muetterties, E.L., see Johnson, A.L. 516
Muetterties, E.L., see Stöhr, J. 520
Muller, K.D. 147

Müller-Dethlefs, K. 332, 379, 450
Mulliken, R. 418
Munkby, L., see Nilsson, P.O. 753
Munro, I.H. 125, 130, 131, 136, 137
Murata, Y., see Daimon, H. 511
Murata, Y., see Kobayashi, K.L.I. 656
Murgai, V., see Mårtensson, N. 707
Murphy, E.L. 125
Murray, J.L., see Korenman, V. 482
Murty, M.V.R.K. 61
Myron, J.R., see Cerrina, F. 478

Nagakura, I. 629
Nagakura, I., see Kakizaki, A. 656
Nagakura, I., see Sato, S. 113
Nagakura, I., see Sugawara, H. 656
Nagel, S.R. 638
Nagoaka, S., see Susuki, S. 379
Naikoka, T., see Maezawa, H. 54, 117
Naito, K., see Hanyu, T. 775
Naito, K., see Sugawara, H. 656
Nakada, I., see Mitani, T. 773, 774
Nakada, I., see Suga, S. 774
Nakai, S., see Maezawa, H. 54, 117
Nakai, S., see Nakamura, M. 336, 442
Nakai, Y., see Nakamura, M. 336, 442
Nakajima, T., see Sasanuma, M. 306
Nakamura, M. 248, 305, 336, 439, 442
Nakamura, M., see Akahori, T. 403
Nakamura, M., see Hayaishi, T. 263, 264, 297, 298, 303, 340
Nakamura, M., see Itikawa, Y. 275
Nakamura, M., see Morioka, Y. 337, 439, 442
Nakamura, M., see Sasanuma, M. 305, 306, 338
Nakano, T., see Kotani, A. 656
Namioka, T. 54, 55, 64, 66
Namioka, T., see Noda, H. 64
Nannarone, S. 490
Nannarone, S., see Chiarello, G. 759
Nannarone, S., see Rossi, G. 554, 557
Narita, S., see Taniguchi, M. 656
Natalis, P. 245, 264
Natalis, P., see Caprace, G. 245, 264
Natalis, P., see Collin, J.E. 244, 245, 264
Neave, J.H., see van der Veen, J.F. 562
Nefedov, V.I. 248, 336
Neilsen, M., see Bohr, J. 496, 497
Nelin, C.J., see Bagus, P.S. 506
Nenner, I. 292, 341, 379, 402, 405, 406, 442, 455
Nenner, I., see Adam, M.Y. 295, 296, 425, 426
Nenner, I., see Baer, T. 244, 264, 333, 334, 375, 398, 400, 402
Nenner, I., see Bodeur, S. 442
Nenner, I., see de Souza, G. 443–447
Nenner, I., see Delwiche, J. 398, 428
Nenner, I., see Dutuit, O. 382–385
Nenner, I., see Eland, J.H.D. 381, 438
Nenner, I., see Guyon, P.M. 333, 334, 379, 382, 398, 402, 403, 408, 410, 411, 428–430
Nenner, I., see Hitchcock, A.P. 455
Nenner, I., see Hubin-Franskin, M.-J. 244, 264, 331
Nenner, I., see Ito, K. 407
Nenner, I., see Lablanquie, P. 380, 381, 407, 434–438
Nenner, I., see Millié, Ph. 436, 438
Nenner, I., see Morin, P. 244, 263, 264, 275, 292, 331, 332, 339, 340, 370–372, 376, 398, 399, 402, 418–420, 428, 444, 450, 452, 454
Nenner, I., see Roy, P. 329, 341, 371, 400–402, 413, 414, 422–427
Nenner, I., see Swensson, S. 402
Nenner, I., see Tabché-Fouhailé, A. 264, 331, 332, 339, 382, 418
Nestell, J.E. 738, 750, 764
Neufeld, J.D. 753
Neukammer, J., see Rinneberg, H. 193
Neviere, M. 82, 83
Neviere, M., see Flamand, J. 82
Neviere, M., see Jark, W. 82
Névot, L. 51, 52
Newman, J., see Grover, J. 456
Newman, N. 588–591
Newman, N., see Kendelewicz, T. 561, 562, 584, 585, 592
Newman, N., see Spicer, W.E. 588
Newstead, D.A., see Binns, C. 521
Newton, R.G. 256, 257, 410
Ng, C.Y., see Ono, Y. 263
Ng, C.Y., see Wu, C.Y.R. 263
Nicholls, J.M. 490, 491
Nicholls, J.M., see Uhrberg, R.I.G. 490
Nicklaus, E. 773
Nicolet, M.A., see Hiraki, A. 546, 576
Niedermayer, R., see Wehking, F. 525
Niehaus, A. 245, 264
Niehaus, A., see Morgenstern, R. 245, 264
Niehus, H., see Engel, T. 700
Nielsen, S.E., see Berry, R.S. 264, 274, 291
Nielsen, U., see Blechschmidt, D. 246–248, 305, 331, 336
Nielsen, U., see Brodmann, R.R. 365
Nielsen, U., see Klucker, R. 753
Nilsson, A. 675, 676, 685, 707

Nilsson, P.O. 725, 726, 750, 753, 770
Nilsson, P.O., see Fäldt, A. 773
Nilsson, P.O., see Kanski, J. 655
Ninomiya, K. 339, 439, 441
Niwano, M., see Koide, T. 108
Nizhnikova, G.P., see Farberovich, O.V. 774
Noda, H. 64
Noda, H., see Maezawa, H. 54, 117
Noda, H., see Namioka, T. 54, 55
Nodgren, J., see Ågren, H. 305, 309
Noeldeke, G., see Baig, M.A. 181, 182, 201
Nogami, J., see Abbati, I. 476, 477
Nogami, J., see Rossi, G. 554, 557
Noggle, T.S., see Larson, B.C. 474
Nomerovannaya, L.V., see Gluskin, E.S. 774
Nomura, O., see Iwata, S. 338
Nordberg, R., see Siegbahn, K. 613, 665, 666
Nordling, C., see Ågren, H. 305, 309
Nordling, C., see Baer, Y. 650
Nordling, C., see Siegbahn, K. 401, 443, 613, 665, 666
Norman, D. 471, 472, 518
Norman, D., see Howells, M.R. 61
Norman, D., see Lamble, G.M. 513
Norman, D., see MacDowell, A.A. 472
Norman, D., see McGrath, R. 508
Norman, D., see Owen, I.W. 473, 506, 507
Norman, D., see Stöhr, J. 472
Norris, C., see Binns, C. 482, 521, 522
Norrish, R., see Basco, N. 394
Northrup, J.E. 490
Northrup, J.E., see Uhrberg, R.I.G. 522, 523
Nosenzo, L. 773
Nowik, I., see Schneider, W.-D. 639
Nozières, P. 614, 626
Nozières, P., see Combescot, M. 635
Nussenzweig, H.M., see Altarelli, M. 753
Nyberg, G.L., see Kibel, M.H. 245, 264
Nye, J.F. 727
Nyholm, R. 674, 688, 689
Nyholm, R., see Hörnström, S.E. 705–707

Ochoa, J.G., see Armaly, B.F. 746
Oepen, H.D., see Eyers, A. 479
Oertel, H. 368, 457
Ogata, S., see Hara, S. 275
Ogawa, M. 294
Ogawa, M., see Lee, L.C. 382
Ogawa, T., see Katoh, H. 775
Oh, S.-J. 636
Oh, S.-J., see Allen, J.W. 656
Oh, S.-J., see Loubriel, G. 506
O'Halloran, M.A., see Guest, J.A. 333, 432
O'Handley, R.C. 757

Ohkuma, H., see Hanyu, T. 775
Ohkuma, H., see Miyahara, T. 774
Okamoto, K. 589
Olson, C.G. 741, 767
Olson, C.G., see Aspnes, D.E. 741
Olson, C.G., see Bartlett, R.J. 774
Olson, C.G., see Leveque, G. 749, 774
Olson, C.G., see Lynch, D.W. 765, 766, 773
Olson, C.G., see Weaver, J.H. 764, 765
Olson, C.G., see Wieliczka, D. 484
Onaka, R., see Kinno, S. 773, 774
O'Neill, D.G., see Franciosi, A. 557, 579
O'Neill, D.G., see Grioni, M. 552–554
Ono, Y. 263
Onuki, H. 774
Oowaki, Y., see Tokura, Y. 775
Orchard, A.F., see Burroughs, P. 638
Ordal, M.A., see Bell, R.J. 727
Orders, P.J. 510, 598
Orders, P.J., see Sinkovic, B. 511
Orel, A.E. 339
Orel, A.E., see Langhoff, P.W. 338
Oren, L., see Lee, L.C. 382
Orudzhev, G.S., see Valiukonis, G. 774
Osantowski, J.F., see Hunter, W.R. 108
Oshio, T., see Nakamura, M. 336, 442
Osland, C.D. 160
Osmun, J.W. 767, 768
Osterheld, B., see Heinzmann, U. 333, 427
Osuch, E.A., see Ono, Y. 263
Ottaviani, G. 546
Ottaviani, G., see Calandra, C. 557
Otto, A. 747
Otto, A., see Koch, E.-E. 748, 749
Oura, K., see Saitoh, M. 525
Outka, D., see Stöhr, J. 246, 254, 331, 517, 518
Owen, I.W. 473, 506, 507
Owen, I.W., see Norman, D. 518
Ozenne, J., see Moseley, J. 402

Padial, N. 329, 338, 339, 413, 414
Padial, N., see Langhoff, P.W. 339
Padmore, H.A. 61, 62, 82
Padmore, H.A., see Binns, C. 482
Pahas, P., see Pajasova, L. 775
Paigné, J., see Guillot, C. 652–654
Painter, L.R., see Arakawa, E.T. 773
Pajasova, L. 775
Palik, E.D. 770, 772
Palmberg, P.W. 548
Palmer, K.F. 753
Palmer, see Palik, E.D. 770, 772
Pan, S.H., see Newman, N. 588–591

Pan, S.H., see Spicer, W.E. 588
Pandey, K.C. 478, 490
Pantelouris, M., see Connerade, J.P. 212
Pantos, E. 158
Pantos, E., see Hubbard, D.J. 73
Paolucci, G., see McConville, C.F. 511
Park, A., see McKinley, A. 573
Parker, V.B., see Wagman, D.D. 561, 577, 578
Parks, R.D. 675, 686
Parks, R.D., see Mårtensson, N. 484, 707
Parlebas, J.C. 656
Parlebas, J.C., see Jo, T. 653–656
Parlebas, J.C., see Kotani, A. 638
Parr, A.C. 129, 244, 264, 268, 275–279, 287, 290, 291, 297–299, 303–305, 331, 332, 340, 370, 372
Parr, A.C., see Butler, J.J. 134
Parr, A.C., see Codling, K. 244, 264, 331
Parr, A.C., see Cole, B.E. 338
Parr, A.C., see Dehmer, J.L. 263, 287, 288, 295, 306, 308–315, 318, 319, 321–329, 340–342, 409, 410, 417, 422, 425, 439, 446
Parr, A.C., see Ederer, D.L. 244, 264, 331
Parr, A.C., see Holland, D.M.P. 279, 340, 341, 372
Parr, A.C., see Poliakoff, E.D. 333, 432, 433, 457
Parr, A.C., see Rosenstock, H. 402
Parr, A.C., see Southworth, S.H. 294, 296, 297
Parr, A.C., see Stockbauer, R. 332, 338
Parr, A.C., see West, J.B. 264, 268, 280, 285–287, 292, 293, 331, 339, 417, 418, 421, 422
Parratt, L.G. 50
Pashuk, I.P., see Vishnevskii, V.N. 773
Pastori Parravicini, G., see Bassani, F. 725
Patella, F., see Perfetti, P. 529, 530
Paton, A., see Duke, C.B. 558, 559
Paton, A., see Kahn, A. 558, 559
Paulick, T.C. 750
Paulikas, A.P., see Veal, B.W. 752, 764, 765
Pauling, L. 513, 568, 580, 590
Paulus, T.J., see Bedwell, M.O. 131
Pavinich, V.F. 748
Pavinich, V.F., see Belousov, M.V. 748
Pavlovic, Z. 262
Pearson, G.L., see Schockley, W. 545
Peatman, W. 377, 398
Peatman, W., see Frey, R. 381
Peatman, W., see Schlag, E. 402
Peatman, W.B. 333, 334
Pelletier, E., see Arndt, D.P. 762, 763
Pelletier, E., see Borgogno, J.P. 738, 762

Pendry, J.B. 666, 717
Pendry, J.B., see Moore, J.D. 652
Pendry, J.B., see Norman, D. 472
Penn, D.R. 651, 653
Penn, D.R., see Tersoff, J. 641
Penney, T., see Kaindl, G. 483, 484, 675, 676, 707
Pepper, S.V. 755
Perfetti, P. 529, 530, 557
Perfetti, P., see Franciosi, A. 553
Perlman, M.L., see Davenport, J.W. 692
Perlman, M.L., see Krinsky, S. 3, 6, 10
Persson, P.E.S., see Uhrberg, R.I.G. 490
Pescia, D., see Hezaveh, A.A. 482
Peterman, D.J. 565, 656
Peterman, D.J., see Lynch, D.W. 773
Peterman, D.J., see Phillip, P. 565
Petersen, H. 60
Petersen, H., see Bauer, R.S. 519
Petersen, H., see Bianconi, A. 338, 439
Peterson, C.M. 755
Peterson, D.T., see Peterman, D.J. 656
Petersson, L.-G., see Johansson, L.I. 482
Petit, R. 82
Petro, W.G. 561, 562, 584
Petro, W.G., see Babalola, I.A. 562
Petro, W.G., see Kendelewicz, T. 559–562, 584
Petro, W.G., see Newman, N. 588, 589
Petroff, Y. 364
Petroff, Y., see Chandesris, D. 656
Petroff, Y., see Guillot, C. 652–654
Petroff, Y., see Louie, S.G. 486, 487
Petroff, Y., see Miranda, R. 482
Pettifor, D.G., see Duthie, J.C. 688
Pettit, G.D., see Woodall, J.M. 593
Peyerimhoff, S., see Marian, C. 403, 404
Pflüger, J. 775
Phillip, P. 565
Phillip, P., see Chang, S. 519
Phillip, P., see Franciosi, A. 519, 520
Phillips, E., see Lee, L.C. 305, 307, 382
Phillips, J.C. 581
Phillips, J.C., see Andrews, J.M. 546
Phillips, J.C., see Pandey, K.C. 478
Piancastelli, M.N. 341, 443
Piancastelli, M.N., see Adam, M.Y. 402, 424, 425
Piancastelli, M.N., see Carlson, T.A. 341, 423, 425
Pianetta, P. 514
Pianetta, P., see Chye, P.W. 572, 576, 577
Pianetta, P., see Lindau, I. 577, 583, 589, 590, 599

Pianetta, P., *see* Spicer, W.E. 493, 502, 519, 545, 583, 589
Pianetta, P., *see* Su, C.Y. 519
Picozzi, P., *see* Balerna, A. 498
Pidzyrailo, M.S., *see* Vishnevskii, V.N. 773
Pijolat, M. 671, 709
Pincelli, U., *see* Cadioli, B. 245, 336
Pinchaux, R., *see* Gross, J.G. 773
Pinchaux, R., *see* Guillot, C. 652–654
Pinchaux, R., *see* Louie, S.G. 486, 487
Pindor, A.J. 482
Pittel, B., *see* Friedrich, H. 338
Piuzzi, F., *see* Castex, M.C. 456
Plaskett, T.S., *see* Kaindl, G. 483, 484, 675, 676, 707
Platau, A., *see* Fuggle, J.C. 639
Platzman, R.L. 359
Plummer, E.W. 337, 372, 407, 416, 470, 471, 478, 486, 506
Plummer, E.W., *see* Albert, M.R. 520
Plummer, E.W., *see* Allyn, C.L. 516
Plummer, E.W., *see* Bartynski, R.A. 486, 488
Plummer, E.W., *see* Chen, C.T. 78
Plummer, E.W., *see* Didio, R.A. 508, 509
Plummer, E.W., *see* Eberhardt, W. 340, 443, 450, 482, 483, 501
Plummer, E.W., *see* Freund, H.-J. 505, 506
Plummer, E.W., *see* Greuter, F. 502, 504, 505
Plummer, E.W., *see* Gustafsson, T. 246, 254, 329, 331, 338, 414
Plummer, E.W., *see* Heskett, D. 519
Plummer, E.W., *see* Levinson, H.J. 486
Plummer, E.W., *see* Schmeisser, D. 511
Poate, J.M., *see* Franciosi, A. 557
Poate, J.M., *see* Sinha, A.K. 546, 577
Pogarev, D.E., *see* Belousov, M.V. 752
Poliakoff, E.D. 333, 366, 415, 419, 432, 433, 457
Poliakoff, E.D., *see* Cole, B.E. 338
Poliakoff, E.D., *see* Pratt, S.T. 332
Poliakoff, E.D., *see* White, M.G. 132, 244, 275, 375
Pollak, F.H., *see* Goel, A.K. 773
Pollak, H., *see* Frey, R. 381
Pollak, H., *see* Schlag, E. 402
Pollak, R.A., *see* Kaindl, G. 483, 484, 675, 676, 707
Pollak, R.A., *see* Kowalczyk, S.P. 650
Pollak, R.A., *see* Ley, L. 666
Pollard, J., *see* Trevor, D. 456
Pollini, I. 774, 775
Polozhentsev, V.E., *see* Sachenko, V.P. 305, 337
Poole, J.H. 30

Poon, H.C., *see* Tong, S.Y. 717
Popper, P. 106
Porteus, J.O. 739, 754
Porteus, J.O., *see* Bennett, H.E. 50, 739
Potts, A.W. 305–307, 309, 320
Pouey, M., *see* Gürtler, P. 366
Praet, M.-T., *see* Natalis, P. 245, 264
Prange, R.E., *see* Korenman, V. 482
Pratt, R.M. 771
Pratt, S.T. 332
Pregenzer, A.L., *see* Barrus, D.M. 338
Preil, M.E. 775
Preil, M.E., *see* Fisher, J.E. 775
Press, W., *see* Grimm, F.A. 424, 426
Price, W.C. 227, 244, 264
Price, W.C., *see* Potts, A.W. 305–307, 309, 320
Prietsch, M., *see* Domke, M. 484
Prietsch, M., *see* Mandel, T. 512
Prince, K., *see* Hezaveh, A.A. 482
Prince, K.C., *see* McConville, C.F. 511
Priol, M. 726, 770
Pruett, C.H., *see* Brown, F.C. 548
Prutton, M., *see* Jones, R.G. 503
Puester, G. 137
Pullen, B.P., *see* Carlson, T.A. 264, 423, 425
Pulm, H. 246
Purcell, K.G., *see* Grider, D.E. 506
Purcell, K.G., *see* Jupille, J. 515, 675, 697
Purvis, M.K., *see* Arndt, D.P. 762, 763
Puschmann, A. 514, 520
Puschmann, A., *see* Döbler, U. 514
Pyle, I.C. 154, 157

Quaresima, C., *see* Perfetti, P. 529, 530
Quemerais, A. 774
Querry, M.R. 751
Querry, M.R., *see* Hale, G.M. 751
Quinn, D.M., *see* Arndt, D.P. 762, 763
Quinn, F.M., *see* Owen, I.W. 473, 506, 507

Rabalais, J. 396
Rabe, P., *see* Friedrich, H. 338, 439
Rabinovitch, K. 45
Radtke, E., *see* Connerade, J.P. 235
Raether, H. 740, 757
Raether, H., *see* Daniels, J. 757, 758
Rafeev, G., *see* Raoult, M. 230
Raisin, C., *see* Bertrand, Y. 773
Rake, F.M., *see* Tasker, R. 168
Rakes, T.R., *see* Franz, L.S. 164
Ramsey, J.B., *see* Cox, J.T. 108
Randall, K., *see* Frasinski, L. 379, 381, 402, 403, 438

Raoult, M. 230, 244, 264, 265, 268, 271–273, 290–294, 331, 412, 413, 419, 421, 422, 433
Raoult, M., see Jungen, Ch. 244, 264, 268
Raseev, G. 243, 244, 246, 264, 268, 275, 280, 285, 331, 338, 339, 409, 410, 419, 422, 423
Raseev, G., see Lefebvre-Brion, H. 244, 264, 268, 331
Raseev, G., see Lucchese, R.R. 243, 246, 296, 318, 340, 425, 426
Raseev, G., see Raoult, M. 244, 264, 268, 290, 292–294, 331, 412, 413, 419, 421, 422, 433
Rau, R. 201
Ravet, M.F., see Krill, G. 635
Read, F.H. 441
Read, F.H., see Harting, E. 278
Read, F.H., see King, G.C. 337, 365, 441
Read, F.H., see Shaw, D.A. 340, 342, 450
Read, F.H., see Tronc, M. 337–339
Reale, A., see Balerna, A. 498
Reguzzoni, E., see Nosenzo, L. 773
Rehfeld, N. 89, 90
Rehn, V. 48, 49, 105, 106, 740, 746, 761
Rehn, V., see Monahan, K. 365
Reifenberger, R., see Franciosi, A. 565
Reihl, B. 482
Reihl, B., see Kaindl, G. 483, 484, 675, 676, 685, 686, 707, 708
Reihl, B., see Mårtensson, N. 484, 707
Reihl, B., see Parks, R.D. 675, 686
Reihl, B., see Schmeisser, D. 511, 512
Reihl, B., see Schneider, W.D. 688
Reilly, J.P., see Wilson, W.G. 332
Reimer, A. 455
Renhorn, I., see Lagerqvist, A. 214
Rescigno, T.N. 337, 338
Rescigno, T.N., see Diercksen, G. 384
Rescigno, T.N., see Langhoff, P.W. 263, 338, 339
Rescigno, T.N., see Orel, A.E. 339
Rhoderick, E.H., see Turner, M.J. 583
Rhodin, T.N., see Dowben, P.A. 503
Ribarsky, M.W., see Wall, W.E. 766, 767
Richard-Viard, M. 380, 402–405
Richard-Viard, M., see Dujardin, G. 380, 427, 435, 436
Richards, J.A. 243, 246, 275
Richardson, C.H., see Norman, D. 472, 518
Richardson, C.H., see Owen, I.W. 473, 506, 507
Richardson, D. 643
Richardson, N.V., see Grider, D.E. 506
Richardson, N.V., see Unwin, R. 264, 297, 298, 303, 332, 340

Richardson, P. 438
Richter, H.W. 560, 561
Riedel, R.A., see Davis, G.D. 564, 565
Rieger, D. 137, 150
Rieger, D., see Himpsel, F.J. 532, 533
Rieger, D., see Schnell, R.D. 137, 150, 503, 674
Riera, A., see Borondo, F. 390
Rife, J.C., see Bedford, K.L. 774
Rife, J.C., see Hunter, W.R. 60, 61
Rife, J.C., see Onuki, H. 774
Rigeissen, J., see Gross, J.G. 773
Riley, C.E., see Puschmann, A. 520
Riley, S.J., see Glownia, J.H. 332
Rimmer, E.M. 147
Rimmer, M.P. 752
Rinneberg, H. 193
Ritchie, B. 339
Ritchie, R.M. 758, 759
Ritchie, R.M., see Elson, J.M. 740
Ritsko, J.J. 773, 774
Ritsko, J.J., see Preil, M.E. 775
Ritz, V.H., see Bermudez, V.M. 757
Rivory, J. 755
Riwan, R., see Soukiassian, P. 667, 702, 703
Rizzi, A., see Chang, S. 519
Robaux, O. 185
Robey, S.W., see Barton, J.J. 510, 598
Robey, S.W., see Tobin, J.G. 521
Robin, M.B. 324, 328, 337
Robin, M.B., see Ishii, I. 443
Robin, S., see Priol, M. 726, 770
Robin, S., see Stephan, G. 770
Robinson, G.Y. 546, 577
Robinson, I.K. 495, 496
Robinson, I.K., see Bohr, J. 496, 497
Roche, A.L. 410
Roche, A.L., see Raseev, G. 338
Roche, M. 424, 425
Roche, P., see Borgogno, J.P. 738, 762
Roessler, D.M. 751, 752
Rohr, K. 3–6
Rohrer, H., see Binnig, G. 495
Roick, E., see Gürtler, P. 366
Romand, J., see Vodar, B. 44
Ron, A., see Pratt, R.M. 771
Rooy, T.L., see Huijser, A. 558, 559
Rose, S.J. 205, 208
Rose, S.J., see Connerade, J.P. 208
Rosei, R., see Olson, C.G. 741
Rosenberg, R.A. 208
Rosenberg, R.A., see White, M.G. 132, 244, 275, 375
Rosenblatt, D.H. 510, 513

Rosenblatt, D.H., see Kevan, S.D. 508, 510
Rosencwaig, A. 644, 645
Rosencwaig, A., see Cohen, R.L. 627, 628
Rosengren, A. 488, 666, 680, 681, 684, 703, 704
Rosenstock, H.M. 402
Rosenstock, H.M., see Botter, R. 298
Rosenwinkel, E., see Miller, T. 674, 691, 692
Rosner, R.A. 162, 166
Rossi, G. 519, 554–557
Rossi, G., see Abbati, I. 519
Rossi, G., see Comin, F. 498
Rossi, G., see Perfetti, P. 557
Rossinelli, M. 774
Rothberg, G.M., see Clauberg, R. 655
Roubin, P., see Chauveau, D. 675, 694
Roubin, P., see Guillot, C. 488, 674, 702
Rowe, E.M. 28, 548
Rowe, E.M., see Snyder, P.A. 230
Rowe, J.E. 546, 583, 588
Rowe, J.E., see Brennan, S. 674, 690, 691
Rowe, J.E., see Citrin, P.H. 503, 514
Rowe, J.E., see Comin, F. 595, 596
Rowe, J.E., see Franciosi, A. 557
Rowe, J.E., see Margaritondo, G. 583, 588
Rowe, J.E., see Schlüter, M. 503
Rowe, J.E., see Weeks, S.P. 137
Roy, D., see Roy, P. 329, 341, 371, 400–402, 413, 414, 422–427
Roy, P. 329, 341, 371, 400–402, 413, 414, 422–427
Roy, P., see Hubin-Franskin, M.-J. 244, 264, 331
Ruben, G.C., see Nestell, J.E. 738, 764
Rubloff, G.W. 557
Ruf, M.W., see Lewandowski, B. 208
Ruf, M.W., see Morgenstern, R. 245, 264
Ruf, M.W., see Niehaus, A. 245, 264
Rühl, E., see Beckmann, O. 368
Rumble, J.R. 263
Ruščić, B. 202

Sachenko, V.P. 305, 337
Sachs, M.W. 146, 153
Sadan, B., see Ahituv, N. 164
Sadovskii, A., see Mazalov, L. 439
Sagawa, T., see Blechschmidt, D. 246–248, 305, 331, 336
Sagawa, T., see Nagakura, I. 629
Sagawa, T., see Nakamura, M. 336, 442
Sagawa, T., see Suzuki, S. 633
Saile, V. 23
Saile, V., see Gurtler, P. 263
Saile, V., see Peatman, W.B. 333, 334, 398

Saile, V., see Rieger, D. 137, 150
Saile, V., see Sprüssel, G. 773
Saitoh, M. 525
Sakamoto, H., see Taniguchi, M. 656
Sakata, S., see Kajiyama, K. 590
Sakisaka, Y., see Dowben, P.A. 503
Salahub, D., see Roche, M. 424, 425
Salaneck, W.R., see Liang, K.S. 572
Samson, J.A.R. 44, 45, 278, 337, 372, 414
Samson, J.A.R., see Gardner, J.L. 245, 264, 286, 287, 372, 418
Samson, J.A.R., see Hancock, W.H. 245, 264
Samson, J.A.R., see Masuoka, T. 434
Sander, M., see Müller-Dethlefs, K. 332, 379, 450
Sands, M. 6
Sang Jr, H.W., see Chiaradia, P. 591
Sang Jr, H.W., see Katnani, A.D. 530, 591
Santucci, S., see Balerna, A. 498
Saprykhina, E.A., see Kondratenko, A.V. 337
Saris, F.W., see van der Veen, J.F. 515
Sasaki, F., see Bagus, P.S. 649
Sasaki, T., see Haensel, R. 750
Sasaki, T., see Ishiguro, E. 319, 324, 328, 340, 439, 441
Sasaki, T., see Kanamori, H. 341
Sasaki, T., see Maezawa, H. 54, 117
Sasaki, T., see Ninomiya, K. 339, 439, 441
Sasaki, T., see Shiles, E. 731–735, 764
Sasanuma, M. 305, 306, 338
Sasanuma, M., see Hayaishi, T. 263, 264, 297, 298, 303, 340
Sasanuma, M., see Morioka, Y. 337, 439, 442
Sasanuma, M., see Nakamura, M. 248, 305, 336, 439, 442
Sassaroli, P., see Nannarone, S. 490
Sato, H., see Itikawa, Y. 275
Sato, K. 332
Sato, K., see Achiba, Y. 332
Sato, K., see Suga, S. 774
Sato, S. 113, 773
Sato, S., see Koide, T. 108
Sato, S., see Miyahara, T. 773
Sato, S., see Nakamura, M. 336, 442
Saunders, V.R., see Delaney, J.J. 332, 340
Savoia, A., see Perfetti, P. 529, 530
Sawatzky, G.A., see Haak, H. 450
Sawatzky, G.A., see Van der Laan, G. 644, 647
Sawatzky, G.A., see Zaanen, J. 648
Schaefer, H.F., see Bagus, P.S. 251
Schafer, W., see Heidrich, K. 774
Schäfers, F. 236, 333, 428
Schäfers, F., see Eyers, A. 35, 479

Schäfers, F., see Heckenkamp, Ch. 35, 375
Schäfers, F., see Heinzmann, U. 333, 427
Schalpbach, L., see Erbudak, M. 674, 675
Schatz, P.N., see Snyder, P.A. 230
Scheer, J.J., see van Laar, J. 545, 586, 589
Schenk, H., see Oertel, H. 368, 457
Schevchik, N.Y., see Kemeney, P.C. 639, 650
Schiller, J., see Hormes, J. 137, 230
Schirmer, J. 263, 332, 714
Schirmer, J., see Cederbaum, L.S. 263, 332, 400
Schirmer, J., see Langhoff, P.W. 263, 339
Schirmer, J., see Reimer, A. 455
Schlag, E.W. 402
Schlag, E.W., see Frey, R. 381
Schlag, E.W., see Müller-Dethlefs, K. 332, 379, 450
Schlag, E.W., see Peatman, W. 377
Schledermann, M. 757
Schlegel, A. 773
Schlüter, M. 490, 503
Schlüter, M., see Larsen, P.K. 503
Schlüter, M., see Zhang, H.I. 546, 593
Schmeisser, D. 506, 507, 511, 512
Schmidt, F.A., see Franciosi, A. 557
Schmidt, F.A., see Weaver, J.H. 557
Schmidt, P.H., see Meixner, A.E. 774
Schmidt, V., see Derenbach, H. 244, 275, 370, 371
Schmidt, V., see Krummacher, S. 263, 332, 339, 340, 402
Schmidt-May, J., see Barth, J. 484
Schmidt-May, J., see Hörnström, S.E. 705–707
Schmidt-May, J., see Nyholm, R. 674, 688, 689
Schmitt, B., see Borgogno, J.P. 738, 762
Schnatterly, S.E. 757
Schnatterly, S.E., see Jasperson, S.N. 757
Schnatterly, S.E., see Nagel, S.R. 638
Schnatterly, S.E., see Treu, J.I. 757
Schneider, B.I. 243, 246, 251, 285, 341
Schneider, B.I., see Collins, L.A. 243, 246, 263, 264, 331, 332, 341, 413, 414, 423, 424
Schneider, H.A., see Boudriot, H. 773
Schneider, W.D. 639, 688
Schneider, W.D., see Domke, M. 484
Schneider, W.D., see Kaindl, G. 676, 685, 686, 707, 708
Schneider, W.D., see Mandel, T. 512
Schneider, W.D., see Mårtensson, N. 707
Schneider, W.D., see Wuilloud, E. 638
Schnell, R.D. 137, 150, 503, 674
Schnell, R.D., see Rieger, D. 137, 150

Schockley, W. 545
Schoen, R.I., see Doolittle, P.H. 244, 264
Schoenes, J. 775
Schollhorn, R., see Manzke, R. 764, 765, 774
Schönhammer, K. 635
Schönhammer, K., see Fuggle, J.C. 629, 635, 637, 638
Schönhammer, K., see Gunnarsson, O. 636–638
Schönhense, G. 333
Schönhense, G., see Eyers, A. 35, 479
Schönhense, G., see Heckenkamp, Ch. 35, 375
Schönhense, G., see Heinzmann, U. 333, 427
Schönhense, G., see Kaesdorf, S. 335
Schott, G.A. 3, 230
Schreiber, M., see Heidrich, K. 774
Schrieffer, J.R., see Davenport, J.W. 246
Schröder, K., see Kisker, E. 481
Schröder, K., see Weller, D. 482
Schulz, G.J. 245, 246, 252, 261, 262, 285, 442
Schulz, G.J., see Nenner, I. 442
Schulz, G.J., see Pavlovic, Z. 262
Schumm, R.H., see Wagman, D.D. 561, 577, 578
Schwartz, B., see Schwartz, G.P. 593
Schwartz, G.P. 593
Schwarz, S., see Braicovich, L. 550
Schwarz, W.H.E. 337, 341, 442
Schwarz, W.H.E., see Friedrich, H. 338, 439
Schweig, A., see Kreile, J. 245, 264, 279, 297, 298, 303, 339–341
Schweitzer, G.K., see Carver, J.C. 649
Schwentner, N. 457
Schwentner, N., see Chergui, M. 418, 457, 547
Schwentner, N., see Hahn, U. 365
Schwentner, N., see Munro, I.H. 125, 130, 131, 136, 137
Schwentner, N., see Wilcke, H. 365, 367
Schwinger, J. 28
Scouler, W.J. 752
Seah, M.P. 547, 548, 572, 709
Seaton, M.J. 182, 183, 186, 227
Seaton, M.J., see Baig, M.A. 181, 182, 201
Seaton, M.J., see Dubau, J. 186
Seaver, M., see White, M.G. 332
Seger, G., see Marian, C. 403, 404
Segev, E. 383, 385
Segmuller, A., see Haelbich, R.P. 50
Seki, K., see Nicholls, J.M. 490, 491
Seki, K., see Pulm, H. 246
Seki, M., see Suga, S. 774
Seki, M., see Taniguchi, M. 656, 674, 692

Seki, S., see Suga, S. 774
Sekreta, E., see Wilson, W.G. 332
Selander, L., see Ågren, H. 305, 309
Selci, S., see Nannarone, S. 490
Sell, J.A. 245, 264, 305-307, 309
Sen, A., see Franz, L.S. 164
Sen, S.K., see Gupta, R.P. 714
Sendall, D.M. 147, 164
Senn, P., see Keller, P.R. 340
Sergeeva, V.M., see Zhuze, V.P. 773
Sette, F. 342, 438, 443
Sette, F., see Eberhardt, W. 340, 443, 450
Sette, F., see Hitchcock, A.P. 341
Sette, F., see Norman, D. 472
Sette, F., see Perfetti, P. 529, 530
Sette, F., see Stöhr, J. 246, 254, 331, 517, 518, 520, 521
Seya, M., see Noda, H. 64
Sham, T.K., see Carr, R.G. 246, 254
Sham, T.K., see Davenport, J.W. 692
Sham, T.K., see Eberhardt, W. 340, 450
Shapira, Y. 559
Shapiro, M. 383, 395
Shapiro, M., see Segev, E. 383, 385
Sharp, T. 386
Sharpless, R., see Black, G. 456
Shaw, D.A. 340, 342, 450
Shelykh, A.I., see Bagdasarov, K.S. 775
Shelykh, A.I., see Farberovich, O.V. 774
Shelykh, A.I., see Gurin, V.N. 773
Shelykh, A.I., see Zhuze, V.P. 773, 774
Sher, A., see Spicer, W.E. 565
Shibuya, S., see Suga, S. 774
Shidara, T., see Koide, T. 108
Shieh, C.C., see Fisher, J.E. 775
Shigeyasu, T., see Kitagawa, K. 235
Shiles, E. 731-735, 764
Shiles, E., see Smith, D.Y. 753
Shimabukuro, R.L., see Henke, B.L. 42, 47, 49, 752, 772
Shimakura, H., see Katoh, H. 775
Shimamura, I. 245, 246, 252, 261, 262, 285
Shin, S., see Daimon, H. 511
Shin, S., see Mitani, T. 773
Shin, S., see Suga, S. 774
Shin, S., see Taniguchi, M. 674, 692
Shinozaki, S.S., see Feldkamp, L.A. 773
Shirley, D.A. 649
Shirley, D.A., see Barton, J.J. 510, 598
Shirley, D.A., see Fadley, C.S. 643, 649
Shirley, D.A., see Ferrett, T.A. 306, 331, 342, 444-447, 455
Shirley, D.A., see Hussain, Z. 472
Shirley, D.A., see Kevan, S.D. 508, 510

Shirley, D.A., see Klebanoff, L.E. 482
Shirley, D.A., see Kowalczyk, S.P. 650
Shirley, D.A., see Lindle, D.W. 331, 341, 446
Shirley, D.A., see Martin, R.L. 666
Shirley, D.A., see McFeeley, F.R. 628
Shirley, D.A., see Piancastelli, M.N. 443
Shirley, D.A., see Poliakoff, E.D. 433
Shirley, D.A., see Rosenberg, R.A. 208
Shirley, D.A., see Rosenblatt, D.H. 510, 513
Shirley, D.A., see Southworth, S.H. 274, 275, 332, 340
Shirley, D.A., see Suzer, S. 205
Shirley, D.A., see Tobin, J.G. 521
Shirley, D.A., see Trevor, D. 456
Shirley, D.A., see Truesdale, C.M. 244, 264, 331, 332, 340, 342, 407, 443, 445-447, 507
Shirley, D.A., see White, M.G. 132, 244, 275, 375
Shlarbaum, B., see Vinogradov, A.S. 337
Shobatake, K., see Achiba, Y. 332
Shoch, J.F. 168
Shockley, W. 486
Shoji, F., see Saitoh, M. 525
Shore, H.B., see Liebermann, L.N. 482
Shustorovich, E., see Baetzold, R.C. 675
Shuto, K., see Hiraki, A. 546, 576
Siddons, D.P. 761
Siddons, D.P., see Hart, M. 761
Sidorin, K.K., see Abramov, V.N. 772
Sidorin, K.K., see Bagdasarov, K.S. 775
Sidorin, K.K., see Farberovich, O.V. 774
Sidorin, K.K., see Gurin, V.N. 773
Sidorin, K.K., see Zhuze, V.P. 773, 774
Siegbahn, K. 401, 443, 613, 665, 666
Siegbahn, K., see Ågren, H. 305, 309
Siegbahn, K., see Baer, Y. 650
Siegel, J., see Dehmer, J.L. 261, 263, 289, 305
Siegel, J., see Dill, D. 246, 261, 331, 337, 338
Siegmann, H.C., see Bänninger, U. 651
Signorelli, A.J. 633
Silberman, J.A. 565
Silberman, J.A., see Kendelewicz, T. 559-562
Silberman, J.A., see Spicer, W.E. 565
Silcox, J., see Batson, P.E. 774
Silcox, J., see Chen, C.H. 760
Sileika, A., see Valiukonis, G. 774
Simon, M., see Nenner, I. 442, 455
Simpson, J. 306
Sinha, A.K. 546, 577
Sinkovic, B. 482, 511
Skeath, P.R. 558, 559, 572
Skeath, P.R., see Abbati, I. 556
Skeath, P.R., see Braicovich, L. 550, 557
Skeath, P.R., see Petro, W.G. 584

Skeath, P.R., see Spicer, W.E.   583, 584, 589
Skibowski, M., see Klucker, R.   744, 748, 749
Skibowski, M., see Schledermann, M.   757
Skibowski, M., see Sprüssel, G.   773
Skillman, S., see Herman, F.   716
Skinner, H.B., see Norman, D.   518
Skorinko, G., see Goel, A.K.   773
Slanger, T.   382
Slowik, J., see Brillson, L.J.   562, 579
Smeenk, R.G., see van der Veen, J.F.   515
Smirnov, I.A., see Gurin, V.N.   773
Smirnov, L.A.   52
Smit, L., see van der Veen, J.F.   562
Smith, A.   414, 416
Smith, C.S.   727
Smith, D.L., see Daw, M.S.   590
Smith, D.L., see Zur, A.   588
Smith, D.Y.   571, 731, 751, 753, 764, 771
Smith, D.Y., see Altarelli, M.   753
Smith, D.Y., see Shiles, E.   731–735, 764
Smith, G.C., see Binns, C.   482
Smith, J.R.   685
Smith, M.E.   332, 341
Smith, N.V.   125, 438, 470, 485, 486, 510, 666
Smith, N.V., see Kevan, S.D.   486, 487
Smith, N.V., see Larsen, P.K.   503
Smith, N.V., see Stoffel, N.G.   521
Smith, R.J.   504
Sneddon, G., see Albert, M.R.   520
Snider, D.R., see Tong, S.Y.   717
Snyder, P.A.   230
So, E., see Kahn, A.   692
Sobolev, V.V.   774
Sochtig, J., see Heidrich, K.   774
Sodhi, R.N.S.   342, 364, 365, 441, 447, 449
Sokolov, A.A.   28
Sokolov, A.V.   726, 728
Solgadi, D., see Lahmani, F.   390, 392–394, 396
Somerton, C., see Campuzano, J.-C.   496
Somerton, C., see Jones, R.G.   503
Sommer, K.   200
Sommer, K., see Baig, M.A.   201, 205
Sonntag, B., see Bruhn, R.   773
Sonntag, B., see Friedrich, H.   338, 439
Sonntag, B., see Haensel, R.   750
Sonntag, B.F., see Koch, E.-E.   243, 263, 398, 438, 441, 726, 770
Sorokin, I.A., see Filatova, E.O.   774, 775
Sotnikova, T.D., see Smirnov, L.A.   52
Soukiassian, P.   667, 702, 703
Southworth, S.H.   274, 275, 294, 296, 297, 332, 340
Southworth, S.H., see Dehmer, J.L.   263, 287, 288, 318, 319, 321–328, 341, 409, 410, 425, 439, 446
Southworth, S.H., see Parr, A.C.   244, 275–278, 370, 372
Southworth, S.H., see Poliakoff, E.D.   433
Southworth, S.H., see Trevor, D.   456
Southworth, S.H., see Truesdale, C.M.   244, 264, 332, 340, 407, 507
Southworth, S.H., see White, M.G.   132, 244, 275, 375
Soven, P., see Levine, Z.H.   243, 246, 264, 297, 298, 303–305, 328, 329, 331, 340, 341
Soven, P., see Levinson, H.   305, 309, 317, 329, 338
Spanier, R.F., see Cahan, B.D.   756
Spanjaard, D.   489, 666, 685
Spanjaard, D., see Chauveau, D.   675, 694
Spanjaard, D., see Desjonquères, M.C.   488, 666, 673, 677
Spanjaard, D., see Guillot, C.   488, 496, 674, 675, 695–697, 702
Spanjaard, D., see Tran Minh Duc   666
Spanjaard, D., see Tréglia, G.   651, 700, 702, 6651
Speer, R.J., see Franks, A.   83, 105
Spicer, W.E.   493, 502, 519, 545, 562, 565, 583, 584, 588, 589, 713
Spicer, W.E., see Abbati, I.   519, 556
Spicer, W.E., see Babalola, I.A.   562
Spicer, W.E., see Bauer, R.S.   754
Spicer, W.E., see Braicovich, L.   550, 557
Spicer, W.E., see Chye, P.W.   572, 576, 577
Spicer, W.E., see Helms, C.R.   567
Spicer, W.E., see Kendelewicz, T.   559–562, 584, 585, 592
Spicer, W.E., see Lindau, I.   577, 583, 589, 590, 599
Spicer, W.E., see List, R.S.   562, 588
Spicer, W.E., see Newman, N.   588–591
Spicer, W.E., see Petro, W.G.   561, 562, 584
Spicer, W.E., see Pianetta, P.   514
Spicer, W.E., see Rossi, G.   519, 554–557
Spicer, W.E., see Silberman, J.A.   565
Spicer, W.E., see Skeath, P.R.   558, 559, 572
Spicer, W.E., see Su, C.Y.   519
Spicer, W.E., see Wagner, L.F.   545
Spieker, P., see Bonse, U.   761
Spiller, E.   762
Spiller, E., see Haelbich, R.P.   50
Spiller, E., see Tatchyn, R.   750
Spizzichino, A., see Beckmann, P.   50
Spohr, R.   377
Sprüssel, G.   773
Srivastava, G.P., see Williams, R.H.   471, 476

Sroka, W., see Hertz, H. 407
Stallings, W. 166, 168, 169
Stanford, J.L., see Rehn, V. 106
Stankiewicz, M., see Frasinski, L. 381, 438
Stankovic, J.A. 162
Stapelfeldt, J., see Moeller, T. 455
Starace, A.F. 295, 304, 331
Starace, A.F., see Dehmer, J.L. 304
Starace, A.F., see Samson, J.A.R. 278
Stark, J.B., see Robinson, I.K. 495
Staunton, J., see Pindor, A.J. 482
Stavrakas, T.A. 233
Stearns, M.B., see Feldkamp, L.A. 759, 773
Stedman, M., see Franks, A. 106
Steel, T., see Hitchcock, A.P. 341
Steele, G. 364, 365
Stefani, G., see Giardini-Guidoni, A. 306
Steglich, F., see Lässer, R. 635
Stegun, I., see Abramowitz, M. 30
Steinmann, W., see Klucker, R. 744, 748, 749
Steinmann, W., see Rieger, D. 137, 150
Steinmann, W., see Schnell, R.D. 137, 150, 503, 674
Stephan, G. 770
Stephens, A. 425
Stephens, J.A. 263, 294–297, 310, 318, 329, 332, 339, 342
Stephenson, P.C., see Binns, C. 482, 521
Stern, E.A. 472
Stern, F. 725, 726, 728, 751
Stevenson, J.R., see Ellis, H.W. 753
Stevenson, J.R., see Hinson, D.C. 753
Stevenson, J.R., see Wall, W.E. 149, 766, 767
Stik, B.J.M., see Tong, S.Y. 692, 717
Stobie, R.W., see Dignam, M.J. 731
Stockbauer, R. 332–334, 338, 428, 507
Stockbauer, R., see Bertel, E. 445
Stockbauer, R., see Butler, J.J. 134
Stockbauer, R., see Codling, K. 244, 264, 331
Stockbauer, R., see Cole, B.E. 338
Stockbauer, R., see Ederer, D.L. 244, 264, 331
Stockbauer, R., see Parr, A.C. 129, 244, 264, 268, 275, 276, 278, 279, 287, 290, 291, 331
Stockbauer, R., see Rosenstock, H. 402
Stockbauer, R., see West, J.B. 264, 268, 280, 285–287, 292, 293, 331, 339, 417, 418, 421, 422
Stocks, G.M., see Pindor, A.J. 482
Stoffel, N.G. 521, 562, 572–574, 599
Stoffel, N.G., see Brillson, L.J. 546, 547, 559, 561–563, 578–580, 591, 592
Stoffel, N.G., see Brucker, C.F. 567, 587, 588, 591

Stoffel, N.G., see Kevan, S.D. 486, 487
Stöhr, J. 246, 254, 331, 471, 472, 505, 516–518, 520, 521, 523, 524, 526, 595
Stöhr, J., see Apai, G. 498
Stöhr, J., see Brennan, S. 513, 674, 690, 691
Stöhr, J., see Cerino, J. 94, 137
Stöhr, J., see Eberhardt, W. 340, 443, 450
Stöhr, J., see Hitchcock, A.P. 341, 443
Stöhr, J., see Hussain, Z. 472
Stöhr, J., see Johannson, L.I. 567
Stöhr, J., see Johnson, A.L. 516
Stöhr, J., see Koestner, R.J. 246, 254, 331
Stöhr, J., see Norman, D. 472
Stöhr, J., see Sette, F. 342, 438, 443
Stoll, W., see Hamnett, A. 337
Stolz, H.J., see Heidrich, K. 774
Storm, E. 770, 771
Straley, J.W., see White, J.J. 752
Strathy, I., see Greuter, F. 502
Strathy, I., see Heskett, D. 519
Streets, D.G., see Potts, A.W. 305–307, 309, 320
Strome, D.H., see Arndt, D.P. 762, 763
Strongin, M., see Eberhardt, W. 340, 450
Studna, A.A., see Aspnes, D.E. 756, 764
Su, C.Y. 519
Su, C.Y., see Abbati, I. 556
Su, C.Y., see Braicovich, L. 550, 557
Su, C.Y., see Chye, P.W. 572, 576, 577
Su, C.Y., see Lindau, I. 577, 583, 589, 590, 599
Su, C.Y., see Petro, W.G. 584
Su, C.Y., see Skeath, P.R. 558, 559, 572
Su, C.Y., see Spicer, W.E. 583, 584, 589
Suassuna, J.F., see Allen, J.W. 656
Subtil, J.L., see Laporte, P. 774
Suga, S. 774
Suga, S., see Mitani, T. 773
Suga, S., see Taniguchi, M. 656, 674, 692
Suga, T. 181
Sugano, S., see Asada, S. 644, 647–649
Sugar, J., see Dehmer, J.L. 304
Sugar, J., see Hill, W.T. 227
Sugawara, H. 656
Sugawara, H., see Kakizaki, A. 656
Sugawara, H., see Sato, S. 113
Sugeno, K., see Kakizaki, A. 656
Sunil, K., see Gregory, A. 406, 407
Šunjić, M.J., see Doniach, S. 626
Surman, M., see Hezaveh, A.A. 482
Surman, M., see McConville, C.F. 511
Susuki, S. 379
Susuki, Y., see Ishiguro, E. 439, 441
Suto, M. 382

Suttcliffe, D.C., see Hopgood, F.R.A. 160
Sutton, M., see Dimon, P. 513
Suzer, S. 205
Suzuki, S. 633
Suzuki, Y., see Ishiguro, E. 319, 324, 328, 340
Suzuki, Y., see Maezawa, H. 117
Svenson, A., see Åsbrink, L. 320
Svensson, W.A., see Carlson, T.A. 341, 423, 425
Swanson, J.R. 254, 255, 318–330, 339, 414, 423, 424
Swanson, L.W., see Chambers, S.A. 553
Swanson, R., see Dill, D. 245, 331, 338
Swartz, C.A. 558
Swenson, R., see Arndt, D.P. 762, 763
Swensson, S. 402
Szargan, R., see Schwarz, W.H.E. 341
Sze, S.M. 544, 545
Szostak, D., see Becker, U. 450
Szostak, D., see Reimer, A. 455

Tabché-Fouhailé, A. 264, 331, 332, 339, 382, 409, 418
Tabché-Fouhailé, A., see Baer, T. 244, 264, 333, 334, 375, 398, 400, 402
Tabché-Fouhailé, A., see Dutuit, O. 382–385
Tabché-Fouhailé, A., see Guyon, P.M. 333, 334, 379, 402
Tabché-Fouhailé, A., see Ito, K. 407
Taibin, B.Z., see Smirnov, L.A. 52
Takagi, H., see Itikawa, Y. 275
Takahashi, M., see Takayanagi, K. 495
Takahashi, S., see Takayanagi, K. 495
Takayanagi, K. 495
Takayanagi, K., see Shimamura, I. 245, 246, 252, 261, 262, 285
Takeuchi, T., see Kitagawa, K. 235
Tal, J., see Kuo, B.C. 140
Tam, W.-C. 337
Tambe, B.R., see Ritchie, B. 339
Tan, K.H., see Brion, C.E. 332, 338, 339, 414
Tanaka, I., see Tanaka, K. 245, 264
Tanaka, K. 245, 264, 403
Tanaka, K., see Susuki, S. 379
Tanaka, T.J., see Henke, B.L. 42, 47, 49, 752, 772
Tanaka, Y., see Ogawa, M. 294
Tanenbaum, A.S. 166
Tang, J.C., see Rosenblatt, D.H. 510
Taniguchi, M. 656, 674, 692
Taniguchi, M., see Suga, S. 774
Tanishiro, Y., see Takayanagi, K. 495
Tasker, R. 168
Tatchyn, R. 750

Taylor, A.J., see Miller, R.F. 746
Taylor, J.A., see Batten, C.F. 320, 333, 334
Taylor, J.W., see Carlson, T.A. 244, 264, 280, 284, 286, 287, 329, 331, 338–341, 407, 422–425
Taylor, J.W., see Grimm, F.A. 339, 341, 424
Taylor, J.W., see Keller, P.R. 264, 297, 298, 303, 305, 332, 340, 341
Taylor, J.W., see Piancastelli, M.N. 341
Tazzari, S., see Buras, B. 5
Teal, G.K. 544
Temkin, A., see Chandra, N. 285
Temple, P.A., see Arndt, D.P. 762, 763
Teo, B.K. 253
Terada, S., see Taniguchi, M. 656
Teranishi, T., see Suga, S. 774
Terentex, N.I., see Tulvinskii, V.B. 752
Ternov, I.M., see Sokolov, A.A. 28
Tersoff, J. 529, 532, 593, 641
Tews, W. 767–769
Thanailakis, A. 583
Theeten, J.B., see Aspnes, D.E. 567, 738, 763
Thevenon, A., see Flamand, J. 82
Theye, M.L., see Abelès, F. 750
Thiel, W. 293, 339, 340, 420, 421, 423
Thiel, W., see Kreile, J. 297, 298, 303, 339–341
Thimm, K., see Heinzmann, U. 427
Thimm, K., see Puester, G. 137
Thiry, P., see Guillot, C. 652–654
Thiry, P., see Hitchcock, A.P. 455
Thiry, P., see Louie, S.G. 486, 487
Thomas, J., see Pollini, I. 774, 775
Thomas, J., see Quemerais, A. 774
Thomas, T.D., see Carroll, T. 450
Thomas, T.D., see Haak, H. 450
Thomas, T.D., see Ungier, L. 450
Thompson, A., see Apai, G. 498
Thomson, D., see Newman, N. 588, 590, 591
Thomson, J.O., see Grimm, F.A. 329, 339, 424, 426
Thomson, J.P., see Brion, C.E. 341, 407
Thonn, T.F., see Arndt, D.P. 762, 763
Thornton, G., see Burroughs, P. 638
Thornton, G., see McGrath, R. 508
Thornton, G., see Norman, D. 472, 518
Thornton, G., see Owen, I.W. 473, 506, 507
Thornton, G., see Rosenberg, R.A. 208
Thornton, G., see Truesdale, C.M. 507
Thornton, G., see White, M.G. 132, 244, 275, 375
Thuault, C., see Guillot, C. 675, 695, 696, 702
Thuler, M.R. 656
Thulstrup, E., see Beebe, N. 403, 404

Thundat, T., see Dev, B.N. 599
Tilford, S.G., see Ginter, M.L. 214
Timothy, J.G. 125, 134
Tiribelli, R., see Giardini-Guidoni, A. 306
Tobin, J.G. 521
Tobin, J.G., see Barton, J.J. 510, 598
Tobin, J.G., see Rosenblatt, D.H. 510, 513
Toigo, F. 740
Toigo, F., see Celli, V. 740
Toigo, F., see Marvin, A. 740
Tokarivskii, M.V., see Vishnevskii, V.N. 773
Tokura, Y. 775
Tomanek, D. 685, 698, 699
Tomanek, D., see Grunze, M. 698, 699
Tomanek, D., see Kumar, V. 680, 681, 704
Tomboulian, D.H. 28
Toney, M., see Bohr, J. 496, 497
Tong, S.Y. 496, 692, 717
Tong, S.Y., see Rosenblatt, D.H. 510, 513
Tosatti, E. 760
Tosatti, E., see Cadioli, B. 245, 336
Tosatti, E., see Greenaway, D.L. 748
Tossell, J.A. 342
Touzet, B., see Flamand, J. 82
Toyozawa, Y., see Kaga, H. 644
Toyozawa, Y., see Kotani, A. 614, 629, 634, 647, 649
Tracy, J.C., see Palmberg, P.W. 548
Trajmar, S. 306
Trajmar, S., see Truhlar, D.G. 262, 263
Trakhtenberg, E.M., see Gluskin, E.S. 49
Tramer, A., see Castex, M.C. 456
Tran Minh Duc 665, 666, 669, 670, 672, 675, 689, 712–714
Tran Minh Duc, see Guillot, C. 496, 675, 695–697, 702
Tran Minh Duc, see Jugnet, Y. 478
Tran Minh Duc, see Tréglia, G. 700, 702
Tréglia, G. 651, 700, 702, 6651
Tréglia, G., see Chauveau, D. 675, 694
Tréglia, G., see Guillot, C. 488, 496, 674, 675, 695–697, 702
Tréglia, G., see Spanjaard, D. 489, 666, 685
Tréglia, G., see Tran Minh Duc 666
Trehan, R., see Sinkovic, B. 511
Treichler, R., see Stöhr, J. 505, 516, 517
Trendel, G., see Heidrich, K. 774
Trepakov, V.A. 774
Treu, J.I. 757
Treusch, J.H., see Heidrich, K. 774
Trevor, D. 456
Tromp, R.M. 495
Tromp, R.M., see Hamers, R.J. 495
Tromp, R.M., see van der Veen, J.F. 515

Tronc, M. 337–339, 442
Tronc, M., see King, G.C. 337, 441
Truesdale, C.M. 244, 264, 331, 332, 340, 342, 407, 443, 445–447, 507
Truesdale, C.M., see Lindle, D.W. 331, 341, 446
Truesdale, C.M., see Southworth, S.H. 274, 275, 332, 340
Truesdale, C.M., see Trevor, D. 456
Truhlar, D.G. 262, 263
Truhlar, D.G., see Rumble, J.R. 263
Truong, V.-V., see Girouard, F.E. 766, 775
Trykozko, R. 774
Trzhasleovskaya, M.B., see Band, I.M. 752, 771
Tsai, B.P. 435
Tsai, B.P., see Batten, C.F. 320, 333
Tsai, B.P., see Werner, A.S. 333, 334
Tseng, M.K., see Pratt, R.M. 771
Tu, K.N., see Ottaviani, G. 546
Tuck, R.A., see Norman, D. 518
Tulvinskii, V.B. 752
Turko, B. 134
Turner, A.M. 480, 482, 486
Turner, D., see Brundle, C. 425
Turner, D., see Franks, A. 83, 105
Turner, D.W., see Baker, C. 298, 299
Turner, M.J. 583
Turowski, M., see Stoffel, N.G. 572, 573
Tuttle Hart, T., see Arndt, D.P. 762, 763

Ueda, A., see Inagaki, T. 753
Uhrberg, R.I.G. 490, 522, 523
Uhrberg, R.I.G., see Nicholls, J.M. 490, 491
Umbach, E. 506
Umbach, E., see Hussain, Z. 472
Umezawa, K., see Katoh, H. 775
Underwood, J.H., see Barbee, T.W. 762
Ungier, L. 450
Ungier, L., see Carroll, T. 450
Ungier, L., see Haak, H. 450
Unwin, R. 264, 297, 298, 303, 332, 340
Unwin, R., see Langhoff, P.W. 297, 298, 303, 339
Unwin, R., see Peatman, W. 398
Uzan, E. 747

Valeri, S., see Abbati, I. 554, 556
Valiukonis, G. 774
Van der Laan, G. 644, 647
Van der Laan, G., see MacDowell, A.A. 97
van der Leeuw, Ph.E., see Brion, C.E. 338
van der Leeuw, Ph.E., see Kay, R.B. 251, 262, 337, 447, 455

van der Veen, J.F. 488, 494, 515, 562, 674, 675, 694, 695, 697, 702
van der Veen, J.F., *see* Eastman, D.E. 488, 666
van der Veen, J.F., *see* Heimann, P. 675, 697
van der Wiel, M.J. 336, 339, 439
van der Wiel, M.J., *see* Brion, C.E. 338
van der Wiel, M.J., *see* El-Sherbini, Th.M. 336
van der Wiel, M.J., *see* Granneman, E.H.A. 125, 133, 134
van der Wiel, M.J., *see* Hitchcock, A.P. 306, 307, 338, 380, 450, 455
van der Wiel, M.J., *see* Kay, R.B. 251, 262, 337, 447, 455
van der Wiel, M.J., *see* Kimman, J. 332
van der Wiel, M.J., *see* Wight, G.R. 251, 262, 336, 337, 442
Van Hove, M.A., *see* Tong, S.Y. 692, 717
van Laar, J. 545, 586, 589
van Laar, J., *see* Huijser, A. 545, 558, 559, 586, 589
Van Vechten, J.A., *see* Phillips, J.C. 581
VanDoren, A.H. 123
Varma, R.R., *see* Williams, R.H. 545, 561, 591, 592
Veal, B.W. 752, 764, 765
Vedrine, J.C., *see* Tran Minh Duc 665, 669, 670, 672, 675, 689, 712, 714
Vedrinski, R.V., *see* Sachenko, V.P. 305, 337
Veigele, W.J. 770, 771
Velicky, B. 751
Velzel, C.H.F. 64
Venghaus, M. 760
Verleur, H.M. 754, 755
Viescas, A.J., *see* Apai, G. 675, 694, 702, 703
Vietzke, K., *see* Rinneberg, H. 193
Viinikka, E.K., *see* Bagus, P.S. 263, 332
Viinikka, W., *see* Bagus, P.S. 400
Vileika, A., *see* Ambrazevivius, G. 772
Vilesov, F.I., *see* Kleimenov, V.I. 245, 264
Villani, A., *see* Furuya, K. 753
Vinciguerra, D., *see* Giardini-Guidoni, A. 306
Vinogradov, A.S. 337, 774
Vinogradov, A.S., *see* Filatova, E.O. 774, 775
Vinogradov, A.S., *see* Gluskin, E.S. 49
Vinogradov, A.S., *see* Zimkina, T.M. 246, 247, 305, 336
Vishnevskii, V.N. 773
Viswanathan, K.S., *see* Wilson, W.G. 332
Vo, Ky Lan, *see* Schneider, B.I. 285
Vodar, B. 44
von Festenberg, C., *see* Daniels, J. 757, 758
von Niessen, W. 305, 308, 309

von Niessen, W., *see* Åsbrink, L. 320
von Niessen, W., *see* Cederbaum, L.S. 263, 332, 400
von Niessen, W., *see* Haller, E. 320, 321, 329
von Niessen, W., *see* Langhoff, P.W. 263, 339
von Niessen, W., *see* Pulm, H. 246
von Niessen, W., *see* Schirmer, J. 263, 332
von Ruden, W. 147
Vorburger, T.V. 757

Wachter, P., *see* Schlegel, A. 773
Wadsworth, B.F. 147
Wadsworth, P.J., *see* Norman, D. 518
Wagman, D.D. 561, 577, 578
Wagner, L.F. 545
Wahlgren, U.I., *see* Lozes, R.L. 251
Waldrop, J.R. 528, 529
Waldrop, J.R., *see* Grant, R.W. 590
Walker, J.A., *see* Botter, R. 298
Walker, J.A., *see* Dibeler, V.H. 263, 303, 306
Walker, R.P. 27, 28
Walker, W.C., *see* Martzen, P.D. 773
Wall, A., *see* Chang, S. 519
Wall, W.E. 149, 766, 767
Wallace, S. 282, 295, 296, 303, 305, 306, 308, 316, 317, 329, 332, 334, 339, 425, 426
Wallace, S., *see* Dehmer, J.L. 263, 280, 282, 285–287, 295, 306, 308–315, 318, 329, 338, 340, 417, 418
Wallace, S., *see* Dill, D. 245, 331, 338
Wallace, S., *see* Kosman, W.M. 342
Wallace, S., *see* Loomba, D. 255, 258, 259, 339
Wallenstein, R. 177
Walter, O., *see* Schirmer, J. 263, 332
Walters, E. 456
Walters, E., *see* Grover, J. 456
Walters, G., *see* Black, G. 456
Wandelt, K., *see* Schnell, R.D. 503
Wang, Y.-T. 164
Wannier, G. 437
Warburton, D.R., *see* McGrath, R. 508
Warburton, D.R., *see* Norman, D. 472
Warburton, D.R., *see* Owen, I.W. 473, 506, 507
Warburton, W.K., *see* Barbee, T.W. 762
Warburton, W.K., *see* Fontaine, A. 762
Waskiewicz, W.K., *see* Robinson, I.K. 495
Watanabe, M., *see* Akahori, T. 403
Watanabe, M., *see* Nakamura, M. 336, 442
Watson, D.K., *see* Lucchese, R.R. 243, 246, 295
Watson, R.E., *see* Davenport, J.W. 692
Watson, R.E., *see* Krinsky, S. 3, 6, 10

Watson, R.E., see Winick, H. 19
Weaver, J.H. 548, 555–557, 562, 764, 765, 771
Weaver, J.H., see Franciosi, A. 478, 553, 556, 557, 579
Weaver, J.H., see Grioni, M. 552–554, 562
Weaver, J.H., see Lynch, D.W. 765, 766, 773
Weaver, J.H., see Peterman, D.J. 656
Weaver, J.H., see Wieliczka, D. 484
Webber, P.R. 674
Weber, J., see Bedford, K.L. 774
Weber, W., see Pflüger, J. 775
Weeks, S.P. 137
Wehenkel, C. 758, 759
Wehenkel, C., see Cukier, M. 773
Wehking, F. 525
Wehlirz, K., see Reimer, A. 455
Wehlirz, R., see Becker, U. 450
Weibel, E., see Binnig, G. 495
Weigold, E. 401
Weinert, M., see Wimmer, E. 689, 690
Weisenbloom, J.F., see Williams, G.P. 30
Welch, J., see Dehmer, J.L. 263, 289
Welch, J., see Dill, D. 261, 338
Weller, D. 482
Welton-Cook, M.R. 498
Wendin, G. 263, 332, 416
Wendin, G., see Eastman, D.E. 650
Weng, S., see Eberhardt, W. 450
Wergand, S.L., see Eberhardt, W. 340
Werme, L., see Siegbahn, K. 401, 443
Werner, A.S. 333, 334
Wertheim, G.K. 483, 500, 629, 649, 667
Wertheim, G.K., see Campagna, M. 483
Wertheim, G.K., see Citrin, P.H. 499, 626, 665, 666, 673–675, 678, 683
Wertheim, G.K., see Cohen, R.L. 627, 628
Wertheim, G.K., see Crecelius, G. 629
Wertheim, G.K., see DiCenzo, S.B. 499, 500
Wertheim, G.K., see Hüfner, S. 639, 640, 643, 650
Wertheim, G.K., see Rosencwaig, A. 644, 645
Wesner, D., see Eberhardt, W. 340, 450
West, J.B. 61, 264, 268, 280, 285–287, 292, 293, 331, 339, 417, 418, 421, 422
West, J.B., see Codling, K. 244, 264, 331
West, J.B., see Cole, B.E. 338
West, J.B., see Ederer, D.L. 85, 244, 264, 276, 331
West, J.B., see Holland, D.M.P. 279, 340, 372
West, J.B., see Howells, M.R. 61
West, J.B., see Kelly, M.M. 106
West, J.B., see MacDowell, A.A. 97, 102, 106, 107, 134, 135, 472

West, J.B., see Marr, G.V. 279
West, J.B., see Parr, A.C. 129, 244, 264, 268, 275, 276, 278, 279, 287, 290, 291, 297–299, 303–305, 331, 332, 340
West, J.B., see Saile, V. 23
West, J.B., see Stockbauer, R. 332, 338
Westra, C., see Van der Laan, G. 644, 647
Westra, C., see Zaanen, J. 648
White, C.W., see Larson, B.C. 474
White, H.W. 201
White, J.J. 752
White, J.J., see Bauer, R.S. 754
White, M.G. 132, 244, 275, 332, 340, 370, 375, 415, 416
White, M.G., see Carnovale, F. 339
White, M.G., see Grover, J. 456
White, M.G., see Poliakoff, E.D. 366, 415, 419
White, M.G., see Rosenberg, R.A. 208
White, M.G., see Walters, E. 456
White, R., see Carlson, T.A. 450
Whitehouse, C.R., see Mackey, K.J. 531, 532
Whitley, T.A., see Carlson, T.A. 244, 264, 331, 340, 341, 407, 423, 425
Whitley, T.A., see Grimm, F.A. 341
Wieder, H.H. 589, 590
Wiegandt, D., see Lee, G. 155
Wieliczka, D. 484
Wight, G.R. 251, 262, 336, 337, 439, 440, 442
Wilcke, H. 365, 367
Wilkins, J.W., see Herbst, J.F. 639
Wilkinson, P.G. 213
Willers, I. 160, 163
Williams, A.R. 673
Williams, A.R., see Eastman, D.E. 650
Williams, A.R., see Moruzzi, V.L. 652
Williams, G.M., see Mackey, K.J. 531, 532
Williams, G.P. 30, 87, 111
Williams, G.P., see Binns, C. 482
Williams, G.P., see Howells, M.R. 61
Williams, G.R.J. 340
Williams, G.R.J., see Hitchcock, A.P. 340
Williams, M.D., see List, R.S. 562, 588
Williams, M.D., see Newman, N. 588, 589
Williams, M.W., see Arakawa, E.T. 767–769, 773, 775
Williams, M.W., see Horton, V.G. 278
Williams, M.W., see Inagaki, T. 772
Williams, M.W., see MacRae, R.A. 752
Williams, M.Z., see Palmer, K.F. 753
Williams, R.H. 471, 476, 526, 543, 545, 546, 561, 562, 565–567, 572, 573, 583, 589–592
Williams, R.H., see Hughes, G.J. 565, 567, 573, 574

Williams, R.H., see Mackey, K.J. 531, 532
Williams, R.H., see McGovern, I.T. 476
Williams, R.H., see McKinley, A. 526, 560, 573
Williams, R.T., see Bedford, K.L. 774
Williams, R.T., see Hunter, W.R. 60, 61
Williams, W., see Truhlar, D.G. 262, 263
Willis, R.F., see Feuerbacher, B. 470, 471, 666, 672
Willis, R.F., see Hezaveh, A.A. 482
Wilmshurst, J.K., see Bowlden, H.J. 753
Wilson, D., see Peatman, W. 377
Wilson, J.A., see Silberman, J.A. 565
Wilson, J.A., see Spicer, W.E. 565
Wilson, K., see Busch, G. 394
Wilson, W.G. 332
Wimmer, E. 689, 690
Winchell, R., see Ajello, J.M. 334
Winick, H. 19, 28, 244, 253
Winick, H., see Brown, G. 116
Winkoun, D., see Dujardin, G. 407, 435, 436
Winter, H., see Pindor, A.J. 482
Wittmaack, K., see Geiger, J. 759
Wiza, J. 134
Wolcke, A., see Heinzmann, U. 236, 427
Wolf, E., see Born, M. 728, 730
Wolf, F., see Peatman, W. 398
Wolf, R., see Birken, H.-G. 755, 767
Wolff, H.W., see Bruhn, R. 773
Wolter, H. 728, 730
Wood, C.E.C., see Okamoto, K. 589
Wood, J.H., see Koelling, D.D. 638
Woodall, J.M. 593
Woodall, J.M., see Freeouf, J.L. 593
Woodruff, D.P. 511
Woodruff, D.P., see Jones, R.G. 503
Woodruff, D.P., see McConville, C.F. 511
Woodruff, D.P., see Puschmann, A. 520
Woodruff, P.R. 244, 264, 290, 407
Woodruff, P.R., see Krause, M.O. 244, 275, 371
Woodruff, P.R., see Marr, G.V. 244, 264, 290
Woodward, A., see White, M.G. 332
Wooten, F. 725, 728, 748, 758
Wooten, F., see Mosteller, L.P. 748
Wort, S., see Eland, J.H.D. 381, 438
Wöste, L., see Bréchignac, C. 456
Wu, C. 382
Wu, C.-K. 752
Wu, C.Y.R. 263
Wuilleumier, F.J., see Krummacher, S. 263, 332, 339, 340, 402
Wuilloud, E. 638

Xu, G., see Tong, S.Y. 496

Yamada, A., see Koide, T. 108
Yamaguchi, S., see Hanyu, T. 775
Yamaguchi, S., see Miyahara, T. 773, 774
Yamaguchi, S., see Nakamura, M. 336, 442
Yamaguchi, S., see Sato, S. 773
Yamaguchi, T., see Suga, S. 774
Yamashita, H., see Miyake, K.P. 60, 61
Yamashita, H., see Nakamura, M. 336, 442
Yamazaki, T., see Kimura, K. 320, 334, 396
Yanajihara, M., see Koide, T. 108
Yeh, J.J., see Abbati, I. 476, 477
Yeh, J.L., see Duke, C.B. 558, 559
Yeh, J.L., see Kahn, A. 558, 559
Yeh, P. 730
Yndurain, F., see Miranda, R. 482
Yoshimine, M., see Tanaka, K. 403
Young, R.H. 751

Zaanen, J. 648
Zacharov, B. 148, 162
Zaera, F., see Stöhr, J. 521
Zakharov, V.M., see Pajasova, L. 775
Zare, R.N., see Anderson, S.L. 332
Zare, R.N., see Greene, C.H. 333, 357, 432
Zare, R.N., see Guest, J.A. 333, 432, 433
Zare, R.N., see Poliakoff, E.D. 333, 432, 433
Zeiger, M.J., see Henrich, V.E. 759
Zemanski, M.W., see Mitchell, A.C.G. 232
Zeppenfeld, K. 748, 759
Zeppenfeld, K., see Daniels, J. 757, 758
Zewail, A., see Nenner, I. 442, 455
Zhang, H.I. 546, 593
Zhao, T.-X. 525, 560, 576
Zhao, T.-X., see Cerrina, F. 567
Zhao, T.-X., see Daniels, R.R. 526, 575, 583, 584
Zhao, T.-X., see Perfetti, P. 557
Zhuze, V.P. 773, 774
Zhuze, V.P., see Bagdasarov, K.S. 775
Zimkina, T.M. 246, 247, 305, 336
Zimkina, T.M., see Filatova, E.O. 774, 775
Zimkina, T.M., see Vinogradov, A.S. 337, 774
Zimmerer, G. 366, 367, 455
Zimmerer, G., see Brodmann, R.R. 365
Zimmerer, G., see Castex, M.C. 455
Zimmerer, G., see Gürtler, P. 366
Zimmerer, G., see Hahn, U. 365
Zimmerer, G., see Jordan, B. 455
Zimmerer, G., see Le Calvé, J. 455
Zimmerer, G., see Moeller, T. 418, 455

Zimmerman, A.M., see Furuya, K. 753
Zimmermann, H., see Day, J.D. 162
Zolnierek, Z., see Fuggle, J.C. 629, 635, 637, 639
Zolnierek, Z., see Hillebrecht, F.U. 635, 643
Zscheile, H., see Kubier, B. 774
Zunger, A. 526, 572–574
Zunger, A., see Daniels, R.R. 526, 575
Zunger, A., see Zhao, T.-X. 525, 560, 576
Zur, A. 588

SUBJECT INDEX

aberration   66, 67, 72, 77, 78, 98, 117
—coefficient   69
—components   73
—-limited resolution   67, 70, 71, 77–79
—theory   24, 63–65
absorbance   764
absorbates   469
absorption coefficient   731, 738, 740, 749, 761
—data   754
—measurements   760
absorption threshold   738
acetylene   520
ACO (LURE, Orsay)   362, 363, 365, 366, 371, 375, 376, 379, 385, 670
acoustic delay line   114
address path   142
ADONE (Frascati)   12
Advanced Light Source (Berkeley), see ALS
AES (Auger electron spectroscopy)   549, 560, 562, 576, 590
Ag   50, 236, 482, 513, 556, 562, 567, 571, 572, 574, 589, 591, 595, 598, 707
—clusters   499
AgCl   754
air-bearing pump   112
Al   49, 50, 107, 108, 486, 550, 551, 557, 558, 560, 562–565, 567, 568, 574–576, 578, 580, 581, 583, 584, 586, 591, 597, 598, 688–690, 730, 731, 740, 755, 764, 769
Al–CdS   588
ALADDIN (Wisconsin)   13
AlF   217, 219
alkali atoms   200, 202
alkaline earth atoms   205
alloys   704–708
Al_2O_3 49, 567, 730, 731, 767, 769
alphanumeric   154
alphanumeric terminal   160
ALS (Advanced Light Source, Berkeley)   13, 116
amplifier   128, 129
amplitude-to-frequency converter   125, 126
analogue meter   116
angle-resolved AES studies   553
angle-resolved photoelectron spectra   299

angle-resolved photoemission extended fine structure, see ARPEFS
angle-resolved techniques   596
angular dispersion   62
angular distribution   281, 292, 361, 368, 397, 427, 432
—anisotropies   331
—of Auger electrons   445, 446
—of photocurrent   253
—of photoelectrons   445
—of photoions   333
—of scattered electrons   246
angular divergence   25, 28, 89, 98
angular-resolved partial photoelectron spectroscopy   417
angular-resolved photoelectron spectroscopy, see ARPES
ANLS (Argonne)   13
ANSI/IEEE-488   146
apparatus function   270
"apple core" toroid   39, 40, 68
Ar   189, 194, 195, 381
—discharge continuum source   213
—matrix   457
area detector   149, 279
ARPANET   162
ARPEFS (angle-resolved photoemission extended fine structure)   510, 598
ARPES (angular-resolved photoelectron spectroscopy)   90, 137, 407, 412, 413, 443, 457, 490, 500, 504, 506, 508, 511, 573, 596, 598, 652
artificial channel method   383
As   492, 573, 577, 581
aspheric component   101
aspheric element   36
aspheric mirror   56
assembler   153
astigmatic coma   66, 67, 69–78, 80
astigmatism   53, 66, 70, 71, 89
astigmatism correction   90
atrophysics   357
asymmetric absorption lines   230
asymmetric photoemission line shape   614
asymmetric top   220, 223
asymmetric top molecule   221

asymmetry parameter β   273–275, 278, 279,
  283, 284, 287–289, 293, 295, 296, 302, 304,
  307, 308, 311–315, 320–327, 329, 370–374,
  408, 420–427, 445, 446, 716
atmospheric chemistry   357
atomic experiments   112
atomic f-values   231
atomic layer engineering   522
atomic photoabsorption cross section   42
atomic relaxation   673
atomic scattering factor   42, 43, 47, 49, 760
attenuation cross section   770
Au   44–46, 48–51, 108, 374, 496, 550, 556,
  560, 562, 565, 567, 569, 570, 572, 577, 578,
  580, 588, 589, 591, 599, 600, 665, 695, 697,
  706, 707, 763
—clusters   499
Au–CdS   588
Au–GaAs   581
Au–Si   570, 579
Auger cascade   444
Auger electron   432, 439, 498, 512
—angular distribution   245, 445, 446
—emission   595
—peak   652–656
Auger electron spectroscopy, see AES
Auger process   360, 361, 434, 439, 443–445,
  448, 450, 452, 453, 455, 472, 554, 594
—decay   473
—emission   599
—energy shift   685
—method   522
—photoemission   525
—spectroscopy   573
—transition   622, 623
Auger spectrum   450, 766
autionization   418
autoabsorption   709, 710
autocollimating mirror   111
autocollimator   110
autocorrelation length of roughness   739
autoionization   186, 200, 243–245, 263–266,
  276, 280, 287, 289, 293, 294, 298, 330, 331,
  359, 364, 377, 384, 385, 398, 405, 409, 410,
  412, 415–417, 427–429, 431, 433, 437, 444,
  446, 447, 449, 456
—profile   185, 194, 304
—resonance   191, 202, 230, 332, 334, 421
—rotational   265
—Rydberg series   229, 332
—Rydberg states   316
—spectrum   227
—state   327
—structure   316

—vibrational   265, 268
automatic ellipsometers   757
autoregressive Fourier transform analysis   510
Avogadro's number   42

B   689
B K-shell X-ray absorption spectrum   324
Ba   15, 208–210, 233
Ba^{2+} 227
backlash   140
backscattering   712
backstreaming   112
backup copies of programs   151
band bending   582, 583, 585, 587–589, 593,
  601, 602
bandbending depth   527
barrier penetration   250
BASIC   144, 154
Be   49, 486, 688, 689
BE shift   689
beam dimension   25
beam direction diagnostics   109
beam divergence   6, 8
beam line   4, 19, 23, 24, 26, 31, 46, 47, 54,
  82, 84–88, 94, 99, 109, 112, 113, 116, 117,
  362, 366, 371
—mirror   102
—optics   116
—shielding   86
beam shutter   114, 115
beam stability   4
beam stop   114
bending magnet   85, 86
benzene (C_6H_6)   230, 516
BEPC (China)   12
beryllium window   24, 94, 111, 114
Bessel function   29
BESSY I, II (Berlin)   13, 32, 33, 60, 78, 375,
  379
β-curve   329
BF_3 249, 254, 255, 261, 318–327, 330
"bicycle tyre" toroid   39, 40, 68
binding energy   470, 506
birefringence   233
birefringent materials   757
bit   144, 151, 152
"bit-slice" microprocessor controller   129
blaze   61, 82, 89
—angle   59
Bloch theorem   717
bond angle   222, 226
bonding configuration   504
Born–Haber cycle   679, 680, 698, 700, 703
Born–Oppenheimer approximation   269

borosilicate glass 107
Br 503, 504, 648, 699
Br_2 253, 699
Br–C bond 455
$Br-CF_2-CF_2-I$ 455
Bragg angle 94–97, 504
Bragg diffraction 761
Bragg peak 96, 598
Bragg plane 96, 762
Bragg reflection 599
Bragg scattering 474
branching ratio 266, 279, 284, 303, 311–315, 317, 320–330, 413, 415, 417, 428, 431, 455
bremsstrahlung radiation 114
Brewster angle 44, 731, 737, 746, 747
brightness 18
Brillouin zone 485, 596
bulk chemical shift 708
bulk electric properties 476
buses 141, 142, 144, 146, 147, 151, 152
byte 144, 151

C 49, 50, 749
C–C stretch mode 299–301, 303
C–I bond 455
Ca 205–207
CAMAC 129, 134, 141–146, 164, 279
CAMAC modules 145
CAMAC–FASTBUS 143
CAMAC–IEEE 488 143
capillary light guide 98
capillary tube 277
carbon 49, 50, 749
—contamination 86, 88, 108
—film 93
—vane pump 112, 113
—window 94
CARS/AFRODITE (Frascati) 12
cascade autoionization 449
catalysis 469
catalysis connection 500
catalyst 473
catalytic science 498
catalytic system 521
CCl_4 330, 457
Cd 198, 200, 235, 446, 563–565, 577
Cd–Te 562, 569, 580, 581
CD_3ONO 390
CdS 562, 570, 574, 580, 581, 588, 599
CdSe 562, 567, 580, 581, 588
Ce 484, 551–554, 562, 629, 635, 637, 639, 686, 687
$CeAl_2$ 635
$CeCu_2Si_2$ 635

$CeNi_2$ 637
centrifugal barrier 186, 202, 204, 245, 246, 248, 249, 251, 252, 254, 281, 358, 359, 410, 439, 446, 478
centrifugal effects 201
centrifugal potential 194, 255
centrifugal repulsion 258
centrifugal term 204, 211
CeO_2 638, 639
CeSe 635
CESR (Ithaca) 13
CF_4 330, 439, 443, 445, 446
CF_3NO 390
CF_3X 382
CH_4 435, 436, 443
C_2H_2 297, 298, 301–304, 330, 332, 334, 402
chalcogen sulphur 500
channeltron 365
CH_3Br 427, 428
chemical effects 331
chemical shift 529, 613, 707
chemical trapping 579, 580, 601
chemical vapour deposition (CVD) 106
chemisorption 500, 502, 504, 507, 508, 511, 522, 525, 533, 573, 698, 702
—geometry 718
—satellite 506
—sites 503
chemisorption-induced phase transition 695
CH_3I 227
CH_4I 450
C_2H_5I 227, 230
chiral molecules 455
CH_3O 394
CH_3ONO 390–393
C_2H_5ONO 390
circular dichroism 457
circularly polarized radiation 232
circulating beam current 4
CIS 600
Cl 442, 453, 513, 514, 648
Cl_2 418, 457, 500, 503
cleaning by glow discharge 108
ClNO 390, 394–396
ClO_2 442
clock pulse generator 127
closed-cycle helium compressor 112
cluster 336, 456, 498, 499, 511, 549, 560, 562, 570, 572–575, 577, 578, 584, 602, 644, 646, 648, 717
—formation 526, 559
—model 634, 638
—volume 500
clustering 571

C_2N_2 331, 334
CO 214, 261, 303, 325, 330–332, 334, 398, 399, 402, 443, 445, 446, 450, 457, 500, 504–506, 511, 516–520, 699, 702
CO_2 261, 262, 329–331, 401, 402, 413, 422, 423, 425–428, 433–435, 443, 445, 446
CO_2^+ 424
Co 480, 482, 557, 648
coherence 6, 18
coherence length, l_c 18, 19
cohesive energy 681, 682, 688, 705
coincidence 378, 379
—counting system 132
—detection technique 377
—gates 133
—technique 362, 368
coma 36, 54–56, 66, 67, 69–77, 80, 89, 90
—aberration 78
—tail 55, 80
combustion 357
communication board 148
compiler 153
complex conductivity 726
complex dielectric function 726, 742, 755
complex refractive index 727, 729, 742, 751
compression layer 52
computer backplane bus 162
computer controlled translation of exit slits 80
computer interfacing 123
computer systems 123
conductivity 96, 106
constant final state spectroscopy (CFS) 599
constant initial state spectroscopy (CIS) 599
constant ionic state method 372
contamination 94, 113, 116
continuum 252
—absorption cross section 247
—coupling 263, 294, 295, 297, 310, 329, 334
—interaction 293
—structure 246, 247
Controller Bus 144
conventional X-ray sources 5
conversion 132
—of protocol 162
converter
—amplitude-to-frequency 125, 126
—digital-to-analogue 129
—time-to-amplitude 125, 126
—voltage-to-frequency 125, 126
cooled absorber 115
cooled block 102
cooling circuit failing 115
Cooper minimum 408, 423–425, 476, 554, 555

Cooper zeros 331
core excitations 46
core level binding energy 678, 699
core level binding energy shift 667, 679, 699, 701
core level shift 690, 693, 697, 698
—of surface atoms 718
core resonance 442
core state diffraction pattern 712
coriolis coupling 388
correlation function 616, 617
correlation time 620
Coster–Kronig process 478
COSY (Berlin) 13
Coulomb attraction 264
Coulomb coupling 310
Coulomb explosion 438, 473
Coulomb function 183
Coulomb gauge 714, 715
Coulomb interaction 183, 450, 622, 635, 643, 650, 667
Coulomb phase shift 420, 421
Coulomb potential 249, 255, 457, 616, 668, 691
Coulomb–non-Coulomb phase shift 716
counting efficiency 129
coupled channels 246
covalency parameter 645
Cr 557, 562, 579, 764
Cr–Si 579
crate controller 143, 144
critical angle 49, 51, 52
critical glancing angle 48
critical wavelength, λ_c 9, 10, 12, 29, 30, 32
cross section 251, 279, 361
—total 261
"crossed" cylindrical mirrors 99, 103
"crotch" area between beam lines 85
cryopump 112, 115, 277
crystal-field pair 255
crystal heating 94
crystal spectrometer 94, 99
crystallographic orientation 683
crystallographic unit cell 727
Cs 201, 202, 204, 513, 514, 702, 703
CS_2 330, 402, 427, 434–440, 442
CU 482
Cu 50, 107, 474, 486, 506, 556, 560, 562, 567, 574, 581, 589, 591, 599, 648, 650, 656, 685
—on glass 52
$CuBr_2$ 647
$CuCl_2$ 647
Curie temperature 481, 643
current amplifier 129

Subject index 833

CVD SiC premirror 78
cylindrical mirror analyser (CMA) 548
cylindrical premirror 90

DAC (digital-to-analogue converter) 129
Daresbury Laboratory 144, 145, 153
Daresbury SRS 7, 8, 10, 11, 13–15, 18, 27, 32, 33, 82, 84, 86, 87, 89, 94, 97, 98, 102, 105–107, 109, 111, 113–115, 117, 129, 134, 136, 379
data acquisition 148, 149, 155
—industrial 123
—scientific 123
—efficiency 150
—program 144, 152, 153, 156
—system 123, 157
data processing 159
data reduction 279
data transfer rate 142
DBr 214–216
DC servo system 141
"dead-time" 128, 129, 132
Debye temperature 741
Debye–Waller factor 741
Debye–Waller form 51
decay channel 364, 365
decay time 618
dedicated SR source 149
deep inner-shell photoionization 331
deflection parameter, K 16, 17
defocus 66, 67, 70, 74, 76–78, 80
—aberration 92
defocusing term 71, 72
"delayed coincidence" 130, 131, 134
demagnification 56, 77, 78, 85, 89–92, 98, 99
DESY (Hamburg) 365–367, 457
detector 123
—high-voltage power supply 126
—signal processing 123, 125
device driver 155
dial gauge indicator 101
diamond 568, 767
—polishing with diamond paste 767, 769
diatomic molecules 219
dichroism 233
dielectric constant 41–43, 57
dielectric function 727, 731, 743, 748, 749, 751, 753–755, 757–760, 763–765
—spectra 753
—tensor 728, 745
dielectric tensor 760
—spectra 749
differential partial cross section 372
differential photoionization cross section 445
diffraction efficiency 83, 84

diffusely scattered radiation 739
digital plotter 150
digital to analogue converter 129
dimers 456
dipole-allowed channels 251
dipole magnet 4, 5, 8, 10, 12, 15, 17, 362, 14, 364
dipole strengths 269
Dirac–Slater approximation 715
direction sensors 4
disc interface 148
"discrete" shape resonance 252, 262
discrete–continuum coupling 329
discrimination 129
discriminator 125, 128
dispersion integral 771
dispersive aberration 74–77, 80
dissocation energy of HO 221
dissociative adsorption 507
dissociative inner-shell ionization 455
dissociative ionization 450, 453
dissolution energy 681
divergence 6, 11, 15, 16
D_2O 220, 221
Doppler broadening 181
Doppler effect 17
DORIS (Hamburg) 13
double ionization 335
double resonant Auger process 453
drift tube 131
dual electron spectrometer 277
duty cycle 14
Dy 687
dye laser 181
DyF_3 627
dynamics of multiphoton process 332, 333

$e-N_2$ scattering 252
eccentricity 38
EELS (electron energy loss spectroscopy) 438, 439, 441, 442, 514, 520, 742, 757–759, 765, 767, 769
effect of temperature 741
effusive beam 279
eigenchannel contour maps 255
eigenchannels 256–261
elastic cross section 262
elastic electron scattering 245, 671
electric dipole interaction 252
electron
—affinity 544
—analyser 278, 369, 372
—bunches 14
—core interaction 243
—correlation 445

—correlation effect 628
——ion coincidence experiments 132–134
——ion coincidence technique 405
——molecule scattering 256
——optics of molecular fields 331
——phonon interaction 741
—refraction 671
—scattering 245, 246
—spectrometer 278
—spin polarization spectroscopy 230, 235
——stimulated desorption 473
electron energy loss spectroscopy, see EELS
electron loss spectroscopy, see ELS
electron spectroscopy for chemical analysis, see ESCA
electronic autoionization 294
electronic channels 292, 293
electronic configuration assignments 219
electronic mail 162
electronic polarizability 498
electronic relaxation 358
electronic states 220
electronics 124
—availability 125
—format 125
—modularity 124
—sensitivity 124
electronless nickel process 105
ellipse 37, 38
ellipsoid 36, 38, 68, 101
ellipsoidal mirror 78
ellipsometric measurements 740, 752, 763, 764
ellipsometry 741, 742, 756, 759, 762
elliptic 54
elliptically polarized light 278
ELS (electron loss spectroscopy) 560
emittance 6, 11, 12, 15, 18, 24–27, 85
encoder 141
energy optimization 73
Energy Research and Development Association 143
epitaxial Al 572
epitaxial growth 530, 532, 571
epitaxial interface 602
epitaxial layer 556, 579
epitaxial system 532
Er 687
ESCA (electron spectroscopy for chemical analysis) 665, 666
escape depth 498, 521, 527, 548, 551, 569, 570, 586, 588
—aberration-limited 78, 79

ESRF (European Synchrotron Radiation Facility) 4, 12, 19, 116
etching 769
Ethernet 167, 168
—LAN 166
ethylene 230, 520
Eu 484, 629, 687, 707
Eu_3O_4 484
European Standards on Nuclear Electronics 143
European Synchrotron Radiation Facility, see ESRF
evaporated film 738, 764
EXAFS (extended X-ray absorption fine structure) 253, 438, 472, 498, 502, 508, 511, 519, 717, 760, 761, 770
excimers 455
excited autoionizing states 334
excited molecular states 333
excited oxygen 502
excitonic states 691
extended X-ray absorption fine structure, see EXAFS
external electric field 457

F 255, 648
F_2 439
f-value 233
factorial difference relation 29
Fano configurational mixing 409
Fano effect 654
Fano parameters 193
Fano profile 212, 416
Fano q-parameter 190, 413, 416
Fano theory 414
Fano–Beutler profile 263, 269, 359, 385
Faraday rotation 230, 233, 235
fast closing valve 113
FASTBUS 147
Fe 480–482, 557, 643, 647–649, 699
Fermat's principle 66
Fermi distribution function 620
Fermi energy 626, 650
Fermi Golden Rule 672
Fermi level 480, 491, 527, 544, 547, 562, 565, 575, 581, 588, 624, 670, 678, 680
—pinning 582, 589, 593, 599, 603
—shift 741
Fermi surface 550
—instability mechanism 492
Fermi vacuum 636
FES (fluorescence excitation spectroscopy) 407, 408, 428, 455
Feshbach resonances 439

fibre-optic cable 167
figure of merit 72
file server 159
film inhomogeneity 738
fixed paraboloid 58
flexural pivots 105
float glass 52, 107
—etched 52
—platinum coated 52
floating-point arithmetic 152
fluorescence 333, 365, 375, 445, 446, 455, 472, 474
—excitation spectrum 383, 385, 390
—lifetime 434
—profile 503
—spectroscopy 90
—yield 504
fluorescence excitation spectroscopy, see FES
focusing mirror 57
formate (HCO_2) 517, 520
FORTRAN 144, 145, 153, 155, 161, 164
Fourier analysis 757
Fourier coefficient 741
Fourier inversion 474
Fourier map 496
Fourier series 755
Fourier spectrum 598
Fourier transform 161, 511, 525, 615, 617, 619, 755
Fourier transformation 728
Fourier transformed data 595
fragmentation of molecular ions 334
Franck–Condon-allowed channels 303
Franck–Condon distribution 280
Franck–Condon effect 279, 298, 301, 422
Franck–Condon factor 274, 284, 291, 397, 398, 409, 416
Franck–Condon factorization 285
Franck–Condon gap 406, 415
Franck–Condon model 396
Franck–Condon overlap 389, 456
Franck–Condon principle 219, 226, 429, 620
Franck–Condon regime 384
Franck–Condon region 366, 402–404, 434, 437
Franck–Condon separation 281
Franck–Condon transitions 416
Franck–Condon value 417
free-atom X-ray absorption cross section 770
free-electron gas 738
—model 752
free-electron laser 12, 335, 364
free jet expansion 456

free radicals 335
free surface 521
Fresnel equations 42, 43, 50
Fresnel relations 728, 730, 743, 746, 747, 755, 761
Fröhlich-type interaction 620
front surface location 100
full-screen editor 153
fused quartz 746
fused silica 97, 105–107
FUTUREBUS 147

Ga 492, 557, 558, 573, 574, 577, 578, 581, 583
GaAlAs 565
GaAs 479, 490, 496, 502, 503, 559, 562, 571, 572, 575, 578, 579, 583, 584, 586, 588–591, 593, 692, 693, 702
GaInAs 565
gallium interface 96
GaP 562, 692
gas jet 279
gas load 277
GaSb 562, 583, 584, 599, 692
GaSe 567
GaTe 749, 754
Gaussian distribution 739
Gaussian form 618
Gaussian function 619, 620
Gaussian image point 66, 67
Gaussian path function 66
Gaussian spectrum 626
Gd 211, 212, 687
GdF_3 627
Ge 96, 490, 496, 503, 504, 530, 690, 691
$GeCl_4$ 253
generalized spherical aberration 66, 69
generalized valence-bond 309
GeO 214
giant resonances 210–212, 269
grain boundary 739
graphics kernel system 160
"graphics metafiles" 160
graphics terminal 150
graphite 744, 748
grating
—blazed 59, 83, 84
—efficiency 82
—groove depth 82, 83
—groove profile 59, 60, 82
—groove shape 57
—groove spacing 82, 83
—groove width 83
—holographic 64, 177, 181, 212, 236

—holographic toroidal 68
—lamella 59, 82–84
—pitch ratio 83
—profile 82
—sinusoidal 59, 83, 84
—spherical 70
—toroidal 68, 70
grazing-angle X-ray diffraction 495
grazing incidence optics 108
Green function 309, 332
grey-level 160

H_2 213, 214, 266–268, 272, 275, 291, 331, 333, 385, 386, 389, 418, 423, 425, 432, 500, 502, 695, 702
H_2 reconstructed surface 697
Hamiltonian 623, 630, 640, 644, 646, 648, 657–659
hard dissociative double ionization 453
"hard" double-ionization process 434
hardware 145
hardware standard 142, 145
Hartree–Fock approximation 613, 628
Hartree–Fock calculation 649, 668, 691
Hartree–Fock single configuration calculation 186, 189
Hartree–Fock theory 627
Hartree–Fock–Slater approximation 715
HASYLAB (Hamburg) 13, 364, 367
HBr 214–216, 428, 450, 451
HCl 415, 416
HCN 330, 334
He 112, 178, 181, 182, 200, 201
heat loading 87, 97
heat of segregation 488, 685
heat of solution 704
heat of vaporization 707
Heidenhain linear encoder 94
helium 24
HELP text 151
"herringbone" structure 511
HESYRL (China) 12
HF 439, 443
Hg 180, 181, 564, 565
Hg–Te 569
HgCdTe 565, 569
HI 227
high-brightness lattice 27
high-energy bremsstrahlung intensity 86
high resolution molecular spectroscopy 90
high resolution normal incidence instruments 93
high Rydberg states 181
high-temperature molecules 335
histogramming memory 134, 145, 146, 153

Ho 687
H_2O 220–223, 226, 382–385, 443, 508
$H_2O(A)$ 395
H_2O_2 382
HoF_3 627
hole localization 331
hole–hole interaction 445
holographic grating 64, 177, 181, 212, 236
holographic microscopy 19
holographic techniques 64
Hopfield absorption 287, 288, 292, 294
Hopfield emission 288, 292, 294
Hopfield series 412, 421, 422
H_2S 220, 222, 224, 226, 247, 424, 425, 442, 443
H_2Se 220, 222, 224, 226
H_2Te 220, 224
Hund's coupling cases 269
H_2X, H_2^+X 225, 269
hybridization broadening 508
hydrocarbon contamination 111
hydrodesulphurization 521
hydrogen background source 213
hydrogen halides 214

I 699
I_2 699
i-C_3H_7ONO 390
ICN 433
IEEE-488 General Purpose Interface Bus 126
"image potential" state 487
In 564, 567, 572, 575
InAs 96, 562
independent-electron methods 246
indirect autoionization mechanisms 331
Indore (India) 12
inhibition of corrosion 469
inner-shell photoabsorption 261
inner-shell spectra 251
InP 560, 562, 574, 576, 581, 583–585, 588, 590, 591
InSb 496, 562, 692
insertion devices 12, 15, 18
INSOR (Japan) 12
instrument function 272
integrated circuit devices 19
integrated cross section 278
interchannel mixing 184
interferometry 237, 760
"interloper" structure 292
International Electrotechnical Commision 143
interpreter 154
intrachannel coupling 329

Invar 107
ion pump 112, 115
ion scattering spectroscopy 494
ion yield 367
ion–ion coincidence technique 435, 452
ionic crystals 246
ionic state 397, 398
ionic threshold 280
ionization
—chamber 365
—channel 269, 278
—efficiency 307
—gauge 113
—potential 181, 213
—threshold 42
—yield measurements 306
Ir 108, 694, 695, 697
itinerant magnetism 482
itinerant surface states 492

Jahn–Teller effect 309
JANET 162
Joule heating 5
JUMBO (crystal spectrometer) 94

K 201–204, 702
K-edge 80, 81
—carbon 80, 81, 88
—nitrogen 80
—oxygen 80
K matrix 256, 257, 295
K-shell of N_2 253
K-shell photoionization 250–252
Kelvin probe 549
kinematic relocation 101
Koopmans' theorem 436, 613, 668
Kr 189
Kramers–Kronig analysis 758, 761, 764, 771, 774
Kramers–Kronig consistency 764
Kramers–Kronig integrals 753
Kramers–Kronig methods 742
Kramers–Kronig relation 42, 727
Kramers–Kronig techniques 748, 749, 751
Kramers–Kronig treatment 754, 755, 767

La 629, 633, 635, 639, 643, 647, 686, 688
Laboratoire d'Utilisation du Rayonnement Electromagnétique, see LURE
$LaBr_3$ 633
$LaCl_3$ 633
LaF_3 633
Lamb shift 182
LAN (local-area network) 160, 165
Landau's principle 482

lanthanides 685–688, 707
La_2O_3 633
$LaPd_3$ 629, 633
laser 18, 110, 233
—alignment 101, 103, 104
——system 87, 88, 109
—annealing 474
—beam 91
—calorimetry 741
—design 357
—-induced fluorescence 455
—system 111
LCAO, calculation 529
lead cladding 114
lead glass 114
LEED (low energy electron diffraction) 490, 491, 494, 496, 498, 511, 513–515, 521, 522, 525, 549, 553, 558, 560, 571, 573, 575, 597, 602, 695, 700, 711, 717, 766
Li 201, 202
LiF pile-of-plates polarizer 757
lifetime 367
light flux 84
linear combination of atomic orbitals, see LCAO
linked-cluster theorem 631
linker 154
liquid gallium 102, 106
lithography 5, 94
lithography station 88
LNRS (Brazil) 12
local-area network, see LAN
longitudinal excitation mode 621
longitudinal excitons 759
Lorentz doublet 232
Lorentz oscillator 754
Lorentzian broadening 633
Lorentzian form 619
Lorentzian function 622, 623, 633
Lorentzian oscillators 752
Lorentzian spectrum 626
low energy electron diffraction, see LEED
low-pass filter 47
low-pass mirror filter 49
Lu 686–688
Lu–Fano analysis 194
Lu–Fano graph 183, 184, 227, 228
luminescence lifetime 743
LURE (Laboratoire d'Utilisation du Rayonnement Electromagnétique, Orsay) 12, 361, 719
Lyman absorption bands 213

Maclaurin's series 69
Madelung energy 494, 691

Madelung potential 668, 690–692
Madelung summation 668, 690
magic angle 369, 716
magnet dipoles 14
magnetic-bearing pump 112
magnetic circular dichroism spectroscopy, *see* MCD
magnetic "dead layers" 482
magnetic field effects 230
magnetic shielding 278
magnetic tape drive 159
magnetism 480
magneto-optical patterns 231, 233, 235
magneto-optical rotation pattern 232
magneto-optical spectra 231
magneto-optical spectroscopy 177
magneto-optical Vernier method 231, 233, 235
magnification 40, 53, 55, 62, 63, 78
mainframe 148
many-body effect 613
many-body response 615
many-electron effects 295
mask etching 533
masks 108
mass spectrometer 368
mass spectrometry 380, 434, 435, 453, 454
matrix experiments 457
matrix-isolated molecules 457
matrix-isolated species 230
MAX (Lund) 13
Maxwell's equations 82
MBE (molecular beam epitaxy) 522, 533, 549, 565, 590, 593, 693
—spectroscopy 137, 230
MCD (magnetic circular dichroism) 230
mean free path for electrons in solids 470
mean free path of electron 709, 717
memory management unit 152
meridian focus 39–41, 53, 54
metal-induced gap state (MIG) 593, 594
metal–anion bonding 562
metal–cation bonding 562
metallic mirror 757
metallic surface state 490, 491
metastable states 335
methoxy (CH_3O) 517
methyl bromide 236
methyl iodide 228, 229
metrology 36
Mflop 161
Mg 179, 181, 191, 688, 689
MgF_2 108
MgO 498
microcomputer 148

microelectronics 521
microfocus 6
microstepping 138, 139
mirror 47, 49, 52, 54
—"apple core" toroid 39, 40
—autocollimating 111
—"bicycle tyre" toroid 39, 40
—box 85, 86, 88, 93, 102, 109
—copper 106
—cylindrical ("crossed") 99, 103
—design 87
—ellipsoidal 91, 98, 104
—elliptical 37, 38
—glancing incidence 35, 54
—materials and coatings 99
—metal 105
——thermoelectrically cooled 105
——water cooled 105
—paraboloid 37
—SiC 102
——CVD 106
——REFEL 107
—surface 50
—toroidal 98
Mn 648, 649
mnemonic 153
MnF_2 644, 650
Mo 107, 764, 766
modulation limit 618, 620
modulation techniques 741
module 141–143, 145
molecular autoionization 265
—dynamics 268
molecular axis 254
molecular beam epitaxy, *see* MBE
molecular experiments 112
molecular orbital 214, 264, 307, 397
molecular photoionization 245, 248, 250, 256, 259, 260, 268
—dynamics 243, 263, 297, 335
molecular quantum defect plot 227
molecular solids 246, 331
momentum of electron states 470
monatomic resolution 581
monochromator 3, 23, 24, 28, 46, 47, 49, 61, 62, 64, 67, 68, 70, 73, 77, 85–87, 91, 98, 99, 104, 111, 114, 123, 126, 276, 277, 362, 364, 366, 367, 372, 438
—crystal 94, 99
—optimized 82
—slitless 82
monolayer 502, 545, 547, 549, 550, 553, 554, 558, 559, 563, 564, 569, 571, 572, 574, 577, 578, 581, 583, 585, 586, 589, 595, 599, 601, 697, 700, 709, 730, 738, 765

Subject index 839

MoS 476
MoSi$_2$ 556
motor driving 123
Mott detector 373, 374
mount (mirror or grating) 100
MQDT (multichannel quantum defect theory) 183–186, 227, 228, 230, 244, 268–271, 274, 292–294, 332, 333, 335, 412, 418, 421, 431, 433
multi-electron shake-up 505
multi-layer mirror 50
multi-user "host" computer system 154
MULTIBUS-I, II 147, 148
multicenter field 243
multichannel analyser 130
multichannel plates 365
multichannel quantum defect theory, see MQDT
multicoincidence experiments 381
multiphoton ionization 332
multiple-angle reflectance measurements 756
multiple scattering 758
—calculations 306
—model 260, 318, 319, 329
multiply excited states 246
multiply reflected beam 729, 749
multiply reflected fields 730
multipole wiggler 8, 15, 16, 18
multiscaling 134, 136
multistepping 138
multitasking 156, 157

N 49, 252, 699
N$_2$ 219, 229, 246, 251, 252, 255, 258–262, 268, 281–283, 288, 295, 296, 303, 304, 310, 325, 330–333, 382, 398, 402, 413, 416, 421–423, 425, 426, 432, 446, 455, 699
N$_2$, π_0 resonance in electron scattering 285
N$_2$-induced reconstruction 697
N$_2$ photoionization 249, 250, 252
Na 201, 203, 204, 626, 688, 689
nanosecond domain 366
Nb 557, 764
NBS SURF-II Facility 98, 134, 245, 276, 277
Nd 639, 687
Ne 112, 186, 187, 189, 201, 204
negative factorials 29
negative-ion system 261
NEXAFS (Near EXAFS) 503, 505, 516–518, 520, 770
NF$_3$ 330
NH 699
NH$_2$ 699
NH$_3$ 382, 443, 699, 702
Ni 480–482, 486, 556, 557, 560, 562, 567, 581, 591, 595, 597, 614, 637, 639, 643, 647, 648, 650, 652, 700
NIM "bin" 125
NIM system 125
NIM units 126
NiSi 595
NiSi$_2$ 595
NO 330–334, 390, 391, 418, 431, 446, 699, 700
NO(A) 393
N$_2$O 330, 331, 382, 398, 400, 406, 409, 415, 427–429
N$_2$O$^+$ 405, 434
NO$_2$ 390
noise 128
—interference 124, 125
—pickup 125
non-bonding orbital 226
normal emission photoelectron diffraction, see NPD
normal incidence 91
normal-incidence monochromator 56, 88, 97, 276
normal-incidence spectrometer 90
normal modes 257
NPD (normal emission photoelectron diffraction) 508, 510
NPS (National Physics Laboratory, UK) 105
NSLS I, II (Brookhaven) 13, 111
nuclear-spin angular momentum 433
null ellipsometry 756
NWC (Naval Weapons Center, USA) 105

O$_2$ 49, 330–332, 334, 382, 398, 402–404, 410, 415, 418, 500, 502, 700, 702
^{18}O$_2$ 405
—in films 749
OCS 330, 382, 398, 428, 435, 442, 443, 445, 446
off-axis aberrations 91
oil contamination 112
on-line analysis 158, 159
one-electron approximation 613
optic axis 748
optical coatings 24, 108
optical constants 45, 46, 108, 742, 747, 750, 755, 763
—data 44
optical data 726
optical disc cartridge 159
—slope error 80, 81
optical elements 24
optical filter 365, 367
optical path function 68
optical properties 739

—data collections, literature 772–777
—of solids 725
—temperature dependence 740
optical surface 105
oriented molecules 246
orthogonal mirror system 103
orthogonality catastrophy 614, 621, 623, 633, 634
Os 108
oscillator model 753
oscillator parameter 755
oscillator strength 49, 200, 211, 247, 248, 250, 251, 258, 272, 304, 328, 384, 400, 401, 443, 754
outgassing 86
"overflow" signal 127, 128
overheating 116
overlaid program 152
oxidation 702

P 572, 656
P_2 218, 219
paging 157
parabola 36, 37
paraboloid 36, 37, 67, 101
paraboloidal collimating premirror 61
paraboloidal mirror 60
partial cross section 251–253, 292–294, 307, 308, 311–315, 317, 318, 321–328, 330, 361, 413, 416, 427, 446
—measurements 309, 329
partial photoionization cross section 305, 306, 369, 372, 408, 412, 417
partial pressure of hydrocarbons 87
partial yield spectroscopy 599
Patterson function 496
Pauling radius 513
Pb 474, 567
PbTe 567
PCI (post-collision interaction) 447
Pd 482, 556, 591, 595, 633
—clusters 499
Pd_2Si 556, 595
Peierls distortion 522
penetration depth of photon field 49–51
Penning gauge 115
PEP (Stanford) 13
PES (photoelectron spectroscopy) 150, 177, 368, 381, 396–398, 408, 414, 434, 443, 450, 725
phase ellipse 26
phase-sensitive lock-in amplifier 137
phase shift 185, 227
—term 427

phase space 24–26
phase transition 484
phonon broadening 689
phonon population 740
phosphor bronze 105
photionization efficiency curve 457
photo-double ionization 208
photo-induced covalency 622
photoabsorption cross section 247
photoabsorption spectra 262
photochemistry 357
photodissociation 357
photoelectric cross section 770
photoelectron 248, 249, 447, 500
—angular distribution 243–246, 250, 254, 263, 266, 268, 269, 278, 280, 285, 305, 306, 369, 419, 432, 445, 446
—angular distribution measurements 370
—backscattering 670, 672
—bands 371
—branching ratio 245, 278
—channels 333
—diffraction 513
—motion in molecular fields 255
——photoion coincidence 334
—spectra of BF_3 319
—spectrum 425
—wave function 258
—yield 429
—yield of gold 372
photoelectron spectroscopy, see PES
photoemission 49
—cross-section resonances 504
—spectroscopy
——angle-integrated 476
——angle-resolved 476
photofragmentation 357
photoinduced covalency 647
photoion–fluorescence coincidence, see PIFCO
photoion–photoion coincidence, see PIPICO
photoion–photoion coincidence technique 434, 453
photoionization 243, 245, 246, 255, 357
—branching ratio 243, 263
—channels 285
——in N_2 280, 282
—cross section 236, 243, 263, 266, 267, 360, 407, 416, 437, 771
—dynamics 244, 271
—dynamics of BF_3 321
—in N_2 329
—in SF_6 329

—mass spectrometry 268, 270, 305, 306, 320
—of H_2 265, 268, 270
—of laser-excited molecular states 335
—of SF_6 305
—process 248
photoionization spectrum (PIS) 244, 261, 292, 410
photoionizing molecules adsorbed on surfaces 254
photometric ellipsometry 756
photometric measurements 763
photometric methods 742
photomultiplier 364
photon bandpass 369
Photon Factory (Japan) 12
photon flux, $N(\lambda)$ 10, 11, 17, 26, 30–33, 116
photon-stimulated desorption 473, 506, 507
photoyield 755
physisorption 500, 511–513
π-bonded chain model 490
picosecond domain 366
PIFCO (photoion–fluorescence coincidence) 381, 382
pinning 584, 590, 601, 602
—level 583
—position 584, 585
—value 584
PIPICO (photoion–photoion coincidence) 380, 381, 438
—spectrum 435
Pirani gauge 115
Pirani/Penning gauge 113
plane grating 57, 58
—monochromator 56, 57
plane polarization 90
plane-polarized light 472
plasma frequency 620
plasma physics 357
plasmon 621, 650, 759
—energy 754
—satellite 623, 626, 627
Pm 686
PN 219
Poisson distribution 618
polar orientation of adsorbate molecule 504
polarization 6, 8, 9, 11, 14, 30, 31, 34, 35, 42–48, 54, 98, 177, 230, 237, 253, 254, 278, 333, 361, 362, 369, 370, 372, 407, 419, 427, 432, 445, 446, 470, 508, 512, 516, 613, 616, 619, 623, 651, 652, 655–657, 669, 670, 672, 689, 701, 710, 712, 715, 725, 728, 729, 731, 740, 742–751, 753, 756, 757, 767
—circular, properties 479
—-dependent data 523

—elliptical 46
—selection rules 503, 504
—TE 43
—TM 43
polarized light 373–375, 478
polarized optical elements 738
polarized radiation 728
polarized synchrotron radiation 496
polishing with diamond paste 767, 769
polypropylene 365
ports 19
post-collision interaction, see PCI
post-focusing mirror 98
post-focusing optics 97
potential barrier 248, 249, 252, 253, 255
—effect 201, 305
potential vector 714, 716
Pr 639, 687
pre-ionization 359
pre-optical system 77, 78
preamplifier 128
precision lead screw 100
predissociation 280, 331, 358, 359
prefocusing mirror 85
prefocusing system 92
premirror 47, 80, 97
processor board 148
proportional counters 365
protocol 168, 169
protocol conversion 162
PSD SEXAFS 508
"pseudo magic angle" 370
pseudocolor 160
pseudocolor display system 134
pseudopotential calculations 478
Pt 108, 499, 506, 556, 693–695, 702, 703, 706, 707
— –Au alloy 705
—clusters 499
pulse amplifier 125
pulse-height analyser 130
pump failure 115
Pyrex 107
Pyrex glass 98
pyridine (C_5H_5N) 516

q-reversal effect 193
quadruply differential measurements 333
quantum defect 182, 183, 185, 202, 204, 227, 268, 431
quartz 44
—crystal oscillator 549
quasichemical theory of solids 681
quenching by hydrogen 501

radial potential 211
—circularly polarized 232
radiation damage 105, 116
radiation hazard 114
radiative decay 358
radius of electron 42, 760
rapid modulation case 622
rare gases 201, 202, 204
ratemeter 125
ray tracing 55, 56, 73, 75, 80
Rb 201, 202, 204
RBS (Rutherford backscattering spectrometry) 546, 562, 576, 602
reactive outdiffusion 601
reciprocal linear dispersion 67
REFEL 97, 106
—SiC mirrors 106
reference channel 129
reference detector 128
reflectance 730, 738–740, 746, 747, 749, 750, 752, 755, 764, 767, 769
—measurements 748, 753, 754, 759, 760, 765, 766
—ratio 744, 745
—spectra 749
reflection analyser 45
reflection efficiency 90, 98
reflection polarizer 44, 45
reflectivity 42, 44–46, 50–52, 504
—calculated 42
—polarized 48
reflectometer 746
refractive index 41, 43, 51, 728, 738, 752, 760, 762
relaxation effect 331, 445, 685
relaxation energy 618, 669
relaxation shake-off 444
relaxation shift 685
remote exit slit 98
residual gas molecules 86
resolution 62, 73, 82, 84, 85, 92
—aberration limited 70, 71, 77
—slit-width limited 62, 63
—source-size limited 77, 78
resolving power 24, 177, 181
"reverse tangent" recess 86
reverse tangent window 111
Rh 108, 637, 762, 763
Ritz formula 182
rms roughness 51
robustness of coatings 108
rolled foils 761
rotary encoder 94
rotary vacuum feedthrough 105

rotation constants 212, 219
rotational structure 224
rotational–vibrational autoionization 270
—structure 269
rotationally resolved spectra of molecules 213
rough surface 757
roughness 738
—-induced absorption 740
—-induced scattering 740
—plane 52
—profile 50
rovibrational ionization channels 265
rovibrational spacing 267
rovibrational threshold 267
(ro)vibronic ionization channels 244
rovibronic levels 398
rovibronic modes 243
rovibronic structure 219, 220, 222, 227, 268
Rowland circle 80
—monochromator 56
RPA (random phase approximation) calculations 620
RRPA (relativistic random phase approximation) calculations 715
Ru 637
ruling alignment in monochromator 101
"rumpling" 498
Rutherford backscattering spectrometry, see RBS
Rydberg constant 182
Rydberg electron 264
Rydberg members 186
Rydberg orbital 264
Rydberg series 177, 183, 186, 191, 194, 225, 252, 267, 268, 303, 383, 409
—in molecules 226
Rydberg states 226, 248, 250, 264, 384, 397, 398
Rydberg structure 246–248
"Rydbergization" 418

S 246–248, 255, 508
S matrix 256
sagittal focus 39–41, 53, 54, 70, 72, 78, 80, 81, 99, 101
sapphire 107
satellite 655, 656
—bands 449
—intensity 654
—lines 400, 401
Sb 562, 572
scalar 128
scalar scattering theory 50

scanning monochromator 90
scanning tunnelling microscopy, see STM
scattered radiation 740, 742
scattering factor 774
Schottky barrier 521, 522, 526, 528, 543, 544, 546, 557, 560–562, 565, 567, 570, 572, 582, 583, 589–591, 601, 602
—formation 589, 594, 599
Schottky relation 593
Schrödinger equation 716
Science and Engineering Research Council 147
scintillator 364
Sc_2O_3 762
—magnitude 705
—measurements 674–676
SCS (surface core level shift) 484, 488–490, 492, 665, 666, 669, 672, 673, 678, 679, 681, 683, 685–693, 703–708, 710, 718
Se 565
second- and higher-order radiation 742
semi-insulator 478
semiconductor 478
semiconductor device fabrication 469
separated function focusing system 80
separated function grating monochromator 56
separated function mirror systems 52, 53
servo motor 137, 140
SEXAFS 49, 469, 471–473, 498, 503, 508, 513–516, 518, 520, 523, 526, 544, 556, 594, 595, 598, 601, 602
—fluorescent 49
Seya monochromator 54, 88–90
SF_6 247–249, 254, 255, 261, 262, 308, 311–316, 318, 325, 329, 330, 449, 457
SF_6^+ 310
SF_6 photoionization 306
—channels 295
SF_5CF_3 330
SF_2O 330
SF_2O_2 330
SGM (spherical grating monochromator) 78–82
shaft encoder 140
shake-up satellite 622, 647
shape parameters 211
shape resonance 243–248, 250, 252, 255, 256, 259–263, 276, 280, 282–285, 295, 297, 306, 309, 310, 318, 319, 325, 327–332, 334–342, 359, 364, 409, 410, 413, 416, 417, 421, 423, 427, 439, 442, 443, 446, 447, 455, 457, 516
—effects in molecules 305

—-enhanced absorption 254
—behaviour 251
—enhanced antibonding 316
—enhancement 295
—feature 248, 254, 298
—states 257
shielded hutches 24
shift of surface band 677
Shockley state 485, 487
short-range interaction 183
Si 96, 252, 446, 447, 490, 495, 498, 503, 551–554, 557, 569, 570, 581, 595, 598–600, 689–691, 761, 762, 764
SiC 97, 106, 117
—CVD 107
—hot pressed 107
—mirror 102
—REFEL 107
$Si(CH_3)_4$ (TMS) 443, 444
$SiCl_4$ 330, 442
SiF_4 330, 442
SiF_6^{2-} 330
signal 124
—analogue 124, 126
—channel 128
—fast rise time 125
—linear 125
—"logic" 125
—preferred levels 125
signal-to-noise ratio 150
SiH_4 450
silicon "islands" 106
Si_3N_4 107
sine bar 100
—drive 58
—scanning mechanism 78
single bunch 15
"single-bunch" mode 130
single-channel analyser 125
single-channel counting 126, 127
single-crystal surface orientation 683
single-hole state 332
single-user system 150
SiO 214
SiO_2 330, 502, 533, 567, 568
Si/SiO_2 interface 502
SLAC (Stanford) 13
Slater exchange 628
Sm 483, 562, 629, 639, 686
small-scale roughness 739
Sn 499, 562
—clusters 499
SO_2 407, 435, 439, 441, 442
SO_4^- 330

soft dissociative double ionization 450
soft double ionization 453
"soft" process 434
soft X-ray photoemission spectroscopy, see SXPS
soft X-ray spectrometer 88, 90, 93, 95, 96
software standard 142, 145
sorption pump 112, 113
source brilliance 6, 8
source size 6, 11, 27, 28
—limited resolution 77, 78
SPEAR (Stanford) 13, 365
spectator channels 450
spectral brilliance 15
—of available storage rings 236
—of available synchrotrons 236
spectral function $J(\theta)$ 617
spectrographs 177
specular reflectance 51
spherical focusing mirror 62
spherical grating 70
spherical grating monochromator, see SGM
spherical top 219
spin polarization 373, 374, 427, 428, 457
—analysis 480
—measurements 481
—of photoelectrons 333, 470
—spectroscopy 236
spin-polarized photoemission 471, 479, 482
spin-resolved diffraction studies 482
spin-resolved photoelectron spectroscopy 375, 457
spontaneous magnetization 482
Sr 179, 181, 185, 207, 232, 233
SRS BL1 HESGM 79–81
SRS program library 158
stacking fault 495
stainless steel 105
—vacuum envelope 114
standard highway 144
Stark potential 457
stepped surface 693
stepping motor 129, 137–141
—driving 125
—translator 139
sticking coefficient 549
stigmatic image 39
Stirling's approximation 29
STM (scanning tunnelling microscopy) 495, 697
stoichiometry 476, 532
streak cameras 136
substrate 67, 105
—ellipsoidal 67

—fused silica 105
—normal toroidal 68
—paraboloidal 67
—toroidal 67
sum rule 753, 764, 767, 771
super Coster–Kronig transition 653, 656
superexcited state 359
SUPERLUMI (at DESY) 366, 367, 457
supernumerary Rydberg series 201
supersonic expansion 433, 456
supersonic jet 279
supersonic molecular beam 369
SURF II, NBS 13
surface anisotropy 764
surface band narrowing 688
surface core level shift, see SCS
surface core shift, see SCS
surface crystallography 488, 666
—relaxation 718
surface energy 704
surface extended X-ray absorption fine structure, see EXAFS
surface magnetism 486, 489
surface mobility 570–572
surface oxidation 764
surface plasmons 739
surface reflectance 728
surface roughness 50–52, 105, 740, 754, 761, 763
—effects 671
surface segregation heat 679
surface shift 713
surface space charge layer 489
surface state 484, 485, 489
—spectroscopy 486
surface X-ray diffraction 496
surface X-ray scattering studies 474
surface–bulk core level shift 694
SXPS (soft X-ray photoemission spectroscopy) 543, 544, 547, 550, 551, 553, 557, 558, 560–565, 569, 571, 576, 577 580, 581, 586–591, 594, 595, 600–603
SXRL (FEL ring Stanford) 13
symmetric top 220
—oblate 220, 222
—prolate 220
synchrotron bunch repetition rate 132
Synchrotron Radiation Center, University of Wisconsin 287
synthetic microstructures 762
system software 145

t-matrix method 651
Ta 199, 488, 702

tachometer 141
Tamm state 485–487
TANTALUS (Wisconsin) 13, 15, 548
Tb 687
TbF$_3$ 627
Te 564, 565
TEM (transmission electron microscopy) 495, 556
tensile strength 106
—spectra 406
TEPICO (threshold electron–photoion coincidence) 402–404
TERAS (Japan) 12
text editor 153
TGM (toroidal grating monochromator) 53, 54, 56–58, 70, 71, 75, 77, 78, 81, 88, 91, 92, 97
thermal load 117
thiophene 520, 521
—ring 520
three-photon laser spectroscopy 220
threshold electron coincidence spectra 406
threshold electron–fluorescence coincidence, see TPEFCO
threshold electron–photoion coincidence, see TEPICO
threshold electrons 334, 377
threshold photoelectron spectrum, see TPES
threshold photoelectron–photoion coincidence, see TPEPICO
Ti 557, 560, 562, 586, 591, 765, 766
time constant 129
time-dependent local density approximation 304
time of flight, see TOF
time-of-flight spectrometer, see TOF spectrometer
time-of-flight spectroscopy, see TOF spectroscopy 298
time-resolved fluorescence 362
—spectra 367
time slice 156
time-to-amplitude converter 125
TiO$_2$ 762
Tl 193, 200, 562
TLS (Taiwan) 13
Tm 687
TMS 446
TMS [Si(CH$_3$)$_4$] 443, 444
TOF 368, 377, 381, 508
TOF spectrometer 374, 376, 380
TOF spectroscopy 131, 132, 415
Token Bus 168, 169

Token Ring 168, 169
toroid 64, 92, 94, 101
toroidal grating monochromator, see TGM
total absorption cross section of BF$_3$ 320
total cross section 271–273, 317, 770
total photoionization cross section 269
total photoionization spectrum 268
total reflection 49
TPEFCO (threshold electron–fluorescence coincidence) 381
TPEPICO (threshold photoelectron–photoion coincidence) 378
—for N$_2$O 429
TPES (threshold photoelectron spectrum) 376, 377, 398, 403, 410, 428, 429
transition metal surface 506
transmission 729, 731
—data 755
—electron energy loss measurements 767
—function 279
—grating 750
—measurements 728, 750, 767
—window 230
transmission electron microscopy, see TEM
triply differential cross section measurements 271, 274, 276, 277
triply differential photoelectron measurements 244, 245, 275, 287, 294, 298, 330, 331, 334
Tristan (Japan) 12
tunable lasers 181
tungsten photocathode 278
tungsten photodiode 278
turbo-molecular pump 112, 115, 277
two-hole one-particle state 332
two-step autoionization 210

UHV surface-science experiment 98
ultraviolet photoelectron spectroscopy, see UPS
uncertainty principle 212
undulator 5, 6, 8, 12, 17–19, 82, 116, 117, 149, 335, 364, 375, 602
—helical 237
unit magnification 99
UNIX operating system 155
UPS (ultraviolet photoelectron spectroscopy) 470, 590, 665, 666, 698
UVSOR (Kyoto, Japan) 12, 377

V 557, 562, 572, 764
vacuum data logging 116
vacuum equipment 111

valence
—orbitals 396
—photoemission spectroscopy 506
—states of SF_6 306
—transition 247
valence-shell spectra 251, 261, 263
valve modules 115
van der Waals bonding 456, 565
van der Waals complexes 456
VEPP-2M (USSR) 13
VEPP-3 (USSR) 13
vibrational autoionization 273, 274
vibrational autoionizing states 274
vibrational branching ratio 244, 268, 269, 273, 274, 292
vibrational channels 291
vibrational data 212
vibrational ionization channels 292
vibrational levels 219
vibrational partial cross section 271–273, 285
vibrational relaxation 358
vibrational–rotational autoionization 265
vibrationally elastic scattering cross section 261
vibronic coupling 310, 332
video camera 137
video monitor 150
virtual orbital 255
virtual storage capability 152
VME-bus 147
VME–CAMAC 143
voltage-to-frequency converter 125
VSi_2 555, 556
VUV ring, Berlin 27
VUV ring, Brookhaven 27

W 488, 496, 670, 694, 695, 702, 712, 764
WAN (wide-area network) 162, 165
Wannier exciton model 457
water-cooled copper block 96, 102
water-cooled mask 116
water cooling 94
water vapour 500, 506, 507
wavefunction collapse 204
wavelength shifter 5, 15, 16
waviness 738
WC 107
weak channels 246
—interaction 248
wide-area network, see WAN
wide bandpass monochromator 94
wiggler 5, 12, 375, 602

Wigner–Seitz cell 667
Winchester disc drive 150
window resonance 194, 292
WO_2 701
WO_3 701
word 152
word length 152
work function 544, 549, 593, 678

X-ray absorption near-edge structure, see XANES
X-ray core emission 530
X-ray crystallographic data 514
X-ray diffraction 496, 544
—technique 512, 598, 601
X-ray interferometers 761
X-ray mass spectrometry 450
X-ray optical spectroscopy 725
X-ray PES, see XPS
X-ray photoelectron intensity 310
X-ray photoelectron spectroscopy, see XPS
X-ray photoelectron spectrum 309
X-ray photoemission spectroscopy 502
X-ray reflectivity 35
X-ray standing wave analysis 503
X-ray standing wave interference spectrometry (SWXIS) 599
X-ray standing wave techniques 598
XANES (X-ray absorption near-edge structure) 438, 469, 471, 472
Xe 186–188, 228
XeCl excimer laser 560
XeF_2 456
XPS BE 690
XPS (X-ray photoelectron spectroscopy) 401, 483, 665–667, 686
XYZ manipulator 279

"yaw" 104
Yb 196, 554, 686, 687, 707
$YbAl_2$ 483

zeolite traps 112
zero-order trap baffle 92
Zerodur 107
Zn 197, 200, 485, 577, 648
ZnF_2 644
ZnS 562, 580
ZnSe 562, 580, 581
ZnTe 562, 580, 581
"zoom" lens 278

COLOPHON

Typeset by ICPC, Dublin, Ireland
Printed by ICG-Printing, Dordrecht, The Netherlands
Bound by Boekbinderij Stokkink, Amsterdam, The Netherlands
Cover design by Jan de Boer, Utrecht, The Netherlands
Desk editor: Jan ten Have
Producer: Mari Plooy

Values for component f_2 of the atomic scattering factor for selected values of atomic number Z at energies between 100 and 1000 eV. Values in the table are recorded in the form $1.42(-)02$ to denote the value $1.42 \times 10^{(-)2}$. (From B.L. Henke, 1985, Scattering factors and mass absorption coefficients, in: X-Ray Data Booklet, Publ. 490, Center for X-Ray Optics, Lawrence Berkeley Laboratory, University of California, Berkeley, CA. Data from Atomic Data and Nuclear Data Tables, Vol. 27 (Academic Press, New York, 1982).)

Z	Energy (eV)									
	100	200	300	400	500	600	700	800	900	1000
3	2.75 00	3.33 00	3.26 00	3.20 00	3.15 00	3.12 00	3.10 00	3.08 00	3.07 00	3.06 00
6	4.08 00	3.67 00	2.88 00	5.50 00	6.10 00	6.28 00	6.35 00	6.37 00	6.36 00	6.34 00
9	6.71 00	7.63 00	7.54 00	7.33 00	6.95 00	6.30 00	5.01 00	7.50 00	8.35 00	8.75 00
12	3.73 00	1.02 01	1.10 01	1.11 01	1.10 01	1.09 01	1.07 01	1.06 01	1.04 01	1.02 01
15	2.93 00	7.00 00	1.22 01	1.38 01	1.43 01	1.44 01	1.44 01	1.43 01	1.42 01	1.41 01
18	6.21 00	4.67 00	7.64 00	1.28 01	1.57 01	1.68 01	1.73 01	1.76 01	1.77 01	1.77 01
21	8.50 00	9.92 00	8.86 00		1.24 01	1.69 01	1.90 01	1.98 01	2.03 01	2.06 01
24	1.06 01	1.35 01	1.42 01	1.37 01	1.17 01		1.56 01	1.92 01	2.12 01	2.23 01
27	9.54 00	1.52 01	1.71 01	1.74 01	1.70 01	1.60 01	1.29 01	7.92 00	1.70 01	2.10 01
30	5.99 00	1.54 01	1.94 01	2.07 01	2.12 01	2.10 01	2.04 01	1.95 01	1.74 01	1.00 01
33	2.69 00	1.27 01	1.98 01	2.24 01	2.40 01	2.45 01	2.46 01	2.44 01	2.40 01	2.33 01

K

L_{III}